THE OXFORD HANDBOOK OF

RELIGION AND SCIENCE

THE OXFORD HANDBOOK OF

RELIGION AND SCIENCE

Edited by

PHILIP CLAYTON

AND

ZACHARY SIMPSON

ASSOCIATE EDITOR

OXFORD

UNIVERSITY PRESS

OXFORD
UNIVERSITY PRESS

Great Clarendon Street, Oxford OX2 6DP

Oxford University Press is a department of the University of Oxford.
It furthers the University's objective of excellence in research, scholarship,
and education by publishing worldwide in

Oxford New York

Auckland Cape Town Dar es Salaam Hong Kong Karachi
Kuala Lumpur Madrid Melbourne Mexico City Nairobi
New Delhi Shanghai Taipei Toronto

With offices in

Argentina Austria Brazil Chile Czech Republic France Greece
Guatemala Hungary Italy Japan Poland Portugal Singapore
South Korea Switzerland Thailand Turkey Ukraine Vietnam

Oxford is a registered trade mark of Oxford University Press
in the UK and in certain other countries

Published in the United States
by Oxford University Press Inc., New York

British Library Cataloguing in Publication Data

Data available

Library of Congress Cataloguing in Publication Data

The Oxford handbook of religion and science
edited by Philip Clayton and Zachary Simpson, associate editor.
Includes bibliographical references.
ISBN–13: 978–0–19–927927–2 (alk. paper)
ISBN–10: 0–19–927927–6 (alk. paper)
1. Religion and science. I. Clayton, Philip, 1956–
II. Simpson, Zachary R.
BL240.3.094 2006
201'.65—dc22   2006019811

Typeset by SPI Publisher Services, Pondicherry, India
Printed in Great Britain
on acid-free paper by
Biddles Ltd., King's Lynn

ISBN 0–19–927927–6   978–0–19–927927–2

1 3 5 7 9 10 8 6 4 2

ACKNOWLEDGEMENTS

During its three-year gestation period the Handbook has profited greatly from the expertise, wisdom, and ongoing support of the Editorial Committee: John Polkinghorne, Arthur Peacocke, Ian Barbour, Nancey Murphy, and Jeffrey Schloss. Because each of them is an acknowledged expert in the field of science and religion—and not least because they did not always agree among themselves on what should and should not be included—the Handbook is more comprehensive and more even-handed than it would otherwise have been.

We have also benefited immensely from the advice of dozens of other scholars around the world, who have invested significant energy and effort in directing us toward persons to ask, debates to include, and mistakes to avoid. I am particularly grateful for the role that my naturalist friends have played in 'balancing out' my theistic perspective and ensuring a uniformly fair treatment of religious traditions, philosophical schools, and the naturalism–theism debate itself.

The staff at Oxford University Press in Oxford have again demonstrated why they are known as the most professional team in academic publishing worldwide. The initial structuring of the project occurred under the direction of Hilary O'Shea. The project came to fruition under the skilled hand and sometimes firm guidance of Lucy Qureshi. We are grateful also to Jean van Altena, Dorothy McCarthy, and Elizabeth Robottom for their invaluable assistance during the copy-editing and production process.

Producing a text of fifty-six chapters would have been impossible without the competent and dedicated help of a whole team of graduate students at Claremont. Andrea Zimmerman worked for six months researching and contacting authors in the early phase of the project; Fay Ellwood took over for the next six months as drafts began to come in; and Emily Bennett brought her editing and organizational skills to the hectic final phase of assembling the Handbook. We gratefully acknowledge the professional help of Casey Crosbie-Nell and Jason Stevens in the final formatting and preparation of the typescript. Joining the team with her usual high standards was my long-time transcriptionist in Santa Rosa, Jheri Cravens.

Above all, I wish to thank the book's Associate Editor and my fellow scholar at Claremont, Zachary Simpson. Not only did he spearhead the correspondence with the dozens of authors, logging in some 2,800 letters sent and received, but he has also worked as a full colleague in making editorial decisions and commenting on chapter drafts. Editing a Handbook of this size is a 'trial by fire' even for the most experienced scholar; if Zachary Simpson's performance on this project is any indication, he has some great books ahead of him.

We gratefully acknowledge permission from Professor Seyyed Hossein Nasr and the Editor of the *Islamic Quarterly* for permission to republish a modified version of his article, 'The Islamic World-view and Modern Science'.

Contents

PART II. CONCEIVING RELIGION IN LIGHT OF THE CONTEMPORARY SCIENCES

PART III. THE MAJOR FIELDS OF RELIGION/SCIENCE

PART IV. METHODOLOGICAL APPROACHES TO THE STUDY OF RELIGION AND SCIENCE

LIST OF CONTRIBUTORS

Peter Atkins is SmithKline Beecham Fellow and Tutor in Physical Chemistry at Lincoln College, University of Oxford.

Scott Atran is Director de Recherche (CNRS) at the Institut Jean Nicod in Paris, and Adjunct Professor of Psychology at the University of Michigan.

Joseph A. Bracken, SJ, is Professor of Theology and Director of the Brueggeman Center for Interreligious Dialogue at Xavier University.

Susan Power Bratton is Professor and Chair of Environmental Studies at Baylor University.

Michael W. Brierley is Chaplain to the Bishop, at Oxford University.

John Hedley Brooke is Andreas Idreos Professor of Science and Religion at Oxford University.

Bernard Carr is Professor of Mathematics and Astronomy at Queen Mary College, London.

Philip Clayton is Ingraham Professor of Theology at Claremont School of Theology and Professor of Philosophy and Religion at Claremont Graduate University.

Ronald Cole-Turner is H. Parker Sharp Professor of Theology and Ethics at Pittsburgh Theological Seminary.

Robin Collins is Associate Professor of Philosophy at Messiah College, Grantham, Pennsylvania.

Terrence W. Deacon is Professor of Biological Anthropology and Linguistics at the University of California, Berkeley.

Celia Deane-Drummond is Professor of Theology and Biological Sciences at University College Chester.

William A. Dembski is Carl F. H. Henry Professor of Theology and Science and Director of the Center for Science and Theology at Southern Baptist Theological Seminary.

Willem B. Drees is Chair of Philosophy of Religion, Ethics, and Encyclopaedia of Religion, University of Leiden, The Netherlands.

George F. R. Ellis is Distinguished Professor of Complex Systems at the University of Cape Town.

Sean Esbjörn-Hargens is co-director and founding member of the Integral Ecology Center of Integral University and Assistant Professor in the Integral Studies Department at John F. Kennedy University.

Richard Fenn is Professor of Christianity and Society at Princeton Theological Seminary.

Owen Flanagan is James B. Duke Professor of Neurobiology at Duke University.

Carl Gillett is Professor of Philosophy at Illinois Wesleyan University.

Ursula Goodenough is Professor of Biology at Washington University.

Niels Henrik Gregersen is Professor of Systematic Theology at the University of Copenhagen.

David Ray Griffin is Professor Emeritus of Philosophy of Religion and Theology at Claremont School of Theology and co-director of the Center for Process Studies.

John Grim is Coordinator of the Forum on Religion and Ecology and Professor in the Department of Religion at Bucknell University.

John F. Haught is Landegger Distinguished Professor of Theology at Georgetown University.

Philip Hefner is Professor Emeritus of Systematic Theology at the Lutheran School of Theology in Chicago.

Martinez Hewlett is Professor Emeritus, Department of Molecular and Cellular Biology at the University of Arizona.

Nancy R. Howell is Academic Dean and Associate Professor of Theology and Philosophy of Religion at Saint Paul School of Theology.

William B. Hurlbut is Consulting Professor at the Neuroscience Institute at Stanford University.

Michael Lambek is Professor of Anthropology at the University of Toronto.

Alister E. McGrath is Professor of Historical Theology at Oxford University.

Sangeetha Menon is a Fellow in the School of Humanities at the National Institute of Advanced Studies in Bangalore.

Mary Midgley is Retired Professor of Philosophy, formerly of the University of Newcastle.

Nancey Murphy is Professor of Christian Philosophy at Fuller Seminary.

Seyyed Hossein Nasr is University Professor of Islamic Studies at George Washington University.

Raymond F. Paloutzian is Professor of Psychology at Westmont College.

Wolfhart Pannenberg is Professor Emeritus of Systematic Theology at the University of Munich.

Ann Pederson is Professor of Religion at Augustana College, Rock Island, Illinois.

Robert T. Pennock is Associate Professor of Philosophy at Michigan State University.

Ted Peters is Professor of Systematic Theology at Pacific Lutheran Theological Seminary and the Graduate Theological Union.

John Polkinghorne is President of Queens' College, Cambridge University, and former Professor of Mathematical Physics, Cambridge University.

William B. Provine is Charles A. Alexander Professor of Biological Sciences at Cornell University.

Holmes Rolston III is University Distinguished Professor of Philosophy at Colorado State University.

Pauline M. Rudd is University Research Lecturer and Senior Research Fellow at the Glycobiology Institute of the University of Oxford.

Robert J. Russell is founder and director of the Center for Theology and the Natural Sciences and Professor of Theology and Science in Residence at the Graduate Theological Union.

Norbert M. Samuelson is Grossman Chair of Jewish Studies at Arizona State University.

Jeffrey P. Schloss is Professor of Biology at Westmont College, Santa Barbara, California.

Robert A. Segal is Professor of Religious Studies at the University of Aberdeen.

LeRon Shults is Professor of Systematic Theology at Agder University, Kristiansand, Norway.

Michael Silberstein is Associate Professor of Philosophy at Elizabethtown College, Elizabethtown, Pennsylvania.

Lisa L. Stenmark is Lecturer in Comparative Religious Studies at San Jose State University and co-chair of the American Academy of Religion's Science, Technology and Religion Session.

Owen C. Thomas is Associate Professor in the Center for Process Biotechnology at the Technical University of Denmark.

Evan Thompson is Associate Professor of Philosophy at the University of Toronto.

Thomas F. Tracy is Professor of Religion at Bates College, Lewiston, Maine.

B. Alan Wallace is founder and president of the Santa Barbara Institute for Consciousness Studies.

Kirk Wegter-McNelly is Assistant Professor of Theology at Boston University.

Michael Welker is Professor of Systematic Theology at the University of Heidelberg.

Phillip H. Wiebe is Professor of Philosophy and Dean of Arts and Religious Studies at Trinity Western University in Canada.

Ken Wilber is director of the Integral Institute, Boulder, Colorado.

Wesley J. Wildman is Associate Professor of Theology and Ethics and Chair of the Philosophy, Theology, and Ethics Department at Boston University's School of Theology.

INTRODUCTION

PHILIP CLAYTON

There was a time when scholars disputed whether the discussion of science and religion could ever be viewed as a specialized field on its own. Admittedly, some attention was always devoted to the relations—and especially the tensions—between these two key areas of human experience, and educated persons generally held strong opinions about whether they could be harmonized. Still, attempts to make progress on questions of science and religion, much less to resolve them fully, were viewed as exercises in futility. Devoting good scholarship to such questions would at best create an impression of rigour and rationality where none could be had.

Of course, from another perspective, *any* discussion of the possibility of 'science and religion' as a distinct field of study represented a clear step forward from the dominant prejudice of an earlier age. After all, prior to such discussions it was common knowledge that science and religion were at war with one another—a warfare so bloody and of such great import that no Geneva Convention could ever regulate its battles.

Today, by contrast, it seems hard to deny that a new area of study has emerged, one devoted to the study of the complex and multifaceted relationships between science and religion. The chapters in this Handbook, and the thousands of references provided here to other bodies of literature, surely testify to the existence of a distinct field of inquiry. Scores of monographs and hundreds of articles appear each year; dozens of conferences are being convened annually on specialized research topics; and refereed journals are springing up to publish important results in the field. Not only scholars of religion, but now more and more scientists are finding that they wish to explore the lines of relationship between the two domains.

The Oxford Handbook of Science and Religion seeks to provide both an introduction to this burgeoning field and a snapshot of the state of the art across its various sub-fields. Detailed typologies of religion–science relations exist, and sophisticated methodological proposals offer rigorous means for comparing them (Part IV). A specialized literature exists for relating each of the major scientific disciplines to religious questions and concerns (Part II). Like other fields of study, Religion/Science is divided into specific sub-fields of study, each of which employs its own methods and relies on established work within religious studies, theology, comparative philosophy, the social scientific study of religion, and other disciplines (Part III). Perhaps most exciting, new discussions are now under way to reinterpret the science–religion relationship from the perspective of each of the major world religions (Part I). One can hope that carefully studying how differently the various religions conceive their relationship with the sciences will help to overcome the tendency, until recently dominant in Western scholarship, to equate 'religion and science' with 'Christianity and science'.

Nine central questions underlie the Handbook project and emerge as leading themes in many of the chapters:

- In what ways are the goals, methods, and results of religious practice and reflection similar to science, and in what ways are they different? If different, are they compatible or incompatible? If compatible, are the two complementary, and if so, exactly how and in what respects?
- How does the relationship between science and religion appear differently when one views it from the perspective of the various religious traditions?
- How does this relationship appear differently when one views it from the perspective of the various particular sciences?
- Which methodologies or standpoints are most helpful in comprehending the relationship between religion and science, and which offer too limited a perspective or distort the subject-matter?
- What light is shed on the core questions of 'science and religion' by the various specific disciplines that study it: history, philosophy of science, religious studies, theology, metaphysics, ethics, and spirituality? In particular, can the often diverging conclusions of the social scientific study of religion and faith-based perspectives be reconciled?
- What is the 'naturalism' with which science is often associated? Does science presuppose the truth of naturalism, or merely the usefulness of naturalistic methodologies for making progress in empirical research? If methodological naturalism is indeed essential to the practice of science, what does this say about naturalism as a metaphysical position? For example, does the success of science provide evidence that naturalistic explanations are ultimately truer than non-naturalistic explanations?
- To what extent (if at all) is the practice of science, or at least the interpretation of its results, affected by one's culture, one's gender, or one's religious presuppositions? To what extent are religious beliefs and practices affected by historical and cultural location, and by scientific beliefs? In the face of so great a diversity on both sides, what shared results can we achieve, and on what basis can we come to such agreement?

- Do science and religion represent massively different ways of knowing, or is there a common definition of knowledge that both share, and perhaps even some common criteria? Are these two spheres of activity necessarily competitors in the human quest for knowledge, or can they function as partners in a multilateral quest?
- What are the *implications* of the field of science-and-religion studies? How can the debates covered in this Handbook shed new light on the fundamental values issues that confront humankind and our planet today?

Just as genetic diversity is crucial for the survival of a community of organisms, and biodiversity is indispensable for the flourishing of an ecosystem, so also a diversity of approaches is crucial if 'religion and science' is to flourish and to progress as a distinct field of study. The careful reader of this Handbook will discern recurring themes and questions. There is consensus that certain theories are inconsistent or have proved less useful, just as there is widespread agreement that other topics, debates, and approaches are particularly important within the contemporary scientific and religious context. Readers will nonetheless also discern crucial differences on key questions. If progress is to be made, the differences will be as critical as the agreements; we have thus sought to foreground them rather than to hide them from sight.

No one person can define a field—if this is true of standard disciplines within the academy, it is all the more true of a massively interdisciplinary field such as the study of science and religion. For this reason, the policy of the Editorial Committee has been emphatically and boldly pluralistic. To the extent that readers find that we have emphasized physics to the exclusion of other sciences, or theism to the exclusion of other world-views, or supernaturalism to the exclusion of naturalism, or Christianity to the exclusion of other religious perspectives, we will have failed at actualizing our central editorial policy. The Handbook does not presuppose that there is a single right relationship between 'religion' and 'science', nor even that religion is necessarily a good thing—as the chapters by a number of the authors will make clear. Most fundamentally, we have sought to represent the field of science and religion not as a series of conclusions that students are to learn and memorize, but as a series of questions and topics that scholars are researching and debating. The goal of the Handbook is to invite readers to join in this debate, to add to its rigour, and to help it advance toward more sustainable conclusions.

This goal should be most clear in Part V. The twenty chapters of this part have been gathered together not as individual presentations of the right answers on each topic, but rather as paired debates between experts focusing on the most hotly contested issues in each field. Although these 'hot topics' chapters are research-based and written by leading scholars, the authors were asked not to pretend to the neutrality and objectivity of an encyclopaedia article. However much the natural sciences may consist of dispassionate theories grounded in objective facts and data (the degree to which this occurs in science being a matter of heated contention among the Handbook authors), any pretence to encyclopaedic objectivity must surely flounder given the interpretative intricacies of religion–science debate. Instead, the authors have made their standpoints and 'locations' explicit—not in order to force readers to adopt the same standpoint, but in order to encourage each one to formulate, and to argue for, his or her own positions on the burning questions of the field. The authors having defended their responses to the most contentious issues, readers are now encouraged to evaluate for themselves the merits of each individual standpoint, interpretation, and argument.

In short, we have aimed not to push a programme but to model a form of dialogue. Underlying the Handbook's editorial policy is at least one value commitment that should be stated as clearly as possible. We have assumed that the scientific and the religious quests are likely to be permanent features of human existence, and that humanity will be much better positioned successfully to navigate the threats that it faces if it draws constructively on *both* dimensions. If ways can be found for science and religion to work together in a complementary and productive fashion, perhaps humankind will have a better chance of overcoming the momentous challenges of the twenty-first century than otherwise.

PART I

RELIGION AND SCIENCE ACROSS THE WORLD'S TRADITIONS

CHAPTER 1

HINDUISM AND SCIENCE

SANGEETHA MENON

INTRODUCTION

Hinduism represents the religion and philosophy that originated in India, and has a historical past covering the experiences of 'thousands of different religious groups that have evolved in India since 1500 BCE' (Levinson 1998). It is the religion of 16 per cent of the world's population, and India is home to more than 90 per cent of the world's Hindus.

It would be incorrect to say that Hinduism is a monolithic religion, owing to its diverse theological traditions and its warm embrace of pluralistic thinking. However, the foundational textual sources can be traced to the corpus of *Sruti*, *Smrti*, and *Darsana* literature. This set consists of the Veda, Purana, Dharmasastra, and the six systems of Hindu philosophy. Vedas are collections of hymns and incantations often guiding worrisome thoughts about the origin of the world and natural forces to gods and goddesses. Purana forms the mythopoetic literature, and the Dharmasastras the code of ethics and moral laws for the individual and the society. *Darsana* forms systematic discussions on metaphysics, epistemology, and ways of living. To a contemporary Hindu the names of the *Bhagavad gītā*, Upanishads, and Puranas are significant sources for her thinking, believing, and understanding.

Today many historians and philosophers of science have started reviewing the dynamic events and historical processes that led to what is called the European Enlightenment and modern science. Monolithic and Eurocentric views about science are being challenged from the context of Eastern and Islamic contributions to world science. The role of China, India, and the Islamic culture in developing the bed for the origin of Western science is a theme being widely pursued. The discussion in this chapter does not trace these origins and theories. Rather, the focus will be on how Hinduism as a religion has coexisted with scientific pursuits, the underpinnings of such partnerships, and the significant contributions of such dialogues to the current engagements between science and spirituality.

The origins of Hinduism take us to a time when human experience and its possibilities were the central issues. Science, ethics, and laws were all in the context of the primary experiences of a culture and a society that saw subjectivity as an essential factor in creating objectivity. Knowledge and experience both contributed to humanization. We will see in the discussion to follow how apparently different enterprises of experience (in other words, culture) and reporting of experience (in other words, a systematic body of knowledge) were given a common space, as well as what are the areas of convergence that Hinduism posits for dialogues between and within science and spirituality. In this chapter, I will particularly look at the Vedantic (Upanishads) tradition.

GUIDELINES AND FUNDAMENTALS

As one of the oldest religions, which originated thousands of years ago, and as a religion that has assimilated the changing urges and attitudes of the Indian mind, Hinduism represents primarily a pluralistic philosophy. The religion of Hinduism, however segregated its factions are, is founded on a tolerant philosophy that is ready to include and integrate. It is pertinent to look at the fundamental principles of Hinduism that guide the contemporary Hindu to practice her religion and also place her in a global context for engagements with issues concerning humankind in general.

The fundamentals employed are acceptance of diverse views in metaphysics, faith, and belief systems; the ideal of *ahimsa*—non-violence; the ideal of *satya*—Truth; the emphasis on ways of living guided by reflection, detachment (*sakshibhava*), and meditation.

The pluralistic philosophy of Hinduism is founded on these fundamentals, which are also the bedrock for fresh and periodic additions of gods and practices to the religion. A striking feature that greets the eyes of a non-Hindu traveller in India is the number and form of deities who become part of the religion and day-to-day life. A tree, a stone, or even an anthill could suddenly be elevated to a divine status enriching the lives of people. Hinduism best explained is a living and growing religion, with an ability to extend the dimension of the divine and integrate new forms of the divine without disturbing the order of the religious system. The fluid face and form of the Hindu god makes sure that there is more inclusion of ideas, practices, and beliefs into the system of a living religion—what might be called a theology with a systems approach that helps integrate knowledge processes with pluralistic coexistence.

Non-violence

Tolerance and non-violence are the basic identities of Hinduism. *Ahimsa* is non-injury and non-violent disagreement. The ideology behind *ahimsa* is 'to agree to disagree' and 'respect for differences'. For a leader like Mahatma Gandhi, the concept of non-violence proved to be not only a political tool but also a value that touched daily life. Such is the power of this value that Einstein wrote to Gandhiji: 'You have shown through your works, that it is possible to succeed without violence even with those who have not discarded the method of violence. We may hope that your example will spread beyond the borders of your country, and will help to establish an international authority, respected by all, that will take decisions and replace war conflicts.'[1]

Gandhiji translated *ahimsa* into positive interpretations of equity and peaceful coexistence. The famous mantra that influenced Gandhiji in a significant manner says that 'the whole world is pervaded by the divine; therefore take what you are given and never covet what belongs to another' (Isavasya Upanishad, 1). The fact that *ahimsa* is one of the preconditions for a person aspiring to Yogic excellence tells us that Hinduism views non-violence not merely as an ethical concept, but as a practice capable of leading to transformation and transcendence. India's freedom movement led by Mahatma Gandhi is a testimonial for this.

The ideal of *ahimsa*, which is hailed as the foundation of religion by *Mahabharata*, is considered the supreme virtue by Hindu teachings. In a religious and metaphysical environment of contending systems of thought and faith, Hinduism, which is virtually a confederation of faiths, had to develop rules of thumb to ensure peaceful and creative interaction between them. We could say that the concept of *ahimsa* thus originated as a response to the plurality of movements within Hinduism. Though *ahimsa* literally means non-injury to living beings, it can be interpreted in different ways, as 'respect for difference', 'coexistence', 'peaceful resolution of conflicts', 'multi-dimensional perspectives', 'learning from each others' experiences', 'humility', or 'ecological harmony of all life forms'.[2]

The Fluid Face of Truth

The concept of Truth has implications for epistemic, ethical, metaphysical, and spiritual definitions in Hinduism. *Satya* is the pursuit of Truth as well as practising Truth in word and deed. The uncompromising connection between what is thought and practised makes Truth a hard value to live, a difficult epistemic concept to define, and an experiential ideal to fulfil. Isavasya Upanishad's mantra says that the 'face of the Truth is concealed by a golden disc' (mantra 14). Ramayana's theory is to 'tell

---

[1] See Einstein's letter to Gandhi, courtesy of Saraswati Albano-Müller; <http://streams.gandhiserve.org/einstein.html>.

[2] Swami Bodhananda in an email to this author, August 2005.

Truth, but say it in a manner that is helpful to others'. Mundaka Upanishad says 'Truth alone prevails'—*satyameva jayate*—(3. 1. 6). If there is conflict between ideals of God and Truth, then it is Truth that a Hindu will choose.

In her search for Truth, the Hindu is ready to undergo self-purification and transformation to refine her means of knowledge. Reason and experiments are therefore not the only valid means of knowing. Depending on the domain of study, reflection, inner transformation, and ontological insights also are means of knowledge. The intricate diagrams of altars in the Vedic literature, such as the symbolic representation of the bird with outspread wings in the Sulvasutra that represents the complex geometry of pre-Euclidian times, indicate the perceptions of a culture that didn't view 'doing science' as a pure objective enterprise but as part of ritualistic experience.[3] The Truth that was pursued demanded a means that is a blend of personal and social engagement, ecological awareness, and advanced mathematics.

The Vedic people performed rituals to harness forces of nature in their favour to gain victory in battle, to receive timely rain, for healthy children, for more cattle, for a good harvest, and for a place in heaven after death. According to their world-view the world was created by the sacrifice of gods, and gods subsisted by the sacrifice of humans, as described in the *Purushasukta* of *Rig Veda*. They developed mathematics, linguistics, astronomy, and metaphysics to aid their rituals, and conceived of a complex system known as the six limbs of Veda (*vedanga*), such as grammar (*vyakarana*), etymology (*nirukta*), manual of rituals (*kalpa*), prosody (*chandas*), and astronomy/astrology (*jyotisha*). For them, gods, nature, and humans formed a web of existence.

The One in the Many

The inclusive identity of Hindu religion is to be seen in the context of the rigour and clarity of its philosophy, exemplified by, for instance, the Jaina view of 'many-sided' and the 'perhaps' view of the real, the Buddhist theory of dependent origination, the Upanishadic concept of two kinds of knowledge, and the Nyaya proposal of empirical criteria to serve as a test for truth, such as the concordance with observed data and affirmation by practical utility. Knowledge at no point can be seen as exclusive of its contextual, relative, and tentative nature. The pragmatic view of knowledge that Indian epistemologists developed coexists with their inclusive and tolerant views about ways of living. Adherence to laws of conduct and social rules (*dharma*), contributing and partaking from the collective memories (*smrti*) of the community, taking the lead to search for the ultimate reality and its identity with the self (*moksha*)—for the Hindu, knowledge meant all this. On non-verifiable transcendental issues Hinduism is open-ended. On practical and ethical issues Hinduism encourages consensus building based on principles of honesty and non-violence.

---

[3] For more on the Sulvasutra see Narasimha (2003).

The famous Vedic statement, one of the foundational propositions of Hinduism, that says 'truth is one but expounded in many ways by the wise' (*Rig Veda*: *ekam sat vipra bahudha vadanti*), tries to relate Truth to reality, and the pursuit of Truth as the pursuit of the real. This dictum maintains a radical Hindu view that in essence existence is one, though the wise describe it differently. Swami Vivekananda, the representative of neo-Hinduism, quoted another immortal verse to signify this idea in his speech in the Parliament of Religions held in 1893 in Chicago: 'As the different streams having their sources in different places all mingle their water in the sea, so, O Lord, the different paths which men take through different tendencies, various though they appear, crooked or straight, all lead to Thee' (Swami Vivekananda 1991: 3).

The later Upanishads conceived of a world with two realities—the higher and the lower—*para* and *apara*, the world of the spirit and of matter. This division continued to exist, and got incorporated in metaphysical theories in the later Vedic and early Upanishadic thought of *brahman* and *maya*, *purusha* and *prakrti*, *kshetrajna* and *kshetra*, and *prana* and *rayi*. The later Upanishads and the *Bhagavad gītā* resolved this dichotomy by positing a higher reality, *parabrahman* or *purushottama* that transcends and includes the duality of matter and spirit, *prakrti* and *purusha*. The Vedantin perceives seer–seen duality as an expression of *parabrahman*, which is realized in the wake of enlightenment. Different Hindu schools, the *Dvaita*, *Visishtadvaita*, *Yoga*, *Tantra*, and *Sakta*, uphold the same idea, with slightly different metaphysics. The thread that runs through all forms of Hindu metaphysics and belief systems is the idea that God, ultimate reality, permeates all material manifestations, and hence that there is no fundamental antagonism between matter and spirit, world and God. Hindu enlightenment is seeing God in every bit of the world and experiencing the harmony of dualities.

In Hinduism and Indian philosophy, pluralism is not limited to the form and nature of the god of one's belief. The Hindu mind conceives pluralism as a method for thinking and experiencing the multidimensionality of reality. Indian epistemology and metaphysics are rich sources of the thinking–experiencing paradigm. Philosophy according to the Hindu view cannot be an alienated rational process, though much discussion goes into theories of knowledge. Why Hindu philosophy is primarily a wisdom tradition is explained by its idea of identity between knowledge and existence, *cit* and *sat*. Many dimensions of Truth, many ways of knowing it, and many modes of being it, are built into the Hindu psyche. Ethical priorities, logical efficacies, and metaphysical theories are all finally supposed to lead to a way of living, and transformation of attitudes, approaches, and experiences. Different sects of Hinduism more or less believe that the world is an extension of God, and hence worldly activity is not opposed to religious life. Faith is not opposed to reason, since they deal with different domains. Tools to realize the Truth can also be different, such as knowledge and devotion. At the same time, knowledge about God, self, and their relation (*jnana*) is complementary to one's practice of religion through love of God and humanity (*bhakti*). The striking feature of Hinduism as a religion is its ability to devise tools for coexistence and conflict resolution, to accommodate diverse views and practices, to see a joint enterprise between scientific thinking and spiritual living.

CROSSROADS

The four crossroads that lead to significant dialogues between Hinduism as a religion and current thinking in science are (i) the Hindu concept of knowledge and method, (ii) theories of causation, (iii) theories of consciousness, and (iv) conflict resolution. The crux of the science and spirituality dialogue seen from a Hindu perspective and the knowing–being nexus in Hinduism can both be explained with the help of these four points of intersection.

Beginnings of Knowledge and Method

Nature was the intimate setting for the Vedic people to do their 'science' and experiments in the wilderness. Knowledge, its origin and nature, dominates several discussions in the Upanishadic literature. This trend continues in the other schools of Hindu philosophy as well. But what unites the Vedic, Upanishadic, and classical schools of Indian thought in their concept of knowledge is that equal importance is given to a scientific pursuit of the knowledge of the 'outer' and 'inner' worlds. The outer and the inner are seen as twin realities of life, and progress for the Vedic people depended on how well they could include each other. On one side the Hindus presented pioneering findings in the field of physical sciences—the outer—such as astronomy, mathematics, chemistry, medicine, and metallurgy—and on the other they developed—the inner—a wisdom tradition. The Hindu ideals of love, compassion, and personal well-being make avenues for 'material' developments to meet with 'spiritual' progress in a common space for optimal development of the person.

First Signs

The earliest signs of an inquiry to trace the 'real', or the basic stuff which things are made of, can be found in expressions like, 'Who has seen that the boneless bears the bony when being born first? Where may be the breath, the blood, the soul of the earth? Who would approach the wise to make this inquiry?' (*Rig Veda*, 1. 164. 4).[4] This and similar verses indicate distinct metaphysical and epistemological routes to reality.

We could also find an inquiry that leads to linguistic, psychological, and transpersonal issues with certain finesse in some of the hymns. The Vedic concept of *rta*, close to the present-day English word 'rhythm', is a result of the recognition of a comprehensive and unifying principle by the Vedic people. Vedic sages recognize *rta* as the rhythm behind the structuring of the dynamic aspects of the universe. In the later part of the Vedic literature (*Samhitas* and *Brahmanas*), the superior nature of

---

[4] Quotations from the Vedas and the *Taittiriya Samhita* are taken from Muller (1978).

mind in relation to speech is recognized. Further, *Taittiriya Samhita* recognizes the limitations of both speech and mind to define the first principle. It says, 'Finite are the hymns, finite the chants, finite the ritual formulae, to what constitute *Brahman* however there is no end' (VII. 3. 1. 4).

Bending Knowledge to Realism and Tentativeness

The Upanishadic theories of creation, the Jaina theory of multiple possibilities of existence, and the Nyaya theory of action fulfilment try to bend a structured concept of knowledge with realistic caution. Major schools of Hindu thinking deal extensively with the means (*pramaana*) of knowledge (*prama*) and validation (*praamaanya*). The concept of *pramaana* could initially be interpreted as a theory of knowledge, of ascertaining knowledge. But its function will not be completely understood without taking into consideration two characteristic features of knowledge as perceived by many of the classical schools of Indian thinking. These two characteristics *abhadhitatva*, of non-contradiction, and *anadhigatatva*, of novelty, lay down the condition for validating knowledge (Hiriyanna 1975). A knowledge statement is of questionable validation if there is another knowledge statement that contradicts the claim of the previous statement. Not being contradicted by another statement alone does not perform the role of validation. The characteristic of non-contradiction is also to be followed by the feature of novelty. Newness of knowledge is as important as non-contradiction in the ascertaining of knowledge. The feature of 'novelty' implies once again the epistemological openness evident in Indian thought.

Beginning and End of Creation

The two creation myths in the Vedic literature are: (i) time-space-event creation is an illusory projection of the transcendental Truth, and (ii) the experience of the world as 'other' is the result of self-ignorance. A significant hymn called *Nasadiya Sukta* says: 'Verily, in the beginning this (universe) was, as it were. Neither non-existent nor existent, in the beginning this (universe), indeed, as it were, existed and did not exist: there was then only that Mind' (*Sathapatha Brahmana*, x. 5. 3. 1).

This hymn tries to mark the boundaries of a conceptual categorization of creation in terms of cause and effect. It says that the wholeness which can only be pointed to as 'That one' is the ground of all existence—*sat*—and, by negation, non-existence is given the designation—*asat*. In the first verse of *Purusha Sukta*, reality is depicted as the *Virat Purusha*, or Cosmic Person, pervading the whole universe but still beyond it.

The Upanishads do not speak of a unitary, divine principle that is opposed to the multiplicity of creation. The 'transcendence' that the Upanishads highlight does not signify an aloofness or exclusion. The Upanishadic ideas of immanence and transcendence, creation and creator, can be understood only in the context of theories of consciousness and self. The Rishis expound consciousness as the ultimate reality, and identify it with the Self. The Upanishadic Rishi considers any doctrine on creation or causality as a myth created to explain mystery. The captivating narratives, mostly based on ecopsychological principles, try to break the built-in linearity in causality theories. Carl Sagan says:

The Hindu religion is the only one of the world's great faiths dedicated to the idea that the Cosmos itself undergoes an immense, indeed an infinite, number of deaths and rebirths. It is the only religion in which the time scales correspond to those of modern scientific cosmology. Its cycles run from our ordinary day and night to a day and night of Brahma, 8.64 billion years long. Longer than the age of the Earth or the Sun and about half the time since the Big Bang. And there are much longer time scales still.... The most elegant and sublime of these is a representation of the creation of the universe at the beginning of each cosmic cycle, a motif known as the cosmic dance of Lord Shiva. The God, called in this manifestation Nataraja, the Dance King. In the upper right hand is a drum whose sound is the sound of creation. In the upper left hand is a tongue of flame, a reminder that the universe, now newly created, with billions of years from now will be utterly destroyed. (Sagan 1980: 32)

From one angle of explanation—that is, the non-existence of anything prior to creation—we find Upanishadic references like 'There was nothing whatsoever here in the beginning' (*Brhadaranyaka Upanishad*, I. 2. 1); 'Non-existence verily was this (world) in the beginning' (*Taittiriya Upanishad*, II. 7. 1); 'In the beginning this (world) was non-existent' (*Chandogya Upanishad*, III. 19. 1). From another angle of explaining creation, *Aitareya Upanishad* speaks of the creator's entrance into the body by the opening in the skull—*vidriti*. *Brhadaranyaka Upanishad* adds that the creator entered up to the very tip of the nails. The purpose of this entry was 'to become everything that there is' and for 'assigning names into the objects and the evolution of their functions' (*Chandogya Upanishad*, VI. 2. 1). *Taittiriya Upanishad* says, 'Having entered it, He became both the actual (*Sat*) and the beyond (*Tyat*), the defined and the undefined, both the founded and the non-founded, the intelligent and the non-intelligent, the real and the un-true. As the real It became whatever there is here' (II. 6. 1). The declaration of the student in *Taittiriya Upanishad*, 'I am food and the eater of food, too'—*aham annam...aham annaadah*—indicates that the chain of existence is essentially cyclic.

From the *Brahmanas* to the Upanishads we find a cosmology that, with a more consistent analysis of creation, reaches a psychology identifying the First Principle with consciousness and the Self. Ranade says: 'Existence is not existence, if it does not mean self-consciousness. Reality is not reality if it does not express throughout its structure the marks of pure self-consciousness. Self-consciousness thus constitutes the ultimate category of existence to the Upanishadic philosophers' (Ranade 1926: 270). A. L. Basham says in his book *The Wonder that was India*: 'The great and saving knowledge which the *Upanishads* claim to impart lies not in the mere recognition of the existence of Brahman, but in continual consciousness of it...Brahman is the human soul, is Atman, the Self' (Basham 1967: 252).

The later Hindu schools of philosophy approach the problem of causation, and creation in particular, in interestingly different ways, yet tied together by a common emphasis on the transformation of experience. The naturalistic tradition of *Samkhya*, the oldest Indian thought, is the basis for many developments in Hindu religion. This tradition avoids the problem of the independent existence of creator and creation by positing a somewhat complex existence of reality that has on the one side dynamic matter, and, on the other, passive spirit. For *Samkhya*, the universe owes its existence to the interaction of *prakrti* and *purusha*, the principles of materiality and consciousness. It is the presence of *purusha* that upsets the equilibrium of a yet unmanifest *prakrti* and kick-starts the evolutionary process of the world. *Samkhya* recognizes the mutual association of consciousness and matter as essential for creation. It is said in *Sarva-siddhanta Samgraha* that 'Through the association (of *prakrti*) with that (*atman*) possessed of consciousness there arises creation' (*Sarva-siddhanta Samgraha*, x, 15–16).[5] Metaphorically illustrated, the 'lame *purusha* cannot operate without the blind *prakrti*'. 'The association of the two, which is like that of a lame man and a blind woman, is for the purpose of Primal Nature being contemplated (as such) by the Spirit' (*Samkhya Karika*, 21).[6] To the question of how long creation subsists, *Samkhya* answers with the help of the famous analogy: When *purusha* has enjoyed all manifestations of *prakrti*, *prakrti* ceases to act. It is like 'a dancer [who] desists from dancing, having exhibited herself to the audience' (*Samkhya Karika*, 59).

Consciousness Leading Back to Self

Hindu theories of creation and cosmology are founded on certain central ideas concerning the self and consciousness. Hindu ideas not only about mind and matter, but also about God, self, death, well-being, and spiritual progress, bring a radically different perspective to the current discussions on consciousness.

A prominent contention in consciousness studies, which is popular as the NCC (neural correlate of consciousness), is that experience is much too complex to be comprehended by 'building-block' approaches. It is possible that segregated explanations of specific sensory functions would give us path-breaking knowledge about the working of some aspects of human mind and consciousness. But then, whether these explanations together will be sufficient to understand the intricacy and integral wholeness of human self and experience is a question that demands considerable attention. The 'binding problem'[7] of consciousness, which scholars are never tired of discussing, is not only the 'puzzle of conscious experience' (Chalmers 1995) but also the most evasive problem of the subjective self, the 'harder problem' (Menon 2001). The Hindu theories of consciousness focus on the subjective self.

---

[5] Quotations from the *Sarva-siddhanta Samgraha* are taken from Rangacharya (1983).

[6] Quotations from the *Samkhya Karika* are taken from Sastri (1973).

[7] 'Binding experiences' are how physical, discrete, quantitative neural processes and functions give rise to experiences that are non-physical, subjective, unitary, and qualitative.

Distinction between Mind and Consciousness

In the *Aitareya Upanishad* a distinction is drawn between consciousness as the real knower and mind as just another sense organ. The various functions that can be classified under the three categories of cognition, affection, and conation are enumerated. 'Perception, discrimination, intelligence, wisdom, insight, steadfastness, thought, thoughtfulness, impulse, memory, purpose, life, desire, control' (III. 1. 2) — all these are identified as the operative names of consciousness. A further point of psychological interest is the analysis of the cognitive act into the knower (*praajna*), the intellect (*prajna*), and cognition (*prajnana*).

Kena Upanishad starts with the psychological inquiry as to what must be regarded as behind the psychophysical functions: namely, thinking, breathing, speech, vision, and action. Why is it that the mind is able to think? Who regulates the vital breath? How is it that the mouth, eye, and ear enable one to speak, see, and hear? Are sense perceptions autonomous, or is there an entity that underlies them? To these queries, the Rishis reply that it is consciousness that underlies all the psychic functions. But sense organs or the mind cannot know it. Consciousness is beyond not only what is known, but also what is unknown. It is beyond the reach of knowledge as well as ignorance. 'Upanishadic knowledge is renunciation of inferiority along with its vessel; it is transcendence of the very condition of inferiority' (Grinshpon 2003). Self-knowledge is a key concept in the Upanishads. We find a frequently occurring refrain, '*yo evam veda*—'one who knows thus'. Upanishadic psychology presents the Self as the pure subject which never becomes an object.

A definition of consciousness is 'that which reveals by itself through every act of cognition'—*prati bodha viditam* (*Kena Upanishad*, II. 4). Consciousness is the innermost subject, which makes cognitions and experiences possible and hence by itself cannot be explained using these. *Taittiriya Upanishad* and *Kena Upanishad* say, 'Whence words return along with the mind, not attaining it' (II. 4. 1); 'There the eye cannot go, nor can speech reach' (I. 3). *Brhadaranyaka Upanishad* says, 'You cannot see the seer of seeing, you cannot hear the hearer of hearing, you cannot think of the thinker of thinking, you cannot understand one who understands understanding' (III. 4. 2). The Upanishads desist a categorical definition of consciousness. On this Upanishadic style, Deussen remarks: 'The opposite predicates of nearness and distance, of repose and movement are ascribed to Brahman in such a manner that they mutually cancel one another and serve only to illustrate the impossibility of conceiving Brahman by means of empirical definitions' (Deussen 1906: 149).

The Inside World of Experiences

There are several verses in the *Brahmanas* that imply the quest for the source of knowledge and experience (Menon 2002*a*). From the origins of Indian philosophy to the classical schools and the works of later savants, the focus in Indian philosophy and wisdom traditions has not been on outside diversity, which one then artificially works to bring into a unity. Rather, the goal has been to discover intuitively a unity and then work towards diversity. This is the case even if we consider the most realist schools.

Epistemological analysis, in Indian thought, is subservient to experiential paradigms. Indian schools of thought, in general, have one common thread—to relate to a larger, deeper, and holistic concept or entity called 'self'. Whether it is for affirmation or denial, they expend considerable reflection and analysis in order to form a philosophy about 'self'. Both analysis (which is a structured, 'leading-to-the-next-step' kind of hierarchical thinking) and experience (as a set of 'given' or self-evident data) are used as epistemological tools in an integral manner to form distinct but interrelated ontologies. Metaphors and images are used as epistemological tools for creating transcendence in thinking, and thereby experiencing. The aim is not to arrive at structured and classified/listed knowledge of the 'other' object or phenomenon, but to gain understanding in relation to an abiding entity whether it be the inner self/no-self/or outer matter.

Neuroscience, Meditation, and Spiritual Altruism

Recently the discussions on synaesthesia[8] and meditation have gained a renewed interest, and journals are devoting many pages to them. Ideas from the corpus of Hindu philosophical and psychological literature which led to 'transcendental meditation' and meditation research two decades ago are gaining centre stage today in the effort to understand the nature of synaesthesia and how much of synaesthetic experience can be simulated. Another area focuses on the current discussions on the role of altruism in sociobiology. The major discussions on altruism that we follow today (especially in the context of sociobiology) give exclusive attention to altruism *as an act* favouring evolutionary or social benefits. What is almost always neglected is the fact that altruism is a phenomenon exhibited by *a self*. It is important to understand what exactly constitutes the 'self-space' that links the various levels of altruistic behaviour in order to know why altruism is discussed at all. We have at least three questions from the Hindu point of view: (i) Is there a rationale for altruistic expressions and behaviours? (ii) Is there an 'emotionale' for altruistic expressions and behaviours? And (iii), What drives altruistic perceptions and behaviours?

The methodological exclusivity granted to altruistic *acts* will not only land us in an artificial epistemology; it also supplies too limited a framework for one to identify

---

[8] Synaesthesia (also spelled synesthesia); from the Greek *syn-* 'union', and *aesthēsis*, 'sensation', is the neurological mixing of the senses. A synaesthete may, for example, hear colours, see sounds, and taste tactile sensations. Although considered a symptom of autism, it is by no means exclusive to those with autism. Synaesthesia is a common effect of some hallucinogenic drugs such as LSD or mescaline. See <http://en.wikipedia.org/wiki/Synaesthesia> and the 2005 issue of *Journal of Consciousness Studies* (12/4–5).

qualitatively the advantages of altruism. To limit altruism to psychological hedonism is to blindfold oneself with respect to the complexity of this phenomenon. It is essential to look at the psychological process whereby we are able to go from a drive for our own pleasure to a belief that we will gain this pleasure by benefiting others (Ablondi 1996). For, if all humans are naturally self-preservationists, then each person's possibility for the fulfilment of his or her self-interests is reduced by the very act of altruism. This could offer a response to Schlick, who said that the processes whereby the general welfare becomes a pleasant goal are 'complicated . . . [and] take place chiefly in the absence of thinking' (Schlick 1939: 417, 424). Even from a utilitarian point of view, the only way to maximize the possibility for fulfilment of self-interest is by being altruistic to another. The Hindu ideal of *lokasamgraha*—the uplifting of all—initiates such a meaning. In the famous dialogue between Maitreyi and her husband and Saint Yajnavalkya, Maitreyi asks 'what is that is most endeared and for whom'? Yajnavalkya, connecting a bare nerve of individual survival to a larger concept of self-identity, responds that 'it is for oneself that everything becomes endearing',[9] meaning that one cannot have purist theories of either self-survival *or* Self-survival.[10]

Hindu ways of thinking and notions about self focus on altruism in the context of *selflessness* and *self-space*. The reason I wish to qualify the nature of selflessness and self-space defined by the Hindu psyche as *spiritual altruism* is essentially related to the emphasis on selflessness in Hinduism as a *state of being*. It is directly connected with the transformation of consciousness, but also influences compassion, empathy, and ideas of the social good.

Conflict Resolution

There are two major paradigms in Hindu ways of thinking, in spite of otherwise great differences among metaphysical and epistemological positions. These are (i) what we actually see and experience, which is constituted by the given and the immanent, and (ii) what we could see and experience, which is constituted by future possibilities and the transcendent (Menon 2003). It is within these two paradigms that the elaborate and detailed discussion of fundamental experiences such as pain and pleasure, sorrow and happiness, selfishness and selflessness, freedom and bondage, the given and the possible, etc., takes place. Hinduism and its philosophy are an attempt to bridge these seemingly two contradicting paradigms, through an exploration of the self, based on systematic discussions of (i) theoretical, (ii) experiential, and (iii) transcendental issues.

---

[9] See <http://www.swami-krishnananda.org/brdup/brhad_II-04.html> for the dialogue between Yajnavalkya and Maitreyi on the absolute Self.

[10] *self* is a well-defined and exclusive identity (*jiva*), and *Self* is a growing, inclusive identity (*atma*).

In the history of humankind, and especially in the new century, the dominant forces that will lead and guide humanity will be the pursuit of knowledge and the need for the coexistence of all life forms. More than at any time in the past, religion and science will be the major collaborators as well as competitors in this common pursuit. The primary question will therefore be how to leverage each other's strengths and minimize weaknesses for the sake of humanity. The Hindu mind, which recognizes the given and the possible as two complementary paradigms for progress, can help to resolve the conflict by introducing a third powerful yet 'unassuming' force: that is, spiritual well-being. What is central to the Hindu concept of such well-being is its soul-centredness and the ability to give up what will be proved to be detrimental and inadequate for healthy coexistence. The much misinterpreted Hindu concept of *maya* proposes a new avatar, so as to provide a sound metaphysical and phenomenological explanation for the passing and apparently real world. *Maya* has often been castigated as a pessimistic concept describing the spatio-temporal world as worthless and illusory. The growing interest in the ideas of quantum entanglement and multiple possible worlds by quantum physicists might provide a welcome note for the dynamic and positive interpretations of maya, which hold that the world is 'real while experiencing, but not independently'.

The junctions and meeting points of discussions on theoretical, experiential, and transcendental issues, I think, will be at centre stage for Hinduism in the emerging science and religion dialogues. The key Hindu pointers will include an emphasis on ethical and spiritual discipline as a prerequisite for deeper self-knowledge and new experiences, a change in self-knowledge which changes the understanding of the given, and a reorientation of experience in order to allow for new responses to emerging situations.

What distinguishes the Indian way of thinking from what we today call the Western way of thinking is the wholesome connection present in the Hindu world between theoretical, experiential, and transcendental issues. It is also this distinguishing feature of Indian thinking that is often dismissed as 'mystic' and 'otherworldly'. The important point missed by such dismissals is that what interested Indian thinking and its guiding principle in the ancient past, classical period, and modern times, is not the linearity and immediate convenience that is provided by rigid, reductionistic structures of knowledge. Rather, Indian thinking has been characterized by an open-endedness in which experience and reflection can together bring about a reorientation of how we construe our self-identities, so that we can respond innovatively to changing situations. This has turned out to be a most difficult challenge for science in particular. Because of their implications for self-identity, the new findings in the sciences of nanomaterials, string theory, stem cell research, brain studies, and evolutionary biology will face religious and cultural disapproval.

How to understand and reorient emerging human identity will rely, to a great extent, on the theories of well-being and transformation of consciousness that we affirm and practise. The works of Swami Vivekananda (1863–1902) and Sri Aurobindo (1872–1950) reflect some of these concerns and offer plausible responses.

Just as we have physical and moral laws, so there are spiritual laws as well. Swami Bodhananda enumerates seven Hindu spiritual laws, 'in order to organize our lives for greater good and to achieve greater happiness' (2004: p. xiii). These laws deal with consciousness/*brahman*, the world of experiences/*maya*, individual identity/*dharma*, individual responsibility/*karma*, social interactions/*yajna*, detached engagement/*yoga*, and effortless work/*lila*.

Guidelines for Discourse and Dialogue

Tarkasamgraha, a widely used premier on Indian logic and dialectics, talks about the fourfold preliminary (*anubandha catushtaya*), the set of requirements for a discourse that may bring about meaningful results. The fourfold preliminary for any discourse are elucidations of *vishaya* (the theme of the discourse), *prayojana* (its major goal), *sambandha* (the relation between the theme of the discourse and its goal), and *adhikari* (the qualified participant).

The practice of specifying the objective and subjective guidelines for a discourse is also found in the foundational texts of Vedanta and *Mimamsa*. The starting verse of the text specifies the nature of inquiry such as for *brahman*, *dharma*, etc. Specifying the defining characteristic of a discourse avoids any doubt that might later ensue about what it is that guides the discourse. The thematic specification of the discourse also helps the student to have a clear picture of what the discourse will *not* talk about and what themes are restricted. Even if the theme of the discourse is known prior to entering into the discourse, the discussion could at some point raise the question of teleology in the mind of the student. Hence the theme, as well as the purpose of a discussion on such a theme, is specified initially. Though it could be a meta-question outside the scope of the discourse, it is essential also to anticipate at least to some extent the relationship between the discourse and the theme of the discourse itself. This would enable one to understand how far the treatise or discourse is representative of the theme.

The final and most important preliminary factor for any discourse is to specify who is qualified to enter that particular discourse. This is a major rule for meta-discourse, which, I think, is almost forgotten in the current discussions of a complex theme like 'consciousness'. The recognition of the aptitude of the person as playing a vital role in the success of discourse and understanding implies the ever-present subjective factor involved in epistemological enterprises. It also implies that under-standing is always finally related to the basic aptitude of the scientist or philosopher, which once again anticipates the essential relationship between epistemology and phenomenology, knowledge of something, and its experience. One instance of expounding the nature of *adhikari* can be seen in the primal text of *Advaita*, *Tattvabodha*, where Sankaracarya talks about the four qualifications required for a pursuit of self-knowledge—'*sadhana catushtaya*'. This includes a distinction between the ontologies of the real and the unreal, a detached view of worldly pleasures, practise of ethical and psychological discipline, and an intense yearning to know the beyond and to be it. These guidelines are expected to resolve competing interests and clarify foundational issues. Most importantly, they serve to achieve 'transcendence in and while thinking' (Menon 2002*a*: 26).

The foundational issues, crossing the rigidity of being theoretical, experiential, or transcendental, which are embedded in the Hindu religion and philosophy are (i) about human mind, consciousness, and experience, and (ii) about self-identity. The guidelines for the exploration of these embedded issues are (i) abstraction: to identify the unitary in the discrete; (ii) placeability: to have an ontological meaning for any experience, its object, and its experiencer; (iii) practice: to have values and discipline as essential guidelines for self-exploration. The role of Hinduism in fostering science and spirituality dialogues is to be placed in this context.

Saints, Science, and Spiritual Quest

The dawn of neo-Hinduism, inspired by saints and social leaders of India like Raja Ram Mohan Roy, Dayananda Saraswati, Swami Vivekananda, Mahatma Gandhi, Sri Aurobindo, Ramana Maharshi, and others has brought to light a uniting force of spiritual quest. Their teachings reiterated the connections between theories of creation, cosmology, and consciousness and theories of self, human identity, and spiritual well-being.

The dominant thoughts and views of Hinduism as a religion and philosophy time and again imbibe the ideas and visions of its savants, who appear at different historical times as poets, spiritual gurus, political leaders, mystics, and so on. The influence of Rabindranath Tagore on Bengal renaissance, his dialogues with Einstein; dialogue between David Bohm and J. Krishnamurthy, and Ramana Maharshi and Paul Brunton, are some instances of this process. In contemporary times, Swami Sivananda, Swami Tapovan, Swami Chinmayananda, Swami Ranganathananda, Maharshi Mahesh Yogi, Swami Bodhananda, and others, are particularly significant in initiating and contributing to significant dialogues and exchanges between science and Hinduism. Their teachings and views demonstrate that spiritual exploration is the midway between science and religion, especially evidenced by the past and present of Hindu religion.

Self-oriented thinking, nested narratives that complement rational processes by serving functions of complex explanations, the systems approach of Upanishadic Rishis and Hindu philosophers, and their tryst with temptations and death—all point to a dimension that could be metaphorically addressed as the 'inner'. The 'inner' chooses to reveal best at the junction points of science and religion. Hindu philosophy identifies such meeting points as points of transcendence and inclusion. Hinduism as a living and growing religion is therefore based primarily on an active and positive interpretation of karma theory, of interpreting and integrating challenges of the present.

A great danger in the golden age of dialogues based on spiritual meeting points is the hasty and immature appropriations of different domains of which science and religion are both sometimes guilty. When religion tries to claim that the current trends, theories, and findings of science existed already in ancient thought, forgetting the historicity and cultural specificity of religion, the way is unfortunately paved for primitive competitions rather than for the healthy pursuit of knowledge. Likewise, when science tries to dispossess religion of its central role in defining new meanings for human identity and spiritual well-being, it critically impinges on our ability to conceptualize the very meaning of existence and survival—that is, human imagination and the pursuit of something still beyond. To see, understand, experience, and be one with the beyond is what inspires the Hindu mind. The points of junction that Hinduism identifies for dialogues that lead to advancement in knowledge and well-being are consciousness, agency, and self-identity. Today these three domains are central to both science and spirituality, and each seeks points of convergence in order to contribute to the greater meaning and enrichment of survival and evolutionary advance. It is hoped that such a joint exploration will ensure human progress, healthy coexistence, and spiritual well-being.

REFERENCES AND SUGGESTED READING

ABLONDI, FRED (1996). 'Schlick, Altruism and Psychological Hedonism', *Indian Philosophical Quarterly*, 23/3–4: 417–24.

BASHAM, A. L. (1967). *The Wonder that was India*. Calcutta: Rupa and Co.

CHALMERS, D. (1995). 'The Puzzle of Conscious Experience', *Scientific American*, December 1995: 62–8; <http://consc.net/papers/puzzle.html>.

DEUSSEN, PAUL (1906). *The Philosophy of the Upanishads*, trans. A. S. Geden. Edinburgh: Morrison and Gibb Ltd.

GRINSHPON, YOHANAN (2003). *Crisis and Knowledge: The Upanishadic Experience and Storytelling*. Delhi: Oxford University Press.

HIRIYANNA, M. (1975). *Indian Conception of Values*. Mysore: Kavyalaya Publishers.

LEVINSON, DAVID (1998). *Religion: A Cross-Cultural Dictionary*. Oxford: Oxford University Press.

MENON, S. (2001). 'Towards a Sankarite Approach to Consciousness Studies: A Discussion in the Context of Recent Interdisciplinary Scientific Perspectives', *Journal of Indian Council of Philosophical Research*, 18/1: 90–117.

——(2002*a*). *Binding Experiences: Looking at the Contributions of Adi Sankaracarya, Tuncettu Ezuttacchan and Sri Narayana Guru in the Context of Recent Discussions on Consciousness Studies*. Shimla: Indian Institute of Advanced Study, Monograph published under the program Knowledge Dissemination Series.

——(2002*b*). 'The Selfish Meme and the Selfless Atma', *Sophia*, 41/1(May): 83–8.

——(2003). 'Binding Experiences for a First Person Approach: Looking at Indian Ways of Thinking (*darsana*) and Acting (*natya*) in the Context of Current Discussions on Consciousness', in Chhanda Chakraborti, Manas K. Mandal, and Rimi B. Chatterjee (eds.), *On*

Mind and Consciousness, Shimla: Indian Institute of Advanced Study, and Kharagpur: Indian Institute of Technology, 90–117.

MULLER, F. MAX (1978) (ed.). *Sacred Books of the East.* Delhi: Motilal Banarsidass Publishers.

NARASIMHA, R. (2003). *About the NIAS Emblem.* Bangalore: National Institute of Advanced Studies.

RANADE, R. D. (1926). *A Constructive Survey of Upanishadic Philosophy.* Poona: Bilvakunja Publishing House.

RANGACARYA, M. (1983) (trans.). *The Sarva Siddhanta-Sangraha of Samkaracharya.* New Delhi: Ajay Book Service.

SAGAN, CARL (1980). *Cosmos.* New York: Random House Inc.

SASTRI, S. KUPPUSWAMI (1932). *A Primer of Indian Logic: According to Annambhatta's Tarka-samgraha.* Mylapore: Madras Law Journal Press.

SASTRI, S. S. SURYANARAYANA (1973) (ed. and trans.). *The Sankhya Karika of Isvara Krsna,* 2nd edn. Madras: University of Madras.

SCHLICK, MORITZ (1939). *Problems of Ethics,* trans. David Ryan. New York: Prentice-Hall.

SWAMI BODHANANDA (2004). *The Seven Hindu Spiritual Laws.* Delhi: BlueJay Books.

SWAMI VIVEKANANDA (1991). *Chicago Addresses.* Calcutta: Advaita Ashrama Calcutta.

CHAPTER 2

BUDDHISM AND SCIENCE

B. ALAN WALLACE

INTRODUCTION

When reading an essay on Buddhism and science, it is natural to assume that Buddhism is a religion, together with Judaism, Christianity, and Islam, because our Western concept of religion has been modelled primarily on the basis of the three Abrahamic traditions. In the West we have developed separately the constructs of *science* and *philosophy*, as initially inspired by Greek and Roman modes of inquiry. Since Buddhism is one among many traditions of inquiry that arose outside the Mediterranean basin, there is no reason to expect it to fit neatly into any of the categories of religion, science, and philosophy that have been forged in the West. To understand what Buddhism brings to the dialogue between religion and science, it should be met on its own terms, without insisting that it conform to Western conceptual categories. Buddhism is both more and less than the sum of these three Western traditions of inquiry.

While Buddhism is often referred to as a 'non-theistic religion', a problematic characterization in many ways, it does have the potential to play a unique mediating role between theistic religions, with their emphasis on faith and divine revelation, and the natural sciences, with their ideals of empiricism, rationality, and scepticism. It may serve as a catalyst for reintroducing the spirit of empiricism in religion with respect to the natural world and in science with respect to spiritual realities and subjective experience in general. This might even lead to a science of religions that would earn the respect and trust of religious believers, scientists, and the public at large.

Religion is often regarded as addressing questions concerning the meaning and purpose of life, our ultimate origins and destiny, and the experiences of our inner life.

Moreover, we commonly deem a system of belief and practice to be religious if it is concerned primarily with universal and elemental features of existence as they bear on the human desire for liberation and authentic existence (Harvey 1981: ch. 8; Gilkey 1985: 108–16; Gould 1999: 93). Stated in such broad terms, Buddhism can certainly be classified as a religion.

Science may be defined as an organized, systematic enterprise that gathers knowledge about the world and condenses that knowledge into testable laws and principles. In short, it addresses questions of what the universe is composed of and how it works (Wilson 1998: 58; Gould 1999: 93). Buddhism is an organized, systematic enterprise aimed at understanding reality, and it presents a wide range of testable laws and principles, such as the propositions set forth in the Four Noble Truths (Dalai Lama 1997). Although Buddhism has not developed historically along the lines of Western science, it is a time-tested discipline of rational and empirical inquiry that could further evolve in ways more closely resembling science as we have currently come to understand it.

Furthermore, philosophy, as it is defined primarily within the context of Western civilization, consists of theories and modes of logical analysis of the principles underlying conduct, thought, knowledge, and the nature of the universe, and it includes such branches as ethics, aesthetics, logic, epistemology, and metaphysics. While there is a general consensus that scientific theories must be testable, at least in principle, by empirical observation or experiment, no such stipulation is made for philosophical theories. They may be evaluated on the basis of reason alone. Buddhism has from its origins included theories and modes of logical analysis of the principles underlying conduct, thought, knowledge, and the nature of the universe. So in this regard, Buddhism may be viewed as a philosophy, or—given the great range of theories within the Buddhist tradition—as a diverse array of philosophies.

While theistic religions are centrally concerned with transcendental realities, such as God, Buddhism is naturalistic in the sense that it is centrally concerned with the causality within the world of experience (Sanskrit: *loka*). Its fundamental framework is the Four Noble Truths, pertaining to the reality of suffering, its necessary and sufficient causes, the possibility of freedom from suffering and its causes, and the practical means for achieving such freedom. This basic structure of the Buddhist enterprise is pragmatic, rather than supernatural or metaphysical, so it bears only some of the family resemblances of Western religions.

While science has overwhelmingly focused on understanding the objective, quantifiable, physical universe in order to gain power over the natural world (Bacon 2004), Buddhism is primarily focused on understanding subjective, qualitative states of consciousness as a means to liberate the mind from its afflictive tendencies (*klesha*) and obscurations (*avarana*). Given the scientific focus on the outer world, the Western scientific study of the mind did not begin until more than 300 years after the time of Copernicus, whereas the rigorous, experiential examination of the mind has been central to Buddhism from the start. Buddhist theories are not confined to the Buddha's inquiries alone, but have been rationally analysed and experientially tested by generations of Buddhist scholars and contemplatives over the past 2,500 years (Wallace 2000: 103–18). Buddhist insights into the nature of the mind and related phenomena are presented as genuine discoveries in the sense that any competent practitioner with sufficient training can replicate them (though different kinds of training pursued within different conceptual contexts do lead to different, and sometimes conflicting, insights). They could thus be said to be *empirical* in the sense that they are based on immediate experience, but that experience consists primarily of first-person, introspective observations, not the third-person externalist observations more commonly associated with science.

In addition, many Buddhist writings are clearly philosophical in nature and can be cross-culturally evaluated as such (Bronkhorst 1999; Tillemans 1999). However, empirical or intellectual inquiry motivated simply by curiosity or knowledge for its own sake has never been a widespread Buddhist ideal. Unlike both Western science and philosophy, the Buddhist pursuit of knowledge occurs within the framework of ethics (*shila*), focused attention (*samadhi*), and wisdom (*prajña*). These comprise the essence of the Four Noble Truths, the path to liberation.

The main body of this chapter focuses on Buddhist approaches to cultivating eudaimonic well-being, probing the nature of consciousness, and understanding reality at large. In each case, religious, scientific, and philosophical elements are blended in ways that may not only lend themselves to dialogue with Western science, but push forward the frontiers of scientific research as well as interdisciplinary and cross-cultural inquiry.

The Buddhist Pursuit of *Eudaimonic* Well-being

Buddhist tradition identifies itself not in terms of the Western constructs of religion, science, and philosophy, but with the Indian notion of *dharma*. While this word takes on a wide variety of meanings within different contexts, 'Buddhadharma' refers to the Buddhist world-view and way of life that lead to the elimination of suffering and the realization of a lasting state of well-being. Such 'sublime dharma' (*saddharma*) is presented in contrast to mundane dharmas (*lokadharma*), which include the classic set of 'eight mundane concerns': namely, material gain and loss, stimulus-driven pleasure and pain, praise and ridicule, and fame and ill repute (Wallace 1993: ch. 1).

These two types of dharma correspond closely to two approaches to well-being studied in psychology today: hedonic and eudaimonic (Ryan and Deci 2001). The hedonic approach, corresponding to mundane dharma, is defined in terms of the pursuit of mental and physical pleasure and the avoidance of pain, whereas the eudaimonic approach, corresponding to sublime dharma, focuses on striving for the perfection that represents the realization of one's true potential (Ryff 1995: 100; Kahneman, Diener, and Schwarz 1999). Hedonic well-being includes pleasurable emotions and moods aroused by agreeable stimuli. I would argue that the evolutionary process of natural selection facilitates such happiness in the course of modifying living organisms so that they can survive and procreate. Eudaimonic well-being, on the other hand, appears to arise not as a result of natural selection, but primarily from practices of the kind Buddhists call *sublime dharma*.

A BUDDHIST MODEL OF SUFFERING

The sublime dharmas taught in Buddhism as a whole have as their principal aim the decrease and eventual complete liberation from suffering (*duhkha*), of which three levels are commonly identified: explicit suffering, the suffering of change, and ubiquitous suffering of conditionality (Tsong-kha-pa 2000: 289–92). *Explicit suffering* refers to all physical and mental feelings of pain and distress. The *suffering of change* refers not to unpleasant feelings, but to *pleasurable* feelings and mental states aroused by pleasant stimuli, as well as the stimuli themselves. It is so called because when the stimulus is removed, the resultant happiness fades, revealing the underlying dissatisfaction that was only temporarily veiled by the pleasant stimulus. The *ubiquitous suffering of conditionality* refers to the state of existence in which one is constantly vulnerable to all kinds of suffering due to the mind's afflictive tendencies. These include the 'three mental toxins' of craving, hostility, and delusion, which are fundamental sources of dissatisfaction. In short, the ground state of such an afflicted mind is suffering, even when one is experiencing hedonic well-being, and this is overcome only through the pursuit of eudaimonic well-being, in which all forms of suffering are ultimately severed from their root.

A BUDDHIST MODEL OF HAPPINESS

As a remedy to the above three-tiered model of suffering, Buddhists aim toward a similarly three-tiered model of happiness (*sukha*). The most superficial level of *sukha* consists of all forms of explicit pleasure that arise from pleasant chemical, sensory, intellectual, aesthetic, and interpersonal stimuli. Some of these are ethically neutral, such as the pleasure of eating sweets; some are ethically positive, such as the joy of performing an act of altruistic service, or taking delight in one's children's success;

and some are ethically malignant, such as taking satisfaction in another's misery. A second level of *sukha* consists of traits of eudaimonic well-being that arise from an ethical way of life and from exceptional states of mental health and balance. The highest level of *sukha* consists of the eudaimonic well-being resulting from freedom from all mental afflictions and obscurations and the complete realization of one's potentials for virtue. One who experiences such total freedom and realization is known as a *buddha*, literally 'one who is awake'.

Hedonic psychology is concerned with the avoidance of explicit suffering and the accomplishment of explicit happiness, and it measures the success of that approach in terms of the amount of happiness and suffering one experiences from day to day. The Buddhist pursuit of eudaimonic well-being, on the other hand, is primarily concerned with gaining freedom from the second and third levels of *duhkha* and realizing the second two levels of *sukha*. However, there is an asymmetry in the causal relation between hedonic and eudaimonic well-being. While the hedonic pursuit of stimulus-driven pleasures may or may not contribute to eudaimonic well-being and may actually interfere with it, the eudaimonic approach enables one to derive increasing pleasure from life in the midst of both adversity and felicity. The hedonic approach focuses on the short-term causes of stimulus-driven happiness, whereas the eudaimonic approach focuses on the long-term causes of well-being that arise from mental balance.

While hedonic well-being is contingent upon outer and inner pleasant stimuli, and is often pursued with no regard for ethics, the Buddhist eudaimonic approach begins with ethics, then focuses on the cultivation of mental balance, and finally centres on the cultivation of wisdom, particularly that stemming from insight into one's own nature. In this regard, eudaimonic well-being may be characterized as having three levels: social and environmental well-being stemming from ethical behaviour in relation to other living beings and the environment, psychological well-being stemming from mental balance, and spiritual well-being stemming from wisdom. These three elements—ethics, mental balance derived from the cultivation of focused attention, and wisdom—are the three 'higher trainings' that comprise the essence of the Buddhist path to awakening.

Ethics

The essence of the first training in ethics consists of the avoidance of injurious behaviour and the cultivation of behaviour that is conducive to one's own and others' well-being. While the topic of ethics in Western civilization is commonly a matter of religious belief or philosophical analysis, and has not been a focus of psychology, in the Buddhist tradition as a whole it is a practical, experiential matter that is at the very core of well-being. All of us are called upon to examine our own physical, verbal, and mental behaviour, noting both short-term as well as long-term consequences of our actions. Although some activity may yield immediate pleasure, if over time it results in unrest, conflict, and misery, it is deemed unwholesome (*akushala*). On the other hand, even if a choice of behaviour involves difficulties in the short term, it is regarded as wholesome (*kushala*) if it leads eventually to contentment, harmony, and eudaimonic well-being for oneself and others. This raises the possibility of ecological, sociological, and psychological research into the role of ethics, not in terms of religious doctrines or societal contracts, but with respect to the types of behaviours that impede and nurture our own and others' genuine well-being.

In Buddhist tradition, ethics is taught before introducing the second kind of training—meditative practices designed to reduce mental afflictions and enhance mental balance—for it has been found that without this foundation, such practices will be of little or no value. Indeed, they may aggravate pre-existing neuroses and other mental imbalances. Likewise, we can reflect upon the limited benefits of teaching people sophisticated therapeutic techniques to reduce depression, anxiety, or rage without exploring the effects of how they are leading their lives.

Mental Balance

While many environmental problems and social conflicts stem from unethical behaviour, according to Buddhism most mental suffering is due to imbalances of the mind to which virtually all of us are prone. A person whose mind is severely imbalanced is highly vulnerable to all forms of *duhkha*, including anxiety, frustration, boredom, restlessness, and depression. These are some of the symptoms of an unhealthy mind, and Buddhists claim that the underlying problems can be remedied through skilful, sustained mental training (Gethin 2001). On the other hand, just as a healthy, uninjured body is relatively free of pain, so a healthy, balanced mind is relatively free of psychological distress.

This is the point of the second phase of Buddhist practice, a key element of which is the cultivation of focused attention (*samadhi*). The training in *samadhi*, however, refers to much more than the development of attentional skills. More broadly, it includes (1) *conative balance*, or the cultivation of desires and intentions conducive to eudaimonic well-being (Tsong-kha-pa 2000); (2) *attentional balance*, including the development of exceptional attentional stability and vividness (Gunaratana 1991; Lamrimpa 1995; Wallace 2005a); (3) *cognitive balance*, including the application of mindfulness to one's own and others' bodies, minds, and the environment at large (Nyanaponika Thera 1973; Gunaratana 1991); and (4) *affective balance*, in which one's emotional responses are appropriately measured and conducive to one's own and others' well-being (Goleman 1997, 2002; Davidson *et al.* 2005; Nauriyal 2005; Wallace 2005b).

A basic hypothesis of Buddhism is that to the extent that the mind loses its balance of any of the above four kinds, its ground state, prior to any chemical, sensory, or conceptual stimulation, is one of *duhkha*, or dis-ease. In response to such dissatisfaction, there are two major options: (1) to follow the hedonic approach of smothering the unpleasant symptoms of these fundamental imbalances; (2) to adopt the eudaimonic approach of getting to the root of these symptoms by cultivating mental balance. Modern society has provided us with a plethora of means to stifle unhappiness, from mood-altering drugs to sensory bombardment, to extreme sports. The more the mind is in a state of imbalance, the more intense the stimuli it requires to smother its internal unrest.

According to Buddhism, no pleasurable stimuli are true sources of happiness in the sense that an artesian well is a source of water and the sun is a source of heat. If they were, we should experience happiness whenever we encounter pleasurable stimuli, and the degree of our happiness should be directly correlated to the intensity and duration of our contact with those stimuli. While sensory experiences, attitudes, other people, and situations seem to 'make us happy', in fact the most they can do is *contribute* to our well-being; they cannot literally deliver happiness to us. The only way, according to the Buddhist hypothesis, to achieve eudaimonic well-being is to balance the mind, and to the extent that this occurs, one discovers a sense of well-being from within, which lingers whether one is alone or with others, active or still.

We turn now to the third element of Buddhist practice, the cultivation of wisdom, particularly through the investigation of consciousness.

Buddhist Science of Consciousness

In his classic work *Science and Civilization in China* (1956) Joseph Needham explored the historical reasons why the civilizations of China and India never developed science as we understand it in the modern West: namely, a quantitative, technologically driven science of the outer, physical world. Similarly, one may ask why Western civilization has never developed a science of consciousness (Whitehead 2004), in which a consensus is reached regarding the definition of consciousness and means are devised to examine directly the nature of consciousness, as well as its necessary and sufficient causes and its causal influences. Buddhism, I maintain, has developed such a rational and empirical discipline of inquiry.

I shall begin by outlining a hierarchy among the natural sciences, showing both the strengths and weaknesses of modern science. While the physical sciences rely heavily on quantitative analysis, axioms of mathematics do not define, predict, or explain the emergence of the physical universe. Isaac Newton modelled the physical laws presented in his *Mathematical Principles of Natural Philosophy* on the axioms of geometry, but his discoveries would have been impossible without careful observations of celestial and terrestrial physical phenomena. Likewise, the current laws of physics alone do not define, predict, or explain the emergence of life in the universe. Biologists needed to develop their own unique modes of observing living organisms, such as Darwin's studies on the Galapagos Islands, as a basis for defining and explaining the emergence and evolution of life in the universe. Similarly, the laws of biology alone do not define, predict, or explain the emergence of consciousness in living organisms; nor is consciousness detected by the instruments of biology. Given the pattern of the physical and life sciences, it follows that cognitive scientists must also devise sophisticated, rigorous means of directly observing mental phenomena as a basis for defining and explaining the origins and nature of consciousness. Galileo refined the telescope and used it to make precise observations of celestial phenomena, and Van Leeuwenhoek used the microscope to make precise observations of minute living organisms. But cognitive scientists have failed to devise a methodology for making reliable, direct observations of the whole spectrum of mental phenomena themselves, which can be made only from a first-person perspective, as I shall discuss below.

William James, a great pioneer of American psychology, proposed that psychology should consist of the study of subjective mental phenomena, their relations to their objects, to the brain, and to the rest of the world. To develop this scientific study of the mind, he proposed a threefold strategy: mental phenomena should be studied *indirectly* through the careful observation of behaviour and of the brain, and they should be examined *directly* by means of introspection. Among these three approaches, he declared that for the study of the mind, '*Introspective Observation is what we have to rely on first and foremost and always*' (James 1890/1950: i. 185). Much as the theories of Copernicus, Darwin, and Mendel were largely ignored for decades after their deaths, so this threefold strategy of James has been discarded for the most part, while behaviourism, cognitive psychology, and neuroscience have dominated the cognitive sciences. The current means of observing mental phenomena directly has not achieved the level of sophistication of the behavioural and neurosciences, so, in this regard, James's comment that psychology today is hardly more than what physics was before Galileo still retains a high degree of validity (James 1892).

There are certainly problems in incorporating introspection—a first-person, qualitative mode of inquiry—into the framework of science, which is centred upon third-person, quantitative methods. Indeed, there have been examples in Western psychology of employing inadequately developed methods of self-reporting that were never able to clarify general principles for understanding mental functions (Danziger 1980). However, these problems may be surmounted by improving the necessary skills for making precise, reliable, introspective observations. Another reason why first-person observation has been so neglected since the time of James is the neuroscientific interest in identifying the *mechanisms* underlying mental processes. Despite this focus, cognitive scientists have yet to identify *any* mechanism that explains how neural processes generate or even influence subjectively experienced mental processes, or, conversely, how mental events influence the brain. They *have* succeeded in identifying the neural *correlates* to specific perceptual and conceptual processes, but the exact *nature* of those correlations remains a mystery. A widespread assumption among cognitive scientists is that neural and mental processes are actually flip sides of the same coin, but this belief has yet to be validated by either empirical evidence or rational argument. All we really know is that specific kinds of neural events are necessary for the generation of specific kinds of mental processes. That hardly amounts to a proof of identity.

In light of the history of science, this insistence on identifying the mechanisms of mental processes may be premature. From the time when Newton identified the natural laws of gravity in 1687, it was 228 years before the mechanism of gravity was explained in Einstein's General Theory of Relativity. Likewise, for the laws of natural selection, a century passed after the publication of Darwin's *On the Origins of Species* in 1859 and Gregor Mendel's formulation of his theories of genetics in 1865 before James Watson and Francis Crick were able to model the structure of DNA. And according to quantum theory, which is commonly cited as the most successful of all scientific theories, no mechanisms have yet been found to explain such phenomena as non-locality, the uncertainty principle, or the collapse of probabilistic wave functions.

It is quite possible that no mechanisms will ever be found to explain the causal interactions between neural and mental events, but this should not deter scientists from developing rigorous methods for observing mental phenomena in the only way possible: through first-person, introspective observation in conjunction with careful observation of behaviour and of the brain, as James proposed more than a century ago.

The physical sciences have undergone two revolutions: the Copernican revolution and the twentieth-century revolution of relativity and quantum theory. The biological sciences have witnessed one revolution, beginning with Darwin and culminating in the Human Genome Project. The cognitive sciences have achieved no similar radical shift in their understanding of mind or consciousness. The basic assumptions about the mind and its relation to the brain that were common in the late nineteenth century remain unchanged and largely unchallenged to this day. Although great advances have been made recently in measuring neural correlates of mental phenomena, it is far from clear whether these objective measures will ever reveal the nature of those correlations and therefore the nature of mind–brain interactions or consciousness itself.

Despite the West's failure to bring about a revolution in the cognitive sciences, it would be hasty to assume that no other civilization has revolutionized the scientific study of the mind. Much as Galileo refined the telescope and used it in unprecedented ways to directly observe celestial phenomena, so the Buddha refined the practice of *samadhi* and used it in unprecedented ways to explore states of consciousness and their objects (Ñāṇamoli 1992). As a result of his own experiential explorations, he came to the conclusion: 'The mind that is established in equipoise comes to know reality as it is' (Kamalaśīla 1958: 205). While such introspective inquiry may seem more philosophical than scientific, consider the definition of the scientific method as 'principles and procedures for the systematic pursuit of knowledge involving the recognition and formulation of a problem, the collection of data through observation and experiment, and the formulation and testing of hypotheses' (*Webster's Ninth New Collegiate Dictionary*). There is nothing in that definition that insists on third-person observation or quantitative analysis, especially for phenomena that are irreducibly first-person in nature (Searle 1994).

THE PSYCHE

Derived from exactly this kind of exploration, three dimensions of consciousness may be posited on the basis of contemplative writings common to the Mahayana Buddhist tradition (which emerged around the beginning of the Christian era). The first of these is the psyche (*chitta*)—the whole array of conscious and unconscious mental processes that occur from birth to death. In Buddhism the primary reason for exploring the psyche is to identify and learn to overcome the afflictive mental processes that generate suffering internally. This is the central theme of the Four Noble Truths and the Buddhist pursuit of liberation.

A thorough understanding of the human psyche must include insight into its origins. The vast majority of contemporary cognitive scientists assume, often unquestioningly, that the brain is solely responsible for producing all mental processes. The uniformity of this view is remarkable in light of the fact that scientists have yet to identify the neural correlates of consciousness or its necessary and sufficient causes (Searle 2002: 49–50; Searle 2004: 119). Researchers in the field of artificial intelligence question whether a carbon-based brain is necessary for the generation of consciousness, and there is no scientific consensus regarding its sufficient causes. The belief that the brain is solely responsible for all states of consciousness stems immediately from the metaphysical principles of scientific materialism, which dominate most scientific thinking today, much as Roman Catholic theology dominated and constrained intellectual life during the time of Galileo (Wallace 2000).

SUBSTRATE CONSCIOUSNESS

Through the development and utilization of highly advanced stages of *samadhi*, which remain unexplored by science, contemplatives in the 'Great Perfection' (Dzogchen) tradition of Indo-Tibetan Buddhism claim to have discovered a second dimension of consciousness: a continuum of individual mental awareness that precedes this life and continues on beyond death, which they call the *substrate consciousness* (*alayavijñana*) (Wallace 1996: ch. 23; Düdjom Lingpa 2004: 31 and 68; Wallace 2000a: 77–8 and 164–6). This relative ground state of the mind is characterized by three qualities: bliss, luminosity, and non-conceptuality. It is most vividly apprehended by meditatively enhancing the stability and vividness of attention, but it naturally manifests in deep sleep and in the dying process.

The human psyche, the first dimension of consciousness mentioned above, emerges, they conclude, not from the body but from this underlying stream of consciousness that precedes species differentiation. While the body *conditions* the mind and is necessary for specific mental processes to arise as long as the substrate consciousness is embodied, the psyche *emerges* from this underlying stream of consciousness that is embodied in life after life. This theory is compatible with all current scientific knowledge of the mind and the brain, so there is nothing illogical about it; nor is it simply a faith-based proposition as far as advanced Buddhist contemplatives are concerned. Scientific materialists, however, insist that mental phenomena emerge solely from the brain, much as bile is secreted from the gall bladder (Searle 2002: 115). What they commonly overlook, though, is that mental phenomena, unlike all other emergent phenomena known to science, cannot be observed by any objective, scientific means. So this assertion is a metaphysical assumption, not an established scientific fact. Something that is purely a matter of religious faith or philosophical speculation as far as scientists in the West are concerned may be an experientially confirmed hypothesis for contemplatives in the East. The demarcation between science and metaphysics—between theories that can and cannot be tested empirically—is determined by the limits of experiential inquiry, not Nature or God.

Thus far, experiential inquiry in science has been confined largely to the exploration of the objective world by way of our five physical senses and the instruments of technology. Mental phenomena themselves, as opposed to their neural and behavioural correlates, are invisible to such objective modes of observation. So, to this day, cognitive scientists have yet to come to a consensus regarding the definition of consciousness; they have no objective means of detecting the presence of consciousness in anything; they have failed to identify even the neural correlates of consciousness, and therefore remain in the dark regarding the necessary and sufficient causes of consciousness. All this suggests that mental phenomena are irreducibly first-person phenomena, and that the only way to restore a true sense of empiricism to the scientific study of the mind is to acknowledge the primary role of introspective observation.

A major reason for the resistance on the part of many scientists to including introspection as a legitimate method of empirical inquiry is that it is quintessentially a private, first-person kind of experience. Scientific inquiry, on the other hand, has achieved its great successes by way of public, third-person observations. It is important to note that these advances in scientific knowledge have focused primarily on objective, quantifiable, physical processes, while conscious mental processes are subjective, qualitative, and invisible to the physical means of observation developed by science. But now, with the recent development of sophisticated psychological and neurophysiological methods of inquiry, the first-person methods of introspection (based on the development of advanced stages of *samadhi*) may be cross-checked with the third-person methods of the cognitive sciences in ways that may expand the horizons of both scientific and contemplative inquiry.

The Buddha claimed to have gained direct knowledge of this continuity of individual consciousness beyond death, as well as direct knowledge of the patterns of causal relationships connecting multiple lifetimes (Ñāṇamoli 1992: 23–6). Many generations of Buddhist contemplatives throughout Asia claim to have replicated his discoveries, so such reports are not confined to the testimony of one individual. From a third-person perspective, all such discoveries based on introspective inquiry remain anecdotal, so only practitioners have 'proof' of their validity. As such, they are accessible only to a privileged few, but this has always been true of many of the most profound scientific truths. It takes years of training to become a qualified 'third person' capable of testing others' alleged discoveries in any advanced field of science. They have never been testable by the general public, who often take them on faith, much as religious believers take on faith the claims of their church. The Buddhist training in *samadhi* required to gain experiential access to the substrate may easily take 10,000–20,000 hours—comparable to the time required for graduate work in science—and until now, such professional training has never been available to cognitive scientists.

Particularly in the Tibetan Buddhist tradition, for centuries there has been keen interest in identifying children who were allegedly accomplished meditators and teachers in their past lives. This has commonly been done by seeking out children who appear to remember their past-life experiences, and scientific research into such instances has also begun (Stevenson 1997). Most cognitive scientists have refused to consider any theory of reincarnation, insisting that it cannot belong in a scientific dialogue *per se.*

While there does not appear to be any neuroscientific means of disproving the hypothesis that the brain is necessary for all states of consciousness, few scientists have expressed concern over the non-scientific nature of their fundamental assumptions about the mind–body problem. Similarly, the Buddhist hypothesis of the substrate consciousness does not easily lend itself to scientific repudiation; but scientific inquiry, with a suspension of disbelief, should first be directed to examining whether any positive evidence exists, before worrying about whether it can be repudiated.

Indirect evidence may be provided by third-person methods, such as the field studies of Ian Stevenson and his scientific successor Jim Tucker (2005). The quantitative, objective tools of observation of science provide no immediate access to any kind of mental phenomena, so they are not likely to reveal any evidence for the substrate consciousness. This can come only from rigorous, first-person methods such as those proposed by the Buddhist tradition. Just as the existence of the moons of Jupiter can be verified only by those who gaze through a telescope, so the existence of subtle dimensions of consciousness can be verified experientially only by those willing to devote themselves to years of rigorous attentional training. And dedication to such refinement of attention is not contingent on accepting the hypotheses of Buddhism or any other contemplative tradition beforehand.

Primordial Consciousness

There is yet a third dimension of consciousness, known as primordial consciousness (*jñana*), or the Buddha-nature (*buddhadhatu*) (Ruegg 1989; Thrangu Rinpoche 1993;

Dalai Lama 2000). This is regarded in the Mahayana Buddhist tradition as the ultimate ground state of consciousness, prior to the conceptual dichotomies of subject and object, mind and matter, and even existence and non-existence. This realm of consciousness is described metaphorically as being space-like and luminous, forever unsullied by mental afflictions or obscurations of any kind. The realization of this state of consciousness is said to yield a state of well-being that represents the culmination of the Buddhist pursuit of eudaimonic well-being, knowledge, and virtue. With such insight, it is said that one comes to understand not only the nature of consciousness, but also its relation to reality as a whole. This raises the truly astonishing Buddhist hypothesis: 'All phenomena are preceded by the mind. When the mind is comprehended, all phenomena are comprehended. By bringing the mind under control, all things are brought under control' (Śāntideva 1961: 68).

This primordial consciousness is, then, the ultimate basis for the other two dimensions of awareness. While each human psyche emerges from its individual substrate consciousness, all streams of substrate consciousness emerge ultimately from primordial consciousness, which transcends individuality. The substrate consciousness can allegedly be ascertained with the achievement of advanced stages of *samadhi*, whereas primordial consciousness can be realized only through the cultivation of contemplative insight (*vipashyana*) (Bielefeldt 1988; Karma Chagmé 1998; Padmasambhava 1998; Wallace 2005*b*: ch. 14). Thus, Buddhism postulates this dimension of awareness not as a mystical theology, but as a hypothesis that can be put to the test of immediate experience through advanced contemplative training open to anyone, without any leap of faith that violates reason.

The above theory of the multiple levels of emergence of consciousness flies in the face of the widespread assumption of cognitive scientists that the brain alone produces all states of consciousness. Such scientists commonly assume that they already *know* that consciousness has no existence apart from the brain, so the only question to be solved is *how* the brain produces conscious states. Neurologist Antonio Damasio, for instance, while acknowledging that scientists have yet to understand consciousness, declares, 'Understanding consciousness says little or nothing about the origins of the universe, the meaning of life, or the likely destiny of both' (Damasio 1999: 28). This assumption is an instance of what historian Daniel Boorstin calls 'an illusion of knowledge'. It is such illusions, he proposes, and not mere ignorance, that have historically acted as the greatest impediments to scientific discovery (Boorstin 1985: p. xv).

Prospectively, were the Buddhist theories of the substrate consciousness and primordial consciousness and the practices for realizing eudaimonic well-being to be introduced into the realm of scientific inquiry, radical changes might occur in both traditions. Buddhism, like all other religions, philosophies, and sciences, is prone to dogmatism. As they encounter the empiricism and scepticism of modern science and philosophy, contemporary Buddhists may be encouraged to take a fresh look at their own beliefs and assumptions, putting them to the test, wherever possible, of rigorous third-person inquiry. Buddhist societies have never developed a science of the brain, nor any quantitative science of behaviour or the physical world, so its understanding of the human mind may be enhanced by close collaboration with various branches of modern science.

The encounter between the cognitive sciences and Buddhism and other contemplative traditions may also bring about deep changes in the scientific understanding of the mind. One possibility is that the first revolution in the cognitive sciences may result from the long-delayed synthesis of rigorous first-person and third-person means of investigating a wide range of mental phenomena. This would be the fulfilment of William James's strategy for the scientific study of the mind, which has been marginalized over the past century. This revolution could be analogous to the emergence of classical physics, culminating in the discoveries of Isaac Newton. If we speculate further into the future, we may envision a second revolution in the cognitive sciences emerging from the study *of* and *with* individuals with exceptional mental skills and insights acquired through sophisticated, sustained contemplative training. This might parallel the revolution in physics in the early twentieth century, which challenged many of our deepest assumptions about the nature of space, time, mass, and energy. Such revolutions in the cognitive sciences may equally challenge current scientific assumptions about the nature of consciousness and its relation to the brain and the rest of the world.

A Return to Empiricism

A reasonable scientific response to the above presentation of Buddhist views on the nature of eudaimonic well-being and the three dimensions of consciousness is one of open-minded scepticism. But such scepticism should be equally directed to one's own beliefs, which may be 'illusions of knowledge' masquerading as scientific facts. Richard Feynman wonderfully expressed this ideal of scientific scepticism thus:

One of the ways of stopping science would be only to do experiments in the region where you know the law. But experimenters search most diligently, and with the greatest effort, in exactly those places where it seems most likely that we can prove our theories wrong. In other words we are trying to prove ourselves wrong as quickly as possible, because only in that way can we find progress. (Feynman 1983: 158)

Buddhism, too, expresses a comparable ideal of scepticism. The Buddha is recorded as having said: 'Monks, just as the wise accept gold after testing it by heating, cutting, and rubbing it, so are my words to be accepted after examining them, but not out of respect for me' (Shastri 1968: k. 3587). The Dalai Lama maintains this self-reflective spirit of scepticism when he writes: 'A general basic stance of Buddhism is that it is inappropriate to hold a view that is logically inconsistent. This is taboo. But even more taboo than holding a view that is logically inconsistent is holding a view that goes against direct experience' (Varela and Hayward 1992: 37).

I have argued in this chapter that Buddhism has developed a science of consciousness, but serious objections may be raised. It may be pointed out that science is characterized by controlled experiments, repeated iterative evolving cycles of hypothesis formation, controlled testing, hypothesis revision, and prediction. However, these traits are not common to all branches of science. Astronomy, geology, meteorology, and ecology are some examples that do not lend themselves to all the above methods. Buddhist rational and experiential inquiry into the nature of consciousness and the world at large bears some qualities in common with modern science, but not all. This opens the possibility of a new contemplative science emerging from the interface between Buddhism and the cognitive sciences, in which rigorous first-person and third-person methodologies are integrated in unprecedented ways. Such a science may serve to bring together spiritual and scientific modes of inquiry, to the enrichment of everyone.

Buddhism is also poised to serve as a mediator between theistic religions, which regard God (existing independently of human experience) as their ultimate authority, and science, which takes Nature (existing independently of human experience) as its ultimate authority. While many theologians claim that God can be known only through faith or reason (versus direct experience), and many scientists claim that the mind can be scientifically studied only inferentially by examining the brain and behaviour, Buddhist contemplatives claim that the potential range of immediate experience is far greater than is commonly assumed. The Buddhist challenge here is to retrieve spiritual realities and physical realities from their respective black boxes and return them to the world of experience, where they rightfully belong.

Such a move accords with William James's proposal of a science of religion that differs from philosophical theology by drawing inferences and devising imperatives based on the scrutiny of 'the immediate content of religious consciousness' (James 1902/1985: 12). Such a science of religions, he suggested, might offer mediation between scientists and religious believers, and might eventually command public adherence comparable to that presently granted to the natural sciences. I conclude this chapter with James's challenge to restore a true spirit of empiricism to both religion and science:

Let empiricism once become associated with religion, as hitherto, through some strange misunderstanding, it has been associated with irreligion, and I believe that a new era of religion as well as philosophy will be ready to begin . . . I fully believe that such an empiricism is a more natural ally than dialectics ever were, or can be, of the religious life. (James 1909/1977: 142)

REFERENCES AND SUGGESTED READING

BIELEFELDT, C. (1988). *Dōgen's Manuals of Zen Meditation*. Berkeley: University of California Press.
BOORSTIN, D. J. (1985). *The Discoverers: A History of Man's Search to Know His World and Himself*. New York: Vintage Books.

BRONKHORST, J. (1999). *Why is there Philosophy in India*. Amsterdam: Royal Netherlands Academy of Arts and Sciences.

H. H. THE DALAI LAMA (1997). *The Four Noble Truths: Fundamentals of Buddhist Teachings*. London: Thorsons.

—— (2000). *Dzogchen: The Heart Essence of the Great Perfection*, trans. Geshe Thupten Jinpa and Richard Barron. Ithaca, N.Y.: Snow Lion Publications.

DAMASIO, A. (1999). *The Feeling of What Happens: Body and Emotion in the Making of Consciousness*. New York: Harcourt, Inc.

DANZIGER, K. (1980). 'The History of Introspection Reconsidered', *Journal of the History of the Behavioural Sciences*, 16: 241–62.

DAVIDSON, R., EKMAN, P., RICARD, R., and WALLACE, B. A. (2005). 'Buddhist and Psychological Perspectives on Emotions and Well-Being', *Current Directions in Psychological Science*, 14/2: 59–63.

DÜDJOM LINGPA (2004). *The Vajra Essence: From the Matrix of Pure Appearances and Primordial Consciousness, a Tantra on the Self-Originating Nature of Existence*, trans. B. A. Wallace. Alameda, Calif.: Mirror of Wisdom Publications.

FEYNMAN, R. P. (1983). *The Character of Physical Law*. Cambridge, Mass.: MIT Press.

GETHIN, R. M. L. (2001). *The Buddhist Path to Awakening*. Oxford: Oneworld Publications.

GILKEY, L. (1985). *Creationism on Trial*. Minneapolis: Winston Press.

GOLEMAN, D. (1997) (ed.). *Healing Emotions: Conversations with the Dalai Lama on Mindfulness, Emotions, and Health*. Boston: Shambhala Publications.

—— (2002). *Destructive Emotions: A Scientific Dialogue with the Dalai Lama*. New York: Bantam Doubleday.

GOULD, S. J. (1999). *Rocks of Ages: Science and Religion in the Fullness of Life*. New York: Ballantine Publishing Group.

GUNARATANA, H. (1991). *Mindfulness in Plain English*. Boston: Wisdom Publications.

HARVEY, V. (1981). *The Historian and the Believer*. Philadelphia: Westminster Press.

JAMES, W. (1890/1950). *The Principles of Psychology*. New York: Dover Publications.

—— (1892). 'A Plea for Psychology as a Science', *Philosophical Review*, 1: 146–53.

—— (1902/1985). *The Varieties of Religious Experience: A Study in Human Nature*. New York: Penguin.

—— (1909/1977). *A Pluralistic Universe*. Cambridge, Mass.: Harvard University Press.

KAHNEMAN, D., DIENER, E., and SCHWARZ, N. (1999) (eds.). *Well-being: The Foundations of Hedonic Psychology*. New York: Russell Sage Foundation.

KAMALAŚĪLA (1958). *First Bhāvanākrama*, in G. Tucci (ed.), *Minor Buddhist Texts, Part II*, Rome: Istituto italiano per il Medio ed Estremo oriente.

KARMA CHAGMÉ (1998). *A Spacious Path to Freedom: Practical Instructions on the Union of Mahāmudrā and Atiyoga*, comm. Gyatrul Rinpoche, trans. B. A. Wallace. Ithaca, N.Y.: Snow Lion Publications.

LAMRIMPA, G. (1995). *Calming the Mind: Tibetan Buddhist Teachings on the Cultivation of Meditative Quiescence*, trans. B. A. Wallace. Ithaca, N.Y.: Snow Lion Publications.

NAURIYAL, D. K. (2006) (ed.). *Buddhist Thought and Applied Psychology: Transcending the Boundaries*. London: Routledge-Curzon.

NEEDHAM, J. (1956). *Science and Civilisation in China*, i: *Introductory Orientations*. Cambridge: Cambridge University Press.

ÑĀṆAMOLI, B. (1992). *The Life of the Buddha: According to the Pali Canon*. Kandy, Sri Lanka: Buddhist Publication Society.

NYANAPONIKA THERA (1973). *The Heart of Buddhist Meditation*. New York: Samuel Weiser.

Padmasambhava (1998). *Natural Liberation: Padmasambhava's Teachings on the Six Bardos*, comm. Gyatrul Rinpoche, trans. B. A. Wallace. Boston: Wisdom Publications.

Ruegg, D. S. (1989). *Buddha-Nature, Mind and the Problem of Gradualism in a Comparative Perspective: On the Transmission and Reception of Buddhism in India and Tibet.* London: School of Oriental and African Studies.

Ryan, R. M., and Deci, E. L. (2001). 'On Happiness and Human Potentials: A Review of Research on Hedonic and Eudaimonic Well-being', *Annual Review of Psychology*, 52: 141–66.

Ryff, C. D. (1995). 'Psychological Well-being in Adult Life', *Current Directions of Psychological Science*, 4: 99–104.

Śāntideva (1961). *Śik asamuccaya*, ed. P. D. Vaidya. Darbhanga: Mithila Institute.

Searle, J. R. (1994). *The Rediscovery of the Mind.* Cambridge, Mass.: MIT Press.

—— (2002). *Consciousness and Language.* Cambridge: Cambridge University Press.

—— (2004). *Mind: A Brief Introduction.* New York: Oxford University Press.

Shastri, D. (1968). *Tattvasaṃgraha.* Varanasi: Bauddhabharati.

Stevenson, I. (1997). *Where Reincarnation and Biology Intersect.* Westport, Conn.: Praeger.

Thrangu Rinpoche (1993). *Buddha Nature.* Hong Kong: Rangjung Yeshi Publications.

Tillemans, T. (1999). *Scripture, Logic, Language: Essays on Dharmakīrti and his Tibetan Successors.* Boston: Wisdom Publications.

Tsong-kha-pa (2000). *The Great Treatise on the Stages of the Path to Enlightenment*, i. Ithaca, N.Y.: Snow Lion Publications.

Tucker, J. (2005). *Life Before Life: A Scientific Investigation of Children's Memories of Previous Lives.* New York: St Martin's Press.

Varela, F. J. and Hayward, J. (1992) (eds.). *Gentle Bridges: Conversations with the Dalai Lama on the Sciences of Mind*, 2nd edn. Boston: Shambhala Publications, 2001.

Wallace, B. A. (1993). *Tibetan Buddhism from the Ground up: A Practical Approach for Modern Life.* Boston: Wisdom Publications.

—— (1996). *Choosing Reality: A Buddhist View of Physics and the Mind.* Ithaca, N.Y.: Snow Lion Publications.

—— (2000). *The Taboo of Subjectivity: Toward a New Science of Consciousness.* New York: Oxford University Press.

—— (2005a). *Balancing the Mind: A Tibetan Buddhist Approach to Refining Attention.* Ithaca, N.Y.: Snow Lion Publications.

—— (2005b). *Genuine Happiness: Meditation as a Path to Fulfillment.* Hoboken, N.J.: John Wiley & Sons.

Whitehead, C. (2004). 'Everything I Believe Might be a Delusion. Whoa! Tucson 2004: Ten Years On and Are We Any Nearer to a Science of Consciousness?', *Journal of Consciousness Studies*, 11/12: 68–88.

Wilson, E. O. (1998). *Consilience: The Unity of Knowledge.* New York: Alfred A. Knopf.

CHAPTER 3

....

JUDAISM AND SCIENCE

....

NORBERT M. SAMUELSON

BACKGROUND

....

This essay on the role of science in Judaism will use some familiar terms in unfamiliar ways. They include 'creation', 'revelation', 'redemption', 'God', 'world', 'humanity', 'belief', 'wisdom', 'philosophy', and 'science'. Let me briefly explain why.

Judaism is the formal expression of the faith of the Jewish people from its origins in the land of Israel almost 4,000 years ago to the present. Throughout that history the people have lived in different places as a minority within a dominant civilization. As a minority, every aspect of its life, including its beliefs, has been influenced both by its own past and the past of its dominant host. Since times, places, and peoples constantly change, so do the beliefs expressed in the worshipping community of the Jewish people. Hence, Jewish thought must always be seen as a history of ideas.

The earliest part of that history—the life of the Jewish people in the Ancient Near East—resulted in the production of a sacred text (the Hebrew Scriptures), which is foundational for Christianity no less than for Judaism. Hence, since Judaism and Christianity share a common textual origin, much of what they believe at least shares a common religious terminology. Furthermore, the centre for Jewish intellectual life for at least the past 500 years has been in Western European Christian civilization. Hence, rabbis and Christian clerics share a common tradition of science. However, the classical development of Jewish philosophy and theology did not occur in Christian lands. Jewish intellectual development took place over more than 1,000 years in lands where the dominant religion became Islam, and where Christianity consequently had little importance. Hence, the meanings of many key terms in this essay are anchored in a tradition of religious thought shared by believing Jews and Muslims which is unfamiliar to Christians. But these meanings are not secrets reserved for a spiritual élite. All that a reader need do is be sensitive to the somewhat distinctive way in which the most common terms in this essay (including the words 'science' and 'religion') are being used.

There are three historical periods which encapsulate the relationship between Judaism and science. Each period is marked by major changes in the religion of Israel and/or in the science of the host civilization. The first period is the one recorded in the canonized Hebrew Scriptures and the civilization of Rome. The term 'religion' here refers to the 'philosophy' of the early rabbis of Judea, and the term 'science' refers to a mixture of Hellenistic philosophies of which the dominant one is Stoicism. The second period is the one recorded in the philosophical biblical commentaries and apologetic treatises of the rabbis who lived from the tenth century on in Muslim lands and continues through the rabbis who lived in Western Christian Europe until the end of the middle of the nineteenth century. In this period the term 'religion' is used generally in more or less its modern Western sense, designating an area of life and belief that is roughly (but not precisely) separate from a comparable area of life and belief called 'secular'. However, its more specific use by the rabbis designates a way of life and belief, one that transcends the secular/religious dichotomy, that is pre-scribed in detail in a set of rabbinic texts of which one (but only one) of the most important is the *Mishneh Torah* by Moses Maimonides (twelfth century CE). In this period the term 'science' refers to a complete philosophy of life known as 'Aristote-lianism', which includes the writings of Aristotle as well as the writings of other Greek and Roman schools (notably Platonists, Epicureans, and Stoics), as those texts were interpreted by late medieval rabbis whose own interpretations used the commentaries of earlier medieval Muslims, who themselves used translations and commentaries by late Hellenistic pagans and Christians. Finally, the third period is the modern period, in which the use of the term 'religion' has been determined primarily by English and Germanic Christian protesters against the Roman Catholic Church, and the term 'science' reflects a process of specialization marked by natural philosophies that are called 'mechanistic' (notably with reference to René Descartes), 'mathematical' (not-ably with reference to the interpreters of Isaac Newton), and 'materialist' (notably with reference to the followers of Darwin and to the continental Positivists). In all three periods the terms 'religion' and 'science' have different meanings, but the new meanings in later periods presuppose the old meanings in earlier periods.

THE COMPLEMENTARY-CONFRONTATIONAL MODEL OF SCIENCE AND RELIGION

The very first pair of commandments discussed by Maimonides at the beginning of the *Mishneh Torah* is the positive commandment to worship God and the negative commandment to avoid idolatry. The entire system of 613 commandments has only a single purpose. It is to create the kind of psychological, physical, and political conditions that make fulfilment of the first pair of commandments possible. Different people will succeed at different degrees in fulfilling them. (The possible levels of fulfilment are enumerable.) At the highest level Jews will be able to live life fully involved in the affairs of this physical world, but they will do so at all times with their bodies and minds completely focused in meditation on and with God. Achieving this end requires the highest level of success in self-discipline, but moral perfection itself is an instrumental value whose end is intellectual. The intellectual goal is a fully adequate understanding of who and what God is. This perfect knowledge is in fact the ultimate object of all knowledge, and any lesser knowledge runs the constant risk of being in reality the worst of all sins—idolatry. Idolatry is the worship of any deity other than the true God, and the positive and literal attribution to God of any characteristics that are not true of him constitutes idolatry.

This radical negative judgement about theological error is a consequence of God's oneness. Oneness entails radical simplicity, such that uniquely in God's case what God is and what God does are the same thing, and that thing can be only a single thing. Hence, anything attributed to God constitutes who God is, so that if that thing is not true of God, then the speakers have unwittingly committed themselves to idolatry. In knowing an object, the knower achieves a form of unity with the object known. Hence, the pursuit of knowledge of God is an effort to achieve unity with God. Therefore, to know God is to worship God. However, if the deity known is the wrong deity, then the speakers not only misspeak God's name; unwittingly or not, they worship a false deity.

Maimonides urges all who read his *Mishneh Torah*, which is his practical guide to those who follow the path of Torah to unity with God, to learn science. God is identified by three primary acts: he creates the universe at its origin; he redeems the world at its end; and between the beginning and the end he reveals the path for humanity to follow. It is a consequence of God's absolute simplicity that these three acts are a single, timeless act that is identical with God. Of the three, the one act that is most accessible to human, natural knowledge (which here just means a knowledge that is independent of revelation) is creation. Through the use of the senses and reason all human beings have the ability to achieve knowledge of the natural laws that govern all creatures in the universe. Since the universe exists by the will of God, then those laws are expressions of God's will. And since in God's case there is no distinction between who God is, what God wills, and what God does, knowledge of the sciences is knowledge of God.

Hence, the terms that Maimonides uses (and the entire medieval tradition used) for 'science' and 'scientist' respectively are 'wisdom' (*chachmah*) and 'sage' (*chacham*). The earliest rabbis were also called 'sages' more than 'rabbis'. A sage is a 'wise man', a person who has acquired wisdom, and 'wisdom' for Maimonides is what it was for the author of Proverbs and the Greek philosophers: that knowledge necessary for the realization of 'happiness', where happiness is understood to be the attainment of the highest moral end of human perfection.

What Maimonides presents here is what in the title of this section I call the 'complementary' model of Judaism and science. The sciences that Maimonides emphasizes are astronomy and physics, with the consequence that on Maimonides' schema, all Jews are commanded to learn physics and astronomy to the maximum level of which they are capable. However, without any violence to Maimonides' original intent, we may extend his ruling to apply to the study of any and all sciences.

Note that Maimonides' model was not the only one proposed by the rabbis. There were other medieval Jewish philosophers—notably Judah Halevi in the twelfth century and Hasdai Crescas in the fourteenth century—who defended, in contrast to Maimonides, a confrontational model in which Judaism and philosophy were seen to be in tension. However, the confrontational model does not have many rabbinic exponents until the late Middle Ages (viz. the fifteenth and sixteenth centuries), and it did not become dominant in rabbinic thinking until modern times (viz. the nineteenth and twentieth centuries). We will not in this essay discuss the reasons for the change. Rather, we will limit ourselves to describing the change within modern Judaism and modern science.

MODERNITY AND THE MUTUAL CHALLENGES OF SCIENCE AND JUDAISM

The way in which Maimonides describes creation, revelation, and redemption—that is, how God acts in relationship to the world, to the Jewish people in particular, and to humanity in general—presupposes that in principle there can be no disagreement between what the Torah says when it is interpreted correctly and what good (meaning true) science affirms. The unity is such that in principle it is legitimate to use good science to discover the correct meanings of the Torah. In demonstrating his method of interpretation Maimonides used the best science of his day. Hence, for someone today to do what Maimonides did would not be to say exactly what Maimonides said. The difference is that were he living today Maimonides' language of science would not be Platonic-Aristotelian-Stoic. With respect to creation and the nature of the physical world, it would be the language of modern physics and astrophysics following in the tradition of the texts of Galileo, Kepler, Newton, Leibniz, Huygens, Maxwell, Einstein, Bohr, Heisenberg, and Schrödinger. The same can be said with respect to revelation and the nature of humanity, which would include at least the language of the biological discoveries of Darwin, Mendel, and Watson and Crick. That few rabbis today engage in such Maimonidean speculation is in part a sign of the dominance of the confrontational model of science and Judaism. But that is not the whole story. In part it is a consequence of the fact that in modern life rabbis no longer function as sages. It is also a consequence of the modern separationist model, which excludes from the domain of religion any subject over which scientists pronounce judgement, in the hope that scientists will grant the same courtesy to them on questions of theology and ethics.

Today the separation model seems dominant among rabbis, even as it is breaking down in the academic world of evolutionary psychologists and genetic engineers. However, this sketch is slightly too simplistic. While rabbis tend to leave questions of scientific theory alone, that is not always the case with the applications of science. For example, in general, traditional rabbis support almost anything that genetic engineers want to do to help human beings produce children. But that position is not a consequence of attributing any inherent value to science. Rather, it is a consequence of a highly liberal stance among traditional rabbis on producing Jewish children in response to the numerical devastation of the Jewish people in the Holocaust. Post-Holocaust the number of living Jews is dangerously low. Hence, to ensure the future survival of the people, help is welcome from every corner. But that does not mean that traditional Jews are also open to what geologists say about the age of the earth, or what evolutionary palaeo-anthropologists have to say about the origins of the human species. Quite the contrary: contemporary traditional rabbis increasingly seem to be attracted to the very same kinds of defences of biblical literalism (despite the fundamental principle of rabbinic interpretation of the Hebrew Scriptures that 'the Torah speaks in human language' rather than saying things precisely as they are) and to the so-called young earth accounts of the American Christian Right (despite the fact that Maimonides and other medieval commentators on Genesis insist that the 'account of creation' has a 'deeper meaning' than its 'literal interpretation'). That these rabbis can adopt biblical literalism as a standard in such cases is, once again, a symptom of the decline of the rabbinate as a source for present and future *chachamin* (sages).

POST-NEWTONIAN PHYSICS AND JEWISH CREATION: NECESSITY, ACCIDENT, AND PURPOSE

If we read as literally as possible what the opening chapters of the Book of Genesis say about the origin of the universe, then the following picture emerges. At its origin, the universe is a sphere of 'earth', surrounded by a ring of 'water', above which hovers the 'wind of God' in a background of darkness. The universe in this initial state is described as 'chaotic and unintelligible'. God 'creates light', which is set over against the dark. He names his creation 'day', and the spatial darkness from which it is set off he names 'night'. He calls this first insertion of division in space 'the first day'. Then follow six more so-called days in which further distinctions or separations are made, all of which he calls 'good'. On the second and third day the space of the universe is differentiated. On the second day God forms a ring of material that he calls 'sky', which is stretched out in the middle of the water to divide it into two separate rings. Then, on day three he moves the globe of the earth slightly into the ring of sky to distinguish what he calls 'dry land' from the rest of the earth, and commands the earth to generate vegetation. The dry land is that part of the earth globe that reaches through the ring of water into the sky. On the next three days God creates the entities that move in the space differentiated on the first three days. On day four (which parallels the first day when the sky was created), God sets things that emit light and designates the largest of them as his governor over the sky. In addition, God commands the light emitters to move in such a way as to determine liturgical seasons. On day five (which parallels the second day when the sky was created in the water) God creates forms of movable life that reside in water and in the sky, and he commands them to 'be fruitful and multiply'. Then, on day six (which parallels the formation of the dry land with its vegetation) God commands the earth to generate entities that live on its surface, all of which are commanded to be fruitful and multiply. Then he forms, with someone's help (the text does not explicitly say whose), the 'human', which he makes male and female. Just as 'the large light emitter' was put in charge of the sky on day four, so the human on day six is ordered to govern all the forms of life on the earth. Finally, on day seven, God's work of creation comes to an end with the creation of rest, as a day of rest.

Every word or phrase set in quotation marks is not explained in the Scriptures themselves. This task is left for the subsequent rabbinic commentators on the text. These commentaries are the main basis for the classical dogma of creation in Judaism. In constructing their interpretations, these rabbis used what to them were the best texts of physics and cosmology. By at least the twelfth century and no later than the fourteenth century, the premier of these books was Plato's *Timaeus*. Using this scientific text to guide them, the critical words and phrases in quotation marks above were said to mean the following.

Since God does not change, his act of creation, with which he is identical, is a single act without beginning or end. Hence, the stages of creation cannot constitute distinct acts in time. Rather, they are logical designators of the distinct levels in God's creation. At the base is the differentiation of space into regions, followed by the creatures who occupy the regions. Earth exists for the sake of vegetation, which exists to feed the living things, which exist to provide food for the species of fish, birds, and animals that occupy the different regions of space. The living things in turn exist to provide food for the humans, whose purpose on earth is to worship God in appropriate ways at appropriate times. The creation is called good because it constitutes the formation of a cosmic political entity whose constitution is morally ideal. The ideal is a constitutional monarchy, where all creatures are citizens governed by a law code that emanates from the supreme ruler in his wisdom. The image presented is of a complex hierarchy of social strata in which each social group knows its unique place in the universe and obeys the appropriate laws. Physical space is unholy and holy; languages as well are holy and unholy; and so are nations. At the top of the hierarchy of the world below the moon are humans, and at the top of the hierarchy of nations of the earth are the Jewish people, who are governed by an oligarchy of rabbis, who themselves are governed ultimately by the highest subspecies of human beings (the prophets) whose experiences of divine will are recorded in the Hebrew Scriptures.

The cosmos is imagined as a series of spheres within spheres. Each sphere is a living thing, whose life principle is a non-physical (or spiritual) entity with which it is intimately associated. The spiritual entity of a sphere is called a 'separate intellect' (in the language of science) and an 'angel' (in the language of the Scriptures). The body of a sphere consists of all the spatially extended entities within its domain, which includes the lesser spheres. The lowest of these spheres contains within it the globe of the earth. Its associated intellect is called 'the active intellect' (in the language of science) and the Shekinah (in the language of the liturgy, which translates into English as God's 'indwelling presence'). The highest level of creatures within the Shekinah's domain is the prophet, who is understood to be a transitional being between humans (who have practical wisdom) and angels (who have theoretical wisdom). Below them are the other nations of humans, below which are the various kinds of living things, below which are the various forms of non-living entities, which are ultimately formed from the four elements of science. These elements—active fire and air ('God's wind') and passive water and earth—constitute the stuff that is formed into entities through a combination of laws of mechanical necessity and divine purpose. All of this is understood to be morally good, and knowledge of it is a primary path to knowledge of God, which itself is a moral obligation.

The cosmology of the rabbinic doctrine of creation draws heavily on the cosmology of Plato and Aristotle, but more from the former than the latter. The difference is critical. Both claim that the universe has no beginning in time. However, as we have seen, neither do the classical rabbis. This rejection has more to do with theology than astronomy. The commitment to a non-temporal creation is a consequence of the rabbis' radical interpretation of divine oneness; God can have only one unchanging act, and that act cannot be subject to temporal origin or end. However, this judgement is also dependent on science. In this case the science is semantics, not physics. On the rabbis' understanding of language, the meaning of sentences consists of ideas, and ideas are spiritual entities. These entities describe substances that are modified by states. In medieval semantics the substance is the reference of the subject of the associated declarative sentence, and the modification of the substance is the referent of the predicate of the sentence. On this theory of language, a predicate names a thing that is contained within another thing that the subject names. Hence, if there can be univocal positive attributions of God, God would necessarily be complex, which would deny his unity.

Modern semantics is no longer Aristotelian. Hence, it is no longer obvious why the statements about God's nature in the Hebrew Scriptures cannot be taken (at least initially) at face value (which would include the ascription to God of physical characteristics). In contemporary liberal Jewish theology anthropomorphic descriptions of God have been reintroduced, especially in feminist theology and in a revival of Jewish romanticism through the study of Cabbalah.

However, this return to theological primitivism does not mean that there are no problems in affirming a literal interpretation of the biblical account of creation as true. It means only that the problems are different. The critical issue for a rabbinic understanding of the dogma of creation in the light of contemporary astrophysics turns on the notion of cosmic causation.

What is consistent in every professed scientific account of the origin is the claim that the universe is the way it is either for purely mechanical reasons or because of chance. Our universe is defined as a domain in which (at least in principle) any thing located in its space may have influence on another thing located in this space. By influence I mean any kind of change in what the thing is or what the thing does. That influence occurs either through some kind of force (of attraction or repulsion) or through physical contact. These influences happen either by chance, which here means without any cause, or by a cause, which in some sense necessitates that the thing changed changes. But nothing within the domain of physics happens for a purpose. Hence, for no reason or purpose whatsoever, our universe is at its origin a single entity whose size is infinitely small but whose density and temperature are infinitely great. The very state of this origin, which occurred at some point in time many millions of years ago, causes the universe to implode as pure energy that disperses through the surrounding empty space. As the energy expands in all directions, it becomes less dense and its temperature lowers. As the temperature lowers, units of energy adhere to each other to become distinct entities, which, with continued cooling form into other distinct entities, all organized in a hierarchy of elementary particles, nuclear particles, material entities, planets, and galaxies.

There is no part of the above story that cannot be incorporated into a Jewish understanding of creation except one—that all of this happens for no reason or purpose whatsoever. Now it is not necessary that the reality of purpose as a mover in our physical world be proved scientifically. On the contrary, all that is needed from science is for the possibility of purpose to be something judged to be reasonable. To my knowledge, most contemporary scientists will grant that the exclusion of purpose from an account of the origin of the universe is only methodological.

The methods used by physicists do not admit any kind of teleological cause, but this does not mean that things do not happen for a purpose. When scientists can explain rabbinic claims, committed Jews value those explanations, because they make a contribution to the faith understanding of the doctrines. (There is no virtue in maintaining some affirmation as a dictate of faith, which here means not provable, when it can be known, which here means provable.) Still, there is no obligation that every religious doctrine be so explained. On the contrary, few rabbis today (contrary to many medieval rabbis) believe that scientific and 'Torah' truths are merely two different ways of thinking about the same thing. So, it is by revelation (which means the tradition of rabbinic interpretation of the Hebrew Scriptures) that we know the purpose of the creation: viz. to generate entities who may worship God, each in its own way. Science tells us nothing about that, but that is the limitation of science. Its strength is that it can tell us in a depth that far exceeds revelation the means by which God so acts.

POST-DARWINIAN BIOLOGY AND JEWISH ETHICS: MATERIALISM, VOLUNTARIANISM, AND RELATIONALITY

Teleology, Life, and Humanity

The operative principle of peaceful coexistence between the sciences and the religions in Western civilization has been separation. Religion has its domain of values into which scientists, as scientists, are not to enter because of their lack of expertise. Similarly, science has its domain of physical events into which clergymen, as clergymen, are not to enter because of their lack of expertise. This is not the classical position of rabbinic philosophy. For Maimonides, the prophet combined both the talent of the seer, who could perceive with skill divine directives, and the talent of the philosopher, who could reason with skill from common-sense experience. In contrast, the first modern Jewish philosopher, Baruch Spinoza, asserted the modern separation. Prophets are persons of great imagination who are skilled in political leadership, but that skill does not constitute knowledge. Knowledge is the exclusive domain of scientists. Until recently, most liberal clergymen, including rabbis, agreed with this modern judgement, and they granted to the scientists the right to judge physical facts while they reserved for themselves the exclusive right to judge morality. Undoubtedly, the unstated premiss upon which the scientists agreed to this restriction was their acceptance of the philosophical judgement that no scientific basis for moral judgements is possible. However, in recent years there have arisen a number of scientists who have rejected this epistemic judgement. They claim that there is one science, evolutionary psychology, which is capable as a science of making moral judgements. As the argument goes, the physical existence of genes does have a purpose: viz. to replicate its own kind. And the existence of all the living things designed by these genes also has a purpose: viz. to provide a good host to enable the genes to replicate. It is this contemporary move of the life sciences into the realm of values that poses the first serious conflict between Judaism and (at least the life) sciences.

The opening chapter of Genesis distinguishes the creation of the human from other life forms on the planet Earth in two ways. First, in addition to being commanded to procreate (a commandment shared with all other living things), the human is commanded to govern the world beneath the sky. Second, whereas all other living things come into existence by means of a direct command from God to the element earth, in the case of the human God says 'Let us make him in our image.' The human, and the human alone, is said to be in 'our' image. The immediate question is, of course, to whom besides God the 'us' and the 'our' refer. The medieval Jewish philosophers answered that it refers to the angels. Furthermore, the 'image' that the human shares in common with God and with the angels is the ability to be consciously aware of God, who is the proper ultimate object of all theoretical rational knowledge.

Modern evolutionary science tells a quite different story about what it means to be a living thing. The story begins in much the same way as Genesis begins. In Genesis, first there is earth (viz. inorganic matter), out of which arise all forms of life (viz. organic matter), including the human. However, just what is this 'life' is different. For the Hebrew Scriptures the life is heated breath (*neshamah*) that is located within the blood of the creature. For the classical rabbis, this breath becomes associated with a 'soul' (*nefesh*), which, as it was for the Greek and Roman philosophers, is a non-material principle of life. In this respect the modern life scientists are closer to the Hebrew Scriptures (where living is a quality of the body) than they are to the rabbinic philosophers (who associate life with an entity that is separate from the body). For the geneticists the principle of life is chemical.

As their story is now told, life begins with cells. These life-creating entities contain chromosomes, which are thread-like structures within the nuclei of the cells. The chromosomes are the bearers of genes. Genes, in turn, are molecules composed of strings of chemicals. The chemicals in this case are deoxyribonucleic acid (DNA), which are held together in a double helix geometrical formation by groups of sugar and of phosphate. The chemical bases of DNA are adenine (A), guanine (G), cytosine (C), and thymine (T). Chemical enzymes (which are proteins, yet another kind of chemical compound) activate the DNA into becoming a new kind of entity (just as yeast, also an enzyme, activates dough into becoming a new kind of substance such as bread). The new creature in this case is ribonucleic acid (RNA), whose bases are adenine, guanine, cytosine, and uracil. In this new state the cell nucleus activates proteins, and the proteins direct embryonic differentiations in their host cells as the cells divide and multiply. It is in this way that all of life develops, and in this respect humans are no different from anything else.

If there is any issue here at all between Judaism and science, it is the same as the issue summarized above with respect to physics: viz. whether any of these processes exhibit divine purpose or whether they occur by chance and/or physical necessity. Until recently the answer of the biologists would have been like the answer of the physicists. However, today some biologists would agree that what occurs in life does have a purpose, but the purpose is directed by the genes, not by God.

Besides the general question of purpose, the life view adopted in modern science is problematic for Jewish philosophers for two reasons. First, the life sciences seem to deny that human beings are distinct from other life forms. Second, they deny that the power to reason has any special moral value. For the classical Jewish philosophers, rationality is a unique human trait that serves the members of this species to fulfil their ultimate moral end of unity with God. In contrast, for the evolutionary psychologist and palaeo-anthropologist, reasoning is only one kind of survival technique that species improvise in response to their environment, and as these techniques go, it is far from the best. (Compare, for example, the slow and clumsy survival activities of the mighty human beings who have to reason out solutions to their problems with the speed and endless flexibility of lowly viruses in adapting to environmental changes without any intellect whatsoever.)

It is interesting to note that the modern scientific position has more in common with the biblical life view than it does with rabbinic philosophy. It is the philosophy, not the Scriptures, that asserts that the life principle is a separate, non-physical or spiritual entity called a soul, and it is the philosophy, not the Scriptures, that makes the end of human morality rational thought. For the Scriptures the term 'soul' means only a principle of life in a material organism, and the end of life is defined by a system of commandments that directs human beings into community with each other and with God.

Practical Issues: 'Be fruitful and multiply'

So far we have considered science in relationship to Judaism only at the level of theory. However, the reality of the encounter of science with Judaism is far more at the practical level, and from this perspective there is even less conflict. Scientific knowledge, particularly in the life sciences, is viewed primarily as medicine. Medicine has always been a preferred field of study and practice by the rabbis. (Maimonides, for example, was a physician.) It was through the study and teaching of this art that Jews first entered the Christian universities of Europe, and it continues to be a field in which Jews are prominent. However, there are a number of challenges to traditional understandings of Jewish law that modern medicine introduces.

For the most part, contemporary discoveries about the nature of heredity and the mechanics of human reproduction have been of great help to the Jewish people. Concerning reproduction, no issue has been more important in collective Jewish life than the physical survival of the Jewish people. The issue of numerical loss is that as the numbers of Jews involved in the Jewish community declines, it becomes increasingly difficult for committed religious Jews to continue the kind of activities that their faith requires. (Below a certain number, kosher shops, for example, cannot stay in business, and synagogues cannot afford to hire educators.) Furthermore, Jewish daily life focuses primarily on the home, and the operation of the home as a centre of spiritual activity requires strong Jewish commitment from both parents. Hence, it is critical for Jews who want to be Jewish and to raise children who want to be Jewish to marry partners who share this spiritual and communal commitment. Usually this means that Jews need to marry Jews. However, after the Holocaust the number of Ashkenazi (European descent) Jews is now so dangerously low that on purely numerical terms future Jewish collective survival is threatened. The threat takes two forms.

First, Jews do not have enough children to enable the population of the community to grow. Hence, anything that science can do to assist Jewish parents in procreation is highly valued. For this reason even the most politically conservative rabbis have adopted the most liberal positions on reproduction, which is the sole issue in the American political spectrum where the moral sensitivity of the 'Jewish Right' is out of step with the leadership of the American 'Christian Right'.

Second, the reproductive pool of Jews is so small that there are some genetic diseases to which Jews are especially prone. Once again, the contribution of contemporary medical science, especially in genetics, is highly valued for two reasons: one, genetic identifications warn prospective parents of a threat in order to prepare them to treat it; two, genetic research holds open the possibility that some day the threat will be completely eliminated.

Of course these benefits also involve problems. Concerning genetic identification, there is always the possibility that parents will choose to abort the foetus, and the morality of abortion is as complicated for religious Jews as it is for committed Christians. However, because of the inherently pro-life stance of traditional Jewish law, and, once again, because of the Holocaust, committed Jews in general will be less likely than other Americans to opt for abortion.

The dramatic increase in scientific knowledge of the nature of inheritance and reproduction also raises distinct sets of problems for Jewish law and ethics. In the past the decision that a person was dead was fairly simple. A feather was placed under the person's nose, and if (but only if) there was no evidence of breath, the person would be declared dead. Today the lack of breathing is only one marker of the end of life, and it is by far the weakest one. Far more significant is the cessation of brain functions. But even brain death does not necessarily mean that the person is dead. It simply is no longer clear at what point it can be said that someone is dead, because it is no longer clear just what it means to be alive. In many cases some form of life can be maintained indefinitely, but it is not clear whether this form is life or is merely existence. Furthermore, where life is only existence, should the life be allowed to continue, and just who has the right to make such a decision?

Genetics has provided a composite blessing with its own problems in yet another way. It is now possible to prove that many Jews who are Cohanim and Levites, which are two categories of biblical priests, have in fact maintained (to many people's surprise) their identity as priests. In general, for traditional religious Jews, inheritance runs through the line of the mother and not through the father. One exception to this rule is priesthood, where the line runs through the father. The father of a priest must be a priest for the son to be a priest, and today, because of a distinctive mutation in the Y chromosome that is passed on only through the father to his son, it is possible to determine a pure line of priestly descent.

Once again, with the positives also come the problems. There are significant groups of peoples, such as the Lembas in Africa, who seem to be Jewish genetically, but who otherwise would never have been considered Jewish before. However, it is far from clear what are legitimate or illegitimate political uses of this kind of genetic data. Should, for example, genetic evidence be used to decide who may or may not qualify for citizenship in the State of Israel under the Law of Return? (On this basis many Iraqis and Syrians might qualify.) Furthermore, should a Jewish state take proactive steps to 'improve' the genetic pool of its citizens? For anyone who remembers the 1920s and 1930s in the United States and in Western Europe, there must be some fear that genetic engineering may simply be a resurrection of the eugenics of the past century.

POST-MARXIST HISTORY AND JEWISH REDEMPTION: 'THE LIGHT UNTO THE NATIONS'

The hope that lies behind genetic engineering is that science enables human beings to improve the quality of human existence, and there are no limits on what science in the service of humanity may accomplish. Many think that in the not too distant future we will have the means to extend indefinitely a qualitatively high level of human life in a world where medicine has conquered disease and the social sciences have cured political injustice. Such a hope is a secular expression of what in Judaism is the expectation of the eventual coming of the messianic age with the hope that it should happen 'speedily in our own day'.

Ours is not the first age to have such messianic hopes for science. The seventeenth- and eighteenth-century confidence in the achievements of the new mathematical science of Newton and Leibniz fuelled the passion of the intellectual élite who tried to change the nature of absolutely everything in the French Revolution. In fact, the world at the beginning of the nineteenth century was a radically different place from what it was at the beginning of the seventeenth century. But the dark side of the Revolution was its economic expression as the Industrial Revolution, in which a new class of people called 'workers' found the conditions of their lives reduced by another new class of people called 'management' to a level of poverty and inhumanity never before known in history. In response to the failure of the nineteenth-century liberal and commercial state spawned by the ideals of the European Enlightenment, there arose a host of new social sciences, including economics, and a serious attempt to transform the arts of history and philosophy into sciences. The new sciences of the late nineteenth and early twentieth centuries in turn produced a new form of this worldly messianism called 'Socialism', which again promised through the political domain to bring about a utopia. However, its two concrete expressions in the early to mid-twentieth century—the Fascist national socialism of Germany, Italy, and Spain, and the Communist international socialism of the Soviet Union and the People's Republic of China—produced brave new worlds that did anything but improve the quality of human life on this planet.

The daily prayers of Jews are all attempts to invoke the coming of the messianic age, and though Jews coming from the perspective of their traditional faith remain sceptical about every new innovation that imaginatively exaggerates the benefits it will bring to humanity, they never give up the hope that this is the time of the coming, and clearly science will play a role in the fulfilment.

The first thing God created was light, but the light of creation is held in reserve for the messianic age. Its light is not the light of our world. In the revolutionary fervour of the seventeenth century, both physical and spiritual light became the central conception in the transformation of the Aristotelian philosophy of the Middle

Ages into the mechanical philosophies of the early modern period and then into the mathematical sciences of our own age. For most of Judaism, this change is viewed as progressive. But the progress is only a process. We are not at the hoped-for end, for the light of this world—both physical and intellectual—is still far from being God's initial light of creation.

Postscript: Intellectual History and Constructive Theology

This essay has been for me no less an exercise in constructive Jewish theology than an exhibition of the intellectual history of Judaism and of the Jewish people. Some readers may have difficulty in seeing how a work that in its appearance is about history could also be theological, but this is the way in which constructive Jewish thought has always been written. Nor is the use of a historical mode to express theology something exclusively Jewish. Non-Jewish theologians of no less stature than Aristotle (in his *Metaphysics*) and Hegel (in his *Encyclopaedia*) have used the same procedure, as have most Muslim and Roman Catholic theologians as well. In fact, my impression is that the notion that theology can proceed in a non-historical way seems culture-specific to a certain kind of Protestant, Western European, modern tradition of both theology and philosophy, where experience and reason are taken to be epistemically authoritative to the exclusion of tradition. (Learning from history is learning through a tradition.)

In my own thinking about both religious and scientific matters, tradition plays a significant but not uncritical role. First, I can think about what I think about only because I am a product of a certain culture, and how that culture impacts upon my thought has to do with its history. Everything I do and think is a product of that history. Second, I live in multiple communities. I am an American as well as a Jew, and I am a product of a certain kind of academic education that shaped the way I examine both my American and my Jewish cultural inheritance. Third, these different cultures are not independent. Jews have lived everywhere at every time as a minority culture within at least one dominant culture, and the dominant cultures have affected how Jews think about everything, including Judaism. At the same time, these dominant cultures have changed because the Jewish people have become part of them. (American Jewish life is distinctively American; America would be a significantly different place if it were not for the Jews; and I am in every aspect of my life an American-Jew, a Jewish-American.) Because I am a product of different cultures, I have a place to stand to look critically at all of them. But I am not a divinity who is free of external influence. Even when I critique where I come from intellectually, my critique remains itself a part of the critiqued.

There is no way to rise above culture. The closest we can come to doing so is through the study of its history, for in the act of looking critically at that history we contribute to its advancement by finding ways to move beyond it. This essay has been such an exercise. Let me end this postscript with two highlights from the body of the essay (making the implicit explicit).

First, the models of complement and confrontation are not mutually exclusive. They are simply two elements of the way in which we learn and grow in our pursuit of knowledge and the improvement of the world. Both are always present. Differences are only a matter of emphasis. Post-Newtonian physics and rabbinic creation doctrine are examples of where the dominant interactions of traditional rabbinic sources and contemporary Jewish philosophical theology with academic studies of elementary and cosmological physics are deeply complementary. However, there still remains an element of tension in the synthesis. (For example, is moral value something that human beings exclusively inject into nature or, despite what our school texts say, does physical nature express moral values?) On the other hand, the interaction between post-Darwinian biology and rabbinic revelation doctrine is an example of where the dominant interaction is pervasively confrontational. At the core of the engagement is the conception of the distinctive nature of being human. (For example, is a human some form of machine whose function is to produce more copies of itself, or an entity who forms communities whose function is to worship God?) However, even in the conflict with biology there remain elements of common endeavour. Certainly for contemporary Jewish thought, especially in areas of genetic engineering, modern biology is viewed as a blessing, especially so after the devastating losses suffered by the Jewish people in the Holocaust.

Second, while the Jewish people have much to gain from the engineering consequences of modern scientific knowledge, the same cannot be said for modern scientific theory. The basic tension, and challenge, is ultimately ontological. At least methodologically, modern science presupposes that everything that is can be reduced to what is physical, and the dynamics of the physical can be understood in solely mathematical and mechanical terms. Jewish thinking has to challenge this reductionism, for its most fundamental insight is that the so-called plastic world (i.e. the world of what our sense organs present to our consciousness as reality) is only one small part of all of reality, and it is far from being reality's most valued realm. However, Jewish thinking must respond to this challenge with knowledge rather than ignorance, and in many respects ours is the most ignorant Jewish community in history.

I do not mean to say that Jews are ignorant. Quite the opposite is the case, for Jews have been at the forefront of all modern scientific advancements in the past 200 years. However, they have pursued science not as Jews; in fact, they have pursued science in conscious opposition to their Jewish inheritance. (Sigmund Freud is a paradigmatic but far from unique case.) The source of the problem is a traditional European rabbinate that in the late Middle Ages and early modern period, in marked opposition to the tradition of Jewish rationalism that culminated in the writings of Maimonides, separated what we recognize to be the study of science from the standard curriculum of the learned Jew. Whereas intellectual energy in the pre-modern Jewish world had been directed into what we call science no less than into meditation and community action, the rabbis who created the modern world of the Jews excluded any continued pursuit of science. The result has been a largely assimilated contemporary Jewry who place little value on their Jewish heritage and their Jewish identity.

In the past century the spiritual leaders of the Jewish people resisted assimilation with considerable creativity—most notably with the development of liberal religious Judaism and secular nationalist Zionism. Both of these movements have been the primary conceptual and pragmatic means by which the Jewish people as a people have sought to survive in modernity. However, there is an inherent weakness in both approaches. Liberal Judaism, modelled on liberal Christianity, is a nineteenth-century response to the mechanistic philosophies of the seventeenth century as it became enacted in the political reforms of the French Revolution in the eighteenth century. As such, liberal Judaism is a development that is at least 100 years too late, for it came just at the time when scientific thinking had finally absorbed the humanist lessons of the Age of Reason and was moving beyond it into a world of uncertainty and non-determinism in the most modern science. In other words, liberal Jewish thinkers embraced scientific certainty at just the time when cutting-edge scientists were discovering that nothing about the world is certain.

Similarly, the political leaders of the Jewish people adopted nationalism as the desired form for Jewish collective identity precisely at the time when the develop-ment of modern technology promised (or threatened, as the case may be) to produce a world of collective identity that transcends national boundaries in the formation of what we today call 'globalism'. It was the ideal of the eighteenth century to create national communities based solely on shared economic and cultural identity in order to replace the failed cosmological forms of religious identity that had prevailed in medieval Europe and led its people into the horrendous religious wars of early modernity—of Christians against Muslims and, finally, Christians against Christians. However, the failure of nationalism became apparent to many European intellectuals in the battlefields of the First World War that initiated us into the twentieth century, at whose end we were using our technological advances to move into a world of both economic and cultural relations that transcend in every respect the increasingly archaic idea of the nation. It is precisely at this time of the obsolescence of the nation-state that both Jews and Muslims have decided to create nation-states and vest their futures in them. Clearly the twenty-first century must show us better, more reasonable alternatives for our present age of technology.

The source of both failures of Jewish life at the beginning of the twenty-first century has its source in the now centuries-old decisions of the community's religious leaders to marginalize science from spiritual religious life. They did so not out of a strategy of conflict, but rather because they simply ceased to see the relevance of science to the future existence and prosperity of the Jewish people. My hope is that in this new century this leadership will revive its more classical approach of har-monization, and promote a communal leadership that is informed of both its religious and its scientific heritage, as a community of 'God-fearers' dedicated to the pursuit of all forms of 'wisdom'.

CHAPTER 4

CHRISTIANITY AND SCIENCE

JOHN POLKINGHORNE

INTRODUCTION

Christian theology has always resisted a Manichaean opposition between God and the world, believing that the universe is God's creation and that, in the Incarnation of the Word made flesh, the One by whom all things were made became a participant in the history of the world (John 1: 3, 14). As a consequence, Christian thinking at its best has sought to be in a positive relationship to all forms of human knowledge, including science, without allowing itself to become distorted by an improper submission to the restricted protocols of purely secular argument. All forms of rational inquiry into aspects of reality have their own particular motivating experiences and indispensable concepts. Therefore, neither science nor theology should make the mistake of supposing that it can answer the other's proper questions. Nevertheless, there has to be a consonance between the answers that each gives, if it is indeed the case that there is a fundamental unity of knowledge about the one world of created reality.

Already in the second century, apologists such as Justin Martyr sought to give a reasoned defence of Christianity in the intellectual context of the later Roman Empire. When Augustine came to write his *Literal Commentary on Genesis* (early fifth century), he was not concerned with some kind of naïve biblicism; rather, he acknowledged that if well-established secular knowledge seemed to conflict with a customary interpretation of Scripture, then the latter might need to be reconsidered. (Much later, Galileo would appeal to this dictum in his controversy with Cardinal Bellarmine about the relationship of Copernican theory to the Bible.) Augustine himself had been persuaded to abandon his early adherence to Manichaeism partly by a recognition that the predictions of contemporary astronomers concerning coming eclipses were more accurate than those offered by the sages.

The later Middle Ages was a period of intense intellectual activity, involving thinkers drawn from all three Abrahamic faiths: Judaism, Christianity, and Islam. On the Christian side, an important figure was Thomas Aquinas. He had been greatly influenced by his teacher, Albert the Great, a student of nature and an early proto-scientist, and by the newly recovered writings of Aristotle. Thomistic think- ing does not oppose faith and reason, but it sees secular and sacred knowledge as complementary to each other in the quest for truth. The distinction between them lies in the one appealing to argument based on general experience, while the other relies on insights given in unique revelatory disclosures. The common quest for truth was well expressed by Anselm when he spoke of theology as 'faith seeking understanding'.

It was in seventeenth-century Europe that science in a recognizably modern form came to birth. Galileo pioneered the combination of mathematical reasoning with skilful observation and experiment. This initiating phase attained its greatest achievement with Isaac Newton's publication in the *Principia* of the theory of universal gravitation, and in the *Opticks* the account of his researches into the nature of light. Scholars have debated why it was that these developments took place there and then, rather than in ancient Greece (with its philosophical acumen and math- ematical discoveries) or in medieval China (in many ways a civilization significantly in advance of contemporary Europe). Counterfactual speculations about alternative courses of history can never establish certain conclusions. Nevertheless, a significant case can be made that it was the doctrine of creation that supplied a supportive ideological setting for the development of modern science. If the world is God's creation, it may be expected to display a rational order expressive of the Mind of its Creator. The ancient Greeks also believed in cosmic order, but they thought that its pattern derived from the necessary form imposed on the creative activity of the demiurge, a pattern to which one might gain access by pure thought. However, Christian theology believed that the order of the world was the free choice of its Creator, and so one had actually to look and see what God had chosen to do—hence the need to combine mathematical reasoning (order) with observation (inspection) in the manner that Galileo employed so successfully. Moreover, if the world is God's creation, it is a worthy object of study, a point that the medieval Chinese may not have appreciated. In the seventeenth century it was a popular saying that God had written two books: the Book of Nature and the Book of Scripture. Both should be read, and if this was done aright, they could not contradict each other, since they had the same Author.

It is certainly the case that the early pioneers of science were mostly people of Christian faith, even if some of them had their problems with the religious author- ities (Galileo) or with Christian orthodoxy (Newton). Another influence that may have been at work was a shift, starting in the late Middle Ages and intensified considerably in Reformation times, to reading the Bible less symbolically and more in a matter-of-fact manner. The adoption of a similar attitude to nature meant that it was regarded no longer as a source of inspiring spiritual images, in the style of the medieval bestiaries, but as significant in its own right. The pelican came to be seen simply as a bird, and not as a symbol of the Eucharist.

Newton had been deeply impressed by the order of the solar system, seeing it as reflecting the power of the Lord of the universe. Early exploitation of the resources of microscopy by Anthony van Leeuwenhoek and Robert Hooke revealed a world of tiny but exquisitely structured life forms. This led to Christian investment in what came to be known as physico-theology, an admiration for the wonderful order of nature, understood as testifying to the character of its Creator. The Cambridge naturalist John Ray, a pioneer of scientific taxonomy, wrote an influential book, *The Wisdom of God Manifested in the Works of Creation* (1691), which ran through many editions. This kind of argument from nature to God reached a peak in William Paley's *Natural Theology* (1802), surveying a wide range of scientific data, both physical and biological. These Christian discussions tended to underplay the ambiguity of the evidence, paying insufficient attention to the darker side of nature, with its malformations and disasters—a point of criticism made trenchantly by David Hume in his *Dialogues on Natural Religion* (1779). There was also insufficient recognition of the logical uncertainty of attempts at a kind of inductive theology, a point emphasized by Immanuel Kant in his insistence on an agnostic division between accessible appearances (phenomena) and the inaccessible nature of things in themselves (noumena).

The development that put an end to Christian reliance on a Paleyesque style of natural theology was not, however, philosophical critique, but a scientific discovery. The publication in 1859 of Charles Darwin's *On the Origin of Species* showed how the patient accumulation and sifting of small differences, taking place over very long periods of time, could give rise to life forms adapted to their environment without the need for the direct intervention of a divine Designer to bring this about. There is an ill-judged interpretation of this seminal event, frequently repeated, that assigns to it the mythical status of a final parting of the ways between science and religion, leading to the triumph of the former and the defeat of the latter. The idea is based on a historically inaccurate notion that Darwin's ideas were immediately and unanimously accepted within the scientific community, while an obscurantist religious community equally unanimously rejected them. This is just untrue. In fact, there were a variety of reactions on both sides. In the scientific community there was a degree of resistance to Darwin that persisted until the discoveries concerning genetics, made by the Moravian monk Gregor Mendel, were recovered at the beginning of the twentieth century. On the religious side, responses were equally varied. Some Christian thinkers, notably Asa Gray in North America and Charles Kingsley and Frederick Temple in Britain, welcomed Darwin's insights from the first. Both of the latter thinkers used a phrase that neatly encapsulates a theological understanding of biological evolution. They said that while no doubt God could have brought into being a ready-made world, it had turned out that the Creator had chosen to do something cleverer than that in making a world in which creatures 'could make themselves'.

The year 1859 certainly did not bring the dialogue between Christianity and science to an end, though it did direct that conversation in new directions. The interaction with science has continued to be a matter of particular concern to Christian thinkers. After this brief historical introduction, it is time to survey the contemporary scene. Five specific topics characterize the present-day discussion.

CREATION

The picture of a static, ready-made world constructed by the divine Artificer had to give way to the dynamic picture of an unfolding creative process. Christian thinking today speaks both of *creatio ex nihilo* and of *creatio continua*. The former concept is understood to express the ontological dependence of all that is on the sustaining will of the Creator; it does not refer simply to the temporal initiation of cosmic history. God is as much the Creator today as God was 13.7 billion years ago, when the universe as we know it emerged from the singular state of the big bang. The cosmos is not understood by Christian theology to possess existence of a kind that would enable it to continue independently of a God who was no more than a deistic spectator of its history, for Christians believe that without the divine sustaining will the world would cease to be. Yet the creation that God continues to hold in existence is not a divine puppet-theatre in which the great Puppet-Master pulls every string. Creatures are allowed to be themselves and to make themselves, in the course of an evolving process that has, over billions of years, turned an initially expanding ball of matter-energy into an arena of complexity, the home of saints and scientists. This unfolding process of *creatio continua* is the way in which creatures explore and bring to birth the new possibilities that emerge from the inherent fruitfulness with which the creation has been endowed.

Cosmic process is pictured as a kind of developing improvisation in which Creator and creatures are both involved, an expression of the divine purpose for creation, a concept that science, of course, brackets out of its self-limited discourse. Scientifically, evolution (understood as a general process as relevant to the formation of galaxies and stars as to the development of terrestrial life) may be conceived as an interaction between two contrasting principles: chance and necessity. Both words need careful explanation. By 'chance' is meant not a capricious randomness, but simply the particularity of historical contingency. The scope of possible happenings very greatly exceeds the range of actual events, so that only a limited set of conceivable options has occurred. A particular genetic mutation happened and turned the stream of life in a particular direction. Had a different mutation occurred, the consequences would have been different. Contemporary Christian theology acknowledges this contingency. It does not picture the history of creation as the inexorable performance of a pre-existent divinely written score. While it can regard the coming-to-be of self-conscious,

God-conscious, beings as a fulfilment of the Creator's intentions, it does not have to believe that specifically five-fingered *Homo sapiens* had been decreed from all eternity.

'Necessity' refers to the lawful regularity of the world. In the next section we shall see that this had to take a very specific form if the evolution of carbon-based life were to be a possibility anywhere at all in the universe. Atheist writers who look to evolutionary thinking as the great principle of universal explanation usually pay scant attention to this vital requirement of fine-tuned specificity. On the other hand, Christian theologians see it as the Creator's gift of fertile potentiality.

Important theological understandings flow from these modern insights into the processes of creation. One of the most significant is a recognition that creation is a kenotic act on the part of the Creator, a self-limiting of the exercise of divine power. Creatures are truly given the liberty to be themselves and to make themselves. Although all that happens depends upon God's permissive will in holding the world in being, not every event that takes place will be in accordance with the divine positive will. A Christian theologian can believe that God wills neither the act of a murderer nor the destruction wrought by an earthquake, but that both are permitted to happen within a creation that has been given a degree of creaturely independence. The great good of a world in which creatures can make themselves has an inescapable shadow side. Genetic mutation has been the means that has both driven the fruitful history of terrestrial life and also been a source of malignancy. The one cannot be had without the other in a non-magical world. This appeal to the letting-be of creative process, recognized as the source of the ambiguities of a cosmic history at once both fertile and destructive, has been a major component in contemporary Christian attempts to wrestle with the problems of theodicy. Few would claim that it removes all perplexities. Christians also have recourse to a further unique and more profound insight into God's relationship to the problem of evil. The Christian concept of God is not one of a deity who is simply a compassionate spectator looking down on the travail of creation. The doctrine of the Incarnation, and in particular the darkness and dereliction of the cross of Christ, imply that the Christian God has also been a fellow-participant in suffering, a sharer on the inside in the bitterness of the world, and not just an onlooker on the outside.

NATURAL THEOLOGY

After 1859 the old-style natural theology, appealing to 'design' held to be visible in the forms of nature, fell into disrepute. Barthian emphasis on the Word of God as the sole source of Christian revelation further marginalized that kind of thinking. Yet in recent years there has been an important revival of Christian interest in the possibility of natural theology. However, this new natural theology adopts a significantly revised strategy of argument compared to that employed by its predecessor.

First, it is more modest in its intellectual ambition. It does not speak of 'proofs' of God, but its concern is with the insightfulness offered by a theistic view of reality. The claim is not that atheism is incoherent, but that it explains less than theism can. Second, the emphasis is not on individual occurrences, such as the optical aptness of the eye, but on the character of the laws of nature themselves, which are the very basis for the possibility of any form of occurrence. The particularities of historical development, such as details in the evolutionary history of life, are acknowledged to be the proper concern of science, and no claim is being made that scientifically statable questions should receive theologically expressed answers. Yet the laws of nature, which are the assumed ground for the scientific explanation of particulars, and which science itself has simply to treat as given brute facts, are held to display a character that makes it intellectually unsatisfying to terminate the search for understanding at this point. Meta-questions arise concerning these laws, whose answering will inevitably take the inquirer beyond science itself. The new natural theology proposes that this further pursuit of understanding through and through will take the seeker after truth in a theistic direction. Thus the revised natural theology does not attempt to rival science on its own ground, in the way that hindsight suggests that Paley's arguments were in danger of attempting to do, but seeks to complement science by setting its insights within a broader and deeper context of intelligibility.

The new natural theology centres principally on two meta-questions. The first asks, Why is science possible at all? Obviously, survival necessity requires some kind of rough-and-ready understanding of what is going on at the level of everyday experience; but this scarcely explains human ability to understand regimes, such as those of subatomic quantum physics or physical cosmology, that are remote from direct impact on human living and whose understanding calls for modes of thought (such as quantum superposition or curved spacetime) that are counterintuitive to everyday reasoning. Moreover, it turns out that it is mathematics—surely the most abstract form of human thought—that affords the key to unlock the deep secrets of the physical universe. It is an actual technique of discovery in fundamental physics to seek theories that are expressed in terms of equations possessing the unmistakable character of mathematical beauty, since time and again it has been found that only such theories have the long-term explanatory fruitfulness that persuades us of their validity. A Nobel Prize-winner in physics, Eugene Wigner, once asked, 'Why is mathematics so unreasonably effective'? Albert Einstein once said that the only incomprehensible thing about the universe is that it is comprehensible.

Science depends for its success upon the world being rationally transparent in this remarkable manner, and scientists feel genuine wonder at the rational beauty thus revealed to their inquiry, an experience that comes as the reward for the long labours of their research. Science itself can offer no explanation of why the universe should be like this, but the fact of deep and satisfying cosmic intelligibility does not seem to be something that should be treated as just a happy accident. Belief in God the Creator makes the rational transparency and rational beauty of the universe comprehensible. Science surveys a world whose order makes it appear shot through with signs of mind, and the religious believer can affirm that this is so because it is indeed the

Mind of God that is revealed in the works of creation. On this view, science is seen to be possible because the world is a creation and human persons are beings who are made in the image of their Creator.

The second meta-question that natural theology addresses is more specific in its character. It asks, Why is the universe so special? Scientists do not like things to be special, for they prefer generality. Their natural inclination would be to suppose that the universe is just a typical specimen of what a cosmos might be like. Yet, as we have come to understand more and more about the processes that turned that initial ball of energy into the home of life, we have come to see that they depended critically on the precise form that lawful necessity takes in our world. The strengths and characteristics of the laws of physics had to be 'finely tuned' to what they actually are for the evolution of life to be possible. While it took ten billion years for any kind of life to appear, and a further almost four billion years before self-conscious beings came on the scene, there is a real sense in which the universe was pregnant with the possibility of life from the immediate aftermath of the big bang onwards, because of the form then taken by the physical fabric of the world. The set of scientific insights expressing this conclusion is called 'the Anthropic Principle', though 'the carbon principle' would have been a better choice of terminology since it is, of course, the general possibility of carbon-based life that is at issue, and not the detailed specificity of *Homo sapiens*.

Many considerations lead to this conclusion. One of the most interesting refers to the manner in which the chemical elements necessary for life came to be made. The very early universe was too simple to produce anything more complex than the two simplest elements, hydrogen and helium. The many heavier elements necessary for life, including carbon itself, could only be made later in the interior nuclear furnaces of the stars and in the supernova explosions that scatter the resulting material out into the cosmic environment. All life is made from the ashes of dead stars. Unravelling this delicate chain of reactions was one of the great astrophysical triumphs of the second half of the twentieth century. It was soon realized that the processes of nucleogenesis depend critically on the nuclear forces being essentially what they are and no different. Small changes would have removed all possibility of carbon and of carbon-based life.

The Anthropic Principle represents an entirely unexpected anti-Copernican turn in scientific thought. Of course, the Earth is not at the centre of the universe, but the fact of life is a profound constraint on what the physical character of the universe can be like. To take another example, the vast size of the observable universe, with its 10^{22} stars, is an anthropic necessity. Only a world at least as big as ours could have lasted the fourteen billion years that correspond to the natural timescale for the coming-to-be of self-conscious beings.

All agree that the observed fine tuning of the constants of nature is necessary for the possibility of carbon-based life. So remarkable a fact does not seem to be something adequately treated as just a fortunate coincidence. However, there are disagreements about what would be the most satisfying form of meta-scientific explanation. Two quite contrasting responses have been proposed. One is the multiverse explanation, the hypothesis that this universe is just one in an enormous portfolio of universes, all existing in detachment 'alongside' each other, and all with different laws of nature. If that multiverse were sufficiently vast and diverse, then one of its members might well 'by chance' be suitable for carbon-based life and, of course, that must be our universe, since we *are* carbon-based life. Sober scientific thinking, as opposed to exuberant speculation, does not offer reliable grounds for belief in the existence of such a multiverse, so the proposal is a metaphysical guess, and one of considerable onto-logical prodigality. An alternative metaphysical guess, and one that many might consider to be more ontologically economical, is that there is just one universe, which is the way it is in its fine-tuned potentiality because it is the creation of a God who purposefully endowed it with just those properties that would enable it to have a fruitful history. Either explanation—multiverse or creation—is trans-scientific, but the theologians can claim that belief in God offers further explanatory insights (for example, into the deep intelligibility of the universe and into the source of religious experience), while the only explanatory work done by the multiverse hypothesis seems to be to defuse the threat of theism. There is little doubt that anthropic considerations have been an important stimulus to work among Christian thinkers on a new revived and revised natural theology.

STRUCTURES OF REALITY

A number of other aspects of the human encounter with reality are regarded by Christian theology as being of significance for theological exploration. One of these is the considerable emphasis placed in contemporary science on the relational character of the physical world. Newtonian physics had pictured physical processes as involving the collisions of individual atoms moving in the container of absolute space and in the course of the unfolding of absolute time. Einstein's great discoveries in relativistic physics put an end to that separable picture, principally through his formulation of general relativity, the modern theory of gravitation. This ties together space, time, and matter in a single package deal of mutual influence. Matter curves space and time, and the curvature of spacetime shapes the paths of matter. Another great twentieth-century discovery, with which Einstein was also associated, even if some-what reluctantly, was the phenomenon of the mutual entanglement of two quantum entities that have once interacted with each other (the so-called EPR effect). This counterintuitive togetherness-in-separation implies a non-local connection by means of which the two retain a power of instantaneous causal influence on each other, however far they may move apart. Einstein himself thought that this was so 'spooky' an idea that it must indicate some incompleteness in quantum theory, but many subsequent experiments have confirmed that this quantum entanglement is indeed a property of nature. When this result is coupled with the extreme sensitivity displayed by macroscopic chaotic systems to the slightest influence coming from their environment, so that they are never truly isolatable from their surroundings, it becomes clear that our common-sense notion of separable entities is far from being unproblematic.

Thus science has discovered that reality is relational to an unexpected degree. The characteristic form of Christian theology is trinitarian. Its fundamental concept of the life of God is the mutual interpenetration and exchange of love between the three divine Persons within the unity of the Godhead, taking place in a ceaseless process that theologians call 'perichoresis'. Trinitarian thinking is fundamentally relational. Of course, general relativity and the EPR effect cannot 'prove' the Trinity, but they are strikingly consistent with what one might expect of the creative work of the triune God.

Another aspect of the human encounter with reality that in different ways is significant for all the faith traditions is the existence of consciousness and the human discernment of value. Perhaps the most astonishing event in the 14-billion-year history of the cosmos of which we are aware has been the dawning of self-consciousness here on planet Earth. In that event, the universe became aware of itself, and as an eventual by-product there emerged the possibility of the scientific understanding of cosmic process and history. The nature of consciousness remains an unsolved mystery. It is clear that it is related to brain activity—a blow to the head will establish as much—but despite the very interesting advances being made in neuroscience, mostly in identifying the neural pathways by which information is received and processed, there is still no real progress in understanding the origin of awareness. A great gap yawns between the most sophisticated accounts of neuronal networking and the simplest mental experiences, such as seeing red and feeling hungry. Triumphalist claims that consciousness is the 'last frontier', soon to be crossed by the victorious armies of a reductionist science, are totally overblown and unjustified. Christian theology does not rejoice at any form of contemporary ignorance, but equally it must resist Procrustean attempts to cut down the richness of reality to make it fit into a bed of diminished scientific explanation. The significant fact of the dawning of consciousness strongly encourages the expectation that the fundamental categories necessary for deep understanding of reality must give appropriate recognition to the personal, and not suppose that the kind of impersonal discourse natural to science is adequate on its own. In common with the other two Abrahamic faiths, Christianity recognizes that the mystery of the divine nature transcends simple notions of personality, but it also recognizes that in recourse to analogical discourse about the divine it is better to call God 'Father' than to call God 'Force'.

The personal mode of encounter with reality involves the acknowledgement of value, both moral and aesthetic. The physical world that is described by science is also the arena of moral imperative and decision. Christian thinking recognizes the existence of ethical knowledge, possessing a certainty at least equal to that of other forms of insight. Our conviction that children are not to be abused or the poor oppressed is not a convention of our society; nor is it some curiously disguised strategy for superior genetic propagation. These moral convictions are insights into the nature of reality itself. Christian theology understands our ethical intuitions to be intimations of the good and perfect will of the world's Creator.

The physical universe is also the carrier of beauty. In purely scientific terms, human experience of music is simply neural response to the impact of air waves on the eardrum. It is clear, however, that this flat manner of speaking does not begin to do justice to the mysterious truth and power of music. Nor is a great painting just a collection of spots of paint of known chemical composition. In aesthetic experience we encounter another rich and undeniable dimension of reality that altogether resists reduction to merely scientific categories. Theology once more comes to our meta-physical aid as it interprets these human experiences of beauty as sharing in the Creator's joy in creation.

These theological insights into the richness of reality are further variations on a theme already discussed in relation to natural theology: theistic belief is more comprehensive and fully explanatory than atheism can manage to be.

God and the World: Divine Action

Christian theology has always resisted a pantheistic identification of God and the world. Recently, however, there has been considerable interest in a panentheistic account of the divine relationship with creation, a view different from pantheism, since it pictures the world as being in God, though the divine being exceeds the world. In part this interest has arisen from the perception of a need to correct an unfortunate tendency in classical Christian thinking. The latter had so strongly emphasized the transcendence of God as to seem to imply too great a distance between Creator and creation. For example, Aquinas believed that God acts upon creatures, but that creatures cannot act upon an impassible God. There are obvious difficulties in reconciling this view with the fundamental Christian conviction that God is love (1 John 4: 16). However, it is not clear that remedying this defect requires embracing panentheism. Many Christian theologians think that simply recovering a stronger account of divine immanence is sufficient to act as a balancing factor in relation to divine transcendence. Those who take this latter view emphasize the importance of maintaining a clear distinction between Creator and creation in order to avoid identifying God too closely with all the evil in the world, and in order to make sure that the One who is believed to be the ground of the hope of a destiny beyond death is not caught up in the eventual futility that is the predicted fate of the present universe.

Panentheists tend to think of divine action in creation after the analogy of human intentional agency exercised within embodiment, often envisaging some form of divine top-down action on the cosmos as a whole. Others seek a different way of conceiving how the divine energies might be at work within the world. All have to make some shift to understand how their ideas might relate to scientific accounts of cosmic process. The topic of divine action dominated the agenda of the Christian discussion of science and religion in the 1990s. The proposals made were diverse, but a common theme in many was some form of appeal to the demise of a merely mechanical account of physical process that had resulted from the twentieth-century discoveries of the widespread presence of *intrinsic* unpredictabilities in science's account of nature. These unpredictabilities are present both at the subatomic level of quantum theory and at the macroscopic level of chaotic dynamics.

Unpredictability is an epistemological property, and there is no necessary entailment from epistemology to ontology. What connection should be proposed is not determined by scientific considerations alone (the existence of both indeterministic and deterministic interpretations of quantum mechanics makes that point clearly enough), for all issues of causal structure are ultimately matters for metaphysical decision, constrained by physics but not settled by it. Those who take a realist view, believing that what we know is a reliable guide to what is the case, will incline to interpret unpredictabilities as signs of some form of ontological openness. The scientific account, based on the exchange of energy between constituents, then need not be considered to be a total account, so that there is room for the action of further causal principles—for example, a top-down influence of the whole on the parts, conceivably relating to the input of information that serves to specify overall patterns of dynamic behaviour.

In view of the complexity of these causal issues, it is scarcely surprising that a variety of different specific proposals were made for how one might consider divine providential action to be exercised, none of which has received universal endorsement. After all, there are equally disputed questions about how human intentional agency is exercised through the instrumentality of our bodies, and there is no reason to suppose that it should be easier to understand divine agency than human agency. However, one important result did stem from this intense activity in the field of science and religion: namely, 'the defeat of the defeaters'. It is clear that physical closure of the causal nexus of the world has not been established, so that claims that science has disproved the possibility of providential agency can be seen to be false. Belief in divine action is no more necessarily negated by an honest science than is belief in free human agency.

The general concept of divine action that has been explored is that the process of the world has an open grain within which God can interact providentially with history in a non-interventionist way, perhaps by the input of pattern-forming information. The universe is pictured as a world of true becoming in which the future is not yet fixed and awaiting our arrival. A number of Christian theologians have concluded from this that since God knows things as they truly are, God will know such a world in its becomingness. God will not only know that events are successive, but will know them in their succession. This implies a genuine divine engagement with time, the presence of a temporal pole in divinity, complementary to the divine eternal pole. Process theology believes in such a dipolar God, but considers it to be a matter of metaphysical necessity. Other Christian theologians who embrace divine dipolarity regard it as a kenotic act of the Creator freely to have accepted such a relationship to a temporal creation. The idea seems to imply a kenosis of divine omniscience, resulting in God's possession of a current omniscience (knowing all that is now knowable) rather than an absolute omniscience (knowing all that will ever be knowable). It would not be a divine imperfection not to know the future, if the future is not yet there to be known.

Other theological consequences are also associated with the picture of an open universe. While there are intrinsic unpredictabilities in nature, there are also processes whose outcomes are reliable—clocks as well as clouds, one might say. The clock-like processes may be interpreted theologically as reflecting the faithfulness of the Creator. In consequence, there are some things that it is not sensible to pray for, a point recognized by Origen in third-century Alexandria, when he said that one should not pray for the cool of spring in the heat of summer. In regimes of cloudy unpredictability it is not possible to disentangle all the threads of causality, itemizing this as due to nature, that as due to human agency, and a third thing as due to divine action. Providence may be discernible by faith, but it will not be demonstrable by analysis.

An account of God providentially at work within the open grain of the universe is one that is free from any notion of occasional and arbitrary interventions. But what about claims of the miraculous, events so totally contrary to common expectation that they cannot be supposed to have been the results of a general providence of this non-interventionist kind? The question of miracle is essential to Christianity, since the resurrection of Jesus lies at the heart of its belief. A miracle of that kind could only be a singular great act of God. Science is concerned with what usually happens, so it cannot categorically deny the possibility of unique events. The real problem of miracle is not scientific, but theological. It is theologically incredible that God should act as a capricious magician, doing something today that God did not think of doing yesterday and won't be bothered to do tomorrow. There must be a deep divine consistency, but that does not condemn the deity to an unvarying regularity of an impersonal kind, like the uniform action of gravity. God's consistency lies rather in a perfectly appropriate relationship to actual circumstances. When those circumstances change, divine response may also change. A transpersonal agent may fittingly be believed to act in unprecedented ways in unprecedented circumstances. The role of theology in relation to miracle is to discern this deeper kind of consistency, a task which has to be undertaken on a case-by-case basis, since there can be no general theory of unique events.

ESCHATOLOGY

Science predicts that, after immense periods of time, the universe will end in futility, either through collapse or (the currently favoured expectation) through long-drawn-out decay. Christian theology is challenged to say how it responds to this prognostication of a dismal fate for creation. The issue is the cosmic version of the equally pointed question posed by the even more certain knowledge that every human life ends in death. In regard to the latter, Jesus pointed to the faithfulness of God as the ground for the hope of a destiny beyond death, affirming his trust in the God of Abraham, the God of Isaac, and the God of Jacob who is the God 'not of the dead but of the living' (Mark 12: 18–27). The purely naturalistic story that science can tell does indeed end in futility, but there is a further theological story that can go beyond the demise of this present universe. God's ultimate purpose is that this world of transience, in the course of whose evolutionary process each generation must give way to the next, will be transformed into the new creation in which death will be no more (Revelation 21: 3–4). Christians believe that this new creation has already begun to grow from the seminal event of Christ's resurrection.

In recent years there has been some serious discussion in the forum of Christian thinking about science and religion, centring on how one might begin to make sense of such an eschatological hope. The key necessity is to find a balance between continuity and discontinuity in the relationship of the new creation to the old. There must be sufficient continuity to ensure that it really is Abraham, Isaac, and Jacob who live again in the kingdom of God, and not just new characters who have been given the old names. Yet the patriarchs cannot be made alive again simply to die again. There must be sufficient discontinuity to ensure that the world to come is freed from the transience and death of this world.

In Christian thinking, the conventional carrier of continuity between this world and the next has been the human soul. It has often been conceived after the Platonic pattern of a detachable spiritual component, released from the body at death. Such a dualist picture of human nature is not essential for Christianity. Many Christian theologians take what is in fact the predominant biblical view, that human beings are psychosomatic unities, animated bodies rather than incarnated souls. But if that is so, what has happened to the human soul? It has not been lost, but it needs to be reconceived. The human person is certainly not to be identified simply with the collection of atoms at any one time making up the body. Those atoms are changing all the time through wear and tear, eating and drinking. What carries personal continuity in this life is the almost infinitely complex, information-bearing pattern in which those atoms are organized. This pattern is the human soul, an insight that is a revival in modern dress of the Aristotelian-Thomistic idea of the soul as the form of the body. This pattern will be dissolved at death with the decay of the body, but it is a perfectly coherent hope that the faithful God will preserve it in the divine memory and ultimately reconstitute the soul's embodiment in an eschatological act of resurrection.

This re-embodiment will have to be in the different 'matter' of the new creation. Again it is a perfectly coherent hope that God will endow this transformed 'matter' with such strong self-organizing principles that it will not be subject to the thermo-dynamic drift to disorder that is the source of the transience of this world. Christian belief in the empty tomb implies that Jesus' risen and glorified body was the transform of his dead body, so that in Christ there is the hope of a destiny for matter beyond the death of this universe. The new creation is not a second act of *creatio ex nihilo*; it arises *ex vetere*, by the redemption of the old from futility. Its different properties result from its being in a more intimate relationship with the life and energies of God than is possible in a world in which creatures are given the evolutionary freedom to make themselves.

The Christian hope is formulated in terms of death and resurrection, not spiritual survival, because Christian thinking considers that it is intrinsic to true human nature to be embodied in some form. We are not apprentice angels. Humans are also intrinsically temporal. Our destiny is not a timeless eternity, but an unending life within the 'time' of the new creation. Everlasting life will not be boring, for fulfilment lies in the ceaseless exploration of the inexhaustible riches of the divine nature that will be progressively revealed to the redeemed in the life of the world to come.

References and Suggested Reading

Barbour, I. G. (1998). *Religion and Science*. San Franscisco: Harper Collins Publishing.

Clayton, P. (1997). *God and Contemporary Science*. Edinburgh: Edinburgh University Press.

Gregersen, N. H., and van Huyssteen, J. W. (1998) (eds). *Rethinking Theology and Science*. Grand Rapids, Mich.: Eerdmans Publishing Co.

Peacocke, A .R. (1993). *Theology for a Scientific Age*. New York: SCM Press.

Polkinghorne, J. C. (1998). *Belief in God in an Age of Science*. New Haven: Yale University Press.

—— (2002). *The God of Hope and the End of the World*. New Haven: Yale University Press.

—— (2001) (ed.). *The Work of Love*. Grand Rapids, Mich.: Eerdmans Publishing Co.

Saunders, N. (2002). *Divine Action and Modern Science*. Cambridge: Cambridge University Press.

CHAPTER 5

ISLAM AND SCIENCE

SEYYED HOSSEIN NASR

INTRODUCTION

The issue of Islam and modern science along with its progeny, modern technology, continues today as one of the most crucial faced by the Islamic community. It has been, and continues to be, addressed by numerous scholars and thinkers, covering nearly the whole gamut of the spectrum of Islamic intellectual activity since the last century. Far from being recent, this intense interest in the subject goes back, in fact, to the beginnings of serious intellectual encounter between the Islamic world and the modern West, in the early thirteenth/nineteenth century, and embraces the *nahḍah* movement in the Arab world, as well as similar movements among Persians, Turks, and the Muslims of the Indo-Pakistan subcontinent at that time. It has also attracted figures as different as Jamāl al-Dīn Asadābādī (al-Afghānī) and Sir Syed Ahmad Khan, Zia Gökalp, Bediüzzam Said Nursi, Muḥammad Iqbal, and the followers of the *salafiyyah* movement, and in our times, the various shaykhs of al-Azhar, as well as practitioners of modern science, such as Abd al-Salam.

While in the earlier days the interest of Muslim thinkers in Western science and, to some extent, technology was due to their intellectual and theological challenge to political independence, at least on paper, of Muslim lands, the interest of Muslim governments in science and technology today is almost always because of what they

I am grateful to the editors of *The Islamic Quarterly* for permission to republish this paper. (The text has been revised by the author.)

feel is their need to gain power, whether it be economic or military, and not wisdom. In fact, an army of modernists have joined the so-called fundamentalists in their blind praise of modern science and technology, equating blindly science in its current English sense with al-'ilm of the Noble Qur'an, Ḥadīth, and the whole Islamic intellectual tradition (Nasr 1990b). Most have remained impervious to the difference between the goal of knowledge as the gaining of wisdom, increasing the depth of understanding of God's signs (āyāt), and his creation in relation to Him and the perfecting of the human soul, on the one hand, and its goal of gaining power to dominate over God's creation and other human beings, especially those belonging to another nation, ethnic or religious group, as the history of the modern West exemplifies so readily, on the other. Of course, one can understand why the blind praise and almost 'worship' of modern science and technology are manifested everywhere among sectors of Islamic society with profound political, theological, and judicial differences when one sees a war in which some sixty people are killed on the other side, and nearly 300,000 on the Muslim side, or observes that in every encounter of Muslims with others from Bosnia to Chechnya to Kashmir to the Philippines, the Muslims are always on the defensive and the ones to be massacred and 'cleansed'. Yet this desire to adopt Western science and technology blindly and without critical appraisal, no matter how understandable on the emotional level or that of political expediency or even necessity, cannot remain heedless of truths that underlie the profound issues of the relationship between the Islamic world-view and modern science. And, Islamically speaking, it is always the truth (al-Ḥaqīqah) that must prevail, and we must always think as Muslims in terms of the truth, rather than expediency, whether it be political or otherwise, never forgetting the Qur'anic verse, 'the truth has come and falsehood has vanished away; verily falsehood is bound to perish' (18. 81).

It is, therefore, necessary before anything else to analyse modern science and subject it to an in-depth criticism from the Islamic view, by which is meant not just any view that claims to be Islamic by combining the external meanings of some verse of the Noble Qur'an with all kinds of concepts and 'isms' imported from the modern West, but the view drawn from the Islamic intellectual tradition, including all its branches, and understood traditionally and in the most universal perspective of Islam, rather than through theological or judicial sectarianism.

A Critique of Modern Science

In this short presentation, it is not possible to do justice to a full criticism of modern science from the traditional Islamic point of view, a task which has been carried out to some extent by other scholars and the author of this chapter in other contexts, although much still remains to be accomplished.[1] Some of the essential points need, however, to be mentioned here.

The first, which has even reached pulpits throughout the Islamic world, is a negative one. It is the refusal to even study Western science critically, often as a result of a kind of intellectual inferiority complex that simply equates Western science with the continuation of Islamic science without any consideration of the shift of paradigm and the establishment of a new philosophy of nature and science during the Scientific Revolution, events which distinguish modern science sharply not only from Islamic science, but also from its own medieval and early Renaissance past. It is astounding that some not only simply equate modern science with Islamic science, but also try to apply the modern philosophy of science, based upon an agnostic science of nature and often in a mode already out of fashion in the West, to judge the veracity or lack thereof of Islamic positions.[2]

The second point concerns the relationship between a value system and modern science. Instead of criticizing the implicit value system inherent in modern science from the Islamic point of view, many of the champions of the blind emulation of modern science and technology claim that it is value-free, displaying their ignorance of a whole generation of Western philosophers and critics of modern science who have displayed with irrefutable arguments the fact that modern science, like any other science, is based on a particular value system and a *specific* world-view rooted in specific assumptions concerning the nature of physical reality, the subject who knows this external reality, and the relationship between the two.

Modern science must be studied in its philosophical foundations from the Islamic point of view, in order to reveal for Muslims exactly what the value system is upon which it is based and how this value system opposes, complements, or threatens the Islamic value system, which for Muslims, comes from God and not merely human forms of knowledge which are based by definition upon human reason and the five external senses, and specifically deny any other possible avenue for authentic knowledge. Muslim thinkers must stop speaking of modern physics as not being Western but international, while hiding its provincial foundations grounded in a particular philosophy and value system related to a specific period of not global, but European history. Even a 747 Boeing jet is not global simply because it is now landing in Samoa as well as Tokyo, Beijing as well as Islamabad or Tehran. Rather, it is the result of a technology derived from a particular view of man's relationship with the forces of nature and the environment, as well as an understanding of man himself, a view which many forces in the modern and even post-modern West are trying to

---

[1] See, e.g., Burckhardt (1972, 1987) and Bakar (1991). As for my own writings, see Nasr (1987, 1990*a*, 1993*a*, 1993*b*, 1995); see also the *MAAS Journal of Islamic Science*, ed. M. Zaki Kirmani, which contains numerous studies on this issue; Rais and Naseem (1984); and Naquib al-Attas (1981).

[2] A case in point is the positivistic and rationalistic philosophy of Karl Popper, which, already seriously criticized in the West, is adopted by a number of people, especially in Iran and Pakistan, to evaluate and criticize the traditional Islamic epistemologies and philosophies of knowledge, including science.

globalize, by eliminating other views of the world of nature and man's relationship to it, including of course, the Islamic one. Modern science is a direct challenge to other world-views, including the Islamic, which claims knowledge of reality based on not reason alone, but also on revelation and inspiration. In any case, a serious criticism of modern science includes the value system upon which it is based and which it propagates. It remains, therefore, a major duty to make clear what these values are and to evaluate and criticize them from the authentic Islamic perspective.

Closely associated with the general question of values is that of ethics. Numerous writers have spoken of how ethical individual scientists are and how the unethical use of modern science is not their fault. There are, of course, many devout Christian and Jewish scientists in the West and Muslim, Buddhist, and Hindu ones in the East. But this fact has nothing to do with the non-ethical character of modern science itself. Some of the most ethical scientists, whom we met in our youth, helped to make a bomb which killed more than 200,000 people in two days in Japan half a century ago, not to speak of the tragedies brought about by German scientists during the Nazi period. Moreover, the results of the work of the most humble scientists who would never put their foot on a marching ant have helped to destroy numerous species in God's creation. As a matter of fact, knowledge and its implications cannot evade ethical implications. Modern science has helped to destroy all other perspectives on nature, including the religious, by relegating their claim to knowledge of the world to poetry, myth, or, even worse, superstition, thus barring them from the citadel of officially accepted knowledge of nature. Yet what remains of ethics in the West is essentially from the Abrahamic tradition, and therefore close in many ways to the ethical principles and practices of Islam. By rejecting the Abrahamic traditions' claim to a knowledge of nature, modern science has helped to create a condition in which this ethical heritage is being eroded more and more every day, since it does not correspond to any objectively accepted knowledge of reality in the modern world. Nor should Muslims ever think that this situation is due only to the weakness of Christianity, and that negative ethical consequences seen in the West would not occur in the Islamic world. Such a conclusion would be nothing other than the result of that superficial and shoddy thinking based usually upon scanty knowledge of the West's intellectual, philosophical, and scientific history, which unfortunately has characterized much of the Muslim response to the modern West since the last century.

What is needed is a positive Islamic critique of modern science, based on knowledge and not slogans, not only concerning what it is, but also concerning what it is not, but which many of its exponents and popularizers claim it to be. Modern science is not the only legitimate science of the natural order, but *a* science of nature, legitimate only within the premises of its assumptions about the nature of both the known object and the thinking subject. Muslims must be able to maintain the traditional Islamic intellectual space for the legitimate continuation of the Islamic view of the nature of reality to which Islamic ethics corresponds, without denying the legitimacy of modern sciences within their own confines. Otherwise, no matter how many times Muslim scientists pray, the displacing of the Islamic intellectual universe by one drawn from modern science, while it may make Islamic countries rich and powerful, will destroy the hold of Islamic ethics upon the larger Islamic community, as one observes not only in the case of the Christian West, but also among those modernized Muslims who have abandoned most of their spiritual and ethical heritage in the name of 'the scientific world-view', propagated on the one hand by the now mostly defunct Marxism as a pseudo-religious slogan, on the other hand foisted as the flag which unifies so many secularists, humanists, and other anti-religious forces in the West.

And finally there is the most essential criticism concerning the at best neutral attitude of modern science concerning religion and the paramount role of science in creating a mental ambience from which God and the eschatological realities are absent and, therefore, finally 'unreal'. Numerous Western writers have tried to show that science is not *against* religion and does not necessarily deny God, and many Muslims have claimed the same. But during most of these debates, religion has always been asked to reform itself or change its doctrines, especially its claims regarding knowledge of both supernatural and natural reality, with the result that four centuries after the rise of modern science, it is religion that is now marginalized, and not science. Occasionally Western theologians, usually in awe of modern science, point to this or that scientific discovery as conforming to a particular religious teaching, unaware of how dangerous it is to correlate that which possesses the character of absoluteness with a form of knowledge which is by definition, transient, although it does reflect certain metaphysical truths if seen from the metaphysical, and not simply scientific, point of view (Smith 1984). This type of shallow correlation is especially evident these days in what is now called cosmology in the West, which is nothing but the extrapolation of astrophysics, and which has nothing to do with cosmology as traditionally understood (Burckhardt 1987). For years, many theologians have been excited by the big bang theory which, they claim, accords with the biblical and even Qur'anic understanding of creation, and many symposia have been held on this matter. Meanwhile, many cosmologists are now beginning to deny the reality of the big bang theory itself.

The significant point here is that there must be a profound analysis and critique of modern science in its relation to religion from the Islamic point of view, and totally opposed to that enfeebled intellectual reaction which first accepts the theories, hypotheses, and even conjectures of modern science as being absolutely true, and then tries to torture this or that verse of the Noble Qur'an or a particular *ḥadīth* to prove Islam's conformity to this most transient form of knowledge, whose prestige emanates not from the illumination that it provides of the nature of reality, but from the fact that it leads to the acquiring of wealth and power over nature, as claimed by one of its founders, Francis Bacon. What is essential to show is that in modern science, the very 'hypothesis' of the existence of God is redundant to the system. One can be a famous physicist who is a devout Catholic, Jew, or Muslim, but also a renowned physicist who is an agnostic or atheist. The reality of God has had nothing to do with the system of modern science as seen by that science, and God has been called by some 'an unnecessary hypothesis'. Today, the world-view of physics itself is changing, and some speak of the necessity of considering God and consciousness as fundamental to even quantum mechanics; but this view is still far from being acceptable to the majority of modern physicists, and the prevalent view taught as *the* only correct and legitimate form of knowledge is one from which God is simply absent, no matter how many modern scientists believe individually in him.

How can Islam accept any form of knowledge that is not rooted in God and does not necessarily lead to Him? How can it explain the universe without even referring to the Transcendent Cause of all things, of which the Noble Qur'an speaks on almost every page? Traditional Islamic thought has provided many profound answers to such quandaries, while the contemporary Islamic world can be characterized as being particularly bereft of responses in the light of existing challenges which would come even close in depth to the answers provided by our ancestors. That is why, in fact, those scientists and thinkers in the West interested seriously in alternative views of reality are much more interested in the Muslim thinkers of old than in contemporary Muslim thinkers. There is no way to find a path toward a healthy relationship between Islam and modern science other than an in-depth critique of any system which claims knowledge of God's creation, including modern science, from the Islamic point of view.

The Question of Absorption of Modern Western Science

For over a century, various Muslim leaders, whether they are religious or political, have pleaded for a rapid and complete absorption of modern Western science, occasionally adding a few pious remarks that this act must be combined with the preservation of Islamic ethics. There is, however, no possibility of absorbing modern science blindly without the greatest consequence for Islamic society. There can, of course, be the type of adoption based on blind emulation rather than judicious selection that one observes in many Islamic countries today, without such attempts to accept modern science *in toto* having led to any notable scientific activity and creativity which are Islamic, or to the complete absorption of the modern sciences. At best, it has led to contributions to the prevalent modern science by men and women who are Muslims, but whose Islamicity has had little to do with the science to which they have contributed. If there were to be a successful total absorption, however, the impact upon the very fibre of Islamic society would be much greater than what one sees today, precisely as a result of the current lack of total success in the carrying out of such a process.

The adoption of Western science can be carried out completely only by absorbing also its world-view, in which case the consequences for the Islamic view of reality, both cosmic and meta-cosmic, cannot be anything but catastrophic. Nor has it been otherwise for other religions. Those who keep mentioning the case of Japan should look not only at the success of that nation scientifically and technologically, while some of its traditional institutions such as that of the Emperor have been preserved and people still continue to wear kimonos and use chop-sticks. The situation must be seen from the perspective of Buddhism and Shintoism and the spiritual havoc wreaked upon the Japanese religious tradition, causing a major social crisis to the extent that now some in Japan are speaking about the 're-Asianization' of their country.

What, then, is to be done? Digestion of any external substance for any living being involves both absorption and rejection. If we were to absorb all that we eat without rejecting some of it, we would die in a short time. The case of a living religion and civilization are similar, and, lest one forget, Islam is still a vital and living religion, and even the great civilization created by this religion, although partly destroyed not only by Europeans, but also by modernized Muslims themselves, is far from being defunct. Islam and Islamic civilization cannot adopt modern science seriously without rejection, as well as absorption, without what one might call judicious adaptation and absorption, based on the principles and nature of the living reality which is performing the act of adopting and absorbing.

If proof be necessary for such an obvious assertion, one needs only turn to the history of the Islamic world during the past century. Rabid modernism, blind adoration, and emulation of modern science have certainly not brought about a major scientific renaissance in the Islamic world. A kind of shallow scientism has produced a large number of scientists, and especially engineers, in the Islamic world, without spawning a scientific activity which would spring from the heart of Islamic civilization itself, from which many Western-trained Muslim students of modern science find themselves alienated. Those with personal piety take refuge in the great gift of faith (*al-īmān*) and continue to pray and recite the Noble Qur'an, but intellectually they feel exiled from the traditional Islamic intellectual universe, which they then begin to criticize as not being really Islamic, thereby creating the cleavage in the Islamic intellectual world conspicuous today in so many Islamic countries. Moreover, this attitude toward Western science has helped to destroy much of the Islamic humanities, thereby creating a vacuum whose consequences are evident in many parts of *dār al-islām*.

What is needed is the rediscovery and reformulation in a contemporary language of the Islamic world-view, within whose matrix alone can any foreign body of knowledge such as modern science be studied, criticized, and digested, and the elements alien to that world-view rejected. Moreover, this world-view, as far as the cosmos and the whole question of various scientific epistemologies are concerned, cannot simply be extracted from the Sacred Law, or *al-Sharīᶜa*, which embodies God's Will for our actions in this world, nor even from *Kalām*, whose role has always been to protect the citadel of faith from rationalistic attacks, nor still from jurisprudence (*al-fiqh*) understood in its current sense rather than in its Qur'anic meaning. Rather, it must be drawn from the *al-Ḥaqīqah*, which lies at the heart of the Noble Qur'an and *Ḥadīth* as expounded and formulated by the traditional commentators, as well as Islamic metaphysics, cosmology, the doctrinal and intellectual aspects of Sufism, and the Islamic sciences themselves.[3] Only in this intellectual tradition, shunned by both the modernists and many of the so-called fundamentalists, can one rediscover the authentic Islamic world-view insofar as it pertains to the knowledge of nature, and in fact, the whole question of the levels of knowledge. Only with its help can Muslims really make their own any kind of foreign body of knowledge which claims to correspond to some aspect of reality without either being destroyed in the process or remaining simple copiers and emulators. The history of the Islamic world during the past two centuries offers many lessons if only one were to study it with the eye possessing correct vision or what the Qur'an calls *al-baṣīrah*.

Steps in the Creation of an Authentic Islamic Science

The road to the achievement of an authentically Islamic science[4] is a long one, yet a road that must be traversed if the Islamic world is to remain Islamic and also create a science in conformity with its own ethos as it did 1,200 years ago. I will mention a few of the milestones on this long and arduous journey.

(1) The first necessary step is to stop the worship-like attitude towards modern science and technology which is prevalent today in much of the Islamic world, where one can detect a crass scientism, long rejected by many of the leading physicists and philosophers of science in the West itself, not only among modernists but also among certain of the juridically most conservative elements of Islamic society who, while rejecting all use of reason within Islamic thought, accept scientism with hardly any questioning. In fact, the very attack of such groups on the Islamic intellectual tradition during the past two centuries has done much to create a vacuum filled

---

[3] Many of my writings, including those mentioned in n. 1 above, have been dedicated to the achievement of this task.

[4] A word must be said about the criticism made in certain circles about the defence of Islamic science as being a form of fundamentalism. This is a crass criticism based on a journalistic term created by the Western media, loaded with pejorative connotations, and then employed by whoever the powers that be in the West do not like at the moment. Any group of people photographed praying together can be and have been used to demonstrate the danger of 'fundamentalism' and its rise. By that logic, your grandmother and mine, who never missed a prayer in their lives, were fundamentalist grandmothers. Furthermore, to draw examples from the modern West and to ask why there must be an Islamic science when there is no Christian science, is to misunderstand the whole of modern Western intellectual history and, in a sense, to place the secularization of the cosmos and the separation of the knowledge of the world from religion as an example and ideal for the Islamic world, as if the religion of Islam based on unity (*al-Tawḥīd*) could ever accept any form of irreducible dualism.

quickly by Western positivism and scientism, creating tensions between external piety and submission to scientism that are bound to have even more catastrophic consequences in the future than those we observe today.

This trend must be reversed, and the whole of modern science and technology be seen not with a sense of inferiority as if a frog were looking into the eyes of a viper, but from an independent Islamic world-view whose roots are sunk in Allah's revelation and which could be compared to the case of an eagle who roams the horizons and studies the movements of the viper without being mesmerized by it. In light of this world-view, the whole notion of decadence in Islamic civilization, especially as far as it concerns the sciences, must be re-examined. The West must no longer dictate the criteria for renaissance, decadence, etc., on its own terms and identify scientific prowess purely and simply with civilization, conveniently forgetting that one can go to the Moon during the same time as teenagers are killing each other in the streets of the country which has sent the astronauts into outer space. Only by basing oneself on the authentic Islamic perspective can the inferiority complex so widespread among the so-called Muslim intelligentsia today be overcome, and the ground readied for creative scientific activity related to the Islamic world-view.

(2) There must be an in-depth study of the traditional Islamic sources, from the Noble Qur'an and *Ḥadīth* to all the traditional works on the sciences, philosophy, theology, cosmology, and the like, to formulate the Islamic world-view and especially the Islamic concept of nature and the sciences of nature. This arduous and yet necessary task must be carried out within the framework of the Islamic intellectual tradition itself, and not simply by going to certain verses of the Sacred Book, often taken out of context, and interpreting them by ourselves, by a mind usually cluttered by ideas, issues, and ideologies as far removed from Islam as possible. Surely this is one of the reasons why the Noble Qur'an refers to guidance in these terms: 'He leadeth astray whom He willeth and guideth whom He willeth' (16. 93).

Men can be misled even in the reading of God's Word if they are not guided by Him. How quaint at best, and worthless at worst, appear those interpretations of the Noble Qur'an and *Ḥadīth* so prevalent today among a number of Muslims in the West, as well as among an army of modernized Muslims in the Islamic world itself. Only the revival of the traditional Islamic world-view can provide for Muslims an authentic alternative to the current Western world-view which is itself now undergoing profound transformations, and in a sense, dissolution rather than being itself a second-rate imitation of the Western view with a few Qur'anic verses interspersed to give such types of interpretations a ring of Islamic authenticity.

The rediscovery of the authentic Islamic world-view, especially as it concerns the sciences of nature, also necessitates a deep study and understanding of the history of Islamic science to which any future authentic Islamic science must graft itself to become a new branch of a tree that has its roots in the Islamic revelation and a trunk and earlier branches which cover the span of fourteen centuries of Islamic history. Unfortunately, Muslims have not been as active as Western scholars in clarifying the history of Islamic science, and those who have done serious work have usually been influenced by the Western understanding of the role of Islamic science in the history of Western science, and the positivistic understanding of the history of science in general as it developed in the West early this century, an interpretation which has been challenged even by some Western scholars of this discipline (for example, Jaki 1978: 3 ff.). As in so many other fields, the agenda in this domain for Muslims has been prepared by Western sources, even in the crucial field of Islamic science. Whatever the interest of the West might be in the history of Islamic science as a chapter in the development of its own science, the interest of Muslims cannot but be to understand the long development of science in Islamic civilization in relation to the Islamic revelation and other elements of the Islamic intellectual tradition. Islamic science must be seen and studied from the Islamic point of view, and appreciated for its achievements and not simply because of its role in the sciences of another civilization, no matter how important that role is and how much it needs to be emphasized today. In the same way that there must be a proper Islamic philosophy of science, so must there be also an *Islamic* history of Islamic science and even history of science in general with its own methodologies, definitions, and purposes while integrating all non-Muslim scholarship on this subject which is of a positive quality from the scholarly, historical, and philosophical points of view.[5]

(3) A larger number of Muslim students should be allowed to study at the highest level the modern sciences, especially the basic sciences or what the West calls pure science. In the Islamic world today we have a large number of doctors and engineers in comparison to those who have studied the pure and theoretical aspects of sciences, such as physics, chemistry, and biology, and who also work at the frontiers of these disciplines where alone profound theoretical transformations can be brought about. Those who speak with such gusto about the necessity of cultivating science and technology do not always realize that what is important is to also train scientists to practise the kind of science that has no immediate utility, without which, even in the field of modern science and technology, Muslims will always be at the receiving end and must of necessity remain satisfied with a few crumbs from the table of Western science and technology.

In contrast to what many have said about our being opposed to the cultivation of Western science, we have never advocated something which, in any case, is not a possibility at this moment of history. Rather, our proposal has been to master in the best manner modern science *while* criticizing its theoretical and philosophical bases, and then through the mastery of these sciences, to seek to Islamicize science by taking future steps within the Islamic world-view and distinguishing what is based upon scientific 'facts' from how that is interpreted philosophically, such as

---

[5] I sought to lay the foundation for such an approach almost forty years ago when I first began to conceive my *Science and Civilization in Islam* (1987); see also Ford (1978–9). In recent years, a number of works have appeared by Muslim scholars following this line of thought. See e.g. the works of Osman Bakar and Syed Numanul Haq.

the stratigraphical structure of the Himalayas or the geometry of a crystal from hypotheses parading as scientific facts, such as Darwinian evolution. We have never preached ignorance, especially for a religion such as Islam which, based upon knowledge, must confront any other school or mode of thought which lays claim to the knowledge of reality. How sad it is, in fact, that in many Islamic countries ruled by powers which claim to be great patrons of modern science, the general quality of education has declined in so many fields during the twentieth century, as a cursory study of such major universities of the earlier days as the University of the Punjab, Aligarh, and Cairo University reveals. It is impossible to understand, criticize, integrate, or transcend any form of science without deep knowledge of it. No amount of sloganeering and emotional outbursts can replace knowledge, whose primacy the Noble Qur'an confirms in the famous verse, 'Are they equal—those who know and those who do not know?' (39. 9).

No one working in an inorganic chemistry lab can follow a formula of doing Islamic science rather than pursuing the chemistry established by Boyle, Lavoisier, and later modern chemists. But physicists and chemists working at the boundaries of those sciences and imbued with not only piety but also knowledge of the Islamic world-view, could transform this science in the direction of an Islamic science of materials in the same way that with a new world-view or paradigm based on rationalism, empiricism, and secularism, seventeenth-century chemists created the new chemistry upon the cadaver of the long tradition of alchemy, whose inner meaning they did not even comprehend. In any case, any hope of opening a new chapter in the history of Islamic science which could integrate what can be integrated from modern science without causing the death of the Islamic view of the cosmos, must rely upon those who, being deeply rooted in the Islamic world-view, also know the modern sciences at their highest level without having become swallowed up and absorbed by the philosophical presumptions and secularist outlook of these sciences.

As for those Muslims who are scientists but not functioning at the boundary of their science, they can at least point out the theoretical limitations of their science, the danger of scientism, the divorce of modern science from ethics and the necessity for Muslims to emphasize the significance of ethics as much as possible, and the crisis that the blind applications of their and other branches of modern science have caused in man's relationship with nature and in the harmony of nature itself. They can also point to all phenomena as *āyāt* of God, with a significance beyond their material reality (see Kirmani 1992, where he deals with this and certain other issues raised in this paper; see also his 1989). How paradoxical it is that while many Muslim political thinkers decry the fact that Muslims are so-called behind Western science and technology, the Islamic world is catching up and even equals or surpasses many areas of the West in its destruction of the natural environment, which is the direct consequence of the application of modern technology. Although this is a story for another day, one can hardly refrain from mentioning how important it is for Muslim thinkers, including scientists, to revive the authentic Islamic view of nature even before the act of the complete integration of modern science into the Islamic world-view takes place (see my 'Sacred Science and the Environmental Crisis—An Islamic Perspective', in 1993*b*; see also Khalid and O'Brien 1992). It is the duty of Muslim scientists dedicated to Islam to point out to Islamic society at large the real dangers which the whole globe faces as a result of the application of a science divorced from ethics, on the one hand, and forgetful of God, on the other.

It is they who, along with other Islamic thinkers, must help open up a place in the current Islamic mental space for the realization of other possibilities for the studying of nature, especially the Islamic way, by pointing to the fact that modern science is a double-edged sword. It performs 'miracles' in medicine, while facilitating the death of millions of the unborn to which such a Western authority as the late Pope was referring when he spoke of the spread of the 'culture of death' in the West. It has brought about many successes in agricultural production, while being directly or indirectly the cause of many more mouths to feed. It has purified the water of the cities in the West, while causing the pollution of their air. And this is not even a question of a balancing act, for if things continue as they are today, there may be no human beings around during the next century to praise the glories of modern science. Such a tragedy could occur not only from science enabling man to transform his dagger into a nuclear bomb without having gained any greater control over his passions, the *nafs al-ammārah* of the Noble Qur'an, but even more so from the so-called peaceful use of the applications of modern science which, combined with human greed, that is not the monopoly of any one people, are waging a most relentless war upon the very natural environment created by God which nurtures us and which has enabled us to live as human beings until now.

(4) Another important step is to revive traditional Islamic sciences whenever and wherever possible, especially in such fields as medicine, pharmacology, agriculture, and architecture. Such an act would not only give greater confidence to Muslims in their own culture, but would also have immense social and economic consequences. Moreover, it would remove to some extent the monopolistic claims of modern science and technology. Even in North America and Europe, non-Western forms of medicine are spreading rapidly, including acupuncture and Ayurvedic medicine. How strange it is that Islamic medicine, one of the richest in the world, is still absent from the scene, disdained most of all by the majority of physicians from the Islamic world itself! But certain realities cannot be denied forever. At last, an association for Islamic medicine is being created in North America by Muslim physicians, and one hopes that this will cause those who are forever mesmerized by what is going on in the West to begin to take the traditional Islamic sciences more seriously. In this domain, in fact, Pakistan and the Muslims of India, who have kept Islamic or so-called *Yūnānī* medicine alive to this day, and such successful institutions as the Hamdard centres in Karachi and Delhi should serve as models for other Muslim lands. One can hardly overemphasize the importance of the role of the revival of the traditional Islamic sciences in *dār al-islām* in creating a bridge between the traditionally learned scholars and practitioners of modern science and as one of the major means of reviving the function of science within the Islamic intellectual universe.

(5) One of the most important steps that must be taken to create a veritable Islamic science is to re-wed science and ethics, not through the person of the scientist, but through the very theoretical structures and philosophical foundations of science. As already mentioned, there is no logical link in the modern world between science and ethics, because the prevalent ethic, which is primarily Christian, corresponds to a world-view that has been supplanted by that of modern science. As for scientific humanism anchored in ever changing human nature, its impact even in the secularistic West is not extensive, as far as creating a viable ethical alternative is concerned. The results of this situation can be seen in the environmental crisis in which all attempts to create and apply an environmental ethics, while simply accepting the scientific view of nature as being alone real, have had little effect in the continuous destruction of the natural environment. What is needed is knowledge of the cosmos that is congruous and shares the same universe of meaning with ethical norms, which are drawn in all civilizations from the religions which have founded them (Nasr 1996). This is, of course, a task for Muslim theologians, philosophers, and ethicists, but must also be shared by scientists themselves. There is, in fact, no possibility of creating an authentic Islamic science which is not wed to Islamic ethics in its world-view and philosophy but is related to ethics only by practitioners of science who may or may not be ethical personally, without, in any case, their ethical concerns having anything to do with the science they produce. These questions become, in fact, even more urgent as new research in bio-medicine and bio-engineering now challenge the very foundations of all religious ethics, Islamic or otherwise.

A WORD ABOUT TECHNOLOGY

The question of technology is a vast one which, although of course related to modern science, poses distinct problems of its own, and is concerned with other philosophical, psychological, religious, social, and economic issues with which we cannot deal here save to emphasize its relationship and at the same time distinction from the pure sciences. It is of course through its technology, rather than pure science, that the West has been able to dominate the world, and it is the power inherent in this technology that at once enriches and causes misery in which most Muslim governments are interested. But in the process of chasing and so-called catching up with Western technology, as if that technology were stationary, Muslim countries have in fact made themselves even more dependent upon the West. With all the discord existing among Muslims, one manoeuvre of the tanks in Iraq can bring more wealth to certain countries in the West from the treasury of the Islamic countries than would be necessary to repair and renovate a city as large as Cairo, along with a few other places. Such a fact is the result of the consequences of attempted reliance by Muslims upon one kind of technology: namely, the modern military one. But the same is true in one way or another of many other forms of technology, ranging from the pharmaceutical to the nuclear. Muslim countries remain receivers of whatever the creators of modern technology choose to sell them, including left-over pharmaceutical products, and they have no choice when certain technologies, such as the nuclear, are denied to them as a result of political considerations. The result is that most Muslim countries are today more dependent upon the West than the people of the north-western provinces in Pakistan were upon the British during the last century when they were fighting against their direct domination.

In this domain, as in science, it is necessary to develop an Islamic understanding of the parameters and factors involved in technology and its use. It must be recalled that nothing can be more anti-Islamic than an economics without ethics, which for Islam is directly linked to the *Sharīʿa*, than a consideration of immediate problems without a vision of what their proposed solution would evolve in the longer period and as it concerns Islamic values as far as man's relationship with God, society, and also God's creation are concerned. Where are the Islamic critics of modern technology similar to those in the West itself who have realized the profound dehumanizing effect of modern technology, from Heidegger and Louis Mumford to Ivan Illich and Theodore Roszak.[6] Even when voices arise here and there, they are hardly considered seriously as the catastrophic situation of the environmental crisis in many areas of the Islamic world bears witness.

What is needed is an evaluation of modern technologies even if all the choices are bad, environmentally and socially speaking, and to select in every situation what is least disruptive of the patterns of Islamic social and individual life and least destructive of the environment. Whenever possible in fields ranging from agriculture to architecture, traditional technologies of the Islamic peoples, which are often less costly and more culturally integrated into Islamic patterns of life, must be preserved and revived without any sense of shame before the technological onslaught of the modern world, which, if continued, cannot but result in the collapse of the very system that supports human life here on earth. There must also be as much effort exerted as possible to develop less disruptive alternative technologies, such as solar energy, and not wait for the day when solar cells made in the West for a few dollars will be sold to Muslim countries for hundreds or thousands of dollars. In this difficult and treacherous minefield of the use and application of modern technology, where organizations such as the World Bank and the IMF followed policies during the past decade that were directly related to the destruction of the Amazon forest and the ozone layer in the atmosphere, Muslims must have their gaze always fixed upon the Islamic teachings

---

[6] This is not to say that no Muslims have addressed this question, but one can hardly consider it as a main concern of the present-day Muslim intelligentsia or as having serious effect upon government activities, despite the writings of a number of Muslim scholars, such as Hasan Hanafi and Parvez Manzoor, and several Muslim scholars who have been associated with the Third World Network (see e.g. Hanafi and Manzoor 1988). See also *Proceedings of the International Symposium Science, Technology, and Spiritual Values—On the Asian Approach to Modernization* (1987), which is concerned with Asia in general, but includes a number of essays on the relationship between technology and Islamic society by Muslim scholars.

concerning the trust (*amānah*) for the protection of not only other human beings, but the whole of the Earth, which God has placed upon our shoulders. Islamic environmental ethics must be revived in the context of the *Sharīʿah* and the Islamic view of nature on the basis of the Noble Qur'an and numerous writings of Islamic sages and seers over the ages, and the two be made the guiding principles and the framework for all technological adaptations and development beyond blind emulation and even immediate human interests, not to speak of the tragic demands of the greed that casts its shadow so strongly in this debate.

CONCLUDING COMMENTS

In conclusion, it is necessary to repeat that any science that could legitimately be called Islamic science, and not be disruptive of the whole Islamic order, must be one that remains aware of the 'vertical cause' of all things, along with the horizontal, a science that issues from and returns to the Real (*al-Ḥaqq*), Who is the Cause of all things. Such a science has been cultivated by Muslims for over a millennium. It must now be resurrected, the Islamic philosophy of science and the Islamic world-view reformulated in a language understandable to contemporaries and, in their light, modern science both critically appraised and judicially absorbed into the Islamic intellectual universe, after which a new chapter could be added to the already illustrious history of Islamic science.

If an authentic Islamic science could be created, upon the basis of the traditional Islamic science, while absorbing those elements of modern science which correspond to some element of reality, be it only the physical, a major step would be taken for the authentic revival of Islamic civilization itself. Moreover, Islam's refusal to accept the divorce between religion, science, and philosophy, as well as science and ethics, could have the deepest consequences for the whole of humanity now standing before the abyss of annihilation caused by the application of a science based upon the forgetting of God by humans who have forgotten their role of protector and steward of His creation. Only a science that issues from the source of all knowledge, from the Knower (*al-ʿAlīm*), and that is cultivated in an intellectual universe in which the spiritual and the ethical are not mere subjectivisms but fundamental features of the cosmic, as well as the meta-cosmic Reality, can save humanity today from this mass suicide that parades as human progress. Let us hope that in these dark hours of human history, the Islamic world, as the bearer of the message of God's last planary revelation, can rise to the occasion to create a veritable Islamic science which would not only resuscitate this civilization, but also act as a major support for those all over the globe who seek a science of nature and a technology which could help men and women to live at peace with themselves, with the natural environment, and above all, with that Divine Reality Who is the ontological source of both man and the cosmos.

REFERENCES AND SUGGESTED READING

RAIS, AHMAD, and NASEEM, AHMAD (1984) (eds.). *Quest for New Science*. Aligarh: Center for Studies in Science.

BAKAR, OSMAN (1991). *Tawhid and Science: Essays on the History and Philosophy of Islamic Science*. Kuala Lumpur: Secretariat for Islamic Philosophy and Science.

BURCKHARDT, TITUS (1972). *Moorish Culture in Spain*, trans. Alisa Jaffa. London: George Allen & Unwin.

—— (1987). *Mirror of the Intellect*, trans. William Stoddart. Albany, N.Y.: State University of New York Press.

FORD, J. G. (1978–9). 'A Framework for a New View of Islamic Science', *ʿAdiyat Halab*, 4–5: 68 ff.

HANAFI, HASAN, and MANZOOR, PARVEZ (1988). *Modern Science in Crisis*. Penang, Malaysia: Third World Network.

JAKI, S. (1978). *The Road of Science and the Way of God*. Chicago: University of Chicago Press.

KHALID, F., and O'BRIEN, JOANNE (1992) (eds.). *Islam and Ecology*. London and New York: Cassell.

KIRMANI, M. ZAKI (ed.). *MAAS Journal of Islamic Science*. Aligarh University.

KIRMANI, M. ZAKI (1989). 'Moving Towards a New Paradigm', in Ziauddin Sardar (ed.), *An Early Crescent*, London: Mansell, 140–62.

—— (1992). 'An Outline of Islamic Framework for a Contemporary Science', *MAAS Journal of Islamic Science*, 8/2: 55–75.

NAQUIB AL-ATTAS, SYED MUHAMMAD (1981). *The Positive Aspects of Tasawwuf: Preliminary Thoughts on an Islamic Philosophy of Science*. Kuala Lumpur: Islamic Academy of Science.

NASR, S. H. (1987). *Science and Civilization in Islam*. Cambridge: Islamic Text Society.

—— (1990a). *Man and Nature: The Spiritual Crisis of Modern Man*. London: Harper Collins; and Chicago: ABC International.

—— (1990b). *Traditional Islam in the Modern World*. London: Kegan Paul.

—— (1993a). *An Introduction to Islamic Cosmological Doctrines*. Albany, N.Y.: State University of New York Press.

—— (1993b). *The Need for a Sacred Science*. Albany, NY: State University of New York Press.

—— (1995). *Islamic Science: An Illustrated Study*. Chicago: Kazi Publications.

—— (1996). *Religion and the Order of Nature*. London and New York: Oxford University Press.

Proceedings of the International Symposium Science, Technology, and Spiritual Values—On the Asian Approach to Modernization. Tokyo: Sophia University and the United Nations University, 1987.

SMITH, W. (1984). *Cosmos & Transcendence: Breaking Through the Barrier of Scientific Belief*. La Salle, Ill.: Sherwood Sugden & Co.

CHAPTER 6

INDIGENOUS LIFEWAYS AND KNOWING THE WORLD

JOHN GRIM

INTRODUCTION

The diversity of peoples and cultures indicated by the term 'indigenous' makes it somewhat ambiguous. However, the local and international struggles for survival of diverse tribal, folk, local, native, and traditional peoples has given focus to the usage of this term. Indigenous knowledge is increasingly used in development, political, and academic settings by indigenous individuals and communities, or those who support them, with regard to authenticating and controlling the ways of knowing created by these distinctive societies (see e.g. Sanders 1977; Jaimes 1992; Jhappan 1992; Wilmer 1993; Alfred 1995; and Smith 1999). Moreover, the appearance in the consultations and documents of international bodies affirms the usage here of the term 'indigenous'. For example, article 1 of the International Labour Organization's Convention 169 regards people 'as indigenous on account of their descent from the populations which inhabited the country, or a geographical region to which the country belongs, at the time of conquest or colonization or the establishment of present state boundaries and who, irrespective of their legal status, retain some or all of their own social, economic, cultural, and political institutions' (see website for the Office of the High Commissioner for Human Rights at www.unhchr.ch/html/menu3/b/62.htm).

The strong social and political emphases in this definition are important, and will be explored in this work, but other dimensions of this term emphasized by indigenous communities deserve mention, such as contexts, territories, cultures, traditions, histories, languages, institutions, and beliefs. Indigenous spokespeople have described these themes as not only inseparable from knowledge but as interwoven into the fabric, or lifeway, of their existence as a people. 'Indigenous' thus refers to small-scale societies around the planet who share and preserve ways of knowing the world embedded in particular languages, story cycles, kinship systems, world-view dispositions, and integrated relationships with the land on which they live. (See also, Aga Khan *et al.* (1987).)

Lifeway is an interrogative concept that raises questions about the ways in which diverse indigenous communities celebrate, work towards, and reflect on their wholeness as a people. Indigenous knowledge is a key component in this communal reflection. In their diverse ways of knowing the world, indigenous peoples draw out their identity and meaning-in-the-world in both the presence of ecosystems and the authority of cosmology. These reciprocal ways of knowing in indigenous lifeways manifest differences in expression and underlie the wisdom and the specificity of indigenous knowledge. Thus, the Maori scholar Linda Tuhiwai Te Rina Smith (2000: 234–5) observes of *whakapapa,* her people's concept of knowing, '*Whakapapa* is a [Maori] way of thinking, a way of learning, a way of storing knowledge, and a way of debating.... *Whakapapa* also relates us to all other things that exist in the world. We are linked through our *whakapapa* to insects, fishes, trees, stones, and other life forms.' The wisdom transmitted here seems to speak across its specific Maori context, but who can determine if that wisdom is transferable as more than environmental poetry?

As a descriptive term, 'indigenous' is inextricably tied to an engaged knowledge that is actively pursued in relationships through the natural world. Indigenous ways of knowing are not simply expressions of an instrumental rationality, or a functional, specialized knowledge framed exclusively for accomplishing specific tasks. As the Santa Clara Pueblo educator Gregory Cajete says (2004: 5), native science is 'a metaphor for a wide range of tribal processes of perceiving, thinking, acting, and "coming to know" that have evolved through human experience with the natural world. Native science is born of a lived and storied participation with the natural landscape. In its core experience, Native science is based on the perception gained from using the entire body of our senses in direct participation with the natural world.' Yet, although the term 'Native science' expresses the process character of indigenous knowledge, its emergence in the context of the whole life of a people, and its sense-based relationship with the non-human world, the general terms 'knowing' and 'knowledge' are preferred here.

The intention is to distinguish indigenous knowledge from Western usages of the term 'science'. As Ellen and Harris (2000: 6–20) point out, science has its own historical roots in local, embedded, experiential, indigenous, and Asian knowledge. However varied the historical roots of science, as a way of knowing it gradually disengaged from any particular community to establish the ideal of constant questioning and re-examination of the assumptions that ground knowledge. Science has no 'lifeway' concept that draws it together in commitment to a people, their ecologies of meaningful place, and their cosmologies of identity. Science makes no commitment to a search for 'balance', either as the compromise between views or as a knowing that positions one in a final religious or metaphysical value. Science strives for ongoing change in which better explanations displace ones that are understood as problematic. Science emphasizes experimentation within an objective, materialist world over attentive presence to any inherent dignity or meaning of reality in itself.

Often scientific observation, the presentation of theories, and the intense debate and competition in science are human-centred. The test of scientific conclusions about the real world is grounded in facts observed in an objectified reality, but facts also flow from wholly human perspectives and technical instrumentation. Science also flows back into the pragmatic concerns of human communities. The feedback loop in science begins and ends with the human, whereas with indigenous knowledge the other-than-human world may find voice in community considerations. Actually, the environmental crises have activated policy considerations on behalf of species and bio-regions that present new and troubling questions for science. Indigenous knowledge, on the other hand, cultivates a deep empathy in relation to biodiversity, in which humans, within their own communities, often stand for voices in nature. For both science and indigenous knowledge the relationships with personal, social, and political modes of power are crucial components of any evaluation of those ways of knowing. That is, just as science has at times been misused and co-opted by corporate power, so also indigenous knowledge has been diverted into forms of personal and social aggrandizement.

Finally, the use of scientific categories to record indigenous knowledge tends to recast it as a global and universal mode of knowing; but indigenous knowledge tends to be more local and specific. There is a further problem in that applied science has become quite closely tied to commercial interests. Thus, the use of any scientific categories to describe indigenous knowledge suggests that those ways of knowing can easily be transmitted within a corporate governance structure that values exploitation over nurturance, and profit over distribution. Two orientations are presented here for understanding knowledge among indigenous peoples. First, the relevance of indigenous knowledge and its ongoing fragmentation in the neocolonialist era of global capital is asserted. Second, this assertion calls for a closer understanding of the interface of the forms of indigenous knowledge, including their acquisition and transmission, within the context of indigenous lifeways.

THE RELEVANCE OF INDIGENOUS KNOWLEDGE

While often thought of as remote minorities, indigenous peoples are a significant and diverse population of more than 500 million peoples in Africa, South

Asia, South-East Asia, Central Asia, Australia, the Pacific region, Northern Eurasia, and the Americas (see Berger (1990); and Barnes *et al.* (1995)). Often these indigenous peoples are marginalized even today within their own nation-states and in international economic globalization. This continues to the point that their existence is threatened in some settings by mining, logging, and other extractive activities in which they have little voice. Typically, indigenous peoples have been positioned in the worldwide hierarchy of nations as wards of individual states. Scholars such as Vine Deloria and Clifford Lytle (1983) offer a more nuanced term: namely, 'the nations within', which deflects the vestiges of colonialism and emphasizes connections to nation-states as well as separateness from them. More recently, international economic and financial bodies have negotiated agreements with governments that mask energy production on native lands and privatization of water and social services imposed on indigenous peoples as 'poverty eradication' and 'sustainable development'. Development projects motivated by these perspectives tend to dismiss any sense of multiple-use of land and resources proposed by indigenous peoples.

Knowledge in the context of indigenous lifeways is centred on the land. Often, indigenous peoples have been the subject of centuries of oppression, so that their lifeways and the ways of knowing embedded within relationships with the land became fragmented. The taking of indigenous homelands resulted in serious challenges to the transmission of indigenous knowledge and the maintenance of diverse indigenous world-views. Typically one of the major moves of colonial domination over indigenous peoples was to denigrate or deny that any form of systematic knowledge could be found in native lifeways. Obviously, by denying indigenous knowledge, arguments could be made that eroded native people's claims to the land on which they lived.

Similarly, an earlier colonial penchant for positivistic science ranked the ways of knowing the world, or epistemologies, of indigenous peoples as inferior, failed, or non-existent. The term 'animism' encapsulated the earlier colonial dismissal of indigenous knowledge. Interestingly, this term has been reappropriated by some native peoples as descriptive of their experiences of a relational exchange with spiritual presences in the world. Bird-David (1999) suggests that the concept of animism needs to be revisited for what it has to tell us of persons-in-relationship as emerging and maturing exchanges that result in indigenous knowledge. That is, a knowing that recognizes personhood in both the human and the presences in the world from which mutual privileges and obligations emerge. Indigenous knowledge from this standpoint is described as a relational epistemology. Tuhiwai Smith (1999) further explores how indigenous knowledge not only flows inward, informing a people of its relationships with self, society, land, and cosmos, but also flows outward affecting relationships with dominant societies. Indigenous knowledge, then, can result in contemporary 'decolonizing methodologies'. These are contemporary and emerging forms of indigenous knowledge that guide the work of indigenous scholars in reasserting the wholeness of fragmented indigenous communities. They do this by means of research and development from the standpoint of indigenous ways of knowing.

Contemporary indigenous societies and the ecosystems in which they reside as vital, interactive wholes are described here by the term 'lifeway'. The close connections between territory and society, religion and politics, cultural and economic life, are the intellectual and emotional basis whereby indigenous peoples maintain and recuperate their knowledge systems. Indigenous lifeways as ways of knowing the world are presented as both descriptive of enduring modes of sustainable livelihood and prescriptive of what Peet and Watts call 'ecological imaginaries' (1996: 7). These are deep, attractor relationships between place and people that activate sensing, minding, and creating at the heart of cultural life.

Lifeways establish the patterns and the ways of perceiving, or sensing, the world. Moreover, indigenous knowledge as minding flows from the conscious conceptual exchange with the world in conjunction with sensing. This felt experience of indigenous knowledge is that of beings-in-the world who are mutually related and dependent on one another for survival, for the knowledge needed to survive, and for the assertion of power that enables survival. Knowledge resides in that 'place' that indigenous people speak of (Cajete 2000) which is not 'out there', as in Western notions of separable knowledge. Rather, the place-based knowledge of indigenous lifeways is embodied in dynamic relationships often expressed in oral narratives and as kinship relations extending through individuals to the whole life community. These dynamic relationships in turn create results in the complementary flow of sensing and minding in lifeways which enable innovation and creativity in the face of life's challenges.

Indigenous ways of knowing orient communities to adapt to change, heal sickness, and respond to the numbing reality of death. In the West there are deep motivations, latent in applied science, that give urgency and ethical force to radically eliminating the dreaded limits of the human condition. Overcoming death by any possible means is given a high priority in the West, even when this involves intrusive technologies that might change the organic nature or body composition of the human or reality itself. These millennial drives relate to biblical notions of a perfection reached in the end-times and do not appear in this symbolic form in indigenous traditions. Rather, there is in indigenous knowledge broadly conceived an orientation to survival that fosters attention to creativity in the whole process of existence, rather than in human science and technology exclusively. In this sense, unlike the anthropocentric character of science, indigenous knowledge is more anthropocosmic. That is, the human has co-creative roles in the cosmic process. Other-than-human 'persons' or places may be more powerful than humans but there is a greater work that they undertake together.

As Leroy Little Bear observes (2000: 81), the interface between indigenous ways of knowing and lifeways establishes an educational context that is co-creative for individuals and societies, for adults and children, for land and cosmos. He writes:

The function of Aboriginal values and customs is to maintain the relationships that hold creation together. If creation manifests itself in terms of cyclical patterns and repetitions, then the maintenance and renewal of those patterns is all-important. Values and customs are the participatory part that Aboriginal people play in the maintenance of creation.

This place-based knowledge of indigenous peoples often results in meaningful stories of reality, or cosmologies, that emphasize the personal, the specific, and the contextual. Affirming indigenous knowledge need not be understood as promoting what Ellen (1993: 126) called the 'myth of primitive environmental wisdom'. Rather, it can be said that indigenous knowledge affirms local peoples and ecologies while meaningfully orienting them in ceremonies and oral narratives to larger cosmic realities. Therefore, in the wider context of the environmental crises facing humanity, indigenous knowledge provides viable alternative visions of human–Earth relations. This knowledge is primarily for indigenous communities themselves, but it also bears significantly on an emerging environmental awareness in any multiform planetary civilization that promotes sustainable life.

An Overview of this Project and Questions Raised by this Topic

This chapter seeks to explore selected examples of these diverse indigenous ways of knowing the world. In its discussions this project acknowledges differences not only among indigenous ways of knowing but also between indigenous knowledge and systems of knowing within industrial-technological societies. This latter difference is especially evident with regard to the presentation and organization of indigenous knowledge using the ideas and methods of Western, Enlightenment thought. Typically, indigenous ways of knowing are framed by such Western template ideas as monotheism, social contract theory, private property and individual rights, unilateral views of democratic governance, and scientific views of the objectivity of reality. This has had the effect of decontextualizing indigenous knowledge, so that some aspects are adapted to scientific categories while other significant native domains, logics, and epistemologies are rejected as unassimilable.

Still, as Arun Agrawal (1995: 415) points out, it is helpful to remember that constructing a 'sterile dichotomy between indigenous and western' may simply obscure ideas and practices that unnecessarily constrict peoples' considerations of potential knowledge transfers. Indigenous knowledge shares with Western science an ethical injunction to know and to describe the world as it appears in both its local and its cosmological manifestations. Both guard against presenting themselves as cosmology in themselves—that is, as standing in place of the world itself. However, their different approaches and concerns tend to bring these ways of knowing to entirely different positions in relationship to reality. Science requires a critical distance from the object of research to find principles of explanation, whereas indigenous knowledge establishes the means for individuals to search for transformative meaning and for communities to find their place in the larger community of life.

Questions linger, then, about which approach to understanding indigenous knowledge is most appropriate. Certainly, positing a 'deep structure' for indigenous knowledge reifies a shared resemblance of indigenous lifeways that is expressed so spontaneously and differently in diverse cultures. Richards (1993: 62) challenges the idea of indigenous practices such as farming being grouped in such a 'misplaced abstraction' as indigenous knowledge. Using the analogy of musicians who train on their instruments and then adapt to particular needs, Richards likens indigenous knowledge to an adaptation of agricultural resource skills and techniques. This emphasis on performance knowledge is similar to the composite approach suggested below in which time, space, authority, and spiritual presences are proposed as coalescing in engaged knowing. But an improvisational performance tailored to each usage and decision-making situation hardly accounts for the cultural depth of many forms of indigenous knowledge.

Questions surface when indigenous knowledge is described and discussed in the language, ideas, and values of a dominant society. Is there an indigenous knowledge theory in the same way that there is scientific theory? In what ways have the great traditions of Chinese and Ayurvedic medicine interacted with Asian indigenous knowledge systems? How does the ideological work of academics, supporting the struggle for recognition of indigenous knowledge, relate to the development community's efforts to build infrastructure for local communities who rely on local resources? What motivates politicized obstructionists who consistently attempt to block national and international recognition of indigenous knowledge practitioners as having valid claims to land and livelihood? Have indigenous knowledge systems been understood, or even heard, in their own languages, voices, values, and epistemological positions?

This brief overview chapter explores several of these issues by considering the organic relationality of lifeway, land, and indigenous knowledge as mutually interactive processes. While differently described by diverse native peoples, indigenous ways of knowing are not simply about creating systems of knowledge; rather, they bring into possibility the lifeway itself. Instead of being understood as an abstraction, indigenous knowledge is relevant, experiential engagement by a people with ecosystems and biodiversity.

Lifeway and Land as Pervasive, Mutual Contexts for Knowing

The significance of lifeway as a concept for understanding indigenous knowledge can be both insightful and limited. It is insightful insofar as it helps an investigator understand the broad cosmology-cum-economy context in which native knowledge is generated, implemented, and transmitted. As Cajete (2004) suggests, indigenous ways of knowing are a 'coming to know' in shared and lived participation with the local landscape. Thus, indigenous knowledge is an experiential form of personal knowledge transmitted in the context of community that relates a task at hand to the larger realities of the lifeway. It is in this context that Johannes Wilbert (1993) describes the planning and construction of a Warao canoe as not simply a personal task and technology. Rather, these indigenous peoples of the Orinoco River delta in Venezuela engage, as they construct the canoe, with vital presences in the environment, and activate a cosmic narrative which gives a larger meaning to their work.

Lifeway can be a limiting concept if it leads to romantic images of frozen, timeless, ahistorical societies who preserve cultural norms outside the world of change. Similarly, the lifeway concept can be misleading if understood as a unit of analysis that separates out fundamentals when, in fact, indigenous knowledge orients towards a field patterned by beings and forces manifesting both spontaneity and coherence. In this sense ceremonies, economic transactions, and kinship systems as singular expressions of lifeways, creatively emerge at the interface with indigenous knowledge and land. As a way into understanding the pervasive and mutual organic relationality of indigenous knowledge, lifeway, and land, consider the statement by the Gitksan and Wets'uwetén elders Gisday Wa and Delgam Uukw (1987: 7, 26), before the Supreme Court of British Columbia, Canada.

Each Gitksan house is the proud heir and owner of an *adáox*. This is a body of orally transmitted songs and stories that act as the house's sacred archives and as its living, millennia-long memory of important events of the past. This irreplaceable verbal repository of knowledge consists in part of sacred songs believed to have arisen literally from the breath of the ancestors. Far more than musical representations of history, these songs serve as vital time-traversing vehicles. They can transport members across the immediate reaches of space and time into the dim mythic past of Gitksan creation by the very quality of their music and the emotions they convey.

Taken together, these sacred possessions—the stories, the crests, the songs—provide a solid foundation for each Gitksan house and for the larger clan of which it is a part. According to living Gitksan elders, each house's holdings confirm its ancient title to its territory and the legitimacy of its authority over it.

In fact, so vital is the relationship between each house and the lands allotted to it for fishing, hunting, and food-gathering that the *daxgyet*, or spirit power, of each house and the land that sustain it are one.

Here the song cycles, sounded as *adáox*, are presented as a body of Gitksan and Wets'uwetén knowledge. Rather than an abstract body of knowledge, however, they are described as living, 'breathed' archives connected to social organization, political leadership, and subsistence practices. This 'social body' is a way of knowing the world that activates the 'breaths of the ancestors'. Moreover, these embodied songs meet the personal somatic intention of a practitioner in the larger field of Gitksan and Wets'uwetén knowledge. Because they are oral, these musical transmissions can be adjusted to the maturity of the learner, his or her level of relatedness to the land, and experiential knowledge of the holistic order. Thus, the song cycles are not fixed as in a literate system, but accommodate different personal learning and teaching styles as well as spiritual accomplishment.

In a comparative aside, it is interesting to observe that social science methods can accommodate and understand the objective forms of organization and political structures embedded in indigenous knowledge. However, the affectivity and meta-phoric embodiment expressing this knowledge are beyond the purview of scientific method. Realization gained through personal experience of ancestral breath, for example, is not reducible to exact units of analysis; nor is it falsifiable by means of experimental method. Rather, the millennia-long transmission of indigenous ways of knowing is grounded in personal accomplishment and responsibility, community awareness and approval, and ecological response and sustainability.

The Gitksan and Wets'uwetén elders state that the ceremonial settings in which the songs are performed involve crests and stories of the lineage of ancestors. These serve to substantiate and affirm individual and group claims to subsistence rights, trad-itional sanctions, and relationships with land and spirit powers. Moreover, the songs transmit epistemological insight into the nature of time, space, authority, and spiritual presence that are not simply objective, reified, and abstract topics. Rather, the passage above strongly suggests sensory participation in ancestral knowledge by means of synaesthetic experiences of audial, visual, emotional, and social ways of knowing. This interweaving of sensing, awareness, and creativity appears to be a recurring mode of indigenous knowing that has virtually no parallel in Western natural sciences, social sciences, or humanities.

TRADITIONAL KNOWLEDGE: TIME, SPACE, AUTHORITY, AND SPIRITUAL PRESENCES

In the above quote the Gitksan and Wets'uwetén elders suggest that the synaesthetic character of the songs transports members into mythic time. This leads to reflection on the ways in which multivalent modes of indigenous knowledge not only express particular dimensions of an indigenous lifeway but actually serve to structure occasions for the active creation of that lifeway. Even brief considerations of these topics, however, help to illuminate the interface of indigenous knowledge, lifeways, and land.

In indigenous ways of knowing notions of time are evident which are strikingly different from the linear arrow image of time, or the metaphor of a flow of reality, or the absolute character of past, present, and future. Obviously, the knowledge em-bedded in the Gitksan and Wets'uwetén songs brings practitioners into deep partici-pation with time. By 'deep' is meant more than simply a mystical unknowable. Rather, indigenous knowledge in its diverse expressions seeks to integrate an authenticity and originality of knowing by exploring the pervasive, interactive, and process-punctuated characters of time, space, authority, and spiritual presences. Like the pervasive darkness of night, a depth experience of the world is an all-encompassing presence that can be punctuated by arresting experiences which afford their own original ways of knowing. The ceremonial space may create correspondences as microcosmic places of re-enactment for depth experiences of the macrocosm transmitted by an indigenous people. Often, when it is imaged in the immediacy of land, a sacred site may be more like a portal leading to that depth than simply the site of the sacred in and of itself. Place-based analyses of indigenous thought have only begun to explore the experiential authority of indigenous knowledge arising from an original and embodied knowing of place. Authority for Gitksan and Wets'uwetén leaders, for example, is established in a world-view linking land, house, affect, ancestors, and spirit powers. How shall we interpret this type of knowledge that evokes powerful emotions just as it establishes social organization and political authority and opens practitioners to complex contemplations on time?

Many indigenous peoples transmit their knowledge in songs, oral narratives, and symbolic actions that relate to specific places and events. For example, Stewart and Strathern (2004) describe how the Duna people of the Lake Kopiago district in the Southern Highlands province of Papua New Guinea assert social identity through narratives called *malu*. The Duna recognize the rights of individuals by means of *malu* narratives to claim precedence for garden locations or for the gathering of medicinal plants. Prerogatives with regard to narrating *malu* are believed to have been transmitted from the time of the ancestor animals. *Malu* operate at at least three levels for the Duna: at the intellectual level, by providing explanations of the world; at the affective level of ceremonies that keep powerful ancestors benevolent; and on the level of contemporary policy formation. That is, the performance of *malu* assists Duna people in framing reactions and responses to the current development of their land. Referring to or reciting *malu* simultaneously generates social stability in mythic-ceremonial contexts as well as symbol-making creativity to meet the Duna peoples' needs for contemporary, adaptive change. This way of knowing is a behaviour, then, that does not simply filter out pragmatic time in favour of a mythic time, but seems to bring both to bear upon one another.

Indigenous knowledge embedded within complex ceremonials is not exclusively liminal—that is, it does not exclusively involve an entry into extraordinary space and time outside the ordinary. In fact, the opposite is the case; indigenous knowledge manifested at peak ceremonial moments has deep and abiding connections both with extraordinary presences such as ancestors and spirit powers, as well as with ordinary events such as canoe making, gardening, gender roles, and healing practices. Time is a key, it seems, to an interpretation of ways in which space, authority, and spirit presences manifest one another in the ceremonial and symbol-making contexts of indigenous knowledge.

Another example is provided by the Dogon peoples of Mali. The Dogon of sub-Saharan Africa continue to celebrate a masked festival (*dama*) whose explicit focus is reburial of the dead, and whose implicit emphasis is on affirming the inherent power and responsibilities integral to speech. Having originated from the animals of the bush, the ceremonial, *dama*, was acquired by women over time. Yet, by means of skilful deceit, it eventually came to be a male privilege with extensive gender and age separations. These strict divisions between women, children, and the dancers suggest intergenerational and gendered facets of the interface between Dogon knowledge, land, and lifeway. Moreover, the songs, masked dances of adolescent Dogon, and the speeches of Dogon elders at *dama* are understood as the context for a 'second burial' of the dead. Fischer's analysis (2004) suggests an interpretation in which the masked festival of *dama*, the reburial of the dead, the concerns about speech, and the gender exclusions all involve aspects of Dogon knowledge. That is, all three are about the lifeway, as well as being forms of indigenous knowledge that must be enacted in order for the lifeway to exist.

The male masked dancers, the elders' speeches, the gender prohibitions on Dogon society as a whole—these all coalesce as entry into mythic time at *dama*. In this way the Dogon create a new existence for the dead—that of being an ancestor—which asserts male control over fertility and reaffirms the spiritual character of speech. This complicated and ever-changing ceremonial brings together Dogon knowledge of spiritual presences in the bush with speech that also comes from the wild bush. The return of the no-longer-remembered dead as interactive presences in the bush are enacted by the Dogon during *dama* in a complex weave of time, space, authority, and spiritual presences. Rather than a simple progression of past, present, and future, time is a self-similarity, a fractal logic, in which the authority of elders, the power-producing land, and the animal spirit presences are all imaged as aligned bodies with inner speech forms. Amidst the gender conflicts, the competition among age grades, and the wrenching grief of mourning, the Dogon assert forms of knowledge that can be spoken in efficacious ways, and at the right time, so as to affect temporary cosmological harmony among those bodies.[1]

The discussions of 'Dreaming' in Australian Aboriginal traditions are also relevant in a discussion of the interface of knowledge, land, and lifeway. The Walpiri, Kaytej, and Pintupi peoples of Central Australia use a variant of *jukurrpa* to sound this complex concept of spiritual presences in the environment-as-cosmology. W. E. H. Stanner gave expression to his understanding of Aboriginal knowledge and these indigenous reflections on *jukurrpa* through his descriptions of Dreaming as 'one possibility thing' and 'everywhen'.[2] His eliptic interpretations suggest the unific and cosmological character of Dreaming, as well as point towards the integral differentiations of time, space, authority, and spiritual presences found within these indigenous lifeways.

---

[1] For Dogon ethnography related to the *dama* see Griaule (1948) and Calame-Griaule (1986). A critique of Griaule's method and resulting work in that period can be found in van Beek (1991). See also van Beek (1993); van Beek and Hollyman (2001); and van Beek and Banga (1992).

[2] Stanner (1966, 1957). Stanner was the first ethnographer to assert that the mythology of Aboriginal peoples of Australia proposed a narrative about the reality of the world that constitutes valid knowledge and a basis for moral law. For Pintupi peoples see Myers (1986). For Walpiri and Kaytej peoples see Bell (1983).

Geographically closer to the Getksan and Wets'uwetén are the sub-Arctic Koyukon peoples of Alaska, who are reported by Nelson (1983: 10) as narrating stories of the mythic 'Distant Time'. In these stories are ethical prescriptions that the Koyukon call *hutlaane*, which are described as regulated relationships with their bio-region. *Hutlaane* are an assemblage of gendered prohibitions that result, according to Nelson, in a conservation system for managing and protecting the bio-region from over-exploitation by humans. Based on restrictions set down in the Koyukon mythic stories of the primal period, or 'Distant Time', *hutlaane* are transmitted as teachings with authority, or moral force, which come from the hunted animals and gathered plants themselves. *Hutlaane* act as strong approval and prohibitions for responsible hunting and gathering. The constraints and allowances of *hutlaane* connect Koyukon views of land, biodiversity, and cosmology. Moreover, they demonstrate an engaged knowledge that sustains both the lifeway and the bio-region. This is a mode of indigenous knowledge in which ethical reflections and moral practices emerge in a sense of time that is not merely linear or chronological. Koyukon time seems more attuned to experiential knowing of the land as punctuated by mythic time that opens to sustenance and responsibility.

Another insightful interpretation is provided by Sharp (2001: 63) in his observations among the Dene peoples of North Central Canada. Acknowledging continuities with Western modes of time, he relates the special character of Dene reflections on *inkoze* as a form of indigenous knowledge treating space, authority, and spiritual presences. *Inkoze* refers to the Dene sense of causality that comes from the power of dreams and the ways in which animals mediate these synaesthetic relationships of person and land. He observed that:

Dene time usage, though it uses linear, directional, and cyclical time, includes usages that do not correspond to any of these. Animal/persons are not limited by our perceived constraints of the physical universe. Their license to suspend the restrictions of what the West considers physical reality is embedded in their inkoze and is particularly conspicuous in their dispensation with the restrictions imposed by time. Dene culture is not dominated by the idea of a now, and time is not seen as a flow between a no-longer-existing past and a not-yet-existing future. There is a sense, and there are circumstances, in which the Dene conceive of reality as effectively being 'simultaneous', that is, time is treated as a dimension that is independent of any flow or directional change within that dimension. The past and the future are as real as the present. Communication and connection between past, now, and future are all possible. The connections between time and the three ordinary dimensions of physical reality are uncoupled. Time sometimes becomes a thing independent of motion within it. All places in time become equally accessible. It is possible for some beings to move anywhere in time rather than having to move in only one direction in time. The effect of this is to make it seem as if the Dene sometimes use time the way Western culture uses place.

Here Sharp insightfully draws attention to the ways in which Dene indigenous knowledge weaves time, space, authority, and spiritual presences into an experiential tapestry of deep, cosmological meaning for a community.

Moreover, Dene perspectives appear in ways to parallel Albert Einstein's strategy in his theory of general relativity. In that theory Einstein described space and time as relative to an observer's motion, and hypothesized flexible and dynamic rather than rigid, unchanging structures, which was contrary to how they been understood in classical Newtonian cosmology. Interestingly, in Einstein's view of a 'block' universe he came up against the challenge of the now-moment of time (Pais 1982). He insisted that, though time and space change, the now-moment must be utterly resistant to change despite our common-sense feelings of time's ceaseless now-moments as like a flow of time. A philosopher remembered a conversation he had with Einstein on this subject: 'Einstein said that the problem of the now worried him seriously. He explained that the experience of the now means something special for man, something essentially different from the past and the future, but that this important difference does not and cannot occur within physics. That this experience cannot be grasped by science seemed to him a matter of painful but inevitable resignation' (from Carnap (1963), cited in Greene (2004: 141)).

While aware of the paradoxical nature of the simultaneity of time, the Dene seem neither seriously pained nor resigned to the constancy of the *inkoze* now-moment. In fact the now-moment does not seem as dominant in their knowledge system as the fact of *inkoze* itself as a simultaneity of time. Dene experiences of *inkoze* are not scientific observations; nor are they a theory like general relativity. But they do seem to be depth insights of Dene indigenous knowledge as these people also struggle to understand an embodied, or organic, mutuality of knowledge, land, and lifeway.

CONCLUDING COMMENTS REGARDING THE ACQUISITION OF INDIGENOUS KNOWLEDGE

Three concluding observations to this overview are drawn from Marlene Brant Castellano (2000: 23–4) regarding the acquisition of indigenous knowledge. She observes that, first, *traditional knowledge* has been handed down by indigenous peoples more or less intact from previous generations. Second, *empirical knowledge* is gained through careful observation of the natural and built environments. Third, *revealed knowledge* is acquired through dreams, visions, and intuitions that are understood to be spiritual in origin. Briefly considering examples of these points allows us to conclude with timely observations on several epistemological characteristics of indigenous knowledge as well as the decolonized responses to the attempts to terminate aboriginal knowledge.

First, *traditional knowledge* is transmitted within an indigenous community especially by elders. This rich concept of elders is another example of an indeterminate, spontaneous, community-generated role that may or may not settle on an older person. Rather, an 'elder' is that person whom an indigenous community recognizes as imbued with the knowledge, responsibility, and spiritual awareness for actively living the role requested of them. In the contemporary world, indigenous elders, women or men, often embody knowledge of language, world-view, and ecology.

Transmission of language is widely recognized as central to the authentic acquisition of that deeper knowledge to which indigenous elders refer when they speak of their identity and meaning as an integral people. Cree elders of the northern sub-Artic regions say, 'The Cree language is our identity (*kinêhiyâwiwininaw nêhiyawê-win*) (Wolfart and Anenakew 1993: 5). Henderson (2000: 263) comments on this claim: 'Aboriginal consciousness and language are structured according to Aboriginal people's understanding of the forces of the particular ecosystem in which they live. They derive most of the linguistic notions by which they describe the forces of an ecology from experience and from reflections on the forces of nature.' Thus, just as indigenous languages derive from experiences of the land, so also world-views emerge, in indigenous lifeways, from visionary experiences of the depth of origins, the beginnings of life, and the unfolding of the universe.

Just as language is not an abstract entity in the indigenous context, so also world-views, and the values that flow from these experiential ways of knowing the world generate a complex of interrelated activities. As has been suggested, rituals, art forms, cults, organizations, and experiences are indeterminate, fluid connectors at the interface of indigenous knowledge, land, and lifeway. Ntuli (2002: 58) describes the crucial role that world-views have in the African setting, saying:

We must now turn our attention to the examination of African value systems that we seek to see reborn. Contrary to Western thought, African thought sees life as a cycle; the world as an interconnected reality; human beings, plants, animals, and the universe as one interconnected whole, and that our survival depends on how these forces interact with each other. In all societies, the beginning and meaning of life lie within the world of myth, and these myths give form through rituals. For these rituals to be effective, dances and other cultic acts are performed, and art objects are created to give form and potency to the ritual. In other words, songs are composed, dances performed, and sculptures and other art objects are created to support rituals. . . . Traditional Africa provided us with a world-view that recognised our sanctity as people and sought to secure our place in the wider spheres of life; to help give meaning and form to our strivings for oneness with the cosmic spirit that guides us.

Just as elders embody an engaged knowledge involving language and world-view values, so too they live their connection to ecosystems in ways of knowing that exemplify *empirical knowledge.*

There is no doubt that 'traditional environmental knowledge', or TEK, has captured the public imagination more than any other aspect of indigenous knowledge. In entering the public sphere, it also reminds us of ways that indigenous knowledge has been commodified. The allure of pharmaceutical panaceas gathered from indigenous knowledge sources tends to mask potentially exploitative activities such as genetic piracy. In these contemporary extractive agendas, biological materials are gathered from native peoples and attempts made to patent the distinctive genetic heritage of indigenous groups. Furthermore, efforts to protect particular expressions of indigenous knowledge using international intellectual property rights have proved ineffective at best, because they entangle indigenous elders and spokespeople in the bureaucratic procedures of multiple nation-states. More significantly, they situate the empirical knowledge of indigenous peoples in an epistemological context that is wholly inconsistent with the knowledge being protected. That is, the languages, ideas, and values supposedly protected by the concept of intellectual property rights are set within a context that is more similar to a colonial mind-set that situates indigenous peoples as subservient clients rather than careful observers, users, and co-creators of their environments. It is also evident that local knowledge can be used by indigenous leaders to oppress their own people (see the argument by David Harvey (1996)). But the historical gravity of responsibility between peoples and lands over centuries of relatedness and use, evident, for example, in oral narratives, precludes dismissal of indigenous knowledge by labelling it as potentially abusive of the rights of its practitioners. Moreover, indigenous knowledge is often an empirical source for science itself.

Indigenous knowledge records orally empirical observations of local ecosystems in complex, particular forms. Werner Wilbert (2001: 400) pointedly distinguishes the empirical knowledge of Warao peoples of the Orinoco River delta in South America from that of Western science. He writes:

The world of the Warao is a manifestation of the supernatural, experienced through life not understood through scientific thought. The product of longitudinal empirical observation and interpretation, it enfolds human society in a mythologized landscape of primordial origin. Rather than self-centered atomistic individualism, it promotes a moral bond of society, restrained by principles of pluralistic coexistence. Principles of restraint are encoded in the world order by holistic design and perpetuated through enculturative learning both private and public.

Perhaps the central accompanying feature of empirical knowledge among indigenous peoples is that of responsibility. Rather then romanticizing traditional environmental knowledge, it is more instructive to learn that many indigenous myths warn of the overuse and misuse of natural goods. Thus, Guss (1989) describes the intricate restrictions and prohibitions on the gathering of materials for basket making among the Yekuana of Venezuela. Basket making remains one of the signal accomplishments whereby Yekuana adults achieve and maintain social status. They also sell some baskets for cash income. Yet, their empirical attention to, and use of, these diverse fibers, roots, and plants is not driven by the goal of limitless achievement. Rather, knowledge transmitted in mythologies about the culture hero Wanadi restrains usage, promotes mental discipline, and binds a community to a deeper vision of itself. Empirical knowledge of the bush by indigenous peoples also raises significant questions about different ways in which the wild is known and the ways in which that knowledge is transmitted.

From many indigenous perspectives, 'wildness' and 'wilderness', as areas in which the human is absent, are puzzling concepts. For example, Robert Jarvenpa (1998: 8–9) reports of the Chipewyan and Han, Dene/Athapaskan peoples of the North American sub-Artic, that a number of relationships link humans to these open spaces. He writes:

To the outsider, much of the subartic landscape may appear 'empty' or 'unoccupied'. This notion holds little meaning for Athapaskans who see virtually all their surroundings as active, alive or occupied in some fashion. Almost any space, whether within or beyond the confines of currently inhabited settlements or camps, may have functioned previously as a culturally meaningful landscape where events transpired and activities occurred.

A fine-grained understanding of the landscape is symbolically codified in language. Athapaskan place name terminologies recognise a myriad of geographical features over extensive regions. Many of these are trenchant descriptions of environmental features and processes. The Chipewyan expression *ts'ankwi ttheba* ('old woman rapids') not only denotes a particularly turbulent section of the Mudjatik River but also evokes the circumstances of a tragic death at that location generations ago. In this way, conventional language and discourse continually situate the topographical landscape in terms of peoples' history and lore.

From a Dene perspective, then, wilderness as the absence of the human is not a working concept. Rather, the traditional environmental knowledge (TEK) of open spaces is transmitted in a complex narrative, mnemonic system based on place-names. (For a classic study see Basso (1983), and his more recent work (1996).) These place-names may or may not refer to what outsiders call sacred sites, but they typically have embedded within them an indeterminacy of time that is a signal feature of the knowing they transmit.

Finally, the character of *revealed knowledge* in dreams, visions, and spiritual intuitions among indigenous peoples is manifest in many of the examples above. Some of the most striking case studies of visionary experiences that led to indigenous knowledge concern the ayahuasca complex of South America, rice agriculture in South-East Asia, and corn agriculture in Mesoamerica. Davis (1998) and Narby (1998) discuss the sophisticated processes involved in making the ecstasy-inducing drink ayahuasca. Its preparation involves not simply the *banisteriopsis* vine, but delicate infusions of several other prepared substances. More important, however, is the widespread understanding among the native peoples who work with this substance that the plants themselves revealed in dreams and visions the complex processes required to produce ayahuasca. Victoria Tauli-Corpuz (2001) discusses Igorot values involved in the terracing, irrigation, and complex rice agriculture among these indigenous peoples of northern Luzon in the Philippines. She speaks of *gawisi*, or responsibility, of the people to land and ancestors, *innayan*, or limits and relations with the spirit world of nature, and *lawa*, the depth experiences of the holy in these relationships.

The mythical origin of maize among Nahua peoples of central Mexico presents a major example of revealed knowledge in which the gods 'robbed' maize for the people and transmitted the knowledge of its production and sustenance in the context of lifeway and land. Javier Galicia Silva (2001: 304) gives a relevant passage from the *Florentine Codex* that encapsulates the fourfold embodiment of this revealed knowledge. That is, the profound teachings of indigenous knowledge typically open insight into the embodied, spiritual presences of the holistic order: namely, the human body, the social body, the ecological body, and the cosmo-logical body. This passage relates maize to the sustenance of these embodiments, saying:

Listen: *Tonacayotl* [maize], Our Sustenance, is for us, all-deserving. Who was it who called maize our flesh and our bones? For it is Our Sustenance, our life, and our being.

It is to walk, move, enjoy and rejoice. Because Our Sustenance is truly alive, it is correctly said that it is he who rules, governs and conquers....

Only for Our Sustenance, *Tonacayotl*, the maize, does our soil subsist, does the world live, and do we populate the world. The maize, *Tonacayotl*, is the true value of our existence.

This revealed knowledge among the Mesoamerican Nahua peoples is related to the distinctly different understandings of the neighbouring Mayan peoples regarding the origin, production, and deeper implications of maize agriculture. Both of these major Mesoamerican civilizations in their knowledge systems reflect on lifeway as a fourfold embodiment of their sacred food, corn. Knowledge flows, then, as a vitality shared across bodies in the cosmos.

CONCLUSION

Western claims of universal knowledge articulated from the eighteenth-century Enlightenment period validated colonialist domination as a divine right or the inevitable tide of progress over the benighted peoples of the Earth. As Noël observed (1994: 79), the logic of universal claims eventually came up against the resistance of indigenous peoples and the assertion of their own forms of knowing. He writes:

After long endorsing the logic of a discourse taught to them as the only one that was valid, the dominated began to feel doubts. At first vague and fleeting, these doubts were aroused by the oppressor's own failure to live up to his idealised model of humanity. As the oppressed became more actively aware of their own worth, their doubts grew more insistent. Gradually, the dominated ceased to see the oppressor's defense of his special interests as the inevitable tribute owed to a superior being. Divine, natural, or historical laws that espoused such narrow designs became suspect. It eventually came to mind that these laws were pure creations of a group wishing to legitimize its privileges.

The current regeneration of indigenous knowledge by native peoples themselves is a testimony to their resistance to ongoing forms of contemporary colonization. Resistance in this sense does not point to a fossilized indigenous knowledge as a desperate coping mechanism. Rather, indigenous ways of knowing actively seek to nourish both the cultural lifeways and the biodiversity of the land.

Increasingly, efforts to decolonize indigenous knowledge have led to programmes that preserve traditional lifeways. Indigenous projects for restoring traditional ways of knowing often seek to harmonize with selected social and subsistence changes from outside communities, and accommodate paradigms and practices such as Western science. Encouraging indigenous youth to enter into scientific knowledge while holding to community lifeways are ongoing challenges for native communities.

Contemporary indigenous projects that draw on both scientific and indigenous knowledge are numerous and diverse. Relevant examples are the indigenous environmental network, the Maori educational reform called Kura Kaupapa Maori, and the United Nations Working Group on Indigenous Peoples. These and other indigenous organizations increasingly report on the current period of globalization-as-colonization and the resurgent decolonizing responses from indigenous peoples themselves.

What this resurgence of indigenous knowledge reveals today is not simply the ashes of an extinct, failed way of knowing, but the embers that indigenous elders are rekindling to confront their peoples with awareness of deeper purpose. The metaphor of fire evokes that renewed understanding of indigenous knowledge whose wisdom may not be available to other peoples. Still, it calls to mind the social and political vitality of peoples informed by these ways of knowing, and the challenges facing the human to acknowledge multiple ways of knowing.

References and Suggested Reading

Aga Khan, Sadruddin, and Hassan bin Talal (1987). *Indigenous Peoples, a Global Quest for Justice: A Report for the Independent Commission on International Humanitarian Affairs.* London: Zed Books.

Agrawal, A. (1995). 'Dismantling the Divide between Indigenous and Scientific Knowledge', *Development and Change*, 26: 413–39.

Alfred, G. R. (1995). *Heeding the Voices of our Ancestors: Kahnawake Mohawks Politics and the Rise of Nationalism.* Oxford: Oxford University Press.

Antweiler, C. (1998). 'Local Knowledge and Local Knowing: An Anthropological Analysis of Contested "Cultural Products" in the Corner of Development', *Anthropos*, 93: 469–94.

Apffell-Marglin, F., and Marglin, S. A. (1990). *Dominating Knowledge: Development, Culture and Resistance.* Oxford: Oxford University Press.

Barnes, R. H., Gray, A., and Kingsbury, B. (1995) (eds.). *Indigenous Peoples of Asia.* Ann Arbor: Association for Asian Studies Monograph no. 48.

Basso, Keith H. (1983). '"Stalking with Stories": Names, Places, and Moral Narratives among the Western Apache', in Edward M. Brunner (ed.), *Text, Play, and Story: The Construction and Reconstruction of Self and Society,* Proceedings of the American Ethnological Society, Washington: The American Ethnological Society, 21–55.

—— (1996). *Wisdom Sits in Places: Landscape and Language among the Western Apache.* Albuquerque, N.M.: University of New Mexico Press.

Battiste, M. (2000) (ed.). *Reclaiming Indigenous Voice and Vision.* Vancouver and Toronto: UBC Press.

Bell, Diane. (1983). *Daughters of the Dreaming.* Melbourne: McPhee Gribble Publishers.

Berger, Julian (1990). *The Gaia Atlas of First Peoples.* London: Gaia Books.

Bird-David, N. (1999). '"Animism" Revisited: Personhood, Environment, and Relational Epistemology', *Current Anthropology*, 40, suppl. (Feb.): S67–91.

Cajete, G. (2000). *Look to the Mountain: An Ecology of Indigenous Education.* Skyland, N.C.: Kivaki Press.

—— (2004). *Native Science: Natural Laws of Interdependence*. Santa Fe, N.M.: Clear Light Publishers.

CALAME-GRIAULE, G. (1986). *Words and the Dogon World*. Philadelphia: Institute for the Study of Human Issues.

CARNAP, RUDOLF (1963). 'Autobiography', in P. A. Schilpp (ed.), *The Philosophy of Rudolf Carnap*, Chicago: Library of Living Philosophers, 37.

CASTELLANO, M. B. (2000). 'Updating Aboriginal Traditions of Knowledge', in G. Dei, B. Hall, and D. Rosenberg (eds.), *Indigenous Knowledge in Global Contexts*, Toronto: University of Toronto Press, 21–36.

Codice Florentino: Book VI y X da la Colección Palatina da la Biblioteca Medicea Laurenziana, 3 vols. (1979). Mexico: Secretaria da Gobernación, Archivo general da la Nación.

COLORADO, P. (1988). 'Bridging Native and Western Science', *Convergence*, 21/2–3: 49–68.

DAVIS, W. (1998). *Shadows in the Sun: Travels to Landscapes of Spirit and Desire*. New York: Broadway Books.

DELORIA, V., and LYTLE, C. (1983). *American Indians, American Justice*. Austin, Tex.: University of Texas Press.

ELLEN, R. F. (1993). 'Rhetoric, Practice and Incentive in the Face of the Changing Times: A Case Study of Nuaulu Attitudes to Conservation and Deforestation', in K. Milton (ed.), *Environmentalism: The View from Anthropology*, ASA Monograph 32, London: Routledge, 126–43.

—— and HARRIS, H. (2000). 'Introduction', in *Indigenous Environmental Knowledge and Its Transformations: Critical Anthropological Perspectives*, Amsterdam: Harwood Academic Publishers, 1–33.

FISCHER, M. (2004). 'Powerful Knowledge: Applications in a Cultural Context', in A. Bicker, P. Sillitoe, and J. Pottier (eds.), *Development and Local Knowledge: New Approaches to Issues in Natural Resources Management, Conservation, and Agriculture*, London and New York: Routledge, 19–30.

GISDAY WA and DELGAM UUKW (1987). *The Spirit of the Land: The Opening Statement of the Gitksan and Wets'uwetén Hereditary Chiefs in the Supreme Court of British Columbia*, Gabriola, B.C.: Reflections.

GREENE, B. (2004). *The Fabric of the Cosmos: Space, Time, and the Texture of Reality*. New York: Vintage.

GRIAULE, MARCEL (1948). *Dieu d'eau: Entretiens avec Ogotommêli*. Paris: Éditions du Chene; published as *Conversations with Ogotemmeli: An Introduction to Dogon Religious Ideas*. London: Oxford University Press for the International African Institute, 1965.

GRIM, J. (2001) (ed.). *Indigenous Traditions and Ecology: The Interbeing of Cosmology and Community*. Cambridge, Mass.: Harvard Divinity School Center for the Study of World Religions.

GUSS, D. M. (1989). *To Weave and Sing: Art, Symbol, and Narrative in the South American Rain Forest*. Berkeley: University of California Press.

HARVEY, DAVID (1996). *Justice, Nature and the Geography of Difference*. London: Blackwell.

HENDERSON, J. (SÁKÉJ) Y. (2000). 'Ayukpachi: Empowering Aboriginal Thought', in Battiste (2000), 248–78.

JAIMES, M. A. (1992) (ed.). *The State of Native America: Genocide, Colonization and Resistance*. Boston: South End Press.

JARVENPA, R. (1998). *Northern Passage: Ethnography and Apprenticeship among the Subartic Dene*. Prospect Heights, Ill.: Waveland Press.

JHAPPAN, RADHA (1992). 'Global Community? Supranational Strategies of Canada's Aboriginal Peoples', *Journal of Indigenous Studies*, 3/1: 59–97.

LITTLE BEAR, L. (2000). 'Jagged Worldviews Colliding', in Battiste (2000), 77–85.

MYERS, FRED (1986). *Pintupi Country, Pintupi Self: Sentiment, Place, and Politics among Western Desert Aborigines*. Washington: Smithsonian Institution Press; Canberra: Australian Institute of Aboriginal Studies.

NARBY, J. (1998). *The Cosmic Serpent: DNA and the Origins of Knowledge*. New York: Jeremy P. Tarcher/Putnam.

NELSON, R. (1983). *Make Prayers to the Raven: A Koyukon View of the Northern Forests*. Chicago: University of Chicago Press.

NOËL, L. (1994). *Intolerance: A General Survey*. Montreal and Kingston: McGill–Queen's University Press.

NTULI, P. P. (2002). 'Indigenous Knowledge Systems and the African Renaissance', in C. O. Hoppers (ed.), *Indigenous Knowledge and the Integration of Knowledge Systems*, Claremont, South Africa: New Africa Books, 53–66.

PAIS, A. (1982). *Subtle is the Lord: The Science and Life of Albert Einstein*. Oxford: Oxford University Press.

PEET, R., and WATTS, M. (1996). *Liberation Ecologies: Environment, Development, Social Movements*. London and New York: Routledge.

RICHARDS, P. (1993). 'Cultivation: Knowledge or Performance', in M. Hobart (ed.), *An Anthropological Critique of Development*, London: Routledge, 61–78.

SANDERS, D. E. (1977). 'The Formation of the World Council for Indigenous Peoples'. International Working Group for Indigenous Affairs Document 29. Copenhagen.

SEMALI, L. M., and KINCHELOE, J. L. (1999). *What is Indigenous Knowledge? Voices from the Academy*. New York: Falmer Press.

SHARP, H. S. (2001). *Loon: Memory, Meaning, and Reality in a Northern Dene Community*. Lincoln, Nebr., and London: University of Nebraska Press.

SILVA, J. G. (2001). 'Religion, Ritual, and Agriculture among the Present-Day Nahua of Mesoamerica', in Grim (2001), 308–23.

SMITH, L. T. (1999). *Decolonizing Methodologies: Research and Indigenous Peoples*. London and New York: Zed Books.

—— (2000). 'Kaupapa Maori Research', in Battiste (2000), 225–47.

STANNER, W. E. H. (1957). 'The Australian Aboriginal Dreaming as an Ideological System', *Proceedings of the North Pacific Science Congress*, 3: 116–23.

—— (1966). *On Aboriginal Religion*. The Oceania Monographs, 11. Sydney: University of Sydney.

STEWART, P. J., and STRATHERN, A. (2004). 'Indigenous Knowledge Confronts Development among the Duna of Papua New Guinea', in A. Bicker, P. Sillitoe, and J. Pottier (eds.), *Development and Local Knowledge: New Approaches to Issues in Natural Resources Management, Conservation, and Agriculture*, London and New York: Routledge, 51–63.

SUZUKI, D., and KNUDTSON, P. (1992) (eds.). *Wisdom of the Elders: Sacred Native Stories of Nature*. New York and Toronto: Bantam Books.

TAULI-CORPUZ, V. (2001). 'Interface between Traditional Religion and Ecology among the Igorots', in Grim (2001), 281–302.

VAN BEEK, WALTER E. A. (1991). 'Dogon Restudied: A Field Evaluation of the Work of Marcel Griaule', *Current Anthropology*, 32/2 (April): 139–67.

—— (1993). 'Processes and Limitations of Dogon Agricultural Knowledge', in Mark Hobart (ed.), *An Anthropological Critique of Development: The Growth of Ignorance*, London and New York: Routledge, 43–60.

—— and HOLLYMAN, STEPHANIE (2001). *Dogon: Africa's People of the Cliffs*. New York: Harry N. Abrams.

—— and PIETEKE BANGA (1992). 'The Dogon and their Trees', in Elisabeth Croll and David Parkin (eds.), *Bush Base*, London and New York: Routledge, 57–75.

WILBERT, J. (1987). *Tobacco and Shamanism in South America*. New Haven: Yale University Press.

—— (1993). *Mystic Endowment: Religious Ethnography of the Warao Indians*. Cambridge, Mass.: Harvard Divinity School Center for the Study of World Religions.

WILBERT, W. (2001). 'Warao Spiritual Ecology', in Grim (2001), 377–407.

WILMER, F. (1993). *The Indigenous Voice in World Politics*. Newbury Park, Calif.: Sage.

WILSON, W. A. (2004) (ed.). 'Special Issue: The Recovery of Indigenous Knowledge', *American Indian Quarterly*, 28/3–4 (Summer & Fall): 359–633.

WOLFART, H. C., and ANENAKEW, F. (1993). *kinêhiyâwiwininaw nêhiyawêwin: The Cree Language is Our Identity*. Winnipeg: University of Manitoba Press.

CHAPTER 7

RELIGIOUS NATURALISM AND SCIENCE

WILLEM B. DREES

INTRODUCTION

Religious belief seems to be about the conviction that there is more than nature. Naturalism is perceived as the claim that there is nothing but nature. Is naturalism therefore an atheistic position, that which should be disproved in 'religion and science'? That seems to be the religious agenda of quite a few in 'religion and science'. My main thesis in this chapter is that it may well be more fruitful to accept 'naturalism' as an understanding of reality with which we have to live, and to explore religious options within the context of naturalism, which may turn out to be more hospitable to religious motives than antagonists suggest.

We ought to accept a naturalistic view, since it is the position that is most respectful of the epistemic success of the natural sciences, and thus cognitively preferable. It is also morally preferable, as it incites us to work with our knowledge. For those who accept theistic considerations, naturalism should be the preferred view of reality as God's creation, since it does not locate God's role in our ignorance or limitations, but in what we know and what we are able to do. The costs of rejecting naturalism are high. The solution is to live with naturalism.

To explicate what it means to live with naturalism, I will first clarify how I understand naturalism. This initial exploration may show why naturalism might be less harmful than some take naturalism to be. Some reasons for and against a

The present chapter reuses some paragraphs and phrases from Drees, 1996, 1997, 1998, 2000.

naturalistic position will be considered. Subsequent sections will explore some of the religious and theological options in association with naturalism. Basically, these will be treated as pertaining to two kinds of naturalism: namely, *theistic naturalism* and *religious naturalism*.

This chapter has its place in a part entitled 'Religion and Science Across the World's Traditions'. But is Religious Naturalism 'a tradition'? Is it one of the world's traditions, alongside Buddhism and Christianity? Perhaps it is a tradition as old as science. We'll get to this towards the end, but first, we have to become more familiar with naturalism in its relations with the sciences, and with religious forms of naturalism.

What Might Naturalism Mean?

Nature

In the present context, 'nature' is not just about wilderness. It also refers to nature as domesticated by humans, to technological artefacts, to humans and their creations such as languages and political institutions. It refers to material reality, but also to that which is done in and through material reality. Music, for instance, does not exist in a vacuum—it needs vibrations, material movements. At least as vibrations, music is part of nature. As we will see, 'naturalism' takes a further step, treating music as a natural phenomenon, not only in its expression as vibrations, but also in its origin in human creativity and intentionality. One might also substitute for nature 'the whole of empirical reality', if that would not prematurely decide certain issues of theology or metaphysics.

Such an encompassing concept of nature need not imply explanatory or valuational reductionism, as if by being natural, an entity would be of less interest. If, for instance, humans are material beings, this does not downgrade humans. It should rather lead to a high esteem for matter, since matter is capable not only of being sand or rock, but also of being Rembrandt, Einstein, Gautama (Buddha), and Jesus.

The Rejection of Dualism and the Success of Science

In arguments for naturalism, there are at least two different motives at work. Naturalism is a response to the success of the sciences. The sciences provide an increasingly integrated and unified understanding of reality, resulting in precise predictions which correspond to empirical results. Success may also be understood practically, since scientific understanding allows us to manipulate parts of reality with enormous precision. Electrons are not merely hypothetical entities, but have become instruments for further research (Hacking 1984: 154). Inspired by the success of the sciences, naturalism is a 'low-level' metaphysics, seeking to follow as closely as is philosophically feasible the insights that the sciences offer (Drees 1996: 11). The theologian Gordon Kaufman writes that he accepts naturalism, taken to mean 'simply that so far as I can see all human (and other) life is to be found within what we call nature, and the whole of human meaning and value, personality and spirituality, has emerged within the complex natural processes of life on Earth and is not induced from outside the natural order' (Kaufman 2003: 96).

The success of the natural sciences is, however, not the sole inspiration for naturalistic positions. For some authors, naturalism begins with the rejection of dualism. Some would consider theism with its supernatural deity to be incoherent or religiously unacceptable. If one allows for divine interventions (miracles), God would go against his own laws. Besides, God might become morally blameworthy for not intervening on many other occasions. Whatever the precise considerations, for those who dislike dualism, naturalism might be attractive as a form of monism. Thus, Jerome Stone defines naturalism as follows:

Negatively, it asserts that there seems to be no ontologically distinct and superior realm (such as God, soul or heaven) to ground, explain, or give meaning to this world. Positively, it affirms that attention should be focussed on this world to provide whatever explanation and meaning are possible in life. (Stone 2003b: 89)

And Charley Hardwick (1996: 5–6) defines naturalism in contrast to supernaturalism as follows:

(1) that only the world of nature is real; (2) that nature is necessary in the sense of requiring no sufficient reason beyond itself to account for its origin or ontological ground; (3) that nature as a whole may be understood without appeal to any kind of intelligence or purposive agent; and (4) that all causes are natural causes so that every natural event is itself a product of other natural events.

Naturalism as Ontology, Epistemology, and History

A relevant approach to naturalism in the context of 'science and religion' is, in my opinion, to use it as a label for a world-view that follows the natural sciences as its major guide for understanding the world we live in and are a part of. Such a naturalism is not formally implied by the sciences, since other logically coherent constructions may be possible; but it is a view of the world that stays as close as possible to mainstream consolidated science when it articulates its ideas about the ontology and history of reality.

With respect to *ontology*, naturalism assumes that all objects around us, including ourselves, consist of the stuff described by chemists in the periodic table of the elements. This stuff is further understood by physicists to consist of elementary particles and forces, and beyond that is assumed to consist of quantum fields, superstrings, or whatever. As the 'whatever' indicates, our knowledge has not yet reached rock bottom. Hence, naturalism cannot be articulated from a fundamental ontology upwards. Nor does it imply that all phenomena can be described in terms of physics and chemistry. A conceptual and explanatory non-reductionism is tenable (Drees 1996: 15 f.; 1997: 531). Even more, a naturalist can also defend the idea that there are genuinely new objects with new properties, even though they have arisen out of other objects. Higher-level properties are not just combinatorial consequences of lower-level properties (Humphreys 1997). Scientists and philosophers of science may clarify how we can understand emergent entities and properties as real and causally efficacious, even if produced by (and 'consisting of' in a material sense) simpler ones, just as future entities will be real and causally efficacious even though produced by present ones (see also Goodenough and Deacon, Ch. 50 below).

With respect to *history*, naturalism understands living beings, humans included, as the current stage in a bundle of Darwinian evolutionary histories on our planet, which itself is a transient phenomenon in a universe that has been expanding for some fifteen billion years. These insights do not commit one to a particular view on processes 'within the first fraction of the first second'; it may be that 'first second' is not an adequate reference at all. It is with history as with ontology: the most fundamental issues about the beginning of our universe and the nature of time, space, and substance are not settled for the naturalist.

Naturalism sees *social and mental life* as one of the fruits of the long evolutionary process. Science is one facet of this, even when it studies our own emergence. Naturalism holds that this is not a vicious circularity. Rather, science and other intellectual enterprises can be seen as building upon human capacities for dealing with our environment, improved piecemeal over many generations; science is a social phenomenon which is cognitively reliable, and increasingly so (Kitcher 1993). Among the social phenomena are also the *religious attitudes, beliefs, and traditions*. They can be studied by cultural anthropologists, historians, and the like. The processes of emergence, development, change, continuation, and extinction of various religions are to be understood within a natural framework, comparable to some extent to the emergence, change, and disappearance of languages and legal systems.

When we come to speak of *religious* naturalism, the issue is not merely whether one can understand the history of religions naturalistically. The issue is what one's own religious or non-religious position is, and how well it copes with a naturalistic self-understanding. This is especially relevant when religious traditions are not merely pre-scientific propositional beliefs, but also powerful motivators, embodying moral and existential convictions. Before we discuss theological and religious options in relation to naturalism, we will first consider some challenges to naturalism as such.

The Wildness of Experience

Is reality not more complex and intractable than naturalism takes it to be? As the novelist John Fowles wrote in his reflective essay *The Tree* (1979): 'Ordinary experience, from waking second to waking second, is...hopelessly beyond science's power to analyse. It is quintessentially "wild", in the sense my father disliked so much: unphilosophical, uncontrollable, incalculable' (1979: 40. f). I agree that such wildness is a genuine feature of reality. That should make us modest with respect to the claim that scientific theories can elucidate all experience. Irreducibility should be part of our understanding of reality. And it is. Chaos theory has made clear what should have been obvious to students of history: we never have sufficient knowledge of all the details to provide a full account of the course of events, and this lack of knowledge results in a lack of predictability. We do not monitor all processes; we cannot even monitor our own inner states. Higher-level disciplines use concepts that are not adequately expressible in terms of those of lower levels; all forms of money are physical, but there is no general identification of money with a particular type of matter (no type–type identity, even though there is token–token identity).

When it comes to humans, beings who attach meanings to their world, processes become even more complex. The humanities are far more difficult than physics. Thus, typically humanistic approaches such as hermeneutics may well be needed; there is no reason why a naturalist should exclude these.

Of course, much more could be written about the complexity of reality and the variety of methods needed to acquire a more or less complete understanding. The point is that there are unpredictable phenomena—for instance, in chaotic systems— but that we have scientific theories and mathematical models that seem to describe and explain such systems, including the unpredictability of their behaviour. Thus, let me suggest the conclusion that the wildness of reality can be understood scientifically, and thus does not count against a naturalistic view.

Pan-experientialist Naturalism?

There is a standard order from lower to higher among the major sciences, something like the following: physics—chemistry—biology—psychology. Atoms gave rise to complex molecules typical of processes within living organisms; some of these organisms have developed in various degrees awareness of their environment and of themselves, an inner life. Atoms don't have sentience, but sentience has arisen in material beings. Most naturalists accept this order, and thus opt for some form of physicalism as the more fundamental ontology, without thereby denying the significance of phenomena such as sentience at higher levels.

Within the naturalist family, however, a minority position reverses the order, and considers human experience as typical of the fundamental ontology. According to process philosophy in the tradition of Whitehead, all actual entities have freedom, sentience, and subjectivity. A most vocal advocate of such a naturalism is David R. Griffin (2001), who is the author of Chapter 27 below. Any naturalism that sticks to the usual order of disciplines he dismisses as materialist and atheistic.

This is not the place for an extensive argument regarding the richer naturalism that Griffin advocates. His more relaxed naturalism might make life easier for religious thought. However, it seems to mistrust the power of the process, as it denies the emergence of sentience and subjectivity. If they are emergent properties, they would not have to be among the basic ingredients at the most fundamental ontological level. Furthermore, the approach is substantially at odds with current science, where the disciplinary order indicated above does seem to express insights about the layered character of reality (see also Peacocke 1993: 217; Drees 1996: 257–9). Thus, not willing to turn scientific understanding upside down, I will not consider such a modified naturalism in the remainder of this chapter.

Normativity and Naturalism

In epistemology, the philosophy of science, and moral philosophy, there seems to be an opposition between naturalism and normativism. That naturalism might be unable to articulate normative aspects of existence seems to lead to the rejection of naturalism by the philosopher Charles Taylor in his *Sources of the Self* (1989: 19–22). Can a naturalist distinguish epistemology from psychology, truth from belief, and moral norms from evolved preferences? That is, if practices such as science and considered moral judgement are human practices, and as such fully natural phenomena rooted in our existence as primates, why should we take these as fundamentally different from pseudo-science or prejudice?

Naturalists tend to deny that there is an absolute demarcation between science and non-scientific activities. However, at the same time they do prefer science over pseudo-science, and thus live by such a distinction. Is this not self-referentially incoherent? The point is not that normative positions in ethics and epistemology don't have somewhat similar problems, but that 'naturalism' faces a problem when it comes to justifying its own criteria.

Naturalists will have to do without absolute norms and procedures. Public justification and individual reflection do strengthen the credibility of rules and norms; piecemeal improvement of morality and scientific methods and criteria makes a real difference (e.g. Kitcher 1985, 1993). In a naturalistic approach, arguing for a normative position, whether in morality or in epistemology, will always be an unfinished project. It is a project in which naturalism can benefit from other philosophical styles, such as pragmatism (with its sensitivity to the way in which our norms are rooted in human practices) and Kantianism (with its reflection on never fully accessible, always elusive, transcendent regulative ideals).

Scientism?

Are naturalists falling prey to scientism (Stenmark 2001)? That is, are they expecting too much from the natural sciences? Of course, some naturalists will have used methods and arguments from a particular domain elsewhere, where they turned out not to be adequate. However, as I indicated above, naturalists should allow for multiple domains or layers of reality, with a variety of methods and approaches. To further one's health, physical exercises may be more useful than exercises in physics.

In social life, the best solution may be the outcome not of calculation but of listening and looking for a consensus. Thus, 'scientism' may be a valid criticism, if one makes a case as to why a certain approach is misplaced in a particular context.

However, the charge of 'scientism' can also become an easy excuse. 'Used inappropriately, accusations of scientism may not clarify debate but muddle it by serving a more general antiscientific agenda. Scientism thus becomes a way of "crying wolf," of creating alarm about a given scientific claim without adequately considering the merits of the methods and evidence in question' (Peterson 2003: 758). It can be a useful, dismissive term for lazy people who do not make a well-focused argument. The charge of 'scientism' can also be brought forward at the expense of limiting science to the instrumental or empirical domain, robbing it of its theoretical dimension, which is where science reaches beyond what has been measured and observed so far. In the context of naturalism, the tendency is to push science as far as possible. Whether the use of science has been pushed too far will always need a specific argument, and it should be an argument that does justice to the multilayered understanding of reality that reasonable naturalists have come to acknowledge.

Four Arguments for Naturalism

So far we have considered various objections and alternatives. What are the main reasons in favour of naturalism? I would suggest that there are four clusters of reasons.

(1) The epistemic success of the natural sciences as it developed in the last century or two, resulting in corroborated theories that have a wide scope, unifying the understanding of phenomena in various contexts, in combination with remarkable precision, is totally without equal with any cognitive understanding offered in previous human history, whether in religious myths, theological systems, or philosophical speculations. This success makes it urgent to take these theories as our best available guides to the understanding of reality.

(2) The natural sciences, in conjunction with technology, have provided us with unprecedented means to act in reality. Actions can be laudable or despicable. However, it is morally preferable to use one's knowledge than to wilfully close one's eyes. Some advocates of other positions in 'religion and science' seek to play down the epistemic significance of science, arguing that it is as much about dogmatic frameworks and prejudices as is religion (as if such a *tu quoque* argument would help religion). In one example, the authors treated the theory that HIV was the cause of AIDS as just a theory that fitted the interests of the pharmaceutical industry and served discrimination (Stahl *et al.* 2002: 110). In their rejection of consolidated science, the authors were explicitly voicing support for the South African leader Mbeki, who by his denial of the viral background of AIDS deprived tens of thousands, if not more than a million people, of effective treatment. There is a major moral risk in playing down established science. This is related to 'the ethics of belief',

though the demands articulated by William K. Clifford in his original contribution (1886) were unrealistic. Playing down established science, even if for morally lofty purposes, may have immoral consequences. Working with the best knowledge available, rather than playing down such knowledge, may well be morally required.

(3) Any theist has good reasons to be a naturalist, maybe not in the ultimate sense of denying God's transcendence, but in the sense of welcoming the insight that nature has an impressive integrity and coherence. If this world is God's creation, any knowledge we have of this world is knowledge of God's creation. God is not to be found so much in the lacuna in our current knowledge, in the gaps, but rather in what we have uncovered. If our skills and powers are gifts of God, we should not look for God when we fail, but rather appreciate God for all that has become possible. Nature, religiously spoken of as creation, is not opposed to God, but rather God's gift. We'll come back to theism and naturalism below.

(4) For the naturalist, there is no way we can keep our religious convictions aloft from critical consideration. Our beliefs are human, just like the wide range of beliefs we come across around us, including those we consider superstitious. Our beliefs are not isolated as revealed insight, but are an intrinsic part of the human heritage. In the end, only those beliefs that can be integrated with the results of the best research and consideration deserve adherence. Research regarding one's own religion was quite a challenge for Western Christianity in the nineteenth century, when historical-critical study of the Bible in itself and in relation to new historical knowledge, e.g. regarding Mesopotamia, was to many traditional Christians far more challenging than Darwin's evolutionary biology. Some seek safety by closing their eyes to such studies, but the intellectual price of such wilful blindness is high. The alternative is to reconsider one's understanding of the significance of the biblical narratives, treating these as human responses, shaped by circumstances and theologies. The human dimension thus acquired by the sources of the tradition invited a hermeneutical questioning as to whether we recognize their experiences and thereby come to share their convictions, or not.

THEISTIC OR RELIGIOUS NATURALISM

If one accepts 'naturalism', what might be *religious* naturalism? Does naturalism not exclude what is typical of religious life? The answer depends not only on the understanding of naturalism, but also on one's concept of religion. If one accepts naturalism, what might be left, and would that be enough to qualify as religious? Or is this question phrased too much with mainstream religion as a yardstick, as if naturalism is chipping away at traditional belief, a watered-down version of religion light? We'll begin at the theistic end of the spectrum and gradually come to some examples of more purely naturalistic positions.

Naturalistic Theism

There is a traditional way to articulate theism that avoids a confrontation with the natural sciences, and that is to emphasize the uniqueness of God's mode of being and activity (e.g. Kaufman 1972; Wiles 1986; Stoeger 1995; Heller 2003). This is articulated in the notion of *creatio ex nihilo*, which does not apply to any natural causality. God creates and sustains all things as their primary cause; all natural causes are real, just as are all entities and events, but they are so because they have been created by God. Such real natural causes are 'secondary causes'. This distinction between primary and secondary causality was developed in the European Middle Ages—for instance, by Thomas Aquinas—but its roots can be traced back at least to Augustine (Thomas 1983; McMullin 1985, 1988; Hebblethwaite and Henderson 1990; Burrell 1993). God creates everything—past, present, and future events—and creates them not as an amorphous bag of events but with their temporal, spatial, and causal relations. The distinction between God and God's activity, on the one hand, and creatures and creaturely activity, on the other, is often articulated as a difference with respect to time: creatures are temporal, whereas God, as conceived in this view, is not temporal. God's eternity is not everlastingness but timelessness. Accepting the whole natural world as the creation of a timeless, transcendent God may be consistent with a naturalistic view of the world, since it accepts the world as understood by the natural sciences as God's creation. There is no need for particular gaps within the world.

On such an understanding of God, theology and the natural sciences relate to each other with respect to the explanation of the natural world as a whole (rather than the explanation of phenomena in the natural world). Or at least, the naturalistically minded theist would claim that the sciences are explanatory within the world, but not explanatory of the world as such. This, in my opinion, is consistent with naturalism. Science offers explanations, but every explanation assumes an initial state and laws. Thus, science explains within a framework, but does not explain the framework as such; limit questions persist (Drees 1996: 17–18, 266–9).

The answers of theists and naturalists to limit questions may be quite different—at least for naturalists driven by a dislike of dualism. If naturalism is defined as including the assumption 'that nature is necessary in the sense of requiring no sufficient reason beyond itself to account either for its origin or ontological ground' (Hardwick 1996: 5–6), a naturalist cannot accept the suggestion that limit questions might allow for a transcendent ground of reality. In my opinion, however, the naturalist should not be too ideological with respect to limit questions. (Thus, there may be tension over this question among naturalists.)

A science-inspired naturalism need not imply the dismissal of such limit questions regarding the existence, structure, and intelligibility of the world. I find *naturalistic theism* a genuine, and attractive, possibility—that is, a fully science-inspired naturalism with respect to the world we live in and experience, combined with openness to the possibility that this remarkable reality is continuously created by a transcendent God.

Naturalistic theism has one major problem, as I see it. It is hard to give reasons, once one accepts a naturalist understanding of created reality, why one would hold such a theological position; 'since there are no real "gaps" to fill, we may be left without an argument for God's existence of the kind that would convince a science-minded generation' (McMullin 1988: 74). Limit questions may exist, but they do not point to a specific answer. *Agnostic naturalism* might be epistemically more appropriate. Theism and naturalism with respect to the world may be reconcilable, but *naturalistic theism* with a concept of a transcendent God would be a species of theism rather than of naturalism.

Theistic Naturalism: God as Ground

Less dualistic, though still within the theistic tradition broadly conceived, is the position of those who speak of God as Ground of Being (see Wildman, Chapter 36 below). A major figure in the articulation of such a theological position has been Paul Tillich. In the religion and science dialogue, Arthur Peacocke (1993) might be the most prominent advocate. This view has come to be formulated often in panentheistic terms—understanding the world to be in God, even though God surpasses the world (Clayton and Peacocke 2003). I found a most inspiring poetic expression among aphorisms in *The Aristos* of John Fowles (1980: 27): 'The white paper that contains a drawing; the space that contains a building; the silence that contains a sonata; the passage of time that prevents a sensation or object continuing forever; all these are "God".' A creative development of such a position has been the suggestion of the theologian and scholar of the New Testament Gerd Theissen (1985) that we understand religious history as adaptation to ultimate reality. Whatever the precise formulation, this position has more deeply ingrained naturalistic presuppositions by seeking to avoid the dualism of a transcendent God and a natural world, even though it maintains a concept of God as surpassing the world—and thus I would prefer to consider this not as *naturalistic theism*, but as *theistic naturalism*.

Religious Naturalism

There are some positions that may be further removed from theism, as they do not understand God as an entity (or as the all-encompassing entity) but rather as a symbol used to speak of our existence and the world we live in, referring to 'the sacred' rather than to God. Are such positions still religious? Stone describes religious naturalism as 'a variety of naturalism whose beliefs and attitudes assume there are religious aspects of this world that can be appreciated within a naturalistic framework. Occasions within our experience elicit responses that are analogous enough to the paradigm cases of religion that they can appropriately be called religious' (Stone 2003b: 89). Speaking of *religious* naturalism may thus be justified if the attitudes and responses are sufficiently analogous.

Let me briefly introduce some varieties of such forms of religious naturalism. Gordon Kaufman interprets the Christian symbolism of God in the context of our existential concerns and our responsibility in a time of ecological and nuclear threats, but also as a figure of speech to speak of an overwhelmingly significant characteristic of processes in the universe: namely, their 'serendipitous creativity' (1993, 2003). He connects such aspects with traditional understandings of God as Creator and the moral call and vision as understood in the Christian tradition. Stone's version of naturalism denies absolute transcendence, but speaks of 'the sacred' in relation to '*situationally transcendent resources and continually challenging ideals in the universe*' (1992: 17; 2003a: 798).

Charley Hardwick gives up on ontology, while seeking to articulate theological content. 'Theological content can break free of ontology if this content is valuational rather than ontological. Such a valuational theism becomes possible when Rudolf Bultmann's and Fritz Buri's method of existentialist interpretation is wedded to Henry Nelson Wieman's naturalistic conception of God' (Hardwick 2003: 111). He quotes approvingly Bultmann, who has said: 'the real meaning of myth does not present an objective world picture but instead expresses our understanding of ourselves in our world' (quoted in Hardwick 2003: 113). 'Although *God* does not refer (any more than rights, duties, values, or point masses need have ontological references), *God* or *God exists* can serve as a complex meta-expression for a form of life that is expressed as theistic seeing-as' (Hardwick 2003: 114). In his book *Events of Grace* (1996), Hardwick reconstructs classical Christian conceptions, such as sin and grace. This is not accidental, but part of his understanding of a religious naturalistic agenda. He does not seek a religious Esperanto without roots in a particular tradition. 'I am constantly reminded here of Santayana's dictum that "the attempt to speak without speaking any particular language is not more hopeless than the attempt to have a religion that shall be no religion in particular"' (Hardwick 2003: 115).

Other religious naturalists are less engaged with a particular tradition. Perhaps 'the evolutionary epic' might become *Everybody's Story* (Rue 1999). Some such more universal inspirations come from the side of scientists with a broader engagement with humanistic interests, such as Ursula Goodenough in *The Sacred Depths of Nature* (1998), or are focused on worldwide challenges such as ecology.

Last but not least, there are also religious naturalists who dispense with all God-language. Donald Crosby speaks in this context of *naturism* (rather than naturalism), 'to distinguish it from conceptions of religious naturalism that make fundamental appeal to some idea of deity, deities, or the divine, however immanental, functional, nonontological, or purely valuational or existential such notions may be claimed to be. The focus of naturism is on nature itself as both metaphysically and religiously ultimate' (Crosby 2003: 117).

RELIGIOUS NATURALISM: A TRADITION AS OLD AS SCIENCE?

Is religious naturalism a tradition, or even one of the world's traditions? That is the context in which this contribution has been invited. There is no explicit institutionalization, as in some religions. There is no clear set of rituals that mark religious naturalists. However, 'religious naturalism' seems a subculture with an identity of its own. Michael Cavanaugh (2000) describes some contemporary contributions, but this subculture has a history that, often unconsciously and occasionally consciously, might be a formative part of its identity. In 1998 *Zygon* devoted a series of contributions to the legacy of Ralph Burhoe. Beyond the history of this specific journal, one may refer to philosophers, scientists, and theologians such as Henry Nelson Wieman, George Santayana, John Dewey, Charles Sanders Peirce, Mordecai Kaplan, and Jack J. Cohen, and to some extent even Alfred N. Whitehead and William James, as forerunners.

Jeffrey Stout argues, in response to exclusionary ways of defining religion and democracy, that in the United States we have 'Emersonian piety' alongside an 'Augustinian' one. 'Emersonian piety' refers to Waldo Emerson, but is used here as a label for a religious attitude that is much more prevalent. Piety is understood not as deference to higher powers, to theological truth as a given, or as reverence for authority, but is rather characterized as self-reliance, taking responsibility for one's thinking. This is not self-reliance as if our achievements are ours in isolation from a tradition shaped by earlier generations. Rather, it is gratitude to earlier generations and the whole of nature, the sources of our existence, but gratitude that is honoured not by receptivity alone, but by moving on, by further explorations. A similar attitude could be articulated by referring to various thinkers of the European Enlightenment (Stout 2004: 20–31). There is in many respects a huge overlap between religious naturalism and American pragmatism.

One may go back further in time, beyond the last century and a half, and claim to be an heir of Spinoza, of his liberal Christian and Unitarian friends and the subsequent Spinozists of various stripes, and of some of the German philosophers (Clayton 2000), as well as of Dissenters and British scientists who became Unitarians (Joseph Priestley) or pantheists (Humphry Davy; see Knight 2000). Of course, every figure is to be seen in the context of his time. Claiming them as ancestors is appropriation out of context, but that is precisely the intellectually ambivalent practice that strengthens identity. These exemplary figures are individuals who are perceived as somewhat heretical by the traditional religious community of their time, while standing in close contact with, if not being part of, the scientific community— precisely the mix that may fit contemporary religious naturalists.

There are Christian, Jewish, and humanist dialects of religious naturalism, as well as biological, psychological, and physicalist ones, reflecting upbringing, training, and heritage as well as needs and situation. Some dialects are dialects of another tradition as well, just as a local dialect near the border of my country may be considered by some as a dialect of Dutch, whereas others might treat it as a dialect of German. Thus, essays such as those of Arthur Peacocke (2000) and David Pailin (2000) may well be read as liberal Christian essays as well as naturalistic ones. There is a wide range of styles, from the sober and minimalist (Stone, Hardwick) to the ecstatic and exuberant (Corrington 1997), from the analytical to the evocative (Goodenough 1998). Religious naturalism is an umbrella which covers a variety of dialects, of which some are revisionary articulations of existing traditions whereas others may be more purely naturalistic religions indebted almost exclusively to the sciences. There is family resemblance, with affinities and disagreements, not unity.

Religious naturalism also takes shape through stories. The evolutionary epic serves as a master narrative, but there are also smaller stories that evoke attitudes and feelings alongside philosophical essays that convey intellectual claims. Ursula Good-enough's *The Sacred Depths of Nature* is an example; my own retelling of a creation story in *Creation: From Nothing until Now* (2002) and Peacocke's 'Genesis for the Third Millennium' (Peacocke 2001: 1–2) might be mentioned here too. Naturalism has its romantic side as well. And it offers a setting for understanding the darker aspects of one's own existence: 'death is the price to be paid . . . My somatic life is the wondrous gift wrought by my forthcoming death' (Goodenough 1998: 151). There is also plenty of work on more systematic theological elaborations (e.g. Stone 1992; Hefner 1993; Kaufman 1993; Hardwick 1996; Peters 2002).

A Tradition as Old as Science?

If one envisions religious naturalism as a tradition as old as science, definition in terms of particular scientific content will not do. Our picture of reality has become far more dynamic, both in cosmology and in biology, compared with the best available understanding in Spinoza's day. It might perhaps be characterized in its continuity by certain attitudes.

One characteristic is the *openness* to research and the readiness to challenge authority focused on persons or ancient books; its concept of piety is not submissive (see Stout 2004: ch. 1). Not just scientific knowledge, but also historical research regarding other cultures and the sources of our own (e.g. biblical criticism) is appreciated. Such a tradition would be, by nature, not communitarian but individu-alistic—and thus its continuity would always be under pressure, since individualisms reproduce with difficulty; if one does not need the church or the community to be saved, children may get that message and do without.

Another feature is a positive *appreciation of this world*, not necessarily naïve, sometimes even renouncing materialism *qua* lifestyle, but in contrast with investing hope in a different world to come.

A third, related feature is an *activist* attitude, as redemption is not expected to happen to us; improvement is to be brought about by human activity. Naturalists appreciate reality, but include in this human activity. The historian John Brooke (Brooke and Cantor 1998; Brooke 2003) has observed that the discourse of improving nature has a long history, e.g. in alchemy, and continues in chemistry (Priestley again) and other transformative disciplines, independently of, or even at odds with, natural theology, which served more to support traditional theology with arguments from design—that is, from the world as observed.

Furthermore, even though some religious naturalists build upon a particular religious tradition, there seems to be a *universalist* intention, in that the religious naturalist expects his approach to be open in principle to persons from all walks of life, of all cultures, and of all faiths. With this universalism, the religious naturalists are *moralists*, who are not just interested in understanding nature, but who seek to articulate humanist values in relation to their understanding of reality.

I am not sure whether it is helpful to understand religious naturalism as a tradition as old as science, and in many ways intertwined with it in its development, or as a response to science. Thus, in addition to reflection on the sciences and the philosophical clarification of various forms of religious naturalism, there is also work to be done by historians of religion and of culture by studying such more diffuse forms of religion, whether related to traditions or as 'something-ism', agnosticism and religious humanism. These could be studied historically and systematically, for motives and arguments, as well as for dynamics.

Am I a religious naturalist? Others have used the label of me. I am not sure that I like the label, as it seems to constrain, whereas I want to explore. I also have some sympathy for the naturalistic theism described above. But certainly, precisely in this attitude of exploring, I fit the tradition referred to above. Or at least, I hope I do. Even if I am not sure whether I am a religious naturalist, I am most interested in understanding what religious naturalism might mean, may become, and will offer.

REFERENCES AND SUGGESTED READING

BROOKE, JOHN HEDLEY (2003). 'Improvable Nature?', in Willem B. Drees (ed.), *Is Nature Ever Evil? Religion, Science and Value*, London: Routledge, 149–69.

—— and CANTOR, GEOFFREY (1998). *Reconstructing Nature: The Engagement of Science and Religion*. Edinburgh: T & T Clark.

BURRELL, DAVID B. (1993). *Freedom and Creation in Three Traditions*. Notre Dame, Ind.: University of Notre Dame Press.

CAVANAUGH, MICHAEL (2000). 'What is Religious Naturalism? A Preliminary Report of an Ongoing Conversation', *Zygon*, 35/2 (June): 241–52.

CLAYTON, PHILIP (2000). *The Problem of God in Modern Thought*. Grand Rapids, Mich.: Eerdmans.

—— and PEACOCKE, ARTHUR (2003) (eds.). *In Whom We Live and Move and Have Our Being: Panentheistic Reflections on God's Presence in a Scientific World*. Grand Rapids, Mich.: Eerdmans.

CLIFFORD, WILLIAM K. (1886). 'The Ethics of Belief', in *Lectures and Essays*, 2nd edn., London: Macmillan, 339–63.

CORRINGTON, ROBERT S. (1997). *Nature's Religion*. Lanham, Md.: Rowman & Littlefield.

CROSBY, DONALD (2002). *A Religion of Nature*. Albany, N.Y.: SUNY Press.

—— (2003). 'Naturism as a Form of Religious Naturalism', *Zygon*, 38/1 (March): 117–20.

DREES, WILLEM B. (1996). *Religion, Science, and Naturalism*. Cambridge: Cambridge University Press.

—— (1997). 'Naturalisms and Religion', *Zygon*, 32: 525–41.

—— (1998). 'Should Religious Naturalists Promote a Naturalistic Religion?', *Zygon*, 33/4 (December): 617–33.

—— (2000). 'Thick Naturalism: Comments on *Zygon* 2000', *Zygon*, 35/4 (December): 849–60.

—— (2002). *Creation: From Nothing until Now*. London: Routledge.

FOWLES, JOHN (1979). *The Tree.* St Alban's: Sumach Press.

—— (1980). *The Aristos,* Rev. edn. Falmouth: Triad/Granada.

GOODENOUGH, URSULA (1998). *The Sacred Depths of Nature.* New York: Oxford University Press.

GRIFFIN, DAVID R. (2001). *Reenchantment without Supernaturalism: A Process Philosophy of Religion.* Ithaca, N.Y.: Cornell University Press.

HACKING, IAN (1984). 'Experimentation and Scientific Realism', repr. in J. Leplin (ed.), *Scientific Realism,* Berkeley: University of California Press, 154–72 (orig. in *Philosophical Topics,* 13 (1982): 71–87).

HARDWICK, CHARLEY D. (1996). *Events of Grace: Naturalism, Existentialism, and Theology.* Cambridge: Cambridge University Press.

—— (2003). 'Religious Naturalism Today', *Zygon,* 38/1 (March): 111–16.

HEBBLETHWAITE, B., and HENDERSON, E. (1990) (eds.). *Divine Action: Studies Inspired by the Philosophical Theology of Austin Farrer.* Edinburgh: T & T Clark.

HEFNER, PHILIP (1993). *The Human Factor: Evolution, Culture, and Religion.* Minneapolis: Fortress Press.

HELLER, MICHAEL (2003). *Creative Tension: Essays on Science and Religion.* Philadelphia: Templeton Foundation Press.

HUMPHREYS, PAUL (1997). 'How Properties Emerge', *Philosophy of Science,* 64 (March): 1–17.

KAUFMAN, GORDON D. (1972). *God the Problem.* Cambridge, Mass.: Harvard University Press.

—— (1993). *In Face of Mystery: A Constructive Theology.* Cambridge, Mass.: Harvard University Press.

—— (2003). 'Biohistorical Naturalism and the Symbol "God" ', *Zygon,* 38/1 (March): 95–100.

KITCHER, PHILIP. (1985). *Vaulting Ambition: Sociobiology and the Quest for Human Nature.* Cambridge, Mass.: MIT Press.

—— (1993). *The Advancement of Science: Science without Legend, Objectivity without Illusions.* New York: Oxford University Press.

KNIGHT, DAVID. (2000). 'Higher Pantheism', *Zygon,* 35/3 (September): 603–12.

McMULLIN, ERNAN (1985). 'Introduction: Evolution and Creation', in E. McMullin (ed.), *Evolution and Creation,* Notre Dame, Ind.: University of Notre Dame Press, 1–56.

—— (1988). 'Natural Science and Belief in a Creator: Historical Notes', in Robert J. Russell, William R. Stoeger, and George V. Coyne (eds.), *Physics, Philosophy, and Theology: A Common Quest for Understanding,* Vatican City State: Vatican Observatory; distributed by University of Notre Dame Press, 49–79.

PAILIN, DAVID A. (2000). 'What Game is Being Played? The Need for Clarity about the Relationships between Science and Theological Understanding', *Zygon,* 35/1 (March): 141–63.

PEACOCKE, A. R. (1993). *Theology for a Scientific Age: Being and Becoming—Natural, Divine and Human,* enlarged edn. London: SCM Press; Minneapolis: Fortress Press.

—— (2000). 'Science and the Future of Theology: Critical Issues', *Zygon,* 35/1 (March): 119–40.

—— (2001). *Paths from Science Towards God: The End of All Our Exploring.* Oxford: Oneworld Publications.

PETERS, KARL E. (2002). *Dancing with the Sacred: Evolution, Ecology, and God.* Harrisburg, Pa.: Trinity Press International.

PETERSON, GREGORY R. (2003). 'Demarcation and the Scientistic Fallacy', *Zygon,* 38/4: 751–61.

RUE, LOYAL R. (1999). *Everybody's Story: Wising Up to the Epic of Evolution.* Albany, N.Y.: SUNY Press.

SELLARS, W. (1963). *Science, Perception, and Reality.* London: Routledge & Kegan Paul.

STAHL, WILLIAM A., CAMPBELL, ROBERT A., PETRY, YVONNE, and DRIVER, GARY (2002). *Webs of Reality: Social Perspectives on Science and Religion.* New Brunswick, N.J.: Rutgers University Press.

STENMARK, MIKAEL (2001). *Scientism: Science, Ethics, and Religion.* Aldershot: Ashgate.

STOEGER, WILLIAM R. (1995). 'Describing God's Action in the World in Light of Scientific Knowledge of Reality', in Robert J. Russell *et al.* (eds.), *Chaos and Complexity: Scientific Perspectives on Divine Action,* Vatican City State: Vatican Observatory; Berkeley: Center for Theology and the Natural Sciences, 239–61.

STONE, JEROME A. (1992). *The Minimalist Vision of Transcendence: A Naturalist Philosophy of Religion.* Albany, N.Y.: SUNY Press.

—— (2003a). 'Is Nature Enough? Yes', *Zygon,* 38/4 (December): 783–800.

—— (2003b). 'Varieties of Religious Naturalism', *Zygon,* 38/1 (March): 89–93.

STOUT, JEFFREY (2004). *Democracy and Tradition.* Princeton: Princeton University Press.

STRAWSON, P. F. (1985). *Skepticism and Naturalism: Some Varieties.* New York: Columbia University Press.

TAYLOR, CHARLES (1989). *Sources of the Self: The Making of Modern Identity.* Cambridge, Mass.: Harvard University Press.

THEISSEN, GERD (1985). *Biblical Faith: An Evolutionary Approach.* Philadelphia: Fortress Press. (Translation of *Biblischer Glaube in evolutionärer Sicht* (Munich: Kaiser, 1984).)

THOMAS, OWEN C. (1983) (ed.). *God's Activity in the World: The Contemporary Problem.* Chico, Calif.: Scholars Press.

WILES, MAURICE (1986). *God's Action in the World.* London: SCM Press.

CHAPTER 8

ATHEISM AND SCIENCE

PETER ATKINS

INTRODUCTION

Science is the only path to understanding. It would be contaminated rather than enriched by any alliance with religion. Such should be (in my view) the attitude of a scientifically alert atheist (a 'scientific atheist'). I shall elaborate and justify this core attitude in what follows.

There are those who consider that the domain of science is restricted to some kind of 'physical world' whereas religion deals with the 'spiritual'. In other words, while religion deals with the great questions of being, science deals with the little ones. A scientific atheist will agree that the domain of science is the physical world, but will hold the view that there is no other variety of world, and that the 'spiritual' is an illusion generated by a physical brain. The same scientific atheist will also hold the view that religion has manifestly failed to answer the big questions, despite claiming them as its own, and—in so far as they are real questions—believe that only science will succeed as well with them as it is succeeding with the little questions. Clearly, at some point it will be important to discuss the nature of this belief and distinguish it from religious belief.

THE CONTRAST OF TECHNIQUE

There are two central features of science that distinguish it from religion. One is its mode of action: its reliance on publicly accessible experimentation, in contrast to religion's private introspection. The other is its attitude: that the ultimate fabric of reality is determinable and in a certain sense comprehensible, in contrast to the ultimate indeterminability and incomprehensibility of the explanations offered by religion. Whereas science is meticulous in its objectivity, and false observation is soon exposed by parading data on public platforms, religion grasps at wisps of observation, and if they strike a sentimental chord, readily and enthusiastically absorbs them into the fabric of belief. In short, whereas science relies on experiment, religion relies on sentiment.

An example that might at first seem perverse is the purported discovery towards the end of the twentieth century of 'cold fusion', the achievement of the fusion of atomic nuclei in a simple pot where huge international effort using tons of equipment had virtually failed. The reports were, if true, wonderful, and represented what mankind was longing for—inexhaustible clean energy. Immediately, the world's scientific community sought to replicate the experiments, but failed, in due course discovering what had misled the original observers. The scientific procedure of public experimentation had overcome the longing for the achievement of a fantastic goal. How different that is from the report of a Virgin Mary on a church steeple (or, increasingly, on various pieces of toast). The religious swarm to see it, and driven by what is essentially sentimentality driven by mass hysteria and cultural suggestion, do purport to see it, and then weld it into their various ecosystems of belief. The whole of the scientific endeavour is based on scepticism of the out-of-the-way; in contrast, the whole of the religious endeavour is based on the rapturous embracing of the bizarre.

The distinction between science and religion can be expressed in a variety of other ways. Thus, scientists are hewers of simplicity out of complexity. They perceive (and enjoy) the awesomely complex (and often stunningly beautiful) attributes of the world around them, but dig deep into its foundations to discover the seeds from which that complexity has sprung. They are awed, but not *overawed*: they acknowledge the intricacy, harshness, and beauty of the world, and especially the intricacy of the activity of the human brain, but then doggedly pursue the sources of that complexity.

The quest downwards in the search for the underlying simplicity is very difficult, and needs to be done with cautious imagination, imagination to identify the path and caution to refer incessantly to observation. The journey back up from the discovered simplicity up to the world of appearance is fraught with difficulty, for the simplicities are not concatenated into a single thread that leads from source to appearance but gang together in a hugely complex web such that an event somewhere can have essentially unpredictable consequences elsewhere. In short, science is really hard work.

In contrast, instead of hewing simplicity from complexity, religion heaps complexity on simplicity: its unconscious goal seems to be to conceal the emptiness of its approach by generating obfuscation. It seeks complexity as the cause and explanation, and instead of ascending its ladder stepwise towards that ultimately unattainable comprehension, ascends by wild leaps of reason that are often closely and scholarly argued but are nothing but opinions guided by prejudice. In short, religion (in its arguments if not its practice), where it seeks comprehension, is really easy.

But should simplicity be the aim? And what is simplicity anyway? Simplicity is the end point of the search for properties that require no further explanation. The attainment of simplicity represents the atomization of explanation, the discovery of the components of an explanation. In other words, the achievement of simplicity is the core aim of reductionism. Reductionism, however, is more than dissection into parts, for it is certainly not a part of the scientific attitude that independent parts account for the properties of a whole. A part of the reductionist programme is in fact its very opposite, the *assemblist* programme, of seeking to understand how feedback and interaction of the components to which an entity has been reduced can result in complex and sometimes unpredictable behaviour of the reassembled entity. Religion, on the other hand, effectively ignores the richness of reductionist explanation and aims in a single gulp to ingest the whole, eschewing and dismissing both the reductionist and assemblist components of reductionism.

This insistence on the discovery of underlying simplicity as the ultimate aim of the scientific endeavour raises the question of how simplicity is recognized, and science knows when to stop. First, it must be stressed that the simplicity sought by science must be a potent simplicity, a simplicity that can account for the complexity of the world. This too is in stark contrast to religious pursuits of knowledge, where the desire is to come to know, in a sentimental sense at least, the potent *complexity* that is asserted to be the fount of all. A God is the ultimate anti-simplicity: a complexity beyond understanding, an entity that is outside understanding. In other words, a God is a synonym of intellectual defeat, the ultimate pessimism, the antithesis of the hopeful, optimistic driving force of science.

One sign of the achievement of a simple explanation is the elimination of a law of behaviour. Thus, much of science consists in examining an entity, identifying a pattern of behaviour, and summarizing that behaviour in terms of a law (in science, a law is a summary of observed behaviour, as in Newton's laws of motion). However, a step towards simplicity is achieved if it can be shown that the law is a natural consequence of the intrinsic nature of the entity, for then the combination 'entity + law' is replaced by 'entity' alone.

As a greatly simplified example of how an intrinsic property accounts for a perceived law, take the propagation of light. Light travels in straight lines. More precisely, the path of a light ray through a medium is such that its time of passage is least (this is a bowdlerized version of Fermat's principle of least time). How, though, before it starts out (or at least in the first moments of its journey) does a ray of light know the path that will, once it has completed its journey, turn out to have been the shortest of all possible paths? Once we realize that the intrinsic character of light is a wave, this behaviour falls into place. In brief, light takes *all* paths between the starting-point and the end-point; however, all but a few paths have neighbours that interfere with each other destructively. That is, when they arrive at the terminal point, the peak of one wave is likely to coincide with the trough of another, so they average to zero at that point. The few paths that do not interfere in this way all lie close to a straight line, for waves travelling along such paths all arrive with their peaks and troughs almost in step. (This result can be expressed precisely mathematically by drawing on the properties of waves.) Thus, because all non-straight line paths interfere destructively, and cancel each other, but straight line paths do not, and survive, an observer is led to conclude that light travels in a straight line. The important point is that a law that seems to govern behaviour turns out to be the natural outcome of complete anarchy. This is an example where casual observation would seem to require both an entity and a law to govern its behaviour, and perhaps even a lawgiver, but science demonstrates that only the entity is necessary, for the law emerges without further imposition. In this case, anarchy is the governor of behaviour.

This example reduces the complexity of the world and diminishes the need for a creative and workaday God. One religious view is that God needed to impose laws of behaviour on the universe at least at its creation. That laws in fact emerge from, and are manifestations of, an underlying anarchy rather does away with that role. Another religious view is that God is the universal proctor of behaviour, ceaselessly and ubiquitously ensuring that laws are obeyed, except where a flamboyant miracle is required. Once again, this busybody notion of God is shown by quiet scientific reflection to be wholly unnecessary, and perceived by an atheist as a fantasy of busybody minds anxious to find a cosmic role for their invention.

THE SCOPE OF SCIENCE

Science is limitless in its scope. Through the discovery of this rather straightforward technique (the scientific method is by no means difficult—it is essentially the application of common sense, going out into the world to make controlled observations of it, making sure that one's results can be replicated by another, and establishing how any discoveries fit into the matrix of other discoveries, and being honest), mankind appears to have stumbled upon a rather obvious way of reaching a true understanding of *anything* of interest, including those aspects of existence that religions have regarded as peculiarly their own. However, in the exercise of its power to answer deeply troubling questions, it has had to distinguish apparently real questions from the merely invented. Among the latter, of course, lie a number closely considered and regarded as deeply significant by the religious.

A general point that it is appropriate to make at this point is that science is widely considered to be concerned with questions of *how* rather than *why*. That is true. However, it is possible, and appropriate within scientific exegesis, to adopt the strong position that there is no legitimate 'why' question. All 'why' questions (including, to take the strongest possible position, in the special field of daily life) are actually convenient shorthand for congregations of 'how' questions. In other words, fully to understand a 'why' question (and its ineluctable flavour of purpose), we need to deconstruct it into its component (and purposeless) 'how' questions. From an atheist's point of view, all cosmic 'why' questions are either congregations of 'how' questions or meaningless inventions (in some cases, of course, both).

What are the great questions of being, and how far can we expect God-free science to go to answer them? My concern at this initial point is simply to separate the wheat from the chaff; I shall return later to the wheat.

One question concerns the origin of everything, which the religious have sought to answer, or at least to provide allegorical entertainments, in the form of creation myths. This question is perhaps real, because there certainly does seem to be something here the origin of which seems to require explanation. It will be classified as wheat and dealt with at length shortly.

A second great question fondly considered by numerous religions is the problem of why we are here at all, our cosmic purpose. This question is chaff. There is no evidence that such a question has any meaningful content. This question has been invented by those who prejudge the issue by considering that the universe must have a purpose if it is here and who cannot come to terms with the possibility that it has no purpose whatsoever. A purposeless universe is just as viable as one with a purpose, and, for some of us, an engagingly amusing accident of cosmic enormity.

It is best to regard questions that arise simply from surmise as empty amusements. Unfortunately, the religious give such surmises centrality. Science, however, has no need of purpose, has detected no sign of it, and finds that it can go about its business in its absence. (It is important, of course, to distinguish cosmic purpose from private purpose, which science does not deny and, through its various branches, such as biology and psychology, seeks to illuminate.)

Some will consider that dismissals such as this are frivolously superficial and that the fact that science dismisses a deep question as nonsense is a sign that the question is outside its reach. This would be a valid criticism if there were the slightest evidence that the universe did in fact have a purpose. There is not the merest whiff of such evidence. Moreover, it is easy to understand why many might think that such a complex and vast entity must be here for a purpose, for most complex and vast terrestrial entities have been made for one purpose or another, even if that purpose is entirely frivolous. However, there are plenty of examples of purposeless constructions (the Moon, for instance), and without even a hint of evidence that the universe does in fact have a purpose, the only rational position to adopt is that it is purpose-free. It is most important for a scientifically alert atheist not to be brow-beaten by those who impose a preconception on the universe and then cry 'Superficial!' when science declines to waste its time on their preconception.

Another great question regarded by many religions as central and their private reserve of information is the nature of the afterlife. Science denies that there is an afterlife. First and foremost, there is no evidence for such a state of existence. There is, of course, a great deal of desperate longing that there should be an afterlife, but longing is one thing, and reality another. Our current understanding of the physical operation of the brain and its ability to generate the intricate and currently moderately mysterious property known as 'consciousness', and in particular the sense of self, rules out without question the fantasy that some kind of function (that is, a soul) can persist in the absence of the physical substrate of the brain. The entire adornment of the debate about the afterlife and the associated speculative and evidence-free fantasies such as reincarnation, transmigration, purgatory, heaven and hell, resurrections (as distinct from recovery from comas), and ghosts falls aside once it is accepted that there is no such thing.

The absurdity of presuming that some sense of self (or whatever is the favoured theological flavour of the persistence corresponding to an afterlife) is not only supported by the view that the physical brain pumps ideas like the heart pumps blood, but is reinforced by the psychological basis of this belief. All that a scientific analysis of the proposition exposes is the psychology of control and the psychology of fear. The relevant psychology of control is the weapon that belief in an afterlife puts into the hands of those whose aim is to intrude into the private lives of others with threats that are feared and cannot be verified. The relevant psychology of fear concerns the inability of individuals to come to terms with the prospect of their own annihilation. It should perhaps be added that of all the fantasies propagated in their own self-interest by the religious, belief in an afterlife is perhaps the most pernicious; for not only does it restrict enjoyment of the pre-afterlife (that is, life itself), but it proves to be a potent source of inspiration and reward for those who wish to kill in religion's name or merely satisfy their blood-lust.

The closest that science can come to studying the afterlife is its investigation of near-death experiences, a kind of interlife, where those on the brink of physical death report a variety of experiences that afterlife enthusiasts leap on to support their otherwise totally unsupported belief. In all cases that have been investigated meticulously, the phenomena reported (tunnels ending in bright lights, and the like) have turned out to be well-established physiological consequences of restricted supplies of blood to the brain, not glimpses of the blissful life to come.

Another great question of being is the nature of God, and in particular his, her, or its existence. Can science illuminate this question, central as it is to most religions?

There are, of course, several types of challenge in this question. One is to invite science to *prove* that there is no God. Of course it cannot do this, for omnipotence can spin out contrivances without restriction. However, the challenge is hardly fair, on a variety of grounds. Bertrand Russell's teapot is a well-known allegory on the question, where he asserts that it is impossible to disprove the assertion that there is a teapot in orbit around Mars. But the core point is why a scientist should be asked to provide the disproof of the assertion. An Occam-like view is that a more primitive assertion (that is, the absence of a positive assertion) should take precedence over one that is less primitive. That is, the atheist's primitive world-view (not needing to assert the existence of a God) should be the starting-point for any argument, and if a theist wishes to bring about a change of mind, then it is up to him or her to supply the evidence. A scientist should not be required to prove a negative: a religious person should be required to prove a positive.

The religious will not accept this argument, of course, not merely because it puts them in an awkward position, but because they hold the view (here I have to presume, not being familiar with the experience) that their evidence is revelation (that is, some kind of hallucination, more often than not one that is culturally conditioned or at least reinforced), and that the certain evidence they have cannot be conveyed at second hand and must be experienced directly and personally. A scientist, even one who is not an atheist, should view such a statement with the greatest suspicion, for it is well known how prejudice, social pressure, and general cultural conditioning can influence a person's judgement. An atheist will be adamant that evidence must be shareable if it is to be acceptable, and will not accept intuition as sufficient.

It goes without saying that there is no scientifically acceptable evidence for the existence of God. That billions of people believe that there is a God, and that many have ruined their life and that of others by martyrdom in the belief's cause, is of course, no substitute for evidence.

That miracles have been reported and ascribed to the hand of God does not amount to evidence. Miracles are of three varieties. There are seemingly miraculous happenings (most commonly cures) that on closer inspection have straightforward, natural explanations. There are seemingly miraculous happenings that turn out to be false reports and require no further explanation (except perhaps to inquire into the state of mind of the active deceiver or the passive deceived). David Hume comes to mind here, of course, and his remark that there is always more reason to believe that the report of a miracle is false than to believe that it actually occurred. The third variety of miracle—what one might call 'hard miracles'—have no natural explanation and, because they involve transgression of natural laws, are truly miraculous. No such hard miracles have ever occurred.

Religions also consider that they have sole suzerainty over questions of ethics. That personal and social deportment is an actual problem cannot be denied, and so we classify it as wheat and, in so far as science has anything to say on the matter, consider it later.

At this point of the harvest, we have two wheaten questions (the origin of everything and ethics) and several mountains of chaff (purpose, the afterlife, the existence of God, miracles). It is time to ignore the chaff and to consider the wheat.

Cosmogenesis

Religions have long been concerned with the problem of cosmogenesis, but apart from the entertainment value of some delightful allegories have provided no insight whatsoever. Some deny that religion stands or falls by its ability to contribute to this major question of being, but others see it as perhaps the ultimate exercise of a God's omnipotence, the creation of an entirely new universe from, presumably, absolutely nothing. Science, too, cannot explain the incipience of a universe without external intervention (or even with intervention), but it is edging ever closer to resolving what is perhaps the biggest question of all.

A sign of the progress that has been made by science within a span of 300 years, in contrast to the total lack of progress stemming from religious speculations in at least ten times as long, is the closest to the moment of incipience that science can reach. Thus, with the formulation of Einstein's theory of general relativity (his theory of gravitation), it has proved possible to trace backwards with confidence our current universe to within milliseconds of the event taken to mark its origin. We can even make experimental observations on the universe back to about a million years after it was formed (by looking out to great distances, from which light has taken billions of years to reach us). In fact, even contemporary observations, such as detailed investigation of the microwave background radiation, the remnants of the big bang of inception, can be used to infer the nature of the events accompanying the formation of the universe.

With decreasing confidence, we can even trace the history of the universe back to within about 10^{-40} seconds of its inception (without, needless to say, discovering any sign of the finger of God), although events this close to the origin are highly speculative because the closer we reach to the origin, the more doubtful it becomes to extrapolate current physical theories. That last remark, however, should not be construed as indicating that science is failing and reaching beyond its grasp. All it means is that science is proceeding cautiously and drawing on its ever increasing toolkit of theories and information. A striking and crucially important aspect of science is its patience. Scientists are conservative revolutionaries: they build bridgeheads out into the sea of ignorance using imagination and cunning, but their constructions are firmly rooted in the known and tested at every stage. Not for them the fiery leaps of sometimes poetic, emotionally charged imagination so characteristic of religions, leaps that are merely imagination however elaborately dressed they are in the trappings of scholarship.

In fact, and although it is not strictly relevant to this discussion, one of the most remarkable aspects of cosmology is that observations made in terrestrial laboratories are found to be applicable to the entire universe. Although scientists have speculated that the fundamental constants and, more broadly, properties might be different in realms far off in space and time, all the observational evidence points to the structural homogeneity of the universe. Moreover, cosmological theories depend crucially on bringing together theories and observations relating to both the extremely small (the fundamental particles) and the extremely large (whole universes): there can be no stronger indication that science is a reliable route to knowledge than that these two great rivers of knowledge that spring from such disparate sources mingle constructively. On the whole, the great rivers of religion mingle destructively where they meet, generating war rather than enhancing understanding.

There is, of course, a further component of this great cosmogenetical question: not merely what happened at the beginning but *how* it all began. On this point, science currently has very little to say. There have been speculations, but they are little more than musings, and barely distinguishable from religion's impotent own. However, there are a few points worth making. First, these speculations all focus on achieving a universe autonomously, without the hand of a creator. Second, although the speculations are only that, they are all cast within a scientific matrix, which suggests—it is no more than a suggestion—that science will be able to direct its attention to this

great problem. Third, we are becoming increasingly aware of the events and composition of the early universe just this side of zero, so the interface between the two
sides of zero (that is, the interface between when there was nothing and when there
was at least a very primitive something) is becoming clearer. Fourth, a careful analysis
of what there is in this universe reveals that it is actually much simpler than casual
inspection suggests, and that the incipience of the universe is less of a problem than it
might at first seem.

This, the simple composition of the universe, is such an important point that it
deserves a little air. When we think scientifically about the universe, we discover that
in a certain sense it is a combination of opposites (the religious, even the God-free
religious, will leap on the analogy of yin and yang; it is undeniable that certain human
thoughts may in due course be found to have a lucky retrospective applicability; many
more, of course, have not). Thus, there is positive and negative electric charge, and the
total quantity of positive charge in the entire observable universe exactly matches the
total quantity of negative charge. A naïve view could be, therefore, that the Creation
did not involve the creation of electric charge but merely the separation of no charge
(one aspect of nothing) into its opposites. The same is true of angular momentum (a
measure of the extent of circular motion). The total angular momentum of the
universe appears to be zero, but there are plenty of examples of local angular
momentum (a spinning planet, for instance). Thus, before the Creation there was
no angular momentum (another aspect of nothing), and after it there is still no
angular momentum overall: the clockwise angular momentum has simply become
separated from the counterclockwise, and appears to us a local rotational motion.

This argument can be extended to other attributes: energy, for instance. There is,
to the casual eye, a lot of energy around, so a particularly big task for any Creator
would be to furnish the cosmos accordingly and munificently. A scientist, however,
looks at the world with a beadier eye. There is certainly a great deal of energy due to
the motion of things, and a vast amount locked up as mass (according to Einstein,
mass and energy are equivalent, mass in a sense being merely a measure of energy
content in a region), including the almost incalculable total mass of all the galaxies.
However, there is also a negative contribution to the total energy, that arising from
the gravitational attraction between all the planets, stars, and galaxies. There is some
reason to suppose that this vast total negative contribution to the energy almost
(perhaps completely) cancels the vast positive contribution to the energy, and that
the total energy of the universe might be close (or actually equal) to zero. Thus, if this
speculation proves correct, the task of the Creator was not to supply vast amounts of
energy but merely to separate no energy (another aspect of nothing) into positive
and negative contributions.

Such speculations might be nonsense. Moreover, there are major constituents of the
universe that are currently almost completely unknown (such as the enigmatic dark
matter that seems to pervade all space). Nevertheless, they point the way to the fact
that science is in the process of simplifying the task of accounting for the incipience of
the universe and giving some hope that one day it will be possible to achieve an
authoritative account of its autonomous inception from absolutely nothing.

In the process of achieving this elucidation, science is already enlarging our conception of what is. Thus, current (and therefore fragile) theories of the early universe, some aspects of which are supported by observation, already indicate somewhat tentatively that this universe is but one of many. If that is so, it provides a possible answer to another vexing question: why our universe appears to be so well-tuned for life. This so-called fine-tuning problem notes that even small deviations of the fundamental constants (such as the charge of an electron) from their actual values would have catastrophic effects on matter, in the sense that stars would burn too fast to produce the elements of life, planetary systems would not survive long enough for evolution to act to produce conscious beings, and so on. There are four possible resolutions of this problem. The religious (that is, lazy) explanation is that it demonstrates the benign hand of God, who chose the fundamental constants with awesome foresight and benevolence (to our kind, at least). Then there are the more intellectually demanding scientific speculations. One is that any universe has to come into being with our mix of constants, and that it is simply a happy coincidence that their values suited our emergence. An alternative is that a universe can come into existence with a ragbag of values of fundamental constants. However, because there are so many universes, perhaps an infinite number, there is sure (as we have found) to be at least one within the ensemble that has the mix of values benign to life. Although countless universes might seem profligate and more demanding than the concept of a single God, do not be deceived: a random pile of countless particles of dust is still less organized than an omnipotent deity (or even a deity with almost no potency).

This important question of whether a Creator was involved in the creation has occupied a lot of space, and although many will regard it as peripheral to the spiritual dimension of religion, many remain puzzled by the simple fact that there is a universe, and religions have sought to provide answers. The scientific position is that it is working hard to provide an observationally verifiable account of the very early universe, and can see that one day it may be possible to account for its incipience without having to invoke active creation.

SPIRITUALITY

Now for the second wheaten question. The core of the argument directed against the universality of science is its abject failure to accommodate the spiritual. That religion has been of overwhelming importance in stimulating great art cannot be denied, but this is hardly an argument in support of the existence of the supernatural. That great art (like great suffering) has sprung from religions is a sign of the deep impact that religious concepts have had on minds, an impact that may be so great because brains have been disposed by evolution to believe, or merely because the cultural impact of religion is so great. Moreover, we accept that there is more to the spiritual than the enjoyment of artistic creation.

Does science fail when it turns its eye inwards? There are several problems here, and we need to distinguish them carefully. Perhaps the simplest is the question of whether science can investigate beauty, or, more generally, whether it is possible to achieve a scientifically based theory of aesthetics. Such a theory would, presumably, be a component of a broader theory of consciousness, so by arguing that consciousness can be understood, we simultaneously argue that the lesser problem of the perception of beauty can be understood too.

The problem of consciousness is of a kind quite different from the problem of cosmogenesis, and almost certainly will be elucidated quite differently too. Whereas cosmogenesis will be expressed in terms of a theory that can be formulated mathematically, our comprehension of consciousness will probably be in terms of its simulation by some kind of computational device (not necessarily a digital computer, but something of a computer-like kind). It is probably misleading to think that there will ever be a 'theory of consciousness' in the same way that there will be a theory of cosmogenesis (or even a 'theory of everything' for the whole of physical reality). There will, of course, be understanding of local issues, such as are already being achieved in relation to perception, cognition, and memory, but there is no need to seek nor want a global 'theory' of consciousness, whatever such a thing would be. Moreover, it might prove difficult to simulate the extra-neuronal consequences of the global, fleeting, modulating chemical environment of neurons, and there might always be doubt that the form of consciousness that has been simulated is in fact the analogue of our own consciousness. Once a form of consciousness has been achieved, there is the possibility of enhancing it; however, it might be that we would not recognize it as a greater form of consciousness and mistake it for stupidity and write off the construction as a failure.

Be that as it may, as far as we are concerned for the present discussion, once a form of consciousness has been simulated, we can do a variety of experiments to explore aspects of the appreciation of beauty, and all the other attributes that we consider to be especially human. For instance, it may be that our appreciation of beauty can in fact be atomized (in the reductionist-assemblist sense of reductionism) into elementary components.

The notion of spirituality extends, of course, beyond the boundaries of beauty, and may be taken to include the sense of moral behaviour. Can science illuminate morality, or must we leave that to religion? To those who dislike the thought that personal freedoms should be infringed by the deliberations of some variety of tribal elder and circumscribed by appeal to the compilations of ancient folk-tales and myths known as 'holy scripture' of one brand or another, it would be helpful if certain aspects of human behaviour and the notion of 'good' could be illuminated by science. We would then know what is intrinsic to our nature and what has been imposed by those whose desire is to control.

We are far from understanding our own nature, but the scientific investigation of the origins of behaviour, as represented broadly speaking by anthropology and psychology, is increasingly illuminating. In short, there are genetically evolved contributions that represent transcripts of our long journey through evolutionary history and show the scars of the struggles of our predecessors for survival. There are also the reflective components of behaviour, where our big brains allow us to consider the consequences of our actions (for good or ill). Such deep understanding of the interplay of these two aspects will provide a much deeper insight into our and others' actions than an appeal to ancient written authority.

Finally, and an aspect of this last remark, that perception and reflection in some cases result in religious beliefs is hardly surprising. One of the most fascinating realms of scientific inquiry is why otherwise intelligent people still believe in gods in general and God in particular. It is easy to understand why some great thinkers, and scientists, of the past were devout; for they were under considerable social pressure and had not been fully exposed—how could they be?—to the enormous advances in scientific understanding of the past century. That Isaac Newton was intensely religious is not an argument for modern scientists to be so, for he was alive while religion was still firmly in control of minds. Many intellectual descendants of Newton have also been fervent believers, and even today some noted scientists remain convinced of the existence of God and consider that the various holy books are in some a reliable guide to understanding. It remains a mystery, but one that could be resolved by the appropriate psychological investigation, why some scientists still adhere to a religious belief. Presumably the explanations will include a sense of personal insecurity. To a non-believing scientist, it remains a source of regret that other scientists should not have the intellectual courage to accept that the natural does not need the supernatural.

BELIEF

Underpinning this entire discussion, of course, is the question of belief. I accept that it is my undemonstrated *belief* that science can illuminate all the great questions of being, and that I believe in its omnipotence and universal competence. These beliefs are of a less demanding kind than the beliefs characteristic of religion, where the justification of the belief will be found on the other side of the grave.

Not to believe in a God is a more primitive view (in the sense of being simpler) than to believe in one, just as not believing that a teapot is in orbit around Mars is more primitive than believing that there is in fact one. The challenge must be for those who believe in a more elaborate explanation of all there is (that it was all provided and is currently overseen by God) rather than those who consider that, given time, and in the light of its current success, science will come up with a simple explanation (that the universe, all its attributes and processes, all the aspects of human behaviour including the belief that there is a god, and the emergence of the universe are all a natural working out of causally connected events), to demonstrate that their more elaborate hypothesis is essential. So far, nothing in science has required the intrusion of any flavour of God.

There is a final point worth making. Atheism, and its justification through science, is the apotheosis of the Enlightenment. Religion implicitly scorns humanity by asserting that it is intellectually simply too puny. Scientifically alert atheism, on the other hand, respects the power of the human intellect to strive for, and in due course achieve, understanding.

PART II

CONCEIVING
RELIGION IN LIGHT
OF THE
CONTEMPORARY
SCIENCES

CHAPTER 9

COSMOLOGY AND RELIGION

BERNARD CARR

INTRODUCTION

The aim of cosmology is to understand the large-scale structure and overall evolution of the universe. It involves both observations—classifying and cataloguing the various contents of the universe—and models to explain these observations. Cosmologists are particularly interested in the origin of the universe and what initial conditions could have led to a world like the one we inhabit. Most adopt the hot big bang model, which supposes that the universe started in a state of great compression some ten billion years ago, but whether the universe will recollapse or continue to expand forever, or whether it is finite or infinite in extent, remain open questions.

Although cosmology uses input from various branches of science, it is primarily concerned with structures on the scale of galaxies and above. The crucial point about such large structures is that they are dominated by gravity, electrical interactions being insignificant because the universe has no net charge. The other two forces in nature—the weak and the strong force (involved in nuclear interactions)—are short-range, and therefore unimportant on cosmological scales. Nevertheless, these forces were very important in the early universe, because the temperature and density were then high enough for the associated interactions to be significant. Indeed, it is clear that many features of the universe must have resulted from processes which occurred in the first moments of the big bang. For this reason, early universe studies have led to an exciting collaboration between particle physicists and cosmologists. This chapter will therefore also refer to recent developments in particle physics.

Some of the questions addressed by cosmologists are traditionally regarded as being in the domain of religion. All human cultures have their creation myths, and the earliest (pre-scientific) cosmological models explicitly reflected religious views. Modern cosmologists prefer to emphasize their links with science rather than religion, but this is a relatively recent development. Indeed, cosmology attained the status of a proper science only in 1915, when the advent of general relativity—the modern theory of gravity—gave the subject a secure mathematical basis. The discovery of the cosmological expansion in the 1920s then gave it a firm empirical foundation, and the detection of the microwave background radiation in 1965 established the hot big bang theory as a branch of mainstream physics. Nevertheless, cosmology is still different from most other branches of science: one cannot experiment with the universe—it may be unique—and speculations about processes at very early and very late times depend upon theories of physics which may never be directly testable. Because of this, more conservative physicists still tend to regard cosmological speculations as going beyond the domain of legitimate science, although another view is that one must change one's concept of what constitutes proper science.

In order to discuss the relationship between cosmology and religion, one must first specify *which* cosmology and *which* religion. This is complicated, because religious views are space-dependent (reflecting the culture and history of the particular part of the world where they originated) but not necessarily time-dependent (adherence to scriptures tending to freeze beliefs), whereas the cosmological view tends to be time-dependent (the favoured model evolving surprisingly rapidly) but not space-dependent (science by its very nature being a global enterprise). In this chapter, I will try to avoid the first problem by referring to religion only in very broad terms; where the discussion is specific, it will refer mainly to the three major monotheistic Western religions. (For further reading, a broader perspective is provided by Butchins (2002).) The second problem (that of time-dependence) is more fundamental, because—as we will see—the relationship between cosmology and religion may change as the cosmological view evolves.

The first part of this chapter is historical in emphasis: it shows how astronomical progress has generally drawn cosmology and religion ever further apart, and ends by summarizing various anti-divine arguments. The second part suggests that recent developments in cosmology may have reversed this trend, and it ends with a summary of the pro-divine arguments. The third part discusses some current topics which are likely to have an impact upon the future relationship between cosmology and religion.

How Progress in Cosmology Seems to Remove God

Early humans started with a 'geocentric' and 'anthropocentric' view, in which the heavens were the domain of the divine and the universe was very much 'alive'.

Humans were the focus of creation, with a direct link to the god (or gods) who sustained the world. However, this perspective was shattered once science started to expand its domain of interest beyond the human scale. By developing new instruments, like the telescope and the microscope, it was possible to extend observations *outwards* to scales much bigger than humans and *inwards* to much smaller scales. Modern cosmology might be regarded as the culmination of this process. While the journey has been intellectually gratifying, it has also entailed a humbling of humanity and a diminishing role for God. The extent of physical space is now so all-encompassing that there seems to be nowhere left for the soul (Wertheim 1999).

The Outward Journey

The seeds of modern cosmology were sown during the Renaissance period by three crucial steps on the 'outer' front. In 1542 Nicolaus Copernicus argued that the heliocentric picture provides a simpler explanation of planetary motions than the geocentric one. Today the Copernican principle is taken to mean that the universe has no centre and looks the same everywhere, although this idea had also been proposed by Nicholas de Cusa in 1444. The advent of this view immediately set astronomy at odds with the major monotheistic traditions, all of which assumed the Earth to be the centre of the universe. The next step occurred when Galileo used the newly invented telescope to show that not even the Sun is special. His observations of sunspots showed that the Sun changes, and in 1610 he speculated that the Milky Way—then known as a band of light in the sky—consists of stars like the Sun but at such a great distance that they cannot be resolved. This not only cast doubt on the heliocentric view, but also vastly increased the size of the known universe.

The third step was Newton's discovery of universal gravity which—by linking astronomical phenomena to those on Earth—removed the special status of the heavens. Furthermore, the publication of his *Principia* in 1687 led to the 'mechanistic' view according to which the universe is regarded as a giant machine. For Newton himself, this testified to the existence of God (second letter to Bentley, 10 Dec. 1662, in Newton 1959–77: 233): 'Blind fate could never produce the wonderful uniformity of planetary movements. Gravity may put the planets into motion but without the divine power, it could never put them into such circulating motions as they have.' However, this blend of science and theism was not to persist with most of his successors.

In the following century astronomers began to map the stars and nebulae in ever greater detail, and were able to use Newton's equations to explain their motions and configurations. In 1750 Thomas Wright proposed that the Milky Way is a disc of stars, and in 1755 Immanuel Kant speculated that some nebulae are 'island universes', similar to the Milky Way but outside it. Nevertheless, even at the start of the twentieth century most astronomers still adopted a Galactocentric view and assumed that the Milky Way comprised the whole universe. For example, this was Einstein's belief when he published his theory of general relativity in 1916. Then in the 1920s the idea that some of the nebulae are outside the Milky Way began to take hold. For some time this was a matter of intense debate, until the controversy was finally resolved in 1924 when Edwin Hubble discovered the distance to the Andromeda galaxy. An even more dramatic revelation came in 1929, when he obtained velocity and distance estimates for several dozen nearby galaxies and found that they are moving away from us with a speed proportional to their distance. Today this is known as Hubble's law, and it has been shown to apply to a distance of 10 billion light-years, a region containing 100 billion galaxies.

The most natural interpretation of Hubble's law is that space itself is expanding, as indeed had been predicted by Alexander Friedmann in 1920 on the basis of general relativity. Einstein rejected this model at the time because he believed that the universe (i.e. the Milky Way) was static, and he even introduced an extra repulsive term into his equations—the cosmological constant—to allow this possibility. After Hubble's discovery, he described this as his 'biggest blunder', although it transpired to be one of his most profound insights. Friedmann's model suggested that the universe began in a state of great compression at a time in the past now known to be about 14 billion years ago, with all galaxies receding under the impetus of an initial explosion. It is interesting that the person who did most to champion this picture was a priest, Georges Lemaître.

Although the discovery of the cosmological expansion gave the big bang theory a secure empirical foundation, it was still some time before it gained full recognition. For example, when they were working on cosmological nucleosynthesis in the 1940s, Ralph Alpher and Robert Herman (1988) recall: 'Cosmology was then a sceptically regarded discipline, not worked in by sensible scientists.' Throughout the 1950s there was also the competing steady state theory, which accepted the expansion of the universe but assumed that there is continuous creation of matter, so that it always looks the same. However, in the 1960s astronomers obtained increasing evidence that the universe is evolving: first, from radio source counts and the discovery of quasars, and then, most decisively, from the discovery that the universe is bathed in a sea of background radiation. This radiation is found to have the same temperature in every direction and to have a black-body spectrum, implying that the universe must once have been sufficiently compressed for it to have interacted with the matter. Subsequent studies of this radiation, most recently by satellites such as COBE and WMAP, have revealed the tiny temperature fluctuations associated with the density ripples which eventually led to the formation of galaxies and clusters of galaxies. George Smoot, principal investigator of the COBE project, described the picture of these ripples as the 'face of God'; today one of the prime aims of cosmologists is to study this 'face' in ever greater detail.

The last decade has seen even more dramatic developments. Although one would expect the expansion of the universe to slow down because of gravity, recent observations suggest that the expansion is actually *accelerating*. We do not know for sure what is causing this, but it must be some exotic form of 'dark energy', most probably related to the cosmological repulsive term introduced by Einstein to make the universe static. Another idea which has become popular is that our entire universe may be just one member of a huge ensemble of universes called the 'multiverse'. More conservative cosmologists would prefer to maintain the *cosmocentric* view that ours is the only universe, but perhaps the tide of history is against them.

This brief historical review of progress on the outer front illustrates three points. First, the expanding vistas opened up by cosmological progress have come at a price: the bigger the universe has grown, the more insignificant humans have become. Second, the heavens have been progressively stripped of their divinity, so we can no longer delude ourselves into thinking that we have some special or singular connection with a Creator. Third, cosmology has had to strive constantly to maintain its scientific respectability, battling not only religious but also scientific orthodoxy.

The Inward Journey

Progress on the inner front has also been unsettling. With the advent of atomic theory in the eighteenth century came the first hints that our experience of the small is just as limited as our experience of the large. While the discovery yielded crucial insights into chemistry, as a first step towards reductionism, it was as disturbing to religious orthodoxy as the Copernican principle. Atomic theory was also linked to statistical mechanics, and this put another cloud on the horizon; for the second law of thermodynamics suggested that the universe must eventually undergo a 'heat death', with life and all other forms of order inevitably deteriorating.

At first atoms could be viewed as solid objects, like billiard balls, but further dramatic developments came early in the twentieth century. The 'billiard ball' picture was demolished by the realization that an atom is mainly empty space, with electrons in orbit around a nucleus comprised of protons and neutrons. Even the solidity of the atomic constituents was soon removed with the discovery of quantum theory: elementary particles became fuzzy, ephemeral entities, described by a 'wave function' which is smeared out everywhere, and the classical deterministic laws were replaced by probabilistic ones. Quantum mechanics shattered our view of the micro-world just as much as relativity theory shattered our perspective of the macro-world. The probing of the micro-world also emphasized our vulnerability. For we discovered the strong and weak nuclear forces, and thereby unleashed an awesome new source of destruction.

On the other hand, the inward journey has reaped huge intellectual rewards: it has revealed that everything is made up of a small number of fundamental particles (e.g. protons and nucleons are made of quarks) interacting through just four forces: gravity, electromagnetism, the weak force, and the strong force. These interactions have different strengths and characteristics, but it is now thought that the last three can be unified as part of a 'Grand Unified Theory' (GUT). It may also be possible to incorporate gravity into the unification using 'superstrings', which has led some physicists to proclaim that we are on the verge of obtaining a 'Theory of Everything'.

The Macro–Micro Connection

Taken together, scientific progress on both the outer and the inner fronts can certainly be regarded as a triumph. In particular, physics has revealed a unity to the universe which makes it clear that everything is connected in a way which would have seemed inconceivable a few decades ago. This unity is succinctly encapsulated in the image of the *Uroborus*. This is shown in Figure 9.1, and demonstrates the intimate link between the macroscopic domain (on the left) and the microscopic domain (on the right). The numbers at the edge indicate the scale of these structures in centimetres. As one moves clockwise from the tail to the head, the scale increases through sixty decades: from the smallest meaningful scale allowed by quantum gravity (10^{-33} cm) to the scale of the visible universe (10^{27} cm). The scales are also given in units of 10^{-33} cm in parentheses.

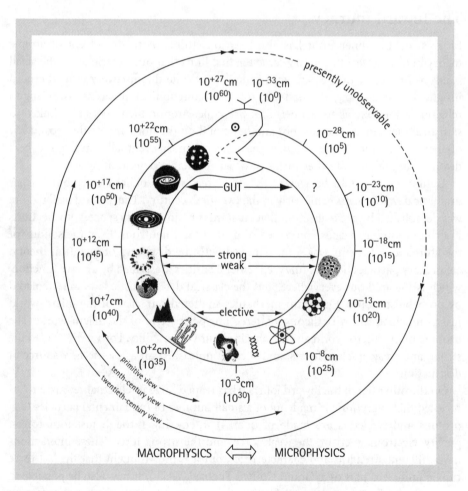

Fig. 9.1.

A further aspect of the Uroborus is indicated by the horizontal lines. These correspond to the four interactions and illustrate the subtle connection between microphysics and macrophysics. For example, the 'electric' line connects an atom to a mountain, because the structure of a solid object is determined by atomic and intermolecular forces, both of which are electrical in origin. The 'strong' and 'weak' lines connect a nucleus to a star, because the strong force which holds nuclei together also provides the energy released in the nuclear reactions which power a star, and the weak force which causes nuclei to decay also prevents stars from burning out too soon. The 'GUT' line connects the grand unification scale with galaxies and clusters because the density fluctuations which led to these objects originated when the temperature of the universe was high enough for GUT interactions to be important.

The significance of the head meeting the tail is that the entire universe was once compressed to a point of infinite density. Since light travels at a finite speed, we can never see further than the distance light has travelled since the big bang, about 10^{10} light-years: more powerful telescopes merely probe to earlier times. Cosmologists now have a fairly complete picture of the history of universe. As one goes back in time, galaxy formation occurred at a billion years after the big bang, the background radiation last interacted with matter at about a million years, the universe's energy was dominated by its radiation content before about 10,000 years, light elements were generated through cosmological nucleosynthesis at around three minutes, antimatter was created at about a microsecond (before which there was just a tiny excess of matter over antimatter), electroweak unification occurred at a billionth of a second (the highest energy which can be probed experimentally), grand unification and 'inflation' (an extra rapid expansion phase) occurred at 10^{-35} seconds, and the quantum gravity era (the smallest meaningful time) was at 10^{-43} seconds.

Arguments for the Absence of God

At this point cosmology and particle physics might appear to collude to diminish the status of humans and, by implication, the role of God. The heavens have been deprived of their divinity, and the more we understand the universe, from the vast expanses of the cosmos to the tiny world of particle physics, the more soulless it seems to become. As Steven Weinberg (1977) says: 'The more the Universe seems comprehensible, the more it seems pointless.' Let us summarize the reasons for gloom.

Humans are insignificant   The steady progress from the geocentric to heliocentric to galactocentric to cosmocentric view shows that humans—as judged by scale—are completely insignificant. We are equally insignificant as judged by duration: the lifetime of an individual—and even an entire civilization—is utterly negligible compared to the timescale on which the cosmos functions. If the history of the universe were compressed into a year, *Homo sapiens* would have persisted for only a few minutes. Nor is it clear how long humans will persist in the future, since we are prone to dangers from both without (asteroids, marauding black holes, exploding stars) and within (nuclear destruction, some devastating new virus). Everything in the universe falls apart, and the presence of humans, it seems, merely accelerates the process! Darwin's theory of evolution also shows that there is nothing special about humans from a biological perspective. *Homo sapiens*, far from being a unique creation in the Garden of Eden, is just the latest stage of development in a series of biological mutations. For a while one might still think of God as guiding evolution or initiating it. However, the discovery of DNA in the 1950s made it clear that what steers evolution is not some divine being 'outside', but a tiny molecular strand inside. But if humans are so insignificant and ephemeral, how can one believe in the existence of a God in whose image we are supposed to have been created and who cares about us?

The big bang removes the need for a divine creator   One traditional argument for God is that he is required to create the universe. Even though few people now interpret religious creation myths—like the biblical account in Genesis—literally, the idea that the universe must have some first cause still has appeal. Although the big bang provides a very cogent picture for the creation of the universe, this does not necessarily preclude God, since one can still ask, 'Who lit the fuse?' Indeed, the fact that the universe had a finite beginning was claimed by Pope Pius XII in 1952 to support Genesis. Implicit here is the notion that the physical description of creation is incomplete, since it must break down at sufficiently early times. On the other hand, as time proceeds, cosmology seems to have provided an ever more complete description. For example, one could envisage the following catechism-style dialogue:

How did the universe originate? The universe started as a state of compressed matter. But where did the matter come from? The matter arose from radiation as a result of GUT processes occurring when the universe had the size of a grapefruit. But where did the radiation come from? The radiation was generated from empty space as a result of a vacuum phase transition. But where did space come from? Space appeared from nowhere as a result of quantum gravity effects. But where did the laws of quantum gravity come from? The laws of quantum gravity are probably no more than logical necessities.

Each step in this dialogue represents many years of painstaking theoretical work, but the upshot is clear. No first cause is needed because the universe contains its own explanation, a view propounded by Stephen Hawking (2001). Even if God does exist, it is not clear that he could have created the universe differently.

The universe, including mind, is just a machine   Since the Enlightenment, the prevailing scientific view has been that the universe—and everything within it—is just a machine. Indeed, every technological innovation is based on this assumption. But if the material content of the universe does not reflect the existence of God, how about our minds? Perhaps consciousness is the 'ghost in the machine' which testifies to his existence. Unfortunately, recent advances in brain research and artificial intelligence suggest that even the mind is a machine. We may appear to have free will, but this could just be an illusion, consciousness being the mere excretion of brains. Machines already think more quickly, remember more precisely, and decide more intelligently than mere human minds, and it has even been claimed that they may eventually develop consciousness.

The anti-divine arguments are summed up very cogently by Peter Atkins (1993, also Chapter 8 above), who claims that the broad features of the world are uniquely specified by the fact that it has emerged from nothing and must permit the development of complexity. There is no Creator, no purpose, and we ourselves are merely the product of chance. Indeed, it seems that science has expunged the need for any divine element in the world so completely that Richard Dawkins can now dismiss believers in a Creator as 'scientifically illiterate'.

HAS FURTHER PROGRESS IN COSMOLOGY REINSTATED GOD?

Curiously, in recent decades cosmology has brought about a reversal in this trend. This is mostly related to the suggestion that mind may be a fundamental rather than an incidental feature of the universe, although linking this notion to God is not inevitable.

The Unity, Beauty, and Comprehensibility of the Universe

The unity of creation and the intimate link between the macroscopic and the microscopic, so aptly encapsulated in the Uroborus, has led some scientists to see evidence of a great intelligence at work in the universe. For example, James Jeans (1931) famously remarked that 'The Universe is more like a great thought than a great machine'. This impression derives from the fact that the world is so cleverly constructed. At the very least, the coherence of the laws which regulate it seems to point to the existence of some underlying organizing principle (Davies 1993). This also relates to the question of why the universe is comprehensible at all. It seems remarkable that, after just a few millennia, we are already on the verge of a 'Theory of Everything'. As Roger Penrose (1997) has emphasized, there seems to be a closed circle: the laws of physics lead to complexity, complexity culminates in mind, mind leads to mathematics, and mathematics allows an understanding of physics. Why should the structure of the world reflect the structure of our minds, and why should our brains have the ability to generate the required mathematics?

There is also an inherent beauty in the universe. The nature of this beauty is hard to define, but it involves mathematical elegance, simplicity, and inevitability. In particular, all the laws of nature seem to be a consequence of a simple set of symmetry principles. For example, symmetrizing electricity and magnetism gives Maxwell's equations; symmetrizing space and time gives special relativity; and invoking gauge symmetries leads to the unification of the forces of nature. Such symmetries can be appreciated only intellectually, but they are profoundly elegant and can be very moving for physicists. The importance of beauty was appreciated by Paul Dirac, who claimed that 'Beauty in equations is more important than fitting experiments' (Cole 1985), and by John Wheeler (1977), who said: 'One day a door will surely open and expose the glittering central mechanism of the world in all its beauty and simplicity.'

The Anthropic Principle

In the last forty years there has developed a reaction to the mechanistic view, which is termed the Anthropic Principle (Barrow and Tipler 1986). This claims that, in some respects, the Universe has to be the way it is because otherwise it could not produce life, and we would not be here speculating about it. Although the term 'anthropic' derives from the Greek word for 'man', it should be stressed that this is really a misnomer, since most of the arguments pertain to life in general.

As a simple example of an anthropic argument, consider the question, Why is the universe as big as it is? The mechanistic answer is that, at any particular time, the size of the observable universe is the distance travelled by light since the big bang, which is now about 10^{10} light-years. There is no compelling reason why the universe has the size it does; it just happens to be 10^{10} years old. There is, however, another answer to this question, which Robert Dicke (1961) first gave. In order for life to exist, there must be carbon, and this is produced by cooking inside stars. The process takes about 10^{10} years, so only after this time can stars explode as supernovae, scattering the newly baked elements throughout space, where they may eventually become part of life-evolving planets. On the other hand, the universe cannot be much older than 10^{10} years, or else all the material would have been processed into stellar remnants. Since all the forms of life we can envisage require stars, this suggests that it can only exist when the universe is aged about 10^{10} years.

So the very hugeness of the universe, which seems at first to point to humanity's insignificance, is actually a prerequisite for our existence. This is not to say that the universe itself could not exist with a different size, only that we could not be aware of it. Dicke's argument is an example of what is called the 'Weak Anthropic Principle', and is no more than a logical necessity. This accepts the constants of nature as given, and then shows that our existence imposes a selection effect on when (and where) we observe the universe. Much more controversial is the 'Strong Anthropic Principle', which says that there are connections between the coupling constants (the dimensionless numbers which characterize the strengths of the four interactions), in order that observers can arise (Carr and Rees 1979). For example, the existence of convective and radiative stars (both needed for life) requires that there be a 'tuning' between the electric and gravitational coupling constants; heavy elements like carbon can be ejected from stars in supernova explosions only because there is a tuning between the weak and gravitational coupling constants; and an interesting variety of chemical elements exists only because there is a tuning between the electric and strong coupling constants. There are also anthropic constraints on various cosmological parameters, including the cosmological constant and the amplitude of the initial density fluctuations, which need to be finely tuned for galaxies to form.

As far as we know, the relationships discussed above are not predicted by any unified theory, and even if they were, it would be remarkable that the theory should yield exactly the coincidences required for life. Cosmologists have therefore turned to more radical interpretations of the anthropic coincidences. The first possibility—clearly relevant to the science–religion debate—is that they reflect the existence of a 'beneficent being' who tailor-made the universe for our convenience. Such an interpretation is logically possible, but most physicists are uncomfortable with it. Another possibility, proposed by Wheeler (1977), is that the universe does not properly *exist* until consciousness has arisen. This is based on the notion that the universe is described by a quantum-mechanical wave function, and that consciousness is required to collapse this wave function. Once the universe has evolved consciousness, one might think of it as reflecting back on its big bang origin, thereby forming a closed circuit which brings the world into existence. Even if consciousness really does collapse the wave function (which is far from certain), this explanation is also somewhat metaphysical.

The third possibility—discussed in more detail later—is that there is not just one universe but lots of them, all with different, randomly distributed coupling constants. In this 'multiverse' proposal, we just happen to be in one of the small fraction of universes which satisfy the anthropic constraints. Of course, invoking many universes is highly speculative, especially since the other universes may never be directly detectable, so some cosmologists remain uncomfortable with the idea. Nevertheless, many anthropically-inclined physicists are attracted to the multiverse because it seems to dispense with God as the explanation of cosmic design. The Strong Anthropic Principle—rather than having teleological significance—just becomes an aspect of the Weak Anthropic Principle.

The Anthropic Principle also explains why the history of the big bang allows an increasing degree of organization to develop. As the universe expands and cools, a 'Pyramid of Complexity' arises, with different levels of structure as one goes from quarks (at the bottom) to nucleons to atoms to molecules to cells and finally to living organisms (at the top). Despite the earlier pessimistic notion of heat death, no violation of the second law of thermodynamics is involved, because local pockets of order can be purchased at the expense of a global increase in entropy. These structures arise because processes cannot occur fast enough in an expanding universe to maintain equilibrium. However, disequilibrium is only possible because of the anthropic fine-tuning of the coupling constants. This suggests that the Anthropic Principle should really be interpreted as a Complexity Principle. This also throws light on the dilemma of what qualifies as an observer in anthropic considerations (i.e. what minimum threshold of awareness is required). If one regards minds (i.e. brains?) as the culmination of complexity, this does not matter much because—at least on Earth—the development of brains seems to have occurred very quickly once the first signs of life arose.

Arguments for the Presence of God

We now summarize various pro-divine arguments to counteract the anti-divine ones at the end of the last part. None of these is decisive—science can probably neither prove nor disprove the existence of God—but they offer what John Polkinghorne describes as 'nudge factors'.

Humans—or at least complexity—are central   The Uroborus symbol suggests that humans may be important after all, in that we occupy a special symmetry point near the bottom. We are not at the centre of the universe geographically, but we are central in the *scale* of things. Of course, this argument would also apply to any other life form or 'container of mind' which may have evolved in the universe. Many people have emphasized that there seems to be a 'life principle' at work in the cosmos, but the presence of life still requires the special anthropic conditions necessary for the Pyramid of Complexity to arise. Whether these conditions reflect the existence of a Creator or a multiverse, our presence is no longer irrelevant.

The unity, beauty, and comprehensibility of the universe point to a Creator   Even though the big bang explains many features of the universe, we have seen that modern physics has revealed other aspects of nature which seem to point to some underlying intelligence. It is possible (though not obligatory) to link this to the notion of God, at least in a deistic non-personal sense (Davies 1993). This might seem reminiscent of Paley's argument—and subject to the same criticism—but it is more subtle. For it is the laws themselves, rather than the structures to which they give rise—the underlying simplicity rather than the complexity—which are so striking. To those of a religious disposition, the miracle of matter may therefore provide some intimation of the divine.

Mind is fundamental to the cosmos   The Uroborus also represents the blossoming of consciousness. The physical evolution of the universe from the big bang (at the top) through the Pyramid of Complexity to humans (at the bottom) is just the start of a phase of *intellectual* evolution, in which mind—through scientific progress—works its way up both sides to the top again. Indeed, as indicated by the outer arcs, the Uroborus can be interpreted historically as representing how humans have systematically expanded the outermost and innermost limits of their awareness. In terms of scale, it is striking that science has already expanded the macroscopic frontier as far as possible, although experimentally we may never get much below the electroweak scale in the microscopic direction. The fact that science really provides a sequence of mental models, each progressively removed from common-sense reality, also emphasizes the importance of mind.

RECENT DEVELOPMENTS

Most of the arguments presented so far could have been made a decade ago. In this section, I will highlight some current cosmological issues which may influence the future relationship between science and religion, although the direction of that influence is still unclear.

Extraterrestrial Life

The anthropic coincidences suggest that life is a fundamental rather than an incidental feature of the universe but this does not resolve the issue of how pervasive life is. If the universe is conducive to life, either because it was designed that way or because it is a privileged member of a multiverse, one might anticipate that life would exist in other places besides Earth. Since the Western monotheistic religions place so much emphasis on the uniqueness of human beings, this is very crucial to the science–religion debate.

From a theoretical point of view, the abundance of life in the universe depends upon the product of two numbers, one very large and the other very small. The very large number is the number of stars in the observable universe, while the very small number is the probability (P) that life will be associated with any particular star. The first number is well known: there are 10^{11} stars in our galaxy and 10^{11} galaxies in the observable universe, so there are roughly 10^{22} stars in total. The value of P is much less definite, the main uncertainty being the likelihood that self-replicating cells (which one assumes are necessary for life) arise on any life-conducive planet. Once this happens, it seems plausible that life will evolve fairly rapidly to form creatures like us, but it is the first step which is problematic.

There are various points of view, depending on how small one believes P to be. For the Earth to be the only site of life in the Galaxy, P would need to be less than 10^{-11}. An empirical argument for this relates to the so-called Fermi–Hart paradox: if advanced life forms have evolved elsewhere in the Galaxy, then one would expect at least some of these to have spread throughout the Galaxy and made contact with us. However, even with one civilization per galaxy, the cosmos would still in a sense be teeming with life. For Earth to be the only site of life in the entire observable universe, which might be regarded as the pessimistic view, P would need to be less than 10^{-22}. The argument for this is theoretical, and stems from low estimates of the probability that self-replicating cells can arise from biological molecules by random processes in the 10^{10} years permitted by the big bang.

Neither of these arguments is compelling. The first might be regarded as self-defeating since, if every civilization accepted it, none would ever embark on the programme of space exploration presumed at the outset. The second is unconvincing because estimates of P are too uncertain; recent developments suggest that life might arise much more easily than a simplistic analysis would suggest (Kauffman 1995). We may therefore turn to the more optimistic view. The number of stars at a particular time expected to be associated with life intelligent enough to be able to communicate is given by the Drake equation. This is the product of a number of factors, some of which can be determined by astronomy; but the result is that the number of communicating civilizations in the Galaxy at any time is roughly the lifetime of a typical civilization in years. (The discrepancy with the earlier view hinges on the value of P.) At this stage, it is impossible to assert with confidence whether the pessimists or the optimists are correct.

Astronomers are now actively looking for radio signals from extraterrestrial civilizations, but have not yet found any. Whatever the outcome of their search, it is likely to have an important impact on humanity. If we *don't* detect a signal, the universe will be a lonelier place, but at least life on Earth will have become more precious, and we will have been restored to our pre-Copernican status. If we *do* detect a signal, the sociological, technological, and perhaps even spiritual consequences will be immense. There is a sense in which this would raise our level of consciousness from a planetary to a Galactic level. Both situations would have important religious implications.

God and the Multiverse

There has been a fundamental change in the epistemological status of the Anthropic Principle in the last decade because of the realization that most models of the origin of the universe involve some form of multiverse (Carr 2006). Admittedly, cosmologists have widely varying views on how different universes might arise. Some invoke models in which the universe undergoes cycles of expansion and recollapse, with the constants undergoing change at each bounce. Others invoke the inflationary scenario, in which our universe is just one of many bubbles, all with different laws of low-energy physics and different coupling constants. An interesting variant of this is *eternal inflation* (Linde 1986), in which the universe is continually self-reproducing. The suggestion that our universe is a 'brane' in a higher-dimensional 'bulk' leads to another multiverse scenario, with collisions between branes producing multiple big bangs. A further possibility is associated with the 'many worlds' interpretation of quantum mechanics, since this seems to be the only sensible context in which to discuss quantum cosmology. A comprehensive review of these scenarios has been provided by Tegmark (1998).

In assessing the multiverse hypothesis, a key issue is whether some of the physical constants are contingent on accidental features of symmetry breaking and the initial conditions of our universe, or whether some fundamental theory will determine all of them uniquely. Only in the first case would there be room for the Anthropic Principle, hence many physicists prefer the latter view. But how likely is this? The current favourite candidate for a fundamental theory is the superstring proposal. This posits that spacetime is either ten-dimensional (supergravity) or eleven-dimensional (M-theory), with four-dimensional physics emerging from the compactification of the extra dimensions. Some people have argued that M-theory may predict all the fundamental constants uniquely, the only input being the size of the eleventh-dimension. However, other people claim that M-theory could permit a huge number of vacuum states; the constants may be uniquely determined within each one, but could be different across the states themselves. This corresponds to what has been termed the 'string landscape' scenario (Susskind 2005). The crucial issue is whether the number of vacuum states is sufficiently large and their spacing sufficiently small to allow room for anthropic constraints.

A very different multiverse proposal comes from Lee Smolin (1997), who argues that the physical constants have *evolved* to their present values through a process akin to mutation and natural selection. The underlying physical assumption is that whenever matter gets sufficiently compressed to undergo gravitational collapse into a black hole, it gives birth to another expanding universe, in which the fundamental constants are slightly mutated. Since our own universe began in a state of great density (i.e. with a big bang), it may itself have been generated in this way (i.e. via gravitational collapse in some parent universe). Cosmological models with constants permitting the formation of black holes will therefore produce progeny (which may each produce further black holes, since the constants are nearly the same), whereas those with the wrong constants will be infertile. Through successive generations of universes, the physical constants will then naturally evolve to have the values for which black hole (and hence baby universe) production is maximized. Smolin's proposal involves very speculative physics, since we have no understanding of how the baby universes are born, but it has the virtue of being testable, since one can calculate how many black holes would form if the parameters were different.

The multiverse proposal certainly poses a serious challenge to the theological view and it is not surprising that atheists find it a more plausible explanation of the anthropic fine-tunings. However, the dichotomy between God and multiverse may be too simplistic. While the fine-tunings certainly do not provide unequivocal evidence for God, nor would the existence of a multiverse preclude God since there is no reason why a Creator should not act through the multiverse. The theological case has been defended by Holder (2004).

The Origin of the Universe and the Nature of Time

Classical physics breaks down at the big bang itself because the density of the universe becomes infinite. Also, the notion of spacetime as a smooth continuum fails because of quantum gravity effects; there must be a 'singularity' where general relativity breaks down. Some cosmologists try to avoid this problem by having the universe bounce at some finite density (e.g. due to a cosmological constant). In this case, the present expansion would have been preceded by a collapsing phase, and the universe may even have existed eternally. However, if there was a big bang singularity, it means that time itself must have been created there. Quantum cosmology purports to describe this process, and even claims that the universe can be created as a quantum fluctuation out of nothing (Isham 1988).

In standard quantum theory, probabilities are associated with physical states. In quantum cosmology, these states are three-dimensional spatial hypersurfaces, with time being defined internally in terms of their curvature. To determine the evolution of the universe, one then uses a 'sum-over-histories' calculation which allows for all possible paths from some initial spatial hyperface to another final one. The crucial development in this approach was the 'no boundary' proposal of Hartle and Hawking (1983). This removes the initial surface by making time imaginary there, thus adroitly sidestepping the usual philosophical problems associated with a moment of creation. Drees (1990) has explored the theological implications of this picture in some detail, but it is unclear to what extent quantum cosmology really does impinge on theology at this particular interface. There is still a basic distinction between the question of *how* the universe came into existence and *why* it did so. Also, if one wants to invoke God, he could probably work through any model of quantum cosmology. Nevertheless, there is an obvious connection here with the theological concept of *creatio ex nihilo*.

CONCLUDING REMARKS

Cosmology addresses fundamental questions about the origin of matter, mind, and life, which are clearly relevant to religion, so theologians need to be aware of the answers it provides. In a sense, cosmology should provide some of the raw material from which religious belief is fashioned. Of course, the remit of religion goes well beyond the materialistic issues which are the focus of cosmology. Nevertheless, inasmuch as religious and cosmological truths overlap, they must be compatible. This has been stressed by George Ellis (1993), who distinguishes between Cosmology (with a big C)—which takes into account 'the magnificent gestures of humanity'— and cosmology (with a small c), which just focuses on physical aspects of the universe. In his view, morality is embedded in the cosmos in some fundamental way (cf. Leslie 1989), and although science cannot deal with such issues, individuals count.

The requirement that cosmology and religion should be compatible has two important implications. First, it makes a 'God of the gaps' view very unattractive from a theological perspective, since it implies that religion is always on the retreat as cosmology advances. Second, since science progresses by a series of paradigm shifts, each providing a better approximation to reality than the previous one, but none representing ultimate truth (Kuhn 1970), one should be wary of religious claims to possess absolute (God-given) truths. Indeed, one might expect religion—like science—to undergo paradigm shifts, so that even some questions addressed in the past (e.g. the location of heaven and hell) become meaningless. For example, such a shift would presumably be triggered by contact with extraterrestrials.

Clearly none of the pro-divine or anti-divine arguments presented in this essay are decisive, and the evidence provided by the study of the physical world will probably always be equivocal. Even those cosmologists who are mystically inclined have not usually based their faith on scientific revelations (Wilbur 2001). More relevant perhaps are studies of consciousness and the world of mind. Such investigations certainly go beyond *classical* physics, since there is a basic incompatibility between the localized features of mechanism and the unity of conscious experience. On the other hand, the classical picture has now been replaced by a more holistic quantum one, and there are some indications that this *can* include consciousness. In any case, many people are sceptical of attempts to formulate a 'Theory of Everything' which neglects such a conspicuous aspect of the world. It is even conceivable that some future paradigm of physics will incorporate mind explicitly in some way. If so, this will surely transform the nature of the science–religion connection in years to come.

References and Suggested Reading

ALPHER, R. A., and HERMAN, R. (1988). 'Reflections on Early Work on Big Bang Cosmology', *Physics Today*, 41: 24–34.

ATKINS, P. (1993). *Creation Revisited*. Oxford: Freeman.

BARROW, J. D., and TIPLER, F. J. (1986). *The Anthropic Cosmological Principle*. Oxford and New York: Oxford University Press.

BUTCHINS, A. (2002). *The Numinous Legacy*. Aldershot: Albatross Press.

CARR, B. J. (2006). *Universe or Multiverse?* Cambridge: Cambridge University Press.

——and REES, M. J. (1979). 'The Anthropic Principle and the Structure of the Physical World', *Nature*, 278: 605–12.

COLE, K. C. (1985). *Sympathetic Vibrations*. New York: Bantam, 225.

DAVIES, P. C. W. (1993). *The Mind of God*. New York: Touchstone/Simon & Schuster.

DICKE, R. H. (1961). 'Dirac's Cosmology and Mach's Principle', *Nature*, 192: 440.

DREES, W. E. (1990). *Beyond the Big Bang*. La Salle, Ill.: Open Court.

ELLIS, G. F. R. (1993). *Before the Beginning: Cosmology Explained*. London: Boyars/Bowerdean.

HARTLE, J. B., and HAWKING, S. W. (1983). 'Wave Function of the Universe', *Physical Review D*, 28: 2960–75.

HAWKING, S. W. (2001). *The Universe in a Nutshell*. New York: Bantam Press.

HOLDER, R. (2004). *God, the Multiverse and Everthing: Modern Cosmology and the Argument from Design*. Aldershot, Ashgate.

ISHAM, C. (1988). 'Creation of the Universe as a Quantum Tunnelling Process', in R. J. Russell *et al.* (eds.), *Our Knowledge of God and Nature: Physics, Philosophy and Theology*, Nôtre Dame, Ind.: University of Nôtre Dame Press, 374–408.

JEANS, J. (1931). *The Mysterious Universe*. Cambridge: Cambridge University Press.

KAUFFMAN, S. A. (1995). *At Home in the Universe*. Harmondsworth: Penguin.

KUHN, T. S. (1970). *The Structure of Scientific Revolutions*, 2nd edn. Chicago: University of Chicago Press.

LESLIE, J. (1989). *Universes*. London: Routledge.

LINDE, A. D. (1986). 'Eternally Existing Self-reproducing Chaotic Inflationary Universe', *Physics Letters B*, 175: 395–400.

NEWTON, I. (1959–77). *The Correspondence of Sir Isaac Newton*, ed. H. W. Turnbull, J. F. Scott, A. Rupert Hall, and L. Tilling, iii. Cambridge: Cambridge University Press.

PENROSE, R. (1997). *The Large, the Small and the Human Mind*. Cambridge: Cambridge University Press.

SMOLIN, L. (1997). *The Life of the Cosmos*. London: Weidenfeld & Nicolson.

SUSSKIND, L. (2005). *The Cosmic Landscape: String Theory and the Illusion of Intelligent Design*. Boston: Little, Brown & Co.

TEGMARK, M. (1998). 'Is the Theory of Everything Merely the Ultimate Ensemble Theory?', *Annals of Physics*, 270: 1–51.

WEINBERG, S. (1977). *The First Three Minutes*. New York: Basic Books.

WERTHEIM, M. (1999). *The Pearly Gates of Cyberspace*. London: Virago.

WHEELER, J. (1977). 'Genesis and Observership', in R. Butts and J. Hintikka (eds.), *Foundational Problems in the Special Sciences*, Dordrecht: Reidel, 3.

WILBUR, K. (2001). *Quantum Questions: Mystical Writings of the World's Greatest Physicists*. Boston: Shambhala Publications.

CHAPTER 10

FUNDAMENTAL PHYSICS AND RELIGION

KIRK WEGTER-McNELLY

INTRODUCTION

What are nature's deepest secrets? Why is there something rather than nothing? The eagerness with which many in modern cultures look to physicists for answers to these and other ultimate life-questions, not to mention the frequency and ease with which some of today's physicists speak of such things as the 'mind of God', testifies to the aura of prestige and power that surrounds the oldest of the modern natural sciences.

Although it is not the purpose of this chapter to ask why, from a political, sociological, or psychological perspective, physics has come to command such authority across cultures (a fascinating question, to be sure), it is worth noting at the outset the assumption in much contemporary religious reflection on the nature and significance of the physical world that fundamental physics is important because it gets to the 'bottom' of things. The reductionism implicit in this assumption often manifests itself even in more explicitly holistic religious perspectives as a vague though powerful sense that physics must be 'deeply' important to religious problems, even if it is incapable of solving them on its own. I begin with such a comment because this assumption, which lies at the root of much contemporary religious interest in physics, is rarely analysed or defended. It is not uncommon among those familiar with the practice of physics to have a sense that it is charged with much holy and unholy potential; but much more needs to be done to account for this feeling in a way that goes beyond the reductionistic impulse and thereby allows for the integrity of the religious as well as the scientific perspective.

Fundamental physics has come to be interesting religiously in at least three different ways, depending on what kind of authority it is granted. Physics becomes *ontologically* interesting for religious reflection when one takes it to reveal something that is both true and unobtainable by other means about the structures and processes of the world. It can also become *epistemologically* interesting when one takes its remarkable predictive power to reveal something significant about the ways in which human rationality can be effectively harnessed to produce reliable knowledge about the world. And if, for whatever reason (perhaps a concern with the issue of reductionism mentioned above), one rejects the ontological and epistemological relevance of physics to religious reflection, it can still be *conceptually* interesting as a rich source of images and ideas which, when extracted from their original context and allowed to take on a life of their own, can extend the range of concepts available to religious reflection for talking about the meaning of human life.

Each of these different modes of religious engagement with physics can provide various sorts of raw material for religious accounts of the nature of existence and transcendence; but clearly the first two modes provide the greatest opportunity for establishing consonance between religious and scientific world-views in their shared quest for understanding. Appropriating images and ideas emerging from modern physics simply as analogies or metaphors can enrich and enliven religious accounts of reality, but the greatest potential for learning and discovery lies in a synthetic approach that embraces all three modes of engagement.

The basic goal of modern fundamental physics has been to provide a theoretical description of the constituents and processes of the physical world on the basis of repeatable experiments using general concepts such as mass, energy, space, and time. As the earliest of the modern scientific disciplines, physics has played a central role in establishing the general methodological approach of the natural sciences, which aims to formulate theories in terms of precise mathematical 'laws' that can be used to generate empirically testable predictions.[1]

Scientific laws in general, and the laws of physics in particular, are commonly understood by scientists and non-scientists alike to carry prescriptive force; i.e. they are presumed to have some sort of quasi-Platonic existence by which they effectively steer natural processes, however blindly, from the present into the future. In common parlance, natural processes are typically said to 'obey' the laws of physics. Among contemporary philosophers of science, however, this robustly realist view of the ontological status and power of physical laws has been subjected to serious scrutiny (Hilgevoord 1994; Russell *et al.* 1996). A more modest and careful, but still realist, view of the laws of physics counts them as relatively accurate descriptions of nature's own regularities, but not as the cause of these regularities. This view, often linked to an epistemological perspective labelled 'critical realism', has attracted the

---

[1] Although this ideal still animates most of the natural sciences, some have begun to question whether we have reached the point of diminishing returns; see Wolfram (2002).

interest of a number of religious thinkers because of the perception that it provides room for religious reflection and theological (re)description to exist alongside scientific accounts. Beyond this mediating view, at the far end of the spectrum one finds the minimalist view that physical laws and the theories of which they are a part amount to nothing more than calculational devices, and should be understood to carry no ontological weight. This view, which in fact is a family of views commonly associated with various labels such as 'anti-realism', 'instrumentalism', and 'empiricism', has been adopted by some philosophers of science in order to avoid various difficulties that attend realist views.[2]

It is perhaps worth noting, even if the point has little bearing on philosophical analysis, that the minimalist view is a relative rarity among working physicists, who generally take the theories they work with to be saying something true, however imperfectly, about the world. Clearly, one's view of the status of physical theories—however one tailors this view to fit different theories—lies at the root of judgements about whether physics is interesting religiously. Rich and varied conceptual possibilities can emerge from any of these various perspectives on the ontological status of physical laws, though the minimalist view would appear to work most strongly against ontological and epistemological connections.

The historical roots of modern physics lie in the musings of the ancient Greeks, most notably those of Pythagoras and his followers. The writings of the Pythagorean tradition, as well as those of Aristotle and other Greek philosophers, were reintroduced to late medieval Europe by Islamic scholars such as Ibn Sīnā (Avicenna), who gave the West not just renewed access to its own Greek heritage but also the insights of a genuinely Islamic scientific and mathematical tradition (Iqbal 2002). These writings and traditions profoundly shaped the subsequent development of European thought, precipitating many of the intellectual shifts that accompanied and in some cases fostered the rise of modern physics during the European Renaissance and, subsequently, the Protestant Reformation and the Enlightenment. Because of this particular historical association, the relevance of modern physics to religious thought was first felt in the context of Western Christian debates over the nature of human and divine agency, at the centre of which was a growing understanding that the newly emerging paradigm of Newtonian physics posed serious challenges to various aspects of the received medieval theological world-view.

[2] For example, the multiple interpretability of the mathematical formalism of quantum theory (Cushing 1994; Norris 2000), or the so-called pessimistic meta-induction: i.e. that because yesterday's reigning theories inevitably turned out to be false in some important respect, the same is likely to be the case with currently accepted theories.

The Rise of Newtonian Physics

Mechanistic Determinism

Early European scientists, including the English physicist Isaac Newton, were at least nominally Christian, though many held views at odds with the official doctrines of their respective churches. Some, like the German astronomer Johannes Kepler, understood themselves to be 'thinking God's thoughts after Him' and saw their work as a kind of hymn to the Creator (Brooke 1991: 22). Others, like the Italian polymath Galileo, attempted to distance the world of scientific ideas from religion by characterizing science and the Church as two distinct authorities operating within separate spheres of knowledge. Quoting the powerful Italian cardinal and Counter-Reformation historian Césare Baronio, Galileo argued famously in his own defence against church authorities that 'the intention of the Holy Ghost is to teach us how one goes to heaven, not how heaven goes' (1957: 186). But this way of distinguishing science from religion initially obscured the far-reaching consequences of the replacement of the medieval view of the world as a rich, multilayered, integrated reality open to divine interaction with a view of the world as a lifeless, autonomous clock-like mechanism closed to any 'external' influence.[3]

Although the Roman Catholic Church is still commonly perceived to have lodged theological objections against the heliocentric, or Sun-centred, account of planetary motion first advanced by the Polish astronomer Nicolaus Copernicus and later promoted by Galileo, historians have recently argued that the Church's hostility toward heliocentrism had less to do with theological objections to the Earth's motion than with the personalities involved and with prevailing controversies over magisterial authority stemming from the Reformation, particularly those relating to matters of biblical interpretation (Fantoli 2003). It is also worth noting that the Church had long allowed its cathedrals to be used as solar observatories, chiefly for the sake of tracking and, eventually, reforming the calendar (Heilbron 1999). The magisterium of the Roman Catholic Church, then, cannot simply be said to have opposed the rise of modern science. It struggled in Galileo's time, as it continues to do in ours, with how best to critically interpret the latest findings of science for the sake of mutual interaction with the world. In the words of Pope John Paul II, 'Science can purify religion from error and superstition; religion can purify science from idolatry and false absolutes. Each can draw the other into a wider world, a world in which both can flourish' (Russell et al. 1990).

It has often been noted that Copernicanism displaced the Earth, and humanity with it, from its privileged position at the centre of the medieval cosmos. (This

---

[3] The negative social and intellectual impact of this shift has been thoroughly analysed in Merchant (1990).

process of dethronement was completed, in fact, only in the twentieth century, when modern scientific cosmology finally discarded the notion of a universal 'centre'.) There is no denying that many felt the decentring effect of Copernicanism to pose a serious challenge to theology. In retrospect, however, the deeper religious significance of the move from geocentrism to heliocentrism was that it marked the first step toward a scientifically comprehensive and *deterministic* account of physical processes. Cartoons still appear noting the risk of gazing through a telescope to one's sense of self-importance, but reconceptualizing the vastness of the universe in terms of religious awe and wonder has not proved as difficult as facing the challenge of physical determinism. The lasting legacy of Copernicanism is best understood in terms of the role it played in giving birth to a deterministic world-view within which the notions of human and divine agency have become enduring puzzles.

The determinism of early modern physics solidified around the grand synthesis effected by Isaac Newton, which united celestial and terrestrial motion into a single conceptual scheme. The plausibility of the heavens as the abode of spiritual beings steadily diminished as the celestial realm came increasingly to be seen and understood as another part of the physical world which was scientifically (i.e. mathematically) analysable in terms of the orderly and deterministic motion of its parts. In this key respect, Newton's unifying account of physical motion, his 'mechanics', shaped the generally deterministic character of early modern science. The phrase 'Newtonian determinism' has since become synonymous with a lack of any genuine novelty or openness in natural processes. As the French mathematician Pierre-Simon de Laplace famously stated, 'Given for one instant an intelligence which could comprehend all the forces by which nature is animated and the respective situation of the beings who compose it . . . for [this intelligence] nothing would be uncertain and the future, as the past, would be present to its eyes' (de Laplace 1917: 4). The real and lasting force of this comment, of course, is not that a superintelligence in a deterministic world could know everything that would ever happen, but that such an intelligence would have nothing to do—thus rendering superfluous the notion of a divine overseer of the world's processes.

Newtonian physics also relied crucially upon the strategy of reductionism—i.e. explaining an object's behaviour solely in terms of the behaviour of its parts—and thus elevated this strategy to the status of a central methodological principle of modern science. Embracing both the determinism and the reductionism of Newtonian science, early modern scientists quickly distanced themselves from modes of explanations that invoked purpose, or *telos*. Increasingly they sought explanations couched exclusively in terms of efficient (i.e. mechanical) causes. It was physics' characterization of the world within this new framework of mechanistic reductionism that led to a significant theological crisis in Christian thought, for if the state of the natural world was completely determined by the relevant physical laws acting upon the prior configuration of its various parts in each preceding moment, could one still conceive of human beings as thinking and acting in the world with genuine freedom? And, equally important, could one still affirm God's ongoing activity in such a world?

Human and Divine Freedom

In response to the mechanistic determinism and reductionism of Newtonian physics, Western philosophers and theologians of the Enlightenment attempted to insulate human freedom from natural-scientific considerations. One of the first to do so was the French philosopher René Descartes, who divided reality into two realms: the material world of mechanical necessity (*res extensa*) and the world of mental free willing (*res cogitans*). The German philosopher Immanuel Kant subsequently advanced a more nuanced form of dualism in which he distinguished between the causal deterministic framework we impose upon the world through our cognitive apparatus in the very act of perceiving it (what he called the *phenomenal* realm) and the world as it exists in and for itself (what he called the *noumenal* realm), positing human freedom as a part of the latter. Following Descartes and Kant, liberal Protestant theologians increasingly ignored the physical dimension of human existence and retreated instead to the 'inner' world of the human spirit. One of the first to move in this direction was the German theologian Friedrich Schleiermacher, who moved religion out of the realm of knowledge and resituated it within the realm of feeling. By the end of the nineteenth century, another German Protestant, Albrecht Ritschl, could write: 'theology has to do, not with natural objects, but with states and movements of man's spiritual life' (1902: 20). In their interactions with early modern physics, these Christian thinkers perceived the twin problematic of determinism and reductionism, and responded by trying to protect human freedom. However, their means of achieving this goal—namely, by isolating the theological account of human existence and freedom from any relevant physical considerations—proved untenable in the long run. What has become clear in recent decades is the importance of the issue of physical embodiment to a full understanding of the human person as a religious and scientific being (Coakley 2000; Lawrence and Shapin 1998).

Newtonian physics also posed a serious challenge to inherited notions of God's presence and activity in the world. In response to this framework, Christian thinkers developed three strikingly different views of whether and how God acts at particular times and places in the world (henceforth 'special divine action'). Some were willing to countenance the idea that the universe lacks the causal powers necessary for bringing about everything that occurs within it. Newton, for example, could see no way of securing the stability of planetary orbits, and thus claimed that it must result from occasional divine adjustment. It seemed reasonable to him and others to locate God's activity in this unexplained, and presumably unexplainable, aspect of the universe. Laplace later demonstrated the self-stabilizing tendency of planetary systems, thereby exposing the chief limitation of what has come to be called the 'God of the gaps' approach: because it relies on ignorance, religion must retreat whenever science fills any explanatory gap. Others advocated a stronger version of the general view that the universe lacks the causal powers necessary for bringing about everything that occurs within it. Commonly called *interventionism*, this view argues that God, as transcendent Creator of the world and its laws, simply breaks the laws of nature whenever God wishes to alter the course of the world by acting in a specific event. God simply creates space in an otherwise deterministic causal structure to make 'room' for particular divine acts. This view presumes, in opposition to Newton's early view, that nature's causal powers are sufficient to account for regular physical processes, but that God also does things in the world which it could not do on its own (e.g. miracles). Eighteenth-century deists rejected this approach on the grounds that the more honest and reasonable response to Newtonian determinism was to abandon the notion of special divine action altogether in favour of a God who brings the world into existence and then refrains from any further interaction with it. German theologian Wolfhart Pannenberg (1993) has also pressed the point that Newton's view of inertial or self-sustaining motion helped to discredit the idea of the world depending upon divine conservation for its continued existence. Throughout the nineteenth and twentieth centuries and into the present, liberal Protestants have continued to maintain the idea that God is present to creation, but only insofar as creation is itself God's one great act, which amounts to relinquishing the notion of 'objectively' special divine acts, including miracles. On the liberal account, one might *perceive* God as acting specially in some particular physical event, but this would be merely a matter of one's own subjective perception (for a notable contemporary proponent of this view, see Wiles 1986).

These three views of special divine action developed by Christian thinkers in response to the rise of Newtonian physics—interventionism, deism, and liberalism—differ sharply from one another, yet they brook a common theological constraint. Each accepts the idea that a God who is understood to alter the course of events in the world must be treated on a par with any other object or causal process in the world. Thus, each accepts the claim that in a Newtonian world of strict determinism there is no 'room' in the physical world for God to act in individual events— a viewpoint I call 'theophysical incompatibilism'. On the basis of this construal of the relationship between special divine action and natural processes, deists and liberals have judged that God does *not* act specially at particular moments in history (even though liberals maintain the ongoing presence of God in creation through the idea of God enacting history itself). Interventionists, on the other hand, have clung to the notion of special divine action by understanding God to override or 'break' the physical world's laws in a special divine act. The far-reaching consequences of this common willingness to accept a 'no-inherent-room-for-God' constraint coming from Newtonian physics cannot be overemphasized. Prior to the rise of Newtonian physics, Christian thinkers simply did not perceive the logical difficulty of asserting simultaneously that God acts at specific times and places and that the world retains its own causal efficacy and integrity. However, the supposed compatibility of these two ideas dissolved in the face of Newtonian determinism, which left in its wake human and divine agency as newly felt problems.

Jumping ahead for a moment to the twentieth century, it is important to note that recent developments in physics have allowed for new, but still theologically incompatibilist, approaches to the problem of special divine action, now commonly referred to as 'non-interventionist' strategies. Those who adopt these strategies accept, on the one hand, the liberal theological view that God must be understood to act with and not against the grain of natural processes—after all, it is argued, God is the one who has established these processes in the first place. They agree, on the other hand, with the interventionist view that God can and ought to be thought of as acting objectively at particular times and places in the world. Non-interventionism attempts to straddle the traditional divide between these two views by locating within nature room for special divine action—which, recall, is a necessary condition for objectively special divine action within the perspective of theophysical incompatibilism—variously within its infinite sensitivity to initial conditions (à la chaos theory; see Polkinghorne 1991), the under-determined character of natural processes (à la quantum theory; see Russell et al. 2001), and higher levels of novelty and freedom (à la complexity and emergence theory; see Peacocke 1993; Clayton 2004), to name the three most widely discussed strategies. In these different types of physical processes, so the non-interventionist argument goes, God can be understood to act objectively in the world without needing to violate its laws.

In light of this recent development, what is presently needed is a more thorough analysis of the strengths and weaknesses of the widely shared assumption of theophysical incompatibilism: i.e. that objectively special divine action is incompatible with physical determinism. The implications of this theological assumption have not yet been adequately scrutinized in the contemporary religion-and-science literature. This need stems in the first instance not from the possibility that the world might turn out, after all, to be a truly closed causal mechanism, but from the observation that accepting natural science as a constraint on theological accounts of God's activity in the world presupposes a competitive, or 'zero-sum', view of the relationship between divine and creaturely activity. The assumption of theophysical incompatibilism appears to lead to a kind of 'domestication of transcendence', to borrow a phrase from American Reformed theologian William Placher (1996). Might it be possible to recover a view of divine activity as the source and guarantor of the integrity of natural processes and creaturely freedom—as one finds it, say, in pre-modern thinkers such as Aquinas, Luther, and Calvin—without giving up the task of serious religious engagement with the natural sciences?

The rise of non-interventionist accounts of special divine action reveals the significant and lasting impact of Newtonian determinism on Christian thought, which continues into the present era despite the fact that numerous developments over the past century have seriously called into question the deterministic and reductionistic assumptions of the classical Newtonian world-view. These new developments have themselves led to a wide variety of new perspectives on the religious significance of physics, to which I now turn.

THE TWENTIETH-CENTURY REVOLUTION

As the nineteenth century drew to a close, many theoretical physicists judged that their labours were nearly concluded. Newtonian mechanics had provided a comprehensive framework for understanding the motion of physical masses under the influence of mechanical forces; electromagnetic theory had provided a basis for understanding the interaction of and relation between electric and magnetic phenomena; and thermodynamics had provided a mechanical account of the phenomena of temperature and heat. The Victorian physicist Lord Kelvin (William Thompson), who was instrumental in the development of thermodynamics, saw nothing but a few inconsequential clouds obscuring the 'beauty and clearness' of physics' horizon. In truth, though, behind these clouds lay deep conceptual and theoretical puzzles regarding the nature of light and the behaviour of atoms. Contrary to Kelvin's expectations, attempts to solve these puzzles ushered in the greatest revolution in the Western conception of the physical world since the time of Galileo and Newton. This exciting era dawned in the form of two new theoretical frameworks, both of which were quickly understood to be deeply at odds with various aspects of the Newtonian world-view: the so-called *relativity theory* developed single-handedly by the Swiss physicist Albert Einstein during the first two decades of the twentieth century and the so-called *quantum theory* of atomic behaviour developed by a host of scientists in the 1920s. The new views of space, time, and causation represented by these frameworks have since led to a remarkable amount of rethinking of the nature of transcendence, the world, and humanity from a variety of religious perspectives and traditions.[4]

Special and General Theories of Relativity

Whereas Newton had conceived of space as God's means of experiencing the world and of time as having infinite extension as well as a uniformly moving present, Einstein in his 1905 *special theory of relativity* (SR) construed space and time as a single reality, *spacetime,* and postulated that the speed of light, not space or time, is the true 'absolute' of the universe (for Einstein's own views on religion, see Jammer 1999). Shorn of their own absoluteness, measurements of the extension and duration of any given object or event will vary, according to SR, when measured by different observers in relative motion. SR thus appears to point to the demise of a universal 'now', and has brought about a reassessment of traditional views of the relationship between divine eternity and creaturely temporality (Russell *et al.* 1988; Peters 1989; Matt 1996; Craig 2001). Additionally, the characterization of time as a fourth dimension has led some to interpret SR as hostile to the very idea of temporal flow. According to proponents of the *block universe* interpretation, the spacetime manifold exists timelessly as a four-dimensional whole. This interpretation of SR calls into question the reality of human freedom and our phenomenological sense of temporal becoming, though it is difficult to know how to take this question seriously in the

---

[4] General reflections on the religious implications of the so-called new physics include Russell *et al.* (1988); Hilgevoord (1994); Heller (1996); Worthing (1996); Toolan (2001); Barr (2003); Wallace (2003); Hodgson (2005).

face of the seemingly essentially temporal character of human experience (Fagg 1995; Isham and Polkinghorne 1996). Is this an instance where the phenomenology of human experience should function as a guide to the interpretation of physical theories?

After the publication of SR, Einstein turned to the problem of developing a theory of gravity based on his relativistic account of spacetime. In his 1915 *general theory of relativity* (GR) he treated gravity geometrically as the curvature of spacetime rather than classically (i.e. in Newtonian terms) as a force acting on masses. According to one common formulation, in GR matter tells spacetime how to curve, and curved spacetime tells matter how to move (Taylor and Wheeler 1992: 275). Within this broad conceptual framework physicists have developed various theories of the origin, structure, and development of the universe under the umbrella of 'modern scientific cosmology'. Extrapolating backwards from the present expansion of the universe, they soon arrived at the notion of a primordial explosion, or 'big bang', which led Pope Pius XII in 1951 to suggest that physics had finally confirmed the Christian doctrine of creation (1972). Much subsequent debate has ensued as to whether the inference to a divine creative act is quite as straightforward as the Pope claimed and whether the concept of *creation* actually entails an absolute beginning to creation or only the more general notion of creation's ontological dependence (see e.g. Jaki 1989; Peters 1989; Drees 1990; Van Till *et al.* 1990; Russell *et al.* 1996; Sobosan 1999). Recent scientific (though highly speculative) proposals such as 'eternal inflation' and 'quantum cosmology' make it possible to perceive the beginning of this universe as one event in a longer (even infinite) series of similar events. Consequently, the big bang now looks less and less like an absolute beginning. Is the lesson here that religion should interact with modern physics only at points of long-standing consensus among physicists, since physics changes quickly around the edges? While it must be said that the Pope's pronouncement was hasty, it is also possible to see in this example hints of a kind of religious engagement with physics, and with science generally, that is more fluid and open to change, not just on the scientific side but on the religious side as well. Christian theologians, at least, have lived too long under the illusion that their pronouncements must transcend time and space (not to mention culture and gender) to be authentic and valuable.

Contemporary scientific cosmology has also reinvigorated the design argument for God's existence. Earlier forms of this argument focused on the intricate order of natural processes manifested in the complex structure of living organisms, but Darwin's powerful case against biological design—that it was merely 'apparent'—successfully shifted the debate to the realm of physics in terms of the so-called Anthropic Principle (though the debate has shifted back toward the realm of biology recently, with the emergence of the Intelligent Design Movement). The Anthropic Principle is based upon the observation that the structure and processes of the universe appear finely tuned for the requirements of our own existence; change any one of the basic parameters governing the physical interactions of the universe ever so slightly, and something about its structure or contents suddenly becomes inhospitable to life as we know it. In its strongest form, this principle has been used to support an inference to a divine 'tuner'. In its weaker form, however, our existence can be seen merely as the result of a kind of cosmic Darwinism (a loose analogy at best) in which there needn't be any surprise at finding ourselves in a particular domain of the universe—or, using more recent terminology, in a branch of a 'multiverse' (Rees 2003)— whose structures and processes are hospitable to life like ours (for a review of the debate, see Barrow and Tipler 1985; Leslie 1989; Murphy and Ellis 1996). This weaker version avoids the theistic conclusion, but much disagreement remains as to whether or not it amounts to a scientific explanation. The debate over the significance of fine-tuning provides an interesting example of how what is judged to count as 'scientific' shifts with theoretical and technological advances. Whereas the anthropic argument was initially taken to be quite compelling to a good number of religiously minded physicists (or, alternatively, a serious problem for physics to solve as a way of avoiding religious implications), this number has fallen as the idea of a multiverse has moved slowly away from the realm of science fiction and closer to the mainstream of cosmological theory.

The design argument runs into further difficulties with the far future of the universe, which appears doomed either to endless expansion and cooling, the so-called *freeze* scenario, or to eventual recollapse and implosion, the so-called *fry* scenario. Neither offers much comfort for those wanting to maintain a more robust and traditional notion of future fulfilment (Polkinghorne and Welker 2000; Ellis 2002). And while it is at least conceivable that life, suitably transformed, could extend itself far into the future, this kind of pseudo-immortalization similarly offers little hope for a more traditional eschatological vision (Dyson 1988; Tipler 1994). Much more likely, humanity and all life on earth will be extinguished in a solar supernova—if, that is, we manage to survive the end of fossil fuels—long before the 'end' of our universe arrives. The religious implications of this piece of the modern cosmological story have only begun to be addressed.

Quantum Theory

As Einstein was rewriting Newton's account of space and time and reshaping our understanding of the universe at the largest scales, another similarly (arguably, even more) radical revolution was taking place at the very smallest scales. In 1900 the German physicist Max Planck turned his attention to one of the most puzzling of the few remaining 'clouds' on the horizon, a problem having to do with the emission and absorption of electromagnetic radiation by atoms. He solved this problem by introducing the curious notion that energy can come only in discrete units, subsequently dubbed *quanta*, rather than in continuously varying amounts, as classical physicists had supposed. This and other breakthroughs led other European physicists such as Niels Bohr, Werner Heisenberg, Erwin Schrödinger, and Paul Dirac to develop quantum theory, which has since achieved spectacular successes in its ability to describe the behaviour of atoms and their components. These successes, however, have come with a price: many of physicists' classical and common-sense intuitions regarding basic concepts such as causality, determinism, separability, and the wave–particle distinction have been called into question along the way. At the quantum level objects appear to change their state over time without any sufficient mechanical cause, evolving in a purely random manner ('indeterminism'); they appear to remain connected to one another even when separated across large distances ('entanglement'); and they appear to behave like waves in some settings and particles in others ('complementarity')—to name only a few of the more bizarre consequences of this new theoretical framework.

Quantum theory has attracted the attention of religious thinkers from a wide variety of traditions (Capra 1975; At-Tarjumana 1980; O'Murchu 1997; Thuan 2001). Some have explored ways of extending Bohr's notion of 'complementarity'—the idea that mutually incompatible descriptions like 'wave' and 'particle' are necessary for a complete description of a single quantum phenomenon—to issues such as the relationship between religion and science and the character of religious language (see e.g. Barbour 1974; Losee 1992). Others, whom I mentioned above when discussing recent developments in the concept of special divine action, have appealed to quantum indeterminism (Russell *et al.* 2001). Still others have interpreted quantum entanglement to be a hint that the universe is a place of subtle interconnection in the midst of bewildering diversity (O'Murchu 1997; Jungerman 2000; Sharpe 2000). Whether these connections will take hold and be seen as a valuable addition to contemporary religious reflection has a great deal to do with the degree to which quantum concepts ever migrate, or fail to do so, into contemporary cultures. Despite the fact that numerous contemporary technologies operate on principles that make sense only within the framework of quantum theory, the atomic and subatomic scales at which these principles apply effectively hides them from general cultural awareness.

Chaos Theory

The remarkable subtlety of physical processes is additionally highlighted by chaos theory, a third significant theoretical development within twentieth-century physics. Strictly speaking, chaos theory resides within the deterministic framework of Newtonian physics. Even so, it suggests that certain processes thought to be describable in principle by deterministic laws, such as weather patterns, can develop in unpredictable and seemingly random ways. But because chaos theory is deterministic, it does not offer any straightforward opportunities for a non-interventionist (and theophysically incompatibilist) account of divine action. Some have argued that despite its presently deterministic form the theory points to a genuine openness in nature's processes; this openness, they argue, will eventually be reflected in some future version of the theory (Polkinghorne 1991; and see the subsequent debate in Russell *et al.* 1995). If such a shift were to occur, chaos theory would provide yet another powerful example of physics moving beyond its Newtonian origins.

String Theory

One final piece of contemporary physics, 'string theory', deserves mention, though it lies on the speculative edge of the discipline—so near the edge, in fact, that physicists fight amongst themselves as to whether it ought to be counted as legitimate physics. Various difficulties with the relativistic and quantum frameworks—especially the challenge of combining the two—have led to the search for a single overarching framework that could incorporate both. String theory, the most widely publicized candidate among a number of competing approaches, treats the fundamental units of the physical world not as mathematical points but as vibrating 'loops' or 'strings'. These strings vibrate not only in the familiar $3 + 1$ dimensions of space and time, but in additional dimensions as well. These extra dimensions, rather than being extended, are 'rolled-up' at the subatomic scale such that we do not experience them.[5] At present neither string theory nor its competitors can be tested empirically. Its speculative nature renders it of little ontological interest from a religious perspective at this point, but it offers a rich array of complex concepts that might prove to be interesting religiously, such as the notion that reality has dimensions beyond those present to our immediate experience.

CONCLUSION

Physicists are currently struggling to unite the various theoretical developments surveyed here under one conceptual framework. At present, relativity theory, quantum theory, and chaos theory each provide quite distinct lenses onto the world's physical structures and processes. Although both relativity theory and chaos theory transform various aspects of Newton's account of space, time, and causation, they also essentially sustain the determinism of the classical framework. Quantum theory, on the other hand, at least according to the most widely held interpretation, dramatically overturns this tradition. Whether string theory or some other new framework will be rich enough to unite these multiple perspectives on the nature of the physical processes is anyone's guess. Physics is at present a scientific discipline deeply at odds with itself, presenting us with remarkable but fractured insights into the physical world that forms the substrate of human existence. The struggle to resolve these tensions will no doubt lead to further opportunities for conversation with religious perspectives. The human quest for meaning and transcendence cannot

---

[5] A helpful though crude analogy is the difference between viewing a drinking straw from a distance and thinking it to be a one-dimensional line and seeing it up close as a two-dimensional surface with one of its two dimensions rolled up on itself.

be reduced to physical explanation, but it can be enriched and enlivened by the remarkable account of the world's natural processes obtained through physics.

REFERENCES AND SUGGESTED READING

AT-TARJUMANA, AISHA (1980). *The Subatomic World in the Qur'an.* Norwich: Diwan Press.

BARBOUR, IAN G. (1974). *Myths, Models, and Paradigms.* San Francisco: Harper & Row.

BARR, STEPHEN M. (2003). *Modern Physics and Ancient Faith.* Notre Dame, Ind.: University of Notre Dame Press.

BARROW, JOHN D., and TIPLER, FRANK J. (1985). *The Anthropic Cosmological Principle.* Oxford: Oxford University Press.

BROOKE, JOHN HEDLEY (1991). *Science and Religion: Some Historical Perspectives.* Cambridge: Cambridge University Press.

CAPRA, FRITJOF (1975). *The Tao of Physics,* 3rd edn. Boston: Shambhala.

CLAYTON, PHILIP (2004). *Mind and Emergence: From Quantum to Consciousness.* Oxford: Oxford University Press.

COAKLEY, SARAH (2000). *Religion and the Body.* Cambridge: Cambridge University Press.

CRAIG, WILLIAM LANE (2001). *Time and Eternity: Exploring God's Relationship to Time.* Wheaton, Ill.: Crossway Books.

CUSHING, JAMES T. (1994). *Quantum Mechanics: Historical Contingency and the Copenhagen Hegemony.* Chicago: University of Chicago Press.

DE LAPLACE, PIERRE-SIMON (1917). *A Philosophical Essay on Probabilities,* 2nd edn, ed. F. W. Truscott and F. L. Emory. New York: John Wiley & Sons.

DREES, WILLEM B. (1990). *Beyond the Big Bang: Quantum Cosmologies and God.* La Salle, Ill.: Open Court.

DYSON, FREEMAN J. (1988). *Infinite in All Directions.* New York: Harper & Row.

ELLIS, GEORGE F. R. (2002) (ed.). *The Far-Future Universe: Eschatology from a Cosmic Perspective.* Philadelphia: Templeton Foundation Press.

FAGG, LAWRENCE W. (1995). *The Becoming of Time: Integrating Physical and Religious Time,* Scholars Press Studies in the Humanities Series, 20. Atlanta: Scholars Press.

FANTOLI, ANNIBALE (2003). *Galileo: For Copernicanism and for the Church,* 3rd edn, trans. G. V. Coyne, SJ. Notre Dame, Ind.: University of Notre Dame Press.

GALILEO (1957). 'Letter to the Grand Duchess Christina (1615)', in S. Drake (ed.), *Discoveries and Opinions of Galileo,* New York: Anchor Books, 175–216.

HEILBRON, J. L. (1999). *The Sun in the Church: Cathedrals as Solar Observatories.* Cambridge, Mass.: Harvard University Press.

HELLER, MICHAEL (1996). *The New Physics and a New Theology,* trans. G. V. Coyne, SJ, S. Giovannini, and T. M. Sierotowicz. Vatican City State: Vatican Observatory Foundation.

HILGEVOORD, JAN (1994) (ed.). *Physics and Our View of the World.* Cambridge: Cambridge University Press.

HODGSON, PETER E. (2005). *Theology and Modern Physics.* Aldershot: Ashgate.

IQBAL, MUZAFFAR (2002). *Islam and Science.* Burlington, Vt.: Ashgate.

ISHAM, CHRIS J., and POLKINGHORNE, JOHN C. (1996). 'The Debate over the Block Universe', in Russell *et al.* (1996), 139–47.

JAKI, STANLEY L. (1989). *God and the Cosmologists.* Edinburgh: Scottish Academic Press.

JAMMER, MAX (1999). *Einstein and Religion: Physics and Theology.* Princeton: Princeton University Press.

JUNGERMAN, JOHN A. (2000). *World in Process: Creativity and Interconnection in the New Physics.* Albany, N.Y.: SUNY Press.

LAWRENCE, CHRISTOPHER, and SHAPIN, STEVEN (1998) (eds.). *Science Incarnate: Historical Embodiments of Natural Knowledge.* Chicago: University of Chicago Press.

LESLIE, JOHN (1989). *Universes.* London: Routledge.

LOSEE, JOHN (1992). *Religious Language and Complementarity.* Lanham, Md.: University Press of America.

MATT, DANIEL C. (1996). *God & the Big Bang: Discovering Harmony between Science & Spirituality.* Woodstock, Vt.: Jewish Lights Publishing.

MERCHANT, CAROLYN (1990). *The Death of Nature: Women, Ecology, and the Scientific Revolution.* San Francisco: Harper San Francisco.

MURPHY, NANCEY, and ELLIS, GEORGE F. R. (1996). *On the Moral Nature of the Universe: Theology, Cosmology, and Ethics,* Theology and the Sciences Series. Minneapolis: Fortress Press.

NORRIS, CHRISTOPHER (2000). *Quantum Theory and the Flight from Realism: Philosophical Responses to Quantum Mechanics.* London: Routledge.

O'MURCHU, DIARMUID (1997). *Quantum Theology: Spiritual Implications of the New Physics.* New York: Crossroad.

PANNENBERG, WOLFHART (1993). *Toward a Theology of Nature: Essays on Science and Faith,* ed. T. Peters. Louisville, Ky.: Westminster/John Knox Press.

PEACOCKE, ARTHUR (1993). *Theology for a Scientific Age: Being and Becoming—Natural, Divine and Human.* Minneapolis: Fortress Press.

PETERS, TED (1989) (ed.). *Cosmos as Creation: Theology and Science in Consonance.* Nashville: Abingdon Press.

PIUS XII (1972). 'Modern Science and the Existence of God', *The Catholic Mind,* 49: 182–92.

PLACHER, WILLIAM C. (1996). *The Domestication of Transcendence: How Modern Thinking About God Went Wrong.* Louisville, Ky.: Westminster/John Knox Press.

POLKINGHORNE, JOHN C. (1991). *Reason and Reality: The Relationship between Science and Theology.* London: SPCK.

—— and WELKER, MICHAEL (2000) (eds.). *The End of the World and the Ends of God: Science and Theology on Eschatology.* Harrisburg, Pa.: Trinity Press.

REES, MARTIN (2003). *Our Cosmic Habitat.* Princeton: Princeton University Press.

RITSCHL, ALBRECHT (1902). *Justification and Reconciliation,* iii: *The Positive Development of the Doctrine,* H. R. Mackintosh, trans. London: T & T Clark.

RUSSELL, ROBERT J., CLAYTON, PHILIP, WEGTER-MCNELLY, KIRK, and POLKINGHORNE, JOHN C. (2001) (eds.). *Quantum Mechanics: Scientific Perspectives on Divine Action.* Vatican City State: Vatican Observatory; Berkeley: Center for Theology and the Natural Sciences.

—— MURPHY, NANCEY, and ISHAM, CHRIS J. (1996) (eds.). *Quantum Cosmology and the Laws of Nature: Scientific Perspectives on Divine Action,* 2nd edn. Vatican City State: Vatican Observatory Foundation; Berkeley: Center for Theology and the Natural Sciences.

—— —— and PEACOCKE, ARTHUR (1995) (eds.). *Chaos and Complexity: Scientific Perspectives on Divine Action.* Vatican City State: Vatican Observatory Foundation; Berkeley: Center for Theology and the Natural Sciences.

—— STOEGER, WILLIAM R., SJ, and COYNE, GEORGE V., SJ (1988) (eds.). *Physics, Philosophy, and Theology: A Common Quest for Understanding.* Vatican City State: Vatican Observatory Foundation.

—— —— —— (1990) (eds.). *John Paul II on Science and Religion: Reflections on the New View from Rome*. Vatican City State: Vatican Observatory Foundation.

SHARPE, KEVIN J. (2000). *Sleuthing the Divine: The Nexus of Science and Spirit*. Minneapolis: Fortress Press.

SOBOSAN, JEFFREY G. (1999). *Romancing the Universe: Theology, Science, and Cosmology*. Grand Rapids, Mich.: Eerdmans.

TAYLOR, EDWIN F., and WHEELER, JOHN ARCHIBALD (1992). *Spacetime Physics: Introduction to Special Relativity*, 2nd edn. New York: Freeman.

THUAN, TRINH XUAN (2001). *The Quantum and the Lotus: A Journey to the Frontiers where Science and Buddhism Meet*. New York: Three Rivers Press.

TIPLER, FRANK J. (1994). *The Physics of Immortality: Modern Cosmology, God, and the Resurrection of the Dead*. New York: Doubleday.

TOOLAN, DAVID S. (2001). *At Home in the Cosmos*. Maryknoll, N.Y.: Orbis.

VAN TILL, HOWARD J., SNOW, ROBERT E., STEK, JOHN H., and YOUNG, DAVIS A. (1990). *Portraits of Creation: Biblical and Scientific Perspectives on the World's Formation*. Grand Rapids, Mich.: Eerdmans.

WALLACE, B. ALAN (2003). *Choosing Reality: A Buddhist View of Physics and the Mind*. Ithaca, N.Y.: Snow Lion Publications.

WILES, MAURICE F. (1986). *God's Action in the World*. London: SCM Press.

WOLFRAM, STEPHEN (2002). *A New Kind of Science*. Champaign, Ill.: Wolfram Media.

WORTHING, MARK W. (1996). *God, Creation, and Contemporary Physics*. Minneapolis: Fortress Press.

CHAPTER 11

MOLECULAR BIOLOGY AND RELIGION

MARTINEZ HEWLETT

INTRODUCTION

At first blush, these seem to be completely disparate ways of thinking: the study of the molecular structure and function of the genetic information in the cell, on the one hand, and a systematized set of beliefs about God, the transcendent, and our relationship to the supernatural, on the other hand. However, like all human activities, both of these have areas of congruence and, as would be expected, areas of controversy. In this chapter, I will introduce you to the background, assumptions, and kinds of descriptive models used in molecular biology. My intention is to provide a way to navigate through the jargon-spiked waters of this field so that you can see how and where these places of overlap between this field and religion exist, as well as where the controversies lie. In doing this, I will not so much focus on defining religion or theology, but will assume that the reader has a reasonably sophisticated understanding of what these terms mean. In addition, since my own religious path is that of Roman Catholicism, some of my theological comments may be flavoured by that particular perspective. For this, I ask your indulgence.

A BRIEF HISTORY OF MOLECULAR BIOLOGY

When Charles Darwin proposed his model of descent with modification as a naturalistic explanation for how the amazing variation in the living world arose, he did not have a clear idea of how variation is actually inherited. The prevailing notion in the mid to late nineteenth century was that inheritance involved some kind of blending of characteristics from two parents into offspring. It was the work of an Augustinian monk, living in Brün, Austria, that was to change this notion and provide us with our current understanding of genetics. Gregor Mendel's work was, however, ahead of his times, and while Darwin's book was widely read immediately upon its publication, Mendel's papers were ignored for nearly fifty years.

Mendel defined the particulate nature of inheritance—that is, characteristics are passed from parent to offspring in units that persist and are not diluted by blending. In this way, traits that might appear in one generation may later reappear, even though intervening generations did not display them. Mendel did not coin the word 'gene' for these traits, but he derived the laws of genetics by which we still understand the behaviour of these units of inheritance.

By the beginning of the twentieth century, Mendel's work had been rediscovered by DeVries and others, and the science of genetics was born. In the spirit of the reductionist methodology of modern science, it became the goal to determine just what structure in the cell might be identified with the gene. The choices were becoming clarified. Genes were physically associated with structures in the cell called chromosomes. Chromosomes consist of large molecules—so-called macromolecules—and these were the prime suspects for what might be the gene. Two of these were receiving particular attention: proteins, long linear chains of amino acids; and nucleic acids, linear molecules consisting of sugar, phosphates, and nitrogen-containing, ring-shaped structures called purines and pyrimidines.

Whatever the chemical nature of the gene might be, it seemed to physicists like Irwin Schrödinger that genes must obey different physical principles. In a short book called *What Is Life?* (1944), Schrödinger, one of the founders of quantum theory, proposed that genes must indeed be unique, since they seem to resist the normal physical processes that lead to dissipation. Genes survive apparently intact over generations. Schrödinger suggested that the study of genes might lead to the discovery of new laws of physics.

This challenge intrigued one young physicist in Germany, Max Delbrück. He determined to pursue this idea, and launched himself away from both physics and the decaying pre-Second World War German Reich to follow Schrödinger's call in the United States. Being a physicist, he reasoned that what was needed was a simple system to study, one in which all of the variables could be controlled and which had only a few easily managed features. He settled on using the bacteriophage or, as it is familiarly known, the phage (the word rhymes with 'cage'), viruses that infect bacterial cells. He was the founding father of what has come to be called the Phage

Group, working during the academic year at the California Institute of Technology in Pasadena, California, and migrating cross-country for the summer to the Cold Spring Harbor Laboratories on Long Island, New York (Cairns *et al.* 1992).

The Phage Group, under the leadership of Delbrück, attracted a wide variety of scientists, including Salvador Luria, a physician, and Leo Szilard, a physicist who had left the Manhattan Project in protest over the continued development of the bomb after the fall of Germany. The group had a large number of young scientists— chemists, geneticists, virologists, physicians, and others who were attracted to this new way of searching for the gene. These men and a few women were, in fact, the first molecular biologists.

At the heart of the new discipline was the focus on the physical and chemical nature of the gene. It had become clear, through the work of Oswald Avery and his colleagues at Rockefeller University, that DNA was the likely candidate macromolecule for the gene. Two members of the Phage Group—Alfred Hershey and Martha Chase—finally demonstrated this in an elegant experiment, using a Waring blender and the viruses that had become one of the central organisms of study for the new field.

Now that the object of their inquiry had been identified, these new biological pioneers set out in earnest. In relatively rapid order a structural model for DNA was proposed (Watson and Crick), a method worked our for how DNA could be reproduced and genes passed from generation to generation (Meselson, Stahl, Kornberg, and others), and, finally, the genetic code broken to reveal how the information that resides in DNA comes to be unfolded into the myriad of structures that make up a cell (Crick, Nirenberg, Khorana, and others). This was the golden age of molecular biology, and hardly a week passed during the decade of the 1950s and 1960s when a new window into the molecular world of the cell wasn't opened.

THE NEO-DARWINIAN SYNTHESIS

When Darwin offered his evolutionary model to the world in 1859, objections did not only arise from the religious community. Members of the scientific community were also not completely ready to accept this idea, given that there was no mechanism for inheritance that fit descent with modification. In fact, some objectors pointed out that if blending inheritance, the prevailing model, were in fact true, there would be no expectation that advantageous traits would survive over generations so that selection could work.

With the rediscovery of Mendel's work at the start of the twentieth century, this problem was solved. By 1942, Julian Huxley, grandson of Darwin's defender Thomas Huxley, could tout what he called the 'modern synthesis', in which the Darwinian model was merged with Mendelian genetics and ideas about populations to produce an overarching paradigm that subsumes the entire field of biology to the present. The so-called neo-Darwinian synthesis now models all of life on the following principles:

1. Genes are information in the form of the linear array of bases that make up the DNA molecules of chromosomes.
2. The traits of an organism (the phenotype) are the direct expression of the information found in the genes (the genotype).
3. Variations in traits are the result of subtle differences in this information (changes in nitrogen-containing base pairs of the DNA: namely, adenine, cytosine, thymine, and guanine).
4. Changes in genes are mutational events that occur in a 'random' way. The word 'random' is used, but by this we really mean 'unpredictable'.
5. A population of entities will have variations in traits that are the result of these mutational events (a process called genetic drift).
6. The force of natural selection operates on this pool of genetic variants, allowing those with greater reproductive fitness to be represented in succeeding generations.

The new discipline of molecular biology quickly became the vanguard of neo-Darwinian thought. After all, this was where the details of how genetic variation takes place and how the information in the DNA is ultimately expressed as traits would be learned. The golden age became nothing less than a quest to understand the entire living world in terms of the informational molecules themselves. As Francis Crick wrote in this oft-quoted passage: 'The ultimate aim of the modern movement in biology is in fact to explain all biology in terms of physics and chemistry... Eventually one may hope to have the whole of biology explained in terms of the level below it, and so on right down to the atomic level' (Crick 1966).

Neo-Darwinism and Religion

It is certainly true that the initial reactions to Darwin's model of the origin of species had much to do with the philosophical implications, as opposed to the science itself. The strain of reductionist epistemology and ontology that had overtaken the modern scientific enterprise was essentially complete by the middle of the nineteenth century. It is not surprising, therefore, that twentieth-century scientists had inherited this same approach. While not every biologist could be called an agnostic or an atheist, it is true that some of the more important interpreters of the new biology, as it was coming to be called, subscribed to the materialist agenda, expressed in the quote from Crick. *Chance and Necessity*, written by the French Nobel laureate Jacques Monod (1971), became an influential book, championing the idea that the universe is, indeed, simply the product of a random (read 'unpredictable') set of events, of which we are just a lucky outcome.

This is not, in and of itself, a religious or even an anti-religious view, although the secular humanism that pervades the book could certainly be called such. However, a belief in the ultimate power of science in general, and molecular biology in particular, to explain everything about life is certainly an assumption in these works. Even for scientists who were not specifically agnostic or atheistic, the idea of science having ultimate explanatory value was assumed, if not stated.

The religious nature of this scientism became focused in the proposal by Crick in 1958 that the flow of information in the biological world constituted the 'central dogma' of molecular biology. It is clear that, in some sense, Crick was making a joke by using this frankly religious language. However, the fact that he and others saw it as a joke is telling. It speaks to an unconscious faith in science and a concomitant disdain for religion. In any case, the central dogma is picked up as the defining statement of the new discipline, such that every modern textbook of biology uses this phrase to describe biological information.

Before he died, Crick attempted to downplay his use of the word 'dogma', calling it a 'poor choice of words'. However, historians of the discipline were quick to adopt the religious implication. Horace Judson wrote a masterful book, describing the birth of molecular biology, based on interviews with those who were around at the beginning. The title of this book is *The Eighth Day of Creation* (Judson 1996).

THE DOMINANCE OF THE MOLECULAR PARADIGM

Carl Woese, a molecular biologist, has contributed a great deal to our models of the natural world, including his insights into how relationships between organisms can be defined. Recently he published a review paper in which he looks at the history of molecular biology in relation to other life sciences. In this paper, entitled 'A New Biology for a New Century', he writes: 'The most pernicious aspect of the new molecular biology was its reductionist perspective, which came to permeate biology, completely changing its concept of living systems and leading to a change in society's concept thereof' (Woese 2004: 174). It was certainly true that, by the later part of the twentieth century, it was impossible to think of any aspect of biology without invoking the neo-Darwinian paradigm. As the tools available to the molecular biologist were perfected, the hold over the other sciences tightened. It became standard procedure to ask about the genes that were involved, whether one was investigating some aspect of cellular metabolism or observing mating behaviour. All of biology, to paraphrase Crick, could be reduced to DNA.

The other life science disciplines, such as developmental biology, organismic biology, and ecology, maintained their separate existence, but increasingly looked to molecular biology for the new tools needed to analyse the living world as stretches of nucleic acid sequence. This move was prompted both by the power of technology to yield new models, as well as by the desire of funding sources to have the latest and best approach. As a result, cherished ideas about whole systems and emergent properties were at odds with the new, atomistic epistemology. Woese encapsulates this conflict when he writes: 'The intuitive disparity between atomic reality and the "biological reality" inherent in direct experience became the dialectic that underlay the development of 20th century biology' (Woese 2004: 174).

FROM SCIENCE TO ENGINEERING

Any science makes progress by the development of better tools for observation. Examples include the telescope for astronomy, the cyclotron for particle physics, and the microscope for cell biology and microbiology. This is no less true for molecular biology. It is especially significant for a discipline whose object of study cannot be directly appreciated with the human senses. The molecular biologist cannot 'see' DNA.

The golden age culminated with the building of models that described how the information in the gene becomes the functionality within the cell. All types of life, from viruses to humans, were assumed and shown to use these same basic methods of genetic expression. These models were built using data obtained with a variety of sophisticated instruments. Each advance in understanding led to a corresponding improvement of the methods. When molecular biologists came to appreciate the structure of DNA as defined by the Watson–Crick model, this suggested not only the mechanism by which the molecule could be duplicated in the cell, but also the ways in which to demonstrate this duplication.

During the heyday of discovery in the 1960s, new molecular features of the living world came to be appreciated in finer and finer detail, leading to ever more elaborate models. Towards the end of that decade, molecular biologists turned their attention towards the cells of higher life forms, the so-called eukaryotic world, which includes the cells that make up humans. Earlier, it was assumed that the information gathered from the simple bacterial cell *Escherichia coli* would be identical for all life forms. As Jacques Monod wrote, 'What's true for *E. coli* is true for the elephant' (Monod 1972).

Most interestingly, this turned out not to be the entire story. As the molecular complexity of eukaryotic cells began to unfold, it became clear that the basic features of all life were the same: all cells use DNA as the genetic material; all cells use RNA as an intermediate informational molecule; and all cells translate the code into protein in essentially the same way. The 'central dogma' is, in effect, universal. However, the details of the processes revealed a myriad of differences, many of which turned out to be truly significant for modelling how the genetics of these higher cells function.

By the end of the 1970s, significant inroads were being made into the workings of the eukaryotic cell. But simpler life forms had not been abandoned. Instead, molecular biologists turned to them as a source of the tools they would need to probe even deeper into the molecules of life, into the very code of life itself. These tools included a variety of enzymes, the biological catalysts that drive the chemical reactions of the cell. These enzymes could be isolated from simple life forms such as bacteria or viruses and then turned into very precise instruments that could manipulate DNA, cutting it and re-splicing it in specific ways. These efforts led to the transition of the molecular biologist from a scientist who could only observe living systems to an engineer who could, in effect, alter and design genes in the test tube.

The era of the genetic engineer began in the mid-1970s, when a number of prominent scientists did something remarkable. They proposed a voluntary moratorium on certain kinds of engineering experiments, until a meeting could be held to discuss the implications. This unprecedented step was taken because the scientists realized what might be at stake. If genes could be manipulated and moved from one organism to another in combinations that did not exist in the natural world, then the potential for designing and producing something dangerous was very real.

A meeting was held at Asilomar Conference Center outside Monterey, California, in February 1975. The conference was attended by molecular biologists, geneticists, legal experts, and bioethicists. The question before the assembly was not whether such experiments should ever be done, but rather what limits to place on the technology. It is very important to note that, in spite of a few dissenting voices, most molecular biologists supported the application of this technology if it could be used safely and with appropriate guidelines in place. As a result, the conference did not really debate the philosophical or ethical issues raised by the technology, but rather went about creating the regulatory agency that would oversee the proper use of the new methods.

And so began the most recent era in the history of molecular biology: the age of the genome and of recombinant DNA.

The Short, Happy Life of Molecular Biology?

Things went swimmingly at first. Universities set up committees, as required by the governing agency, that would review and pass on research using the new technology. The US federal agency responsible for regulation was the Recombinant DNA Advisory Committee, or RAC. Each unit requesting federal funds had to establish a local version of this group, consisting of both researchers and community representatives. I served for much of my career at the University of Arizona as a member of our Institutional Biosafety Committee.

With guidelines in place, the technology was exploited in the service of modelling the functional aspects of gene expression. However, it was clear from the outset that these methods of genetic manipulation also lent themselves to industrial applications. After all, once you could isolate the gene that encodes human insulin or human growth hormone and move it into a bacterial cell such that the protein could be produced in commercially important quantities, it is easy to see how these tools became much more than merely avenues of fruitful research.

Biotechnology companies flourished, and their public stock offerings were wildly anticipated and traded. By the end of the 1980s there was hardly an academic molecular biologist who did not have some tie to a biotech company, whether it be as a founder, a member of the board of directors, or a grant recipient. True, some of the most innovative work was done in these companies, and major improvements in the technology led to even more fanciful but achievable goals. Nonetheless, molecular biology seemed to be passing from a science that asked basic questions and built explanatory models to an applied discipline which provided the tools for this new (pardon the pun) growth industry.

Molecular biology could have been heading for oblivion, remembered only for its storied past. What was needed was a new challenge to set it back on track. That challenge came along in the form of the Human Genome Project. The quest to sequence our own genetic material proved to be both the salvation of the field and the driving force behind a major paradigm shift through which we are currently living.

The Human Genome Project: Headlong into the Wall of Anomaly

When the US Department of Energy (DOE) approached leading molecular biologists in 1984, at a meeting it co-sponsored in Alta, Utah, and asked, 'With the current technology and large sums of money, would it be possible to sequence all of human DNA?', the answer, after little hesitation, was a resounding 'Of course' (Cook-Deegan 1989).[1] This was, for biology, the first attempt at what physicists call 'big science'. This

---

[1] An interesting aside concerns the role of the DOE in the project. Why the DOE? The history of this goes back to the Manhattan Project. After the bombs were dropped on Hiroshima and Nagasaki, President Truman signed a directive establishing the Atomic Bomb Casualty Commission (ABCC), which later became the Radiation Effects Research Foundation (RERF) (2005). The effort was placed within the purview of the Atomic Energy Commission (AEC), which inherited the Manhattan Project facilities at Los Alamos. The purpose of the ABCC and the RERF was and remains the detection of mutagenic effects of low-level radiation on human DNA. Initial observations of survivors of the two blasts were not sensitive enough to reveal fine changes. As a result, the RERF was always looking for better

would be a massive effort. At first, it was suggested that the entire project be housed at the Los Alamos National Laboratories, where some of the preliminary work had already begun. In the end, however, a decision was made to distribute the work to a number of centres around the country, and, as both the European and Japanese scientific community joined in, to centres around the world.

Here, then, was a project worthy of the discipline that had, as its principle tenet, the central dogma. Sequencing the human genome would allow humans to understand themselves right down to the atomic level, as Crick had predicted. In fact, the most reductionist interpretation would be that the genomic sequence was all that was necessary to completely define what it means to be human. The *imago Dei* was retired in favour of the *imago DNA*.

The project was launched in 1986, with the stated timeline of finishing the sequence in about twenty years. It became obvious early on that much of human DNA is not involved in defining the sequence of a protein. In fact, large regions of human DNA don't seem to contain any genes whatsoever. The term 'junk DNA' was coined to describe these parts of the genome, although the hubris of this appellation did not seem to strike many of those involved. After all, dogma is dogma, and whatever lies outside belief is either heresy or junk.

In order to speed up the process, it was decided that the only DNA that needed to be sequenced were those regions that encode a protein. These so-called expressed sequences then became the target of a massive push, resulting in the announcement in February of 2001 that a working draft of the sequence of the human genome had been completed. The millions of bases were deposited in large, public databases, maintained at the National Institutes of Health and Los Alamos National Laboratories. Now the next phase, some would say, is where the real work began. What do these sequences mean?

The first inkling of a problem with this scenario was the realization that humans had many fewer genes than predicted. In fact, rather than the 100,000 or so genes predicted, it turns out that we have somewhere between 20,000 and 25,000, not very much different from the fruit-fly. If this is the case, then what makes us different from the fly?

A closer examination of the data revealed an even greater problem. The long-cherished notion of the gene was now being challenged. Just what is a gene? What seemed to be obvious at the start was no longer clear. In April 2003, Michael Snyder and Mark Gerstein published a paper entitled 'Defining Genes in the Genomic Era'. They came to the remarkable conclusion that 'Ultimately, we believe that defining genes based solely on the human genome sequence, while possible in principle, will not be practical in the foreseeable future' (Snyder and Gerstein 2003: 260).

analytical tools. What finer tool could there be for the analysis of changes in the human genome than the sequence of the genome itself? The AEC morphed into the Nuclear Regulatory Commission (NRC), which is under the DOE. Thus, the DOE was a co-sponsor of this meeting, along with the International Commission for Protection against Environmental Mutagens and Carcinogens (ICPEMC).

What might this mean? If, in fact, humans have roughly the same number of genes as the fruit-fly, and if the long-cherished paradigm of the gene is no longer definable at the level of the DNA sequence itself, then what is to become of Crick's goal of explaining everything in terms of chemistry and physics?

It was, in fact, a physical chemist who said it best. Michael Polanyi argued in 1968 that, in principle, life cannot be reduced to the chemistry of DNA. While chemistry explains the ability of a G (guanine) to pair with a C (cytosine), for instance, it does not explain the nature of biological information, of which a GC base pair is only one small bit. It is, rather, the context of that GC pair that leads to the emergent property of information. As a result, he stated, life cannot be reduced to the chemistry of the base pairs. In fact, he contended that the information content of DNA must be insensitive to the chemistry of the base pairs. Otherwise a GC pair could not be found within different contexts meaning different things. In effect, a GC base pair in a dog's DNA would mean exactly the same thing as it does in the fruit-fly. Since this is not the case, the information is not reducible to chemistry.

It is perhaps not surprising, then, that when molecular biology finally completed its *magnum opus*, it ran headlong into this Polanyian issue. Molecular biology, a methodologically reductionist science, was faced with an epistemic problem. Reductionism as a way of knowing something was no longer fully tenable in practice, though it had methodological advantages. The solution to this problem came from an unexpected direction, and resulted in what may be one of the major paradigm shifts in science.

BACK TO THE FUTURE: A RETURN TO HOLISTIC SCIENCE

In his wonderful philosophical essay, Carl Woese points out that the problems in biology had always, until the advent of the molecular approach, been considered within the context of broader biological systems. Evolution, he argues, is a holistic problem, as is the issue of developmental biology. However, the dominance of molecular biology, beginning in the middle of the twentieth century, put all these other views either on the back burner or into the historical archives.

When the dominant paradigm of a discipline is challenged by either recurring anomalies or unexplainable phenomena, that discipline is ready for a shift. Thomas Kuhn used as his predominant exemplar of this effect the Copernican revolution that supplanted the Ptolemaic model. In our recent history we have also seen the shift in physics from the Newtonian to the quantum model. I would argue that we are in the midst of this same kind of shift in biology.

The massive database of the Human Genome Project is accompanied by similar databases of sequences for other organisms, including the fruit-fly, the mouse, the round worm, and several other experimental organisms. The worldwide database collection recently celebrated the milestone of reaching 100 gigabases. How to manage all of this data? Enter two saving disciplines: complexity theory and network analysis.

In 1967 Stanley Milgram, a Yale sociologist, published the results of a remarkable study. He asked a simple question: how many steps or links are there between any two people on the planet? In order to answer this question, he picked, at random, names from the telephone directory in a Midwestern town. He sent each of them a letter, asking that they mail a card to a specific person in New Haven, Connecticut. If they knew this person, they were to mail the card right away. If they did not, they were to send the card to someone whom they believed might know the person. At each stage, a report would be sent back to Milgram. From this experiment, he calculated that there are, on average, about 5.6 steps between any two people on the Earth, from which we derive the 'six degrees of separation' dictum.

It turns out that this is an example of a scale-free, or small-world, network. It also turns out that such networks are descriptive of organization at all levels, from the social networks such as those that Milgram investigated, to the worldwide web, from food chains to the Hollywood acting community, and from whole organisms to the network of reactions that take place inside the cell.

A number of books have been published recently that describe the features of such networks, including those by Albert-László Barabási (2002) and Duncan Watts (2003). For our purposes, it is only important to realize that the properties of the network are not simply derived by summing up the individual parts. Rather, the properties are emergent—that is, the network has features that can be understood only in a holistic manner, and not simply as the supervenient properties of its constitutive parts. It is the connections and interactions between the nodes, the points or centres that make up the network, that are important, not the nodes in and of themselves, in this emergence.

When this model is applied to genomic databases, it turns out that it really isn't the genes alone, or even the proteins encoded by those genes (the proteome, as it is called), but rather the interaction patterns between the proteins and various other macromolecular components in each cell that constitute the network. As a result of this realization of increasing complexity and interactivity, we are now seeing the beginnings of a holistic or network approach to living systems, where interaction maps, or interactomes, are produced, describing the ways in which all of the known proteins in a cell are linked in a network.

This change in molecular biology happened quite suddenly. One of the first major indicators that a substantive alteration of the basic assumptions had taken place was the publication of an issue of *Science*, the journal of the American Association for the Advancement of Science, devoted to networks and biology (*Science*, 2003). The papers contained in this issue examined the network approach to understanding everything from intracellular metabolic reactions to neural systems to social insects. During the succeeding years, rarely a month has gone by without the publication of interaction maps, for the fruit-fly (Giot *et al.* 2003), the roundworm (Li *et al.* 2004), or yeast (Hood *et al.* 2004). For now, much attention is being given to producing such maps.

And how do these maps appear? How are they generated? The accomplishment is a merging of microarray technology, large-scale computation, and complexity analysis. In effect, the experimenter determines all possible interactions of the proteins expressed from the genome of a particular organism. This is done by asking, in pair-wise tests, which proteins do or do not interact. 'Interact' in this case means that the two proteins bind together with sufficient strength that their pairing can send off a signal that is detected in a computer scan. This so-called two-hybrid system can be used to analyse the relationships of all of the predicted protein products from a genetic sequence.

Once the interactions have been determined, a computer is used to generate a network map that shows how every protein connects to every other protein. These maps take on the appearance of massive, three-dimensional networks, with some proteins having connections to a myriad other proteins. The presence of these highly connected nodes is emblematic of a scale-free, or small-world, network, where the properties of the network are more than the sum of the parts of that network. Such networks therefore have features that can best be described as emergent.

Most molecular biologists in this area have been content with making such maps and publishing them as such. However, in a recent paper Lee Hood, a pre-eminent molecular biologist, attempted to determine just how the emergent properties of such maps would affect the behaviour of a cell (Hood *et al.* 2004). In these experiments, the effect of a mutation of one of these key nodal proteins was examined. When the mutant was produced and its interactome determined, the ripple effect on the network was pronounced, and went far beyond what might have been predicted. Functions far removed from the mutated protein were affected. It was apparent that the mutation altered much more than the function of the altered protein. These kinds of experiments may represent the new direction that the discipline is taking: the use of emergent network properties as a model for understanding living systems at all levels.

It is certainly true that most molecular biologists are blissfully unaware of this ongoing change in paradigms. If I may speak for others in my field, we are lovers of new tools. Having the ability to use sophisticated molecular methods coupled with powerful computational software to produce interaction maps certainly qualifies as a new tool, one that comes with all sorts of bells and whistles. We biologists are busy learning how to use the tool and finding out what it can and cannot do. As a result, the notion of a paradigm shift does not even occur to us. Or perhaps it may be too soon to identify this as a true shift in the underlying approach. In the early stages of the quantum revolution, it was not at all clear that the new quantum physics was more than just a clever trick to get around insoluble problems. In fact, Einstein never accepted quantum mechanics, claiming that the model was unnecessary and that everything could be well modelled by proper use of statistical methods. This may be the case with network biology. However, it appears to me that when molecular biologists are talking about emergent properties as real rather than epiphenomenal, we are heading in a very different direction.

New Directions, New Reflections:
A Religious Reaction

Ian Barbour (1990) and John Haught (1995) have argued that the interaction between science and theology fall into one of four typologies: conflict, contrast, conversation, and confirmation. The conflict model is, of course, the one popularized by the media. The contrast model was favoured by biologists like the late Stephen Jay Gould, who proposed the idea of non-overlapping magisteria (NOMA).

Most people involved in the theology–science discourse favour a conversational model, if not one that results in mutual confirmation of some kind. Considering the previous orientation of molecular biology, it is difficult to imagine a conversation when both participants are facing in opposite directions. This does not take into account that many working scientists have presupposed, as their professional position, at least, that science has everything to say, and that theology is, at best, a 'soft' discipline.

It is my contention that science in general, and molecular biology in particular, has reached the time for it to take its proper place at the table of discourse. Science sometimes needs to sit and listen in humility. Indeed, it is the experience of paradigm shift that can humble a discipline enough to quiet the intellectual self-talk and allow other voices to be heard.

One such voice is that of Beatrice Bruteau. In *God's Ecstasy* (1997), she writes of a trinitarian reaction to science. Her purpose in this book is to show contemplative Christians why they should be excited about science. She says at the very beginning: 'My hope is to show religious readers that scientific knowledge of the natural world (which includes people and people's cultures) is important, is part of our religious life, our practice, the way we live divine life' (Bruteau 1997: 9).

As her chief metaphor she uses the idea that the Trinity is, in fact, an interacting system, a symbiotic system, and a community. All of these features are, of course, part of living organisms. In the previous version of molecular biology, the reductionist dictum would consider these features to be epiphenomena of genome-encoded information that specified everything about that organism, whether it be yeast, fruit-fly, or human. In the new approach to biology, however, the language of networks and systems is perfectly intelligible. This does not mean that a molecular biologist reading Bruteau's book will come to an agreement with the Christian contemplative life. However, it does mean that both molecular biologists and theologians in this case are, for the first time, using very similar language. With her focus on trinitarian theology, Bruteau writes, for instance: 'From elementary particles in the atom, through atoms in molecules, molecules in cells, cells in organisms, organisms in societies, to social actions and even ideas—all of them being organised as systems—the trinitarian image, as a Many–One, as a Community, has been present and growing' (Bruteau 1997: 9).

This book, published in 1997, presages the change in molecular biology that did not really become evident until six years later. One could ask, in agreement with

Bruteau, if it is now possible to envision a coherent anthropological description, a new version of the *imago Dei*, that begins with networks of interacting molecules and emerges into humankind's interactions with the divine presence. It may be possible, I suggest, to use a kind of language at every level that, instead of rejecting the description that went before, instead incorporates that description into a richer fabric of understanding. Could it be, for example, that the interaction of humans with the Divinity and with each other is consistent with, but not reducible to, the interactome of the human cell?

CONCLUDING REMARKS

Molecular biology as a discipline has had a short and tumultuous history. It has risen to become the pre-eminent paradigm of the life sciences and, from that position, has dominated the research agenda of the last half of the twentieth century in disciplines far removed from the pure study of genes. As an example, molecular medicine is based firmly on this paradigm. Throughout this rapid rise, the goal had always been the finer and finer description of genes, leading, as Crick and others argued, to an understanding of everything at the atomic level. With the massive complexity issues that arise from the ultimate reductionist project, the Human Genome Project, molecular biology has been offered a chance for humility and introspection. The change from atomism to systems may ultimately be profound, and may result in a way not only to move through the morass of sequence data, but also to overcome the philosophically impoverished position of ontological reductionism.

REFERENCES AND SUGGESTED READING

BARABÁSI, A-L. (2002). *Linked: The New Science of Networks*. New York: Perseus Books Group.

BARBOUR, I. (1990). *Religion in an Age of Science*. San Francisco: Harper San Francisco.

BRUTEAU, B. (1997). *God's Ecstasy: The Creation of a Self-Creating World*. New York: Crossroad Classic Publishing Company.

CAIRNS, J., STENT, G., WATSON, J. (1992). *Phage and the Origins of Molecular Biology*. Cold Spring Harbor, N.Y.: Cold Spring Harbor Laboratory Press.

COOK-DEEGAN, R. (1989). 'The Alta Summit, December 1984', *Genomics*, 5: 661–6.

CRICK, F. (1966). *Of Molecules and Men*. Seattle: University of Washington Press.

GIOT, L., *et al.* (2003). 'A Protein Interaction Map of *Drosophila melanogaster*', *Science*, 302: 1727–36.

HAUGHT, J. (1995). *Science and Religion: From Conflict to Conversation*. Mahwah, N.J.: Paulist Press.

HOOD, L., HEATH, J. R., PHELPS, M. E., and LIN, B. (2004). 'Systems Biology and New Technologies Enable Predictive and Preventative Medicine', *Science*, 306: 640–3.

JUDSON, H. (1996). *The Eighth Day of Creation: Makers of the Revolution in Biology*, expanded edn. Cold Spring Harbor, N.Y.: Cold Spring Harbor Laboratory Press.

LI, S., *et al.* (2004). 'A Map of the Interactome Network of the Metazoan *C. elegans*', *Science*, 303: 540–3.

MILGRAM, S. (1967). 'The Small World Problem', *Psychology Today*, 1: 60–7.

MONOD, J. (1971). *Chance and Necessity*. New York: Alfred Knopf.

——(1972). 'Tout ce qui est vrai pour le Colibacille est vrai pour l'éléphant'; <http://www.pasteur.fr/infosci/archives/mon/im_ele.html>.

POLANYI, M. (1968). 'Life's Irreducible Structure', *Science*, 160: 1308–12.

Radiation Effects Research Foundation (2005). <http://www.rerf.or.jp/>.

SCHRÖDINGER, E. (1944). *What is Life? The Physical Aspects of the Living Cell*. Cambridge: Cambridge University Press.

Science (2003). *Science*, 302.

SNYDER, M., and GERSTEIN, M. (2003). 'Defining Genes in the Genomic Era', *Science*, 300: 258–60.

WATTS, D. (2003). *Six Degrees: The Science of a Connected Age*. New York: W. W. Norton and Company.

WOESE, C. (2004). 'A New Biology for a New Century', *Microbiology and Molecular Biology Reviews*, 68: 173–86.

CHAPTER 12

EVOLUTIONARY THEORY AND RELIGIOUS BELIEF

JEFFREY P. SCHLOSS

If there is ever a time in which we must make profession of two opposite truths, it is when we are reproached for omitting one.

Pascal, *Pensées*

INTRODUCTION

'There is a grandeur in this view of life', Darwin famously mused at the close of the *Origin*, 'with its several powers, having been originally breathed by the Creator into a few forms or into one; and that . . . from so simple a beginning, endless forms most wonderful have been, and are being evolved' (Darwin 1963: 445). Although it has been suggested that this benediction was a disingenuous sop to Victorian sensibilities, the undeniable fact remains, as biologists soundly affirm, that there *is* a grandeur to the fecund creativity of life, and its emergent forms are indeed most wonderful. For many, the very 'breath of the Creator' is evident in the wondrously endowed cosmos Darwin helped elucidate, and faith in that Creator has been correspondingly magnified (van Till 1996; Haught 2000).

But things are not that easy. Around the time Darwin wrote the above, he mentioned in a letter to Hooker, 'What a book a devil's chaplain might write on the clumsy, wasteful, blundering, low, and horribly cruel work of nature!' (Darwin

1990: 178), and lamented more seriously to Asa Gray, 'I cannot persuade myself that a beneficent and omnipotent God would have designedly created' the contrivances of parasitoids for consuming their hosts alive (Darwin 1993: 224, cited in Ruse 2001: 130). This apparent double-mindedness is not shallow duplicity but profound ambivalence, reflecting interpretive ambiguity with the power to both rend internal coherence and polarize community. Almost immediately, there was a radical division not only between, but also within, scientific and religious communities over the implications of Darwin's ideas. Contemporary historiographic critiques of science–religion warfare notwithstanding, there is no denying that Darwinism has been seen by both advocates and critics as challenging important theological beliefs of the Abrahamic traditions. Writing for the *Encyclopaedia Britannica*, Gavin de Beer claims that 'Darwin did two things: he showed that evolution was a fact contradicting scriptural legends of creation and that its cause, natural selection, was automatic, with no room for divine guidance or design. Furthermore, if there had been design it must have been very maleficent to cause all the suffering and pain that befall animals and men' (de Beer 1974: 23). Actually, the above offers two, but suggests three impacts of Darwin—contradicting Scripture, obviating divine design, and magnifying the problem of evil. In this chapter I want to explore these three issues, at the heart of what for the past 150 years has been referred to as the Evolution–Creation Struggle (Ruse 2005), Debate (Numbers 1995), Controversy (Larson 2003), or Battle (Ratzsch 1996).

To begin with, such a depiction of dichotomous contest is a threefold misnomer. First, the 'debate' is not just over scientific and religious understandings of origins, but over the very character of the natural world. And it entails deep ambiguities in the relationship between nature and religious belief that have always existed, which are not so much caused as further illuminated by evolutionary theory. Moreover, these interpretive ambiguities exist very much *within* evolutionary theory as well. Second—and this will be a major focus of this chapter—there is not just one 'controversy' or 'struggle', but many. And almost every issue involves not a two-sided debate—evolution versus creation—but a continuum, often an entire landscape, of nuanced positions. Third, the very nomenclature of evolution and creation fails to represent not only the variable landscape of topics and positions, but also the complex explanatory hierarchy in scientific and religious understanding. Evolutionary biology and theology entail, and therefore interact at, various *levels* of interpretive scale.

Eschewing dichotomous caricatures of the issues, there is nevertheless one feud about which it is appropriate to cry, 'A plague on both your houses!' The polar extremes are not 'creationists' versus 'evolutionists', which are neither homogeneous nor discrete, much less orthogonal taxa to begin with. Rather, *within each* variable domain there are, on the one hand, those who are persuaded that the two perspectives are utterly irreconcilable (Dawkins 2003; Johnson 1993, 2000), and on the other hand, those who sanguinely believe that there is neither serious tension nor potentially fruitful interaction, because there is no substantial overlap (Miller 1999; Gould 2002). The premise of this chapter is that there is profound intersection between theology and evolutionary theory, entailing both unresolved ambiguity and under-exploited opportunity.

LEVELS OF INQUIRY

Evolutionary Scale

There is an epistemically ascending scale of theologically significant propositions in evolutionary theory. First is the set of observational data, or the 'text' of nature. In the fossil record, exploration of the tropics, and other ways, the eighteenth and nineteenth centuries discovered, in a sense, previously hidden manuscripts that forever changed our sense of the authorized version of nature's text. Notwithstanding that 'facts' are seen through the interpretive lens of theory, several startling and theologically significant propositions have come to be accepted as fact. The idea of deep time, or a 14-billion-year-old cosmos, calls into question certain readings of Genesis and the very idea of a Designer instantaneously creating life through direct causes (Haught 2000). It also entails a temporal humbling of humanity's place in creation, analogous to the Copernican spatial humbling. The primordial nature of death and suffering challenges dominant Augustinian, if not biblical, notions of an initially perfect creation. The observation of continuing creation raises questions of initial perfection and God's rest, and the massive cataclysms and possible progressive nature of such change harbour significant theological implications. Last, and most significant, is the notion of common descent. While in Darwin's time this proposition had the status of a theoretical inference from the data, it is now itself regarded as a historical 'fact'. The implications of common descent for anthropology and ethics are profound and widely debated not only between, but also within, scientific and theological communities (Rachels 1990).

The next level involves 'exegesis' of what the text is telling us about origins. Of course the radical exegetical claim of Darwinism is that there is an entirely adequate naturalistic explanation for the origin of organic adaptation and biotic diversity: variation and selective replication are understood as fully competent to do the work previously believed to require divine design. With the removal of this last impediment, biology is 'scientized', and naturalistic mechanisms achieve the status of not only warranted inference but also requisite starting assumptions for explanations of life (Dewey 1910). Indeed, not only is natural selection universally accepted as a mechanism of evolutionary diversification, it is also widely presumed to explain the historical 'alpha and omega' of evolution: life's origin and the recent emergence and nature of self-aware, rational, moral creatures. This is the case even though we presently lack comprehensively accepted proposals and there is some criticism within the scholarly community about whether these phenomena lie within the bounds of

Darwinian (though not naturalistic) explanation (e.g., see Gould 1991: 455 for biogenesis; Dupré, 2001 for human nature).

Third, it is not just generic naturalism, but the interpretive level involving the specific *kind* of naturalism in neo-Darwinism, that may be most significant. The former makes a Designer-God unnecessary, the latter seems to make one untenable—not just in the eyes of anti-evolutionists (Johnson 1995) or evolutionary materialists (Dawkins 2003; Dennett 1995), but even in the eyes of evolutionary theologians (Haught 2000, 2003). Why so? The integration of selection theory with genetics by the Modern Synthesis is taken by (ambiguously entitled) neo-Darwinian interpretations to entail three distinctives. First, the nature of variation and differential replication is construed to entail random mutation and blind selection, or contingency. Therefore the process of evolution is not just undirected, but disteleological, much like a drunk stumbling down the street (Gould 1996). Second, competitive struggle is not only the outcome, but the driving force behind creation. And even if co-operation and symbiosis play a larger role in evolution than some allow, their ultimate effect is to confer competitive reproductive advantage. Third, there is a particular kind of reductionism involved in neo-Darwinian naturalism. It is not just the explanatory reductionism of viewing genes as the units of variation and selective transmission, but organic life itself is understood as serving the *telos* of the gene, or being 'survival machines for genes' (Dennett 1990: 59; Dawkins 1995). 'DNA neither knows nor cares. DNA just is. And we dance to its music' (Dawkins 1995: 133). Indeed, for some, these three interpretive perspectives combine to yield an ultimate meta-interpretive deciphering of organismic reality: 'What's in it for me is the ancient refrain of all life' (Barash 1977: 167). There is considerable scientific disagreement about the first three points and substantial debate over the last, as well as a wide range of possible theological responses. The theological implications of neo-Darwinism and its internal controversies are therefore far more profound than just the issues of common descent or naturalistic origins.

Fourth, there is an epistemology that is not strictly entailed by, but is made possible by, evolutionary biology. This is evolutionary scientism—the view that only scientific understanding constitutes genuine knowledge, and given Darwinism, questions of ultimate purpose have no answers that qualify as knowledge. Richard Dawkins points out that we may rightly ask the temperature or colour of many things, 'but you may not ask the temperature question or the colour question of, say, jealousy or prayer'. Similarly, then, we may ask the purpose of a bicycle or another clearly designed artefact, but 'the Why question...when posed about a boulder, a misfortune, Mt. Everest, or the universe...can be simply inappropriate, however heartfelt' (1995: 97). John Haught refers to this refusal or inability to ask what it all *means* as 'a kind of "cosmic literalism" stuck on the surface of nature', analogous to a biblical literalism that misses the depths of Scripture (2003: p. xiv). Quite ironically, and quite revealingly, Dawkins's very examples are dramatically illustrative of this point, as one wonders in what sense it is 'simply inappropriate' to speak of being green with jealousy or lukewarm in prayer. Such metaphors of interiority make no scientific sense; but that only reflects the limits of science, and those who circumscribe the domain of appropriate questions to what is answerable by science.

Theological Scale

Theological characterizations of the origin and nature of life scale in an analogous fashion to the above evolutionary perspectives. Because I will explore a number of these issues more deeply in subsequent sections, I will provide only a brief outline here. First—and most prominent in many American religious responses to evolutionary theory—is the interpretation of the biblical account(s) of origins. Notwithstanding the profound differences between constituencies who read the text quite literally and those who encounter it largely theologically, even a modest reading of historical (e.g. a Fall) or anthropological (e.g. *imago Dei*) content in Genesis will not avoid interaction with evolutionary theory.

Second, there are several issues that emerge not from the biblical account of origins, but merely from what might be considered generic understandings of God as Creator. The issue of design, and more generally of creation's testimony to divine purpose, the problem of evil, and the question of divine action are all provoked by evolutionary theory. Third, there is a range of issues in biblical and systematic theology that are not strictly coupled to views of origins, but are nevertheless related to evolutionary theory. Theological anthropology (human uniqueness, original sin), biblical ethics (agapic love, naturalistic morality), and the grounding of salvific and eschatological hope may be challenged or enriched, but may not escape the implications of evolutionary theory.

Finally, there is the issue of religious epistemology. This involves more than just proposing a counterpoint to evolutionary scientism, i.e. asserting the importance to human flourishing of religious in addition to scientific understanding. Indeed, a number of evolutionary materialists are happy to concede that religious ideas make important contributions to human well-being, even though such ideas are false—'useful fictions' (Ruse 1994) or practical untruths (Wilson 2003). The point is not the metaphysical assertion that religious belief is fictitious (which is not an entailment of evolution). The deeper issue is the epistemological question of what the mind is for, what it is equipped to 'know'. From a strict—one might say fundamentalist—Darwinian perspective, the brain, like every other organ, exists to serve the gonads (Dennett 1990; Wright 1997). The brain functions so that we *tend to construe as true* what is reproductively efficacious—true or not. Traditional religious epistemology affirms the ostensible opposite: the mind, indeed the person, exists for the very purpose of knowing and loving the truth, itself personal, even at the possible expense of radical reproductive relinquishment. Of course there is currently a provocative flourishing of scientific and philosophical debate over reductive and adaptationist explanations of mind, and this constitutes both a need and an opportunity for engagement by theistic metaphysics (Plantinga 2002; Haught 2003).

BIBLICAL VIEWS OF ORIGINS

Although approaches to interpreting scriptural texts on origins and relating them to science vary along a continuum, there are, if not two distinct poles, at least two kinds of issues involving interaction between scriptural and evolutionary accounts of origins.

Historiography

One issue involves the relationship between the Genesis story and the *actual events* that occurred when earth, life, and humanity came to be. There is a continuum here between young earth creationism's commitment to a 10,000-year-old cosmos, special creation of individual taxa, Edenic Adam, and Noachian flood (with versions that do and do not assert confirmation by science), to 'progressive creationism's ancient earth and evolved biota, punctuated by episodes of creative intervention and the special creation of humanity directly from dust, to the more modest and recent papal affirmations of Darwinian evolution, along with divine involvement in creation of the human spirit, to several fully evolutionary understandings with varying accounts of divine influence (see Table 12.1).

Table 12.1. Major historiographic approaches to creation and evolution

	Age of cosmos	Noachic flood	Common descent	Historic Adam	Divine influence	Scientific detection
Ascientific YEC	<10k	Yes	No	Yes	Yes	No
Scientific YEC	<10k	Yes	No	Yes	Yes	Yes
Progressive Creationism	>13 B	No	No	Yes	Yes	Yes
Intelligent Design	–	–	–	–	Yes	Yes
Evolutionary Creationism	>13 B	No	Yes	Yes	Yes	No
Theistic Evolution	>13 B	No	Yes	No	Yes	No
Evolutionary Theism	>13 B	No	Yes	No	No	No

Notes: YEC: Young Earth Creationism.
Ascientific YEC, i.e. 'appearance of age' or gap theories.
Scientific Creationism, i.e. 'flood geology'.
Divine Influence: Evolutionary history has been providentially guided by means other than establishment of initial conditions, though not necessarily by interventionist or detectable action.
Scientific Detection: posits empirical evidence confirming a reading of Scripture or understanding of God's action in creation, typically coupled with interventionist understandings of divine action.
Nomenclature for the last three positions is variable.

Clearly, *any* interpretation that involves an attribution of divine action (e.g. *creatio ex nihilo*) or a historical fact (e.g. monophyletic origin of humanity) opens the door to concord or discord with evolutionary theory. This violates the terms of Galileo's truce for science and religion, which asserts that the Scriptures teach 'how to go to heaven, not how the heavens go'. But any proposal for peace by separation is not adequate for an understanding of the biblical tradition that posits, to continue the play on words, a sacred history of 'how heaven came to us'. The question for the Abrahamic traditions is not whether, but in what way, the Scriptures teach that God has interacted with history. Clearly, the more literal the readings and the more ancient the history, the greater the discord with evolution.

In light of this, and given the rich history of allegorical theological interpretation, it is tempting to view biblical literalism about origins using the Weberian concept of religious routinization—that is, religion that has lost contact with profound, sacred meanings and substituted reified but nominally religious understanding (Schloss 2005). Haught (2003) insightfully suggests that both creationism and evolutionary materialism represent shallow literalisms that fail to see deeper meanings. Although these criticisms may have merit, they fail to represent the complexity of the issues, and Haught's point even involves a patent dis-analogy. While evolutionary materialism clearly entails a rejection of deeper interpretations of life, religious commitment to literalist historiography—involving a gradient from affirming the Resurrection, to the Davidic reign, to wilderness wanderings, Abrahamic historicity, or an Edenic Adam—is often predicated precisely on *accepting* the deeper significance of an event as allegory, type, example, sign, or down payment on a promise with crucial theological significance. Historical literalism(s) may involve scriptural misinterpretation, but it is not necessarily shallow interpretation, unresponsive to allegorical or theological meanings.

Perhaps partly in this context Alvin Plantinga suggests that accepting a young earth on the basis of scriptural testimony, in the face of scientific evidence to the contrary, ought not be automatically judged as 'pathological or irrational or irresponsible or stupid' (Plantinga 1991: 15). Michael Ruse responds that even if a religious believer fully accepts Darwinism, if she could allow the possibility that someone else could reasonably reject Darwinism on the basis of Scripture, that belief system is 'irresponsible and stupid' and 'ought to be rejected' (2001: 59). Plantinga is not suggesting that avoiding evidential ambiguity by substituting bad science for legitimate science could be warranted by religious belief. The more nuanced claim is that in cases of conflict between properly conducted empirical science and religious belief, given the fallibility of all human understanding, and an adequate internal coherence and epistemic foundation of religious belief, it may not be irrational to accept religious understanding. This entails living with the tension of unresolved conflict in evidence.

My point is not to emphasize, much less to authorize, extreme literalism, but to locate this issue—along with the range of more moderate historiographies involving evolution and religion—within an epistemology that promotes, rather than forecloses, dialogue. First, it is neither productive nor warranted to withhold the presumption of rationality: in two decades of interacting with those holding wildly varying beliefs about evolution and creation, it is clear that the more generously concerns and reasons are respected, the more genuine is discussion and assessment of one's own and other positions in light of both theology and science. Inclusion in such dialogue constitutes an essential opportunity for interpretive checks and balances.

Second, the life of faith cannot avoid encountering dissonance between the observed world and religious hope or revelatory promise. Honestly facing tension, and on some occasions affirming one's understanding of revelation—not reflexively, but deliberatively and in the midst of 'fear and trembling'—is risky, but not intrinsically irresponsible. Indeed, wrestling with ambiguity and, like Jacob, perhaps finding deepened faith in the very experience of failure to prevail, may constitute a crucial portal to religious maturity.

It is the lack of this mature willingness to walk with an epistemic limp, so to speak, that may account not so much for biblical literalism as for the more injurious attempt to compress ambiguity out of experience by reconstructing science to fit a reading of Scripture. Ironically, an entirely analogous error is made in Ruse's (and some theologians') insistence that religious beliefs must be entirely trimmed to fit prevailing scientific understanding or circumscribed to avoid any overlap. Thus the problem shared by evolutionary materialism and creationism is not so much strict literalism, as naïve concordism: preferring wooden harmonization to the ongoing quest for dynamic coherence. This insistence on easy concord in the face of theological questions raised by the natural world, from evolutionary naturalists and creationists to Job's wife and friends, typically results in two extremes: exhortations to 'curse God and die', or attempts to defend God with 'proverbs of ashes'.

Finally, not all is ambiguous. Clearly the Scriptures view creation as a historical process, whether or not they provide an account of its history. In the beginning all things came to be through the *Logos*; creation was not instantaneous but involved God's progressive ordering, blessing, and empowering; there have been disruptively profound changes in humanity's relationship to God and nature since humans arose. These claims are neither modest nor incoherent in evolution's light. The more abiding problems with evolution and the biblical text end up connecting this history to anthropology.

Anthropology

There are three fundamental issues in the relationship between evolutionary anthropology and biblical (primarily Genesis) accounts of humanity: the identity and origin of Earth's first humans, human uniqueness or *imago Dei*, and the Fall. Surveying, much less assessing, the varied theological reflections on these topics is not remotely within the scope of this chapter, but I will comment briefly on recent evolutionary ideas that have special relevance.

First, after considerable debate over monophyletic versus polyphyletic and multiregional versus African accounts of origins, there is substantial agreement that modern humans arose from a single ancestral population in Africa. Indeed, humans are genetically quite closely related, and mitochondrial DNA and Y-chromosome studies have even been represented as indicating that humanity may have arisen from individual male and female progenitors, though this is a misinterpretation of the data (Ayala 1998). What about the divine instantiation of a soul at the origin of humanity? In one sense, this is entirely contrary to evolutionary theory and the scientific enterprise itself, which invokes no miracles for explanation. In another sense, it entails a metaphysical question which the methodological naturalism employed by science claims not to answer, but merely to disregard in seeking explanations. Would instantiation of a soul in a fertilized egg be contrary to embryology? Not unless one tried to make explanatory use of it in embryological development. Similarly, one could posit the divine implantation of a soul in the course of evolution, as long as it did not represent a causal explanation *for* evolution. However, such a causally inert postulate is not very interesting, reflecting the criticism that 'science does not contradict religion; but it makes it increasingly improbable that religious discourse has any subject matter' (Dupré 2003: 60).

The issue of human uniqueness has been fascinatingly revisited by recent evolutionary theory. Qualitatively distinctive characteristics attributed to humans include language, morality, and reputationally mediated indirect reciprocity; culture and extra-somatic storage of information; degrees of social co-operation—even altruism—unique to mammalian biology. This has provoked the development of a hierarchical theory of selection by which behavioural phenotypes are understood to reflect not merely genes by environmental interaction, but also the top-down influence of emergent information that is irreducible to genes. Richard Dawkins makes the startling claim that 'we, alone on earth', are capable of 'cultivating and nurturing pure, disinterested altruism—something that has no place in nature, something that has never existed before in the whole history of the world' (Dawkins 1989: 201). While being an emphatic affirmation of human uniqueness, this naïve statement also raises profound scientific and philosophical, not to mention theological, questions. It reflects a neo-dualism that takes humanity out of nature, rather than impressing God's image on nature. Nevertheless, there is surely new fuel to stoke the fire of theological reflection here (Schloss 2002).

Finally, the question of a Fall remains perhaps one of the most significant sources of tension between evolution and traditional Christian theology. There are several emerging ideas in evolutionary theory that constitute fertile, though ambiguous, resources for theological integration. One is the notion of conflicting legacies of group and individual selection, resulting in a profound behavioural and affective ambivalence between egalitarian co-operation and self-seeking (Boehm 1999; Schloss 2002). Another involves genetic lag, a mismatch between fundamental desires and cognitive structures that arose in the Pleistocene period and the radically different social and physical environments we now inhabit. Finally, there is the proposal of an abrupt and recent explosion of cognition, which other brain processes have not kept up with and which entail significant dissonance (Mithen 1999). All of these ideas provide ways of understanding the profound sense of interior division, motivational ambivalence, and social estrangement that humans interpret as having lost something, reflecting radical estrangement from an idyllic, innocent past or paradisaical home that was more whole and hospitable to flourishing. However, none of these approaches is easily reconcilable with the traditional notion that what has been lost is unbroken communion with God and one another, fractured by a singular, historic wilful disobedience. The 'brokenness' or incompleteness of human life is historically embedded, but is not sinful. Nevertheless, there are proposals that affirm original sin (Hefner 1993), which apart from God's grace may subvert the eschatological healing of brokenness that our history burdens us with (Haught 2003; Edwards 1999).

DESIGN AND PURPOSE

It is widely contended that 'Darwin's great contribution was the final demolition of the idea that nature is the product of intelligent design' (Rachels 1990: 110). Moreover, in eliminating the need to invoke God as a causal explanation, Darwinism is viewed as eradicating the warrant for believing in God at all: 'Without the argument from design there is nothing very credible left of theism generally, and Christianity in particular' (Dupré 2003: 56). This strong scientistic claim as to what constitutes adequate evidential warrant for belief in God is accepted by many materialists and creationists. Ironically, though, design arguments are conspicuously underemphasized in the biblical tradition (Hebrews 11: 3), and in Christian theology there is an ongoing understanding of faith as not resting on such arguments, from Augustine to Anselm, Pascal, Kierkegaard, and Barth.

Beyond its implications for the existence of nature's Designer, evolution is regarded by some to make implausible the very notion of purpose in or for nature itself. Dawkins famously asserts that when the world is viewed through an evolutionary lens, 'there is, at bottom, no design, no purpose, no evil and no good, nothing but blind, pitiless indifference' (1995: 133). The assertion of purposelessness can mean a number of different things, often conflated. At one level 'purpose' can simply mean the target-regulated behaviour of a functional system, and the legitimacy of such teleological language in the absence of a Designer is currently debated, even though biologists speak of a trait's proximal function and an organ's ultimate fitness-enhancing purpose. At a deeper level, the purpose can entail not just the functionality of evolution's organismic actors, but the aesthetic, moral, or sacred meaning—the *telos* in the play itself. Dawkins views this notion as a 'universal delusion' (1995: 96), and Hans Jonas observes that the Darwinian 'combination of chance variation and natural selection, completed the extrusion of teleology from nature. Having become redundant even in the story of life, purpose retired wholly into subjectivity' (1982: 46).

At a deeper level still, beyond purpose *in* the products or process of evolution, there is the question of purpose *for* evolution. As using a rock for a paperweight entails but is not inferable from properties of the rock, apart from knowledge of the agent employing it, so the question of purpose *for* the cosmos includes, but may not be answerable by, its evolutionary character alone. If God's purpose for nature is to promote his pleasure, or glory, or that it should culminate in beings who love and are loved by him, this remains a theological question. It may be informed by science, but is subverted by scientistic concordism of both evolutionary materialists and creationists. *Contra* Dawkins, it may indeed make sense to ask if the cosmos has purpose, though it does not make sense for science to provide the answer. Nor, therefore, does it make sense to defend purpose at the beachhead of design.

Contemporary responses to the above issues tend to entail three kinds of approaches.

Supernatural Design: The Actors

One approach involves the attempt to rehabilitate traditional design arguments by rejecting naturalistic evolutionary accounts for the origin of living organisms. Called 'intelligent design theory' by its proponents and 'intelligent design creationism' by its critics, it is strictly speaking neither a theory (by virtue of explanatory consilience or fruitfulness), nor creationism (by virtue of affirming the historicity of Genesis or other scriptural accounts of origins). In light of its historiographically minimalist but crucial commitment to interventionist accounts of origins, it might best be regarded as intelligent design supernaturalism.

There are three fundamental components to the intelligent design (ID) programme. First, it argues that the question of whether something has been designed is wholly legitimate and that it can be posed in a way that is rigorously formalizable. Second, in light of their claim that design hypotheses are falsifiable by empirical data, and because disciplines such as anthropology, cryptography, and the Search for Extraterrestrial Intelligence (SETI) routinely do so, ID proponents maintain that their endeavour ought rightly be called 'design science'. Third, ID asserts that there is adequate evidence for a design inference in the origin of life, based on the explanatory failure of Darwinian accounts.

These issues have been amply argued, including in this volume, but because so much of the debate has been primarily philosophical and scientific in focus, and so polemical in tone, I will make several mediating comments related to science–religion dialogue. First, the *question* of design has for centuries been, and surely still is, a legitimate one to ask. In-principle attempts to rule out the very question from serious intellectual consideration not only entail an appropriate opposition to anti-evolutionism, but may also reflect a rejection by scientific reductionism of any kind of agency: there are those who strenuously argue that scientific naturalism in general, and the Darwinian algorithm in particular, demands rejecting the notion that intelligent agency exists *at all* in nature, or if it does, that it may be employed as an explanation (Cziko 1997). On the other hand, ID's method for detecting design—an 'explanatory filter' that concludes that if something is not explainable by law or chance, then it is designed—is flawed. What this eliminative filter rules out is material causes, hence detecting not intelligent design, but miracles or supernatural (at least immaterial) agency. Even if we were certain that something belonged at the last level (and Hume's critiques apply to such certainty), the eliminative rather than attributive operation of the filter does not enable it to distinguish between genuine design and the undesigned products of a drunken demon (Ratzsch 2001). In addition to scientifically problematic false positives, there are theologically troublesome false negatives. Because a design inference requires interventionist abridgement of natural processes, it is incapable of detecting design *in* natural processes. If there is a cosmic Designer, he must occasionally break the rules by pulling cards from his pocket and insinuating them into the deck of nature, rather than providentially arranging the deck for winning hands in a fairly dealt game (Dembski 2001; Murray 2003, 2006; see Chapters 42 and 43 below).

Second, insisting that the question of design be called science—given the contentious polarization by the politics of what is taught in American public schools—ironically distracts and sabotages open discussion of the question itself. There are no a priori reasons for rejecting the possibility that the cosmos contains divinely created entities that are incapable of having arisen by natural means alone. If so, it is reasonable to expect that it would be distinguishable from a counterpart universe without such things (Dawkins 2003). Whatever the enterprise is called (and there is appreciable warrant for just calling it metaphysics), the outcome of such discussion is presumably important to both science and theology (Dawkins 2003; Plantinga 1991). Insistence that it either be authorized as science or delegitimated as religion is counterproductive to both enterprises.

Moreover, proposing a Creator as a scientific conclusion may actually constrain the potential contribution of theology to science. If 'design' were posited not as a theologically untainted inference from gaps in natural regularities, but as a starting assumption explicitly informed by theological understandings of purpose, it could serve as a well-spring of investigable hypotheses about aspects of the natural world that themselves have been compressed to fit reductionistic pre-commitments: e.g. the nature of altruism, morality, religious belief, and evolutionary progress.

Third, and most contested, is the claim that irreducible complexity (IC)—a functional system that requires all its parts in order to function at all—cannot be explained by Darwinian mechanisms and therefore attests to ID. Other than identifying the seminal logical error of conflating a system that loses all function by any subtraction with a system that cannot have arisen by successive functional addition, I will not review the debates over the IC claim.

Of course the ID agenda provokes standard debates over God-of-the-gaps thinking and interventionist understandings of divine action, debates which, intellectual fashion notwithstanding, themselves reflect metaphysical questions of causal continuity that are far from fully settled. But underlying the IC $\rightarrow$ ID debate is also a significant issue of divine historiography. On the one hand, IC posits God leaping over historical constraints to create things fully formed. On the other hand, numerous critics posit no need for divine interaction in history, given the fundamental theological premiss of creation's fully endowed formational resources (van Till 1996). What both extremes do not address is the profound ambiguity of a creation that is *not* sufficiently endowed for historical development that averts tragic incompleteness, but does not magically leap over finitude to attain its destiny by divine transmutation. The persistent biblical notion is that God's grace transforms history redemptively, yet at every point is embedded in and somehow subject to the constraints of history. This mystery is profoundly enriched, though far from fully funded, by the evolutionary perspective of an unfolding cosmic history.

Fine-Tuning: The Stage

An alternative response to design has been to affirm fully that living organisms, the 'actors' in the evolutionary play, have arisen through natural regularities described by Darwin. But the prerequisites for biological evolution—the extraordinary precision of the physical-chemical 'stage'—suggest that the cosmos has been designed to support the evolutionary drama. Such arguments do not posit abridgements of scientific laws, but maintain that the finely tuned character of nature's endowments requires, or at least suggests, an underlying intelligence.

There are two approaches to this line of thinking. One emphasizes fine-tuning of the *necessary* conditions for the origin of life. Cosmological fine-tuning maintains that the fundamental physical constants of the universe could not be minutely different from what they are for life to be possible (Barrow and Tipler 1988). Biochemical arguments emphasize the unusual fitness of the pre-biotic chemical environment to support life (Barrow *et al.* 2006). Geological perspectives claim that the planetary conditions necessary to support intelligent life are unusually rare (Ward and Brownlee 2003).

A second approach emphasizes the *sufficient* conditions to make likely, even inevitable, something like the evolutionary history we have (Conway Morris 2004). Such a perspective challenges assertions of radical contingency that depict evolutionary history as a drunken stumble, which, if repeated innumerable times, would never wander into this point again (Gould 1996). While 'sufficiency' arguments are consistent with the notion that a historical trajectory for evolution is providentially 'built in' to nature, this does not entail the stronger claim that evolution itself is purposive. Directional inevitability is not teleology.

All of these ideas involve providential 'deck-stacking' rather than interventionist 'double-dealing' approaches to design. And for all, design represents a possible metaphysical implication, but is not invoked as a scientific explanation. All are also vigorously debated, though disagreement tends to emphasize the plausibility of divergent accounts rather than reciprocal charges of science's conspiratorial hijacking by ideology. The ambiguous regress of claims and counterclaims may itself reflect the Pascalian affirmation of 'two opposite truths' with which this chapter began: 'All appearance indicates neither a total exclusion nor a manifest presence of divinity, but the presence of a God who hides himself. Everything bears this character' (2002: 90).

Evolutionary Theology: The Play

The above approaches focus on design of the actors or stage, but do not address the plot or purpose of the play itself. Many post-Darwinian theologies attempt just that. Michael Ruse (1997) observes that Darwin did not actually demolish natural theology, but that his work resulted in the focus shifting from the ostensibly designed *products* of evolution to the progressive *process* of evolution. But there was also a general shift in approach, away from natural theology to theology of nature: interpreting the theological meaning of natural history rather than inferring God's existence or attributes as a causal explanation for natural history.

There is a continuum even within theology of nature. On the one hand, the Barthian eschewal of natural theology modestly affirms that creation can provide 'parables of the Kingdom'. A homely example may be C. S. Lewis's observation that nature cannot teach us that God is glorious, but that, having learned that elsewhere, it can furnish images of glory (Lewis 1960). On the other hand, others—including recent evolutionary cognitive theories of supernatural agent attribution (Atran 2004)—affirm that nature does 'teach' or directly evoke religious intuitions. The primordial experience is, 'The heavens declare the glory of God', by way not of design argument, but of fundamental perception.

This continuum exists in interpreting the evolutionary process. As mentioned at the beginning of the chapter (and unrelated to fine-tuning), the very fact that the cosmos has the marvellous capacity to generate life is taken to reflect the immense intelligence and benevolence of a God who extravagantly endows creation. Richard Dawkins observes that if there were an intelligence underlying the algorithmic process that generates biological complexity, it would have to be massively more complex than the sum total of all the products themselves (Dawkins 1986: 141). For this very reason he parsimoniously rejects God, as explaining 'precisely nothing'. But evolutionary theology does not *conclude* God's existence as an explanation of nature's endowments. Having 'learned that elsewhere', it views nature as reflecting his providence and wisdom. ID advocate Philip Johnson criticizes this as merely putting 'a theistic spin on the story provided by materialism' (2000: 100). Such a view ironically affirms the scientist premiss that meaning is just a gloss over the ultimate, reductive reality of how matter behaves. Evolutionary theology in general (in fact, common experience) rejects this in recognizing meaning as a *deeper* understanding of, not a mere varnish on, the cosmos.

The discussion doesn't stop at nature's fecund production of actors, though. It goes on to reflect theologically on the plot of the play, primarily in two ways. First, there is a tradition that proposes theological significance for evolutionary progress (Ruse 1997). Progress is more than mere directionality, entailing change in a valued direction. Discerning progress thus involves empirical questions of evolutionary trends and subjective questions of valued ends. Given the theological implications and undeniable social misuse of the progress idea, its very framing has been vigorously opposed: 'Progress is a noxious, culturally embedded, untestable, nonoperational, intractable idea that must be replaced if we wish to understand evolutionary history' (Gould 1988: 319). While the subjective criteria for progress do not contribute to evolutionary explanation, given sufficient constraints on possibility space and a specified set of evaluative criteria, it is not true that progress is intractable. It turns out empirically that a constellation of taxonomic and life history traits has increased over evolutionary time: species diversity, trophic depth, homeostatic control, sensory acuity, behavioural and locomotor freedom, various measures of complexity at cellular, organismal, and social levels, body mass and lifespan, per capita parental investment, and capacity for intersubjective awareness and inter-organismal attachment. Hans Jonas (1982) refers to this as an increase in biotic potency involving a deepening of organismal teleology, or an 'ascent of soul'. Even without the interpretive metaphors, by widespread standards of value—including those posited by Darwin himself—this trajectory constitutes progress. Moreover, it is not just narrowly anthropomorphic or ethnocentric, but truly biocentric: one could say that evolution entails a process by which the cosmos 'might have life, and have it more abundantly'.

True and even beautiful though this is, the observation can be theologically saccharine if it fails to recognize the profound ambiguity of this kind of 'progress'. What increases is potency, not goodness. Conflation of the two may even reflect the demonic. With increased potency or deepened organismal *telos* comes heightened capacity for fulfilment *and* tragedy, for (moral or natural) goodness *and* evil.

A further, riskier step is to suggest that the ambivalent contrasts of such progress reflect a larger purpose, a reason behind or beneath the evolutionary drama itself. That reason, or *telos*, according to many evolutionary theologians working in the process tradition, is 'the maximizing of cosmic beauty' (Haught 2000: 130). This *telos* is emphatically consonant with the evolutionary pageant, while comprehending its disparate contrasts within a unifying rubric of beauty. But it entails a comparable if not graver moral ambiguity, being a mere reformulation of the progress ethic in aesthetic terms, along with the concomitant magnification of the hideous. Moreover, it risks deriving notions of divine purpose and human hope from the extrapolation of nature's trajectory, rather than revelatory promise of its redemption.

An alternative approach, to invoke a distinction made at the beginning of this section, affirms that there is a purpose *for* evolution, but no discernible purpose *in* evolution. The evolutionary story or 'narrative character of nature' (Haught 2003: 67) is not, in the last analysis, a drama that can be read for theological instruction. The nature or existence of a divine purpose for the cosmos, involving the reconciling of all things to God, is not discernible by science or by theological reflection on the evolving world that science illuminates. But this does not mean that, 'having learned purpose elsewhere', we cannot both understand this world in terms of God's purposes and enrich our understanding of these purposes in light of the evolving world they engage. One way to view this is that what is being maximized as evolution amplifies life is precisely ambiguity—the unresolved intensity of life's precarious teleonomic perch between destiny and tragedy (Jonas 1982). This entails an intrinsic escalation in neither goodness, nor beauty, nor love, but in the capacity—and the need—to be saved by and for all three.

EVOLUTIONARY EVIL

Ambiguity is one thing. Straight out, unmitigated evil is another.

The evolutionary process is rife with happenstance, contingency, incredible waste, death, pain, and horror.... Whatever the God implied by evolutionary theory and the data of natural history may be like, He is not...a loving God.... Careless, wasteful, indifferent, almost diabolical. He is certainly not the sort of God to whom anyone would be inclined to pray. (Hull 1991: 486)

Of course, two things are going on in this prototypic passage: interpretive natural history and theological reflection. It is not entirely clear which drives which, but I want to comment briefly on each insofar as they represent crucial issues in evolution and religion.

With respect to natural history, three things are salient. First, what a mysterious, anguishing but undeniable irony it is that the same world that gives rise to this invective also inspires 'a grandeur in this view of life'—often in the same person. Nature red in tooth and claw is also green in bloom and bough: fecund, vital, extravagant, tender, beautiful. We are back to ambiguity again, and Pascal's opposite truths. Second, in what demonstrable sense are the above descriptors 'true'? A number of them—contingency, waste, horror—are judgements that reflect assumptions about, rather than suggest answers to, metaphysical questions. To choose one, distinguishing between waste and, on the one hand lavishness, and on the other hand efficiency, is a question of teleology—requiring knowledge of what something is 'for'. 'Think of all those waves, for all those centuries, just going to waste', comments the surf movie *Endless Summer* upon discovering the 'perfect wave' in South Africa. Every work of art is every engineer's waste. But even from an engineer's perspective, nature's economy wastes virtually nothing, and selection can be seen as an entrepreneurial innovator, continually replacing slackers. (Ironically, who would want to pray to an efficiency expert God either? Elusive opposites: 'We played the flute and you did not dance; we sang a dirge and you did not mourn.' No God read directly from the text of nature is the loving God of biblical faith.)

Third, and most important, is the issue of natural evil actually posed or exacerbated by *evolution*, as opposed to being a feature of the world independent of evolutionary process? This is not to minimize the profound theological problem of natural evil, but to question in light of this volume's focus whether it is coupled in any meaningful way to the scientific theory of evolution. In fact, it may represent an attempt by atheology to piggyback on the cultural authority of science, analogous to creationism. At one level the answer is easy: the sum total of all the death in the world is precisely one per creature. This is the case whether evolution is true or false, and it did not take Darwin to make it into a theological issue. At a deeper level, though, it is not the magnitude but rather the role of death and competition that seems to change profoundly in light of Darwinian theory. The vexation is that they represent not *post hoc* impositions on, but the very driving force of, creation. Dark art indeed. But not quite so fast. Death actually is *post hoc*—temporally and ontologically subsequent to life. And evolution's driving force is not death, but variation and differential repro-duction, which would exist even for immortal creatures with unlimited resources. In the last analysis, the problem is the ancient enemy—not evolution, but death itself. And it is possible that evolution may even help furnish a theodicy, by revealing a creative grace that brings value out of struggle.

This then brings us to the theological question of a loving God behind evolution. On the one hand, evolution does not seem to raise the stakes of natural evil. In fact, coupling the issue to the dilemma of extinct biota, parasitoid wasps, and 'millions of sperm' that never get to fertilize an egg (Hull 1991) does seem to be straining at gnats, or rather gametes and hymenoptera, while camels of monstrous twentieth-century evil are on the hoof. On the other hand, does evolution challenge existing theodicies? It clearly obviates the idea that death entered the world by Adam's sin, but that Augustinian notion has hardly been a central theodicy in biblical tradition. Along those lines, one can still invoke demonic evil, and the carnage of evolution could even be viewed as the casualities of epic spiritual warfare. Other more profound theodicies of soul making, necessary evil, divine kenosis or hiddenness, eschatological redemp-tion, incarnational co-suffering—however adequate or inadequate—are all coherent in light of evolution. In fact, each one is enriched by the notion of a historically constrained yet indeterminate, incrementally evolving biotic intensification.

Finally, evolution may itself be something of a resource for theodicy. First, the Gospel's affirmation is that in God's cruciform economy he graciously turns death to life. This is true of the Resurrection, of the redemption that comes from the cross, even of biogeochemical cycling and of the evolutionary process. Not that we learn the principle of redemption from evolution, but having learned it elsewhere, we see it there—as literal and allegorical history. Second, while evolution does not increase the amount of suffering, in another sense it certainly increases its depth. To the very extent that living beings have ends, and seek and desire their attainment, and sense (and in humans know) their loss, there is suffering. The capacity for such suffering increases with the evolutionary intensification of organismic teleonomy or biotic depth. But to be able to choose such a loss, on behalf of a more valued end, ushers in an ascendance of regulatory and then volitional potency—the more agency, the more capacity to suffer loss for greater gain—that culminates over the evolutionary process in precisely the capacity to love. It is not clear what love would mean if there could never be a costly gift on behalf of the beloved. The same is true of worship. David says, 'I will not offer to the Lord that which costs me nothing' (1 Chron. 21: 24). It is, at the same time, this evolved, autonomous capacity to relinquish or experience cost, that also allows us to receive: Nicholas of Cusa (1988: 692) gratefully ponders, 'How could you give yourself to me, unless you had first given me to myself?'

REFERENCES AND SUGGESTED READING

ATRAN, SCOTT (2004). *In Gods We Trust: The Evolutionary Landscape of Religion.* New York: Oxford University Press.

AYALA, FRANCISCO (1998). 'Biology Precedes, Culture Transcends: An Evolutionist's View of Human Nature', *Zygon*, 33/4: 507–23.

BARASH, DAVID (1977). *Sociobiology and Behavior.* New York: Elsevier.

BARROW, JOHN, and TIPLER, FRANK (1988). *The Anthropic Cosmological Principle.* Oxford: Oxford University Press.

BOEHM, CHRISTOPHER (1999). *Hierarchy in the Forest: The Evolution of Egalitarian Behavior.* Cambridge, Mass.: Harvard University Press.

CONWAY MORRIS, SIMON (2004). *Life's Solution: Inevitable Humans in a Lonely Universe.* Cambridge: Cambridge University Press.

——FREELAND, STEPHEN, and HARPER, CHARLES (2006, forthcoming). *Fitness of the Cosmos for Life: Biochemistry and Fine-Tuning.* Cambridge: Cambridge University Press.

CUSA, NICHOLAS (1988). *Nicholas of Cusa's Dialectical Mysticism: Text, Translation, and Interpretive Study of* De Visione Dei, trans. Jasper Hopkins. Minneapolis: Arthur J. Banning Press.

CZIKO, GARY (1997). *Without Miracles: Universal Selection Theory and the Second Darwinian Revolution.* Cambridge, Mass.: MIT Press.

DARWIN, CHARLES (1963). *On the Origin of Species.* Norwalk, Conn.: The Heritage Press.

——(1990). *The Correspondence of Charles Darwin,* vi: *1856–1857.* Cambridge: Cambridge University Press.

——(1993). *The Correspondence of Charles Darwin,* viii: *1860.* Cambridge: Cambridge University Press.

DAWKINS, RICHARD (1986). *The Blind Watchmaker: Why the Evidence of Evolution Reveals a Universe without Design.* New York: W. W. Norton.

——(1989). *The Selfish Gene.* Oxford: Oxford University Press.

——(1995). *River Out of Eden.* New York: Basic Books.

——(2003). *A Devil's Chaplain: Reflections on Hope, Lies, Science and Love.* New York: Houghton Mifflin.

DE BEER, GAVIN (1973–4). 'Evolution', in *The New Encyclopaedia Britannica*, 15th edn., London: Encyclopaedia Britannica, vii. 7–23.

DEMBSKI, WILLIAM (2001). *No Free Lunch: Why Specified Complexity Cannot Be Purchased Without Intelligence.* New York: Rowman & Littlefield.

DENNETT, DANIEL (1990). 'The Myth of Original Intentionality', in K. A. Mohveldin Said, R. Viale, and William Newton-Smith (eds.), *Modelling the Mind*, Oxford: Clarendon Press, 48–62.

——(1995). *Darwin's Dangerous Idea: Evolution and the Meaning of Life.* New York: Simon & Schuster.

DEWEY, J. (1910). *The Influence of Darwin on Philosophy, and Other Essays.* New York: P. Smith.

DUPRÉ, JOHN (2001). *Human Nature and the Limits of Science*. Oxford: Clarendon Press.

—— (2003). *Darwin's Legacy: What Evolution Means Today*. Oxford: Oxford University Press.

EDWARDS, DENIS (1999). *The God of Evolution*. Mahwah, N.J.: Paulist Press.

GOULD, S. J. (1988). 'On Replacing the Idea of Progress with an Operational Notion of Directionality', in Matthew Nitecki (ed.), *Evolutionary Progress*, Chicago: University of Chicago Press, 319–38.

—— (1991). *Bully for Brontosaurus*. New York: Norton.

—— (1996). *Full House: The Spread of Excellence from Plato to Darwin*. New York: Harmony.

—— (2002). *Rocks of Ages: Science and Religion in the Fullness of Life*. New York: Ballantine.

HAUGHT, JOHN (2000). *God after Darwin: A Theology of Evolution*. Boulder, Colo.: Westview Press.

—— (2003). *Deeper than Darwin: The Prospect for Religion in the Age of Evolution*. Boulder, Colo.: Westview Press.

HEFNER, PHILIP (1993). *The Human Factor: Evolution, Culture, and Religion*. Minneapolis: Fortress Press.

HULL, DAVID (1991). 'The God of the Galapagos', *Nature*, 352: 485–6.

JOHNSON, PHILLIP (1993). *Darwin on Trial*. Downers Grove, Ill.: InterVarsity Press.

—— (2000). *Wedge of Truth: Splitting the Foundations of Naturalism*. Downers Grove, Ill.: InterVarsity Press.

JONAS, HANS (1982). *The Phenomenon of Life: Toward a Philosophical Biology*. Chicago: University of Chicago Press.

LARSON, EDWARD (2003). *Trial and Error: The American Controversy over Creation and Evolution*. New York: Oxford University Press.

LEWIS, C. S. (1960). *The Four Loves*. New York: Harcourt Brace.

MILLER, KENNETH (1999). *Finding Darwin's God: A Scientist's Search for Common Ground Between God and Evolution*. New York: Cliff Street Books.

MITHEN, STEVEN (1999). *The Prehistory of Mind*. London: Thames & Hudson.

MURRAY, MICHAEL (2003). 'Natural Providence (or Design Trouble)', *Faith and Philosophy*, 20/2: 307–27.

—— (2006). 'Natural Providence: Reply to Dembski', *Faith and Philosophy*, 23/3 (forthcoming).

NUMBERS, RONALD (1995) (ed.). *Creation–Evolution Debates*. New York: Taylor & Francis.

PASCAL, BLAISE (2002). *Pensées*. Grand Rapids, Mich.: Christian Classics Ethereal Library.

PLANTINGA, ALVIN (1991). 'When Faith and Reason Clash: Evolution and the Bible', *Christian Scholar's Review*, 21/1: 8–32.

—— (2002). 'An Evolutionary Argument against Naturalism', in James Beilby (ed.), *Naturalism Defeated? Essays on Plantinga's Evolutionary Argument against Naturalism*. Ithaca, N.Y.: Cornell University Press, 1–12.

RACHELS, JAMES (1990). *Created from Animals: The Moral Implications of Darwinism*. New York: Oxford University Press.

RATZSCH, DELVIN (1996). *Battle of Beginnings*. Downers Grove, Ill.: InterVarsity Press.

—— (2001). *Nature, Design, and Science*. Albany, NY: SUNY Press.

RUSE, MICHAEL (1994). 'Evolutionary Theory and Christian Ethics: Are They in Harmony?', *Zygon*, 29/1: 5–24.

—— (1997). *Monad to Man: The Concept of Progress in Evolutionary Biology*. Cambridge, Mass.: Harvard University Press.

—— (2001). *Can a Darwinian be a Christian?: The Relationship between Science and Religion*. New York: Cambridge University Press.

—— (2005). *The Evolution–Creation Struggle*. Cambridge, Mass.: Harvard University Press.

SCHLOSS, JEFFREY P. (2002). '"Love Creation's Final Law?": Emerging Evolutionary Accounts of Altruism', in S. Post, L. Underwood, J. Schloss, and W. Hurlbut (eds.), *Altruism and Altruistic Love: Science, Philosophy, and Religion in Dialogue*, New York: Oxford University Press, 212–42.

—— (2005). 'Hath Darwin Suffered a Prophet's Scorn? Evolutionary Theory and the Scandal of Unconditional Love', in Charles Harper (ed.), *Spiritual Information*, Philadelphia: Templeton Press, 291–9.

VAN TILL, HOWARD J. (1996). 'Basil, Augustine, and the Doctrine of Creation's Functional Integrity', *Science and Christian Belief*, 8: 21–38.

WARD, PETER, and BROWNLEE, DONALD (2003). *Rare Earth: Why Complex Life is Uncommon in the Universe*. New York: Springer-Verlag.

WILSON, DAVID SLOAN (2003). *Darwin's Cathedral: Evolution, Religion, and the Nature of Society*. Chicago: University of Chicago Press.

WRIGHT, ROBERT (1997). *The Moral Animal: Evolutionary Psychology and Everyday Life*. New York: Vintage.

CHAPTER 13

ECOLOGY AND RELIGION

SUSAN POWER BRATTON

NEW SCIENCE, OLD RELATIONSHIPS

Ecology is one of the newer branches of the life sciences. Along with genetics, its origin as a formal field of study lies in the nineteenth century, but its academic development and expansion are products of the twentieth century, particularly the period following the Second World War. Such biological giants as Charles Darwin, Alfred Wallace, and Alexander von Humboldt investigated the relationships between organisms and their environments well before the German zoologist Ernst Haeckel coined the term 'ecology' in 1866. Haeckel adopted the Greek word *oikos*, meaning economy, household or dwelling (Smith and Smith 2001: 3) A staunch defender of Darwin, Haeckel originally described ecology as:

the economy of nature—the investigation of the total relations of the animal to both its inorganic and organic environment; including above all, its friendly and inimical relations with those animals and plants with which it comes directly and indirectly in contact—in a word, ecology is the study of all those complex interrelations referred to by Darwin as the conditions of the struggle for existence. (Quoted in Golley 1993: ii. 207)

Today, ecology may be defined as 'the study of the relationship between organisms and their physical and biological environments' (Ehrlich and Roughgarden, 1987: 3), or in terms of its scientific function: 'Ecology works at characterizing the patterns seen in nature, studying the complex interactions among organisms and their environments, and understanding the mechanisms involved in biological diversity' (Smith and Smith 2001: 3). Contemporary ecology incorporates numerous sub-fields such as ecosystem, population, community, landscape, chemical, behavioural, aquatic, and terrestrial ecology.

Non-specialists often confuse ecology with environmental science, or even with environmental management and politics. Ecology explicitly investigates the interaction of biotic systems with their environment. Many professional volumes on 'ecology and religion' are actually framed around environmental science—an applied discipline, integrating biology with geology, chemistry, oceanography, and atmospheric science, in the study of human impacts on the earth's physical and biotic systems. This chapter will first cover religious interaction with ecological phenomena prior to or independently of the rise of modern life science. The second half will investigate the past two centuries of religious interaction with the development of ecological thought in the Euro-American context, and will conclude with a brief survey of the last quarter-century of ecological and environmental science dialogue with world religions.

RELIGION AS A RESERVOIR OF ECOLOGICAL KNOWLEDGE

The world's religions have historically served as important cultural reservoirs of ecological understanding. Religious myths and rituals order and convey information about the geography and availability of natural resources, the behaviour of important food species or of predators, and the habitats and properties of plants. Although religious traditions often emphasize species or environmental features with consumptive value, such as salmon, bison, or dependable springs, the world's diversity of religious art and myth contains myriad accounts of species not directly utilized by humans. Religion appreciates the ecological roles of organisms of little caloric importance, such as ravens and eagles, and utilizes these creatures to symbolize natural processes. Religious rituals often imitate nature. The buffalo dances of the Native American peoples depict bison behaviour and movements. Myths about salmon, with their associated religious art including totem and mortuary poles, describe not just the migratory patterns, but also the relationships of salmon to other species, such as seals, bears, and orcas. A first area of interaction between religion and ecological science, therefore, is the study of how religions investigate, process, and express ecological reality.

Creation myths and religious cosmologies frequently incorporate detailed accounts of regional environments and ecological processes. The first two chapters of Genesis are relatively short and abstract, for example, in comparison to a Hawaiian creation chant called the *Kumulipo*, which begins with a hot earth, in recognition of volcanism as the source of island formation. In the *Kumulipo*, the first creature born of the divine pair of primary deities, *Kumulipo* and *Po'ele*, is the coral polyp. This expresses not evolutionary order, but the role of coral as a keystone species, critical to the diversity of sea life around the islands. The chant continues to devote lines to pairs of invertebrates, including starfish, sea cucumbers, barnacles, oysters, mussels, limpets, and mother-of-pearl. The sea urchins, identified as a tribe, divide into short-spiked, smooth, long-spiked, ring-shaped, and thin-spiked. The second individual song in the chant names the major fishes, and the third names the birds and flying insects, the fourth the crawlers, such as turtles and lobsters, the fifth the diggers who cultivate the land, and the sixth the nibblers, including the Polynesian rat (Beckwith 1951: 58–88). The whole is an impressive species inventory, particularly of the shallow reef. The chant classifies organisms by similarity in form and habitat. The audience for the creation story could presumably recognize far more species than the chant honours. The *Kumulipo*, however, reinforces the importance of distinguishing among related species, encourages learning the composition of each habitat, and announces that the biodiversity of the islands is primal and basic.

INDIGENOUS RELIGION REGULATING ENVIRONMENTAL MANAGEMENT

Religious coda or laws and taboos can dictate ecologically sound management strategies for regional ecosystems. The Oregon country Indians of the American Pacific Northwest moderated salmon harvest via strict observance of usufruct rights to designated stretches of streams and rivers and limited fishing in these territories to specific kin groups. Myths 'extolled the necessity of respect'. Examples are the story of a magic trap, which filled itself so effectively that its owner, often the trickster Coyote, could not keep up, and, unable to cook the salmon, cursed the trap. Both the trap and the salmon disappeared. Narratives of murdered salmon regenerated from a single egg or from bones emphasize the importance of reproduction and the natural cycles replenishing the salmon populations (Lee 1993; Taylor 1999: 30–1; Lichatowich 1999; Scarce 2000).

Many of the Oregon country tribes held first-salmon ceremonies, as the salmon runs began. During a period which might extend for more than two weeks, religious ritual and taboo constrained salmon harvest to only those fish that could be immediately consumed during the festival. The Tillamook prescribed lengthwise cutting to prepare a single fish for the headman, who ate it all except the bones and the blood. They then burned the remains and returned them to the river in a disposal ceremony. The ritual sequence recognizes the importance of salmon reproduction and the regenerative properties of the salmon carcasses. Functionally, the first-fish ceremony allows for escapement and spawning for a portion of the run

(Taylor 1999: 27–38). Scientific ecological study has found that the decaying Pacific salmon, dying after spawning, provide critical inputs to support aquatic food chains, which in turn feed juvenile salmon. The migratory salmon transport nutrients such as phosphorus from the sea and actually fertilize the flood plain forests of coastal rivers (McClain *et al.* 1998).

The Indians of Oregon country were technologically sophisticated fishermen, who deployed poisons, weirs, seines, and traps as well as gillnets. Hardly benign, their methods were as potentially effective as those of the industrialized fishing industry, and they harvested a high proportion of the salmon runs. Yet, unlike the 'scientific fisheries management' of the industrial era, which has been impotent in the face of intense exploitation of the salmon and other riparian resources, such as hydropower, aboriginal stewardship effectively balanced harvest with natural production. Environmental historian Joseph Taylor (1999) concludes that the Indians' superior spatial organization with families and salmon chiefs monitoring individual watersheds and breeding sites, in combination with religions effective in constraining capture and consumption, resulted in a sustainable fishery. Shared myths and ceremonies teaching practical ecology to entire societies encouraged communal co-operation and environmental restraint, and also co-ordinated entire villages into simultaneous activity, including critical abstinence from fishing for a set time or in specific locales.

DEITIES AS PERSONIFICATIONS OF ECOLOGICAL PROCESSES

Even when ecologically astute, myths and religious explanations do not function as scientific reporting does. Myths are a reservoir of insights about human behaviour, and often serve as templates for cultural responses to social issues. Compared to scientific reporting, myth and religious explanations for ecological phenomena do change through time, but they are often conservative, incorporating new knowledge or circumstances slowly. Myths frequently retain images whose meaning is lost or modified, and may attribute natural disasters to human causes, such as violations of community taboos, which modern science would hold to be unrelated to natural phenomena. Despite the potential limitations of conveying complex and sometimes obscure cultural meanings, myth and religious ritual can be accurate metaphors for environmental processes of importance to humans.

Deities, spirits, and mythic beings are frequently personifications of natural phenomena or summarize the ecological dynamics of culturally important flora or fauna. Kamapua'a, the Hawaiian supernatural being who is half pig and half god, is a shape-shifter capable of turning himself into not just a boar, but into a fish and various plants. His legends describe him rooting like a wild hog, while also conveying protocols for ritual sacrifice of hogs and other species. He courts the volcano goddess Pele, who rebuffs him with fountains of flame. Associated with water, Kamapua'a threatens to douse Pele's fires with precipitation, and seeks the aid of his sister, who sends fog and rain. An army of hogs overrun Pele's domain, and the fiery crater fills with water. The pig-god then has his way with Pele, and they divide the island of Hawaii into two regions: she takes the leeward or dry side, and he takes the windward or wet side, with its rainforest and prime hog habitat (Beckwith 1970: 203–13). Embedded in the legend is an ecologically perceptive description of vegetation recovery after volcanic disturbance and of the relationship of the trade winds and topography to vegetation. Human-style courtship becomes a metaphor for the interactions between volcanism and the oceanic climate, and vice versa. The myth also contains a lesson about keeping potentially destructive pig populations at bay.

The legends of Kamapua'a and Pele describe the behaviour of natural hazards, such as lava flows. In an oral culture, this is a dependable way to 'remember' less common but high-intensity environmental disturbances. When pursued by his stepfather Olopana and a band of armed warriors, Kamapua'a jams his koa-wood canoe into a stream above a waterfall in one of the steep valleys of the pali, or central mountain ridge of O'ahu. The water stops flowing downstream and builds up behind the canoe, and Olopana's puzzled followers, not grasping the danger, continue pursuit up the stream course. When Kamapua'a removes the canoe, a flash flood runs down the narrow gorge and destroys the oblivious warriors (Thompson 1966: 46–52; Beckwith 1970: 203–13). In ecological reality, natural log-jams and landslides may hold back water after heavy rains. The channelling of precipitation in the narrow valleys can also send a lethal torrent raging downstream. Hiking tourists, unfamiliar with the legends of Kamapua'a, have been killed in the resulting flash floods, whereas a Hawaiian would be wary of the steep-sided valleys. Kamapua'a the trickster reflects the capricious aspect of day-to-day environmental processes.

Today field ecologists, ethnographers, and environmental planners are giving greater consideration to the role of regional and indigenous religions in interpreting and managing environmental resources. Traditional knowledge embedded in religion, such as an understanding of the pharmaceutical properties of plant species or the long-term fluctuations in salmon runs, may be difficult to replace. Further, indigenous environmental management informed by religious tradition is usually sustainable, whereas modern scientific management often fails to balance harvest with ecosystem or population productivity, thereby resulting in serious environmental degradation.

ECOLOGICAL INSTRUCTION IN SACRED TEXTS

Although sometimes inaccurately viewed as 'unecological', world religions valuing preservation and study of sacred texts also document and interpret ecological phenomena. The Hebrew Scriptures, foundational to Judaism and Christianity, record the interactions of a culture based on herding and tillage with an upland landscape characterized by unpredictable rains and easily erodible soils. In order to farm productively, the ancient Hebrews constructed cisterns to retain precious water and built erosion-control terraces from the loose stone of the shallow soils. If one terrace failed, the resulting slump of soil and rock could wash out the terraces below it. This method of farming thus relied on a high degree of community co-operation, and was vulnerable to disruption during periods of social disorder, such as wars and invasions. Although the Hebrew Scriptures do not contain explicit directions for maintaining these erosion-control structures, perhaps because the methods were common knowledge, the texts mention the terraces numerous times and have explicit regulations for care of fields and vineyards. Isaiah 5 describes the destruction transpiring in the vineyards when the hedges are removed and the fences are broken (Hillel 1991: 98–102).

Leviticus and Exodus outline the laws of agricultural 'neighbourliness', including instructions to leave the corners of the fields untilled so as to provide for widows, orphans, and wildlife, and allow a seventh-year rest for agricultural fields as a practical response to the need for renewal in shallow upland soils (Waskow 2000; Wirzba 2001; Brueggemann 2002). The biblical laws, not unlike the first-fish festival, call for community co-operation to prevent ecological disaster. Numerous passages in the Book of Proverbs admonish individuals, including farmers, not to be greedy and to begin new enterprises slowly in order to discourage expanding flocks or farm fields in response to years of high rainfall, in anticipation of the inevitable years of little rain and much lower productivity (Bratton 2003). The Genesis story of Joseph and his brothers warns of the potential consequences of extended drought, and teaches the value of leaving supplies in reserve, as well as of sharing. The Hebrew Scriptures rarely anthropomorphize animals, yet their celestial god, who rides the storm clouds and releases the rains, personifies the ever-present issue of inadequate and fluctuating precipitation. Today, religious practitioners as diverse as Jewish vintners, Mennonite dairy farmers, Dutch Reformed truck gardeners, and Catholic nuns engage in 'gleaning' for the poor turn to the Hebrew Scriptures for guidance concerning sustainable management of agricultural ecosystems.

Islam, as a religion guided by sacred texts, has carefully considered the management of different classes of land and water resources. Islamic law not only distinguishes between rivers, springs, and wells, but also recognizes that the usage of large natural and perpetually flowing rivers should be differentiated from that of smaller rivers, those rivers that must be dammed to supply water, and artificial canals and irrigation ditches (Dutton 1992). The concept of *hima* regulates unowned lands, conserving them for the common good. Aside from the basic ethical principles set forth in the Qur'an, the Muslim tradition of environmental regulation relies on the opinions and analysis of imams and other forms of Islamic jurisprudence (Llewellyn 2003). The efficacy of the Islamic management may be observed in the rich and diverse Muslim cultures developing in regions where lack of precipitation is a constant threat to agriculture, as well as in moister tropical climates where fields may be flooded periodically.

HISTORIC WESTERN RELIGIOUS VALUES AND ECOLOGY

Scientific ecology arose in a Euro-American context in which Christianity was the dominant religion. Classical Greek and Roman, Islamic and Jewish scholars have, however, all left their mark on ecology's precursor, natural history. Through the medieval period Christians imitated the approaches of ancient authors such as Aristotle and Pliny the Elder, who classified biological phenomena such as the feeding habits of sea urchins. Influenced by Islamic science following the Christian conquest of Muslim Spain in the eleventh century, Christian proto-science began to rely increasingly on careful observation (Lindberg 1992: 180–2; Grant 1996: 23–4). The fascination with natural form is evident in the great outpouring of organic realism in Gothic sculpture. At Rheims Cathedral, built around 1230, for example, a diverse local flora of more than thirty recognizable species crowns the capitals and doorways and implies that peaceful, organic growth supports the great projects of God (Camille 1996: 134–5).

From its beginnings in the first century CE, Christianity valued natural metaphor in its art and literature. Paul Santmire (1985: 31–53), however, documents an early tension in Christian cosmology. He argues that Irenaeus (c.130–200) was very affirming of the Christian relationship with nature by his invocation of creation history as one work of God, whereas Origen (c.185–254), by emphasizing 'the unchanging One, God, dwelling above in eternity, surrounded by a world of rational spirits', demotes the cosmos to a secondary theological concern and prompts Christian alienation from nature. As this dialogue continued into the Middle Ages, St Bonaventure (1221–74) presented a 'fecund' triune God who diffuses eternal goodness and divine life into the creation, while St Francis of Assisi (1182–1226) extended Christian charity and love to the 'lowly creatures', treating them as brothers and sisters before God (Santmire 1985: 97–119). Historian of ideas Clarence Glacken (1967: 353–428) identifies the early modern period as dominated by physico-theology, which assumes that the Earth was planned or designed by God. The scientific investigation of nature was therefore a beneficial exercise in exposing and admiring God's perfect order. At the same time, museums with zoological and botanical collections began to appear in European cities and, along with the budding universities, stimulated the formal study of nature (Moore 1993: 69–76). These activities were congruent with the theology of Reformers such as John Calvin, who praised the cosmos as God's 'most beautiful theatre' in which every natural object had an appointed place and function. The environment demonstrated God's 'power, goodness, wisdom and eternity' (Santmire 1985: 128).

These concepts of divine order emigrated with Christians to the Americas. Colonial Puritan leader Cotton Mather found that natural theology supported the practice and the advance of science. The student of nature followed the 'Footsteps of a Deity in all the Works of Nature', and could 'by the Scale of Nature ascend to the

God of Nature'. Discovering God's laws and observing natural features and processes strengthened rather than threatened one's faith (Stoll 1997: 74–5). For perhaps the greatest of all colonial American theologians, Jonathan Edwards, 'the infinite fullness of God led inevitably to the emanation of excellency, beauty, happiness and know-ledge of himself [God] in the creation of the world'. Edwards himself took long meditative walks in natural settings, thus engaging in the practice of reading the symbolic language of nature and the emanation of the divine in the cosmos to better grasp the universal truth of God (Stoll 1997: 81–3).

Like doctors, lawyers, and teachers, Enlightenment clergy composed nature journals, pressed plants, or listed regional bird species. Gilbert White, an eighteenth-century Anglican curate kept a nature journal recording the daily events in the southern English countryside. In *The Natural History of Selborne* (1789) White observed and recorded foraging and nesting behaviour of birds such as the blue titmouse and the ring-ousel and determined the feeding preferences of hedgehogs. He accumulated species lists utilizing Carolus Linnaeus's newly developed Latin binomial nomenclature (G. White 1985). Although this is not yet scientific ecology, the careful notation of range and habitat, the recognition of the differential roles or niches of individual species, and the explanation of regional biotic diversity are necessary precursors to the extraction of ecological principles. William Bartram, a botanist and eighteenth-century Friend, or Quaker, explored the Cherokee and Creek territories of the southern Appalachians. His sophisticated descriptions of vegetation in relation to topography still provide useful historic documentation of biota. Bartram wrote in the introduction to his journals (1928: 15): 'This world, as a glorious apartment of the boundless palace of the sovereign Creator, is furnished with an infinite variety of animated scenes, inexpressibly beautiful and pleasing, equally free to the inspection and enjoyment of all his creatures.'

ROMANTICISM AND TRANSCENDENTALISM

With its emphasis on the contemplation of nature as providing religious or philosophical insights, nineteenth-century Romanticism countered Enlightenment reductionism and perception of nature as mechanical or comprised of dissociated elements. Although many Romantics were relatively mainstream Christians or Jews, others investigated pantheism or sought a unified spirit in nature. During the nineteenth century, Hindu and Buddhist religious texts began to appear in European-language editions. Henry David Thoreau was familiar with the *Bhag-avadgītā*, and, with Ralph Waldo Emerson, translated Eugene Burnouf's French-language scholarship on Buddhism into English (Hodder 2001: 143–5, 177). The Transcendentalist world-view was based on the more ancient concept of correspondence. The human community is a small microcosm within a much greater macrocosm or the transcendent domain of nature. The ultimacy of the great cosmos infers that there is 'a law implicit in the scheme of things, a controlling providence, a natural or moral law that unfolded in the very order that the cosmos represented' (Albanese 1988: 21).

Romanticism and Transcendentalism perceived nature as highly dynamic and subject to change through deep time. The English painter J. M. W. Turner chose avalanches in the Alps and converging currents at sea as his subjects, invoking the chaos of nature to reflect the deeper meaning of human life. Thoreau was an accomplished field naturalist and the first person to fully describe the process of plant succession, an important ecological concept. His late work *Faith in a Seed*, suggesting the spiritual value of ecological process in its title, describes revegetation of exhausted farm soil and the strategies of different native plant species for dispersal (Thoreau 1993). In *My First Summer in the Sierra*, John Muir, founder of the Sierra Club, enthusiastically describes the effects of spring floods on the stream channels and the origin of the moss-covered boulders, once transported by glacial melt water. He then contrasts these major geological forces to the 'small low tones of the current gliding past the side of the boulder-island, and glinting against a thousand smaller stones down the ferny channel!' Muir (1911: 47–9) concludes: 'The place seemed holy, where one might hope to see God.'

Scientific ecology's dedication to comprehending the whole as more than the sum of its parts and to describing natural process in terms of embedded cycles has intellectual precursors in nineteenth-century philosophy and theology. Ernst Haeckel, for example, was a founder of the Monist League, which seeks a unified all-encompassing reality (Bramwell 1989: 39–56). As K. S. Shrader-Frechette and E. D. McCoy (1993: 279) have argued, however, the complexity and stochastic nature of ecological process in a forest or coral reef or other living community have actually inhibited the development of 'exceptionless empirical laws and a deterministic general theory' in ecological science. Today's community ecology, for example, is still largely based in the analysis of specific, natural history-based cases, albeit a far more mathematical form of research than its nineteenth-century counterpart. Philosophical holism has presupposed a greater order, which has proved very difficult to define scientifically.

A second realm of subtle religious influence is in the early ecological emphasis on the climax concept, advocated by American ecologist Fredrick Clements, who proposed that ecological succession ended in a stable biota community with relatively predictable components. Early ecologists maintained a strong dichotomy between natural and human-influenced ecosystems. The Transcendentalist interest in the primal wilderness reflects nineteenth-century valuation of undisturbed nature as Edenic, and therefore existing in an eternally unchanging and perfect state (Williams 1987). The Hudson River School of landscape painting saturated wilderness views with 'divine light' and utilized inaccessible mountains as metaphors for God. Today's ecologists have curtailed the definition of an ideal climax community and, taking a more post-modern perspective, treat the outcomes of disturbance and succession as relative to specific conditions and the history of a locale, if subject to rules of community assemblage.

Examples of direct religious influence on ecological paradigms are few. A potential case is the work of W. C. Allee, a Quaker who, holding communitarian values, believed that ecology had placed an excessive emphasis on competition and should further research co-operation between species and individuals (pers. comm., R. H. Whittaker; McIntosh 1985). His finding that co-operation, often in the form of flocks or herds, was necessary to the survival of individuals (now termed the 'Allee effect') is important to conservation biology and the management of endangered species. Reduction of packs of predators, such as African wild dogs, below a critical size may, for example, precipitate the demise of the remaining individuals and accelerate an extinction vortex.

Historian Donald Worster (1993: 189) has argued that the values of progressive Protestantism, particularly the Reformed denominations such as Presbyterians, Quakers, and evangelical Methodists, have 'provided an important spawning ground for environmental reform movements'. Worster lists John Wesley Powell, early explorer of the Grand Canyon and advocate of water conservation; Stephen Mather, Director of the US National Park Service; Mary Austin, natural history writer; William O. Douglas, Supreme Court Justice; and John Muir, as examples of environmental leaders from strong Protestant backgrounds. Important additions to Worster's list are Rachel Carson, a Presbyterian (Lear 1997) and Howard Zahniser, a framer of the Wilderness Act and a Free Methodist (pers. comm., Zahniser family). Worster (1993: 196–9) proposes that Reformed Protestantism has left a legacy of moral activism, ascetic discipline, egalitarian individualism, aesthetic spirituality, and support for applied science, which also characterize American environmentalism, and thereby influence the trajectories and research priorities of environmental science.

THE POST-MODERN DIALOGUE BETWEEN ECOLOGY AND RELIGION

Through the second half of the twentieth century to the present day, political response to environmental degradation and the impact of industrialization has driven most religious dialogue with ecological science. In an article published in *Science* in 1969, Lynn White Jr. concluded that the Christian emphasis on a transcendent deity who is exterior to and above nature was a major root of techno-industrial culture's disregard for nature. White pointed to St Francis of Assisi as a possible ecological role model and suggested that non-Western religions such as Buddhism might be inherently more environmentally sensitive and accountable.

In the vigorous dialogue that followed, religious environmentalists have chosen from among three basic strategies: reviving or extracting ecologically sound beliefs from historic religious traditions, replacing 'unecological' religions with those believed to be more ecologically sound, or reforming the world's religions to grapple more effectively with planetary environmental degradation.

Attempts at recovering an ecologically friendly religious heritage include that of Lutheran theologian Paul Santmire (1970), who in an early response to Lynn White Jr. constructed *Brother Earth: Nature, God and Ecology in a Time of Crisis*, as an environmental theology based on the life and example of the medieval St Francis. Christians have also rediscovered the environmental values of desert ascetics following in the footsteps of St Antony the Great of Egypt and of 'Celtic Christian spirituality', originating with saints such as Brigit and Columba of Ireland (Bratton 1993). Muslims have studied the living universe so beautifully celebrated by the poet Rumi (Clarke 2003). Hindus have sought inspiration for environmental practice in the myths of Krishna the cowherd (Prime 1992). Looking for more environmentally compatible religious alternatives, Westerners have studied Zen Buddhist contemplative practice and aesthetics as an entrée to a better understanding of nature. Western Buddhists see themselves as 'promoting a different vision' which can influence the balance of acts that add to or subtract from the earth's burden (Timmerman 1992: 74–5). The New Age movement, in syncretizing elements of different religions, has advocated Native American beliefs as guides to earth-friendly living, invoked the Eightfold Path of Buddhism, and studied the creation-centred spirituality of former Dominican friar Mathew Fox (Peters 1991: 52–93, 120–31).

Intellectual and community leaders of all the world's religions have articulated ethical and cosmological rationales for better environmental care. These include the following:

The belief that the cosmos or living creatures have inherent or intrinsic worth   Genesis 1, the first book of the Torah or Pentateuch, proclaims that God created and blessed all life and declared that even the creeping things and the sea monsters are 'good'. The Hebrew word for 'good' is *tob*, implying that creation is both worthy and beautiful (Bratton 1993). A Pagan might argue for inherent worth based on deities emanating from nature and an understanding of nature awe. Zen Buddhists, in contrast, may find Western theological definitions of intrinsic value incompatible with Buddhist concepts of emptiness. A Buddhist might instead advocate 'an emphatic identification with all life' or a transpersonal approach to all things (James 2004: 102–3).

Encouragement of individual and community responsibility toward the environment   Buddhists teach that uncontrolled craving (*tanha*) and suffering (*dukkha*) not only harm the human who clings to them, but also prompt poor environmental decision making (Harvey 2000; Tucker and Williams 1997). Jews have invoked principles from the Torah, such as *bal tashchit*, or 'do not destroy', forbidding wanton destruction of nature (Schwartz 2001). Religions in general have generated strategies for community organization, denominational statements, petitions for environmental accountability, educational literature, congregation- and school-based environmental programmes, grass-roots resistance to environmental health threats, and restoration and sustainable management projects.

A belief that the world is infused with Spirit   Indigenous religions of the eastern North American woodlands limit wanton destruction of game or environmental resources out of respect for *manito*, or a supernatural power infusing not just other species, but also inanimate objects and landscape features as well (Tooker 1979: 11–30). Japanese care for nature is encouraged by Shinto, where *kami* inhabit trees, rocks, and mountains (Earhart 1982). Christian theologian Jürgen Moltmann (1985) has advocated a panentheist interpretation of the universe which emphasizes divine immanence in the physical and biotic environment, while Sallie McFague (1993) has proposed an organic model of the Earth as God's body.

A belief that humanity is spiritually linked to the universe or cosmos   For Daoists, right relationship arises from religious experience, rather than from theological dogma or formulations. Chinese religion emphasizes *xiao*, or filial piety, which is extended, not just from the family to the state, but to the entire natural world—a realization 'arising from the ecstatic religious experience of union with the entire environment' (Paper 2001: 118; cf. Girardot *et al.* 2001). Roman Catholics recognize the sacramentality of creation, and reflect on the passage of life's seasons of growth and dormancy in the liturgical calendar (Irwin 1996). According to the Dalai Lama of Tibet (2000: 16–17): 'All material objects can be understood in terms of how the parts compose the whole, and how the very idea of whole depends upon the existence of the parts. . . . So when we consider the universe in these terms . . . we also understand that the entire phenomenal world arises according to the principle of dependent origin.' For the Buddhist, things, including humans, 'do not have an independent autonomous reality'.

A belief that religious and cultural preservation and ecological preservation are integrally linked   For indigenous peoples the native species, natural events, and regional landscapes which are threatened by environmental degradation are critical to religious rituals and traditions (Grim 2001). The Navaho have struggled to prevent Glen Canyon dam from flooding sacred Rainbow Bridge, a natural rock arch, among other ritual locales (Kelly and Francis 1994). Hindu women of the Chipko movement have placed themselves between logging crews and sacred groves in order to protect both religious traditions and the important ecological services the trees provide, such as firewood and watershed protection (Prime 1992). Ladakh Buddhism, from the high mountains of central Asia, emphasizes rituals linked to seasonal rhythms, and in response to a difficult physical environment values frugality and recycling so completely that 'there is literally no waste'. In Ladakh, where the populace faces an influx of tourists, conversion to a cash economy, and the disintegration of the indigenous culture, Buddhist organizers have begun organizations to promote environmentally sound development (Norberg-Hodge 1992: 46). Religions may also share environmental concerns such as the preservation of Mount Kailas in western Tibet, which is a sacred locale for Hindu, Buddhist, Jain, and Bonpo practitioners (Johnson and Moran 1989).

Ecology's and Environmental Science's Impact on Religion

Since the Second World War, ecology has increasingly influenced religion. Scholarly investigation of the structure and function of religions has invoked ecological models and concepts, including those of energy and nutrient flow through ecosystems, adaptation to habitat, interspecific competition, population regulation, and resilience to disturbance. Roy Rappaport in *Pigs for Ancestors* (1968) investigated the relationship of ritual feasts to management and consumption of New Guinea wild pig herds, and thereby the distribution of calories and protein. Richard Nelson in *Make Prayers to the Raven* (1983) correlated Koyukon taboos and regulation of hunting to the carrying capacities and population dynamics of northern game species. Religions, particularly new or 'alternative' religions, have directly invoked ecological terminology and concepts. Today's pagans believe themselves to be 'fundamentally "Green" in philosophy and practice', and borrow scientific language in defining humanity as 'one part of an elaborate and evolving community of beings, a web of life, an ecosystem' (Harvey 1997: 126, 131).

World religions have applied ecology to cosmology. Beginning in the 1970s, Christians began to explicate environmental theologies, dubbed 'ecotheologies', as the efforts expanded in the 1980s (Bratton 1983). Christians often retain traditional vocabulary and the trinitarian Godhead, while attributing the biodiversity resulting from evolutionary process to the ultimate character of the Creator or the Holy Spirit. Similarly, Hindus have explored Vedic cosmology, where *karma* is the source of the planet's productivity. Hindu sacrifice and food offerings establish a reciprocal relationship with the ecological energetics of the Earth (Prime 1992: 31). Hindu *dharma*, caring for the welfare of all living beings, is an ethos 'giving rise to harmony and understanding' in human relationships with all of creation (Dwivedi 2000). Although God is independent of the Five Great Elements, the Elements are not separate from God and support the function of the universe. The earth is Dharani and the Mother of all living; thus life has value above that of property or inherent worth. The Hindu practice of *Yajna*, incorporating both the sacrifice of the individual ego and the ritual burning of impurities, removes human greed and lust from the environment (Rao 2000). These concepts in turn inform scientific process. Apffel-Marglin and Parajuli (2000) have contrasted supposed scientific separation of fact and value in nature and notions of disembeddedness versus the embeddedness of ethnic Hindu communities protecting 'sacred groves', which serve as wildlife and forest preserves.

The impact of a single theory from environmental science may be demonstrated by investigating the interface of religion and chemist James Lovelock's concept of Gaia, where life interacts with and may partially regulate planetary geophysical dynamics such as the composition of the atmosphere. Buddhists have responded by analysing the connection between the cosmic order (*dharma*) and Gaia (Badiner 1990).

Pantheists conclude that Gaia unifies life and non-life (Harrison 1999). Christian feminist Rosemary Ruether (1992) has compared Teilhard de Chardin's Omega Point to Gaia, as both assume a living and evolving planet. Ruether uses Gaia to model planetary healing. Eric Rosenblum (2001) suggests that *tikkun olam*, or ecosystem restoration, is consistent with the Jewish mystical concept of redemption, which is analogous to mending the broken 'blood vessels' of creation. By presenting the Earth as an integrated, dynamic entity, the science encourages monotheism or pantheism, rather than a portrait of nature personified by multiple deities forming an anthropomorphic assemblage. Scientific holism and systems modelling have fuelled a conceptualization of the great goddess for today's feminists, which is more universal than the classical Greek portrait of the earth as γι, or as a female deity in a complex pantheon.

RELIGIOUS CRITIQUE OF ECOLOGICAL AND ENVIRONMENTAL SCIENCE

Although ecology is an inherently evolutionary branch of the life sciences, fundamentalist Christian rejection of evolutionary paradigms has not been a major component of Christian interaction with ecology. Advocates for intelligent design of living organisms have shown little interest in debating ecological principles, even though the historical understanding of God as Designer was a key motive for Christian study of ecological process. Challenges from the major world religions to environmental science primarily concern the social interpretation of models predicting environmental stresses, such as demographic projections of the growth of the Earth's human population or computer-generated estimates of increases in regional surface temperatures that are driven by climbing proportions of greenhouse gases in the atmosphere. More conservative Christians and Muslims, for example, may be troubled by advocacy for regulation of human population growth based on predictions of future natural resource shortages.

Adherents to new or 'renewed' religions, such as Pagans and Wiccans, are often sceptical of science's influence in post-modern economic and political culture, and perceive their own return to respecting spirit in nature as healing or countering the ill effects of techno-industrial societies' excessively rational, unfeeling, and manipulative approach to the environment. While ecofeminists, including those committed to goddess-centred spirituality, are seeking to dialogue with scientific ecology, they contest its assumptions, research priorities, and disregard for the perspectives of women and indigenous peoples. Ecofeminists have critiqued the application of hierarchical structures to ecosystem analysis and the ecologists' preference for studying competition in nature rather than mutualism and co-operation among organisms (Warren 1994; Zabinski 1997).

The last area of religious critique concerns the use of 'ecological' rationales to exclude or disenfranchise ethnic minorities. During the nineteenth and first half of the twentieth century, National Socialists (Nazis) and others with racist or ultra-nationalist agendas touted social Darwinism and idealized concepts of human relationship to the ecological landscape as evidence of the inadequacies of 'foreign' or supposedly 'non-native' ethnicities such as Jews. Daniel Gasman (1971) and Anna Bramwell (1989) have implicated Ernst Haeckel, whose writings were admired by Nazis involved in alternative religions, as a scholar forwarding ecological racism. The US National Park system was established to preserve natural environments, but originally excluded Native Americans. Indigenous peoples continue to have major concerns about the economic abuse of scientific expertise to rationalize environmentally damaging development endeavours, remove control of natural resources from regional cultures, and denigrate the legitimacy of religious rationales for environmental management.

THE GLOBALIZATION OF THE DIALOGUE BETWEEN ECOLOGY AND RELIGION

As the Earth enters an era of globalization of national economies, there has been a call from both academics and environmentalists to encourage all religions to respond to planet-wide environmental issues in terms of their own traditions, rituals, coda, and cultures. In 1986 philosophical environmental ethicist Eugene Hargrove edited *Religion and Environmental Crisis*, which emphasized Christian perspectives, but also included articles representing world religions such as Judaism, Native American, Islam, Daoism, and classical polytheism. In 1992 the World Wide Fund for Nature, aware that religious values could either help or hinder public understanding of environmental science and politics, sponsored a series of short, accessible volumes on 'World Religions and Ecology' summarizing the positive environmental values of Christianity, Islam, Judaism, Buddhism, and Hinduism. Between 1996 and 1998, the Harvard School Center for the Study of World Religion hosted a series of environmental conferences. Mary Evelyn Tucker (2003: 9), who spearheaded the project, believes that: 'the environmental crisis calls the religions of the world to respond by finding their voice within the larger Earth community. In so doing, the religions are now entering their ecological phase and finding their planetary expression.' In addition to the five religions covered by the World Wide Fund for Nature, the Harvard conferences sponsored commentaries on Confucianism, Jainism, Daoism, and indigenous religions, thereby greatly expanding the representation of Asian, southern hemisphere, and oral traditions.

As ecological and environmental science enter the twenty-first century, their dialogue with world religions is indeed continuing to expand. Deep ecologists, who promote a biocentric ethic considering both spirituality and egalitarian political values have encouraged responses from a variety of religious perspectives (Barnhill and Gottlieb 2001). University and seminary course and programme offerings are becoming more common, as are the textbooks supporting them (Kinsley 1995; Foltz 2003). John Carroll and Keith Warner (1998) have forwarded the scientific voice in *Ecology and Religion; Scientists Speak*, where commentators include sociobiologist E. O. Wilson on the topic of natural philosophy; Eliot Norse, a Jewish conservation biologist, on the meaning of death in nature; and ecologist William Gregg on Baha'i values and environmental management for a sustainable biosphere. Rather than merely offering a summary of some idealized human interaction with nature, religious interaction with environmental science is becoming more specific. It is now engaging individual environmental issues, including human population growth, global warming, the preservation of endangered species, and the management of specific ecosystems such as oceans, forests, and rivers. Among other benefits, these environmental interactions among religions and denominations represent a peaceful and thoughtful common attempt to solve urgent practical problems affecting the quality of human life and the fate of the Earth's diverse species and ecosystems.

REFERENCES AND SUGGESTED READING

ALBANESE, C. (1988). *The Spirituality of the American Transcendentalists: Selected Writings of Ralph Waldo Emerson, Amos Bronson Alcott, Theodore Parker, and Henry David Thoreau.* Macon, Ga.: Mercer University Press.

APFFEL-MARGLIN, F., and PARAJULI, P. (2000). ' "Sacred Grove" and Ecology: Ritual and Science', in C. K. Chapple and M. E. Tucker (eds.), *Hinduism and Ecology,* Cambridge, Mass.: Harvard University Press, 291–316.

BADINER, A. H. (1990) (ed.). *Dharma Gaia: A Harvest of Essays in Buddhism and Ecology.* Berkeley: Parallax Press.

BARNHILL, D. L., and GOTTLIEB, R. S. (2001) (eds.). *Deep Ecology and World Religions: New Essays on Sacred Ground.* Albany, N.Y.: State University of New York Press.

BARTRAM, W. (1928). *Travels of William Bartram.* New York: Dover Publications; originally published 1791.

BECKWITH, M. (1951). *The Kumulipo: A Hawaiian Creation Chant.* Chicago: University of Chicago Press.

—— (1970). *Hawaiian Mythology.* Honolulu: University of Hawaii Press.

BRAMWELL, A. (1989). *Ecology in the 20ᵗʰ Century: A History.* New Haven: Yale University Press.

BRATTON, S. (1983). 'The Ecotheology of James Watt', *Environmental Ethics,* 5: 225–36.

—— (1993). *Christianity, Wilderness and Wildlife: The Original Desert Solitaire.* Scranton, Pa.: University of Scranton Press.

—— (2003). 'The Precautionary Principle and the Book of Proverbs: Toward an Ethic of Ecological Prudence in Ocean Management', *Worldviews,* 7(1–2): 54–91.

BRUEGGEMANN, W. (2002). *The Land: Place as Gift, Promise, and Challenge in Biblical Faith.* Minneapolis: Fortress Press.

CAMILLE, M. (1996). *Gothic Art: Glorious Visions.* New York: Harry N. Abrams.

CARROLL, J. E., and WARNER, K. (1998) (eds.). *Ecology and Religion: Scientists Speak*. Quincy, Ill.: Franciscan Press.

CLARKE, L. (2003). 'The Universe Alive: Nature in the *Masnavi* of Jalal al-Din Rumi', in R. C. Folts, F. M. Denny and A. Baharuddin (eds.), *Islam and Ecology*, Cambridge, Mass.: Harvard University Press, 39–65.

DALAI LAMA, HIS HOLINESS (2000). *A Simple Path: Basic Buddhist Teachings by His Holiness the Dalai Lama*. London: Thorsens.

DUTTON, Y. (1992). 'Natural Resources in Islam', in F. Khalid and J. O'Brien (eds.), *Islam and Ecology*, London: Cassell, 51–67.

DWIVEDI, O. P. (2000). 'Dharmic Ecology', in C. K. Chapple and M. E. Tucker (eds.), *Hinduism and Ecology*, Cambridge, Mass.: Harvard University Press, 3–22.

EARHART, H. B. (1982). *Japanese Religion: Unity and Diversity*. Belmont, Calif.: Wadsworth Publishing.

EHRLICH, P. R., and ROUGHGARDEN, J. (1987). *The Science of Ecology*. New York: Macmillan.

FOLTZ, R. C. (2003) (ed.). *Worldviews, Religion and the Environment*. Belmont, Calif.: Wadsworth/Thompson Learning.

GASMAN, D. (1971). *The Scientific Origins of National Socialism: Social Darwinism in Ernst Haeckel and the Monist League*. London: Macdonald.

GIRADOT, N. J., MILLER, J., and XIAOGAN, L. (2001) (eds.). *Daoism and Ecology: Ways within a Cosmic Landscape*. Cambridge, Mass.: Harvard University Press.

GLACKEN, C. (1967). *Traces on the Rhodian Shore: Nature and Culture in Western Thought from Ancient Times to the End of the Eighteenth Century*. Berkeley: University of California Press.

GOLLEY, F. B. (1993). *A History of the Ecosystem Concept in Ecology: More than the Sum of the Parts*. New Haven: Yale University Press.

GRANT, E. (1996). *The Foundations of Modern Science in the Middle Ages: Their Religious, Institutional and Intellectual Contexts*. Cambridge: Cambridge University Press.

GRIM, J. A. (2001) (ed.). *Indigenous Traditions and Ecology: The Interbeing of Cosmology and Community*. Cambridge, Mass.: Harvard University Press.

HARGROVE, E. (1986) (ed.). *Religion and Environmental Crisis*. Athens, Ga.: University of Georgia Press.

HARRISON, P. (1999). *Pantheism: Understanding the Divinity in Nature and the Universe*. Boston: Element.

HARVEY, G. (1997). *Contemporary Paganism: Listening People, Speaking Earth*. New York: New York University Press.

——(2000). *An Introduction to Buddhist Ethics*. Cambridge: Cambridge University Press.

HILLEL, D. (1991). *Out of the Earth: Civilization and the Life of the Soil*. Berkeley: University of California Press.

HODDER, A. D. (2001). *Thoreau's Ecstatic Witness*. New Haven: Yale University Press.

IRWIN, K. W. (1996). 'The Sacramentality of Creation and the Role of Creation in Liturgy and Sacraments', in D. Christiansen and W. Glazer (eds.), *'And God Saw that it Was Good': Catholic Theology and the Environment*, Washington: United States Catholic Conference, 103–46.

JAMES, S. P. (2004). *Zen Buddhism and Environmental Ethics*. Aldershot: Ashgate.

JOHNSON, R., and MORAN, K. (1989). *Tibet's Sacred Mountain: The Extraordinary Pilgrimage to Mount Kailas*. Rochester, Vt.: Park Street Press.

KELLY, K. B., and FRANCIS, H. (1994). *Navaho Sacred Places*. Bloomington, Ind.: Indiana University Press.

KINSLEY, D. (1995). *Ecology and Religion: Ecological Spirituality in Cross-Cultural Perspective*. Upper Saddle River, N.J.: Prentice-Hall.

LEAR, L. (1997). *Rachel Carson: Witness for Nature*. New York: Henry Holt and Co.

LEE, K. N. (1993). *Compass and Gyroscope: Integrating Science and Politics for the Environment*. Washington: Island Press.

LICHATOWICH, J. (1999). *Salmon without Rivers: A History of the Pacific Salmon Crisis*. Washington: Island Press.

LINDBERG, D. (1992). *The Beginnings of Western Science: The European Scientific Tradition in Philosophical, Religious and Institutional Context, 600 B.C. to A.D. 1450*. Chicago: University of Chicago Press.

LLEWELLYN, O. A. (2003). 'The Basis for a Discipline of Islamic Environmental Law', in R. C. Folts, F. M. Denny, and A. Baharuddin (eds.), *Islam and Ecology*, Cambridge, Mass.: Harvard University Press, 185–247.

MCCLAIN, M., BILBY, R., and TRISKA, F. (1998). 'Nutrient Cycles and Responses to Disturbance', in S. Kantor (ed.), *River Ecology and Management: Lessons from the Pacific Coastal Ecoregion*, New York: Springer, 347–72.

MCFAGUE, S. (1993). *The Body of God: An Ecological Theology*. Minneapolis: Fortress Press.

MCINTOSH, R. (1985). *The Background of Ecology: Concept and Theory*. Cambridge: Cambridge University Press.

MOLTMANN, J. (1985). *God in Creation: A New Theology of Creation and the Spirit of God*. San Francisco: Harper & Row.

MOORE, J. A. (1993). *Science as a Way of Knowing: The Foundations of Modern Biology*. Cambridge, Mass.: Harvard University Press.

MUIR, J. (1911). *My First Summer in the Sierra*. New York: Houghton Mifflin.

NELSON, R. K. (1983). *Make Prayers to the Raven: A Koyukon View of the Northern Forest*. Chicago: University of Chicago Press.

NORBERG-HODGE, H. (1992). 'May a Hundred Plants Grow from One Seed: The Ecological Tradition of Ladakh Meets the Future', in M. Batchelor and K. Brown (eds.), *Buddhism and Ecology*, London: Cassell, 41–54.

PAPER, J. (2001). 'Chinese Religion, "Daoism", and Deep Ecology', in D. L. Branhill and R. S. Gottlieb (eds.), *Deep Ecology and World Religions: New Essays on Sacred Ground*, Albany: State University of New York Press, 107–26.

PETERS, T. (1991). *The Cosmic Self: A Penetrating Look at Today's New Age Movement*. San Francisco: Harper San Francisco.

PRIME, R. (1992). *Hinduism and Ecology: Seeds of Truth*. London: Cassell.

RAO, K. L. S. (2000). 'The Five Great Elements (*Pancamahabhuta*): An Ecological Perspective', in C. K. Chapple and M. E. Tucker (eds.), *Hinduism and Ecology*, Cambridge, Mass.: Harvard University Press, 23–38.

RAPPAPORT, R. A. (1968). *Pigs for Ancestors*. New Haven: Yale University Press.

ROSENBLUM, E. (2001). 'Is Gaia Jewish? Finding a Framework for Radical Ecology in Traditional Judaism', in M. Yaffe (ed.), *Judaism and Environmental Ethics*, Lanham, Md.: Lexington Books, 183–205.

RUETHER, R. R. (1992). *Gaia & God: An Ecofeminist Theology of Earth Healing*. San Francisco: Harper San Francisco.

SANTMIRE, P. H. (1970). *Brother Earth: Nature, God and Ecology in a Time of Crisis*. New York: Thomas Nelson.

——(1985). *The Travail of Nature: The Ambiguous Ecological Promise of Christian Theology*. Minneapolis: Fortress Press.

SCARCE, R. (2000). *Fishy Business: Salmon, Biology and the Social Construction of Nature*. Philadelphia: Temple University Press.

SCHWARTZ, E. (2001). '*Bal tashchit*: A Jewish Environmental Precept', in M. Yaffe (ed.), *Judaism and Environmental Ethics*, Lanham, Md.: Lexington Books, 230–49.

SHRADER-FRECHETTE, K. S., and McCOY, E. D. (1993). *Method in Ecology: Strategies for Conservation*. Cambridge: Cambridge University Press.

SMITH, R. L., and SMITH, T. M. (2001). *Ecology & Field Biology*. San Francisco: Benjamin Cummings.

STOLL, M. (1997). *Protestantism, Capitalism, and Nature in America*. Albuquerque, N.M.: University of New Mexico Press.

TAYLOR, J. E. (1999). *Making Salmon: An Environmental History of the Northwest Fisheries Crisis*. Seattle: University of Washington Press.

THOMPSON, V. L. (1966). *Hawaiian Myths of Earth, Sea and Sky*. Honolulu: University of Hawaii Press.

THOREAU, H. D. (1993). *Faith in a Seed*. Washington: Island Press.

TIMMERMAN, P. (1992). 'It is Dark Outside: Western Buddhism from Enlightenment to the Global Crisis', in M. Batchelor and K. Brown (eds.), *Buddhism and Ecology*, London: Cassell, 65–76.

TOOKER, E. (1979) (ed.). *Native North American Spirituality of the Eastern Woodlands*. Mahwah, N.J.: Paulist Press.

TUCKER, M. E. (2003). *Worldly Wonder: Religions Enter their Ecological Phase*. Chicago: Open Court Publishing.

—— and WILLIAMS, D. R. (1997) (eds.). *Buddhism and Ecology*. Cambridge, Mass.: Harvard University Press.

WARREN, K. (1994). *Ecological Feminism*. London: Routledge.

WASKOW, A. (2000). *Torah of the Earth: Exploring 4,000 Years of Ecology in Jewish Thought*. Woodstock, Vt.: Jewish Lights Publishing.

WHITE, G. (1985). *The Essential Gilbert White of Selborne*. Boston: David R. Godine.

WHITE, L. Jr. (1967). 'The Historic Roots of our Ecological Crisis', *Science*, 155: 1203–7.

WILLIAMS, D. R. (1987). *Wilderness Lost: The Religious Origins of the American Mind*. Cranbury, N.J.: Associated University Presses, Susquehanna University Press.

WIRZBA, N. (2001). *The Paradise of God*. Oxford: Oxford University Press.

WORSTER, D. (1993). *The Wealth of Nature: Environmental History and the Ecological Imagination*. New York: Oxford University Press.

ZABINSKI, C. (1997). 'Scientific Ecology and Ecological Feminism', in K. Warren (ed.), *Ecofeminism: Women, Culture, Nature*, Bloomington, Ind.: Indiana University Press, 314–24.

CHAPTER 14

NEUROPHENOMENOLOGY AND CONTEMPLATIVE EXPERIENCE

EVAN THOMPSON

INTRODUCTION

Scientific investigation of the mind, known since the 1970s as 'cognitive science', is an interdisciplinary field of research comprising psychology, neuroscience, linguistics, computer science, artificial intelligence, and philosophy of mind. The presence of philosophy in this list is telling. Cognitive science, although institutionally well established, is not a theoretically settled field like molecular biology or high-energy physics. Rather, it includes a variety of competing research programmes—the computational theory of mind (also known as classical cognitive science), connectionism, and dynamical and embodied approaches—whose underlying conceptions of mentality and its relation to biology, on the one hand, and to culture, on the other, are often strikingly different (see Clark 2001, for a useful overview).

It is important to keep this situation in mind in any discussion of the relationship between cognitive science and religion, for different theoretical perspectives in cognitive science can combine with different scientific approaches to religion. Rather than review these possibilities here, however, I shall describe one recent approach, known as neurophenomenology (Lutz and Thompson 2003; Varela 1996). Although neurophenomenology is not directly concerned with the cognitive science of religion, it is highly relevant to this field, especially the psychology and biology of religious experience.

Neurophenomenology is an offshoot of the embodied approach in cognitive science (Varela *et al.* 1991). The central idea of the embodied approach is that cognition is the exercise of skilful know-how in situated action. The most important feature of this approach, for our purposes here, is that experience is not seen as an epiphenomenal side issue, but is considered central to any adequate understanding of the mind, and accordingly needs to be investigated in a careful phenomenological manner. Phenomenology and experimental cognitive science are thus seen as complementary and mutually informing modes of investigation. Neurophenomenology builds on this view with the specific aim of understanding the nature of consciousness and subjectivity and their relation to the brain and body.

The working hypothesis of neurophenomenology is that phenomenological accounts of the structure of human experience and scientific accounts of cognitive processes can be mutually informative and enriching (Thompson 2006; Varela *et al.* 1991; Varela 1996). The term 'phenomenology' in this context refers to disciplined, first-person ways of investigating and analysing experience, as exemplified by the Western philosophical tradition of phenomenology (Moran 2000) and Asian contemplative philosophies, especially (though not exclusively) Buddhism. The reason why the Buddhist tradition is particularly relevant in this context is that its cornerstone is contemplative mental training and critical phenomenological and philosophical analysis of the mind based on such training (Dreyfus and Thompson 2007; Lutz *et al.* 2007). Thus, neurophenomenology intersects with religion not so much as an object of scientific study, as it is for the cognitive science of religious beliefs and behaviours (e.g. Boyer 2001, 2003, 2005), but rather as a repository of contemplative and phenomenological expertise. According to neurophenomenology, such expertise could play an active and creative role in the scientific investigation of consciousness (Lutz *et al.* 2007; Lutz and Thompson 2003; Thompson 2005).

Religion includes many other things besides contemplative experience, and many religions have little or no place for contemplative experience. Conversely, contemplative experience is found in other contexts besides religion, such as philosophy (McGee 2000). For these reasons, the term 'religion', at least as it is generally used in the West, is not a good designation for the kind of practice and experience that neurophenomenology seeks to bring into constructive engagement with cognitive science. A better description might be the kind of self-cultivation and self-knowledge cultivated by the world's contemplative 'wisdom traditions' (Depraz *et al.* 2003). Nor does the term 'science–religion dialogue' describe the motivation for neurophenomenology, for the aim is not to compare, evaluate, or adjudicate between the claims of science and religion, but to gain a deeper understanding of human experience by making contemplative phenomenology a partner in the scientific investigation of consciousness.

Of course, if 'science–religion dialogue' were understood as this sort of task—and many, especially in the Asian traditions, do understand it in this way—then the gap between neurophenomenology and religion–science discussions would not be so great. Similarly, if the goal of gaining a deeper understanding of human experience is taken as a religious practice—as it certainly is in Buddhism—then neurophenomenology might be seen as part of, or at least parallel to, religious practice.

The Jamesian Heritage

Over 100 years ago, William James, in his *Principles of Psychology*, wrote that in the study of subjective mental phenomena, 'Introspective Observation is what we have to rely on first and foremost and always' (James 1981: 185). Psychology, as James presented it in this landmark book, is the study of subjective mental phenomena—mental events as experienced in the first person—as well as the study of how mental states are related to their objects, to brain states, and to the environment. Whereas physiological psychology studies the relation of mind and brain, including the naturally evolved 'mutual fit' of mental faculties and the environment, introspection studies mental states in their subjective manifestations. Yet, what exactly is introspection? James continued: 'The word introspection need hardly be defined—it means, of course, the looking into our own minds and reporting what we there discover. *Everyone agrees that we there discover states of consciousness*' (James 1981: 185).

This passage is often quoted, but less often remarked is that James hardly thought introspection to be easy or an infallible guide to subjective mental life. Later in his book, when discussing sensed moments of transition in the subjective stream of thought and feeling, he wrote:

Let anyone try to cut a thought across in the middle and get a look at its section, and he will see how difficult the introspective observation of the transitive tracts is.... The attempt at introspective analysis in these cases is in fact like seizing a spinning top to catch its motion, or trying to turn up the gas quickly enough to see how the darkness looks. (James 1981: 236–7)

James clearly did not think that we already know the nature and full range of thought and feeling simply because we are able to look into our own minds. In 1904 James heard the Theravada Buddhist renunciate Anagarika Dharmapala lecture at Harvard on the Buddhist conception of mind. According to the Buddhist view, there is no single, permanent, enduring self underlying the stream of mental and physical events. Afterwards, James rose and proclaimed to the audience, 'This is the psychology everybody will be studying twenty-five years from now.' He apparently meant not so much Buddhist psychology *per se*, but a psychology of the full developmental range of human consciousness, pursued with the kind of phenomenological precision exemplified by Buddhism (Taylor 1996: 146).

James's prediction, of course, was too optimistic. The words of another founding father of American psychology, James McKeen Cattell, also from 1904, indicate the path that much of psychology took in the years to come: 'It is usually no more necessary for the subject in a psychology experiment to be a psychologist than it is for the vivisected frog to be a physiologist' (Cattell 1904, as quoted by Lyons 1986: 23). The strategy that psychology pursued was to objectify the mind as much as possible, either as behavioural performance, physiological response, or, with the rise of cybernetics and then cognitive science, as non-conscious information processing. 'Consciousness' became a taboo term; introspection was rejected as a method for investigating the mind; and it was no longer necessary for the psychologist to have any disciplined first-person expertise in the subjectivity of mental life. Although there were notable exceptions to this trend, such as Gestalt psychology and phenomenological psychology, this 'taboo of subjectivity' (Wallace 2000) has influenced the scientific study of the mind for decades.

It has taken over a century, not a quarter of one, for the science of mind to begin to find its way back to James's vision of a science of mental life, including 'the varieties of religious experience' (James 1997), which integrates experimental psychology, neuroscience, and phenomenology. In recent years, a small but growing number of cognitive scientists have come to accept that there cannot be a complete science of the mind without understanding subjectivity and consciousness, and that cognitive science accordingly needs to make systematic use of introspective first-person reports about subjective experience (Jack and Roepstorff 2002, 2003). As cognitive neuroscientist Chris Frith recently stated: 'A major programme for 21st century science will be to discover how an experience can be translated into a report, thus enabling our experiences to be shared' (Frith 2002: 374).

CONTEMPLATIVE MENTAL TRAINING AND COGNITIVE SCIENCE

This renewed appreciation of the first-person perspective raises the question of how to obtain precise and detailed first-person accounts of experience. On the one hand, it stands to reason that people vary in their abilities as observers and reporters of their own mental lives, and that these abilities can be enhanced through mental training of attention, emotion, and metacognition. Contemplative practice is a vehicle for precisely this sort of cognitive and emotional training. On the other hand, it stands to reason that mental training should be reflected in changes to brain structure, function, and dynamics. Hence, contemplative practice could become a research tool for developing better phenomenologies of subjective experience and for investigating the neural correlates of consciousness.

The potential importance of contemplative mental training for scientific research on consciousness is central to neurophenomenology (Lutz *et al.* 2007). Concretely, neurophenomenology proposes to incorporate 'first-person methods' of examining experience into experimental research on subjectivity and consciousness. First-person methods sensitize individuals to their own mental lives through the systematic training of attention, emotion regulation, and metacognitive awareness (awareness of cognition) (Varela and Shear 1999). Such methods and training have been central to the Buddhist tradition since its inception (Wallace 1998, 1999). In Tibetan Buddhism, contemplative mental training is often described as a systematic process of 'familiarizing oneself' with the moment-to-moment character of mental events (Lutz *et al.* 2007). This description points towards the relevance of contemplative mental training to neurophenomenology: contemplative training cultivates a capacity for sustained, attentive awareness of the moment-to-moment flux of experience, or what James famously called 'the stream of consciousness'. For this reason, the Buddhist tradition holds special interest for neurophenomenology (Lutz *et al.* 2007; Varela *et al.* 1991).

It is worth reconsidering, from this vantage-point of contemplative mental training, how psychology came to reject introspection shortly after James. According to the standard history, introspection was given a fair try but failed. It allegedly failed because the two rival schools of introspectionist psychology were unable to agree whether there was such a thing as imageless thought. James had already observed, however, that the form of introspection practised by these schools was stilted and tedious, because it focused on the sensations caused by impoverished sensory stimuli (James 1981: 191–2). It is not surprising that introspection of this sort turned out to be so unilluminating, as Gestalt psychologists and phenomenologists also later remarked (Köhler 1947: 67–99; Merleau-Ponty 1962: 3–12). Furthermore, the textbook history neglects to mention that the rival schools did agree with each other at the descriptive level of introspective phenomenology; their disagreement was instead at the level of theoretical or causal interpretation. One lesson to be learned from this debate, therefore, is not that introspection is a useless method for obtaining descriptive accounts of subjective experience, but rather that psychology needs to discriminate carefully between the description of subjective phenomena and causal-explanatory theorizing (Hurlbert and Heavey 2001). A similar lesson should be drawn from the famous studies of Nisbett and Wilson in 1977: they observed that subjects often said that their behaviour was caused by mental events when it was really the result of external manipulation. Yet these inaccurate subjective reports were causal-explanatory in form, not rigorously descriptive and phenomenological. Again the lesson to be learned is that experimental participants need to be coached to pay strict attention to their felt cognitive processes and to avoid causal-explanatory conjectures (Hurlbert and Heavey 2001).

Yet how is such attention to be cultivated? First-person methods of examining experience are concerned with precisely this question (Varela and Shear 1999). What makes Buddhist contemplative mental discipline exemplary in this context is its pragmatic refinement and theoretical sophistication (Depraz *et al.* 2003). Whereas James described introspection as simply 'looking into our own minds and reporting what we there discover', Buddhism speaks of sustained attention to, and analytic discernment of, one's own mental processes. Buddhist phenomenology distinguishes between attentional stability and instability due to mental excitation, and between attentional vividness and dullness due to mental laxity (Wallace 1999). Buddhist phenomenology also discusses the metacognitive monitoring of these qualities of attention, and Buddhist epistemology discusses the degree to which a mental cognition ascertains or fails to ascertain its mental object, according to various conditions (Dreyfus 1997). According to this perspective, if the stream of thought and feeling is lucid, rather than turbulent and murky, then introspection in James's sense will be much richer in its discoveries and reports.

The working hypothesis of neurophenomenology appeals to this notion of refined first-person observation and description of subjective mental events. In an experimental context, this working hypothesis is twofold. First, phenomenologically precise first-person reports produced through mental training can provide important information about endogenous and externally uncontrollable fluctuations of moment-to-moment experience, such as quality of attention (Lutz *et al.* 2002). In addition, individuals who can generate and sustain a particular type of contemplative state cultivated in the Buddhist tradition—a state in which one's mind reposes, awake and alert, in the sheer 'luminosity' of consciousness (its quality of non-reflective and open awareness), without attending exclusively to any particular object or content—could provide important information about subjective aspects of consciousness not readily apparent or accessible to ordinary introspection or reflection (Lutz *et al.* 2007).

Second, the refined first-person reports produced through mental training can help to detect and interpret physiological processes relevant to consciousness, such as large-scale dynamical patterns of synchronous oscillatory activity in neural assemblies. Experimental studies following this approach have already cast light on the neurodynamics of conscious visual perception (Cosmelli *et al.* 2004; Lutz *et al.* 2002), epileptic activity and associated subjective mental events (Le Van Quyen and Petitmengin 2002), pain experience (Price *et al.* 2002; Rainville 2005), and the neurodynamical correlates of meditative states in highly trained Tibetan Buddhist practitioners (Lutz *et al.* 2004).

A further conjecture regarding contemplative mental training and experience is also important. Individuals who can generate and sustain specific sorts of mental states, and report on those states with a high degree of phenomenological precision, could provide a route into studying the causal efficacy of mental processes—how mental processes may modify the structure and dynamics of the brain and body. According to a neurodynamical perspective, mental states are embodied in large-scale dynamical patterns of brain activity (Thompson and Varela 2001), and these patterns both emerge from distributed, local activities and also globally shape or constrain those local activities. One can thus conjecture that in intentionally generating a mental state, large-scale brain activity shifts from one coherent global pattern to another, and thereby entrains local neural processes (Freeman 1999; Thompson and Varela 2001). Thus, individuals who can intentionally generate, sustain, and report on distinct types of mental states could provide a way of testing and developing this idea.

Neurophenomenological research based on the foregoing hypotheses has potentially profound implications for both cognitive science and contemplative wisdom traditions. Were such research to prove fruitful, adept contemplatives could become a new kind of scientific collaborator, rather than simply a new type of experimental participant, for their first-person expertise would be directly mobilized within scientific research on the mind. To conclude this chapter, I would like to relate this idea to the overall theme of this Handbook.

Towards a Contemplative Science of Mind

At the outset of this chapter, I stated that the aim of neurophenomenology is not to adjudicate between the claims of science and religion with regard to human experience, but to gain a deeper understanding of experience by making contemplative phenomenology a partner in the scientific investigation of consciousness. Varela, Thompson, and Rosch (1991) have described this approach as one of 'mutual circulation' between science and experience. According to the logic of mutual circulation, each domain of cognitive science, phenomenological philosophy, and contemplative mental training is distinct and has its own degree of autonomy—its own proper methods, motivations, and concerns—but they also overlap and share common areas. Thus, instead of being juxtaposed, either in opposition or as separate but equal, these domains can flow into and out of each other, and so be mutually enlightening.

This vision of mutual circulation does not fit easily within the established frameworks of the science–religion dialogue. We can appreciate this point by distinguishing the mutual circulation perspective from some of the main representative positions staked out in the science–religion dialogue, particularly as this dialogue touches on the nature of the human mind.

First, exploring the mutual circulation of mind science and contemplative experience is different from viewing science and religion as 'non-overlapping magisteria' (Gould 1999). This separate-but-equal strategy of insulating science and religion is highly problematic. It divides science and religion along the lines of a subject–object dualism: science addresses the empirical world conceived as a realm of objectivity, whereas religion addresses the subjective realm of human purposes, meaning, and value. Yet this subject–object dualism breaks down in the face of the intersubjectivity of human experience (Thompson 2005). Intersubjective experience is the common terrain of both science and religion, and it is poorly understood when fractured along the lines of a subject–object (or fact–value) dichotomy (Wallace 2005).

Second, the mutual circulation approach is different from looking for the physiological correlates of religious experiences (e.g. Newberg et al. 2001). The key difference is that adept contemplatives, as mentioned above, are considered not simply as experimental participants, but as scientific collaborators. Thus, the mutual circulation approach enables us to envision future cognitive scientists being schooled in contemplative mental training and phenomenology, as well as brain-imaging techniques and mathematical modelling, and future contemplative practitioners being knowledgeable in neuroscience and experimental psychology. Science and contemplative knowledge could thus mutually constrain and enrich each other. James envisioned this sort of prospect over a century ago in his writings on scientific psychology and religious experience (see Taylor 1996).

Third, the mutual circulation approach is different from the cognitive science of religion, especially evolutionary psychology explanations of religious thought and behaviour (Boyer 2001, 2003, 2005). Although these explanations are illuminating in linking religious concepts to our intuitive understandings of agency, social relations, and misfortune (see Boyer 2005), they neglect the contemplative aspect of certain religious traditions. Whereas evolutionary psychology takes religious notions and norms as objects of explanation for evolutionary and functionalist cognitive science, neurophenomenology looks to the role that contemplative mental training and experience can play in a phenomenologically enriched cognitive science.

A common feature of the three approaches to science and religion I have contrasted with the mutual circulation approach is that they take the concepts of 'science' and 'religion' largely for granted. These concepts, however, are deeply problematic. They are European intellectual categories that have been shaped in recent Western history by the science–religion conflicts of the European Enlightenment and modernity. As such, they do not map in any clear way on to the knowledge formations and social practices of certain other cultural traditions, in particular those of Asian contemplative wisdom traditions (see Hut 2003). As Wallace has recently written in his introduction to a volume on Buddhism and science:

> The assertion that Buddhism includes scientific elements by no means overlooks or dismisses the many explicitly religious elements within this tradition.... Buddhism is very much concerned with human purposes, meaning, and value. But, like science, it is also concerned with understanding the realms of sensory and mental experience, and it addresses the questions of what the universe, including both objective and subjective phenomena, is composed of and how it works.... Buddhism does address questions concerning the meaning and purpose of life, our ultimate origins and destiny, and the experiences of our inner life. But the mere fact that Buddhism includes elements of religion is not sufficient for singularly categorizing it as a religion, any more than it can be classified on the whole as a science. To study this discipline objectively requires our loosening the grip on familiar conceptual categories and preparing to confront something radically unfamiliar that may challenge our deepest assumptions. In the process we may review the status of science itself, in relation to the metaphysical axioms on which it is based. (Wallace 2003a: 9–10)

In this chapter, I have proposed that certain contemplative wisdom traditions— Buddhism most notably, though not exclusively—and certain approaches in cognitive science—the embodied approach and neurophenomenology—are not simply compatible, but mutually informative and enlightening. Through back-and-forth circulation, each approach can reshape the other, leading to new conceptual and practical understandings for both. At stake in this possibility is nothing less than the prospect of a mature science of the mind that can begin to do justice to the rich and diverse traditions of human contemplative experience.

REFERENCES AND SUGGESTED READING

BOYER, P. (2001). *Religion Explained: The Evolutionary Origins of Religious Thought.* New York: Basic Books.

—— (2003). 'Religious Thought and Behaviour as By-products of Brain Function', *Trends in Cognitive Sciences,* 3: 119–24.

—— (2005). 'Gods, Spirits, and the Mental Instincts that Create Them', in J. Proctor (ed.), *Science, Religion, and the Human Experience,* New York and Oxford: Oxford University Press, 237–60.

CATTELL, J. M. (1904). 'The Conceptions and Methods of Psychology', *Popular Science Monthly,* 60.

CLARK, A. (2001). *Mindware: An Introduction to the Philosophy of Cognitive Science.* New York and Oxford: Oxford University Press.

COSMELLI, D., DAVID, O., LACHAUX, J.-P., MARTINERIE, J., GARNERO, L., RENAULT, B., and VARELA, F. J. (2004). 'Waves of Consciousness: Ongoing Cortical Patterns during Binocular Rivalry', *Neuroimage,* 23: 128–40.

DEPRAZ, N., VARELA, F. J., and VERMERSCH, P. (2003). *On Becoming Aware: A Pragmatics of Experiencing.* Amsterdam and Philadelphia: John Benjamins Press.

DREYFUS, G. (1997). *Recognizing Reality: Dharmakirti's Philosophy and its Tibetan Interpretations.* Albany, N.Y.: State University of New York Press.

—— and THOMPSON, E. (2007). 'Indian Theories of Mind', in P. D. Zelazo, M. Moscovitch, and E. Thompson (eds.), *The Cambridge Handbook of Consciousness,* New York and Cambridge: Cambridge University Press.

FREEMAN, W. J. (1999). *How Brains Make up their Minds.* London: Weidenfeld & Nicolson.

FRITH, C. (2002). 'How Can We Share Experiences?', *Trends in Cognitive Sciences,* 6: 374.

GOULD, S. J. (1999). *Rocks of Ages: Science and Religion in the Fullness of Life.* New York: Ballantine.

HURLBERT, R. T., and HEAVEY, C. L. (2001). 'Telling What We Know: Describing Inner Experience', *Trends in Cognitive Sciences,* 9: 400–3.

HUT, P. (2003). 'Conclusion: Life as a Laboratory', in B. Alan Wallace (ed.), *Buddhism and Science: Breaking New Ground,* New York: Columbia University Press, 399–416.

JACK, A. I., and ROEPSTORFF, A. (2002). 'Introspection and Cognitive Brain Mapping: From Stimulus-Response to Script-Report', *Trends in Cognitive Sciences,* 6: 333–9.

—— —— (2003) (eds.). *Trusting the Subject? The Use of Introspective Evidence in Cognitive Science,* i. Thorverton, UK: Imprint Academic.

JAMES, W. (1981). *The Principles of Psychology.* Cambridge, Mass.: Harvard University Press.

—— (1997). *The Varieties of Religious Experience.* New York: Touchstone Press.

KÖHLER, W. (1947). *Gestalt Psychology.* New York: Liveright.

LE VAN QUYEN, M., and PETITMENGIN, C. (2002). 'Neuronal Dynamics and Conscious Experience: An Example of Reciprocal Causation before Epileptic Seizures', *Phenomenology and the Cognitive Sciences,* 1: 169–80.

LUTZ, A., and THOMPSON, E. (2003). 'Neurophenomenology: Integrating Subjective Experience and Brain Dynamics in the Neuroscience of Consciousness', *Journal of Consciousness Studies,* 10: 31–52.

—— DUNNE, J., and DAVIDSON, R. J. (2007). 'Meditation and the Neuroscience of Consciousness: An Introduction', in P. D. Zelazo, M. Moscovitch, and E. Thompson (eds.), *The Cambridge Handbook of Consciousness,* New York and Cambridge: Cambridge University Press.

—— LACHAUX, J.-P., MARTINERIE, J., and VARELA, F. J. (2002). 'Guiding the Study of Brain Dynamics by Using First-Person Data: Synchrony Patterns Correlate with Ongoing Conscious States during a Simple Visual Task', *Proceedings of the National Academy of Sciences* (USA), 99: 1586–91.

—— GREISCHAR, L. L., RAWLINGS, N. B., RICARD, M., and DAVIDSON, R. J. (2004). 'Long-Term Meditators Self-Induce High-Amplitude Gamma Synchrony during Mental Practice', *Proceedings of the National Academy of Sciences* (USA), 101: 16369–73.

LYONS, W. (1986). *The Disappearance of Introspection*. Cambridge, Mass.: MIT Press.

McGEE, M. (2000). *Transformations of Mind: Philosophy as Spiritual Practice*. Cambridge: Cambridge University Press.

MERLEAU-PONTY, M. (1962). *Phenomenology of Perception*, trans. Colin Smith. London: Routledge.

MORAN, D. (2000). *Introduction to Phenomenology*. London: Routledge.

NEWBERG, A., D'AQUILI, E., and RAUSE, V. (2001). *Why God Won't Go Away: Brain Science and the Biology of Belief*. New York: Ballantine Books.

NISBETT, R. E., and WILSON, T. D. (1977). 'Telling More Than We Can Know: Verbal Reports on Mental Processes', *Psychological Review*, 84: 231–59.

PRICE, D., BARRELL, J., and RAINVILLE, P. (2002). 'Integrating Experiential-Phenomenological Methods and Neuroscience to Study Neural Mechanisms of Pain and Consciousness', *Consciousness and Cognition*, 11: 593–608.

RAINVILLE, P. (2005). 'Neurophénoménologie des états et des contenus de conscience dans l'hypnose et l'analgésie hypnotique', *Théologique*, 12: 15–38.

TAYLOR, E. (1996). *William James, on Consciousness beyond the Margin*. Princeton: Princeton University Press.

THOMPSON, E. (2005). 'Empathy and Human Experience', in J. Proctor (ed.), *Science, Religion, and the Human Experience*, New York and Oxford: Oxford University Press, 261–86.

—— (2006). *Mind in Life: Biology, Phenomenology, and the Sciences of Mind*. Cambridge, Mass.: Harvard University Press.

—— and VARELA, F. J. (2001). 'Radical Embodiment: Neural Dynamics and Consciousness', *Trends in Cognitive Sciences*, 5: 418–25.

VARELA, F. J. (1996). 'Neurophenomenology: A Methodological Remedy for the Hard Problem', *Journal of Consciousness Studies*, 3: 330–50.

—— and SHEAR, J. (1999) (eds.). *The View from Within: First-Person Approaches to the Study of Consciousness*. Thorverton, UK: Imprint Academic.

—— THOMPSON, E., and ROSCH, E. (1991). *The Embodied Mind: Cognitive Science and Human Experience*. Cambridge, Mass.: MIT Press.

WALLACE, B. A. (1998). *The Bridge of Quiescence: Experiencing Tibetan Buddhist Meditation*. La Salle, Ill.: Open Court.

—— (1999). 'The Buddhist Tradition of *shamatha*: Methods for Refining and Examining Consciousness', in Varela and Shear (1999), 175–88.

—— (2000). *The Taboo of Subjectivity: Toward a New Science of Consciousness*. New York: Oxford University Press.

—— (2003a) (ed.). *Buddhism and Science: Breaking New Ground*. New York: Columbia University Press.

—— (2003b). 'Introduction: Buddhism and Science', in Wallace (2003), 1–29.

—— (2005). 'The Intersubjective Worlds of Science and Religion', in J. Proctor (ed.), *Science, Religion, and the Human Experience*, New York and Oxford: Oxford University Press, 309–27.

CHAPTER 15

PSYCHOLOGY, THE HUMAN SCIENCES, AND RELIGION

RAYMOND F. PALOUTZIAN

INTRODUCTION

It is an interesting puzzle that two broad areas of human endeavour and inquiry, the human sciences and religion, both have among their ultimate goals to understand and help human beings. It is hard to find an aim in all of human scholarship that is more lofty and more important than this one. This is because, whether through religious avenues or through scientific ones, if we are able to fully understand how human beings function—think, feel, and act—we will have struck upon the single most important piece of knowledge of all time. This knowledge is the intellectual golden ring. He or she who has knowledge has power. He or she who has knowledge of the workings of human beings, whether it is implemented through religious or through secular media, has ability akin to that held by the possessor of the golden ring of ancient tales—the power to influence, guide, mould, and yes, even control the behaviour of other people.

I wish to thank Erica Swenson for her help in the preparation of this chapter, and the Catlin Foundation, whose grant supported her research assistantship. Portions of this chapter were presented as part of a symposium entitled 'Integrative Themes in the Current Science of the Psychology of Religion' (Crystal L. Park, chair) at the American Psychological Association Convention, Washington DC, August 2005.

Those who aspire to liberal, democratic, and humanistic ideals would rightfully cringe at the thought of such knowledge being placed in the charge of only a specialized group of authorities, whether they be religious leaders or scientists of human nature. It would be preferred that the ring stay in its resting place in the bottom of the sea. This is because, as the hard march from the past shows us, individuals or small groups of persons entrusted with a great storehouse of treasures cannot be trusted to guard and use those treasures with wisdom and compassion for the good of the whole. Yet it is precisely the kind of knowledge that would make this possible, at least in principle, that those in the human sciences of religion, and in particular in the psychology of religion, are trying to discover or, more accurately, to create. We intend to know the ring.

IT's ALL PSYCHOLOGICAL

Let us explore the picture that is taking shape as we begin to see what this knowledge looks like and begin to understand the human creature described by it. The psychology of religion is a rich storehouse of treasures that offers a compelling argument to the science–religion dialogue. This will become clear as we examine research approaches and what they can and cannot tell us about religion, present the scope of scholarly areas of inquiry and the interface of each one with religion, and identify the fundamental, core issues that prevail across the myriad specific areas of research.

I am convinced that the psychology of religion and its companion human sciences of religion are at a threshold. Research and theoretical advances have occurred during the past twenty-five years that are sufficiently powerful to guide scholarship in both intradisciplinary and interdisciplinary ways as far as the eye can see. The fields look poised, ready to begin the work that will become their biggest contribution combined with its biggest risk: i.e. the complete understanding of the psychological processes involved in religion. This progress has been documented in the *Handbook of the Psychology of Religion and Spirituality* (Paloutzian and Park 2005a). The research base is vast, and ranges from the micro (e.g. neurological factors in religious or spiritual experience) to the macro (religious factors in international terrorism and peace). Park and I (Paloutzian and Park 2005b) have proposed that the integration of the vast body of research and that of the future may be facilitated by two pivotal ideas: the multilevel interdisciplinary paradigm (Emmons and Paloutzian 2003) and the model of religion as a meaning system (Park and Folkman 1997; Park 2005a; Silberman 2005a, 2005b). These two ideas, respectively, provide the overarching umbrella under which integrative research and theory is encouraged and a common language that can apply to religion as it is studied across all topic areas (Park and Paloutzian 2005).

PSYCHOLOGY OF RELIGION OR RELIGIOUS PSYCHOLOGY?

I take it as a given that in order for this large science–religion conversation to yield benefits for the world it is necessary for all participants to engage in it with a truly open mind. In the ideal case, all persons would listen and speak to the issues with detached objectivity. But it is questionable whether 'objectivity' exists, and even if it does, psychological research consistently shows that people approach tasks of perception and interaction from the point of view of their own biases (Alicke *et al.* 1995). Scholars are no different.

Nevertheless, the importance of enabling all minds to be genuinely open is made clear when we examine the difficulty of finding common ground when one or more of the participants believes that he or she has the truth and that it is absolute. For example, those who espouse so-called Creation Science insist that the earth is between 6,000 and 10,000 years old, deriving this view from one interpretation of the Book of Genesis. This 'young earth' view is arrived at logically prior to and independent of an examination of the data. Thus when the data are examined, they are attributed to a process that conforms to the pre-held view. No other conclusion is acceptable. It would seem, therefore, that if a point of view is already held in a sufficiently fixed, absolute way, there is little real capacity for dialogue. The danger is that there may be only combat.

Similar difficulties can be seen among psychologists attempting to dialogue who presuppose either a secular Western epistemology or a Muslim religion-based epistemology. Khalili and Colleagues (2002) and Murken and Shah (2002) present a dialogue between a Western psychologist of religion and an Islamic psychologist. The epistemological assumptions of the two participants differ in the extreme: they are incompatible. In the secular Western mind, religion is an aspect of culture similar to other aspects. By contrast, in the strict Islamic mind, religion defines the culture, so that all other aspects of it are subsumed within the religion (see the remarks by Murken in Khalili *et al.* 2002). This would include knowledge and the conduct of science. The published dialogue illustrates the long way we have to go to bring others into this conversation. Sooner or later it is necessary for people on opposite sides of an issue to collaborate. In the psychology of religion this would be exemplified by psychological researchers conducting projects on problems that concern them both and that cannot be addressed without the full mutual participation of both sides. It is the co-operation of people who think in different ways, working on research projects of mutual benefit, that automatically engages them in dialogue with each other. Fundamentally, this means doing psychology of religion rather than religious psychology.

Religion is not one thing, however. It is a multidimensional variable whose facets include beliefs, knowledge, practices, feelings, effects, myths or foundational stories, and ethics (see texts such as Paloutzian 1996; Wulff 1997; Spilka *et al.* 2003, for discussions of these dimensions; see Smart 1989, for an application of them).

Furthermore, many people in the world have either an expanded or a different concept of that which transcends them and guides their lives. For some people it is a personal God but with less connection to traditional religious institutions; for others it is an alternative form of supreme being; and for others it is a purpose, principle, state of being, or other construction of ultimate values or concern (Emmons 1999). Although it is probably psychologically functionally equivalent to being religious, many people today prefer to call themselves 'spiritual' instead of or in addition to 'religious'. Therefore, dialogue between science and religion will probably have to embrace the span of meanings included within both religion and spirituality (Zinnbauer and Pargament 2005).

RELIGION AND SPIRITUALITY

Over the course of the twentieth century the use of the terms 'religion' and 'spirituality' evolved on somewhat parallel but partially overlapping tracks. It seems commonly assumed that prior to mid-century these two terms were understood to mean more or less the same thing. A devoutly religious person understood him- or herself to be spiritual, and someone who was seen as a spiritual person would be so labelled because he or she demonstrated genuineness or consistency in the practice of a faith. Roughly speaking, this is how these two concepts would have been understood for approximately the first half of the last century.

What has 'religion' come to mean?   Somewhere around mid-century, however, the usage of these two terms began to diverge. 'Religion' gradually came to have reference primarily to traditional, established faith traditions. These have known histories, organizations, and outlets for routine activities such as worship, ministry, or outreach. In simple categories, religion meant Protestant, Catholic, or Jewish. In psychological and sociological research on religion, this three-group categorization was the operational definition of religion for many studies (Argyle and Beit-Hallahmi 1975). Some additional precision was attained in studies that further subdivided Protestants into denominations and subdivided Jews into the three main branches of Judaism.

At the same time, people began to look outside established church structures for an additional resource or as an alternative route to enhance their spirituality and meaning in life. The traditional church was not working for many people, so they searched for other ways to connect with something larger than themselves. This emerging search for spirituality expressed itself in both religious and non-religious forms.

Many new religious movements (NRMs) emerged (Melton 1986), many of whose adherents were attracted to the group they were in because of a search for genuine spirituality (Richardson 1995; Paloutzian, Richardson, and Rambo 1999). Some NRM converts were searching for a spirituality different from (some would say more 'authentic' than) the religion in which they were raised. For them, participation in their new religion was still a matter of finding spirituality within a group that defined itself in religious terms.

What has spirituality come to mean? For others the search to satisfy spiritual needs led them away from identifiably religious groups. Their alternative was a non-religious form of spirituality. Whatever aspects of traditionally religious institutions these departing souls were responding to (e.g. emphasis on ritual, formality, trad-itionalism, doctrine, creeds, and codes), the search was on for a meaning, purpose, and fulfilment that would transcend them; they found it unsatisfactory when it was defined in the traditional terms of God and church. For these persons, the conver-sation changed from which doctrines to believe and which religious practices to engage in to which values to hold and which experiences to enhance.

How are spirituality and religion similar and different? During the past fifteen years research has empirically teased apart what *religion* and *spirituality* mean to people (Pargament 1997; Zinnbauer *et al.* 1997; Hill *et al.* 2000; Zinnbauer and Pargament 2002, 2005; Hill and Pargament 2003). The two concepts overlap but are not synonym-ous. Both religion and spirituality tend to be associated with frequency of prayer, church attendance, and intrinsic religious orientation (see Allport and Ross 1967 and Hill and Hood 1999), and both terms connote attempts to connect with that which is perceived to be sacred (Hill *et al.* 2000). However, spirituality is more associated with mystical experiences and being hurt by clergy, and connotes more concern with personal growth and existential issues. In contrast, religion is more associated with authoritarianism, church attendance, and self-righteousness, and connotes more personal and institu-tional practices and commitment to church or denominational beliefs (Zinnbauer *et al.* 1997). Thus, the way people describe their spirituality and/or religiousness takes differ-ent forms. For example, people say that they are both religious *and* spiritual, *neither* religious *nor* spiritual, spiritual *but not* religious, religious *but not* spiritual, or that they embrace a peculiar combination of religious spirituality combined with non-religion: as was said by one of my students, 'I am a spiritual Christian *but not* religious' (Paloutzian and Park 2005*b*).

Implications for the Science–Religion Dialogue

Two implications follow from the above considerations about what religion is. First, and most important for psychology of religion research, is that it may not matter whether a person says that he or she is religious or applies traditional God or church language to him or herself. More fundamental than the words that someone uses is the psychological process that underlies and mediates people's commitment to that which lies beyond themselves (Frankl 1963; Emmons 1999, 2000). Psychologically speaking, it does not matter whether this is called religion, spirituality, or something else. The terminology seems to be a matter of personal preference (Paloutzian and

Park 2005b). Modern research reflects this. For example, Hill and Hood's (1999) compendium of measures includes many that are phrased to assess people's religiousness, spirituality, and related processes expressed in a variety of ways. Future research will need to expand to encompass the increasingly diverse ways in which people talk about what may be functionally the same process.

Second, and equally important for the broader science–religion dialogue, it is essential that the focus of the conversation be understood as primarily psychological. A variety of disciplines and a number of religions are participants in the dialogue, as is evident from the contents of this Handbook. Each of them plays a crucial role, whether historical, religiously positional, scientific, or philosophical. But at the most fundamental level our concern is not with, for example, the history of the relation between science and the church or with a 'real' conflict between one approach to Christianity or Islam and the conduct of science. Our fundamental concern is with the meanings that these things have to people today. This is because it is to those current constructions of meaning that people respond (Park 2005a; Silberman 2005b), even if those have roots in the past or in one's religious teaching. The historical path of an issue or the pronouncements that a religion makes in relation to science matter insofar as they are processed psychologically in the here and now. Thus, the contribution of the psychology of religion to the dialogue is fundamental, because our task is enhanced when we understand the unique role that religiousness plays functionally in the human mind. Therefore, scholarship by others in the science–religion field would be enhanced in two important, perhaps indispensable ways by incorporating research in the psychology of religion. First, the exploratory (or 'dialoguing') process is itself a matter of the perception, processing, and interpretation of information within a person's meaning system. This means that whether someone understands or accepts a point of view different from his or her own depends not only on logical arguments or evidence, but also heavily on the psychological processes involved in the construction and maintenance of meaning. Psychologically, truth is in the meaning system of the beholder. Second, in the end people may respond to what a new idea means to them no matter how logically compelling it might be. Thus the goal of significantly fostering the well-being of humans requires that psychological knowledge be integral to the process. To have a good impact, our logic needs psycho-logic.

Religion as Unique and Non-unique

The issue of whether religion is like or unlike other human activities is foundational to the psychology of religion (Dittes 1969), and perhaps to the larger science–religion dialogue. If religion operates in the same way as any other human activity operates, then it is non-unique, an instance of behaviour in general, and there is no compelling reason other than the practical importance of religion in the world for psychology or any other science to engage it (McCrae 1999). If, on the other hand, religion is an activity that is intrinsically different, that plays a role or operates in ways that nothing else does, then it is unique and warrants this dialogue because scientific knowledge cannot be complete without it, and solutions to problems may be possible that could not be had any other way (McCrae 1999). The science–religion dialogue seems largely to assume the latter position—that there is something about religion that is unique and that drives human life to such a degree that science must engage it.

The broad science–religion conversation, as well as the psychological study of individual people, will contribute the most by adopting the approach that the unique and the non-unique assumptions are both true. Looking at religion from the point of view of a psychologist, it is obvious that much religious belief and behaviour and many religious emotions and cognitions operate by the same processes by which any other beliefs, behaviour, emotions, and cognitions operate. This should neither surprise nor threaten anyone, including the strict religious believer. But also, there is a mounting body of evidence (Paloutzian and Park 2005a; Pargament 2002) that there are aspects of religion not found elsewhere. One often noted aspect of this uniqueness is religion's ability to draw commitment to that which the person perceives as sacred (Pargament 1997; Pargament, Magyar-Russell, and Murray-Swank 2005; Silberman 2005b). Whatever this unique aspect of religion is, however, it does not seem to manifest itself at a psychological level only. Its expression is evident across various levels of analysis and multiple disciplines.

INTEGRATING THE RESEARCH

One of the most daunting tasks in an enterprise as grand as the science–religion dialogue, whose charge includes pooling together ideas from many fields of science and many religions, is how to integrate the vastly diverse database and ideas. The closer we come to identifying elements common to a number of areas, the more likely we are to settle upon ways of synthesizing them. No field has come upon the single best schema for how to do this, but Crystal Park and I recently proposed that five themes may be sufficient for this integrative function for psychology of religion research (Paloutzian and Park 2005b; Park and Paloutzian 2005). The psychology of religion is 100 years old and has never had an idea, theory, principle, or assembly of them that could integrate the diverse approaches and types of data in the field. James Dittes (1969) correctly and pointedly informed us that we had large reams of data, various species of theory big and small, and each had little to do with any other. There were two psychologies of religion: one of theories (Freud, Jung, and variations) and one of numbers (questionnaire responses on myriad religious beliefs, practices, experiences, etc.). Each went forward on its own track, either incapable of, or uninterested in, relating to the other. Meanwhile, the rest of psychology was moving step by step from being an unintegrated field with an array of disconnected topics to its present state with extraordinary cross-theoretical and cross-empirical work among its subdisciplines. Sigmund Koch, in his mid-century six-volume review (1959–63), concluded that psychology could not be a coherent science. But the opposite came to be true. I watched psychology evolve into a synthetic discipline with an increasingly unified and unifying idea and database integrated by a broad cognitive orientation that connects other approaches. Perhaps these themes can facilitate integration across the broad range of issues that make up the larger science–religion dialogue.

The five integrative themes that we think are sufficient to enable us to dialogue about the whole field are (1) the paradigm issue, (2) methods and theory, (3) the question of meaning, (4) the path of the psychology of religion, and (5) the role of the psychology of religion.

Multilevel Interdisciplinary Paradigm

The psychology of religion has almost always been pre-paradigmatic. Richard Gorsuch (1988) took a look in the first-ever *Annual Review of Psychology* chapter on this topic; he concluded that the field did not yet have an integrative idea or common language. The truth of the time: psychologists of religion were still trying to measure what their topic was about. The good news is that from 1988 to the publication of the second-ever *Annual Review* chapter (Emmons and Paloutzian 2003) the field was transformed in the richness of the data collected, the range of methods used, and the ideas driving the research and used to interpret the data (Hood and Belzen 2005). The field is doing what the rest of psychology has been doing: i.e. gradually moving towards a synthesis of the varied data around common ideas (Paloutzian and Park 2005b; Park and Paloutzian 2005). As a reflection of this trend and as a stimulus to further it, we argue for the use of the *multilevel interdisciplinary paradigm* (Emmons and Paloutzian 2003). *Multilevel* means that research in the specialized areas within the psychology of religion can be interrelated and brought together in a common theory; *interdisciplinary* means cross-fertilization of research in the psychology of religion with that of allied fields such as anthropology, neuroscience, biology, sociology, and so forth. This paradigm is secondarily a description of how the field has been in the recent past, and is primarily a concept that we hope guides all of our thinking in the future.

Method and Theory

For much of the field's history, theory and empirical research in the psychology of religion had little to do with each other. The well-known theories about the psychological bases of religion written by Freud (1927) and Jung (1938), for example, had no major counterpart in empirical research. Likewise, the mostly questionnaire data empirical studies were done with seemingly no concern for theoretical relevance. But it is precisely the creation of good theory to which good data contribute, due to a method–theory–method feedback loop that is inherent in all sciences. Recent scholarship seems to have changed this, however (Corveleyn and Luyten 2005; Kirkpatrick 2005a, 2005b), so that the interaction between the numbers and the ideas can cut across all topic areas and help bring them together in a more unified picture.

Theory

Most gratifying in the area of theory are recent advances in empirical approaches to testing modern psychoanalytic theories of religion. Corveleyn and Luyten (2005) have documented a growing body of psychoanalytic research that uses methods other than the traditional case-study method (Luyten, Blatt, and Corveleyn 2006). These recent approaches enable data collected to test psychoanalytic ideas to interact with their counterpart material from mainstream psychology. They include techniques such as experience sampling, diary methods that allow for a more in-depth look at psychological processes in real life, and longitudinal methods enhanced by advanced statistical techniques, including structural equation modelling, growth curve modelling, and survival analysis (Willett, Singer, and Martin 1988). Corveleyn and Luyten (2005) explain how approaches such as these make it possible to examine empirically the complexity of psychodynamic hypotheses.

Equally gratifying is the recent introduction of evolutionary psychology to the psychology of religion. Lee Kirkpatrick (2005a, 2005b) has made a strong case that evolutionary psychology can serve as an overarching meta-theory that, in addition to already having demonstrated its power in the biological sciences, has the capability of subsuming under its large umbrella a number of psychological models and theories that are smaller in scope. For example, he has explained how a number of the well-known ideas in recent psychology of religion research such as intrinsic versus extrinsic-quest religious orientation (Batson, Schoenrade, and Ventis 1993), fundamentalism (Altemeyer and Hunsberger, 2005), the tendency to make supernatural attributions (see Spilka et al. 2003, for review), and spiritual intelligence (Emmons 1999, 2000), each of which has its own database and explanatory reach, may be able to stand both time and test and fit within a larger system of mini-models as well as mid-level theories such as attachment theory (Kirkpatrick 2005a; Granqvist 2006) by meeting a set of criteria for an adaptive process. Key to his argument is that it is not sufficient merely to say that a process is adaptive. It is necessary to explain why it is so. The system of logic offered by Kirkpatrick combined with recent psychodynamic, mid-level, and mini-model theories, may be able to accomplish the integrative vision so long as the right sorts of data are available.

Methods

The recent methodological advances may be rich enough to allow this. No longer are empirical studies in the psychology of religion carried out only by giving questionnaires to people and calculating zero-order correlations among all possible combinations of variables (Hunsberger 1991). Instead, we now have a menu of both quantitative and qualitative approaches for looking at religion, both from the point of view of the outside observer and from the point of view of inside the mind of the religious person. The quantitative approaches reflect the traditionally understood objective approach in the nature of the data collected. The qualitative approaches (also called hermeneutical, because they are heavily interpretive procedures) are more recent, and allow for detailed examination of the phenomenology of religion, an attempt to see it as it is seen by the experiencing person. Such studies are often based on the researcher's interpretation of a description of religious experiences or meanings. The criticism of this approach is that the researcher's own hermeneutical bias can determine the categories into which bits and pieces of information are placed.

Hood and Belzen (2005) illustrate how it is the combination of these methods that gives us the rich blend of data that we need in order to get a glimpse of the complexity of religious meaning. One line of Hood's research is illustrative. For a number of years Hood and others have examined religious serpent handling by people in the rural US southern states. What has been learned comes from studies with open-ended interviews and phenomenological methods to identify the experience of handling serpents from the handler's perspective (Williamson, Polio, and Hood 2000), electrophysiological measures of handlers in the laboratory (Burton 1993), an oral history of the handlers' tradition (Hood, in press), videotaping of a number of serpent-handling services of the same group over an extended period of time (Hood-Williamson Archive for Serpent-Handling Sects of Appalachia), which allows for longitudinal studies to be conducted, and participant observation (Hood and Kimbrough 1995). Add to this the possibility of time-sampling subjects' mental events and behaviours both *in vivo* and under random assignment experimental manipulations, invoking emotional states and/or memories mimicking those before, during, and after real services by either role-play or hypnotic induction procedures, and the picture that emerges of the psychological meanings involved in religious serpent handling becomes much more intricate and complete. Further, the use of multiple methods facilitates multilevel intradisciplinary and interdisciplinary research (Silberman 2005*b*).

Methodological Pluralism

Arguments have been made in favour of one or the other approach. I think, however, that the issue of whether the quantitative or the qualitative approach gets us closer to the truth about the psychological meaning of religion is akin to the problem of whether the chicken or the egg came first. Fortunately, Corveleyn and Luyten (2005) argue for a methodological pluralism whereby those who use one approach would see the other as complementary and cross-fertilizing. They state their case this way.

[W]e believe that the existing divide ... between a hermeneutic, interpretive approach ... and a (neo-)positivistic approach ... is not only to a large extent artificial, but also unfruitful. ... There is no (quasi-)experimental research without previous theorizing and subsequent interpretation. Likewise, interpretations can and should be empirically tested. ... Whereas it can be said that much (quasi-)experimental research in the psychology of religion concerns 'impeccable studies of nothing very much,' many interpretive studies are vulnerable to the critique that 'anything goes' in such studies. Hence, instead of seeing these approaches as conflicting, they should rather be seen as completing each other, with much possibility of mutual enrichment. (Corveleyn and Luyten 2005: 87–8)

The Question of Meaning

Scholars have recently proposed that the concept of *religion as a meaning system* provides a common language capable of connecting diverse areas of psychology of religion research (Park and Folkman 1997; Park 2005a; Silberman 2005a). Questions of meaning are typically construed as theological or philosophical, but they are also and, perhaps fundamentally, psychological questions. When we ask what something means, we are asking what it stands for, what its implications are, what its representations and connections are in the human mind (Baumeister 1991; Wong and Fry 1998; Park 2005a; Silberman 2005b). Meaning is always in relation to something else. Thus, to create a theory of the psychological processes in religiousness that captures the heart and soul of what it is about, we need to answer the question of meaning's meaning in religion, because religion is essentially about meaning. By extension, then, whether we're talking about religious experience (Hood 2005), religious development (Boyatzis 2005; Levenson, Aldwin, and D'Mello 2005; McFadden, 2005), religious actions (Donahue and Nielsen 2005; Spilka 2005), the neurological processes involved in religious experience (Newberg and Newberg 2005), coping (Park 2005b), religious processes in physical and mental health (Miller and Kelley 2005; Oman and Thoresen 2005), or the role of religion in international terrorism and peace efforts (Silberman 2005c), we are invoking questions about meaning.

I can but sketch a brief picture of what a meaning system is. Since there are different accounts of meaning systems (e.g. Park and Folkman 1997; Park 2005a; Silberman 2005a), even if they share a common core (diagrammed well by Park 2005a), I will present my own social-psychological way of construing it (Paloutzian 2005). A meaning system is a structure within a human cognitive system that includes attitudes and beliefs, values, focused goal orientations, more general overall purposes, self-definition, and some locus of ultimate concern. Each element affects the others, so that when pressure is imposed on one component of the system (e.g. a person who holds one set of religious beliefs is introduced to beliefs of a different religion and is encouraged to convert), the beliefs that are under pressure confront information already in the other elements of the system, and if changed in the direction of the pressure may become inconsistent with them. If the initial beliefs are resistant enough, the ties among the elements of the system sustain belief and repel the pressure to change. But if resistance capabilities have not developed, the pressure on the beliefs may be strong enough to dent or topple other elements of the system. Thus a meaning system can be modified in one or more aspect(s). Any modification constitutes some amount of transformation of the meaning system. When the degree of change reaches a certain threshold, we call it religious conversion.

If the whole system is replaced by a completely different one, then we consider it a dramatic spiritual transformation, the relatively rare example of the radical convert.

Using this brief portrait of how the model of religion as a meaning system helps us understand religious conversions and spiritual transformations, it is easy to extrapolate and apply this model to all aspects of religiousness, from the micro (neuropsychology of religious experience) to the macro (religious motivation to or justification of violence and terrorism). It may be possible that an extrapolation of this model might serve some of the needs within the science–religion dialogue.

The Path and the Role of the Psychology of Religion

Inherent in what we do are the fourth and fifth integrative themes, i.e. the path and the role of the psychology of religion. The very process of doing this science means that we either are, or hope to be, on a path that goes somewhere worth our time. We hope that this includes development of the science itself, and in a way that feeds and draws from other fields. This has often been said, but now we have a label for a paradigmatic idea (the multilevel interdisciplinary paradigm) and a common language for religion that is capable of facilitating cross-fertilization of diverse research areas (religion as a meaning system) to help us see better where to go. Closely related to this, the role of the psychology of religion is usually said to include the contribution of something unique to general psychology and the generation of knowledge that can be translated into application for human good. These goals seem to be fostered by these recent disciplinary advances.

Summary

Three of the five integrative themes seem straightforward: the path of the research, the role of the psychology of religion, and the methods–theory–methods feedback loop. No matter what topic area we work in, we are somehow engaged in these processes. The other two are pivotal. Upon them the success of the others hinges. The multilevel interdisciplinary paradigm and the model of religion as a meaning system may be intellectual devices that can foster more expansive research programmes and more visionary theoretical integration. If so, they will also have enhanced the contribution of the psychology of religion to the science–religion dialogue.

WHAT LIES BEYOND THE BOUNDARY?

A conversation about the relationship between science and religion properly has boundaries that would place some questions or activities within the scope of the dialogue and other questions or activities outside of it. All of the above discussion falls properly within the dialogue. However, there is one kind of question that lies outside the boundary, and although asking it is honest, and people have their beliefs or disbeliefs about it, it is, psychologically and scientifically speaking, not answerable—we will never reach closure on it. I am referring to questions like whether prayer cures disease or makes people psychologically healthier under conditions in which the one being prayed for has no knowledge that he or she is being prayed for. Activities of this sort, e.g. conducting experiments to test for long distance prayer effects, reflect an 'experimental theology of miracles' that is logically doomed from the start. This amounts to trying to conduct an experiment to test whether God does what one asks.

I think that no outcome of such an experiment is scientifically meaningful. This is because, unless I am missing something, it is not possible to state a valid theological process that might mediate any 'effects' that might occur. Another illustration is evident in a paper I rejected for publication in *The International Journal for the Psychology of Religion*. (I'll change the particulars, but the gist of the story is true.) The authors had collected data on whether praying for someone to behave and feel differently on the job (without the target person knowing that he or she was being prayed for, using a double-blind method) produced changes in behaviour and feelings in that person, compared to a control group that was not prayed for. Besides the differences between groups being non-significant, the paper could have been scientifically interesting, except that there was no way the authors could write a non-miracle-laden psychological model for why such effects would be hypothesized. Were effects to be found, what social-psychological processes would explain them? If no effects were found, what theory is falsified? Science is a game of creating good theory, and it does not seem possible to create a good scientific theory about the effects of prayer (i.e. in this instance, testing to see if God intervenes upon request) because of the nature of what is prayed for (i.e. something that is God's decision), what prayer is theologically, and to whom the prayers are addressed.

To clarify: central to the idea that good science requires at least the possibility of creating a good theory to explain phenomena is the idea that predictions must be falsifiable. This means that if the data come out in a way opposite to one's hypothesis, one must be able and willing to say, 'My idea about the process was wrong.' This means, for example, that if heart surgery patients randomly assigned to be prayed for are hypothesized to get well better or faster or in greater numbers than a non-prayer control group (e.g. current research by H. Benson, reported in Myers 2005), it has to be possible to produce data that would discount the model of the process (God's decision) said to be involved. The problem with experiments of the type noted above is that the God to whom the prayers are addressed (who, the 'model' hypothesizes, produces the outcomes in the prayed-for group) can do anything it wants and is presumably never wrong. Simply put, God decides.

The uninterpretability becomes crystal clear when we realize that when people pray to God to heal their sick loved-one, for example, they almost always include saying 'if it be your will' or words similar to those. But when this is added to the prayer, all bets (and scientific tests) are off. All psychologically valid bases for hypothesizing any particular outcome are gone. God presumably is able to answer prayers by either granting or denying the request, with or without the pray-er's awareness (i.e. God doesn't have to inform someone of the answer or whether there is one), at any time (now, many years from now), over any time period (two days, fifteen years), via any method, through any vehicle, and with or without anyone's knowledge. In other words, by the very nature of the being to whom the prayers are addressed, one gives up having any basis (scientific or otherwise) for presuming to empirically test for an outcome. No matter how the data come out, they are not interpretable within the framework of any scientific model that I am able to think of, because the process cannot be assessed or discounted.

The psychology of religion and the larger science–religion dialogue may have a number of boundaries, and one of them is whether the validity of an idea or a claim about a process can be assessed. A foundation for doing good science is whether we are able to assess the process. Doing good science–religion dialogue would seem to rest upon the same foundation.

Conclusion

The psychology of religion has developed to such a degree during the past quarter-century that its contributions to the science–religion dialogue are compelling. The boundaries within the subdisciplines of psychology are diminishing, and in their place we see the rise of research that pulls together ideas from previously isolated lines of work. Similarly, cross-fertilization of psychological research with allied fields has increased, and this trend will continue. It is as if the golden ring is beginning to take shape sufficiently well that we may soon be able to recognize its beauty.

An important picture of human beings is inserted into the dialogue by the most recent research that comes out of this area. It is not a particularly rational picture. For example, the research in clinical, personality, and social psychology that leads to the development of the model of religion as a meaning system suggests that perhaps we do not arrive at our conclusions about reality or our pictures of the world by means of a purely rational process following the steps of Aristotelian logic. Perhaps the questions we are posing cannot be answered in the form in which we are asking them. Where there are only parts, our perceptual system wants to see a whole. Thus, humans have a mental structure that guides our perception of the elements that feed this dialogue, and our capacity to attribute meaning helps us understand how we come to see certain conclusions or inferences. But the research that instructs us about such workings of the human mind says that the process is psychological, not necessarily (purely) logical. Instead, we may think what we think, conclude what we will, and *then* construct a rationale and basis for holding it, and perhaps a meaning behind it. This is logically independent of whether the meaning is 'actually there'.

REFERENCES AND SUGGESTED READING

ALICKE, M. D., KLOTZ, M. L., BRIETENBECHER, D. L., YURAK, T. J., and VREDENBURG, D. S. (1995). 'Personal Contact, Individuation and the Better than Average Effect', *Journal of Personality and Social Psychology*, 68: 804–25.

ALLPORT, G. W., and ROSS, M. J. (1967). 'Personal Religious Orientation and Prejudice', *Journal of Personality and Social Psychology*, 5: 432–43.

ALTEMEYER, B., and HUNSBERGER, B. (2005). 'Fundamentalism and Authoritarianism', in Paloutzian and Park (2005a), 378–93.

ARGYLE, M., and BEIT-HALLAHMI, B. (1975). *The Social Psychology of Religion*. London: Routledge.

BATSON, C. D., SCHOENRADE, P., and VENTIS, W. L. (1993). *Religion and the Individual: A Social Psychological Perspective*. Oxford: Oxford University Press.

BAUMEISTER, R. F. (1991). *Meanings of Life*. New York: Guilford Press.

BOYATZIS, C. J. (2005). 'Religious and Spiritual Development in Childhood', in Paloutzian and Park (2005a), 123–43.

BURTON, T. (1993). *Serpent-handling Believers*. Knoxville, Tenn.: University of Tennessee Press.

CORVELEYN, J., and LUYTEN, P. (2005). 'Psychodynamic Psychologies and Religion: Past, Present, and Future', in Paloutzian and Park (2005a), 80–100.

DITTES, J. E. (1969). 'Psychology of Religion', in G. Lindzey and E. Aronson (eds.), *The Handbook of Social Psychology*, 2nd edn., Reading, Mass.: Addison-Wesley, v. 602–59.

DONAHUE, M. J., and NIELSEN, M. A. (2005). 'Religion, Attitudes, and Social Behaviour', in Paloutzian and Park (2005a), 274–91.

EMMONS, R. A. (1999). *The Psychology of Ultimate Concerns: Motivation and Spirituality in Personality*. New York: Guilford Press.

—— (2000). 'Is Spirituality an Intelligence?: Motivation, Cognition, and the Psychology of Ultimate Concern', *The International Journal for the Psychology of Religion*, 10: 3–26.

—— and PALOUTZIAN, R. F. (2003). 'The Psychology of Religion', *Annual Review of Psychology*, 54: 377–402.

FRANKL, V. E. (1963). *Man's Search for Meaning*. New York: Washington Square Press.

FREUD, S. (1927/1961). *The Future of an Illusion*, trans. J. Strachey. New York: Norton; originally published, 1927.

GORSUCH, R. L. (1988). 'Psychology of Religion', *Annual Review of Psychology*, 39: 201–21.

GRANQVIST, P. (2006). 'On the Relation between Secular and Divine Relationships: An Emerging Attachment Perspective and a Critique of the "Depth" Approaches', *The International Journal for the Psychology of Religion*, 16/1: 1–18.

HILL, P. C., and HOOD, R. W. Jr. (1999). *Measures of Religiosity*. Birmingham, Ala.: Religious Education Press.

—— and PARGAMENT, K. I. (2003). 'Advances in the Conceptualization and Measurement of Religion and Spirituality: Implications for Physical and Mental Health Research', *American Psychologist*, 58: 64–74.

—— —— HOOD, R. W. Jr., McCULLOUGH, M. E., SWYERS, J. P., LARSON, D. B., and ZINNBAUER, B. J. (2000). 'Conceptualizing Religion and Spirituality: Points of Commonality, Points of Departure', *Journal for the Theory of Social Behaviour*, 30: 51–77.

HOOD, R. W. Jr. (2005). 'Mystical, Spiritual, and Religious Experiences', in Paloutzian and Park (2005a), 348–64.

—— (in press) (ed.). *Handling Serpents: Pastor Jimmy Morrow's History of the Jesus' Name Tradition*. Mercer, Ga.: Mercer University Press.

—— and BELZEN, J. A. (2005). 'Research Methods in the Psychology of Religion', in Paloutzian and Park (2005a), 62–79.

—— and KIMBROUGH, D. (1995). 'Serpent-handling Holiness Sects: Theoretical Considerations', *Journal for the Scientific Study of Religion*, 34: 311–22.

Hood-Williamson Archive for Serpent-Handling Sects of Appalachia. Lupton Library. University of Tennessee at Chattanooga.

HUNSBERGER, B. (1991). 'Empirical Work in the Psychology of Religion', *Canadian Psychology*, 32: 497–504.

JUNG, C. G. (1938). *Psychology and Religion*. New Haven: Yale University Press.

KHALILI, S., MURKEN, S., REICH, K. H., SHAH, A. A., and VAHABZADEH, A. (2002). 'Religion and Mental Health in Cultural Perspective: Observations and Reflections after the First International Congress on Religion and Mental Health', Tehran, 16–19 April 2001, *The International Journal for the Psychology of Religion*, 12/4: 217–37.

KIRKPATRICK, L. A. (2005a). *Attachment, Evolution, and the Psychology of Religion*. New York: Guilford Press.

—— (2005b). 'Evolutionary Psychology: An Emerging New Foundation for the Psychology of Religion', in Paloutzian and Park (2005a), 101–19.

KOCH, S. (1959–63) (ed.). *Psychology: A Study of Science*, 6 vols. New York: McGraw-Hill.

LEVENSON, M. R., ALDWIN, C. M., and D'MELLO, M. D. (2005). 'Religious Development from Adolescence to Middle Adulthood', in Paloutzian and Park (2005a), 144–61.

LUYTEN, P., BLATT, S. J., and CORVELEYN, J. (2006). 'Minding the Gap between Positivism and Hermeneutics in Psychoanalytic Research', *Journal of the American Psychoanalytic Association*, 54/2: 571–610.

McCRAE, R. R. (1999). 'Mainstream Personality Psychology and the Study of Religion', *Journal of Personality*, 67: 1209–18.

McFADDEN, S. H. (2005). 'Points of Connection: Gerontology and the Psychology of Religion', in Paloutzian and Park (2005a), 162–76.

MELTON, J. G. (1986). *Encyclopedic Handbook of Cults in America*. New York: Garland.

MILLER, L., and KELLEY, B. S. (2005). 'Relationships of Religiosity and Spirituality with Mental Health and Psychopathology', in Paloutzian and Park (2005a), 460–78.

MURKEN, S., and SHAH, A. A. (2002). 'Naturalistic and Islamic Approaches to Psychology, Psychotherapy, and Religion: Metaphysical Assumptions and Methodology—A Discussion', *The International Journal for the Psychology of Religion*, 12: 239–54.

MYERS, D. G. (2005). 'Psychological Science Meets the World of Faith', *APS Observer*, 18/10: 14–18.

NEWBERG, A. B., and NEWBERG, S. K. (2005). 'The Neuropsychology of Religious and Spiritual Experience', in Paloutzian and Park (2005a), 199–215.

OMAN, D., and THORESEN, C. E. (2005). 'Religion and Spirituality: Do They Influence Health?', in Paloutzian and Park (2005a), 435–59.

PALOUTZIAN, R. F. (1996). *Invitation to the Psychology of Religion*, 2nd edn. Boston: Allyn and Bacon.

—— (2005). 'Religious Conversion and Spiritual Transformation: A Meaning-system Analysis', in Paloutzian and Park (2005a), 331–47.

—— and PARK, C. L. (2005a). (eds.). *Handbook of the Psychology of Religion and Spirituality*. New York: Guilford Press.

—— —— (2005b). 'Integrative Themes in the Current Science of the Psychology of Religion', in Paloutzian and Park (2005a), 3–20.

—— RICHARDSON, J. R., and RAMBO, L. R. (1999). 'Religious Conversion and Personality Change', *Journal of Personality*, 67: 1047–79.

PARGAMENT, K. I. (1997). *The Psychology of Religion and Coping*. New York: Guilford Press.

PARGAMENT, K. I. (2002). 'Is religion nothing but . . . ?: Explaining Religion versus Explaining Religion Away', *Psychological Inquiry*, 13/3: 239–44.

—— MAGYAR-RUSSELL, G. M., and MURRAY-SWANK, N. A. (2005). 'The Sacred and the Search for Significance: Religion as a Unique Process', *Journal of Social Issues*, 61/4: 665–88.

PARK, C. L. (2005a). 'Religion and Meaning', in Paloutzian and Park (2005a), 295–314.

—— (2005b). 'Religion as a Meaning Making Framework in Coping with Life Stress', *Journal of Social Issues*, 61/4: 707–30.

—— and FOLKMAN, S. (1997). 'Meaning in the Context of Stress and Coping', *Review of General Psychology*, 1: 115–44.

—— and PALOUTZIAN, R. F. (2005). 'One Step toward Integration and an Expansive Future', in Paloutzian and Park (2005a), 550–64.

RICHARDSON, J. T. (1995). 'Clinical and Personality Assessment of Participants in New Religions', *The International Journal for the Psychology of Religion*, 5: 145–70.

SILBERMAN, I. (2005a) (ed.). 'Religion as a Meaning-System', *Journal of Social Issues*, 61/4 (Special issue).

—— (2005b). 'Religion as a Meaning-System: Implications for the New Millennium', *Journal of Social Issues*, 61/4: 641–64.

—— (2005c). 'Religious Violence, Terrorism, and Peace: A Meaning System Analysis', in Paloutzian and Park (2005a), 529–49.

SMART, N. (1989). *The World's Religions*. Englewood Cliffs, N.J.: Prentice-Hall.

SPILKA, B. (2005). 'Religious Practice, Ritual, and Prayer', in Paloutzian and Park (2005a), 365–77.

—— HOOD, R. W. Jr., HUNSBERGER, B., and GORSUCH, R. (2003). *The Psychology of Religion: An Empirical Approach*, 3rd edn. New York: Guilford Press.

WILLETT, J. B., SINGER, J. D., and MARTIN, N. C. (1988). 'The Design and Analysis of Longitudinal Studies of Development and Psychopathology in Context: Statistical Models and Methodological Recommendations', *Development and Psychopathology*, 10: 395–426.

WILLIAMSON, P. W., POLIO, H. R., and HOOD, R. W. JR. (2000). 'A Phenomenological Analysis of Anointing among Serpent Handlers', *Journal for the Scientific Study of Religion*, 10: 221–40.

WONG, P. T. P., and FRY, P. S. (1998) (eds.). *The Human Quest for Meaning: A Handbook of Psychological Research and Clinical Applications*. Mahwah, N.J.: Erlbaum.

WULFF, D. (1997). *Psychology of Religion: Classic, Contemporary*, 2nd edn. New York: Wiley.

ZINNBAUER, B. J., and PARGAMENT, K. I. (2002). 'Capturing the Meanings of Religiousness and Spirituality: One Way Down from a Definitional Tower of Babel', *Research in the Social Scientific Study of Religion*, 13: 23–54.

—— —— (2005). 'Religiousness and Spirituality', in Paloutzian and Park (2005a), 21–42.

—— —— COLE, B., RYE, M. S., BUTTER, E. M., BELAVICH, T. G., HIPP, K. M., SCOTT, A. B., and KADAR, J. L. (1997). 'Religion and Spirituality: Unfuzzying the Fuzzy', *Journal for the Scientific Study of Religion*, 36: 549–64.

CHAPTER 16

SOCIOLOGY AND RELIGION

RICHARD FENN

SEARCHING FOR THE SACRED

Social order is precarious, and novelty, inspiration, and desire can be especially subversive or even destructive. Sociology has long focused, therefore, on the problem of order: why there is so much of it or, under other conditions, so little? In search of answers to both of these questions, sociologists of religion investigate the sacred. When a social system begins to emerge from the flux of everyday life, it develops boundaries that separate it from those outside. To signify the continuity of certain relationships over time, to give them an identity, and to mark their difference from other relationships, social systems define themselves in symbols, construct themselves through rituals, and realize themselves through practices. When friendships and families, communities and ethnic groups, clans and entire peoples, institutions, corporations, and nation-states, claim for themselves a capacity to transcend the passage of time, their symbols, rites, and practices take on the quality of the sacred. Whether it is called the numinous or the holy, charisma or mana, the sacred is always somewhat mysterious, even when it becomes embodied in a stone or a plant, a person or a word, a practice or a way of life, an institution or an entire society. Always by implication, and often explicitly, those who remain outside the sacralized social system are regarded as belonging to a world that will not transcend the passage of time. That external world, beyond the pale of the sacred, is thus relegated to the merely temporal; unable to stand the test of time, the world of the secular is headed irrevocably toward death. No one is more aware of the fatal implications of being excluded from the sacred than the Muslims in Jerusalem who, as this essay is being written, witness evangelical Christians in Jerusalem displaying two large stones inscribed as 'cornerstones of the third temple'. To initiate the building of this temple on the Temple Mount is their goal, and such an initiative will inevitably result in an Armageddon-sized battle in the Middle East. That is the effect of the sacred: to consign all that is beyond its boundaries to extinction. Of course, there are often less violent manifestations of the sacred that reflect the fault lines in any society that distinguish men from women, the old from the young, the relatively noble and powerful from the relatively powerless and ignoble, the ordained from the lay, the collective from the merely individual, the magnificent from the mundane. Wherever the line is drawn between the sacred and the secular, however, it is only those within the precincts of the sacred who have a purchase on eternity and will therefore transcend the passage of time. Time is always and everywhere already running out on the remainder.

In all of its manifestations, whether personal or social, natural or supernatural, the sacred provides only *a limited embodiment of unfulfilled possibilities*. What might well be called the upper-case Sacred, then, or the Sacred itself, comprises the set of all possibilities that any social system must necessarily exclude. To the extent that a society defines itself over and against nature, for instance, the Sacred may include notions about animal spirits or the supernatural. Angels and demons have long qualified for inclusion in the Sacred, but they tend to escape direct sociological observation. To the extent that a social system defines itself over and against the remainder of humanity, then, the Sacred will include all the subhuman and the superhuman. Centaurs and messiahs or kings returning from the dead have also long found a place in the Sacred, but these too tend to elude the attention of sociologists of religion, who confine themselves to the study of understanding and explaining the ways in which social worlds take on a life of their own, become mysterious and authoritative enough to ask for tribute or even sacrifice, and consign those outside their parameters to a world that is at best temporary and is always destined to disappear or be destroyed. How do religious institutions, clearly 'man-made', take on sacred authority?

To ask how a humanly constructed social universe takes on the aura of the Sacred, inspires devotion, and requires sacrifice is still necessary, whether one lives under a constitutional, divine-right monarchy or in a messianic nation with a volunteer army. It is even more pressing when those who claim for themselves a monopoly on the sacred relegate even their co-religionists to a secular world that is passing away and call for a holy war, an Armageddon, that will purify the world once and for all.

THE SACRED AS SUBVERSIVE

Not every society, of course, enshrines traditional authority or calls for sacrifice. In some societies, people tend to leave the dead behind relatively easily, gather nuts and berries, hunt for the occasional small animal, treat women reasonably well, stand on very little ceremony, and leave little behind in the way of monuments. On their travels they may be accompanied by shamans with a gift for seeing who or what is coming; these specialists in the Sacred know how to investigate and explain premature or abrupt departures. Like shamans with direct access to a world of possibility that may or may not be fulfilled in any given time or place, prophets and seers suggest ways of freeing a traditional society from its past for the sake of a future in which the first may very likely be the last. Where social order is moribund or scarce, charismatic leaders may suggest that it is time for the dead to bury the dead, and may announce a nascent society emerging in the midst of the old (Wilson 1973). Certainly charismatic leaders may pose a threat even to a social order that is fairly coherent and intact by mobilizing resentment and by activating a residue of perennially unrealistic aspiration for satisfactions that most civilizations are unable or unwilling to provide.

When sociologists of religion inquire into the ways in which the sacred disrupts a society or enables it to make effective claims on the loyalty and affections of its members, they are seeking to understand religion by investigating its effects. A functionalist analysis of this sort is implied whenever the sacred is discussed as a polar opposite of the secular or the profane. The effect of sacralizing a way of life or an institution, a community or an entire nation, is to relegate whatever is excluded from the sacred to the secular world that is always and everywhere already passing away. It is also to take the mystery from the profane world outside the sacred, since its possibilities are everywhere open to inspection: no Sunday, then no Monday; no temple (*fanus*), then no profane (pro-*fanus*). As societies create the sacred, so they also create its opposite in a world that is passing away and deserves no reverence whatsoever. The function of the sacred, then, is to create that which will not stand the test of time and is therefore expendable, but the purely secular or profane then becomes not only a by-product of the sacred, but its enemy.

For sociologists of religion, then, the problem of order requires an investigation into the tension between the sacred and the secular or profane. If sociologists tend to regard the social order itself as the epitome or origin of the sacred, they are likely to see the individual as a derivative of social order that can become its antithesis. If it is manhood that is sacralized, then it is womanhood that becomes the derivative and potential negation of manhood, and so on for nobility and commonness, elders and the young, the collective and the merely personal, society and the individual. That is why, in searching for threats to social order, sociologists have long focused on the varieties of individualism that in any society may foster deviant and subversive convictions (Lukes 1985). The sacred is thus the source of social order and yet sows the seeds of disorder and negation.

Conversely, to explain why there is so much order, sociologists may investigate the ways in which rituals of conversion or exorcism separate individuals from their own most intimate convictions and desires. No rituals are able fully or permanently to replace the individual's psyche with a prescribed mentality. That is why sociologists also investigate the longings, perceptions, and viewpoints that remain outside the realm of ritualized self-understanding, experience, and speech (Bell 1997; Rappaport 1999). Because language frames a sense of possibility, sociologists of religion have studied the ways in which religious institutions authorize some speakers at the expense of others, set the limits on what can be said, remembered, or hoped for, suppress awareness of desire, postpone or reject certain possibilities for satisfaction, define collective memory and anticipation, and create a community defined by the reach of the word.

THE SACRED IN TENSION WITH RELIGION

There is thus another reason why the sacred is always sowing the seeds of its own destruction, or, to put it more formally, why any functionalist analysis of the sacred is likely to end up in a dialectical argument. Because the sacred always points beyond itself to possibilities that the social order can scarcely acknowledge or include, the sacred is always potentially subversive or antinomian. That is why it needs to be contained, and it is one function of religious institutions to provide such containment. Thus every social order is haunted by excluded possibility, and every form of language has meanings that point to suppressed desires, aspirations, and longings. Because sacred speech evokes the presence of invisible authorities and gives voice to suppressed or excluded possibilities, religious institutions authorize some utterances and place others in a barbarian limbo beyond the reach of known language. Some have the right to speak on behalf of spirits of the dead, while others intone as if it were the dead themselves or a god who is speaking. Those who have ears to hear but cannot hear, or whose presence is uncalled for, are beyond the pale of communication (Moore 2000). However, when religion loses its monopoly on and control over the sacred, hitherto unheard of possibilities become common parlance.

To control language is thus to control an awareness of possibility. In traditional societies, where religious beliefs and practices determine who can talk about specific subjects at various times and places, sociologists must study religion if they are to understand power and the rules for linguistic engagement. In highly ritualized societies, those rules place severe limits on who can say what, and words themselves operate within relatively narrow semantic limits. The possibilities enshrined in a sense of the sacred may be very limited, while pointing beyond themselves to a world of possibility that is intended to remain permanently enshrined in mystery beyond the range of speech and thought. Whatever religion defines to be unutterable remains in a world of possibility beyond the reach of all except an élite with access to secret or esoteric knowledge. Especially where religious language controls public awareness, there is much that necessarily goes without saying, whether because it is implicit or taken for granted, or because it is prohibited (Bloch 1989).

Wherever religion permeates all other aspects of the social system, anyone who breaks the rules governing sacred speech and language may therefore pose a serious political or economic threat and destabilize received notions of affinity and kinship. In the 1960s a younger generation in Western democracies burned flags or wore them on the seats of their pants, refused pledges, sat in prohibited public places, interrupted liturgies, consumed illegal substances, and developed their own forms of subversive celebration. Those with unauthorized knowledge, who speak with uncommon authority or in a strange tongue, undermine any system based on the ownership and control of sacred speech; Pentecostals are always a threat to those who seek to maintain a monopoly on sacred language. Less threatening but still potentially subversive are those who, like liberal or secular interpreters of the Bible, go beyond the literal or original meaning of sacred language and trade in allegory or metaphor (Martin 2002). Whoever pushes the limits of the possible and enables people to imagine what had been unimaginable may be accused of using magic to overthrow religion. Jesus, after all, was accused of having a demon. Certainly in modern societies religious institutions have lost their monopoly on the sacred, which is now found in New Age religiosity, science fiction sagas, in films like *The Lord of the Rings* or the Harry Potter series, in sports, and in political claims to religious authenticity.

As many social systems develop cybernetic characteristics, they make themselves up as they go along, and constitute themselves by acts of communication over great distances among people who are relative strangers to each other. The sacred has escaped from the control of religious authorities and, in cybernetic social systems especially, expands the range of what people are able or allowed to imagine or conceive. With the sacred now roaming at large, it is very difficult for any modern social system to keep the genie of unfulfilled possibility confined within the limits of traditional religion. That is why, in modernizing societies, charismatic speech of the sort associated with Pentecostalism becomes widespread and carries within it a potential for accelerated social change. That may also be why charismatic speech seems to its critics to be a form of magical thinking at odds with a reality long circumscribed by traditional religion.

The possibilities contained in the sacred increasingly escape the control of religious institutions. In literature, the media, and the arts, as well as through popular psychotherapeutic practices, individuals increasingly have access to a mythical realm of ghosts and demons, to sublime or ecstatic experience, and to the symbols of the unconscious. Even when sacred speech is highly ritualized and is uttered only in the right way and at the right times and places by the right people, however, it can only point beyond itself to a world of possibility that will never, until the last word is spoken, be completely realized. Even the Christian Eucharist is a meal that looks forward to a final, eschatological feast.

THE SACRED AS A CODE FOR VIOLENCE

At some level, societies know that they are based on the foreclosure and postponed fulfilment of possibilities for both life and death. Every social system has to eliminate or control animal spirits, create an index of impossible or prohibited satisfactions, deflect violence on to safe domestic or foreign targets, and control affection and hatred for both the living and the dead. Sociologists of religion investigate the ways in which these possibilities are coded and concealed in religious belief and practices. Sacred rites contain a sign that violence has been done. Even when the horns of a slain beast are placed on the altar, where they hold out hope that life may yet return and food again be abundant for a faithful people, they inevitably signify that life has come to a violent end. The sacred always offers only a very limited embodiment of unfulfilled possibility (Fenn 2001).

Thus religious rituals may defer to the past or project into the future whatever may be too much for any community or society fully to experience or acknowledge in the present. That is why sociologists have long been interested in the ways in which, through implicit knowledge, a people may keep a conspiracy of silence over the violence in their history. Communities and societies suppress knowledge of the violence, or the threat of violence, on which they are based. The potential for fratricidal hatred may be coded as the memory of an act of violence: a Romulus killing a Remus, a Cain slaying an Abel. Regicidal and patricidal passions may be coded in the myth or memory of a slain king whose death initiates a kingdom that awaits his perennial return. Regardless of the historicity of these myths, they signify that the sacred has a guilty secret: e.g. undying hatred, longing for a prohibited alliance, or the memory of an ancestor, for instance, who killed an Egyptian. Through religious myth, opposing passions are coded as possibilities that have been regretted, postponed, or foregone and remain perennially unfulfilled. Religious myths may acknowledge a history of infanticide in stories about aborted or prohibited child sacrifice, just as religious rites of initiation may enact the symbolic drowning or castration of a child, and revivals or exorcisms may enact spiritual attacks on the psyche of deviant individuals. Religious rites may therefore conceal as much as they convey. Unfulfilled possibility lends the aura of the sacred to the rite, while concealing or distracting attention from the actuality of violence.

There is something that remains hidden or incomplete about many rites: an element of unknowing or misrecognition that allows the participants to take part in what some have called 'a comedy of innocence'. Thus, among some peoples, the ingestion of bitter herbs may well code the memory of cannibalism, just as ritualized lynching, the killing and burning of African Americans, has impressed some observers as having had cannibalistic associations and overtones (Patterson 1998). The same comedy of innocence may be enacted in rites of initiation in which not only symbols and words but also gestures make references to the death of the young *initiand*, and in at least one initiatory rite, both the men and the women of the village chase the young through the streets shouting 'Kill, kill, kill'. Many rites of initiation, in opening the way to a new life, also enact the symbolic death of the *initiand*. Some participants may be less aware than others that they are targets of partially disguised resentment, even when they engage in public displays of their unworthiness, as in the shaming rituals of a new king or, in modern electoral campaigns, of a candidate for high public office. Of many such sacrifices it may be said that 'they know not what they do'.

THE SACRED AS CODE FOR POSSIBILITY

In all of its manifestations, whether personal or social, natural or supernatural, the sacred provides *a limited embodiment of unfulfilled possibilities*, especially those associated with past or future violence. Some of these possibilities preserve and enhance life, or, in religious terms, constitute 'salvation': good health, knowledge of the future, satisfaction of grievances, the fulfilment of desires, and even a measure of justice. However, the sacred also offers a limited embodiment of other possibilities that are threatening to life itself: protection from violence, demonic influences, disease, a reversal of fortunes, the protracted disappointment of old grievances, possession by malign spirits, the frustration of desire, resentment over rank injustice, and sudden death. Thus the sacred always offers only a *limited embodiment* of these as yet unfulfilled possibilities for both death and life, salvation and destruction—indeed, for the fulfilment or devastation of every human need and aspiration.

The devotee is given only a foretaste of the benefits and a promise of more to come as a reward for devotion, fidelity, and obedience. Similarly, the sacred offers only a limited embodiment of the possibilities that are inimical to life itself; otherwise contact with the sacred would be fatal or lethal. That is why, of course, only a priest who has attained the highest levels of purity would have been allowed to approach the Holy of Holies in the Temple at Jerusalem, contact with which was thought inevitably to be fatal to the unworthy and impure. As religious institutions become less able to contain the sacred within the constraints of ritual and orthodox belief and practice, demands increase for the reversal of the social order, for the satisfaction of longings and grievances, for the cancellation of old debts, and for the beginning of a new age in which the impossible and the unspeakable become ordinary speech and aspiration. What is often decried as a widespread and popular tendency to feel victimized and to pour out resentment on uncaring, usually 'liberal' or 'secular', élites is part of this release of pent-up religious resentments no longer contained within religious rites.

Because the sacred embodies only *unfulfilled* possibilities, it always points beyond itself to the full range of possibilities for either salvation or destruction. This set of all possibilities, both actual and hypothetical, I call the Sacred, and it is to the sociology of religion what dark matter is to astrophysicists, or 'the god above the god of theism' is to theologians. Direct exposure to all the possibilities contained in the Sacred would of course stagger the mind and the imagination. Who can stand in the presence of the deity, the Sacred itself? That is why the deity is best approached in the form of the lower-case sacred, since it is crucial that the entire range of possibilities for both life and death itself remains unfulfilled if even limited access to them is to be granted. Otherwise, as the apocalyptic imagination has long known, were these possibilities for salvation or destruction to be fulfilled, the end would have come. That is why the sacred is the object of awe, fascination, fear, devotion, and allegiance: in Durkheim's view *la vie serieuse*. By postponing the fulfilment of the most extreme possibilities, the sacred thus 'buys time' for the social system; by making a small down payment of sacrifice, it gets a purchase on eternity for the social order. However, when apocalyptic movements gain widespread opportunity in Jewish, Islamic, and Christian societies, they signify that the social system may be increasingly unable to buy time by keeping people waiting for their various satisfactions.

The failure of any system to fulfil aspirations for security and comfort, health and honour, inevitably creates resentment. So does the tendency of any social system to be based on real or symbolic violence to the individual. It is the function of ritual, as I have suggested, both to disguise and express that violence and to inure a population to the incurably partial fulfilment of their desires. Put more simply, ritual expresses, disguises, deflects, and offers partial satisfaction of the resentment caused by the social order itself. When rituals fail to buy time for a social order, as in the decades prior to the civil war in Palestine in 66–73 CE, apocalyptic enthusiasms become exceedingly difficult to contain. Contemporary apocalyptic movements signify that formal, traditional rites are less able than ever to persuade many to sacrifice their desires, their longings for power and recognition, and their claims on life itself. The failure of ritual to offer temporary relief for resentment over unsatisfied longings and grievances increases demand for a final turning of the tables, when the last will be first.

Rituals and the Reality Principle

The sacred thus provides a code for the social system that defines the limits of legitimate aspiration. In providing partial and selective access to limited embodiments of possibilities that must remain unfulfilled lest they in fact dissolve the social order, rituals perform an important social function: what used to be called maintenance of the reality-principle. When rituals work, they redirect intergenerational hostility to safer targets within a generation, such as rivals for offices that represent patriarchal spiritual or political authority. Without adequate ritualization, these conflicts take to the streets, shed blood, import reinforcements from outside the social system, and degenerate into civil war (Lincoln 1985). Therefore it is not stretching a point to suggest that rituals are an evolutionary universal that have helped some societies to avoid self-destruction.

The less religion is able to monopolize the sacred, the more a society will be haunted by an awareness of unfulfilled possibility. Even in traditional societies, the sacred is very difficult to contain or demarcate: difficult, that is, to 'institutionalize'. Because the living so often have unfinished emotional or financial business with persons who have died, the dead become objects of fear or devotion. The less religion is able to contain and constrain the sacred, the more difficult it becomes for formal rituals to contain or fulfil all the sentiments attached by the living to the dead, or to confine apparitions of the dead to only certain times and places during the year. Even in societies where religion does monopolize the sacred, affections for those who die young are especially difficult to contain in formal rites; that is why their spirits are so often experienced as dangerous or demonic, and why so many practices have been devised to assuage their unfulfilled desires or obligations. However, as rituals fail to assuage and contain resentment over unfulfilled longings, whether for relief from fear and doubt or for a more abundant life, the sacred may take a wide range of unauthorized shapes, from ghosts and demons to individuals claiming supernatural powers of their own. As covenants are 'written on the heart' rather than on tablets of stone or in sacred texts, the social system's 'reality principle' becomes increasingly difficult to locate or maintain.

As more formal rituals lose adherents and efficacy, wide areas of social life may become 'serious', i.e. require devotion, elicit sacrifice, and promise great benefits to those who comply. Weber's interest in 'the Protestant ethic' reflects the development of a character type and of social practices that required religious self-discipline in the midst of otherwise mundane social and economic activities. Contrast societies that institutionalize the sacred in formal rituals confined to certain times and places; in these societies much of everyday life remains unritualized, relatively 'unserious', and therefore more open to utilitarian, pragmatic, or self-interested activities. Where rituals are highly formal and sharply distinguished from everyday life, some dramatic displays supporting the reality principle will be highly lethal, like the *autos-da-fé* in which humans were indeed sacrificed; other public celebrations that attack the reality principle will be more orgiastic, like that of Mardi Gras. However, in societies where more areas of social life are highly ritualized, much of the mundane becomes serious and subject to social discipline over both aggressive and erotic impulses.

It is always difficult for religion to institutionalize the sacred, to protect such areas of everyday life as the family, work, and politics from erratic and potent intrusions of unruly spirits. In societies where ritual offers highly limited and selective access to the sacred only at very specific times and places and under tightly controlled conditions, the sacred is highly institutionalized. Even under these conditions, however, the sacred may take a variety of uninstitutionalized forms: ghosts and demonic spirits that inhabit wild places on the margins of the social order, individuals with extra-ordinary powers or whose impulses are demonstrably resistant to social control, and ascetics who prefer the desert or the forest to the monastery and the local church. The more a society relies on tightly bounded, formal rituals to institutionalize the sacred, the less disturbance the sacred may cause to everyday life and the mundane.

DE-INSTITUTIONALIZING THE SACRED AND RITUALIZING EVERYDAY LIFE

The more a society institutionalizes the sacred under professional control in limited times and places, the more control and discretion individuals have over their own actions in mundane areas of social life. Where the sacred is highly institutionalized, i.e. restricted to formal rituals alone, the more freedom is available in other areas of social life for people to act as their own agents under their own authority, for their own purposes, and in accordance with their own needs. Activities become individualistic rather than corporate; goals become more personal than collective; standards become more instrumental than ethical; and orientations become far more secular than religious. Vast areas of social life are therefore regarded as secular and of merely temporal concern. Under these conditions it is more difficult for the larger society to imbue areas of work and politics, family life and education, play and contests, with the values of the sacred and to elicit sacrificial motives and self-disciplined performances in the conduct of everyday life. What a society therefore gains in delimiting the sacred to particular times and places, and in embodying the sacred in specific persons and performances, it loses in the scope of the sacred.

When the sacred is highly institutionalized in formal rituals performed under professional control in specific and limited times and places, economic activity is subject to chance, to competition, and to the play of personal interest. This is Weber's point about various forms of capitalism uninformed by the Protestant ethic. These sorts of capitalism depended on winning battles or on risky voyages, and they suited societies where the sacred was highly institutionalized in certain formal rituals performed only at certain times and places. In these societies wide areas of social life in work and play, in war and politics, were freed from ascetic disciplines, open to risk taking, and subject therefore to chance and the play of extraordinary personalities. Under the impact of Protestantism, however, more aspects of everyday life were ritualized and thus brought under the control of ascetic self-disciplines for inquiry and investment. Scientists could explore the mysteries of creation by investigating the table of elements or inquiring into the movement of bodies or the stars, just as the laity could investigate their own sacred mysteries by becoming able to read basic texts in the vernacular or by working out their faith not only in fear and trembling but by the punctual and productive discharge of their debts and duties.

So long as the sacred is highly institutionalized in formal rituals sharply demarcated from everyday life, the flow of capital is likely to be difficult to predict or control, high interest rates may be charged, and with social trust relatively low, economic goals will be focused on the near future rather than the long term. Under these conditions, economic or political activity is likely to be controlled by families or ethnic fraternities that place constraints on opportunities to invest or to engage in novel forms of work and politics. However, when the sacred escapes from the tight control of institutions like the church, it legitimates and disciplines innovation, entrepreneurship, and experimentation in a wide range of areas: science and economic activity during the sixteenth century, the interrelationships of races, genders, and generations in the late twentieth century.

Where the sacred is less well institutionalized, less contained in formal rituals performed only under professional control, the sacred becomes an aspect of mundane activities and everyday life: so much so that even the human person may come to be regarded as sacred. To encompass the sacred in so many and complex settings, religious beliefs and values become relatively abstract: witness current attempts in the United States to distinguish 'people of faith' from others, Christian or not, who do not share a set of values abstractedly defined as a 'culture of life'. Abstractions like these may be code words for more highly limited notions, but the rhetoric itself, being so highly generalized, fails to define the sacred it seeks to encompass. Thus the semantic range of code words becomes contested in the public arena. In the same way, the notion of a 'culture of life' may be intended to demonize as death loving all those who are in favour of a woman's right to an abortion, but it can also be used to attack those who favour an evangelical war in foreign territory. What a religious culture gains in scope and flexibility under these conditions, it loses in relevance and specificity.

As the sacred becomes less subject to institutional control, wide areas of social life may become more highly ritualized, in the sense that work becomes more subject to a collective discipline, more responsive to chains of command, its language more limited in the range of meaning, and its demands on the individual's identity, motivation, and allegiance more rigorous. It is increasingly difficult to determine what aspects of social discipline are intended simply to professionalize workers and which are intended to socialize wider areas of the psyche, such as moods and motivations, attitudes, internal moral standards, and world-views. If at the same time a society is undergoing a process of differentiation, in which the economy, work, and politics become relatively free from direct religious control and supervision, each area of social life may develop its own forms of the sacred. Bureaucracies sacralize their own forms of organization regardless of the impact of this process on their ability to achieve their stated goals; the nation also becomes an object of sacred allegiance; control over esoteric knowledge in science and industry gives technocracy the aura of the sacred.

The Emergence of the Individual

In describing this process, in which areas of social life have become not only increasingly differentiated but also sacralized, sociologists of religion have used the notion of secularization in ways that imply both necessary restrictions on the authority and influence of religious institutions over the individual and the blurring of the boundary between the sacred and the profane (the de-institutionalization of the sacred). In this process, individuals are able to take a wider range of both religious and secular roles; the meaning of sacred symbols is stretched to cover a wider range of mundane situations; sacred stories are embellished or replaced by narratives of secular provenance; the range of orthodox belief is expanded to include personal opinion; and the meaning of particular words and symbols is more likely to become complex and multivalent.

Societies differ in the extent to which the rituals that define and modify the lives of individuals over the course of a lifetime and in the various aspects of everyday life also link them symbolically to the larger society. In some societies, individuals take part in a wide range of rituals that may have little or no connection with the rituals that provide collective self-definition. Sometimes rather disparagingly, sociologists of religion have labelled these rituals as magic and have reserved the notion of religion for those rituals that are more specifically collective in their scope and identity. Some have made a distinction between rites that evoke a highly transcendent deity and those that evoke spirits that are more immediately accessible and more useful for responding to a wide range of individual wants and complaints. Here again, one finds the notion that what sanctifies the agency of individuals and legitimates their demands is relatively secular, temporal, mundane, instrumental, or merely individualistic, whereas the sacred pertains to what expresses or is conducive to the legitimacy and effectiveness of the social system as a whole. Thus New Age religiosity, or the transfer of religious authority and authenticity to individuals through such doctrines as the priesthood of all believers, has been considered by some sociologists to reflect the process of secularization.

The Differentiation of Religion from Social Life

Along with the de-institutionalization of the sacred, sociologists have investigated the extent to which religion has become differentiated from the rest of the larger society. In highly differentiated societies, religion no longer provides integration into the society as a whole and becomes more visibly a special interest: one part lobbying along with all the others for a share of public discourse, political influence, and social control. Religious rhetoric becomes increasingly double-coded, contestable, subject to a variety of interpretations, and of no obvious relevance to decision making under specific conditions. Under these conditions, religion becomes one of many voices in the public arena, one of many subsystems in a society, where it must compete for influence over the economy, politics, the family, and education, to name only a few other subsystems.

In seeking to determine the extent to which religion is differentiated from the rest of the larger society, sociologists raise a number of questions concerning the extent to which religion is entangled in the business of everyday life: with the way people work or fight; with the way business gets done and money made; with the way that power is created, shared, and imposed; with the way children are taught how to obey and conform; with the way in which people learn to defer to authority or to acquire it for themselves; with the way power is exercised and justified; with the extent to which women are treated as a form of investment rather than as autonomous agents; with the way that death is understood, depicted, and managed; with the way outsiders and other societies are imagined and treated; with the way that violence is disguised, imposed, and justified; with the way that the past and the future are conceived and imagined to be more or less distant from and irrelevant to the present.

The more highly differentiated religion becomes from other subsystems, the more difficult it will be for religion directly to shape a nation's identity: to enshrine the collective memory of old battles and injuries and entertain visions of future satisfaction, triumph, and redemption; to control its law or polity; to shape the limits and extent of social trust; to have a role in determining who is available for a loan or other forms of social investment; to qualify individuals and groups for elective office or for other positions of trust; to legitimate the dominance of one gender over the other; to legitimate the call for individuals to sacrifice themselves for the sake of the community or larger society.

The more that religion is differentiated from other aspects of the social system, the more religion can put its own doctrinal, symbolic, and ritual house in order without confusion caused by outside influences, but the less direct control it has over social institutions, policies, and practices. What religion gains in autonomy and self-control, it tends to lose in its capacity to influence the larger society. Similarly, the more systematic and rational religion becomes, the more it is able to recognize and punish discrepant beliefs, deviant practices, and unauthorized sources of inspiration, but the less it is able to cope with innovation and change.

As a religious system becomes more highly differentiated from other subsystems, its beliefs may become more clearly orthodox, its practitioners more disciplined, its followers more compliant. On the other hand, what a more highly rationalized religious system gains in integrity and completeness, it loses in mystery. Thus, the more differentiated religion becomes from the larger society, the more objective and formulaic become its beliefs, and the more professional and practised become its practitioners. What religion gains in internal mastery and control, it loses not only in mystery but in spontaneity and in access to novel sources of inspiration and authority that could enable religion to respond to unfamiliar situations and challenges.

THE DE-DIFFERENTIATION OF RELIGION FROM THE LARGER SOCIETY

Some societies that have operated at relatively high levels of differentiation between religion and the rest of the social system may then become somewhat less differentiated. The boundaries become blurred between scientists and politicians, the news media and business, insurance companies and the practice of medicine, lobbyists and legislators. Under such conditions of increasing de-differentiation, it becomes more difficult to determine whether activities that may be apparently religious on the surface are really political or economic. It becomes increasingly important to resolve a number of questions: whether there is one set of rules for everyone who does business; whether contracts are given and interest charged on the basis of a person's political or religious affiliation; whether social policies are intended to benefit particular groups, institutions, or categories of the population based on religious motivations; whether notions of need and merit are constructed on the basis of religious identity and affiliation or are relatively secular. During periods of de-differentiation, then, it is increasingly difficult to know whether an activity is fundamentally religious or secular, just as it is difficult to know whether a doctor is following protocols set by the medical profession or those set by an insurance company, or whether a politician's vote is determined more by the public than by a special interest. Doctors may be disciplined who, on religious grounds, refuse to provide medical care to an otherwise deserving patient.

When de-differentiation occurs, professionals in specialized fields who mix religion with their practices raise questions about the nature of the practice itself; is the doctor practising medicine or his or her faith? Is the jurist interpreting the law or covertly applying the Bible to a particular case? Even the Bible itself may be taken to offer information about creation that is as reliable, and ideas about the origin of species that are as credible, as those of natural scientists. It is this opening up of relevant stories and of semantic ranges that jeopardizes the disciplines in a society undergoing de-differentiation and creates a demand for univocal, simple, literal meanings, based on hard, incontrovertible facts, that can stand the test of time. This demand for fixed, plain, and reliable sources of authority is one indicator of a deflation in social trust.

DEFLATIONARY PERIODS AND A REACTIONARY RETURN TO THE SACRED

In such deflationary periods, a society is less able to export moral outrage and to give its converts and subjects external fields to conquer or colonize. Under these conditions resentment can turn inward against domestic institutions and authorities. Citizens then expect less of their politicians, want the law strictly interpreted and enforced, place their trust in fixed assets like real estate and the family, and return to a more literal interpretation of sacred texts. As resentment turns inward, rather being exported outside the social system, people in high places become targets of suspicion and accused of disloyalty, and those who seem outwardly to be decent and upstanding citizens are suspected of secret moral and religious failings. Public trust is withdrawn from major institutions, as people develop low opinions of experts and intellectuals, scientists and politicians. In deflationary periods, public trust is re-invested in institutions like the family and the homestead, in traditional ways of life and religious belief. There is a move back to basics in everyday life: to deeds rather than words, to the original meaning of the Bible or the Constitution rather than later interpretations, to precedent rather than novelty, to knowledge that is accessible rather than the esoteric or scholastic, to art that is more representative than abstract, and to the individual's own religious commitments as opposed to more authoritative or customary credentials. Clearly there are affinities in deflationary periods between conservative or even reactionary developments in the polity or the economy and religious fundamentalism.

In periods during which trust is radically withdrawn from public institutions, and resentment is focused on aspects of the society itself rather than directed to peoples and places outside the social system, a society will seem to its members to be moribund and therefore running out of time. That is, in extremely deflationary periods, there will be tendencies not only to fundamentalist but also to apocalyptic religious belief, sentiment, and practice.

Deflationary periods see conservative or even reactionary developments in religion, in the polity, and in the economy. There will increasingly be demanding calls for loyalty and even sacrifice in order to restore the hard assets of the social system in the commitments of the individual.

The De-institutionalization of the Sacred and Longings for Betterment

Even in a society in which there are strong deflationary tendencies toward stricter and more literal interpretations of authoritative texts like the US Constitution or the Bible, and toward far less trust in and expectations of dominant social institutions and élites, there may also be contrary tendencies. Alongside segments of the population carrying deflationary tendencies, others may tend to envisage a future brighter than the present and to promote the notion that the culmination of history will await the success of efforts to bring social betterment to entire classes of people and purification to the soul. Among Pentecostal communities that are now as transnational as they are local, religious belief and practice foster not only personal morality but social disciplines that tend in the long run to produce economic and political progress.

As perhaps unwitting agents of the process of secularization, Pentecostal communities within one or two generations do produce individuals with high levels of education and competence in a wide range of occupations and professions. As the sacred becomes increasingly disentangled from a local web of familial, economic, and political associations, salvation becomes less a matter of participating in formal rites and more a process of sacralizing other aspects of social life. The de-institutionalization of the sacred fosters longings for 'betterment' marked by self-discipline, social and economic progress, and high levels of responsibility for social welfare. Gertrude

Himmelfarb (2005) notes that in the eighteenth-century conservative social and political theorists like Edmund Burke drew on precisely such religious developments to support their theories about the popular bases of social solidarity and moral progress, just as more recently David Martin (2002) has commented on comparable developments in worldwide Pentecostalism.

If sociologists of religion wish to study the extent to which the sacred has become de-institutionalized, they will need to investigate the ways in which communities and societies experience, imagine, and construct the passage of time. To what extent do people believe that the present offers an opportunity to realize the unfulfilled potentials of the past? To what extent does a particular society interpret new situations in the light of precedents rather than as novel? To what extent does a society believe that actions are unrepeatable and their effects irreversible, rather than being open to future revision and redemption? Is there a sacred history or national myth that gives the semblance of continuity and development to the sequence of events and to the passage of time? How are moments perceived: as simply fleeting or potentially everlasting, as possibly opportune or actually critical, as mundane or potentially sublime, as calling for endurance or for decisive action? Is either the past or the future pressing on the moment, so that old scores have to be settled and visions realized? Or is the present relatively open and indefinitely extended and filled with new departures?

A PREDICTION

I would like to venture a prediction: that as the sacred becomes increasingly more diffuse and therefore more difficult to institutionalize, there will be a reaction against the involvement of religion in government. As the sacred increasingly becomes a dimension of work, education, and politics, countervailing tendencies will seek to keep these areas of social life at a safe distance from centres of religious control. In this process, furthermore, religion as a subsystem will again become more highly differentiated from other subsystems like the polity or the economy, and its beliefs and values will become more generalized and abstract, to cover a wider and more complex set of contingencies and situations. What religion will lose in specific and immediate relevance to or control over particular situations, it will gain in the appearance of transcendence. Because religious symbols will be too abstract to provide specific guidance in particular situations, however, ethicists will continue to develop an increasingly situational and pragmatic casuistry for the guidance of everyday life.

As it becomes more difficult to contain the sacred within formal rituals controlled by religious professionals, so political leaders will continue to assume the mantle of the prophet, just as bureaucracies will continue to develop routines that are sacred, regardless of the goals of the organization, and a secular priesthood of scientists may continue to control access to esoteric knowledge and authority. Lawyers may make renewed attempts to define their function as making their clients 'whole', while doctors may seek to resuscitate the 'sacred doctor–patient' relationship, and both the military and the business world may renew their emphasis on their 'missions'. Only if new attempts are made to differentiate religious institutions from positions of control in politics, education, and the economy, will each sphere of activity, from work and play to religion and warfare, be able to develop and maintain its own more informal ways of symbolizing and addressing the sacred. Without such an attempt to re-differentiate religion from control positions, especially in the political and the judicial systems, there will be increasingly deflationary tendencies, radical distrust of essential institutions, and renewed demands to limit the meaning of religious and judicial texts to narrow and strict constructions.

Sociologists have further work to do in assessing the extent to which religion, in any community, institution, or society that they are studying, expresses and embodies such inflationary or deflationary tendencies. Under what conditions could the escape of the sacred from the constraints imposed by religious institutions become conducive to 'inflationary' trends? As aspirations for the satisfaction of old longings and grievances become increasingly plausible, will individuals become more likely to believe in the promises of political candidates, to expect the future to be better than the present, and to allow the value of money to be determined not by fixed assets, like gold, but by exchange rates? Is there a correlation between the use of credit cards, metaphoric language, trust in public rhetoric, a relatively broad construction of law and the US Constitution, liberal politics, and a willingness to revise liturgies, to reinterpret sacred texts and doctrines, and lower standards for clerical behaviour and church membership? Do these inflationary trends depend on the degree to which religion itself is differentiated from the polity and the judicial system?

On the basis of the argument in this essay, I would predict that, as religious institutions lose their control over the sacred and become more differentiated from other aspects of the larger society such as government and the courts, individuals will take a more flexible approach to language and the interpretation of texts, prefer a metaphoric to literal usage of words, and free such documents as the Constitution of the United States and the Bible from the limits of strict construction; and only then will politics be liberal and religion progressive.

REFERENCES AND SUGGESTED READING

BELL, CATHERINE (1997). *Ritual: Perspectives and Dimensions.* New York and Oxford: Oxford University Press.

BLOCH, MAURICE (1989). *Ritual, History, and Power: Selected Papers in Anthropology.* London and Atlantic Highlands, N.J.: Athlone Press.

—— (1991). *Prey into Hunter.* Cambridge: Cambridge University Press.

FENN, RICHARD K. (2001). *Beyond Idols: The Shape of a Secular Society.* Oxford: Oxford University Press.

HIMMELFARB, GERTRUDE (2005). *The Roads to Modernity: The British, French, and American Enlightenments.* New York: Vintage.

LINCOLN, BRUCE (1985). *Religion, Rebellion, Revolution: An Interdisciplinary and Cross-Cultural Collection of Essays.* New York: St Martin's Press.

LUKES, STEVEN (1985). *Individualism.* Oxford: Blackwell.

MARTIN, DAVID (2002). *Pentecostalism: The World their Parish.* Oxford: Blackwell.

MOORE, BARRINGTON Jr. (2000). *Moral Purity and Persecution in History.* Princeton: Princeton University Press.

PATTERSON, ORLANDO (1998). *Rituals of Blood: Consequences of Slavery in Two American Centuries.* Washington: Civitas/Counterpoint.

RAPPAPORT, ROY A. (1999). *Ritual and Religion in the Making of Humanity.* Cambridge: Cambridge University Press.

WILSON, BRYAN (1973). *Magic and the Millennium.* Oxford: Heinemann Educational Books.

CHAPTER 17

ANTHROPOLOGY AND RELIGION

MICHAEL LAMBEK

INTRODUCTION

Ever since Plato there has been a binary distinction in Western thought between reason and its other, described, respectively, as standing outside or inside the object of thought. Plato called these positions 'philosophy' and 'poetry'; in recent centuries they have sometimes been referred to or refracted as enlightenment and enchantment, modernity and tradition, or science and religion, and they find applications in a wide range of analogies concerning the West and the rest, male and female, and so forth. They are generally the terms of those who consider themselves to hold the position characterized by the former member of each set of oppositions. Despite the historical success of this way of looking at the world, Plato's own pupil, Aristotle, offered a distinct alternative. Aristotle begins not with a binary opposition but with a triad of complementary modes: contemplative thought, practical reason, and creative production (*poiēsis*). These modes—thinking, doing, making—are universal and pervasive to humankind. In the Aristotelian tradition, humans—philosophers and scientists included—are always speaking or thinking from within some kind of practice. Further, such practices are historically, not transcendentally, located: contingent, not absolute. An implication is that social practices, modes of thought,

My thanks to Joshua Barker, Wendy James, Eva Keller, Tanya Luhrmann, Philip Clayton, and Zach Simpson for their careful and very helpful responses to a first draft, and to Jackie Solway for stimulating discussion. Naturally, I take responsibility for all errors and infelicities. This chapter was written with the benefit of a research leave supported by the Social Science and Humanities Research Council of Canada and the Department of Social Sciences, University of Toronto at Scarborough.

and the acts and products of human creation are not necessarily strictly comparable to each other along a single or simple axis of value ('truth'); instead, they are incommensurable to one another. Incommensurability poses a severe challenge to unilineal narratives of progress, and also means that often one cannot make an exclusive choice between alternatives; that the exercise of judgement—practical reason (*phronēsis*)—entails finding the right balance, generally a path of moderation.[1]

While Aristotle's thought was congenial to the medieval Catholic Church and influenced many thinkers subsequently, including both Marx and Durkheim, it has not so easily found a home in either social science or the public imagination. Yet, it is the argument underlying this essay (an argument which has not escaped a binary opposition of its own) that the Aristotelian framework affords a superior orientation for the anthropology of religion and for discerning the relationship (or, to be more precise, the ongoing history of the relationships) between religion and science. This entails moving beyond a strictly 'intellectualist' appreciation of religion and science as comparable forms of knowledge or reasoning and embracing their ethical and aesthetic dimensions as well.[2]

STRUCTURE AND HISTORY
IN AND OF ANTHROPOLOGY

Grappling with the relationship between religion and science has been—and will doubtless remain—central to anthropology. Indeed, one could almost say that it is intrinsic to the field, that religion/science stands, like nature/culture in Lévi-Strauss's (1963) theory of myth, as the irresolvable opposition around which anthropological thought builds itself. This is so for two main reasons. First, the religion/science opposition has stood synecdochally in anthropology for even larger questions: notably, the debate between relativism and rationalism, and the contrast and transition between holistic but diverse 'primitive' or 'traditional' worlds and the disenchanted, fragmented, but ultimately singular 'modern' one. Second, much as anthropology would like to see itself as an objective observer of human institutions and transformations, it is itself situated within the broad discursive field constituted

---

[1] In the Kuhnian model of science, however, successive incommensurable paradigms do more or less fully replace one another.

[2] A number of the points condensed in these first two paragraphs are developed in Lambek (2000). My depiction of Plato is strongly influenced by Havelock (1963), while that of Aristotle owes a good deal to Bernstein (1983) and beyond him to Gadamer (1985) as well as to MacIntyre (1984). Lloyd (1990) offers a penetrating account of the polemical emergence of science as a self-conscious style of inquiry in ancient Greece.

by science and religion, and it has always also been an interested party in the debate between them, pulled between explanation and interpretation or, in the terms of its methodological injunction, between observation and participation. As a result, despite many insightful contributions and developments, clarifying the relationship between science and religion remains an ongoing therapeutic task or feature of anthropology, rather than a fully realized or realizable scientific goal. It is internal as well as external to the practice of anthropology.

This is not to say that anthropologists hold a unified position on these issues or that they have not contributed both evidence and arguments to wider public and scholarly debates. Unlike a Lévi-Straussian opposition, the tension between religion and science is thoroughly shaped by the historical conditions of its time and has changed for anthropology over the course of its own history, each phase leaving significant traces in successive layers of theoretical debate. I paint the picture with extremely broad brush strokes as a series of three phases of anthropological thinking in conjunction with wider intellectual and sociopolitical processes. In its inception as an academic discipline in the late nineteenth century, anthropology was, to draw on MacIntyre's (1990) term, 'encyclopaedist'. Anthropology saw itself as an objective, neutral science of reason that challenged the obfuscations and misapprehensions of religion and tried to locate religion's place in human history. The birth was not easy, as manifest in the career of Robertson-Smith (Beidelman 1974), but Edward Tylor had a kind of assurance in stating that religion was rational but grounded in error.[3] The general project, too easily denigrated today for its affinity with, if not actual modality as, colonial governmentality, was also part of the radical Enlightenment programme of locating humankind ('man') as creatures of nature, rather than God.

The surpassing of evolutionism by functionalism, cultural particularism, and structuralism gave the argument a new direction. There was a general deconstruction of the overly objectified typologies and reified categories of the earlier period and a relocating of humans as a product less of nature than of themselves ('culture'). For much of the twentieth century the progressive task of anthropology was to show the order, logic, morality, and beauty in what seemed to the majority of Europeans and North Americans to be uninteresting, primitive, backward, disorderly, disappearing, and generally unworthy in societies and systems of thought. The role of anthropology was no longer to critique religion but to appreciate it from a distance, i.e. indirectly, by means of 'other' societies. Lévi-Strauss's title *Tristes Tropiques* (1969) conveys the sensibility of the period. Anthropology's subjects were for the most part conceived as distant, quiescent, and relatively powerless, and there was an ethical imperative to represent them in the face of the onslaught of change, whether one saw it as 'modernization' or exploitation.

During both these phases, the present (since called 'modernity') was identified with the growth of secularism, and anthropology understood itself as a secular discipline, sometimes concealing from itself its strong romanticist tendencies. But

---

[3] I have not provided references for the early work. For excerpts of some classic texts, as well as reprints of a number of the articles referenced below, see Lambek (2002a).

by the end of the twentieth century, with the resurgence of religion in the United States and within nationalist, transnational, and global politics (a recognition epitomized by the general surprise occasioned by the Iranian revolution), but also with the rise of scepticism within the academy about the nature of science and secularism themselves (as phrased by diverse strands of post-structuralist, post-modernist, and post-colonial thought) and the concomitant affirmation of history as the master paradigm, anthropology finds itself squirming, no longer content or able simply to champion the richness of religion against science and modernization narratives, or to return to an ostensibly value-free objectivist science of religion.

The tension between anthropology's scientific rationalism and its humanistic relativism, perspectivism, and historical particularism is particularly acute at the time of writing this chapter, when the President of the United States advocates the teaching of 'evolution by design' alongside natural selection (the primary area of science targeted by fundamentalist religion) in American schools and continues to refuse subsidies for AIDS prevention programmes that promote the use of condoms. Do anthropologists simply interpret the coherence of conservative Christianity and analyse the power of its rhetoric? Or do we try to fight for the naturalist and evolutionary premises on which anthropology and the life sciences are built? If there is a compass to the anthropological direction, perhaps it lies in unmasking or decentring hegemonic assumptions, undue power, unfairness, and dogmatic or absolutist thinking. These are, of course, not the special province of either religion or science *per se*, but only of certain manifestations and invocations thereof.

If, during the first two phases of anthropological thought delineated above, science itself was unproblematic, and if in the first phase anthropology saw itself unproblematically *as* a science, these facts are not true of the present age. A number of things have changed beyond the political fortunes of religion. First, anthropology has increasingly questioned its own status as a science;[4] and second, science itself has become an object of anthropological inquiry alongside and roughly equivalent to that of religion. Both religion and science can be described as systems of human thought and practice, each with strengths and weaknesses, neither perfect according to their own standards, the practitioners of each struggling with issues of moral judgement, creativity versus iteration, and the prejudices entailed by their own means of production and reproduction and modes of seeing the world.

Advocates and practitioners of both science and religion must consider how their subject articulates with politics and the economy, the industry of war, environmental issues, the mobility of people, the spread or containment of disease, and the new genetic and reproductive frontiers. Each must face up to its commitments to forms of geo- and biopolitics and to the contradictions therein. Anthropology, again, is not simply an observer and explicator of these trends and processes, but is increasingly self-conscious regarding its own agency and lack thereof.

---

[4] For lack of sufficient space and knowledge I forgo discussion of explicitly scientific contemporary anthropological perspectives on religion, notably those drawing on cognitive—and now neuro—science or calling themselves 'neo-Darwinian'. See Ch. 25 below.

In general, then, the current phase of anthropological thought has witnessed a move away from claiming a particular expertise or understanding of the nature or essence of religion—and from participation in the theological or scientific debates that such claims entail—toward analysing the politics of 'religion', as though from outside (and perhaps, outside 'science') but thereby necessarily 'inside' some other, at times seemingly inchoate, mode of practice. The current articulation of the recurrent theoretical fault line of anthropology is that between the ironic or sceptical genealogical observer and the complicit, but possibly critical, hermeneutic participant (i.e. the person who accepts Gadamer's (1985) argument that we are all located within traditions and that all traditions entail their prejudices, that anthropology shares horizons with both religion and science). Between these (non-binary) positions anthropologists must construct both their research programmes and their politics.

RELIGION AND SCIENCE DISTINGUISHED AND RELATED

Of course, the task of determining the relationship between science and religion is also irresolvable for reasons that do not pertain specifically to anthropology: notably the fact that religion and science themselves are not fixed or homogeneous entities but dynamic, heterogeneous bodies of thought, practice, and production. As anthropologists turned for inspiration from Durkheim to Weber and from non-Western religions to Islam, Christianity, and various post-colonial transformations of other religious traditions, among the observations they have come increasingly (and belatedly) to make is that specific religious traditions encompass diverse ('internal') debates and arguments, their practitioners and advocates struggling to be more or less closely connected or committed to immediate political issues, including responses to scientific arguments and discoveries. Indeed, a major question for the anthropology of religion, both theoretical and empirical, is the degree to which religion is able to constitute itself outside or at arm's length from the political sphere and thereby reproduce itself on a longer historical trajectory and at a slower pace than those of current affairs and scientific discovery (Rappaport 1999).

Conversely, an anthropology of science, informed by developments and debates in the philosophy of science, worries over social constructionism and its limits, and observes the practical exigencies and political conflicts that shape 'laboratory life' and the production and reception of ostensibly value-free scientific facts, theories, and goods. A study of the production and reception of scientific 'goods' must entail consideration of temporality (e.g. speed, duration, and turnover) and of spatiality and authorization, and the ways in which these processes are determined by or articulated with the capitalist marketplace and its ideological offshoots such as neo-liberal policy. If science could once be considered value-free in contrast to religion, its market value and subsumption into the military-industrial complex are all too evident. Moreover, aesthetic and ascetic bodily practices that were once the province of religion have come increasingly within the scope of science and the consumer market, recent fashions in cosmetic surgery and mood-altering pharmaceutical drugs being only the latest examples.

Perhaps no social scientist has considered the relationship between religion and science as carefully as Max Weber. In his essay 'Science as a Vocation' (1946), Weber argues for a distinction between fact and value. Insofar as science addresses the former and religion the latter, the two fields are complementary yet necessarily at arm's length. However, insofar as science does not treat of value, Weber argues, it cannot itself comprehend why its adherents choose it as a calling rather than, say, the church, and, insofar as it constitutes a specific calling (as he understood the term) distinct from religion, so it is in a kind of competition with religion. Under conditions of rationalization, that is, since the rise of science and the increasing disenchantment of the world, religion impels a choice, not only 'for' religion instead of or possibly alongside science, but for commitment to one religion or set of values as opposed to another. Science, by contrast, although it is characterized by great specificity, does not require a choice between competing value orientations within itself. Science cannot on internal grounds decide which areas of research (nuclear fission, stem cells, ecological versus physiological research on cancer, etc.) should be funded, how far to proceed, what to do with the results, and so on—all are areas of value (ethics) rather than fact *per se*.

In contrast to the evolutionists, who saw science arising sequentially after or out of religion, Weber, and Robert Merton (1949) after him, also noted affinities between a scientific outlook and specific kinds of religious perspectives and social positions rather than others. Perhaps one could summarize Weber's position by saying that he viewed science and religion as incommensurable, and hence inclined towards a nominalist position with respect to theorizing their relationship. At any rate, that is not unlike the conclusion of the present essay as a whole.

It is by no means clear that either religion or science constitutes a distinct type ('natural' or otherwise) for which there are respectively a number of discrete, commensurate tokens. They are each more likely to constitute polythetic sets. Although this issue has bedevilled the anthropological study of kinship to the point that some theorists have suggested that there is no such thing as kinship, the critique has not been carried so far with respect to either religion (but see Southwold 1978) or science. There is a quiet assumption in each case that type/token relations hold. Perhaps the strongest challenge to this view of religion comes from Talal Asad (1993), who criticized Clifford Geertz (1966) (and the whole encyclopaedist tradition, Geertz's Weberian and hermeneutic position being hardly typical of objectivist social science notwithstanding) for the assumptions entailed in attempting to define religion in the first place. Asad (2003) has since attempted to theorize and document the rise of 'religion' (as a discursive subject and self-conscious set of disciplinary practices) in tandem with secularism with reference to both state power and colonial and transnational relations in classifying and apportioning domains and forms of discipline.

Asad's emphasis on the role of power and discourse in authorizing what may count at any period as religious practice or religious truth obviously holds for science as well. However, if science and religion can be viewed by anthropology within similar analytic frameworks, and if their respective adherents are faced with similar political and practical challenges, this is not to conclude that science and religion are equivalent or to be collapsed into one another, either by practitioners or by observers.

EXPLANATION, ACCOUNTABILITY, AND CLOSED SYSTEMS

It is evident that contemporary religion does not dispense with science (or at least technology) *per se*; fundamentalism condemns only what it interprets to contradict a correct reading of its sacred texts. Both the Islamic government of Iran and the *de facto* Christian government of the United States pursue programmes of nuclear power. The primary area of science targeted by religion in the USA is evolutionary theory, particularly the evolution of the human species, but extending to all life forms and even to geology insofar as religion advocates either intelligent design or creation along the lines—and with the timing—of a literal interpretation of the Genesis story. It is perhaps a paradox that evolution is challenged in theory at the very moment in history when intelligent—or not so intelligent—design is in fact tampering with its very elements and mechanisms by means of fields like genetic engineering that are enabled by the state. But it is no wonder that anxious political and military leaders might wish to place ultimate responsibility outside their own hands and into God's.

The question of responsibility—or accountability—has been seen by some anthropologists as a critical dividing line between the spheres of science and religion. Whereas science explains by means of so-called natural cause, religion attributes action more often to personal cause, be it a personified God, gods, spirits, or humans acting as saints, witches, sorcerers, shamans, or magicians. Religion thereby fills a function from which science abstains: namely, providing a theodicy. The ability—indeed, necessity—of religion to maintain interpretability—that is, to explain bafflement, suffering, and ethical paradox, rather than evade, postpone, or dismiss answers to these questions—is for Geertz (1966) one of the critical markers for distinguishing religion from science.[5] It is likewise central to E. E. Evans-Pritchard's famous analysis (1937) of how, within multiple levels of causality, witchcraft serves the Azande as an explanation of misfortune. A granary can collapse if its wooden posts are eaten by

---

[5] Insofar as Geertz views religion, science, and common sense as shifts in perspective rather than discrete objects (a position on which he is not consistent), the main thrust of Asad's critique is vitiated.

termites, but only witchcraft can explain why that moment of collapse coincides with my neighbour's decision to sit beneath it. (At the same time, of course, the system produces its own set of 'victims': namely, those discovered to be witches—as well as general anxiety and mistrust.)

Evans-Pritchard's argument also points in the other direction, towards a similarity between science and religion as both tokens of a type that might be referred to as a coherent or comprehensive 'system of thought'. Building in a somewhat unacknow-ledged way on Malinowski (1922), Evans-Pritchard's *tour de force* is memorable for his demonstration of the practical logic of witchcraft beliefs, how witchcraft and the ways to counter it form a relatively closed, internally consistent system for the Azande. His famous remark that Azande 'reason excellently in the idiom of their beliefs, but...cannot reason outside, or against, their beliefs because they have no other idiom in which to express their thoughts' (1937: 338) was taken by certain philosophers as a feature also of scientific paradigms. As Stanley Tambiah summar-izes Wittgenstein's *On Certainty* (written 1949–51):

A mistake is something which can be tested and shown to be wrong. But the idea of testing already implies some particular system which has as its foundation a set of presuppositions and propositions which cannot themselves be tested or doubted. These propositions make the activity of testing possible by determining what will count as evidence for arguments and verification.... Says Wittgenstein: 'Whether a proposition can turn out false after all depends on what I make count as determinants for that proposition'. 'The *truth* of certain empirical propositions belongs to our frame of reference'. 'All testing, all confirmation and disconfirma-tion of a hypothesis takes place within a system...The system is not so much the point of departure as the element in which arguments have their life'. (Tambiah 1990: 64)

And just as Evans-Pritchard showed how secondary elaborations conserve the basic premises of the system, so too, some have argued, scientific paradigms preserve themselves from contrary evidence.

Of course, Wittgenstein's description does not preclude comparison of different systems or presuppositions, and some of anthropology's most exciting discoveries come when we penetrate to truly distinctive ones, as in Eduardo Viveiros-de-Castro's depiction of Amerindian perspectivism (1998). His analysis of the way in which many Amerindian groups presume that humans, animals, and spirits each see both them-selves and one another differently from within different kinds of bodies not only opens up a whole new understanding of Amerindian shamanism, but provides an astonishing alternative to Western categories of nature, culture, and 'supernature' from which so much theorizing about religion and science begins.

Evans-Pritchard's work has always been taken as a paradigm.[6] But in hindsight, and noting how little appears to have changed in Zande witchcraft, it may be suggested that the system was and is too tight, does its job too well, and that some comparison of systems of thought as more or less closed or open is appropriate. Thus, in contrast to the Azande, my ethnographic research on the western Indian

---

[6] But not exactly as a Kuhnian paradigm; his mode of analysis in *Witchcraft, Oracles and Magic* was widely lauded, but rarely taken up by successors.

Ocean island of Mayotte showed that during the 1970s and 1980s a diagnosis of sorcery was not inevitable there, that there were alternative modes of explanation, sorcerers were rarely named or accused, and narratives of causality frequently remained open to further events and interpretations (Lambek 1993). At the other extreme is the horrific scenario of child witches that has recently arisen in Kinshasa. Filip De Boeck (De Boeck and Plissart 2004) is at considerable pains to describe, by means of a Lacanian analytic vocabulary, how nightmare (fanned by Christian congregations) can become reality. The comparative question is how such systems are regulated or become deregulated—in other words, to look not only at the internal logic but at modes of authorization (as indicated by Asad) and at how they articulate with wider political processes and events.

In sum, Evans-Pritchard's analysis remains significant, but not quite for the reasons often claimed. A negative consequence of a common reading of the Zande paradigm, and one reinforced by the Boasian attention to cultural coherence and the Durkheimian and structuralist legacies of synchronic holism, was the picture of traditional, static, closed systems, contrasted negatively with the positive, dynamic, open system of modern science, idealized by Popper, that could prove them wrong and thereby succeed them. This is patently false because, as noted, openness and closure do not map onto any line of historical or evolutionary progress. As articulated by Wittgenstein, science too must have its limits to falsifiability; conversely, a fully bounded static system is equally inconceivable. Moreover, as Evans-Pritchard himself noted, 'mystical and scientific thought could be compared as normative ideational systems in the same society', and, moreover, we can observe people switching between modes of thought or frames of mind as the context changes (Tambiah 1990: 92).

Drawing on Alfred Schutz, Tambiah (like Geertz) prefers to speak of alternating orientations to reality. Tambiah revives Lévy-Bruhl's idea of religion as participation, and manages to link it with developments in semiotics and various ideas in psychoanalysis, feminist psychology, and philosophy. Tambiah probably overstates his case in setting up two basic orientations to the world that he calls, respectively, 'causality' and 'participation'. This is a dual opposition remarkably reminiscent of Plato's distinction between philosophy and poetry, and, as noted above, recurrent in its modern refraction as enlightenment versus enchantment; and like these, it omits most ordinary language and life. Moreover, the opposition risks replication by those tempted to offer scientistic explanations for specific religious practices.[7]

---

[7] It is interesting that the two extreme ends of the spectrum often come together in the hands of specific thinkers about religion, who might, e.g., try to combine work on neurotransmitters (causality) with their own ostensibly mystical experiences in the field (participation). The Society for the Anthropology of Consciousness is full of this kind of thing. My own position could not be further removed from such facile mediation of binary oppositions, emphasizing instead the sociocultural (rather than either the natural or the transcendental) reality of religious phenomena, the rigorous hermeneutic and reflexive stance required of interpreters, and the significance of a tripartite Aristotelian rather than a dualist model (Lambek 2000, 2002b).

Empirically (and logically) it is the case that people always have recourse to incommensurable ideas and practices (Lambek 1993). These are ones for which a clear algorithm of binary choice is not possible; rather, judgement must be continuously activated. Furthermore, outside the Abrahamic legacy, it is by no means clear that religion has meant exclusive loyalty to one god or prophet. The sorts of heterogeneity of religious traditions characteristic of many Asian social fields (e.g. Gellner 2001) and the polytheism of Hinduism or Yoruba religion have arguably demonstrated more openness and flexibility (though the Abrahamic religions all also have their internal streams of debate and conversations between branches of tradition). Contrary to the ideas of many Christian and Muslim scholars, polytheism is not intrinsically either rationally or ethically inferior to monotheism; indeed, one could conceive of arguments that it is superior. Certainly, any religion that claims exclusive access to the truth must have at least that assumption regarded as false by a comparative anthropology.

A comparison of common forms of science and religion suggests that the former are generally narrower in scope than the latter. Not only does science generally abdicate from Weberian questions of theodicy, it is largely unable to address the social functions that Durkheim and his British disciples attributed to religion. That is to say, science is not as richly expressive of society in the way that religious symbols or rituals may be (but see, e.g. Martin 1994); nor is it as directly contributory to social solidarity or the enrichment of moral life. As recent events have shown, religion continues to form a powerful vehicle for drawing together collective sentiments and for providing a source of collective identity within mass or global society. Religion provides an alternative, and stands in some contrast to (and possibly mystification of) impersonal bureaucracy, the alienation characteristic of capitalist production, the amorality of capitalist exchange, and the anomie characteristic of capitalist consumption. With the partial exception of dedicated working scientists, who view their practice as a Weberian vocation (or the effects achieved through the prescription of anti-depressants), science cannot offer these advantages. A similar comparison of the functional attributions and limitations of religion and science could be made along Freudian lines with respect to the unconscious and the elaboration of fantasy. However, interesting new work on the 'technoscientific imaginaries' of, say, communications engineers (Barker 2005) suggests that science may connect to society and psyche more broadly than these remarks imply.

MAGIC AND MEDICINE

The opposition between science and religion has often and tellingly been mediated by a third term: namely, magic. *Magic, Science and Religion* was the title of a set of influential essays by Malinowski (1954); the triad reappears in an extremely useful overview by Tambiah (1990), which I have already cited; and magic is provocatively paired with modernity in a recent collection edited by Birgit Meyer and Peter Pels

(2003). In the early formulations magic serves as a way to displace what theorists assume that science is successfully able to criticize in religion. The ostensibly instrumental, mechanical, and simple-minded elements of belief and practice are classified as magic, leaving religion as a domain characterized by soteriological or transcendental concerns, purified from error and illusion and clearly distinguishable from science. Under the magic carpet are thus swept beliefs and practices that might blur the boundaries between science and religion or threaten to subsume one in the other. By more clearly distinguishing the respective functions of religion and science from one another, the model thereby serves to reduce the sense of their competition. It also, implicitly, serves to distinguish the world of 'ethical' religions from the practices of smaller-scale societies which are ostensibly more easily assimilated to the category of magic. Whereas numerous ethnographers show the vacuity of this view for understanding practices in smaller-scale societies, the essays in Meyer and Pels (2003) demonstrate, conversely, the 'magic' inherent within 'modernity'. Here 'magic' embraces the religious, irrational but persuasive impulse within science. Several of the authors, especially Michael Taussig (2003), go further in demonstrating the way in which magic always entails—and plays with—scepticism alongside credulity.

Robin Horton (1967) also challenged the attribution of magical thinking to smaller-scale societies by arguing in neo-Tylorean fashion that 'African religion' is actually much closer to science than is generally assumed. His argument, however, fails to acknowledge that in the end the beliefs and practices of smaller-scale societies are not strictly comparable to science, if only because they are not discrete institutions on the order of science. Tambiah (1990) offers a sharp rebuttal to this kind of category error (but cf. Appiah 1992).

For many years, then, the question was how to define religion and science so as to distinguish them but also compare them, and to use lessons drawn from the one, either positively or negatively, to help understand the other. The project was considerably refined by means of the development of structural analysis. Borrowing the term from William James, Mary Douglas (1966) offered a salutary attack on what she called 'medical materialism': namely, the explanation of specific religious practices, such as the Jewish taboo against pork, contrary to what religious practitioners themselves might say, as 'really' based on material criteria, such as the health dangers of poorly cooked meat. Her argument was elaborated in the bracing critique by Marshall Sahlins (1976) of various forms of Malinowskian functionalism or American cultural evolutionism that explain religion in terms of what it does, thus ultimately in terms of a materialist science. Douglas and Sahlins also went a long way towards unpacking the cultural logic of science itself, a project that has been furthered by developments in medical anthropology—for example, in Margaret Lock's (2002) demonstration that even death is differentially defined and regulated in Japanese and North American forms of biomedicine. In decisions over organ transplants it is not so easy to distinguish fact from value.[8]

---

[8] Limits or border areas of science and religion are also evident in psychiatry, in questions like the displacement of the soul by memory (Hacking 1995); in science fiction understood as myth (Moisseef 2005); and so forth.

Yet, insofar as structuralism as practised by the master (Lévi-Strauss) held out the goal of discovering 'laws of the mind', the structuralist comparison of religion and science was not necessarily a rejection of scientific explanation of religious facts *per se*, but rather a call to widen the boundaries of what science, Anglo-American science in particular, could conceive as valid forms of explanation. Thus these critiques also demonstrated implicitly the way in which science, too, is defined by culture and by specific intellectual traditions. For a certain kind of cultural structuralist both the science and the religion of a given society build from the same structural roots even if, on the surface level, they prove very different from one another. This is perhaps a source of what, from a different intellectual tradition, Weber described as 'elective affinities'.

Perhaps the most decisive advance was that initiated in the philosophy of language. Once it was understood that language can do other things than represent, it was possible to understand religious practices in a new light. Whereas science works to explain or represent the world, and does so by means of propositions that can then be judged as accurate or inaccurate, true or false (or at least understands itself to be operating in this manner), much of what falls under religion is now more clearly seen as non-representational and non-explanatory but rather as constitutive (hence surpassing the 'intellectualist' arguments of Tylor, Evans-Pritchard, and Geertz, as well as the 'symbolic' arguments of the structure-functionalist school). Religious utterances are as often illocutionary (performative) or perlocutionary (rhetorical) speech acts as they are locutionary (descriptive) statements. The success of these acts is not to be judged by criteria of correspondence truth or falsity but, in the case of illocutionary acts, by means of what J. L. Austin (1962) termed felicity conditions, having to do with how well and appropriately they are performed. Roy Rappaport phrased the consequences succinctly:

> *The state of affairs is the criterion by which the truth, accuracy or adequacy of a statement is assessed.* In the case of performatives there is an inversion. If, for instance, a man is properly dubbed to knighthood and then proceeds to violate all the canons of chivalry... we do not say that the dubbing [was] faulty, but that the subsequent states of affairs are faulty. *We judge the state of affairs by the degree to which it conforms to the stipulations of the performative act.* (1999: 133; italics original)

Rappaport's analysis illuminates both the question of religious facticity and truth and Durkheimian insights concerning ritual and religion as the moral foundation of society.

On the perlocutionary side, religion incorporates multiple sensory media and has an aesthetic dimension largely missing from science (despite the advances of PowerPoint). Music, art, dance, and other forms of experience play a significant role (James 2003). This is beautifully illuminated in works such as Turner (1967), Witherspoon (1977), Fernandez (1983), Kapferer (1983), Daniel (1984), and Hirschkind (2001). Together, the illocutionary and the perlocutionary go a long way to explain the success of non-Western healing systems (but also apply to biomedicine) (e.g. Lévi-Strauss 1963; Tambiah 1973; Lambek 1993; Antze 2002).[9]

---

[9] This is not to say that particular forms of healing—e.g. Buddhist 'mindfulness'—may not also be efficacious in other respects.

Of course, the performative and rhetorical effects are often mystified in religion, such that practitioners often do operate at least partially by means of a correspondence theory of truth. And in a surprising way this returns us to science. Because while science claims that its own speech is not illocutionary or perlocutionary, and that its truth is a correspondence truth, a Foucauldian analysis would lead us to think quite otherwise. Insofar as science is discourse, so it 'produces' the objects of which it speaks, as in Foucault's analysis of sexuality (1978). The ontological status of such objects of science has been the subject of much lively and heated debate by philosophers (e.g. Hacking 1998), and is great territory for anthropologists interested in exploring contemporary practices of rationality and ethics (e.g. Rabinow 1996) or for returning with new analytic tools to the practices of small-scale societies (James 1988). Here, where it was least expected, science and religion appear to approach each other once again.

CONTEMPORARY FERMENT

The task now is not to paint global pictures of 'science' and 'religion', but to study instances of each—i.e. sets of practices that claim or are claimed to be either science or religion—and describe their discursive foundations, actions, and effects, and their relationship to other discursive claims and practices within the same social milieu. We now accept that there is no single, unidimensional comparison possible between 'science' and 'religion', but lots of micro-comparisons, on the order of Foucault's capillary relations of knowledge and power. How do these diverse sets of practices shape the ethical landscape or orient people within it, and how do they constitute or 'colonize' various 'lifeworlds'? Conversely, how do ordinary people respond to cultural fragmentation and the 'disembedding of religion' (to import Karl Polanyi's idiom developed with respect to the economy; cf. Taylor 2004) and draw on the array of alternatives available to them, producing various bricolages or intensifications and attempting to transform situations of imposed power or ostensible but shallow choice into ones of moral integrity, dignity, and serious judgement? How, in sum, are specific regimes of value transcended and transfigured?

How, in particular, are we to explain the growing interest in religion, often with a concomitant ostensible rejection of science? Anthropological answers are forged through the practice of ethnography, i.e. immersion within specific communities of practice and careful listening. Rather than review a range of studies or attempt to build a comprehensive explanation, I illustrate a single, provocative case.

Eva Keller makes the useful point that even if anthropologists give up the systematic comparison of 'religion' and 'science' as a false problem, it remains a salient issue for many people trying to make sense of their own positions. Thus an anthropological contribution is to understand the many practical, imaginative, and theoretical ways in which various people (societies, congregations) conceptualize and try to resolve the issue. Keller (2005a) describes Malagasy Seventh-Day Adventists, who 'are not concerned with defending religion against science, but rather, from their point of view, with debating one scientific theory against another [creationism against evolutionism]'.[10]

The Christians Keller describes apparently attempt to draw on the authority of science in order to say that they, too, are approaching the world scientifically, seeking 'proof'. Keller's point is not to claim that Adventism is scientific, but rather that Malagasy Adventists want to understand the world by means of rational thought and evidence. Whereas Keller compares the Adventists to Kuhnian scientists, I would suggest that from the perspective of science, for which its distinctiveness as a rigorous mode of inquiry is acute, the proof seeking of the Adventists could only be a faulty kind of mimesis, a sort of unintentional parody. When the challenge is closer to home, as in legal battles over the right to teach 'evolution by design', such mimesis can only appear disingenuous.[11]

What the Adventists are doing is claiming the right to appropriate authority and interpret the world for themselves. They are not against science *per se*, but only against the discursive authority that would exclude their own voices from its practice. Many modern religious movements can similarly be understood (in part) as attempts to reappropriate knowledge and truth from distant experts and to provide alternatives to the authoritarian regimes of church or state and to the diffuse but pervasive capillary systems of power/knowledge characteristic of modernity.[12] In this respect the Adventists are not so different from the advocates of complementary medicine. The scientific and medical establishment finds itself challenged on the one side by religious fundamentalism and on the other by New Ageism.

Keller's Adventists operate like good intellectualists, and it may be that the diversity of anthropological theories of religion merely replicates the possible kinds of orientations of local religious congregations, some of whom function like Tyloreans and others like Lévy-Bruhlians. Thus, whereas Keller concludes that Tambiah's dualism does not apply to her Protestant fundamentalists, who appear to be entirely on the side of 'causality', Tanya Luhrmann (2005) reaches virtually the opposite conclusion in her ethnography of an American evangelical group, the Vineyard Christian Fellowship, arguing that their cultivation of practices that enable them to hear God speak directly to them validates Lévy-Bruhl's idea of participation.

If religion has been shrunk by the authority of science, the circumscription and regulation of the state, and the demands of capitalist production and consumption, it attempts to return the favour. Charismatic and evangelical revivals may be seen as attempts to restore the centrality and comprehensiveness of religious practice, to

[10] See also Keller 2005b.

[11] Contemporary North American creationist arguments illustrate not the proto-science characterized in the writings of intellectualists like Horton (1967) so much as a kind of pseudo-science.

[12] Of course, this is not to be taken as an exclusive explanation or the whole picture. A focus on power can be linked to matters of ethical disquiet and redemption (Burridge 1969) as well as to missionary funding and many other factors. For a masterly overview of recent historical trends and theories to account for them, see Hefner (1998).

reseal the rift between disposition and cosmos, and to shift the reduction of the content of religion from precarious and fragmented 'beliefs' back to holistic, substantive 'knowledge'. This is not so much to de-privatize the religion of a secular age as to remove the very boundary between public and private so that religious ethics, practice, and comportment pervade everyday life and the lifeworld.[13]

Rather than compare or evaluate religious alternatives according to criteria established by science—that is, by discourses external to and different from religion—we should ask whether we can come up with evaluative criteria internal or intrinsic to religion. This is what Rappaport offers us in what is the single most ambitious and important anthropological attempt to provide a unified analysis of religion and science. Rappaport begins his masterwork by inquiring about 'the nature of religion and of religion in nature' (1999: 1), and carefully distinguishes between what he calls the scientifically lawful and the religiously meaningful as well as the kinds of truth appropriate to each. He offers a coherent formal model of religion, in which 'ultimate sacred postulates'—utterances that are deeply meaningful but informationally empty—have a central place. Having elaborated the significance of religion for social life, human evolution, and even the future of the planet, he concludes some 450 pages later by discerning pathologies of religion. These occur when the most sanctified utterances or forms of authority take on political, social, or material specificity. Such overspecification of sacred postulates, or oversanctification of the specific, leads to religious conflict and inflexibility. When, as in the tacit sanctification of profit and consumption under capitalism, it raises 'relative, contingent, and material values to the status of ultimacy', it similarly 'relativize[s] the absolute, for it identifies the absolute with the status quo and the material . . . [and] vulgarizes, profanes, and degrades the ultimate' (p. 443). Rappaport, after Tillich, refers to this as 'idolatry'. Overly literal interpretations of specific sacred texts are instances of overspecification that not only support immediate political and social conservatism but (paradoxically) risk both social breakdown and exposing the sacred to general invalidation (pp. 444–5). Similar pathologies may be discerned in science, as, for example, when evolutionary theory is over-extended into social Darwinism, sociobiology, or even the shape of ideas (Lewontin 2005). What Rappaport calls idolatry in religion is akin to reductionism in science.[14]

Conclusion

I sum up with several points. First, once we accept the historical emergence and spread of science as a unique discursive formation (or set of related formations), it becomes (anachronistic) nonsense to talk about the relationship *between* religion

---

[13] See e.g. Csordas (1997) on Catholics, and Hirschkind (2006) and Mahmood (2005) on Muslims.

[14] This argument was developed together with Jackie Solway.

and science, or religion *as* a kind of science in societies that have not yet encountered or internalized this development. (For some, the same would hold for the historical emergence of 'religion' as an explicit category and subject of political discourse.) This does not preclude us from examining the rationality of practices and discourses in these societies, but merely from trying to fit them into a mould that is not theirs. Within 'modern' societies that do comprehend science—and this includes the entire world today—one must realize the diversity of possible relationships between science and religion, beginning with the unequal education and formation that individuals, classes, or status groups of various kinds receive in one or the other, but examining also the dominant or hegemonic public formulations of their relationship as well as diverse counter-hegemonic alternatives.

Second, the view of at least this anthropologist is that although religion has been challenged by the rise and success of science, and has had to accommodate itself to science, and although science may now find itself challenged by religion; although religion and science have emerged and defined themselves (or been defined by the state) in relationship to one another, found themselves in direct opposition on certain issues, or attempted to imitate each other; although they can each be made to temper (or inflame) the excesses of the other; and although both religion and science risk the dangers of provincialism but also afford a means to escape it—at base they are incommensurable. This means that religion and science cannot be judged or compared along a single axis of measurement, and therefore that they will continue to irritate or complement each other without either one being able to fully subsume, displace, vanquish, or eliminate the other (unless, in conflagration, they eliminate us all). However, if their fault lines and borderlands are problematic, they are also extremely interesting, characterized by vibrant creativity and intense ethical questioning. As to their relationships in the future, even their very natures, much will depend on the politics of the state and the regulation and fate of capital.

Third, the very pairing of 'science-and-religion' invites intellectualist accounts of religion, and I have doubtless given them excessive attention in this essay. But just as abstract thought is not the best or only lens through which to compare religion with science, so too, reason is not restricted to science, nor, conversely, is science to be exclusively identified with or understood with respect to abstract reason. Both religion and science are characterized by combinations of contemplative thought, practical judgement, and creative performance. Thinking, doing, and making are a part of everyday life as well. Conviction, disposition, and gracefulness derive not only from facility *with* the objective vehicles or instruments of larger powers but also from agility *as* their vehicle, instrument, or subject. The stark Platonic opposition between dispassionate reason and impassioned participation is misleading.

Finally, the *de facto* identification of science with abstract reason and religion with engaged performance, the incommensurability of science and religion in the modern world, the destabilization of the transcendent or foundational claims of each, and the ultimate uncertainty that their conjunction or opposition imposes, all beg for triangulation with a third construct: namely, ethics.[15]

REFERENCES AND SUGGESTED READING

ANTZE, PAUL (2002). 'Memory and the Pragmatics of Transference in Psychoanalysis'. Paper presented to the Eighth Annual Conference in the Human Sciences at George Washington University.

APPIAH, KWAME ANTHONY (1992). *Old Gods, New Worlds: In My Father's House: Africa in the Philosophy of Culture.* New York: Oxford University Press.

ASAD, TALAL (1993). *Genealogies of Religion.* Baltimore: Johns Hopkins University Press.

—— (2003). *Formations of the Secular.* Stanford, Calif.: Stanford University Press.

AUSTIN, J. L. (1962). *How to Do Things with Words.* Oxford: Oxford University Press.

BARKER, JOSHUA (2005). 'Engineers and Political Dreams: Indonesia in the Satellite Age', *Current Anthropology,* 46/5: 703–27.

BEIDELMAN, THOMAS (1974). *William Robertson-Smith and the Sociological Study of Religion.* Chicago: University of Chicago Press.

BERNSTEIN, RICHARD J. (1983). *Beyond Objectivism and Relativism: Science, Hermeneutics, and Praxis.* Philadelphia: University of Pennsylvania Press.

BURRIDGE, K. O. L. (1969). *New Heaven New Earth.* Oxford: Oxford University Press.

CSORDAS, THOMAS (1997). *Language, Charisma, and Creativity.* Berkeley: University of California Press.

DANIEL, VALENTINE (1984). *Fluid Signs: Being a Person the Tamil Way.* Berkeley: University of California Press.

DE BOECK, FILIP, and PLISSART, MARIE-FRANÇOISE (2004). *Kinshasa: Tales of the Invisible City.* Brussels/Tervuren: Ludion.

DOUGLAS, MARY (1966). *Purity and Danger.* New York: Praeger.

EVANS-PRITCHARD, E. E. (1937). *Witchcraft, Oracles and Magic among the Azande.* Oxford: Clarendon Press.

FERNANDEZ, JAMES (1983). *Bwiti: An Ethnography of the Religious Imagination in Africa.* Princeton: Princeton University Press.

FOUCAULT, MICHEL (1978). *The History of Sexuality,* i. New York: Vintage Books.

GADAMER, HANS-GEORG (1985). *Truth and Method.* New York: Seabury.

GEERTZ, CLIFFORD (1966). 'Religion as a Cultural System', in Michael Banton (ed.), *Anthropological Approaches to the Study of Religion,* London: Routledge, 1–46.

GELLNER, DAVID (2001). *The Anthropology of Buddhism and Hinduism.* Oxford: Oxford University Press.

HACKING, IAN (1995). *Rewriting the Soul.* Princeton: Princeton University Press.

—— (1998). *The Social Construction of What?* Cambridge, Mass.: Harvard University Press.

HAVELOCK, ERIC A (1963). *Preface to Plato.* Oxford: Blackwell.

---

[15] Ethics here serves as a kind of obverse to the other triadic term invoked: i.e. magic. Insofar as magic identifies instrumental practices within religion and mystifying practices within science or its public invocations, it serves to mediate between science and religion rather than to offer a viable alternative to dualism.

HEFNER, ROBERT (1998). 'Multiple Modernities: Christianity, Islam, and Hinduism in a Globalizing Age', *Annual Review of Anthropology*, 27: 83–104.

HIRSCHKIND, CHARLES (2001). 'Passional Preaching, Aural Sensibility, and the Islamic Revival in Cairo', *American Anthropologist*, 28/3: 623–49.

—— (2006). *Ethics of Listening: Affect, Media, and the Islamic Counter-public*. New York: Columbia University Press. Forthcoming.

HORTON, ROBIN (1967). 'African Traditional Thought and Western Science', *Africa*, 37: 1–2.

JAMES, WENDY (1988). *The Listening Ebony: Moral Knowledge, Religion and Power among the Uduk of Sudan*. Oxford: Oxford University Press.

—— (2003). *The Ceremonial Animal: A New Portrait of Anthropology*. Oxford: Oxford University Press.

KAPFERER, BRUCE (1983). *A Celebration of Demons: Exorcism and the Aesthetics of Healing in Sri Lanka*. Bloomington, Ind.: Indiana University Press.

KELLER, EVA (2005a). 'How, Exactly, Does the World Work?' Paper prepared for the workshop on Zafimaniry Anthropology in honour of Maurice Bloch, London School of Economics, June.

—— (2005b). *The Road to Clarity: Seventh-Day Adventism in Madagascar*. New York: Palgrave-Macmillan.

LAMBEK, MICHAEL (1993). *Knowledge and Practice in Mayotte*. Toronto: University of Toronto Press.

—— (2000). 'The Anthropology of Religion and the Quarrel between Poetry and Philosophy', *Current Anthropology*, 41: 309–20.

—— (2002a) (ed.). *A Reader in the Anthropology of Religion*. Malden, Mass.: Blackwell.

—— (2002b). *The Weight of the Past*. New York: Palgrave-Macmillan.

LEWONTIN, RICHARD (2005). 'The Wars Over Evolution', *New York Review of Books*, 16 (20 Oct.): 51–4.

LÉVI-STRAUSS, CLAUDE (1963). *Structural Anthropology*. New York: Basic Books.

—— (1969). *Tristes Tropiques*. New York: Atheneum.

LLOYD, G. E. R. (1990). *Demystifying Mentalities*. Cambridge: Cambridge University Press.

LOCK, MARGARET (2002). *Twice Dead: Organ Transplants and the Reinvention of Death*. Berkeley: University of California Press.

LUHRMANN, TANYA (2005). 'Learning to Hear God Speak'. Paper presented at the symposium 'Learning Religion: Anthropological Perspectives', Lisbon, September.

MACINTYRE, ALASDAIR (1984). *After Virtue*. Notre Dame, Ind.: University of Notre Dame Press.

—— (1990). *Three Rival Versions of Moral Enquiry*. London: Duckworth.

MAHMOOD, SABA (2005). *Politics of Piety: The Islamic Revival and the Feminist Subject*. Princeton: Princeton University Press.

MALINOWSKI, BRONISLAW (1922). *Argonauts of the Western Pacific*. London: Routledge & Kegan Paul.

—— (1954). *Magic, Science, and Religion and Other Essays*. New York: Doubleday.

MARTIN, EMILY (1994). *Flexible Bodies: The Role of Immunity in American Culture from the Days of Polio to the Age of AIDS*. Boston: Beacon.

MERTON, ROBERT (1949). *Social Theory and Social Structure*. Glencoe, Ill.: Free Press.

MEYER, BIRGIT, and PELS, PETER (2003) (eds.). *Magic and Modernity: Interfaces of Revelation and Concealment*. Stanford, Calif.: Stanford University Press.

MOISSEEF, MARIKA (2005). 'La Procréation dans les Mythes Contemporains: Une Histoire de Science-Fiction', *Anthropologie et Sociétés*, 29/2: 69–94.

RABINOW, PAUL (1996). *Essays on the Anthropology of Reason*. Princeton: Princeton University Press.

RAPPAPORT, ROY (1999). *Ritual and Religion in the Making of Humanity.* Cambridge: Cambridge University Press.

SAHLINS, MARSHALL (1976). *Culture and Practical Reason.* Chicago: University of Chicago Press.

SOUTHWOLD, MARTIN (1978). 'Buddhism and the Definition of Religion', *Man,* 13: 362–79.

TAMBIAH, STANLEY (1973). 'Form and Meaning of Magical Acts', in Robin Horton and Ruth Finnegan (eds.), *Modes of Thought,* London: Faber, 199–229.

—— (1990). *Magic, Science, Religion, and the Scope of Rationality.* Cambridge: Cambridge University Press.

TAUSSIG, MICHAEL (2003). 'Viscerality, Faith, and Skepticism: Another Theory of Magic', in Meyer and Pels (2003), 272–306.

TAYLOR, CHARLES (2004). *Modern Social Imaginaries.* Durham, NC: Duke University Press.

TURNER, VICTOR (1967). *The Forest of Symbols.* Ithaca, N.Y.: Cornell University Press.

VIVEIROS DE CASTRO, EDUARDO (1998). 'Cosmological Deixis and Amerindian Perspectivism', *Journal of the Royal Anthropological Institute,* 4/3: 469–88.

WEBER, MAX (1946). 'Science as a Vocation', in H. H. Gerth and C. Wright Mills (eds.), *From Max Weber: Essays in Sociology,* New York: Oxford University Press, 129–56.

WITHERSPOON, GARY (1977). *Language and Art in the Navajo Universe.* Ann Arbor: University of Michigan Press.

PART III

THE MAJOR FIELDS OF RELIGION/ SCIENCE

CHAPTER 18

CONTRIBUTIONS FROM THE HISTORY OF SCIENCE AND RELIGION

JOHN HEDLEY BROOKE

INTRODUCTION

It is sometimes assumed that a simple story can be told about the historical relationship between science and religion. On one overview, 'science' and 'religion' existed in harmony for centuries, conflicting only in the modern period. On this reading, even the debate over Charles Darwin's *Origin of Species* (1859) was nothing more than a storm in a Victorian teacup (Raven 1943). By contrast, the converse is often assumed: 'science' and 'religion' have existed in more or less perpetual warfare, until recently, when the potential for peace has supervened. This second view is attractive to those who believe that twentieth-century physics, in particular, has given unprecedented access to the mind of God (Davies 1990, 1993).

How can both accounts be true? Each story is problematic because of the level of generality assumed. When science is placed in opposition to religion, as it so often is, it is easy to conclude that the one must exclude the other. For Darwin's 'bulldog', Thomas Henry Huxley, it was simply impossible to be a soldier for science and a loyal son of the church. Or, as Darwin's cousin Francis Galton put it, the practice of science does not accord with the priestly temperament (Turner 1978). Yet one of the lessons of professional historical scholarship is that too much of our thinking about the mutual bearings of science and religion is governed by an uncritical acceptance of the dichotomies and antitheses that pervade popular discussion. In short, serious history enriches and complicates the picture (Brooke 1991). For example, an Oxford theologian of the late nineteenth century, Aubrey Moore, when reflecting on evolutionary theory, famously wrote that under the guise of a foe Darwin had done the work of a friend (Peacocke 1985: 111). Such arresting remarks of this kind immediately suggest a more nuanced picture. Darwin, still seen as a foe in some constituencies, had done the work of a friend by liberating Christianity from bondage to a *deus ex machina*, a magician who intervened to conjure new species into existence. For Moore, a creative process of evolution was consonant with a theology of Incarnation in which divine immanence was restored. The subtlety of his remark stands in judgement over simplistic models of how science and religion should be related. In this respect historical scholarship both informs and supports the claim that more sophisticated taxonomies are required for capturing the multifaceted relations between science and religion (Stenmark 2004; see also this volume).

TOTAL DISPARITY?

One way of collapsing a rigid dichotomy might be to show that science and religion have more in common than images of hostility suggest. There is a limit to how far one can press this case because, in their respective practices, scientists and religious devotees engage in very different exercises. Prayer, meditation, and worship contrast with the testing of hypotheses through controlled experimentation. Nevertheless, the historical record is enlightening on this point, because perceptive commentators have noted that there is no complete disparity. For the theologian Thomas Chalmers, who led the Disruption of the Scottish Church in 1843, science and religion had this in common, that both were 'refined abstractions from the grossness of the familiar and ordinary world'. In his view there obtained 'a very close affinity between a taste for science, and a taste for sacredness'. The two resemble each other in this, he wrote, 'that they make man a more reflective and less sensual being, than before' (Topham 1992: 406). Such sentiments are not to be dismissed as mere effusions from the past. Comparable things are said about science in contemporary scientific literature, as in a recent editorial in the journal *Science* where editor-in-chief Donald Kennedy explains that 'my love of science has much to do with its mystery' (Kennedy 2005: 19). Research may be about answers, but science, he continues, is about questions. Science, in other words, is as much about the lure of the mysterious as about its elimination. Historical work on Albert Einstein has shown that what he experienced as a 'cosmic religious feeling' was associated with an ultimate mystery. The very success of science in demonstrating the intelligibility and comprehensibility of the universe served to underline the mystery as to *why* nature should be intelligible and comprehensible (Jammer 1999: 52, 73). It was also Einstein who observed that the emotional state that enables great scientific achievements to be made is 'similar to that of the religious person or the person in love' (Brooke and Cantor 1998: 227). Science and religion may, then, have elements in common, and historical research has exposed them. Einstein was convinced that beauty was a guiding principle in theoretical physics, and would instinctively reject formulations he perceived to be ugly. The epistemic virtues of simplicity, elegance, symmetry, and beauty are visible in many scientific theories of the past and have often contributed to their success (Polanyi 1958: 133; Brooke and Cantor 1998: 207–43). Precisely how much they have contributed in specific cases, such as the acceptability of Copernican astronomy, has given rise to sophisticated debate (McMullin 1988).

To tease out common elements is not, however, the only way of breaking down barriers. Biographies of eminent scientists have often revealed religious leanings, as with the biblical, Sandemanian Christianity of Michael Faraday (Cantor 1991). To imagine that there could only be a compartmentalizing of the scientific and religious commitments, even when this might sometimes be implied by the subject, can be to miss subtle interconnections.

Science as a Religious Activity

Despite sensationalist novels such as Dan Brown's *Angels and Demons*, most science historians would deny that scientific and religious sensibilities have always clashed. Scientific thinkers may not always have been the most orthodox in their theology (Brooke and Maclean 2005), but among the pioneers of modern science there was often a fusion of scientific and theological interests (Funkenstein 1986). Isaac Newton was not an orthodox Christian, since he denied the doctrine of the Trinity. Yet Newton described the task of the natural philosopher as the deduction of causes from their effects until one arrived at the 'first cause', which was not mechanical. As he wrote to Richard Bentley in 1692, the fact that the planets orbited the sun in the same direction and in the same plane could not spring from any natural cause alone. What Newton described as the business of natural philosophy involved the discussion of divine attributes and God's relation to the world of nature.

It has often been said that, with Newton, the 'scientific revolution' reached its apotheosis, in which science emancipated itself from religious concerns. It would be more accurate to say that Newton reinterpreted theological concepts, such as divine providence and divine omnipresence, through the categories of his science. Because space was constituted by God's omnipresence, Newton could be confident of its homogeneity and of the universality of his laws of motion and gravitation (Brooke 1991: 139). The unity of nature presupposed the unity of the divine mind (Snobelen 2001).

Newton nicely shows how history complicates the picture. By some contemporaries his science was perceived as friendly to religion. William Whiston, his successor to the Lucasian Chair of Mathematics at Cambridge, rejoiced that Newton's *Principia* (1687) had smashed the mechanistic materialism of Descartes's natural philosophy (Force 1985: 33–7). But at the same time Newton was deeply suspected of heresy, since his science appeared to be associated with Socinian tendencies (Snobelen 1999). High Church Anglicans in the eighteenth century were sometimes severe critics of Newton's science (Stewart 1996). Consequently there is no simple answer to the question of whether Newton's science was friendly or hostile to 'religion'. It depended on where you were coming from.

One conclusion, however, is inescapable. The coexistence and interpenetration of theistic and scientific concepts in a mind as celebrated as that of Newton must destroy the popular stereotypes. Historical examples can be suggestive in other ways too. They destroy dichotomies by revealing the possibility of mediating positions.

MIDDLE POSITIONS

When discussing the great revolutions in scientific thought that have changed perceptions of our place in the universe, it is tempting to contrast new with old in black-and-white terms, missing the subtlety of middle positions. There is a telling example in the Galileo affair. The publication of Galileo's book on the 'Two Chief World Systems' (1632) precipitated the famous trial of 1633. Routinely, his 'correct' Copernican science is juxtaposed with the incorrect geocentric system of Aristotle and Ptolemy, the Roman Catholic Church suffering opprobrium for having silenced him and for suppressing the truth. That is how it looks in retrospect. At the time, the issues were much less clear (McMullin 2005). Were there only two world systems worth debating? To present the matter that way was undoubtedly attractive to Galileo, since to disprove one system was to promote the other. But there was a middle position. This was the system proposed by Tycho Brahe, in which the Earth remained stationary, around which the sun still revolved. The crucial difference from the traditional model was that the planets (excepting Earth) orbited the Sun (Gingerich 2004: 75–6). Mathematically equivalent to the Copernican system and difficult to disprove, it was perceived, certainly by some Jesuit astronomers, as a satisfactory alternative. It was relevant to the way in which relations between astronomy and religion were perceived at the time, because this 'compromise' proposal was compatible with a geocentric reading of the salient biblical passages such as Joshua 10: 12–13. The situation looked rather different if the mathematical modelling of the heavens were taken to represent a physical reality. The Tychonic system could then appear less elegant than the Copernican, since it required two centres of motion and the combined physical motions of the planets would be cumbersome. This, however, serves to underline the important point that one of the central issues in the Galileo affair was whether the models of mathematical astronomy were merely instruments for making planetary predictions or whether they could be elevated to the status of physical hypotheses. The debate was not simply about biblical authority; it also involved the status of astronomy as a discipline. In fact, recent scholarship has shown how close Galileo's principles of biblical interpretation were to those adopted by the theologians of the Inquisition (Carroll 2005).

Middle positions have usually been possible in scientific controversies where religious interests have also been at stake. It might be objected that the Tychonic system disappeared eventually, and that it therefore constitutes an inauspicious example. But middle positions are not always so vulnerable, and have often eased the assimilation by religious bodies of otherwise threatening innovations. An example from the nineteenth century would be the work of the British anatomist and palae-ontologist Richard Owen. On the question of the origin of species Owen was neither a 'creationist' in the narrow modern sense nor a Darwinian. Owen emphatically believed that new species arose through what we would call natural causes. But he also considered it premature to regard Darwin's mechanism of natural selection as the agency of transformation (Richards 1987). The persistence of the same basic bone structure in the vertebrate kingdom encouraged him to abstract from it an archetypal skeleton, which he could present as an idea in the mind of God. The evolutionary process could then be understood as an unfolding of a divine plan, a kind of continuous creation, in which the archetype had successive instanti-ations (Rupke 1994). Owen's position was one of many competing accounts of biological evolution in which the sufficiency of natural selection was questioned and a theistic integration achieved (Bowler 2001: 122–59). This was important because it meant that Christians did not have to choose directly between natural selection and miraculous intervention. And there are modern equivalents. One can be a neo-Darwinian and still point to constraints in the evolutionary process that have led to convergent trends, such as the independent reappearance of the same kind of eye (Conway Morris 2003: 328–30).

So far in this discussion the terms 'science' and 'religion' have been used infor-mally, as if they are unproblematic. But this is unsatisfactory, because it assumes that there is a thing called science, of which we can specify the essence, and a thing called religion, of which the essence can also be distilled. Historical scholarship can help us here, because the meaning of these words has changed with time. It also reminds us that it is more appropriate to speak of 'sciences' and 'religions'. When we do, any simple dichotomy loses its rigidity.

SCIENCES AND RELIGIONS

One of the most fascinating subjects for historical investigation is the manner in which boundaries between disciplines are constructed and how they change with time. The proliferation of specialized sciences in Europe did not begin to accelerate until the late eighteenth and early nineteenth centuries, but distinctions had previously been drawn between different sciences according to their subject-matter and methods. It would have been obvious to Newton's contemporaries that the study of astronomy involved very different practices from the study of chemistry, natural history, and natural philosophy. The last had been associated with the study of motion and eventually graduated into what was later called physics. But in the seventeenth century, natural philosophy, as its name implies, was studied as a branch of philosophy. This meant that it was broader in scope than would now be permitted within the culture of natural science. As we have already seen, natural philosophy for Newton embraced theological questions; it did not exclude them. Reputable discussions of the transformation from natural philosophy to natural science are now available (Knight and Eddy 2005), and they stand as a warning against the imposition of supposedly timeless categories of 'science' and 'religion' onto historical realities. It has even been argued that natural philosophy in the seventeenth century was primarily a form of theology (Cunningham 1991), a view that risks conflating natural philosophy with natural theology and which has prompted a vigorous debate that can be followed in volume 5 of the journal *Early Science and Medicine* (2000: 258–300). Just as the word 'science' has changed its meaning from any organized body of knowledge (including, therefore, theology) to modern research-based and highly specialized activities (deliberately excluding theological interference), so the word 'religion' is also an artefact, finding new application in the Enlightenment, when comparative approaches were needed for the analysis of different cultures, their practices and rituals (Harrison 1990). Revealingly, the word 'scientist' is a relatively recent invention, first coined by the Cambridge polymath William Whewell in the 1830s.

Whewell himself used historical examples to accentuate differences between the sciences. Such sensitivity has important implications for the discourse of 'science and religion'. At a given point in time one science might be posing a challenge to a particular religion, when another might be offering support. Such disparity can occur even when the sciences have much in common. For example, during the seventeenth century the telescopic discoveries of Galileo undoubtedly posed a problem for traditional Christian theology. The vast number of stars invisible to the naked eye, yet revealed by the new instrument, raised the question of why a rational Creator would have created such a plethora of useless objects. They might of course shine on other worlds, and that became a standard response. But that, too, was not without theological repercussions, since embarrassing questions could be asked about the relevance of Christ's redemptive role to the denizens of other worlds (Dick 1982; Crowe 1986). Another scientific instrument, however, the microscope, opened up a hidden world that was both beautiful and awesome: divine craftsmanship was almost immediately discernible in the intricate structures of living things, from the scales of a fish to the eye of a fly. That a Creator had instilled life into the minutest mite deeply impressed Robert Boyle as he reflected on the meticulous and ingenious craftsmanship visible in the microscopic world. Even the proboscis of the flea, observers decreed, had its uses, since it provided a free blood-letting service for those unable to afford a physician (Harrison 1998: 171–6).

To emphasize a plurality of sciences is not to overlook the fact that scientific communities, in their interface with the public, have tried to articulate a formalism usually called 'the' scientific method. A sophisticated literature has existed for some time on the rhetoric and politics of such discourses in this supposedly unitary, but often protean, method (Schuster and Yeo 1986). Historical investigation exposes not only a differentiation of sciences and their methods, but also a diversity in the metaphysical conclusions drawn by different protagonists. For example, during the second half of the nineteenth century, physicists and geologists reached contrary conclusions on the age of the Earth—not a trivial matter when the mechanism of evolutionary transformation was at issue (Burchfield 1975). Similarly, in the early years of the twentieth century, we find astronomy and biology at cross purposes on the subject of extraterrestrial life. Whereas astronomers gleefully did their probability calculations concerning the likely number of planets at similar distances from their suns as we are from ours, the co-founder of the theory of natural selection, Alfred Russel Wallace, actually used evolutionary theory to exterminate their inhabitants. Writing rather as Stephen J. Gould would do in his *Wonderful Life* (1991), Wallace stressed that at each point of evolutionary divergence, so great were the contingencies that it was inconceivable that the same path leading to intelligence akin to ours could have been followed elsewhere in the universe (Crowe 1986: 531).

Rather than multiply examples, I will consider another plurality: the diversification of faith traditions. Even within one and the same religion, there can be disparate evaluation of the sciences. There is, for example, an extensive literature on the role of Christian Dissenters in promoting a culture of science and technology—a literature that has recently undergone renewed scrutiny (Wood 2004). In one of the most famous historical theses concerning the relevance of religious beliefs to the expansion of the sciences, Robert Merton (1938) argued that Puritan values were more conducive to scientific activity than the more contemplative Catholic spiritualities—a conclusion spawning cogent rejoinders (Morgan 1999). But it is also clear that different world faiths have had their own distinctive attitudes towards the sciences, rendering monolithic treatments suspect. A comparative study of the responses of Quakers and Jews to science and modernity has been a valuable recent addition to the literature (Cantor 2005). It is sometimes said that certain Hindu traditions have been more amenable than the classical theism of mainstream Christianity to ecological sensitivity and evolutionary perspectives (Gosling 2001: 41; Edelmann 2005). In another influential master narrative Christianity has itself been blamed for a set of attitudes conducive to our environmental crises (White 1967), a thesis that has been co-opted by apologists from other religious traditions, but one that has also been overstated. White's contention was that the biblical injunction to exercise dominion over nature could easily lead to an exploitative domination of nature. An important qualification is that, until the seventeenth century, there was nothing intrinsic in Christian biblical exegesis to generate such exploitation, but once nature became a more conspicuous resource for human benefit, as it did for Francis Bacon, the Genesis text was reinterpreted to justify the appropriation (Harrison 1999).

Contrasts between different cultural traditions have been drawn in many ways that impinge on the evaluation and characterization of the sciences. Seeking to understand why, despite their impressive technological achievements, Chinese societies did not produce the abstract and mathematically expressed 'laws' of nature that featured in the physical science of the Christian West, Joseph Needham supposed that the absence of the idea of a personal God, legislating for nature, might partly explain the deficit (Needham 1969: 301–27). There is a sense, however, in which to focus on alleged deficiencies can itself be chauvinistic. Not surprisingly, when histories of science are written from the standpoint of a particular religious tradition, they tend to be constructed so as to display the religion concerned in the best possible light. Given the manner in which many Western historians neglected the contributions of Muslim philosophers, it is perfectly understandable why the scientific achievements of the 'golden age' of Islam—especially in the well-documented fields of astronomy, optics, mathematics, and medicine—should now be reaffirmed. Examples of reputable scholarship on the subject of science and philosophy in medieval Islamic theology include King (1993), Sabra (1994), Saliba (1994), and Ragep and Ragep (1996). The achievements of Ottoman science have been thoroughly documented by Ihsanoğlu (2004).

But there are also concerns that playing the game of 'Who discovered X first?' can lead to a naïve kind of apologetic history in which too much is claimed for precursors of later science (Ragep 2001: 49). Warnings have also been issued against exaggerating the degree to which the shape and content of medieval science were determined by religious presuppositions, whether Jewish, Christian, or Muslim (North 2005). Serious historical study raises the additional problem that just because a particular piece of scientific research was conducted within a specified religious culture, it does not follow that it was either inspired or opposed on the ground of theological doctrine. The same has been said of the Graeco-Arabic translation movement in ninth-century Baghdad (Gutas 1998: 191). When examining the preconditions of the possibility of particular scientific initiatives, it can be a mistake to privilege the 'religious' parameter as if it could be completely abstracted from social, economic, and political realities. To understand the Galileo affair, with all its many nuances, it must not be isolated from the political realities of Counter-Reformation Rome, when Pope Urban VIII was under pressure from Catholic Spain to show less leniency towards heretics, and when Galileo, of whom he had earlier been supportive, was perceived to have betrayed him (Biagioli 1993). Historical approaches to 'science and religion' remind us of the importance of political power and the constitution of authority in contexts where scientific innovations have impinged on religious sensibilities. The many different reinterpretations of the Galileo episode in subsequent generations underline the point (Finocchiaro 2005). Where there is sensitivity to the importance of time, place, and culture, questions about the relationship between science and religion invite the immediate response: whose science and whose religion? (Brooke and Cantor 1998: 43–72).

ENTAILMENT OR CONSONANCE?

Even if there were only one science and only one religion, the dichotomies could still be challenged. This is because one and the same science could simultaneously both underpin and threaten a religious perspective. This was manifestly true of geology in the decades preceding Darwin's *Origin of Species*. The fossil record was used to underpin a doctrine of creation and to undermine atheism. One of Darwin's mentors at Cambridge, Adam Sedgwick, pointed out that species had not existed from eternity, as older forms of atheism had suggested. Species had come into existence that had not been there before (Brooke 1991: 194). In the period before evolutionary theories became the norm, this appeared to be a cogent argument in favour of progressive creation. But that same fossil record also pointed to the extinction of species, with all the melancholy consequences that were addressed in Tennyson's *In Memoriam*.

From such historical examples we learn that great care is needed when assessing what scientific discoveries entail for religious belief. In the literature of popular science, there is a tendency to make the science more portentous in its implications than a more sober estimate would allow. To claim that a new discovery has major implications for religious belief is a well-tried way of catching public attention. But the idea that scientific data or even scientific theories *entail* certain consequences for metaphysical or theological positions is greatly overstated. One might even suggest that scientific knowledge is neutral with respect to theism, since it can be made consonant with either theism or atheism. *How* we interpret its cultural meaning will largely depend on our presuppositions. Many treatments of science and religion are vitiated precisely because they are looking for implications and entailments that are simply not there. This is not such a radical observation. Einstein said of his theory of relativity that it carried no implications for religion (Jammer 1999: 155). T. H. Huxley said of Darwin's theory that it had no more to do with theism than had the first book of Euclid (Huxley 1887: ii. 202). Of course there is opposition between a literal reading of Genesis and our modern understanding of cosmology and evolution, and for many that is not a trivial matter. Why it is not, and why there should be such an unmitigated struggle in the USA, are themselves questions that call for historical understanding (Numbers 1992; Moore 1993; Larson 1997; Ruse 2005).

Another way to deconstruct the dichotomies is to switch attention from time to space. Some of the finest work in the field of science and religion has been accomplished by scholars sensitive to the effect of local geographies on receptivity to new ideas.

THE PLACE OF PLACES

Suppose one asks the innocent question, How did Presbyterian Christians respond to Darwin's theory of evolution in the thirty years or so following its publication? The answer turns out to be that much depends on where you were. If you were in Belfast in 1874, you might have heard the physicist John Tyndall deliver his Presidential Address before the British Association for the Advancement of Science. Annoyed by the neglect of the sciences in Christian colleges, Tyndall went on the offensive, declaring that the scientific world would claim and wrest from theology the entire domain of cosmological theory. In a historical survey designed to show the triumph of science over superstition, Tyndall associated Darwin's theory with his own secular monism, thereby alienating a religious constituency that would henceforth regard Darwinism with dismay. At Princeton, by contrast, despite the rejection of Darwinism by Charles Hodge, Presbyterians were for the most part less troubled, with the more tolerant view epitomized by James McCosh (Livingstone 1992). In Scotland it was a different story again: Presbyterian voices were initially concerned; yet, according to David Livingstone, the Darwinian theory was far less a worry than the radical biblical criticism of Robertson Smith (Livingstone 2003: 117–19). A contrast has also been drawn between the reception of Darwin's theory in New Zealand and in the southern states of America. In New Zealand, the theory, with its emphasis on competition and on what Darwin in his subtitle called the 'preservation of favoured races in the struggle for existence', was attractive to settlers who could use it to justify their ruthless treatment of the Maori (Stenhouse 1999). For different racial reasons Darwin's theory did not appeal to the polygenists of the American South, for the important reason that, on the interpretation of both Darwin and Wallace, it was actually supportive of monogenism, since all races were ultimately derived from a common ancestor (Livingstone 2003: 4; Brooke 2005: 176–7).

The critical point is that whether a particular piece of science is perceived as friend or enemy may crucially depend on local events and circumstances. The lesson is no less relevant today. Discussions of the kind 'Is "creation science" really science? Or is it religion?' continue to feature in American courts of law in a manner that has not (yet) been experienced in Britain. Is the appeal to intelligent design 'science', or is it 'religion'? The either/or dichotomy, so prevalent in North American contexts, has much to do with the protection of a Constitution different from the British. When testing legalities, the teaching of ideas about creation or design is perceived to be either the teaching of science or the teaching of religion. No other option, it seems, is to be entertained. But are 'science' and 'religion' the only meaningful categories when confronting issues that involve competing metaphysics and world-views?

ONLY TWO POSSIBILITIES?

It would surely be a move in the right direction if young people were helped to see that discourse about 'creation' or about 'design' may belong neither to science nor to religion in any straightforward way. These discourses belong to metaphysics and to the philosophy of religion—to the philosophy of science, also. Too often there is reference to the 'dialogue between science and religion' as if there were only two parties. Conversations between scientists and theologians are set up as if the immediate juxtaposition might produce original insights and a *modus vivendi*. Without the mediation of the humanities and social sciences, however, the conclusions drawn from such conversations can sometimes be naïve. In this essay I have tried to show how the mediation of historical study can be instructive. The mediation of metaphysics is also fundamental. Would there be so much misunderstanding of Christian doctrines of creation if there were a greater familiarity with creation understood as a relationship of ultimate, ontological dependence of the world on God, rather than as the magic of an anthropomorphic deity? There are resources here in the shape of specific histories that have addressed the interweaving of creation doctrines and science (Kaiser 1997) and the interdependencies between natural theology and the natural sciences (Brooke and Cantor 1998: 141–243). Although William Paley's *Natural Theology* (1802) is often taken to typify the genre, historical study reveals the diversification of the design argument during the nineteenth century (Lightman 2001), qualifying the popular conception that Darwin destroyed it.

The assumption that there are only two categories worth talking about has found expression in many modern discussions of 'science and religion'. In *Rocks of Ages* (1999) Stephen J. Gould at least enabled both to survive as long as one accepted his principle of NOMA—that there should be no overlapping magisteria. For Gould, the province of the one is the determination of facts about the world, the proper province of the other the determination of moral values. This NOMA principle may be an ideal; but it is not one to which Gould himself was able to adhere. In his controversy with Simon Conway Morris over the fossils of the Burgess Shale, Gould conceded that his own interpretation, which stressed the accidental and contingent factors in the course of evolution, had been shaped by personal beliefs and preferences (Gould 1998: 55). This raises a tantalizing but difficult question, which only historical research can fully illuminate: have there been contexts, and do such contexts still exist, in which the cognitive content of a scientific theory has been informed (and not merely deformed) by religious commitment (Brooke, Osler and

Van der Meer 2001)? How, for example, might serious historical research deal with the claim that without the Christian doctrine of creation there would have been no modern science?

REVISING A REVISIONIST HISTORY

There were many versions of Christianity competing for attention in seventeenth-century Europe, and many variants of an emerging scientific culture. The suggestion that science (singular) was somehow an offspring of Christianity (singular), with the intimate organic connection that this implies, has not withstood the test of critical scholarship. This does not mean that the old 'warfare' thesis was correct. There were resources within the Christian tradition that, when appropriated and reinterpreted, could be used to justify scientific activity. This is not, however, the same as asserting a simple causal relation between Christianity and scientific endeavour.

A serious and influential case for asserting intimate connections between science and Christian doctrine was made by the philosopher Michael Foster during the 1930s (Foster 1934). If science ultimately depended for its rationality on the existence of an orderly and intelligible world, this was a presupposition that could be supplied by a doctrine of creation, in which the unity of nature also reflected a monotheistic faith. Thinking along similar lines, A. N. Whitehead had already asserted that 'faith in the possibility of science, generated antecedently to the development of modern scientific theory, is an unconscious derivative from medieval theology' (Whitehead 1925: 19). It is also undeniable that prominent scientific thinkers of seventeenth-century Europe presented their scientific insights in language redolent of religious conviction. The astronomer Kepler clearly believed that the order of nature was best captured by exhibiting the geometrical harmonies pervading the cosmos. Mathematics was the language that mediated between the divine and the human mind. Galileo, for whom mathematics was the language of nature, saw human reason as a divine gift, to be used in reading the book of nature. Newton's belief in the universality of nature's laws was, as suggested above, a reflection of his robust monotheism.

Christian beliefs could also be brought to bear on questions of methodology. It was a common refrain among advocates of what became known as the 'experimental philosophy' that the armchair philosophizing of the scholastics was arrogant in its presumption that reason alone would give access to the truth. If God had been free to make whatever world God wished, only an empirical investigation could reveal the nature of the world that had actually been made. For the Dutch Calvinist historian of science Reijer Hooykaas, Protestant Christianity had a special role in making modern science possible. This was not merely by encouraging freedom of thought, though that would be of no small import. Through its emphasis on biblical teaching and a celebration of the sovereignty of God, it cleared the air for a science of nature by eliminating all mediating agents between an omnipotent, sovereign God and the workings of nature. Hooykaas liked to speak of the 'de-deification of nature', a process leading to a world-view in which nature and God were so related that to investigate the laws of nature was to investigate the way in which God normally chose to act in the world (Hooykaas 1973). The 'law' metaphor made sense if the Christian God could be identified as a divine legislator, a point that gained currency even within a Marxist historiography of science (Zilsel 1944).

Such connections may indicate the relevance of Christian doctrine to the rise of science and to a revisionist historiography that turns the warfare thesis on its head. Yet proper caution must be exercised before drawing too strong a conclusion. The pre-eminence that Hooykaas gave to Protestant theologies, and Calvinism in particular, could easily obscure the many scientific contributions made by Catholic scholars, among whom were leading exponents of a mechanical philosophy of nature: Galileo, Mersenne, Gassendi, and Descartes (Ashworth 2003). After examining the philosophies of nature constructed by leading Catholic natural philosophers, William Ashworth concluded that not only was there no unanimity, but that their several positions could equally have been found among Protestant thinkers (Ashworth 1986).

One of the most sophisticated accounts of Protestant initiatives in creating the space for scientific inquiry is now Peter Harrison's *The Bible, Protestantism and the Rise of Natural Science* (1998). His argument is multifaceted, suggesting that new ways of reading the Bible had a spin-off in new ways of reading the book of nature. No longer was the latter seen as a deposit of natural objects, each in some way emblematic of spiritual truths. The more literal interpretations of Scripture, prioritized within Protestant communities, arguably led to understandings of nature in which what was important was not a set of spiritual symbols but the physical connections between things that the sciences could explore. In particular, the Protestant rejection of multiple meanings of specific biblical texts had a parallel in the scientist's quest (visible in Newton) for a single, precise explanation of each natural phenomenon.

Was such a movement specifically Christian? This becomes an important question in light of the fact, stressed by Harrison, that the expansion of an experimentally grounded natural philosophy in seventeenth-century England was accompanied as much by a reformulation of natural theology as by a revision of doctrine. According to exponents of the 'new philosophy', nature was still a book to be read, but it increasingly became a resource to be used for human benefit. As lowly a creature as the silkworm had a raised profile once one focused on its value to humankind. One of the paradoxes of the Scientific Revolution was that, even as new cosmologies had a decentring effect on our place in the universe, the utilitarian thrust of the new science became more resoundingly anthropocentric.

Natural theologies with their appeals to design and beauty in creation could, of course, be supportive of a Christian commitment. But their attraction often consisted in their transcending of religious divisions. Arguments for design could equally appeal to deists, in their drive to dispense with revelation altogether. Images of a clockwork universe could be used to celebrate divine designs; the problem for

Christian orthodoxies was that they could also be used to underwrite the autonomy of nature. This illustrates again that there was no simple entailment either from Christian doctrines and values to a reverence for science, or from new forms of mechanistic science to refashioned models of divine activity. This is one of the most important lessons from historical inquiry. It constitutes a critique of those popular writers today who argue that the conclusions of science entail whatever views about Christianity or about religion in general that they happen to prefer.

For a balanced view it must be recognized that exclusive claims for the dependence of early modern science on a Christian culture are unsustainable. If Christianity was so germane to the rise of science, why did so many centuries elapse before any scientific fruits became visible? Rejoinders that point to the need for other social and economic preconditions to have been in place have the effect of diluting, even trivializing, the claim for a distinctively Christian input (Gruner 1975: 81). A more subtle objection has been that some formulations of Christian doctrine, concerning the Fall for example, were not conducive to scientific inquiry because they gave prominence to concepts of forbidden knowledge (Harrison 2001). Reformulation was necessary to accommodate visions of a scientific utopia. When Francis Bacon promoted the applied sciences with the argument that they would help to restore the dominion over nature that humankind had sacrificed at the Fall (Webster 1975), it was a reinterpretation of Christian theology, not a natural outgrowth. This is why the language of appropriation and reinterpretation seems more realistic when referring to the resources within Christendom that could be used in diplomacy for the sciences.

It must also be obvious that the scientific achievements of the Indian, Chinese, and Muslim philosophers of earlier periods preclude any crude claim for Christian ownership of science, as must the achievements of the Greek philosophers, so respected by the European giants of the seventeenth century (Rashed 1980; Lloyd and Sivin 2002). Galileo was full of praise for Archimedes, and Newton for Pythagoras. Moreover, references to the harmony between scientific and Christian beliefs were often made in self-defence, as when Galileo urged the compatibility of Copernican astronomy with Scripture in his *Letter to the Grand Duchess Christina* (McMullin 2005: 88–116). A recent study draws attention to the discomfort felt by the English clergy in the seventeenth century when devoting time to natural history or natural philosophy that they sensed should rather be given to their pastoral duties (Feingold 2002). There was no smooth passage from Christian conviction to the practice of the sciences.

Such considerations complicate any historical account of a supposed dependence of the modern scientific movement on Christian doctrines and values. The question of location is again of paramount importance when examining how the relations between Christianity and the sciences were constructed. In the German states, Lutherans played a key role in disseminating Copernican astronomy (Barker 2005); in Denmark and Sweden, Lutheran theology was more deeply suspicious of the new science (Brooke 1991: 100). Calvin may have created the space for a new astronomy through his doctrine of biblical accommodation (Hooykaas 1973); but Calvinism in some locations—Scotland, for example—could be distinctly unwelcoming to a heliocentric cosmology (Brooke 1991: 100). Contemporary scholarship is enriching our understanding of this cultural diversity, which cuts across the generalizations of apologetic literature (Brooke and Ihsanoğlu 2005; Brooke and Maclean 2005). This expansion of historical understanding is happily occurring with reference to religious traditions other than those of Christianity. The first major analysis of Jewish responses to the theory of evolution has now been published (Cantor and Swetlitz 2006), setting a precedent for future research and creating further opportunities for comparative study.

REFERENCES AND SUGGESTED READING

ASHWORTH, W. B. (1986). 'Catholicism and Early Modern Science', in D. C. Lindberg and R. L. Numbers (eds.), *God and Nature: Historical Essays on the Encounter between Christianity and Science*, Berkeley and Los Angeles: University of California Press, 136–66.

—— (2003). 'Christianity and the Mechanistic Universe', in D. C. Lindberg and R. L. Numbers (eds.), *When Science and Christianity Meet*, Chicago: University of Chicago Press, 61–84.

BARKER, P. (2005). 'The Lutheran Contribution to the Astronomical Revolution', in Brooke and Ihsanoğlu (2005), 31–62.

BIAGIOLI, M. (1993). *Galileo Courtier*. Chicago: University of Chicago Press.

BOWLER, P. J. (2001). *Reconciling Science and Religion*. Chicago: University of Chicago Press.

BROOKE, J. H. (1991). *Science and Religion: Some Historical Perspectives*. Cambridge: Cambridge University Press.

—— (2005). 'Darwin, Design and the Unification of Nature', in J. D. Proctor (ed.), *Science, Religion, and the Human Experience*, New York: Oxford University Press, 165–83.

—— and CANTOR, G. N. (1998). *Reconstructing Nature: The Engagement of Science and Religion*. Edinburgh: T & T Clark.

—— and IHSANOĞLU, E. (2005) (eds.). *Religious Values and the Rise of Science in Europe*. Istanbul: Research Centre for Islamic History, Art and Culture.

—— and MACLEAN, I. (2005) (eds.). *Heterodoxy in Early Modern Science and Religion*. Oxford: Oxford University Press.

—— OSLER, M. J., and VAN DER MEER, J. (2001) (eds.). *Science in Theistic Contexts: Cognitive Dimensions*, Osiris 16. Chicago: University of Chicago Press.

BURCHFIELD, J. D. (1975). *Lord Kelvin and the Age of the Earth*. London: Macmillan.

CANTOR, G. N. (1991). *Michael Faraday: Sandemanian and Scientist*. London: Macmillan.

—— (2005). *Quakers, Jews, and Science: Religious Responses to Modernity and the Sciences in Britain, 1650–1900*. Oxford: Oxford University Press.

—— and SWETLITZ, M. (2006) (eds.). *The Jews and Evolution*. Chicago: University of Chicago Press.

CARROLL, W. (2005). 'Galileo and the Myth of Heterodoxy', in Brooke and Maclean (2005), 115–44.

CONWAY MORRIS, S. (2003). *Life's Solution: Inevitable Humans in a Lonely Universe*. Cambridge: Cambridge University Press.

CROWE, M. J. (1986). *The Extraterrestrial Life Debate 1750–1900: The Idea of a Plurality of Worlds from Kant to Lowell*. Cambridge: Cambridge University Press.

CUNNINGHAM, A. (1991). 'How the *Principia* Got its Name', *History of Science*, 29: 377–92.

DAVIES, P. (1990). *God and the New Physics*. Harmondsworth: Penguin.

—— (1993). *The Mind of God*. Harmondsworth: Penguin.

DICK, S. J. (1982). *Plurality of Worlds: The Extraterrestrial Life Debate from Democritus to Kant*. Cambridge: Cambridge University Press.

EDELMANN, J. (2005). 'A Meeting with Charles Darwin and the *Bhāgavata Purāna*', *Journal of Vaishnava Studies*, 13: 113–26.

FEINGOLD, M. (2002). 'Science as a Calling? The Early Modern Dilemma', *Science in Context*, 15: 79–119.

FINOCCHIARO, M. (2005). *Retrying Galileo, 1633–1992*. Berkeley and Los Angeles: University of California Press.

FORCE, J. (1985). *William Whiston: Honest Newtonian*. Cambridge: Cambridge University Press.

FOSTER, M. B. (1934). 'The Christian Doctrine of Creation and the Rise of Modern Science', *Mind*, 43: 446–68.

FUNKENSTEIN, A. (1986). *Theology and the Scientific Imagination from the Middle Ages to the Seventeenth Century*. Princeton: Princeton University Press.

GINGERICH, O. (2004). *The Book Nobody Read*. New York: Walker & Co.

GOSLING, D. L. (2001). *Religion and Ecology in India and Southeast Asia*. London: Routledge.

GOULD, S. J. (1991). *Wonderful Life: The Burgess Shale and the Nature of History*. Harmondsworth: Penguin.

—— (1998–9). 'Showdown on the Burgess Shale—A Debate with Simon Conway Morris', *Natural History*, December 1998/January 1999: 48–55.

—— (1999). *Rocks of Ages: Science and Religion in the Fullness of Life*. New York: Ballantine.

GRUNER, R. (1975). 'Science, Nature and Christianity', *Journal of Theological Studies*, 26: 55–81.

GUTAS, D. (1998). *Greek Thought, Arabic Culture*. London: Routledge.

HARRISON, P. (1990). '*Religion' and the Religions in the English Enlightenment*. Cambridge: Cambridge University Press.

—— (1998). *The Bible, Protestantism and the Rise of Natural Science*. Cambridge: Cambridge University Press.

—— (1999). 'Subduing the Earth: Genesis 1, Early Modern Science, and the Exploitation of Nature', *Journal of Religion*, 79: 86–109.

—— (2001). 'Curiosity, Forbidden Knowledge, and the Reformation of Natural Philosophy in Early-Modern England', *Isis*, 92: 265–90.

HOOYKAAS, R. (1973). *Religion and the Rise of Modern Science*. Edinburgh: Scottish Academic Press.

HUXLEY, T. H. (1887). 'On the Reception of the "Origin of Species"', in F. Darwin (ed.), *The Life and Letters of Charles Darwin*, 3 vols., London: Murray, ii. 179–204.

IHSANOĞLU, E. (2004). *Science, Technology and Learning in the Ottoman Empire*. Aldershot: Ashgate Variorum.

JAMMER, M. (1999). *Einstein and Religion*. Princeton: Princeton University Press.

KAISER, C. (1997). *Creational Theology and the History of Physical Science*. Leiden: Brill.

KENNEDY, D. (2005). 'Editorial', *Science*, 309: 19.

KING, D. (1993). *Astronomy in the Service of Islam*. Aldershot: Ashgate Variorum.

KNIGHT, D. M., and EDDY, M. D. (2005). *Science and Beliefs: From Natural Philosophy to Natural Science, 1700–1900*. Aldershot: Ashgate.

LARSON, E. J. (1997). *Summer for the Gods: The Scopes Trial and America's Continuing Debate over Science and Religion*. Cambridge, Mass.: Harvard University Press.

LIGHTMAN, B. (2001). 'Victorian Sciences and Religions: Discordant Harmonies', in Brooke, Osler, and Van der Meer (2001), 343–66.

LIVINGSTONE, D. N. (1992). 'Darwinism and Calvinism: The Belfast–Princeton Connection', *Isis*, 83: 408–28.

—— (2003). *Putting Science in its Place: Geographies of Scientific Knowledge.* Chicago: University of Chicago Press.

LLOYD, G., and SIVIN, N. (2002). *The Way and the Word: Science and Medicine in Early China and Greece.* New Haven: Yale University Press.

MCMULLIN, E. (1988). 'The Shaping of Scientific Rationality: Construction and Constraint', in E. McMullin (ed.), *Construction and Constraint: The Shaping of Scientific Rationality,* Notre Dame, Ind.: University of Notre Dame Press, 1–47.

—— (2005) (ed.). *The Church and Galileo.* Notre Dame, Ind.: University of Notre Dame Press.

MERTON, R. K. (1938). 'Science, Technology and Society in Seventeenth-Century England', *Osiris*, 4: 360–632.

MOORE, J. R. (1993). 'The Creationist Cosmos of Protestant Fundamentalism', in M. E. Marty and R. S. Appleby (eds.), *Fundamentalisms and Society: Reclaiming the Sciences, the Family, and Education,* Chicago: University of Chicago Press, 42–72.

MORGAN, J. (1999). 'The Puritan Thesis Revisited', in D. N. Livingstone, D. G. Hart, and M. A. Noll (eds.), *Evangelicals and Science in Historical Perspective,* New York: Oxford University Press, 43–74.

NEEDHAM, J. (1969). *The Grand Titration.* London: Allen & Unwin.

NORTH, J. D. (2005). 'Western Science, Jews and Muslims', in Brooke and Ihsanoğlu (2005), 15–30.

NUMBERS, R. L. (1992). *The Creationists: The Evolution of Scientific Creationism.* New York: Knopf.

PEACOCKE, A. R. (1985). 'Biological Evolution and Christian Theology—Yesterday and Today', in J. Durant (ed.), *Darwinism and Divinity,* Oxford: Blackwell, 101–30.

POLANYI, M. (1958). *Personal Knowledge: Towards a Post-Critical Philosophy.* London: Routledge & Kegan Paul.

RAGEP, F. J. (2001). 'Freeing Astronomy from Philosophy: An Aspect of Islamic Influence on Science', in Brooke, Osler, and Van der Meer (2001), 49–71.

—— and RAGEP, S. P. (1996) (eds.). *Tradition, Transmission, Transformation.* Leiden: Brill.

RASHED, R. (1980). 'Science as a Western Phenomenon', *Fundamenta Scientiae*, 1: 7–21.

RAVEN, C. E. (1943). *Science, Religion and the Future.* Cambridge: Cambridge University Press.

RICHARDS, E. (1987). 'A Question of Property Rights: Richard Owen's Evolutionism Reassessed', *British Journal for the History of Science*, 20: 129–71.

RUPKE, N. A. (1994). *Richard Owen: Victorian Naturalist.* New Haven: Yale University Press.

RUSE, M. (2005). *The Evolution–Creation Struggle.* Cambridge, Mass.: Harvard University Press.

SABRA, A. I. (1994). *Optics, Astronomy and Logic: Studies in Arabic Science and Philosophy.* Aldershot: Ashgate Variorum.

SALIBA, G. (1994). *A History of Arabic Astronomy: Planetary Theories during the Golden Age of Islam.* New York: New York University Press.

SCHUSTER, J. A., and YEO, R. R. (1986). *The Politics and Rhetoric of Scientific Method.* Dordrecht: Reidel.

SNOBELEN, S. D. (1999). 'Isaac Newton, Heretic: The Strategies of a Nicodemite', *British Journal for the History of Science*, 32: 381–419.

—— (2001). '"God of gods, and Lord of lords": The Theology of Isaac Newton's General Scholium to the *Principia*', in Brooke, Osler, and Van der Meer (2001), 169–208.

STENHOUSE, J. (1999). 'Darwinism in New Zealand, 1859–1900', in R. L. Numbers and J. Stenhouse (eds.), *Disseminating Darwinism: The Role of Place, Race, Religion and Gender,* Cambridge: Cambridge University Press, 61–89.

STENMARK, M. (2004). *How to Relate Science and Religion*. Grand Rapids, Mich.: Eerdmans.

STEWART, L. (1996). 'Seeing through the Scholium: Religion and Reading Newton in the Eighteenth Century', *History of Science*, 34: 123–65.

TOPHAM, J. (1992). 'Science and Popular Education in the 1830s: The Role of the *Bridgewater Treatises*', *British Journal for the History of Science*, 25: 397–430.

TURNER, F. M. (1978). 'The Victorian Conflict between Science and Religion: A Professional Dimension', *Isis*, 69: 356–76.

WEBSTER, C. (1975). *The Great Instauration: Science, Medicine and Reform 1626–1660*. London: Duckworth.

WHITE, L. (1967). 'The Historical Roots of our Ecological Crisis', *Science*, 155/(March) 1203–7.

WHITEHEAD, A. N. (1925). *Science and the Modern World*. New York: Macmillan.

WOOD, P. (2004) (ed.). *Science and Dissent in England, 1688–1945*. Aldershot: Ashgate.

ZILSEL, E. (1944). 'The Genesis of the Concept of the Physical Law', *Philosophical Review*, 51: 245–79.

CHAPTER 19

CONTRIBUTIONS FROM THE SOCIAL SCIENCES

ROBERT A. SEGAL

INTRODUCTION

There has long been tension, not between social scientists and practitioners of religion but between social scientists and scholars of religious studies, whom I dub 'religionists'. Religionists have conventionally assumed that the social sciences are guilty of multiple sins, none of them forgivable: ignoring the proverbial believer's point of view; denying the irreducibly religious nature of religion; analysing religion exclusively functionally, reductively, and explanatorily; analysing religion exclusively materially and behaviourally; and denying the truth of religion.

I maintain that the social sciences—anthropology, sociology, psychology, and economics—are guilty of none of these charges, which rest on misconstruals of the social sciences. Ironically, those religionists who contend that contemporary social scientists have come round to accepting the religionist approach likewise misconstrue the social sciences.

There are genuine philosophical objections to social science, not merely to the social scientific study of religion, and some will be considered. Different from these objections are post-modern ones, which are objections to theorizing *per se*. Post-modern objections, I maintain, rest on a further array of misconstruals or, more exactly, confusions.

THE RELIGIONIST MISCONSTRUAL OF SOCIAL SCIENCE

The contemporary exemplar of the religionist position is Mircea Eliade. According to him, and to a tradition that goes back to the Victorian scholar Friedrich Max Müller, religion originates and functions to bring believers close to God. No other origin or function is allowed. If the sole way to explain religion is religiously, then only those in the field of religious studies are qualified to study religion. Anthropologists, sociologists, economists, and psychologists, all of whom explain religion non-religiously, are excluded. Religious studies thereby deserves the status of an independent discipline.

I maintain that the arguments offered by Eliade and others for a monopoly on the study of religion mischaracterize the social sciences in at least five ways.

A social scientific account of religion ignores the believer's point of view Assume that believers maintain that they become religious and remain religious to get close to God. No social scientist ignores this point of view. On the contrary, it is what social scientists are trying to explain. If they were to ignore it, they would be left with nothing to explain. Their explanation need not, however, be the believer's own.

A social scientific account of religion denies the irreducibly religious nature of religion No social scientist denies that the manifest nature of religion is religious: that believers pray because they believe in God. The question is *why* they believe in God. The question is to be settled by research, not by a priori pronouncements from religionists like Steven Kepnes: 'After all, what we [students of religion] want to understand, what we want to study, is religion, and not society or psychology or brain chemistry' (Kepnes 1986: 509). How does Kepnes, any more than Eliade, know that religion is not sociological or psychological or chemical in nature? His declaration is dogmatic.

Eliade declares not merely that religion must be analysed only religiously but, more, that those who presume to study it any other way thereby cease to study religion: 'A religious phenomenon will only be recognised as such if it is grasped at its own level, that is to say, if it is studied *as* something religious. To try to grasp the essence of such a phenomenon by means of physiology, psychology, sociology, economics, linguistics, art or any other study is false' (Eliade [1958] 1963: p. xiii). Eliade is half right: a sociological account, for example, does partly transform religion into a sociological phenomenon. But it does not do so altogether.

Take Durkheim's account of religion in *The Elementary Forms of the Religious Life* ([1915] 1965). Durkheim asserts that whenever a group gathers regularly, there will be religion, and that without a group, there will be no religion. Nevertheless, Durkheim is not reducing religion to a group. Rather, he is accounting for religion *as* a group activity. Even if the regular gathering of a group guarantees the outcome, there is no religion until that group produces metaphysics, ethics, symbols, rituals, and the experience of God. Undeniably, Durkheim is trying to make religion as fully sociological a phenomenon as he can, but he, like all other social scientists, recognizes a point beyond which he cannot go: the point, wherever it lies, at which a group becomes a distinctively religious group.

A social scientific account of religion is functional, reductive, and explanatory;
a religionist account of religion is substantive, non-reductive, and interpretive   One
religionist strategy for fending off the social sciences is to delimit the nature of a social
scientific inquiry. A social scientific account is thus commonly characterized as
'functional', 'reductive', and 'explanatory'—terms that are used interchangeably.
By contrast, a religionist account is conventionally characterized as 'substantive',
'non-reductive', and 'interpretive'—terms that are likewise used interchangeably. For
example, Kepnes, in an attempt to reconcile a social scientific approach with a
religionist one, refers to 'substantive or nonreductive methodologies' (Kepnes
1986: 504), as if they were identical, and contrasts them to functional and reductive
ones. Conversely, he writes that 'Those who utilise methods of explanation are often [and
rightly] called reductionist' (Kepnes 1986: 508), and he contrasts reductionists to those
who, using methods of interpretation, or 'understanding', are called non-reductionists:

[W]e need not see the study of religion as either a scientific attempt to *explain* religion
[*reductively*] in terms of sociology, psychology or physics, or an intuitive and analogical
attempt to grasp the meaning of religion [*non-reductively*] from the believer's standpoint.
The study of religion...requires *both* methods of *understanding* and *explanation*. Thus, the
so-called *reductionist* and *nonreductionist* approaches to the study of religion are not mutually
exclusive. (Kepnes 1986: 505; italics added)

Neither 'functional', 'reductive', and 'explanation' nor 'substantive', 'non-reductive',
and 'interpretation' ('understanding') are in fact synonymous. 'Functional' and 'substantive' refer to *definitions* of religion. 'Reductive' and 'non-reductive' refer to either
explanations or *interpretations* of religion. 'Explanation' and 'interpretation' refer to
methods of studying religion.

Neither functional and substantive *definitions* nor explanatory and interpretive
methods correspond to reductive and non-reductive *explanations*. A reductive
explanation departs from the believer's own account of the origin and function of
the believer's religiosity. A non-reductive explanation captures that account. But
the reductive *explanations* of Marx, Durkheim, and Freud, among others, employ
substantive *definitions*, and at least theologian Paul Tillich's non-reductive *explanation* employs a quasi-functional *definition*: religion as whatever one values most.

The meaning of religion provided by an *interpretation* need not be conscious, so
that there can be reductive as well as non-reductive interpretations. Even
R. G. Collingwood, the doyen of interpretivism, permits the historian to discover
the meaning of an action of which the agent was unaware: 'The historian...can even
discover what, until he discovered it, no one ever knew to have happened at all'
(Collingwood 1946: 238). Certainly an *explanation* can be either non-reductive or
reductive. It can be either the believer's own account of the origin and function of the
believer's religiosity or that of the social scientist. The accounts can even coincide.
The distinction between reductive and non-reductive does not, then, correspond to
that between explanatory and interpretive.

A social scientific account of religion is materialist and behaviourist; a religionist
account of religion is mentalist   Just as religionists try to fend off the social sciences
by deeming a social scientific account of religion 'functional', 'reductive', and
'explanatory', so they try to fend off the social sciences by deeming a social scientific

account of religion materialist and behaviourist—terms also sometimes used interchangeably. As religionist Robert Fuller states:

The problem, however, is that the particular kind of empiricism insisted upon by our modern
social sciences fates us to remaining in the dark of night. By restricting the scope of reality to
the *material* forces shaping everyday life, the empirical method has shed no light on the great
issues that face humanity both as individuals and as a species. (Fuller 1987: 501; italics added)

In actuality, the social and even the natural sciences allow for mentalist as well as
materialist accounts of human behaviour. Science does not mean materialism. The
relationship of the mind to the brain remains an open scientific question. Philosophers of science such as Carl Hempel allow for mental as well as physical causes of
human behaviour (see Hempel 1965: 463–87). Adolf Grünbaum writes of 'the myth
that the explanatory standards of the natural sciences are intrinsically committed to a
physicalistic reductionism such that psychic states (e.g., intentions, fears, hopes,
beliefs, desires, anticipations, etc.) are held to be, at best, epiphenomena, having
no causal relevance of their own' (Grünbaum 1984: 75).

Moreover, few social scientists are themselves materialists. That is, few deny the
existence of culture and other forms of mental life. Even as resolute a materialist as
the anthropologist Marvin Harris seeks only to *explain* culture materially, not to deny
the reality of it. Moreover, Harris grants that the superstructure 'may achieve a degree
of autonomy' from the infrastructure (Harris [1979] 1980: 56). Marx (in Marx and
Engels 1957) himself scarcely denies the existence of religion or any other part of the
superstructure, which for him, too, can sometimes operate independently of
the infrastructure. Religion is not an illusion. The illusion is the assumption that
the superstructure explains itself: that the origin and function of religion are ultimately, not just directly, religious rather than economic.

Just as few social scientists are materialists, so few are behaviourists. Outside
psychology, the best known is sociologist George Homans (1961). But Homans,
following B. F. Skinner, is a methodological rather than a logical behaviourist. He
sidesteps the issue of the status of the mind rather than, like the philosopher Gilbert
Ryle, reducing the mind to a tendency to behave in a certain way.

A social scientific account of religion denies the truth of religion   It is typically
asserted that a social scientific account of either the origin or, less often, the function
of religion denies the truth of religion. This characterization is triply wrong. First,
most contemporary social scientists, in contrast to earlier ones, shun the issue of
truth as beyond their social scientific ken. They confine themselves to the issues of
origin and function. Rather than seeking to determine whether religion is true, they
seek to determine why religion is *believed* to be true.

Second, those contemporary social scientists who do assess the truth of religion
claim that the social sciences either do or should assert the *truth*, not the falsity, of
religion. Victor Turner (1975: 195–6) berates his fellow social scientists for denying the
truth of religion. As a relativist, Mary Douglas (1975: pp. ix–xxi; 1979) considers
true the beliefs of all cultures. Robert Bellah (1970: 252–3) comes to declare religion
true, though true to the experience of the world rather than true of the world itself.

Peter Berger ([1969] 1970: 52–97; [1979] 1980: 58–60, 114–42) comes to maintain that the social sciences can confirm the truth of religion.

Third, classical social scientists who do pronounce religion false—for example, Tylor, Frazer, Marx, and Freud—do not do so on the basis of their social scientific findings. Rather, they do the reverse: they argue for a secular origin and function on the grounds of the falsity of religion. For them, religion is false on philosophical, not social scientific, grounds.

Take Marx (in Marx and Engels 1957). For him, religion is dysfunctional not because it fails to accomplish its intended function but because the escapist and justificatory functions that it does accomplish are more harmful than helpful. But for Marx, religion would not be escapist if he believed in the place of escape: heaven. Marx, then, deems religion dysfunctional because he deems it false, but not false because dysfunctional. Someone else might invoke economic harm as an argument against the existence of a fair or powerful God, in which case the dysfunctional effect of religion would argue for the falsity of religion. But Marx himself disbelieves in a God of any kind, and does so on philosophical, not social-scientific, grounds.

The charge that the social sciences deny the truth of religion assumes that the social sciences not only *do* deny but also *dare not* deny the truth of religion. They dare not exactly because they are supposedly limited to the issues of origin and function. For them to enlist their conclusions about the origin of religion to assess the truth of religion would be to commit the genetic fallacy. For them to enlist their conclusions about the function of religion to assess the truth of religion would be to commit what I call the functionalist fallacy. In actuality, as Freud argues in *The Future of an Illusion* ([1961] 1964) and as I have argued more generally, it is possible to argue from the origin or the function of religion to either the truth or the falsity of religion without committing either fallacy (see Segal 1980, 2005).

If, as I maintain, all of the objections lodged against the propriety of a social scientific analysis of religion are illegitimate, then social scientists are as entitled to study religion as religionists are.

The Religionist Misconstrual of Contemporary Social Science

Ironically, there has arisen among some religionists, though certainly not Eliade, the opposite view of the social sciences. Social scientists are now considered entitled to study religion as fully as religionists are. Why? Because social scientists have at last become converted to the religionist position.

Religionists who make this claim do not assert that contemporary social scientists, in contrast to classical ones, consider religion true rather than false or helpful rather than harmful. Rather, they assert that 'contemporaries', in contrast to 'classicals', account for religion religiously. But do they?

Peter Berger ([1967] 1969, [1969] 1970, [1979] 1980) contends that religion serves to make human life meaningful. A meaningful life is one not merely explained but also outright justified. Suffering, above all death, mocks the justifications offered by secular society. By ascribing all events to the will of God, religion offers far sturdier justifications. Ritual implants and confirms those justifications. Both because every human being needs meaning, or meaningfulness, and because no secular society can fully provide it, religion is indispensable.

Yet religion is still only a means to a secular end. For Berger, humans crave meaning in general, not religious meaning in particular. They seek not the sacred itself but only the justifications it bestows. Religion serves not the irreducibly religious purpose of providing religious meaning but the secular purpose of providing meaning *per se*. Berger's account of religion is therefore reductive.

Clifford Geertz (1968, 1973, 1983) likewise maintains that religion serves to make life meaningful. By a meaningful life he means one not necessarily justified, as for Berger, but simply explained or endured. Threats to meaning can come not only, as for Berger, from suffering, or what Geertz calls 'unendurable events', but also from merely inexplicable events, among which Geertz locates death, and from outright unjustifiable ones like the Holocaust. Whereas unjustifiable events need to be outright justified, inexplicable ones need only to be explained, and unendurable ones need only to be endured.

For Geertz, as for Berger, religion does not emerge until after the efforts of secular society have failed. Religion arises to provide a far more cogent and powerful explanation, alleviation, or justification of troubling experiences than the common sense of seculardom offers. Geertz calls that explanation, alleviation, or justification a 'world view'. The world view not only reinforces *existing* social life, as for Berger, but also offers a model for creating a *new* social life, or an 'ethos'. Ritual fuses the world view with the ethos.

Geertz maintains that his recognition of an existential need served by religion makes him non-reductive, but that need is, as for Berger, for meaning in general, not for religious meaning in particular. Even if for Geertz, as for Berger, religion is indispensable, it remains only a means to a secular end.

Mary Douglas (1966, 1970, 1975) maintains that religion functions to satisfy the need less for existential than for intellectual meaning. Humans need not merely to explain, endure, or justify their experiences, as for Berger and Geertz, but more fundamentally simply to organize them. Religion, like other domains of culture, organizes experience by categorizing it. Without categories, life would be not merely baffling, painful, or unjust but outright incoherent. For Douglas, as for Berger and especially Geertz, rituals are the means by which order is imposed. To take Douglas's most famous example, Jewish dietary laws serve to prohibit the eating of animals that violate the categories into which living things are divided.

Religion for Douglas is not the only means of organizing experience. On the contrary, she is eager to show that various secular activities like spring cleaning are in actuality ritualistic. Yet even if religion were for her, as for Berger, the sole means of serving its function, that function would still be non-religious.

What is true of Berger, Geertz, and Douglas is also true of the other contemporaries embraced by religionists: Robert Bellah, Victor Turner, and Erik Erikson (1958). Contemporaries and religionists remain far apart. Here Eliade is right: the social sciences still need to be fended off if religious studies is to retain its monopoly on the study of religion.

Trends in the Social Scientific Study of Religion

Even if contemporary social science yields no reconciliation with religious studies, there has been a broadening of the social scientific field. The standard focus on the origin and function of religion has been broadened into a concern as well with, first, the propriety of religion and, second, the truth of religion. Berger and Bellah best represent these dual trends. The third trend, best represented by Geertz, has been the supplementing of an explanatory approach to religion with a hermeneutical, or an interpretive, one.

The Propriety of Religion

In his earlier writings (Berger 1961a, 1961b, [1967] 1969) Berger rails against life made easy by religion. He denounces religion for accepting rather than challenging seculardom. He denounces his own variety of Christianity for supporting rather than questioning such American values and institutions as financial success, class and racial divisions, the Cold War, capital punishment, and the family unit (see Berger 1961a: 116). Religion that seeks to justify society constitutes what Berger, employing Sartre's famous term, calls 'bad faith'.

Earlier Berger deems bad faith not only the use of religion to sanction seculardom but, reciprocally, the use of seculardom to justify religion. Ironically, Berger denounces as improper what Geertz applauds as effective: the meshing of a conception of reality with a way of life. Wherever religion fits society snugly, the affirmation of it requires no effort (see Berger 1961a: 40–1).

In his later writings (Berger [1969] 1970, [1979] 1980; Berger et al. 1973) Berger stresses that the affirmation of at least modern religion demands effort. For the competition from other religions and from seculardom renders any claim to certainty tenuous. Bereft of certainty, one must leap unto faith: 'Faith is no longer socially given, but must be individually achieved' (Berger, Berger, and Kellner 1973: 81). For Berger, religion is proper when the basis of commitment to it is faith.

When, in his later writings, Bellah turns from Japan (1957) and the rest of the world to the United States, he changes his aim from analysing religion to heeding it. His claim that there exists an American 'civil religion' is in part a mere restatement of Durkheim's fundamental claim that every society worships itself (see Bellah 1975, 1976, 2002; Bellah and Hammond 1980). Yet Bellah is interested less in the social function of civil religion than in the obligation that civil religion imposes. Whereas for Durkheim duty to God means duty to society, for Bellah duty to society means duty to God. For Bellah, as for Berger, religion should serve to judge whether society is living up to its ideals.

The Truth of Religion

Earlier Berger views the social sciences as unable to assess the truth of religion. Later Berger, beginning with *A Rumor of Angels* ([1969] 1970), reverses himself and comes to view the social sciences as able to affirm the *truth of* religion. Earlier as well as later Berger is intent on reconciling the social sciences with religious truth, but earlier Berger does so by declaring the issue of truth beyond the social scientific ken: 'it is impossible within the frame of reference of scientific theorizing to make any affirmations, positive *or* negative, about the ultimate ontological status of this alleged reality' (Berger [1967] 1969: 100).

Later Berger reconciles the social sciences with religious truth by arguing that the social sciences can *establish* the existence of God—this by cataloguing 'signals of transcendence', or those experiences of hope, humour, and above all order that entail, because they presuppose, the existence of the transcendent: 'Thus man's ordering propensity implies a transcendent order, and each ordering gesture is a signal of this transcendence' (Berger [1969] 1970: 57). Whereas earlier Berger spurns any evidence for belief as bad faith, later Berger solicits evidence.

Bellah calls his later approach to religion 'symbolic realism'. In contrast to his own earlier 'consequential reductionism', symbolic realism is concerned not with the effect of religion on society but with the 'meaning' of religion for believers themselves. Religion for believers is not a scientific-like account of the world but an encounter with it: 'If we define religion as that symbol system that serves to evoke ... the totality that includes subject and object and provides the context in which life and action finally have meaning, then I am prepared to claim that ... religion is true' (Bellah [1969] 1970: 252–3).

The Hermeneutical Approach to Religion

Like both Berger and Bellah, Geertz (1973, 1983, 1995) shifts his focus—in his case from an explanatory approach to religion to an interpretive one. What he means by interpretation fluctuates. When he follows the philosopher Paul Ricoeur (1983), all of culture, including religion, is considered akin to a literary text, which therefore requires the equivalent of exegesis. Here interpretation refers to the theme of the text, explanation to the origin and function of the text.

When Geertz follows Gilbert Ryle, the distinction between interpretation and explanation is more technical. Geertz uses Ryle's own example of the difference between twitching and winking (see Geertz 1973: 6). In the familiar sense of the terms 'cause' and 'meaning', a twitch is causal, or meaningless, because it has no purpose. It is involuntary and therefore unintentional. It is not inexplicable, for its cause explains it, but it is purposeless. A wink is meaningful because it has a purpose as well as a cause—more accurately, a purpose rather than a cause. It is voluntary and therefore intentional.

In Ryle's sense of the terms, a wink is meaningful rather than causal not just because it is purposeful, or intentional, but also because the purpose is inseparable from the behaviour: in winking, one does not first contract one's eyelids and then wink but, rather, intentionally contracts one's eyelids. The purpose cannot therefore be the cause of the behaviour, for cause and effect must be distinct. The purpose and the behaviour are two aspects of a single action rather than, as in causal explanation, the cause and the effect. If the contraction were the effect of winking, winking would be the cause. But because the contraction is the expression of winking, winking is the meaning.

To describe only the behaviour would be to give what Geertz, following Ryle, calls a 'thin description'. To describe the meaning expressed by the behaviour is to give what Geertz, again following Ryle, calls a 'thick description', or an interpretation.

In stating that humans strive to make sense of life, earlier Geertz is stating that that striving *causes* them to engage in religious and other sense-making activities. In continuing to state that humans strive to make sense of life, later Geertz is simply stating *that* they engage in religious and other sense-making activities, which *express* rather than *effect* humanity's sense-making character.

One of the many other things that Geertz means by 'interpretation' is the primacy of the particular over the general. By their nature, generalizations disregard the distinctiveness of the particular, where, Geertz tells us, the significance of any cultural phenomenon lies: 'the notion that the essence of what it means to be human is most clearly revealed in those features of human culture that are universal rather than in those that are distinctive to this people or that is a prejudice we are not necessarily obliged to share' (Geertz 1973: 43).

Geertz's insistence on the superiority of the particular to the general helped pioneer the post-modern approach to culture, including religion. Yet Geertz himself has never gone so far as his avowedly post-modernist successors. He does not reject generalizations altogether. He does not dismiss the possibility of an objective analysis of culture. Most important, he does not pit an interpretive approach against a scientific one. In even 'Thick Description' (Geertz 1973: ch. 1), his programmatic interpretive statement, he asserts that interpretation should supplement, not supplant, explanation. Anthropology should be an 'interpretive science'.

OBJECTIONS TO THE SOCIAL SCIENCES AS SCIENTIFIC

The strategy of religionists has been to claim that the social sciences are irrelevant to the study of religion. Rarely has the scientific status of the social sciences been challenged by them. Yet surely one can do so. If the theories applied to religion fall short of their own purported standard, then the application must fall short, and religionists can continue to ignore the social scientific challenge.

Many of the objections to social science as science rest, however, on an erroneous view of what makes science science. Erroneous, to begin with, is the assumption that, to be scientific, the social sciences dare not go beyond observables—behaviour—to unobservables—beliefs and emotions. The natural sciences begin with the observable world but venture beyond it to account for it. True, there are philosophies of science, notably instrumentalism and constructive empiricism, that refuse to go beyond the observable realm, but the sciences themselves do not.

Equally erroneous is the assumption that, to be scientific, the social sciences must be predictive. John Stuart Mill, defender *par excellence* of the scientific status of the study of humans, cites 'tidology' as an indisputable science that nevertheless makes inaccurate predictions (see Mill [1872] 1988: 31). Mill simply distinguishes between the exact and the inexact sciences, which means presently, not inherently, inexact. Astronomy was once an inexact science that has since become an exact one, as likely has tidology, and as may yet the social, or 'moral', sciences.

Alasdair MacIntyre has argued that 'law-like generalizations' in the social sciences fall short of their counterparts in the natural sciences (see MacIntyre 1981: 84–102). And for him fall short they must. For whereas science predicts, there is 'systematic unpredictability in human affairs' (MacIntyre 1981: 89). He contrasts statistical generalizations in the social sciences to those in the natural sciences. But his contrast in fact rests on a misunderstanding of statistical laws (see Salmon 1992: 416–17) and, more, on an exaggerated view of laws in the natural sciences, where, notably, *ceteris paribus* clauses are also to be found (see Hempel 1988: 150–1). As ever more precise instruments have been created, the accuracy of predictions in exact sciences such as astronomy and physics has actually decreased (see Salmon 1992: 406). The most fundamental laws of physics may turn out to be inherently probabilistic, in which case there will never be perfect prediction.

OBJECTIONS TO THE STUDY OF HUMAN BEINGS AS SCIENTIFIC

If one set of objections to social scientific theorizing is that the social sciences do not qualify as scientific, another set of objections is that the human world, including religion, cannot be studied scientifically. One of these objections has already been considered: the claim that the cause of human behaviour is mental rather than physical and that science deals with only the physical world.

Another objection is the assumption that humans are free, so that their behaviour cannot be predicted. But this objection confuses prediction with control. While free will versus determinism remains a perennial issue, the notion of compatibilism distinguishes between *knowing* a person's future behaviour and *causing* it. A psychiatrist who knows a patient well enough to be able to predict the patient's behaviour is not causing the patient to behave in that way.

For Hempel (1965: 463–87) and others, reasons, or meanings, are simply mental causes and can therefore be studied scientifically. For 'interpretivists' like Collingwood (1946), meanings are distinct from causes. By 'interpretivism', or interpretation, one may mean simply the attribution of human behaviour to mental causes, with explanation referring to the attribution of human behaviour to physical causes. By this tame usage, which is found in Max Weber ([1947] 1964), Hempel would qualify as an interpretivist, though he himself would subsume interpretation under explanation. Alternatively, by interpretation one may mean the message, or theme, of behaviour, with explanation referring to the account of the behaviour. This usage is found in Ricoeur (1983) and at times in the inconsistent Geertz (1973). Here explanation and interpretation are answers to different questions rather than different answers to the same question—in our case, Why did or do persons become religious?

By contrast to these usages, Collingwood deems interpretation and explanation incompatible ways of accounting for behaviour. Causes, which explanation provides, are separate from the behaviour they bring about. Meanings, which interpretation provides, are not separate from behaviour, in which case what they bring about are their *expressions* rather than their *effects*.

To use Collingwood's own example, to say that the *cause* of Brutus's stabbing of Caesar was a hunger for power would be to say that power hungering was a trait in Brutus separate from the stabbing, which was the effect of that trait. To say that the *meaning* of Brutus's stabbing was a hunger for power would be to say that power hungering was a trait in Brutus expressed outwardly in the stabbing. The deed becomes part of the meaning, or *definition*, of power hungering. As a cause, power hungering says only what brought about the deed. As a meaning, power hungering in addition categorizes the deed *as* power hungering. Indeed, to say *what* Brutus did— sought power—is to say *why* he did it. As Collingwood famously puts it, 'When he [the historian] knows what happened, he already knows why it happened' (Collingwood 1946: 214). This distinction between interpretation and explanation is to be found not only in Collingwood but also in the philosophers William Dray (1957) and Peter Winch (1958)—and insofar as he follows Ryle, Geertz. For all four, the distinction is that between what Collingwood and Dray call 'history', a term used broadly to include much of social science, and natural science. (On the varying ways of distinguishing between interpretation and explanation, see Segal 1992.)

Donald Davidson (1963) has argued against the interpretivist claim that reasons cannot be causes because they are part of the behaviour they spur. Davidson notes that causal relationships between events in the world, such as Brutus's stabbing of Caesar, can be described in more than one way. Brutus's stabbing of Caesar is describable as either the expression or the effect of a hunger for power. Davidson accepts the interpretivist view that an interpretation not only accounts for a behaviour but also classifies and thereby redescribes it, but he argues that there can be a causal as well as a meaningful redescription. Moreover, a redescription that is not also an explanation fails to connect the classification to the behaviour that it is supposed to classify. Unless Brutus's hunger for power *caused* him to stab Caesar, the stabbing can scarcely be *classified* as power hungering.

The issue is not whether all of the philosophical objections to social scientific theories have been met. Davidson himself stresses the difficulty of formulating laws that connect reasons to actions. The issue is that the objections rest on a proper grasp of theorizing. Post-modern objections, I proceed to argue, do not.

POST-MODERN OBJECTIONS
TO SOCIAL SCIENCE

By no coincidence, post-modernism has been embraced by religionists for its un-compromising rejection of science, natural and social alike. A post-modern approach to religion spurns a scientific approach as outdated because modern. Indeed, 'modern' and 'scientific' are used interchangeably.

In his introduction to *Critical Terms for Religious Studies* Mark C. Taylor unhesi-tatingly pronounces the post-modern approach to religion superior to the modern one:

For interpreters schooled in postmodernism and poststructuralism, the seemingly innocent question 'What is...?' is fraught with ontological and epistemological presuppositions that are deeply problematic. To ask, for example, 'What is religion?' assumes that religion has something like a general or even universal essence that can be discovered through disciplined investigation....But what if religion has no such essential identity? What if religion is not a universal phenomenon?...Investigators create—sometimes unknowingly—the objects and truths they profess to discover. Some critics claim that appearances to the contrary notwith-standing, religion is a *modern Western invention*....

[M]any critics schooled in poststructuralism insist that the very effort to establish similar-ities where there appear to be differences is, in the final analysis, intellectually misleading and politically misguided. (Taylor 1998: 6–7, 15)

This statement evinces most of the confusions that constitute post-modernism. I take them up one by one.

The Confusion of Universality with Essence

Contrary to Taylor, theories of religion do not claim to have uncovered the 'essence' of religion. Essence is a metaphysical issue. Theorizing is a merely empirical enter-prise. Theories claim to have discovered only the conditions for the emergence (origin) and perpetuation (function) of religion. Few theories claim to have discovered even the necessary and sufficient conditions for the emergence and perpetuation of religion. For example, Max Weber offers only necessary, not sufficient, conditions for religion: religion—more precisely, religion beyond the stage of magic—requires a cult, but not every cult produces the next, priestly stage of religion. By contrast, Emile Durkheim does offer necessary as well as sufficient conditions for religion: whenever a group amasses regularly, there will be religion. But the explanations offered by most theories claim to be merely probabilistic, so not even sufficient. Most theories modestly claim that if certain conditions exist, religion will probably, but not always, arise and persist.

The Confusion of Invention with Discovery

Even grant that 'religion', itself an ancient term, is somehow a modern Western invention. That claim is made relentlessly nowadays—most exhaustively by Daniel Dubuisson in his *The Western Construction of Religion* (2003), though, to be sure, he deems religion a hoary Christian invention rather than a modern secular one. But there is a difference between asserting that 'religion' arose in a particular setting and asserting that it thereby fails to apply to other settings. To the extent that Dubuisson strives to show, not simply to declare, that the concept does not apply beyond its point of origin and should therefore be abandoned, his procedure is unimpeachably modern.

The post-modern contention of Taylor, or of 'some critics', is that no testing is necessary. Religion, because created by the modern West, not merely *may* prove but *must* prove a misfit anywhere else. But how can post-modernists be so confident? They can be so only by confusing invention with discovery. To quote Taylor again: 'Investigators create ... the objects and truths they profess to discover' (Taylor 1998: 7). But how does Taylor know? It can only be that knowledge of the time and place of the origin of definitions and theories reveals this insight to him. But to deny the possibility of discovery on the grounds that one knows the point of origin is to commit the genetic fallacy. Theories in the natural sciences, no less than those in the social sciences, arise at some time and place.

The Confusion of Students with Natives

Against those who, like Frazer, assumed that 'primitives' possessed magic or religion *rather than* science, Bronislaw Malinowski argued that in fact 'primitives' have science as well as magic and religion. That 'primitives' do not themselves use the term 'science' did not faze Malinowski. By contrast, post-moderns would be apoplectic. For them, the issue is not whether the term 'religion' fits but who concocted it. Taylor thus praises Jonathan Z. Smith for alerting us to the danger of mistaking 'our' categories for 'theirs'. As Smith declares in his contribution to the Taylor collection, '"Religion" is not a native term; it is a term that is created by scholars for their intellectual purposes and therefore is theirs to define' (Smith 1998: 281; repr. in Smith 2004: 193–4). What Smith cautions about 'religion', he also cautions about other standard terms in religious studies.

Are we to believe that modern theorists did not realize that the vocabularies they created were unknown to the cultures to which those vocabularies were applied? Did Jung naïvely assume that 'archaic man' chatted away about his 'archetypes'? If not, then what are we being told? That the terms *may* not fit cultures other than ours? Who would demur? Only one option is left: that our terms *cannot* fit, and cannot fit precisely because they are ours, not theirs.

To make this claim is to confuse roles. No doctor defers to a patient in making a diagnosis. The patient may harbour the ailment, but the doctor is trained to identify it. Deferring to the patient confuses the subject—the patient—with the scholar—the doctor. Medical terms are not assumed to be the subject's. The issue is whether they fit.

The Confusion of Explanation with Effect

Theories of religion are accounts of religion. They are not policy statements. Some theorists like religion, others dislike it—this because they like or dislike the effects of religion. Post-modernism introduces a meta-level of assessment: now one likes or, usually, dislikes not religion but *theories* of religion, and this for their effect on religion and in turn the effect *of* religion. The inspiration here comes from Michel Foucault, whom Russell McCutcheon, in his *Manufacturing Religion* (1997), applies to theories of religion.

Rather than attacking all theories of religion, McCutcheon attacks only the religionist theory epitomized by Eliade. He objects to Eliade's ignoring the non-religious origin and function of religion, and thereby supposedly sanctioning whatever political effect religion in fact has. The religionist theory, or 'discourse', 'manufactures' the theory of religion not merely to give religiosity autonomy, as has long been argued, but even more to deflect attention away from the political origin and function of religion. Whereas Marx and Engels maintained that religion serves to perpetuate inequality, McCutcheon tells us that a *theory* of religion serves to do the same: 'By overlooking the importance of these additional [i.e. material] aspects of human existence, scholars [of religion] may not necessarily be promoting these imbalanced distributions of wealth or influence, but they certainly minimalize the significance of such factors' (McCutcheon 1997: 13).

Who would ever have imagined that capitalists needed any further spur to their avarice? But even if one grants the religionist theory, or discourse, its effect, the theory can still be true. To say otherwise is to commit the functionalist fallacy. If true, the theory should be retained, and retained for doing what theories do, which, to reverse Marx's *Theses on Feuerbach*, is merely to interpret the world, not to change it.

The Confusion of Failed Theories with All Theories

Theories of religion can fail for multiple reasons. For example, they may turn out to account for some religions, but not all. Or they may account inadequately for all religions. For the brand of post-modernism inspired by Jacques Derrida, theories contradict themselves: they are undermined by the presence of contrary currents in the texts that present the theories.

The most brilliant application of Derridean deconstructionism to theories of religion is Tomoko Masuzawa's *In Search of Dreamtime* (1993), the subtitle of which is *The Quest for the Origin of Religion*. Masuzawa assumes that classical theories of religion sought above all the historical, one-time origin of religion. She lumps religionist theories with social scientific ones and takes as her prime targets Durkheim, Freud, Eliade, and Müller. Against them all, she argues that their own texts undermine their intentions.

For example, Durkheim's definition of the sacred as the ideal society is supposedly undermined by the continual appearance in *The Elementary Forms of the Religious Life* of another definition: the sacred as the opposite of the profane. Freud's attribution, in *Totem and Taboo* (1950), of the origin of religion to the sons' rebellion against their tyrannical father is supposedly undercut by Freud's own characterization of this would-be historical deed as fantasy.

Masuzawa's argument is tenuous. Classical theorists sought the recurrent more than the historical origin of religion (see Durkheim [1915] 1965: 20). Even Freud, who in *Totem and Taboo* comes closest to seeking the historical origin of religion, seeks only the first *stage* of religion. Moreover, classical theorists were concerned as much with the function of religion as with the origin, recurrent or historical. Above all, the presence in theories of inconsistencies argues merely for the provisional state of the theorizing rather than, as for Masuzawa, for any systematic undermining of the effort.

Masuzawa's approach is post-modern in the conclusion she draws: that the quest for historical origin, for her the key concern of at least classical theorizing, must be abandoned. But even suppose that all classical theorists failed in a common quest for the historical origin of religion. Dare subsequent theorists not try? Why does the failure of some theories spell the failure of all?

The Confusion of the Quest for Similarities with the Denial of Differences

Taylor declares that 'the very effort to establish similarities where there appear to be differences is . . . intellectually misleading' (Taylor 1998: 15). Who would disagree? But the only way to determine whether similarities mask differences is by research.

Moreover, why are differences more significant than similarities? Similarities, not just differences, often lie beneath the surface. An emigré to a foreign country is typically struck first by the differences and only later notices the similarities. The 'privileging' of differences means the rejection of theories, for theories are generalizations. But theories do not deny differences. They deny the *importance* of differences. Theories do not bar the quest for differences. They simply do not undertake it themselves. Their goal is precisely to account for similarities. At the same time that quest is unavoidable even for those seeking differences, for differences begin only where similarities end (see Segal 2001: 348–9).

I maintain that the post-modern challenge to the social scientific study of religion is as dubious as the traditional religionist challenge has been. Post-modernism in no way fends off the social sciences, which remain free to study religion as fully as they choose.

REFERENCES AND SUGGESTED READING

BELLAH, R. N. (1957). *Tokugawa Religion*. Glencoe, Ill.: Free Press.

—— (1970). *Beyond Belief*. New York: Harper & Row.

—— (1975). *The Broken Covenant*. New York: Seabury.

—— (1976). 'The Revolution and Symbolic Realism', in J. C. Brauer (ed.), *Religion and the American Revolution*, Philadelphia: Fortress Press, 55–73.

—— (2002). 'Meaning and Modernity: America and the World', in R. Madsen *et al.* (eds.), *Meaning and Modernity*, Berkeley: University of California Press, 255–76.

—— and HAMMOND, P. E. (1980). *Varieties of Civil Religion*. San Francisco: Harper.

BERGER, P. L. (1961*a*). *The Noise of Solemn Assemblies*. Garden City, N.Y.: Doubleday.

—— (1961*b*). *The Precarious Vision*. Garden City, N.Y.: Doubleday.

—— ([1967] 1969). *The Sacred Canopy*. Garden City, N.Y.: Doubleday Anchor Books.

—— ([1969] 1970). *A Rumor of Angels*. Garden City, N.Y.: Doubleday Anchor Books.

—— ([1979] 1980). *The Heretical Imperative*. Garden City, N.Y.: Doubleday Anchor Books.

—— BERGER, B., and KELLNER, H. (1973). *The Homeless Mind*. New York: Random House.

COLLINGWOOD, R. G. (1946). *The Idea of History*, ed. T. M. Knox. New York: Oxford University Press.

DAVIDSON, D. (1963). 'Actions, Reasons, and Causes', *Journal of Philosophy*, 40: 685–700.

DOUGLAS, M. (1966). *Purity and Danger*. London: Routledge & Kegan Paul.

—— (1970). *Natural Symbols*, 1st edn. New York: Pantheon.

—— (1975). *Implicit Meanings*, 1st edn. London and Boston: Routledge & Kegan Paul.

—— (1979). 'World View and the Core', in S. C. Brown (ed.), *Philosophical Disputes in the Social Sciences*, Brighton: Harvester Press; Atlantic Highlands, N.J.: Humanities Press, 177–87.

DRAY, W. H. (1957). *Laws and Explanation in History*. Oxford: Clarendon Press.

DUBUISSON, D. (2003). *The Western Construction of Religion*, trans. W. Sayers. Baltimore: Johns Hopkins University Press.

DURKHEIM, E. ([1915] 1965). *The Elementary Forms of the Religious Life*, trans. J. W. Swain. New York: Free Press.

ELIADE, M. ([1958] 1963). *Patterns in Comparative Religion*, trans. R. Sheed. Cleveland: Meridian Books.

ERIKSON, E. (1958). *Young Man Luther*. New York: Norton.

FREUD, S. (1950). *Totem and Taboo*, trans. J. Strachey. New York: Norton.

—— ([1961] 1964). *The Future of an Illusion*, trans. W. D. Robson-Scott, rev. J. Strachey. Garden City, N.Y.: Doubleday Anchor Books.

FULLER, R. C. (1987). 'Religion and Empiricism in the Works of Peter Berger', *Zygon*, 22: 497–510.

GEERTZ, C. (1968). *Islam Observed*. New Haven: Yale University Press.

—— (1973). *The Interpretation of Cultures*. New York: Basic Books.

—— (1983). *Local Knowledge*. New York: Basic Books.

—— (1995). *After the Fact*. Cambridge, Mass.: Harvard University Press.

GRÜNBAUM, A. (1984). *The Foundations of Psychoanalysis*. Berkeley: University of California Press.

HARRIS, M. ([1979] 1980). *Cultural Materialism*. New York: Vintage Books.

HEMPEL, C. G. (1965). *Aspects of Scientific Explanation and Other Essays in the Philosophy of Science*. New York: Free Press.

—— (1988). 'Provisoes: A Problem Concerning the Inferential Function of Scientific Theories', *Erkenntnis*, 28: 147–64.

HOMANS, G. C. (1961). *Social Behavior*, 1st edn. New York: Harcourt, Brace.

KEPNES, S. D. (1986). 'Bridging the Gap between Understanding and Explanation Approaches to the Study of Religion', *Journal for the Scientific Study of Religion*, 25: 504–12.

MACINTYRE, A. (1981). *After Virtue*, 1st edn. London: Duckworth.

MARX, K., and ENGELS, F. (1957). *On Religion*. Moscow: Foreign Languages Publishing.

MASUZAWA, T. (1993). *In Search of Dreamtime*. Chicago: University of Chicago Press.

MCCUTCHEON, R. (1997). *Manufacturing Religion*. New York and Oxford: Oxford University Press.

MILL, J. S. ([1872] 1988). *The Logic of the Moral Sciences: A System of Logic*, Book VI, 8th edn. (1st edn. 1843). La Salle, Ill.: Open Court.

RICOEUR, P. (1983). *Hermeneutics and the Human Sciences*, trans. and ed. J. B. Thompson. Cambridge: Cambridge University Press.

SALMON, M. H. (1992). 'Philosophy of the Social Sciences', in M. H. Salmon *et al.*, *Introduction to the Philosophy of Science*, Englewood Cliffs, N.J.: Prentice-Hall, 404–25.

SEGAL, R. A. (1980). 'The Social Sciences and the Truth of Religious Belief', *Journal of the American Academy of Religion*, 48: 403–13.

—— (1989). *Religion and the Social Sciences*. Atlanta: Scholars Press.

—— (1992). *Explaining and Interpreting Religion*. New York: Peter Lang.

—— (2001). 'In Defense of the Comparative Method', *Numen*, 48: 339–74.

—— (2005). 'James and Freud on Mysticism', in J. Carrette (ed.), *William James and 'The Varieties of Religious Experience'*, London and New York: Routledge, 124–32.

SMITH, J. Z. (1998). 'Religion, Religions, Religious', in Taylor (1998), 269–84; repr. in Smith (2004), ch. 8.

—— (2004). *Relating Religion*. Chicago: University of Chicago Press.

TAYLOR, M. C. (1998) (ed.). *Critical Terms for Religious Studies*. Chicago: University of Chicago Press.

TURNER, V. (1975). *Revelation and Divination in Ndembu Ritual*. Ithaca, N.Y.: Cornell University Press.

WEBER, M. ([1947] 1964). *The Theory of Social and Economic Organization*, ed. T. Parsons, trans. A. M. Henderson and T. Parsons. Glencoe, Ill.: Free Press.

WINCH, P. (1958). *The Idea of a Social Science and its Relation to Philosophy*. London: Routledge & Kegan Paul.

CHAPTER 20

CONTRIBUTIONS FROM THE PHILOSOPHY OF SCIENCE

ROBIN COLLINS

INTRODUCTION

Other than ethical issues surrounding science, most work in the philosophy of science centres on two main issues: the ontology given by science—that is, what science is telling us about the nature of the world—and the methodology and epistemology of science. Since it is such a vast field, philosophy of science makes many contributions to understanding the interaction between science and religion, far more than can be discussed here. One of the most important contributions that philosophy of science potentially could make, however, is helping provide a framework for understanding what the sciences are telling us about the world that is friendly to religion yet true to science. One of the most important questions in this regard, I believe, is the issue of reductionism and its alternatives.

Various forms of reductionism have been mainstays of those scientists and philosophers who are the most vocal opponents of religion, such as Peter Atkins, Richard Dawkins, Edward Wilson, and Steven Weinberg. It is not that reductionism is itself incompatible with religious belief: one could consistently view the world as a giant Newtonian clock and still believe that God created and sustains it, as many in the past did. Reductionism, however, paints a sparse picture of nature and human beings that, at least for many, seems opposed to our religious sensibilities. God's relation to nature, for instance, becomes largely external, with divine action consisting of a sort of mechanical intervention, insofar as it exists at all. And, in the case of the human person, it leaves little room for the integration of mind with body: either one claims that the mind is merely some complex biochemical process in the brain, or one claims that the relation between mind and brain involves some sort of intervention in the otherwise rigid mechanical order of the brain. In fact, Alister McGrath, following many others, claims that the desacralization of nature characteristic of both mechanical philosophy and much Protestantism is partially responsible for the rise of atheism, since this desacralization evacuated God from the natural world and hence from much of everyday experience (2004b: 200–6). Further, as Simon Critchley, a leading expert in Continental philosophy, has pointed out, this reductionist view has resulted in a deep alienation of human beings from the world, since human values and meaning seem to have no place in nature. Indeed, Critchley claims that one of the primary motivations of Continental philosophy—such as the philosophy of existentialism—has been to address this alienation, without questioning the claim that the 'scientific world-view' is the reductionist world-view (2001: 8–9, 49–53, 71, 111–13, 115–17, 120–1, 126).

Of course one could accept this, and then find meaning largely either in some transcendent realm, or in some philosophy like existentialism. A better way, however, would be to develop an alternative to reductionism that is more friendly to human values and religion. One such widely discussed alternative is the view of *emergent complexity*—the idea that the universe is arranged in hierarchical levels of complexity with new properties and causal powers emerging at each level. This view is discussed in Chapters 45–7 of this Handbook and elsewhere.[1] In this essay, I will develop another alternative to reductionism, and briefly indicate its potential for providing a powerful ontological, epistemological, and methodological framework for the science and religion dialogue. This alternative grows out of considering what I believe is the strongest reason for rejecting the reductionist world-view: namely, quantum mechanics—that extraordinarily paradoxical theory that is widely either misunderstood or ignored. I will call the alternative I develop the *non-reductive intelligibility view*, developing a proposal suggested by scientist/theologian John Polkinghorne and theologian Eric Mascall, though perhaps not in the way they would (Mascall 1956: 82–6, 174–7; Polkinghorne 2002: 86). Although I will critique emergent complexity and compare it with my view, I think that at this pre-paradigm stage, all alternatives to reductionism should continue to be rigorously developed. We will begin by explicating a common form of reductionism, what I call 'compositional reductionism' (CR), and then seeing why quantum mechanics presents serious difficulties for it.

---

[1] For a current book-length treatment of emergence, see Clayton 2004.

Compositional Reductionism Explained

CR is a strategy of explanation pursued from the Scientific Revolution in the seventeenth century until today, and still the framework of thought in which most scientists work. According to CR, the properties and behaviour of a whole can *in principle* be explained in terms of the properties, spatial relations, and external causal interactions of its parts, in conjunction with their causal and spatial relations with the environment. This programme was very successful until the beginning of the twentieth century, with the advent of general relativity and quantum mechanics. Even the electric and magnetic fields introduced in the nineteenth century were still understood in terms of CR, with their parts now being an infinity of points in space with various field values being assigned to them. CR has no doubt been a powerful view of the world, bringing an ontological and explanatory unity to the physical world and the sciences, which is one of the reasons why it has held such sway over people's minds. Below, we will show how quantum mechanics (QM) presents almost fatal difficulties for CR.

It should be noted, however, that reductionism need not always be CR. Another important kind of reductionism involves claiming that one domain of phenomena, such as psychology, is reducible to another domain, such as neurology, without subscribing to CR. In its more global form, it claims that there is a single theory, such as string theory, which can account for all phenomena, even if that theory is non-CR. Obviously, we cannot take on these other forms of reductionism here, though I will argue that QM also raises problems for non-CR versions of global reductionism. I believe, however, that the failure of CR casts significant doubt on these other forms of reductionism.

Quantum Mechanics and Compositional Reductionism

Quantum Holism and CR

The first problem that QM presents for CR is QM's inherent holism. Within the standard formulation of QM, a physical system—such as that involving a single particle—is assigned a state vector, which is a mathematical entity that purportedly tells us everything there is to know about the system.[2] For states consisting of more

---

[2] The standard formulation of QM is the formulation actually used by physicists. It consists of a set of formal rules for assigning states to physical systems, determining how the states evolve, and drawing statistical predictions regarding the results of various measurements of the system—such as a measurement of the energy, momentum, or position of a

than one particle, typically QM does not allow one to assign each particle its own state vector, and hence it is impossible to represent each particle as being in a distinct state. Such systems are said to be in an *entangled* state. In such cases, one can speak meaningfully only of the state of the whole system. Thus, in QM, the state of the whole is not fixed by the intrinsic properties of the parts—such as the particles 'composing' a system—and the structural (spatial) relations between those parts. Indeed, in the case of identical particles—such as a system of many electrons—there are no such relations, since the parts cannot be represented as separate entities—in the jargon of QM, they are *indistinguishable*.

Entanglement shows that QM represents the world in a way that is inconsistent with CR, since CR would require that each particle have a definite state. Various interpretations of QM have been given, but none yields the CR picture described above (an interpretation of QM is an attempt to understand what QM is telling us about the world). To illustrate this conflict with CR, we will look at David Bohm's *hidden variable interpretation*, an interpretation of QM that comes closest to the classical view of the world. In fact, the famous theorem due to John Bell, called 'Bell's theorem', implies an even stronger claim than this: no account of reality in which underlying reality is modelled as consisting solely of localized parts causally interacting with one another can reproduce the (now largely verified) predictions of QM, unless those parts are thought to interact with each other instantaneously in very odd ways. This puts severe restraints on the compatibility of CR with any successor theory to QM. As a leading philosopher of physics, Tim Maudlin, concludes after examining the holism question from the perspective of alternative ways of formalizing QM:

The physical state of a complex whole cannot always be reduced to those of its parts, or to those of its parts together with their spatiotemporal relations, even when the parts inhabit distinct regions of space. Modern science, and modern physics in particular, can hardly be accused of holding [CR] reductionism as a central premise, given that the result of the most intensive scientific investigations in history is a theory that contains an ineliminable holism. (1998: 55)

Maudlin concludes from this that QM refutes CR (1998: 55). Further, it should be noted, quantum field theory and string theory, which both follow the rules of quantum mechanics, present similar problems for CR, though they may take different forms.[3]

particle. (A physical system is simply the physical object, or objects, one is trying to study— e.g. a beam of light going through a small hole or a hydrogen atom in a magnetic field.) In addition to these formal rules, physicists have also developed a set of heuristic rules for assigning states to physical systems.

The standard formulation comes in two varieties, the 'state-vector' formulation and the 'density matrix' formulation. All the main conclusions drawn below apply to both formulations, although I will typically present my arguments in terms of the state-vector formulation since this is the most commonly used and the easiest to follow.

[3] See e.g. Cushing (1988: 27). General relativity also presents problems for CR, as stressed by Roger Penrose (1989: 220–1), though I cannot discuss this here.

Why Physicists Use CR

Why, then, do physicists often present their field as providing a CR picture of physical reality, with talk of things being composed of particles with definite properties? One reason is the difficulty of communicating using quantum concepts, and thus the physics community falls back on classical language. Another is that to a great extent the classical picture is very useful, especially if one adds quantum corrections to one's classical model. For instance, the Bohr model of the atom, in which the atom is pictured as a tiny solar system with electrons whirling around the nucleus, is not consistent with any interpretation of QM (and thus is literally false), though it is very useful. Similarly, physicists steeped in general relativity will still speak of the 'gravitational force', even though in general relativity the concept of gravitational force is replaced by the concept of curvature of spacetime.

A final, and perhaps most important, reason is that a quantum-mechanical model of a system is typically developed by first developing a classical model of the system, and then substituting quantum operators for the classical variables. This procedure is called 'quantization'. Classical models thus provide a useful rung for developing the more correct quantum theory, even in esoteric branches of physics such as string theory. Thus they retain a key conceptual role in physics. Since the quantum operators inhabit an entirely different mathematical space, however, the mathematical relations between the mathematical entities in the quantum model are entirely different. Thus the classical model cannot be said to be even approximately true, though it constitutes an important step to the quantum-mechanical model. We will briefly return to this issue below.

Chemical CR

So, I have argued, CR is inconsistent with both the way in which the QM formalism represents reality and all major interpretations of QM. But, could CR still be valid at a higher level, such as for chemistry? At least for chemical compositional reductionism (CCR), in which atoms and molecules are treated as the fundamental building blocks, the answer appears to be no. The problem with CCR is that atoms and molecules are also inhabitants of the quantum world, even though for most practical applications they can be treated without appeal to quantum mechanics. This means that any non-quantum theory of the behaviour of groups of these atoms and molecules will be seriously incomplete, and thus CCR will fail. This can be most explicitly seen for such phenomena as superconductivity, superfluity, and the operation of the laser, where QM plays a central explanatory role. Since there is no clear boundary between the classical and the quantum realms, any sort of CR is likely to fail.

The Measurement Problem and Reductionism

Although quantum holism presents a fatal problem for CR, it still allows for a more general sort of physical reductionism which builds in some sort of holism, as we will see below when we look at David Bohm's 'interpretation' of QM. Such a general reductionism would claim that the behaviour of all objects can be understood in terms of the equations of QM, though not in the way required by CR. The much discussed *measurement problem* of QM, however, poses a major, though not fatal, challenge to even this form of reductionism. In the standard way in which quantum mechanics is formulated, the result of a measurement interaction—that is, the state in which the physical system ends up after being measured—cannot be explicated in terms of the microphysical causality given by the Schrödinger equation. Yet, the Schrödinger equation is the equation that determines the evolution of the state of individual particles, and groups of particles, even if these groups are understood in a non-CR way. Put in simple terms, in the case of a quantum system being measured, the behaviour of the system is not portrayed in the standard formulation of QM as simply the result of the behaviour of the 'parts' and their interactions with themselves and the measurement apparatus. Rather, in the case of measurement interactions, the standard formulation postulates a second, mysterious process—often called the process of 'state-vector reduction'—in order to obtain the correct experimental results. In more than sixty years, no one has provided an adequate understanding of this process in terms of any underlying physical processes, and there are serious barriers in principle to doing so.

One peculiar consequence of the nature of this postulated measurement process is that one cannot take the quantum models used by physicists, such as a model of an atom as a nucleus with electrons swirling around, literally—that is, as directly corresponding to reality—without running into severe paradoxes. This is nicely illustrated by the so-called delayed-choice experiments.[4] In these experiments, the attribute to be measured (energy, position, spin, etc.) is chosen a significant time after the event, such as the decay of an atom, that is under study. What one finds is that, given that one takes one's models literally, the properties one is forced to ascribe to the system under study at time t_0 depend on choice of measurement at some future time, say $t_0 + 1$ sec. As E. T. Jaynes says in commenting on one of these sorts of experiments,

It is pretty clear why present quantum theory not only does not use—it does not even dare to mention—the notion of "real physical situation.". . . somewhere in this theory the distinction between reality and our knowledge of reality [as exemplified by our choice of what observable to measure] has been lost, and the result has more the character of medieval necromancy than of science. (1980: 42)

The simplest resolution to the measurement problem is to deny that quantum theory provides an account of underlying reality at all, but instead claim that it is merely a useful calculating device. This is a common view among physicists, and the sort of view that Jaynes implicitly criticizes in the above quotation when he says that

---

[4] For a fairly accessible overview of these experiments, see Greenstein and Zajonc 1997: 35–42, 92–105, 193–8.

quantum theory dare not even mention a 'real physical situation'. This view, of course, implicitly denies any reductionist interpretation of QM.

Another resolution is to construct an *interpretation* of quantum mechanics that attempts to circumvent or solve the measurement problem. Such interpretations can either be reductionistic or non-reductionistic. Because of the sort of problems illustrated by the delayed-choice experiments, reductionistic interpretations do not take the models used by physicists literally, but instead offer an alternative (non-CR) model of underlying reality that purportedly reproduces the statistical predictions of quantum mechanics. All such interpretations run into significant problems, however, which is the reason why none of them is widely accepted. But even if a viable reductionist interpretation could be found, it would still be the case that the actual success of the heuristic models and the standard formulation of QM used by physicists offers no support for reductionism, since these models cannot be taken as directly corresponding to reality. Rather, all it would show is that some form of non-CR reductionism is compatible with the empirical predictions of quantum mechanics. These reductionist interpretations, therefore, appear to be driven more by an a priori philosophical commitment to reductionism than by actual scientific practice. I will now illustrate the above points by looking at David Bohm's hidden variable interpretation of QM, a non-CR reductionist interpretation that is one of those most widely discussed among philosophers.

Bohm's Interpretation of Quantum Mechanics

David Bohm's well-known hidden variable theory/interpretation of non-relativistic QM is the interpretation that in many ways stays closest to the CR picture of reality, although, like all viable interpretations, it must deny CR.[5] In Bohm's interpretation, every particle has a definite position and velocity, which is not how the standard formulation of QM represents reality.[6] Further, macroscopic objects—that is, objects much much larger than an atom—are considered to be composed of large aggregates of these particles. To obtain the predictions of quantum mechanics, Bohm must hypothesize that these particles are guided by what he calls the 'pilot wave' or 'quantum potential', which takes a certain specified mathematical form. The quantum potential can be imaginatively thought of as a giant octopus, with tentacles around each particle in the universe, guiding it in such a way that the predictions of QM are duplicated. To account for the observable consequences of quantum holism,

[5] For an accessible and sympathetic presentation and discussion of this theory, see Albert 1992: ch. 7. See also Bohm 1980: 70–110.

[6] Specifically, the standard formulation represents particles as being in superposed states of position, energy, and the like. A single particle, e.g., is typically represented as being in multiple locations at once (i.e. a superposition of position states), with each location being assigned a certain weight given by a complex number—i.e. a number with a real and an imaginary component.

the quantum potential must be non-local, meaning that it cannot be thought of as a field spread throughout space—that is, with spatio-temporal parts. Instead, it is written as a field inhabiting the $3N$-dimensional configuration space of the N particles in the system. (For example, assuming there are at least 10^{80} particles in the universe, it would inhabit at least a 3×10^{80}-dimensional space!) Besides the quantum potential, Bohm must hypothesize a special distribution of particles at the beginning of the universe (or some new underlying randomizing process) to reproduce the statistical predictions of quantum mechanics.

Each of these postulates presents serious difficulties. In Collins (1996), for example, I argue that this postulate of a special initial distribution in itself gives us sufficient reason to reject Bohm's interpretation. Further, in Bohm's theory the particles play no causal role; all the causal work is done by the quantum potential. Being causally inert, it is difficult to say what the idea of a particle even means or how we could ever come to know about particles, since they have no effect on us. Finally, field theory versions of Bohm's theory have run into particular difficulties in accounting for fermions (such as electrons) and in being combined with special relativity (Huggett and Weingard 1994: 386–7).

In any case, because of the non-local character of the quantum potential, Bohm's theory is in conflict with CR: although macroscopic objects are composed of particles, the behaviour of the particles is determined by the non-local quantum potential that cannot be divided in spatial parts. It is still a reductionist (and deterministic) theory, however, in that the behaviour and properties of everything in the universe are determined by the quantum potential. Further, insofar as the (inadequate) relativistic versions of his theory have been developed, the particles are replaced by fields, with macroscopic objects being manifestations of such fields. The behaviour of these fields is in turn determined by a non-local 'superquantum potential', and hence the same conflict with CR arises again. Bohm's is one of the two major (re)interpretations of QM that attempt to retain a classical picture with its particles and fields.[7] It illustrates the minimum elements of the classical CR picture of reality that quantum holism forces one to give up.

It should be noted, however, that, as with virtually all other interpretations, Bohm's interpretation does not take the standard formulation of QM and the models used by physicists as corresponding directly to reality. For example, unlike Bohm's theory, in the standard formulation, particles do not have definite positions and velocities, and there is no quantum potential. Yet, the quantum formalism and heuristic models used by scientists have been both theoretically and experimentally fruitful. As the late physicist/philosopher James Cushing notes, 'In recent high-energy physics one sees how a theorist's intuitions are often led by the mathematics of a formalism rather than by the physics as in a previous era' (1988: 34). Thus, the formalism itself is largely what has fruitful content, a point that has been extensively argued for by Mark Steiner (1998). Although physicists construct models of underlying reality, taking such models literally has not proved fruitful. Neither has

[7] The other is the so-called spontaneous collapse theory. See e.g. Monton (2004).

constructing any sort of reductionist model, such as Bohm's; rather, it is the formalism along with merely heuristic models that have proved fruitful. This is one reason why, unlike philosophers, most physicists have not been attracted to Bohm's theory or other reductionist accounts. As John Polkinghorne has remarked, for physicists 'there is an air of contrivance about it [Bohm's theory] that makes it unappealing [to physicists]' (2002: 55).

AFTER REDUCTIONISM: A NEW VIEW
OF SCIENCE

QM, therefore, gives us almost definitive reasons to reject CR, and poses serious problems for even non-CR forms of microphysical reductionism. If reductionism is false, how do we understand the nature of physical reality? Reductionism at least offers a unified, simple view of physical reality.

At the end of his introductory book on quantum theory John Polkinghorne suggests that 'it is intelligibility (rather than objectivity), that is the clue to reality' (2002: 86). I agree. In light of the problems raised by QM for both CR and non-CR forms of reductionism, I propose a new view of reality that I call 'non-reductive intelligibility' (NRI). This view in turn suggests another view, which I will call 'theistic non-reductive intelligibility' (TNRI), since, I will argue, the sort of intelligibility that we find in the universe suggests theism. At the same time theism allows us to deepen and fill out the conception of NRI.[8] Nonetheless, one could consistently hold NRI without taking the step to TNRI.

To say that nature is intelligible means, among other things, that nature is such that human beings can understand it. One kind of intelligibility is that offered by the reductionist. For the reductionist (whether CR or some other variety), nature is intelligible because we can construct a single model which, to at least a significant degree of approximation, directly corresponds to the underlying physical reality, from which in principle we could explain the behaviour of all material objects. To claim that the intelligibility is non-reductive, in contrast, means that there is no such single model. Rather, it suggests a view of the sciences promoted by, among others, philosopher of science John Dupré (1993) for reasons that are independent of QM. In Dupré's view, each area of science has a certain amount of independence from other domains. A cursory glance at the various sciences reveals a wide range of different domains of inquiry, each with its own explanatory concepts and principles: physics,

---

[8] By theism, I mean the claim that an intelligence created the universe and hence transcends it to some degree. Traditional, or classical theism, both in the West and the East (e.g. major Hindu forms of theism) typically add that this intelligence is all good, omnipotent, omniscient, and the like.

chemistry, biology, psychology, sociology, and then within each science, subdisciplines that also have separate concepts and explanatory principles—such as cell biology and evolutionary biology. Each area is linked in various ways with other domains, with the nature of each linkage determined on a case-by-case basis.

As Dupré notes, this suggests that there is no overarching scientific method that applies to all sciences (1993: 229–43). Rather, each discipline has its own revisable set of what Stephen Toulmin has famously called 'ideals of natural order' (1961: 47–61). These ideals not only concern what we should expect the world to be like, but determine the aim of science and what counts as an explanation. In Aristotelian physics, for instance, the ideals of natural order centred on the four causes, whereas in the mechanical philosophy, it involved explaining phenomena in terms of the interaction of very small bits of matter (atomism). On the other hand, the methods of explanation characteristic of contemporary evolutionary biology—such as those involving natural selection—are different from the Pythagorean sorts of explanation characteristic of fundamental physics that emphasize elegant mathematical structures. This view of science, therefore, suggests a methodological and epistemological pluralism, with one significant factor that brings unity to the various disciplines being nature's intelligibility (and 'discoverability'—see below). It could thus be thought of as combining two key elements of modernist and post-modernist insights with regard to science.[9]

NRI and Other Alternatives to Reductionism

To help better understand NRI, it is useful to consider how it fits in with other alternatives to reductionism, such as emergent complexity and Alfred North Whitehead's process metaphysics. It agrees with them in its stress on the rich, interconnected nature of physical reality and its intelligibility. In contrast to them, however, NRI does not postulate any particular ontology of the physical world to undergird this intelligibility. Rather, it allows each science to provide its own window into what makes the world intelligible, without trying to fit them into a single ontology. To illustrate, consider some specific areas of contrast between the approach of NRI and the programme of emergent complexity. One of the key theses of emergent complexity is that 'the world appears to be hierarchically structured: more complex units are formed out of more simple parts, and they in turn become the "parts" out of which yet more complex entities are formed' (Clayton 2004: 60). As articulated above, the holism of QM (along with the measurement problem) seriously calls into question the very idea that wholes have well-defined parts of which they are composed, at least at the level of microphysics or chemistry.

As mentioned above, according to NRI, each science provides a window into the order of a certain aspect of nature, where the way in which the order uncovered by each science relates to the other sciences is determined on a case-by-case basis. NRI

---

[9] Theologian Alister McGrath (2004a: 139–52), following Roy Bhaskar, endorses a similar approach to NRI, labelling it 'Critical Realism'.

does not postulate any overarching metaphysical ontology—such as the hierarchies of emergence—that tells us how one science relates to another. The window is in turn determined by the multifaceted and complex history of models, heuristics, theories, and experimental practices characteristic of the science in question. The domains of the sciences, however, are arranged in a non-ontological *hierarchy of generality*, with physics being the most general, then chemistry, then biology, and the like. Because the postulated entities and processes in a less general domain fall under more general domains, this means that the laws and principles of a more general domain can often provide a partial explanation for the laws and principles of a less general domain— e.g. physics provides a partial explanation of chemical laws, and chemistry of biological laws, but not vice versa. We must keep in mind, however, that these explanations are relative to the models of each science, with the models of each science providing insight into reality without necessarily directly representing reality, as discussed more below. From the perspective of NRI, the problem with emergent complexity is that it mistakes a hierarchy of generality for an ontological hierarchy.

Further, the form of NRI that I advocate disagrees with emergent complexity in its claim that new properties emerge at higher levels of complexity, at least if complexity is understood as a purely objective feature of the world that can be specified independently of human interests. One problem is that from the perspective of the field theories of current fundamental physics, a non-denumerably infinite amount of information is required to specify any configuration of matter, no matter how small, since the values of the various physical fields would have to be specified at a non-denumerably infinite number of spatial (or spatio-temporal) points. Thus, from an information-theoretic perspective, a field is infinitely complex. Even though all systems are therefore infinitely complex, we consider some systems more complex than others because in the highly complex systems only a very small proportion of possible arrangement of parts results in a property or function that we can easily recognize or find of interest, such as the arrangement of parts in a radio or a living cell. The degree of complexity, therefore, does not seem to be a completely objective feature of the world, contrary to what emergent complexity presupposes.

Scientific Realism and NRI

Now QM implies that the intelligibility that nature exhibits does not require that the constructs in our models directly correspond, even approximately, to physical reality. Many have responded to this fact by adopting some form of instrumentalism, in which the formalism of QM and its heuristic models are seen as merely useful calculational devices, without offering any significant insight into the nature of reality. In philosophy of science, one leading objection to instrumentalism is the so-called 'no-miracles argument'. According to this argument, if the entities postulated by microphysics do not really exist (or if the formalism of microphysics does not correspond even approximately to underlying reality), then it is a 'miracle' that physics has been so successful, in terms of both novel predictions and guiding the development of technology. Although there is something right about this argument against instrumentalism, it does not support realism in the typical sense of our models corresponding approximately to reality, since the formalism and heuristic models used by physicists are very successful; yet they cannot be taken to correspond directly to reality, as argued above—e.g. when we considered the case of delayed-choice experiments.

Nonetheless, one can still contend that our models offer fundamental insights into the nature of reality over and above simply being useful instruments of prediction. Each area of science, with its own rich history of instrumentation, heuristic constructs, metaphors, and theories, can be thought of as providing a window into the order of one aspect of the physical world. The sort of realism that stresses insight (and metaphor), instead of some sort of semantically precise correspondence, is the kind of realism that many leading defenders of scientific realism, such as Ernan McMullin, claim is more true to the actual practice of science (1984: 30–6, esp. 36). This form of realism helps to reinstate the truth-indicating value of non-literal language (such as metaphor and symbol), which appears to be essential to much religious discourse, but with the rise of science and its accompanying reductionism has often been considered to have at best a secondary status as far as revealing the true nature of the world. Accordingly, many religious believers should find this sort of realism congenial, since they must adopt this form of realism in the realm of religion insofar as they believe that religious discourse is revelatory of reality while at the same time to a large extent non-literal.

Theistic Non-Reductive Intelligibility

This form of realism, however, still does not offer an overall explanation of significant facets of the success of scientific methodology, and thus does not adequately satisfy the intuitions underlying the no-miracles argument. A key intuition here is that the success of the scientific enterprise is something that calls for explanation. An important ingredient in the success of science is the 'user-friendliness' of the structure of the universe for gaining insight into that structure, something I call its 'discoverability'. Realism alone, even of the non-literal variety, does not account for this. As I explain below, discoverability takes multiple forms: the degree of success of recognizably false heuristic constructs, of physicist's intuitions, of purely formal mathematical manipulations in developing new theories (as e.g. famously noted by Eugene Wigner (1960)), and of the criterion of mathematical beauty and elegance in forming fundamental physical theories. As Mark Steiner concludes in his book on the topic, the universe appears to be more 'user-friendly' than one would expect under metaphysical naturalism (1998: 176). Theism naturally accounts for each of these ways in which scientific methodology is successful: it makes sense under a theistic perspective for God to have a providential purpose for human beings of coming to partially understand the natural world and develop technology, thus accounting for nature's discoverability; and given that, as theists have typically held, God has a perfect (or at least a significant) aesthetic sense, one would expect creation to manifest beauty and elegance at a fundamental level.[10] This means that a theistic version of NRI (TNRI) helps satisfy the strong intuitions of those who subscribe to the no-miracles argument in a way that realism, as typically construed, cannot, and thus should be a natural step for many of those with realist inclinations.

What are some examples of this discoverability? One example mentioned above is the beauty and elegance of the laws of nature. This has contributed enormously to the development of physics—going all the way back to Newton—as many physicists have commented on. Nobel Prize-winning physicist Steven Weinberg, for instance, devotes a whole chapter of his book *Dreams of a Final Theory* to explaining how the criteria of beauty and elegance are commonly used with great success to guide physicists in formulating laws. For example, as Weinberg points out, 'mathematical structures that confessedly are developed by mathematicians because they seek a sort of beauty are often found later to be extraordinarily valuable by the physicist' (1992: 153). Later, Weinberg comments that 'Physicists generally find the ability of mathematicians to anticipate the mathematics needed in the theories of physics quite uncanny' (1992: 157).

Another example is the quantization procedure discussed above, which is significantly discussed by Steiner (1998: 96–7, 136–75). As mentioned above, the quantization procedure involves constructing false classical models for a physical system, and then substituting quantum operators for the classical variables. Because the mathematical relations between quantum operators are entirely different from those between classical variables, the classical models cannot be thought of as even approximately correct. That this procedure works at all seems like a 'miracle'. As Roger Penrose notes, 'This procedure looks like hocus-pocus! But, it is not just mathematical conjuring! It is genuine magic which works' (1989: 288). Many similar examples are discussed by Steiner (1998).

The idea of discoverability also provides a theistic explanation for why CR and its accompanying mechanistic view of the universe have been so successful, even though the inherent holism and non-locality of QM imply that they are ultimately false. In order to create a world whose underlying order is discoverable, God would have to create a world that is approximately separable, and hence in which CR would be very successful. The reason is that the ability to break a system into parts that can be separated from the rest of the environment to a high degree of accuracy allows us to study the properties of a system in idealized conditions, without having to consider the extremely messy and unknown influences of the environment or other extraneous factors. This allows for 'controlled experiments'. A theistic explanation of the discoverability of the universe, of course, assumes that God would have good reason for wanting human beings to discover the underlying structure of the universe.

---

[10] In saying that God has a perfect aesthetic sense, I am not attempting to endorse any sort of theodicy, such as a Leibnizian theodicy in which the existence of suffering—such as occurs in the evolutionary process—contributes to the overall aesthetic perfection of the world.

TNRI's Implications for Scientific Methodology and Epistemology

The sort of intelligibility that QM and the idea of discoverability suggest, however, not only lends support to theism, but the belief that God created the universe can positively contribute to our understanding and elaboration of this sort of intelligibility, in analogy to theism's much discussed historical role in the rise of science. One way it could do this is by strengthening the case for the ideal of beauty, elegance, and discoverability as a replacement for mere simplicity as an ideal of natural order, something already suggested by physicists' extensive use of the criteria of beauty and elegance. From a theistic perspective, God would have a reason to create a universe that exhibited elegant and beautiful fundamental structures, as argued above. Simplicity, however, does not seem to have any intrinsic value, at least not for an infinite, omniscient being. But simplicity does have value insofar as it is part of discoverability, elegance, and beauty. The simplicity of the equations of physics at each stage of the development of physics—such as Newton's equation of gravity and Einstein's equation of general relativity—have enormously contributed to human beings having discovered them. Further, simplicity is an essential part of the classical conception of beauty and elegance. Simplicity with variety was the defining feature of the classical conception of beauty or elegance as articulated by William Hogarth in his 1753 classic *The Analysis of Beauty*, where he famously used a line drawn around a cone to illustrate this notion. According to Hogarth, simplicity apart from variety, such as a straight line, is boring, not elegant or beautiful. Thus, I suggest, it is because simplicity often contributes to the beauty and discoverability of nature that it has mistakenly been taken to be the premier virtue of a theory.

Theism suggests not only that beauty (instead of mere simplicity) is the appropriate criterion to apply, but also that in many domains nature will exhibit more than the sparse sort of beauty (akin to Greek architecture) that Weinberg and others claim is characteristic of the mathematical structures of fundamental physics (Weinberg 1992: 149). Under a conception of God as infinitely creative and having a perfect and deep aesthetic sense, for example, it would make sense for the fabric of creation to be richly interconnected and interwoven, in clever, deep, subtle, and elegant ways, expressive of many different types of beauty from the sparse classical sort to the more 'post-modern', with its wild extravagance, as characteristic of the evolution of life on the Earth.

Among other things, such a rich view of nature holds out the hope of providing the needed room and subtlety in nature for grounding a truly sacramental view of nature, along with more adequate accounts of divine action and non-reductive accounts of the mind–body relationship. To illustrate, several philosophers have argued that non-reductive accounts of the mental would involve an enormous complexity of laws linking mental states (such as sensations and experience) with the brain. The leading materialist philosopher J. J. C. Smart has taken this as a powerful argument for reductionism (1970: 54), whereas philosopher Robert Adams

(1987) takes this as a strong reason to appeal to God, *instead* of science, to account for the relation of the mind to the body. If elegance and beauty are taken as fundamental ideals of natural order in place of mere simplicity, both arguments are misguided, since they assume that legitimate scientific explanations must be simple. Rather, we would expect some domains of nature to express those sorts of beauty that involve a high degree of complexity at the fundamental level. Similar things could be said concerning speculation about forefront issues such as biocentric laws, higher-level patterns of teleology in evolution, such as explored by Teilhard de Chardin (1955), Simon Conway Morris (2004), and the like. More generally, science should focus on finding intelligible (and in many cases, elegant) patterns in nature, instead of the ideal being explaining reality in terms of the causal powers of a few basic constituents, though certainly this latter form of explanation is of value in some domains.

One might also expect nature to reflect other attributes of God: God's eternity and infinity, and God's mysteriousness. Accordingly, theists should not be surprised that the universe is very old and vast, and perhaps even infinitely large, as some cosmologists speculate. This is why, as I have emphasized elsewhere, theists should not be opposed to new cosmic speculation, particularly that arising out of inflationary cosmology, in which our universe is one of an incredibly large if not infinite number of universes generated by some physical process.[11] Although a finite, single universe is certainly compatible with God's infinite creativity, an infinitely large universe and/ or multiverse arguably makes even more sense from a theistic perspective. The same goes for the depth of nature: in his more philosophical work, for instance, physicist David Bohm has hinted at the idea that nature should be thought of as like an onion, with perhaps an infinite number of layers of more and more subtle orders of operation (1980: 193). It should also be mentioned that the tremendous advance of science in this last century has uncovered a deep rational structure, as exemplified by QM. Yet, at the same time, it has increased our sense of mystery, by not allowing us to fit reality into any neat conceptual scheme, testified to by the puzzles and paradoxes arising out of QM. This also fits well with theistic religious sensibilities, which hold in tension our ability to rationally comprehend reality (since God is the ultimate creator of our minds) and the deep mysteriousness of reality.[12] Thus, I believe, TNRI has great potential for providing a metaphysically, religiously, and scientifically fruitful framework for thinking about physical reality.

[11] See Collins (forthcoming).

[12] Although probably obvious from the foregoing discussion, it should be stressed that, whatever the merits or lack thereof, of the intelligent design movement (which has gained significant attention in the United States), TNRI should not be confused with it. Unlike advocates of intelligent design, I am not proposing to substitute design explanations for purely naturalistic explanations of physical phenomena, but only claiming that theism can have a significant influence on what we take to be the ideals of natural order, especially in forefront areas such as scientific studies of consciousness.

REFERENCES AND SUGGESTED READING

ADAMS, ROBERT (1987). 'Flavors, Colors, and God', in *The Virtue of Faith and Other Essays in Philosophical Theology*, Oxford: Oxford University Press, 243–62.

ALBERT, DAVID (1992). *Quantum Mechanics and Experience*. Cambridge, Mass.: Harvard University Press.

BOHM, DAVID (1980). *Wholeness and Implicate Order*. London: Routledge & Kegan Paul.

CLAYTON, PHILIP (2004). *Mind and Emergence: From Quantum to Consciousness*. Oxford: Oxford University Press.

COLLINS, ROBIN (1996). 'An Epistemological Critique of Bohmian Mechanics', in J. Cushing, A. Fine, and S. Goldstein (eds.), *Bohmian Mechanics and Quantum Theory: An Appraisal*, Dordrecht: Kluwer Academic Publishers, ch. 18.

—— (forthcoming). 'A Theistic Perspective on the Multiverse Hypothesis', in Bernard Carr (ed.), *Universe or Multiverse?*. Cambridge: Cambridge University Press. (Related articles found at <www.fine-tuning.org> or <www.RobinCollins.org>.)

CONWAY MORRIS, SIMON (2004). *Life's Solution: Inevitable Humans in a Lonely Universe*. Cambridge: Cambridge University Press.

CRITCHLEY, SIMON (2001). *Continental Philosophy: A Very Short Introduction*. Oxford: Oxford University Press.

CUSHING, JAMES T. (1988). 'Foundational Problems in Quantum Field Theory', in H. Brown and R. Harré (eds.), *Philosophical Foundations of Quantum Field Theory*, Oxford: Clarendon Press, 25–39.

DUPRÉ, JOHN (1993). *The Disorder of Things: Metaphysical Foundations of the Disunity of Science*. Cambridge, Mass.: Harvard University Press.

GREENSTEIN, GEORGE, and ZAJONC, ARTHUR (1997). *The Quantum Challenge: Modern Research on the Foundations of Quantum Mechanics*. Boston: Jones and Bartlett Publishers.

HOGARTH, WILLIAM ([1753] 1997). *The Analysis of Beauty*. ed., intro., and notes by Ronald Paulson. New Haven, Conn.: published for the Paul Mellon Center for British Art by Yale University Press.

HUGGETT, NICK, and WEINGARD, ROBERT (1994). 'Interpretations of Quantum Field Theory', *Philosophy of Science*, 61: 370–88.

JAYNES, E. T. (1980). 'Quantum Beats', in A. O. Barut (ed.), *Foundations of Radiation Theory and Quantum Electrodynamics*, New York: Plenum Press, 37–43.

MASCALL, E. L. (1956). *Christian Theology and Natural Science: Some Questions on their Relations*. London: Longman's, Green and Co.

MAUDLIN, TIMOTHY (1998). 'Part and Whole in Quantum Mechanics', in E. Castellani (ed.), *Intepreting Bodies: Classical and Quantum Objections in Modern Physics*, Princeton: Princeton University Press, 46–60.

McGRATH, ALISTER (2004a). *The Science of God: An Introduction to Scientific Theology*. Grand Rapids, Mich.: Eerdmans.

—— (2004b). *The Twilight of Atheism: The Rise and Fall of Disbelief in the Modern World*. New York: Doubleday.

McMULLIN, ERNAN (1984). 'The Case for Scientific Realism', in J. Leplin (ed.), *Scientific Realism*, Los Angeles: University of California Press, 8–40.

MONTON, BRADLEY (2004). 'The Problem of Ontology for Spontaneous Collapse Theories', *Studies in the History and Philosophy of Modern Physics*, 35/3: 407–21.

PENROSE, ROGER (1989). *The Emperor's New Mind: Concerning Computers, Minds, and the Laws of Physics*. New York: Oxford University Press.

POLKINGHORNE, JOHN (2002). *Quantum Theory: A Very Short Introduction*. Oxford: Oxford University Press.

SMART, J. J. C. (1970). 'Sensations and Brain Processes', in C. V. Borst (ed.), *The Mind–Brain Identity Theory*, London: Macmillan Press, 52–66.

STEINER, MARK (1998). *The Applicability of Mathematics as a Philosophical Problem*. Cambridge, Mass.: Harvard University Press.

TEILHARD DE CHARDIN, P. (1955). *The Phenomena of Man*. London: Collins.

TOULMIN, STEPHEN (1961). *Foresight and Understanding: An Enquiry into the Aims of Science*. Bloomington, Ind.: Indiana University Press.

WEINBERG, STEVEN (1992). *Dreams of a Final Theory*. New York: Vintage Books.

WIGNER, EUGENE (1960). 'The Unreasonable Effectiveness of Mathematics in the Natural Sciences', *Communications on Pure and Applied Mathematics*, 13: 1–14.

CHAPTER 21

CONTRIBUTIONS FROM PHILOSOPHICAL THEOLOGY AND METAPHYSICS

JOSEPH A. BRACKEN, SJ

Our generation and the two that preceded it have heard almost nothing but talk of the conflict between faith and science, to the point where, at one moment, it decidedly seemed as though science was called on to replace faith. Now the longer the tension between them continues, the more obvious it is that the conflict is to be resolved by some entirely different form of equilibrium—not by elimination or dualism, but synthesis. After almost two centuries of passionate struggle, neither science nor faith has managed to diminish the other; quite the contrary, it becomes clear that they cannot develop normally without each other. (Teilhard de Chardin 1999: 203)

While Pierre Teilhard de Chardin may have been overly optimistic here in his estimate of how soon the synthesis of religion and science might take place, his deep faith in the achievability of the project remains an important motivational factor in the contemporary religion and science discussion. In this chapter, I will first review briefly how the conflict of interests between proponents of religion and science in the modern era arose historically. Secondly, I will indicate how various contemporary writers in the field of religion and science have tried to ease this tension. Finally, I will offer my own vision for the reconciliation of religion and science, based largely upon the philosophy of Alfred North Whitehead, but suitably revised so as to affirm key Christian beliefs.

How the Tension Arose

As Ian Barbour makes clear in the opening chapter of *Religion and Science*, there was a synthesis between religion and science in the medieval period of Western Europe (Barbour 1997: 4–9). The philosophy of Aristotle provided the conceptual common ground for theologians and philosophers of nature at that time. Aristotle was interested in the explanation of physical reality in terms of intelligible forms or essences and their purpose within an overall world-view (Barbour 1997: 5). Hence, final and formal causality took precedence over efficient and material causality in his metaphysics. This coincided nicely with the reflections of theologians like Thomas Aquinas on the God–world relationship and the efforts of philosophers of nature to determine the workings of divine providence in the world of creation. Thus, despite ecclesiastical condemnations of radical Aristotelianism in 1270 and 1277, the latter's philosophy over time became accepted in late medieval Europe as the necessary philosophical basis for the articulation of a comprehensive Christian world-view embracing both religion and science (Lindberg 2002: 65–71).

By the sixteenth century, however, this synthesis of philosophy, science, and religion was increasingly questioned at least in scientific circles. Besides Galileo, Descartes and Newton figured prominently in a new mathematically based approach to the world of Nature. All three were staunch believers in the existence of God, but each conceived God principally as the unipersonal God of natural theology rather than as the tripersonal God of Christian revelation. In his extensive theological writings, Newton was in fact a fierce opponent of traditional Christian belief in the divinity of Jesus and the doctrine of the Trinity (Westfall 2002: 156–7). Thus all three indirectly paved the way for the emergence first of deism and ultimately of atheism in academic circles.

Perhaps the easiest way to trace the movement from theism to deism to atheism is to review the life history of a celebrated French *philosophe* Denis Diderot. In his early years he was a student at the Jesuit College in Langres, France, and even considered the possibility of becoming a Jesuit. But, after studies at the University of Paris, he converted from theism to deism: namely, to belief in a Creator God who never interferes in the operation of the laws of Nature, once having instituted them. But later he became a militant atheist when it occurred to him that matter is capable of self-generation: 'Matter is no longer the inert, geometric extension of Descartes, nor the Newtonian mass identified with inertia and known through its resistance to change. Now matter is the creative source of all change' (Buckley 1987: 249). Even a Creator God in the minimal sense is not needed as an explanation for the way things are in this world.

There was, of course, intense opposition within educated circles in France and elsewhere to the outspoken atheism of Diderot and other *philosophes*. But the long-term effect of their attack on Christianity was to convince practising scientists that religion and science should be kept separate, since neither one can contribute significantly to the growth of knowledge in the other discipline. Pierre Laplace, for example, the leading Newtonian scientist in post-Revolutionary France, corrected the irregularities in the celestial mechanics of Newton without reference to God, and devised a 'nebular' hypothesis for the origin of the solar system (Numbers 2002: 239). Clearly, this was contrary to the account of creation in the Book of Genesis. But, as Michael Buckley comments, Laplace was thereby no more of an atheist than Descartes, who likewise insisted that the world of Nature is governed by strictly mechanical principles (Buckley 1987: 325). Where he differed from Descartes and Newton was in his assumption that theology has nothing to do with physics. That same assumption appears to be operative in the minds of many scientists (and theologians) even to this day.

Searching for the Causal Joint

In the last few decades of the twentieth century and in the first decade of the twenty-first century, however, there has been a tremendous growth of scholarly interest in rethinking the relationship between religion and science. One of the pioneers in this endeavour was certainly Ian Barbour, with his Gifford Lectures in Scotland in 1989, initially published as *Religion in an Age of Science* (1990) and then reissued in expanded form as *Religion and Science: Historical and Contemporary Issues* (1997). In both texts he sets forth four models for the interrelation of religion and science that have been, and continue to be, operative in the minds of individuals active in the field: conflict, independence, dialogue, and integration. The model of conflict he dismisses as the work of fundamentalists on both sides, individuals unwilling to admit the inevitable limitations of their own discipline. The independence model is far more widespread and influential at the present time, but only because both scientists and theologians are reluctant to give time and energy to boundary issues where an alleged conflict of interest between religion and science needs to be resolved. As the growing number of publications and conferences on science and religion makes clear, the model of dialogue between proponents of religion and science is highly regarded at the present time. But Barbour himself seems to favour the fourth model, that of the integration of religion and science within a new overarching world-view. What he has in mind here is the philosophy of Alfred

North Whitehead, although with some reservations. He is critical, for example, of Whitehead's understanding of the human self as simply a series of moments of experience rapidly succeeding one another, and of the seeming inability of his followers to explain the diversity of different levels of existence and activity within the cosmic process (Barbour 1997: 290; 2002: 97–8). But at the same time he believes that Whitehead offers the most promising philosophical conceptuality for the integration of religion and science to date.

Given the aforementioned number of recent books and articles on religion and science, I will limit myself to a single key issue in the present discussion, and cite representative thinkers for the various positions that have emerged so far. The issue in question is that of divine agency within the world of creation. How can one reasonably affirm divine providence over the creative process without violation of the laws of nature as known to natural science? For, if many natural scientists like Laplace have no need for the 'God hypothesis' in their teaching and research, how can there be a fruitful dialogue between theologians and scientists in which each group has something to contribute to the other? Where, in brief, is the 'causal joint' at which God's activity can plausibly be said to impact on the world of creation? (Clayton 1997: 192).[1]

William Stoeger, Jesuit priest and astrophysicist at the Vatican Observatory in Tucson, Arizona, defends the classical Thomistic distinction between God's primary causality within creation and the secondary causality of creatures, but in more nuanced fashion than his predecessors in the Thomistic tradition. After indicating how the laws of nature are nothing more than human approximations of somewhat hidden regularities operative in the world of creation, he concludes:

God can be conceived as acting through the laws, but the ones through which God is acting principally are not 'our laws,' but rather the underlying relationships and regularities in nature itself, of which 'our laws' are but imperfect and idealized models. And these underlying interrelationships and regularities possess aspects which we are unable to represent adequately—for example, the grounds of possibility, of necessity, and of existence itself. (Stoeger 1996: 230–1)

Thus, while we observe the workings of the laws of Nature only 'from the outside', God experiences and knows them 'from within' in all their relationships, 'including those which determine their possibilities and necessities and grounding in God' (Stoeger 1996: 231). Since we have no proper analogy for primary causality in human experience, however, it still must stand as an exception to any philosophical or scientific explanation of causal relationships within Nature.

In an essay entitled 'The Metaphysics of Divine Action', the Anglican priest and scientist John Polkinghorne explores the possibility of using contemporary chaos theory as an explanation of the causal joint. 'For a chaotic system, its strange attractor represents the envelope of possibility within which its future motion will be contained. The infinitely variable paths of exploration of this strange attractor are not discriminated from each other by differences of energy. They represent different

[1] I prescind here from the work of Michael Behe, William Dembski, and other contemporary proponents of 'intelligent design', since with their theories they seem to be conflating religion and science rather than showing their necessary complementarity (see Haught 2000: 1–10).

patterns of behavior, different unfoldings of temporal development' (Polkinghorne 1995: 153). In conventional chaos theory these different patterns of possibility are brought about by minor disturbances within the environment, but Polkinghorne suggests that this unpredictability within Nature provides an opening for God to communicate 'active information' to the chaotic system without interfering with its normal operation. There is no energy transfer here from God to the world of creation, but only an infusion of further information, a form of 'top-down' causality to complement the 'bottom-up' causality of the system itself (Polkinghorne 1998: 71–2).

Robert Russell and Wesley Wildman doubt, however, that chaos theory can be so readily used to provide the causal joint for the interaction of God and creatures in this world. For, strictly speaking, the mathematics of chaos theory seems to support metaphysical determinism more than metaphysical openness or indeterminism (Wildman and Russell 1995: 84). For this reason, Nancey Murphy, Professor of Philosophy at Fuller Theological Seminary, argues that divine agency is more likely operative at the microscopic or quantum level of existence and activity within Nature than at the macroscopic level of chaotic systems. Her claim is that, while entities at the quantum level 'have their distinguishing characteristics and specific possibilities for acting, it is not possible to predict *exactly when* they will do whatever they do' (Murphy 1995: 341). Since there is no 'sufficient reason' for the entities themselves to determine that choice, Murphy concludes that it is either due to complete randomness or divine determination (Murphy 1995: 341). She opts for divine determination, which carries with it the implication that God is active in every event at the quantum level. But this is not total predestination on God's part, since entities at the quantum level are still free to actualize their innate potentialities in their own way (Murphy 1995: 342–3). Furthermore, as Murphy sees it, God can affect the thoughts, feelings, and actions of human beings through stimulation of neurons in the brain: 'God's action on the nervous system would not be from the outside, of course, but by means of bottom-up causation from within' (Murphy 1995: 349). Finally, in co-operation with the free and intelligent activity of human beings, God can exercise a form of top-down causality on the world of Nature.

Arthur Peacocke, former director of the Ian Ramsey Centre in Oxford, is quite sceptical of efforts to find the causal joint in the interaction of God and the world of creation, above all (as Murphy proposes) at the quantum level through bottom-up causality. Instead, he proposes that God works exclusively in creation from the top down:

The world-as-a-whole, the total world system, may be regarded as 'in God,' though ontologically distinct from God. . . . If God interacts with the 'world' as a whole at a supervenient level of totality, then God, by affecting the state of the world-as-a-whole, could, on the model of whole–part constraint relationships in complex systems, be envisaged as able to exercise constraints upon events in the myriad sub-levels of existence that constitute that 'world' without abrogating the laws and regularities that specifically pertain to them. (Peacocke 1995b: 282–3; also 1993: 157–60)

What Peacocke has principally in mind here is the interplay of the mind, the brain, and the rest of the body as a single unitive event at any given moment within human consciousness. If the mind is a 'unitive, unifying, centered constraint on the activity of our human bodies', then God can be analogously conceived as a 'unifying, unitive source and centered influence on events in the world' (Peacocke 1995*b*: 284–5; see also 1993: 160–3). The analogy is imperfect, since God transcends the world in a way that a human being, even in moments of full consciousness, does not transcend her body. But it at least makes clear that God's interaction with the world is more by way of a 'flow of information' than as a transfer of energy, more a type of formal and final causality than efficient causality. In this way, says Peacocke, God acts persuasively rather than forcefully with creatures, respecting the spontaneity and ontological independence of creatures, above all, at the human level (Peacocke 1995*a*: 138–42).

Philip Clayton, Professor of Philosophy and Religion at the Claremont School of Theology and the Claremont Graduate University in Claremont, California, agrees with Peacocke that the best model for the God–world relationship is panentheism (everything in God but ontologically distinct from God), and that divine agency in the world should be seen as analogous to the mind–body relation within human beings (Clayton 1997: 232–5). There are, of course, inevitable limitations to this approach. According to traditional Christian belief, God 'precedes the world, guides its evolution and continues in existence after its end' (Clayton 1997: 239). The same cannot be said of the relationship of the mind to the body within human beings. Likewise, most Christians believe in life after death in union with the risen Christ; but this too seems impossible if the mind–body relationship is too close. Hence, Clayton's strategy is first to challenge a purely materialistic approach to reality: that is, to establish from a scientific perspective the legitimacy of specifically mental properties (e.g. thinking and willing) over and above the activity of neurons in the brain (Clayton 1997: 247–57). Then, from a philosophical perspective, he argues that both scientists and theologians must have recourse to metaphysical assertions about reality which are not empirically testable in order to present a coherent picture of reality within their own discipline (Clayton 1997: 259–60). Accordingly, theologians can legitimately make certain claims about the God–world relationship (e.g. that God is a personal being who transcends the physical world even though immanent within it at all times, that human beings are made in the image of God, and that God can grant immortality and bodily resurrection to human beings after their death), even though these assertions cannot be verified empirically. For they frame the Christian world-view derived from Scripture and church teaching (Clayton 1997: 261–4).

Locating the Causal Joint?

Many other authors could be cited in connection with this discussion on the causal joint for the interaction of God with the world of creation. But the authors cited above seem to cover the basic options. One can appeal with Stoeger to primary causality as qualitatively different from secondary causality operative in causal relations within this world. But there is no analogue for primary causality within human experience. With Polkinghorne one can appeal to the way in which God could conceivably communicate 'active information' to chaotic systems on the macroscopic level to influence their further development, and with Murphy we can make basically the same argument on the quantum level, given the alleged intrinsic indeterminacy of subatomic particles. But one can counter-argue in both cases that the alleged indeterminacy is only a gap in our human knowledge of the laws of Nature which will someday be remedied by further scientific investigation. With Peacocke and Clayton one can appeal to the way in which top-down causality works within hierarchically ordered natural systems and urge that God is operative within the world in a manner akin to the way in which the mind influences the body (and is affected by the body) within human beings. But there are limits to this model of the God–world relationship from a theological perspective. Hence, while all these options shed light on the issue of the causal joint, none of them seems to offer a fully satisfactory solution to the problem of the interaction between God and the world of creation. None of them, for example, offers a supporting metaphysical conceptuality in which the trans-empirical hypotheses of both religion and science could be grounded and thus rendered more plausible.

Ian Barbour, to be sure, grounds his understanding of the God–world relationship in the metaphysics of Alfred North Whitehead. But, as noted above, he has reservations about certain features of Whitehead's scheme from a strictly philosophical perspective: namely, the ongoing ontological identity of the self in human consciousness and the interplay between higher and lower levels of Whiteheadian societies within the overall order of Nature. In my opinion, Barbour's reservations are justified; only a new way of conceiving Whiteheadian societies as more than simply aggregates of analogously constituted actual occasions (momentary subjects of experience) is needed to make Whitehead's philosophy a plausible choice for mediating between the expectations of traditional Christian theology and contemporary natural science. In the following pages, I will set forth such a revised understanding of Whitehead's category of society, and then show its applicability to both theology and natural science.

At the conclusion of *God and Contemporary Science*, Clayton notes that matter is no longer easy to identify, given Albert Einstein's celebrated mathematical equation $E = mc^2$ and its application to the notion of 'force-fields' within contemporary physics (Clayton 1997: 263). Along the same lines, I believe that Whiteheadian societies should be interpreted as structured fields of activity for their constituent actual occasions and that emphasis should be laid upon the character of actual occasions as psychic energy events rather than mini things. Furthermore, there is a textual basis in Whitehead's thought for such an understanding of the reality of societies. In *Process and Reality* Whitehead refers to background societies for any given set of concrescing (or becoming) actual occasions as 'environments' arranged in 'layers of social order' which directly influence the self-constitution of those same actual occasions (Whitehead 1978: 90). Then a few lines later he notes that 'in a society, the members can only exist by reason of the laws which dominate the society, and the laws only come into being by reason of the analogous characters of the members of the society' (Whitehead 1978: 90–1). These few remarks of Whitehead about the interrelationship of societies and their constituent actual occasions are in my judgement key to a better understanding of how bottom-up and top-down causation can be simultaneously operative within Whiteheadian societies along the lines indicated above by Peacocke and Clayton.

Bottom-up causation is easy to understand, since in Whitehead's own words, 'agency belongs exclusively to actual occasions' (Whitehead 1978: 31). Each actual occasion is active in its own self-constitution. Indirectly, of course, it contributes to the structure of the society (or societies) to which it belongs, because analogously constituted actual occasions find themselves as a result of their individual self-constitution grouped together into societies with 'a common element of form' (Whitehead 1978: 34). But is this sufficient to account for the relatively unchanging character of those same societies? Whitehead claims that within his philosophy 'it is not "substance" which is permanent, but "form"' (Whitehead 1978: 29). But how is the 'common element of form' transmitted from one set of actual occasions to the next so as to guarantee continuity of form? In line with his own commitment to metaphysical atomism (Whitehead 1978: 35), Whitehead asserts that each constituent actual occasion within a given society 'transmutes' the multiple physical feelings which it prehends from its predecessors so as to acquire a single physical feeling of its environment as a nexus or community of past actual entities, and then adapts that unified physical feeling to its own purposes in its self-constitution (Whitehead 1978: 250–4). Presumably all the constituent actual occasions will undergo this process of transmutation in basically the same way and thus retain their ontological identity as a society with a definite pattern of self-organization.

My counter-argument for many years now has been that this is much too complicated and chancy. A much simpler explanation is that the society is a field of activity structured by the interplay of its constituent actual occasions from moment to moment which likewise serves as the principle of continuity for the transmission of form from one set of actual occasions to another. The actual occasions, in other words, do not individually have to transmute the common element of form from their predecessors. Rather, they prehend the common element of form already resident in the field as a result of the activity of those same predecessors. Then by their dynamic interrelation here and now they either transmit the structure of the society unchanged to their successors or adapt it very slightly in line with their own somewhat changed existential situation. In either case, the society as a structured field of activity for its constituent actual occasions is what survives as successive generations of actual occasions come into and go out of existence.

This might initially appear to be simply a metaphysical nuance to be debated among Whiteheadians alone. But, once accepted as plausible, it has unexpected explanatory power for understanding the simultaneous operation of bottom-up and top-down causality within Nature. Likewise, it makes clear how 'supervenience' and 'emergence' are factors in the evolution of more complex natural structures along the lines indicated by Peacocke and Clayton. The key factor in both these instances is that unlike 'substances' in the classical sense, fields can be layered within one another and thus serve as respectively the infrastructure or superstructure of their neighbours. Thus, within a given society the inherited structure of the field acts as top-down causation on its constituent actual occasions here and now through what we might call formal causality. Unlike an Aristotelian form, however, the structure of the field is not active but passive; that is, it is simply available for prehension by the currently concrescing actual occasions which then exercise bottom-up causality on the structure of the field for the next set of actual occasions. Likewise, since Whitehead allows for different levels or grades of complexity for actual occasions (Whitehead 1978: 177–8), one can justify the emergence of higher levels of existence and activity within Nature by proposing that, as the structure of a given society becomes more complex, its constituent actual occasions necessarily become more complex until by their dynamic interrelation at a given point they spontaneously generate a new society or structured field of activity with new higher-order actual occasions to populate the field. In this way, without any outside intervention, through a strictly immanent process, one structured field of activity can supervene upon or be emergent from a predecessor structured field of activity. The lower-level field of activity still acts as an indispensable infrastructure for the functioning of the higher-level field, even as the higher-level field of activity conditions the self-constitution of the actual occasions in the lower-level field(s) of activity (Whitehead 1978: 106).

Likewise, applied to the mind–brain correlation, this notion of a Whiteheadian society as a structured field of activity for its constituent actual occasions yields surprising results. In *Process and Reality* Whitehead distinguishes 'entirely living' actual occasions which constitute the 'regnant nexus' of actual occasions within a more complex 'structured society' from its necessary infrastructure, layers of subordinate sub-societies of non-living actual occasions (Whitehead 1978: 103). On the level of the mind–brain correlation this nexus of entirely living actual occasions corresponds to the mind, and the infrastructure of sub-societies of non-living actual occasions to the brain with its elaborate neuronal network. He then adds that while the nexus of entirely living actual occasions is not a society in the strict sense, it still supports 'a thread of personal order along some historical route of its members' (Whitehead, 1978: 104). I argue that this thread of personal order among entirely living actual occasions is best accounted for in terms of a superordinate field of activity which integrates and co-ordinates the activity of all the sub-fields of activity within the brain.[2]

Given the plausibility of this argument, the problem of an ontological dualism between mind and brain disappears. For, entirely living actual occasions constituting

---

[2] 'Finally, the brain is coordinated so that a peculiar richness of inheritance is enjoyed now by this and now by that part; and thus there is produced the presiding personality at that moment in the body. Owing to the delicate organization of the body, there is a returned influence, an inheritance of character from the presiding occasion and modifying the subsequent occasions through the rest of the body' (Whitehead 1978: 109). Given the movement of the 'presiding personality' from part to part of the brain, and yet its wide-ranging influence on the rest of the body through the brain, a field metaphor easily comes to mind.

the mind or soul represent only a higher grade of actual occasion, not a different kind of entity altogether from the non-living actual occasions constituting the neuronal network of the brain. Likewise the mind or soul is emergent out of the interplay of those same non-living actual occasions organized into sub-fields of activity within the brain. Thus it is both dependent on, and yet independent of, its infrastructure of subordinate sub-societies. It is not a higher-order property or function of the brain, as Clayton and others maintain (Clayton 2004: 128–9), but an entity in its own right, the superordinate field of activity within a Whiteheadian structured society of hierarchically ordered sub-societies.

There is, of course, still another objection which Clayton and others lodge against such a scheme for the mind–brain connection. It presupposes panpsychism: namely, 'that every level of reality possesses some sort of mental experience' (Clayton 2004: 130), a claim that cannot be verified empirically and thus will inevitably be rejected or at least held in suspicion by the scientific community. But is this reductively a failure in imagination on the part of both contemporary natural scientists and philosopher/theologians? As already noted, Einstein stipulated that matter and energy are interchangeable. Yet is energy preferably to be associated with inert particles of matter or with self-constituting subjects of experience exhibiting both particle-like and wave-like properties?[3] In any case, I turn now to the final section of this essay, dealing with various theological implications of my scheme, including my own solution to the problem of the causal joint.

THEOLOGICAL IMPLICATIONS

Belief in God as Trinity, three divine Persons who are collectively still one God, is in Clayton's terms a 'trans-empirical' hypothesis (Clayton 1997: 260; 2004: 179–85). That is, it is well grounded in the Christian Bible, above all in the Gospel of John and the epistles of Saint Paul, and has been a consistent factor in church teaching over the centuries; but it cannot be verified simply through appeal to immediate experience or current scientific research. Yet, as I shall make clear below, it is consistent with a field-oriented approach to Whiteheadian societies. This field-oriented approach to reality, moreover, provides a plausible philosophical explanation for the phenomenon of emergence within cosmology and evolutionary biology. One may reasonably ask, therefore, whether a trinitarian understanding of God within the context of a

---

[3] See e.g. Bracken 1991: 57–73; likewise Barbour 2005 and Athearn 2005. Both authors emphasize that contemporary natural scientists may unconsciously be working with an outdated ontology and thus be indirectly guilty of what Whitehead termed the fallacy of misplaced concreteness (cf. Whitehead 1967b: 51; 1966: 154).

field-oriented approach to physical reality provides the long-sought causal joint for divine–human interaction.

For, as noted earlier, unlike classical 'substances', fields can be layered within one another in such a way that they can serve respectively as the infrastructure and superstructure of one another. Thus if the Trinity can be said to constitute the top-most, all-inclusive field of activity within a field-oriented approach to reality, and if the Trinity thereby both influences and is influenced by the vast network of finite fields of activity contained within it, then the problem of the causal joint is in principle solved. In what follows I sketch simply the broad outline of this proposal and refer the reader to previous publications for further details (Bracken 1995: 52–69; 2001: 94–105).

Using Whitehead's terminology, I maintain that the three divine Persons of the classical doctrine of the Trinity are 'personally ordered societies' of actual occasions (Whitehead 1978: 34–5), each with an infinite field of activity proper to its own function within the Godhead. But, since three non-overlapping infinite fields of activity is a logical impossibility, together they constitute a single all-comprehensive field of activity with three distinct foci or perspectives. According to Whitehead, the principle of creativity whereby 'the many become one and are increased by one' is simply a metaphysical given applied to the self-constitution of actual entities (White-head 1978: 21). I argue that it is the nature or principle of existence and activity for the divine Persons, and that it applies not only to each of them individually as a personally ordered society of divine actual occasions, but also to their coexistence as members of one and the same divine field of activity. The field, in other words, is their collective and objective self-expression, that which endures in virtue of their ongoing dynamic interrelations as immaterial subjects of experience.

Furthermore, I suggest that the world of creation as a very large but still finite network of societies of finite actual occasions comes into existence and is sustained at every moment within this divine matrix, or all-comprehensive field of activity. All these sub-societies of actual occasions or subordinate fields of activity contribute their structure or 'common element of form' to the structure of the divine matrix, akin to the way in which Whitehead in *Process and Reality* describes how all the events taking place in this world at every moment are integrated within the 'conse-quent nature' of God (Whitehead 1978: 349–51). Likewise, akin to Whitehead's scheme, I argue that in virtue of their 'primordial nature' or unlimited vision of possibilities for the future, the three divine Persons offer a 'lure' or an 'initial aim' to every concrescing actual occasion within this vast network of finite sub-societies constituting the world of creation (Whitehead 1978: 244). But, since creativity is in the first place their own divine nature, along with the initial aim they communicate to the concrescing actual occasion, a finite share of their own creativity which empowers it to become itself in line with its own self-constituting 'decision'. In this way, the three divine Persons are involved in the creation or self-constitution of the individual actual occasion, but they do not determine how it will happen. Their activity with respect to their creatures is more by way of final causality than efficient causality (as in scholastic metaphysics).

Another tenet of classical Christian belief is that human beings after their death will enjoy eternal life in union with Christ as the risen Lord. As I see it, this too is possible within the God–world relationship presented here, provided that one accepts the implications of the metaphysics of intersubjectivity implicit therein. Whitehead stipulated that 'the final real things of which the world is made up' are actual occasions or momentary subjects of experience (Whitehead 1978: 18). But he could not account for intersubjectivity within his metaphysical scheme, since an actual occasion cannot prehend its predecessors except as 'superjects', devoid of subjectivity or the power of self-constitution (Whitehead 1978: 25–8). Even God prehends finite actual occasions objectively as superjects, not as coexisting subjects of experience. Some years ago Marjorie Suchocki sought to remedy this lacuna in Whitehead's metaphysics, and thereby to legitimate the possibility not simply of objective immortality but also of subjective immortality for finite actual occasions within the divine consequent nature. She proposed that God prehends finite actual occasions in the moment of 'enjoyment' when the actual occasion is subject and superject at the same time (Suchocki 1988: 81–96). Hence, God can integrate them as both subject and superject into the divine consequent nature, and thereby share with them the divine life on an intersubjective basis.

Though ingenious, Suchocki's proposal, in my judgement, still does not meet the classical expectations of Christians about eternal life and resurrection of the body. One must stretch Whitehead's scheme even further to make it suitable for that task. One should begin by stating that intersubjectivity exists not between individual actual occasions or momentary subjects of experience, but between the societies of which they are constituents in terms of their mutually overlapping fields of activity. Whiteheadian societies, to be sure, are agents or subjects of experience not in their own right, but only in virtue of their constituent actual occasions. Yet every society has a subjective focus in terms of a regnant actual occasion or set of co-ordinate actual occasions. Thus, when two or more societies of actual occasions merge to form a common field of activity, intersubjectivity becomes a reality in virtue of that shared field of activity.

Applied to Christian belief in eternal life and the resurrection of the body, this means that the three divine Persons and all their creatures share a common field of activity (in the language of the Bible, the Kingdom of God) which is structured by the decisions of the divine Persons and all their creatures from moment to moment. But, as creaturely societies of actual occasions cease to exist in this world, their final constituent actual occasions are incorporated into a still higher-order intersubjective relationship with the divine Persons than they enjoyed while in this life. I say 'final constituent actual occasions' since all that is needed for a created society of actual occasions to enjoy subjective as well as objective immortality within the field of activity proper to the divine Persons is a subjective focus in terms of a final actual occasion or set of actual occasions. Thus, not individual actual occasions as in Suchocki's proposal (Suchocki 1988: 107–12), but societies of actual occasions, the persons and things of ordinary experience, achieve eternal life in union with the divine Persons.

I say 'persons and things' also quite deliberately, since societies of actual occasions redeemed by divine grace include in the first place the societies of actual occasions constitutive of our minds or souls. But they also include the sub-societies of actual occasions constitutive of our bodies as well as our minds, so that we may enjoy eternal life as a psychosomatic unity. Yet, if our human bodies can be thus redeemed and transformed, then the whole of material creation should likewise somehow experience 'resurrection' within the divine life. Our bodies, after all, are composed of the same basic elements as the rest of the universe. In brief, then, within the divine field of activity nothing that ever existed is lost; everything survives even though the manner of survival will differ notably from creature to creature. Finally, as I see it, 'resurrection' does not have to wait until the projected end of the world. It happens at every moment as societies of finite actual occasions in one way or another cease to exist. One can thus introduce into Whitehead's metaphysical scheme something like Teilhard de Chardin's vision of the Cosmic Christ as the Omega Point or cumulation of the evolutionary process, but present it as it is achieved at every moment rather than only at the end of the world.

Further elaboration of this process-oriented eschatology is presented elsewhere (Bracken 2005). Sufficient for this essay is the basic exposition of a metaphysical scheme which seems apt for mediating between the claims of contemporary natural science and those of Christian theology. That is, it offers a rationale for the emergence of higher-order forms of life from lower-order levels of organization within Nature even as it provides a plausible philosophical argument for the retention of cherished Christian beliefs. As such, it offers a 'common language' which has been missing in discussions between scientists and theologians since the beginning of the modern era. As a system of metaphysics, of course, it is a trans-empirical hypothesis which cannot be fully verified either in terms of common-sense experience or in virtue of contemporary scientific research. But, as Whitehead commented in *Adventures of Ideas*, what is important about a proposition or set of ideas is not in the first place that it be true, but that it be interesting, provocative of further reflection (Whitehead 1967*a*: 244). This chapter was written with this goal in mind.

References and Suggested Reading

ATHEARN, D. (2005). 'Toward an Ontological Explanation of Light', *Process Studies*, 34: 45–61.
BARBOUR, I. G. (1990). *Religion in an Age of Science*. San Francisco: Harper & Row.
—— (1997). *Religion and Science: Historical and Contemporary Issues*. San Francisco: Harper-Collins.
—— (2002). *Nature, Human Nature, and God*. Minneapolis: Fortress Press.
—— (2005). 'Evolution and Process Thought', *Theology and Science*, 3: 161–78.
BRACKEN, J., SJ (1991). *Society and Spirit: A Trinitarian Cosmology*. Toronto: Associated University Presses.
—— (1995). *The Divine Matrix: Creativity as Link between East and West*. Maryknoll, N.Y.: Orbis Books.

BRACKEN, J., SJ (2001). *The One in the Many: A Contemporary Reconstruction of the God–World Relationship*. Grand Rapids, Mich.: Eerdmans.

—— (2005). 'Subjective Immortality in a Neo-Whiteheadian Context', in J. Bracken, SJ (ed.), *World without End: Christian Eschatology from a Process Perspective*, Grand Rapids, Mich.: Eerdmans, 72–90.

BUCKLEY, M., SJ (1987). *At the Origins of Modern Atheism*. New Haven: Yale University Press.

CLAYTON, P. (1997). *God and Contemporary Science*. Grand Rapids, Mich.: Eerdmans.

—— (2004). *Mind and Emergence: From Quantum to Consciousness*. Oxford: Oxford University Press.

HAUGHT, J. (2000). *God after Darwin: A Theology of Evolution*. Boulder, Colo.: Westview Press.

LINDBERG, D. (2002). 'Medieval Science and Religion', in G. Ferngren (ed.), *Science and Religion: A Historical Introduction*, Baltimore: Johns Hopkins University Press, 57–72.

MURPHY, N. (1995). 'Divine Action in the Natural Order: Buridan's Ass and Schroedinger's Cat', in Russell *et al.* (1995), 325–57.

NUMBERS, R. (2002). 'Cosmogonies', in G. Ferngren (ed.), *Science and Religion: A Historical Introduction*, Baltimore: Johns Hopkins University Press, 234–44.

PEACOCKE, A. (1993). *Theology for a Scientific Age: Being and Becoming—Natural, Divine, and Human*. Minneapolis: Fortress Press.

—— (1995a). 'Chance and Law in Irreversible Thermodynamics, Theoretical Biology, and Theology', in Russell *et al.* (1995), 123–43.

—— (1995b). 'God's Interaction with the World: The Implications of Deterministic "Chaos" and of Interconnected and Interdependent Complexity', in Russell *et al.* (1995), 263–87.

POLKINGHORNE, J. (1995). 'The Metaphysics of Divine Action', in Russell *et al.* (1995), 147–56.

—— (1998). *Belief in God in an Age of Science*. New Haven: Yale University Press.

RUSSELL, R., MURPHY N., and PEACOCKE A. (1995) (eds.). *Chaos and Complexity: Scientific Perspectives on Divine Action*, Notre Dame, Ind.: University of Notre Dame Press.

STOEGER, W., SJ (1996). 'Contemporary Physics and the Ontological Status of the Laws of Nature', in R. Russell, N. Murphy, and C. Isham (eds.), *Quantum Cosmology and the Laws of Nature: Scientific Perspectives on Divine Action*, 2nd edn., Notre Dame, Ind.: University of Notre Dame Press, 207–31.

SUCHOCKI, M. (1988). *The End of Evil: Process Eschatology in Historical Context*. Albany, N.Y.: State University of New York Press.

TEILHARD DE CHARDIN, P. (1999). *The Human Phenomenon*, trans. Sarah Appleton-Weber. Brighton: Sussex Academic Press.

WESTFALL, R. (2002). 'Isaac Newton', in G. Ferngren (ed.), *Science and Religion: A Historical Introduction*, Baltimore: John Hopkins University Press, 153–62.

WHITEHEAD, A. H. (1966). *Modes of Thought*. New York: Free Press.

—— (1967a). *Adventures of Ideas*. New York: Free Press.

—— (1967b). *Science and the Modern World*. New York: Free Press.

—— (1978). *Process and Reality*, corrected edn., ed. D. Griffin and D. Sherburne. New York: Free Press.

WILDMAN, W., and RUSSELL, R. (1995). 'Chaos: A Mathematical Introduction with Philosophical Reflections', in Russell *et al.* (1995), 49–90.

CONTRIBUTIONS FROM SYSTEMATIC THEOLOGY

WOLFHART PANNENBERG

THE NEED FOR DIALOGUE

Why should Christian theologians get involved in a dialogue with natural science? Personal reasons aside, they should take an interest in science because they have to account for the world of nature, including human beings, in the context of their existence as God's creation. When Christians confess God as the Creator of the world, it is inevitably the same world that is also the object of scientific descriptions, although the language may be quite different. Therefore, theologians must be concerned with the question of how theological assertions about the world, and about human beings as God's creation, can be related to their descriptions by scientists. After all, there is only one world, and this one world is claimed as God's creation in the Bible and in the faith of the Church.

Now there are obvious differences between the way the Bible, especially in its first chapter, presents the world as God's creation and the account of the reality of the universe, of the stars and the Earth, of vegetation, animals, and the human being, in modern science. In order to do justice to those differences, it is important to be aware of the fact that to a large extent they are due to historical differences between the knowledge about natural forms and processes available in the sixth century before Christ (and presupposed by the biblical authors) and that of modern science. Thus there is not simply a contrast between revelation and empirical knowledge, but between a now obsolete form of empirical knowledge about the world that was once used to articulate in detail the belief in God's creation of the whole world and a modern form of empirical knowledge that might be looked upon in light of the question of whether it could serve a similar function in the modern situation. In asking such a question, one should be aware of the fact that the biblical writers did not simply take over the empirical knowledge of their time, but also adapted it for theological purposes. Is such a thing conceivable and also legitimate with regard to modern science? Certainly the facts cannot be changed at will, but the perspectives on their interpretation can be more or less different. It is desirable, in a dialogue between religion and science, to reach agreement with scientists on issues of this sort.

When scientists talk to the general public about their methods and their results, they use a language different from that of their science. They do not write equations on the blackboard, but rather engage in a sort of philosophical reflection upon what they do as scientists. Similarly, it is on this level of philosophical reflection that the dialogue between theology and science is taking place. Such philosophical reflection, of course, requires appropriate awareness of the methods and results of science in the proper sense. One task of the scientists involved in such a dialogue is to make sure that proper awareness of the methods and results of science is indeed present on the part of the theologians participating in the dialogue. On their side, the appropriate awareness concerning the task of theology and regarding to particular problems connected with the traditional theological language about God and creation is to be safeguarded.

In addition, however, familiarity with the tradition of philosophical thought about nature is also required, especially in connection with the philosophical origins of scientific language itself. Key scientific terms such as 'movement', 'energy', 'atom', 'time', 'space', but also 'law', 'field', 'contingency', and others have had a philosophical prehistory that has influenced and to some extent still influences their 'formalized' use in science, and especially the reflective awareness of their meaning and impact. Attention to the philosophical roots of scientific terminology can also help dialogue partners to notice connections with theological issues and concerns. Therefore, it should be considered a requirement of successful dialogue between theology and science to take into consideration the philosophical framework of the history of science.

CONTINGENCY AND LAW

If general agreement on the conditions of such dialogue is obtained, it is necessary to select for closer study a set of issues that are relevant to science as well as to the theological understanding of the world as creation. In the 1960s a small group of German physicists, philosophers, and theologians selected for such a closer study the concept of natural law, on the one hand, and that of contingency, on the other.[1]

---

[1] See Müller and Pannenberg (1970). My contribution to this volume was published in English translation in Pannenberg (1993: 72–122).

Among the reasons for this selection was the consideration that the idea that contingent events express the freedom of God the Creator and Lord of history is central to theology in a fashion similar to the way that the concept of natural law is to science.

In addition, there are important relationships between contingency and natural law. The application of each natural law presupposes initial conditions and marginal conditions that are contingent relative to the regularity expressed by that particular law. This element of contingency belongs to the logic of the concept of natural law itself and of its application. Contingency in this sense has been aptly called nomo-logical contingency (Russell 1988), in distinction to contingency as an index of real events. The occurrence of the initial and marginal conditions in the applications of a particular law may, of course, be explained by another law, so that the appearance of their contingency seems to be dissolved. But this other law again presupposes contingent conditions for its applications. Thus the problem of contingency—of the contingent occurrence of data for the application of natural law—cannot be discarded in principle.

This seems to suggest that contingency of events may be the basic reality of nature. It is the contingent sequence of such events within which regularities occur and can be observed, such as in the regularities of sequence that can be described by formulas of natural law. Such a conception of reality has the advantage of overcoming the opposition between contingency and determination by law, because a contingent sequence of events does not exclude the possibility of regularities occurring in that sequence, though the regularities that can be described in terms of natural law represent only one aspect of the sequence of events.

Such a view presupposes, of course, that contingency is indeed a basic feature of physical reality, no less than the reality of law is, and is even constitutive of the occurrence of laws themselves, since they are perceived as based on regularities in the sequence of events. It is understandable that scientists at first feel uneasy at this point, since the endeavour of science seems to aim at the reduction of apparent contingency to the operation of laws in the course of events. Therefore, many scientists hesitated to accept the assumption of irremovable contingency in the reality of nature. At the time of the conversations in the Sixties, the most prominent example of such irremovable contingency was provided by quantum physics. Although quantum field theory had restored the possibility of a deterministic description of quantum processes with regard to great numbers of events, in the form of statistical laws, the individual quantum events remained in a state of unpredictability. Notwithstanding Albert Einstein's famous dictum that God does not play dice, on the subatomic level contingency seems irremovable. But later on, the investigation of non-linear thermo-dynamic processes, and especially the work of Ilya Prigogine and Isabelle Stengers, demonstrated that even in macrophysical processes—if they take place far from thermodynamic equilibrium—contingent events occur that are not only unpredict-able and irreversible, but also have the potential of changing the direction of evolutionary processes at 'bifurcation points' (Prigogine 1980 and Prigogine and Stengers 1986). Processes like these have been called 'chaotic'. But such a 'chaos' is not completely irregular. It was certainly correct to speak of a 'deterministic chaos', since even here the concept of natural law does not become inapplicable (Bonting 2002; Ganoczy 1995). Its application is quite obviously reduced, however, to the description of regularities that can be observed in sequences of contingent events.

The idea of natural law as relating to the regularities observed in the sequence of contingent events should not be considered to be opposed to contingency, because the application of natural law itself presupposes that sequences of (contingent) events happen, since otherwise laws could not be observed either. In theology, on the other hand, the search for natural law should not be looked upon as reducing the possibility of speaking of God's action in contingent events of nature and history. Rather, the occurrence of regularities in the course of events is itself a contingent fact. It is generally considered in the Bible as a special disposition by God the Creator, a beneficial arrangement for human beings, because it renders the world they inhabit reliable for them (Genesis 8: 22). In fact, the operation of laws in the course of events is a precondition for every other form of stability in the world of creation, especially for the emergence of all enduring forms of created existence. Therefore, human beings owe gratitude to the Creator for the creation of regularities that obey natural laws. Nor does a proper concept of God's miraculous working conflict with the operation of laws in the course of nature. Augustine defined the concept of miracle in relation to our human subjective experience as an exceptional occurrence, but not as a violation of natural law. Such a violation or break of law would abolish the very concept of natural law, since such a law by definition does not suffer exceptions. Otherwise there would not be a law. On the other hand, no single law and no system of law is exhaustive in governing the course of events. Thus there is always room for the unexpected (Pannenberg 2002).

In the dialogue of Christian theologians with the natural sciences, and also in the Christian doctrine of creation, the concept of contingency deserves a place of basic importance. This was strongly emphasized, a few years after my own article on 'Contingency and Natural Law' (in Müller and Pannenberg 1970) which was written for the above-mentioned discussion group in Germany, by the British pioneer of a new theological reappropriation of natural science, Thomas F. Torrance, in his important book *Divine and Contingent Order* (1981). Torrance argues, in the line of medieval Christian thought and especially of Duns Scotus, that the Christian doctrine of creation basically affirms the contingent existence of the world in distinction from God. Contingency characterizes the existence of the world as a whole, as well as of each of its parts. This contingency of the creatures is the correlate and expression of the freedom of God in his activity as Creator. The concept of contingency had its origin in the philosophy of Aristotle, where it designated an aspect of formless matter, viz. its indeterminacy and mere possibility. But in Christian medieval thought it came to be related to the concept of divine freedom, and especially to the exercise of this freedom toward creatures, and among them specifically to the concept and exercise of human freedom. It is for this reason that the idea of contingency is so extremely important in a Christian doctrine of creation.

HISTORY AND CONTINUOUS CREATION

Contingency is also important for an appropriate account of the biblical concept of history. Histories were conceived as sequences of contingent events. These events could be interpreted in terms of divine as well as human actions. Their contingency was bound up with the element of novelty. The historical process as a whole was also taken to be contingent, particularly in its outcome. Therefore the meaning of each event in the sequence depends on its final outcome, its eschatological future.

It was an important (though indirect) contribution to the dialogue between science and theology that in 1948 the German physicist (and, later on, philosopher) Carl Friedrich von Weizsäcker presented the big bang cosmology under the title of a 'history of nature'. The unity of nature and of the universe was no longer conceived as a timeless order of repetitive processes, but as a unique 'history' beginning with the big bang and moving through the production of increased complexity in natural forms all the way to the emergence of the human race. Weizsäcker emphasized the irreversibility of this temporal process based on the second law of thermodynamics, which he called the 'principle of the historicity of nature', because it opens up the prospect of a process of irreversible changes (Weizsäcker 1948). The reason is that in this process entropy is always increasing in general, though in particular cases novelty may occur (i.e. a decrease of entropy), but only at the price of even greater increases in entropy in the surroundings. This also became the point of departure for the later theses of Prigogine and others on the rise of novelty and its lasting effects in the course of evolution.

There was no difficulty in principle of integrating the evolution of life with the concept of a history of nature. The fight about the Darwinian doctrine of evolution was one of the most unnecessary controversies between Christian theology and natural science in the course of their entire history. It is perhaps understandable that the thesis of evolution through natural selection was considered in the late nineteenth century by friends and foes as the ultimate triumph of a mechanistic description of nature, though in truth it should actually be considered (as with G. Altner) as a breakthrough to a new, historical view of nature (Altner 1986). But the lasting resistance to the evolutionary doctrine was not only the result of a literalist reading of the Bible (in spite of Genesis 1: 11, 24) and of the belief in the constancy of natural species after their first creation. Even more, it was due to the observation that if the theory of natural selection were correct, there must apparently be a breakdown of William Paley's argument from design. The argument from design had, however, acquired a quite inordinate position of importance in the discussion of theology and science in the nineteenth century. Considered from a biblical perspective, it is at best secondary in importance. Much more important in the dialogue between theology and science is the issue of contingency—both in the broad sense of the contingent emergence and existence of everything created and in the more special function of contingency as a source of novelty.

This significance of contingency, however, was taken into account in the later development of evolutionary theory in the twentieth century in terms of emergent evolution and organic evolution. If the evolution of life is looked upon as a process of emergence of ever-new forms, there is no reason for theologians any longer to reject the theory of evolution. Even in the early history of the debate on Darwin's theory, the contributors to the book *Lux Mundi*, edited by Charles Gore in 1899, rightly acclaimed the new theory as overcoming a mechanistic view of nature that left no room for God except as the initial author of the natural order, but not as acting creatively in the further process of nature. The circle around Charles Gore found it possible to integrate the Darwinian theory of evolution in a Christian concept of salvation history. Thus the Incarnation could be interpreted as the culmination of the evolution of life. This approach was continued by some of the most important British theologians, including William Temple and Charles Raven and, more recently, Arthur Peacocke (1979). A similar conception was developed in the famous and influential work of Teilhard de Chardin.

The theological integration of the concept of organic evolution required a revision of the theological concept of creation itself. In the theological tradition, the concept of creation was connected primarily with the beginning of the world, when, according to the first chapter of the Bible, the foundation of the order of nature and of the archetypes of creatures was laid. This conception of creation, however, did not easily admit the emergence of something significantly new in later phases of the world's history. Therefore, in the dispute over Darwinism, the emphasis on the constancy of natural species would inevitably clash with the new ideas on evolution. That idea, in turn, required a vision of a process of continual emergence, through which new and sometimes significantly novel organisms arose.

Evolution fits less easily with an exclusive emphasis on the beginning; it harmonizes more easily with the idea of a Creator who is continuously and creatively active, as expressed, for example, in Psalm 104 and in various prophetic passages. It was therefore with good reason that the idea of a 'continuing creation' has been increasingly emphasized by those involved in the dialogue between theology and science, especially after the dispute over Darwinism.[2] But the emphasis on continuous creation does not of course exclude the importance of a first beginning of everything. Nor is it opposed to the notion of 'creation from nothing', since that notion simply says that the creature did not exist at all before it was created (cf. Romans 4: 17). The notion of *creatio ex nihilo*, therefore, does not only apply to the first beginning of the world; it applies to *each* act of creation of something genuinely new in the history of the world.

When I said earlier that the issue of contingency—both in natural processes and with regard to the existence of the world as a whole—is far more important in the biblical doctrine of creation than the idea of design is, this was not intended to

---

[2] See e.g. Barbour (1966: 348 f.), where, however, the concept of continuing creation is taken to be opposed to *creatio ex nihilo*, whereas the intended opposition is rather 'creation in the beginning'.

exclude design and purpose completely from our understanding of the natural world. One has to be aware, however, of the danger of anthropomorphism in language about God's design of the world and with regard to the course of its history. The notions of design and purpose are rooted in the description of our human actions. As humans, we set purposes for ourselves regarding a future that is different from our present situation, and then we look for means in order to realize our purposes in the process of our action. This image is not appropriate with regard to God's activity, because God is eternal. His presence is not lacking a future. God is not looking ahead from the beginning of time to a distant future so as to set purposes for his activity.

Nevertheless, purposive language can be applied to God's action in the history of his creation insofar as the eternal God looks at his creation as a whole, across the entirety of its history. Here, earlier events can be seen as related not only to their past, but also to the future completion of creation. Present events are oriented not only to the future of their own particular history (and thus to the wholeness of their destiny), but also to the future of the world as a whole, which conditions the significance of each part. With regard to this constellation, one can affirm an element of purpose or 'top-down causation' in the processes of nature and also in the activity of its creator, formulated in terms of God's foreordination and intentions. This element helps to account for the fact of cohesion in the sequence of events, as does also the efficacy of natural law. But the observation of apparent teleology should never be considered as an alternative explanation of phenomena, as occurred in the dispute over Darwin.

Time, Space, and Eternity

If the reality of the actual world is basically and ultimately characterized as a sequence of contingent events, or 'actual occasions' (A. N. Whitehead), then time is also essential—not just in the domesticated form of a reversible sequence, but rather as an irreversible sequence, where each further step is genuinely new. The concept of time as irreversible sequence already involves the contingency of events: if the temporal sequence is unique and irreversible, then each new moment of time is different from all previous ones. It is only within geometric abstraction, where time is 'spatialized' and represented as a fourth dimension added to the three dimensions of Euclidian space, that the temporal sequence may be reversed. In historical processes, time is always irreversible. This applies also to the biblical view of reality as a history that moves from a particular beginning to an eschatological end or consummation. One of the challenges which the modern big bang cosmology (together with the second law of thermodynamics) raised stemmed from the fact that it presented the universe as an irreversible process, a 'history of nature', whereas in earlier phases of modern science natural processes, ruled by laws, were considered to be reversible in principle. Once again, then, it is obvious why modern scientific cosmology has become so important for the dialogue between theologians and scientists.

In the reality of nature, time is closely connected with space. This is evident in the phenomenon of movement. Time is the more basic concept, insofar as spatial relations presuppose some form of simultaneity (even though the theory of relativity told us that there is no absolute simultaneity). Nevertheless, space is indispensable in the measurement of time. From the heavenly movements of Sun and Moon which were crucial for measuring days, weeks, months, and years, to the invention of more artificial clocks, temporal sequences have always been measured by some form of spatialization. This is one of the reasons why the genuine nature of time so easily escaped the attention of theorists.

In the dialogue between theology and natural science, it is very important that space and time cannot be reduced to geometric models or descriptions—neither to Euclidian space nor to the spacetime concept of the theory of relativity. The reason is that all geometrical description works with units of measurement. Each unit is a finite part of space (or time) and, as such, already presupposes some more comprehensive space (or time). Prior to the perception or conception of finite spaces, then, there is always the intuition of space as an infinite whole or of time as an infinite whole, as Immanuel Kant and, before him, Samuel Clarke argued convincingly. Time and space can therefore not be exhaustively represented by geometrical models, important as such models otherwise are. Furthermore, the infinite whole of space and time (in their primordial intuition) is different from the conception of infinite space (or time) that results from the continuous and never-ending addition of more and more finite parts. The whole manifests a priority over every conception of parts.

Why, then, is this issue important in the dialogue between theology and science? The reason is that theology has to speak of God's eternity and omnipresence in his creation, which involves the question of God's relationship to space and time. In the idea of omnipresence, space is clearly implied as the medium within which something is present to somebody. Newton therefore considered space as the medium of God's presence to his creatures, each of whom exists at their particular place in space. It was in this sense that Newton referred to space as the *sensorium Dei* in his *Opticks*. Leibniz accused Newton of pantheism as a result, because he understood Newton to take space to be an attribute of God, which would mean that God himself would be spatial, divisible into parts, and composed out of parts. Against this criticism, Clarke defended Newton by distinguishing the space of God's omnipresence, which is infinite and indivisible, from the space of creatures who are composed of parts and divisible into parts. The space of God's omnipresence, then, is the undivided infinite space, which, according to Kant, is the ultimate condition of conceiving of any finite spaces, their divisions, and compositions. If the space of God's omnipresence were identical with geometrical space, which is composed of units of measurement, pantheism would indeed be an inevitable consequence. One sees this result in the philosophy of Spinoza, with whom Albert Einstein sympathized in the twentieth century.

The theory of relativity demonstrated that space is not an infinite and empty receptacle prior to the existence of the finite bodies that find their places in it. Rather, spatial and temporal relations are dependent on the existence of masses and bodies. Still, always first presupposed is the undivided and infinite space which is prior to every conception of finite spaces and to all measurement. This makes it possible to conceive of the space required for divine omnipresence as distinct from and prior to the spatial relations of bodies and masses. Similarly in the case of time: if the undivided and infinite totality of time, understood as a precondition for conceiving of finite times and of duration, is prior to the times and durations of finite things, then the eternity of God can be distinguished from, but also related to, the time of creatures and also to the time of physicists. Eternity is not timeless, nor is it the ever-increasing continuation of temporal process. Eternity is, as Plotinus said (and Boethius after him), the 'wholeness of life in simultaneous possession'—the infinite and undivided unity of time, which according to Kant is presupposed in any conception of limited and particular times. This also explains why and how the eternal God can be present at the particular times of his creatures and at particular occasions in their lives.

Thus, the connection of the eternity and omnipresence of God with the spaces and times of his creatures is of great importance for understanding his active and creative involvement with them. It explains how God can create and preserve his creatures in their particular spaces and times. It also helps to explain how God can actively intervene in their lives. Affirming such interventions is indispensable in Christian theology. Otherwise theology would end up with a God who may have created the universe and its order in the beginning, but who is not further involved with the natural course of its processes. To scientists, on the other hand, the idea of divine interference with the natural course of events has generally met with suspicion, if not outright rejection, because it seems hard to reconcile with the determination of all natural processes by natural causes according to the laws of nature. This issue of divine interference or intervention in natural processes therefore needs careful clarification.

GOD'S ACTION AND PHYSICAL MOVEMENT

In pre-modern times, before the rise of classical physics, there was no particular paradox involved in conceiving of the action of God the Creator in connection with the movements of his creatures. The creatures were understood to depend on God not only for their existence but also for their preservation, which included their ability to act and the exercise of their actions. In their actions—indeed, even in the free actions of humans—God's co-operation was considered a necessary condition. On the other hand, God was held to be able to intervene in the course of events without the help of secondary causes.

With the rise of classical physics, the conditions for the interpretation of movement changed considerably. The first factor was that the Aristotelian concept of movement, which also included qualitative changes, was reduced to local changes, explicitly (for example) by Pierre Gassendi. The second factor was the new principle of inertia, which no longer assumed rest to be the natural state of bodies, but rather perseverance in either a state of rest or of movement. The consequence was that movement as such no longer required an explanation, but only the alteration of the states of movement or rest. The reason for changes in state (from motion to rest, or vice versa) Descartes attributed to the mutual interactions of bodies and to the transfer of their movements from one to the other, since God was supposed to preserve each creature in its state. God was believed to have created all of his creatures in a state of movement or rest and to preserve them, as far as he is concerned, in this state. All changes in the world of nature from the original state of affairs to the present, then, are due not to divine actions but to the mechanical interactions of bodies. This mechanistic conception was criticized not only by the theologian John F. Buddeus, but also by Isaac Newton, as leading to atheistic consequences. This was one of the reasons why Newton developed his concept of force, which as *vis impressa* is not only a transfer of movement from body to body but could also include immaterial forces through which, Newton believed, God's active presence in his creation was at work. After Newton, however, the spiritual aspect of his concept of force was discarded in favour of the mechanistic model, which did indeed lead to the exclusion of God's activity from the interpretation of natural processes. If all forces are to be attributed to bodies, and only bodies exercise forces upon others, then God can no longer be conceived as a source of movements in the natural world. Whatever God is supposed to be, he certainly cannot be a body.

Such was the impasse in the relationship between theology and natural science with regard to the interpretation of natural movement. To some, an opening out of this impasse seemed to be suggested by the argument from design: many creatures, along with their complexities and beauties, are so subtly organized that their existence appears to be comprehensible only as an effect of the intelligent design of a creator. This teleological argument did not exactly restore belief in the efficient activity of God in the processes of nature, however. The question of how God's design could be realized could not be answered in this way. In any event, the Darwinian theory of evolution ruined this recourse to the argument from design. The impasse caused by mechanistic determinism could be overcome only by a conception of natural force that no longer considered forces in terms of the impact of bodies on other bodies. Such a new concept of force was indeed developed in the origins of field theory, especially by Michael Faraday.

The concept of field has been developed in modern physics by Maxwell and Einstein, and quantum field theory beyond its origins in the 'field of force' concept of Faraday. Nevertheless, Faraday's introduction of the field concept was an epochal step, not only in the history of physics but also with regard to the relationship between science and theology, because the disengagement of the concept of force from its attribution to bodies marked the end of the mechanistic description of natural processes. Field became a new foundational concept. Faraday already conceived of bodily phenomena in terms of field effects—that is, as manifestations of fields. This new paradigm also had the potential to offer a new interpretation of God's continuous activity in his creation, in accordance with the biblical concept of the divine Spirit and his participation in the creative action of God.

Thomas F. Torrance ([1969] 1978) was the first theologian to adapt the field language of physics to the theological description of God's activity in history, and in particular to the event of the Incarnation. He spoke of the 'creator Spirit of God' as constituting the 'force of energy' of this field of divine activity. This move gained plausibility, in my view, through the suggestion of Max Jammer that a historical connection exists between the biblical and early Greek conception of Spirit (*pneuma*) in the sense of 'air in movement' (as in wind, storm, or breath) and the origins of the modern field concept in physics. Indeed, this famous historian of science called the ancient (Stoic) concept of *pneuma* 'the direct precursor of the field concept' of physics. Since the Hebrew concept of spirit (as in Genesis 1: 2 and Psalm 104: 30) is quite similar to that of the Stoic thinkers in this respect, it indeed seems plausible to relate the work of the divine Spirit in God's interaction with his creation to the modern concept of field, and I followed Torrance's approach in volume 2 of my own *Systematic Theology* (1991). When in the biblical creation story it is said that 'the Spirit of God was moving over the face of the waters' (i.e. the primeval waters; cf. Genesis 1: 2), the divine Spirit was conceived as the origin of all physical movement. And when in the Gospel of John it is said that God himself is a Spirit (John 4: 24), and shortly before it is written of the Spirit that his nature is that of the wind that 'blows where it wills, and you hear the sound of it, but you do not know whence it comes or whither it goes' (3: 8), this conception of Spirit is much closer to Faraday's concept of a field of force than to the concept of mind, or *nous*, which was long considered to be the equivalent of the biblical concept of Spirit.

Of course, a theological use of the field concept in order to describe God's interactions with the creatures does not exactly follow any particular field theory within physics, nor even the general form of such theories, the main difference being that physical field effects can be measured by waves while God's activity cannot. Nevertheless, there is more than a superficial similarity. For this reason, one need not speak of the theological use of the field concept as a mere metaphor, insofar as the historical sources of the field concept reveal a common root in the ancient concept of Spirit. A metaphorical element, of course, is present in the scientific field concept itself, if one assumes that the 'ordinary' meaning of the term 'field' is to be found in the field of the farmer.

If the theological adoption of the field concept is admitted, it can help to make understandable how the activity of the Creator in his creation can coexist with the proper activities of the creatures themselves. The key presupposition here is that field effects can be superimposed upon one another. Specific fields can be regarded as manifestations of more comprehensive fields. There is no competition, then, between the creator Spirit of God and created agencies. Rather, as the omnipresence of God permeates all the space of the creatures, so God's Spirit permeates all natural forces and the life of the creatures, and thereby empowers them in their own activities. The

Spirit is at work in all creatures; yet to some of them, according to the biblical witness, even a share in the Spirit is given. This is the case not only with humans, but also with animals (Genesis 1: 30; 7: 22), and perhaps the ability for independent movement provides the reason for this biblical assumption. In any event, consciousness is only a secondary and derivative manifestation of the life-giving activity of the Spirit, even in human beings.

CONCLUDING REMARKS

The issues discussed in this chapter are important for a Christian doctrine of creation that is concerned with its relationship to natural science. They deal with the core presuppositions that are required for any theological use of concepts derived from the natural sciences—just as the biblical account of the creative act of God in Genesis 1 made comparable use of the knowledge about the world of nature that was available at that time. I have attempted such a presentation of the doctrine of creation in the second volume of my *Systematic Theology* (1991). There the idea of a first beginning of the universe is connected with the doctrine of continuing creation, and this is related in turn to the formation of enduring creatures in the process of the universe's expansion, which produced a cooling effect that allowed for the emergence of enduring forms. We can consider the production of such creatures to be the intrinsic aim that was implicit in the act of creation—assuming that a goal of creation was to bring about creatures that enjoy a status of relative independence *vis-à-vis* God. The Christian doctrine of creation strongly affirms the relative independence of creatures—not only with regard to one another, but also with regard to God himself—as essential in the act of creation itself.

The personal difference and self-distinction of the Son in relation to the Father is the model for such independent existence of creatures. It is an independence, however, that finds its satisfaction in the free submission of the Son to the Father and in his communion with his eternal life. This is the special destiny of living creatures, and in particular of human beings. Being destined to independent existence and to a freedom that is fulfilled in communion with God through submission to him is the aim of all creation; it is a determination to which God sticks even in the face of the finitude of his creatures, of their sin and their death. The eschatological accomplishment of a new life in communion with God, a life that will conquer sin and death, is the final accomplishment of creation. It will affirm and ultimately realize the original determination of the Creator to grant his creatures an independent existence in communion with himself. This final future certainly transcends the dialogue of theology with science, of course. Still, according to the Apostle, it concerns not only human beings, but also the world of nature (Romans 8: 19 ff.) which scientists are now exploring.

REFERENCES AND SUGGESTED READING

ALTNER, G. (1986) (ed.). *Die Welt als offenes System.* Frankfurt: Fischer Taschenbuch.

BARBOUR, I. G. (1966). *Issues in Science and Religion.* New York: HarperCollins.

BONTING, L. (2002). *Chaos Theology: A Revised Creation Theology.* Ottawa: Novalis Press.

GANOCZY, A. (1995). 'Chaos, Zufall', in *Schöpfungsglaube: die Chaostheorie als Herausforderung der Theologie,* Mainz: Grünewald.

MÜLLER, A. M. K., and PANNENBERG, WOLFHART (1970). *Erwägungen zu einer Theologie der Natur.* Gütersloh: Mohn.

PANNENBERG, W. (1991). *Systematic Theology,* ii. Grand Rapids, Mich.: Eerdmans.

—— (1993). *Toward a Theology of Nature: Essays on Science and Faith,* ed. Ted Peters. Louisville, Ky.: Westminster/John Knox.

—— (2002). 'The Concept of Miracle', *Zygon,* 37: 759–62.

—— (2005). 'Eternity, Time, and Space', *Zygon,* 40: 97–106.

PEACOCKE, A. (1979). *Creation and the World of Science.* Oxford: Oxford University Press.

—— (1986). *God and the New Biology.* New York: HarperCollins.

PRIGOGINE, I. (1980). *From Being to Becoming: Time and Complexity in the Physical Sciences.* San Francisco: W. H. Freeman.

—— and STENGERS, J. (1986). *Dialogue mit der Natur,* 6th edn. Munich: Piper.

RUSSELL, R. J. (1988). 'Contingency in Physics and Cosmology: A Critique of the Theology of Wolfhart Pannenberg', *Zygon,* 23: 23–43.

TORRANCE, THOMAS F. ([1969] 1978). *Theological Science.* Oxford: Oxford University Press.

—— (1981). *Divine and Contingent Order.* Oxford: Oxford University Press.

WEIZSÄCKER, C. F. VON (1948). *Die Geschichte der Natur.* Göttingen:Vandenhoeck & Ruprecht.

CHAPTER 23

CONTRIBUTIONS FROM PRACTICAL THEOLOGY AND ETHICS

TED PETERS

INTRODUCTION

Practical theology is concerned with the life of the Church and the daily life of the Christian in both church and society. Practical theology is frequently distinguished from speculative or systematic theology, where most of the conversation regarding the relationship of science and faith has been taking place. The parish pastor or priest provides the channel through which practical theology flows into the life of the congregation.

The charge of the parish pastor or priest is threefold: *preaching, teaching,* and *counselling.* Of course, there is more to the weekly ministry than these three alone. Preaching takes place within a planned worship setting with liturgy and the celebration of the sacraments. Teaching involves administering a Christian education programme and overseeing Bible studies. Counselling is only the core of conversation taking place during hospital visitation, home visitation, and numerous other opportunities for interpersonal engagement. A faithful pastor finds precious little time to stop, read, think, meditate, pray, and discern. But this is the inescapable lot of those who feel they have been called by the Holy Spirit to ordained service.

Yet, the revolutionary new rapprochement between science and theology provides the parish pastor with a treasure chest of intellectual jewels that could enrich his or her preaching, teaching, and perhaps even counselling ministries. To leave this treasure chest unopened would be to deny oneself a wealth of resources for parish ministry. In what follows, I will identify some areas of the new dialogue between science and theology that could definitely enhance the effectiveness of preaching and teaching, if not counselling and pastoral care as well.

We will first look at pastoral hermeneutics, which includes the prophetic and constructive tasks of biblical preaching. We will then review eight different models of understanding the relationship between science and faith, half that draw upon the image of warfare and half which advocate peaceful co-operation. We will give special attention to the controversy surrounding Darwinian evolution. Finally, we will turn to the frontier of genetic research and the ethical issues surrounding the phrase 'playing God', recommending that the pastor demythologize science while appropriating science to a theological understanding of the world in which we live.

PASTORAL HERMENEUTICS

Why might interest in modern science be of value to the parish preacher or teacher or counsellor? After all, the task of a church leader is to interpret the Bible, not natural science. Yet, the interpretation of the Holy Scriptures on behalf of listening ears in the context of the modern world requires attention to the modern world-view, which has been determined primarily and extensively by natural science.

Yes, interpreting the Bible is what distinguishes the work of the Christian parish pastor. Yet, the dialogue with natural science can aid the pastor in this task of interpretation, first and foremost in preaching. The preacher will need to deal with science on two fronts: first, and critically, as a myth to be transcended by the proclamation of the gospel; and, second, as an essential component in post-critical world-view construction.

To proclaim the Word of God in a sermon in such a manner that it becomes the Word of *God* and not merely that of the preacher is the unenviable charge to the conscientious pastor. 'When the Gospel is preached, God speaks', wrote Karl Barth (1963: 12); and the presence of the Word of God makes the pulpit a sacred and awesome place. In order to help ensure that the Word of God comes through with impact, the pastor in the pulpit needs to attend to two contradictory or paradoxical thrusts within the sermon's structure: world-view deconstruction and world-view construction. The first is prophetic; the second is pastoral. The first is critical; the second is post-critical. Let me start by saying a few things about the critical task of world-view deconstruction.

God is not of this world. God transcends this world. This does not mean that God is distant; it means that God cannot be controlled by anything we do or anything we think. God cannot be subordinated to anything we believe or imagine. God comes to us from outside our imaginations, challenging and cracking our previously accepted understandings of ourselves and our world. To proclaim the gospel as God's Word, the preacher needs to challenge our this-worldly assumptions and loyalties so that we can open ourselves to a message that comes from beyond.

Cultures and individuals construct in their imaginations world-views or pictures of reality that include the whole world and themselves in it. Such views of reality provide the conceptual framework within which we understand our place in an otherwise immense and mysterious cosmos. Yet the God who revealed himself to the people of ancient Israel is simply not of this world. The God of the first commandment—make no images!—transcends all our images of God. To proclaim the Word of God includes smashing our understanding of how the world works in order to reveal a higher reality, God. The preacher, by example and by message, needs to communicate to his or her listeners: don't trust what you imagine the world to be like; trust God instead.

During the middle third of the twentieth century when neo-orthodox and existentialist theology reigned, most theologians referred to this as preaching the 'kerygma', or preaching the gospel, accompanied by 'demythologizing'. 'Demythologizing' was (and still is) an important term. It did not mean eliminating the Bible's mythological world-view and replacing it with a scientific world-view. Rather, it meant interpreting the message of Jesus Christ distinguishable from any world-view. Rudolf Bultmann wrote, 'Its [de-mythologizing] aim is not to eliminate the mythological statements but to interpret them. It is a method of hermeneutics' (1958: 18).

For this discussion, the word 'myth' is almost equivalent to 'world-view'. The prophetic task or critical task of the preacher is to break through all world-views in order for the kerygma to call listeners to personal trust in God, trust in the God who transcends this world and transcends our imagination of how this world works. Preaching consists of a call to faith, with faith understood as existential trust in a God who remains mysterious and unknowable, even if trustworthy. By 'world' we mean our world-view or world of meaning, *Weltanschauung*. This needs to be destroyed, or in post-modern terms 'deconstructed', so that our attention can be drawn toward the divine grace that comes to us from beyond the world. The preacher would enter the pulpit ready to smash idols—that is, to critically dismember idols in the form of ideas or concepts or beliefs about how the world works. In radical obedience to the first commandment against graven images, the preacher would proclaim that God comes from beyond all our mental images of this world and confronts us with the decision to either have faith in the transcendent or continue to rely idolatrously upon our this-worldly imagination.

World destruction is the prophetic task of the preacher. What I here call the *prophetic* task is built on Paul Tillich's 'Protestant principle', which 'emphasises the infinite distance between God and man' (1959: 68). By criticizing, challenging, and deconstructing the world-view within which we live, the preacher would call us to faith in the infinitely transcendent God who has revealed the Godself as gracious in the cross of Jesus Christ. The limit to this prophetic preaching task is that it cannot on its own tell us how to live in this world, even if it is a world partially constructed by the human imagination. Human beings cannot live on a day-to-day basis on prophetic preaching alone. Criticism is not enough. We all need to envision ourselves within a world that is packed with meaning, existential meaning. This leads to the second of the preacher's tasks: namely, to construct a world-view where all things are oriented toward the God of grace. This is the pastoral element in the sermon.

Since the days of Thomas Aquinas, the job of the systematic theologian has been to describe all things in the world in light of their relationship to God. This is the job of the contemporary preacher (and teacher) as well as the theologian. The meaningfulness of the life of every person sitting in a church pew is conditioned by the picture of the world that the preacher paints. The preacher is charged with the task of painting a picture of reality that orients all things to the God who transcends all things. This is a most difficult task, to be sure; yet nothing less is demanded.

Once the preacher has called us out of the world to listen to the eternal kerygma, or Word of God, then the preacher turns around and calls us back into the world with a new orientation. Our response to the gospel reorients our understanding of ourselves and of the world within which we live. This happens whether the preacher is aware of it or not. It is better to be aware.

Implicit in the sentences and metaphors and images and allusions the preacher employs in the sermon is a description of a single comprehensive reality in which both the listener and God are components. This is as important as it is inescapable. The verbal pictures of the world the preacher draws are the primary vehicle for evoking a sense of meaning, of belonging, of orientation, of welcome, of acceptance.

Language and world-view belong together. 'The gospel does bring change and resulting *challenge*, yet the way we language that *challenge* from the pulpit determines a great deal. The language choices preachers make influence the theological under-standings with which hearers leave' (Rogers 2004: 270; italics original).

Our concern here is this: what role does modern science play in structuring the world-view and the language that frames the sermon? Note what is not being said here. Our concern is not to demand that modern science provide the content of the sermon. The kerygma, or gospel, provides the central content. The role that science plays is found in the framing world-view, whether to be prophetically demytholo-gized or to become a component in world-view reconstruction.

Science is an unavoidable factor in describing persuasively to virtually all modern people just how our world works. It is with this in mind that Philip Clayton writes while discussing physical causality, 'Remember that the question is not how to *prove* that God is active in the world at particular moments, but rather how to think this possibility in a manner that does not conflict with what we now know of the world' (1997: 193; italics original).

Science is the reigning myth, so to speak. Science sets the standards for credibility. No one can live with a sense of truthfulness in a world-view that is irreconcilable with the picture of nature drawn by physicists, chemists, biologists, and ecologists. Whatever the preacher assumes at the level of the conceptual frame will become implicitly audible to his or her listeners; if it is irreconcilable with science, the credibility of everything the preacher says will unnecessarily be doubted.

THE WARFARE WORLD-VIEW

World-view construction will include two overlapping but distinguishable components: the realm of nature and the realm of science. Science is one way to interpret nature; it is distinguishable from nature and worthy of analysis in its own right. In this section we will ask: Just how might the parish pastor include natural science in the world-view he or she seeks to construct?

This larger question is prompted by a set of smaller questions. Do we want to perpetuate the widespread belief that science and faith are at war with one another? Do we want people of faith to fear the sciences as threats to religious belief? Do we want members of congregations to suspect that the scientists who are also members may be secretly committed to atheism? Do we want our young people considering a future career to eschew studying science out of the fear that it may be of the Devil?

To address these questions directly or indirectly in preaching and teaching will require on the part of the parish pastor or priest at least a modicum of sophistication regarding just how science and religion in fact relate to one another. Surprisingly, multiple ways of construing the relationship between science and faith make up the currency of today's exchange of ideas. Here we will look at eight models or patterns of interaction between science and faith, suggesting directions for pastoral assessment and employment (Peters 2003: ch. 1).

Four Warfare Models

The first four models fit the widespread belief—myth, if you will—that science and faith are at war. The first model is *scientism*. In the contemporary West, the term 'scientism' refers to naturalism, reductionism, or secular humanism—that is, to the belief that there exists only one reality: namely, the material world. Further, science provides the only trustworthy method for gaining knowledge about this material reality. Science has an exhaustive monopoly on knowledge. It judges all claims by religion to have knowledge of supernatural realities as fictions, as pseudo-knowledge. All explanations are reducible to secularized material explanations. According to this model, religion loses the war by being declared false knowledge.

Peter Atkins, an Oxford chemistry professor, represents the position. 'My conclusion is stark and uncompromising. Religion is the antithesis of science; science is competent to illuminate all the deep questions of existence, and does so in a manner that makes full use of, and respects the human intellect. I see neither need nor sign of any future reconciliation' (Atkins 1997; also Chapter 8 above).

Against scientism, the preacher needs to speak critically and prophetically. The world of scientism is a sealed-off natural world that has closed its doors and windows to transcendence. Not only is it anti-religious, it only pretends to be scientific. Actual science as a research enterprise does not need the ideology of scientism. One can have perfectly good science without adding this ideology of materialism or reductionism. The preacher can help listeners to distinguish between healthy science and unhealthy scientism.

The second model on our list is *scientific imperialism*, a close ally of scientism. Scientific imperialism does not outrightly dismiss religion. Rather, it uses materialist reductionism to explain religious experience and reassess theological claims. Scientific imperialists grant value to religion and religious contributions to society. They may even grant the existence of God. Yet, scientific imperialists claim that science provides a method for discerning religious truth that is superior to that of traditional theology. In contemporary discussion this approach is taken by some physical cosmologists such as Paul Davies and Frank Tipler when explaining creation or eschatology, and by sociobiologists such as E. O. Wilson and Richard Dawkins in proffering a biological explanation for cultural evolution, including religion and ethics. Here religion is defeated in the war by being conquered and colonized.

Scientific imperialism may appear to the preacher like the whore of Babylon. It appears at first to be attractive, because on the face of it scientists say good things about religion. But its cup of abominations is seductive. Once within its grip, what is precious to the Christian disappears: a transcendent God, an active God, a gracious God, and a God who is capable of redemption. Prophetic critique is called for.

Now we turn to warfare from the point of view of the other army. *Ecclesiastical authoritarianism*, our third model, is what every scientist fears from the Church. According to this model, modern science clashes with religious dogma that is authoritatively supported by ecclesiastical fiat, the Bible, or in Islam by the Qur'an. The best example is the 1864 *Syllabus of Errors*, promulgated by the Vatican. Here it is asserted that scientific claims must be subject to the authority of divine revelation as the Church has discerned it. The Second Vatican Council in 1962–5 reversed this, affirming academic freedom for natural science and other secular disciplines. Ecclesiastical authoritarianism wins the war over science through intellectual intimidation.

The twenty-first-century pastor is not likely to be tempted to defend church dogma against an alleged scientific assault as did the Roman Catholic Church of the nineteenth century, to be sure. Yet, the integrity of theologically derived truths and observations needs to be explored without the anxiety that science will soon explain all such claims away. The parish leader, in this circumstance, needs to trust the truth—that is, trust that open and fair discussion of any theological idea or any scientific idea will lead eventually to edifying truths and not to any embarrassment for either faith or science. Church teaching situations should embrace an atmosphere of confident openness and exploration.

The War over Darwinian Evolution

The fourth of our warfare models is the *battle over Darwinian evolution*. The battle-fields are churches, public school classrooms, school boards, university lecture halls, and the courts in the USA, Australia, and Turkey, with little or no notice in Europe. Before picking sides and leaping into battle with guns blazing, the pastor should pause to see who is fighting with whom about what. For that reason, I will pause here to provide more detail than I have for the other models.

Five positions are discernible, making it much more complicated than the image of a simple war between science and religion might connote. The first position would be that of *evolutionary biology strictly as science* without any attached ideological commitments. The reigning theory is neo-Darwinian. Neo-Darwinism combines Charles Darwin's original nineteenth-century concept of natural selection with the twentieth-century concept of genetic mutation to explain the development of new species over 3.8 billion years. Defenders of quality science education in the public schools most frequently embrace this 'science alone' approach.

The second position combines neo-Darwinism with the scientism mentioned above to formulate a *materialist ideology*. This ideology includes repudiation of any divine influence on the course of evolutionary development. Spokespersons for *sociobiology*, such as E. O. Wilson and Richard Dawkins, are aggressive and vociferous. Evolution here provides apparent scientific justification for scientism, scientific imperialism, and in some cases belligerent atheism. Charles Darwin himself did not draw atheistic implications from his science, writing, 'My views are not at all necessarily atheistical' (1888: 312). But his disciples do. Harvard geneticist Richard Lewontin (1997) contends that '*Science*, as the only begetter of truth.... Materialism is absolute, for we cannot allow a Divine Foot in the door.'

The third position is *scientific creationism*. During the fundamentalist era of the 1920s, biblical creationists appealed to the authority of the Bible to combat the rise in influence of Darwinism. Since the 1960s, creationists have based their arguments not on biblical authority but rather on counter-science—hence their label, *scientific creationists*. They argue, for example, that the fossil record contradicts standard appeals to natural selection over long periods of time. Those known as young earth creationists (YEC), such as the leaders of the Institute for Creation Research near San Diego, California, hold that the planet Earth is less than 10,000 years old and that all species of plants and animals were originally created by God in their present form. They deny macro-evolution—that is, they deny that one species has evolved from prior species—although they affirm micro-evolution—that is, evolution within a species. Key here is that creationists justify their arguments on scientific grounds. 'Creation is true, evolution is false, and real science confirms this', writes Henry Morris (1993: 308–9).

The fourth position is *intelligent design*. Advocates of intelligent design sharply attack neo-Darwinian theory for overstating the role of natural selection in species formation. They argue that slow incremental changes due to mutations combined with natural selection are insufficient to explain the emergence of new and more complex biological systems. Many of the life forms that have evolved are irreducibly complex, and this counts as evidence that they have been intelligently designed. Intelligent design scholars such as Michael Behe, Philip Johnson, and William Dembski posit that appeal to a transcendent designer is necessary for the theory of evolution to successfully explain the development of life forms. Here scientific questions lead to theological answers.

The fifth position is *theistic evolution*, according to which God employs evolutionary processes over deep time to bring about the human race, and perhaps even to carry the natural world to a redemptive future. Theistic evolution first appeared in the late nineteenth and early twentieth centuries, even in the work of conservative Princeton theologian B. B. Warfield, for whom God's *concursus* with nature brought about the human race, just as God's *concursus* wrote the Scripture with human minds and hands. Teilhard de Chardin is perhaps best known for his evolutionary cosmology directed by God toward a future 'Point Omega'. Among contemporary scholars at work in the field of science and religion, the roster of theistic evolutionists includes Arthur Peacocke, Philip Hefner, Robert John Russell, Nancey Murphy, Kenneth Miller, John Haught, Martinez Hewlett, and Howard van Till. This school of thought is not occupied with defending evolution against attacks by advocates of scientific creationism or intelligent design; rather, it seeks to work through questions raised by randomness and chance in natural selection in light of divine purposes and ends (Peters and Hewlett 2003; Russell *et al.* 1998).

It is my view that the best of these alternatives is theistic evolution. Along with my colleague Martinez Hewlett, we argue that the biblical testimony that God creates new things indicates that God's creative activity is ongoing. We affirm *creatio continua* along with *creatio ex nihilo*, setting us apart from the scientific creationists. We further affirm that God has a purpose *for nature* that cannot be discerned scientifically *within nature*; and this sets us off from intelligent design. We believe that the purpose for the long story of evolution will be determined by, and revealed by, the advent of God's promised new creation (Peters and Hewlett 2003: ch. 7).

That a war over evolution is being fought is clear. However, because the actual points at issue deal specifically with the explanatory adequacy of natural selection, it would be misleading to dub this simply a war between science and religion. I here stress the point that what the parish pastor needs to know is this: despite the fact that superficially this looks like a battle between science and faith, deep down it is not. All combatants revere and respect science. It is in essence a battle over what constitutes good science. A parish leader who views this as a battle of science versus faith and who takes the side of faith does so only at grave peril.

FOUR NON-WARFARE MODELS

We have just reviewed four ways of understanding the relationship between science and faith in terms of warfare: scientism, scientific imperialism, ecclesiastical authoritarianism, and the battle over evolution. But warfare is not the only lens through which to look at this relationship. In fact, the image of warfare could be misleading. 'An image of perennial conflict between science and religion is inappropriate as a guiding principle', writes John Hedley Brooke (1991: 33). In the contemporary globe-wide conversations between scientists and religious thinkers, at least four non-warfare or co-operative models have emerged. Any or all of these could become material for the parish pastor to use in constructing a meaningful world-view that includes modern science.

The first and most widely embraced model for relating science and religion is the *two languages* model. According to this model, science speaks one language, the language of facts, and religion speaks a different language, the language of values. The two language model—sometimes referred to as the 'independence' model—is the prevailing view of both scientists and theologians in Western and Asian intellectual life. Science attends to objective knowledge about objects in the penultimate realm, whereas religion attends to subjective knowledge about transcendent dimensions of ultimate concern. Modern persons need both, according to Albert Einstein, who claimed the following: 'Science without religion is lame and religion without science is blind' (1982: 49). Warfare is avoided by establishing a border and keeping science and faith in their respective territories.

This two languages model should not be confused with the classic model of the *two books*, according to which the book of Scripture and the book of Nature each provide an avenue of revelation for God (Hess 2003). The difference is that the two books model sees science as revealing truth about God, whereas the two language model sees science as revealing truth solely about the created world.

Already many preachers rely on the two language model when expositing biblical texts such as the Genesis creation accounts. After grasping for a few possible connections or cross-overs between Genesis and big bang cosmology or Darwinian evolution, the preacher typically elects to say that science tells us what happened, while the Bible tells us what it means. This is perfectly legitimate. It is safe, and satisfactory. It does not risk losing any credibility. It provides an accessible safety zone within which to present the biblical message unencumbered with otherwise difficult-to-answer questions.

For many scholars in the dialogue, however, the two languages model for keeping the two independent of one another is inadequate. It is theoretically inadequate from the point of view of the theologian. The theologian asks: If it is true that the God of Israel is the creator of this world, and if it is true that natural scientists are gaining accurate knowledge of how this world works, then, sooner or later, we would expect to see some convergence, or at least consonance, between the two domains of knowing.

This leads to the next non-warfare model, *hypothetical consonance.* Going beyond the two language view by assuming an overlap between the subject-matter of science and the subject-matter of faith, *consonance* directs inquiry toward areas of correspondence between what can be said scientifically about the natural world and what can be said theologically about God's creation. Even though consonance seems to arise in some areas, such as the apparent correspondence of big bang cosmology with the doctrine of creation out of nothing, consonance has not been confirmed fully in all relevant shared areas. Therefore, the adjective *hypothetical* applies to theology as well as science. The key hypothesis of this model is that there can be only one shared domain of truth regarding the created world, and science at its best and faith at its best both humble themselves before truth. It follows, then, that one should trust that consonance will eventually emerge. Hypothetical consonance provides the basis for what some call 'dialogue between science and theology' and others the 'creative mutual interaction of science and theology'.

It is inappropriate to ask the preacher to engage hypothetical consonance because so little in the typical sermon is of a hypothetical character. Hypotheses are for scholars who explore new ideas; sermons occasionally, though rarely, do this. Perhaps what could inform a sermon would be a background of unsaid trust in the truth, trust in the truth wherever it may be found. The sermon could carry the mood that truth, even if unearthed by a scientist, could be celebrated by a person of faith.

The third non-warfare model, and number seven on our comprehensive list, is *ethical overlap.* Building on the two languages model, wherein mutual respect between scientists and religious leaders is affirmed, some express a strong desire for religious co-operation on public policy issues deriving from science and technology. The ecological crisis and human values questions deriving from advances in genetics both enlist creative co-operation.

Pope John XXIII told us in *Pacem in terris* that Roman Catholics could make partners with 'all persons of good will' when working for world peace. Working with all persons of good will in the community or around the world—even scientists of good will—to make our planet a better place in every respect ought to be the daily diet of any Christian congregation.

This brings us to the final example on our list, number eight, *New Age spirituality.* Having left the conflict or warfare model behind, synthetic spiritualities, such as those found in the New Age movement, seek to construct a world-view that integrates and harmonizes science with religion. Evolution becomes an overarching concept that incorporates the sense of deep time and imbues the development of a global spiritual consciousness as an evolutionary advance for the cosmos. Many here are prompted by the visionary theology of Teilhard de Chardin, although this Jesuit forerunner could not himself be categorized as New Age. Others in the New Age movement seek to integrate the experience of mystery articulated in Hinduism and Buddhism with advanced discoveries in physics, such as indeterminacy and quantum theory (Peters 1991).

The metaphysics of the New Age will attract some pastors and repel others. Although some Western scientists and many Eastern scientists resonate with New

Age spirituality, it is generally held in academic disrepute because it lacks the rigour found in both scientific research and classical Christian theology. Ethical overlap with special emphasis on ecology would be the domain most likely shared by New Agers and Christian pastors.

THE PROBLEM OF SCIENTISTS PLAYING GOD

One dramatic component of the reigning scientific world-view is discernibly mythical: namely, the concept of 'playing God'. In fact, the concept of playing God is itself a survival of the classical Greek myth of Prometheus. In the modern scientific setting, this myth takes the thinly disguised form of Faust, Frankenstein, or *Jurassic Park*. The parish preacher and teacher need to be aware of the actual myths within the worldview, so that demythologizing the scientific myth can become part of the hermeneutical process.

Public policy controversies, as we can all see, invoke abhorrence of 'playing God'. The phrase 'playing God' refers to the power that science confers upon the human race to understand and to control the natural world. Even though the phrase 'playing God' is by no means a theological phrase, it sounds like one; and people within the churches will ask about its meaning and significance (Peters 2002). The pastor needs to be prepared to answer.

Upon close examination, the widely used phrase 'playing God' seems to connote three overlapping meanings. The first is connected to basic scientific research: to play God is to *learn God's awesome secrets*. The second is found in the hospital. It connotes the fact that doctors have gained the *power over life and death*. The third refers to the ability of scientists to *alter life and influence human evolution*. This meaning is best expressed in the story of Frankenstein, the mad scientist who violates an invisible boundary and crosses over into the sacred realm of nature; then nature rises up vengefully and unleashes death-dealing chaos. Our society fears the mythical mad scientist, who by violating nature may cause a backlash that will lead to suffering on the part of us all.

We assume here that scientific understanding leads to technological control. We want control. Yet we doubt our own wisdom to know how to use this control. In our attempt to gain control over nature, we may so violate nature that it will lash back with destructive force. This fear is most associated with genetics and ecology. Genetic engineering, wherein we alter our genome and perhaps alter our own essence, is the primary area of science that provokes fears of playing God. Such fears also arise in ecology, where we worry that civilization may soon pass the point of no return and the environment will poison the human race into extinction.

I have suggested above that we are talking here about myth, about the place of science in the contemporary world-view. In this case, an actual mythical story is at work. Our use of the phrase 'playing God' relies on the ancient Greek myth of Prometheus. According to this myth, when the world was being created, the sky-god Zeus was in a cranky mood. The top god in the Olympiad decided to withhold fire from earth's inhabitants, consigning the nascent human race to relentless cold and darkness. Prometheus the Titan, whose name means 'to think ahead', could foresee the value of fire for warming homes and providing lamplight for reading late at night. He could anticipate how fire could distinguish humanity from the beasts, making it possible to forge tools. So Prometheus craftily snuck up into the heavens where the gods dwell and where the Sun is kept. He lit his torch from the fires of the Sun, then he carried this heavenly gift back to Earth.

The gods on Mount Olympus were outraged that the stronghold of the immortals had been penetrated and robbed. Zeus was particularly angry over Prometheus's *hubris*, so he exacted merciless punishment on the rebel. Zeus chained Prometheus to a rock where an eagle could feast all day long on the Titan's liver. The head of the pantheon cursed the future-oriented Prometheus: 'Forever shall the intolerable present grind you down.' The moral of the story, which is remembered to the present day, is this: human pride, or hubris, that leads us to overestimate ourselves and enter the realm of the sacred will precipitate vengeful destruction. The Bible appears to provide a version of the same point: 'Pride goes before destruction' (Proverbs 16: 18).

For us in the modern world who think scientifically, no longer does Zeus play the role of the sacred. Nature has replaced the Greek gods. It is nature that will strike back in the Frankenstein legend or its more contemporary geneticized version, Michael Crichton's novel *Jurassic Park* (1990) and the subsequent movies. The theme has become a common one: the mad scientist exploits a new discovery, crosses the line between life and death, and then nature strikes back with chaos and destruction.

Interpreting the Gene Myth

How should religious leaders interpret this classic myth as it influences contemporary culture? Should the parish pastor believe the myth? Should the parish pastor act out of the world-view of this myth? Or should the pastor demythologize?

One thing to observe is that the God of 'playing God' is not necessarily the God of the Bible. Rather, it is divinized nature. In Western culture nature has absorbed the qualities of sacredness. Science along with technology risk profaning the sacred.

Despite the fact that this myth deals with a god other than the God of the Bible, numerous religious leaders have embraced it; and they have taken up rhetorical arms against science. A 1980 task force report, *Human Life and the New Genetics*, includes a warning by the US National Council of Churches: 'Human beings have an ability to do Godlike things: to exercise creativity, to direct and redirect processes of nature. But the warnings also imply that these powers may be used rashly, that it may be better for people to remember that they are creatures and not gods.' A United Methodist Church Genetic Science Task Force report to the 1992 General Conference stated similarly: 'The image of God, in which humanity is created, confers both power and responsibility to use power as God does: neither by coercion nor tyranny, but by love. Failure to accept limits by rejecting or ignoring accountability to God and interdependency with the whole of creation is the essence of sin.' In the hands of Christian leaders, the myth tells us that we can sin through science by failing to recognize our limits, and thereby violate the sacred.

However, there is an alternative route one can take. As noted above, genetics (along with ecology) is the field of research that provokes the most anxiety regarding the threat that scientists will play God. This is because DNA has garnered cultural reverence. The human genome has become tacitly identified with the essence of what is human. A person's individuality, identity, and dignity have become connected to his or her individual genome. Therefore, if we have the hubris to intervene in the human genome, we risk violating something sacred. This tacit belief is called by some the 'gene myth', by others 'the strong genetic principle', or 'genetic essentialism'. This mytheme—a mini-myth within the larger Promethean myth within the still larger world-view of science—provides an interpretive framework that includes the assumed sacrality of the human genome combined with the fear of Promethean pride.

The preacher and teacher in the parish setting should critically, if not prophetically, question the gene myth. Doubt should be registered about the equation of DNA with human essence or human personhood. A person is more than his or her genetic code. The National Council of Churches of Singapore put it this way in *A Christian Response to the Life Sciences*: 'It is a fallacy of genetic determinism to equate the genetic makeup of a person with the person' (2002: 81). No person is reducible to his or her genome. No person is a victim of a thoroughgoing genetic determinism. At some level, this cultural myth needs demythologizing, if not descientizing, if the parish pastor is to move people to a reasonable and healthy understanding of human nature in light of our faith in God.

GENETICS, ETHICS, AND WORLD-VIEW CONSTRUCTION

In the pastoral setting, increasingly parishioners will come to their clergy for counsel and advice on genetic issues. Initially, couples planning to bring children into the world will visit their genetic counsellor at the clinic and then show up on the pastor's doorstep for further discussion. Pastors will need to understand the overlap between pastoral and ethical concerns that will come in a single package. Stem cell therapy, selecting genes for future children, altering genomes, aborting defective foetuses, and envisioning a genetic future for children will appear on the list of concerns. The parish pastor needs to be ready.

Here, we will look at one of the issues that might confront the pastor which combines counselling and ethical concerns: namely, the relationship between genetic therapy and genetic enhancement. In addition to appearing as a pastoral matter, it is also a public policy issue.

Employment of advancing genetic technologies to alter human DNA leads to considerations regarding the distinction between therapy and enhancement. At first glance, therapy seems justifiable ethically, whereas enhancement seems Promethean and dangerous. The term 'gene therapy' refers to directed genetic change of human somatic cells to treat a genetic disease or defect in a living person. With 4,000 to 6,000 human diseases traceable to genetic predispositions—cystic fibrosis, Huntington's disease, Alzheimer's, many cancers—the prospects of gene-based therapies are raising hopes for dramatic new medical advances. Few if any find ethical grounds for prohibiting somatic cell therapy via gene manipulation.

The term 'human genetic enhancement' refers to the use of genetic knowledge and technology to bring about improvements in the capacities of living persons, in embryos, or in future generations. Enhancement might be accomplished in one of two ways: either through genetic selection during screening or through directed genetic change. Genetic selection may take place at the gamete stage or, more commonly, as embryo selection during pre-implantation genetic diagnosis (PGD) following *in vitro* fertilization (IVF). Genetic changes could be introduced into early embryos, thereby influencing a living individual, or by altering the germ line, influencing future generations.

Some forms of enhancement are becoming possible. For example, introduction of the gene for IGF-1 into muscle cells results in increased muscle strength and health. Such a procedure would be valuable as a therapy, to be sure; yet it lends itself to availability for enhancement as well. For those who daydream of so-called designer babies, the list of traits to be enhanced would likely include height and intelligence, as well as preferred eye or hair colour. Concerns raised by both secular and religious ethicists focus on economic justice—that is, wealthy families are more likely to take advantage of genetic enhancement services, leading to a gap between the 'genrich' and the 'genpoor'.

The most ethical heat to date has been generated over the possibility of germ line intervention, and this applies to both therapy and enhancement. The term 'germ line intervention' refers to gene selection or gene change in the gametes, which in turn would influence the genomes of future generations. Because the mutant form of the gene that predisposes for cystic fibrosis has been located on chromosome 4, we could imagine a plan to select out this gene and spare future generations the suffering caused by this debilitating disease. This would constitute germ line alteration for therapeutic motives. Similarly, in principle, we could select or even engineer genetic predispositions to favourable traits in the same manner. This would constitute germ line alteration for enhancement motives.

Such efforts at genetic engineering are risky. Too much remains unknown about gene function. It is more than likely that gene expression works in delicate systems, so that it is rare that a single gene is responsible for a single phenotypical expression. If we remove or engineer one or two genes, we may unknowingly upset an entire system of gene interaction that could lead to unfortunate consequences. The prohibition against 'playing God' serves here as a warning to avoid rushing in prematurely with what appears to be an improvement but could turn out to be a disaster. Ethicists frequently appeal to the precautionary principle—that is, to refrain from germ line modification until the scope of our knowledge is adequate to cover all possible contingencies.

It is important to note that the precautionary principle does not rely upon the tacit belief in DNA as sacred. Rather, it relies upon a principle of prudence that respects the complexity of the natural world and the finite limits of human knowledge

CONCLUSION

The parish pastor or priest could be motivated to invest time and energy in the worldwide dialogue between science and theology out of pure curiosity, or even out of a desire to enhance the effectiveness of his or her own ministry. I have tried to show here that three areas of pastoral responsibility—preaching, teaching, and counselling—could all benefit from increased sophistication regarding the scientific picture of the world in which we live. By no means do I suggest that science provide the content of what the preacher or teacher or counsellor says; yet, for the sake of credibility and relevance, what is said should resonate with what is unsaid about the scientific world-view we all share.

REFERENCES AND SUGGESTED READING

ATKINS, P. (1997). 'Religion the Antithesis of Science', *Chemistry and Industry*, 20 (January): 'Comment' section.

BARTH, K. (1963). *The Preaching of the Gospel.* Louisville, Ky.: Westminster/John Knox Press.

BROOKE, J. H. (1991). *Science and Religion: Some Historical Perspectives.* Cambridge: Cambridge University Press.

BULTMANN, R. (1958). *Jesus Christ and Mythology.* New York: Charles Scribner's Sons.

CLAYTON, P. (1997). *God and Contemporary Science.* Grand Rapids, Mich.: Eerdmans.

DARWIN, C. (1888). *The Life and Letters of Charles Darwin, Including an Autobiographical Chapter*, ii, ed. Francis Darwin. London: John Murray.

EINSTEIN, A. (1982). *Ideas and Opinions.* New York: Crown.

HESS, P. M. J. (2003). 'God's Two Books: Special Revelation and Natural Science in the Christian West', in Ted Peters, Gaymon Bennett, and Kang Phee Seng (eds.), *Bridging Science and Religion*, Minneapolis: Fortress Press, 123–40.

LEWONTIN, R. (1997). Review of *The Demon Haunted World* by Carl Sagan, *New York Review of Books* (9 January).

MORRIS, H. (1993). *History of Modern Creationism*, 2nd edn. Santee, Calif.: Institute for Creation Research.

National Council of Churches of Singapore (2002). *A Christian Response to the Life Sciences*. Singapore: Genesis Books.

PETERS, T. (1991). *The Cosmic Self*. San Francisco: HarperCollins.

—— (2002). *Playing God? Genetic Determinism and Human Freedom*. New York and London: Routledge.

—— (2003). *Science, Theology, and Ethics*. Aldershot: Ashgate.

—— and HEWLETT, M. (2003). *Evolution from Creation to New Creation*. Nashville: Abingdon.

ROGERS, T. G. (2004). '*What* and *How* of North American Lutheran Preaching', *Dialog*, 43/4: 263–71.

RUSSELL, R. J., STOEGER, W. R., and AYALA, F. J. (1998). *Evolutionary and Molecular Biology: Scientific Perspectives on Divine Action*. Vatican City State: Vatican Observatory; Berkeley: Center for Theology and the Natural Sciences.

TILLICH, P. (1959). *Theology of Culture*, ed. Robert C. Kimball. London and New York: Oxford University Press.

CHAPTER 24

CONTRIBUTIONS FROM SPIRITUALITY: SIMPLICITY— COMPLEXITY— SIMPLICITY

PAULINE M. RUDD

The outstanding efforts that are being made by scientists, philosophers, and theologians to disseminate the essence of the science and religion dialogue to the general public have raised the awareness of people in the pews to the importance of this debate in the modern world. Nevertheless, it is not easy for a congregation to know how to engage seriously with this field if they have neither a scientific nor a philosophical background. It is not necessarily evident that this debate can have a positive impact on the spiritual growth of us all. Growth in spirituality is important because it is the underlying reason behind the religious practice of most ordinary worshippers. In general, theological treatises become relevant to us only when they relate directly to our personal life experience.

The aim of this chapter is to demonstrate to non-specialists that science itself can provide a profound basis for meditation, using a series of short vignettes that address questions relevant to our everyday lives. Meditation often begins by focusing on something beautiful such as a flower or a shell—with a little information it is equally possible to use a mathematical concept or a molecular interaction to lead us into deep reflection and even contemplation. Reciprocally, the aim is to show that the results of spiritual insight can be directly relevant to the practice of science. Insights from religious experience and praxis, such as enlightenment or the dark night of the soul, have their parallels in the scientific enterprise. Scientific insights do not threaten religious practice, but instead provide a means by which we can re-examine our beliefs and mature in our comprehension of the Almighty.

Although there is a common search for truth, science and religion are not entirely analogous. The methodology by which data are acquired is different and the factors that constitute 'proof' are not the same. Although we can enhance our view of the world by engaging with either or both, it is unhelpful to extrapolate literally from one to the other. It is particularly unadvisable for theologians, because a particular scientific world-view will change with time, and it is essential to develop a theology that is consistent with our current knowledge of the world but not dependent on it. Theology itself has little to say about the scientific method—but it provides a basis for the personal and ethical development of the scientist, and is a means of attaining the maturity required for the practice of science. By the same token, Holy Writ is not a handbook for scientists, who do not, on the whole, conduct their science by taking account of data extracted from Scripture. Science measures reality in terms of reality. Religion deals with the interaction of reality with abstract values and concepts. It addresses the attempt of the created world to interact with that which is not material.

In contrast to a commonly held view, many scientists practise their religion without compromising their scientific thinking. It is certainly challenging to hold in tension insights from both areas of life without compartmentalizing them to avoid conflict; but, then again, becoming an integrated person is one of the great challenges in life. There is no question that there are ideas in science that are inconsistent with the teachings of religion, and vice versa; but such data present an opportunity as well as a challenge, and point to the need for further exploration.

Modern biology, philosophy, and religion pose many questions, including those which are addressed in this reflective chapter.

a. How do humans cope with the enormous complexity of their lives, find purpose and meaning, and make decisions?
b. How do we find an answer to the question, 'What is creation for?'
c. Can the concept of God be valuable to us even if we are not describing a material reality?
d. How far are we controlled by our genes, by the initial conditions of the universe, or by a God who has predetermined everything?
e. Can we understand how to work in partnership with God and nature?
f. How do we move forward decisively when we have only partial information?
g. How does the creative process happen?
h. How do order and disorder relate to creativity?

i. How do we deal with pain and death?

j. What personal qualities are required in order to engage effectively with science or religion?

k. Is searching for answers to questions a way to lose one's faith?

SIMPLICITY AND COMPLEXITY

The universe in which we stand is both elegantly simple and breath-takingly complex. It is simple enough for us to penetrate the mathematics, physics, chemistry, and biology that underpin the existence of everything from the smallest fundamental particle to the most complicated of living organisms. It is complex enough for us to be challenged to describe and explain the seemingly infinite number of ways in which the basic building blocks of matter can be combined to achieve unique, functioning, sentient, emergent forms of life.

Moreover, the universe is both unified by the fundamental constants that constrain all that exists, and diversified by the spatial and temporal location of matter and the relationships that such singularities allow. The simple and the unified give rise to the complex and the diverse. A major undertaking for modern biology is to explore the boundaries where the simple and the complex relate co-operatively with each other so that new properties and functions can emerge. An example of this is the 'immunological synapse', where cells infected with pathogens interact with patrolling T-cells that give instructions to the infected cell to die. Although the properties or 'rules' which govern the behaviour of the complex interactions at the cell junctions do not transgress those of the individual molecules that form the synapse, it is clear that living organisms depend on the integration of 'simple' entities into larger more 'complex' machineries.

DECISION MAKING

Human beings too are both simple and complex. We live in a glorious technicolour world in which we somehow deal decisively with the messiness of everyday life on a minute-by-minute basis despite the complexity of the information that both drives and informs us. Our most basic decisions are guided by the possibilities and limitations imposed by our genes and environment. The metabolic systems that keep our bodies alive, the repair mechanisms that enable us to survive environmental insults, and our immune systems that combat disease operate largely without any conscious intervention by us, according to principles that we can increasingly understand. These systems support our continued existence, whether we like it or not.

However, living a fulfilled human life is far more complicated than simply remaining alive. As individuals we need to become integrated into an even more complex world, so that we find a niche where we can flourish physically and emotionally. Our cultural environment teaches us how to relate to others, and gives us the practical skills and materials for learning what we need to deal with the complex world of work and family. Perhaps these cultural values give us the courage we need to persist in the face of personal failure.

Yet most of us expect to achieve even more in the fleeting moment of consciousness that we are privileged to experience. Expressing our creative ideas and responding to beauty, love, joy, sorrow, death, and loss require even more complex information than our genes and culture alone can provide. This kind of information arises from a synthesis that takes place deep within the human psyche; it is an awareness that we grasp as tenuously as a dream, but which, even as it slips through our fingers, we struggle to articulate so that it can be of lasting value to ourselves and to others.

WHAT IS EVERYTHING FOR?

A scientific world-view is enriching and indispensable in today's world, but we need to do more than just appreciate and exploit the material advantages that it provides. We need to discover how to relate the detailed information about the material world that is now at our fingertips with the deepest yearnings of the human spirit. We need to integrate our scientific world-view into our inner worlds that traditionally have been informed by a multitude of other insights. Science is just one window through which we view the world, and alone it is not enough. If we are to make the complex choices that confront us with good judgement, then we need to complement our scientific base with views from other windows—from art, music, literature, and the great religions of the world. Above all, we need to address the age-old question, 'What is everything for?' To be more specific, if we reject the concept of a god who controls every aspect of our lives either through divine intervention or through genetic determinism, how will we determine the purpose of our lives? For, regardless of our certain knowledge that our lives are finite, that the Earth will eventually be consumed, and in the end all that we know will no longer exist, we continue to imbue our existence with meaning. Our choices are informed by the short- or long-term sense of purpose which we give to our lives. Our choices demonstrate the extent to which we are in touch with the essence of our human nature, which in its positive aspects is at various times creative, courageous, restless, inquiring, austere, competitive, co-operative, generous, rational, ethical, and full of wonder. Our choices also reflect the ways that we have dealt, or not dealt, with the destructive forces within us. Without doubt, the cumulative effect of our choices will spell out death or glory to this planet.

How Does the Creative Process Happen?

The immense creative burst that is the universe which has given birth to us is displayed in endless forms that inspire and delight us. We can only marvel at the vast power of matter and energy that have combined to form everything from the stellar displays in the night sky to the unseen molecular machines that provide our cells with energy. Our universe is the manifestation of this creative force that in most religions provides a foundation for the concept of God.

Innate biological systems control the mechanisms that sustain life, and our cultural inheritance helps us to make the most of our environment, but an individual original creative discovery is novel—it is not programmed or learned. Humans are also filled with creative energy, and we bring this to birth in any number of forms, including art, music, drama, science, and literature. Once we have encountered new insights at the centre of our being, our impulse is to create something which will allow us to articulate our inner experience so that we can integrate it into our own lives and share it with others. The continuous challenge is to bring our informed, intuitive inner knowing to a reality that we can integrate with other forms of knowing and use as a platform of consolidated ideas from which to step further into the unknown.

There is a kind of intuitive wonder which we experience when we look at the natural world and see a glorious sunset or a leopard running across the plains or a perfectly formed crystal. However, there is another aspect to wonder which primarily involves the intellect, because it appreciates the mechanism behind the beauty. Eventually this type of intellectual wonder can give way to the intuitive, for the mechanisms can become so well studied and understood that the distance between the observer and the observed is somehow eclipsed, and then the scientist intuitively understands the sunset, the leopard, or the crystal. We become one with the reality of things.

For example, we often identify with the molecules that we work with. As a scientist, I use this deep and intuitive sense to understand how some molecules malfunction in order to cause a disease such as rheumatoid arthritis. Before we can hope to achieve this, it is necessary to understand all that we can about the molecule which has attracted our attention—properties such as its size, shape, dynamics, the distribution of charge on its surface, its surroundings, and the details of its interactions with other molecules. So we build up a picture in our imaginations, and, as we do more experiments, the picture becomes more detailed, so in our minds the molecule assumes giant proportions (see Figure 24.1). In the end we are able somehow to walk around it; we can explore inside it; we can feel how it moves, how it sounds and feels. We can predict intuitively how it will respond to changes in its environment and finally, when we can consciously express our intuition, we can test our predictions by experiment.

A model that addresses the way in which intuitive understanding culminates in conscious knowing is an interpretation of the Divine as expressed in the painting of the Trinity by the artist Theodore Zeldin (see Figure 24.2). This icon of the three-in-one and one-in-three powerfully portrays the gentleness and strength of the relationships between the three aspects of the Godhead, showing how each gives meaning to the others and sustains the others, yet depends on them to achieve its own wholeness. The first aspect represents that part of ourselves which is the root and ground and source of our being, the Creator God who, when we approach, creates a sense of awe in us as when we look into vast mountain ranges or deep into a starlit sky. The second aspect, the human face of God encompassing justice, mercy, compassion, and truth, represents that part of our being to which we have access, which we have struggled to bring to conscious expression, articulating it first in music, symbol, and the visual arts, and finally in words. The third aspect, wisdom, represents the way by which we synthesize conscious and unconsciously acquired information and then bring understanding to conscious knowing. The Trinity is the basis of unified wholeness. Harmonious complexity, as in the components of a flower, can be manifested in simplicity of form when we view the whole.

These three aspects of our nature are inextricably woven together, for human beings have rational minds and contemplative and intellectual knowledge. Intuitive and rational interpretations of our human experience, whether scientific, artistic, or religious, move through our levels of understanding to nurture and, once integrated, sustain our whole being in harmony.

How do Order and Disorder Relate to Creativity?

In the practice of science, scientists often find inspiration from other aspects of their lives. For example, exploring the relationship between order and chaos is important when thinking about creativity. In a Zen garden in Kyoto it seems that everything is asymmetric, including the tree branches which are deliberately bent. When a raindrop falling from the tip of a twig into the pool below instantly makes a small impact, we can encompass it in our vision. A second later, the circling ripples are too wide for us to take in at a glance, yet we know that we can understand the geometry.

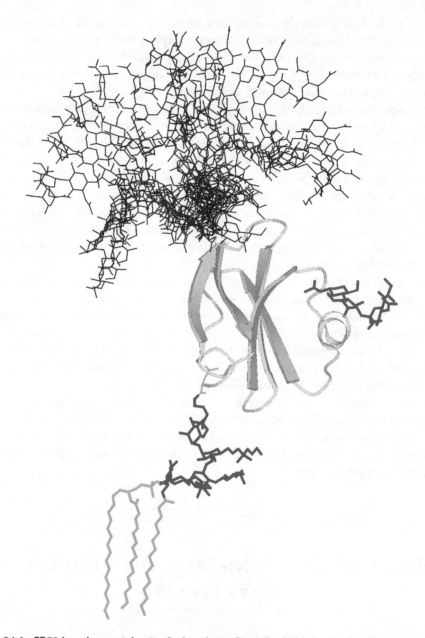

Fig. 24.1. CD59 is a glycoprotein attached to the surface of red blood cells. It protects the cells from damage in inflammatory sites. CD59 has a protein domain, one N–linked sugar shown as a series of conformational overlays to convey the space that it occupies, a smaller O–linked sugar and a glycosylphosphatidyl inositol anchor. *Source*: M. R. Wormald, R. A. Dwek, and P. M. Rudd., 'Review on Glycobiology: Glycosylation: Heterogeneity and the 3D Structure of the Protein' in P. M. Rudd, and R. A. Dwek (eds.), *Critical Reviews in Biochemistry and Molecular Biology*, 32 (1997): 1–100, 40.

Fig. 24.2. Trinity by Theodore Zeldin.

A moment later the first ripples have reached the bank and are reflecting back across the circles, and suddenly we are lost. In a split second we are thrown from a position of complacency and control to one of uncertainty and confusion (see Figure 24.3).

Another unsettling feature of the garden is the raked stone garden which again challenges us to come to terms with complexity. As we sit on the balcony contemplating the moss that grows only on the stones, we still struggle to be in control by trying to do the impossible and see all of the fifteen stones simultaneously. The garden is designed so that only fourteen can be seen at any one time. When, finally, we become still and prepared to accept that we cannot know everything, we lift our eyes to the walls of the garden which are set at less than 90 degrees; we experience an illusion that we are looking into the far distance, and are again acutely aware that the order and control that we crave is not a feature of the natural world. There are no straight lines or right angles in natural forms of life. The creative force in the universe glories in complexity.

Fig. 24.3. Ryōanji temple in Kyoto. The stone garden in the temple of the Silver Pavilion (Ginkakuji, as it is popularly known; the actual temple name is Jishōji).

In the tea ceremony, in the practice of *chanoyu*, we learn many things, one of which is that to live creatively means to live fully in the present moment where we are not constrained by the past nor held in thrall to fears of what lies in the clouds on the horizon.

Indeed, the whole of the garden proclaims the idea that the greatest potential for creativity is when our minds are still free to choose from all the options that lie before us. Once we begin to order things, we are more comfortable, because we feel we are establishing control and gaining insight. Yet, as we continue, our options steadily become narrower. We need to establish order in our efforts to comprehend the physical universe and the heart of humans, yet once we have stabilized an idea with experimental evidence or religious praxis, we need to move forward. If we hope to be truly creative and understand something of the creative spirit of the universe, we need to remain 'uncomfortable' with the tension and continually return to the edge of the unknown and immerse ourselves in complexity until we find another thread to follow. Simplicity and precise function emerges from complexity.

CAN THE CONCEPT OF GOD BE VALUABLE TO US EVEN IF WE ARE NOT DESCRIBING A MATERIAL REALITY?

For millennia people have projected their profound internal awareness onto the concept of a god or of a way, expanding the concept to include such attributes as justice, mercy, steadfast love, truth, compassion, charity, harmony, forgiveness and redemption, eternal life and ultimate reality. A concept is a powerful thing. A concept may not have the properties of matter, but nonetheless it can affect human behaviour as profoundly as a natural disaster. The concept of God provides a means of articulating some of our deepest insights in a way that can be communicated to others through common language, ritual, and myth, and developed as a practical guide to living, finding meaning, and guiding choices. Regardless of whether this God exists apart from human consciousness, the practice of religion has provided a basis for ethics and a purpose for living in almost every part of the world. The value of dealing with a god that represents more than simply naked power is that it provides a context in which we can develop our own value systems in a way that involves both the heart and the intellect.

CAN WE UNDERSTAND HOW TO WORK IN PARTNERSHIP WITH GOD AND NATURE?

Our own small struggles to express our scientific or religious insights allow us to experience the courage and commitment involved in paying the personal costs of creativity. The greater our struggle, the greater is our appreciation that all creations come, like the jewel for which one would give all that one has, at a great price. The greater is our own commitment, perseverance, and humility, the more we can identify with the power that led to our own existence. Millennia of cosmic evolution were required to give rise to the solar system, and further millennia for life, death, rearrangement, and trial and error to culminate in the emergence of our earliest ancestors and, eventually, despite all the odds, ourselves. At last humans evolved as conscious intelligent beings, able not only to comprehend the universe but increasingly to play a significant role in the development of life on Earth. Today we no longer need courage to fight nature to survive. If we are to continue as a species, we now need courage to work in partnership with nature and the spirit recognized by Wordsworth 'that impels all thinking things, all objects of all thought, and rolls through all things'. Partnership is not just one-sided. It involves respect for the other and the courage to relinquish total control. We have the freedom to make informed choices, and with freedom comes the responsibility to discern how best to use it. Science opens many possibilities through which we can change the world; however, we need more than scientific insights to know how to use these opportunities well.

In science we do not control or manipulate data. We try not to manipulate or control each other either, but to work in international partnerships or consortia, learning from each other's history and culture. By the same token, we cannot control God, and if we wish to become responsible individuals, then we do not want God to control us either. As we mature, we recognize that we are in a co-creative partnership. Control by us or by God may satisfy our desire for security, but it doesn't really fit with our observance and experience of nature, or indeed with experience of ourselves at the level of intentional action. It appears, instead, that there is openness, a range of possibilities within nature. If we either envisage God as someone who designed everything once and for all at the beginning of creation, or we envision ourselves as the ultimate controllers of the natural world through science, then we miss the majesty of the possibility of creation itself being creative as it responds in harmony with changes to itself. Perhaps God's involvement with us is such that together we can attain real novelty, contingency, and opportunity that preserve the integrity of life in the process.

How Do We Make Decisions When We Have Only Partial Information?

Much of the time we have to work with partial information. We travel in the dark, dealing with probabilities and weighing up different levels of risk. We see through a glass darkly; we walk in a mist where things that are hidden are clearer to us than we are able to articulate. We need all of the information that comes from our penetrating scientific method, as well as all of the sensitivity that comes from the receptive, responding spiritual depth that lies within us. Most of all we need to find a balance between the two. Mount Sinai is mysteriously beautiful and full of soft blue-velvet shadows when bathed in moonlight and lit by stars and planets hanging like lanterns from the Middle Eastern sky. The small camps of Arabs selling water to the travellers are sprinkled along the winding track up the mountain side, their twinkling lights showing the way to the summit. It is a quiet, gentle, feminine experience inviting silence and reflection and inner stillness from those of all religions (and of no religion) who ascend to the summit to greet the dawn. The Middle Eastern sun rises with unbelievable splendour over the mountain tops, dimming the deep shadows of the night with penetrating shafts of red and gold, bringing heat and light to an awakening world. In a moment of exquisite harmony, half of the mountain is bathed in sunlight while, behind the observers of the rising Sun, the remainder lingers in the pale light of the remaining stars and the fading Moon.

Even science itself sometimes requires us to be receptive and to listen. It is in a profound silence that nature reveals herself through long hours of experimentation, whirring machines, and piles of data. From simple single manipulations on the bench, complex information is generated which then distils into elegant conclusions. The buds that veil the blossom finally open to display the beauty of the flower without our further intervention.

Is Searching for Answers to Questions a Way to Lose One's Faith?

Scientists are driven to ask questions and solve problems. Searching for answers to questions is not a challenge to faith, but a route to it. Religious truth is not about making us comfortable, but about making us strong to work out the purposes we have set for ourselves in the world. If both science and religion are to continue to support our deepest hopes and aspirations, nothing can be thought of as the final conclusion. In both science and religion it is crucial to distinguish clearly between what we know to be true and what we would like to be true, but have neither demonstrated nor experienced. Often it seems that we need to revise our earlier scientific models or develop our concepts of God to accommodate new data or insights. Sometimes the way feels very lonely, for at the cutting edge in both science and religion we are called upon to make our own individual journeys of discovery. In our quest for truth in science and religion, we are responsible for following the path lit by the spirit within us with integrity and also for integrating these findings into the field or community in which we live.

What Personal Qualities are Required to Engage Effectively with Science or Religion?

The personal qualities needed to be a creative scientist are not so very different from those long recognized as being important in the monastic journey. They include a desire for truth, perseverance, detachment from material things and from self-interest, humility, and a level of austerity by which to accept criticism and to sacrifice comfort and pleasure to reach the goal. We have to feel a need, an emptiness, and a passion before we can truly find a place to begin our own scientific journey. We need to exercise judgement when staking out the way, courage to tread a new path, determination to journey onwards in spite of personal failure, and single-mindedness in the struggle to order our thoughts, accomplish our aims, and articulate our discoveries. The ability to continue in spite of the apparent insignificance of our lives and the apparent pointlessness of this creation requires more than the unconscious response of our genes and our social surroundings can teach us. It requires inspiration and humility before nature as we seek to unravel her ways. We move from unknowing to knowing. It is like crossing a river—we cast off from one bank, out of the shallows into the deep. We swim out of our depth, buffeted by the currents of ignorance and self-doubt, until our feet once again touch the ground and we arrive transformed by new insights on the far bank.

How Do We Deal with Pain and Death?

The natural world is also restless and dynamic, constantly rearranging and reinterpreting itself to create new features, new worlds, and new life. It is often a very violent universe, working through fire and tempest, death, destruction, and pain to forge new life. The question of theodicy troubles sentient creatures, but not the rest of the natural world. Pain, like so much that makes us human, is something that evolves with consciousness. The physical forces and systems that have given rise to life feel no pain and no regret. Death is simply a means to a metamorphosis that allows change. Death and decay give the raw materials back into the hands of forces that reshape them here into a butterfly, now into a flower, and there into a dictator. The cost of the beauty and elegant adaptation that we see around us is that of all the species that never made it, all the victims of disease, accident, and predators, all the mutations that led to non-viability. Nature is not economical and does not count the cost of human heartache. Science and religion both attempt to mitigate the pain in their different ways. Science provides us with information and gives us a means to change what can be physically changed. Religion helps us to deal with what cannot be changed and with what can be alleviated only by changing the human heart.

We may wish to live in a secure, unchanging world, but the reality is that stability is an illusion that we can espouse only because of our short lifetimes compared with the geological time frame, and perhaps order is an illusion, too. For we all attempt to classify the natural world—by colour, by size, by species, by sex—but we know that many things do not fall neatly into a single category; there are always overlaps, exceptions, and fuzziness at the boundaries—turquoise is neither blue nor green; theft is wrong, except under some circumstances; a zebra is neither black nor white; and who can determine the exact diameter of a tree trunk?

There was once a time when God was envisioned as the great designer, ordering the natural world in a magnificent hierarchical structure. Eventually we would understand all the laws, and everything would be understood—indeed, physicists still work to develop an elegant 'Theory of Everything' which will unify all current theories, bringing together quantum theory and relativity to enable us to explain and predict all the events in the universe. While such an aim might be achievable at the level of particle physics, in the world that our senses open up to us at least some of the order we perceive exists in our minds and is a consequence of our temporal and spatial location and our use of language. Maybe we impose order on the natural world to fill an essential need we have to make paths and tracks so we can travel in the vast network which constitutes our minds, just as we use roads and rivers to navigate the physical world and make constellations out of the stars.

Reality is hard to define. Concepts as well as scientific proof can affect our lives, and concepts can often form the basis of real choices. Concepts or models may or may not turn out to be based on flimsy evidence, and they may be but a pale shadow of the reality they are seeking to reveal; but in any case they need to be taken seriously.

THE SPIRIT OF TRUTH

Models may not be the reality, but if they are pragmatic and help us to develop our thinking, they are valuable. The models which we use to visualize the molecules with which we work and the words with which we describe them are symbols, not the reality. However, they are powerful images that enable us to describe and grasp the experimental raw data that is beyond the reach of our physical senses. In science our common understanding of systems, symbols, and experimental practice means that we are able to communicate with each other (even across different scientific disciplines) and get to the bottom line quickly. These are both learned but also added to by each individual researcher, so that they form an ever-changing representation of cutting-edge research. Systems, symbols, and experimental practice enable us to bring intuitive knowledge to conscious testable reality. For scientists, as well as those who practise religion, work in a cloud of unknowing where things are not understood, and even the tools to explore the questions do not yet exist. The language of science is austere, sparse, and focused. A sentence can sometimes take hours to construct, for each word can be as laden with significance as that in a poem as we struggle to express new ideas. Writing in such a way is a skill that is hard to learn, pruned as it must be from subjectivity, imprecision, and speculation. It is a challenge to find the right words when we know that to speak at all is to dilute or distort the reality. Words are but the messengers of reality, the verbal is provisional, and only the truth is real.

The words and symbols which we use to describe God are likewise not the complete disclosure of reality, but a means of articulating an even deeper knowledge which is also beyond the reach of our physical senses. We all need to develop some kind of language to express our basic concepts. Religion, music, and art, as much as science, involve learning a language that can convey ideas. For example, we use the language of symbolism and ritual actions to convey the meaning of the Eucharist. At the heart of the Mass, bread and wine symbolize the eternal life-giving Spirit and enable us to open our own lives to this sustaining presence. These symbols provide a way of connecting 'earth' and 'heaven', the 'reality' with the 'potential' of our lives, and our conscious understanding with our intuitive understanding of the world. In religion, doctrine, symbols, and rituals act, as they do in science, to enable communication between individuals and across religions. I once wandered into a tiny Romanian Orthodox chapel in Transylvania one evening at dusk. No matter that I couldn't follow word for word the liturgy. The icons streaming with light and meaning, the sonorous voices, the intense devotion of the people, the familiar symbolic gestures of the clergy, meshed instantly with my own faith and religious experience, so that by the end of the service a real bond had developed between us. One of the most profound experiences in my life took place in Kyoto with Buddhist friends, another on a mountain top in Israel amongst Jews and Muslims, and another in Singapore in a Hindu temple. Such experiences make communication possible between people who speak different languages, come from cultures that are worlds apart, yet share religious insights that are deeper than any of these.

Those who are privileged to enter into such life-changing moments of insight and reflect upon their experience use painting, music, poetry, writing, and symbolism and ritual to attain understanding. These media allow us to continue to draw strength and inspiration from the revelation and also allow others access to these sacred experiences. We have within our world religions, and outside them, too, a huge treasury of first-hand experience from which we can draw inspiration in the depths of sorrow, the heights of joy, or just in the normal run-of-the-mill day. A piece of music or a simple symbol can represent a thousand ideas, maybe a million, maybe as many as the people who engage with it. Art, symbolism, and ritual provide a complex, living, growing representation of that part of ourselves and our communities that helps us to deal with our emotional environment, to communicate with each other, and a way to find meaning and purpose in our lives and, most importantly, the courage to be. Proscriptive dogma that cramp the spirit and would make us believe that everything is known, that nothing remains but for us to put ancient teachings into practice, fails to understand the glory of a dynamic creation where we may act to change things responsibly with the power that created us.

How Far are We Controlled by our Genes, by the Initial Conditions of the Universe, or by a God Who has Predetermined Everything?

The genome gives a framework within which an organism operates. In the human immuno-deficiency virus, which replicates very rapidly, frequent mutations change the structure of the viral coat, allowing the virus to survive in the hostile environment of our immune systems. Virus particles that our immune systems recognize are eliminated, and new clades which can evade the immune system expand to take over.

It is trial and error, where the viral genome randomly mutates and the environment in the host selects which will survive. Absolute determinism and fidelity in the HIV genome would restrict its survival. The coat proteins function mainly to deliver the virus to a new host cell, and as long as the structure is retained at a few crucial sites that mediate this process, much of the rest of the viral coat can change, so that the immune system is continually challenged.

In complex higher organisms that reproduce less often, significant changes in a species happen slowly, directed by recombinations that accompany sexual reproduction and opportunistic mutations. In a single individual the inherited genome is fixed at conception and does not change unless there is damage to the genetic material. DNA repair mechanisms are in place to avoid the consequences of such damage, which are usually disastrous. However, the genes do not operate independently—the proteins they code for are expressed in response to signals which come from the conditions *in utero* and from the environment after birth.

Our genetic make-up is fundamental to the person we are, but it does not constitute all that we are. It represents potential, and imposes constraints on what we can do.

At the physical level sophisticated systems control growth, repair, and infection, and our cells access the genome in their nuclei to generate the particular proteins and lipids at the precise levels in which they are needed. Almost all of this flexible, adaptable 'housekeeping' goes on without any conscious intervention by us. We become aware of these systems only when something goes wrong, such as when we do not have adequate nutrition to keep the systems operating or a disease develops. Even then there are fallback mechanisms that will attempt to keep things running.

At the social level we recognize cues that enable us to live in communities more or less harmoniously. Our genes certainly play a role in determining our personality traits, but as we know very well, we are heavily influenced by our cultural experience. Acquiring the accumulated wisdom of our predecessors enables us to maximize the possibility of survival and improve the quality of our lives. Living dynamically within our constantly changing environment demands appropriate flexible responses.

The genome is an entity that can be structurally described and mapped. The environment has measurable physical parameters. The human person is less easy to categorize. It emerges from physically definable bodies and brains, and yet it is not yet clear how the physical aspects we can observe and measure are linked with our conscious awareness of ourselves as integrated individuals who have abstract ideas of non-material entities, of ourselves, and our world. Yet this very aspect of our personalities contributes to decisions that critically determine our responses to our environment.

Absolute determinism imposed by our genes acting without information from the environment would seriously curtail our ability to respond to novel situations and to select intelligently from the choices before us. Intelligent action requires a conscious decision following the assimilation and integration of numerous diverse pieces of information. Intentional choices require judgement; the decisions we make reflect our priorities. Our priorities reflect our desire to secure our personal and communal survival, for happiness and fulfilment of our personal destiny.

Religious experience is also a complex activity in which the same 'input' or experience may be interpreted in different ways in different cultural environments, giving rise to alternative 'outputs' or practice. In fact, the analogy between these two sets of processes sheds light on the nature of religious experience and religious symbols. Mathematical or biochemical descriptions of a system constitute an attempt to find general statements that can apply to many singularities. In religion, symbolism may fulfil the same purpose and provide a unifying function in inter-religious dialogue.

Neither overly deterministic approaches in science nor extreme fundamentalism in religion takes into account our need to be creative and to select intelligently between the many choices that lie before us. Nor do these philosophies support our conviction that, as mature individuals, we bear responsibility for protecting and enhancing life, perhaps even as co-creators with the Almighty.

For light and truth came into the world that we might have life, and have it more abundantly.

PART IV

METHODOLOGICAL APPROACHES TO THE STUDY OF RELIGION AND SCIENCE

CHAPTER 25

THE SCIENTIFIC LANDSCAPE OF RELIGION: EVOLUTION, CULTURE, AND COGNITION

SCOTT ATRAN

INTRODUCTION

Ever since Edward Gibbon's *Decline and Fall of the Roman Empire*, first published in 1776, scientists and secularly minded scholars have been predicting the ultimate demise of religion (cf. Dawkins 1998). But, if anything, religious fervour is increasing across the world, including in the United States, the world's most economically powerful and scientifically advanced society. An underlying reason is that science treats humans and intentions only as incidental elements in the universe (Russell 1948), whereas for religion they are central. Science is not particularly well suited to deal with people's existential anxieties, including death, deception, sudden catastrophe, loneliness, and the longing for love or justice. It cannot tell us what we ought to do, only what we can do (Sartre 1948). Religion thrives because it addresses people's deepest emotional yearnings and society's foundational moral needs.

Science, then, may never replace religion in the lives of most people and in any society that hopes to survive for very long. But science can help us understand how religions are structured in and across individual minds and societies—or, equivalently for our purposes, brains and cultures—and also, in a strictly material sense, why religions endure. Recent advances in cognitive science, a branch of psychology with roots also in evolutionary biology, focus on religion in general, and on awareness of the supernatural in particular, as a converging by-product of several cognitive and emotional mechanisms that evolved under natural selection for mundane adaptive tasks (Atran 2002).

As human beings routinely interact, they naturally tend to exploit these by-products to solve inescapable existential problems that have no apparent worldly solution, such as the inevitability of death and the ever-present threat of deception by others. Religion involves costly and hard-to-fake commitment to a counterintuitive world of supernatural agents who are believed to master such existential anxieties (Atran and Norenzayan 2004). The greater one's display of costly commitment to that factually absurd world—as in Abraham's willingness to sacrifice his beloved son for nothing palpable save faith in a 'voice' demanding the killing—the greater society's trust in that person's ability and will to help out others with their inescapable problems (Kierkegaard [1843] 1955).[1]

RELIGION AS AN EVOLUTIONARY BY-PRODUCT

Explaining religion is a serious problem for any evolutionary account of human thought and society. All known human societies—past or present—bear the very substantial costs of religion's material, emotional, and cognitive commitments to factually impossible, counterintuitive worlds. From an evolutionary standpoint, the reasons why religion shouldn't exist are patent. Religion is materially expensive, and it is unrelentingly counterfactual and even counterintuitive. Religious practice is costly in terms of material sacrifice (at least one's prayer time), emotional expenditure (inciting fears and hopes), and cognitive effort (maintaining both factual and counterintuitive networks of beliefs).

Summing up the anthropological literature on religious offerings, Raymond Firth (1963: 13–16) concludes that 'sacrifice is giving something up at a cost.... "Afford it or not," the attitude seems to be'. That is why 'sacrifice of wild animals which can be

---

[1] The outlines of the factually preposterous world a person is committed to must be shared by a significant part of society, lest the person be considered a deviant psychopath or sociopath (e.g. lest Abraham's willingness to sacrifice his beloved son Isaac be considered attempted murder or child abuse).

regarded as the free gift of nature is rarely allowable or efficient' (Robertson Smith 1894: 466). As Bill Gates aptly surmised, 'Just in terms of allocation of time resources, religion is not very efficient. There's a lot more I could be doing on a Sunday morning' (cited in Keillor 1999).[2]

Functionalist arguments, including adaptationist accounts, usually attempt to offset the apparent functional disadvantages of religion with even greater functional advantages. There are many different and even contrary explanations for why religion exists in terms of beneficial functions served. These include functions of social (bolstering group solidarity, group competition), economic (sustaining public goods, surplus production), political (mass opiate, stimulant to rebellion), intellectual (explaining mysteries, encouraging credulity), and emotional (allaying terror and anxiety) utility, as well as health and well-being (increasing life expectancy, acceptance of death). Many of these functions have obtained in one cultural context or another; yet all have also been true of cultural phenomena besides religion.

Such descriptions of religion often insightfully help to explain how and why given religious beliefs and practices provide competitive advantages over other sorts of ideologies and behaviours for cultural survival. Still, these accounts provide little explanatory insight into cognitive selection factors responsible for the ease of acquisition of religious concepts by children, or for the facility with which religious practices and beliefs are transmitted across individuals. They have little to say about which beliefs and practices—all things being equal—are most likely to recur

---

[2] In sum, religious sacrifice generally runs counter to calculations of immediate utility, such that future promises are not discounted in favour of present rewards. In some cases, sacrifice is extreme. Although such cases tend to be rare, they are often held by society as religiously ideal: e.g. sacrificing one's own life or nearest kin. Researchers sometimes take such cases as prima facie evidence of 'true' (non-kin) social altruism (Rappaport 1999; Kuper 1996), or group selection, wherein individual fitness decreases so that overall group fitness can increase (relative to the overall fitness of other, competing groups) (Sober and Wilson 1998; Wilson 2002). But this may be an illusion.

Consider suicide terrorism (Atran 2003). The 'Oath to Jihad' taken by recruits to *Harkat al-Ansar*, a Pakistani-based ally of Al-Qaeda, affirms that by their sacrifice they help secure the future of their 'family' of fictive kin: 'Each [martyr] has a special place—among them are brothers, just as there are sons and those even more dear.' In religiously inspired suicide terrorism, these sentiments are purposely manipulated by organizational leaders and trainers to the advantage of the manipulating élites rather than the individual (much as the fast food or soft drink industries manipulate innate desires for naturally scarce commodities like fatty foods and sugar to ends that reduce personal fitness but benefit the manipulating institution). No 'group selection' is involved for the sake of the cultural 'superorganism' (Wilson 2002; cf. Kroeber 1963)—like a bee for its hive—but only cognitive and emotional manipulation of some individuals by others. In evolutionary terms, quest for status and dignity may represent proximate means to the ultimate end of gaining resources, but, as with other proximate means (e.g. passionate love), they may become emotionally manipulated ends in themselves (Tooby and Cosmides 1992).

in different cultures and are most disposed to cultural variation and elaboration. None predicts the cognitive peculiarities of religion, such as:

- Why do agent concepts predominate in religion?
- Why are supernatural agent concepts culturally universal?
- Why are some supernatural agent concepts inherently better candidates for cultural selection than others?
- Why is it necessary, and how is it possible, to validate belief in supernatural agent concepts that are logically and factually inscrutable?
- How is it possible to prevent people from deciding that the existing moral order is simply wrong or arbitrary and from defecting from the social consensus through denial, dismissal, or deception?

This argument does not entail that religious beliefs and practices cannot perform social functions, or that the successful performance of such functions does not contribute to the survival and spread of religious traditions. Indeed, there is substantial evidence that religious beliefs and practices often alleviate potentially dysfunctional stress and anxiety (Ben-Amos 1994; Worthington *et al.* 1996) and maintain social cohesion in the face of real or perceived conflict (Allport 1956; Pyszczynski *et al.* 1999). It does imply that social functions are not phylogenetically responsible for the cognitive structure and cultural recurrence of religion. The claim is that religion is not an evolutionary adaptation *per se*, but a recurring cultural by-product of the complex evolutionary landscape that sets cognitive, emotional, and material conditions for ordinary human interactions (Kirkpatrick 1999; Boyer 2001; Atran 2002; Pinker 2004). Religion exploits ordinary cognitive processes to display passionately costly devotion to counterintuitive worlds governed by supernatural agents. The conceptual foundations of religion are intuitively given by task-specific panhuman cognitive domains, including folk mechanics, folk biology, and folk psychology. Core religious beliefs minimally violate ordinary notions about how the world is, with all of its inescapable problems, thus enabling people to imagine minimally impossible supernatural worlds that solve existential problems, including death and deception.

THE SUPERNATURAL AGENT: HAIR-TRIGGERED FOLK PSYCHOLOGY

Religions invariably centre on supernatural agent concepts, such as gods, goblins, angels, ancestor spirits, and jinns. Granted, non-deistic 'theologies', such as Buddhism and Taoism, doctrinally eschew personifying the supernatural or animating nature with supernatural causes. Nevertheless, common folk who espouse these faiths routinely entertain belief in an array of gods and spirits that behave counter-intuitively in ways that are inscrutable to factual or logical reasoning.[3] Even Buddhist monks ritually ward off malevolent deities by invoking benevolent ones, and conceive altered states of nature as awesome.[4]

Mundane agent concepts are central players in what cognitive and developmental psychologists refer to as 'folk psychology' and 'theory of mind'. A reasonable speculation is that agency evolved as a hair-triggered response in humans, who needed to react 'automatically' under conditions of uncertainty to potential threats (and opportunities) by intelligent predators (and protectors). From this evolutionary perspective, agency is a sort of 'Innate Releasing Mechanism' (Tinbergen 1951) whose original evolutionary domain encompasses animate objects but which inadvertently extends to moving dots on computer screens, voices in the wind, faces in clouds, and virtually any complex design or uncertain situation of unknown origin (Guthrie 1993; Hume [1757] 1956).[5]

Experiments show that children and adults spontaneously interpret the contingent movements of dots and geometrical forms on a screen as interacting agents with

---

[3] Although the Buddha and the buddhas are not regarded as gods, Buddhists clearly conceive of them as 'counter-intuitive agents' (Pyysiäinen 2003). In Sri Lanka, Sinhalese relics of the Buddha have miraculous powers. In India, China, Japan, Thailand, and Vietnam, there are magic mountains and forests associated with the Buddha; and the literature and folklore of every Buddhist tradition recount amazing events surrounding the Buddha and the buddhas.

[4] Experiments with adults in the USA (Barrett and Keil 1996) and India (Barrett 1998) illustrate the gap between theological doctrine and actual psychological processing of religious concepts. When asked to describe their deities, subjects in both cultures produced abstract and consensual theological descriptions of gods as being able to do anything, anticipate and react to everything at once, always know the right thing to do, and able to dispense entirely with perceptual information and calculation. When asked to respond to narratives about these same gods, the same subjects described the deities as being in only one place at a time, puzzling over alternative courses of action, and looking for evidence in order to decide what to do (e.g. to first save Johnny, who's praying for help because his foot is stuck in a river in the USA and the water is rapidly rising; or to first save little Mary, whom he has seen fall on railroad tracks in Australia where a train is fast approaching).

[5] When triggered by a certain range of stimuli, an innate releasing mechanism 'automatically' unleashes a sequence of behaviours that were naturally selected to accomplish some adaptive task in an ancestral environment. Consider food-catching behaviour in frogs. When a flying insect moves across the frog's field of vision, bug-detector cells are activated in the frog's brain. Once activated, these cells in turn massively fire others, in a chain reaction that usually results in the frog shooting out its tongue to catch the insect. The bug-detector is primed to respond to any small dark object that suddenly enters the visual field. For each *natural domain*, there is a proper domain and (possibly empty) actual domain (Sperber 1996). A *proper domain* is information that it is the device's naturally selected function to process. The *actual domain* is any information in the organism's environment that satisfies the device's input conditions, whether or not the information is functionally relevant to ancestral task demands—i.e. whether or not it also belongs to its proper domain. If flying insects belong to the proper domain of frog's food-catching device, then small wads of black paper dangling on a string belong to the actual domain.

distinct goals and internal motivations for reaching those goals (Heider and Simmel 1944; Premack and Premack 1995; Bloom and Veres 1999; Csibra *et al.* 1999). Such a biologically prepared, or 'modular', processing programme would provide a rapid and economical reaction to a wide—but not unlimited—range of stimuli that would have been statistically associated with the presence of agents in ancestral environments. Mistakes, or 'false positives', would usually carry little cost, whereas a true response could provide the margin of survival (Seligman 1971; Geary and Huffman 2002).

Our brains may be trip-wired to spot lurkers, and to seek protectors, where conditions of uncertainty prevail (when startled, at night, in unfamiliar places, during sudden catastrophe, in the face of solitude, illness, prospects of death, etc.). Plausibly, the most dangerous and deceptive predator for the genus *Homo* since the late Pleistocene has been *Homo* itself, which may have engaged in a spiralling behavioural and cognitive arms race of individual and group conflicts (Alexander 1989). Given the constant menace of enemies within and without, concealment, deception, and the ability to generate and recognize false beliefs in others would favour survival. In potentially dangerous or uncertain circumstances, it would be best to anticipate and fear the worst of all likely possibilities: the presence of a deviously intelligent predator.

From an evolutionary perspective, it's better to be safe than sorry regarding the detection of agency under conditions of uncertainty. This cognitive proclivity would favour emergence of malevolent deities in all cultures, just as a countervailing Darwinian propensity to attach to protective caregivers would favour the appearance of benevolent deities. Thus, for the Carajá Indians of central Brazil, intimidating or unsure regions of the local ecology are religiously avoided: 'The earth and underworld are inhabited by supernaturals.... There are two kinds. Many are amiable and beautiful beings who have friendly relations with humans...others are ugly and dangerous monsters who cannot be placated. Their woods are avoided and nobody fishes in their pools' (Lipkind 1940: 249). Similar descriptions of supernaturals appear in ethnographic reports throughout the Americas, Africa, Eurasia, and Oceania (Atran 2002).

In addition, humans conceptually create information to mimic and manipulate conditions in ancestral environments that originally produced and triggered evolved cognitive and emotional dispositions (Sperber 1996). Humans habitually 'fool' their own innate releasing programmes, as when people become sexually aroused by make-up (which artificially highlights sexually appealing attributes), fabricated perfumes, or undulating lines drawn on paper or dots arranged on a computer screen—that is, pornographic pictures.[6] Indeed, much of human culture—for better or

---

[6] An example from ethology offers a parallel. Many bird species have nests parasitized by other species. Thus, the cuckoo deposits eggs in passerine nests, tricking the foster parents into incubating and feeding the cuckoo's young. Nestling European cuckoos often dwarf their host parents (Hamilton and Orians 1965): 'The young cuckoo, with its huge gape and loud begging call, has evidently evolved in exaggerated form the stimuli which elicit the feeding

worse—can be arguably attributed to focused stimulations and manipulations of our species' innate proclivities. Such manipulations can serve cultural ends far removed from the ancestral adaptive tasks that originally gave rise to the cognitive and emotional faculties triggered, although manipulations for religion often centrally involve the collective engagement of existential desires (e.g. wanting security) and anxieties (e.g. fearing death).

Recently, numbers of devout American Catholics eyed the image of Mother Teresa in a cinnamon bun sold in a Tennessee shop. Latinos in Houston prayed before a vision of the Virgin of Guadalupe, whereas Anglos saw only the dried ice cream on a pavement. Cuban exiles in Miami spotted the Virgin in windows, curtains, and television after-images as long as there was hope of keeping young Elian Gonzalez from returning to godless Cuba. And on 9/11, newspapers showed photos of smoke billowing from one of the World Trade Center towers that 'seem[ed] to bring into focus the face of the Evil One, with beard and horns and malignant expression, symbolizing to many the hideous nature of the deed that wreaked horror and terror upon an unsuspecting city' ('Bedeviling: Did Satan Rear His Ugly Face?', *Philadelphia Daily News*, 14 September 2001). In such cases one sees a culturally conditioned emotional priming in anticipation of agency. This priming, in turn, amplifies the information value of otherwise doubtful, poor, and fragmentary agency-relevant stimuli. This enables the stimuli (e.g. cloud formations, pastry, ice-cream conformations) to achieve the mimimal threshold for triggering hyperactive facial recognition and body-movement recognition schemata that humans possess.

In sum, supernatural agents are readily conjured up perhaps because natural selection has trip-wired cognitive schema for agency detection in the face of uncertainty. Uncertainty is omnipresent; so, too, the hair-triggering of an agency detection mechanism that readily promotes supernatural interpretation and is susceptible to various forms of cultural manipulation. Cultural manipulation of this modular mechanism and priming facilitate and direct the process. Because the phenomena that are thereby created readily activate intuitively given modular processes, they are more likely to survive transmission from mind to mind under a wide range of different environments and learning conditions than entities and information that are harder to process (Atran 1990; Boyer 1994). As a result, they are more likely to become enduring aspects of human cultures, such as belief in the supernatural.

'Minimally counterintuitive' worlds allow supernatural agents to resolve existential dilemmas. Supernatural agents, like ghosts and the Abrahamic Deity and Devil are much like human agents, psychologically (having belief, desire, promise, inference, decision, emotion) and biologically (having sight, hearing, feel, taste, smell, coordination), although they lack material substance and some associated physical

response of parent passerine birds. . . . This, like lipstick in the courtship of mankind, demonstrates successful exploitation by means of a "super-stimulus"' (Lack 1968). Late nestling cuckoos have evolved perceptible signals to *manipulate* the passerine nervous system by initiating and then arresting or interrupting normal processing. In this way, cuckoos are able to subvert and co-opt the passerine's modularized survival mechanisms.

constraints. As we shall see in the next section, these imaginary worlds are close enough to factual, everyday worlds to be perceptually compelling and conceptually tractable, but also surprising enough to capture attention and prime memory, and thus 'contagiously' to spread from mind to mind.

CULTURAL SURVIVAL: MEMORY EXPERIMENTS WITH COUNTERINTUITIVE BELIEFS

Many factors are important in determining the extent to which ideas achieve a cultural level of distribution. Some are ecological, including the rate of prior exposure to an idea in a population, physical as well as social facilitators and barriers to communication and imitation, and institutional structures that reinforce or suppress an idea. Of all cognitive factors, however, mnemonic power may be the single most important one at any age (Sperber 1996). In oral traditions that characterize most human cultures throughout history, an idea that is not memorable cannot be transmitted and cannot achieve cultural success (Rubin 1995). Moreover, even if two ideas pass a minimal test of memorability, a more memorable idea has a transmission advantage over a less memorable one (all else being equal). This advantage, even if small at the start, accumulates from generation to generation of transmission, leading to massive differences in cultural success at the end.

One of the earliest accounts of memorability and the transmission of counterintuitive cultural narratives was Bartlett's (1932) classic study of 'The war of the ghosts'. Bartlett examined the ways in which British university students remembered, and then transmitted, a Native American folk tale. Over successive retellings of the story, some culturally unfamiliar items or events were dropped. Perhaps Bartlett's most striking finding was that the very notion of the ghosts—so central to the story—was gradually eliminated from the retellings, suggesting that counterintuitive elements are at a cognitive disadvantage. Bartlett reasoned that items inconsistent with students' cultural expectations were harder to represent and recall, hence less likely to be transmitted than items consistent with expectations.

In recent years, though, there has been growing theoretical and empirical work to suggest that minimally counterintuitive concepts are cognitively optimal: that is, they enjoy a cognitive advantage in memory and transmission in communication. Religious beliefs are counterintuitive because they violate what studies in cognitive anthropology and developmental psychology indicate are universal expectations about the world's everyday structure, including basic categories of 'intuitive ontology', i.e. the ordinary ontology of the everyday world that is built into the language learner's semantic system (e.g. person, animal, plant, and substance; see Atran 1989).

Religious ideas are generally inconsistent with fact-based knowledge, though not randomly. Beliefs about invisible creatures who transform themselves at will or who perceive events that are distant in time or space flatly contradict factual assumptions about physical, biological, and psychological phenomena (Atran and Sperber 1991). Consequently, these beliefs will more likely be retained and transmitted in a population than random departures from common sense, and thus become part of the group's culture. Insofar as category violations shake basic notions of ontology, they are attention-arresting and hence memorable. But only if the resultant impossible worlds remain bridged to the everyday world can information be stored, evoked, and transmitted. As a result, religious concepts need little in the way of overt cultural representation or instruction to be learned and transmitted. A few fragmentary narrative descriptions or episodes suffice to mobilize an enormously rich network of implicit background beliefs (Boyer 1994, 2001).

Basic conceptual modules—naturally selected cognitive faculties—are activated by stimuli that fall into a few intuitive knowledge domains, including folk mechanics (inert object boundaries and movements), folk biology (species configurations and relationships), and folk psychology (interactive and goal-directed behaviour). Ordinary ontological categories are generated when conceptual modules are activated. Among the universal categories of ordinary ontology are person, animal, plant, and substance. The relationship between conceptual modules and ontological categories is represented as a matrix in Table 25.1.

Changing the intuitive relationship expressed in any cell generates what Pascal Boyer (2000) calls a 'minimal counterintuition'. For example, switching the cell (− folk psychology, substance) to (+ folk psychology, substance) yields a thinking talisman, whereas switching (+ folk psychology, person) to (− folk psychology, person) yields an unthinking zombie (cf. Barrett 2000).[7]

In one series of experiments, Barrett and Nyhoff (2001) asked participants to remember and retell Native American folk tales containing natural as well as non-natural events or objects. Content analysis showed that participants remembered 92 per cent of minimally counterintuitive items, but only 71 per cent of intuitive items.[8] These

---

[7] These are general, but not exclusive, conditions on supernatural beings and events. Intervening perceptual, contextual, or psycho-thematic factors, however, can change the odds. Thus, certain natural substances—mountains, seas, clouds, Sun, Moon, planets—are associated with perceptions of great size or distance, and with conceptions of grandeur and continuous or recurring duration. They are, as Freud surmised, psychologically privileged objects for focusing the thoughts and emotions evoked by existential anxieties like death and eternity. Violation of fundamental social norms also readily lends itself to religious interpretation (e.g. ritual incest, fratricide, status reversal). Finally, supernatural agent concepts tend to be emotionally powerful because they trigger evolutionary survival templates. This also makes them attention-arresting and memorable. For example, an all-knowing bloodthirsty deity is a better candidate for cultural survival than a do-nothing deity, however omniscient.

[8] Barrett and Nyhoff (2001: 79) list as common items: 'a being that can see or hear things that are not too far away', 'a species that will die if it doesn't get enough nourishment or if it is severely damaged', 'an object that is easy to see under normal lighting conditions'. Such items

Table 25.1. **Mundane relations between naturally selected conceptual domains and universal categories of ordinary ontology.**

ONTOLOGICAL CATEGORIES	Conceptual Domains (and associated properties)				
	Folk mechanics	Folk biology		Folk psychology	
	(Inert)	(Vegetative)	(Animate)	(Psychophysical, e.g. hunger, thirst)	(Epistemic, e.g. believe, know)
PERSON	+	+	+	+	+
ANIMAL	+	+	+	+	−
PLANT	+	+	−	−	−
SUBSTANCE	+	−	−	−	−

Note: Changing the relation in any one cell (+ to −, or − to +) yields a minimal, supernatural counterintuition.

results, contrary to the findings in Bartlett's classic experiments, seem to indicate that minimally counterintuitive beliefs are better recalled and transmitted than intuitive ones.

Importantly, the effect of counterintuitiveness on recall is not linear. Too many ontological violations render a concept too counterintuitive to be comprehensible and memorable. Boyer and Ramble (2001) demonstrated that concepts with too many violations were recalled less well than those that were minimally counter-intuitive. These results were observed immediately after exposure, as well as after a three-month delay, in cultural samples as diverse as the Midwestern United States, France, Gabon, and Nepal. Consistent with the idea that this memory advantage is related to cultural success, a review of anthropological literature indicates that religious concepts with too many ontological violations are rather rare (Boyer 1994).

Although suggestive, these studies leave several issues unresolved. For example, why don't minimally counterintuitive concepts occupy most of the narrative structure of religions, folk tales, and myths? Even casual perusal of culturally successful materials, like the Bible, Hindu Vedas, or Maya Popul Vuh, suggests that counterintuitive concepts and occurrences are a minority. The Bible consists of a succession of mundane events—walking, eating, sleeping, dreaming, copulating, dying, marrying, fighting, suffering storms and drought—interspersed with a few counterintuitive occurrences, such as miracles and appearances of supernatural agents like God,

fall so far below ordinary expectations that communication should carry some new or salient information that Barrett and Nyhoff (2001: 82–3) report: 'common items were remembered so poorly relative to other items. . . . In some instances of retelling these items, participants tried to make the common property sound exciting or unsusual.' In other words, some subjects tried to meet minimum conditions of relevance (Sperber and Wilson 1995). For the most part, common items failed these minimum standards for successful communication.

angels, and ghosts. One possible explanation for this is that counterintuitive ideas are transmitted in narrative structures. To the extent that narratives with too many counterintuitive elements are at a cognitive disadvantage, cognitive selection at the narrative level would favour minimally counterintuitive narrative structures.

In one study that tested this hypothesis, Norenzayan, Atran, Faulkner, and Schaller, (in press; Atran and Norenzayan 2004) analysed folk tales possessing many of the counterintuitive aspects of religious stories. They examined (1) the cognitive structure of Grimm Brothers' folk tales, and (2) the relative cultural success of each tale measured in terms of stable patterns of retention of story elements over time. The hypothesized non-linear relationship between the frequency of counterintuitive elements and cultural success was confirmed (Figure 25.1). Minimally counterintuitive folk tales (containing two or three supernatural events or objects) constituted 76.5 per cent of the culturally successful sample, whereas stories with fewer counterintuitive elements (less than two), and with excessive numbers of counterintuitive

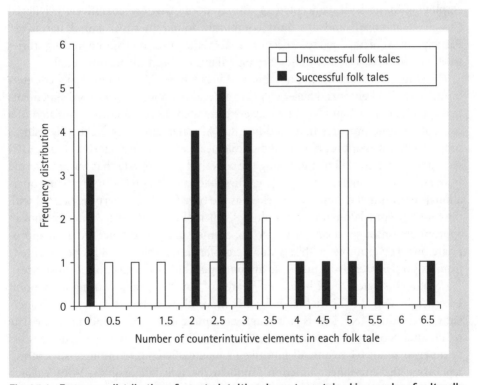

Fig. 25.1. Frequency distribution of counterintuitive elements contained in samples of culturally successful and unsuccessful folktales. The y-axis is the frequency distribution of culturally successful and unsuccessful folk tales. The x-axis is the number of counterintuitives in each folk tale. The number of counterintuitives for some folk tales is expressed as a fraction (1.5, 2.5, etc.). This is because when the two raters who counted disagreed, their ratings were averaged. Successful tales clustered within the range of 2–3 counterintuitives; unsuccessful tales had a flat distribution.

elements (more than three) constituted only 30 per cent and 33 per cent of the culturally successful sample, respectively. Overall, minimal counterintuitiveness predicted cultural success of folk tales accurately 75 per cent of the time. Perceived memorability and ease of transmission, but not other features of the folk tale (e.g. whether the tale contains a moral lesson, interest value to children), partly mediated the relationship between minimal counterintuitiveness and cultural success. While results indicate that cultural success is a non-linear (inverted U-shaped) function of the number of counterintuitive elements, success was not predicted by unusual narrative elements that are otherwise intuitive.

If memorability is the critical variable that mediates the effect of minimal counterintuitiveness on cultural success, than minimally counterintuitive knowledge structures should enjoy superior memory in the long run. To test this hypothesis more directly, in a related study Atran and Norenzayan (2004) examined the short- and long-term memorability of knowledge structures that systematically varied the proportion of counterintuitive elements. Our methodology differed from prior studies by employing 'basic level' concepts (e.g. thirsty door) that are cognitively privileged (Rosch *et al.* 1976), and are most commonly found in supernatural narratives. Participants were not cued to expect unusual events or to transmit interesting stories to others. Instead, a standard memory paradigm was used to measure recall.

The study examined the memorability of intuitive (INT) and minimally counterintuitive (MCI) beliefs and belief sets over a period of a week. Two-word statements that represented INT and MCI items were generated. Each statement consisted of a concept and one property that modified it. INT statements were created by using a property that was appropriate to the ontological category (e.g. closing door). MCI statements were created by modifying the concept by a property that was transferred from another ontological category (e.g. thirsty door). This procedure explicitly operationalizes minimal counterintuitiveness as the transfer of a property associated with the core conceptual domains of folk physics, folk biology, and folk psychology from an appropriate ontological category (person, animal, plant, substance) to an inappropriate one. For example, a 'thirsty door' transfers a folk-biological property (thirst) from its proper category (animal) to an improper category (inert object/substance).

US students rated these beliefs on degree of supernaturalness using a six-point Likert scale, with MCI beliefs significantly more likely to be associated with supernaturalness than INT beliefs. Although no differences were found in immediate recall, after a one-week delay minimally counterintuitive knowledge structures led to superior recall relative to all intuitive or maximally counterintuitive structures,[9] replicating the curvilinear function found in the folk tale analysis. With Yukatek Maya speakers, minimally counterintuitive beliefs were again more resilient than

---

[9] Maximally counterintuitive statements (MXCI) were created by modifying a concept with two properties taken from another ontological category (e.g. squinting wilting brick). To control for memory differences on two- versus three-word items, for each MXCI statement a matching statement was generated, only one of the properties being counterintuitive (e.g. chattering climbing pig).

intuitive ones. A follow-up study revealed no reliable differences between the Yukatek recall pattern after one week and after three months (Atran and Norenzayan 2004), indicating a cultural stabilization of the recall pattern.

In brief, minimally counterintuitive beliefs, as long as they come in small proportions, help people remember and presumably transmit the intuitive statements. A small proportion of minimally counterintuitive beliefs give the story a mnemonic advantage over stories with no counterintuitive beliefs or with too many counterintuitive beliefs, just like moderately spiced-up dishes have a cultural advantage over bland or very spicy dishes. This dual aspect of supernatural beliefs and belief sets—commonsensical and counterintuitive—renders them intuitively compelling yet fantastic, eminently recognizable but surprising. Such beliefs grab attention, activate intuition, and mobilize inference in ways that greatly facilitate their retention, social transmission, cultural selection, and historical survival (cf. Atran 2001).

Meta-representing Counterintuitive Worlds: A Theory of Mind Experiment

If counterintuitive beliefs arise by violating innately given expectations about how the world is built, how can we possibly bypass our own hard-wiring to form counterintuitive religious beliefs? The answer is that we don't entirely bypass common-sense understanding, but conceptually parasitize it to transcend it. This occurs through the cognitive process of meta-representation.

Humans have a meta-representational ability to form representations of representations. This allows people to understand a drawing or picture of someone or something as a drawing or picture, and not the real thing. It lets us imagine fiction and gives us an ability to think about being in different situations and deciding which are best for the purposes at hand, without our having to actually live through (or die in) the situations we imagine. It affords us the capacity to model the world in different ways, and to conscientiously change the world by entertaining new models that we invent, evaluate, and implement. It enables us to become aware of our experienced past and imagined future as past or future events that are distinct from the present that we represent to ourselves, and so permits us to reflect on our own existence. It allows people to comprehend and interact with one another's minds.

Meta-representation also lets people retain half-understood ideas, as when children come to terms with the world in similar ways when they hear a new word. By embedding half-baked (quasi-propositional) ideas in other factual and common-sense beliefs, these ideas can simmer through personal and cultural belief systems and change them (Sperber 1985; Atran and Sperber 1991). A half-understood word or idea is initially retained meta-representationally, as standing in for other ideas we already have in mind. Supernatural ideas always remain meta-representational.

After Dennett (1978), most researchers in folk psychology, or 'theory of mind', maintain that attribution of mental states, such as belief and desire, to other persons requires meta-representational reasoning about false beliefs. Not until the child can understand that other people's beliefs are only representations—and not just recordings of the way things are—can the child entertain and assess other people's representations as veridical or fictional, truly informative or deceptive, exact or exaggerated, worth changing one's own mind for or ignoring. Only then can the child appreciate that God thinks differently from people, in that only God's beliefs are always true.

In one of the few studies to replicate findings on 'theory of mind' in a small-scale society (cf. Avis and Harris 1991), Knight, Sousa, Barrett, and Atran (2004) showed forty-eight Yukatek-speaking children (twenty-six boys and twenty-two girls) a tortilla container and told them, 'Usually tortillas are inside this box, but I ate them and put these shorts inside.' They asked each child in random order what a person, God (*dyoos*), the Sun (*k'in*), principal forest spirits (*yumil k'ax'ob'*, 'Masters of the Forest'), and other minor spirits (*chiichi'*) would think was in the box. As with American children (Barrett *et al.* 2001), the youngest Yukatek (4 years) overwhelmingly attribute true beliefs to both God and people in equal measure. After age 5, the children attribute mostly false beliefs to people but continue to attribute mostly true beliefs to God. Thus, 33 per cent of the 4-year-olds said that people would think tortillas were in the container versus 77 per cent of 7-year-olds. By contrast, no significant correlation was detected between answers for God and age.

Collapsing over ages, Yukatek children attribute true beliefs according to a hierarchy of human and divine minds, one in which humans and minor spirits are seen as easier to deceive. Mental states of humans were perceived as different from those of God, as well as from those of Masters of the Forest and the Sun-God. God is seen as all-knowing, and local religious entities fall somewhere in the middle (Figure 25.2). Lowland Maya believe God and forest spirits to be powerful, knowledgeable agents that punish people who over-exploit forest species. For adults, such beliefs have measurable behavioural consequences for biodiversity, forest sustainability, and so forth (Atran *et al.* 2002, 2005). In brief, from an early age people may reliably attribute to supernaturals cognitive properties that are different from those of parents and other people.

In brief, human meta-representational abilities, which are intimately bound to fully developed cognitions of agency and intention, also allow people to entertain, recognize, and evaluate the differences between true and false beliefs. Given the ever-present menace of enemies within and without, concealment, deception, and the ability both to generate and to recognize false beliefs in others would favour survival. But because human representations of agency and intention include representations of false belief and deception, human society is forever under threat of moral defection. If some better ideology is likely to be available somewhere down the line, then (reasoning backward by induction) there is no more justified reason to accept the current ideology than convenience.

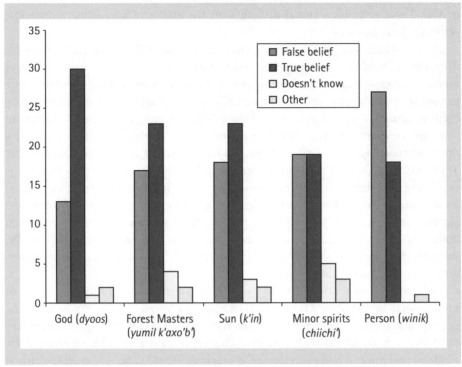

Fig. 25.2. What's in the container? Yukatek Maya children's responses to a false belief task.

As it happens, the very same metacognitive aptitude that initiates this problem also provides a resolution through meta-representation of minimally counterintuitive worlds. Invoking supernatural agents who may have true beliefs that people ordinarily lack creates the rational conditions for people to steadfastly commit to one another in a moral order that goes beyond apparent reason and self-conscious interest. In the limiting case, an omniscient and omnipotent agent (e.g. the supreme deity of the Abrahamic religions) can ultimately detect and punish cheaters, defectors, and free-riders no matter how devious (Frank 1988; Dennett 1997).

EXISTENTIAL ANXIETY: AN EXPERIMENT ON WHAT MOTIVATES RELIGIOUS BELIEF

If supernatural agents are cognitively salient and possess hidden knowledge and powers, then they can be invoked to ease existential anxieties such as about death and deception that forever threaten human life everywhere. To test this, Norenzayan, Hansen, and Atran (2005, reported in Atran and Norenzayan 2004) built on a study by Cahill and colleagues (1994) dealing with the effects of adrenaline (adrenergic activation) on memory.

The hypothesis was that existential anxieties (particularly about death) not only deeply affect how people remember events, but also their propensity to interpret events in terms of supernatural agency. Each of three groups of college students were primed with one of three different stories (Table 25.2): Cahill *et al*.'s uneventful story (neutral prime), Cahill *et al*.'s stressful story (death prime), and another uneventful story whose event structure matched the other two stories but which included a prayer scene (religious prime). Afterwards, each group of subjects read a *New York Times* article (2 October 2001) whose lead ran: 'Researchers at Columbia University, expressing surprise at their own findings, are reporting that women at an *in vitro* fertilization clinic in Korea had a higher pregnancy rate when, unknown to the patients, total strangers were asked to pray for their success.' The article was given under the guise of a story about 'media portrayals of scientific studies'. Finally, students rated strength of their belief in God and the power of supernatural intervention (prayer) on a nine-point scale.

The results show that strength of belief in God's existence and in the efficacy of supernatural intervention (Figure 25.3) are reliably stronger after exposure to the death prime than to either the neutral or the religious prime (no significant differences between either uneventful story). This effect held even after controlling for religious background and prior degree of religious identification. In a cross-cultural follow-up, seventy-five Yukatek-speaking Maya villagers were tested, using stories matched for event structure but modified to fit Maya cultural circumstances. They were also asked to recall the priming events. We found no differences among primes for belief in the existence of God and spirits (near ceiling in this very religious society). However, subjects' belief in the efficacy of prayer for invoking the deities was significantly greater with the death prime than with the religious or neutral primes. Awareness of death more strongly motivates religiosity than mere exposure to emotionally non-stressful religious scenes, like praying. This supports the claim that emotionally eruptive existential anxieties motivate supernatural beliefs.[10]

According to Terror Management Theory (TMT), cultural world-view is a principal buffer against the terror of death. TMT experiments show that thoughts of death function to get people to reinforce their cultural (including religious) world-view and derogate alien world-views (Greenberg *et al*. 1990; Pyszczynski *et al*. 1999). On this view, then, awareness of death should enhance belief in a world-view-consistent deity but diminish belief in a world-view-threatening deity. An alternative view is that the need for belief in supernatural agency overrides world-view defence needs for death-aware subjects.

To test these competing views, Norenzayan, Hansen, and Atran told seventy-three American undergraduates that the prayer groups described in the first experiment

[10] In control conditions, equally anxiety-provoking scenarios (e.g. a visit to the dentist where dental pain is experienced) did not lead to stronger supernatural belief. Moreover, whenever stronger anxiety was found in the mortality salience condition, controlling for self-reported anxiety (measured on the PANAS scale) failed to eliminate the effect (Norenzayan and Hansen, in press).

Table 25.2. Three stories with matching events used to prime feelings of religiosity: neutral (uneventful), death (stressful), religious (prayer scene)

	Neutral	Death	Religious
1	A mother and her son are leaving home in the morning.	A mother and her son are leaving home in the morning.	A mother and her son are leaving home in the morning.
2	She is taking him to visit his father's workplace.	She is taking him to visit his father's workplace.	She is taking him to visit his father's workplace.
3	The father is a laboratory technician at Victory Memorial Hospital.	The father is a laboratory technician at Victory Memorial Hospital.	The father is a laboratory technician at Victory Memorial Hospital.
4	They check before crossing a busy road.	They check before crossing a busy road.	They check before crossing a busy road.
5	While walking along, the boy sees some wrecked cars in a junk yard, which he finds interesting.	While crossing the road, the boy is caught in a terrible accident, which critically injures him.	While walking along, the boy sees a well-dressed man stop by a homeless woman, falling on his knees before her, weeping.
6	At the hospital, the staff are preparing for a practice disaster drill, which the boy will watch.	At the hospital, the staff prepares the emergency room, to which the boy is rushed.	At the hospital, the boy's father shows him around his lab. The boy listens politely, but his thoughts are elsewhere.
7	An image from a brain scan machine used in the drill attracts the boy's interest.	An image from a brain scan machine used in a trauma situation shows severe bleeding in the boy's brain.	An image from a brain scan that he sees reminds him of something in the homeless woman's face.
8	All morning long, a surgical team practices the disaster drill procedures.	All morning long, a surgical team struggles to save the boy's life.	On his way around the hospital, the boy glances into the hospital's chapel, where he sees the well-dressed man sitting alone.
9	Make-up artists are able to create realistic-looking injuries on actors for the drill.	Specialized surgeons are able to re-attach the boy's severed feet, but cannot stop his internal hemorrhaging.	With elbows on his knees, and his head in his hands, the man moves his lips silently. The boy wants to sit beside him, but his father leads him away.

(Continued)

Table 25.2. (*Continued*)

	Neutral	Death	Religious
10	After the drill, while the father watches the boy, the mother leaves to phone her other child's pre-school.	After the surgery, while the father stays by the dead boy, the mother leaves to phone her other child's pre-school.	After a brief tour of the hospital, while the father watches the boy, the mother leaves to phone her other child's pre-school.
11	Running a little late, she phones the pre-school to tell them she will soon pick up her child.	Barely able to talk, she phones the pre-school to tell them she will soon pick up her child.	Running a little late, she phones the pre-school to tell them she will soon pick up her child.
12	Heading to pick up her child, she hails a taxi at the number 9 bus stop.	Heading to pick up her child, she hails a taxi at the number 9 bus stop.	Heading to pick up her child, she hails a taxi at the number 9 bus stop.

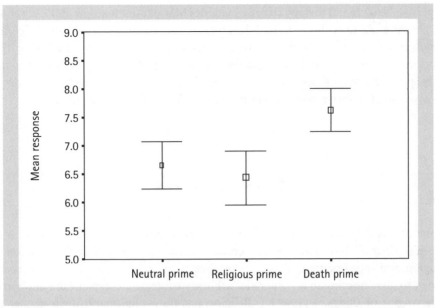

Fig. 25.3. Strength of belief in supernatural power after priming (neutral, religious, or death) and then reading a newspaper article about the effects of prayer on pregnancy. Vertical bars represent margin of error at p = 0.05.

above were Buddhists in Taiwan, Korea, and Japan. Supernatural belief was measured either shortly after the primes, or after a significant delay between the primes and the belief measures. When the primes were recently activated, as expected there was a stronger belief in the power of Buddhist prayer in the death prime than in the control prime. Remarkably, death-primed subjects who previously self-identified as strong believers in Christianity were more likely to believe in the power of Buddhist prayer. In the neutral (control) condition, there was no correlation between Christian identification and belief in Buddhist prayer.[11] Given a choice between supernatural belief versus rejecting an alien world-view (Buddhism), Christians chose the former. This finding is difficult to explain in terms of bolstering a proprietary cultural world-view.

There was no evidence for differences in recall of priming events after subjects rated strength of belief in God and the efficacy of supernatural intervention. With this in mind, note that uncontrollable arousal mediated by adrenergic activation (e.g. subjects chronically exposed to death scenes) can lead to Post-Traumatic Stress Syndrome if there is no lessening of terror and arousal within hours; however, adrenergic blockers (e.g. propranolol, guanfacine, possibly anti-depressants) can interrupt neuronal imprinting for long-term symptoms, as can

[11] Under control conditions, the Christian subjects are unwilling to believe in Buddha or shamanic spirits, and they believe in God far more than they believe in Buddha or spirits.

cognitive-behavioural therapy (work by Charles Marmar discussed in McReady 2002: 9). A plausible hypothesis is that heightened expression of religiosity following exposure to death scenes that provoke existential anxieties may also serve this blocking function. It remains to test the further claim that existential anxieties not only spur supernatural beliefs, but that these beliefs are in turn affectively validated by assuaging the very emotions that motivate belief in the supernatural.

CONCLUSION

None of this is to say that the 'real' function of religion and the supernatural is to promise relief of all outstanding existential anxieties, any more than the function of religion and the supernatural is to neutralize moral relativity and establish social order, to give meaning to an otherwise arbitrary existence, to explain the unobservable origins of things, and so on. Religion has no evolutionary functions *per se*. Rather, existential anxieties and moral sentiments constitute—by virtue of evolution—ineluctable elements of the human condition; and the cognitive invention, cultural selection, and historical survival of religious beliefs in the supernatural owe much to its success in accommodating these elements. Other factors in the persistence of religion as humankind's provisional evolutionary destiny involve naturally selected elements of human cognition. These include the inherent susceptibility of religious beliefs to modularized (innate, universal, domain-specific) conceptual processing systems, such as folk psychology, that favour the survival and recurrence of the supernatural within and across minds and societies.

REFERENCES AND SUGGESTED READING

ALEXANDER, R. (1989). 'Evolution of the Human Psyche', in C. Stringer (ed.), *The Human Revolution*. Edinburgh: University of Edinburgh Press, 455–513.

ALLPORT, G. (1956). *The Nature of Prejudice*. Cambridge, Mass.: Harvard University Press.

ATRAN, S. (1989). 'Basic Conceptual Domains', *Mind and Language*, 4: 7–16.

——(1990). *Cognitive Foundations of Natural History: Towards an Anthropology of Science*. Cambridge: Cambridge University Press.

——(2001). 'The Trouble with Memes: Inference versus Imitation in Cultural Creation', *Human Nature*, 12: 351–81.

——(2002). *In Gods we Trust*. New York: Oxford University Press.

——(2003). 'Genesis of Suicide Terrorism', *Science*, 299: 1534–9.

——and NORENZAYAN, A. (2004). 'Religion's Evolutionary Landscape: Counterintuition, Commitment, Compassion, Communion', *Behavioral and Brain Sciences*, 27: 713–70.

——and Sperber, D. (1991). 'Learning without Teaching: Its Place in Culture', in L. Tolchinsky-Landsmann (ed.), *Culture, Schooling and Psychological Development*, Norwood, N.J.: Ablex, 39–54.

——Medin, D., and Ross, N. (2005). 'The Cultural Mind: Ecological Decision Making and Cultural Modeling within and across Populations', *Psychological Review*, 112: 744–76.

—— ——Ross, N., Lynch, E., Vapnarsky, V., Ucan Ek', E., Coley, J., Timura, C., and Baran, M. (2002). 'Folk Ecology, Cultural Epidemiology, and the Spirit of the Commons: A Garden Experiment in the Maya Lowlands, 1991–2001', *Current Anthropology*, 43: 421–50.

Avis, J., and Harris, P. (1991). 'Belief–Desire Reasoning among Baka Children', *Child Development*, 62: 460–7.

Barrett, J. (1998). 'Cognitive Constraints on Hindu Concepts of the Divine', *Journal for Scientific Study of Religion*, 37: 608–19.

——(2000). 'Exploring the Natural Foundations of Religion', *Trends in Cognitive Science*, 4: 29–34.

——and Keil, F. (1996). 'Conceptualizing a Non-natural Entity', *Cognitive Psychology*, 31: 219–47.

——and Nyhoff, M. (2001). 'Spreading Nonnatural Concepts', *Journal of Cognition and Culture*, 1: 69–100.

——Richert, R., and Driesenga, A. (2001). 'God's Beliefs versus Mother's: The Development of Nonhuman Agent Concepts', *Child Development*, 72: 50–65.

Bartlett, F. (1932). *Remembering*. Cambridge: Cambridge University Press.

Ben-Amos, P. G. (1994). 'The Promise of Greatness: Women and Power in an Edo Spirit Possession Cult', in T. Blakely, W. van Beek, and D. Thomson (eds.), *Religion in Africa*, Portsmouth, N.H.: Heinemann, 119–34.

Bloom, P., and Veres, C. (1999). 'The Perceived Intentionality of Groups', *Cognition*, 71: B1–B9.

Boyer, P. (1994). *The Naturalness of Religious Ideas*. Berkeley: University of California Press.

——(2001). *Religion Explained*. New York: Basic Books.

——and Ramble, C. (2001). 'Cognitive Templates for Religious Concepts', *Cognitive Science*, 25: 535–64.

Cahill, L., Prins, B., Weber, M., and McGaugh, J. (1994). 'Beta-adrenergic Activation and Memory for Emotional Events', *Nature*, 371: 702–4.

Csibra, G., Gergely, G., Bíró, S., Koós, O., and Brockbank, M. (1999). 'Goal Attribution without Agency Cues', *Cognition*, 72: 237–67.

Dawkins, R. (1998). *Unweaving the Rainbow*. Boston: Houghton Mifflin.

Dennett, D. (1978). 'Response to Premack and Woodruff: Does the Chimpanzee have a Theory of Mind?', *Behavioral and Brain Sciences*, 4: 568–70.

——(1997). 'Appraising Grace: What Evolutionary Good is God?', *The Sciences*, 37: 39–44.

Firth, R. (1963). 'Offering and Sacrifice', *Journal of the Royal Anthropological Institute*, 93: 12–24.

Frank, R. (1988). *Passions within Reason*. New York: W. W. Norton.

Geary, D., and Huffman, K. (2002). 'Brain and Cognitive Evolution', *Psychological Bulletin*, 128: 667–98.

Gibbon, E. ([1776] 1845). *Decline and Fall of the Roman Empire*. London: International Book Co.

Greenberg, J., Pyszczynski, T., Solomon, S., Rosenblatt, A., Veeder, M., Kirkland, S., and Lyon, D. (1990). 'Evidence for Terror Management Theory II', *Journal of Personality and Social Psychology*, 58: 308–18.

GUTHRIE, S. (1993). *Faces in the Clouds: A New Theory of Religion*. New York: Oxford University Press.

HAMILTON, W., and ORIANS, G. (1965). 'Evolution of Brood Parasitism in Altricial Birds', *Condor*, 67: 361–82.

HEIDER, F., and SIMMEL, S. (1944). 'An Experimental Study of Apparent Behavior', *American Journal of Psychology*, 57: 243–59.

HUME, D. ([1757] 1956). *The Natural History of Religion*. Stanford, Calif.: Stanford University Press.

KEILLOR, G. (1999). 'Faith at the Speed of Light', *Time*, 14 June.

KIERKEGAARD, S. ([1843] 1955). *Fear and Trembling and the Sickness unto Death*. New York: Doubleday.

KIRKPATRICK, L. (1999). 'Toward an Evolutionary Psychology of Religion and Personality', *Journal of Personality*, 67: 921–52.

KNIGHT, N., SOUSA, P., BARRETT, J., and ATRAN, S. (2004). 'Children's Attributions of Beliefs to Humans and God: Cross-cultural Evidence', *Cognitive Science*, 28: 117–26.

KROEBER, A. L. (1963). *Anthropology: Culture Patterns and Processes*. New York: Harcourt, Brace and World (originally published 1923).

KUPER, A. (1996). *The Chosen Primate*. Cambridge, Mass.: Harvard University Press.

LACK, D. (1968). *Ecological Adaptations for Breeding in Birds*. London: Methuen.

LIPKIND, W. (1940). 'Carajá Cosmography', *Journal of American Folk-Lore*, 53: 248–51.

McREADY, N. (2002). 'Adrenergic Blockers Shortly after Trauma can Block PTSD', *Clinical Psychiatry News*, February, 9.

NORENZAYAN, A., and HANSEN, I. (2006). 'Belief in Supernatural Agents in the Face of Death', *Personality and Social Psychology Bulletin*, 174–87.

—— ATRAN, S., FAULKNER, J., and SCHALLER, M. (in press). 'Memory and Mystery: The Cultural Selection of Minimally Counterintuitive Narratives', *Cognitive Science*.

PINKER, S. (2004). 'The Evolutionary Psychology of Religion'. Paper presented at the annual meeting of the Freedom from Religion Foundation, Madison, Wisconsin, 29 October, 2004; <http://pinker.wjh.harvard.edu/articles/media/2004_10_29_religion.htm>.

PREMACK, D., and PREMACK, A. (1995). 'Origins of Social Competence', in M. Gazzaniga (ed.), *The Cognitive Neurosciences*, Cambridge, Mass.: MIT Press, 205–18.

PYSZCZYNSKI, T., GREENBERG, J., and SOLOMON, S. (1999). 'A Dual Process Model of Defense against Conscious and Unconscious Death-related Thoughts: An Extension of Terror Management Theory', *Psychological Review*, 106: 835–45.

PYYSIÄINEN, I. (2001). *How Religion Works*. Leiden: Brill.

RAPPAPORT, R. (1999). *Ritual and Religion in the Making of Humanity*. New York: Cambridge University Press.

ROBERTSON SMITH, W. (1894). *Lectures on the Religion of the Semites*. London: A. & C. Black.

ROSCH, E., MERVIS, C., GREY, W., JOHNSON, D., and BOYES-BRAEM, P. (1976). 'Basic Objects in Natural Categories', *Cognitive Psychology*, 8: 382–439.

RUBIN, D. (1995). *Memory in Oral Traditions*. New York: Oxford University Press.

RUSSELL, B. (1948). *Human Knowledge: Its Scope and Limits*. New York: Simon & Schuster.

SARTRE, J.-P. (1948). *Being and Nothingness*. New York: Philosophical Library.

SELIGMAN, S. (1971). 'Phobias and Preparedness', *Behavioral Therapy*, 2: 307–20.

SOBER, E., and WILSON, D. S. (1998). *Unto Others*. Cambridge, Mass.: Harvard University Press.

SPERBER, D. (1985). 'Anthropology and Psychology', *Man*, 20: 73–89.

—— (1996). *Explaining Culture*. Oxford: Blackwell.

——and WILSON, D. (1995). *Relevance: Communication and Cognition*, 2nd edn. Oxford: Blackwell.

TINBERGEN, N. (1951). *The Study of Instinct*. London: Oxford University Press.

TOOBY, J., and COSMIDES, L. (1992). 'The Psychological Foundations of Culture', in J. Barkow, L. Cosmides, and J. Tooby (eds.), *The Adapted Mind*, New York: Oxford University Press, 19–36.

WILSON, D. S. (2002). *Darwin's Cathedral*. Chicago: University of Chicago Press.

WORTHINGTON, E., KURUSU, T., McCULLOUGH, M., and SANDAGE, S. (1996). 'Empirical Research on Religion and Psychotherapeutic Processes of Outcomes', *Psychological Bulletin*, 19: 448–87.

CHAPTER 26

VARIETIES OF NATURALISM

OWEN FLANAGAN

THE MANY MEANINGS OF NATURALISM

Here are some things 'naturalism' has been taken to mean or imply.

1. Philosophy should 'respect', 'be informed by', 'wholeheartedly accept' the methods and claims of science.
2. When a well-grounded philosophical claim and an equally well-grounded scientific claim are inconsistent (whatever 'equally well-grounded' means), the scientific claim trumps.
3. Philosophical questions are not distinct from scientific questions—they differ, if they do differ, only in level of generality.

This chapter has had the longest shelf-life, on and off, of any paper I've ever written. The germination stage began in graduate seminars in the late 1970s with W. V. O. Quine and his first Ph.D. student, George Berry, now both deceased. I dedicate this paper to their lives and work. The original ancestral version of this essay was delivered at a conference on 'Naturalism and Anti-naturalism' in Elizabethtown, Pa., almost a decade ago. I am most grateful to Michael Silberstein, Al Plantinga, Stuart Silvers, Susan Haack, and Michael Ruse for helpful discussions during that wonderful summer week in Pennsylvania. Close and not-so-close continuations of the original were delivered to audiences in Athens, Georgia, and Athens, Greece, and Los Angeles, St Louis, Ithaca, New York, and Rethnymo, Crete. I thank those audiences. Special thanks also for the wisdom of Philip Clayton, John McDowell, Michael Tye, David Wong, Hagop Sarkissian, Kevin DeLapp, Andrew Terjesen, Tom Polger, Eddy Nahmias, Alex Rosenberg, and Leonore Fleming all of whom helped me to articulate better what is, I hope, a defensible and robust form of naturalism.

4. Both science and philosophy are licensed only to describe and explain the ways things are.

5. Both philosophy and science are, in addition to the businesses of description and explanation, in the business of giving naturalistic justifications for epistemic and ethical ideals and norms.

6. There is no room, or need, for the invocation of immaterial agents or forces or causes in describing or accounting for things.

7. Mathematics and logic can be understood without invoking a Platonic (non-naturalistic) ontology.

8. Ethics can be done without invoking theological or Platonic foundations. Ethical norms, values, and virtues can be defended naturalistically.

9. Naturalism is another name for materialism or physicalism; what there is, and all there is, is whatever physics says there is.

10. Naturalism is a form of non-reductive physicalism; there are genuine levels of nature above the elemental level.

11. Naturalism is a thesis that rejects both physicalism and materialism; there are natural but 'non-physical' properties, e.g. informational states.

12. Naturalism claims that most knowledge is a posteriori.

13. Naturalism is indifferent to claims about whether knowledge is a priori or a posteriori, so long as whatever kind of knowledge exists can be explained, as it were, naturalistically.

14. Naturalism is, first and foremost, an ontological thesis that tells us about everything that there is.

15. Naturalism is, first and foremost, an epistemic thesis, which explains, among other things, why we should make no pronouncements about 'everything that there is'.

I could go on.[1]

Wittgenstein and Bouwsma

Now there are two objections to almost any kind of naturalism: one due to Wittgenstein, the other to O. K. Bouwsma. In his *Tractatus* of 1922 Wittgenstein writes:

4.111   Philosophy is not one of the natural sciences.

4.112   Philosophy aims at the logical clarification of thoughts.

---

[1] I really mean that I could go on. I have developed lists of the meanings of 'naturalism' that I won't subject the reader to, but which remind me of Paul Simon's song 'Fifty Ways to Leave Your Lover'. Important work on the variety of meanings of 'naturalism' includes Kitcher (1992); Rosenberg (1996); entries on 'Naturalism' in Audi (1999), and in Honderich (1995). A new and excellent work, *Naturalism in Question*, ed. De Caro and Macarthur (2004) contains classic and many new state-of-the-art reflections on naturalism, its meaning, and its prospects as a distinctive and robust philosophical position.

4.1121   Psychology is no more closely related to philosophy than any other natural science [is related to philosophy].

4.1122   Darwin's theory has no more to do with philosophy than any other hypothesis in natural science.

4.111 and 4.112 do not entail anti-naturalism, but 4.1121 and 4.1122 do—at least of the Quinean variety. However, although Wittgenstein's view is anti-naturalistic, it is so pretty much solely in virtue of *stipulating* against naturalism, not by *arguing* against the credibility of any one of its various meanings.

Bouwsma's worry is of greater concern, since it is that naturalism is a glib and promiscuous doctrine. He writes (1948): 'The naturalist is excited about something...the scientific method.... [B]ut math is not knowledge that depends in any way on scientific method.... Has the naturalist ever heard about numbers?' And he adds: 'For the philosophical naturalist the universal applicability of the experimental method is a basic belief.'

He then says this:

Compare to vacuumism, the belief in the universal applicability of the suction nozzle: The vacuum cleaner salesman may argue to himself, 'If I ever give this up, I'll never sell another vacuum cleaner. It is basic.' To the housewife who asks: 'And can it dust books?' he replies, 'of course'! And when he shows her and finds that it does not do so well, does he deny the universal applicability of the nozzle? No such thing.... Since the universal applicability of the nozzle is now the touchstone of dust, either he is not skillful [yet] with the nozzle or there is no dust—it just seems like dust.

I will not here explore Bouwsma's vacuumism analogy, except to say that as a self-proclaimed naturalist, I feel the force of his point. It is something to worry about.

Naturalism: A Common Core?

In any case, with Bouwsma's credible concern in view, the first point is that the terms 'naturalism' and 'naturalist' lack a single determinate meaning. Furthermore, many variants once out on the streets seem open to the vacuumism worry.

The *OED* suggests that the original *philosophical* meaning of the term 'naturalism' dates back to the seventeenth century and meant 'a view of the world, and of man's relation to it, in which only the operation of natural (as opposed to supernatural or spiritual) laws and forces is admitted or assumed'.

Barry Stroud (1996) writes, 'Naturalism on any reading is opposed to supernaturalism...By "supernaturalism" I mean the invocation of an agent or force which somehow stands outside the familiar natural world and so whose doings cannot be understood as part of it. Most metaphysical systems of the past included some such agent. A naturalist conception of the world would be opposed to all of them.' Indeed,

Stroud goes on to suggest that anti-supernaturalism is pretty much the only determinate, contentful meaning of the term 'naturalism'. Assuming he is right, then anti-supernaturalism forms the common core, the common tenet, of 'naturalism' insofar as 'naturalism' is anything like a coherent philosophical doctrine spanning the last four centuries. Let me be clear about a matter of considerable importance: the objectionable form of 'supernaturalism' is one according to which (i) there exists a 'supernatural being or beings' or 'power(s)' outside the natural world; (ii) this 'being' or 'power' has causal commerce with this world; (iii) the grounds for belief in *both* the 'supernatural being' *and* its causal commerce cannot be seen, discovered, or inferred by way of any known and reliable epistemic methods.

Stroud suggests that all reputable philosophers, except for Alvin Plantinga, are naturalists according to the deflated sense of the term. Stroud is, I am sure, wrong in thinking that supernaturalism is extinct among philosophers.[2] That temporarily to one side, suppose, as I think is plausible, that some sort of commitment to the dispensability of supernaturalism is a *necessary* condition for naturalism. Still, five things are worth commenting on. First, the commitment to the dispensability of supernaturalism does not entail a rejection of all forms of spirituality or religion. Theologians and philosophers who are religious naturalists reject the conjunction of (i)–(iii) above, and that amounts to a rejection of the objectionable form of supernaturalism. Furthermore, many forms of embodied spiritual commitment are themselves committed to the dispensability of supernaturalism in the objectionable sense (i)–(iii). Buddhism comes to mind, as do some strands in Hinduism, Jainism, many kinds of shamanism, as well as among many liberal Christian communities in North America (Quakers, Unitarian Universalists). Second, putting the *OED*'s and Stroud's criteria together, naturalism is a very general thesis; neither what is 'natural', 'a natural law', or 'a natural force', nor what is 'non-natural', 'supernatural', or 'spiritual' are remotely specified. All the important details are left out or need to be spelled out.

---

[2] Kitcher (1992), like Stroud, claims that naturalism is pretty much the only game in town. I can't really get into the matter here. But this claim has credibility only to the degree that they intend some version of the idea that for the *purposes* of doing ontology in what I will below call the non-imperialistic sense, or for doing naturalized epistemology or ethics, divine agency does not need to be introduced to play an *explanatory* role. Stroud thinks that Plantinga is a naturalist when it comes to descriptive epistemology, but not when it comes to normative epistemology. As I read Plantinga, he is resolutely dedicated to using the well-accepted tools of modal logic to make the case for theism. But there are other major contemporary philosophers whose views should also give him pause. W. P. Alston, Alasdair MacIntyre, and Charles Taylor come to mind. Taylor expresses the idea at the end of *Sources of the Self* (1992) that perhaps God can play a role in the justification of our ethical norms. MacIntyre holds a two-level view; like Aquinas, he thinks that there are natural and supernatural justifications of norms. The natural justifications are satisfactory, but the divine ones are 'more' ultimate. One way to state my point is that Plantinga, Alston, MacIntyre, and Taylor are theists, but it is not clear that they are philosophical supernaturalists in the objectionable sense that involves commitment to (i), (ii), and (iii). It's (iii), of course, that is the big troublemaker. If the case for (i) and (ii) can be made with acceptable epistemic methods, then we have a form of religious naturalism.

Third, the thesis is ambiguous between an ontological claim and an epistemological claim—one could read the claim as one to the effect that no supernatural entities can be countenanced when saying 'what there is'; or it might be read as a weaker claim to the effect that one should dispense with supernatural stuff in *explaining* things, in particular the things in *this* world, while making no restrictions on what one can believe exists. Fourth, the view is stated a bit more negatively than a proud proponent of a philosophical view should like. We are not told what the naturalist is committed to so much as what the view excludes. Imagine a political leader who, when asked about her political position, says: 'Well, I really can't say what my view is, but, rest assured, it is *not* communism.' The fifth point is that even if the belief in the dispensability of supernaturalism is a necessary condition for naturalism, the *scope* of the thesis has not been specified.

SCOPE

Let me explain this last point about *scope* a bit more. All economists I know believe that one does not need to invoke divine agency in explaining whatever it is that economists allegedly explain. Economics in our time, like the other human sciences, dispenses with supernaturalism for the sake of doing economics. Economists are naturalists in this narrow sense. Economics abides by what I'll call *methodological naturalism*. When it comes to divine agency, economics simply dispenses with it, presumably because economists find no explanatory work for divine agency to do in economics.

Suppose a poll reveals that most economists believe in God; that is, when asked questions, say, about the origin of the universe or life after death, they invoke divine agency.[3] Economists, or these economists, we might say, are *ontological non-naturalists* in the domain or domains comprised by these non-economic questions.

Are economists who are methodological naturalists when doing economics and ontological non-naturalists when doing something else inconsistent? I don't see that they are. If they are, explain to me how.

Imagine further—what I suspect is in fact the case—that many economists when pressed about their views concerning human agency *within* economics turn out to be closet Cartesians; that is, they believe in a view of the mind–body relation—the relation between reason, desire, and action—that involves interaction between a non-physical mind and a physical body. Suppose, further, that despite all the talk about the 'iron force of the invisible hand' of economic rationality, economists at

---

[3] Economists can't think that value is grounded non-naturalistically without great effort. But I leave that to the side for now.

least sometimes speak as if they believe in a libertarian conception of agency. Are these economists courting inconsistency?

I am inclined—but only half-heartedly—to say 'yes', since according to one conception of naturalism, it is supposed to exclude not only immaterial divine agents from doing any explanatory work but immaterial finite agents as well. So imagine, while thinking in this vein, that we bring the economists into group therapy sessions run by naturalistically inclined philosophers. We get them to see their commitments about the mind–body relation and about free will clearly, and as a community they resolve to be compatibilists about free will. But imagine that on the mind–body problem they are only willing to budge a small distance. Suppose the economists say this to the philosophers: 'We've thought a lot about your worries that our methodological naturalism is unstable so long as we don't take a firm stand to the effect that mental events *are* complex brain events. But in fact we don't see that we depend on any view whatsoever about the metaphysics of mind *in doing economics*. Sure, we make assumptions about the motivational structure of mind, and (thanks to you) we see that we assume complex psychophysical laws. But the nature of mind and the nature of the mind–body relation is your problem, not ours. Philosophical therapy has taken us this far: we see that *qua* economists we can be, possibly should be, *neutral* on the metaphysics of mind. Thank you for your services. God bless the intellectual division of labour. Goodbye.'

Are these economists inconsistent? Again, I don't see that they are. The first group of economists I imagined were saved from inconsistency by distributing some commitments *inside* and some *outside* economics. The community of economists now being imagined claims that certain questions considered as litmus tests for membership in the club of naturalists by most contemporary *philosophers of mind* are questions on which they *qua* economists—even, I suspect, *qua* naturalistic philosophers of economics—can remain neutral or agnostic about *inside* economics.[4]

I have been focusing so far mostly on the contrast between naturalism and supernaturalism. The reason is to see if the contrast reveals, as I think it does, the one necessary condition of naturalism as a historical doctrine, the one thing about which all card-carrying naturalists agree, or should agree.[5]

---

[4] Crudely, the *scope* of a view has to do with how many domains it claims to reach across. The *range* of a view would then involve what it is ontologically committed to using: say, some device like Quine's proposal in 'On What There Is' (1948), e.g. 'to be is to be the value of a bound variable'.

[5] However, even if this is true, one implication of what I have said so far is that one cannot infer from a commitment to methodological naturalism within one of the special sciences—even if the methodological naturalism is tied to or results in ontological naturalism within *that* domain—that one is an ontological naturalist across the board. Going in another—and in this paper underdeveloped—direction, one should not judge that, for example, a mathematical Platonist thinks or should think, just on the basis of her mathematical Platonism, that any traditional theological beliefs are worth entertaining.

ONTOLOGY: IMPERIALISTIC
AND NON-IMPERIALISTIC

If I have established anything so far, it is simply these two points. First, in terms of the history of the usage of the term 'naturalism', some kind of exclusion of the supernatural, of the spiritual, is required. The exclusion condition can be stated precisely: any form of supernaturalism that embraces (i)–(iii) above is excluded. But the second point—the one that has to do with what I call 'the scope question'—is that even this sort of minimal core meaning says too little to characterize a robust and positive philosophical doctrine.

We can see this by looking again at the *OED*'s and Stroud's characterizations. Recall that the *OED* states that the philosophical meaning of 'naturalism' is 'a view of the world, and of man's relation to it, in which only the operation of natural (as opposed to supernatural or spiritual) laws and forces is admitted or assumed'. Whereas Stroud (1996) writes: 'Naturalism on any reading is opposed to supernaturalism...By "supernaturalism" I mean the invocation of an agent or force which somehow stands outside the familiar natural world and so whose doings cannot be understood as part of it.'

Examined closely, these two formulations do not say the same thing. The reason is this: the *OED* definition is compatible with a view like that of the French deists—a view according to which a supernatural force is invoked to explain what occurred before the big bang as well as what might occur after the big bust. To be sure, even on the *OED* characterization of naturalism, the naturalist cannot admit or assume spiritual agents or forces in explaining what Stroud calls 'the familiar natural world'. But the *OED* formulation does not exclude, as Stroud's rendition does, belief in agents or forces that stand *outside* the familiar natural world.

I want to be clear that nothing in the analysis so far precludes spirituality or religion. It depends on whether the objectionable form of 'supernaturalism' is espoused. According to that objectionable (and not unfamiliar) form of supernaturalism, (i) there exists a 'supernatural being or beings' or 'power(s)' outside the natural world; (ii) this 'being' or 'power' has causal commerce with this world; and (iii) the grounds for belief in *both* the 'supernatural being' *and* its causal commerce cannot be seen, discovered, or inferred by way of any known and reliable epistemic methods.

Many forms of naturalistic spirituality reject supernaturalism in the objectionable sense described by Ursula Goodenough (1998; also Chapter 50 below). I myself am religious: a Celtic-Catholic-Buddhist. And I see my ethical commitments as supported and enhanced by deep transcendental cognitive convictions and emotions that powerfully ground a conviction that I am a part of the whole, inextricably connected to everything else that there is. But I reject (i)–(iii).

Stroud's formulation invites the interpretation that naturalism, understood positively, is an *ontological* position of the widest scope. Naturalism says that what there is, *and all there is*, is the natural world. A weaker view, one invited by the *OED*, is that naturalism is an ontological position of narrower scope—it is a view about what, to put it crudely, we should be ontologically committed to, given our cognitive capacities and limits when talking about *the* world, 'world' with a small 'w'.[6]

Speaking for myself, I do not hold, and would not recommend holding, an all-encompassing or imperialistic ontological naturalism. My reason is pretty simple: I don't see that we can or do have access to any information that would warrant any all-encompassing claims about everything that there is *or* is not.

Now this might appear to give comfort to the supernaturalist—indeed, to aid and abet his cause. Never fear. The supernaturalist can take no solace from the point about our epistemic limits. It is not as if in saying that the naturalist cannot make warranted claims about what there is, or is not, in the widest possible sense, I make room for the supernaturalist to move in. The reason is simple: my point applies to all *Homo sapiens*. *Qua* finite, historically embedded animals, humans are in no position to make positive or negative assertions about 'everything that there is, and is not, in the widest possible sense'. No variety of imperialist ontologist, naturalist nor non-naturalist, should find a friend in me or in what I have said so far.

What I do think is warranted, all things considered, is a form of ontological naturalism about *this* world—*for all we know and can know, what there is, and all there is, is the natural world*. Truth be told, this is pretty vague. Since the conception of what is 'natural' is not a constant, the central concept in the motto lacks a clear and determinate meaning.[7] Still, vague as it is, the view is not friendly to theism. The epistemological humility called for is not so humble that it tolerates agnosticism. Theological claims do not work and for that reason they are something akin to nonsense, lacking in cognitive significance, as they used to say in the old days.

---

[6] I usually chastise those who promote this sort of argument within the philosophy of mind, e.g. Colin McGinn, calling them 'mysterians' (Flanagan 1992). Here I am simply acknowledging the force of these arguments against ontology in the widest sense. A different point, worth pursuing, is that we have been very successful as a species at overcoming obstacles, e.g. our inability to fly, by inventing prosthetics. Airplanes get us to fly, and microscopes and telescopes get us to see beyond nature's endowment. Past success at overcoming physical or cognitive barriers may partly explain why we are not inclined to accept any limit as insurmountable.

[7] Stroud (1996) calls this 'open-minded or expansive naturalism', and he says this about the view: 'What I am calling more open-minded or expansive naturalism says we must accept everything we find ourselves committed to in accounting for everything we believe is so and want to explain . . . But now it . . . looks as if this expandable or open-minded form of naturalism does not amount to anything very substantive or controversial. . . . If that is still called "naturalism" the term by now is little more than a slogan on a banner raised to attract the admiration of those who agree that no supernatural agents are at work in the world' (Stroud 1996: 54). Stroud goes on to suggest that we might just as well call the view 'open-mindedness, period'.

That said, I want to be clear that nothing I have argued so far suggests that ontology or metaphysics be eliminated. Quite the contrary. I am attracted to the view that the ontological thrust of philosophy originates in the interests and purposes of those individuals who are, depending on one's perspective, blessed with or afflicted with (to paraphrase Sellars) 'the desire to understand how things, in the widest possible sense of the term, hang together, in the widest possible sense of the term'.

Witness the economists discussed earlier, who do not have the thirst for ontology that many philosophers have. For those who have this thirst there is plenty of work to be done extracting the ontological commitments made across the disciplines, evaluating the warrant for these multifarious commitments, and attempting to continually update our picture of what ontologically it makes sense to be committed to.[8] The impulse to make the picture consistent is itself one of the things that the naturalist needs to account for.

So far I have been focusing on such questions as: Are there many meanings of naturalism? Is there some necessary feature required for being a naturalist? The answer to both these questions, I have suggested, is 'yes'. I have also been trying to figure out whether, and to what extent, it is a wise philosophical strategy to be ontologically imperialistic about one's naturalism, and I have suggested that it is not wise. But then again it is not wise, for the very same reasons, to be imperialistic about one's supernaturalism. There is little warrant for *confidence* in imperialistic naturalism or supernaturalism.

For the rest of this chapter I am going to switch gears and address an objection or set of objections directed specifically at the programme of naturalized epistemology and naturalized ethics, focusing specifically on the case of naturalized ethics. The objection has to do with certain alleged incapacities that ethical or epistemological naturalism has when it comes to dealing with norms and normativity, with justification, and with 'ought' talk—key issues for religion-and-science discussions.

The objection is important. The opponents of naturalism are right: if there is no room for normativity, then there is no room for naturalism. Let me explain.

---

[8] William James presents a particularly interesting case, both as someone who struggled all his life with the problem of naturalism and non-naturalism, and as a reminder of how the job description one decides to work under can make it easier or harder to make charges of inconsistent ontological commitments stick. In a famous passage in the 'Epilogue' to *Psychology: The Briefer Course*, James (1892) writes: 'Let psychology frankly admit that *for her scientific purposes* determinism may be *claimed*, and no one can find fault....Now ethics makes a counter-claim; and the present writer, for one, has no hesitation in regarding her claim as the stronger, and in assuming that our wills are "free"...the deterministic assumption of psychology is merely provisional and methodological.' The strategy here is to keep two prima facie incompatible commitments afloat by claiming a scope difference and thus denying the real incompatibility. Because the philosopher, especially the ontologist, has the Sellarsian thirst to understand how 'things considered, in the widest possible sense, hang together, in the widest possible sense', he has far greater demands placed upon himself—purely in virtue of his vocation to draw things together into a coherent 'big picture'.

Epistemology Naturalized
and the Problem of Normativity

In 'Epistemology Naturalised', Quine (1963) suggested that epistemology be assimilated to psychology. Although I personally never read Quine's arguments for naturalization as arguments against a normative role for epistemology, many have. Putnam writes: 'The elimination of the normative is attempted mental suicide... Those who raise the slogan *"epistemology naturalised"*... generally *disparage* the traditional enterprises of epistemology' (1982: 229). And Jaegwon Kim writes: 'If justification drops out of epistemology, knowledge itself drops out of epistemology. For our concept of knowledge is inseparably tied to that of justification... itself a normative notion' (1993: 224–5).[9]

The alleged problem with epistemology naturalized is this: psychology is not in general concerned with *norms* of rational belief, but with the description and explanation of mental performance and mentally mediated performance and capacities.

Assuming for now that this apparent problem is real, then the naturalized epistemologist has two routes open. One is to give up altogether the projects of explaining norms and providing justification—perhaps by explaining all normative speech acts along emotivist lines, as noises people make when trying to get others to think and act as they wish. Since I agree with Putnam and Kim that this is a very bad road to go down, we must take a different road.

Here is the suggestion in broad strokes. The best way to think of epistemology naturalized is not one in which epistemology is a 'chapter of psychology', where psychology is understood merely descriptively, but rather to think of naturalized epistemology as having two components: a *descriptive-genealogical* component *and* a *normative* component. Furthermore, not even the *descriptive-genealogical* component will consist of purely psychological generalizations, for much of the information about actual epistemic practices will come from biology, cognitive neuroscience, sociology, anthropology, and history—from the human sciences broadly construed. More obviously, *normative epistemology* will not be part of psychology, for it involves the gathering together of norms of inference, belief, and knowing that lead to success in ordinary reasoning and in science. And the evolved canons of inductive and

---

[9] Actually Putnam (1982: 229) seems to think that this worry is not one for the naturalist metaphysician, whose job is largely descriptive: 'The materialist metaphysician often uses such traditional metaphysical notions as *causal power* or *nature* quite uncritically. The "physicalist" generally doesn't seek to *clarify* these traditional metaphysical notions, but just to show that science is progressively clarifying the *true* metaphysics. This is why it seems just to [describe] *his* enterprise as "natural metaphysics," in strict analogy to the natural theology of the eighteenth and nineteenth centuries.' These are the words that precede and give the context to the anti-Quinean pronouncement that 'Those who raise the slogan *"epistemology naturalised"*, on the other hand, generally *disparage* the traditional enterprises of epistemology'.

deductive logic, statistics and probability theory, most certainly do not describe actual human reasoning practices. These canons—take, for example, principles governing representative sampling and warnings about affirming the consequent—come from abstracting successful epistemic practices from unsuccessful ones. The database is, as it were, provided by observation of humanity, but the human sciences do not (at least as standardly practised) involve extraction of the norms. So epistemology naturalized is not epistemology psychologized *simpliciter*.[10] But since successful practice—both mental and physical—is the standard by which norms are sorted and raised or lowered in epistemic status, pragmatism reigns.[11]

ETHICS NATURALIZED

The same worries that Putnam and Kim express over Quine's conception of naturalizing epistemology recapitulate Kant's worries over Hume's approach to naturalizing ethics. And I take it that John McDowell's criticism of 'bald naturalism' in favour of 'second nature naturalism' is a way of stating the same concern: namely, that at least some kinds of naturalism are not equipped to explain ethical normativity. Although my approach is different from McDowell's, I don't think I can be charged with bald naturalism. But we'll see. Actually, I'm not sure who McDowell would think is an actual bald naturalist in ethics—unless he is thinking of John Mackie, possibly A. J. Ayer, and perhaps some evolutionary psychologists.

In any case, moral psychology, sociology, and anthropology—what Kant called the 'empirical side of morals'—might tell us what individuals or groups think ought to be done, what they believe is right or wrong, what they think makes a good person, and so on. But all the human scientific facts taken together, including that they are widely and strongly believed, could never justify any of these views.

In the *Groundwork*, Kant writes that a 'worse service cannot be rendered morality than that an attempt be made to derive it from examples'. Trying to derive ethical principles 'from the disgusting mishmash' of psychological, sociological, or anthropological observation, from the insights about human nature that abound 'in the chit-chat of daily life', that delight 'the multitude', and upon which 'the empty headed regale themselves' is not the right way to do moral philosophy.

---

[10] To my mind the best work in naturalized epistemology is that of Alvin Goldman (1986, 1992). Goldman never tries to *derive* normative conclusions from descriptive premises, although in ethics he sees (as I do) the relevance of the empirical to the normative and sensibly uses Peircean abduction to link the two. Furthermore, he continually emphasizes the historical and social dimensions of epistemology in a way that Quine did not.

[11] In epistemology, pragmatic evaluation is done relative to our cognitive aims. These to be sure are themselves norms, and as such are subject to the same sort of requests for rationales and warrant as all other norms.

What is the right way to do moral philosophy? We need 'a completely isolated metaphysics of morals', a pure ethics unmixed with the empirical study of human nature. Once moral philosophy has derived the principles that ought to govern the wills of all rational beings, then, and only then, should we seek 'the extremely rare merit of a truly philosophical popularity'.[12]

Despite his commitment to the project of the Enlightenment, Kant believed in God, the God of pietistic Lutheranism. And he believed that God was, in fact, the ultimate source of morality. Kant's key insight involved seeing that disagreement about theological details could be circumvented so long as God had given us a faculty of pure practical reason in which and through which all conscientious persons could discover the right moral principle—and indeed he (God, that is) had.

Kantian ethics, *qua* philosophical theory, we might say, is not openly supernaturalistic; but it is not naturalistic either. It would be naturalistic if we could give an account of a faculty of pure practical reason that possesses moral principles not gleaned from the observation of human practices and assessments of practices that work differentially well to meet our aims, and which, in addition, fits with the findings of the mental sciences. There is no such faculty that meets these criteria, and thus no faculty to account for.

The point I want to stress first is that we can—indeed, we should—conceive of naturalistic ethics in pretty much the same way as we think of naturalized epistemology. Naturalistic ethics will contain a *descriptive-genealogical* component that will specify certain basic capacities and propensities of *Homo sapiens*, e.g. sympathy, empathy, egoism, and so on, relevant to moral life. It will explain how people come to feel, think, and act about moral matters in the way(s) they do. It will explain how, and in what ways, moral learning, engagement, and response involve the emotions. It will explain what moral disagreement consists in and why it occurs; and it will explain why people sometimes resolve disagreement by recourse to agreements to tolerate each other without, however, approving of each other's beliefs, actions, practices, and institutions. It will tell us what people are doing when they make normative judgements. And finally, or as a consequence of all this, it will try to explain what goes on when people try to educate the young, improve the moral climate, propose moral theories, and so on.[13]

It should be pointed out that every great moral philosopher has put forward certain *descriptive-genealogical* claims in arguing for substantive normative proposals,

[12] Thanks (or no thanks) to Kant, the dominant conception of the intellectual division of labour as I write makes a sharp distinction between moral philosophy and moral psychology. Moral philosophy is in the business of saying what ought to be, what is really right and wrong, good and evil, what the proper moral principles and rules are, what counts as genuine moral motivation, and what types of persons count as genuinely good. Most importantly, the job of moral philosophy is to provide philosophical justification for its 'shoulds' and 'oughts', for its principles and its rules.

[13] I want to make one general observation before I proceed. Most philosophers understand how a theory of meaning can have implications for meta-ethics; e.g. if we discover that ethical terms, like most other terms, lack definitions, it is not surprising that G. E. Moore couldn't

and that although most of these claims suffer from sampling problems and were proposed in a time when the human sciences did not exist to test them, they are almost all testable—indeed, some have been tested (Flanagan 1991).

For example, here are four claims familiar from the history of ethics which fit the bill of testable hypotheses relevant to normative ethics: (i) the person who knows the good does it; (ii) if one (really) has one virtue, one has the rest; (iii) morality breaks down in a roughly linear fashion with breakdowns in the strength and visibility of social constraints; (iv) in a situation of profuse abundance, innate sympathy and benevolence will 'increase tenfold', and the 'cautious jealous virtue of justice will never be thought of'.

Presumably, how the *descriptive-genealogical* claims fare matters to the normative theories, and would have mattered to their proponents.[14]

find a definition of 'good'. Failure to find a definition of 'good' would no more prove that it names a non-natural property than the same failure to find definitions for 'fuzzy' or 'chair' would prove that fuzziness is a non-natural property and chairs are non-natural objects. However, defenders of naturalistic ethics like myself are continually asked to explain how a better picture of moral psychology *can* contribute to our understanding of ethical theory in general, and normative ethics in particular. Moral psychology, cognitive science, cultural anthropology, and the other mental and social sciences can tell us perhaps how people *in fact* think and behave. Ethical theories tell us what the aims of ethics are, where to look to ground morality, and so on, while normative ethics tells us how we *ought* to feel, think, and act. It is hard to see how such factual or descriptive knowledge can contribute to the projects of helping us to understand the aims of ethics, where the sources of moral motivation lie, and how we ought to live. I used to be patient with this sort of hybrid of bewilderment and criticism of the programme of naturalized ethics. I am no longer patient, and I want to see if I can switch the burden of proof here. Read every great moral philosopher: Plato, Aristotle, Hobbes, Hutchinson, Smith, Hume, Kant, Mill, etc. I claim that you will find in each and every one of these philosophers a well worked-out *philosophical psychology* which postulates basic human dispositions that help or hinder morality, mental faculties—e.g. reason or emotion—where moral motivation has its source, and an argument for privileging one set of dispositions or one faculty when it comes to justifying a moral view. Furthermore, and relatedly, I claim that whatever normative theory is promoted by the philosopher in question will be best understood by seeing it in the light of the philosophical psychology he espouses. The point is that one ubiquitous feature of the tradition is that everyone, everyone, thinks that their philosophical psychology (I mean to include philosophical anthropology) has *implications* for ethics.

If this much is right, the question arises as to why the contemporary movement to naturalize ethics raises so many hackles? My guess is this: *philosophical psychology*, the sort that can be done from an armchair and which is an assemblage of virtually every possible view of mind, is now giving way to *scientific psychology*, which may eliminate some of the classical views of mind on empirical grounds. If this happens, then our ethical theories will be framed by better background theories about our natures. What could be wrong with this? Perhaps the fear is that if the background theory is scientific, this makes ethics a science. Or that if the background theory is a science, we can suddenly violate the laws of logic and derive 'oughts' from 'is's'. But no one has suggested these things!

[14] What I am calling the descriptive-genealogical component will itself be normative in one sense, since it will involve descriptions of human actions, etc., and thus traffic in intentional description. But it will not be normatively ethical.

Nonetheless, *no* important moral philosopher, naturalist or non-naturalist, has ever thought that merely gathering together all relevant descriptive truths would yield a full normative ethical theory. Morals are radically underdetermined by the merely descriptive, the observational; but so too, of course, are science and normative epistemology. All three are domains of inquiry where ampliative generalizations and underdetermined norms abound.

The distinctively normative ethical component extends the last aspect of the descriptive-genealogical agenda: it will explain why some norms—including norms governing choosing norms (values, virtues)—are good or better than others. One common rationale for favouring a norm or set of norms is that it is suited to modify, suppress, transform, or amplify some characteristic or capacity belonging to our nature—either our animal nature or our nature as socially situated beings. The normative component may try to systematize at some abstract level the ways of feeling, living, and being that we, as moral creatures, should aspire to. But whether such systematizing is a good idea will itself be evaluated pragmatically.

Overall, the normative component involves the imaginative deployment of information from *any* source useful to criticism, self/social examination, formation of new or improved norms and values, improvements in moral educational practices, training of moral sensibilities, and so on.[15] These sources include psychology, cognitive science, all the human sciences, especially history and anthropology, as well as literature, the arts,[16] and ordinary conversation based on ordinary everyday observations about how individuals, groups, communities, nation-states, the community of persons, or sentient beings are faring.[17]

---

[15] One might wonder what important moral philosopher—not overtly interested in religious grounding—did *not* conceive of his project this way. Kant and Moore come to mind.

[16] Richard Rorty (1991: 207) convincingly suggests that the formulation of *general moral principles* has been less useful to the development of liberal institutions than has the gradual *expansion of the imagination* '[through works] like those of Engels, Harriet Taylor and J. S. Mill, Harriet Beecher Stowe, Malinowski, Martin Luther King Jr., Alexis de Tocqueville, and Catherine MacKinnon'.

[17] Critics of naturalized ethics are quick to point out that notions like 'flourishing', 'how people are faring', 'what works for individuals, groups, nation-states, the world, etc.', are vague, virtually impossible to fix in a non-controversial way. This is true. The pragmatist is committed to the requirement that normative judgements get filled out in conversation and debate. Criteria of flourishing, what works, and so on will be as open to criticism as the initial judgements themselves. It is hard, therefore, to see how the criticism *is* a criticism. The naturalist is open to conversational vindication of normative claims; she admits that her background criteria, cashed out, are open to criticism and reformulation; and she admits that words like 'what works', 'what conduces to flourishing', are superordinate terms. Specificity is gained in more fine-grained discussion of particular issues. But in any case there is no ethical theory ever known, naturalist or non-naturalist, which has not depended on abstract concepts. Thin concepts sometimes yield to thick concepts: 'That's bad.' 'Why'? 'Because it is immodest.' Now, one can and often does stop here. But one can go on in any number of directions: 'Why is it immodest?; Why should I care about immodesty?'

The standard view is that descriptive-genealogical ethics can be naturalized, but that normative ethics cannot. One alleged obstacle is that nothing normative *follows* from any set of descriptive-genealogical generalizations. Another alleged obstacle is that naturalism typically leads to relativism, is deflationary, and/or morally naïve. It makes normativity a matter of power: either the power of benign but less than enlightened socialization forces, or the power of those in charge of the normative order, possibly Fascists or Nazis or moral dunces. A third obstacle is that norms themselves do not fit within a naturalistic picture of things—norms are not, one might say, part of the natural fabric of this world. The fourth obstacle to a naturalized normative ethics is related to the three previous obstacles, but is usefully marked off on its own. It is that the project of normative justification requires us to get *outside* the natural world, including the world of 'all actual, and thus far actualized, human practices', to the world of unactualized possibilities, and to the space of better or worse ways of being—ways of being that have not been actualized and that perhaps are even unactualizable in principle, but that are nonetheless worthy as ideals.

I will not remotely be able to answer all these concerns in a satisfactory manner. But let me sketch out some of the things that the ethical naturalist can say in response to these worries, things that I hope will at least allay any fears that ethics naturalized is simply a non-starter.

NATURALIZING NORMATIVITY

Consistent with my theme in the first part of the paper, I want to be clear that, just as there are many varieties of naturalism, speaking generally, so there are differences among those committed specifically to the programmes of epistemology naturalized and ethics naturalized. Indeed, one difference (which appears, for example, in Quine's later work) is that one can believe that the programme of epistemology naturalized can succeed but believe that ethics naturalized cannot. This is not because normativity cannot be naturalized but because ethics is 'methodologically infirm', and permanently so. It is not a discipline in a position to make 'cognitively significant' claims, as it were. But I can't go into this line of argument here (Flanagan 1982, 1988).

In any case, speaking for myself, here are some tenets of ethics naturalized as I conceive it. First, ethical naturalism is non-transcendental in the following respect: it will not locate the rationale for moral claims in the a priori dictates of a faculty of pure practical reason—there is no such thing—*nor* in divine design, which, even if there is such a thing, is beyond our cognitive limits to speak of. Because it is non-transcendental, ethical naturalism will need to provide an 'error theory' that explains the appeal of transcendental rationales and explains why they are less credible than pragmatic rationales, possibly because they are disguised forms of pragmatic rationales.

Briefly, here is what I have in mind: it may well be a natural cognitive tendency to want reasons for action. Unless one is an eliminativist or a physicalist in the reductive sense—a bald naturalist—reasons exist, as do norms and ideals. Reasons, furthermore, can be causes. But being a reason that causes is not the same as being a reasonable cause; or, better: a motivating reason is not, in virtue of being motivating, something that it is reasonable to believe in or something in terms of which to justify one's (other) thoughts or actions. If I believe that 'Santa Claus will not deliver coal to me unless I behave myself', this will motivate me, but it is not the sort of thing we think a sensible adult should believe, let alone be motivated by.

However, since beliefs that have contents that don't refer are no problem for the naturalist, the causal power and efficacy of beliefs about things that don't exist is not something that worries the naturalist either. It is largely a matter of psychological, sociological, and anthropological inquiry why different sorts of things are motivating reasons: i.e. why certain reasons and not others motivate at different times and places. The role for the normative naturalist is to recommend ways of finding good reasons for belief and action and to indicate why it makes sense to be motivated by such reasons.

Motivational grounds to one side, there is always the interesting question of whether, even if we judge the motivating reasons for some norm or set of norms to be unwarranted, we judge the norm or set of norms themselves unwarranted. There is no strict implication. One is inclined to say that even if one behaves well only because one believes Santa Claus thoughts, and even though there is no Santa Claus, one should still behave well—albeit for non-Santa-Clausy reasons. On the other hand, it may just be the case across multifarious social contexts that things like Santa Claus thoughts motivate as well, if not better, than 'Mom and Dad disapprove of XYZ' thoughts. If this is so, we need an explanation, one that explains how beliefs in certain kinds of non-existent objects can motivate and motivate powerfully. False beliefs that produce goods are an interesting phenomenon, but they create no special problem for the naturalist.

Suppose, as seems plausible, that Kant intended his *grounding* of the categorical imperative in pure practical reason both to rationalize the categorical imperative and to motivate us to abide by it. If one denies, as I do, that there is such thing as pure practical reason, and if we also think that the categorical imperative expresses deep moral insight, then we need to give an alternative account of how Kant came or could have come to express the deep insights he expressed. Likely sources include his own pietistic Lutheranism, his wise observations that many thoughtful people see a distinction between happiness and goodness, and emerging Enlightenment ideals about human equality and respect for persons. I don't mean to be suggesting that Kant's insights are justified, if they are justified, by the full story of the genesis of these insights. My point is that Kant was (a) standing at a certain place in the articulation and development of certain norms in Europe; (b) was heir to a set of critical norms for thinking about norms; and (c) deployed these norms of rationality and criticism when evaluating the practices and opinions revealed in history *and* when imaginatively extrapolating from history. His situation and his smarts situated him nicely to express some of the deepest moral insight ever expressed. However, although Kant was very smart, he lacked insight when it came to telling us what it was that he was consulting in displaying his deep moral insights.[18]

Regarding the challenges to naturalism based on open-question arguments or allegations of fallacious inferences from is to ought, the ethical naturalist has all the resources to meet the challenges effectively. With regard to open-question problems, ethics naturalized need not be reductive, so there is no need to define 'the good' in some unitary way such that one can ask the allegedly devastating question: 'But is that which is said to be "good", good?' To be sure, some of the great naturalists—most utilitarians, for example—can be read as trying to define the good in a unitary way. This turned out not to work well, in part because the goods at which we aim are plural and resist a unifying analysis. But secondly, the force of open-question arguments fizzled with discoveries about failures of synonymy across the board—with discoveries about the lack of reductive definitions for most interesting terms.

With regard to the alleged is—ought problem, the smart naturalist makes no claims to establish moral norms demonstratively; he or she points to certain practices, values, virtues, and principles as reasonable based on inductive and abductive reasoning.[19] Indeed, anyone who thinks that Hume thought that the fallacy of claiming to move demonstratively from is to ought revealed that normative ethics was a non-starter, hasn't read Hume. After the famous passages in the *Treatise* about is–ought, Hume proceeds for several hundred pages to do normative moral philosophy. He simply never claims to *demonstrate* anything. Why should he? Demonstration, Aristotle taught us long ago, is for the mathematical sciences, not for ethics.

I've been arguing that neither epistemology naturalized nor ethics naturalized will be a psychologized discipline *simpliciter*—neither will be a 'chapter of psychology'. Both will make use of information from all the human sciences, and in the case of

---

[18] Hume, I dare say, was doing roughly the same thing, but understood somewhat better than Kant what he was doing when he engaged in espousing certain norms. Surely no one thinks that Hume's arguments against religious institutions and religious belief were based on anything like simple description of the practices of most people. He believed that religious belief and practices led, more often than not, to cruelty and intolerance. Given that fact and that, in addition, such beliefs and practices are based on claims that humans lack the cognitive equipment to make with warrant, we have a two-pronged argument for the adjustment of ordinary epistemic and ethical norms.

[19] Speaking now only for myself: my kind of ethical naturalism implies no position on the question of whether there really are, or are not, moral properties in the universe in the sense debated by moral realists, anti-realists, and quasi-realists. The important thing is that moral claims can be rationally supported, not that all the constituents of such claims refer or fail to refer to 'real' things. Furthermore, in both the realism/anti-realism case and the cognitivist/non-cognitivist case different answers might be given at the descriptive and normative levels. J. L. Mackie (1990) is an example of a philosopher who thought that ordinary people were committed to a form of realism about values, but were wrong. In spite of this, Mackie saw no problem with advocating utilitarianism as the best moral theory, and in that sense was a cognitivist—a cognitivist anti-realist, as it were.

ethics from the arts as well, for the arts are a way we have of expressing insights about our nature and about matters of value and worth. The arts are also—indeed, at the same time and for the same reasons—ways of knowing, forms of knowledge, natural knowledge. Actually I want to say the same for sacred texts as well: Greek, Roman, and Egyptian mythology; the Talmud; the *Bhagavadgita*; the Old and New Testaments; the Analects of Confucius; the sayings of Mencius and Buddha; and numerous others. Such works provide many of the deepest insights ever expressed into our natures and our goods. Naturalists will, however, be fussier than most about 'origin stories', especially ones that court the supernatural. But for reasons suggested earlier, concerns about the ontological commitments revealed in these stories need not—indeed, should not—make the naturalist worry that deep truths are *not* being stated. If it ever becomes important (which it rarely does), then the naturalist will need to explain how and why she is committed to the view absent the origin myth.[20] There are, as we saw above, many ways to conceive these texts as full of spiritual insight.

Overall, norms will be generated, evaluated, and revised by examining all the available information in the light of standards we have evolved about what guides or constitutes successful practice. This will include, of course, practices about identifying, specifying, and defending certain norms as superior to others. First-order, second-order, third-order, and possibly higher-order evaluation of norms is something that natural human minds can do (or capacity that can be developed in certain cultures).

I should say something quick about two more issues, since I raised them: the first is how *and* whether norms—and let us add to the list: ideals, imagination, allegiance to ideals, normative guidance, and so on—can be cashed out naturalistically. Non-reductive naturalistic versions of philosophy of mind and of minds following norms, such as those defended by Allan Gibbard, Simon Blackburn, Alvin Goldman, Alasdair MacIntyre, Mark Johnson, George Lakoff, and myself, among others, have shown that such minds are possible and actual.

The other issue is how ethics naturalized avoids (to pick the worst-case scenarios) extreme relativism or—even worse—nihilism. The answer is simple. The ends of creatures constrain what is good for them. Not all kinds of food, clothing, and shelter suit us animals, us members of the species *Homo sapiens*. Nor do all interpersonal and intrapersonal practices suit us. We are social animals with certain innate capacities

---

[20] I need to state this carefully. Remember that the objectionable form of supernaturalism conjoins (i) belief in supernatural powers with (ii) claims about their causal efficacy, (iii) absent recourse to any intersubjectively certified epistemic methods. (iii) is the main troublemaker. Natural theology claims to establish (i) and (ii) in epistemically respectable ways: e.g. by familiar inductive or abductive arguments, such as the cosmological or design arguments. But I think these arguments fail. Both theistic arguments and naturalist arguments about origins commit themselves to an infinite regress, of spirit or matter, respectively. So it's a draw at that point. But naturalistic arguments can explain much better how this universe came to be in standard causal terms, matter and energy producing more of the same. Theistic arguments have no theory about how that which is not material or energetic can cause anything. This, by the way, is the big problem for 'intelligent design'.

and interests. Although the kinds of play, work, recreation, knowledge, communication, and friendship we seek have much to do with local socialization, the general facts that we like to play, work, recreate, know, communicate, and befriend seems to be part, as we say, of human nature. Even prior to the powerful (natural) effects of culture, we prefer different things when it comes to shelter, play, communication, and friendship than beavers, otters, dolphins, birds, orang-utans, and bonobos. This much constrains extreme relativism.

Nihilism is also not a problem. Humans seek value; we aspire to goods, to things that matter and interest us. Nihilism can be a problem when what I earlier called motivating reasons are exposed as not good 'grounding' or 'justifying' reasons. The loss of faith in parental wisdom and authority during adolescence is an example lately on my mind. Nihilism is also a familiar problem for theists who lose the faith. And for the same reasons. But nihilism is not a special problem for naturalists. Animals like surviving. Reflective animals like living well. Over world-historical time reflective animals develop goals for living: welfare, happiness, love, friendship, respect, personal and interpersonal flourishing. These are not an altogether happy and consistent family of values. Still, even if there are incompatibilities involved among the ends we as animals, socialized animals, seek, the fact remains that there *are* ends we seek, and nihilism is not normally an issue—it is not usually a 'live option'. Nihilism is the view that nothing matters. Things do *matter* for us—certain things matter because of our membership in a certain biological species, and certain things matter to us in virtue of how we have evolved as social beings with a history. That is the way it is.

I close with Dewey's insight that 'Moral science is not something with a separate province. It is physical, biological, and historic knowledge placed in a humane context where it will illuminate and guide the activities of men' (1922: 204–5). What is relevant to ethical reflection is everything we know, everything we can bring to ethical conversation that merits attention—data from the human sciences, from history, from literature and the other arts, from playing with possible worlds in imagination, and from everyday commentary on everyday events.

To repeat a point I have made elsewhere, one lesson such reflection teaches, it seems to me, is that if ethics is like any science or is part of any science, it is part of *human ecology*[21] concerned with saying what contributes to the well-being of

---

[21] Why think of ethics naturalized as a branch of human ecology? Well as I insist in Flanagan (1991) the principle of minimal psychological realism (PMPR) is not sufficient to fix correct theory, because many more theories and person-types are realizable than are good, *and* because many good ones have yet to be realized. Moral theories and moral personalities are fixed (and largely assessed) in relation to *particular environments* and *ecological niches* which change, overlap, etc. Therefore, it is best to think of ethics as part of *human ecology*, i.e. neither as a special philosophical discipline nor as a part of any *particular* human science. Are all ways of life OK? Are the only legitimate standards of criticism 'internal' ones? The answer is *no*. What is good depends a great deal on what is good for a particular community, but when that community interacts with other communities, then these get a say. Furthermore, what can *seem* like a good practice or ideal can, when all the information from history, anthropology, psychology, philosophy, and literature is brought in, turn out to have been tried, tested,

humans, human groups, and human individuals in particular natural and social environments.[22]

Thinking of normative ethical knowledge as something to be gleaned from thinking about human good relative to particular ecological niches will, it seems to me, make it easier for us to see that there are forces of many kinds, operating at many levels as humans seek their good; that individual human good can compete with the good of human groups and of non-human systems; and finally, that only some ethical knowledge is global—most is local, and appropriately so. It might also make it seem less compelling to find ethical agreement where none is needed. Of course, saying what I have said is tantamount to affirming some form of ethical relativism. I intend and welcome this consequence.

SOME FINAL THOUGHTS ON RELATIVISM

My original impulse was to leave things at that. Defending ethics naturalized, a pragmatic ethic conceived as part of human ecology, originally seemed like enough for this essay. However, I know from speaking with others over the years that any kind of relativism needs very elaborate defence, and this despite the fact that almost everyone I know is a moderate relativist in my sense (although they don't always know, see, or admit it). For the sake of the reader who wonders how the argument for moderate relativism would go on from here, I provide this sketch.

One initial observation: to say that one is a moderate relativist or a pluralistic relativist (Wong, in press) is not to say that there will not be consistent consensus on certain big-ticket moral truths. Murder and rape will be judged wrong everywhere. Why? Because universal conditions of human flourishing demand it. Where, then, does the pluralism come in? Well, here are a couple of examples. Among certain Nepalese nomads a certain kind of polyandry is practised. A bride marries all the brothers in a family. Is there anything wrong with this practice? From what I know, it works OK. Everyone is happy with the practice; no one is exploited. If so, the practice is morally acceptable, even though it doesn't appeal to us.

Among certain Hindus a son should never get his hair cut or eat chicken the day after his father dies. This is a serious moral violation among Hindus. What is

and found not to be such a good idea. So if ethics is part human ecology, and I think it is, the norms governing the evaluation of practices and ideals will have to be as broad as possible. To judge ideals, it will not do simply to look and see whether healthy persons and healthy communities are subserved by them in the here and now, but this 'health' must be bought without incorporating practices—slavery, racism, sexism, and the like, which we know can go unnoticed for some time—that can keep persons from flourishing, and eventually poison human relations, if not in the present, at least in nearby generations.

[22] David Wong asks if the method I favour favours consequentialism. The answer is no.

going on here may be hard for us to understand. The best explanation is that the deep structure of the Hindu form of life and the associated conception of filial piety yield a different moral obligation at the surface. No surprise. But it is a perfect example of relativism worth respecting.

The question that invariably arises against even the moderate relativist is how a relativist can make *any* credible value judgements. A relativist might try to *influence* others to adopt certain norms, values, and attitudes, but this couldn't be done *rationally*, since the relativist doesn't believe in rationality.

This is an old debate, one the anti-relativist cannot win. The relativist does believe in rationality. She simply thinks that multifarious social worlds will yield reasons to accept somewhat different practices or ways of instantiating shared values (as in the Hindu case). The charges against relativism are indefensible and invariably strike me as unimaginative, born themselves of some sort of irrational fear that even moderate relativism will cause the social world to come undone. We are to imagine that the relativist could have nothing to say about evil people or practices—about Hitler, for example. Here I can only gesture towards a two-pronged argument that can be spelled out in detail (see Flanagan 2002). The first prong involves emphasizing that relativism is the position that certain things are relative to other things. So 'being a tall person' is relative. Relative to what? Certainly not to everything. It is relative to the average height of persons. It is not relative to the price of tea in China, or to the number of rats in Paris, or to the temperature at the centre of the earth, or to the laws regarding abortion, or to zillions of other things. The relativist is attuned to relations that matter, to relations that have relevance to the matter at hand. Even if there is no such thing as 'transcendent rationality' as some philosophers conceive it, there are perfectly reasonable ways of analysing problems, proposing solutions, and recommending attitudes. This is the essence of pragmatism. Pragmatism is a theory of rationality.

The second prong of the argument involves moving from defence to offence. Here the tactic is to emphasize the contingency of the values we hold dear, while at the same time emphasizing that this contingency is no reason for not holding them dear and constitutive of meaning. If it is true, as I think it is, that whether consciousness of the contingency of life undermines confidence, self-respect, etc., depends on what attitudes one takes towards contingency, then there are some new things to be said in favour of emphasizing 'consciousness of contingency'. Recognition of contingency has the advantage of being historically, sociologically, anthropologically, and psychologically realistic. Realism is a form of authenticity, and authenticity has much to be said in its favour. Furthermore, recognition of contingency engenders respect for human diversity which engenders tolerant attitudes. This has generally positive political consequences. Furthermore, respect for human diversity and tolerant attitudes are fully compatible with deploying our critical capacities in judging the quality and worth of alternative ways of being. There are judgements to be made of the quality and worth of those who are living a certain way, and there are assessments about whether we should try to adopt certain ways of being that are not at present our own. Attunement to contingency, plural values, and the vast array of possible human personalities opens the way for the use of important and under-utilized human capacities: capacities for critical reflection, for seeking deep understanding of alternative ways of being and living, and for deploying our agentic capacities to modify our selves, engage in identity experimentation, and meaning location within the vast space of possibilities that have been and are being tried by our fellows. It is a futile but apparently well-entrenched attitude that one ought to try to discover the single right way to think, live, and be. But there is a great experiment going on. It involves the exploration of multiple alternative possibilities, multifarious ways of living, some better than others and some positively awful from any reasonable perspective. The main point is that the relativist has an attitude conducive to an appreciation of alternative ways of life and to the patient exploration of how to use this exposure in the distinctively human project of reflective work on the self, on self-improvement. The reflective relativist, the pragmatic pluralist, has the right attitude—right for a world in which profitable communication and politics demand respect and tolerance, but in which no one expects a respectful, tolerant person or polity to lose the capacity to identify and resist evil where it exists; and right in terms of the development of our capacities for sympathetic understanding, acuity in judgement, self-modification—and, on occasion, radical transformation.

REFERENCES AND SUGGESTED READING

AUDI, ROBERT (1999). 'Naturalism', in *The Cambridge Dictionary of Philosophy*, Cambridge: Cambridge University Press.

BOUWSMA, O. K. (1948). 'Naturalism', *Journal of Philosophy*, 45: 12–22.

DE CARO, M., and MACARTHUR, D. (2004) (eds.). *Naturalism in Question*. Cambridge, Mass.: Harvard University Press.

DEWEY, JOHN (1922). *Human Nature and Conduct*. New York: Henry Holt.

FLANAGAN, OWEN (1982). 'Quinean Ethics', *Ethics*, 93: 56–74. Repr. in Dagfinn Føllesdal (ed.), *Naturalism and Ethics: The Philosophy of Quine*, ii, New York: Garland, 2001.

—— (1988). 'Pragmatism, Ethics, and Correspondence Truth: Response to Gibson and Quine', *Ethics*, 98: 541–9. Repr. in Dagfinn Føllesdal (ed.), *Naturalism and Ethics: The Philosophy of Quine*, ii, New York: Garland, 2001.

—— (1991). *Varieties of Moral Personality: Ethics and Psychological Realism*. Cambridge, Mass.: Harvard University Press.

—— (1992). *Consciousness Reconsidered*. Cambridge, Mass.: MIT Press.

—— (1996a). 'Ethics Naturalized', in L. May, M. Friedman, and A. Clark. (eds.), *Mind and Morals: Essays on Ethics and Cognitive Science*, Cambridge, Mass.: MIT Press.

—— (1996b). 'Ethics Naturalized: Ethics as Human Ecology'. Revised version in *Self-Expressions: Mind, Morals, and Meaning of Life*, Oxford: Oxford University Press.

—— (2002). *The Problem of the Soul: Two Visions of Mind and How to Reconcile Them*. New York: Basic Books.

—— SARKISSIAN, HAGOP, and WONG, DAVID B. (forthcoming). 'Ethics Naturalized', in Walter Sinnott-Armstrong (ed.), *The Psychology and Biology of Morality*, Oxford University Press.

GOLDMAN, ALVIN I. (1986). *Epistemology and Cognition.* Cambridge, Mass.: Harvard University Press.

—— (1992). *Liaisons: Philosophy Meets the Cognitive and Social Sciences.* Cambridge, Mass.: MIT Press.

GOODENOUGH, URSULA (1998). *The Sacred Depths of Nature.* Oxford: Oxford University Press.

HONDERICH, TED (1995). *The Oxford Companion to Philosophy.* Oxford: Oxford University Press.

JAMES, WILLIAM (1892). *Psychology: The Briefer Course.* New York: Dover Publications.

KIM, JAEGWON (1993). 'What is Naturalized Epistemology', in *Supervenience and Mind: Selected Philosophical Essays,* Cambridge: Cambridge University Press.

KITCHER, PHILIP (1992). 'The Naturalists Return', *Philosophical Review,* 101: 53–114.

MACKIE, J. L. (1990). *Ethics: Inventing Right and Wrong.* New York: Penguin Group.

PUTNAM, HILARY (1982). 'Why Reason Can't Be Naturalized', *Synthese,* 52: 3–24. Repr. in *Philosophical Papers,* i, New York: Cambridge University Press,

QUINE, WILLARD VAN ORMAN (1948). 'On What There Is', *Review of Metaphysics,* 2: 21–8. Repr. in *From a Logical Point of View,* Cambridge, Mass.: Harvard University Press, 1953.

—— (1963). *Ontological Relativity and Other Essays.* New York: Columbia University Press.

—— (2001). 'Reply to Morton White'. Repr. in Dagfinn Føllesdal (ed.), *Naturalism and Ethics: The Philosophy of Quine,* ii, New York: Garland.

RORTY, RICHARD (1991). 'On Ethnocentrism', in *Philosophical Papers,* i, Cambridge: Cambridge University Press.

ROSENBERG, ALEXANDER (1996). 'A Field Guide to Recent Species of Naturalism', *British Journal for the Philosophy of Science,* 47: 1–29.

STROUD, BARRY (1996). 'The Charm of Naturalism', *Proceedings and Addresses of the American Philosophical Association,* 70: 43–55.

TAYLOR, CHARLES (1992). *Sources of the Self: The Making of the Modern Identity.* Cambridge, Mass.: Harvard University Press.

WITTGENSTEIN, LUDWIG ([1921] 2001). *Tractatus,* trans. D. F. Pears and B. F. McGuinness. New York: Routledge Classics.

WONG, DAVID (in press). *Natural Moralities: Pluralistic Relativism.* Oxford: Oxford University Press.

CHAPTER 27

INTERPRETING SCIENCE FROM THE STANDPOINT OF WHITEHEADIAN PROCESS PHILOSOPHY

DAVID RAY GRIFFIN

INTRODUCTION

The task of harmonizing scientific and religious thought is at the heart of the version of process philosophy developed by Alfred North Whitehead (1861–1947). Philosophy 'attains its chief importance', Whitehead said, by fusing science and religion 'into one rational scheme of thought' (1978: 15). This task is especially important in the present age, he observed, because science and religion now 'seem to be set one against the other' (1967b: 182).

The basic method for overcoming the apparent conflicts, he held, is to show that science and religion can be harmonized within a philosophical cosmology that exemplifies the standard philosophical criteria of excellence: adequacy and self-consistency. Achieving self-consistency is not terribly difficult. But cosmologies often fail to approach adequacy because they are not based on 'the whole of the evidence'. Traditionally, cosmologies have drawn upon science, aesthetics, ethics, and religion. But 'each age has its dominant preoccupation', which gives its thinkers a tendency to emphasize some dimension of experience to the neglect of others (1967*b*: p. vii).

This tendency goes far toward explaining why science and religion in the West now seem to be in conflict. On the one hand, the idea of God was developed at a time when the scientific attitude, which had earlier flourished in Greece, was in decline. In the modern period, on the other hand, cosmology has been based almost exclusively on the data from the physical sciences at the expense of the evidence from aesthetics, ethics, and religion (1967*b*: p. vii). In each case, the failure to take into account the whole of the evidence led to 'the chief error in philosophy': namely, 'overstatement' (1978: 7). These overstatements create conflicts, because they lead to 'an exclusion of complementary truths' (1926: 144).

Reasons for the Apparent Conflict of Science and Religion

Supernaturalistic Theism

After having long been an atheist or agnostic, Whitehead decided that both 'the general character of things' and our religious, moral, and aesthetic experience point to the reality of an all-pervasive actuality worthy of the name God (1926: 98–9; 1967*b*: 173–4; 1967*a*: 115; 1978: 40, 46; 1968: 103).

He rejected as a disastrous exaggeration, however, the supernaturalistic version of theism, with its 'theology of a wholly transcendent God creating out of nothing an accidental universe' (1978: 95). One problem with this theology is that it leads to an intolerable problem of evil, because God must be held responsible for 'all evil as well as of all good' (1967*b*: 179).

For our present purposes, the main problem with this theology is that it conflicts with an assumption essential to 'the full scientific mentality': namely, the assumption 'that all things great and small are conceivable as exemplifications of general principles which reign throughout the natural order', so that 'every detailed occurrence can be correlated with its antecedents in a perfectly definite manner, exemplifying general principles' (1967*b*: 5, 12).

A form of theism compatible with this assumption would be a naturalistic theism, according to which God could not occasionally interrupt the causal principles that generally obtain. Divine influence would be conceived as a regular part of, rather than an interruption of, these general causal principles. The creation of our world would not have been the beginning of finite matter of fact, but the introduction of 'a certain type of order' within a previously chaotic realm of finite existence (1978: 96).

This doctrine of creation out of chaos implies that beneath the contingent laws of nature, which apply only to the present cosmic epoch, are some *metaphysical* principles, which are inherent in the very nature of things. Belonging to the very nature of God, these principles cannot be violated, as Whitehead indicated in his dictum 'God is not to be treated as an exception to all metaphysical principles, invoked to save their collapse', but as 'their chief exemplification' (1978: 343).

Whitehead thereby affirmed what we can call naturalism$_{ns}$, with 'ns' standing for 'non-supernaturalist'. His proposal to theistic religious communities is that they can best help overcome the conflict of their doctrines with scientific assumptions by replacing supernaturalistic with naturalistic theism.

Modern Science's One-Sided Cosmology

The appearance of conflict between science and religion has been due equally to the version of non-supernaturalist naturalism (naturalism$_{ns}$) that became embodied in the scientific community's cosmology. Whitehead considered this cosmology one-sided because, as indicated above, it is based solely on evidence from the natural sciences, leaving them unmodified by evidence from moral, aesthetic, and religious experience (1967b: p. vii). Science's truths were thereby exaggerated into falsehoods. One of philosophy's central tasks is, accordingly, 'to challenge the half-truths constituting the scientific first principles' (1978: 10).

Whitehead's term for this one-sided cosmology is 'scientific materialism'. This view 'presupposes the ultimate fact of an irreducible brute matter, or material', which is 'senseless, valueless, purposeless' (1967b: 17). Whitehead's name for this concept of matter was 'vacuous actuality', meaning something that is actual, yet devoid of experience (1978: 29, 167). Although in the seventeenth century this view of matter was combined with supernaturalistic theism, the scientific community eventually extended its materialism to the universe itself, thereby ruling out any form of theism.

A third dimension of today's scientific cosmology, beyond its materialism and atheism, is its reliance upon the sensationist doctrine of perception, according to which sensory perception provides our only source of information about the world.

Given these three dimensions, I call this cosmology 'naturalism$_{sam}$', 'sam' standing for 'sensationist-atheist-materialist naturalism'. Although it is often simply called 'naturalism', with no qualifier, naturalism$_{sam}$ is not the only possible embodiment of naturalism$_{ns}$. Whitehead considered it, in fact, a very inadequate embodiment.

Whitehead criticized naturalism$_{sam}$ partly because it cannot do justice to our moral and religious experiences, but also because it is inadequate for science itself (1967b: 83–4). Various developments within the scientific community since the time Whitehead wrote have reflected agreement with his judgement that reductionist materialism cannot do justice to many empirical facts. Nevertheless, naturalism$_{sam}$

remains the dominant view of the scientific community, as represented by its ideological leadership, as I have argued elsewhere (Griffin 2000).

Underlying Whitehead's twofold critique of naturalism$_{sam}$ is the fact that it fails miserably with respect to the fundamental criterion for judging any type of thought—the criterion I have come to call 'hard-core common sense'.

THE CRITERION OF HARD-CORE COMMON SENSE

The claim that one theory is superior to another must be based upon some criteria for measuring superiority, or the claim is simply an arbitrary assertion of preference. Whitehead rightly insisted on the standard criteria: self-consistency and adequacy to the relevant facts.

Some thinkers have alleged that this appeal does not escape arbitrariness, because these criteria are themselves arbitrary. To renounce these criteria, however, would be to renounce rationality and hence thought itself—as Hilary Putnam came to see with regard to self-consistency. At one time, Putnam denied that there are any a priori truths, thereby implying that even the most fundamental laws of logic might be revisable. But later, Putnam argued that we must regard the principle of non-contradiction as an absolutely unrevisable a priori truth, without which rational thought would be impossible (1983: 98–114).

If a cosmological theory, to be worthy of belief, must be self-consistent, it must also be adequate to the relevant facts. But some thinkers have charged that any attempt to apply it will inevitably involve circularity, because people's judgement as to the 'facts' is always a function of their culturally conditioned belief systems. The criticism that one's system of beliefs does not do justice to X—where X might be evolution, telepathy, or miracles—can be dismissed by saying that there is no such thing as X. The criterion of adequacy to the facts could be applied in a non-circular way, however, if there are some beliefs that, not being a function of cultural conditioning, are universally accepted.

Common Sense and Self-Contradiction

Whitehead aligned himself with a philosophical tradition, known as the Scottish 'common-sense' school, according to which there are some beliefs of this type. Hume's opponent Thomas Reid, the most famous member of this school, said of these common-sense ideas that the very act of denying any of them would entail self-contradiction and hence 'metaphysical lunacy' (1997: 268–9).

It should be noted that to use the term 'common sense' for such notions, and only such notions, is to use it in a restricted way. In common parlance, the term is often used for ideas that are widely held at a certain time and place but are not presupposed in human experience as such, and hence are not universal. Referring to ideas of this type as soft-core common sense, I use the adjective 'hard-core' to refer to those common-sense ideas that are truly common to all human beings in the sense that they are inevitably presupposed in human practice, even by people who verbally deny them.

Whitehead's adherence to this approach is shown by his 'metaphysical rule of evidence', according to which 'we must bow to those presumptions, which, in despite of criticism, we still employ for the regulation of our lives' (1978: 151). 'Such presumptions', Whitehead added, 'are imperative in experience.' They are imperative—in the sense that our theories must include them to avoid irrationality—because we inevitably presuppose them in practice. Any theory that denied them would, therefore, involve 'negations of what in practice is presupposed' (1978: 13)— negations that are now sometimes called 'performative contradictions' (Jay 1993). In relation to hard-core common-sense ideas, accordingly, we have no choice but to affirm their truth. The impossibility of rationally denying them is implied by the law of non-contradiction. As John Passmore put it: 'The proposition p is absolutely self-refuting, if to assert p is equivalent to asserting both p and not-p' (1961: 60).

Whitehead, like Reid before him, criticized Hume for violating this criterion by doubting things in his philosophical theory that he never doubted, as Hume himself pointed out, in practice. For example, Hume said that as a philosopher, he had to be a solipsist, doubting the existence of a real world beyond his sensory impressions, and had to deny causation understood as real influence. These denials followed from his sensationist empiricism.

As an empiricist, he (rightly) held that he could include in his theory only notions rooted in direct perception. Given his sensationism, however, he concluded that we have no direct perception of other actualities or of causation as real influence. Accordingly, he left these ideas out of his philosophical theory, relegating them to 'practice'.

For Whitehead, the whole point of philosophy is to show how all our beliefs can be co-ordinated with each other. Explicitly rejecting Hume's refusal to make the attempt, Whitehead said: 'Whatever is found in "practice" must lie within the scope of the metaphysical description. When the description fails to include the "practice", the metaphysics is inadequate and requires revision. There can be no appeal to practice to supplement metaphysics' (1978: 13). Because Hume did make this appeal, thereby resting content with uncoordinated beliefs, he should be regarded as 'the high watermark of anti-rationalism in philosophy'. Instead of using the inevitable presuppositions of practice to *supplement* his theory, Hume should have employed them to *revise* it (1978: 153, 156).

This common-sense criterion will be employed in the following three sections, in which I demonstrate the inadequacy of naturalism$_{sam}$ by pointing out its failing with regard to three things crucial for science: (1) knowledge of the actual world, (2) knowledge of ideal forms, and (3) an explanation of the mind–body relation.

Knowledge of the Actual World

Science claims to be an *empirical* enterprise, which includes the idea that its basic notions are rooted in direct experience. The *sensationist* version of empiricism, however, does not provide an experiential basis for several hard-core common-sense notions about the actual world that are employed by science as categories for interpreting sensory data.

The External World

One such notion is that of the 'external world', meaning a world that exists independently of our perceptions and conceptions. 'The belief in an external world independent of the perceiving subject', said Albert Einstein, 'is the basis of all natural science' (1931: 66).

But, as we saw, Hume noted that sensory perception does not ground this belief. It delivers only sense-data, such as coloured shapes, not an actual world beyond these data. It thereby provides no evidence against solipsism. The scientist's belief in an actual world beyond his or her own experience cannot, therefore, be a rational belief grounded in evidence, but only a matter, as George Santayana (1955) put it, of 'animal faith'.

This irrationalism, illustrated by Hume and Santayana, was also exemplified by Willard Quine, widely considered America's most important philosopher of science in recent decades. Quine was emphatic about his sensationism, insisting that 'whatever evidence there *is* for science *is* sensory evidence' (1969: 75). Given this sensationist version of empiricism, he said that 'our statements about the external world face the tribunal of sense experience' (1953: 41).

But Quine also said, in agreement with Hume, that this sensory experience provides no knowledge of physical objects, so they are in the same boat as Homer's gods. Nevertheless, said Quine, 'I believe in physical objects and not in Homer's gods' (1953: 44). From Quine's sensationist perspective, therefore, science's affirmation of an external world is entirely arbitrary.

The Past and Time

The past and time are two more hard-core common-sense concepts central to science. But the sensationist version of empiricism can provide no grounding for them. Santayana noted that sense perception provides no knowledge of the existence of a past world. He pointed out, accordingly, that the sensationist doctrine of perception implied not merely solipsism but 'solipsism of the present moment' (1955: 14–15).

The concept of time depends upon the distinction between the past, the present, and the future. Because sensory perception cannot ground any of these distinctions, a sensationist philosophy of science leaves the concept of time with no empirical basis.

Causation and Induction

Another hard-core common-sense idea presupposed in science is efficient causation, meaning the causation of one thing by another. But as we saw, Hume showed that sensory perception provides no basis for affirming the real influence of one thing on another, and hence the idea that there is a necessary connection between an effect and its cause(s).

Causation in this sense is presupposed, moreover, by the concept of induction—the assumption that, for example, an experiment performed today in New York can be repeated tomorrow in Edinburgh. Accordingly, if the idea of a necessary connection between cause and effect is groundless, so is the idea of induction.

This problem threatens the very rationality of science, Hans Reichenbach pointed out, because it suggests that science 'is nothing but a ridiculous self-delusion' (1938: 346). A. J. Ayer's desperate solution—that we should 'abandon the superstition that natural science cannot be regarded as logically respectable until philosophers have solved the problem of induction' (1952: 49)—shows that the problem is, among those who presuppose naturalism$_{sam}$, widely regarded as insoluble.

KNOWLEDGE OF IDEAL ENTITIES

Science, while presupposing knowledge of the actual world, equally presupposes knowledge of entities that have ideal, rather than actual, existence—entities often called 'Platonic Forms'.

Mathematical Objects

The ideal entities most obviously required in science are the objects of mathematics, which are especially central to physics. This centrality is a triple embarrassment to naturalism$_{sam}$, because knowledge of such entities conflicts with its sensationism, its atheism, and its materialism.

The most straightforward conflict is with materialism, understood as the doctrine that nothing exists except physical particles and aggregations and properties thereof. This world-view is nominalistic, having no room for Platonic Forms. For example, Richard Rorty, using Whitehead's term for Platonic Forms, says that there are 'no eternal relations between eternal objects' (1989: 107–8).

Sensationism implies that even if ideal entities somehow existed, we would not be able to know of their existence, because our sensory organs can be excited only by material objects. Reuben Hersh has expressed the conflict thus:

An inarticulate, half-conscious Platonism is nearly universal among mathematicians.... Yet... [h]ow does this immaterial realm relate to material reality? How does it make contact with flesh and blood mathematicians?... Ideal entities independent of human consciousness violate the empiricism of modern science. (1997: 12)

Gilbert Harman says: 'We do not and cannot perceive numbers', because '[r]elations among numbers cannot have any... effect on our perceptual apparatus' (1977: 9–10).

One famous mathematician, Kurt Gödel, suggested that our knowledge of these objects comes through a non-sensory type of perception, which we call mathematical intuition (Gödel 1990: 268). But this proposal was greeted with scorn. C. Chihara asked: 'What empirical scientist would be impressed by an explanation this flabby?' (1982: 217). Hilary Putnam, insisting that 'we think with our brains, and not with immaterial souls', declared: 'We cannot envisage *any* kind of neural process that could even correspond to the "perception of a mathematical object"' (1994: 503).

The atheism of naturalism$_{sam}$ creates two more reasons to deny the possibility of mathematical knowledge. First, ever since Aristotle, in criticizing Plato's doctrine of Forms, pointed out that non-actual things can exist only in actual things, Platonists have usually said that the Forms subsist in God. The rejection of theism, by ruling out this answer, creates what can be called the 'Platonic problem': namely, that there is no place for such forms to exist. As Hersh explained:

For Leibniz and Berkeley, abstractions like numbers are thoughts in the mind of God.... [But] the Mind of God [is] no longer heard of in academic discourse. Yet most mathematicians and philosophers of mathematics continue to believe in an independent, immaterial abstract world—a remnant of Plato's Heaven.... Platonism without God is like the grin on Lewis Carroll's Cheshire cat.... The grin remained without the cat. (1997: 12)

The second problem is that even if thinkers assume that mathematical objects can exist on their own, the rejection of theism creates what can be called the 'Benacerraf problem', after Paul Benacerraf (1988), who argued (rightly) that true beliefs can be considered *knowledge* only if that which makes the belief true is somehow *causally* responsible for the belief. Penelope Maddy has summarized the resulting problem: '[H]ow can entities that don't even inhabit the physical universe take part in any causal interaction whatsoever? Surely to be abstract is to be causally inert. Thus if Platonism is true, we can have no mathematical knowledge' (1990: 37).

The problems created by the sensationism, atheism, and materialism of naturalism$_{sam}$ have produced chaos in the philosophy of mathematics. Some philosophers have decided to take the counterintuitive route of simply denying the reality of mathematical objects, as illustrated by book titles such as *Science without Numbers* (Field 1980) and *Mathematics without Numbers* (Hellman 1989).

Most commentators agree, however, that this 'non-realist' or 'formalist' solution is inadequate to the presuppositions of mathematicians themselves. As Y. N. Moschovakis says, it violates 'the instinctive certainty of most everybody who has ever tried to solve a [mathematical] problem that he is thinking about "real objects"' (1980: 605). Similar observations have been made by other philosophers (Maddy 1990: 2–3; Hersh 1997: 7).

Suggesting a second solution, William Lycan, admitting that his own appeal to mathematical sets was 'an embarrassment to physicalism, since sets et al. are non-spatiotemporal, acausal items', said that unless he was to reject the idea of mathematical sets altogether, he would have to naturalize them (1987: 90). This option has been developed by Penelope Maddy, who has attempted to replace the Platonic version of realism with a materialistic version. We can overcome the problem of 'unobservable Platonic entities', she suggests, by 'bringing [mathematical] sets into the physical world' so that they are no longer 'abstract' but have 'spatio-temporal location' (1990: 44, 59, 78).

This solution is clearly desperate. What could it mean to say that the entire realm of mathematics is embodied in the physical world in such a way as to be observable, especially when only a small portion of the realm explored by pure mathematicians is exemplified in our world?

A third solution, that of Quine, is to affirm the existence of mathematical objects while simply ignoring all the problems. On the basis of his 'physicalism', which declares physics to be the arbiter of what exists, Quine affirmed the existence of mathematical objects solely because of their indispensability to physics. Not flinching from the arbitrary exception that this doctrine introduced, he declared his ontology to be 'materialism, bluntly monistic except for the abstract objects of mathematics' (1995: 14). But Quine never discussed the Platonic problem of *where* and *how* these objects might exist in an otherwise materialistic universe.

Quine's allowance for mathematical objects also introduced arbitrariness into his enforcement of the 'tribunal of sense experience'. While using this tribunal to rule out Platonic moral norms (1986), he simply 'ignore[d] the problem', as Putnam put it, 'as to how we can know that abstract [mathematical] entities exist unless we can interact with them in some way' (1994: 153).

Logical Truths

Science also presupposes the knowledge of logical truths. As Putnam points out, however, 'the nature of mathematical truth' and 'the nature of logical truth' are one and the same problem (1994: 500). Therefore, naturalism$_{sam}$, if true, should equally prevent scientists from knowing logical truths.

Normative Principles

Scientific activity also presupposes the objectivity of normative principles, such as moral ideals. Most medical research, for example, presupposes that some states of affairs—such as the absence of unbearable pain—are better than others, and that it is right to promote those states of affairs. But naturalism$_{sam}$ has been used to deny the possibility of moral truths.

For example, John Mackie, denying that moral values are 'part of the fabric of the world', said that it is *not* an 'objective, intrinsic, requirement of the nature of things' that 'if someone is writhing in agony before your eyes' you should 'do something about it if you can' (1977: 24, 79–80). Essentially the same position has been articulated by Gilbert Harman and Bernard Williams.

These denials of the possibility of moral knowledge are based partly on sensationism (Mackie 1977: 38–9; Harman 1977: 9–10; Williams 1985: 94), partly on materialism—Harman rules out moral truths by declaring that '*all* facts are facts of nature' (1977: 17)—and partly on atheism (Mackie 1977: 48). The Platonic problem created by atheism has been especially emphasized by Harman. Having said that 'our scientific conception of the world has no place for [normative] entities' (1989: 366), he explains this by saying: 'Our scientific conception of the world has no place for gods' (1989: 381).

For science to reject the existence of normative truths, however, is self-defeating. As Charles Larmore points out, the reasons for denying the reality of objective moral values apply equally to cognitive values, which concern 'the way we ought to think'. To deny objective cognitive values would imply the absurd conclusion that the idea that we ought to avoid self-contradiction is merely a preference, with no inherent authority (Larmore 1996: 87, 99). Larmore's well-argued conclusion is that antiPlatonic naturalism, if carried through consistently, 'would destroy the very idea of rationality' (1996: 100–2).

In sum, although science presupposes the capacity of scientists to know various kinds of ideal entities, the possibility of such knowledge is denied by all three dimensions of naturalism$_{sam}$.

THE MIND–BODY RELATION

This version of naturalism also creates an insuperable mind–body problem. Long recognized as the central problem of modern philosophy, it is now seen to be a problem for science as well. Descartes's starting point, that our conscious experience is the thing of which we are most certain, is still valid. But his dualism, by means of which he put human consciousness outside nature and hence beyond the purview of natural science, is now rightly rejected. Accordingly, the scheme of ideas used to discuss the entities of physics, chemistry, and biology cannot be considered adequate unless it can also explain the relationship between our brains and our conscious experiences.

Yet this has proved impossible for naturalism$_{sam}$, primarily because of its materialistic view of nature, according to which the world's ultimate units are devoid of both experience and spontaneity. This view prevents a coherent treatment of three hard-core common-sense notions about the mind–body relation: (1) conscious experience exists; (2) it exerts influence upon the body; and (3) it has a degree of self-determining freedom.

The Reality of Conscious Experience

The hard-core status of the reality of one's own conscious experience is obvious. The impossibility of doubting it was famously emphasized by Descartes. Now Jaakko Hintikka has shown that Descartes's argument involved the notion of a performative contradiction. If I say, 'I doubt herewith, now, that I exist,' explained Hintikka, 'the propositional component contradicts the performative component of the speech act expressed by that self-referential sentence' (1962: 32). Any attempt to avoid the mind–body problem by eliminating all references to conscious experience, as proposed by some 'eliminative materialists', would involve this kind of self-refuting contradiction.

Accordingly, if science endorses a materialist view of nature, it needs an explanation of how beings with conscious experience could have emerged out of entities wholly devoid of experience. From the time of Descartes to the present, however, no such explanation has been produced.

Some philosophers, such as Thomas Nagel, have pointed out that this problem is insoluble in principle. Using *en soi* for a being that is devoid of experience and *pour soi* for one that, having experience, exists 'for itself', Nagel wrote: 'One cannot derive a *pour soi* from an *en soi*. . . . This gap is logically unbridgeable. . . . [A] conscious being [cannot be created] by combining together in organic form a lot of particles with none but physical properties' (1979: 188–9). The failure to surmount this difficulty is now admitted both by dualists, who think of the mind as a full-fledged actuality, and by materialists, who think of consciousness as merely an emergent or supervenient property of the brain.

From the side of dualism, Geoffrey Madell has said that 'the appearance of consciousness in the course of evolution must appear for the dualist to be an utterly inexplicable emergence' (1988: 140–1). Another dualist, Richard Swinburne, has used this difficulty to support the existence of an omnipotent deity, saying that 'science cannot explain the evolution of a mental life', but 'God, being omnipotent, would have the power to produce a soul' (1986: 198–9). This solution is, of course, ruled out by naturalism$_{ns}$, so dualism's problem of consciousness remains unsolved.

From within the materialist camp, Colin McGinn has declared that because it is impossible to understand how 'the aggregation of millions of individually insentient neurons [constituting the brain] generate subjective awareness', the problem is insoluble in principle (1991: 1). In the face of this problem, McGinn adds: '[S]cientific naturalism [by which he means naturalism$_{sam}$] runs out of steam. . . . It would take a supernatural magician to extract consciousness from matter' (1991: 45). McGinn and other materialists cannot, of course, accept that solution, so they are left with a problem seen to be insoluble in principle.

Mental Causation

The idea that conscious experience affects bodily behaviour, usually called 'mental causation', is also widely recognized as a hard-core common-sense notion. Ted Honderich says that its main recommendation is 'the futility of contemplating its denial'. Speaking of epiphenomenalism—the doctrine that conscious experience does *not* exert causal efficacy on the body—Honderich says: 'Off the page, no one believes it' (1987: 447). Jaegwon Kim likewise says that any theory, to be adequate, must have room for psychophysical causation, as when, feeling a pain, one's decision to call the doctor leads one to walk to the telephone and dial (1993: 286). John Searle, in a similar vein, includes 'the reality and causal efficacy of consciousness' among the 'obvious facts' about our minds, endorsing the 'commonsense objection to "eliminative materialism"' that it is "crazy to say that . . . my beliefs and desires don't play any role in my behaviour"' (1992: 54, 48).

Neither dualists nor materialists, however, have been able to explain the possibility of mental causation. For dualists, the problem is understanding how a mental entity could influence physical entities, understood to be completely different in kind. As Madell admits, 'the nature of the causal connection between the mental and the physical, as the Cartesian conceives of it, is utterly mysterious' (1988: 2). Descartes himself was not embarrassed by this mysteriousness, because he could appeal to divine omnipotence (Baker and Morris 1996: 153–4, 167–70). But dualists who reject such appeals have an insoluble problem.

The problem is equally serious for materialists. Searle, for example, asks rhetorically: 'How could something mental make a physical difference . . . in our brains? . . . Are we supposed to think that thoughts can wrap themselves around the axons or shake the dendrites or sneak inside the cell wall and attack the cell nucleus?' (1984: 117). Searle means to be ridiculing dualism. But his rejection of mental causation contradicts his own statement that the causal efficacy of consciousness for our bodily behaviour is an 'obvious fact'.

This problem is even more serious for materialists because of their insistence, in Jaegwon Kim's words, 'that if we trace the causal ancestry of a physical event, we need never go outside the physical domain'. This insistence denies 'the Cartesian idea that some physical events need nonphysical causes' (1993: 280). But this insistence contradicts Kim's affirmation that we walk to the telephone *because* we have decided to make a call. Upon seeing this contradiction, Kim admitted that materialism seems 'to be up against a dead end' (1993: 367).

The aspect of mental causation with which materialism is most obviously in conflict is rational activity. According to the materialist world-view, all causation is *efficient* causation. The rational activity of a scientist is, however, an example of *final* causation, because it is action in terms of some norm, such as the norm of self-consistency. But the materialist world-view has no room for such activity, because the mind is equated with the brain, and the brain's activities are said to be determined by the causal activities of its most elementary parts, assumed to consist entirely of chains of efficient causation.

McGinn raises this problem by asking 'how a physical organism can be subject to the norms of rationality. How, for example, does *modus ponens* get its grip on the causal transitions between mental states?' (1991: 23 n.). McGinn admits that materialism can provide no answer, thereby illustrating Putnam's charge that most science-based philosophies are self-refuting because they 'leave no room for a rational activity of philosophy' (1983: 191).

Freedom

The idea that our consciously directed acts are based on a degree of self-determining freedom is equally recognized as one of our hard-core common-sense beliefs. Searle, after pointing out that people have been able to give up *some* common-sense beliefs, such as the (*soft-core* common-sense) beliefs in a flat Earth and literal 'sunsets', points out that 'we can't similarly give up the conviction of freedom because that conviction is built into every normal, conscious intentional action.... [W]e can't act otherwise than on the assumption of freedom, no matter how much we learn about how the world works as a determined physical system' (1984: 97). Thomas Nagel says: 'I can no more help holding myself and others responsible in ordinary life than I can help feeling that my actions originate with me' (1986: 123).

Some philosophers try to make this *feeling* of freedom compatible with complete determinism by redefining freedom. According to this compatibilist definition, to say that I did X freely is *not* to say that I could have acted otherwise (Lycan 1987: 113–14). But to speak of freedom only in this compatibilist, Pickwickian sense is *not* to speak of freedom as we presuppose it in practice, as seen by both Nagel (1986: 110–17) and Searle (1984: 92–5).

Materialists have not been able to affirm freedom in this genuine (incompatibilist, libertarian) sense. Searle says that science 'allows no place for freedom of the will' (1984: 92). This denial follows from materialistic assumptions, which Searle summarizes thus: 'Since nature consists of particles and their relations with each other, and since everything can be accounted for in terms of those particles and their relations, there is simply no room for freedom of the will' (1984: 86). Searle concludes that, although we necessarily presuppose freedom, it must be an illusion (1984: 86, 97). Explicitly admitting his failure to reconcile his world-view with a (hard-core) common-sense conviction, Searle says that although, 'ideally, I would like to be able to keep both my commonsense conceptions and my scientific beliefs... when it comes to the question of freedom and determinism, I am... unable to reconcile the two' (1984: 86).

Searle's inability to affirm genuine freedom is echoed by many other materialists, such as McGinn (1991: 17 n.), Nagel (1986: 110–23), and Daniel Dennett (1984). Crucial to these denials is the assumption that the mind is not an entity distinct from the brain, with its billions of cells, so there is no conceivable locus for free decisions to be made. As Dennett puts it, the human head contains billions of 'miniagents and microagents (with no single Boss)', and 'that's all that's going on' (1991: 458, 459).

Dualists, by contrast, can affirm freedom by saying that there *is* a 'single Boss', this being the mind, understood as an entity distinct from the brain with the capacity to make decisions. But, given the inability of dualists to explain causal interaction between this mind and its brain, they have failed to explain how this freedom can express itself in bodily practices, such as those involved in scientific research.

The upshot of the previous three sections is that a new version of scientific naturalism is needed, not simply for the sake of overcoming the apparent conflicts between science and religion (even after religion has given up supernaturalism), but also for the sake of science itself. A new version of naturalism, intended as a new framework for science, was developed by Whitehead.

Scientific Naturalism$_{ppp}$

Whitehead's version of naturalism$_{ns}$ can be called 'naturalism$_{ppp}$', with 'ppp' standing for 'prehensive-panentheist-panexperientialist naturalism' (Griffin 2001: 171). In this position, the sensationism of naturalism$_{sam}$ is replaced by a *prehensive* doctrine of perception, according to which sensory perception is derivative from a more fundamental mode of perception, in which there is a non-sensory grasp (prehension) of other things. Atheism is replaced by *panentheism*, according to which the universe of finite things lives within an all-inclusive experience. And the materialistic view of nature is replaced by *panexperientialism*, according to which all genuine individuals, down to the most elementary units of nature, have at least some iota of experience and spontaneity.

This doctrine does *not* attribute experience and spontaneity to all things whatsoever, so it cannot be quickly dismissed by pointing out that rock behaviour manifests no spontaneity suggestive of a unified experience. Rather, experience is attributed only to true individuals—whether simple individuals, such as quarks, or 'compound individuals', such as animals—not to aggregational societies of individuals, such as sticks and stones (Hartshorne 1972). I have emphasized this point by speaking of 'panexperientialism with organizational duality' (Griffin 2001: 6).

I will now indicate how the various failures of naturalism$_{sam}$ discussed above are overcome by naturalism$_{ppp}$.

Knowledge of the Actual World

The fact that we inevitably presuppose the reality of the external world, the past, time, and efficient causation is explained by saying that we have a direct, non-sensory experience of them. Whitehead, focusing especially on sensationism's inability to affirm efficient causation, refers to our direct prehension of prior actualities as 'perception in the mode of causal efficacy'. In these prehensive acts, we are directly aware of the other thing *as actual* and *as exerting causal efficacy* on us. Our awareness of the past, and hence time, arises primarily from prehending prior moments of our own experience, from which memory arises.

We are now in a position to understand one of Whitehead's main criticisms of the sensationist doctrine of perception: that although sensory perception provides the *precise observations* on which science depends, it does not provide the basic *categories* used in scientific theory to interpret the observational data. Those categories are provided by our non-sensory mode of perception (Whitehead 1978: 169–70; 1967a: 225; 1968: 133).

The sensationist doctrine of perception hence involves a disastrous exaggeration—concluding from the fact that sensory perception is the only source of *precise* data that it is the only source of data, *period*.

Knowledge of Ideal Entities

Just as all three dimensions of naturalism$_{sam}$ contributed to making science's knowledge of ideal entities appear impossible, all three dimensions of naturalism$_{ppp}$ are relevant for overcoming this appearance.

The replacement of materialism with panexperientialism means that the existence of ideal entities is not ruled out, so both mathematical objects and normative ideals can be affirmed. Whitehead, in fact, explicitly affirmed eternal objects—his term for Platonic Forms—of both types: eternal objects of the *objective* species, which he equated with 'the mathematical Platonic Forms', and eternal objects of the *subjective* species, which include normative ideals (1978: 291).

Panexperientialism also allows, without any exception to the general principles of the universe, the idea that the experiences constituting the universe of finite things exist within an all-inclusive experience, which can influence the world somewhat as our minds influence our bodies.

This panentheistic view of the universe allows us to reaffirm the old idea that mathematical, logical, moral, aesthetic, and cognitive ideals can both exist and have causal efficacy because they subsist in God. In Whitehead's language, eternal objects subsist in 'the primordial nature of God' (1978: 46), having influence by virtue of being envisaged by God with appetition for their actualization in the world (1978: 32–4). Providing an answer to both the Platonic and the Benacerraf problem, Whitehead says that we experience ideals by virtue of their presence in the divine, non-local agent: 'There are experiences of ideals—of ideals entertained, of ideals aimed at, of ideals achieved, of ideals defaced. This is the experience of the deity of the universe' (1968: 103).

Given the prehensive doctrine of perception, this prehension of ideals, by means of prehending the divine experience in which they subsist, requires no *ad hoc* hypothesis of a special 'moral sense'. Rather, we can apprehend these ideal forms by means of the same mode of perception through which we perceive the past world's causal efficacy.

Explaining the Mind–Body Relation

The change from the materialist to the panexperientialist view of nature means that our hard-core common-sense assumptions about the mind–body relation can be affirmed. If the cells in the brain are centres of experience, then the idea that they give rise to minds is no longer inconceivable. Even McGinn grants this point, saying that if we could suppose neurons to have 'proto-conscious states', it would be 'easy enough to see how neurons could generate consciousness' (1991: 28 n.).

Panexperientialism also overcomes the obstacles to affirming mental causation. Dualists, while verbally affirming interaction between mind and brain, cannot explain how wholly unlike things could interact. Materialists, seeing the impossibility of dualistic interaction, identified the mind and the brain, thereby making inter-action equally impossible. Panexperientialism, by contrast, allows interactionism without dualism. As Charles Hartshorne, the major process philosopher after Whitehead, explained: '[C]ells can influence our human experiences because they have feelings that we can feel. To deal with the influences of human experiences upon cells, one turns this around. *We* have feelings that *cells* can feel' (1962: 229).

Finally, panexperientialism with organizational duality can explain how our apparent freedom is genuine. Like dualism, panexperientialism regards the mind as an actuality that, being distinct from the brain, is a conceivable locus of decision making. But, unlike dualism, this position is not undermined by the inconceivability of dualistic interaction—which led dualism to collapse into materialistic identism, with its explicit denial of freedom.

Panexperientialism does agree with materialism that all actual entities are of the same ontological type. But panexperientialism also agrees with dualism's strong point: the recognition of a structural difference between aggregational entities, such as rocks and typewriters, and compound individuals, such as cats and human beings. Materialistic identism, not allowing this distinction, has to regard intentional human actions as, in Whitehead's words, 'purely governed by the physical laws which lead a stone to roll down a slope and water to boil' (1958: 14).

From the perspective of panexperientialism, this analysis works for a stone because although the molecules have some iota of spontaneity, the various spontaneities 'thwart each other, and average out so as to produce a negligible total effect' (Whitehead 1967a: 207). In a human being, by contrast, there is a dominant member, the mind, which can synthesize the various influences from the body, make a decision, then co-ordinate the body so as to carry out this decision. Whereas the stone's behaviour can be explained entirely in terms of efficient causation, the explanation of human behaviour requires reference to final causation—a purpose freely pursued (Whitehead 1968: 28–9).

Accordingly, although there is no ontological dualism, there is an organizational duality, and 'diverse modes of organization' can produce 'diverse modes of functioning' (1968: 157). We need not suppose, therefore, that because the behaviour of *some* things is wholly determined by efficient causes, science must assume that the behaviour of *all* things is so determined.

IMPLICATIONS FOR EVOLUTIONARY THEORY

I will conclude by indicating the differences that the acceptance of naturalism$_{ppp}$ by both the religious and the scientific communities could make in relation to evolutionary theory, which has long been the major source of conflict.

In its most extreme form, this conflict pits neo-Darwinian evolutionism against 'scientific creationism', which rejects the reality of macro-evolution, and hence common ancestry, in favour of the doctrine that each species is created by God *ex nihilo*. This version of the conflict, which has been especially intense, would obviously be overcome insofar as supernaturalistic theism is replaced by naturalism$_{ppp}$. Much of the conflict between the religious and scientific communities would continue, however, because of neo-Darwinism's rootage in naturalism$_{sam}$. But the remaining conflicts would be overcome if the scientific community adopted a theory of evolution based on naturalism$_{ppp}$.

As I have explained elsewhere (Griffin 2000: ch. 8), the set of doctrines constituting neo-Darwinism, as widely presented, involves a combination of scientific and metaphysical doctrines. Besides accepting the basic scientific doctrines (which assert the reality of macro-evolution as well as micro-evolution), scientists operating in terms of naturalism$_{ppp}$ would also, by definition, accept neo-Darwinism's metaphysical conviction that macro-evolution occurs naturalistically.

A theory of evolution based on naturalism$_{ppp}$ would, however, reject four other metaphysical doctrines: namely, materialism, determinism, nominalism, and the claim, reflecting the atheism of naturalism$_{sam}$, that the evolutionary process is entirely undirected. It would also be open to rejecting, if warranted by empirical evidence, three derivative scientific claims: namely, that evolution is gradualistic, involving no jumps; that macro-evolution involves only the kinds of processes that occur in micro-evolution; and that the only processes involved are random variations and natural selection. Finally, a theory of evolution based on naturalism$_{ppp}$ would definitely reject three moral-religious implications of the neo-Darwinian account: namely, that the universe is non-progressive, amoral, and meaningless.

As this brief sketch indicates, a theory of evolution based on naturalism$_{ppp}$ would provide a middle way between the extremes of neo-Darwinism and creationism. It would endorse the main claims of the scientific community: namely, that evolution has occurred, and in a fully naturalistic way. But it would reject the dimensions of neo-Darwinism that have been the primary sources of the strong moral and religious objections to evolutionism as currently taught.

REFERENCES AND SUGGESTED READING

AYER, A. J. (1952). *Language, Truth, and Logic*. New York: Dover.
BAKER, GORDON, and MORRIS, KATHERINE J. (1996). *Descartes' Dualism*. London and New York: Routledge.

BENACERRAF, PAUL (1983). 'Mathematical Truth', in Paul Benacerraf and Hilary Putnam (eds.), *Philosophy of Mathematics*, 2nd edn., Cambridge: Cambridge University Press, 402–20.

CHIHARA, C. (1982). 'A Gödelian Thesis Regarding Mathematical Objects: Do They Exist? And Can We Perceive Them?', *Philosophical Review*, 91: 211–17.

DENNETT, DANIEL E. (1984). *Elbow Room: The Varieties of Free Will Worth Wanting*. Cambridge, Mass.: MIT Press.

—— (1991). *Consciousness Explained*. Boston: Little, Brown.

EINSTEIN, ALBERT (1931). 'Maxwell's Influence on the Development of the Conception of Physical Reality', in J. J. Thomson *et al.* (eds.), *James Clerk Maxwell: A Commemorative Volume*, Cambridge: Cambridge University Press, 66–73.

FIELD, HARTRY (1980). *Science without Numbers*. Princeton: Princeton University Press.

GÖDEL, KURT (1990). 'What is Cantor's Continuum Problem? Supplement to the Second [1964] Edition', in *Collected Works*, ed. Solomon Feferman et al. New York: Oxford University Press, ii. 266–9.

GRIFFIN, DAVID RAY (2000). *Religion and Scientific Naturalism: Overcoming the Conflicts*. Albany, N.Y.: State University of New York Press.

—— (2001). *Reenchantment without Supernaturalism: A Process Philosophy of Religion*. Ithaca, N.Y.: Cornell University Press.

HARMAN, GILBERT (1977). *The Nature of Morality: An Introduction to Ethics*. New York: Oxford University Press.

—— (1989). 'Is There a Single True Morality?', in Michael Krausz (ed.), *Relativism: Interpretation and Confrontation*, Notre Dame, Ind.: University of Notre Dame Press, 363–86.

HARTSHORNE, CHARLES (1962). *The Logic of Perfection and Other Essays in Neoclassical Metaphysics*. La Salle, Ill.: Open Court.

—— (1972). 'The Compound Individual', in *Whitehead's Philosophy: Selected Essays 1935–1970*, Lincoln, Nebr.: University of Nebraska Press, 41–61.

HELLMAN, G. (1989). *Mathematics without Numbers*. Oxford: Oxford University Press.

HERSH, REUBEN (1997). *What is Mathematics, Really?* New York: Oxford University Press.

HINTIKKA, JAAKKO (1962). 'Cogito, Ergo Sum: Inference or Performance', *Philosophical Review*, 71: 3–32.

HONDERICH, TED (1987). 'Mind, Brain, and Self-Conscious Mind', in Colin Blakemore and Susan Greenfield (eds.), *Mindwaves: Thoughts on Intelligence, Identity, and Consciousness*, Oxford: Blackwell, 445–58.

JAY, MARTIN (1993). 'The Debate over Performative Contradiction: Habermas versus the Poststructuralists', in *Force Fields: Between Intellectual History and Cultural Critique*, New York and London: Routledge, 25–37.

KIM, JAEGWON (1993). *Supervenience and Mind: Selected Philosophical Essays*. Cambridge: Cambridge University Press.

LARMORE, CHARLES (1996). *The Morals of Modernity*. Cambridge: Cambridge University Press.

LYCAN, WILLIAM G. (1987). *Consciousness*. Cambridge, Mass.: MIT Press.

MACKIE, JOHN (1977). *Ethics: Inventing Right and Wrong*. New York: Penguin Books.

MADDY, PENELOPE (1990). *Realism in Mathematics*. Oxford: Clarendon Press.

MADELL, GEOFFREY (1988). *Mind and Materialism*. Edinburgh: Edinburgh University Press.

McGINN, COLIN (1991). *The Problem of Consciousness: Essays Toward a Resolution*. Oxford: Blackwell.

MOSCHOVAKIS, Y. N. (1980). *Descriptive Set Theory*. Amsterdam: North-Holland.

NAGEL, THOMAS (1979). *Mortal Questions*. London: Cambridge University Press.

—— (1986). *The View from Nowhere*. New York: Oxford University Press.

PASSMORE, JOHN (1961). *Philosophical Reasoning*. New York: Basic Books.

PUTNAM, HILARY (1983). *Realism and Reason*. New York: Cambridge University Press.

—— (1994). *Words and Life*, ed. James Conant. Cambridge, Mass.: Harvard University Press.

QUINE, WILLARD VAN (1953). *From a Logical Point of View*. Cambridge, Mass.: Harvard University Press.

—— (1969). *Ontological Relativity and Other Essays*. New York: Columbia University Press.

—— (1986). 'Replies', in Lewis Edwin Hahn and Paul Arthur Schilpp (eds.), *The Philosophy of W. V. Quine*, La Salle, Ill.: Open Court, 663–5.

—— (1995). *From Stimulus to Science*. Cambridge, Mass.: Harvard University Press.

REICHENBACH, HANS (1938). *Experience and Prediction*. Chicago: University of Chicago Press.

REID, THOMAS (1997). *An Inquiry into the Human Mind on the Principles of Common Sense* [1764]: *A Critical Edition*, ed. Derek R. Brookes. University Park, Pa.: Pennsylvania State University Press.

RORTY, RICHARD (1989). *Contingency, Irony, and Solidarity*. Cambridge: Cambridge University Press.

SANTAYANA, GEORGE (1955). *Scepticism and Animal Faith*. New York: Dover.

SEARLE, JOHN R. (1984). *Minds, Brains, and Science*. London: British Broadcasting Corporation.

—— (1992). *The Rediscovery of the Mind*. Cambridge, Mass.: MIT Press.

SWINBURNE, RICHARD (1986). *The Evolution of the Soul*. Oxford: Clarendon Press.

WHITEHEAD, ALFRED NORTH (1926). *Religion in the Making*. New York: Macmillan.

—— (1958). *The Function of Reason*. Boston: Beacon Press.

—— (1967a). *Adventures of Ideas*. New York: Free Press.

—— (1967b). *Science and the Modern World*. New York: Free Press.

—— (1968). *Modes of Thought*. New York: Free Press.

—— (1978). *Process and Reality*, corrected edn., ed. David Ray Griffin and Donald W. Sherburne. New York: Free Press.

WILLIAMS, BERNARD (1985). *Ethics and the Limits of Philosophy*. Cambridge, Mass.: Harvard University Press.

CHAPTER 28

ANGLO-AMERICAN POST-MODERNITY AND THE END OF THEOLOGY– SCIENCE DIALOGUE?

NANCEY MURPHY

WHAT IS ANGLO-AMERICAN POST-MODERNITY?

I have often regretted using the term 'post-modern'; people just *will* take it to mean what they want it to mean. So I begin by sorting through some of its many uses. It was apparently first used to designate styles in art and architecture that were self-consciously post-modernist (Toulmin 1990: 6). It is most often associated with recent moves in Continental philosophy and literary criticism. Since these are largely French, might they not be designated as *après-moderne*, since we need the plain English to refer to significant changes in Anglo-American philosophy? There are also changes in popular culture that are called post-modern; but it is not clear whether they are related to these philosophical changes.

My task in this chapter, then, is to describe emerging patterns in Anglo-American philosophy that represent radical breaks from the thought patterns of Enlightened modernity and to spell out consequences of these changes for the theology–science dialogue. One might ask why the focus on philosophy. The rationale is based on the relation that philosophy bears to the rest of culture. Huston Smith has written that:

The dominant assumptions of an age color the thoughts, beliefs, expectations, and images of the men and women who live within it. Being always with us, these assumptions usually pass unnoticed—like the pair of glasses which, because they are so often on the wearer's nose, simply stop being observed. But this doesn't mean they have no effect. Ultimately, assumptions which underlie our outlooks on life refract the world in ways that condition our art and our institutions, . . . our sense of right and wrong, our criteria of success, . . . what we think it means to be a man or woman, how we worship our God or whether, indeed, we have a God to worship. (Smith 1989: 3–4)

A central task of philosophy is to expose these often invisible assumptions, to critique them, to suggest improvements or replacements; these new theories often become the assumptions on which the next era of scholarship is based.

The terms 'Continental' and 'Anglo-American' are conventional designations for two styles of philosophy that developed in the nineteenth century as different reactions to the work of Immanuel Kant. The Continental tradition has stayed closer to the humanities, while the Anglo-American tradition allies itself as closely as possible with science; the Continental seeks to bridge the gap between wisdom and knowledge, while the Anglo-American focuses on clarifying and solving philosophical problems (Critchley 2001). The most significant movements making up the Anglo-American tradition have been logical positivism, originating with the Vienna Circle in the 1920s (thus Michael Dummett quips that it might better be called 'Anglo-Austrian' (Critchley 2001: 32)); logical atomism, whose heyday was in England in the 1930s; and a wide variety of movements from then until the present falling under the heading of 'analytic philosophy'. A less contentious name for what comes next might therefore be 'post-analytic' philosophy.

In what follows I shall describe new moves in three traditional branches of philosophy: metaphysics, philosophy of language, and epistemology. My contention is that, beginning half a century ago, whole clusters of terms in each of these domains have taken on new uses, and that these changes have radical consequences for all areas of academia. My task here will be to note their actual and potential contributions to the dialogue between theology and science.

CROSSING THE POST-MODERN DIVIDE

My sense of radical discontinuity in the recent history of philosophy came about serendipitously. My late husband, James McClendon, and I were invited to a conference on 'The Church in a Postmodern World'. The speakers included George Lindbeck, historical theologian at Yale; Diogenes Allen, philosopher at Princeton

Theological Seminary; Robert Bellah, sociologist at the University of California, Berkeley; and others. We asked ourselves: if these speakers represent post-modernity, then what is the substance of their differences from modern predecessors? Our sense was that the nature of the questions, the very terms of the arguments, had shifted, and that a handful of philosophers could be credited with the change. In epistemology and philosophy of science, there was the rejection of foundationalism in favour of the holist views of the likes of W. V. O. Quine and Thomas Kuhn. In philosophy of language there was the shift from theories of meaning based on reference or representation to a focus on the social uses of language, found especially in the works of J. L. Austin and Ludwig Wittgenstein. In ethics the shift was the rejection of modern 'generic' individualism in favour of Alasdair MacIntyre's more complex theory of the priority of the social. This rejection of individualism is one aspect of a broader revision of views of the relation of parts and wholes reflected in science and other branches of philosophy—the rejection of modern atomist metaphysics in all its forms.

I said above that the *arguments* have shifted. By this I mean to acknowledge that modern thought was never monolithic. Rather, typical arguments such as those between 'mainline' epistemologists and their sceptical opponents tend to presuppose the same basic theory of knowledge—foundationalism. Thus, modern epistemologists could be thought of as falling along a spectrum or axis, with radical sceptics (despairing foundationalists) at one end and optimistic foundationalists at the other. In philosophy of language, if language ordinarily gets its meaning from what it represents in the world, then some account has to be given of non-factual forms of discourse such as ethics, aesthetics, and religion. The modern answer has generally been some form of expressivist theory, such as that of the emotivists in ethics. So here we have not so much a spectrum of views, but correlative theories of language, both assuming that normative meaning must be referential. Individualists in the modern period have always had to contend with a minority of theorists who favoured the social group as a focus of analysis; however, modern forms of communitarianism have tended to assume the same *generic* view of the individual as their reductionist opponents—the view that individuals are all alike for the purposes of social, political, and ethical theory. These three axes, then, could be thought of as forming a 'Cartesian co-ordinate system' for locating modern 'positions'. Post-modern thinkers are those who have removed themselves from modern intellectual space altogether.

I have since come to think that modern thought cannot be understood or surpassed without also recognizing a fourth feature: a peculiar 'inside-out' conception of mind. René Descartes, the father of modern philosophy, described himself as a thinking thing, distinct from and somehow 'within' the body. Thinking is a process of focusing 'the mind's eye', but focusing on what? On ideas *in* his mind. Thus there arose the modern image of the 'Cartesian theatre': the real 'I' is an observer in the mind, looking at mental representations of what is outside. This image of the real knower as having no direct contact with the world has bedevilled the whole of modern epistemology, and shows up still in worries about whether there is a real world 'out there' for our knowledge to represent.

Finally, a characteristic of modern thought was its ideal of objectivity, of universal knowledge, reflected in an impersonal academic style—a style intentionally flouted here.

THEOLOGY AND SCIENCE AFTER ATOMISM

Atomism entered modern thought when the demise of the Aristotelian system required an entirely new physics, and scientists revived ancient atomist theories of matter. The essence of atomism is not merely the postulation of indivisible particles but also the assumption that the behaviour of the particles determines the behaviour of all complex wholes. The image that holds modern thinkers captive is that of a clock. In a (mechanical) clock, *must* it not be the case that the parts determine the behaviour of the whole?

So the atomism that was for the Greeks pure metaphysics has become embodied in a variety of scientific research programmes; it has become scientific theory. Yet, I suggest, it continues to function metaphysically, though in a looser sense. Modern thought, not only in the sciences, but in ethics, political theory, epistemology, and philosophy of language, has tended to be *atomistic*—that is, to assume the value of analysis, of finding the 'atoms', whether they be the human atoms making up social groups, atomic facts, or atomic propositions.

In addition, modern thought has been *reductionistic* in assuming that the parts take priority over the whole—that they determine, in whatever way is appropriate to the discipline in question, the characteristics of the whole. Thus, the common good is a summation of the goods for individuals; psychological variables explain social phenomena; atomic facts provide the justifying foundation for more general knowledge claims; the meaning of a text is a function of the meaning of its parts.

The development of an alternative to atomist reductionism is still very much a work in progress. An early contribution came from debates in philosophy of biology beginning in the 1920s; 'emergent evolutionism' was proposed as an alternative to both vitalism and mechanistic-reductionist accounts of the origin of life (Stephan 1992: 25). Later, psychologist Roger Sperry (1983) argued for an emergentist account of cognition. Until very recently, emergence theses have been regarded with suspicion by the majority of philosophers. At present, though, there is a growing literature in both science and philosophy attempting to clarify and substantiate the claim that as one goes up the hierarchy of complexity in the natural world, new entities emerge with properties that cannot be reduced to the properties of their parts, and perhaps with new causal powers that are not merely the summation of the causal contributions of their parts (Murphy and Stoeger 2007).

Another contribution has been the development of the concept of downward or top-down causation, beginning with Donald Campbell in the 1970s (Campbell 1974).

There appears not to have been a great deal of attention paid to this development at the time. The concept is now well known, but still hotly debated, in philosophy of mind: do the mental capacities that arise from the brain have any downward effects on the body, or is it really the brain that is doing all of the causal work?

A parallel set of developments occurring since the middle of the twentieth century have been the new disciplines of cybernetics, systems theory, chaos theory, and studies of non-linear and self-organizing systems. All of this together leads me to suggest that there is something of a paradigm or world-view shift in process. So far from Descartes's account of hydraulic animal bodies and Newton's clockwork universe, the universe is now seen to be composed not of objects but of *systems*. The components of the systems are not atoms, but structures defined by their relations to one another, rather than by their primary qualities. Concepts of causation based on mechanical pushing and pulling are being replaced by the concept of attraction in phase space. In such a picture the stark notions of determinism and indeterminism are replaced by the notions of propensity, probability, and constraint (Juarrero 1999).

What impact will the triumph of systems thinking have on the theology–science dialogue? It will merely be the vindication of positions held all along by significant contributors to the contemporary discussion such as Ian Barbour and Arthur Peacocke. The reductionist issue is relevant to the theology–science discussion in at least two ways. First, if human behaviour were reducible to biology, this would amount to an unacceptable account of human nature for most Christians and other religious believers, who see themselves as genuinely accountable for their actions. Most important is the use that has been made of a non-reductive account of the hierarchy of the sciences for understanding the appropriate logical and conceptual relations between theology and the sciences. A particularly fruitful model is the suggestion that theology be viewed as the top-most science in the hierarchy. The notion of theology as the top science in a non-reducible hierarchy originated implicitly in Barbour's and explicitly in Peacocke's writings. It is implicit in *Issues in Science and Religion* in Barbour's claim that 'an interpretation of levels can contribute to a *view of man* which takes both the scientific and the biblical understanding into account' (Barbour 1966: 360). This implies that the religious perspective is an indispensable level of description of human life. This notion was explicit in Peacocke's *Creation and the World of Science*:

It seems to me that no higher level of integration in the hierarchy of natural systems could be envisaged than [worship and other religious activities] and theology is about the conceptual schemes and theories that articulate the content of this activity. Theology therefore refers to the most integrating level we know in the hierarchy of natural relationships of systems and so it should not be surprising if the theories and concepts which are developed to explicate the nature of this activity...are uniquely specific to and characteristic of this level. (1979: 369)

A consequence of this move is that we can expect the relations between theology and the sciences to be analogous to the relations among other sciences. Each science has a subject-matter of its own; it makes use of its own particular language, both theoretical concepts and descriptive terms. And for many purposes, each science has its own integrity—its own autonomy from other levels. If we take theology to be a science in the hierarchy, we can say the same things regarding its special subject-matter, its distinctive language and concepts, its relative autonomy from the other sciences.

However, the sciences have benefited immensely from relations with other sciences. Reductive explanations have greatly strengthened all of the sciences by *explaining* phenomena that could only be *described* at the higher levels. For example, breaking the genetic code (at the biochemical level) allowed for explanations of inherited characteristics that could be described but not explained at the organismic level. If we take theology to be one of the sciences in the hierarchy, this means that while its focus of interest does not overlap directly with that of any other science, neither is it isolated from the others. Findings in both the natural and the human sciences do have relevance for theology.

THE USES OF LANGUAGE IN THEOLOGY AND SCIENCE

By the early decades of the twentieth century, philosophy of language had become a central focus of Anglo-American philosophy. Richard Rorty (1967) speaks of 'the linguistic turn', the judgement that philosophical problems could best be addressed by attending to language. This supplanted the turn to epistemology, which marked the beginning of the modern period. It should not be surprising, then, that a great deal of attention was given in the twentieth century to the development of explicit theories about language, answering especially the question: How does language get its meaning? Or, better: What is the meaning of 'meaning'? The predominant modern answer has been what could be called either the referential or the representative theory: words get their meaning from the things in the world to which they *refer* or to the states of affairs they *represent*.

However, with this clear referential approach to language came the recognition that whole realms of discourse, such as ethics and aesthetics, were not, in this sense, meaningful. This prompted the elaboration of a second theory of language—or, more precisely, the elaboration of a theory of *second-class* language. Here the emphasis is on the function of language to express the emotions, attitudes, or intentions of the speaker. Hence, this is termed the expressivist theory.

I perceive a second revolution in philosophy in the work of philosophers of language who break with this dichotomous account of language in favour of an emphasis on the *varied* uses of language in social practices—to find the meaning, look for the use. Most significant here are J. L. Austin (1962) and Ludwig Wittgenstein, both writing in the mid-twentieth century. My claim that their work represents a post-modern turn is controversial. While Austin and Wittgenstein have a few

followers among professional philosophers of language, most prominent philosophers of language are pursuing projects that this essay would count as modern.
Significant uses of Austin's and Wittgenstein's works are primarily found in other
disciplines. Thus, we might say that a revolution in philosophy of language was
narrowly averted.

Wittgenstein's work is notoriously difficult to summarize, but I shall attempt to
convey something of his criticisms of both referential and expressivist accounts of
language. In *Philosophical Investigations* he tries to imagine a purely referential
language. A characteristic strategy of his philosophical method is to devise highly
simplified languages—'language games' as he calls them. Here is one.

Let us imagine a language for which the description given by Augustine is right. The language
is meant to serve for communication between a builder A and an assistant B. A is building
with building-stones: there are blocks, pillars, slabs and beams. B has to pass the stones, and
that in the order in which A needs them. For this purpose they use a language consisting of the
words "block", "pillar", "slab", beam". A calls them out;—B brings the stone which he has
learnt to bring at such-and-such a call.—Conceive of this as a complete primitive language.
(1953: § 2)

Wittgenstein is content to count this as a purely referential language, but almost
any additional complexity forces us to recognize words that work differently.

Let us now look at an extension of language 1). The builder's man knows by heart the series of
words from one to ten. On being given the order, "Five slabs!", he goes to where the slabs are
kept, says the words from one to five, takes up a slab for each word, and carries them to the
builder. Here both parties use the language by speaking the words. Learning the numerals by
heart will be one of the essential features of learning this language.…

(Remark: We stressed the importance of learning the series of numerals by heart because
there was no feature comparable to this in the learning of language 1). And this shows us that
by introducing numerals we have introduced an entirely different *kind* of instrument into our
language.) (1958: 79)

There are no special objects to which the numbers refer. We can only understand
their meaning by seeing how they are *used*. Notice that these language games are only
intelligible because Wittgenstein has told us what the builders are *doing* with the
language. 'Block' is a command, not a description of any sort.

Wittgenstein has not presented arguments here. In his view philosophy consists in
showing. It is therapy aimed at helping us escape from puzzles created by misleading
mental pictures, or by the theories of other philosophers! Here he has *shown* us how
restricted a purely referential language would have to be.

We can use Wittgenstein's attack on the notion of a 'private language' to show
comparable limitations on a purely expressivist language. With regard to the expression of inner states such as sensations, he asks: 'But could we also imagine a language
in which a person could write down or give vocal expression to his inner experiences—his feelings, moods, and the rest—for his private use?' (1953: § 243).
Wittgenstein does not mean using our existing language (such as English) to record
private experiences. He is asking us to try to imagine the invention of a language for
describing feelings if all that is involved is someone's attaching descriptions to inner

states. Such a language is logically impossible: using language requires regularity, the following of rules. But with no public *criteria* for knowing whether one is getting it right, there is no difference between following a rule and *imagining* that one is following a rule. So a purely expressivist language could never get off the ground. Language is the *ruled* use of speech, and no rules can apply in the case of private associations between signs and inner experiences. Without rules, the 'word' is merely a noise (1953: §§ 256–65).

How, then, is it possible to speak of inner experiences? We learn to apply terms on the basis of shared behaviour. We can teach a child to say 'I have a toothache' because children with toothaches tend to behave in recognizable ways. Due to the way in which language is interwoven with life, we can teach the rules. So both the descriptive and the expressive aspects of language are intelligible only within language games, which involve publicly observable behaviour, social conventions, and action. Language is bound up with 'forms of life'.

Wittgenstein's work has stimulated a number of sophisticated approaches to religious and theological language. For example, Lindbeck, one of the 'post-liberal' theologians who first stimulated my interest in Anglo-American post-modernity, emphasizes the respects in which religions resemble cultures or languages together with their correlative forms of life. Religions are understood as idioms for the construing of reality and the living of life. Doctrines are best treated as second-order discourse—as rules to guide both practice and the use of first-order religious language (praise, preaching, exhortation). Notice the post-modern feature: Lindbeck's first question about doctrines is, What do they *do*? This recognition of the inextricability of language from life recalls the work of both Austin and Wittgenstein.

I have argued (Murphy 1996) that one of the deepest divisions between (modern) liberal and conservative theologians is that, beginning with Friedrich Schleiermacher, liberals have adopted an expressivist theory of religious language, while conservatives hold to a referential-representative theory. This difference has shown itself in their different attitudes toward science. Conservatives are infamous for their claims that modern science, especially evolutionary biology, is in conflict with Christian teaching. Liberals are equally well known for having accommodated theology to the modern world. It is these different theories of religious language that allow for conflict, on the one hand, and easy accommodation, on the other.

Much of the character of liberal theology owes its inspiration to Kant. To protect religion and morality from the consequences of Newtonian determinism, Kant made a sharp distinction between the physical world as known to science and the moral world. Two different kinds of reasoning are involved: pure reason and practical reason, respectively. The (very limited) knowledge we can have of God is not a part of the sphere of pure reason. Newtonian science (with its mechanistic conception of causality) does not contradict religious belief, because it belongs to a sphere of knowledge different from the sphere to which religious knowledge belongs, and one cannot reason from one sphere to the other.

Schleiermacher (1821–2) adopted Kant's general strategy but argued that there are not two spheres but three: the spheres of knowing (science), doing (morality), and feeling (religion). 'The piety which forms the basis of all ecclesiastical communions is, considered purely in itself, neither a Knowing nor a Doing, but a modification of Feeling, or of immediate self-consciousness' ([1928] 1976: 5). So doctrines, strictly speaking, are not knowledge, but rather systematically ordered *expressions* of piety. Nor can they be derived by reasoning from any kind of knowledge.

> Christian dogmas are supra-rational . . . For there is an inner experience to which they may all be traced: they rest upon a given; and apart from this they could not have arisen, by deduction or synthesis, from universally recognised and communicable propositions . . . Therefore this supra-rationality implies that a true appropriation of Christian dogmas cannot be brought about by scientific means, and thus lies outside the realm of reason. ([1928] 1976: 67)

Thus, science and theology have nothing to do with one another. Adopting a term from contemporary philosophy of science, we can say that they are *incommensurable*. This is the strategy that Ian Barbour calls the 'two worlds' approach.

It is probably no accident, then, that most of the theologians involved in the theology–science dialogue hold some version of the referential theory of language: if expressivism and referentialism are the only options, then one will only bother with the attempt to reconcile theology and science if one is *not* an expressivist. Many such scholars hold nuanced referentialist theories, signalled by their calling their position 'critical realism', as opposed to naïve realism, and reflecting their recognition of non-literal, often metaphorical features of theological language.

What would it mean for the theology–science dialogue if participants held post-modern theories of language? First, it would put an end to attempts to argue a priori for a general theory of scientific or theological language, whether instrumentalist or realist. Second, it would lead to more cautious use of concepts drawn from the two sorts of disciplines. Consider the concept(s) of time. Although 'time' is a noun, it clearly does not refer to some *thing*. To know its meaning is to know the rules for its use in a variety of language games. For example, Augustine wrote that time is an aspect of creation, and therefore we ought not to say that God created *in time*. Within the language game of big bang cosmology there is no use for the phrase 'before the big bang'. A Wittgensteinian would be cautious about drawing inferences from one language game in order to attempt to answer questions raised in another.

Added grounds for caution in pursuing metaphysical speculation come from the work of George Lakoff and Mark Johnson (1999). They argue that the concepts of philosophy such as *time, causation, knowledge,* and *mind* have very little literal content, and that the literal content is fleshed out by the structure of complex metaphors and by metaphorical implication. For example, the only literal content in our concept of time has to do with comparison of events. Our ability to reason about things temporal depends on a variety of metaphors such as the spatial orientation metaphor, according to which the observer is 'located' in the present, the future is 'ahead', and the past is 'behind'. We also have the moving time metaphor in which the observer is stationary and time 'passes'. They say that philosophy goes wrong when it fails to notice that metaphors are just that—metaphors. And I would say the same of theology–science dialogue. When we realize this, we may come to the conclusion that some of the questions raised in the theology–science dialogue (such as God's relation to time) may simply be unanswerable. In other cases, failure to recognize the role of metaphors simply hampers discussion. An example is the way in which the metaphor of mind as inner theatre provides an obstacle to the non-reductive account of persons needed for a Christian anthropology.

SCIENCE AND THE COGNITIVE STATUS OF THEOLOGY

The sorts of relationships that are possible between science and theology depend on the cognitive status of the two disciplines. The most interesting current work assumes that both make justifiable claims to knowledge. While the knowledge claims of theology have been highly suspect throughout the modern era, I argue that this is due not to a defect in theology but rather to the lack of adequate theories of knowledge. As epistemology and philosophy of science have become more sophisticated, it has become increasingly possible to make a case for the justification of theological knowledge; the replacement of foundationalist theories of knowledge with holist theories has made the most significant difference. It is widely recognized that the end of foundationalism is the end of modern epistemology, so this is another point at which post-modern philosophy makes a difference to the theology–science dialogue.

James Turner (1985) argues that disbelief was not a live option in Europe and the USA until roughly between 1865 and 1890. This is surprising, because we are all aware of proofs for the existence of God going back through the Middle Ages. However, it has become common to see medieval philosophers and theologians as *not* intending to persuade atheists to believe in God, but rather as engaging in the modest task of showing that reason could justify belief in a God already accepted on other grounds and for other reasons. The medieval synthesis made God so central to all branches of knowledge and all spheres of culture that it was *in*conceivable that he *not* exist.

The interesting story with regard to the surviving theism is how conceptions of knowledge have changed throughout the modern period, and what these changes have meant for the problem of justifying belief in God. The first and most significant change, one inaugurating the modern era itself, is a change in the meaning of the word 'probable'. Medieval thinkers distinguished between *scientia* and *opinio*. *Scientia* was a concept of knowledge modelled on geometry; *opinio* was a lesser but still respectable category of knowledge, not certain but probable. But for them 'probable' meant subject to approbation, theses approved by one or more authorities. Theo-logical knowledge fared well in this system, being that which is approved by the highest authority of all, God himself.

However, the multiplication of authorities that occurred with the Reformation made resort to authority an increasingly inapplicable criterion for settling disputes.

The transition to our modern sense of probable knowledge depended on recognition that the *probity* of an authority could be judged on the basis of *frequency* of past reliability. Here we see one of our modern senses of 'probability' intertwined with the medieval sense. Furthermore, if nature itself has testimony to give, then the testimony of a witness may be compared with the testimony that nature has given in the past. Thus one may distinguish between internal and external facts pertaining to a witness's testimony to the occurrence of an event: external facts have to do with the witness's personal characteristics; internal facts have to do with the character of the event itself—that is, with the frequency of events of that sort. Given the 'problem of many authorities' created by the Reformation, the task increasingly became one of deciding which authorities could be believed, and the new sense of probability—of resorting to internal evidence—gradually came to predominate. The transition from authority to internal evidence was complete.

Princeton philosopher Jeffrey Stout (1981) traces the fate of theism after this epistemological shift. The argument from design was reformulated in such a way that the order of the universe supplies only empirical evidence, not proof, for God's existence. In addition, in an early stage of development it became necessary to provide evidence for the truth of Scripture. If such evidence could be found, then the content of Scripture could be asserted as true. In a later stage it was asked why the new canons of probable reasoning should not be applied to the contents of Scripture as well. Historian Claude Welch writes that by the beginning of the nineteenth century the question was not merely *how* theology is possible, but whether theology is possible *at all* (1972: 59). Stout's prognosis is grim: theologians must either seek some vindication for religion and theology outside the cognitive domain or else pay the price of becoming intellectually isolated from and irrelevant to the host culture.

My own view is much less pessimistic. I have argued that theology's failure to meet modern standards for justification of belief is due not so much to the irrationality of theology as to the fact that modern *theories* of rationality have been too crude to do justice to theological reasoning—and not only to theological but to scientific reasoning as well. Only now do we have theories of human reasoning that are (in Paul Feyerabend's terms) sly and sophisticated enough to do justice to the complexity of scientific reasoning. I would say all the more so with regard to theology. The turning point, as already noted, was the development of post-foundationalist theories of knowledge, and the most significant of these have come from philosophy of science.

Kuhn (1962/1970) set off a revolution in philosophy of science. In his work we find all of the arguments needed to overturn foundationalism in science and replace it with a holist model. First, paradigms (later called 'disciplinary matrices') are accepted or rejected as a whole. Second, data do not have the characteristics required even by weakened versions of foundationalist epistemologies. They are theory-laden—partially dependent on parts of the theoretical structure itself. Thus, if one wishes to retain the building metaphor, we have to recognize not only (as Karl Popper did) that data are more like piles driven into a swamp than like a solid foundation, but also (and this is devastating to the foundationalist picture) that they are partially 'suspended from an upper-story balcony'.

Ian Barbour's is probably the best-known use of Kuhn's work for purposes of philosophy of religion (1974). He argued that religions are like Kuhnian disciplinary matrices; Jesus is the paradigm for Christian life. I have argued that it is a mistake to compare religions as a whole to disciplinary matrices (Murphy 2004). A better approach is Hans Küng's thesis (in Küng and Tracy 1989: 3–33) that there are paradigms in *theology*, such as Augustinianism and Thomism.

I believe that there is still no better account of the structure and justification of scientific reasoning than that of Imre Lakatos (1970/1978). Kuhn's original account of scientific paradigms was ambiguous, and still remains somewhat vague after his clarifications in the second edition of *The Structure of Scientific Revolutions*. Lakatos's account of a scientific research programme has a great deal in common with Kuhn's paradigms but is much more precise. A research programme takes its identity from its 'hard core', a highly abstract, even metaphysical account of the aspect of reality with which the programme is concerned. Surrounding the core theory is a network of auxiliary theories that extend and apply the core thesis. These theories are supported by data, in conjunction with 'theories of instrumentation', which serve to explain why data of these sorts are relevant to the theories they support.

The hard core itself is generally too abstract to be subject to any direct confirmation or falsification. So, while the auxiliaries are adjusted over time under empirical pressure, the core can be falsified only by replacing the entire programme. The criterion for rejection of a programme is based on the recognition that any theory can be saved if enough qualifications are added. Some such additions lead to further discoveries and explanations ('novel facts'), while others are merely *ad hoc*. Programmes can be compared in terms of the extent to which their changes over time are 'progressive' rather than *ad hoc*; progressive programmes should be accepted in place of 'degenerating' rivals.

I have written on the value of Lakatos's work for understanding theological 'research programmes': systematic theologies are differentiated by means of their different core theories regarding what Christianity is essentially about (e.g. salvation from sin, existential orientation, social ethics). Auxiliary theories concern the standard doctrines or theological *loci*. Data come from history, Scripture, and religious experience. The equivalent of theories of instrumentation are 'theories of interpretation'—historiographical principles, hermeneutic theories—and a 'theory of discernment' to allow for recognition of religious experiences that have some degree of intersubjective consensus and genuine theological import (Murphy 1990; see also Hefner 1993; Clayton 1989; and Russell 2002).

THE END OF THEOLOGY–SCIENCE DIALOGUE?

Throughout this chapter I have been speaking of the cognitive status of the academic discipline of *theology*, whereas the title of this book is the Oxford Handbook of *Religion* and Science. I turn now to the larger question of religion, its relation to science, and the contribution that contemporary Anglo-American philosophy can make. I have argued that it makes sense to see MacIntyre's work as the next stage of development of historicist-holist epistemology after Lakatos (Murphy 1998). We can see a rationale for reading MacIntyre in this way in the fact that he offered an early account of his epistemological insights in an article in which he replied to Kuhn's philosophy of science and noted shortcomings in Lakatos's own response to Kuhn (MacIntyre 1977/1989). Lakatos argued that one could choose between competing research programmes on the basis of one being more progressive than its rival. Feyerabend (1970) countered that this criterion is inapplicable, because sometimes degenerative programmes suddenly become progressive again, so one never knows when it is rational to give up a programme. MacIntyre's insight is to point out that there is an asymmetry between the rivals. From the point of view of the superior programme, it is possible to explain *why* the inferior programme failed, and failed at just the point it did. One example is the competition between the Copernican and Ptolemaic programmes. With Copernicus's theory it is possible to explain the (apparent) retrograde motion of the planets—the major stumbling block for Ptolemy. 'What the scientific genius, such as Galileo, achieves in his transition, then, is not only a new way of understanding nature, but also and inescapably a new way of understanding the old science's way of understanding nature' (MacIntyre 1977/1989: 152).

The primary stimulus for further development of these epistemological insights came from MacIntyre's work in philosophical ethics. In *After Virtue* he argued that moral positions could not be evaluated apart from traditions of moral inquiry. Yet, without a means of showing one tradition to be rationally superior to its competitors, moral relativism would follow (1981/1984). In two succeeding books he has elaborated his concept of a tradition and shown by example the possibilities for such comparative judgements (1988, 1989). Traditions generally originate with an authority of some sort, usually a text or a set of texts. The tradition develops by means of successive attempts to interpret and apply the texts in new contexts. Application is essential: traditions are socially embodied in the life stories of the individuals and communities who share them, in institutions and social practices.

One aspect of the adjudication between competing traditions is construction of a narrative account of each: of the crises it has encountered (incoherence, new experience that cannot be explained, etc.) and how it has or has not overcome these crises. Has it been possible to reformulate the tradition in such a way that it overcomes its crises without losing its identity? Comparison of these narratives may show that one tradition is clearly superior to another: it may become apparent that one tradition is making progress while its rival has become sterile.

In addition, if there are participants within the traditions with enough empathy and imagination to understand the rival tradition's point of view in its own terms, then protagonists of each tradition, having considered in what ways their own tradition has by its own standards of achievement in inquiry found it difficult to develop its inquiries beyond a certain point, or has produced in some area insoluble antinomies, ask whether the alternative and rival tradition may not be able to provide resources to characterize and to explain the failings and defects of their own tradition more adequately than they, using the resources of that tradition, have been able to do. (MacIntyre 1988: 166–7)

MacIntyre's work provides the best resources to date for assessing the rational acceptability of a religious tradition. It involves comparison with its live competitors on the basis of the intellectual crises that each has encountered, whether each has succeeded in its own terms in resolving the crises, and, if not, whether one can explain the other's failure. While this methodology could be used to compare religious traditions with one another, my interest here is what I shall call 'the modern scientific materialist tradition'.

Atheism cannot be seen as the simple removal of belief in God from an otherwise unchanged world-view; it is, rather, the development of an alternative world-view alongside the various theistic traditions. Baron d'Holbach, writing in the eighteenth century, could be called the father of modern atheism. His 350-page *System of Nature* was a systematic avowal of materialistic atheism: a treatment of the world as a whole, humanity's place in it, human immortality, the structure of society, and the relation of morality and religion, all as a coherent totality. After circulation of this book, full unbelief became a discussable position for the first time in modern Europe (Gaskin 1989: 88). The rise of atheism required several crucial developments: alternative accounts of the origin of religion; theories explaining the persistence of religion in our own day, Freud's, Marx's, and Nietzsche's being the most powerful; and the development of accounts of the grounding of morality in something other than the will of God.

Many contemporary authors, such as Richard Dawkins and Daniel Dennett, can be seen as attempting to contribute to and update this tradition. One way of describing the project is to say that it is filling in the top position in Peacocke's hierarchy of the sciences. When theology is removed from this position, something else takes its place. Carl Sagan's work is an interesting example. He offers a peculiar mix of science and what can only be called 'naturalistic religion'. He begins with biology and cosmology, but then uses concepts drawn from science to fill in what are essentially religious categories that fall into a pattern surprisingly isomorphic with the Christian conceptual scheme. He has a concept of ultimate reality: 'The Universe is all that is or ever was or ever will be.' He has an account of ultimate origins: Evolution with a capital E. He has an account of the origin of sin: the primitive reptilian structure in the brain, which is responsible for territoriality, sex drive, and aggression (see Ross 1985). E. O. Wilson's work in sociobiology is a striking example of a new contribution to the task of finding a non-theistic grounding for morality.

What has all this to do with the theology–science dialogue? Some of the Christian tradition's major crises have resulted from the rise of modern science. First there was the demise of the Aristotelian-Ptolemaic world-view with which theology had become allied. There was the development of empiricist epistemology, already addressed above. There was the failure of arguments from design, which had become too important a part of Christian apologetics by Darwin's day (Buckley 2004). A current challenge comes from the prestige of science and the fact that proponents of the materialist tradition have done a much better job of incorporating science into its world-view than

Christian scholars have done—until recently. The significance of the theology–science debates is that a significant group of Christian scholars have gone far beyond mere debate in order to incorporate the findings of science into Christian theology. The end of the theology–science dialogue—'end' in the sense of purpose or goal to be achieved—is to meet the challenge of a rival, materialist tradition on its own terms.

REFERENCES AND SUGGESTED READING

AUSTIN, J. L. (1962). *How to Do Things with Words*, ed. J. O. Urmson. Oxford: Clarendon Press.

AYALA, FRANCISCO JOSÉ, and DOBZHANSKY, THEODOSIUS (1974) (eds.). *Studies in the Philosophy of Biology: Reduction and Related Problems.* Berkeley and Los Angeles: University of California.

BARBOUR, IAN G. (1966). *Issues in Science and Religion.* New York and San Francisco: Harper & Row.

——(1974). *Myths, Models, and Paradigms.* New York: Harper & Row.

BECKERMANN, ANSGAR, FLOHR, HANS, and KIM, JAEGWON (1992) (eds.). *Emergence or Reduction?: Essays on the Prospects of Nonreductive Physicalism.* Berlin and New York: Walter de Gruyter.

BUCKLEY, MICHAEL J. (2004). *Denying and Disclosing God: The Ambiguous Progress of Modern Atheism.* New Haven: Yale University Press.

CAMPBELL, DONALD T. (1974). ' "Downward Causation" in Hierarchically Organised Biological Systems', in Ayala and Dobzhansky (1974), 179–86.

CLAYTON, PHILIP (1989). *Explanation from Physics to Theology: An Essay in Rationality and Religion.* New Haven: Yale University Press.

CRITCHLEY, SIMON (2001). *Continental Philosophy: A Very Short Introduction.* Oxford: Oxford University Press.

FEYERABEND, P. K. (1970). 'Consolations for the Specialist', in Imre Lakatos and Alan Musgrave (eds.), *Criticism and the Growth of Knowledge*, Cambridge: Cambridge University Press, 197–230.

GASKIN, J. C. A. (1989). *Varieties of Unbelief: From Epicurus to Sartre.* New York: Macmillan.

HAUERWAS, STANLEY, and JONES, L. GREGORY (1989) (eds.). *Why Narrative? Readings in Narrative Theology.* Grand Rapids, Mich.: Eerdmans.

HEFNER, PHILIP (1993). *The Human Factor: Evolution, Culture, and Religion.* Minneapolis: Fortress Press.

JUARRERO, ALICIA (1999). *Dynamics in Action: Intentional Behavior as a Complex System.* Cambridge, Mass.: MIT Press.

KÜNG, HANS, and TRACY, DAVID (1989) (eds.). *Paradigm Change in Theology.* Edinburgh: T & T Clark.

KUHN, THOMAS (1962/1970). *The Structure of Scientific Revolutions*, 2nd edn. Chicago: University of Chicago Press.

LAKATOS, IMRE (1970/1978). 'Falsification and the Methodology of Scientific Research Programmes', in Worrall and Currie (1978), 8–101.

LAKOFF, GEORGE, and JOHNSON, MARK (1999). *Philosophy in the Flesh: The Embodied Mind and Its Challenge to Western Thought.* New York: Basic Books.

MACINTYRE, ALASDAIR (1977/1989). 'Epistemological Crises, Dramatic Narrative, and the Philosophy of Science', *Monist*, 60/4: 453–72. Repr. in Hauerwas and Jones (1989), 138–57.

—— (1981/1984). *After Virtue: A Study in Moral Theory*, 2nd edn. Notre Dame, Ind.: University of Notre Dame Press.

—— (1988). *Whose Justice? Which Rationality?* Notre Dame, Ind.: University of Notre Dame Press.

—— (1989). *Three Rival Versions of Moral Enquiry: Encyclopaedia, Genealogy, and Tradition.* Notre Dame, Ind.: University of Notre Dame Press.

MURPHY, NANCEY (1990). *Theology in the Age of Scientific Reasoning.* Ithaca, N.Y.: Cornell University Press.

—— (1996). *Beyond Liberalism and Fundamentalism: How Modern and Postmodern Philosophy Set the Theological Agenda.* Valley Forge, Pa.: Trinity Press International.

—— (1998). *Anglo-American Postmodernity: Philosophical Perspectives on Science, Religion, and Ethics.* Boulder, Colo.: Westview Press.

—— (2004). 'Religion, Theology, and the Philosophy of Science: An Appreciation of the Work of Ian Barbour', in Russell (2004), 97–107.

—— and STOEGER, WILLIAM R. (2007). *Evolution and Emergence: Systems, Organisms, Persons.* Oxford: Oxford University Press.

PEACOCKE, ARTHUR R. (1979). *Creation and the World of Science.* Oxford: Oxford University Press.

PETERS, TED, RUSSELL, ROBERT JOHN, and WELKER, MICHAEL (2002) (eds.). *Resurrection: Theological and Scientific Assessments.* Grand Rapids, Mich.: Eerdmans.

RORTY, RICHARD (1967) (ed.). *The Linguistic Turn: Recent Essays in Philosophical Method.* Chicago: University of Chicago Press.

ROSS, THOMAS M. (1985). 'The Implicit Theology of Carl Sagan', *Pacific Theological Review*, 18/ 3: 24–36.

RUSSELL, ROBERT JOHN (2002). 'Bodily Resurrection, Eschatology, and Scientific Cosmology', in Peters, Russell, and Welker (2002), 3–30.

—— (2004) (ed.). *Fifty Years in Science and Religion: Ian G. Barbour and his Legacy.* Aldershot: Ashgate.

SCHLEIERMACHER, FRIEDRICH ([1928] 1976). *The Christian Faith*, ed. and trans. H. R. Mackintosh and J. S. Steward. Edinburgh: T & T Clark.

SMITH, HUSTON (1989). *Beyond the Post-Modern Mind*, 2nd edn. Wheaton, Ill.: Theosophical Publishing House.

SPERRY, ROGER W. (1983). *Science and Moral Priority: Merging Mind, Brain, and Human Values.* New York: Columbia University Press.

STEPHAN, AKIM (1992). 'Emergence—A Systematic View on its Historical Facets', in Beckermann, Flohr, and Kim (1992), 25–48.

STOUT, JEFFREY (1981). *The Flight from Authority: Religion, Morality, and the Quest for Autonomy.* Notre Dame, Ind.: University of Notre Dame Press.

TOULMIN, STEPHEN (1990). *Cosmopolis: The Hidden Agenda of Modernity.* New York: Free Press.

TURNER, JAMES (1985). *Without God, Without Creed: The Origins of Unbelief in America.* Baltimore and London: Johns Hopkins University Press.

WELCH, CLAUDE (1972). *Protestant Thought in the Nineteenth Century.* New Haven: Yale University Press.

WITTGENSTEIN, LUDWIG (1953). *Philosophical Investigations*, trans. G. E. M. Anscombe. Oxford: Blackwell; New York: Macmillan.

—— (1958). *The Blue and Brown Books: Preliminary   Studies for the 'Philosophical Investigations'.* New York: Harper & Row.

WORRALL, JOHN, and CURRIE, GREGORY (1978) (eds.). *The Methodology of Scientific Research Programmes: Philosophical Papers, Volume I.* Cambridge: Cambridge University Press.

CHAPTER 29

TRINITARIAN FAITH SEEKING TRANSFORMATIVE UNDERSTANDING

F. LERON SHULTS

INTRODUCTION

The methodological maxim 'faith seeking understanding' is almost unanimously embraced among Christian theologians, irrespective of the extent to which they have self-consciously engaged in dialogue with contemporary science. This ubiquitous motto is notoriously polyvalent, however, serving as a banner for foundationalists and anti-foundationalists alike. The usefulness of this traditional principle for contemporary interdisciplinary dialogue will depend on the clarity with which we answer the questions: Whose faith? Which understanding? Whither the seeking? My assignment is to treat the question of method, and so this chapter will attend primarily to the philosophical and historical developments that have shaped our current context in ways that facilitate the embrace of particularity and passion within interdisciplinary dialogue. I begin where I am, as a passionate participant in my own Christian tradition. I believe that the best way to conserve the intuitions of this tradition, however, is to liberate them for transformative dialogue as we seek to render intelligible our experience of and with one another.

In this chapter I will argue that integrative developments in late modern philosophy of science and the broader (re)turn to the hermeneutical significance of the category of relationality have opened up conceptual space for the renewal of a *trinitarian* faith that seeks *transformative* understanding as it engages in the discourse among the fields of contemporary science. The first step is to show the connection between two disintegrating tendencies in the early modern period: the separation of the doctrine of the Trinity from 'scientific' discourse, and the compartmentalization of faith and reason. After outlining this link, I will turn in the second part to an exploration of some of the integrative developments in late modern epistemology that contribute to overcoming these dichotomies. Recovering the centrality of the category of relationality can help philosophers of science articulate the reciprocity between commitment and rational inquiry and can help Christian theologians retrieve the illuminative power of trinitarian reflection.

This will set the stage for the third part, which re-examines the intuitions behind the call for 'faith seeking understanding' as it developed in the stream of the Western tradition that flowed through Augustine and Anselm. In some ways these theologians contributed to the marginalization of the doctrine of the Trinity and to the bifurcation of reason from faith. In the wake of late modern philosophical and theological developments, however, we now have an opportunity to retrieve and emphasize at least two aspects of their approach that have often been obscured: apophatic humility and soteriological passion.

Acknowledging the complexity of the relations among particular modes of faith and particular expressions of rationality in the science–religion dialogue may facilitate the recovery of the generative force of some of the categories that have emerged out of the interpreted experience of biblical faith, especially the doctrine of the Trinity. My focus here will be not on the material outworking of such a reconstruction (cf. Shults 2005), but on outlining the conceptual space that makes such a move possible without collapsing back into a naïve fideism or a rigid rationalism.

THE COMPARTMENTALIZATION OF FAITH AND REASON AND THE EXCLUSION OF TRINITARIAN REFLECTION FROM 'SCIENTIFIC' DISCOURSE

The revival of trinitarian doctrine in twentieth-century theology is well known to many (if not most) Christian participants in the religion–science dialogue (for an introduction, cf. Peters 1993; Grenz 2004). With a few notable exceptions, however, trinitarian reflection has not been brought deeply into the interdisciplinary discussion (cf. Shults 2004). The material exclusion of particular relational Christian categories from 'scientific' discourse is intimately tied to the formal separation of faith from reason in the early modern period. I place 'scientific' in scare quotes here because that discourse from which trinitarian categories were excluded was based on an idealized understanding of 'science' as an objective, neutral pursuit of *scientia* that was putatively untainted by a subjective, passionate concern for *sapientia*—wisdom for practical living in the face of the other.

By the mid-seventeenth century Blaise Pascal felt he had to decide between the God of the philosophers and the trinitarian God of the Bible. After his death a note was discovered sewn into his clothes, in which he acknowledged with fiery passion his embrace of the 'God of Jesus Christ' and rejected the philosophical god of the 'scholars' of his day. Pascal felt compelled to choose between living faith and what he saw as a sterile natural philosophy that aimed to prove the necessary existence of a rational causative substance. This generic 'theism' seemed to have little to do with the relational experience of divine love disclosed through the incarnating Logos and the indwelling Pneuma. Conversely, the dynamic experience of the human desire for goodness and beauty had little to do with the work of 'science', which attended primarily to propositional truth.

By the end of the seventeenth century the doctrine of the Trinity was still causing 'nice & hot disputes' (cf. Dixon 2003), but it no longer had the significant illumina-tive or generative force that it had in patristic theology. Developments in Christian theology itself had contributed to this exclusion. In early medieval Europe most theological education involved commenting on the *Sentences* of Peter Lombard, for whom the Unity and Trinity of God were treated together in Book I: *De Dei Unitate et Trinitate*. In the thirteenth century, however, Thomas Aquinas's *Summa Theologiae* split these doctrines apart. Early in his systematic presentation he treats the divine Unity, which he believed could be proved by philosophical (Aristotelian) logic. The treatment of the Triunity of God, however, came significantly later, and was main-tained by appeal to faith in revelation and the authority of the Church. This split was further widened in the fourteenth century by William of Occam, who affirmed the doctrine of the Trinity as articulated by the fathers of the early church, but empha-sized (more strongly than Thomas) that it could not be proved by human logic. His nominalist attention to rational analysis of individual substances facilitated the rise of modern science, but rendered the idea of a trinitarian relational unity increasingly problematic to 'reason'.

Many theologians responded to the perceived challenges of the emergence of modern science by attempting to develop rational proofs for the existence of a divine rational causative substance, rather than by outlining the illuminative power of the relational and dynamic categories that emerged out of religious experience (cf. Buckley 1987). The heuristic categories of Logos and Pneuma, which had long been ingredients of philosophical and scientific reflection, became ciphers for the intellect and will of a *single* divine subject, and 'theism' came to be defined as belief in *a* person outside time who was the necessary First Cause of all effects. As causality came to be explained by means of the mechanistic and mathematical apparatus of post-Newtonian physics, however, the idea of God was further detached from science, and deism followed quite naturally. These categories also intensified the problem of evil, which led to a flurry of 'theodicies' that ultimately failed to stem the tide of atheism, i.e. the reaction against the idea of God as *a* person who is wholly good, whose intellect is all-knowing, and whose will is all-powerful. The 'theistic' projection of Neoplatonic faculty psychology onto God not only tied theology to a particular (now outmoded) concept of person, it also forced theology to leave behind the doctrine of the Trinity as it engaged science.

This marginalization was intensified by the hardening of the distinction between faith and reason in early modern forms of classical foundationalism. Histories of this shift usually begin with René Descartes, whose 1641 *Meditations on First Philosophy* outlined a method of doubting the senses and relying only on clear and distinct ideas in the mind. Later John Locke rejected the notion of innate ideas and insisted that all ideas are formed on the basis of sensations; reason is dependent upon empirical experience. Locke also set the tone for Enlightenment debates over the relationship between faith and reason in his 1689 *Essay concerning Human Understanding*; for him revelation is certainly true, but 'it belongs to Reason to judge of its being a Revelation, and of the signification of the Words, wherein it is delivered' (IV. 18).

The viability of founding knowledge on either innate ideas or sense perception was challenged by David Hume, whose 1739 *Treatise of Human Nature* questioned the very notions of causality, substance, and personality. This scepticism woke Immanuel Kant from his 'dogmatic slumbers' and led him to limit theoretical reason in order to 'make room for faith' in his 1789 *Critique of Pure Reason*. Nevertheless, Kant still presupposed that these are two separate functions; 'faith' is relegated to practical reason (and judgement) and excluded from pure (theoretical) 'reason'.

In the wake of this compartmentalization, Christian theology found itself pulled in two different directions, which we can illustrate by comparing two nineteenth-century theologians: Friedrich Schleiermacher and Charles Hodge. Schleiermacher's approach was to accept the limitation of theoretical reason and to base his presentation of dogmatics on foundational experiences in the realm of 'feeling' rather than 'thinking'. In the introduction to *The Christian Faith* (1830), he explains that the whole of theology is a description and explication of the pious self-consciousness, i.e. the feeling of absolute dependence.

Hodge represents the other extreme. He rejects Kant's limitation of reason and insists that human thought can know God in the same way that scientists know nature. Reflecting the positivism of the late nineteenth century, he believed that 'The Bible is to the theologian what nature is to the man of science. It is his store-house of facts ... the theologian must be guided by the same rules in the collection of facts, as govern the man of science' (1981: 10–11; cf. p. 365). Hodge's foundationalism is implicitly Cartesian: he argues that his idea of an infinite God is 'a perfectly clear and distinct idea' in his mind (p. 359).

Despite their varying responses to the dualism between faith and reason in the early modern philosophical tradition, both Schleiermacher and Hodge are led (in different ways) to further marginalize trinitarian reflection. Schleiermacher left his treatment of the Trinity to a brief section at the end of his *The Christian Faith*, indicating that the 'main pivots' of Christian doctrine are 'independent' of the doctrine of the Trinity ([1830] 1989: 741). Like many of his protestant scholastic predecessors, Hodge spent most of his energy in 'theology proper' defending theism; only after 250 pages of apologetics for the idea of God as a single subject who is an immaterial causal substance does he devote 40 pages to the doctrine of the Trinity, proposing various analogies that (not surprisingly) lean toward modalism.

On both sides of the conservative–liberal continuum today many participants in the science–religion dialogue still feel constrained to exclude trinitarian reflection from interdisciplinary discussion. This hesitancy is due in large part to the incipient dichotomies that structure the continuum itself. Either we base all of our inquiry on absolute foundations accessible to all rational individuals, or we are at sea in relativist rafts with no basis for linking our incommensurable webs of discourse about faith. Either we accede to self-authenticating premises that establish apodictic theological conclusions (foundationalism) or we accept that our theological intuitions have no probative force outside our own idiosyncratic reflection (anti-foundationalism). Those on the extreme poles of this continuum presuppose that these are the only options.

As I have argued elsewhere (Shults 1999), this 'either-or' structuring of the debate can be clarified by noting how four pairs of ideas are methodologically related: experience and belief, truth and knowledge, individual and community, and explanation and understanding. Foundationalism tends to emphasize the first of each dyad: experience as the basis of belief, the unity of truth, reason in the individual, and the universality of explanation. Anti-foundationalism, on the other hand, privileges the web of belief as conditioning experience, the plurality of know-ledge, the rationality of the community, and the particularity of understanding. These methodological decisions have reinforced the exclusion of trinitarian reflection from dialogue with science, either because theology ought to accept the doctrine of Trinity based on faith apart from reason (foundationalism), or because theology ought to confine itself to the linguistic expressions that shape its own community (anti-foundationalism).

Is it possible to bring the resources of the revival of trinitarian doctrine in twentieth-century Christian theology directly into the dialogue with science without reverting to absolutist foundationalism or falling into relativist anti-foundational-ism? The liberation of interdisciplinary discourse from the hegemony of early modern dichotomies can provide space for Christian theology to recover the integral role of trinitarian reflection in its articulation of material issues such as the God–world relation and divine agency. In much of the Western philosophical and theo-logical tradition before the Enlightenment, faith and reason worked together in a reciprocal relation. On this side of the modernist divide, we cannot simply return to naïvely pre-modern ways of embracing their mutuality, but we may search for resources for refiguring the integral relation between *fides* and *ratio* in our contem-porary context.

INTEGRATIVE DYNAMICS IN LATE MODERN EPISTEMOLOGY AND THE PHILOSOPHICAL TURN TO RELATIONALITY

Several developments in late modern philosophy have opened up conceptual space for overcoming many of the various dualisms of the Enlightenment, including the bifurcation between faith and reason. In the last few decades many participants in the science–religion dialogue have struggled to find a middle way between the extremes of foundationalism and anti-foundationalism, while attempting to maintain the valid intuitions of both. Elsewhere I have explored the emergence of this *post-foundationalist* model of rationality, which aims to integrate the concerns of each side of the dyads mentioned above (Shults 1999; cf. van Huyssteen 1997, 1999). This model may be summarized by the following couplets, which attempt to move beyond the either/or impasse by articulating the possibility of a both/and relationality.

- Interpreted experience anchors all beliefs, *and* a network of beliefs informs the interpretation of experience.
- The objective unity of truth is a necessary condition for the intelligible search for knowledge, *and* the subjective multiplicity of knowledge indicates the fallibility of truth claims.
- Rational judgement is an activity of socially situated individuals, *and* the cultural community indeterminately mediates the criteria of rationality.
- Explanation aims for universal, trans-contextual understanding, *and* understanding derives from particular contextualized explanations.

In this context, I will limit myself to a brief exploration of some of the implications of these couplets for the interdisciplinary dialogue between science and religion.

The reciprocity between personal belief and personal experience is operative in both scientific and theological inquiry. All intellectual searching emerges out of 'faith', i.e. out of our trust in the fecundity of a web of beliefs, which itself has been mediated by our experience. Far from inhibiting scientific inquiry, the 'fiduciary component' or 'personal coefficient' (cf. Polanyi 1962) of knowledge is essential, for commitment and intellectual passion are an impetus for the pursuit of knowledge. This does not mean that we must give up on the ideal of the unity of truth (either in theology or in science), even though we may need to acknowledge that this ideal is a *focus imaginarius* that guides and orients our search for knowledge. We ought to begin with our participation within our own particular contexts (where else could we begin?), but we ought also to distance ourselves from our contexts as we face our own fallibility and explore the intersubjective and trans-communal critiques of other persons in other traditions.

As Philip Clayton argues, the long battle for Christian particularity is over, for particularity is precisely what late modern thought celebrates. He notes that unfortunately most theologians 'continue to fight for exactly what is now being offered to them free of charge' (1997: 4). Interdisciplinary dialogue makes it difficult to ignore the particularity of our categories or rely on allegedly universal reasons for our beliefs. The intellectual passion that drives both theology and science operates within and emerges out of a particular paradigm (Kuhn) or tradition (MacIntyre), which guides the criteria for what is reasonable. Being socially located does not entail the impossibility of intersubjective and trans-communal discourse; on the contrary, it is precisely the experience of embodied desire that evokes the longing for knowledge of the other.

The way in which passion forms an inextricable element of rational inquiry has also been demonstrated by findings in the field of neuroscience. Early modern faculty psychology reinforced the separation of reasoning and believing, which were allegedly functions of the soul's powers of intellect and will, respectively. Neuroscientific studies have shown that human thinking cannot be so easily separated from human feeling and doing, because the formation of beliefs is mediated by emotional and psychological experiences that are mediated through neurobiological functioning (cf. Damasio 1999). These developments do challenge the compartmentalization of thinking (reason) from willing (faith), but they also provide an opportunity to recover the more holistic anthropology of the Hebrew Bible and the New Testament (Shults 2003: 165–81).

The division of the sciences between those that aim for explanation (*Erklärung*) and those that aim for understanding (*Verstehen*) has also been challenged by developments in late modern epistemology. This nineteenth-century dichotomy led to the hardening of the separation between the 'natural' sciences (*Naturwissenschaften*) and the 'human' sciences (*Geisteswissenschaften*), which itself is reminiscent of early modern bifurcations between 'extended' and 'thinking' things, fact and value, objective and subjective, etc. The collapse of these dualisms has opened conceptual space for interdisciplinary models that both recognize the overlapping concerns of various fields of inquiry and respect the integrity and focus of particular sciences (cf. Clayton 1989; Murphy and Ellis 1996: 204). These developments make it easier to integrate our explication of concepts like space and time with our understanding of concepts like person and community.

These integrative dynamics have flourished as part of a broader shift in philosophy that has recovered the hermeneutical significance of the category of 'relationality'. In *Reforming Theological Anthropology* (2003), I outline the way in which this conceptual turn has facilitated the growth of interdisciplinary dialogue. The purpose of providing the following brief overview of these developments in this context is to suggest that this shift also opens space for the integration of the relational categories of trinitarian reflection into this dialogue. In Aristotle's philosophy of science, the category of 'substance' was privileged over the category of 'relation'. True knowledge of a thing involved defining its substance, marking off its genus and differentiae. The way in which the thing was related to other things was 'accidental'—not essential for understanding or explaining the reality of the thing itself (e.g. *Categories* II, VII; *Metaphysics* XIV. 1).

Over the centuries, however, the difficulties with a theory of knowledge (and predication) that failed to attend sufficiently to the relationships between things became increasingly evident. The patristic and medieval debates over the reality of universals and their relation to particular things was in large part driven by the clash between Aristotelian and Neoplatonic concepts of 'substance'. The dominance of this concept (as well as its utility and coherence) was challenged by philosophers like Locke and Hume, and by the end of the eighteenth century Kant found it necessary to reverse Aristotle's preference for substance. In his own list of categories in the *Critique of Pure Reason* Kant made 'of relation' a broader category than 'substance and accidents' (B 106).

Hegel privileged relationality even more strongly, challenging the basic separation between the categories of substance and accident in the *Science of Logic* (1812–19). Like Kant, he subsumed these categories under the broader category of 'relation', but in a more radical way. For Hegel, both substantiality and accidentality refer to determinations of the totality or the whole; this 'whole' is neither 'being' nor 'essence', however, but their dialectical unity in the reflective movement of the '*absolute relation*', which is the highest category in the objective logic. This emphasis on relationality played itself out in different ways among other nineteenth-century philosophers. For example, Søren Kierkegaard relied on relational categories in his description of the self as 'a relation that relates itself to itself or is the relation's relating itself to itself in the relation; the self is not the relation but is the relation's relating itself to itself' (1980*b*: 13; cf. 1980*a*: 43, 88–9). C. S. Peirce (1998) developed a new approach to categories that led him to propose three 'classes of relations', which he called 'monadic', 'dyadic', 'triadic'—or Firstness, Secondness, and Thirdness.

The turn to relationality not only shaped the work of twentieth-century philosophers as diverse as A. N. Whitehead and Emmanuel Levinas, but also had ramifications for the scientific endeavours of interpreting humanity and the cosmos. The basic unit of analysis for seventeenth-century psychology was the substance of the individual soul, but contemporary social science attends to the relational systems within which personal consciousness emerges. Whereas Newtonian physics relied on the idea of individual substances colliding mechanically in absolute space and time, contemporary physics uses implicitly relational categories in its hypotheses about field theory, quantum gravity, emergent complexity, and of course Einstein's own theories of 'relativity'.

Hermeneutically privileging the category of relationality may also help us articulate the inherently reciprocal relationship between faith and reason. Instead of asking whether we should begin with rational proofs and 'add' faith when we hit a mystery, or whether we should begin with our fideistic commitments and then 'add' reasonable arguments only when pressed, we might begin with the relationality within which 'faith' and 'reason' are mutually constituted. To believe something or to trust someone requires some knowledge of the thing or person. To know something or someone requires some level of commitment, a fiduciary connection to that which is known. Rationality involves committing oneself to a belief, and faith involves making judgements about what is trustworthy. Both scientific and theological inquiry are intrinsically shaped by this dialectical relation between faith and reason, between dwelling within our interpreted experience of being bound in relation and seeking ever more adequate interpretations of those experienced relations.

I have not meant to imply that all late modern philosophical developments have been integrative. Much twentieth-century philosophy, including aspects of the work of Rorty and Foucault, for example, has not only functioned disintegratively, but has even rejected the very notion of epistemology. These elements represent a *de*-constructive response to the collapse of foundationalism and the compartmentalization of faith and reason. The other extreme would be what we might call a *paleo*-constructive response, which tries to ignore the challenges associated with late modernity. It is precisely in the dialectic between these poles, which pull us toward relativism or absolutism, that we find ourselves faced with the task of responding *re*-constructively, aiming to conserve the intuitions of our traditions by liberating them for transformative interdisciplinary dialogue.

REFIGURING THE *FIDES QUAERENS INTELLECTUM* TRADITION

In light of these late modern developments, how might we critically appropriate this classical methodological maxim for our contemporary context? On the one hand, it is important to admit that the way in which Augustine and Anselm developed and utilized this formula actually contributed to the foundationalist compartmentalization of faith and reason and the exclusion of trinitarian reflection from scientific discourse. On the other hand, I will also argue that some of the deep intuitions that guided their methodological approaches were obscured by the early modern adaptation of their maxim, and that the retrieval and reconstruction of these intuitions may facilitate our integrative efforts in the contemporary interdisciplinary dialogue between science and theology.

The theologian most commonly associated with this theological motto is St Anselm (1033–1109 CE). In his preface to the *Proslogion* Anselm explains that he had originally intended to call it *Faith Seeking Understanding*, which indicates how central this methodological issue was for him. At the end of the first chapter, Anselm confesses to God: 'I do not seek to understand that I may believe, but I believe in order to understand (*credo, ut intelligam*). For this I also believe—that unless I believed I should not understand' (Anselm 1962: 53). Here Anselm was adapting a favourite refrain of St Augustine's which appeared (among other places) in his commentaries on the Gospel of John: *credere, ut intelligas*—'believe, in order that you may understand' (cf. *Tractates*, 15, 27, 69).

Augustine's biblical warrant for this exhortation was Isaiah 7: 9c. The old Latin variant text used by Augustine (and apparently Anselm as well; cf. van Fleteren 1991) read *Nisi credideritis, non intelligetis*, which is usually translated as 'unless you believe, you will not understand'. This passage provides a point of entry for exploring the ambiguous way in which faith and understanding have been related in the Western tradition flowing from Augustine, as well as an insight into the way in which we might relate them today in the dialogue between theology and science. The context of Isaiah 7 is a prophetic call to King Ahaz and the people of Judah to renew their faithfulness to YHWH and a reassurance of divine faithfulness in the face of their enemies. The NRSV translates this phrase as 'If you do not stand firm in faith, you shall not stand at all'.

The Hebrew text involves a play on words. Terms that derive from the root *'mn* appear in both parts of the phrase. In the first case it is in the Hiphil form (*ne'emān*), which means to cleave fast to, or firmly trust in. In the second part of the phrase it is in the Niphal form (*he'emîn*), where it connotes being held fast or confirmed in trust. The way in which the people of Judah are being held in being is linked to their faithfully standing firm in trusting relation to YHWH. Ancient Hebrew does not have an abstract word for 'faith' analogous to the early modern idea of cognitive assent to a proposition; the ancient Israelites were concerned with concrete 'faithfulness' (*'emûnâh*, which is the noun form of the root *'mn*). The focus in the Hebrew Bible in general, and in Isaiah in particular, is on standing in faith, on fidelity in relation to God and neighbour, not simply abstract rational inquiry.

Unfortunately the Greek Septuagint rendered this passage in a way that obscured this idea: *kai ean mē pisteusēte, oude mē sunēte*. The final verb in this translation (from *suniēmi*) indicates intellectual comprehension or understanding, which hardly captures the fiduciary connotation of the Hebrew *he'emîn*. This contributed to the notion that Isaiah's prophecy had to do primarily with mental comprehension ('belief') rather than being upheld by divine faithfulness. The Latin Vulgate translation—*si non credideritis non permanebitis*—is more faithful to the Hebrew text, but this was not the translation upon which Augustine or Anselm depended. As I have argued elsewhere (in Shults and Sandage 2003: 172–8), the New Testament idea of 'faith' (*pistis*) also has to do primarily with the ways in which we bind ourselves to one another in our struggle to know and be known. The biblical idea of faith(fulness) incorporates intellectual understanding but focuses on the way in which persons become wise in their embodied communal relations.

Although it was not their intention, Augustine's and Anselm's misappropriation of Isaiah contributed to the overshadowing of the intimate connection between knowing and being in faithful relation in the biblical witness, and eventually to the bifurcation between fidelity and rationality. The process of compartmentalization had already begun among several Latin theologians in the patristic period (e.g. Tertullian), but the division between rational inquiry and fiduciary commitment had hardened significantly by the time we get to Anselm. In his *Monologion*, for example, he aims to arrive at conclusions about divine being based solely on logical reasoning; his *Cur Deus Homo* attempts to prove the Incarnation by 'absolute reasons', setting aside any prior knowledge or experience of Christ.

Moreover, the Western reliance on the category of substance, inherited from Aristotelian metaphysics, made it difficult for Anselm (and Augustine before him) to articulate the relations between the Father, the Son, and the Holy Spirit while maintaining the unity (and simplicity) of the divine being (cf. Shults 2005). They had confidence that human reason could prove the existence of a single divine substance, but the relationality of the triune life was more problematic. The doctrine of simplicity demanded that there could be no division between substance and accidents in God. However, if relations are accidental, how could the relations among the persons of the Trinity be 'real' in the divine being (substance)? The tendency to spell out the trinitarian relations as modes of the divine substance flowed naturally from their philosophical dependence on the category of substance (cf. Augustine, *On the Trinity*, IX–XI; Anselm, *Monologion*, 78). As we saw above, this problematization of trinitarian relationality only intensified in the late medieval and early modern periods. On this side of the Enlightenment, however, we may be able to recover some of the deeper intuitions of the *fides quaerens intellectum* tradition and refigure them in light of the (re)turn to relationality.

Both Augustine and Anselm emphasized the importance of reason and intellectual searching, but they also recognized their inherent limitations. Even in the *Proslogion* Anselm makes this explicit in his prayer to God: 'thou art a being greater than can be conceived' (ch. XV). Augustine's *Confessions* are replete with acknowledgements that God is beyond the grasp of his human reason. God is not simply one of the objects of human reason, but the One who contains all things and is present to and in all things. Augustine confesses that God is more intimate to him than he is to himself, and that rational desire itself is wholly dependent on the presence of God (cf. *Confessions*, X. 23). We can refer to this aspect of the tradition as *apophatic humility*. The basic insight from the apophatic tradition that I commend for our appropriation is not that we are limited to negative propositions about God, but that the arrogant desire to capture the infinite with finite propositions must itself be negated.

The experience of being limited in relation to the unlimited presence of God, who graciously upholds and orients the search for understanding, has led contemplative theologians throughout the centuries into a humility that is also characterized by a unique kind of confidence. This is not a surreptitious foundationalist hubris that halts inquiry, but a confident humility that accepts its absolute dependence on the presence of an unlimited and illimitable presence in which all things—including all inquiry in its essential limitedness—live and move and have their being (Acts 17: 28). The humble confidence that emerges out of the experience of an infinite fiduciary presence from, through, and to which all things are (Romans 11: 36) intensifies one's desire for understanding. Apophatic humility acknowledges that one's limitation is itself gracious, for it is the condition of one's ongoing delight in the evocative presence of eternity that quickens one's longing to be held fast, to know, and to be known in faithful community.

Augustine illustrates this humility in his commentary on Genesis, where he anticipates that future scientific discoveries may very well challenge his understanding of the nature of light. In such a case, he would be happy to reformulate his understanding. 'If reason should prove that this [new] opinion is unquestionably true', he would return to the text for a new attempt at interpretation. 'It is a disgraceful and dangerous thing for an infidel to hear a Christian, presumably giving the meaning of Holy Scripture, talking nonsense on these [scientific] topics; and we should take all means to prevent such an embarrassing situation, in which people show up vast ignorance in a Christian and laugh it to scorn' (1982: 42–3). In this context, Augustine exemplifies a non-anxious willingness to engage new understandings of the world as he continues trying to articulate his experience of being held fast by the One who evokes all of his intellectual desiring. His confidence is not in his rational ability to come to conceptual closure, but in his experience of being upheld by the unlimited and illimitable presence in whom the intelligibility of the world is ultimately grounded.

A second dimension of this tradition that I commend for late modern appropriation is what we might call the *soteriological passion* that compels faith to seek *transformative* understanding. Christian theology aims not only for 'knowledge' (*scientia*) but for a right relation to divine 'wisdom' (*sapientia*), which rescues us from our broken and painful relations to ourselves and others, and liberates us into joyful relationship with God. Although Augustine often focused on the idea of understanding in terms of *scientia* and theological propositions, he also recognized that the goal of *scientia* was to move through to *sapientia* (*On the Trinity*, XIII. 19). His *Confessions* take the form of a prayer, and his concern is for intimate knowledge of divine wisdom, which saves him from his destructive relations to others and transforms him by orienting him toward faithfulness in his community. Anselm's *Proslogion* is also in the form of a prayer, which begins with his confession of the overwhelming pain of his striving to know and be known (ch. I), and ends with his grateful orientation toward the reception of joy in the trinitarian God (ch. XXVI).

In its dialogue with science, Christian theology ought to acknowledge that its search for understanding is guided by a passion for transformation. Theological inquiry involves *scientia*, rendering intelligible human religious experience in the world, but it is essentially oriented toward *sapientia*, learning to live together wisely and faithfully in redemptive communion. Christian faith humbly (and confidently) seeks more than abstract intellectual propositions; it passionately pursues forms of concrete embodied fellowship that really transform human life. If this passion for salvation is narrowly construed as a proselytizing of the 'unsaved' by compelling their cognitive assent in a prayer of 'salvation', theology will more than likely inhibit rather than transform the dialogue. However, if our passion is guided primarily by our longing for *salus* in community, for health and wholeness through the salutary ordering of our life together, then it is precisely this passion that should make theology interesting to scientists or anyone else interested in salubrious dialogue.

We may very well be critical of some of the substance-dominated trinitarian formulations of the Augustinian tradition, but this should not hinder us from appropriating their willingness to bring the illuminative power of explicitly theological categories into dialogue with science and their emphasis on the transformative dimensions of the search for understanding. In our late modern context we ought to resist the unilateral tendency implicit in Anselm's motto, which fostered early modern foundationalism by suggesting that the movement is unidirectional—from faith to understanding. Faith itself emerges and intensifies within the passionate search for understanding. Intellectual reflection on the experience of being held fast and called into intimate redemptive community ought to transform our fidelity in relation to God and our neighbours.

CONCLUSION

I have argued that the compartmentalization of faith and reason in the early modern period was intimately connected to the exclusion of trinitarian reflection from scientific discourse, narrowly construed as the dispassionate pursuit of objective knowledge. Conversely, the revival of trinitarian doctrine in twentieth-century theology was closely linked to the emergence of the integrative aspects of late modern philosophy of science as well as to the broader turn to relationality. As I have argued elsewhere (Shults 2003), it is more appropriate in theological discourse to speak of a re-turn to relationality. Long before the formal reversal of Aristotle by Kant and Hegel, the intrinsically relational ideas of the incarnating divine Logos and the indwelling divine Pneuma guided the presentation of Christian doctrine in thinkers as diverse as Gregory of Nyssa and Jonathan Edwards. Acknowledging the interplay between methodological and material developments may help us take advantage of new opportunities for reconstructive dialogue that move beyond the level of defending a generic 'theism'.

Christian theology may contribute best to the dialogue precisely by acknowledging its particular passion for the trinitarian God as it seeks transformative understanding. My goal in this chapter has been to address primarily methodological issues, but I bring trinitarian theology into material dialogue with some of the relevant sciences in my *Reforming the Doctrine of God* (2005). Christians interpret their religious experience as a being called into right relation with the incarnating Logos and indwelling Pneuma of God. As lovers of wisdom, philosophers and scientists (*qua* human persons) are also interested in the rational order(ing) of the universe that informs and holds all things together, as well as the principles that guide the emergence and complexification of life within the dynamic fields of energy that constitute the cosmos.

Theology does not compete with other forms of inquiry, but thematizes the ultimate origin, condition, and goal of the phenomena explored by the various fields of science. Christian theology may explicate its understanding of the intensively infinite presence of the triune Creator as that ultimate reality that constitutes finite creatures and evokes their desire to 'grope' after God (Acts 17: 27). A foundationalist rendering of *fides quaerens intellectum* can lead theologians to a unidirectional methodology that begins with privileged pro-legomena and then follows reason as far as it can, but appeals to 'mystery' if the argument leads to a contradiction. I propose a para-legomena approach in which we reflect on our methodology as we go along. This means that trinitarian imagination need not be the immunized 'foundation' of Christian theology; however, the experience of the appealing mystery of the infinite trinitarian God who conditions all of our noetic desiring may very well be the 'fund' or 'fount' that inspires our participation in a redemptive fellowship of knowing and being known, orienting us toward a transformed and transforming future.

In his encyclical *Fides et Ratio*, Pope John Paul II expressed his admiration for the advances of science, but urged scientists not to abandon 'the *sapiential* horizon within which scientific and technological achievements are wedded to the philosophical and ethical values which are the distinctive and indelible mark of the human person' (2003: n. 106). Theologians and scientists share a common humanity and may embrace a common goal—understanding that transforms our lives together. In interdisciplinary dialogue, our interactions may be marked sometimes by fighting together (*confligere*) and sometimes by singing together (*consonare*), but all of our questing together (*conquirere*) ought to be characterized by a passionate desire for wisdom. Our longing for truth can and should be integrated with our longing for goodness and beauty. With all of the other interlocutors in the ongoing human drama, Christian theologians may enter this shared quest with the categories of their own tradition, explicating their illuminative power as they humbly explore the critical and constructive insights of other religions and scientific disciplines.

REFERENCES AND SUGGESTED READING

ANSELM (1962). *Basic Writings*, trans. S. N. Deane, 2nd edn. La Salle, Ill.: Open Court Classics.

AUGUSTINE, ST (1982). *The Literal Meaning of Genesis*, Ancient Christian Writers 41, trans. and annotation J. H. Taylor. New York: Newman Press.

—— (1993). *Confessions*, trans. F. J. Sheed. Indianapolis: Hackett.

BUCKLEY, MICHAEL J., SJ (1987). *At the Origins of Modern Atheism*. New Haven: Yale University Press.

CLAYTON, PHILIP (1989). *Explanation from Physics to Theology: An Essay in Rationality and Religion*. New Haven: Yale University Press.

—— (1997). *God and Contemporary Science*. Grand Rapids, Mich.: Eerdmans.

DAMASIO, A. (1999). *The Feeling for What Happens: Body and Emotion in the Making of Consciousness*. New York: Harcourt, Brace & Co.

DIXON, PHILIP (2003). *Nice & Hot Disputes: The Doctrine of the Trinity in the Seventeenth Century*. New York: Continuum.

GRENZ, STANLEY (2004). *Rediscovering the Triune God: The Trinity in Contemporary Theology*. Minneapolis: Fortress Press.

HODGE, CHARLES (1981). *Systematic Theology*, i. Grand Rapids, Mich.: Eerdmans.

JOHN PAUL II, Pope (2003). *Fides et Ratio: on the Relationship between Faith and Reason*. Sydney: Pauline Books.

KIERKEGAARD, SØREN (1980a). *The Concept of Anxiety: A Simple Psychologically Orienting Deliberation on the Dogmatic Issue of Hereditary Sin*, trans. and ed. Reidar Thomte and Albert Anderson. Princeton: Princeton University Press.

—— (1980b). *The Sickness unto Death*. Princeton: Princeton University Press.

MURPHY, NANCEY, and ELLIS, GEORGE F. R. (1996). *The Moral Nature of the Universe*. Minneapolis: Fortress Press.

PEIRCE, C. S. (1998). *The Essential Peirce*, ii, ed. Peirce Edition Project. Bloomington, Ind.: Indiana University Press.

PETERS, TED (1993). *God as Trinity*. Louisville, Ky.: Westminster/John Knox Press.

POLANYI, MICHAEL (1962). *Personal Knowledge: Towards a Post-Critical Philosophy*. Chicago: University of Chicago Press.

SCHLEIERMACHER, FRIEDRICH ([1830] 1989). The Christian Faith, ed. and trans. H. R. MacKintosh and J. S. Stewart. Edinburgh: T & T Clark.

SCHRAG, CALVIN O. (1992). *The Resources of Rationality: A Response to the Postmodern Challenge*. Indianapolis: Indiana University Press.

SHULTS, F. LERON (1999). *The Postfoundationalist Task of Theology: Wolfhart Pannenberg and the New Theological Rationality*. Grand Rapids, Mich.: Eerdmans.

—— (2003). *Reforming Theological Anthropology: After the Turn to Relationality*. Grand Rapids, Mich.: Eerdmans.

—— (2004). 'The Role of Trinitarian Reflection in the Religion–Science Dialogue', in Niels Henrik Gregersen and Marie Vejrup Nielsen (eds.), *Preparing for the Future: The Role of Theology in the Science–Religion Dialogue*, Aarhus: University of Aarhus Press, 27–40.

—— (2005). *Reforming the Doctrine of God*. Grand Rapids, Mich.: Eerdmans.

—— and SANDAGE, STEVEN J. (2003). *The Faces of Forgiveness: Searching for Wholeness and Salvation*. Grand Rapids, Mich.: Baker Academic.

VAN FLETEREN, FREDERICK (1991). 'Augustine and Anselm: Faith and Reason', in George C. Berthold (ed.), *Faith Seeking Understanding*, Manchester, NH: St Anselm College Press, 57–66.

VAN HUYSSTEEN, J. WENTZEL (1997). *Essays in Postfoundationalist Theology*. Grand Rapids, Mich.: Eerdmans.

—— (1999). *The Shaping of Rationality: Toward Interdisciplinarity in Theology and Science*. Grand Rapids, Mich.: Eerdmans.

CHAPTER 30

RELIGIOUS EXPERIENCE, COGNITIVE SCIENCE, AND THE FUTURE OF RELIGION

PHILLIP H. WIEBE

I

Beliefs in invisible beings capable of displaying their much vaunted powers in the visible world, and interacting with humans, have been present among all the world's peoples, according to written records, folklore, and archaeological remains. The extensive accounts that Sir James Frazer (1890) gave to English readers more than a century ago about gods and goddesses, spirits and demons, from various cultures around the world, suggest that similar experiences might have given rise to similar religious beliefs. In Western culture during the medieval era, transcendent realities were set in a Ptolemaic world of hollow and transparent globes. These globes were thought to be inhabited by three hierarchies of angels, as first described by pseudo-Dionysius (C. S. Lewis 1964: 70 f.), as well as by other creatures variously known as elves, gnomes, trolls, fairies, hags, nymphs, satyrs, and centaurs (cf. Harrison 1903:

379 f. for Greek origins), and by other terms. Enlightenment influences have virtually eliminated beliefs in the latter beings (cf. Bord 1997 on evidence for fairies) and have contributed significantly to misgivings about angels and evil spirits. Frazer poetically observed that 'the army of spirits, once so near, has been receding farther and farther from us, banished by the magic wand of science from hearth and home, from . . . haunted glade and lonely mere, from the riven murky cloud that belches forth the lightning, and from those fairer clouds that pillow the silver moon or fret with flakes of burning red the golden eve' (1890: 546).

Daniel Pals argues that the theories of religion advanced by such prominent theorists as Frazer, Sigmund Freud, Emile Durkheim, Mircea Eliade, E. E. Evans-Pritchard, E. B. Tylor, and Clifford Geertz share the view that 'religion consists of belief and behaviour associated in some way with a supernatural realm, a sphere of divine or spiritual beings' (1996: 270). The entities postulated to exist by religion are widely seen as lacking direct empirical support, as resulting from an inability to identify causes correctly, and as belonging to an age in which claims about fabulous events that they supposedly explained were too readily believed. The familiar naturalistic hierarchy of sciences, according to which chemistry is dependent upon the foundation of physics, biology upon chemistry, psychology upon biology, and other social sciences upon psychology, leaves no obvious place for religion. Its typical claims about a supernatural reality seemingly have no place in contemporary naturalism, and little or no place in the academy. Moreover, when the academy has addressed religion, according to Rodney Stark (2003), it has focused primarily on its rituals (2003; cf. Poloma 1995: 166).

In directing my attention here to the significance and nature of religious experience, I wish to recall the vital link that was once in place between the human soul and religion, and comment on the significance of the contemporary study of mind—now known as cognitive science—for the study of religion, including religion's future in the academy.

II

According to Cambridge classicist John Burnet (1915–16; also Cornford 1932: 50; Taylor 1953: 132; and Guthrie, Introduction to Rohde 1893: p. xi), it was Socrates who gave Western culture the concept of soul in the form that we know it. By encouraging speculation about the nature of soul, the Presocratic Ionian philosophers, who are widely credited with having 'invented' science and philosophy (Rohde 1893: 24 f.), set the stage for the subtle change that Socrates introduced. Greek literature prior to Socrates associated soul with events that were part of unconscious life or semi-conscious experience, e.g. prescient dreams, apparitions of the dead, and trance-like states in which a god was thought to speak—experiences which had, and still have, religious connotations. Burnet considers Socrates' innovation to have consisted in implicating soul also in *ordinary* phenomena that are part of waking consciousness, by characterizing soul as that in virtue of which humans are wise or foolish, good or bad (Plato, *Crito*, 48a). Although subsequent discussion of soul increasingly focused on conscious experience, the concept did not lose its earlier associations with religious phenomena. The Platonic view that the soul was immortal (*Phaedo*, 79b–e) and would be judged in some post-mortem assize (*Phaedo*, 63c) firmly linked the soul with religion, and Christianity's endorsement of this concept of soul during its long hegemony over intellectual life in the West ensured the soul's prominent role in understanding both religious experience and non-religious conscious experience. Since an immortal soul is strikingly different from a mortal body, an ontological distinction between the two was maintained; Descartes is celebrated for having given this distinction its sharpest expression.

The post-Cartesian developments of the concept of soul are well known: dualistic interactionism was questioned; then 'soul' was replaced with 'mind', which does not carry the religious associations of 'soul'; and eventually 'mind' was replaced with 'mental states', which does not suggest a questionable substance but only the properties of something. Conscious mental states and processes are now seen as properties of a person, and the religious experiences first featured in the Greek understanding of soul, such as apparitions and altered states of consciousness, have virtually disappeared from view in philosophy of mind. Naturalistic theories early in the twentieth century, such as behaviourism, explored strategies for eliminating references even to minds and mental states, evidently because of their association with religious questions concerning immortality and also because of the 'spooky stuff' (cf. Churchland and Sejenowski 1990: 227) that mind was thought to be. *Methodological behaviourists* followed B. F. Skinner in arguing that human behaviour could be explained without reference at all to mental events, and even that such mental terms as 'hoping', 'expecting', and 'feeling' were not needed in recording observations (1964: 106). This reductive proposal attempts to eliminate mental states by refusing to use the terms purporting to denote them. *Philosophical behaviourists* such as Rudolf Carnap (1932) and Gilbert Ryle (1949) claimed that apparent references to mental events were really references to human behaviour or to dispositions to act, and that careful analyses would make that apparent. This proposal acknowledges the existence of mental states, but then argues that these can be reduced by finding identities between them and indisputably natural phenomena.

A decisive methodological shift toward mental states took place in the last three or four decades of the twentieth century. Theorists who were broadly sympathetic to physicalism began to look on mental states more favourably (e.g. J. J. C. Smart 1959, 1965; Putnam 1960), arguing that they could be interpreted as 'theoretical entities' postulated by a theory embedded in common-sense thought purporting to explain human behaviour. Wilfred Sellars indulged in some fictional reconstruction in order to show that a language allowing references to thoughts, sense impressions, choices, feelings, memories, and other mental states or processes (folk psychology) could be viewed as expanding the expressive powers of a language containing only descriptive terms that refer to 'public properties of public objects located in space and enduring through time' and logical expressions, however impressive the latter language might be (1963: 178).[1] Sellars does not suppose either that the 'inner' episodes need to be located in a separate substance (1963: 187), which is how classical dualists about mind and body had understood mind, or that they need to be understood as physiological in character, which is how materialists of various kinds insisted on understanding them. They are simply entities postulated by a theory to account for verbal acts and other acts typical of human life.

The work of David Lewis (1966, 1970, and 1972) is particularly illuminating on interpreting mental states as theoretical entities. According to Lewis, terms purporting to denote theoretical entities (T-terms) are given meaning in the context of other terms (O-terms) whose meaning is already established. Unlike the Positivists, who argued that the O-terms needed to denote *observable* objects, properties, or relations, Lewis held that the O-terms needed only to be antecedently understood (1972: 250 f.). He showed that terms purporting to denote mental states have clear meaning in spite of their metaphysical openness. We can speak about anger, for example, by virtue of its *causal relationships* (1966: 20) to various readily observed phenomena that are typically conjectured to cause it, such as rude gestures and verbal abuse, and the phenomena it typically causes, such as acts of retaliation. If verbal abuse, say, is selected as a typical cause in introducing this postulated state, then anger is partially (or contextually) defined by this causal relation. An implication is that statements linking verbal abuse and anger become mere tautologies: e.g. to say that verbal abuse causes anger is to say no more than that verbal abuse causes that which is caused by verbal abuse. Once anger has been introduced as a postulated entity, however, it can be used in descriptions and explanations of other phenomena. If verbal abuse of a person is followed by bouts of depression and self-deprecation in that person, but depression and self-deprecation have not been used to introduce anger, we can offer the conjecture that anger causes (in conjunction with other factors, not all of which might be known) depression and self-deprecation. This statement is not a tautology, but one that purports to add to our understanding of the causal role of anger.

Defining mental states contextually, by identifying the causal roles they are deemed to play in the theory that postulates their existence, has allowed theorists to revive folk psychology as a significant theory and to postpone questions about ultimate realities implicated in human behaviour and experience until a time when the relationships between the levels of inquiry presupposed by scientific research are clearer. The various reductionist positions made popular in the twentieth century—viz. identity theory, functionalism, and eliminativism—can all be articulated with respect to the contention that mental states and processes are 'theoretical entities' postulated by a theory purporting to account for human behaviour and experience. Developments in neuroscience itself suggest that it will discover hitherto unknown,

---

[1] His well-known essay 'Empiricism and the Philosophy of Mind' (1963) was first given in lectures at the University of London in 1956.

and consequently unexpressed, relationships between mental states (Churchland 1992: 64 f.). Inasmuch as mental states are defined in relation to that which is indisputably natural, these states could be deemed natural—the boundary between the natural and the non-natural can be made a *methodological* matter that need not reflect fixed metaphysical realities.

The first important link between minds and supernatural agents can now be described. The methodological approach to mental states I have just sketched can be extended to the 'realities' that religion has addressed, so that the beings that religion has asserted to be real are viewed as entities postulated to explain certain phenomena. God need not initially be considered a being that is defined by way of superlative properties, whose existence is the conclusion of a deductive or an inductive argument, which is how philosophy has traditionally approached religion. God does not even need to be considered non-material, but can be viewed as 'a-something-we-know-not-what' that is postulated to exist, for instance, to account for the big bang.[2] In introducing God in this way, the reference of the term 'God' becomes established, an implication of which is that the statement that God caused the cosmos to exist becomes tautological. Of course, some other phenomenon could be selected for introducing God in the reconstruction of religion that I am proposing, such as the apparitions that form an important part of the histories of both Hebrew and Christian faiths.

Once a being has been introduced, the 'theory of religion' can be expanded to include other phenomena or events of which the postulated entity is conjectured to be the cause. For example, Judaism, Christianity, and Islam all claim that God has acted in various ways in human history—e.g. in encounters alleged to have been experienced by Abraham and Moses as described in the narrative portions of the Pentateuch—and in so doing these faiths can be viewed as extending the causal role(s) that God is deemed to have. Moreover, various other 'spirits' invoked in descriptions and explanations, including evil spirits and holy angels, can also be viewed as introduced by the causal roles these postulated 'entities' are deemed to play. Again, no position on the ultimate nature of the postulated entities need be taken, and questions about the relationships that the postulated entities of religion have to the known natural order can be postponed. Religion could even be seen as an extension of nature inasmuch as the beings postulated to exist are contextually defined by their causal relations to objects or phenomena that are indisputably natural.

The significance of causation in contextually defining unobservable, postulated entities can be illustrated from physics, where baryon-II particles were first introduced in an interpretation of tracks on photographic plates.[3] A straight line was seen

---

[2] This is not the only hypothesis that needs to be considered, of course. Quentin Smith (2002) has advanced a complex probabilistic argument for the view that a timeless, zero-dimensional point is more likely the cause of spacetime's beginning to exist than God.

[3] I have lost the reference for this, but a comparable discussion concerning K-mesons can be found in Wehr, Richards, and Adair (1978: 450 f.).

to emerge from the point of collision of known particles, followed by a blank, and then followed by branching lines in the form of a 'V' that began at a point in the trajectory of the straight line. The straight line indicates that the collision produced a charged particle; its relatively short length indicates that this unstable particle soon disintegrated to produce an uncharged particle that was postulated to exist (baryon-II), which quickly disintegrated in turn to produce two particles of like charge—hence, the 'V' branch, indicating that the particles are repelling each other. Baryon-II is a true unobservable (in this experiment), for it corresponds to the blank on the photographic plate, and is contextually defined in relation to the event that produces it and the effect(s) that it in turn causes. Other relations (besides causation) between known objects and entities postulated to exist by a theory can also help to establish the reference of theoretical terms, e.g. whole–part relations, relations of similarity, and spatio-temporal relations.

Various fields of science and exact inquiry besides physics have been opened up by theories that postulate the existence of theoretical entities, e.g. 'inheritance factors' (genes) in Gregor Mendel's theory, natural selection in Darwin's evolutionary theory, and the unconscious in Freud's theory of the person. Virtually all fields of science appear to postulate entities that are rarely or never observed, as well as properties that do not always admit of direct observation either. The concept of being observable is not as straightforward as it was once considered to be, for electron microscopes, computer technology, and other sophisticated kinds of equipment have expanded the concept (Shapere 1982). Of course, various entities once postulated to exist have later come to be seen as non-existent—phlogiston is one of the better-known examples from the history of chemistry. Unobservable entities and the theories that postulate them are subject to critical scrutiny, although that scrutiny is indirect, and consequently much more complex to execute when those entities are unobservable than when they are observable. Richard Rorty holds that 'empirical evidence is irrelevant to talk about God', remarking that this viewpoint, advanced by both David Hume and Immanuel Kant, applies equally to theism and atheism (Rorty and Vattimo 2004: 33). The approach I am advancing here makes exactly the opposite point: viz. that empirical evidence is relevant to religious claims of all kinds, but that this evidence is a complex matter because of the nature of religious belief systems as theories.

In the reconstructed view of religion I am proposing, its postulated entities are open to reduction in any of its relevant forms (cf. Hooker 1981), as well as to other modifications that would inevitably arise from construing it according to the model of theories that postulate entities that are unobservable or rarely observable. This approach allows the 'theories of religion' to be rescued from the academic waste-bin, much as the theory of mental states was rescued three or four decades ago. Naturally, this approach also brings religious views under critical scrutiny in ways that its proponents might find challenging, but it need not be viewed as inimical to the unique place that God—self-existent Being, Creator, Sustainer of all other existents, and Initiator of special acts—has been given in historic Judaism, Christianity, and Islam.

God and other spirits have often been invoked very uncritically in order to explain experience and phenomena, but we should not be too hard on our ancestors who authored these claims. Theories that postulate unobservables are not subject to the usual strictures that we impose on inquiries concerning observables. For example, John Stuart Mill's five criteria for determining that one thing is a cause of another cannot be used with unobservables, the unfortunate result of which is that theories that posit them can readily be used irresponsibly. One might view the Logical Positivist movement as having wielded very sharp methodological instruments (cf. Hempel 1950) in its efforts to eliminate irresponsible theorizing. Its methodology came to be seen as unworkable, however, inasmuch as it could not admit theoretical physics into the domain of empirically meaningful claims. In readmitting unobservable entities into theories, we need to keep various strictures in place, in order to keep from adopting theories whose links to reality are so remote that claims to truth cannot be plausibly addressed. Some of these strictures can be gleaned from scientific domains in which the structure of theories postulating unobservable (or rarely observed) objects is already well known. The best known of these theories is atomism, around which an extensive body of literature concerning methodological issues has developed. But we should not assume that atomism will supply a model for theorizing that can be routinely imposed on other domains of inquiry also postulating unobservable objects.

The approach to religion that I am advocating here focuses on its truth claims, but of course religion is much more than propositions purporting to assert some hidden truth about the world: it includes rituals, social practices, ethical outlooks, doctrines, and more (N. Smart 1976: 6 f.). However, its truth claims remain an object of great interest as well as a matter of considerable dispute, both in the academy and among the educated public. The 'primitive' people said to have invented religion sometime in human prehistory could be viewed as having anticipated a form of argumentation that has proved to unlock vital fields of science in the modern age. Of course, they were not as parsimonious as they perhaps ought to have been in postulating religious beings, certainly not as parsimonious as William of Occam's methodological descendents have become, who will not make existential claims beyond those that are absolutely necessary.

In the approach that I am advocating, religion can be seen as advancing a complex set of assertions about putative objects and their properties, as well as about supposed causal relationships between these objects and natural objects whose existence is uncontroversial—a feature that allows religion to be considered an extension of 'nature' as this is more narrowly understood. Religion is a conceptual domain in its own right, taking its place within (or alongside) the common-sense domain, from which the relatively autonomous scientific domains have evolved. Religion purports to describe features of human life and experience that go beyond the specialized sciences, even psychology, although nothing about it need be construed as inconsistent with those sciences. It does not depend upon a sharp distinction between matter and spirit (or soul) in order to be understood, and is not unusually dependent upon other conceptual domains. Religion needs to be reconceived so that its place, possibly vital to many other intellectual and practical undertakings, is not overlooked in a comprehensive view of the world.

Religion should not be ignored, as though the concepts and (preliminary) onto-logical commitments implicit in its discourse might be successfully eliminated if we pretend that religion does not exist and consequently stop using the language of religion altogether. Such a stratagem is comparable to that of the methodological behaviourists, who seemed to think that mental states would 'go away' if people stopped reporting them. Neither do we need to insist that every concept having its home in religion must be understood in narrow naturalistic terms in order to be acceptable. This stratagem is comparable to that of the philosophical behaviourists who could not countenance mental states without proposing naturalistic equivalents. Religion is a fact of life, and progress will not readily be made on understanding its place, if it has one, in a *comprehensive* view of the universe (cf. James [1902] 1997: 410) if it is ignored or has some reduction routinely imposed upon it. Religion will exist in general culture whether or not it exists in the academy. Moreover, its implausible and possibly dangerous forms—forms that tend toward superstition or are apt for use in political oppression—will hardly be addressed convincingly if a culture's trained theorists ignore it.

III

As we have seen, the database for the theory of folk psychology is human behaviour in general. We might ask: 'What is the database for another theory also generally embedded in ordinary thought and language, viz. *folk religion*?' A significant portion of the database for folk religion is religious experience, but it also includes the acts alleged to have occurred that the beings typically postulated to exist in religion—so-called spirits—have supposedly caused. Ninian Smart, for one, appears to concur with this approach, construing religious experience as involving 'some kind of "perception" of the *invisible* world or... a perception that some visible person or thing is a manifestation of the invisible world' (1976: 13). He also observes, as many have, that reports of such perceptions are typically set within a specific religious framework, and are 'clothed in the mythological and symbolic forms of the age' (1976: 11).

Caroline Franks Davis has undertaken a broad survey of the domain of religious experience, and drawing her examples primarily from Buddhism, Christianity, Hinduism, and Islam, has identified six categories into which such experience might be placed: viz. visionary, quasi-sensory, revelatory, interpretive, regenerative, and numinous (cf. Swinburne 1979: 249–53 for a similar categorization, and Spilka and McIntosh 1995 for further discussion). She characterizes an experience as 'a roughly datable mental event which is undergone by a subject and of which the subject is to some extent aware' (1989: 19).[4] This characterization of religious experience makes such experiences a proper subset of mental events in general, which is a point to which I will come back.

Some of the categories that Davis identifies are well known from studies that have already been undertaken. William James, for example, is famous for documenting religious conversion, i.e. regenerative experiences (1902), and the numinous is most readily associated with Rudolf Otto (1917). Davis considers quasi-sensory experiences to consist primarily of physical sensations, such as 'dreams, voices and other sounds... the feeling of being touched... and the sensation of rising up (levitation)' (1989: 36). This category also includes visions and apparitions, about which I will say more below. Interpretive experiences are ones that people regard as religious because of their particular religious backgrounds (1989: 33). This category allows for great variety, inasmuch as religious backgrounds vary considerably in different faiths. Among interpretive experiences we need to include thoughts, hunches, spontaneous beliefs, and feelings such as awe, ecstasy, and peace, which have often been seen as events in which invisible agents are causally implicated. The numerous examples that Davis discusses in her book suggest that the classification she has proposed probably captures most of them, but only a detailed and intimate acquaintance with religious experiences could bear out the claim that none have been overlooked.

Religious experience is not the only element in the database of religion, however, for many religions assert that specific acts have occurred that causally implicate God or other beings as having religious significance. Muhammad is said to have been visited by the angel Gabriel in 612 CE, for example, and while we are correct to view this event as involving mental events of various kinds, to those who believe that the visitation actually occurred, the events are more than mental. Mental events are involved in the *experience* of all events, but proponents of religious claims typically construe some of these claims as going beyond that which is experienced (in a narrow sense). The ambiguity of the term 'experience', especially its narrow and broad meanings, needs to be kept in mind when surveying the events that comprise the database for the 'theory of religion'.

We can readily see that certain portions of folk religion have already been substantially eliminated in Western thought. Many physical illnesses were once attributed to evil spirits, for example, but virtually all such attempts at explanation have disappeared. Moreover, most mental illnesses are no longer explained by reference to evil spirits—a theoretical development that is widely viewed as salutary (e.g. Myers and Jeeves 1987: 41). These are examples of reduction by elimination, but some theorists have also suggested that identities between claims concerning spirits and psychological claims can be advanced. For example, Carl Jung once suggested that demon possession could be viewed as 'a correct rendition of his [the schizophrenic's] psychical condition, for he is invaded by autonomous figures and thought-forms'

---

[4] This definition is similar to that found in *The Oxford Dictionary of Philosophy*: 'a stream of private events, known only to their possessor... [which] makes up the conscious life of the possessor' (Blackburn 1994: 130).

deriving from the unconscious (1939: 134). Only a few phenomena are occasionally attributed to evil spirits now, unlike the medieval era in which numerous phenomena were thought to have that origin (cf. Kramer and Sprenger 1486, which identifies seventy), and the few that remain are highly controversial. The topic of evil spirits is frightening to some, and also troublesome, since so much political oppression has been wreaked in its name.

The special acts featured in religious belief systems are commonly dubbed 'miracles', but I cannot digress to discuss this much debated term and all the issues surrounding alleged miracles.[5] The question as to whether any special acts occur that cannot be plausibly placed into a naturalistic framework, as this is presently understood and known, and that appear to require the causal efficacy of some beings or entities generally deemed to belong to religion, is a very difficult one. The domain of religious experience, critically approached, provides a less controversial database for folk religion than special acts, since no doubt surrounds the existence of experiences having religious significance—though of course, their explanation remains a matter of dispute.

Relatively few places are seemingly devoted to the collection of accounts of religious experience and phenomena, but one important centre is that established by Sir Alister Hardy, now located at the University of Wales in Lampeter. This centre has collected more than 6,000 accounts of experiences from those who reported having them (cf. Maxwell and Tschudin 1996 for representative accounts). The recent doctoral research undertaken by Emma Heathcote-James at the University of Birmingham into contemporary reports of encounters with angels (2002) is another hopeful indicator that religious experiences and phenomena might be attracting more attention in the academy. The 800 reports she has collected come from people of all backgrounds: doctors, barristers, teachers, nurses, members of the clergy, homemakers, the unemployed, and prisoners; their religious backgrounds include Christian, Jewish, Hindu, and Muslim faiths, and some were atheistic or agnostic.

Religion derives from what we might call 'an age of experience', but the age in which we in Western culture now live is dominated by knowledge obtained through experimentation. Virtually everything once advanced as knowledge on the basis of shared experience has come under critical scrutiny as experimental work has revealed that much of what was once believed is flawed, mistaken, or even incorrectly conceived. The dominance of experimental evidence over experiential evidence has weakened the capacity of religion to advance its possibly unique insights about the cosmos and human life. Philosopher Stephen Braude has suggested that evidence might be categorized as experimental, semi-experimental, or anecdotal (1986: ch. 1). Semi-experimental evidence, including accounts of lived experience, consists of claims that cannot be readily obtained at will, but are sufficiently numerous to be worthy of being included in serious theorizing; anecdotal evidence consists of

[5] Most theorists construe miracles as events that *violate* natural laws, but they could alternatively be viewed as events caused by beings uniquely featured in religious belief systems. The latter is the view of St Augustine and Karl Rahner (Schwebel 2004: 164).

claims—often one-off claims—that are insufficiently numerous to be rendered plausible, and consequently may not be included in theorizing about the world.

The insights obtained into near-death experiences (NDEs) in the last thirty years are important in several ways to the subject at hand, for the semi-experimental data that have been obtained have reawakened interest in the possibility of post-mortem survival—a phenomenon that cannot easily be dissociated from religion (cf. Fox 2003). Carol Zaleski (1989) has shown that visionary 'journeys' reported from antiquity and medieval times are strikingly similar to NDEs in their phenomeno-logical character. NDEs render these earlier experiences both more understandable and more credible than they once were. The reports that Raymond Moody (1975) first published of near-death experiences were initially met with scepticism (Moody 1977), but the large number of similar NDEs subsequently collected from around the world has evidently convinced sceptics that people actually have the experiences that had been reported for several millennia, and also by Moody. Although theorists are not agreed about how NDEs should best be explained, scepticism about their occurrence, much as Moody first described them, has disappeared. NDEs demonstrate that if an experience is reported widely enough, even though it might be received sceptically at first, it can become credible, thus overturning Hume's famous objection to reports of miracles (cf. Earman 2000: 8 f.). In a critically reconstructed theory of religion, the phenomena that would take central place would be of the semi-experimental kind.

The second important link between cognitive science and religion can now be described. Inasmuch as religious experiences (as narrowly defined by Davis) are mental events, the content of folk religion will eventually be studied by cognitive science as it attempts to complete its task of showing that mental states can be reduced or eliminated without weakening an understanding of human experience (cf. Lancaster 2004: 46 f.). The objectives of cognitive science are evidently *compre-hensive*: i.e. aimed at establishing that the mental can be *fully reduced*. Consequently, cognitive science will not be able to ignore religious experience, and when it finally gets around to examining the mental states that are part of religious experience, it will naturally address the question of their sources. This is how the religious experiences that partly characterized the concept of soul in the Presocratic and medieval eras, but became excluded in the modern era when conscious states became the focus of philosophy of mind, will again come to be given scientific attention. The future of cognitive science ineluctably includes religious experience.

IV

The remarks I have made to this point have been largely methodological; so, to give them more concrete form, I will briefly discuss several examples of religious experi-ences that I personally researched. Among the six categories of religious experience identified by Davis, the most impressive examples for possible implications concerning the reality of some largely invisible order belong to the first two: viz. visionary and quasi-sensory. I will first comment briefly on the four other categories identified by Davis.

Interpretive experiences appear to be capable of being understood within different frameworks, so their impressiveness for even tentative ontological claims is apt to be uncertainly received. Revelatory experiences seem to have significance primarily for those who have them, so their value in arguing that some reality exists that is not part of naturalism (as this is commonly understood) is modest. I grant, however, that this category could include cognitive insights that could add to their epistemic impressiveness; an extensive Christian tradition, whose most famous proponent is perhaps St John of the Cross, claims that 'intuitive knowledge' of God and his ways supersedes anything that empirical evidence might adduce. Experiences that are primarily affective in character, e.g. feelings of peace, absolute dependency (cf. Schleiermacher 1830: Intro., ch. 1[6]), awe, and ecstasy (cf. I. M. Lewis 2003), might impress those who undergo these experiences, but they are not apt to be considered by others, especially those who are inclined to view religious claims suspiciously, to provide impressive grounds for thinking that some obscure form of religious reality has been encountered. For example, William Alston has argued that a person might perceive that God is 'sustaining her in being, filling her with His love, strengthening her, or communicating a certain message to her' (1986: 655); but Kai Nielsen contends that these are only affective qualities, viz. feelings of love, dependency, goodness, or power, which carry no significant ontological import (1994: 6 f.). Spirit possibly permeates *all* of life, but advancing such a claim on the basis of what appear to be rather mundane religious phenomena seems to be an ineffective stratagem in the contemporary intellectual climate, which often is suspicious of religious claims and holds to the adequacy of naturalism.

Among the most important of religious experiences (in the narrow sense) are those that involve human sensory mechanisms, particularly sight and touch, since these are vital to constructing views of the world derived from ordinary (non-religious) experience. All description of religious experience (in the broad sense) is culture-relative, and culture also influences the understanding of that experience and its significance. Some have argued, in addition, that behavioural roles must first be learned from specific religious traditions in order for religious experience to occur (cf. Holm 1995, esp. on the work of Sundén). In spite of these factors, however, certain experiences—whatever the culturally authorized language might be that is used to describe them—more strongly suggest than do others that some obscure form of objective reality has been encountered. Visions of Jesus Christ constitute an

---

[6] He is widely credited with having given experience a central place in Protestant theology, and with having influenced Paul Tillich, who insisted that the 'feeling' of absolute dependency had a significant cognitive component to it that was often overlooked by Schleiermacher's critics (Depoorter 2004). Ralph W. Hood Jr. sees Schleiermacher's work as showing how every religion can be traced back to an experiential source (1995: 571 f. and 595).

important subgroup of visionary experiences in general, for they have been so widely reported in Western history that there is no serious doubt that such experiences occur; of course, people disagree about how they might be explained. Historic Christian tradition claims that appearances of Jesus, sometimes also described as apparitions or visions, began within days of his death. This tradition postulates the 'risen Jesus' to provide a causal account for the perceptual experiences in which he is thought to be seen.[7] The relevant perceptions might turn out to have their sources wholly within the percipient, in which case those who speak of them as hallucinations, as this is conventionally understood, would be correct. On the other hand, these phenomena (some at least) might involve a form of reality whose relationship to better-known and indisputably natural forms is unknown. The 'hallucination theory' is a competitor to the theory that postulates the existence of difficult-to-examine religious realities. A third possibility is that these experiences are indeed hallucinatory, in some sense of this much-used word (cf. Fulford 1991: 230–1 for ten different interpretations), and that their religious significance derives from the manner in which they are interpreted (cf. N. Murphy 1998: 144). For the sake of simplicity, and because interpretative experiences consist of relatively 'soft' data, I will discuss the phenomenon of Christic visions in relation to the first two of these explanations.

One of the most widely embraced explanations of aberrant perceptions among psychiatric researchers is the *perceptual release theory*. It asserts that information obtained through sensory perception is stored, altered, and then 'released into consciousness' at a later time and experienced as a hallucination, much as stored and altered sensory perceptions are thought to be 'released when we sleep' and are experienced as dreams. According to Louis West, this mechanism was advanced as the basis for both dreams and hallucinations more than a century ago by such prominent figures as Jean Esquirol (West 1975: 287), a French psychiatrist who distinguished insanity from mental retardation early in the nineteenth century, and Hughlings Jackson (West 1962: 277), a British psychiatrist who contributed significantly at the end of the nineteenth century to an understanding of epilepsy and of disorders arising from injury to the brain. Sigmund Freud also advanced it (1900: lect. VII, pt. B). This theory has found its way into more complex theories (e.g. Horowitz 1975; cf. Brasic 1998 for a review) that have been proposed to account for aberrant perceptions, including visions, and it might be thought adequate by some. I will briefly discuss it in reference to several Christic visions I investigated (1997, 2000, 2004). The historical context of such experiences is significant for understanding them, and William James has been criticized for ignoring this (Proudfoot 2004:

[7] Roman Catholic discussions of Christic visions generally construe visions that have occurred after the Ascension of Jesus to be a consequence of angels mediating images of him to people on Earth, since he is deemed not to leave his heavenly abode now; cf. *New Catholic Encyclopedia*: 'Apparitions of Christ, Mary, and the blessed are to be considered as representations effected through the instrumentality of angels' (art. 'Visions', xii. 717). Such dogmatic claims are seemingly impossible to address in an empirical way.

43 f.); but I will not go into any of those details here except to say that the thirty visionaries I interviewed from Canada, Great Britain, Australia, and the United States all had contact or familiarity with Christianity. Several were clergy, but none was in monastic life. Moreover, no one seems to have brought an experience about by such mechanisms as excessive fasting, sensory deprivation, minimizing sleep, or taking hallucinogens.

Jim Link reported to me that his first Christic vision occurred as he sat down in his living room one evening to watch television. He said that the television screen suddenly became invisible and the sound inaudible, and that he felt as though he was enveloped in a curtain, although he could not see one. He tried to look out of the large window beside him in the living room, but he could see neither the wall nor the window in the wall. The first thought that hit him was that he had watched so much television that he had lost his sight and hearing, but he knew this to be absurd! Then a figure in regal robes appeared, and beckoned to him. Although the figure wore a hood that prevented its face from being seen, Jim still identified the figure as Jesus. The words from the New Testament came to him: 'Come to me all you who are weak and heavy laden, and I will give you rest.' Jim felt that he was being invited to embrace Christian faith, in which he had no interest—he was attending his wife's church just to please her. The *perceptual release theory* seems to be capable of explaining this perceptual experience, especially because of the fact that the Christic figure was not superimposed upon the visual space that Jim knew himself to be in, but formed a vital part of a whole aberrant perceptual complex. The 'being' in his visual field was not simultaneously observed and reported by others, was not sensed by two of his sensory modalities (such as sight and touch), and did not leave any 'residue' in the spatio temporal causal world, so the theory that the resurrected Jesus appeared to him does not appear to be a superior conjecture than that provided by the *perceptual release theory.* Other reports are more problematic for this theory, however.

Barry Dyck reported an intersubjectively observable phenomenon as an apparent effect of his Christic vision. Barry was hospitalized and in traction, with strict instructions not to move, after having broken three neck vertebrae in a skiing accident on Mount Baker in the State of Washington. Barry says that eight days after the accident Jesus appeared to him at the end of his hospital bed. Barry sat up and grasped the hands of the figure he took to be Jesus, and begged to die because the pain from the accident was excruciating. In wordless communication Barry was informed that this would not be permitted. Barry then fell asleep, and says that when he awakened the next morning, his pain and swelling were gone. He convinced the attending doctor to release him from hospital, and within a week resumed his regimen of exercise and running. Barry said that the tactile and visual elements of his experience meshed with each other, as do our common experiences in which we see our hands touch the objects that we feel them touch. If Barry's experience was hallucinatory, those who wish to defend the *perceptual release theory* as an explanation of it will need to show how these nicely meshed perceptions of several kinds are 'released into conscious experience', and how these phenomena might cause (or be causal concomitants of) physical healing. It might be objected, of course, in the spirit of David Hume, that falsity of the report of a healing is more probable than that a healing occurred. No simple response to this objection is really possible, apart from finding so many similar reports that the objection itself becomes incredible, as with NDEs.

Some visionaries I spoke to said that they were able to conduct a 'reality check' by turning away and then looking back to see the same object in the same place where it was first seen. Helen Bezanson reported this in connection with each of her two Christic visions (at different times of her life). She said that her first experience (in a church service) began with a tactile sensation of someone touching her hand. Her eyes were closed in prayer at the time, so she opened them to see if someone had touched her, but no one was even near enough to do so. She closed them again, and again felt the same touch. When she opened them a second time, she saw a figure standing on a pedestal some nine feet away whom she immediately identified as Jesus. He appeared much as tradition has imagined him, i.e. with a long robe, long hair, and a beard, and was surrounded by radiance, not simply a halo around his head but an oval shape around his entire body. Helen reported that she had the sense that she was looking at God, but she could not explain exactly what gave her that impression. She looked around the room at the other people who were present, to see if any of them gave any indication that they saw the same thing, but none did. Helen was able to look away and back again several times. The figure that she took to be Jesus finally disappeared, but not before communicating the sense to Helen that she was accepted and loved.

Helen's capacity to conduct the described reality check understandably gave her the sense that the object seen was external to her. Again, the *perceptual release theory* seems inadequate as an account of the occurrence of the same visual perceptions each time Helen faced the front of the building. One would think that everyone has a myriad of transformed 'memory clusters' waiting to be released into conscious experience, and that the identical cluster would not be 'released into conscious experience' as a result (or a concomitant) of something as insignificant as how her head was positioned. Moreover, the fact that the event began with a tactile perception, and was then followed by a visual one, suggests that the *perceptual release theory* in its present form is too simplistic to account for the varying phenomenological details (cf. Wulff 1995 for meanings of 'phenomenology') of her experience. Explanations that appeal to psychological states alone, e.g. wishes to have such experiences (G. Murphy 1945), repressed sexual impulses (Carroll 1986: 141 f.), or stress (Jaynes 1976: 91 f.), cannot explain the phenomenological details in such experiences either. Helen's belief that she stumbled upon some supermundane reality is not irrational.

Although reports of collective experiences can be found among Christic visions, they are rare. Collective experiences have also been reported recently in connection with encounters with angels. One report collected by Heathcote-James was subsequently researched by Carol Midgley, a reporter for *The Times* of London, who says that a being considered to have been an angel was seen in a church in England during a baptismal service by about half of the people present, including the rector. Midgley quotes the rector, who insisted on anonymity to prevent visits from curiosity seekers, as saying:

Suddenly there was a man in white standing in front of the [baptismal] font about eighteen inches away. He was a man but he was totally, utterly different from the rest of us. He was wearing something long, like a robe, but it was so white it was almost transparent. . . . He was just looking at us. It was the most wonderful feeling. Not a word was spoken; various people began to touch their arms because it felt like having warm oil poured over you. The children came forward with their mouths wide open. Then all of a sudden—I suppose it was a few seconds, but time seemed to stop—the angel was gone. Everyone who was there was quite convinced that the angel came to encourage us. (*The Vancouver Sun*, 12 December 2000; cf. Heathcote-James 2002: 46–7)

Collective apparitions manage to 'penetrate' the ordinary spacetime-causal continuum, giving the claim that something real has been encountered some plausibility. Of course, most of the events having import for religious claims are not nearly as dramatic as this one.

The approach I am taking here is empirical, broadly speaking, but some religious claims are seemingly impossible to approach in a strictly empirical manner: e.g. the claim that God exists in triune form. An empirical approach, however, should not be seen as interfering with other approaches to religion that theology might advance. Religion has been a source of seemingly profound insights into hidden realities as well as a source of superstition and irresponsible theorizing. Western culture has understandably had an uneasy relationship with religion since the rise of modern science. But the methodology I have outlined in this chapter suggests a way by which we can rethink the relationship between religion and science.

The experiences I have alluded to briefly are an important part of the database of the theory of folk religion, as I envision it. This theory postulates beings—the resurrected Christ or angels in the examples I have described—to account for the experiences reported. More precisely, the postulated entities and the experiences, including the reported mental states, belong to a conceptual domain in which an array of causal connections is conjectured to exist; moreover, because the theory of folk religion is causally linked to uncontroversial objects that belong to other conceptual domains, including the objects of common sense and various sciences, it is 'empirically anchored'. Much about folk religion is not amenable to direct observation or straightforward empirical testing, but it is not unique in that respect, for other theories postulating unobservable entities share this characteristic. If folk religion never made reference to events capable of intersubjective observation, or capable of leaving some 'trace' in the spacetime-causal world, we might be justified in thinking that religion is little more than the imposition of some archaic conceptual scheme on natural phenomena. I submit, however, that religion attempts to do more: viz. it attempts to address a form of reality that is seemingly difficult to access, and about which we probably know very little.

What remains to be done is the hard work of collecting accounts of religious experiences in all their phenomenological detail, as well as of other events that have relevance to religious beliefs. Even if this work is not undertaken out of interest in or sympathy for religion, it should be done for the sake of cognitive science, whose broadly reductive objectives will require the phenomenological detail I am referring to here. Traditional religious views would not escape unscathed in this process, and the limits of empirical inquiry would probably not remain where they are at present, but the unfriendly stand-off that now often exists between religion and science would begin to be overcome.

References and Suggested Reading

ALSTON, WILLIAM (1986). 'Perceiving God', *Journal of Philosophy*, 83: 655–65.

BLACKBURN, SIMON (1994) (ed.). *The Oxford Dictionary of Philosophy*. Oxford and New York: Oxford University Press.

BORD, JANET (1997). *Fairies: Real Encounters with Little People*. New York: Dell.

BRASIC, JAMES R. (1998). 'Hallucinations', *Perceptual and Motor Skills*, 86: 851–77.

BRAUDE, STEPHEN (1986). *The Limits of Influence: Psychokinesis and the Philosophy of Science*. London: Routledge & Kegan Paul.

BURNET, JOHN (1915–16). 'The Socratic Doctrine of the Soul', *British Academy Proceedings*, 8: 235–59.

CARNAP, RUDOLF ([1932] 1990). 'Psychology in Physical Language', repr. in William G. Lycan (ed.), *Mind and Cognition: A Reader*, Oxford: Blackwell, 23–8.

CARROLL, MICHAEL P. (1986). *The Cult of the Virgin Mary: Psychological Origins*. Princeton: Princeton University Press.

CHURCHLAND, PATRICIA (1992). 'Reductionism and Antireductionism in Functionalist Theories of Mind', in Brian Beakley and Peter Ludlow (eds.), *The Philosophy of Mind: Classical Problems/Contemporary Issues*, Cambridge, Mass.: MIT Press, 59–67.

—— and SEJENOWSKI, TERENCE J. (1990). 'Neural Representation and Neural Computation', in William G. Lycan (ed.), *Mind and Cognition: A Reader*, Oxford: Blackwell, 224–52.

CORNFORD, F. M. (1932). *Before and After Socrates*. Cambridge: Cambridge University Press.

DAVIS, CAROLINE FRANKS (1989). *The Evidential Force of Religious Experience*. Oxford: Clarendon Press.

DEPOORTER, ANNEKATRIEN (2004). 'Existential Participation: Religious Experience in Tillich's Method of Correlation', in Lieven Boeve and Laurence P. Hemming (eds.), *Divinising Experience: Essays in the History of Religious Experience from Origen to Ricoeur*, Leuven: Peeters, 146–78.

EARMAN, JOHN (2000). *Hume's Abject Failure*. Oxford: Oxford University Press.

FOX, MARK (2003). *Religion, Spirituality and the Near-Death Experience*. London: Routledge.

FRAZER, SIR JAMES ([1890] 1994). *The Golden Bough: A Study in Magic and Religion*, new abridgment, ed. Robert Fraser. London: Oxford University Press.

FREUD, SIGMUND ([1900] 1952). *The Interpretation of Dreams*. Repr. Chicago: Encyclopaedia Britannica.

FULFORD, K. W. M. (1991). *Moral Theory and Medical Practice*. New York: Cambridge University Press.

HARRISON, JANE ELLEN (1903). *Prolegomena to the Study of Greek Religion*. Princeton: Princeton University Press.

HEATHCOTE-JAMES, EMMA (2002). *Seeing Angels: True Contemporary Accounts of Hundreds of Angelic Experiences*. London: John Blake.

HEMPEL, CARL (1950). 'Problems and Changes in the Empiricist Criterion of Meaning', *Revue Internationale de Philosophie*, 4: 41–63.

HOLM, NILS G. (1995). 'Role Theory and Religious Experience', in Ralph W. Hood Jr. (ed.), *Handbook of Religious Experience*, Birmingham, Ala.: Religious Education Press, 397–420.

HOOD, RALPH W. Jr. (1995). 'The Facilitation of Religious Experience', in Ralph W. Hood Jr. (ed.), *Handbook of Religious Experience*, Birmingham, Ala.: Religious Education Press, 568–97.

HOOKER, C. (1981). 'Towards a General Theory of Reduction', *Dialogue: Canadian Philosophical Review*, 20: 38–59, 201–36, 496–529.

HOROWITZ, M. J. (1975). 'Hallucinations: An Information-processing Approach', in R. K. Siegel and L. J. West (eds.), *Hallucinations: Behaviour, Experience and Theory*, New York: Wiley, 163–95.

JAMES, WILLIAM ([1902] 1997). *Varieties of Religious Experience: A Study in Human Nature*. New York: Book of the Month Club.

JAYNES, JULIAN (1976). *The Origins of Consciousness in the Breakdown of the Bicameral Mind*. Toronto: University of Toronto Press.

JUNG, CARL G. ([1939] 1973). 'On the Psychogenesis of Schizophrenia', *Journal of Mental Science*, 85. Repr. in *Theories of Psychopathology and Personality: Essays and Critiques*, ed. Theodore Millon, 2nd edn., Philadelphia: W. B. Saunders Co., 128–36.

KRAMER, HEINRICH, and SPRENGER JAMES, ([1486] 1971). *Malleus Maleficarum*, trans. Montague Summers. Repr. New York: Dover.

LANCASTER, BRIAN L. (2004). *Approaches to Consciousness: The Marriage of Science and Mysticism*. Houndmills, Basingstoke: Palgrave Macmillan.

LEWIS, C. S. (1964). *The Discarded Image: An Introduction to Medieval and Renaissance Literature*. Cambridge: Cambridge University Press.

LEWIS, DAVID (1966). 'An Argument for the Identity Theory', *Journal of Philosophy*, 63: 17–25.

—— (1970). 'How to Define Theoretical Terms', *Journal of Philosophy*, 67: 427–46.

—— (1972). 'Psychophysical and Theoretical Identifications', *Australasian Journal of Philosophy*, 50: 249–58.

LEWIS, I. M. (2003). *Ecstatic Religion: A Study of Shamanism and Spirit Possession*, 3rd edn. London: Routledge.

MAXWELL, MEG, and TSCHUDIN, VERNENA (1996) (eds.). *Seeing the Invisible: Modern Religious and Other Transcendent Experiences*. Oxford: Westminster College, Religious Experience Research Centre.

MOODY, RAYMOND (1975). *Life after Life*. Covington, Ga.: Mockingbird Books.

—— (1977). *Reflections on Life after Life*. New York: Bantam Books.

MURPHY, GARDNER (1945). 'An Outline of Survival Evidence', *Journal of the American Society for Psychical Research*, 39: 2–34.

MURPHY, NANCEY (1998). 'Nonreductive Physicalism', in Warren S. Brown, Nancey Murphy, and H. Newton Malony (eds.), *Whatever Happened to the Soul: Scientific and Theological Portraits of Human Nature*, Minneapolis: Fortress Press, 127–48.

MYERS, DAVID, and JEEVES, MALCOLM (1987). *Psychology through the Eyes of Faith*. San Francisco: Harper & Row.

New Catholic Encyclopedia, ed. W. J. McDonald *et al.* (18 vols. New York: McGraw-Hill, 1967.

NIELSEN, KAI (1994). 'Perceiving God', in J. J. MacIntosh and H. A. Meynell (eds.), *Faith, Scepticism, and Personal Identity*, Calgary: University of Calgary Press, 1–16.

OTTO, RUDOLF ([1917] 1950). *The Idea of the Holy: An Inquiry into the Non-rational Factor in the Idea of the Divine and its Relation to the Rational*, trans. John W. Harvey, 2nd edn. London: Oxford University Press.

PALS, DANIEL (1996). *Seven Theories of Religion*. New York: Oxford University Press.

POLOMA, MARGARET M. (1995). 'The Sociological Context of Religious Experience', in Ralph W. Hood Jr. (ed.), *Handbook of Religious Experience*, Birmingham, Ala.: Religious Education Press, 161–82.

PROUDFOOT, WAYNE (2004). 'Pragmatism and "an Unseen Order" in *Varieties*', in Wayne Proudfoot (ed.), *William James and a Science of Religions: Reexperiencing the Varieties of Religious Experience*, New York: Columbia University Press, 31–47.

PUTNAM, HILARY (1960). 'Dreaming and Depth Grammar', in R. J. Butler (ed.), *Analytical Philosophy: First Series*, New York: Barnes & Noble, 211–35.

ROHDE, ERWIN ([1893] 1966). *Psyche: The Cult of Souls and Belief in Immortality among the Greeks*, trans. W. B. Hillis, 8th edn., 2 vols. New York: Harper & Row.

RORTY, RICHARD, and VATTIMO, GIANNI (2004). *The Future of Religion*, ed. Santiago Zabala. New York: Columbia University Press.

RYLE, GILBERT (1949). *The Concept of Mind*. London: Hutchinson.

SCHLEIERMACHER, FRIEDRICH ([1830] 1963). *The Christian Faith: English Translation of the Second German Edition*, ed. H. R. Mackintosh and J. S. Stewart, 2 vols. New York: Harper & Row.

SCHWEBEL, LISA (2004). *Apparitions, Healings, and Weeping Madonnas: Christianity and the Paranormal*. New York: Paulist Press.

SELLARS, WILFRED (1963). *Science, Perception and Reality*. London: Routledge & Kegan Paul.

SHAPERE, DUDLEY (1982). 'The Concept of Observation in Science and Philosophy', *Philosophy of Science*, 49: 485–526.

SKINNER, B. F. (1964). 'Behaviorism at Fifty', in T. W. Wann (ed.), *Behaviourism and Phenomenology: Contrasting Bases for Modern Psychology*, Chicago: University of Chicago Press, 79–108.

SMART, J. J. C. (1959). 'Sensations and Brain Processes', *Philosophical Review*, 68: 141–56.

—— (1965). 'Conflicting Views about Explanation', in R. S. Cohen and M. W. Wartofsky (eds.), *Boston Studies in the Philosophy of Science*, ii, New York: Humanities Press, 157–69.

SMART, NINIAN (1976). *The Religious Experience of Mankind*, 2nd edn. New York: Scribner's.

SMITH, QUENTIN (2002). 'Time Was Created by a Timeless Point: An Atheist Explanation of Spacetime', in Gregory E. Ganssle and David M. Woodruff (eds.), *God and Time: Essays on the Divine Nature*, New York: Oxford University Press, 95–128.

SPILKA, BERNARD, and McINTOSH, DANIEL N. (1995). 'Attribution Theory and Religious Experience', in Ralph W. Hood Jr. (ed.), *Handbook of Religious Experience*, Birmingham, Ala.: Religious Education Press, 421–45.

STARK, RODNEY (2003). 'Why Gods Should Matter in Social Science', *Chronicle Review*, 39/49, B7 (6 June).

SWINBURNE, RICHARD (1979). *The Existence of God*. Oxford: Clarendon Press.

TAYLOR, A. E. (1953). *Socrates: The Man and his Thought*. Garden City, N.Y.: Doubleday.

WEHR, M. R., RICHARDS, JAMES A. Jr, and ADAIR, THOMAS W. III. (1978). *Physics of the Atom*, 3rd edn. Reading, Mass.: Addison-Wesley.

WEST, LOUIS J. (1962). 'A General Theory of Hallucinations and Dreams', in L. J. West (ed.), *Hallucinations*, New York: Grune & Stratton, 275–91.

—— (1975). 'A Clinical and Theoretical Overview of Hallucinatory Phenomena', in R. K. Siegel and L. J. West (eds.), *Hallucinations: Behaviour, Experience and Theory*, New York: Wiley, 287–311.

WIEBE, PHILLIP H. (1997). *Visions of Jesus: Direct Encounters from the New Testament to Today.* New York: Oxford University Press.

—— (2000). 'Critical Reflections on Christic Visions', *Journal of Consciousness Studies, Controversies in Science and the Humanities* (Special Issue: *Cognitive Models and Spiritual Maps,* ed. Jensine Andresen and Robert K. C. Forman) 7: 119–44.

—— (2004). 'Degrees of Hallucinatoriness and Christic Visions', *Archiv für Religionspsychologie,* 24: 201–22.

WULFF, DAVID M. (1995). 'Phenomenological Psychology and Religious Experience', in Ralph W. Hood Jr. (ed.), *Handbook of Religious Experience,* Birmingham, Ala.: Religious Education Press, 183–99.

ZALESKI, CAROL (1989). *Otherworld Journeys.* New York: Oxford University Press.

CHAPTER 31

TOWARD A COMPREHENSIVE INTEGRATION OF SCIENCE AND RELIGION: A POST-METAPHYSICAL APPROACH

SEAN ESBJÖRN-HARGENS
AND KEN WILBER

WHICH SCIENCE? WHICH RELIGION?

There are many exciting conversations occurring across the world at cafés, on campuses, in laboratories, during conferences, and at places of worship, but few are as passionate as the conversation about the relationship between science and religion. This conversation is arising in many contexts: neuroscience finding the

The authors would like to thank Annie McQuade and Michael Zimmerman for their feedback on an earlier draft of this chapter. All figures have been prepared by Paul Salamone.

'godspot' in the brain, the Dalai Lama meeting with scientists to discuss consciousness, the debate over teaching intelligent design versus evolutionary theory in US schools, applying quantum physics to 'prove' mysticism, and the conferences sponsored by the John Templeton Foundation.

These different points of contact between science and religion highlight another reason why this discussion is so energized: everyone has a different meaning of 'science' and a different understanding of 'religion'. As we will see, there are *at least* three common though different meanings of the terms 'science' and 'religion'.

We believe an Integral approach can sort through the different definitions and understandings of 'science' and 'religion' and honour the partial truth claims made by every perspective in this crucial exploration. With an Integral approach we can begin to untangle the Gordian knot of traditional religion and modern science in a post-modern world. In other words, the Integral approach provides a way of truly integrating the many aspects and understandings of science with the many facets and perspectives of religion. And it does this in a way that speaks to traditional, modern, and post-modern understandings of both science and religion. The integral approach that we are aware of that is most capable of this task is *Integral Theory*: a post-metaphysical understanding that relies on the AQAL (all-quadrant, all-level) framework and Integral Methodological Pluralism (IMP).

Integral Theory provides a comprehensive means of integrating the four dimension-perspectives of objectivity, interobjectivity, subjectivity, and intersubjectivity (and their respective levels of complexity). Integral Theory also includes the major methodological families in a way that avoids postulating the existence of pre-existing ontological structures of a Platonic, archetypal, Patanjali, or Yogachara-Buddhist variety.

The goal of this chapter is to outline the ways in which Integral Theory, applying IMP, can provide a successful approach to integrating the disciplines of science and religion. We introduce the Integral approach and the application of IMP. Then we draw on IMP to explain some of the important features of both Integral Science and Integral Religion. Finally, we identify some key considerations for integrating science and religion.

AN INTEGRAL APPROACH

Integral Theory is an inclusive approach to today's world, which is often characterized by disciplinary turf wars and clashes between traditional, modern, and post-modern perspectives. Integral Theory offers a framework that is the result of over thirty years of cross-cultural and post-disciplinary scholarship and application (Wilber 1999–2000; 2000a, b). The Integral model is post-disciplinary in that it can be used successfully in the context of approaches considered *disciplinary* (e.g. helping to integrate various schools of psychology into Integral Psychology), *multidisciplinary* (e.g. helping to investigate ecological phenomena from multiple disciplines), *interdisciplinary* (e.g. helping to apply methods from political science to psychological investigation), and *transdisciplinary* (e.g. helping numerous disciplines and their methodologies interface through a content-free framework).[1] As a result of its applicability across and within disciplinary boundaries, Integral Theory has received a wide embrace from individuals associated with a variety of fields: art, business, ecology, medicine, finance, consciousness studies, religion, correctional education, criminology, education, psychology, health care, nursing, politics, sexuality and gender studies, social services, future studies, and sustainability, to name just a few.[2]

Often represented by the acronym AQAL, Integral Theory's signature phrase 'all-quadrants, all-levels' is shorthand for the multiple aspects of reality recognized in an Integral approach. There are at least five recurring elements that comprise an Integral approach: quadrants, levels, lines, states, and types. These five components represent the basic patterns of reality that repeat in multiple contexts. To exclude any element in any given inquiry or exploration is to forgo a truly comprehensive understanding. By including these basic elements, an Integral practitioner ensures that they are considering the main aspects of any phenomenonon: all-quadrants, all-levels, all-lines, all states, and all-types.

The first element, *all-quadrants*, refers to the basic perspectives an individual can take on reality: the interior or exterior of individuals and collectives, which is often summarized as the four dimensions of experience (subjectivity), culture (intersubjectivity), behaviour (objectivity), and systems (interobjectivity).[3] Each of these perspective-dimensions is irreducible; each has its own validity claim (truthfulness, justness, truth, and functional fit) and modes of investigation, as indicated in Figure 31.1.

---

[1] The main distinction between interdisciplinary approaches and transdisciplinary ones is best captured by Julie Klein (1990) when, drawing on Erich Jantsch's work, she argues: 'Whereas "interdisciplinary" signifies the synthesis of two or more disciplines, establishing a new metalevel of discourse, "transdisciplinarity" signifies the interconnectedness of all aspects of reality, transcending the dynamics of a dialectical synthesis to grasp the total dynamics of reality as a whole' (p. 66). For additional information on interdisciplinary and transdisciplinary approaches consult Klein (1990, 1996); Moran (2002); and Nicolescu (2002).

[2] For examples of the many fields that Integral Theory has been applied to, see *AQAL: Journal of Integral Theory and Practice* <www.aqaljournal.org> and Integral University <www.integraluniversity.org>, where more than twenty-five centres (e.g. Integral Art, Integral Medicine, Integral Science, and Integral Religious Studies) are devoted to exploring Integral approaches in their respective disciplines.

[3] The quadrants can represent both the basic perspectives that any individual can take on something (this is called a *quadrivium*—four views) and the basic dimensions of an individual. So while artefacts such as tables and chairs do not have four quadrants (dimensions), they can be looked at from the four quadrants (perspectives).

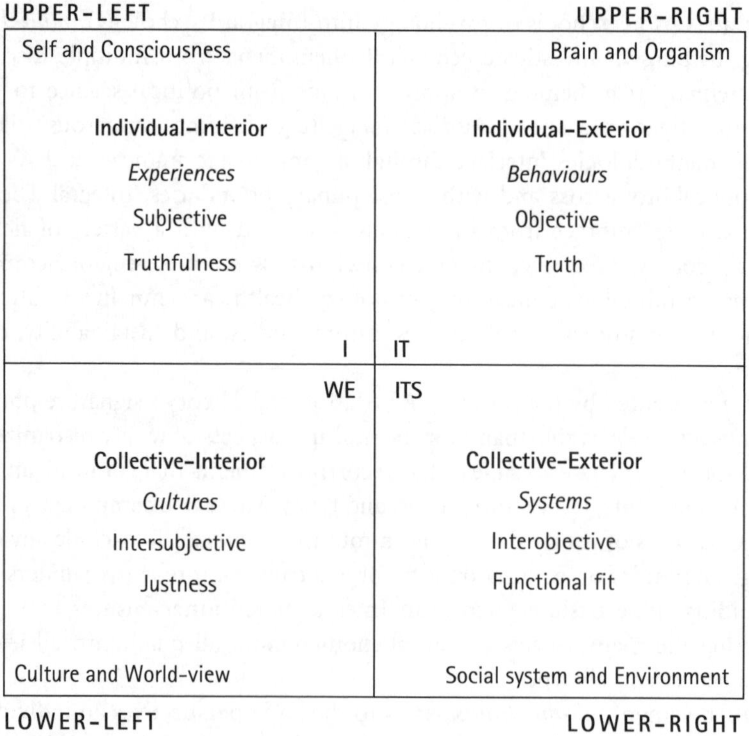

UPPER-LEFT   UPPER-RIGHT

Self and Consciousness	Brain and Organism
Individual–Interior	**Individual–Exterior**
Experiences	*Behaviours*
Subjective	Objective
Truthfulness	Truth
I IT	
WE ITS	
Collective–Interior	**Collective–Exterior**
Cultures	*Systems*
Intersubjective	Interobjective
Justness	Functional fit
Culture and World-view	Social system and Environment

LOWER-LEFT   LOWER-RIGHT

Fig. 31.1. Some aspects of the four quadrants.

The next four elements of the Integral model all arise in each of the four quadrants. *All-levels* are the occurrence of complexity within each dimension (e.g. the levels of physical complexity achieved by evolution in the behaviour quadrant);[4] *all-lines* are the various distinct capacities that develop through each of these levels of complexity (e.g. the developmental features of cognitive, emotional, and moral capacities in the experience quadrant); *all-states* are the temporary occurrence of any aspect of reality within the four quadrants (e.g. the occurrence of weather states in the systems quadrant); and *all-types* are the variety of styles that aspects of reality assume in the various domains (e.g. types of festivals in the cultural quadrant). These five elements are often represented by the AQAL diagram represented in Figure 31.2.

Integral Theory posits that if an approach to science or religion excludes any of these components, it falls short of a truly Integral approach, even if it includes more

---

[4] Within Integral Theory 'levels' are most commonly used to refer to either the general altitude of complexity in any of the quadrants or specific levels within various lines of development. The context will indicate the usage.

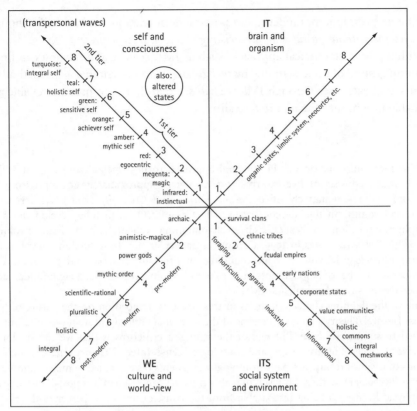

Fig. 31.2. Some aspects of AQAL as they appear in humans.

than other approaches, because each element is understood to be part of each and every moment. Integral Theory assigns no ontological or epistemological priority to any of the elements, because they are understood to co-arise and 'tetra-mesh'.

To integrate these all too often contentious disciplines, Integral Theory uses an Integral Post-metaphysical approach and its corollary, Integral Methodological Pluralism (IMP).[5] This post-metaphysical approach is important for many reasons. First and foremost, any system (scientific or religious) that does not reckon with modern Kantian and post-modern Heideggerian thought cannot survive with any intellectual respectability (agree or disagree, they have to be addressed). That means that any attempt to integrate science and religion must be post-metaphysical in some sense. Second, just as Einsteinian physics applied to objects moving slower than the speed of light collapses into Newtonian physics, so too an Integral Post-metaphysics can

---

[5] For another discussion of the importance of post-metaphysics consult Habermas (1992).

contain all pre-modern, modern, and post-modern religious and scientific thought and systems *without postulating pre-existing ontological structures.*[6]

With a post-metaphysical approach, such as IMP, science and religion can find a common ground of understanding by recognizing the different *and* valid methods of inquiry that each use. Through IMP we can see that they each procure reliable and verifiable insight into the nature of reality.

---

[6] For more information on this Integral approach to post-metaphysics consult Wilber (2003), which consists of five excerpts (for a total of approximately 600 pp.) from the forthcoming book tentatively titled *Kosmic Karma and Creativity* (*KKC*). *KKC* is to be the second volume in the *Kosmos Trilogy*. The task of *KKC* is to fully develop the post-metaphysical position that Wilber has been championing explicitly since the issue of volumes i–iv of his *Collected Works* in 1999. Consult in particular the 'Introduction' of vol. ii and *Integral Psychology* in vol. iv. While building on previous material and positions, *KKC* introduces a number of new concepts, such as tetra-meshing, AQAL space, eight fundamental perspectives of any individual, and IMP.

One of the defining characterizations of this phase is its position on the nature of 'pre-givens'. Integral Theory's major criticisms of the perennial philosophy are numerous and too detailed to summarize here. But one of the strongest criticisms is that we can no longer conceive of 'levels of reality' in a separate ontological sense. Integral Theory rejects entirely the notions of levels of reality as separate ontological existents (as explained in many endnotes in *Integral Psychology*). Rather, any levels of reality must be conceived of in a post-Kantian, post-metaphysical sense, as being inseparable from the consciousness that perceives them. This consciousness is investigated not by metaphysical speculation, but by empirical and phenom-enological research.

To summarize, this post-metaphysical position holds that there are a few involutionary a prioris, which are laid down as Spirit becomes manifest. These include Eros (an impulse towards higher unities, i.e. wider identifications), Agape (an impulse towards embracing all forms, i.e. more inclusion), a morphogenetic field of developmental potential called 'The Great Nest of Being and Knowing' (formerly referred to as the Great Chain of Being when conceived as containing pre-given ontological levels of reality), and a handful of prototypical forms (i.e. the twenty tenets detailed in Wilber (1995)). Everything else in the manifest realm that appears as a pre-given is to be understood as an *evolutionary* a priori, or a 'Kosmic memory habit' (i.e. a probability wave); that is to say, the form or pattern under question was laid down in time and then inherited by subsequent moments. Thus, today's a posteriori is tomorrow's a priori! This implies that today's potentials will become tomorrow's constraints.

As a result of this stance, levels/stages/waves of being and knowing cannot be conceived as involutionary a priori, but rather are evolutionary a priori to the extent that they have been enacted by communities of intersubjects and a Kosmic memory habit or morphogenetic field has been established. The more a particular form has been enacted, the stronger that form becomes, and the more subsequent forms inherit that form. In short, this is a theory of karma: how the past influences the present.

Consequently, the 'lower' levels of psychological development are relatively fixed, while the 'higher' levels, often referred to as soul and spirit, remain as potentials with slight imprints (resulting from the consciousness pioneers of saints, shamans, yogis, sages, and mystics across all traditions). Consequently, the post-rational 'stages' are anyone's 'game'. In other words, the transpersonal realms are understood as potentials and not as fixed realities.

INTEGRAL METHODOLOGICAL PLURALISM

IMP is a collection of practices and injunctions guided by the observation that 'Everyone is partially right!' Each practice or injunction associated with either science or religion enacts and therefore discloses a different aspect of reality. No method discloses reality in its entirety, but each offers some truth and some useful perspective. Integral Theory proposes three principles to uncover and include the partial truths of all perspectives: *non-exclusion* (acceptance of truth claims that pass the validity tests for their own paradigms in their respective fields); *enfoldment* (some sets of practices are more inclusive, holistic, and comprehensive than others); and *enactment* (various types of inquiry will disclose different phenomena, depending in large part on the quadrants, levels, lines, states, and types of the inquirer). These three principles serve to include the greatest number of various forms of truth disclosed by different methodologies.

The essential point is that any truly Integral approach touches bases with as many important areas of research as possible before returning to the specific issues and applications of a given practice. An Integral approach means, in a sense, the 'view from 50,000 feet'. It is a panoramic look at the modes of inquiry (or the tools of knowledge acquisition) that humans use, and have used, for decades, and sometimes centuries. This inclusion of various methodologies and perspectives is based on the idea that no human mind can be 100 per cent wrong. Or, we might say, nobody is smart enough to be wrong all the time. And this means, when it comes to deciding which approaches, methodologies, epistemologies, or ways of knowing are 'correct', the answer can only be, 'All of them'. That is, all of the numerous practices or paradigms of human inquiry—including physics, chemistry, hermeneutics, collaborative inquiry, meditation, neuroscience, vision quest, phenomenology, structuralism, subtle energy research, systems theory, shamanic voyaging, chaos theory, developmental psychology—all of those modes of inquiry have an important piece of the overall puzzle. Since no mind can produce 100 per cent error, this inescapably means that all of these approaches have at least some partial truths to offer an Integral conference, and the only really interesting question is: What type of framework can we devise that finds a place for the important if partial truths of all of these methodologies? To say that none of these alternatives is 100 per cent wrong is *not* to say that any is 100 per cent right. Integral approaches can be very rigorous in standards of evidence and efficacy, a rigour that many holistic approaches let go of too quickly in an attempt to be 'all inclusive'.

One result of the three aforementioned principles is that within each of the four quadrants there are two major types or families of methodologies: those that examine the *inside* aspects of that particular quadrant and those that examine the *outside* aspects of that quadrant.

Consequently, given that the quadrants represent the basic perspectives that an individual can take on any occasion, each individual contains at least eight fundamental perspectives: the inside and the outside view of each of the four

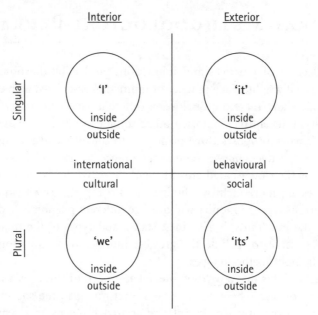

Fig. 31.3. Eight fundamental perspectives.

quadrants of any occasion. Eight fundamental methodologies have arisen out of these eight basic perspectives. They have developed practices, injunctions, and techniques to gain reproducible knowledge (or verifiable repeatable experiences) for each perspective. Some of the better known of these methodologies are summarized in Figure 31.4.

The eight methodological families are *Phenomenology*, which directly explores experience (the insides of individual interiors); *Structuralism*, which explores formal patterns of direct experience (the outsides of individual interiors); *Autopoiesis Theory*, which explores self-regulating behaviour (the insides of individual exteriors); *Empiricism*, which explores observable behaviours (the outsides of individual exteriors); *Social Autopoiesis Theory*, which explores self-regulating dynamics in systems (the insides of collective exteriors); *System Theory*, which explores the functional fit of parts within an observable whole (the outsides of collective exteriors); *Hermeneutics*, which explores intersubjective understanding (the insides of collective interiors); and *Ethnomethodology*, which explores formal patterns of mutual understanding (the outsides of collective interiors). In short, individuals contain all of these dimensions (as disclosed by these respective modes of inquiry) in each and every moment. These methodologies taken together are referred to as 'Integral Methodological Pluralism'.

IMP has important and beneficial consequences for integrating science and religion because it honours each unique approach to reality while recognizing that each uses

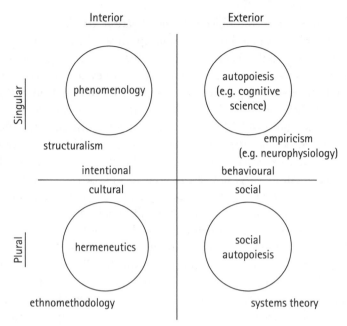

Fig. 31.4. Eight methodologies.

various *partial* perspectives and methodologies to disclose reality. Clearly certain approaches to science or religion prefer different methodological zones. Yet in principle both disciplines can apply all eight methods (all eight perspectives) to investigate reality. Now let's examine how IMP creates an Integral Science and an Integral Religion.

INTEGRAL SCIENCE

As we will demonstrate, science can be defined along a continuum from narrow empiricism to broad or deep empiricism. Integral Theory provides a number of important distinctions useful in defining the multiple meanings of 'science'. What, if anything, is the common denominator between hard, soft, social, life, hermeneutic, and contemplative sciences? In what way are they all concerned with empirical reality? If we begin with the quadrants and IMP, 'science' often means those disciplines that study the outsides of exterior phenomena associated with the Right-Hand quadrants: for example, physics, chemistry, biology, and neurology in the UR and ecology, geology, astronomy, systems theory, chaos, and the complexity sciences in the LR. In all these cases the objects of investigation are the outsides of exterior phenomena described from a third-person perspective, which can be represented by (3p × 3-p × 3p).[7] The only distinction is whether the investigation is aimed at individual or collective occasions. Similarly, disciplines such as cognitive science and social autopoiesis are concerned with describing exteriors from a third-person perspective, but they focus on the insides of those exteriors—providing complex maps of the 'view from inside' (3p × 1-p ×3p). These approaches are often labelled the 'hard' sciences, because they all describe *exteriors*, insides and outsides from a third-person perspective: sciences of exteriors. The life sciences typically focus on a particular level of complexity: namely, biology as opposed to physics, which deals with the laws of matter. Similarly, the social sciences, such as economics, generally focus on the level of human complexity in the LR quadrant.

These 'hard' sciences are often contrasted with the 'soft' sciences, or sciences of interiors that focus on the Left-Hand quadrants from a third-person perspective. For example, there are those disciplines that focus on the UL, such as developmental psychology and developmental structuralism, and those that focus on the LL, such as ethnomethodology and cultural anthropology. These approaches describe the outside of interiors from an objective vantage-point (3p × 3-p × 1p). Presumably, their study of outsides (3-p) of interiors from a third-person perspective (3p) is what constitutes a 'soft' scientific approach. They are not soft in their commitment to third-person description or in their investigation of outsides. Rather, they have been labelled 'soft' because they investigate interiors (1p), which do not manifest in the sensorimotor world. So while the hard sciences examine *exteriors* from a third-person perspective, the soft sciences examine *interiors* from a third-person perspective. They are all 'scientific' according to proponents of the soft sciences, because they all examine objects using third-person descriptors and focus on those objects' outsides (in fact, proponent of the 'soft' sciences have pointed out that their disciplines should be labelled the 'harder' sciences, since their (1p) object of investigation is more elusive than objects sitting around in the external world (3p)). This is often not convincing enough to 'scientists' of the hard sciences. For them it is not enough to provide a third-person perspective, even if it is of the outside of phenomena; one must investigate exterior reality—not interiors.

---

[7] The perspectives of perspectives of perspectives approach of IMP leads to a new type of mathematical notation that replaces traditional variables with perspectives. Using the shorthand of first person (for the inside in general) and third person (for the outside in general), then meditation is 1p × 1-p × 1p (or the inside view of the interior awareness of my first person). Cognitive science is 3p × 1-p × 3p (a third-person conceptualization of a first-person view from within the third-person or 'objective' organism). This 'integral math' can get much more complicated than this, with many more terms, but those are some examples for a start (one can actually build a type of mathematics here, with the equal sign representing 'mutual understanding or resonance'). In this chapter we are using the following three-variable notation: first person (1p) or third person (3p) × inside (1-p) or outside (3-p) × interior (1p) or exterior (3p). Integral math works best with four variables, but for our purposes three will suffice.

If the hard sciences study the insides and outsides of exteriors, and the soft sciences study the outsides of interiors, are there sciences that study the insides of interiors? Not surprisingly, yes! Hermeneutics—often defined as the 'science of interpretation'—studies the LL. Phenomenology and the contemplative sciences study the UL. Edmund Husserl (1970), the founder of phenomenology, was deeply committed to science, and wanted to provide a methodology for disclosing the essential structures of experience, including what is experienced and how it is experienced. These disciplines are characterized by their first-person perspective on the insides of interiors in both individuals and collectives (1p × 1-p × 1p). Unlike the other 'sciences' we have considered, the methods of hermeneutics and phenomenology do not directly involve a third-person perspective, often considered the hallmark of science—hard or

Individual-Interiors	Individual-Exteriors
Soft (Mind) Sciences	Hard (Natural) Sciences
Developmental psychology (1x3x1) Developmental structuralism (1x3x1) Interior phenomenology (1x1x1)	Physics (3x3x3) Chemistry (3x3x3) Cognitive (3x1x3) Molecular biology (3x3x3) Botany (3x3x3) Neurology (3x3x3) Behaviourism (3x3x3)
Collective-Interiors	Collective-Exteriors
Soft (Cultural) Sciences	Hard (Natural) Sciences
Ethnomethodology (3x3x1) Anthropology (3x3x1) Cultural studies (3x3x1)	Astronomy (3x3x3) Geology (3x3x3) Ecology (3x3x3) Environmental (3x3x3) Soft (Social) Sciences Political science (3x3x3) Economics (3x3x3) Sociology (3x3x3) Linguistics (3x3x3)

Fig. 31.5. Some common fields of science.

soft, exterior or interior. As a result, these disciplines have been excluded from so-called scientific investigation. If we include them as scientific enterprises, then we do so based on criteria other than the use of a direct third-person perspective.

Common to all these sciences (e.g. hard, soft, and contemplative) is their drive for repeatable empirical evidence that can be confirmed by other experts in their field. They follow what Integral Theory refers to as the three strands of good science: *instrumental injunction, direct apprehension, communal confirmation or rejection* (Wilber 1983a, 1998). Instrumental injunction refers to an actual practice, an exemplar, a paradigm, an experiment, or an ordinance. It is always of the form 'If you want to know this, do this'. Direct apprehension refers to an immediate experience of the domain brought forth by the injunction: that is, a direct experience or apprehension of data (even if those data are mediated, at the moment of experience they are immediately apprehended). William James pointed out that one of the meanings of 'data' is direct and immediate experience, and science anchors all of its concrete assertions in such data. Communal confirmation or rejection is a checking of the results—the data, the evidence—with others who have completed the injunction and apprehensive strands adequately. Thus all kinds of science are in fact empirical in the broadest sense of experiential. This is a much broader definition of science than the narrow definition of sensory experience usually associated with it.

An Integral approach recognizes both horizontal and vertical empiricism: horizontal, in that researchers can use the three strands of good science (*instrumental injunction, direct apprehension, communal confirmation or rejection*) in any domain explored by the eight methodologies; vertical, in that there are many levels of experience, and therefore many levels of empiricism. In vertical empiricism there is *sensory empiricism* (experience of the sensorimotor world), *mental empiricism* (including logic, mathematics, semiotics, phenomenology, and hermeneutics), and *spiritual empiricism* (experiential mysticism, contemplative spirituality, and transpersonal experiences—confirmed by the community of practitioners who have performed the appropriate injunctions). This means that there is evidence seen by the *eye of flesh* (e.g. intrinsic features of the sensorimotor world), evidence seen by the *eye of mind* (e.g. mathematics and logic and symbolic interpretations), and evidence seen by the *eye of contemplation* (e.g. satori, *nirvikalpa samadi*, gnosis). Each of the three eyes of knowing is natively attuned to its correlative realm of data: sensibilia, intelligibilia, and transcendelia, respectively. However, the eye of mind (or reason) can focus on both the realm of sensibilia and transcendelia. Thus, there are, broadly speaking, at least five different types of empiricism or experientialism (see Figure 31.6).[8]

---

[8] Kurt Koller (2005b) notes that 'There can also be examples of contemplation looking at mind, contemplation looking at body, and likewise flesh looking at both mind and Spirit. Wilber covers these modes briefly when articulating several historical "category errors". A category error is the attempt of one or another eye of knowing to interpret other realms of data in terms of its native realm' (cf. n. 17). See also Koller (2005a, 2005c) for more exploration of Integral Science.

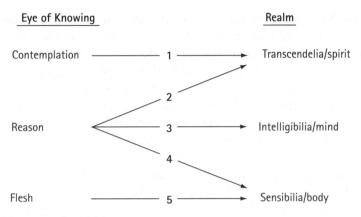

Fig. 31.6. Five types of empiricism.

In addition, an Integral approach to science recognizes that these three strands can be used within various world-views, from magic (impulsive) to mythic (conformist) to rational (conscientious) to systems (autonomous) to transpersonal (ego-aware) (Cook-Greuter 1999). In other words, there are levels of science. Each developmental world-view will define science based on its own perspective. The three principles of Integral post-metaphysics include and honour the context of each level while also judging and discriminating as to the partial value of each. For example, magical science such as various folk sciences (voodoo) or mythic science (creation science) can both follow the three strands even though they are pre-rational, and can therefore properly be considered empirical in the broad sense within their world-view. However, in a larger context the limits of such a naïve empiricism are evident. Likewise, rational sciences like physics and systems sciences like ecology, which are what most people have in mind when they think of science, are also expressions associated with a particular world-view, which are limited from a larger Integral context. Lastly, transpersonal or contemplative sciences such as centring prayer or Mahamudra also follow the three strands and are empirical in a broad sense. One reason why it has been difficult for modern science to accept transpersonal methodologies of investigation is because they appear suspiciously similar to pre-rational forms of science due to their non-rational modes (Wilber 1983a: ch. 8). But as we have explained, non-rational is not anti-empirical when empiricism is understood in the broad sense.

As we have detailed, there are a number of logical movements in shifting from a narrow definition of science to a broad and inclusive definition. Each of these turns is important to understand if we are to integrate science and religion. First, in its most narrow form, science is understood as empirical investigation of exteriors from the outside (3p × 3-p × 3p) or the inside (3p × 1-p × 3p). This definition of science tends to deny the validity of religion, and often pits modern, rational science against traditional, superstitious religion. Next, we see that in some contexts science expands to include interiors, but only those couched in third-person language (3p × 3-p × 1p). Embedded in this definition, scientists investigate religion from a scientific, third-person perspective via disciplines such as cultural anthropology. Third, Integral Theory expands science even further to include the study of interiors from a third-person perspective *and* a first-person perspective as long as we apply the scientific method of the three strands of valid knowledge. At this point religious traditions such as Mahamudra, Zen, and Christian mysticism become scientific insofar as they provide reliable practices for accessing various trans-personal aspects of reality.[9] Thus, the next turn on the road to integrating science and religion is the use of the Integral approach, which answers the concerns of science *and* post-modernism regarding the nature of interiors individually and collectively. The Integral approach recognizes the partial truth in all of these under-standings of science from narrow to broad. The application of IMP gives all of their definitions a place in the science and religion dialogue. We emphasize that we must be clear and concise about what we mean by 'science' and 'religion' for fruitful and generative dialogue. Integral Science recognizes that science can be understood as a *data domain*, such as the Right-Hand quadrants, a *method* such as the three strands of good science, and as a *level* of understanding such as the rational world-view.[10]

Having provided an overview of the many meanings of science, we now turn to the many meanings of religion.

---

[9] Ironically, it is at this point also that interiors come under attack—not by science as much as by post-modernism, which points out that these interiors are shot through with inter-subjective structures and backgrounds and therefore do not exist in any independent sense. Thus Alan Wallace's (2000) defence of subjectivity against the scientific establishment (objectivity) is not as important as defending subjectivity from post-modern deconstruction (intersubjectivity). See also Ch. 2 above.

[10] It can also be understood as a *judgement* such as a third-person cognitive discrimination. Likewise, religion is often associated with the judgement of moral discrimination—and at times aesthetic discrimination. While the eight methodological families reveal phenomena, it is important to realize that they do not determine the type of judgement an individual can take up in relationship to the phenomena disclosed. There are three broad judgements that a person can perform: cognitive ('is it real or true?'), moral ('is it ethical or good?'), and aesthetic ('is it attractive or beautiful?'). In other words, even though science is often associated with the True and religion with the Good, there is the True, the Good, and the Beautiful of both science and religion. Science is usually associated with cognitive judgement, and religion with moral (and to some extent aesthetic) judgement. If one recognizes that all three judgements are important, one is involved in another way of integrating science and religion.

INTEGRAL RELIGION

There are few areas that have as many different associations, connotations, and definitions as religion and spirituality. This diversity of meaning highlights the important role that religion and spirituality play in people's lives and communities and explains why there is so much disagreement in this area. In a general sense 'religion' tends to refer to LL cultures of meaning, symbolism, and theology about God or Spirit. 'Spirituality', on the other hand, usually refers to UL direct felt experiences of insight, love, wisdom or compassion, presence, and grace of the Absolute or the Divine. Interestingly, in terms of levels of psychological development, 'religion' is more often associated with traditional values, whereas 'spirituality' is often connected with post-conventional values. And modern values find both suspect—though there are attempts at providing rational proofs for the existence of God.[11] Additionally, Integral Theory has identified nine different, often exclusive, meanings of 'religion' and five distinct uses of 'spirituality'.[12] Each of these uses is legitimate—we are free to define religion and spirituality any way we wish, and clearly we have—but we *must specify that meaning*. Many scholars and practitioners of both religion/spirituality and science have several implicit but often very different definitions in mind, and they slip between them in a way that generates pseudo-conclusions. The AQAL model recognizes the context in which each definition is accurate and meaningful, and allows each and every one of those definitions to have its place in the interface between science and religion.

In addition to sorting through the multiple uses of terms like 'religion' and 'spirituality', the AQAL model provides a space for Integral Religion (and Spirituality) to emerge. It does so by identifying a number of key issues that have dogged religion for some time. Each quadrant contains phenomena that are crucial for a more comprehensive, balanced, and Integral approach to reality, to the universe, to God and Goddess, and to Spirit. In effect, the four quadrants represent the four hands of God in the manifest realm—leave any one of them out, and one compromises one's relationship with Radiant Spirit.

Let us unpack the different developmental understandings of God, as it is a defining element of Integral Religion. To integrate science and religion, it is necessary to recognize that there is no single God of which religion speaks and which spirituality experiences. A leading developmental theorist, Jean Gebser (1985), found that human beings evolve through at least five major levels of development, which he called archaic, magic, mythic, mental, and integral. If we accept that those stages are more or less right, then there is an archaic God, a magic God, a mythic God, a mental God, and an integral God (with possible higher stages and experiences of God to come).

---

[11] The four most common rational proofs are known as the ontological argument, the cosmological (first cause) argument, the teleological (design) argument, and the moral argument. Consult Rowe and Wainwright (1998).

[12] The nine definitions of 'religion' can be found in Wilber (1983b). The five definitions of 'spirituality' can be found in Wilber (2000a).

An *archaic God* sees divinity in strongly instinctual forces. A *magic God* locates divine power in the human ego and its magical capacity to change the animistic world with rituals and spells. A *mythic God* is located not on this earth but in an other-worldly heavenly paradise, entrance to which is gained by living according to the covenants and rules given by this God to his chosen peoples. A *rational God* is a demythologized Ground of Being that underlies all forms of existence. And an *integral God* is one that transcends and embraces all of the above. Thus Integral Religion recognizes that there are multiple versions of God, and that all of them are worthy of worship and devotion (in their healthy expressions). All of these understandings of God are important because they each capture an irreducible dimension of the Divine in its multidimensional glory. Each 'higher' stage of development actually builds upon and includes the lower, so the lower stages are more fundamental, whereas the higher stages are more significant. Exclude or repress any one of them, however, and one is in trouble. As a result, one ends up with a broken picture of God while claiming that the part one is holding in one's hand is what deserves a nice frame. Tracing that development—while honouring each and every stage as an equally crucial component of that development—is an important part of any Integral approach to religion and spirituality.[13] Moreover, this understanding is crucial for bringing science and religion together under the post-metaphysical umbrella.

Unlike traditional religion's embrace of various metaphysics, Integral Religion embraces an Integral post-metaphysics. This is essential for integrating science and religion, because both science and post-modern theory have produced some devastating critiques of pre-modern metaphysics. As a result, Integral Post-metaphysics replaces *perceptions* with *perspectives*. Thus, for example, the Whiteheadian and Buddhist notion—that each moment is a momentary, discrete, fleeting subject that apprehends dharmas or momentary occasions—is itself a third-person generalization of a first-person view of reality in a first person ($3p \times 1\text{-}p \times 1p$). Each moment is *not* a subject prehending an object; it is a perspective prehending a perspective—with Whitehead's version being a truncated version of that multifaceted occasion, a version that actually has a hidden monological metaphysics (Wilber 1995, 1997, 2000*a*). Integral Post-metaphysics can thus generate the essentials of Whitehead's view, but without assuming Whitehead's hidden metaphysics.

The same is true for the central assertions of the great wisdom traditions: an Integral Post-metaphysics can generate their essential contours *without* assuming their extensive metaphysics. The incredibly important truths of the great traditions could not easily withstand the powerful critiques offered by both modernity and post-modernity. Modernist epistemologies demanded evidence, which the pre-modern traditions were ill prepared to provide, even though traditional contemplative practices offered ample verifiable evidence in favour of claims about Spirit

---

[13] Interestingly, the God of one level often becomes the devil of the next level. For example, the pagan gods of the mythic level become the devil (e.g. Pan) at the mythic level. The mythic God of the Judaeo-Christian religion becomes the devil to the rational God of the Western Enlightenment, and so on.

(contemplation was always a modern epistemology ahead of its time in a pre-modern world). Concluding that no evidence was available to support truth claims about spiritual reality, modernist epistemologies rejected pre-modern religious traditions more or less in their entirety.

Not that it mattered too much, because post-modernity rejected both pre-modernity and modernity. The truth advanced by post-modernist epistemologies is that all perceptions are actually perspectives, and that *all perspectives are embedded in bodies and cultures*, and not just in economic and social systems (which modernist epistemologies from Marx to systems theory had already asserted). If modernity flinched and recoiled in face of these post-modern critiques, one can imagine how the pre-modern traditions fared.

IMP highlights an array of fundamental perspectives, some of which the post-modernist epistemologies would emphasize. In particular, AQAL insists that every occasion has a Lower-Left quadrant (intersubjective, cultural, contextual), and that the quadrants 'go all the way down'.[14] In simpler terms, all knowledge is embedded in cultural or intersubjective dimensions. Even transcendental knowledge is a four-quadrant affair: the quadrants do not just go all the way down; they go all the way up as well.

Modernity focused on the Right-Hand quadrants of objective exterior evidence, while post-modernity focused on the Lower-Left quadrant of intersubjective truth and the social construction of reality. But there was one area that the great traditions specialized in, an area not yet understood, or even recognized, by modernity and post-modernity, and that was the interior of the individual—the Upper-Left quadrant with all its states and stages of consciousness, realization, and spiritual experiences. By situating the great wisdom traditions in an Integral framework, we can salvage their Upper-Left experience and wisdom. Virtually the entire Great Chain of Being fits into the Upper-Left quadrant. Shorn of its metaphysical structures, the wisdom of the pre-modern traditions fits into an Integral framework that allows room for modern and post-modern truths as well.

Just as a Post-metaphysics approach and IMP broaden and deepen narrow science into Integral Science, so they also broaden and deepen narrow religion into Integral Religion, while honouring all the partial truths in between. Like Integral Science, Integral Religion recognizes that religion can be understood as a *data domain* (such as the Left-Hand quadrants), as a *method* (such as those approaches that use the three strands of valid knowledge), and as a *level* of understanding such as a traditional (ethnocentric) world-view or a trans-rational (theocentric) world-view.

Previously we tracked the expansion from narrow science to broad science to Integral science and the ways in which each of those moments contributed (or did not contribute) to an integration with religion. Likewise, when we examine this progression in the context of religion, we see a similar pattern.

---

[14] Whereas the quadrants as perspectives go all the way down (e.g. to the atomic level), the eight methods do not, because they involve a level of self-reflection that is a developmental achievement even among humans.

Narrow religion, often considered religious fundamentalism, is an all too prevalent understanding of religion. This ethnocentric (and sometimes egocentric) expression of religion has the same psychological developmental structure as scientism![15] Integration of science and religion in this context occurs only to the extent that science is placed in service of dogmatic views of understanding divine law. Second, there are rational and world-centric understandings of religion, where someone recognizes that all religious traditions can liberate people from selfishness and provide a context for an intimate relationship with God. It is within this broader understanding of religion that people often attempt to use modern science to prove the Torah, or use brain imaging to map mystical states, and so on. They emphasize the Right-Hand correlates of Left-Hand dimensions. Next, post-modernism interprets religion and science as a series of power/truth claims and places them all on an equal footing (thereby negating development and depth), but does very little to integrate them.

A more inclusive view sees religion as an esoteric core to the great traditions—often called the Great Chain of Being or the perennial philosophy. All too often, in the context of this understanding of religion, science turns to quantum physics to demonstrate the underlying quantum grid of reality. Unfortunately, this is a disaster, a reduction of Spirit in the worst sense (Wilber 1982, 1983a, 1984). Finally, Integral religion recognizes the validity of these previous understandings of religion through a post-metaphysical embrace. In addition to jettisoning the unnecessary ontological pre-givens of traditional metaphysics, this embrace uses IMP to legitimate reproducible spiritual experience and knowledge so that they can be scrutinized by the appropriate community of the adequate (those who have the necessary training in any particular methodology or set of methods). At this level of understanding, science is satisfied that religion is not saddled with unnecessary ontological structures and that religion is following the three strands of valid knowledge.

INTEGRATING SCIENCE WITH RELIGION

Having provided an Integral overview of both science and religion, we can now turn our attention to the salient issues involved in integrating them. One key to understanding these various attempts is to recognize that different world-views have different versions of science and religion, and thus have a different way of trying to

[15] It is important to keep in mind that there are many ethnocentric expressions of science; rationalism, technology, and research can all be appropriated by individuals and organizations with fundamentalist and dogmatic perspectives, using science to further their own ethnocentric goals.

integrate them (Wilber 1998). Thus, each world-view discloses a different valid understanding of both science and religion and their relationship (see Figure 31.7).

With different understandings, different attempts at integration occur. Within a magical world-view science and religion are undifferentiated, and local 'folk' understandings of science, such as causal relationships and taxonomies, support local religious practices (voodoo, witchcraft). The boundary between science and religion is largely absent. A mythic world-view unites science and religion through dogmatism, as in creation science, in which religion accounts for science. In rational world-views, logic and rationality integrate science and religion. Here God becomes a proof. Now science proves religion/God. The post-modern world-view emphasizes plurality in both science and religion, through interdisciplinary research and interfaith dialogue respectively. Transpersonal world-views have not emerged on any large cultural scale, but to the extent that they exist, science and religion are integrated in transrational knowing. It is only with Integral perspectivalism that all these forms of integration are recognized and included, integrating science and religion in their methodological nature (see Figure 31.8).

If we start with traditional religion and modern science and then look at the zones of inquiry of IMP, we notice that all the zones that involve a third-person perspective are represented by science, and the two zones that involve a first-person perspective are often viewed as the domain of religion. In other words, science is often associated with those methods that examine the outsides of the exteriors, and religion is usually associated with those methods that deal with the insides of the interiors. In this sense these two disciplines hold opposite methodological poles. No wonder they are often at odds with one another (see Figure 31.9).

It becomes clear with the IMP approach that while religion has often been confined to the insides of interiors for individuals and collectives (Phenomenology of Religion and Hermeneutics of Religion) there are scientific (i.e. third-person) disciplines that take religion as object of investigation in all the other methodological zones. Thus, all eight methodological families can investigate religion. Let us start

Science	Religion
Transpersonal science: Meditation	Transpersonal religion: Mysticism
Post-modern science: Systems Theory	Post-modern religion: Religious pluralism
Rational science: Physics and Biology	Rational religion: Deism
Mythic science: Scientism	Mythic religion: Fundamentalism
Magic science: Folk science	Magic religion: Voodoo and Paganism

Fig. 31.7. Levels of science and religion and some examples.

| Transpersonal mysticism: Science and religion are always already. |
| Post-modern relativism: Science and religion are equally valid narratives. |
| Modern rationalism: Science proves religion. |
| Traditional Fundamentalism: Religion proves science. |

Fig. 31.8. Levels of integration between science and religion.

with the outsides of exteriors and move toward the insides of interiors. One of the main fields that studies the outside of the individual exteriors is Neurotheology, or what is sometimes called Neuroreligion, which documents the neurological basis of spiritual experience (e.g. McKinney 1994; Austin 1998; Newberg, d'Aquili, and Rause 2001). There are also genetic and biological approaches (e.g. Alper 2001; Pearce 2002;

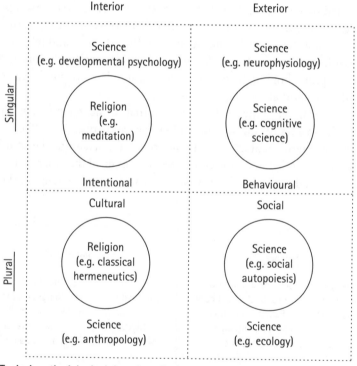

Fig. 31.9. Typical methodological domains of science and religion.

Hamer 2004). The main field associated with the outside of the collective exteriors is Sociology of Religion, which often focuses on institutional dimensions (e.g. Lenski 1963; Wilber 1983*b*; Weber 1993; Durkheim 1995). The inside of individual exteriors is investigated by the field of Cognitive Science of Religion, which looks at the cognitive mechanisms underlying religion (e.g. Andresen 2001; Pyysiainen 2003). Within this approach there are those who situate cognitive mechanisms within an evolutionary context (Boyer 1994; Atran 2004, also Chapter 25 above). At the collective level, Niklas Luhmann's (2000) work on religion and communication explores the inside of collective exteriors. Moving to the Left-Hand quadrants we find the fields of Psychology of Religion (e.g. Fowler 1981; Wilber, Engler, and Brown 1986) and Anthropology of Religion (e.g. Eliade 1958; Lévi-Strauss 1963; Berger 1969; Geertz 1976; Wilber 1981; Foucault 1986), both of which study the outside of individual and collective interiors to identify structural patterns of personal experience and cultural meanings of the Divine. This leaves the fields of Phenomenology of Religion (e.g. Bettis 1969; Twiss 1992; Waardenburg 2001) and the Hermeneutics of Religion (e.g. Osborne 1991; Gadamer 1999; Kearney 2001; Phillips 2001), both of which focus on the insides of the interiors, exploring the individual experience and mutual understanding of the sacred. These last two are empirical in the broad sense of following the three strands of valid knowledge (see Figure 31.10).

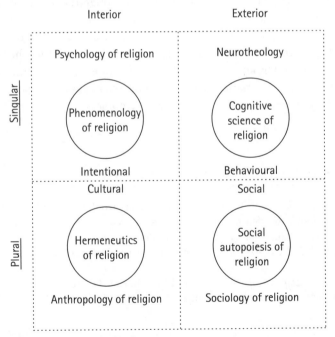

Fig. 31.10. Study of religion scientifically.

Clearly IMP allows science and religion to be integrated by recognizing which methodological zones various approaches are most qualified to inhabit. Those zones inhabited by religion (i.e. phenomenology and hermeneutics) can be understood as scientific, through broad empiricism. It also shows how the third-person zones can be used to study religion to create a more Integral understanding of religion. Again, as we have demonstrated, one reason why the integration of science and religion has been so difficult is that science and religion can be many things to different people.

Not only does the Integral approach recognize the important truths in the many possible ways to integrate science and religion; it also provides a post-disciplinary framework that illustrates their true but partial nature. IMP unlocks the many meanings and reveals the common post-metaphysical language that unites science and religion. Through its guiding principles of *non-exclusion, enfoldment,* and *enactment,* IMP can integrate science and religion regardless of the meaning one has in mind.

If one defines science and religion as *domains of inquiry,* the Right-Hand exterior quadrants versus the Left-Hand interior quadrants, the Integral approach points out that all four quadrants co-arise and are different aspects of the same occasion. Thus, science and religion are inseparable sides of the same Integral coin. If they are defined as *methods,* the Integral approach points out there are eight fundamental methodological families that both science and religion can use to investigate reality: either with disciplines of science being used to study religious phenomena, or with religious practices (broad empiricism) investigating aspects of reality. If they are defined as *levels,* the Integral approach points out that even though religion is often associated with the ethnocentric level and science with a world-centric level, world-views at different developmental levels generate distinct understandings of both science and religion, and therefore take unique approaches to integrating them. Clearly, as a result, science and religion can and must be integrated at multiple levels of understanding.[16] So no matter how we define science and religion, the Integral approach brings them together in an inclusive embrace.

REFERENCES AND SUGGESTED READING

ALPER, M. (2001). *The 'God' Part of the Brain,* 5th edn. Brooklyn, N.Y.: Rouge Press.
ANDRESEN, J. (2001). *Religion in Mind: Cognitive Perspectives on Religious Belief, Ritual, and Experience.* New York: Cambridge University Press.
ATRAN, S. (2004). *In Gods We Trust: The Evolutionary Landscape of Religion.* New York: Oxford University Press.
AUSTIN, J. (1998). *Zen and the Brain.* Cambridge, Mass.: MIT Press.
BERGER, P. L. (1969). *The Sacred Canopy.* Garden City, N.Y.: Anchor Doubleday.

[16] And if science and religion are defined as a *judgement* (i.e. 'What is real' and 'What is good' respectively), the Integral approach points out that the three judgements of the Good, the True, and the Beautiful can each be made in the context of science or religion.

BETTIS, J. D. (1969) (ed.). *Phenomenology of Religion; Eight Modern Descriptions of the Essence of Religion.* New York: Harper & Row.

BOYER, P. (1994). *The Naturalness of Religious Ideas: A Cognitive Theory of Religion.* Berkeley: University of California Press.

COOK-GREUTER, S. (1999). *Postautonomous Ego Development: A Study of its Nature and Measurement.* Dissertation Abstracts International-B 60(06) (no. AAT 9933122).

DURKHEIM, É. (1995). *The Elementary Forms of Religious Life,* trans. Karen Fields. Glencoe, Ill.: Free Press.

ELIADE, M. (1958). *Patterns in Comparative Religion,* trans. R. Sheed. London: Sheed & Ward.

FOUCAULT, M. (1986). *The History of Sexuality,* iii: *The Care of the Self,* trans. R. Hurley. New York: Random House.

FOWLER, J. (1981). *Stages of Faith: The Psychology of Human Development and the Quest for Meaning.* San Francisco: HarperCollins.

GADAMER, H. (1999). *Hermeneutics, Religion, and Ethics,* trans. J. Weinsheimer. New Haven: Yale University Press.

GEBSER, J. (1985). *The Ever-Present Origin,* trans. N. Barstad and A. Mickunas. Athens: Ohio University Press. Original publication, 1953.

GEERTZ, C. (1976). *Religion of Java.* Chicago: University of Chicago Press.

HABERMAS, J. (1992). *Postmetaphysical Thinking: Philosophical Essays,* trans. W. M. Hohengarten. Cambridge, Mass.: MIT Press.

HAMER, D. (2004). *The God Gene: How Faith is Hardwired into our Genes.* New York: Doubleday.

HUSSERL, E. (1970). *The Crisis of European Sciences and Transcendental Phenomenology,* trans. D. Carr. Evanston, Ill.: Northwestern University Press.

KEARNEY, R. (2001). *The God Who May Be: The Hermeneutics of Religion.* Bloomington, Ind.: Indiana University Press.

KLEIN, J. T. (1990). *Interdisciplinarity: History, Theory, and Practice.* Detroit: Wayne State University Press.

—— (1996). *Crossing Boundaries: Knowledge, Disciplinarities, and Interdisciplinarities.* Charlottesville, Va.: University Press of Virginia.

KOLLER, K. (2005a). 'Architecture of an Integral Science', *AQAL,* 1/2.

—— (2005b). 'The Data and Methodologies of Integral Science', *AQAL,* 1/3.

—— (2005c). 'An Introduction to Integral Science', *AQAL,* 1/2.

LENSKI, G. (1963). *The Religious Factor: A Sociological Study of Religion's Impact on Politics, Economics and Family Life.* New York: Doubleday Anchor Book.

LÉVI-STRAUSS, C. (1963). *Totemism.* Boston: Beacon Press.

LUHMANN, N. (2000). *Die Religion der Gesellschaft.* Frankfurt: Suhrkamp.

MCKINNEY, L. (1994). *Neurotheology: Virtual Religion in the 21st Century.* Cambridge, Mass.: American Institute for Mindfulness.

MORAN, J. (2002). *Interdisciplinarity.* London: Routledge.

NEWBERG, A., D'AQUILI, E., and RAUSE, V. (2001). *Why God Won't Go Away: Brain Science and the Biology of Belief.* New York: Ballantine Books.

NICOLESCU, B. (2002). *Manifesto of Transdisciplinarity.* Albany, N.Y.: SUNY Press.

OSBORNE, G. (1991). *The Hermeneutical Spiral: A Comprehensive Introduction to Biblical Interpretation.* Downers Grove, Ill.: InterVarsity Press.

PEARCE, J. (2002). *The Biology of Transcendence: A Blueprint of the Human Spirit.* Rochester, Vt.: Park Street Press.

PHILLIPS, D. Z. (2001). *Religion and the Hermeneutics of Contemplation.* Cambridge: Cambridge University Press.

PYYSIAINEN, I. (2003). *How Religion Works: Towards a New Cognitive Science of Religion.* Leiden: Brill.

ROWE, W. L., and WAINWRIGHT, W. J. (1998) (eds.). *Philosophy of Religion: Selected Readings,* 3rd edn. New York: Harcourt Brace College Publishers.

TWISS, S. (1992). *Experience of the Sacred: Readings in the Phenomenology of Religion.* Hanover, NH: University Press of New England.

WAARDENBURG, J. (2001). *To the Things Themselves: Essays on the Discourse and Practice of the Phenomenology of Religion.* Herdon, Va.: Walter de Gruyter Inc.

WALLACE, B. A. (2000). *The Taboo of Subjectivity: Towards a New Science of Consciousness.* New York: Oxford University Press.

WEBER, M. (1993). *The Sociology of Religion.* Boston: Beacon Press.

WILBER, K. (1981). *Up from Eden.* Garden City, N.Y.: Anchor Press/Doubleday.

—— (1982) (ed.). *The Holographic Paradigm and other Paradoxes: Exploring the Leading Edge of Science.* Boston: Shambhala.

—— (1983*a*). *Eye to Eye: The Quest for the New Paradigm.* New York: Anchor/Doubleday.

—— (1983*b*). *A Sociable God: A Brief Introduction to a Transcendental Sociology.* New York: McGraw-Hill.

—— (1984) (ed.). *Quantum Questions: Mystical Writings of the World's Great Physicists.* Boulder, Colo.: Shambhala.

—— (1995). *Sex, Ecology, Spirituality: The Spirit of Evolution.* Boston: Shambhala.

—— (1996). *A Brief History of Everything.* Boston: Shambhala.

—— (1997). *The Eye of Spirit: An Integral Vision for a World Gone Slightly Mad.* Boston: Shambhala.

—— (1998). *The Marriage of Sense and Soul: Integrating Science and Religion.* New York: Random House.

—— (1999–2000). *The Collected Works,* 8 vols. Boston: Shambhala.

—— (2000*a*). *Integral Psychology.* Boston: Shambhala.

—— (2000*b*). *A Theory of Everything.* Boston: Shambhala.

—— (2003). *Introduction to Excerpts from Volume 2 of the Kosmos Trilogy.* Excerpts A, B, C, D, and G. Retrieved 18 November, 2004 from <http://wilber.shambhala.com>.

——ENGLER, J., and BROWN, D. (1986) (eds.). *Transformations of Consciousness.* Boston: Shambhala.

PART V

CENTRAL THEORETICAL DEBATES IN RELIGION AND SCIENCE

'SCIENCE AND RELIGION' OR 'THEOLOGY AND SCIENCE'?

CHAPTER 32

SCIENCE AND THEOLOGY: THEIR RELATION AT THE BEGINNING OF THE THIRD MILLENIUM

MICHAEL WELKER

'SCIENCE AND THEOLOGY DISCOURSE' OR 'SCIENCE AND RELIGION DISCOURSE'?

Why do organizers of interdisciplinary conferences, sponsors of such events, and editors of academic books prefer the label 'Science and Religion Discourse' to 'Science and Theology Discourse'? From an outside perspective the concentration on 'theology and science' (over against 'religion and science') seems to be narrow and limited. For 'theology' is mostly associated with 'Christian theology', or even with only one discipline of 'Christian studies': namely, 'systematic theology'. Although one-third of the world's population belongs to one of the Christian churches, and although most of the scientifically innovative institutions of teaching and research are located in cultures and environments in which Christianity and Judaism are dominant (Zuckerman 1996, with information beyond the United States), the limitation indicated by 'Science and Theology' seems unattractive if not unbearable.

A second reason for reservations against this label has to do with the opinion that 'theology' seems to require a personal engagement and a commitment of faith which does not agree with 'scientific objectivity'. Despite the fact that this also holds true for the terms 'religion' and 'religious', from an outside perspective the label 'Science and Religion Discourse' appears to allow for more objectivity than the announcement of a theological enterprise. Over against an engaged theological approach, 'religious studies' can be done from a more detached point of view which is often associated with 'objectivity'.

This general outside understanding, however, is not shared by most academic perspectives and those who reflect on the discourse between theology, religious studies, and the sciences. There is no such thing, and there can be no such thing, as a discourse between a religion and a science, not to speak of a discourse between 'religion as such' and 'science as such', whatever that might be. In order to enter into a meaningful dialogue with 'science', it is necessary to bring 'religion' on to an academic, or at least intellectual, level (theology, religious studies, philosophy of religion, etc.). The minimum requirement for a dialogue involving the two areas of knowledge is that religious people or persons with a more or less profound knowledge of one religion or several religions enter into a discussion with scientifically trained people about general or specific topics which are meaningful in each field. If the discourse aims to produce insights that can win academic respect and credibility, religious people should have some academic training and an academically cultivated understanding of the religion(s) brought into play. Only then can 'theology' (or equivalents to it in non-Christian institutions of religious education and research) and 'religious studies' be adequate conversation partners for the sciences.

From this academic perspective the label 'Science and Religion' seems to aim at comparative observations and reflections between the fields rather than a dialogue that aims at producing new insight, a dialogue that enriches both and helps each to develop further. It seems to evoke the juxtaposition of two different types of world-view, which could become a topic in comparative religious studies or in the study of world-views, of different ideologies, or of their clashes and the potentials for connection between them. But such comparative studies, interesting as they might be, are not what most organizers of 'Science and Religion' or 'Science and Theology' discourses have had in mind over the last decades. Their intentions were not directed towards comparative observations in, and reflections on, two areas of insight and knowledge, but rather towards an interactive and productive dialogue between representatives of these two areas. Cf. Peacocke (1993); Rae, Regan, and Stenhouse (1994); Barbour (1997); Polkinghorne (1998); Peters and Bennet (2003). On both sides, this implied a personal commitment to the establishment and testing of truth claims. Although academic rules have had to be negotiated and respected in the discourse, the simple opposition between 'subjective commitment' on the side of theology and 'objective reflection and research' on the side of the sciences was not an option (Polkinghorne and Welker 2000: 1–13). The struggle for insight, certainty, and truth on both sides was not only tolerated, but rather required, for a lively dialogue.

In light of this requirement, the plea for the adequacy of the label 'Science and Theology' was strengthened. The widespread desire, however, to understand 'discourse' between two different areas of knowledge and research as the search for, and the establishment of, a 'meta-level' above them became severely questioned. The danger that comparative religious studies could easily produce nothing but such mere meta-levels was seen. The typically modern move to talk first about methodological issues and metaphysical or epistemological presuppositions in order to provide some 'common ground' before one could turn to specific issues and topical questions was suspected to be an illusion (Welker 2001: 165 ff.). Although general methodological reflections are indispensable in any interdisciplinary discourse, cf. Gregersen *et al.* (1998); Coleman (2001); Heller (2003), they must not become ends in themselves. If the well-meant preparation of the 'common ground' is isolated from the search for specific insights and from raising specific questions relevant to the fields involved, it leads to nothing but general observations and popular-philosophical reflections which hover above both fields. Instead of gaining 'common ground', such a discourse can at best produce 'common horizons', at worst only 'common clouds'.

As became clear at an international conference at which six familiar models of dialogue between theology or religious studies and the sciences in different countries were compared, the intended warming-up phase of a meta-level discourse on issues of presuppositions and methods can easily turn stale and unproductive when the raising of specific issues on both sides is notoriously postponed. One thinks here of Samuel Beckett's oft-quoted phrase, 'Are you merely interested in everything or also in something specific?' The interest in dealing with specific questions and topics is crucial for a fruitful 'Science and Theology Discourse' as well as for successful discourse between topically focused religious studies and scientific thinking.

In a broad-scale overview of the international discourse between science and religion in the last third of the twentieth century, Ted Peters diagnosed shifts from a 'methodological phase' to a 'physics phase' (Russell, Stoeger, and Coyne 1988). The first phase was preoccupied with the preconditions, constellations, and forms of the dialogue in general. The second phase started with cosmological and natural religious questions which originated mostly in the realm of physics. Peters (1998*b*: 3–7) also saw a shift towards a phase in which 'theological questions arising out of evolutionary theory and genetic research' have become more prominent. As many publications document, even in the 'physics phase' (in which physics dominated the shaping of the discourse) there was a growing interest in raising specific theological questions which were taken also to be relevant for scientific minds and scientific thinking.[1] With this phase not yet at its end, a complex 'biology and theology phase' has been developing in which topics of 'divine action' (Russell, Stoeger, and Ayala, 1998) and numerous questions in anthropology (see the exemplary treatments by

[1] Cf. Polkinghorne (1994), as well as the 1998 documentation of a multi-year international and interdisciplinary discourse at the Center of Theological Inquiry, Princeton, published in *Theology Today*, 55; Albright and Haugen (1997); Polkinghorne and Welker (2000); Peters, Russell, and Welker (2002); also cf. the recent books by Polkinghorne (2002, 2004).

Brown, Murphy, and Malony (1998); Gregersen, Drees, and Görman (2000); Jeeves (2004)) and ethics (Peters, 1998*a*) engage both sides of the discourse.

Basic Forms of Theology and Science: Problems of their Interaction

Both 'theology' and 'science'—in history as well as in the contemporary situation—come in an overwhelming wealth of forms, a vast array of methods, topics, interests, and goals. In the midst of all these appearances of 'theology' and 'science', it seems helpful to distinguish three basic and general forms. For the sake of a comfortable overview I should like to speak, first, of maximalist forms of theology and science—THEOLOGY and SCIENCE in capital letters. Second, I will describe minimalist forms of theology and science—theology and science not capitalized. Third, the most common form of academic everyday enterprise is the theological and scientific thinking which moves between maximalist and minimalist theologies and sciences—Theology and Science with a capital 'T' and a capital 'S' respectively (Welker 2000).

By THEOLOGY in capitals we understand an elaborate interconnection of thought and conviction related to God and God's workings. A complex system, or even systemic form, that can be drawn from biblical, historical, and philosophical sources is characteristic of a THEOLOGY. By SCIENCE in capitals we understand an elaborate interconnection of thought, conviction, and tested experience related to nature and its texture. When we speak, for instance, of the THEOLOGY of Augustine, Luther, or Schleiermacher, or the THEOLOGY of the Reformed or Lutheran confessional writings, we refer to such a THEOLOGY. THEOLOGIES can, but need not be, theological classics. We can speak of the THEOLOGY of a work of dogmatics, of a congregation, or of a church tradition, without necessarily praising or even accepting it. There can be THEOLOGIES that we find unconvincing or that may even appear dead to us. The elaborate interconnection of theological thought and conviction is the point of THEOLOGIES. Such an elaborate interconnection can be a blessing or a curse. Schleiermacher warned against attempts to work with theologies that cultivate modes of thought and imagination 'in which nobody really thinks any longer'.

Similar observations can be made with respect to SCIENCE. Relativity theory, quantum theory, the discovery of the structure of DNA, but also great theories of the past which are at best of merely historical interest to contemporary scientists, can stand for SCIENCE as an elaborate interconnection of thought, conviction, and tested experience related to nature and its texture.

A discourse between THEOLOGY and SCIENCE does not occur any longer. The inner complexity and the internal logics of each seem to block all potential endeavours. Geniuses like Newton, and above all thinkers in pre-modern history, tried to connect or fuse complex theories of both sides. But none of these attempts won credibility in theological or scientific communities in the long run. Today, only amateurs dream of relating THEOLOGY and SCIENCE in a complex theory. They will at best produce a private philosophy. Both THEOLOGY and SCIENCE certainly inspire many persons interested in the dialogue. However, the idea that complex theologies and complex scientific theories could directly have an impact on each other is not backed by experience.

What is the constellation like on the minimalist level? We refer to theology (not capitalized) when we claim that every religious seeker or pious person, and every member of a religious community, is enabled to come up with theological utterances and to make theological contributions. As I have argued in more detail elsewhere (Welker 2000; Polkinghorne and Welker 2001), not every remark about God is theological. Not even every pious utterance can be considered theological. The sigh directed to God and the silent prayer are no more theological utterances than the cynical remark about God or the presentation which leaves no doubt that it deals with a religion that is spiritually profoundly foreign to the speaker.

Although a theological utterance about God or about religious matters does not have to evidence a well-developed faith (as would be expected of a THEOLOGY), it must show a minimum of conviction and a minimal degree of having been existentially influenced. A statement about God and religious matters cannot be regarded as theological unless its speaker, regardless of his or her reflective distance or values, shares with others a minimum of certainty. Nor can it be regarded as theological for the speaker unless it signals the search for spiritual reliability and truth, or the need to believe; but if not personal, then at least such a search should be acknowledged minimally in the topics and contexts he or she deals with. This is the first of two conditions to qualify an utterance as theological.

The second presupposition is no less demanding. A theological utterance must be formulated in words and must be comprehensible. Others must be able to follow its logic, and it must be capable of material development. In order to reach the level of theological propositions, religious utterances must express certainties that are communicable, comprehensible, and open to development with respect to their object and content. Although academic theology sometimes finds this side underdeveloped in individual piety and church life, we should not operate with a cliché that leaves conviction and existential involvement to the church and comprehensibility (as well as material development) to the academy. By its very commitment to teaching and research, the academy cannot avoid reaching out for existential commitment, personal certainty, and conviction. Academic theology decays when it simply administers THEOLOGIES, treating them as more or less interesting exhibits in a museum.

The minimalist and elementary operation on the side of a scientific approach can be found in observations of natural entities or events that can be repeated and empirically verified, and in the discovery of regularities which make possible the first steps towards building a theory.[2] These basic activities of scientific thinking and

---

[2] I am grateful to my Heidelberg physicist colleague Jörg Hüfner for a helpful exchange on these issues.

speaking can, of course, be stimulated by SCIENCE in the form of a complex theory. But this need not be the case. Many more or less immediate observations and reflections about nature and natural events can have a scientific character in the minimalist sense described above.

On this minimalist level we can discover some interesting fusions between scientific and theological experiencing and thinking, fusions which have become influential in the history of thought and experience. These fusions, however, should not be regarded as the practice and the result of a dialogue. The fusions lie in the so-called natural awareness of God or in the presentiment of the Divine. Schleiermacher's feeling of 'absolute dependence' was, as he claimed, the result of an empirical observation. It was the observation that human beings cannot escape a twofold feeling in their inner self-consciousness: a feeling of freedom and a feeling of dependence. Even the most dramatic experience of freedom would be connected with some sense of givenness and directedness. And even the most depressing feeling of dependence would, in its very experience, contain an element of spontaneity. Schleiermacher termed this unavoidable and inescapable duality the 'feeling of absolute dependence', and in the 'whither' of this feeling he located God (1960, §§ 3–5).

This theological manoeuvre may invite questions. But it is an example of the attempt to fuse an observation which can generally be repeated and tested experientially with a basic theological thought and argument.[3] Similar reflections are possible in general perspectives on the natural and cultural environments in which we live. John Calvin, in his famous *Institutes* (1559), develops such an argument. In chapter 3 of the first part of his book he says: 'That there exists in the human mind, and indeed by natural instinct, some sense of Deity, we hold to be beyond dispute.' Calvin points towards the sense of a vague 'bearing' or 'grounding' power which surrounds us in such a way that we can neither get a firm grip on it nor avoid and escape it (Welker 1999a: ch. 2). In chapter 5 he argues that this reflection of the encompassing environment in our mind gains relative clarity in the 'beautiful order of the world', in the structures of nature, the order of our bodies, the accessibility of reality through our knowledge, etc. In a more penetrating way, John Polkinghorne has argued that there are 'windows onto reality' that provide a basis for a 'natural theological approach': the order and rational beauty of the universe, the fecundity of the universe, the emergence of consciousness, and the emergence of religious awareness across the globe (Polkinghorne and Welker 2001: chs. 1 and 7).

Both Calvin and Polkinghorne, however, state the deep ambiguity of such an approach. Polkinghorne says that we equally have to acknowledge 'windows onto reality' in the forms of moral evil, natural evil, and the finitude of life (Polkinghorne and Welker 2001: chs. 1 and 7). Calvin argues that on a basis of the 'natural religious awareness', human beings can never attain more than relative, problematic degrees of clarity with regard to the invisible reality of the Divine.

---

[3] Philip Clayton rightly commented that this means that theology as such must always cede ground to the scientific in these discussions.

Thus neither THEOLOGY and SCIENCE, nor theology and science in their basic modes, are themselves candidates for the discourse intended. Rather, for this discourse we need the third form, which moves between elementary authentic experiences, observations, and thoughts, on the one hand, and elaborate theories, on the other. I shall refer to this form as Theology and Science, with capital 'T' and 'S' respectively. These forms are operative in most of our everyday academic work—to the extent that it is open for dialogue and mutual stimulation, mutual challenge, and inspiration—as a rule within one's own field. Theology and Science—and this is characteristic of sound, respected academic enterprises—are pursued in 'truth-seeking communities'. The fact that their practitioners belong to truth-seeking communities provides common grounds for both of them. Although the cultivation of their disciplinary boundaries is, as a rule, part of their ability to stay on track in their search, their opening up to another discipline, or more precisely, to some challenging insights or modes of experiencing and thinking from other fields of discourse, can become desirable and fruitful. Given the specific and systemic limits of each discipline's attempt to gain relevant insights, and in light of the dangers of working with problematic reductionisms, the dialogue can challenge both theological and scientific thinking and possibly draw specific insights from THEOLOGIES and SCIENTIFIC THEORIES without any attempts to synthesize or fuse the different approaches. The mutual respect for systematic and systemic differences of the fields should accompany all Theology and Science discourses.

SCIENCE AND THEOLOGY AS TRUTH-SEEKING COMMUNITIES AND FIGHTERS AGAINST BAD REDUCTIONISMS

'Truth-seeking communities'—I am indebted to John Polkinghorne (1994: 149; 2000: 29 f.; Polkinghorne and Welker 2001: ch. 9) for this expression—are not to be confused with groups which announce that they have found the truth and now possess it. Truth-seeking communities are groups of human beings who indeed raise truth claims, but above all develop and practice open and public forms and procedures in which these truth claims are subjected to critical and self-critical examination. The academy, active in research and education, is one such truth-seeking community. This leads to a two fold definition of truth-seeking communities:

On the one hand, truth-seeking communities advance processes in which certainty and consensus can be developed, interrogated, and heightened. In doing so, they must guard against reducing truth to certainty and consensus.

On the other hand, truth-seeking communities advance processes in which complex states of affairs can be made accessible in repeatable and predictable ways. In doing so, they must guard against reducing truth to the repeatable, predictable, and correct investigation of the subject under consideration.

The path of the search for truth is characterized adequately only by the reciprocal relationship between, on the one hand, the interrogation and heightening of certainty and consensus, and, on the other hand, the repeatable, predictable, and correct investigation of the subject under consideration.[4] This path can be travelled only in open and public critical and self-critical communication. The accomplishment, the value, and the blessing of truth-seeking communities cannot be overestimated, although self-critically and realistically we have to take into account the fact that most often other interests also play a part in these communities. Even if they are neither guided by the search for maximum cultural resonance and for moral and political influence, nor by vanity and the desire for power and control, the seemingly 'innocent' search for pedagogical and technological success may occasionally conflict with the sincere search for truth.

Thus the sober recognition that there are no pure and perfect truth-seeking communities is very helpful indeed. For it helps us to distance ourselves from the blind self-privileging of academic work, and it helps the appreciation of 'justice-seeking communities', and of communities which aim at 'physical and psychic therapy and the restoration of health'. We also have the obligation to respect communities which seek 'political loyalty and a corresponding exercise of influence', communities that seek 'economic and monetary success', and communities that seek to maximize 'public attention and resonance'. It is characteristic of pluralistic societies that truth-seeking communities do not absolutize themselves, but that they recognize and delineate their important and indispensable contributions to the entire society, enabling them to be perceived in other contexts as well. Such side-reflections can help to calibrate the potentials of the Science and Theology discourse and to avoid overestimating or 'ideologizing' its potentials.

As stated before, the great potentials of the Science and Theology discourse lie neither in the establishment of a meta-level above their areas of research, nor in attempts to synthesize both approaches. The great potentials of these dialogues lie instead in raising boundary sensitivities and in gaining specific insights into conceptual limits and the dangers of pernicious reductionisms (Welker 2004 and 2006).

Reductionisms emerge when certain phenomena of an area of possible experience, or certain theoretical or experimental tools and certain figures of thought that can help to disclose this area, are taken to be the *only* phenomena, the *one* guiding principle, or the *sole* key to disclose it. As soon as such an approach convinces a broader group of scholars, or even a broader public, at a certain time, a certain reductionism becomes entrenched. Such a reductionism can become powerful as soon as the researchers working with it come up with astounding new insights which lead to all kinds of successes in theory and praxis. If the research promises new

---

[4] Consensus, coherence, and correspondence concepts of truth thus have to be correlated.

potentials not only to produce further insights but also to enhance political and military power, technological and economic success, or physical welfare and new possibilities of healing, the power of the reductionism increases considerably.

The reductionism now not only gains an impact on scholarly opinions; it makes its way into encyclopaedias and textbooks. It gains an impact on political and economic policies and their readiness to distribute trust and money. It leads to the institutionalization of new academic disciplines, laboratories, and research institutes. It awakens public hopes for better, easier, less endangered, and longer lives, economic expectations of new sources of how to increase industrial income, and political calculations of maximized loyalty and power. By all this it awakens a potentially unlimited willingness to invest money; media attention; personal, public, and political trust; and academic and technological energy into its enterprise. It thus becomes a real power in many areas of social and cultural life.

The enormous success of a reductionistic academic enterprise is not necessarily a danger in itself. As a rule, a successful reductionism in the forms sketched above remains a latent paradigm, and can indeed be stimulating for a while. As soon as its potentials for generating new insights cease, it can be seen as having been a reductionism, or even explicitly proves itself to be stale or boring. It then has to make room for corrected, broader, and more subtle views on the topic it is concerned with—or for other reductionisms. There is, however, the danger that a reductionism will become so powerful that it systematically blocks and distorts other processes of research and potentials of insight.

To name an example, many in the humanities, in philosophy, law, theology, and religious studies are at present afraid of several 'physicalist' anthropological reductionisms that are connected with the recent successes in research on the brain, the genome, and in other fields. But sensitive scholars see the same danger on the 'mentalist' side. Increased sensitivities in the dialogue between the natural sciences and the humanities (including religious studies and theology) to this danger of a powerful reductionism (from either side) have led to a growing dissatisfaction with classical starting points of anthropological research and discourse within social, cultural, psychological, philosophical, and theological studies. Classical starting points in anthropology—such as 'the human being' as a self-referential subject, as a reference point in I–Thou constellations (Torrance 2004), as the typical or the unique member and co-shaper of a moral community or of an environment of sociable interaction (Schleiermacher 1999; Welker 1999b)—have come under scrutiny. Deeper, more realistic, and more subtle forms that might help researchers to grasp the 'koinonial' and embodied nature of human beings are being sought out (Torrance 2004, following John Zizioulas). Anthropologies that only work within a (post-)Cartesian or a (post-)transcendental approach no longer seem able to provide such forms and frameworks.

On the other hand, naturalistic reductionisms became obvious when questions were asked about whether the anthropologies based on brain research or on the 'genetic view' (Brown, Murphy, and Malony 1998: chs. 3, 4, and 9) would be able to integrate diagnostic insights and certainties, and whether they could reach the level of consensus which theology, philosophy, religious, legal, and political studies have enjoyed on the basis of a clearly reductionistic and outdated anthropological model—the model of the free, self-conscious, self-referential subject and its 'dignity'.

A clear perspective on the reductionisms on both sides has been gained through a Theology and Science discourse that has made anthropological research reconsider the relation of God to human beings, revealing that this 'relation' is constituted not merely by God's sustenance of human creatures or by their feeling of 'utter dependence' on God. The divine creativity has to be differentiated into God's (1) sustaining, (2) rescuing and saving, and, finally, (3) elevating and ennobling work on creation. Naturalistic reductionisms—of all kinds—tend to be found on the first level. Anthropologies shaped by social, political, and moral concerns easily come up with reductionisms located on the second level. Typically religious and spiritual perspectives are in danger of creating reductionisms on the third level. Various tensions, both obvious and latent, between religiously, ethically, and scientifically concerned approaches, result from these different concentrations. The attempt in anthropological research to avoid one-sided reductionisms and to appreciate the full spectrum of human existence, as well as the commitment to a critical and self-critical realism which must not focus strictly on one dimension at the expense of the others—all this has united scientists and theologians in fruitful co-operation. No big solutions have been proclaimed, but many small bridges have been built on the way to a more encompassing anthropological paradigm (which is still in the making). At the beginning of the third millenium, this example can demonstrate the challenge to truth-seeking communities in Theology and Science to become aware of the great potentials of their dialogue—but also the patience and modesty required if it is to come up with fruitful results.

REFERENCES AND SUGGESTED READING

ALBRIGHT, C. R., and HAUGEN, J. (1997) (eds.). *Beginning with the End: God, Science and Wolfhart Pannenberg.* Chicago: Open Court.

BARBOUR, I. (1997). *Religion and Science: Historical and Contemporary Issues,* rev. edn. San Francisco: Harper.

BROWN, W., MURPHY, N., and MALONY, N. (1998) (eds.). *Whatever Happened to the Soul? Scientific and Theological Portraits of Human Nature.* Minneapolis: Fortress Press.

CALVIN, J. ([1559] 1957). *Institutes of the Christian Religion,* i, trans. H. Beveridge. Grand Rapids, Mich.: Eerdmans.

COLEMAN, R. (2001). *Competing Truths: Theology and Science as Sibling Rivals.* Harrisburg, Pa.: Trinity International.

GREGERSEN, N., DREES, W., and GÖRMAN, U. (2000) (eds.). *The Human Person in Science and Theology.* Edinburgh: T & T Clark.

—— —— —— and VAN HUYSSTEEN, W. (1998) (eds.). *Rethinking Theology and Science: Six Models for the Current Dialogue.* Grand Rapids, Mich.: Eerdmans.

HELLER, M. (2003). *Creative Tension: Essays on Science and Religion.* Philadelphia and London: Templeton Foundation Press.

JEEVES, M. (2004) (ed.). *From Cells to Souls—and Beyond: Changing Portraits of Human Nature.* Grand Rapids, Mich.: Eerdmans.

PEACOCKE, A. (1993). *Theology for a Scientific Age.* London: SCM Press.

PETERS, T. (1998a) (ed.). *Genetics: Issues of Social Justice.* Cleveland: Pilgrim Press.

—— (1998b) (ed.). *Science and Theology: A New Consonance.* Boulder, Colo.: Westview Press.

—— and BENNET, G. (2003) (eds.). *Bridging Sciences and Religion.* Minneapolis: Fortress Press.

—— RUSSELL, R., and WELKER, M. (2002) (eds.). *Resurrection: Theological and Scientific Assessments,* 2nd edn. Grand Rapids, Mich.: Eerdmans.

POLKINGHORNE, J. (1994). *The Faith of a Physicist: Reflections of a Bottom-Up Thinker,* The Gifford Lectures, 1993–4. Princeton: Princeton University Press.

—— (1998). *Science and Theology: An Introduction.* Minneapolis: Fortress Press.

—— (2000). *Faith, Science and Understanding.* London: SPCK.

—— (2002). *The God of Hope and the Ends of the World.* London: SPCK.

—— (2004). *Science and the Trinity: The Christian Encounter with Reality.* London: SPCK.

—— and WELKER, M. (2000) (eds.). *The End of the World and the Ends of God: Science and Theology on Eschatology.* Harrisburg, Pa.: Trinity Press.

—— —— (2001). *Faith in the Living God.* London: SPCK; Philadelphia: Fortress Press.

RAE, M., REGAN, H., and STENHOUSE, J. (1994) (eds.). *Science and Theology: Questions at the Interface.* Grand Rapids, Mich.: Eerdmans.

RUSSELL, R. J., STOEGER, W., and AYALA, F. (1998) (eds.). *Evolutionary and Molecular Biology,* Scientific Perspectives on Divine Action, 3. Vatican Observatory and Berkeley: CTNS.

—— —— and COYNE, G. (1988) (eds.). *Physics, Philosophy, and Theology: A Common Quest for Understanding.* Vatican Observatory and Berkeley: CTNS.

SCHLEIERMACHER, F. (1960). *Der christliche Glaube,* 7th edn., ed. M. Redeker. Berlin: de Gruyter.

—— (1999). 'Notes on Aristotle: Nichomachean Ethics 8–9', trans. J. Hoffmeyer, *Theology Today,* 56: 164–8.

TORRANCE, A. J. (2004). 'What is a Person?', in Jeeves (2004), 199–222.

WELKER, M. (1999a). *Creation and Reality.* Minneapolis: Fortress Press.

—— (1999b). '"We Live Deeper Than We Think": The Genius of Schleiermacher's Earliest Ethics', *Theology Today,* 56: 169–79.

—— (2000). 'Theology in Public Discourse Outside Communities of Faith?', in Luis E. Lego (ed.), *Religion, Pluralism, and Public Life: Abraham Kuyper's Legacy for the Twenty-first Century,* Grand Rapids, Mich.: Eerdmans, 110–22.

—— (2001). 'Springing Cultural Traps: The Science-and-Theology Discourse on Eschatology and the Common Good', *Theology Today,* 58: 165–76.

—— (2004). 'The Addressee of Divine Sustenance, Rescue, Salvation and Elevation: Towards a Non-Reductive Understanding of Human Personhood', in Jeeves (2004), 223–32.

—— (2006). 'Theological Anthropology Versus Anthropological Reductionism', in R. Kendall Soulen and Linda Woodhead (eds.), *God and Human Dignity,* Grand Rapids, Mich.: Eerdmans, 317–42.

ZUCKERMAN, H. (1996). *Scientific Elite: Nobel Laureates in the United States.* New Brunswick, N.J.: Transaction Publishers.

CHAPTER 33

RELIGION-AND-SCIENCE

PHILIP HEFNER

INTRODUCTION

'Religion and science' has become a cliché, at times scarcely more than an empty cipher, a moniker for a wide range of activities, including any organizational activity, research, and writing that in some way qualify as attempts to relate religion and science. This cliché character masks the fact that bringing religion and science into some kind of concourse, even if it is only linguistic, advances a proposal for our consideration. As a proposal, it is far more than a cliché; rather, it gives rise to serious thinking about how we conduct this interaction between these two powerful movements.

The proposal advanced by religion-and-science is this: If we engage science from a stance within religious experience, we will seek above all to forge links of meaning between the world that science describes and that which is most important to us. If we are actively involved in what has come to be called the 'religion-and-science dialogue' or the 'religion-and-science field', this effort to forge links to what is most important constitutes our primary agenda. Since discerning what is most important is inherently a religious task, religion-and-science in effect becomes a spiritual-religious undertaking, even if one approaches it from the perspective of a scientist; or whether, to recall Ian Barbour's well-known typology (1990), we consider it to be a dialogue, an effort at integration, or a kind of warfare.

At the outset, we recognize that 'religion-and-science' is both an undertaking in its own right and also an object of study, just as both religion and science separately are carried out by practitioners and at the same time are objects for study. The distinction here is between the actual 'doing' and the study of the 'doing'.

Philosophy and the social sciences (including history) study religion as well as science, and they may also study 'religion-and-science'. Such study is meta-analysis that distances itself from the actual doing of religion-and-science; it is in fact the scientific or philosophical study of religion-and-science. This study is important and worthy of attention, but the emphasis here is on the actual conduct of the undertaking we call religion-and-science. Practitioners are planted *within* the domain of religion-and-science; they do not stand at a remove from it. It should be clear, however, that this present essay is meta-analysis, an exercise in philosophical reflection on religion-and-science. I am engaging in thinking about religion-and-science thinking, I am not actually 'doing' it. I refer readers who are familiar with my work to my proposals concerning humans as created co-creators. These proposals are examples of 'doing' religion-and-science.

By way of preliminary comment, it is to be noted that religion-and-science is distinguished from theology-and-science, which receives a separate discussion in these pages. Furthermore, I am regarding the term religion-and-science as a collective noun—not as referring to two separate entities, but as the designation of an attempt that is variously called 'dialogue', 'encounter', 'conflict', 'interface', or 'integrating'. Hence, I use hyphens to connect the words: religion-and-science.

Forging Meaning and the Processes of Discernment

Forging links that relate our lives to the world around us is equivalent to establishing meaningfulness between ourselves and the world. Since, as I describe below, our experience of the world, to a significant degree, is mediated through scientific knowledge, the work of forging links of meaning is itself our most significant engagement with science. The term 'meaning' is complex and ambiguous. In this discussion, I define meaning as the establishing of a link or a 'fit' between what is important to us in our own lives and the world in which we live. The 'world in which we live' includes our own human nature, as well as larger streams of terrestrial and cosmic nature. Meaning is established, for example, if a person can take into account one or more natural processes—say, the evolution of life on planet Earth or big bang cosmology—and come to the conclusion either that those processes are supportive of her or his deepest values or that they are hostile to those values. In the first case, a positive meaning is achieved, in which the person feels as if her or his life is in sync with the world; in the second case, the results are negative, in which the person feels that meaning is to be found in defiance of natural processes. Some form of naturalist philosophy may accompany the first case, whereas an Existentialist or even a supernaturalist philosophy may well correlate with the second case. Naturalistic philosophies have difficulties conceptualizing negation and tragedy, whereas Existentialist and supernaturalist systems of thought are able to do so. However, in actual instances of meaning formation, persons are often naïve about their philosophical assumptions, and when they are aware, many different permutations are possible.

Discernment is not often cited as a constituent of religion-and-science, even though it is fundamental to the forging of meaning. Discernment permeates the practice of religion and spirituality; discernment zeroes in on what is actually reflected as most important in our lived-out lives, and it informs our efforts to be critical of those values and redirect our lives. Since values are in fact embodiments of what we consider most important, spiritual discernment could be designated by the more mundane term 'values clarification'. Scientific knowledge, particularly when coupled with technology and placed in the context of a busy, complex modern civilization, may not alter perennial values—love, family, and justice, for example—but it does radically transform the context in which those values embed themselves and the forms in which they affect our lives. Sometimes, as well, science and technology form a screen or a haze that can veil fundamental values from our awareness. Discernment enables us to take the measure of the ever new forms in which values are to be found, to pierce the veil, and to get a fresh sense of what is important and the shape the 'important' assumes.

Discernment is an embodied process of knowing and judging. It emerges in the depths of our person, in our biology and our neurobiology, as well as in our emotions and our reason. It proceeds in a multilayered fashion, through intuition, trial and error, and thinking; it is embodied in the richness of our ambient world. Discernment is shaped by our culture and our personal relations. Even in retrospect, we cannot know fully how it works or what causes it to come upon discoveries or pass them by. Subliminally, as well as rationally, we sort through what William James called the 'big blooming buzzing confusion' that assaults our senses as we live our lives. We seek to discern the contours of the significant, the 'most important', and sift it out from what is trivial and secondary. On occasion, we do encounter the world at first hand, through naked and immediate sense perception—the smile of a baby, the touch of a loved one, the split second crashing of one car into another, or the pain of a sniper's bullet in battle. But just as often our meeting with the world is mediated by our knowledge. Richard Wilbur describes this encounter in his poem 'In the Field' (1969). A group of young people on an overnight jaunt in the country walks in a field under the starry sky. Their first reaction is to identify the zodiac constellations and recount the myths that accompany them. This is great fun, as it breeds a familiarity with the stars; the sky is a comforting canopy. Then they suddenly recall their school learnings of scientific cosmology. The universe is 15 billion years old and still expanding. The stars are not mythic figures with stories to tell; they are the products of random impersonal processes—and so are the happy wanderers in the field. A chill falls over the group, and they retire for the night. Their experience of the world—clothed in scientific knowledge—has brought them to despair. They encountered the stars as incarnated bits of human knowledge—at first the knowledge conveyed by ancient myth and then by contemporary physics and astronomy. They did not rationalize their comfort or the chill of their depression; they felt it in their very being—Wilbur calls it the 'nip of fear' (Wilbur 1969: 18). They felt the billions of years, the impersonal cosmic laws, and what physicist Steven Weinberg (1993) calls 'the pointlessness' of the cosmos.

Science and scientific knowledge possess an integrity that stands apart from religion-and-science. However, within the discerning process they take on a spiritual dimension, because they are indigenous to our structures of meaning. Scientific knowledge is not an appendage to some essential knowledge of the world and self that we hold in our minds. That knowledge is not an 'accident' that pertains to a deeper, changeless 'substance'. Given the capabilities of our minds and how they work, scientific knowledge is, rather, of a piece with the embodiments in which the world presents itself to our equally embodied intuition and rationality, for our discerning and our valuing. We cannot scratch beneath the surface of cosmological science and find some 'more real' universe in which to live our lives. Nor can we peel off the patina of our neurobiology in order to come upon a 'more real' brain and mind with which to think and feel. The nature of the external world and our own human nature are intrinsically shaped by the science we are taught, from our earliest school years through our adulthood. We do not decide that it will be this way—this is the given territory of our discernment processes.

Consequently, if our world and our lives in the world are charged with meaning, and if they can be conceived as religiously significant, their scientific incarnations must be charged with that same meaning and significance. This is not intended to exalt science, as if to raise it to some dominating position; rather, it is the acknowledgement of the place that scientific knowledge plays in our perception of nature. Let me give an example. The sciences of ecology, primatology, and genetics have provided ever more knowledge that human beings are closely related to the rest of the natural world. Not only are we dependent on the natural environment, we share so much of our genetic and developmental nature with certain other species—chimpanzees, to name one such species—that we can say we are 'kin' to these other animals. The force of this closeness with other animals affects our self-consciousness so powerfully because of the scientific descriptions in which it comes to us. Before the work of primatology, genetic comparison, and mitochondrial DNA studies, our kinship with other animals was at most an intuition tinged with romanticism. The knowledge of this kinship cannot but weigh on our spiritual-religious processes of discernment. What is the meaning of the links between us and the other animals? What does this kinship say about our place in the natural order of the world? What is its import for discerning the purposes of humans? Do we, in the light of this knowledge, have specific responsibilities toward our newly acknowledged kinsfolk? Most religions will bring God into this process of questioning. What does God intend by creating us through the evolution of species? If we are created in God's image, do these kin share in that 'image'? Must God's laws now be interpreted to include our moral responsibilities toward these kin? We owe much of our human capability to our genetic possibilities. If we share so much of our genome with other animals, what does this sharing imply concerning the values and purposes to which we should direct our capabilities? These questions that arise in the processes of spiritual discernment are by no means trivial; they aim directly at what is most important to us—our nature, our purposes, our relationships to transcendence.

In this example, it is clear that we are seeking insight—let us call it 'revelation'—into the meaning of nature, the nature of the world and our own human nature. But this 'nature' whose meaning we seek to fashion is inseparable from the scientific forms under which we engage it. Although genetics, primatology, and ecological science in and of themselves are thoroughly materialist enterprises with no spiritual intentions, they enter of necessity into the domain of discernment and the quest for meaning, and when they do, they become spiritually charged. The process of meaning formation in this example does not arise from a timeless abstract view of 'nature', but rather from the concrete scientifically arrived-at datum that we share our DNA with chimpanzees and with creatures more distant from us in the evolutionary track, like dogs and even earthworms. We might say that elements in the process of spiritual discernment have become 'scientized', not because of theological or philosophical assumptions about science or even our spiritual propensities, but rather because in our current situation when we engage nature, that nature is embodied in scientific forms.

To say that 'religion-and-science' is a spiritual-religious undertaking is to remind ourselves that 'religion-and-science' is a domain of interactive engagement. We do not simply observe religion-and-science or study its components; the grappling with meaning *vis-à-vis* that which is most important to us requires our full engagement, because it is the meaning of our lives that is at stake, not just the mapping of an academic territory.

RELIGION DEFINED AS THE QUEST FOR THE 'MOST IMPORTANT'

We may be clear in our understanding of science, but we face serious ambiguities in our attempts to define religion. I intend no normative definition of religion here; I underscore my thesis that religion-and-science focuses on discerning what is 'most important' and establishing meaning for our lives and our world within the ambience of the scientific knowledge. 'What we consider most important' becomes a defining mark of religion in this context, and from this perspective religion can apply equally to the explicit religious ideas that are rooted in particular religious communities and also to the more general human quest for meaning that avoids religious terminology and may be tied to no particular community. Encompassing both what Paul Tillich called latent and manifest expressions of religion is essential for religion-and-science. God is that which is most important for some, while others, like Buddhists, speak in non-theistic modes. Narrow concepts of religion do not serve us well in considering the human quest for meaning. Tillich is of particular help to us with his insistence that religion is expressed wherever persons relate to that which is most important to them, whether the symbols of religion are overtly present or not; he used the term 'ultimate concern' to designate that which is most important ([1960] 1988*a*). He retrieves a well-established theological tradition. Martin Luther understood that what matters most to a person is in fact that person's god. Such a concept of religion is important for this discussion, because it recognizes the broad range of engagement with science that occurs within the universal human quest for meaning.

'RELIGION-AND-SCIENCE' HAS ITS ROOTS IN PRIMORDIAL HUMAN HISTORY

The effort to link the known world meaningfully to what matters most to us is not a recent thing in human history. Rather, it is a primal human quest that is as old as the human species itself. Anthropologists point out that humans have long devoted enormous attention to understanding the world and their place in it, and also to interpreting the world in terms of what is most important to them. Émile Durkheim emphasized this perennial human quest in his classic work, first published in 1912, *The Elementary Forms of Religious Life* (1912). With his immediate focus on the aborigines of Australia, Durkheim argued that the totemic system was a precursor of modern science in its attempt to order the observed world in a systematic fashion. The totemic classification was, of course, integrated into a religious meaning system. Durkheim believed that in concentrating on the Australian aborigines, he was dealing with the simplest known religion, and thereby uncovering the fundamental laws of all religions. In his view, therefore, it was clear that the quest for discovering order and meaning was primal to human nature, and that this quest goes forward by means of an integral synthesis, what today we call religion-and-science.

Durkheim was a pioneering scientific theorist of religion, and many advances in such theorizing have taken place in the intervening decades since he did his work. Nevertheless, the insight I have detailed here stands the test of time: religion-and-science cannot be separated at a foundational level. Other scholars (e.g. E. B. Tylor [1873] 1958) considered religion to be engaged in 'primitive' explanations of the world. In Durkheim's view, religion also reflected on the meaning of those explanations; the totemic system, for example, not only described, but also provided meaning by relating its descriptions of the world to the life of the community and its individuals. Since the social reality was primary for Durkheim, he considered religion to be a construction of the community; at the same time, he ascribed significance to the social encounter with reality in terms that could suggest a transcendent dimension to the community and its constructions. His insight is comparable to our thesis that if links are to be established between the world and what we consider most important, we must gain knowledge of the world, which is the domain of science, and we must be aware of what is most important or meaningful to us, which is the domain of religion.

THE TERM 'RELIGION'

The term 'religion' in religion-and-science is not a formality. Certain groups and publications have chosen to identify themselves with 'theology' and science. 'Religion' is a troublesome term to theologians for several reasons. For the past three centuries in which the academic scientific study of religion has become a prominent field of university study, the academy has increasingly pushed Christian theology to the periphery, and in not a few cases out of the university altogether. This movement began as early as Giambattista Vico and David Hume in the eighteenth century and continues with the work of Auguste Comte, E. B. Tylor, Sigmund Freud, Émile Durkheim, and their successors up to the present day. Theology was a constituent element of universities in Western Europe and the Americas from the beginning of higher education in those societies. It came under criticism, however, as a sectarian enterprise. As great scholars, many of them in the fields of ethnography and the developing disciplines of sociology and psychology, subjected religion to their methods of study, theologians charged that religion could be understood adequately only from the 'inside', that a religious interpretation of religion is the only adequate approach. This appeared obscurantist, since the university must study all human endeavours if it is to fulfil its mission. Departments of theology became isolated from the larger stream of university studies. In addition, in a period in which we became increasingly aware of the plurality of world religions and in which students engaged in objective study of the religions of the world, it seemed incongruous that Christianity led by its theologians would resist such study of its own religion. To this day, the scientific study of religion and Christian theology are not reconciled, except in a very few instances.

Arguments have been advanced against the idea of 'religion'. Within the theological discipline itself, in the mid-twentieth century, Karl Barth, whose impact on Christian theology was monumental in his own time and continues even today, held that religion is an attempt to domesticate God by means of ritual, organization, and moral code; hence religion is a product of human hubris. Christianity, he insisted, is not a religion; rather, it is a 'faith' (Barth 1936: 314). Religion also has a record of being co-opted by societies, ideologies, and social classes. To some it brings associations of rigid organization and enforced practices; it seems to stifle the intellectual activity that seems appropriate and necessary for engaging science.

Theology, on the other hand, is primarily an intellectual activity; it probes questions and has traditionally challenged religious hierarchy and organization.

Other considerations, however, favoured using the term 'religion' in the engagement with science and scientists. For scientists as well as for both the academic and the general public, religion is a familiar idea, even to the uninformed, whereas theology seems esoteric and sectarian, since it is closely associated in the public's mind with the organizational church. Further, in the current situation of religious pluralism, the Christian association with theology is even more evident. No other religious community calls its thinkers 'theologians' or their body of reflection 'theology'. Jews, Hindus, Buddhists, and Muslims, for example, recognize the term 'thinker', 'philosopher', or 'teacher' when applied to their intellectual reflection, but they do not call themselves theologians. For Buddhists, the term can be offensive, since etymologically theology refers to 'discourse about God'.

Perhaps the very nature of the term 'religion' that gave rise to tensions between theologians and other scholars in the university context—namely, that it must be the object of scientific study—is a reason why it is suitable for the 'religion-and-science' discussion. This fact gives religion-and-science parity in the academic context, and also in the public context where scientists and religious thinkers meet. The term 'religion' gives entrée to the public character of the engagement, which seems not to be the case with 'theology'.

RELIGION-AND-SCIENCE AND THEORETICAL THINKING

Religion is essentially a *practice*; religious thinking must be oriented to that practice. As a consequence, religion-and-science is also a practical undertaking. Theoretical thinking is certainly inseparable from the religious-spiritual quest for meaning. All of the world's major religions have, over the centuries, left a legacy in philosophy of such excellence and significance that it has had a lasting impact on the cultures in which those religions have existed. The religious context requires thinking of the highest order. Nevertheless, within the context of religion-and-science theoretical thinking is not accorded a stand-alone autonomy. Such thinking, be it philosophical or theological, is not carried out for its own sake, but rather for another purpose: namely, to inform the discernment that seeks meaning.

Ironically, even though this role of theoretical thinking in religion-and-science comports well with science, it is not widely accepted in the academic world. To be sure, science does engage in theoretical thinking for its own sake: theorizing is exciting. But the theorist is always monitored by the experimentalist, who challenges theories to make sense of the experimental data and advance knowledge thereby. Even when theory gets ahead of experimental data, data finally determine the usefulness of the theory. Steady-state cosmological theories were superseded not because their proponents were not competent and brilliant, but because big bang theories accounted better for certain data, such as the Hubble measurement of the universe's expansion and the background radiation discovered by Penzias and Wilson. Since data also play an important role in the domain of religion-and-science, establishing relationships of meaning between our scientific knowledge of the world and what we deem most important plays the final hand, so to speak, in monitoring the usefulness of theoretical thinking.

Two examples of interesting theoretical solutions to issues on the interface of religion-and-science are instructive. The first is the so-called Anthropic Principle. This argument focuses on the observation that life on planet Earth is possible only because of a relatively narrow range of physical conditions, some of which are the speed of the universe's expansion after the initial singularity (the big bang), to allow enough time for life to evolve; the distance of the Earth from the Sun, to provide optimum temperatures; the formation of a suitable atmosphere, which in turn requires a delicate balance of constituent chemicals. At each stage of the universe's history the trajectory was timed to pass through the window of opportunity necessary for the emergence of life. From this, proponents argue, we can conclude that the universe is 'fine-tuned' for the emergence of life. Although many scientists in the field find this argument unpersuasive, it has gained a following among some scientists and many theologians. Quite apart from the scientific controversy, the Anthropic Principle proves not to be useful to religion-and-science as a quest for meaning, on several counts. Clearly, it would be of great significance for forging meaning of the world if it could be demonstrated that there is a discernible trajectory in the universe's development that leads directly from the initial singularity to the emergence of human beings on this planet. However, the anthropic argument cannot claim this trajectory. At most, it argues for the formation of life—not intelligent life, and not human life. That simpler forms of life—say, one-celled organisms—are the result of cosmic fine-tuning contributes little to the meaningfulness of human life. Furthermore, while proponents of this argument usually claim that this fine-tuning suggests an intentionality in cosmic evolution, the Anthropic Principle does not touch the so-called theodicy problem; that is, it offers no answer to the question why this imputed cosmic intentionality resulted in human life that is so fragile and vulnerable to inherent defects, with the resultant misery that follows. Since this problem of unmerited evil and ensuing tragedy is one of the major challenges to the forging of meaning, we must conclude that the anthropic argument is not very helpful.

A similar verdict is drawn concerning the related argument for 'intelligent design'. This argument insists that there are phenomena in the natural world that are too complex to be the product of natural processes alone, including evolution. These phenomena must be the result of a design, and presumably a designer. The cilia in the lungs are often mentioned by proponents of this theory as an example of an 'irreducibly complex' entity that must be the product of a designer. Again we meet here an argument that is interesting and intellectually challenging. Like the anthropic argument, however, it fails to address the question of defective design and the evil and tragedy that result. Consequently, in the final analysis we must judge that this theoretical argument is inadequate for the quest for meaning. These arguments in fact pose obstacles to the search for meaning, because they lead us to the conclusion that if there is intentionality in the universe that does not preclude evil and tragedy for humans, we are left with a negative meaning for human life: the cosmic intentionality is indifferent to humans and to what they hold as most important. For those who believe in God, this would be tantamount to concluding that God is either not powerful enough to overcome the defect in divine design or else does not intend to preserve humans from that defect.

We see a different use of theoretical thinking in the so-called kenosis theory. 'Kenosis' is translated from Greek as 'self-emptying', as it appears in the Christian New Testament (Philippians 2: 7), speaking of Jesus Christ, who although he was 'in the form of God' 'emptied (*ekenosa*) himself, taking the form of a slave, being born in human likeness'. This idea of kenosis is elaborated to assert that God is not distanced from the evolutionary processes by which we have emerged, nor from the twists and turns of evolution, with its gains and losses in natural selection, the emergence of novelty and also extinctions. Rather, God enters into the processes and is thus present to all creatures. One might question whether this kenosis theory is adequate to the task that its proponents assign to it, and whether it is elaborated fully enough at the present time; nevertheless, it is clearly supportive of the effort to forge meaning, because of its assertion that we can count on a divine presence in the evolutionary processes. In itself, it does not resolve the theodicy question, but it does not ignore the problems; it brings God into the dialectic between redemption and tragedy.

This kenosis theory is no less theoretical than theories of the Anthropic Principle and intelligent design, but its intention and 'cash value' are practical in a way that is not immediately clear in the case of anthropic and intelligent design thinking. Evolutionary theory is an important scientific contribution to our knowledge of the world, and God is the ultimately important reality. The kenosis theory attempts to bind these two factors together to forge meaning.

RELIGION-AND-SCIENCE AND APOLOGETIC STRATEGIES

The religion-and-science field includes, from time to time, argumentation that seeks to defend statements that grow out of specific attempts to forge meaning. The Anthropic Principle can be employed, for example, to defend the belief in a creator God. This apologetic strategy is almost always polemical in its style and its intentions, and it is frequently a polemic waged against materialist reductionism. Such polemic has its own reward, but it is not really consistent with religion-and-science as a religious-spiritual undertaking, and it is not very useful. Spiritual discernment, the heart of religion-and-science, is not a polemical activity. Rather, it is wholly constructive in its search for what is most important and in its struggle to forge links of meaning. Spiritual discernment is undertaken in humility; it is well aware of its uncertainties and the provisional nature of the conclusions it reaches. Discernment is always open to correction, and it is always testing its insights, laying aside inadequate conclusions, and searching for better ones. Spiritual discernment is not the sort of thing that one argues about and defends—not because it is not fallible, subjective, and open to error, but because it is rooted in very personal interior dynamics that are tied to individual and communal experience. This experience includes critical thinking, based on study and learning, but it is finally determined by the authenticity of the experience of the persons who generate the discerning processes. If a person comes to the conviction that the love of God is present in the cosmic processes of evolution, for example, and finds that a symbolic concept of divine kenosis supports the conviction, then that person will be eager to explore whether the idea of kenosis is properly understood and applied; but the conviction itself will not become a matter of polemic, any more than the conviction, say, that one ought to show kindness in personal relationships. This conviction will not be used to attack other positions; nor is it likely to assume a defensive posture against criticisms.

The fact that authenticity of experience plays such a critical role in forging meaning and defining what is most important throws light on two important aspects of proposals for meaning that emerge from the discernment process of religion-and-science: their status and the importance of community.

The Status of Religion-and-Science Proposals

The objective component of religion-and-science proposals may be instructive, and even persuasive, but because of the subjective component, the conclusions may be received with ambivalence. The experiential factor that is inherent in religion-and-science brings with it a variety of possibilities that in the final analysis defy our comprehension. Even when the array of knowledge is formidable and the analysis impeccable, the proposal will be persuasive only within the bounds of the individual's or community's canons of what is meaningful.

All proposals for meaning share this rootedness in the authenticity of experience, but it is especially prominent in religion. When we recognize this factor, we see that we must acknowledge a fundamental and inescapable plurality, both of discernment and of conviction. It is here that religion-and-science differs from science, and even from most philosophy and much of theology. In their public proposals science and philosophy distance themselves from the experience of the proposer. The value of a scientific hypothesis lies precisely in the fact that it transcends the experience of the proposer(s). Philosophy shares this intention, even if it cannot match the distance and replicability of scientific proposals. For the most part philosophy speaks from the position of 'meta-language', in ways that religion-and-science speech does not. Proposals from religion-and-science possess a confessional character that is muted in science and philosophy. To be sure, contemporary philosophical discussions underscore the social and psychological 'location' of our intellectual work; it all bears a confessional dimension. This is accentuated in religion-and-science, because the proposals grow out of the explicit and poignant effort to establish the meaning of life. The concern for what is most important reinforces the confessional dimension of religion-and-science. Particularly in the United States, we find it difficult to reach a consensus in society even on penultimate values that pertain to survival; when the discussion is intensified by introducing a concern for ultimate values, that which is most important, consensus is impossible to reach—we must settle for a plurality of perspectives that will be reflected in the discussion of how science relates to the meaning of the world. In considering these factors, I have proposed (Hefner 1993: 25–7) that the criterion most applicable to proposals emanating from religion-and-science is fruitfulness.

The philosopher of science Imre Lakatos (1978) established fruitfulness as a criterion in the context of his concept of scientific 'research programmes'. Fruitfulness comes into play in his suggestion that while a research programme cannot be verified or falsified, it can be judged as to whether it is progressive or degenerating, i.e. whether it possesses or lacks the capacity to stimulate new insights. A progressive research programme, according to Lakatos, throws light on 'new facts' and 'dramatic, stunning, and unexpected' interpretations of the world. Such a progressive programme is to be preferred over a programme that no longer engenders novel facts and interpretations, and hence may be considered to be degenerating.

Lakatos in effect circumvented the potential epistemological and metaphysical impasse created by the confrontation between philosophical realist positions that insist on some form of referentiality between our proposals and 'something real', on the one hand, and post-modern and post-foundationalist positions, that put the emphasis on our 'constructing' the real, on the other hand.

When the proposals of religion-and-science are understood as explorations of the meaning of the world in the light of scientific knowledge, the appropriateness of this concept of fruitfulness as criterion is underscored. In a scientific perspective the meaning of the world cannot be static or timeless, and our explorations share this ever enlarging dynamic. Under the rubric of fruitfulness, religion-and-science is a future-oriented effort, pushing the frontier of our sensibility and discernment and putting its world together in dramatic and unexpected ways. Religion-and-science is a restless probing, which is at the same time unsettling and critical. Its proposals deconstruct established meanings—always enlarging received traditions and often overturning them. This is the way of discernment, which is at pains not only to retrieve the past, but even more, to envision the future. Neither religion nor science countenances indifference to the future. The religions of the world, particularly

Judaism, Christianity, and Islam, focus on what the world can become; the sciences assume a world that is woven on the loom of evolution and emergence. Accordingly, the criterion of adequacy for any proposal cannot be some idea of full and unchanging 'truth'. Rather, we look for fresh interpretations that prove reliable for exploring the world that confronts us in ever new ways.

Applying the criterion of fruitfulness in religion-and-science proposals reaches beyond the scientific use, however, in that the process of discernment in the former is more holistic, extending fruitfulness beyond the causalities of the empirical world into personal and moral dimensions that are inherent in the search for meaning. Thus, in religion-and-science fruitful proposals speak not only of personal authenticity, but also of possibilities that are larger than the experience of the individual proposer. Fruitfulness does not claim to be right or final—perhaps that is why theologians frequently do not hold it in high regard as a criterion. Scientists, on the other hand, do work with this criterion. As an expression of epistemological modesty, this criterion sets limits on our expectations of 'proof' or 'verification', suggesting that the value of a proposal rests, rather, on the possibility that others may find it useful for further thinking. At the same time, offering its proposals as fruitful hypotheses serves as invitation to others to share in exploring future possibilities.

The Role of Community in Religion-and-Science

Precisely because religion-and-science is so significantly confessional and relies on spiritual discernment, it requires a community in which to work. Science also requires a community, since it relies so heavily on peer group evaluation of its proposals. Religion-and-science needs community for a different reason, though. Its proposals are rendered vulnerable to distortion and obscurantism by the very factors that constitute its *raison d'être*: discernment and authenticity of experience. Religion-and-science cannot proceed with confidence unless it can be resourced by the imagination and critical interaction of a community. This situation poses a difficulty, since the two primary communities available to religion-and-science are the religious community and the academic community. Academia fosters the critical thinking that is essential, but it is ambivalent about religion-and-science, because discernment, authenticity of experience, and confessional thinking are not its criteria of success and advancement. Academia prizes the distanciation that is only one part of the religion-and-science enterprise. In light of these considerations, there is good reason to doubt that religion-and-science will ever become an academic 'field' in any conventional sense of that term. Religious communities welcome processes of spiritual discernment and the confessional stance, but by and large they consider distanciation to be a defect. Furthermore, they do not naturally consider scientific knowledge to be an important factor in the processes of discernment. Religion-and-science is forced, therefore, to do its work in a boundary position, dependent on both academic and religious communities, but fully at home in neither.

Consequently, religion-and-science must form its own community by drawing together members from diverse intellectual, religious, and professional regions, even though doing so renders it even more unconventional to both academic and religious communities. In the long run, this position on the boundary is a great advantage and resource to religion-and-science, not only because it ensures a richer set of spiritual and intellectual resources, as well as distinctive perspectives, but also because it serves as a kind of 'reality test' for religion-and-science that will never allow it to forget the fragility and unconventionality of its enterprise.

RELIGION-AND-SCIENCE REQUIRES A PRACTICAL, MORAL EXPRESSION

Because religion is a set of embodied practices, borne by embodied communities, moral behaviour is inherent to its nature. Religions and their communities are embedded in culture and its social expressions. Religion's success throughout human history in surviving the selective pressures of evolution—both biological and cultural—demonstrates that it is an element of culture. Its evolutionary success suggests, furthermore, that religion has performed a service to culture. Indeed, scientific studies of religion frequently make this point, most often emphasizing that religion is co-opted by culture for its ends. However true this judgement may be, the intentionality of religion itself aims at something different: promoting the common good. Moral behaviour is an inherent mark of religion, as is often expressed in religious texts. This behaviour is intended to benefit both individuals and the larger group; its motivations may be described in terms of obedience to divine command or enhancement of personal holiness, but the religious community clearly believes that it will serve the good of all, whatever terminology is used to express its motivation.

In light of this basic character of religion, it is unthinkable that religion would not engender moral behaviour. As it forges links of meaning between what is most important and the scientifically understood world, religion-and-science will concern itself with the common good. The meaning it achieves is not exclusively cerebral, and its consequences are more than personal satisfaction. Meaning is discerned and expressed through reaching out into the world in which we live and contributing to it. Religion's search to discern what is most important and science's struggle to depict the world adequately come together in religion-and-science in a focus on knowledge that benefits the human community. Action in the world is an expression of commitment to what is held to be most important, and it is also an articulation of what is meaningful to us. Religion-and-science is neither complete nor faithful to its own nature as a practical discipline if it does not include reflection on the common good and what behaviours are required to maintain it.

The view that I elaborate here holds that religion-and-science is more than either religion alone, as it is commonly understood, or science alone, as it is commonly practised. In this respect, religion-and-science is a *tertium quid*; it is a novel emergent in our time, expressing the human struggle for the meaning of the world in relation to that which is most important. For some that meaning is a discovery or a revelation that is grounded in transcendence, while for others it is a human construction of meaning. Religion-and-science privileges science, in that it believes that the world presents itself to us in terms of embodied scientific understandings. At the same time, religion-and-science is itself religious in nature, in that it is always oriented on what we deem most important (in Tillich's term, *ultimate concern*). This essential work of religion-and-science includes different types of activity, some of them preliminary to the essential task—gathering knowledge, analysing concepts, arguing the validity of theories—but in the final analysis the essential work of forging meaning is what is at stake. When all is said and done, the success or failure of the religion-and-science undertaking is determined by how well it expresses the meaning of the world, scientifically understood, and relates it to what is most important for the long-term future.

References and Suggested Reading

Barbour, Ian G. (1990). *Religion in an Age of Science*. San Francisco: Harper & Row.

Barth, Karl (1936). *Church Dogmatics*, I, 2. Edinburgh: T & T Clark.

Durkheim, Emile ([1912] 1995). *The Elementary Forms of Religious Life*, trans. Karen Fields. New York: Free Press.

Hefner, Philip (1993). *The Human Factor: Evolution, Culture, Religion*. Minneapolis: Fortress Press.

Lakatos, Imre (1978). *The Methodology of Scientific Research Programmes*. Cambridge: Cambridge University Press.

Tillich, Paul ([1960] 1988a). 'The Relationship Today between Religion and Science', in J. Mark Thomas (ed.), *The Spiritual Situation in Our Technical Society*, Macon, Ga.: Mercer University Press, 151–8.

—— ([1960] 1988b). 'Religion, Science, and Philosophy', in J. Mark Thomas (ed.), *The Spiritual Situation in Our Technical Society*, Macon, Ga.: Mercer University Press, 159–72.

Tylor, E. B. ([1873] 1958). *Religion in Primitive Culture*. New York: Harper & Brothers.

Weinberg, Steven (1993). *The First Three Minutes: A Modern View of the Origin of the Universe*, 2nd edn. New York: Basic Books.

Wilbur, Richard (1969). 'In the Field', in *Walking to Sleep: New Poems and Translations*, New York: Harcourt Brace Jovanovich, 18–21.

SCIENCE, THEOLOGY, AND DIVINE ACTION

CHAPTER 34

QUANTUM PHYSICS AND THE THEOLOGY OF NON-INTERVENTIONIST OBJECTIVE DIVINE ACTION

ROBERT JOHN RUSSELL

INTRODUCTION TO NIODA

Since the rise of modern science in the seventeenth century, Christians have hungered for an intellectually and spiritually satisfying view of non-interventionist objective divine action (NIODA) that can support a theology of special providence located between general providence, on the one hand, and miracle, on the other. Such a view must take the natural sciences seriously and work with an incompatibilist view of human, and thus by rough analogy divine, freedom to act in nature as well as history. Can we at last overcome the 'forced option' between conservative preferences for objective divine action that is interventionist and liberal tendencies to prefer non-interventionist but merely subjective divine action? (Murphy 1997: esp. ch. 5).

The task of this chapter is to evaluate a specific proposal for NIODA based on an indeterministic interpretation of quantum mechanics (QM). To begin, we need working definitions of the terms involved and the assumptions brought to the conversation. We can then stipulate the criteria which *any* potential candidate for NIODA must satisfy. Finally, before specifically turning to QM-NIODA, I will respond briefly to six frequent misconceptions (FAQs) regarding NIODA that tend to obscure the conversation and delay progress in assessing serious candidates for NIODA.

Terms and Assumptions

(1) *Laws of nature*   By 'laws of nature' I mean the regularities of natural processes as subsumed into scientific theories, most often through mathematical formulation. Examples would be Newton's law of gravity and the Dirac equation in relativistic quantum mechanics. Some scholars view the laws of nature *philosophically* as having an ontological status, typically in the sense of Platonic realism. They are then often said to 'govern' natural processes prescriptively, and such processes are said to 'obey' the laws of nature. Others view them as descriptions of simple natural regularities or, perhaps underlying these regularities, the causal efficacy inherent in nature. In either case, from a *theological* perspective the laws of nature and the kinds of causal efficacies they represent are due ultimately to God's faithful and trustworthy action in creating the world *ex nihilo*, both as a whole and at each moment, and in giving the world its natural regularities as described by these laws. I tend to view the laws of nature in the latter, descriptive sense, although I sometimes say that nature 'obeys' them because this encourages me to take seriously the consequences of scientific theories even if these mitigate against my theological position.

(2) *Ontological indeterminism*   By 'ontological indeterminism' I mean that nature does not always provide a sufficient efficient cause for a specific effect. The decision to regard nature as (in)deterministic is a philosophical interpretation based on the best-known scientific theories and the laws they incorporate.

In classical physics, the fundamental laws were deterministic and implied, philosophically, that nature itself is deterministic, a closed causal system of forces rigidly determining the motion of matter. This mechanistic view of nature challenged human free will and divine action. Statistical laws, as in the kinetic theory of gases, were used merely for practical purposes because the underlying forces and the relevant boundary and initial conditions were too complicated to make explicit calculations possible. Accordingly, chance is really 'epistemic ignorance' and points to our lack of detailed knowledge of the underlying causes. It includes chance events affecting a single trajectory, such as tossing a coin, and chance events consisting of the random juxtaposition of two trajectories, such as a car crash (Peacocke 1998: 360–4).

Twentieth-century natural science opens the possibility of interpreting chance as a sign of ontological indeterminism in nature. Scholars in theology and science have made powerful cases that various fields, including cosmology, thermodynamics, chaos theory, the neurosciences, and quantum mechanics, do indeed point to ontological indeterminism. If this is correct, it would mean that the presence of statistics in these fields arises not from our ignorance of the underlying deterministic forces but from the fact that there are, in reality, no sufficient underlying forces or causes that fully determine particular physical processes, events, or outcomes. Chance as indicative of 'ontological indeterminism' is radically different from chance as mere 'epistemic ignorance'. NIODA is a search for scientific theories that support ontological indeterminism.

(3) *Objective versus subjective acts of God*   Conservatives stress the possibility of objective acts of God. Put in counterfactual terms, events are considered the result of an 'objective act of God' if they would not have occurred in precisely the way they did had God not acted in a distinctive or special way in bringing them about. Conversely, liberals believe that God acts uniformly in all events, even though some may be viewed as 'subjectively' special when the religious believer attributes to them specific revelatory meaning or distinctive divine agency (Tracy 1995: 294–6).

(4) *'Direct' or 'indirect' acts*   The distinction between *direct* and *indirect* acts comes from the philosophy of action regarding human agency. By a 'direct act' or a 'basic act' I mean an act which an agent accomplishes without having to perform any prior act. By an 'indirect act' I mean an act which an agent eventually accomplishes by setting into motion a sequence of events stemming from a direct act which the agent performs.

In turning to divine action I will use the distinction between direct and indirect acts analogously, recognizing the severe apophatic limitations on any such analogy. An objective act of God may be either a direct act of God or an indirect act resulting from God's direct act elsewhere in nature. Every event in the universe, including (but not limited to) the absolute beginning of the universe at '$t = 0$' (if it had such a beginning), is a direct act in the sense of its sheer existence—that is, of its being created *ex nihilo*. Each event in nature exists *per se* because it is created directly by God. That is, God doesn't create event A by acting through event B. To exist is to exist by the direction action of being created or held in being by God.

(5) *Mediated and immediate divine action*   By 'mediated', I mean that God acts in, with, and through the existing processes of nature without thereby becoming a secondary, or natural, cause. By 'unmediated' or 'immediate' I mean God's action of creation *ex nihilo* which accounts for the ontological existence of the world as a whole and at every moment of time. The act through which every event in nature exists *per se* is an unmediated divine act; it is not mediated by anything that exists prior to it in time. To underscore this we can say that God's direct act of creating an event is not even mediated by the event itself, since its existence *per se* is ontologically prior to its capabilities as an existing event to mediate God's action of creation *ex nihilo*. In short, every event, in that it exists, is the direct result of the unmediated creative act of God *ex nihilo*. At the same time the character of all events in the world excepting $t = 0$ (if there is such an event) is also the result of God's mediated action—that is, God's action

mediated in, with, together, and through prior events, and this action is mediated through the secondary causal processes of nature.

Note that, by combining (4) and (5) we can delineate the following possibilities. Events may be considered as the result of God's immediate and direct action (i.e. the event $t = 0$, where the existence and the nature of the event are the direct result of God's act), as embodying God's mediated and direct action (where the existence of the event is God's direct, immediate act, but its character is mediated by the nature of the event), and as representing God's mediated and indirect action (where the character of the event is also mediated by God's previous action via the processes of nature). Note that in this scheme it would make no sense to talk about God's immediate and indirect acts; all indirect divine acts are mediated (i.e., even if God acts indirectly through the mediation of natural processes, the existence *per se* of every event is never bestowed by God through these processes; it is God's direct gift to each and every event).

Note too that we simply cannot answer the question of 'how' God acts, and these comments are definitely not meant to be understood in that way. God's causality is radically different from any of the kinds of causality we know about, just as God's nature as necessary being is ontologically different from ours as contingent being. An 'apophatic epistemic aura' surrounds our entire thinking about divine action, and must not be forgotten, lest we seem to be 'explaining' it or 'answering' the 'causal joint' problem, etc. Because divine agency is radically different from natural agency, it would probably be more circumspect everywhere in the preceding comments to refer to an event as 'the locus of the effect of God's action' rather than as the effect of God's action.

(6) 'Top-down', 'whole–part', 'lateral', and 'bottom-up' causality, and their combinations   Proposals for NIODA take several forms. 'Top-down' causality refers to God's action at a higher epistemic and phenomenological level than the level of the effects (e.g. the 'mind/brain' problem). 'Whole–part' causality or constraint refers to the way the boundary of a system affects the specific state of the system. 'Lateral' refers to effects lying in the same epistemic level (e.g. physics, biology, etc.) as their causes, but greatly amplified by the long causal chain (e.g. chaos theory and the 'butterfly' effect) which produces them. 'Bottom-up' causality refers to the way in which the lower levels affect higher, more complex levels (e.g. quantum mechanics). Most scholars want to combine all four types of causality when it comes to human agency in the world and to God's action in human life and history. The challenge, however, is to conceive of God as acting in the processes of biological evolution or physical cosmology long before the arrival of any kind of complex biological organism (let along humanity). Here bottom-up causality may be the only approach available.

(7) (In)compatibilist views of God's action in relation to nature as (non)-intervertionist   The term '(in)compatibilism' arises in the philosophy of mind and concerns the problem of free will. Roughly speaking, an incompatibilist asserts that human freedom requires physical indeterminism; a compatibilist (such as Kant) asserts that human freedom is consistent with physical determinism.

These terms can be extended provisionally, and by analogy, to the problem of God's action in nature. Both compatibilists and incompatibilists usually agree that the laws of nature ultimately describe God's regular action working in, with, and through natural processes. For a compatibilist, divine action is consistent with a deterministic world, since what God does in bringing about special events is exactly what God does in bringing about ordinary events, and nothing more. What we consider to be special events are only subjectively special.

Conversely, for an incompatibilist, objective divine action is inconsistent with a deterministic world, since what God does in bringing about special events is more than what God does in bringing about ordinary events. What we consider to be special divine acts are objectively special. It is crucial to note that a compatibilist will label God's special objective action 'interventionist', whether or not the world is deterministic, because God's special objective action goes beyond what the laws of nature describe, *whether or not God's action contradicts God's ordinary action in and with nature.* Contrary to this, an incompatibilist will view God's special objective action as 'interventionist' *only if the world is deterministic.* If the world is indeterministic, then God's special objective action is *non-interventionist* when it brings about events which go beyond those described by the laws of nature without contravening or disproving them, because natural efficient causality, as described by these laws, is created by God *ex nihilo,* to be insufficient to bring these particular events about.

Criteria for a Successful Proposal for NIODA

I am now prepared to state the criteria for deciding whether I have a successful proposal for NIODA:

For non-interventionist objective divine action to be intelligible in light of science from an *incompatibilist* perspective, the events that result from God's action must occur within a domain of nature in which the appropriate scientific theory can be interpreted philosophically in terms of *ontological indeterminism.* The events themselves must be considered as *direct, mediated,* and *objective acts of God.*

'FAQs': Responses to Six Frequently Asked Questions (Typically Misconceptions) about NIODA

A number of misconceptions about NIODA frequently arise, which tend to prevent or detract from a substantive assessment of the proposals. My hope is to minimize any future confusion over just what the NIODA project is and is not about, so that serious discussion of specific proposals can be more efficient and constructive.

(1) *NIODA is not 'physico-theology', nor is it meant to prove that or 'explain' how God acts in nature* NIODA is not a form of natural theology, or physico-theology, and

is most certainly not an argument from design. Instead, it is part of a general constructive theology pursued in the tradition of *fides quaerens intellectum*, whose warrant and justification lie elsewhere, such as in Scripture, reason, and experience, and which incorporates the results of science and the concerns for nature into its broader framework mediated by philosophy. Science should not include reference to God's action in nature as part of its explanation of the world. Theology, however, in *its* explanation of the world should do so. This is as it should be for the mutual integrity of, and distinction between, the two fields of inquiry, and for the order of containment entailed by emergence views of epistemology which requires that theology include and be constrained by, while irreducibly transcending, science.

(2) *NIODA is not a gaps argument in either the epistemic or the ontological sense of gaps*   Let us consider the two senses separately.

Type I: Epistemic gaps. An *epistemic* gaps argument is based on what we don't know about the world, and invokes God to explain it. But many gaps in our current understanding of nature will eventually be filled by new discoveries or changing paradigms in science. We ought not to stake our theological ground on transitory scientific puzzles: candidates for a successful NIODA must *not* be based on epistemic gaps. Instead, they must be based on what is *known* by one branch of science within a reasonable interpretation of it: namely, ontological indeterminism.

Type II: Ontological gaps. An *ontological* gaps argument assumes that natural processes are ontologically deterministic; God must create gaps in order to act in nature. An ontological gaps argument is therefore *interventionist*. The problem with interventionism is that it suggests that God is normally absent from the web of natural processes, acting only in the gaps that God causes. Furthermore, since God's intervention breaks the very processes of nature which God created and constantly maintains, it pits God's special acts against God's regular action, which underlies and ultimately causes nature's regularities. Finally, it undermines the very integrity and autonomy of science which the 'theology and science' interaction seek to uphold because it implies that God's action is equivalent to a natural, or secondary, cause which science on its own should include (e.g. intelligent design). I agree that we *should* avoid an ontological gaps argument when these gaps are viewed as disruptions of nature by God's intervention. Instead, a successful approach to NIODA must claim that the processes of nature are created by God *ex nihilo* with intrinsic, naturally occurring gaps.

(3) *NIODA is not undermined by the fact that scientific theories can be given multiple and mutually contradictory interpretations*   I agree that multiple interpretability is a real problem for NIODA, but this is not particularly surprising or unavoidable, since multiple interpretability is a real problem for any theology seeking to engage with scientific theories. In short, every scientific theory is multiply interpretable! Clearly we cannot avoid the reality of multiple interpretability. What each scholar must do instead is to build a response to it directly into her or his methodology for relating theology and science. In my view, the best response is to take a 'what if' stance to this problem: be rigorously clear in acknowledging the multiple interpretability of a given theory, in choosing one particular interpretation, and in stressing that this approach to NIODA is hypothetical and tentative. With this stated up front, one can proceed to

be as clear as possible about what this interpretation would tell us about the world *if it were true, which it might in fact be.*

(4) *God's action is not reduced to a natural cause*   NIODA does *not* reduce God to a natural cause because, according to the philosophical interpretation of the candidate theory in science, there are no efficient natural causes for the specific events in question.

(5) *God's action is hidden from science*   NIODA is entirely consistent with the basis of science in methodological naturalism. Neither theology nor science will view 'God' as a proper part of a scientific explanation of the world, such as when an explicit reference to God is hidden behind the rhetoric of 'intelligent design'. More sharply, God's direct action according to NIODA will be hidden in principle from science, because, according to ontological indeterminism, there is no natural cause for each event in question for science to discover.

(6) *NIODA is not meant to address 'miracles'*   Objectively special divine acts support and fulfil the meaning of God's general acts that provide for the regularities of nature even as they go beyond their meaning in surprising and novel ways. Still, they are not 'miracles' in the Humean sense: they are not interventions by God which suspend the ordinary regularities of nature or violate the laws of nature that we construct to describe these regularities. Nor are they what theologians for millennia have meant by miracles, namely the nature miracles, the healing miracles, and the central threefold miracle of the Incarnation, Resurrection, and Ascension of Christ—events which involve the transformation of nature as a whole and with it the transformation of the laws of nature (Russell 2002).

QM-NIODA: The Proposal and its Assessment

A variety of proposals have been explored by diverse scholars in recent years. Many creative results can be found in the series of collaborative publications by the Vatican Observatory (VO) and the Center for Theology and the Natural Sciences (CTNS).[1] In my opinion the most promising approach is to base NIODA on quantum mechanics, with the specific philosophical interpretation that nature's ontology at the subatomic level is at least partially indeterministic. This approach has roots in the early 1950s with scholars such as Karl Heim (1953), E. L. Mascall (1956), and William Pollard (1958), and includes very recent work by George Ellis, Nancey Murphy, and Thomas Tracy.

---

[1] See the first five entries in the Bibliography under Russell *et al.* with the indication 'CTNS/VO Series'. For convenience, these volumes are referred to in the remainder of the Bibliography by the initials given here.

Quantum mechanics, c.1900–30, describes the behaviour of atomic and subatomic particles with extraordinary accuracy. It is a foundational theory in contemporary physics, which, when combined with special relativity, leads to quantum field theory and, eventually, to the gamut of current supersymmetry and string theories. Yet QM can be interpreted in a variety of competing and conflicting ways which date back to its formation and which remain highly debated today (Herbert 1985; Shimony 2001).[2] Here I will adopt the view first championed by Werner Heisenberg as a form of the Copenhagen interpretation: namely, that quantum mechanics depicts nature as ontologically indeterministic. By 'ontologically indeterministic' I mean, again, that nature provides the necessary but not the sufficient causes for quantum events to occur. While the Schrödinger equation applies deterministically to the propagation of the wave function and includes efficient causes in the form of potential energies (representing forces at work in nature), during a quantum event, or 'collapse of the wave function', the Schrödinger equation does not apply, and there is no efficient natural cause that brings about this event. It is this interpretation which forms a promising basis for what I will call 'QM-NIODA'. My central thesis is that God acts objectively and directly in and through (mediated by) quantum events to actualize one of several potential outcomes; in short, the collapse of the wave function occurs because of divine and natural causality working together even while God's action remains ontologically different from natural agency.

A variety of theological issues now emerge in the relationship between divine action and the Heisenberg interpretation of quantum physics. I will separate them into general issues and crucial issues.

General Theological Issues

(1) *How QM-NIODA responds to six key FAQs regarding NIODA proposals*

(a) *Is QM-NIODA an epistemic or an ontological gaps argument?* No! QM-NIODA is *not* an epistemic gaps argument; instead, it relies on what we *do* know about nature, assuming that quantum physics is the correct theory and ontological indeterminism its correct interpretation. Therefore, it is *not* an ontological gaps argument; it does *not* require God to 'break into' the causally closed processes in nature. Instead, God has created the universe *ex nihilo* such that some natural processes at the quantum level are insufficiently determined by prior natural events. Because nature is indeterministic, God acts as continuous creator together with nature, which supplies the material and formal causes, to bring about quantum events. In such a non-interventionist account of divine action, we are relating God's action in the world to our knowledge of the world based on quantum mechanics, not to our ignorance about the world. With it objective special

---

[2] In this brief chapter I will not treat such critical issues as non-locality, entanglement, challenges to realism, etc. See Russell (2001).

providence is achieved without contradicting general providence, since God's particular acts, being non-interventionist, do not violate or suspend God's ordinary action.

(b) *Is God's action at the quantum level in effect a natural cause?* No! QM-NIODA does *not* reduce God to a natural cause, because, according to the philosophical interpretation of quantum mechanics deployed here, there are no efficient natural causes for a specific quantum event. If God acts together with nature to produce the event in which a radioactive nucleus decays, God is not acting as a natural, efficient cause.

(c) *Is God's action at the quantum level hidden from science?* Here my response is Yes! for several reasons. First, as stated above, all proposals for NIODA are entirely consistent with methodological naturalism. Second, God's direct action at the quantum level will be hidden in principle from science because, given this philosophical interpretation of QM, there is no natural cause for each specific quantum event for science to discover. Third, and alternatively, God's action will remain hidden from science because it will take the form of realizing one of several potentials in the quantum system, not of manipulating subatomic particles as a quasi-physical force.

(d) *Can even God know the outcome of a quantum process given the underlying ontological indeterminism?* Here my response is Yes!, contrary to Peacocke, who rejected the relevance of quantum indeterminacy to the problem of divine action. According to Peacocke, even God cannot know which potential state will become actual during a quantum event, and therefore what the future trajectory will be for that state, because of quantum indeterminism (Peacocke 1995: 279–81). But I am claiming that God acts together with nature to determine which quantum outcome becomes actual; God can know which potential state will become actual, since *God causes it to become actual!* In essence, quantum indeterminism is the result of it being God, not nature, which determines the outcome (Russell 2001: 314; see also Ellis 1999: 471–2). Moreover, once the system is in a definite state, its future state is predictable to God, because it is determined by the Schrödinger equation (although it would be better to claim that God 'knows' the future state in its own present than 'foreknows' it by predicting it from the present).

(e) *Does QM make divine action 'episodic'?* Contrary to Polkinghorne, I do not believe that QM-NIODA makes divine action 'episodic' (Polkinghorne 1995: 1523; 2001: 186–90). Polkinghorne has actually given several arguments against QM-NIODA. (i) The first focuses on chaos theory, where small changes in the initial conditions are amplified rapidly into large changes as the system develops in time. Some have speculated that quantum physics may be the ultimate source of these initial changes, but to move forward, we need an explicit theory referred to as 'quantum chaology' that unites quantum mechanics and chaos theory and thus accounts for how quantum indeterminacies are amplified by chaotic processes. But the search for such a theory has floundered so far on a host of technical problems. According to Polkinghorne, we must first solve these problems before treating QM as a reliable basis for NIODA. (ii) Quantum physics is subject to competing interpretations, including deterministic ones. It is unwise to base NIODA on a specific interpretation of QM, since this interpretation may turn out to be invalid. (iii) Then there's the measurement problem: How can a piece of apparatus yield exact measurements on a quantum system if it is composed of elementary particles obeying the indeterminacy principle? (iv) Finally, divine action related to quantum mechanics would be 'episodic', because the indeterminacies in quantum behaviour arise only in 'those (occasional) events which qualify, by the irreversible registration of their effects in the macro-world, to be described as measurements'. Such episodic divine action is far too limited an account in light of the general, continuous, multilayered character of God's action in the world.

With respect to (i), I disagree with Polkinghorne about the relevance of quantum chaology to QM-NIODA. There are numerous ways in which quantum processes both underlie and give rise to specific effects in the classical, macroscopic world that do not depend on chaos to amplify them (see below). Quantum chaology is clearly a problem for Polkinghorne's chaos-based approach to NIODA, but it is not a problem for QM-NIODA. (ii) The problem of multiple interpretability was addressed above. Finally, point (iii) is the standard criticism of Bohr's 'two worlds solution' to the measurement problem. It is not a criticism of the claim being made here which is based, instead, on Heisenberg's advocacy of ontological indeterminism.

(iv) What about the charge that the measurement problem makes God's actions 'episodic'? First, Polkinghorne asserts that indeterminacies in quantum behaviour arise *only* when an irreversible registration of their effects occurs in the macro-world. I have argued instead that they occur not only at the micro-macro level of irreversible interactions but also in irreversible interactions at the micro-meso and the micro-micro levels. This leaves his second claim that measurements occur only occasionally. In fact, however, they can occur at any time and place in the universe when the conditions are right for micro-micro, micro-meso, as well as micro-macro, irreversible interactions. This suggests a God who is *acting providentially everywhere and at all times* in and through all of nature—a God whose agency is hardly 'episodic' (Russell 1998: 211–12; 2001: 310).

(f) *Why the Saunders/Wildman 'tetralemma' argument against QM-NIODA fails in principle* Nicolas Saunders has offered a lengthy criticism of the special divine action project in general ('SDA') and the QM-NIODA project in particular ('QSDA'). He concludes that the case being made for 'the "traditional understanding" of God's activity in the world (is) extremely bleak ... (and that) *contemporary theology is in crisis*'—a judgement which has been quoted frequently (Saunders 2002: esp. chs. 5, 6). A detailed rebuttal to Saunders would require much more space than is available here. Wesley Wildman, however, has produced a careful, and I believe fair, summary of Saunders's arguments (Wildman 2004), and I shall use it here to challenge both Saunders and Wildman (see n. 3 below).

According to Wildman, the argument advanced by Saunders depends on four propositions whose conjunction provides 'the most demanding criterion for an adequate theory of SDA'. Indeed, it constitutes 'the criterion of success' for SDA proposals. Wildman poses the conjunction as a 'tetralemma': (1) objectivity, (2) incompatibilism, (3) non-interventionism, and (4) the 'strong ontological view of the laws of nature'. By the latter Wildman means Saunders's view that the stochastic laws in quantum mechanics refer to 'principles or deep structures of nature that statistically govern *each individual event* within an ensemble of events'. Wildman claims that 'all theories of SDA fail to meet this criterion'—the tetralemma. Recognizing this failure in advance, each advocate in the QM-NIODA field intentionally 'protects' his or her version of SDA by 'weakening or rejecting one of the four propositions defining the criterion for success'. The failure of SDA proposals to meet the tetralemma accounts, according to Wildman, for Saunders's dismissal of these proposals (Wildman 2004: 57, 41, 43/table 2, 56).

What, then, of the tetralemma? Is it a valid criterion of success for SDA proposals? Hardly. As Wildman himself caustically states, if the strong ontological interpretation were correct, 'we do not need . . . two chapters of [Saunders's] book or a bunch of conferences to conclude that a non-interventionist account of QSDA is impossible'. I would add that none of the scholars searching for QSDA (QM-NIODA) view the tetralemma as representing the 'criterion for success' for their proposals. None worked at 'weakening or rejecting one of the four propositions' in the tetralemma in order to avoid its fateful verdict. The reason for this is quite simple: *the tetralemma is intrinsically self-contradictory*, as should be obvious: an incompatibilist account of non-interventionist objective, special divine action (NIODA) requires that nature is causally *indeterministic*; but a strong ontological interpretation of the stochastic laws of nature (if such an interpretation is even cogent) means that nature is *deterministic*, governed event by event by these stochastic laws.

It is not surprising that Tom Tracy dismisses the tetralemma out of hand, writing that 'these four assertions are logically incompatible' and that the tetralemma 'cannot possibly define "the criterion for success"' for SDA proposals. Not affirming the criterion is not a 'weakening of such proposals', as Wildman claims, because 'a theory is not compromised by its inability to accomplish a logically impossible task'. Indeed, Tracy (2004) calls support for the tetralemma a 'flatfooted mistake'.[3]

---

[3] Does Wildman agree with Saunders? Actually he criticizes Saunders for not stating why we must accept a strong ontological interpretation of the laws of nature in the first place; it is a 'key lapse' in Saunders's book. He also criticizes proponents of QM-NIODA (Ellis, Murphy, Tracy, and me) for not providing reasons for rejecting it. More significantly, however, he provides his own reason for why SDA proposals must fail the tetralemma, viz. Kant's insight rooted in the antinomy of reason: causality in nature and human freedom can never be reconciled unless we presuppose a *compatibilist* view of freedom. According to Wildman, Kant's argument 'applies equally well to divine freedom to act'. Hence SDA proposals must inevitably fail (Wildman 2004: 58). My response to Wildman is to disagree with his endorsement of Kant. Instead, I believe that the search for an indeterministic interpretation of natural causality, particularly in light of QM, represents a new conception of nature, to which Kant's metaphysics on this point at least is inapplicable. Philip Clayton (2004), in response to Wildman's article, lists five powerful arguments that seek to refute Kant's view.

(2) *Divine action at the quantum level and general providence*

The quantum-mechanical properties of fundamental particles ultimately account for many of the classical properties of the ordinary world of nature. For example, the statistics associated with protons, electrons, and other 'fermions' give rise to such features as the impenetrability and electrical conductivity of matter, while the statistics of photons, gravitons, and other 'bosons' produce phenomena such as superconductivity and the attractive forces in nature. It is to this world of ordinary experience that we attribute God's general providence (or continuous creation): namely, the ongoing creation and sustenance of the general features of the classical world of physics, geology, chemistry, meteorology, evolutionary biology, and so on. Thus, what we routinely take as general providence arises *indirectly* from God's *direct* action of sustaining in existence quantum systems and their properties during both their time evolution and their irreversible interactions (Russell 1988: 344–6; 1998: 200–1; Murphy 1995: 340–3; Ellis 2001: 259–60).

(3) *Divine action at the quantum level and its relation to special providence and theistic evolution*

QM-NIODA also views the domain of quantum mechanics as giving rise to particular events in nature (i.e. special providence). While it is widely asserted that individual quantum events always 'average out' at the macroscopic level, thus making quantum mechanics irrelevant to special providence, it is actually quite clear that quantum processes underlie and give rise to specific effects in the macroscopic world in several ways (which, to repeat, do not involve chaotic phenomena and thus 'quantum chaology').

One way is through those phenomena, such as superfluidity and superconductivity, which, though found in the ordinary world, are really 'bulk' quantum states—what Ellis calls 'essentially quantum effects at the macro level' (Ellis 2001: 261–2). Another, quite different way is through specific quantum processes, which, when amplified correctly, result in particular effects in the classical world. Obvious examples range from such jury-rigged situations as 'Schrödinger's cat' to such routine measurement devices such as a Geiger counter or a photo-multiplier. In fact, the production of specific effects at the macroscopic level from quantum processes includes a whole range of phenomena *in nature* such as the animal eye responding to a single photon, mental states resulting from quantum events at neural junctions, or the eventual phenotypic expression of a single genetic mutation in an organism (Russell 2001: 299, 306). Consequently, I claim that a quantum-based NIODA is enormously relevant to deploying a robust account of 'theistic evolution' in which God's non-interventionist objective divine action works in and with nature at the physical and biological levels of complexity, resulting in the neo-Darwinian evolution of life on Earth (Russell 1998).

In previous writings, I pointed to a watershed accomplishment in theology and science when, in the 1970s, Arthur Peacocke shifted the discussion of chance from a conflict model, 'law versus chance', as urged by atheists such as Jacques Monod (unfortunately, a formulation all too often accepted by Christians who reject evolution), to an integrative framework, 'law and chance'. As a result of this shift, Christians could claim that God acts through both law and chance to create physical, chemical, and biological novelty in nature. Still, the meaning of chance in this context may not be adequate for a genuine sense of non-interventionist divine action in specific events in time. I suggest that we now face a more fundamental shift in our discussion of 'law and chance' in light of quantum physics: a shift from chance in classical physics (where chance as mere epistemic ignorance of underlying causal processes precludes NIODA) to chance in quantum physics (where chance as onto-logical indeterminism is open to NIODA). Rather than saying that God deistically watches the endless unfolding of the potentialities built into nature at the beginning, as the early proponents of theistic evolution seemed to imply, we can now say that God indirectly creates order in the classical realm by (1) directly creating a quantum-mechanical universe with the properties that give rise to many of the phenomena in the classical world and (2) by acting directly in time as the continuous creator in, with, and through the indeterminism of quantum events to bring about novelty in the classical world. God is thus truly the God of both order and novelty in the physical and biological realms (Russell 1998: 344–6).

In summary, then, God's action at the quantum level can be seen as bringing about, in a non-interventionist mode, both the general features of the world we describe in terms of *general providence* (or continuous creation) and those specific events in the world to which *special providence* refers.

Crucial Theological Issues

We are now ready to move directly to the key questions in the debate on divine action and (non-relativistic) quantum physics.

(1) *Does God act providentially (general and/or special) in all, or only in some, quantum events?*

Murphy supports the claim that God acts intentionally in *all* quantum events. In her view, all quantum events involve a combination of natural and divine causality; they are determined, though only in part and not solely, by God (Murphy 1995: 340–3). Tracy explores the option that God acts in some but not all quantum events (Tracy 1995: 321–2).

On the one hand, I find Murphy's approach helpful for several reasons. The idea of God acting in all quantum events supports the theological claim that God does more than sustain the existence of all events and processes; in fact, God governs and co-operates with all that nature does. This idea also offers us a subtle but compelling way of interpreting God's action as leading to both general and special providence. Tracy's option seems to violate the principle of sufficient reason, since some quantum events would occur without sufficient prior conditions, constraints, or causes. Yet on the other hand it underscores the 'special' character of 'special providence': God's direct acts in key quantum events are special, not only because their indirect outcome is special, but also because God normally does not act in other quantum events beyond creating them and sustaining them in being.

Actually we can combine Murphy's pervasiveness of divine causality with Tracy's concern for the event to be objectively special because of the nature of quantum statistics: God acts in all events (God's action is never 'more' or 'less', but always equally causative). Still, on certain occasions, God will choose to actualize one state in particular, and not the other, because that state, and not the other, promotes life, thus conveying God's intentionality in this particular event.

In sum, QM-NIODA delivers just what is needed for non-interventionist object-ive, special providence. It involves *objective* special providence, for it involves a difference in what actually happens; it is objective special *providence*, since it truly conveys God's intentions through events that nurture life and wellbeing in the world; and it is *special* providence, because it is that event that we use to refer to God's providence against the assumed backdrop of the general situation itself: a wonderful outcome, a healing, a renewal of hope. Most importantly, it is *non-interventionist* objective special providence, because it is an act of objective special providence that God achieves without violating or suspending the ongoing processes of nature and the laws that describe them. So in short, God causes all the processes of the ordinary world (general providence), but a few of them genuinely convey special meaning because the choices God makes in causing *them*, and not the other options available to God, bring them about (Russell 2001: 315–17).

I also want to reiterate that I am not proposing an explanation of *how* God acts in nature (i.e. the 'causal joint' problem or the relation between primary and secondary causality); in addition this is, at most, a proposal about one of many domains in nature where the effects of God's acts arise. Hence, for all that has been said here, my proposal is fundamentally circumscribed and moderated by the profoundly apo-phatic nature of theological language.

(2) *Quantum physics, divine action, and the problem of human freedom*

The problem of free will, as formulated in the modern period, is the following: how are we able to act freely in the world if, as in the classical science picture, deterministic laws govern us somatically? Actually the problem arises only on an *incompatibilist/ libertarian* account of free will, which I adopt here. Many scholars have seen quantum indeterminism as a way out of the impasse: perhaps the human mind, through some form of 'top-down' causality (viz. the mind–brain problem), can objectively influence the brain through a direct or basic act which then indirectly affects the movements of the body via the central nervous system, making the enactment of free choices possible because the body is not determined mechanistically. This, however, raises a concern: how do we allow God's action to determine the quantum events that occur in my body and still allow for my own mind/brain to determine them? I will call this the problem of 'somatic overdetermination' (Russell 2001: 317–18).

My suggestion is that God acts in all quantum events in the universe until the evolution of organisms capable of even primitive levels of consciousness. God then increasingly refrains from determining the neurophysiological outcomes we associate with conscious choices, leaving room for top-down, mind–brain causality in conscious and self-conscious creatures. This would be one version of the standard 'solution' to the problem of free will: namely, God's voluntary or metaphysically necessary self-limitation, but seen now as a temporal development of the limitations, from minimum to maximum. God also abstains from acting in those quantum events underlying bodily dispositions resulting from indirect, central nervous system triggering, thereby allowing the developing levels of consciousness to act out their intentions somatically. Hence God bequeaths us not only the capacity for mental experience via God's special action in evolution and the resulting rise of the central nervous system, but God also bequeaths to us the capacity for free will and the capacity to enact our choices by providing at least one domain of genuine indeterminacy in terms of our somatic dispositions.[4] This approach suggests a rough analogy between the mind–brain problem, in which the mind acts as a source of influence on neurophysiologically located quantum events, and God's action in bringing about particular outcomes ('measurements') out of quantum indeterminacies (even though—and this is where the analogy fails completely—the means by which the brain does this is radically different from the means by which God does it).

(3) Quantum physics, divine action, and the challenge of theodicy

The problem of theodicy, of course, is a perennial issue for theism. If God is purely good, and if God can really act in history, why doesn't God minimize the evil done by humanity (i.e. 'moral evil')? When we expand the scope of divine action to include the evolutionary history of life on Earth, the question becomes: Why doesn't God act to minimize suffering, disease, death of individual organisms, and extinction of species (i.e. 'natural evil')? Theodicy has been discussed extensively in the 'theology and science' literature, where its subtle connection to the problem of human freedom has frequently been stressed. But theodicy becomes a particularly intense issue in light of the present thesis regarding a non-interventionist approach to objective, special divine action. George Ellis put the problem eloquently: '[T]here has to be a cast-iron reason why a merciful and loving God does not alleviate a lot more of the suffering in the world, if he/she has indeed the power to do so' (Ellis 1995: 360; see also 384 and Tracy 1998, 2001).

In response to the challenge of theodicy, Murphy calls on her notion of God's respect for the integrity or 'natural rights' of all creatures. This certainly works for humanity. Being non-coercive, God's action is consistent with human freedom and thus addresses, in part, the issue of theodicy as 'moral evil'. But what of 'natural evil':

---

[4] This discussion needs further refinement to take into account the claim *that God's grace, active within the human person, makes free will possible,* including the way grace liberates the will from the bondage of sin. The metaphor of divine 'self-limitation' mitigates against this insight. In essence, I am in agreement here with Ted Peters, who rejects a 'zero-sum' view of divine/human agency and, in particular, its excessive deployment in current 'kenotic' theologies of creation.

why does not God act to prevent suffering in nature in those cases where human freedom is unaffected, including the vast sweep of pre-human and even pre-sentient evolution? I believe that the search for an acceptable response to theodicy should be sought not within the doctrine of creation but within a fully developed theology of redemption as Resurrection-based new creation. I believe that it is only here that we will find the 'cast-iron reasons' that Ellis so rightly demands—reasons that will have the form of the cross and the empty tomb (Russell 2002). In any case, the problem of theodicy is stunningly exacerbated by all the NIODA proposals, including my own. The development of an adequate theological response is an overarching goal for future theological research.

REFERENCES AND SUGGESTED READING

CLAYTON, PHILIP (2004). 'Wildman's Kantian Skepticism: A Rubicon for the Divine Action Debate', *Theology and Science*, 2/2: 186–90.

ELLIS, GEORGE F. (1995). 'Ordinary and Extraordinary Divine Action: The Nexus of Inter-action', in *CC* 359–96.

—— (1999). 'Intimations of Transcendence: Relations of the Mind and God', in *NP* 449–74.

—— (2001). 'Quantum Theory and the Macroscopic World', in *QM* 259–92.

HEIM, KARL (1953). *The Transformation of the Scientific World*. London: SCM Press.

HERBERT, NICK (1985). *Quantum Reality: Beyond the New Physics*. Garden City, N.Y.: Anchor Books, Doubleday.

MASCALL, ERIC L. (1956). *Christian Theology and Natural Science: Some Questions in Their Relations*, Bampton Lecture Series. New York: Longmans, Green & Co.

MURPHY, NANCEY (1995). 'Divine Action in the Natural Order: Buridan's Ass and Schrödin-ger's Cat', in *CC* 325–57.

—— (1997). *Anglo-American Postmodernity: Philosophical Perspectives on Science, Religion, and Ethics*. Boulder, Colo.: Westview Press.

PEACOCKE, ARTHUR (1995). 'God's Interaction with the World: The Implications of Deter-ministic "Chaos" and of Interconnected and Interdependent Reality', in *CC* 263–87.

—— (1998). 'Biological Evolution—A Positive Theological Appraisal', in *EMB* 357–76.

POLKINGHORNE, JOHN C. (1995). 'The Metaphysics of Divine Action', in *CC* 147–56.

—— (2001). 'Physical Process, Quantum Events, and Divine Agency', in *QM* 181–90.

POLLARD, WILLIAM G. (1958). *Chance and Providence: God's Action in a World Governed by Scientific Law*. London: Faber & Faber.

RUSSELL, ROBERT JOHN (1988). 'Quantum Physics in Philosophical and Theological Perspec-tive', in *PPT* 343–74.

—— (1998). 'Special Providence and Genetic Mutation: A New Defense of Theistic Evolution', in *EMB* 191–223.

—— (2001). 'Divine Action and Quantum Mechanics: A Fresh Assessment', in *QM* 293–328.

—— (2002). 'Bodily Resurrection, Eschatology and Scientific Cosmology: The Mutual Inter-action of Christian Theology and Science', in Ted Peters, Robert John Russell, and Michael Welker (eds.), *Resurrection: Theological and Scientific Assessments*, Grand Rapids, Mich.: Eerdmans Publishing Company, 3–30.

—— STOEGER, WILLIAM R., SJ, and COYNE, GEORGE V., SJ (1988) (eds.). *Physics, Philosophy, and Theology: A Common Quest for Understanding*, CTNS/VO Series. Vatican City State: Vatican Observatory Publications (*PPT*).

—— MURPHY, NANCEY, and ISHAM, C. J. (1993) (eds.). *Quantum Cosmology and the Laws of Nature: Scientific Perspectives on Divine Action*, CTNS/VO Series, 1. Vatican City State: Vatican Observatory Publications (*QCNL*).

—— —— and PEACOCKE, ARTHUR R. (1995) (eds.). *Chaos and Complexity: Scientific Perspectives on Divine Action*. CTNS/VO Series, 2. Vatican City State: Vatican Observatory Publications (*CC*).

—— STOEGER, WILLIAM R., SJ, and AYALA, FRANCISCO J. (1998) (eds.). *Evolutionary and Molecular Biology: Scientific Perspectives on Divine Action*, CTNS/VO Series, 3. Vatican City State: Vatican Observatory Publications (*EMB*).

—— MURPHY, NANCEY, *et al.* (1999) (eds.). *Neuroscience and the Person: Scientific Perspectives on Divine Action*, CTNS/VO Series, 4. Vatican City State: Vatican Observatory Publications (*NP*).

—— CLAYTON, PHILIP, *et al.* (2001) (eds.). *Quantum Mechanics: Scientific Perspectives on Divine Action*, CTNS/VO Series, 5. Vatican City State: Vatican Observatory Publications (*QM*).

SAUNDERS, NICHOLAS (2002). *Divine Action and Modern Science*. Cambridge: Cambridge University Press.

SHIMONY, ABNER (2001). 'The Reality of the Quantum World', in *QM* 3–16.

TRACY, THOMAS F. (1995). 'Particular Providence and the God of the Gaps', in *CC* 289–324.

—— (1998). 'Evolution, Divine Action, and the Problem of Evil', in *EMB* 511–30.

—— (2001). 'Creation, Providence, and Quantum Chance', in *QM* 235–58.

—— (2004). 'Scientific Perspectives on Divine Action? Mapping the Options', *Theology and Science*, 2/2 (Oct.): 196–201.

WILDMAN, WESLEY J. (2004). 'The Divine Action Project, 1988–2003', *Theology and Science*, 2/1: 31–75.

CHAPTER 35

THEOLOGIES OF DIVINE ACTION

THOMAS F. TRACY

MODERN MISGIVINGS

One familiar and contentious response to the growth of scientific knowledge insists that long established ways of thinking about God and God's relation to the world can no longer be sustained if we take seriously emerging new understandings of the world. Steven Weinberg, for example, remarks that what we are learning in physics and cosmology makes it difficult to persist in the idea that the world is produced by an 'interested' deity who has purposes for its history and is concerned with good and evil in the lives of human beings. Yet, if theologians retreat from this robust conception of God, they are left with a nebulous Ultimate 'so abstract and unengaged that He is hardly to be distinguished from the laws of nature', and this 'makes the concept of God not so much wrong as unimportant' (1992: 244–5). The net effect of our expanding scientific understanding, Weinberg has said, is that 'The more the universe seems comprehensible, the more it seems pointless' (1977: 149). Weinberg has plenty of distinguished company in making claims of this kind (e.g. see Monod 1971; Wilson 1978, 1988; and Dawkins 1986). These large-scale conclusions stretch well beyond the bounds of physics into metaphysics, and they ought not to go unchallenged when they claim for themselves the lofty authority of science. Nonetheless, views of this kind reflect a widespread response to the picture of the world being pieced together in the sciences, and they vividly illustrate the interpretive challenges facing contemporary thinkers who wish to show the coherence of their religious and scientific beliefs.

It is not only a subset of scientists, writing in a polemical mood for the general public, who proclaim the end of religion as we have known it. A significant number of theologians, not to be entirely outdone by atheists, have concluded that thinking about God must fundamentally change if theistic religious traditions are to participate fully in the intellectual world shaped by the natural sciences. Writing in the first half of the twentieth century, Rudolph Bultmann sounded several themes that recur again and again. The defining mark of myth, says Bultmann, is that it conceives of God acting in a way that 'breaks into and disrupts the continuum of natural, historical, or psychical events—in short, as a "miracle"' (1951: 196–7, as translated and quoted by Ogden 1961: 91–2). If we accept the scientific understanding of the world as a unified network of cause and effect, then there is 'no room for God's working' (Bultmann 1958: 65), and we must give up the idea that God acts within the world's history to affect the course of events. This general pattern of analysis is echoed by a succession of modern theologians. Gordon Kaufman, for example, notes that the sciences approach the world as 'a web of interrelated events that must be understood as a self-contained whole' (1972: 132; also 1993: 271), and he concludes that 'in such a world acts of God (in the traditional sense) are not merely improbable or difficult to believe: they are literally inconceivable' (1972: 134).

If these diagnoses of the intellectual situation are correct, then the prognosis is very poor indeed for one of the central ideas in the theistic religious traditions. Both scientific critics and worried theologians acknowledge the historic importance of the concept of God as one who acts purposefully to call the world into being and to guide its history. Judaism, Christianity, and Islam all include in their sacred scriptures a rich collection of stories about divine action. The God depicted in these texts not only creates and sustains all things, but also acts within the world's unfolding history, affecting the lives of individuals and communities. This understanding of God has intimately shaped religious thought and practice for millennia.

Theologians have long recognized, however, that the language of divine action presents difficult interpretive challenges. Representations of God as an intentional agent are a central, but not the only, way of thinking about God in these traditions, and the narratives of divine action are themselves diverse in their depiction of the character and purposes of the divine agent. Modern literary and historical scholarship makes evident the complex relations between story and history; the stories are typically theology in narrative form, rather than anything we would think of as a historical report. In unpacking their theological content, what claims about divine action should we make? If, for example, we do not treat the story of the escape of the Hebrew people from slavery in Egypt as a straightforward historical description, then what (if anything) should we say about what God did? Communities of religious practice recite these stories and proclaim God's mighty acts of liberation. But are theologians prepared to explain, even in quite general terms, what the God who acts actually does? This difficulty was vividly illustrated in the plight of the 'biblical theologians' who recognized the importance of stories about divine action in identifying God's nature and purposes, but who found themselves unable to explain what they meant by an act of God (Wright 1952; Gilkey 1961). One important reason for their puzzlement and reticence was precisely the worry we have noted about the viability of the idea of divine action in an age of science.

These misgivings about the idea of divine action are not, of course, the only challenges inspired by reflection on new scientific understandings of the world. Weinberg's remarks, for example, reflect a sense that human longings for purpose and meaning are undercut by a recognition of the vanishingly small place we occupy in the enormities of space and time revealed by modern cosmology. We must also wrestle with questions about God's goodness, given the long history of struggle, suffering, and death involved in the evolution of life. I will focus here just on the question of divine action; but it is worth noting that much of what theists have said about the problems of meaning and suffering have been intimately linked to the idea that God acts on our behalf. If talk of particular divine action in the world has ceased to be intelligible, then this will have implications that reverberate throughout the entire network of theological ideas; it will, for example, decisively constrain the options at the heart of Christian doctrine in understanding the person and work of Jesus Christ. It is important, therefore, to take a look at the reasons for concluding that the sciences rule out this idea, and to see whether it might be defended in the face of these challenges.[1]

The Argument Against Divine Action in the World

How compelling are arguments that appeal to the natural sciences in contending that we must abandon the idea of divine action in the world? It is worth noting a minor

---

[1] Wesley Wildman, Ch. 36 below, presses a rather different case against theologies of divine action. Although he contends that ground-of-being theologies have some characteristic advantages in the dialogue with the natural sciences, his primary objection is that the ascription of intentional agency to God, even when all the necessary qualifications are added, is incoherent. At its root, this is because 'determinate entity theism' of any sort contradicts itself when it affirms that God is infinite yet also asserts that God is distinct from the world and acts purposefully to call it into being. 'To be infinite is to lack determining contrasts' (p. 620). There is a more than superficial paradox here, however, since the property of 'lacking contrasts' itself establishes a contrast with things that stand in relations of contrast. This paradox can be escaped by recognizing that one of the conditions for thought about anything is that it be marked out from what it is not. Any subject of predication will be a determinate entity in this restricted sense, and this alone does not entail finitude, at least in any theologically problematic respect. But the tag 'determinate entity theism' carries with it additional connotations of limitation and finitude, so is likely to be resisted by most theists, who affirm some or all of the classical metaphysical attributes of God. There are, as Wildman notes, deep conceptual tensions between the metaphysical and the personal attributes of God, but the long debate over whether these tensions generate contradictions remains open and lively, and cannot readily be collapsed into just two options: viz. either a finite personal God or an infinite non-personal Ground of Being. This is a point, in fact, that Wildman acknowledges in his remarks on Aquinas.

irony at the outset. A number of scientists have supposed that this religious idea has been decisively undercut by the discovery that chance plays so critical a role in the structures of nature, especially in the history of life (e.g. Monod 1971). This startling contingency appears to be at odds with theological claims about divine purpose in nature. Theologians, on the other hand, have tended to locate the crucial problem in the tightly woven web of causal necessities described by the sciences. In their view, the universal reign of causal law appears to push divine action out of nature, unless that action takes the unpalatable form of a miraculous intervention. I will suggest below that the seeds of a reply to both difficulties may be found in the combination of chance and law.

Consider the way in which Bultmann makes his case against the continued viability of the idea of divine action in the world. His position involves the following pattern of argument:

1. The idea of miraculous divine intervention is no longer acceptable to modern human beings whose understanding of the world has been shaped by the sciences.
2. Science commits us to understanding events as occurring within an unbroken continuum of natural causes.
3. Any act of God that alters the course of events in the world will disrupt the causal continuum of nature: i.e. it will be a miraculous intervention.
4. So the idea of particular divine action in the world must be given up altogether.

Many theologians have found some version of this argument to be highly persuasive, and they have looked for ways to interpret the idea of divine action without saying that God acts on particular occasions to affect the course of history. Bultmann himself sought to preserve the idea of divine action by moving it out of the domain of 'objective' events and into the (putatively) scientifically inaccessible depths of existential encounter between the divine and human subjects (1958: 67–70). This strategy implicitly draws upon a neo-Kantian distinction between the phenomenal and noumenal realms, and those who do not wish to join Bultmann in this flight to the noumenal self have had to look elsewhere for a strategy of interpretation. Typically, this involves absorbing the idea of providence into the doctrine of creation. Deists identified all of God's purposive activity with the original creative act that established the lawful structures and initial conditions of the universe. Liberal theologians have followed Schleiermacher, who rejected miracles yet did not want to be left with the inactive God of deism, and so proposed that God acts in every event uniformly as the ontological source on which all things immediately depend (1830/1963: 170 ff.). On this view, God does not so much act *in* history as *enact* history as a whole.

If the argument sketched above is correct, then this may be the best interpretive strategy available to contemporary theologians. So what should we say about this argument? All three premises raise complex issues, and it will be helpful to consider each in turn.

Miracles

Understood in the truncated modern sense as acts of God outside the lawful structure of nature, miracles have for some time been out of favour, at least among those who count themselves (or wish to be counted by others) as critically minded. It is surely correct to say that this is due at least in part to the influence of the natural sciences; but we need to think with some care about why the sciences have this effect. Do the natural sciences rule out the very possibility of miraculous divine intervention? It is not easy to see how this claim could be defended; certainly the actual results of scientific inquiry to date do not justify it. It does appear, however, that the natural sciences are committed in principle to explaining events in terms of immanent causal relationships rather than transcendent intentional agency. It follows that scientists, faced with a tough case, will conclude that they cannot explain the event in question, rather than explain it as an act of God. But the sciences provide no basis for ruling out in advance the possibility of encountering a tough case of this sort.

The trouble with miracles, then, may not be that modern science rules them out in principle, but rather that we do not have good evidence that they occur. David Hume, of course, presented the classic version of this argument, contending that we can never be in an epistemic position rationally to accept a miracle report (Hume [1748] 2000: sect. 10). Hume's elegant argument has stimulated a long debate, but even if we conclude that his argument fails, it seems clear that an attitude of initial dubiousness about miracle claims is justified. The tremendous success of the sciences has taught us to look for naturalistic explanations, and although Bultmann significantly underestimated the capacity of human beings to persist in believing things that he himself found unacceptable, educated folk generally do not move through the world anticipating interventions by supernatural powers.

In addition to these epistemic problems, there are also theological reasons to resist tying one's account of divine action too closely to miracle claims. One might hold that the idea of persistent divine intervention is at odds with God's wise foresight and moral goodness in creation. It is important to add, however, that theists have good reasons not to rule out this form of divine action altogether. If God is the free creator of the entire system of natural law, then God surely could choose to act in the world in ways that exceed the causal powers of creatures. It appears that some of the classical claims of the Christian faith (e.g. about the Resurrection) involve divine action that transcends the present limits of the created world, transforming the original creation into something new.

Scientific Explanation and Causal Completeness

The second premise of the argument can be read in more than one way, and its initial plausibility owes something to this ambiguity. First, it can be taken as asserting that the sciences are committed to offering complete causal explanations whenever they can. This is simply a remark about the strategy and goals of the sciences; it does not claim that this explanatory programme will always be successful. Second, it may be read as asserting that the natural sciences are committed to the view that every event has sufficient causal conditions in the prior history of the world. It is this stronger reading of the second premiss that is needed in order for the argument to reach its conclusion. If nature forms a closed system of finite causes, then it appears that God can act in the world only by breaking in upon it and disrupting its internal structures.

Should we accept this strong version of the premiss? The truth of universal causal determinism need not be assumed by the natural sciences in order to get on with their explanatory work. It is sufficient to adopt the merely methodological and provisional commitment to causal completeness expressed by the first reading of this premiss. Nor have the sciences established the truth of universal determinism as one of the results of their inquiries. The question of whether we live in a deterministic universe is left unsettled in the current state of science, and it is likely to remain open for some time. Quantum theory is relevant here to questions both of scientific method and of scientific findings. On the one hand, quantum mechanics shows that physics can generate a powerful explanatory and predictive system without assuming (and indeed, while giving up) causal completeness. On the other hand, quantum mechanics can be given an ontological interpretation in terms either of indeterminism (e.g. Copenhagen approaches) or determinism (e.g. the Bohmian approach).

Divine Action and the Structures of Nature

Without the second premiss, the third premiss does not follow, and the argument as a whole collapses. But that is hardly the end of the matter. Our interest is in mapping possibilities for conceiving of divine action, and if we reject universal determinism, then a new possibility arises that this argument did not countenance. In an indeterministic world of the right sort, it would be possible for God to act through the structures of nature, yet leave those structures entirely undisturbed. God's action would realize one of the alternative possibilities generated within, but left open by, the causal history of the world. This would alter the direction of the world's development so that events evolve differently from how they would have had God not so acted; but it would do so without displacing natural causes. The result is a non-interventionist and non-miraculous particular divine action.

Whether this is a viable and valuable option depends on a number of critical issues both in science and in theology. On the scientific side, there is of course the question of whether current scientific theories permit a plausible indeterministic interpretation. It is important to note that not just any sort of indeterminism will do. The system described by the theory must include branching pathways that are structured in such a way that the 'choice' between them (i) will have necessary but not sufficient conditions in the prior history of the system, and (ii) can make a significant difference to the subsequent course of events. On the theological side, we must ask whether these conceptual possibilities turn out to be theologically useful. Does this way of conceiving of divine action have a theological role to play, or can theology get along just as well without it? I want to start by considering some of the theological issues, since this will determine whether it is worthwhile to pursue this option at all.

The Theological Context: Multiple Modes of Divine Action

Creation as God's Fundamental Act

It might be argued that theology does not need to claim that God acts to affect the course of events (whether in an interventionist or a non-interventionist way) once the world's history is under way. Suppose for a moment that the argument we just rebutted had succeeded. In that case we would grant that the world must be understood as a causally complete structure and that God does not intervene within that structure. This view entails that God ordains the entire history of the universe in the act of creation. By establishing the laws of nature and specifying the initial conditions under which those laws operate, God determines the subsequent course of events. In a world of this sort, everything that happens can be regarded as an act of God in at least two respects.

First, God gives being to all finite things. As classically understood, this divine act of creation, unlike our own creativity, does not merely change what already is, but rather brings about the very existence of the creature *ex nihilo*. Further, this act of radical creativity is not a one-time event, but is, rather, the act of sustaining, or conserving, the creature's existence at every moment. God therefore acts universally, continuously, and intimately in all events as their ontological ground, or primary cause.

Second, God may choose to structure the world in such a way that creatures function as derived, or secondary, causes in a network of natural relationships. Rather than directly causing a stone to become warm, God may heat it by means of the Sun. Or switching from a medieval to a modern example, rather than directly creating rational animals, God may establish a system of lawful relationships in nature that brings about the emergence of such creatures over time. It will be perfectly correct to say that God warms the stone or creates rational animals, though God does so by means of created causes. Indeed, in a deterministic universe, everything that follows from God's creative act will be an intentional action of God carried out by means of the system of nature.[2]

---

[2] This obviously deepens the already profoundly difficult problem of evil. Partly for this reason, many modern theists reject universal determinism in favour of a view that grants a place in God's creation for indeterministic free action.

In sum, even if God never acts within the stream of events once the world's history is under way, God can be said to act in every event both directly, as its ontological ground, and indirectly, as the designer of the network of secondary causal relationships. If we deny determinism and contend that the universe includes chance events and/or indeterministic free decisions, then the account of divine action will become more complex. In a causally open universe it will remain the case, of course, that God acts in every event as its absolute ontological ground. It will also be true that God establishes the laws (now both deterministic and probabilistic) which set the boundaries of possibility within which causally open processes operate. But we will have to qualify the claim that every effect of secondary causes and finite agents is an indirect intentional act of God; chance and choice will have causal consequences in the events that follow from them, and these will radiate outward in the world's history. This has implications for divine action to which I will return in a moment.

Three Senses of Special Divine Action

The hallmark of the view I just sketched is the universality and uniformity of divine action. Is there any basis within this account for singling out particular events as God's acts? There are two ways in which this can be done.

First, an event may be distinguished from other events because it particularly discloses to an individual or a community God's presence and purposes in the world. The escape of the Hebrew people from servitude in Egypt may have been an outcome of the normal (if surprising) processes of nature and human history; yet it can have special significance as the occasion for this community's recognition of God's liberating purposes. What makes this event special is not the mode of God's action within it, which is no different here than elsewhere, but rather its epistemic role in revealing to us something of the character or direction of the universal activity of God. An event that plays this type of epistemic role we will call a *subjectively special* act of God.

Second, an event may be distinguished from other events because it realizes or advances God's purposes in an especially significant way. The escape across the swampy shallows of the Red Sea may have involved nothing other than the ordinary workings of wind and water; yet it can mark an important advance or turning point in the unfolding course of history. The fact that this event was built into history from the outset in no way undercuts either its status as an act of God or its special role in fulfilling God's purposes. We can call such an event a *materially*, or *functionally*, *special* act of God.

In addition to subjectively and materially special divine action, there is a third possibility, which takes a crucial step beyond the 'uniformitarian' view of divine action. We might single out an event as special because God acts directly at a particular time and place within the world's history to create the conditions for its occurrence. This event will be distinguished from other events by virtue of its causal history. It will have an extensive network of necessary conditions in the prior history of the universe, but that history would not produce this event without God's specific action to alter the course of its development. We can call this an *objectively*, or *causally, special* divine action.

It is of course this third type of special divine action that has provoked so many misgivings in modern theology. If we accept a deterministic picture of nature, then it appears that any such action must necessarily take the form of a miraculous intervention, as Bultmann's argument contended.[3] We have seen, however, that if we move beyond determinism, the possibility arises of non-interventionist, objectively special divine action. But given the concepts of subjectively and materially special divine action, is there any theological need to invoke this third form of special divine action?

This question is linked to an extensive network of theological considerations, just one feature of which I will briefly note here. The biblical narratives and the forms of religious life they have fostered give a significant role to images of God acting in response to human actions in an ongoing dynamic relationship. If God's providential governance of history is contained entirely in God's creative act, then can we say that God responds to the prayers of the faithful or to the cry of the oppressed or to the restless human heart? It appears that in a uniformitarian account, all of God's responses to the actions of creatures will need to be built in from the outset. It is fairly easy to see how God might do this in a thoroughly deterministic world; as we have seen, every event, including all the actions of finite agents, will be specified in the design of creation, so there is no reason why God's responses could not be also. But, as we have also seen, if we introduce indeterminism into our picture of the universe, whether as chance or as (incompatibilist) freedom or both, then the world's history in its entirety will no longer follow deductively from its laws and boundary conditions. This provides a reason to think that God might act within the unfolding course of events in order to respond to developments that were not built into the plan of creation. There are ways to avoid this conclusion. We could accommodate the unpredictable results of indeterministic freedom, for example, by arguing that God, in choosing which possible world to make actual, has knowledge of what every possible free creature would in fact choose to do in every possible circumstance.[4] This would once again make it possible for God to design into the causal history of

---

[3] A number of authors have argued that this is not the case: i.e. that God can bring about particular changes in the course of events in a closed causal order without interrupting the finite causal series. The most fully developed proposal is offered by Arthur Peacocke (1993), who argues that God acts as a structuring constraint on the finite natural system as a whole. This proposal is appealing, but I have argued elsewhere that in the end it does not avoid the need for causal openness in the structures of nature (Tracy, forthcoming).

[4] This is known as God's 'middle knowledge': i.e. a knowledge of contingent truths that depend not on God's sovereign will but rather on the choices that possible free creatures would make if given the chance. The idea has its origins in the thought of the sixteenth-century theologian Louis de Molina ([1594] 1988), and has returned to prominence in recent philosophy of religion (e.g. see Flint 1998). For additional discussion of the bearing of this idea on responsive divine action, see Alston (1993: 191–9).

the world responses to the actions of free creatures. But the philosophical machinery required by this line of argument is sufficiently problematic that it makes sense to explore the conceptual options for non-interventionist, objectively special divine action.

THE SCIENTIFIC CONTEXT: LAW, CHANCE, AND DIVINE ACTION WITHOUT INTERVENTION

We turn, then, to the question of whether the contemporary natural sciences can plausibly be interpreted in a way that would make possible an account of divine action without intervention. Two scientific preconditions must be met in order to launch such a proposal. First, it must be plausible to offer an indeterministic interpretation of the theory: i.e. an interpretation according to which some events in the theory's domain have necessary but not sufficient causal conditions. Second, at least some of these underdetermined events must have the potential to make a difference in the subsequent development of the world's history. In recent discussions of divine action, a number of areas of science have been explored with these desiderata in mind. Let me briefly mention two of the proposals that have emerged.

Chaos Theory

John Polkinghorne has given careful consideration to chaos theory, where we find systems whose behaviour depends with exquisite sensitivity on their initial conditions, producing widely different results from infinitely small differences in starting points. Given our limited ability to specify the initial conditions, the future development of chaotic systems is unpredictable in principle. This leads Polkinghorne to a metaphysical conjecture: perhaps, he contends, the deterministic mathematics we use to describe these systems is an abstraction from a more supple, or flexible, underlying structure in nature. We might then conceive of God acting through this inherent openness to shape the development of the world's history. Specifically, Polkinghorne notes that some chaotic systems display pathways of development so finely differentiated that the mathematics describing them approaches the limiting case of infinite density, where there is no energy difference between nearby alternatives. He suggests that we might think of God choosing between these alternatives, making an input of information but not of energy, yet significantly affecting the overall development of the system.

This ingenious proposal faces some important difficulties. Chaotic unpredictability arises in mathematics strictly as a function of deterministic non-linear equations. There is no basis in chaos theory itself for affirming the unpredictability of the system described by these equations while denying its deterministic dynamics; part of what is so remarkable about chaos theory is precisely that determinism gives rise to unpredictability. Polkinghorne advances an appealing metaphysical conjecture about the open structures of nature 'beneath' the simplifications and idealizations of scientific theory. But it is not clear that chaos theory supports or invites this general metaphysical view. On the contrary, to the degree that we affirm a cautious realism about chaos theory, adopting the maxim that 'what we know is a reliable guide to what there is' (Polkinghorne 1995: 148), the conclusion would seem to be that the systems in nature described by this theory are in fact deterministic.[5]

Quantum Theory

In exploring the possibility of indeterministic openness in the structures of nature, the other obvious place to look is quantum theory. Here too there are open questions and significant difficulties. Does quantum theory meet the two criteria I noted above for presenting a viable resource for theology? It is clear that quantum theory permits, though it does not require, an indeterministic interpretation. Unlike chaos theory, the quantum formalism is resolutely probabilistic at crucial points. This feature of quantum theory can be interpreted in a number of different ways. David Bohm (1952), for example, has produced a version of the theory in which the properties of quantum entities are always determinate, and uncertainty about those properties is merely epistemic. There is a conceptual price to be paid for this recovery of determinacy and causal completeness (e.g. the postulation of non-local hidden variables). Most physicists prefer variants of the interpretations offered by Niels Bohr (1958) and Werner Heisenberg (1958), although this 'Copenhagen' approach carries its own conceptual costs. Interpretations of this sort accept that some of the properties of quantum entities (e.g. the position, momentum, and spin of an electron) do not have a determinate value, but can be described only as a set of coexisting probabilities. This superposition state evolves deterministically through time according to the Schrödinger wave equation until there is an interaction (whether in the laboratory or in nature) that brings about the irreversible collapse of the wave function to a determinate value for the 'measured' property. The quantum formalism spells out the probability of realizing each of the values that the property can display, but it does not specify which value will be obtained on any given occasion.

This opens up the possibility that we might conceive of God as acting in some or all of these transitions to designate which of the probabilistically permitted possibilities is actualized. A divine action of this sort would not displace natural causes, since

[5] For further discussion of Polkinghorne's proposals see e.g. Polkinghorne (1998) and Saunders (2002: ch. 7).

ex hypothesi these events have necessary but not sufficient causal conditions in the prior history of the universe. Rather, God would act to determine what the system of nature leaves undetermined. Divine action at the quantum level, therefore, is not an 'intervention' in nature, if by that term we mean an action that breaks a chain of finite causes. Some critics have contended that this action 'occurs by means of God "ignoring" or intervening against the measurement probabilities "predicted" by the orthodox theory' (Saunders 2002: 155). The obvious reply is that as long as God's determination of quantum events preserves the probability distributions spelled out by the theory, no natural law has been ignored or overridden; indeed, if God determines all quantum transitions, then the probabilistic laws of quantum mechanics will just be a description of the pattern of God's action (Murphy 1995: 344–8). The notion of a 'violation' of a probabilistic law is problematic in any case, since no single event will fall outside the law unless the lawful probability of its occurrence is zero.

The more worrisome difficulty for this form of divine action concerns the second scientific issue I noted initially: namely, the question of whether the underdetermined events allowed by the theory can make a significant difference to the world's unfolding history. If quantum chance is altogether averaged out in higher-level regularities described by deterministic laws, then divine action at the quantum level will do nothing more than underwrite the lawful structures of nature. We need to ask whether selection between alternative possibilities at the quantum level can make a difference at the macroscopic level. It is clear that the 'amplification' of quantum effects is possible, since we now routinely construct devices for just this purpose. But are there structures in nature that have this effect? It may be tempting to appeal to chaotic processes in this connection, given the capacity of such systems to generate widely divergent outcomes from minute differences in initial conditions. The difficulty here is that the relation between chaos theory and quantum mechanics is as yet unclear (e.g. the Schrödinger equation is linear). Quite apart from a theory of quantum chaos, however, it appears that there are structures in nature that can amplify quantum effects. A number of authors have noted that vision involves an amplification of the interaction between photons and the biochemical structures of the eye (e.g. Ellis 2001: 260). Robert Russell has carefully explored the role of quantum effects in genetic mutation; the effects of these mutations are perpetuated or extinguished by natural selection, and they significantly shape the history of life (Russell 1998: 205–8).

It appears, therefore, that an indeterministic interpretation of quantum theory holds promise of being helpful in constructing an account of non-interventionist special divine action. Any proposal of this sort, of course, will be tied to developments in quantum theory and to new approaches to the interpretive problems it presents. There are fundamental unsettled questions in this field: e.g. puzzles about measurement and wave function collapse and about the relationship of quantum mechanics to other scientific theories such as relativity or chaos theory. If theological interpretations of quantum theory are to be attempted at all, they must be offered with a tentativeness that reflects the open-ended and uncertain character of the inquiry. The idea of divine action at the quantum level presents an ongoing direction for research, rather than a settled position.

More Misgivings: The Return
of the God of the Gaps?

Perhaps, then, it would be wisest to avoid such speculation altogether. That certainly has been the dominant strategy in modern theology, as it has sought to reduce its empirical 'exposure' by giving up claims that are vulnerable to changing scientific understandings of the world. It is clear that theology has not benefited from seizing upon lacunae in scientific explanation and, with momentary triumph, inserting God as the missing explanatory term. That is the familiar, and justly maligned, manoeuvre of turning to a 'God of the gaps'. But note that the idea of divine action at the quantum level differs from this strategy in two crucial ways. First, according to leading contemporary interpretations of quantum mechanics, gaps in causal explanation at the quantum level do not simply reflect the incompleteness of the theory, but instead reveal openness in the structure of the natural world. In exploring indeterministic understandings of nature, we are not exploiting a failure of current theory, but rather are grappling with the consequences of its success. It is precisely what we now claim to know about the world that creates the opportunity for this constructive theological interpretation. Of course there are, as we have seen, alternatives to indeterministic interpretations of quantum theory. But there can be no theological discussion of quantum theory that does not make use of some interpretation, and it is perfectly legitimate for theological interests to play a role in deciding which of several defensible interpretations to pursue.

Second, theological proposals about divine action at the quantum level do not claim that the sciences, for their own explanatory purposes, must invoke the concept of God. This conception of divine action is not put forward as a rival to naturalistic scientific hypotheses. On the contrary, it is part of a candidly theological project of interpretation that sets out to integrate central theistic affirmations with current scientific theories. This enterprise of faith seeking understanding is perpetually challenged and renewed by changing understandings of the world; if we are to form some conception of God's activity in the world, we need to attend with care to what we know (or think we know) about the world in which God acts. When a highly successful scientific theory gives rise to a remarkable new picture of the natural order as a subtle interplay of chance and law, then this is something that a theology of nature will need to take into account.

It may be contended, nonetheless, that proposals about divine action at points of underdetermination in nature make the mistake of supposing that God requires gaps in the structure of creation in order to realize the divine purposes. This, the objector may claim, assumes that God acts on the same level as creatures, and the result of this excessively anthropomorphic conception of God is that the divine agent must compete with created causes, either pushing them aside in interventionist miracles or delicately bringing divine influence to bear at points where the system of finite causes is incomplete.

It would indeed be a mistake to adopt this as a general view of divine action. But in pursuing the possibility that God might act through open structures in nature, we are not committed to this wider theological error. Recall that we have been exploring the idea of non-interventionist, objectively special divine action. I have argued that this is a conceptually viable option for a critically minded theology that takes the sciences seriously. But clearly this is just one of a range of ways in which we can understand God to act, and it is by no means the most basic. The fundamental form of divine action is the creative act that grounds and sustains the existence of all finite things. Here there is no possibility of competition between divine and created causalities, for apart from God's creative activity, there would be no finite causes or agents. As we have seen, God can also be said to act by means of the system of created (secondary) causes, so that events which follow deterministically within this structure can be regarded as indirect divine acts. This too permits no trade-off between God's activity and that of creatures. The situation is different, however, if we also say that God chooses to act at a particular time and place within the series of finite causes to bring about a development that would not have occurred had God not so acted. In this case, God's action among created causes will supply one of the necessary conditions for the resulting event, and any description of the secondary causal antecedents of this event must necessarily be incomplete; there will be a gap in its causal history.[6] A theologian may decide for any number of reasons not to include this form of divine action in his or her account of God's activity in the world. But theists who affirm God's freedom as Creator have no basis for denying that God *could* act this way should God choose to do so. Nor does the affirmation of this form of divine action compromise the transcendence of the divine agent. The God who calls the universe into being out of nothing and who sustains it in existence at every moment cannot be just one cause among others, whatever the particular mode of divine action in any instance.

CONCLUSION

In sum, there are several ways of thinking about divine action available to contemporary theologians.

- First, we may understand God to act as creator in every event as its absolute ontological ground.
- Second, we may understand God to act indirectly through the order of created causes (the system of natural law) that God has established.
- Third, we may understand God to act indirectly through the actions of free agents whose decisions have been shaped by the rest of God's activity in the world.

---

[6] For a more detailed discussion of these questions about divine action and the openness of nature, see Tracy (1995) and (forthcoming: sect. 2.1).

- Fourth, we may understand God to act to determine some or all of what is left underdetermined in the order of created causes (i.e. non-interventionist, objectively special action).
- Fifth, we may understand God to act within the world's history to bring about effects that modify or exceed the causal powers of creatures (i.e. interventionist, objectively special action).

All of these ways of conceiving of divine action can be called upon in giving an account of God's purposive activity in the world, and theological proposals will be differentiated in part by how they marshal and deploy them. There are good reasons to resist the demand (or temptation) to offer a detailed explanation of exactly which of these options provides the means by which God acts in any instance; theological reflection on divine action must begin and end in a recognition of our epistemic limits. Modern theologians have compelling reasons to be cautious in their appeals to the fifth mode of divine action: i.e. interventionist miracles. But this does not mean that the idea of objectively special divine action must be abandoned altogether. There are promising prospects for developing the fourth option in an account of non-interventionist divine action that is consonant with widely accepted interpretations of contemporary science. This possibility has been overlooked, in part, because so many theologians have been in the grip of the flawed argument we considered initially, an argument that persists in treating universal causal determinism as a presupposition of scientific inquiry even as the sciences have developed beyond such claims. I have argued, on the contrary, that we need not pre-emptively exclude any of these ways of thinking about divine action on the grounds that they have been ruled out by modern science.

References and Suggested Reading

Alston, William P. (1993). 'Divine Action, Human Freedom, and the Laws of Nature', in Robert Russell, Nancey Murphy, and C. J. Isham (eds.), *Quantum Cosmology and the Laws of Nature: Scientific Perspectives on Divine Action*, Vatican City State: Vatican Observatory Publications; Berkeley: Center for Theology and the Natural Sciences, 185–207.

Bohm, David (1952). 'A Suggested Interpretation of Quantum Theory in Terms of Hidden Variables, I & II', *Physical Review*, 85: 166–93.

Bohr, Niels (1958). *Atomic Physics and Human Knowledge*. New York: Wiley.

Bultmann, Rudolph (1951). *Kerygma und Mythos*, ed. H. W. Bartsch. Hamburg: Herbert Reich-Evangelischer Verlag.

——(1958). *Jesus Christ and Mythology*. New York: Charles Scribner's Sons.

Dawkins, Richard (1986). *The Blind Watchmaker*. New York: Norton.

Ellis, George (2001). 'Quantum Theory and the Macroscopic World', in Robert Russell, Philip Clayton, Kirk Wegter-McNelly, and John Polkinghorne (eds.), *Quantum Mechanics: Scientific Perspectives on Divine Action*, Vatican City State: Vatican Observatory Publications; Berkeley: Center for Theology and the Natural Sciences, 259–91.

FLINT, THOMAS (1998). *Divine Providence: The Molinist Account*. Ithaca, N.Y.: Cornell University Press.

GILKEY, LANGDON (1961). 'Cosmology, Ontology, and the Travail of Biblical Language', *Journal of Religion*, 41: 194–205.

HEISENBERG, WERNER (1958). *Physics and Philosophy: The Revolution in Modern Science*. New York: Harper & Row.

HUME, DAVID ([1748] 2000). *An Enquiry concerning Human Understanding*, The Clarendon Edition of the Works of David Hume, ed. Thomas L. Beauchamp. New York: Oxford University Press.

KAUFMAN, GORDON (1972). *God the Problem*. Cambridge, Mass.: Harvard University Press.

—— (1993). *In the Face of Mystery: A Constructive Theology*. Cambridge, Mass.: Harvard University Press.

MOLINA, LUIS DE ([1594] 1988). *On Divine Foreknowledge: Part IV of the Concordia*, trans. Alfred J. Freddoso. Ithaca, N.Y.: Cornell University Press.

MONOD, JACQUES (1971). *Chance and Necessity: An Essay on the Natural Philosophy of Modern Biology*, trans. Austryn Wainhouse. New York: Alfred A. Knopf.

MURPHY, NANCEY (1995). 'Divine Action in the Natural Order: Buridan's Ass and Schrödinger's Cat', in Robert Russell, Nancey Murphy, and Arthur Reacocke (eds.), *Chaos and Complexity: Scientific Perspectives on Divine Action*, Vatican City State: Vatican Observatory Publications; Berkeley: Center for Theology and the Natural Sciences, 325–57.

OGDEN, SCHUBERT (1961). *Christ without Myth*. New York: Harper & Brothers.

PEACOCKE, ARTHUR (1993). *Theology for a Scientific Age*, 2nd edn. London: SCM Press.

POLKINGHORNE, JOHN (1995). 'The Metaphysics of Divine Action', in Russell *et al.* (eds.), *Chaos and Complexity: Scientific Perspectives on Divine Action*, Vatican City State: Vatican Observatory Publications; Berkeley: Center for Theology and the Natural Sciences, 147–56.

—— (1998). *Belief in God in an Age of Science*. New Haven: Yale University Press.

RUSSELL, ROBERT JOHN (1998). 'Special Providence and Genetic Mutation: A New Defense of Theistic Evolution', in Robert Russell, William Stoeger, SJ, and Francisco Ayala (eds.), *Evolutionary and Molecular Biology: Scientific Perspectives on Divine Action*, Vatican City State: Vatican Observatory Publications; Berkeley: Center for Theology and the Natural Sciences, 191–223.

SAUNDERS, NICHOLAS (2002). *Divine Action and Modern Science*. Cambridge: Cambridge University Press.

SCHLEIERMACHER, FRIEDRICH (1830/1963). *The Christian Faith*, 2nd edn., i, ed. H. R. Mackintosh and J. S. Stewart. New York: Harper & Row.

TRACY, THOMAS (1995). 'Particular Providence and the God of the Gaps', in Russell *et al.* (eds.), *Chaos and Complexity*.

—— (forthcoming). 'Special Divine Action and the Laws of Nature', in Robert John Russell et al., *Scientific Perspectives on Divine Action: Fourteen Years of Problems and Progress*, Vatican City State: Vatican Observatory Publications; Berkeley: Center for Theology and the Natural Sciences.

WEINBERG, STEVEN (1977). *The First Three Minutes*. New York: Basic Books.

—— (1992). *Dreams of a Final Theory*. New York: Pantheon Books.

WILSON, EDWARD O. (1978). *On Human Nature*. Cambridge, Mass.: Harvard University Press.

—— (1988). *Consilience: The Unity of Knowledge*. New York: Alfred A. Knopf.

WRIGHT, G. ERNEST (1952). *God Who Acts: Biblical Theology as Recital*. London: SCM Press.

CHAPTER 36

GROUND-
OF-BEING
THEOLOGIES

WESLEY J. WILDMAN

INTRODUCTION

This chapter concerns a complex family of theological viewpoints, collected under the name 'ground-of-being' theologies. It discusses their relationship to various forms of theism, their shared themes, and their connections with the natural sciences. Ground-of-being theologies have in common two important negations: they *deny that ultimate reality is a determinate entity*, and they *deny that the universe is ontologically self-explanatory*. The positive formulations of ground-of-being theologies vary. Some stay within theism, and others not; some embrace ontological categories, and others repudiate them; some use conceptualities of substance, and others categories of process; some are fundamentally monistic, and others pluralistic; some are indistinguishable from religious naturalism, and others are nurtured within hierarchical cosmologies containing supernatural entities and events.

Ground-of-being theologies are important, because their denial that ultimate reality is a determinate entity establishes a valuable theological contrast with determinate entity theisms such as personal theism and process theism—two ideas of God prominent both in modern theology and in the contemporary science–religion dialogue. Determinate entity views assert that God is an existent entity with determinate features including intentions, plans, and capacities to act, though the various views interpret these features quite differently. By contrast, ground-of-being theologies challenge the very vocabulary of divine existence or non-existence. They interpret symbolically the application to ultimate realities of personal categories such as intentions and actions, and regard literalized metaphysical use of such ideas as a category mistake. They are wary of the *analogia entis* (analogies controlled by the contrast between divine and human being) because, even if the idea of divine being were intelligible, we do not know how to compare human and divine being. They regard determinate entity theisms as excessively vulnerable to anthropomorphic distortion and, in this way, continue the resistance to anthropomorphic idolatry evident in many of the world's sacred religious texts, including the Bible and the Qur'an, the Daodejing and the *Bhagavadgītā*. Of course, many determinate entity theologians are acutely sensitive to the problem of anthropomorphic distortion, and try to build in safeguards, but ground-of-being theology has better intrinsic resistance.

Ground-of-being theologies are also important because their denial that the universe is ontologically self-explanatory resists a flattened-out kind of atheism, affirms that all of reality is ultimately dependent for its very being on an ontological ground, and articulates an authentic basis for religion and value in human affairs. The necessary intimacy between all of being and its ultimate ontological source means that ground-of-being theologies have fascinating interactions with the natural and social sciences. They are complementary to determinate entity theisms in this respect, with the two families of views often responsive to different evidence, and either seeking answers to different questions or else seeking different answers to the same question, as we shall see. Moreover, ground-of-being theologies have impressive intellectual lineages in all large religious and philosophical traditions; they have articulate defenders in all eras; and they are intrinsically interesting both in themselves and as dialogue partners with the natural and social sciences. The relevance of these theologies to the question of divine action will become clear in due course.

ULTIMATE REALITIES AND THEOLOGICAL MODELLING STRATEGIES

Religion is often concerned with ultimacy in various modes, such as ultimate realities, ultimate ways of life, ultimate authorities, ultimate wisdoms, ultimate truths, and ultimate concerns. A given religious context tends to subordinate some modes of ultimacy to others, thereby creating a distinctive style of ultimacy speech, one that may not be easily translatable into other styles. Yet it is possible to focus on one mode of ultimacy for the sake of investigation, taking care to avoid inappropriate generalizations about religion. The focus here is on ultimate realities, which is a serviceable cross-cultural comparative category.[1] It does not distort what we describe

---

[1] This reflects the results of the Comparative Religious Ideas Project, which sought to identify through a rigorous process of comparison and analysis which categories work best to

by means of it, provided that we remember that ultimate realities are secondary features in some religions and even within certain theological traditions.[2]

The words 'ultimate' and 'ultimacy' suggest finality, and thus the phrase 'ultimate realities' denotes our bold attempts to express what is most profound and definitive about the whole of reality. Most religious intellectuals—I call them theologians advisedly[3]—acknowledge the difficulty of speech about ultimate matters. It would be unsurprising if there were a cognitive mismatch between human beings and the ultimate realities they seek to describe, or if ultimate realities were so dense with meaning and power that any portrayal is necessarily a fragmented perspective rather than a comprehensive and consistent description. This likelihood is reinforced by the diversity of renderings of ultimate realities, the intractable disagreements among theologians, and the testimony of mystics and religious adepts. Some theologians limit themselves to poetic and rhetorically potent modes of discourse as a result, and there is real value in the indirection and grace of such speech. If we are to speak of ultimate realities at all, however, there must be a role for those who attempt to bring all their rational powers to bear on the task and, properly wary of intellectual hubris, approach the challenge as rigorously as possible. This is theological inquiry: perpetually tentative, yet imaginative, disciplined, and systematic.

There is ample evidence in both the history of science and the history of metaphysics that human inquirers build conceptual models in order to understand and explain, and that primary metaphors and analogies support model construction. Mathematical models of the physical world routinely involve tropic elements in their interpretation and application, as when we say that forces are vectors in vector spaces and forces combine as vectors add. The role of primary metaphors is even more prominent in metaphysics and theology, where formal languages such as mathematics are not available to aid modelling. Theologians with systematic

---

describe what is important about the ideas of world religious traditions, minimizing distortion and arbitrariness. See Neville 2001a, 2001b, 2001c.

[2] In relation to religions, most forms of Buddhism stress an ultimate way of life as a path to enlightenment. Some Buddhist traditions even regard thinking about ultimate realities as a kind of distraction from which we must detach ourselves if we are to achieve enlightenment. An example for theological traditions would be the youthful tradition of process theology, which does not focus much at all on ultimate reality, but only on God, which is an actual entity with a special role within the whole of reality.

[3] All large religious traditions have an intellectual wing whose members, often organized into numerous sub-traditions, concern themselves with the credibility of their religion's beliefs and practices, and who try to construct compelling rational formulations of them. This activity lacks a common name across traditions, because familiar names such as 'theology' or 'philosophy of religion' involve confusing baggage. Most obviously, the word 'theology' suggests theism, which misses what intellectuals in non-theistic traditions are about. Despite this difficulty, and because of the need for a label, the word 'theology' has been catching on; today we have Jewish theology and Islamic theology and even Buddhist theology. I shall use 'theology' to denote this activity, acknowledging the difficulties and also the diversity in theological styles.

inclinations are sharply aware of the limitations of the tropes on which they rely for modelling ultimate realities and try to compensate. They skilfully juxtapose tropes; they strive to be clear about the senses in which metaphors and analogies apply and do not apply; and they regulate the conceptual tensions inherent in using multiple metaphors. This is an intellectually disciplined form of balancing akin to the practical balancing of popular religion, which typically is a riot of images regulated through narrative structures and ritual practices. It is usually only theologians and religious adepts who strive for optimal consistency in conceptual modelling of ultimate realities.

Given this understanding of theology as model construction, a fundamental choice in framing an idea of ultimate realities is the source of primary metaphors. After that, a great deal depends on skilfully combining and regulating tropic elements. Reflection on ultimacy draws our attention to the most profound features of our experience as sources of primary metaphors. Thus, it makes sense to appeal to the most complex and intense form of being that we know to have emerged from the history of nature: namely, ourselves. This is the basis for anthropomorphism in theology, which is inevitable and religiously useful to some degree, and only religiously dangerous or intellectually defective in extreme or rigidly literalized forms. The appeal to human beings as the source of theological metaphors is also the inspiration for the critiques of religion as an illusion serving human existential and social interests. Yet using ourselves as models for God is a meaningful strategy for expressing what is ultimately important and ultimately real. This is clearest in divine creation theories, where the creator seemingly ought to surpass in dignity and complexity the most intense forms of created being. In these views, the ways in which reality is ultimately personal vary considerably, depending on the features of human being that predominate in the metaphysical model.

The most common personal modelling strategy is to draw primary metaphors from human beings as determinate entities that contrast with their environment, that possess intentions and plans and powers to influence that environment, and that change in response to their environment. This leads to determinate entity theism, which holds that ultimate reality is a determinate divine entity that contrasts with the created world, is influenced by the created world, possesses intentions and plans, and acts providentially within history and nature. This view is prominent within some important sacred religious texts. Because of its humanly comprehensible dimensions, it is the source of many narrative structures that regulate popular religions and their expansive explorations of symbolic material. Some determinate entity modelling strategies make all of reality a kind of home for human beings by rendering reality as the creation of a personal deity who cares about human destiny and each individual human being. These approaches interpret cosmic and cultural history alike as exhibiting the plans of this divine being, weaving the moral and spiritual disasters of human life into a sacred narrative that recounts the historic interactions between God and the world. They make prayer an intimately personal, two-way encounter that is full of expectation for divine conversation and answers to petitions. An everlasting divine person makes an afterlife for human persons easy to conceive and, given the intimate bond between God and the believer, almost inevitable. The harmony between the narrative elements of such theological models and some traditions of practical religion means that theologians who seek to interpret the convictions and activities of religious groups take determinate entity theism extremely seriously.

Theologians are also well aware of the problems with determinate entity theism. Metaphors and analogies are both like and unlike the objects described by means of them. Human beings are often morally confused and weak in will; they are subject to disease and decay and death; they are profoundly dependent on a natural environment for their survival; they are a social species with an evolutionary history; and they are limited in intelligence and wisdom, patience and power. These features of determinate human beings are typically denied of God because of the role God plays in key religious narratives. For example, if God is the final goal of human relational questing and the source of salvation and liberation, God must be all-wise and morally perfect, by contrast with the moral ambiguity of nature and human life. If God acts everywhere and always in the universe, then God must somehow be present to every part of space and time, by contrast with the local character of the information-conveying causal interactions with which we are more familiar. If God is creator of everything, then God had better not be subject to some cosmic clock. The anthropocentric character of the reasoning here is not necessarily problematic, because this may be the best rational way to grasp ultimate reality of which human beings are capable. But the vocal critics of determinate entity theism can only be held at bay if sympathetic theologians mount a sound articulation and defence of the idea that God acts. Realizing this, numerous supporters of determinate entity theism have given intense attention to this topic in recent decades, with some success.[4]

My task is not to delve into the varieties of determinate entity theism, which run from process theism to deism, and from the philosophical subtleties of Boston Personalism to the dualistic hypostatization of human experiences of pleasure and pain in Zoroastrianism and Manichaeism. Nor can I review the fascinating details surrounding the question of divine action in the context of determinate entity theism (though I will comment briefly on the issue below).[5] Rather, with this description of the modelling strategies of determinate entity theism in place, let us note that theologians have made different decisions about the source of primary metaphors for modelling ultimate realities. Most of these lead to ground-of-being theologies along one of two paths.

---

[4] The conferences and volumes of the so-called Divine Action Project are a leading example. See Russell *et al.* 1993, 1995, 1998, 1999, 2001.

[5] See Thomas Tracy's contribution to this *Handbook* (Ch. 35) for a review of the topic of divine action in relation to determinate entity theism. For my review of the Divine Action Project see Wildman (2004). Responses to this paper from Philip Clayton, John Polkinghorne, William R. Stoeger, and Thomas Tracy appear in *Theology and Science*, 2/2 (Oct. 2004): 173–204. For my reply, see Wildman (2005).

On one path, we stay close to the stream of modelling strategies that flows from the fount of human nature, generalizing a selected feature of human beings to the whole of reality. Instead of focusing on humans as determinate beings with intentions and plans and powers to act, however, we might model ultimate realities in terms of the highest human virtues of goodness, beauty, and truth—understood with Plato as the deep valuational structures that permeate the form of everything real. We might centralize the most mysterious and least understood feature of humans, which surely is consciousness, and use it to model ultimate realities as a kind of ultimate consciousness pervading reality in which we participate in our own way—a common strategy within Hindu philosophy. We might concentrate on human relationality and model ultimate realities as relationality itself, distinct from any related entities—a path that leads to the *pratītya-samutpāda* doctrine of Buddhist philosophy, in which relations constitute entities, rather than the other way around. We might concentrate on the human ability to create novelty, notice that all of nature seems to share this characteristic across the various degrees and dimensions of complexity, and then model ultimate realities as creativity itself—an alternative represented by process metaphysics.[6]

On the other path, we follow alongside a stream with a different source. Instead of appealing to the most intense form of being that we know—human being—we might turn to the most pervasive and general features of reality, insofar as we can cognitively grasp them. This makes as much sense as drawing primary metaphors from human beings, and perhaps it makes more sense insofar as human beings are exceptional rather than typical, so long as the resulting view of ultimate realities can accommodate exceptional phenomena such as human beings. The most direct way to do this is simply to identify ultimate reality with everything there is, which is pantheism. This view in its strict form is quite rare in the history of philosophy, because it explains nothing and offers no moral orientation to the world. It merely proposes a lexical equivalence between the word 'God' and everything we already experience. For this reason, it has been more common to interpret ultimate realities in terms of one or more universal characteristics of the whole of reality as we experience it. For example, everything has being and is being; so, with Aristotle, we might say that ultimate reality is Being Itself. Everything realizes its potential as it acts according to its nature; so we might say, again with Aristotle, that ultimate reality is Pure Act, free of any potentiality, and thus also immutable. The fact of having features in common is similarly universal and fundamental, and the basis for speaking, with Plotinus and the Neoplatonists, of ultimate reality as One. Reality seems to be an entanglement of law-like, ordering forces and chance-like, chaotic processes, so we might regard ultimate reality as the co-primal entanglement of fundamental principles of order and chaos, which has been an important option

---

[6] This has been an important strategy in modern Western theology, with Alfred North Whitehead espousing it in his view of ultimate reality (though not in his view of God, which is an actual entity within the entire creative flux of the universe, and thus a species of determinate entity theism); see Whitehead (1978). Gordon Kaufman (1993, 2004) also articulates it.

pervading Chinese philosophy. These are all ground-of-being theologies. It is possible to combine such ground-of-being views with a world of determinate supernatural entities. For example, the variety of forces and powers in nature and human experience might lead us to hypostatize all of them in a glorious pantheon of personal deities loosely organized by a High God, or not organized at all—there are many such examples in the history of human cultures. As with process theology, this is a picture of determinate divine entities in a wider cosmological environment whose ultimate origins and meaning often remain unexplained. If they were explained, however, a ground-of-being view would usually result.

Along either of these two paths, the resulting models of ultimate realities describe not a determinate entity, but the ontological deep structures of reality itself. It follows that ground-of-being theologies can support the possibility of a God with some determinate characteristics. This determinate character in all cases would be more akin to a principle (such as Dao or Being Itself or Pure Act or the Good or a symbiosis of law-like and chance-like processes) than a personality. Thus, the denial of determinate entity theism remains.

Ground-of-being theologies posit a close relationship between the whole of reality, as human experience discloses it and makes it available for inquiry, and its ultimate metaphysical and religious character. This closeness makes ground-of-being theologies heavily indebted to forms of inquiry that produce our understanding of the world, including especially the social and natural sciences, but also the humanities, the fine arts, and the crafts of politics and economics. Ideas of ultimate realities as a determinate entity have metaphysical leverage for more of a disjunction between the character of the world and the character of God than the ground-of-being family of views can sustain. This can be helpful for producing a hopeful intellectual response to the pervasive realities of evil and suffering; some theologians find it reassuring to imagine that a determinate entity divinity has a moral character different from, higher, and certainly less ambiguous that that of the world as we experience it. Others (including me, I must confess) find this prospect even more disturbing than the morally ambiguous world of our experience. By contrast with this possibility of difference in character between God and the world, ground-of-being theologies model ultimate realities in such a way that the moral ambiguity of reality is a natural outcome deriving from the character of ultimacy itself.

It is important to acknowledge here another way of speaking of ultimate realities: namely, *apophasis*, or saying of ultimate realities that we simply cannot describe them. This is not a modelling strategy, in one respect, because apophatic theology declines modelling for the sake of testimony to a reality that utterly transcends human understanding. Yet the negation techniques of apophatic theologians are well defined.[7] They recur across traditions and languages and religious contexts. They give structure and meaning to communities of religious adepts with their

---

[7] Michael Sells (1994) argues that mystical techniques of negation and unsaying are regular and describable, and constitute a definite mode of speech. The doctoral dissertation of Timothy Knepper (Boston University, 2005) significantly extends Sell's research.

mystical practices. Thus, there is a kind of modelling at work among apophatic mystics and their theological kin, and certainly a definite kind of lifeworld and language game construction. While apophatic testimony to the incomprehensibility of ultimate realities is neither ground-of-being theology nor determinate entity theism in itself, it has far stronger affinities with ground-of-being theologies because they explicitly place ultimate realities further from human conceptuality by turning away from determinate entity modelling strategies. Thus, it is common to find theologians treating ground-of-being theology as a provisional theoretical discourse on the way to an apophatic destination.[8]

THREE SHARED FEATURES OF GROUND-OF-BEING THEOLOGIES

I have sketched a diverse range of ground-of-being alternatives to both determinate entity theism and the apophatic refusal to model ultimate realities. Whether their primary metaphors derive from profound features of human beings generalized to all of reality or from universal and fundamental features of the whole of nature, ground-of-being theologies have in common three important characteristics.

First, they throw down the gauntlet to determinate entity theism in a different way than do the anthropomorphic projection critiques of Ludwig Feuerbach and Sigmund Freud and the associated social control analyses of Karl Marx and Émile Durkheim (Friedrich Nietzsche, of course, makes both critiques).[9] The ground-of-being critiques charge many forms of determinate entity theism with philosophical inconsistency on the grounds that the idea of an infinite determinate entity is incoherent.[10] To be determinate is to have features in contrast with an

---

[8] Among recent Christian theologians this is amply evident in Karl Rahner and Hans Urs von Balthasar on the Catholic side, and Paul Tillich on the Protestant side. But the most consistent exponents of this approach are the Madhyamaka philosophers of Mahayana Buddhism, such as Nāgārjuna and Bhāvaviveka, who self-consciously frame theoretical talk about ultimate matters (for them, especially the quest for enlightenment) as a kind of middle way between the chattering noise of conventional reality and the blessed silence of emptiness. There are parallels to this in both Vedanta (Śaṅkara) and Daoist philosophy (even in the Daodejing).

[9] These well-known critiques of religion have classical status, but there is an emerging family of critiques from evolutionary biology and neuroscience that equally powerfully support critiques of religion as anthropomorphic misunderstanding. For one of each, see Boyer (2001) and Newberg and D'Aquili (2001).

[10] This is the basis for Hegel's distinction between the bad infinite, which merely extends finite characteristics to an infinite degree, and the good infinite, which transcends finite contrast. If it can be saved at all, the idea of an infinite determinate entity must have recourse

environment, whereas to be infinite is to lack determining contrasts. Thus, it is important for a theological model not to assert determinateness and infinitude of ultimate realities in the same respect. But some forms of determinate entity theism do exactly this. For example, they assert that God is infinite in respect of any temporal view, yet forms specific intentions, which necessarily are temporal conceptions, which give rise to a theological version of the famous metaphysical problem of time and eternity. Again, they assert that God is infinite in power, yet can only act in limited ways to achieve the divine purposes and to alleviate suffering and judge evil, which gives rise to the equally famous problem of theodicy, with associated conundrums of kenoticism and eschatology. The impressive debates over such theological problems show the extent to which theologians have tried to make good on the claim of some forms of determinate entity theism that God can be infinite and determinate in the same respect. Such theologians refuse the critique of incoherence, but often they do not indicate how the concept of an infinite determinate entity is philosophically coherent, the paradox having become taken for granted and so invisible within their local theological communities.[11]

By contrast, modern religious philosophers from Johann Fichte, Friedrich Schelling, and Georg Hegel to Alfred North Whitehead, Charles Hartshorne, and Robert Neville take the critique with complete seriousness, but respond differently. Whitehead and Hartshorne accept that, to have the determinate character they want to assign God in their process cosmologies, God must be subject to determinate contrasts with other features of the determinate world, and cannot be infinite in those respects (Hartshorne 1984). Philosophically consistent forms of determinate entity theism are the result, though the price paid for this is no substantive theory of

---

to the bad infinite. A classic expression of this critique is that of Johann Fichte, who wrote, 'Only from our idea of duty, and our faith in the inevitable consequences of moral action, arises the belief in a principle of moral order in the world—and this principle is God. But this living principle of a living universe must be Infinite; while all our ideas and conceptions are finite, and applicable only to finite beings—not to the Infinite. Thus we cannot, without inconsistency, apply to the Divinity the common predicates borrowed from finite existence. Consciousness, personality, and even substance, carry with them the idea of necessary limitation, and are the attributes of relative and limited beings; to affirm these of God is to bring Him down to the rank of relative and limited being. The Divinity can thus only be thought of by us as pure Intelligence, spiritual life and energy;—but to comprehend this Intelligence in a conception, or to describe it in words, is manifestly impossible. All attempts to embrace the Infinite in the conceptions of the Finite are, and must be, only accommodations to the frailties of man.' See J. Carl Mickelsen (ed.), 'Memoir of Johann Gottlieb Fichte', in Fichte (1889). Notice that Fichte himself assigns to God determinate characteristics such as pure intelligence, spiritual life, and energy, which begs the question of how he understands divine infinitude in relation to these.

[11] Karl Barth is an important exception to this trend: he developed a complex theological hermeneutic of the cognitive priority of revelation whereby we do not truly know God in Godself even through revelation, and even though faith is properly grounded in revelation. This affirms the paradox but ventures a kind of theological agnosticism about divine infinity that at least generates an answer in kind to the challenge.

ultimate realities and no solution to the classic philosophical problem of the one and the many.[12] Unlike Whitehead and Hartshorne, Neville insists on God being creator as the way definitively to solve the problem of the one and the many, and so he must protect divine infinitude. But he refuses to allow God to have a determinate character logically prior to creation—after all, what could 'determinate character of God' mean when there is nothing to contrast with God's character? Neville's uniquely consistent theory of creation has both God's nature and the world's becoming determinate in a singular primordial event of creation, accordingly.[13]

This instance of contrasting philosophical and theological intuitions continues a long-standing conflict within Western religious thought, and it is important to approach it with sympathy and understanding. Consider the era of doctrinal formation within Christianity, culminating in the epic Ecumenical Councils of Nicaea (325), Constantinople (381), and Chalcedon (451). Despite the unattractive and sometimes deadly politicking of this era, serious intellectual questions were at stake, particularly concerning how to conceive of God. Theologians of the time had inherited two streams of wisdom that were in tension with one another. On the one hand, biblical religion, present especially in the Septuagint and the emerging compilation of early Christian writings that came to be called the New Testament, portrayed God mostly as a divine person who made plans and covenants and plainly acted in history and nature. This portrayal is accompanied especially in the Hebrew Bible by stern warnings about idolatry and occasional comprehension-defying theophanies, but there is no question that the dominant biblical picture of God is as a determinate entity. On the other hand, Greek philosophy, present in this era early on as late Stoicism and Middle Platonism, and later as Neoplatonism, portrayed ultimate reality as a transcendent One from which all determinate things have their life and purpose and value. This God has characteristics such as aseity, immutability, impassibility, and transcendence—all negations conveying essentially the same point that God is not a determinate entity. Arguably, the biblical writings register such features in their critiques of idolatry and in hints that God is beyond human comprehension, but in general the Bible subordinates these abstract features to personal characteristics. The resulting tension established the famous Athens–Jerusalem conflict that has rumbled on throughout the history of Christian theology, and in parallel forms in Jewish and Muslim theology, all of which are inheritors of Greek philosophical traditions as well as sacred textual testimony to God as a determinate—indeed, a personal—entity.

Christian theologians of the doctrinal formation period boldly attempted to forge a synthesis between the two perspectives. This involved fending off extremes on both

---

[12] Whitehead dramatically reframes the one-and-many problem so that it is about the emergence and dissolution of societies of actual occasions, but offers no answer to the classical, large-scale version of the problem, which has typically received answers in terms of creation within theistic contexts.

[13] This creation is not the act of a determinate divine being. Rather, it is logically prior to determination of anything divine or human (Neville 1968).

sides: unqualified personal theism and unqualified Gnostic philosophies. The former seemed plainly pagan and mythological, while the latter seemed opposed to the intimately personal portrayal of God in the Hebrew Bible and in the teaching and example of Jesus Christ. St Augustine's *Confessions* is a glorious example of this synthesis, combining philosophical acuity with spiritual intimacy. Only a harshly positivist critic would reject this synthesis as a futile jumble of contradictions; these theologians were grappling with two vast intuitions and trying to affirm both, rather than reducing the doctrine of God to one side or the other. In this way, personal characteristics of divine reality were typically set within a ground-of-being philosophical framework, while ground-of-being conceptualities were interpreted with an eye to spiritual vibrancy and salvific relevance. The most systematic examples of striking this balance are the majestic *Summas* of St Thomas Aquinas, for whom God was Being Itself and also Holy Trinity, miraculous actor, and source of grace and salvation.[14] Thomas's view cannot easily be called determinate entity theism, despite supportive textual evidence, because it adopts an *Esse Ipsum, Actus Purus* (Being Itself, Pure Act) conceptual model. But neither is it consistent ground-of-being theology, because Thomas plainly also attributes to God characteristics that properly belong only to determinate entities. It is, rather, the epitome of the Athens–Jerusalem paradox of Christian theology. While determinate entity theisms and ground-of-being theologies are clearly distinguishable from one another in our era, they both have some claim on the classical synthetic theological models of medieval and patristic Christian theology.

Today, the goal of synthesizing Jerusalem and Athens remains, but contemporary Christian theologians have also rehabilitated more one-sided projects in search of

---

[14] See *Summa Contra Gentiles* and *Summa Theologica*. Retrievals of Aristotelian philosophy long lost in the West, these Christian theological works were initially controversial, but have proved enormously influential. For example, Thomas's influence is evident today in a contrast between Catholic and Protestant approaches to divine action. While post-Reformation biblicism tended to separate theology from its philosophical sources, Catholic theologians continued the synthetic heritage of patristic theology. Contemporary Protestant theologians think of special divine action as natural-law-abrogating miracle or natural-law-conforming causal entries into history and nature, or else they relegate God's action to a single eternal act of creation (with no special acts, as in Friedrich Schleiermacher), or reduce it to purely subjective encounter (as in Rudolf Bultmann—as if human subjectivity were not tied to the brain and its causal interactions). The views affirming special objective divine action presume and entail determinate entity models of God. But these models of God are generally unappealing to most Catholic theologians, and so the models of divine action that imply determinate entity theism are similarly unappealing. Rather, they tend to hold Thomas's ground-of-being view instead, rejecting the idea of God as an actor alongside other finite actors, and affirming that, as Being Itself, God acts in all events as primary cause beneath the flux of ordinary secondary causes with which we are utterly familiar. Making this Thomistic conception of divine action amenable to the idea of special divine acts is famously difficult in our time, in a way that it was not for Thomas himself. The primary–secondary causation model seems to require a distinction in degrees of divine focus of attention to make sense of any distinction between ordinary divine support of all causes and special divine acts. This difficulty recapitulates the more fundamental Athens–Jerusalem paradox of the Christian idea of God.

consistent models of God, freer of the conceptual tensions inherent in the synthetic project. Thus, determinate entity theisms have achieved honourable standing—even in process forms that reject traditional Christian teachings such as creation *ex nihilo*, divine infinitude, and omnipotence. Personal theists routinely reject or radically reinterpret the classical doctrines of divine aseity, immutability, and impassibility as inappropriate incursions of philosophical conceptualities into the biblical portrayal of God as a divine person.[15] For their part, some ground-of-being theologians have also interpreted the synthetic project as impressive but ultimately doomed to futility. They resolve the paradoxical tension by treating personal attributes of God as non-literal symbolic affirmations and privileging the conceptual framework of a plausible metaphysics over the narrative framework of biblical theism.[16] Paul Tillich, a ground-of-being theist, rightly affirmed the continuity of his view with biblical theism (1964), but this continuity was with the theocentric, prophetic, and iconoclastic elements of the Bible, not with the portrayal of God as a divine person.

In summary, ground-of-being models of ultimate realities—whether as Thomas's Being Itself or Plotinus's One or Plato's Good or Hegel's *Geist* or Tillich's Power of Being or Neville's Creator—present a serious alternative to determinate entity theism. They press critiques against determinate entity theism: religious critiques of anthropomorphism, philosophical critiques of incoherence, and theological critiques of excessive innovation relative to the synthetic heritage of Christian theology, which generally interpreted personal symbolism for God in a ground-of-being framework, despite the insoluble paradox that seems to result.

The second feature that ground-of-being theologies have in common is a rich appreciation of the symbolic life of practical religion. Within a ground-of-being theology, many religious beliefs, including those characteristic of determinate entity theism such as special divine action, make sense only if understood non-literally as symbolic expressions of the religious significance of the world we experience. A cynical interpreter might see this as hostility to practical religious concerns, but ground-of-being theologians take symbolism more seriously than this. To recall one of Paul Tillich's most pointed instructions to his students, we should never say 'merely a symbol'.

Consider the belief, common among all religions, that God (or a given supernatural entity in non-theistic contexts) is a personal being who acts specially to answer prayers. A determinate entity theist typically is willing to affirm this sort of divine responsiveness and intentional activity literally: there really is a divine being who hears, and the world is different from how it would otherwise be if this divine entity chooses to act in response. Ground-of-being theologies regard this belief as mistaken if we interpret it at the level of literalized metaphysics, but as profoundly meaningful if we interpret it non-literally, as a symbolic expression of human dependence on a

---

[15] A seminal example of this is Harnack (1976), with a more accessible presentation in Harnack (1978). Among contemporary theologians, the standard example is Moltmann (1974).

[16] An excellent example of this is Neville (2001*d*).

ground of being and the sometimes happy way that the creative flux of events sometimes works out for human beings. Metaphysically, this is action without intentional agency and without causal joints between God and nature. Spiritually, ground-of-being theologies treat divine action as religious symbolism that engages people in spiritually transformative praxis within the flow of life-giving and life-threatening events.

Religious symbols are not merely targets for demythologization or remythologization, on this view, but means of engaging the ground of being in our lives. We might imagine that we could live without religious symbols that need to be reinterpreted in a metaphysical framework more plausible than the narrative framework that gave birth to them. I think this is unlikely. Without religious symbols to help us conceive our world and orient ourselves to it, the moral character of human life would be perpetually superficial and localized. The world would remain a terrifying jumble rather than becoming a kind of cosmic home. Other people around us would reinforce our fears rather than being opportunities for deepening understanding through compassion. Our connection to the divine depths of our experience would remain undeveloped. Religious symbols are always broken and of uncertain parentage, but they also enable us to recognize the depths of our world around us and approach life with courage, civility, and creativity.[17]

Symbolizing ultimate realities in personal terms is common, and indeed enormously popular, in all religions. The Bible, the Qur'an, and some Vedic literature encourage this personal view of ultimate reality even while resolutely resisting the anthropomorphic mistakes that so often accompany such symbolism. Even putatively non-theistic religions such as Buddhism are, in popular forms, replete with gods and monsters, bodhisattvas, and discarnate entities that form intentions and act freely in the world to get or give what they want—all relative to the narrative structures that guide daily religious life for Buddhists. The popularity of such symbolism may derive from hard-wired propensities to picture the world in anthropomorphic terms—hence the social success of groups that nurture such symbols.[18] Theologians who believe that they must take the first-order symbolism of religious practice at face value as much as they possibly can would be hard pressed to adopt a theological view other than determinate entity theism. But this approach risks reducing the critical task of theological reflection merely to serving the ideological needs of religious institutions, which properly include rationalizing and legitimating the beliefs that make psychological comfort, corporate identity, and social power possible.

Meanwhile, ground-of-being theologians face their own unappetizing challenge. By refusing the literalized metaphysics of a personal, intentional, active divine entity, they must go beyond generalized approval of the engaging power of religious

[17] One of the most important theories of religious symbolism from a ground-of-being perspective is Neville (1996).

[18] This is one contention of numerous recent works on the religious implications of neuroscientific understandings of human cognition; see Newberg and D'Aquili (2001).

symbols to explain how engagement works and to give careful interpretations of symbols whose literalized metaphysical sense, they hold, is mistaken. The explanatory task is sometimes neglected, as if engagement (or its equivalents) were a self-evident or foundational concept.[19] The interpretative task has been standard fare in the intellectual traditions of all religions with sacred literatures, from ancient times until now. This is not a novel or newly difficult task, therefore, so much as the continued discharging of a perpetual theological obligation.

The third common feature of ground-of-being theologies is that they embrace the whole of reality in all its complexity and ambiguity and speak of ultimate realities as the fundamental reason why things are this way. This is the basis for the analogy of 'ground' in the phrase 'ground of being': ultimate realities are the most basic ontological condition of reality. Insofar as we can know anything at all about ultimate realities, we will gain this knowledge by examining not putative supernaturally delivered divine revelation but nature and experience. Indeed, ground-of-being theologians interpret nature and human experience as the primal spring of all revelation, all divine disclosure, all insight, and all transformative understanding.[20] Sacred texts testify to these fecund origins, and theological traditions formulate and reflect on the wisdom encountered there.

This amounts to a collapse of the traditional distinction between revealed and natural theology, which depends on two modes of obtaining knowledge about ultimate realities—a distinction ventured in all theistic religions. But if there is one mode only, then revealed theology is not only naturalized, but natural ways of knowing must also be revelation, whether inside or outside theology. If we understand ultimate reality as the ground of being, in any of the various senses that this has been tried, then all knowledge, regardless of subject-matter, is the result of engagement with a reality that grounds and transcends us, that we encounter as given rather than simply at our cognitive disposal, that resists our ideas about it and forces us to adapt. In respect of its givenness, or equivalently our thrownness (to use Martin Heidegger's term), the world is revelation. In respect of its cognitive penetrability and receptiveness, the world is an object of knowledge and a means of theological inquiry.

This understanding of revelation indicates the sense in which ground-of-being theologies resist supernaturalism: supernatural entities are not problematic in themselves, but supernatural modes of gaining information about ultimate realities contradict the very idea of God as ground of being. Such supernatural knowledge reconstitutes the idea of God as a determinate entity that possesses and conveys information otherwise unavailable to human beings. The naturalized understanding

---

[19] Aristotle may have complained that his teacher, Plato, had not properly explained what participation was or how it worked, but not all Platonists have failed in this task. Neville (1996) explains what engagement means by conceiving truth not as correspondence between propositions and states of affairs but in semiotic and axiological terms as the carry-over of value from something interpreted to the interpreter. This dynamic, causal interpretation of truth means that engagement is present in nature wherever there is interpretation.

[20] See e.g. the differently angled approaches of Heidegger (1962) and Hart (1968).

of revelation also indicates the sense in which ground-of-being theologies are tightly knitted into the whole fabric of human knowledge, and dependent on the wisdom and skills of human inquirers for their content and plausibility. Unsurprisingly, therefore, ground-of-being theologies have a lot at stake in the natural and social sciences, as well as other modes of studying the ways in which human beings experience the world.

The closeness between nature and its ontological ground explains why ground-of-being theologies have close affinities with religious forms of naturalism. Indeed, does not ground-of-being theology finally reduce to ground-of-being (or religious) naturalism? Because of pervasive suspicion in many naturalist quarters toward religious traditions, sympathy for theological terminology is often absent there, and the need for distinctive language pronounced. Yet, at the conceptual level, religious naturalism typically is also ground-of-being theology. In fact, naturalism and ground-of-being theologies are overlapping classes of metaphysical views whose common territory can be equally well called 'religious naturalism' or 'ground-of-being naturalism'. Consider the two non-overlapping territories.

On the one hand, some forms of naturalism (not 'religious naturalism') are bluntly opposed to ground-of-being theologies. These are strictly positivistic, physicalist forms of empiricism, entertaining no questions that cannot be answered from within the scope of the physical sciences. From the point of view of ground-of-being theologies, these forms of naturalism are arbitrarily truncated metaphysical theories. They refuse to consider legitimate questions about the ultimate origins and meaning of nature, the reality of aesthetic and moral values, and the ontological basis for the mysterious applicability of mathematics to modelling nature. All of these issues press the question of an ontological ground for nature, and non-religious naturalists cannot answer them without reconstituting an ultimate ontological basis for reality. The better, more adequate way to refuse such questions is explicitly to adopt a kind of ascetic spiritual discipline that refuses speculative theorizing, in a manner akin to some versions of Madhyamaka and Zen Buddhist philosophy, to austere versions of post-modernist deconstruction, and of course to apophatic mysticism in all traditions. Engaged or not, however, the questions remain, and even the refusal to consider them is mute testimony to their importance. I consider it one of the great discoveries of modern philosophy of religion that consistent non-religious naturalism, or equivalently, ontologically and axiologically flattened-out atheism, is intellectually untenable.

On the other hand, there are forms of ground-of-being theology that cannot be given a religious naturalist translation. For example, some ground-of-being theologies propose grand cosmological schemes that are utterly indigestible to a naturalist of any kind, such as the perennial philosophy's Great Chain of Being, with its hierarchically organized gods and angels and demons, its spirits and discarnate entities, its human beings, and its lesser animals and plants and inert matter (Smith 1976). The perennial philosophy belongs to the ground-of-being family because its picture of ultimate reality is God beyond God, God without attributes (*nirguna Brahman*). And religious naturalism belongs for the same reason. But the latter's relatively sparse ontological inventory is incompatible with the perennial philosophy. It follows that there is a meaningful distinction between naturalism and ground-of-being theologies, even while the two families of views overlap.

GROUND-OF-BEING THEOLOGIES AND THE NATURAL SCIENCES

Unsurprisingly, given the necessary closeness between nature and its ontological basis, ground-of-being theologies connect to the natural sciences differently than do determinate entity theisms. To appreciate this difference, consider a concrete example: the design argument. Despite significant differences of detail and context, the design argument works roughly the same way in its medieval form (say, in Thomas Aquinas's Fifth Way in *Summa Theologica*, Part I, Question 2, Article 3), in its modern form (say, in William Paley 1802), and in its contemporary form (say, in the intelligent design movement—see Dembski 1998): it seeks to infer the activity of a designer from apparent design in natural objects and processes. An eye, complex cellular mechanisms, an ecosystem, our solar system—each has seemed designed to some people, a state of affairs begging for an explanation. The natural impulse to propose that some intelligence actually designed what seems to be designed has historically seemed compelling because no other explanation was available.

David Hume (1780) famously pointed out that even if the inference from apparent design to a designer is a sound one, we can conclude nothing about the character of this designer, or even the number of designers.[21] This was a sound logical point and dented the design argument's usefulness for specifically Christian apologetics, but had little effect on the core of the design argument, which is the inference from apparent design in nature to a designer. Charles Darwin's (1859) theory of evolution, by contrast, had a profound impact on the plausibility of the design argument in its biological version simply by articulating an alternative explanation for apparent design. Now the question became: which hypothesis is the better explanation of apparent design in nature: an intelligent designer (in any number of forms) or biological evolution?

Much the same transformation occurred in the cosmological version of the design argument. In its strongest form, the cosmological design argument takes off from the fine-tuning of the fundamental constants of physics, without which life would be inconceivable (Barrow and Tipler 1986). The apparent contingency of well co-ordinated and fine-tuned fundamental constants has seemed beyond the ability of

---

[21] Hume argued through an elegant dialogue form, among other things, that no confirmation of specifically Christian theological claims about God is possible.

science to explain, and thus to be persuasive evidence for intelligent design. Alternative explanations of fine-tuning depend on scenarios with many universes, each with different sets of fundamental constants. A mother universe with daughter universes having different constants was proposed, as was an everlasting cycle of cosmic expansions and contractions with different constants for each big bang, and numerous other variations on the multiverses theme.[22] One version of string theory now supports a rich mathematical model of a multiverse with non-overlapping, expanding regimes, each with distinct settings for basic physical constants. The model provides a quantum-mechanical explanation for the way the set of constants applying in a new section of the universe migrate from the set in place when the new expanding section is born.[23] While still speculative, these mathematical models are robust enough to have an impact on the fine-tuning wing of the design argument. Moreover, these multiverse explanations for fine-tuning will probably become more detailed, and perhaps empirically distinguishable from competitor theories in the years to come. So the question in this case is which hypothesis is the better explanation of apparent design in nature: an intelligent designer (in any number of forms) setting physical conditions for life or a scenario in which a much vaster universe automatically explores countless settings of physical constants, some producing life and others not?

Determinate entity theisms and ground-of-being theologies have quite different responses to these developments in the biological and cosmological versions of the design argument. To put the difference succinctly, ground-of-being theologies have a lot to lose if the design argument succeeds. The success of the design argument would strongly suggest that God is, after all, an entity capable of planning ahead and acting so as to set physical constants or assemble macromolecules and cellular machinery, and this would be powerful support for determinate entity theism. To the extent that this is a genuine possibility, faint though it may be given the vicissitudes of scientific theories, ground-of-being theologies are vulnerable to falsification—though surely this is a great virtue as far as their intelligibility is concerned. By contrast, determinate entity theisms coexist relatively easily with either the success or the failure of the design argument. Consider the alternatives.

On the one hand, if the design argument succeeds, then otherwise unexplainable contingencies of nature are credited to divine action, including possibly special divine interventions at crucial moments in cosmic and evolutionary history. This outcome would delight some personal theists, including especially theists within the intelligent design movement, for whom this would represent political as well as intellectual triumph. Other personal theists and perhaps most process theists

---

[22] John Wheeler developed the original version of the oscillating universe in the 1960s, but this proposal was ruled out in work done by Roger Penrose and Stephen Hawking. Paul J. Steinhardt and Neil Turok (2002) have revised the original idea in a more credible way. Andrei Linde (1994) proposed the creation of child universes from a parent universe's 'quantum foam', describing this as the self-reproducing inflationary universe.

[23] The leading version of this view is Susskind (2005).

would probably be less delighted to see God's action on such blunt display, because this would suggest that perhaps God ought to have intervened more decisively at moments of human civilizational mayhem. Despite this complication for theodicy, determinate entity theism would manage quite well if the design argument succeeds.

On the other hand, the design argument may fail, perhaps because contingency in the fundamental constants evaporates with the development of quantum cosmologies and string theory, and because complexity theory succeeds with biochemistry and evolutionary theory in explaining cellular assemblage, organ development, and similar challenges in biology. In that case, determinate entity theism happily falls back on the less aggressive position of theistic evolution. This view holds that God made (and possibly makes) the world in whatever way current science suggests is probably the case, without any need deliberately to set contingencies such as physical constants or miraculously to assemble molecules and cells and organs and ecosystems. Intelligent design theists would not be happy with this at all, because it would confirm their worst fears: namely, that evolution is divinely created to run automatically, without subsequent intervention, or else is just an aspect of the divine depths of nature. But most determinate entity theists would not be perturbed in the least by this outcome.[24]

In either case, therefore, most forms of determinate entity theism fare splendidly. Only extreme types of personal theism are tripped up by the failure of the design argument. By contrast, ground-of-being theologies are inevitably committed to theistic evolution in a particular sense, or to its non-theistic, religious naturalist equivalent. They cannot tolerate the success of the design argument if it suggests deliberative divine action. Indeed, they predict its failure, and correspondingly expect the success of attempts to explain cellular complexity and contingency of fundamental constants without any need to invoke intelligent design. Ground-of-being theologies hypothesize that the contingent elements of the universe do not include physical constants and biological complexity, but merely the vast and bare fact that the universe itself exists.

This difference between the two families of theologies illustrates the sense in which ground-of-being theologies are more tightly knitted into the character of physical reality than determinate entity theisms. In the latter, the particularity of the divine nature allows for the possibility that a determinate entity, God, could make the world in such a way that it contained no hints about the divine character save the sheer fact that God must have the power to create. Ground-of-being theologies do not allow for this possibility, because they frame ultimate reality metaphysically as the ground of the world as we encounter it. This gives ground-of-being theologians strong incentive to support inquiry of every kind into the worlds of nature and value, of experience and consciousness. They all illumine the ground of being through detailing the character of being in all its richness. This also imposes on ground-of-being

---

[24] Indeed, most or all of the theologians in the Divine Action Project affirm theistic evolution in something like this sense (Russell *et al.* 1993, 1995, 1998, 1999, 2001).

theologians a strong obligation to ensure that what they say about ultimate reality applies to the world in all its dimensions as human inquiry unfolds it.

Two important questions now arise. First, are there renewed prospects for natural theology in the presence of ground-of-being theologies? Critiques of natural theology derive from the insufficiency of human reasoning powers, the intrinsic ambiguity of nature in the face of metaphysical questions, or the transcendence of God such that we cannot expect creation to be informative about the divine nature or existence. Ground-of-being theologies directly affect the third of these critiques. In theistic formulations, the ground of being transcends the world as its mystery and ontological condition; the divine nature is precisely what the world discloses in the depths of its physical and valuational structures and processes. If the first two difficulties can be managed, therefore—and this is a complex matter in itself— ground-of-being theologies do make natural theology more promising.

Second, can the natural sciences or other types of human inquiry leverage a judgement in favour of ground-of-being theologies over determinate entity theisms, or vice versa? It is difficult to assess entire classes of theologies, but their common features permit some room for comparative judgements. We have seen that determinate entity theisms have considerable flexibility in relation to scientific discoveries, because of the way they construct transcendence between God and the world. We have also seen that ground-of-being theories are theological interpretations of the world precisely as the sciences and other forms of inquiry discover it to be. These two considerations entail that the sciences cannot directly discriminate between the two families of theological views. Yet we also saw that the two families have different postures in relation to the question of intelligent design, with the ground-of-being views increasingly vulnerable should scientific inquiry fail to explain the contingency of physical constants and apparent design in the biological realm. This suggests that the *extended failure* of scientific inquiry could have an impact on the decision between ground-of-being theologies and determinate entity theisms in a way that the *success* of science cannot.

Moreover, science and other types of inquiry could have an indirect influence on this metaphysical contest by changing criteria for plausibility. For example, it is conceivable that the natural sciences could strongly reinforce (without directly entailing) a particular philosophy of nature, as Aristotle's science strongly supported his teleological, organismic philosophy of nature. The basic ontological principles of such a philosophy of nature—in Aristotle's case, there must be an ontological basis for purposive phenomena in nature that harmonizes these natural purposes in a kind of teleological ecology—would then function as plausibility conditions for any theological interpretation of nature. In this way—which is to say by means of a philosophy of nature with ontological principles serving as plausibility criteria for theological models—a comparative form of natural theological argumentation may be possible (Wildman 2006). This would not be traditional natural theology, with its pretensions to infer theological truths directly from nature. Nor could it lead to decisive conclusions in favour of one and against another theological proposal. But this mediated and comparative type of natural theology reasoning can put pressure on some theological views more than others, depending on how poorly or well they harmonize with the plausibility conditions deriving from the philosophy of nature. And this, in turn, may be the most that the natural sciences and other forms of inquiry can offer to the theological competition between ground-of-being theologies and their determinate entity rivals. The rest of the debate remains internal to theology and metaphysics.

References and Suggested Reading

Barrow, John D., and Tipler, Frank J. (1986). *The Anthropic Cosmological Principle*. Oxford: Clarendon Press; New York: Oxford University Press.

Boyer, Pascal (2001). *Religion Explained: The Evolutionary Origins of Religious Thought*. New York: Basic Books.

Darwin, Charles (1859). *On the Origin of Species by Means of Natural Selection, or the Preservation of Favoured Races in the Struggle for Life*. London: John Murray.

Dembski, William A. (1998). *The Design Inference: Eliminating Chance through Small Probabilities*. New York: Cambridge University Press.

Fichte, Johann (1889). *The Popular Works of Johann Gottlieb Fichte*, i. trans. William Smith, 4th edn. London: Trübner & Co.

Harnack, Adolf von (1976). *History of Dogma*. Gloucester, Mass.: Peter Smith.

—— (1978). *What is Christianity?* Gloucester, Mass.: Peter Smith.

Hart, Ray L. (1968). *Unfinished Man and the Imagination: Toward an Ontology and a Rhetoric of Revelation*. New York: Herder & Herder.

Hartshorne, Charles (1984). *Omnipotence and Other Theological Mistakes*. Albany, N.Y.: State University of New York Press.

Heidegger, Martin (1962). *Being and Time*, trans. E. Robinson and J. MacQuarrie. New York: Harper.

Hume, David (1780). *Dialogues Concerning Natural Religion*. London.

Kaufman, Gordon (1993). *In Face of Mystery: A Constructive Theology*. Cambridge, Mass.: Harvard University Press.

—— (2004). *In the Beginning ... Creativity*. Minneapolis: Fortress Press.

Knepper, Timothy D. (2005). 'How to Say What Can't be Said: Techniques and Rules of Ineffability in the Dionysian Corpus' (Ph.D. diss., Boston University).

Linde, Andrei (1994). 'The Self-Reproducing Inflationary Universe', *Scientific American*, 271/ 5: 48–55.

Moltmann, Jürgen (1974). *The Crucified God: The Cross of Christ as the Foundation and Criticism of Christian Theology*. New York: Harper & Row.

Neville, Robert Cummings (1968). *God the Creator: On the Transcendence and Presence of God*. Chicago: University of Chicago Press.

—— (1996). *The Truth of Broken Symbols*. Albany, N.Y.: State University of New York Press.

—— (2001*a*) (ed.). The Human Condition *The Comparative Religious Ideas Project*, i: Albany, N.Y.: State University of New York Press.

—— (2001*b*) (ed.). Ultimate Realities *The Comparative Religious Ideas Project*, ii: Albany, N.Y.: State University of New York Press.

—— (2001*c*) (ed.). Religious Truth *The Comparative Religious Ideas Project*, iii: Albany, N.Y.: State University of New York Press.

NEVILLE, ROBERT CUMMINGS (2001d). *Symbols of Jesus: A Christology of Symbolic Engagement.* Cambridge and New York: Cambridge University Press.

NEWBERG, ANDREW, and D'AQUILI, EUGENE (2001). *Why God Won't Go Away: Brain Science and the Biology of Belief.* New York: Ballantine Books.

PALEY, WILLIAM (1802). *Natural Theology: Or, Evidences of the Existence and Attributes of the Deity Collected from the Appearances of Nature.* London: Faulder.

RUSSELL, ROBERT J., MURPHY, NANCEY, and ISHAM, C. J. (1993) (eds.). *Quantum Cosmology and the Laws of Nature: Scientific Perspectives on Divine Action.* Vatican City State: Vatican Observatory Publications; Berkeley: Center for Theology and the Natural Sciences.

—— —— and PEACOCKE, ARTHUR (1995) (eds.). *Chaos and Complexity: Scientific Perspectives on Divine Action.* Vatican City State: Vatican Observatory Publications; Berkeley: Center for Theology and the Natural Sciences.

—— STOEGER, WILLIAM R., SJ, and AYALA, FRANCISCO J. (1998) (eds.). *Evolutionary and Molecular Biology: Scientific Perspectives on Divine Action.* Vatican City State: Vatican Observatory Publications; Berkeley: Center for Theology and the Natural Sciences.

—— MURPHY, NANCEY, MEYERING, THEO C., and ARBIB, MICHAEL A. (1999) (eds.). *Neuroscience and the Person: Scientific Perspectives on Divine Action.* Vatican City State: Vatican Observatory Publications; Berkeley: Center for Theology and the Natural Sciences.

—— CLAYTON, PHILIP, WEGTER-McNELLY, KIRK, and POLKINGHORNE, JOHN (2001) (eds.). *Quantum Mechanics: Scientific Perspectives on Divine Action.* Vatican City State: Vatican Observatory Publications; Berkeley: Center for Theology and the Natural Sciences.

SELLS, MICHAEL (1994). *Mystical Languages of Unsaying.* Chicago: University of Chicago Press.

SMITH, HUSTON (1976). *Forgotten Truth: The Primordial Tradition.* New York: Harper & Row.

STEINHARDT, PAUL J., and TUROK, NEIL (2002). 'A Cycle Model of the Universe', *Science*, 296: 1436.

SUSSKIND, LEONARD (2005). *The Cosmic Landscape: String Theory and the Illusion of Intelligent Design.* London: Little, Brown.

TILLICH, PAUL (1964). *Biblical Religion and the Search for Ultimate Reality.* Chicago: University of Chicago Press.

WHITEHEAD, ALFRED NORTH (1978). *Process and Reality: An Essay in Cosmology*, corrected edn. ed. David Ray Griffin and Donald W. Sherburne. New York: Free Press.

WILDMAN, WESLEY J. (2004). 'The Divine Action Project, 1988–2003', *Theology and Science*, 2/1: 31–75.

—— (2005). 'Further Reflections on the Divine Action Project', *Theology and Science*, 3/1: 71–83.

—— (2006). 'Comparative Natural Theology', *American Journal of Theology and Philosophy*, 26: 173–90.

PANENTHEISM AND ITS CRITICS

CHAPTER 37

THE POTENTIAL OF PANENTHEISM FOR DIALOGUE BETWEEN SCIENCE AND RELIGION

MICHAEL W. BRIERLEY

Go for a walk. Get wet. Dig the earth.

(Rowan Williams)

Thus Rowan Williams concludes the foreword to a report about the environment published by the Church of England's Mission and Public Affairs Council in 2005 (Williams 2005: p. viii). Williams is more classical theist than panentheist; but his sentiment, urging people to re-experience and re-learn their connections with the natural world, finds a strong echo within panentheism. Indeed, panentheists would contend not only that panentheism is better able than classical theism to support and resource this sort of environmental activity and respect, but, furthermore, that there are many areas, of which ecology is just one, in which panentheism both corresponds

Some of the material in this chapter is drawn from Ph.D. research at the University of Birmingham. I am grateful to Marilyn McCord Adams, Philip Clayton, Richard Harries, and Keith Ward for their comments.

more closely to people's experience and also is more sensitive to ethical concerns than alternative forms of theism. Among these other areas is dialogue between science and religion.

'Dialogue' between science and religion—the very act of using this volume— presupposes participants or readers who believe that they have something to gain from engagement between the two disciplines. All but the most hardline positions or attitudes that scientists or theologians may adopt in relation to the possibility of their disciplines' interaction suggest that such interaction might be fruitful (see e.g. the range of positions categorized in Clayton 2001c: 214–15). This essay, written from the perspective of theology and advocating a certain kind of theism over and above two main rivals, assumes that scientists will be willing to consider that the theism under consideration—panentheism—has something to offer: namely, the overcoming of some of the difficulties, both theological and scientific, under which classical theism, the hitherto dominant form of theism, has laboured. This chapter is not the place to highlight the theological advantages of panentheism; rather, after exploring the meaning of panentheism, it will look at divine action as an example of current debate in science and religion where panentheism makes important contributions. But the essay will then also draw attention to the areas of experience and ethics implied by Williams's words, in the belief that panentheism is able to aid the connection of science to the subjective realities of religious experience and moral decision making.

THE MEANING OF 'PANENTHEISM'

'Panentheism', from the Greek *pan en theos*, means 'all in God', and is commonly distinguished from (and conceived as the median between) classical theism, in which God is essentially separate from the cosmos, and pantheism ('all [*is*] God'). The 'in' is thus the critical term. According to the second (1989) edition of the *Oxford English Dictionary*, 'in' expresses 'the relation of inclusion, situation, position, existence, or action, within limits of space, time, condition, circumstances, etc.'. Its use in theology is, in common with other theological language, metaphorical (McFague 1982). As a metaphor, the 'in' might be most helpfully explicated through analysis of a continuum of further metaphors (cf. Clayton 2004d: 251–2) as set out below.

1. God is separate from the cosmos.
2. The cosmos will be in God.
3. God is present to the cosmos.
4. God contains the cosmos.
5. God is affected by the cosmos (e.g. God suffers).
6. God acts in and through the cosmos.

7. The cosmos is a sacrament, or sacramental.
8. God penetrates the cosmos.
9. God is the ground of the cosmos.
10. The cosmos is God's body.
11. God includes the cosmos, as a whole includes a part.
12. God and the cosmos are inextricably intertwined.
13. God is dependent on the cosmos.
14. God is dipolar.
15. God is totally dependent on, or coterminous with, the cosmos.

Points (1) and (15) lie outside panentheism, and represent classical theism and pantheism respectively. Point (2) is a classical position, while also a form of panentheism known as 'eschatological panentheism', held for example by Keith Ward (2004: 72) and John Polkinghorne (1994: 64, 168; 1996: 55; 2000: 90–1, 94–5; 2002: 115; 2004: 166). On this view, while all is not yet 'in God', this is the situation for which God is (and the cosmos should be) aiming, and which will ultimately pertain. Niels Gregersen has called this position 'soteriological panentheism' (2004: 21, 24–7). It differs from the eschatology of panentheism 'proper', which would hold that while all is 'in God', God is not yet '*all* in all' (cf. 1 Corinthians 15: 28), which can happen only when evil is finally eliminated from the cosmos.

Point (4) might be thought to provide the most straightforward meaning of panentheism's 'in'—the sense that something is contained by something else, such as water 'in' a saucepan. It is possible, however, on a container model, still to conceive of God as essentially outside the cosmos. The saucepan, after all, is 'outside' the water. In the notion of *zimzum*, for example, explored by Jürgen Moltmann (1981: 109–11; 1985: 86–93, 156–7; 2001: 145–8; 2003: 62, 119–20), God makes a space within Godself in which the cosmos exists. Panentheists wish to establish a closer relationship than that implied by God as a mere 'container'; the meaning of 'in' that is required is one that articulates mutual coinherence between God and the cosmos.

To avoid the implication that God remains outside the cosmos, and to establish mutual coinherence, panentheists often invert the meaning 'all in God' to state that God is also 'in' everything. This notion attempts to supplement, with a closer relationship, the transcendence safeguarded by 'all in God' and implied by God as a simple 'container'. The classic definition of panentheism in the *Oxford Dictionary of the Christian Church*, for example, specifies that God's 'Being is more than, and is not exhausted by, the universe', but also touches on point (8) (and point (11)) in suggesting that 'the being of God includes and penetrates the whole universe, so that every part of it exists in Him' (Cross and Livingstone 2005: 1221). Ironically, however, 'God in all' can be interpreted in ways that are compatible with classical theism, such as divine omnipresence (point (3)). On its own, then, the phrase 'all in God' may not take understanding of panentheism very far; but 'God in all' does not take it much further, being capable of both classical and panentheistic interpretations.

Hence points (4)–(8) can be common to both classical theism and panentheism. They represent modifications of classical theism (cf. Griffin 2005); even if one maintains point (1), it is possible to stretch classical theism to these points. Classical theism is sometimes stretched even to point (9), though this is arguably inconsistent; it has also been stretched, for example by Ramanuja and Austin Farrer, to point (10). As long as point (1) is held, the resulting outlook is classical. That is why possibility (point (5)) is not intrinsically panentheistic: it can be maintained under both classical theism and panentheism, even though it has a bias towards the latter on account of its closeness to points (12) and (13) (cf. Brierley 2001: 230–1). Similarly, the language of point (6) (on which point (7) is logically dependent) often implies a panentheistic doctrine, while it is also susceptible to classical interpretation.

Panentheism *starts* with point (4), but points (8) and (9) begin to signify a distinctively panentheistic doctrine. Point (9) was championed above all by Paul Tillich, and is a crucial aspect of panentheism, for it is not possible for something to be entirely separate from its ground (hence panentheism is a negation of point (1)). It is possible for panentheism to stop with points (8) and (9), a position that one might call 'basic' panentheism. Arthur Peacocke, for example, is guarded about point (10) and resists point (11) (1993: 371; 2004a: 145–6). Point (10) marks the beginning of 'full-blown' or 'advanced' panentheism (I prefer the terms 'basic' and 'advanced' to 'weak' and 'strong', which have also been used to describe different panentheisms (Peterson 2001: 399), as less value-laden[1]). Philip Clayton, who of contemporary theologians has the clearest grasp of panentheism, has called point (10) the 'panentheistic analogy' (1997a: 100–2; 2000a: 702–3; 2003: 209–11; 2004c: 83–4) and has shown that it lies at the very root of the whole principle of analogy. It has had a mixed reception by panentheists, however, because of the points where the analogy breaks down (Brierley 2004a: 6–7; 2005: 23–4). Not least, 'spirit' and 'matter' are susceptible to dualist interpretations as well as psychosomatic 'monist' interpretations; also, science regards the person as emergent from the body, while most panentheists would not wish to assert that God is emergent from the cosmos (Gregersen 2004: 20; Griffin 2004: 44).

This raises the question of appropriate analogies for the relationship between God and cosmos in panentheism. Peacocke (1993: 173–7; 1998: 359–60; 2004a: 144; 2004b: 105–6) uses the analogy of the relationship between a composer and her or his music, but the difficulty with this, as with other analogies of artist and artwork, is that the expression of art is physically *separate* from the composer (cf. Temple 1934: 265–6). Similarly, the difficulty with Augustine's analogy of the cosmos as a 'sponge' in water (Augustine 1991: 115) is that it is possible for the sponge to exist out of water. A stronger analogy, which does not involve such separation and which is in effect a variation on Clayton's panentheistic analogy, is the cosmos as an unborn baby within (and always within) its mother's womb—dependent on the mother for every aspect of its existence, yet with a limited freedom and beyond the mother's total control (cf. Case-Winters 1990: 220–7; Clayton 1999: 291; Peacocke 2001: 139, 142; 2004a: 147, 151–2; 2004b: 142; and McDaniel 2005). Julian of Norwich, in one of her revelations,

---

[1] I would not now regard the distinctions between 'weak' and 'strong' panentheism as 'superficial', as I did in Brierley 2004a: 6.

exclaimed that Christ 'is our Mother for we are for ever being born of him, and shall never be delivered!' (Julian of Norwich 1966: 164).[2]

Point (13) has two stages, which could be categorized as (13a) and (13b). The first of these would hold that God is *freely* dependent on the cosmos, the second that God is *necessarily* dependent on the cosmos. Under the first option, God chooses to be dependent; under the second, the dependence is part of God's nature. The difference between these two positions is discussed further in Brierley (2004a: 10) and Clayton (2004d: 254). Point (14) is process theism: this is perhaps the most well-known form of panentheism, but it would be unfortunate to deduce from the fact that it has been of considerable influence on doctrine in the twentieth century, and from the fact that some expositions of panentheism have misguidedly equated panentheism with process thought, that process theism is generally representative of panentheism or occupies the 'middle ground' (a mistake made in Thomas 1999: 286); the spectrum of points given above clearly indicates that it is not, and does not. Process theism represents the most advanced form of panentheism, with its marked limitations of the divine, to the extent that God represents one influence among many in the cosmos, suggesting a degree of separation that, ironically, would be characteristic of more 'basic' points on the spectrum.

I have shown elsewhere how some of the points above—(5)–(7), (10), (12), and (13)—are logical entailments of panentheism (Brierley 2004a: 5–12). As is clear from the fact that some of these points are shared with classical theism, not all the points are on their own necessarily panentheistic, and not all of them are developed in every expression of the doctrine. Panentheism also has logical effects on other Christian doctrines: there are distinctive panentheistic interpretations, for example, of christology, ecclesiology, miracles, and eschatology (Brierley 2006).

Overlaid on the spectrum of beliefs analysed above is the question of whether or not evil is privative. Some panentheists, such as Tillich, following the scholastic tradition, have held that evil has no existence in itself, so that God is ('only') in the 'good' of the cosmos. For panentheisms of this variety, the cosmos has intrinsic positive value. Other panentheists, such as Clayton, regard evil as an existing constituent of the cosmos in God, with the result that the cosmos is morally neutral. Such panentheists are able to state, as do Clayton and Peacocke, that cosmic events are divine actions *per se* (Clayton 1999: 293 ('Every physical event is an act of God'); 2001b: 209, 212; cf. Peacocke 1998: 359; 1999: 235; 2001: 146; 2004a: 144). Those who hold evil to be privative, and who identify God with the good, do not ascribe all cosmic actions to God in this way because of those actions' moral ambiguity.

The spectrum above helps to emphasize that there are different varieties of panentheism (Brierley 2004a: n. 132). It would be as foolish to criticize 'panentheism' for existing in different forms as it would be to criticize classical theism for the same characteristic. Having now analysed this spectrum, it is possible to identify panentheism's distinctiveness. This can be expressed in terms of three premises: first, that God

---

[2] There are of course disanalogies with this as well: a mother does not *need* a baby as points (12) and (13) would have God in relation to the cosmos.

is not separate from the cosmos (panentheism as a negation of point (1)); second, that God is affected by the cosmos (either in the 'basic' sense of point (5) or the 'advanced' sense of points (12) and (13)); and third, that God is more than the cosmos (panentheism as a negation of point (15)). These premisses distinguish panentheism from the following:

- classical theisms in which God is separate from the cosmos, although perhaps present to it (either in a general way or in a specific isolatable way) and even affected by it (a position that asserts points (1), (3) and (5));
- classical theisms in which God is 'in' the cosmos but not fundamentally affected thereby (that is to say, a denial of points (5), (12) and (13));
- pantheisms in which God is not so much 'in' the cosmos as the cosmos itself (point (15)).

These three premisses—that God is not essentially independent or separate from the cosmos, is affected by the cosmos, and is more than the cosmos—are specific enough to be distinctive, but general enough to include most of the detailed forms that panentheism has taken. They consist of a denial of points (1) and (15) and an adherence to either point (5) ('basic' panentheism) or points (12) and (13) ('advanced' panentheism). They amount in fact to a definition of panentheism, and demonstrate that panentheism is more than a mere 'emphasis' on divine 'immanence' of the sort that classical theism could fully accommodate (as some critics have alleged). The three premisses are also very close to Gregersen's two-pronged definition of 'generic panentheism': namely, that the cosmos is in some sense 'in' God and that the relations between God and cosmos are 'in some sense bilateral' (Gregersen 2004: 22; cf. Clayton 2004c: 83; 2004d: 252).

These premisses help to articulate the 'mutual coinherence' that is the fundamental panentheistic conviction. The second premiss, that God is affected by the cosmos in addition to the cosmos being affected by God, accounts for the fact that the words 'relational ontology' and 'reciprocity' are often used in connection with panentheism. The distinctiveness of panentheism here is well illustrated by the classical theistic and panentheistic understandings of divine love. Classical theism views God's love as *agapē*, or benevolence, needing no love in return; panentheism regards God's love as an inextricable mix (cf. point (12)) of *agapē* and *erōs*, 'gift-love' and 'need-love', so that something of God's self is fulfilled in the act of loving: God 'needs somebody to love' in order for the divine love to be complete. The first premiss, that God is not separate, is of equal importance, however. The dangerous implication of relational language is that the cosmos is perceived as an independent entity (Peterson 2001: 402–3). The cosmos is not 'independent' of God in any way; indeed, it is radically dependent on God at every moment and at every level. Even the (relative) freedom of the cosmos is grounded in God and dependent on God.

Herein lies the importance of point (9) and the panentheism of Paul Tillich (Nikkel 1995: 29–82; see also 199–217). Tillich's great aphorism was that God is not *a* being, but Being Itself, or the Ground or Power of Being. God as *a* being or person implies that God is beside or alongside the cosmos, and thus essentially separate from or external to the cosmos (Nikkel 1995: 40, 42–4, 46, 69 n. 87). This is precisely what Tillich wished to avoid. He regarded externality as characteristic of finitude. God, however, is infinite, and nothing is external to the infinite, so the infinite includes (point (11)) the finite; God includes the cosmos (Nikkel 1995: 55). In this way, Tillich stands at the heart of advanced panentheism, on a line stretching from Nicholas of Cusa through Hegel to Clayton, for all of whom the infinite includes the finite (Clayton 2000*b*: 147–51; 2001*a*: 195; cf. 2004*c*: 81; 2004*d*: 253).

Advanced panentheism developed in the wake of the Enlightenment, and the principal theological movements that have embodied the doctrine have been idealism, modernism (Brierley 2004*b*), process theism, and some other liberal theologies, including theologies of 'protest' against classical theism and some of the 'South Bank religion' of the 1960s. Nevertheless, it is clear that aspects of panentheism can be found within the Christian tradition stretching back to the Bible (Brierley 2004*a*: 4 n. 116; cf. Borg 2000: 42–3; 2004: 66, 69), as well as within other ancient religions (cf. Hartshorne and Reese 2000: 294–7, 306–10; Whittemore 1956; Gregersen 2004: 34 n. 37; and Clayton 2004*d*: 250 n. 1). Panentheistic sentiments have particularly been voiced by mystics throughout Christian history (Brierley 2004*a*: 4 n. 109; cf. Lee 1946: 252–3; Fox 1974: 361; and Dombrowski 1996: 164–71).

PANENTHEISM AND DEBATE ON DIVINE ACTION

Scientists can find panentheism resonating with many aspects of their endeavours. There are points of connection between the interrelation of matter and energy in contemporary physics and Joseph Bracken's 'field' version of process panentheism (Bracken 1991, 1992, 1995, 1997, 2004); between concepts of emergence and panentheist 'internalism'; and between theories of vitalism or life force and the interrelationship that panentheism allows between Spirit and spirit.

Panentheism has also illuminated recent debates on divine action, an area of particularly thorough interdisciplinary research and an area that historically has been one of the chief causes of science's distancing of itself from religion.[3] Recent theories of divine action range from theistic naturalism to objectively special divine action without 'gaps' in the causal order, to objectively special divine action *with* 'gaps' in the causal order (for typologies, see Russell 1995, 2001, and Southgate 2005: 269–82). Panentheists argue for the first two positions, as exemplified in the work of Christopher Knight, Clayton, and Peacocke.

---

[3] See esp. Russell *et al.* 1995, 1996, 1998, 1999, 2001. For an introduction to the discussion, see esp. Russell 1995, 2001 and Clayton (forthcoming).

An exponent of theistic naturalism, Knight suggests that there is no 'special' action of God in the cosmos beyond the 'single act' of God's general providence, infusing the cosmos with grace and sustaining it in being through the laws of nature and their intrinsic potentials (which are not of course fully known) (Knight 2001: 11–22; 2004; 2005). The problem of the 'causal joint' is thus solved by grounding all causal joints in divine love. Maurice Wiles, whose panentheism was less explicit than that of Knight, was an earlier exponent of a similar view. Knight defends his view from the charge of deism by holding that God, as ground, participates, and thus is still intimately involved, in every event. Theistic naturalism has the advantage of not conceiving of God as 'a cause among causes'.

Clayton's theory of divine action rests on his panentheistic analogy: God's relation to the cosmos is like a person's relation to her or his body. Some of the body's functions happen automatically and unconsciously, such as breathing and the beating of the heart. While Clayton would wish to assert that God is nevertheless conscious of all action, this 'automatic' action corresponds to the general law-like regularities of the cosmos and God's general providence, equivalent to Knight's naturalism. Clayton also wishes to assert, however, that there is special, or focused, divine action in the cosmos, corresponding to the conscious and intentional acts of a person on her or his body. He uses the scientific concept of 'supervenient' properties, which 'emerge' from lower levels but are not reducible to those levels' causal influence, and which them-selves exercise a 'top-down' causal influence on lower levels, to suggest that God as a 'whole' influences or lures the 'parts' of the cosmos (Clayton 1997a: 232–69; 2001c: 231; 2004d: 263–4). The doctrine can be called *strong emergence*: 'God could guide the process of emergence through the introduction of new information (formal causality) and by holding out an ideal or image that could influence development without altering the mechanisms and structures that constrain evolution from the bottom up (final causality)' (Clayton 2004b: 632–3; cf. 2004a: 187–99). This influence seems to be at the mental level of 'integrated persons', by God as (the unitive category of) Spirit (Clayton 2004d: 264; cf. 2001a: 194 and Bracken 2004: 219).

The position of Peacocke, who has been an important influence on both Knight and Clayton, seems to hover in between their two views. He is clear that God acts in the processes of the world and is not an 'additional' influence in any way; moreover, he uses the term 'sacramental panentheism' as an equivalent to theistic naturalism. Like Clayton, however, he also uses the language of emergence and supervenience, and indeed pioneered the suggestion that divine action can occur through 'whole–part' influence, which is 'special' in its effects (Peacocke 1993: 159–60; 1995: 282–5, 287; 2001: 110; for commentary, see Clayton 1997a: 220–7). Knight suggests that Peacocke has become decreasingly naturalistic (2001: 15–17), while Clayton suggests that Peacocke has become increasingly naturalistic (2004d: 262–3, though cf. 2002: 272 n. 34). The fact that Peacocke regards divine action as intended in its *effects* suggests that he might be nearer to Knight's theory of divine action than to Clayton's.

For centuries religion has suffered from the concept that God acts by intervening in the laws of nature and the affairs of the world. Science has recoiled from religion partly because interventionist classical notions of divine action simply are not compatible with what science has revealed, just as society has generally reacted against theism, towards agnosticism and atheism, in philosophical and moral reaction against the excesses of classical theism. The three theories just summarized offer a form of theism without interventionism or irruptionism; panentheism removes the *need* for distance, and thus can bring scientists and theologians closer together.

PANENTHEISM AND RELIGIOUS EXPERIENCE

Clayton has outlined seven possible reasons for adopting panentheism as a doctrine (Clayton, 2004*c*: 73–4):

- belief that classical theism is no longer viable;
- belief that panentheism is more compatible than classical theism with science;
- belief in a metaphysic which panentheism underlies;
- belief that panentheism is better able than classical theism to support religious doctrines (a development of the first reason);
- belief that panentheism can mediate between different faiths;
- belief that panentheism can respond better than classical theism to objections to religious belief (this seems to be a negative version of the positive, fourth reason);
- belief that panentheism has better ethical or political implications than classical theism.

My own assent to panentheism has stemmed from a gradual convergence of convictions: first, from the philosophy of religion, that forms of theism other than classical can better respond to certain common objections to religious belief, and in particular to the problem of theodicy (the sixth of Clayton's reasons); and second, from the theological conclusion that authentic love entails limitations of the divine that classical theism cannot accommodate (approximating most closely to the fourth of Clayton's reasons), a conviction that corresponds to the human subjective experience of love.

This correspondence with subjective experience would not be significant unless it correlated with broad patterns of religious experience. As it turns out, religious experiences have indeed often been panentheistic in orientation. This is characteristic of the intense religious experiences that are categorized as mysticism, but is also true of religious experiences more generally, as seen not least in the New Age movement and in the counterpart interest, within Christian faith, in Celtic spirituality (Brierley 2004*a*: 14 n. 204). Such experiences are invariably of something transcendent, or 'more', which comes in or through something everyday, ordinary, or routine. Marcus Borg relates an experience of the British theologian Leslie Weatherhead, who, in the dingy corner of a train leaving London one murky November evening, felt the compartment filled with light, and himself overwhelmed and possessed by a sense of love, humility, and joy, particularly towards his fellow passengers (Borg 1997: 44).

Many other individuals have felt a particular environment suffused with light or love, and themselves at peace with the world or united with the world. Their experience is of something more than what the world offers on its own, yet at the same time that something is intimately connected or bound up with the particular part of the cosmos in which they find themselves. It is something 'more' than, yet also organically 'in and through', the cosmos as they know it; it is transcendence 'within' immanence. It is not separate from the cosmos, yet neither is it simply the resources of the cosmos itself. It is a feeling of mutual coinherence; it is panentheistic. Borg has testified to his own experience of being brought up under a classical theistic model, only to discover panentheism as a framework that made more sense: 'becoming aware of panentheism made it possible for me to be a Christian again' (Borg 1997: 12, 37–48; cf. 2000: 42–4).

Hence to Clayton's seven reasons can be added an eighth:

• belief that panentheism represents more accurately than classical theism people's (religious) experience.

If asked why panentheism has come to the fore in modern theology, one need go no further than the fact that it articulates a metaphysical position which is consonant with much, and often popular, religious experience, instances of which are increasingly articulated.

Given the foundational nature of experience for human being, it is unlikely that scientists or theologians will be attracted to panentheism *purely* on the basis of Clayton's second reason above, that panentheism is more compatible with science than classical theism. Scientists are human beings, after all, often with religious experiences. The attraction of scientists to panentheism is therefore likely to involve, in addition to theoretical consonance, convergence between theory and any personal subjective experience of 'God'. Part of what panentheism offers is a holistic integrity between 'head' and 'heart'.

The Moral Potential of Panentheism

Clayton's seventh reason above for adopting panentheism is the belief that panentheism has better ethical or political implications than classical theism. Herein, I believe, lies panentheism's greatest potential. However close God and the cosmos are made to be, there is in classical theism a basic separation that panentheism overcomes. 'Despite the goodwill of its defenders, classical theism is fatally mired in imagery of divine aloofness and detachment' (Cowdell 2000: 21). Conceiving of God as independent or separate from the cosmos has a cost; each version of classical theism has a cost. The cost is moral, because the relationship of God to the cosmos sets the tone or example for internal relationships within the cosmos.

Sophisticated forms of classical theism employ a variety of means to try to establish intimacy between God and the cosmos (see e.g., the analytical theists challenged in Dombrowski 1996). Even Aquinas held that God was 'in' the cosmos as an agent is in that in which its action takes place (Gregersen 2004: 23). Norris Clarke suggests that 'outside' God in classical terminology means simply 'not identical with God; they *are not* God' (Clarke 1990: 108, Clarke's emphasis). But being a 'part' of God in no way suggests identity with God; there are ways of distinguishing between God and the cosmos, without the nuances of separation and independence (and their attendant moral difficulties) that the word 'outside' involves. Being a 'part' of God does not compromise distinction. Similarly, in an essay of characteristic depth, Rowan Williams rejects panentheism because of God's character as sheer gift, of being for others, which needs or receives nothing in return (Williams 2000: 63–78). But does a lack of receipt necessarily follow from sheer gift? Is that not a presupposition? Conceiving of God as not separate from the cosmos and needing the cosmos does not reduce the cosmos to the level of functionality for God; God's receipt need neither detract from nor undermine the unconditionality of divine love. God's receipt can be a consequence, without being the purpose, of loving. Indeed, if God is not to be the *object* of the cosmos, then God's indwelling grounding of it would seem to be the only possible alternative.

The moral edge of panentheism stems from the fact that it corresponds more closely than other forms of theism to the basic religious conviction of 'mutual coinherence' or human connectedness with the divine—the conviction that God is 'more inward than [our] most inward part' (Augustine 1991: 43), 'closer [to us] than breathing' (Alfred Tennyson, 'The Higher Pantheism', in Ricks 1987: 706). Panentheism seeks to stress that 'the infinite God is ontologically *as close to finite things as can possibly be thought without dissolving the distinction of Creator and created altogether*' (Clayton 1999: 290, Clayton's emphasis). Relations under panentheism are always, as David Nikkel observes, 'internal' (Nikkel 2003: 641). Inclusion is understood 'organically'. The 'in' of panentheism is a fundamental conviction about the closeness of God and the cosmos, in contradistinction to any concept of God as separate or entirely independent. Panentheism is able to say that whatever the situation in the cosmos, however small or large, however catastrophic, it is still 'in' God. God is not just present to it, but inseparable from it. God is not so much a compassionate friend as a pregnant mother who bears the pain (cf. Romans 8) with infinite capacity to redeem, restore, and renew. This is not necessarily to say that every action in the cosmos is divine action, even if the cosmos is 'part' of God. One would not say that the action of the foetus is the mother's action, even though the foetus is entirely 'in' the mother and 'part of' the mother. It is important to recognize the moral ambiguity of the cosmos (Page 2004), and that this ambiguity extends to the cosmos's deepest level. The point is that all takes place 'within' God, and that the goodness of God is 'in' everything to some extent, however morally ambiguous. The 'within' of panentheism is required if only—and perhaps only—to avoid the negative connotations of having God 'without'.

The moral implications of conceiving of God as 'in' the cosmos and as necessarily affected by it are manifest in many different fields of theology: pastoral theology, feminist theology, ecotheology, 'economic' liberation theology, sexual theology, theology of religions, and so on. Why is it that much ecotheology is panentheist in orientation? It is because respect for the cosmos is greater when God is conceived as 'in' it than when God is regarded as independent of it. Why is it that feminism finds a natural ally in panentheism? It is because patriarchy and its ill effects are associated with 'independence', whereas feminism celebrates the closest possible relatedness—which is precisely how panentheism characterizes the relationship of the cosmos with the divine. Why is it that interfaith dialogue develops further when participants subscribe to panentheism? It is because God in these cases can readily be seen 'in' different traditions and religious figures (while panentheism is capable of trinitarian interpretation, it is not necessarily so, *contra* Clayton (1997*b*: 133–5, 137–8; 1998: 202, 207–8; 2005: 252)). Similarly, it is no coincidence that a key theologian for the acceptance of gay and lesbian relationships, Norman Pittenger, was a panentheist (Brierley 2006). There are also connections between panentheism and 'economic' liberation theology.

Panentheism has moral potential in pastoral theology, particularly in the context of suffering. Its concept of a non-interventionist God is a natural home for theodicean strategies such as the limitation or reconception of divine omnipotence, and the free-will defence. The notion of a passible God with the closest possible relationship to the cosmos can (but not must) bring pastoral benefit. To be of assistance here, however, it is necessary for panentheism either to adopt the scholastic and Tillichian notion of evil as privative, rather than having actual existence itself, in order to keep God unsullied, or to adopt the process conception of God's dipolarity, whereby the evil of the cosmos exists as part of God, but not at God's absolute 'pole'. I prefer the Tillichian view as avoiding the 'inconsistency' of the process dipolar God; it is also interesting to note the connection between the doctrine of evil as privative and the 'absence of God' sometimes experienced by those in situations of suffering.

Panentheism shows how contemporary work in science-and-religion should not be detached from questions of practical ethics, 'earthly, everyday concerns', as such work has been criticized for being (Deane-Drummond 2004: 234).[4] Panentheism can provide a metaphysical basis for challenging science's assumption that the biosphere is endlessly malleable, 'to our pleasure or whim' and to the benefit of only a few (Conway Morris 2003: 311–30). All these practical implications are moral dimensions of panentheism, and they derive from the closest possible association of God and the cosmos, which panentheism represents. Clayton is wary of attributing too much value to a metaphysical position because of its political consequences (Clayton 2004*d*: 259), but 'moral truth' and 'metaphysical truth' are by no means mutually exclusive (Harris 2004), and indeed may themselves be 'inextricably intertwined'.

---

[4] Ironically, Deane-Drummond believes the answer to lie in the relationship of 'otherness' provided by classical theism.

CONCLUSION

Enlightenment science presented data which were incompatible with classical theism and required that alternative theistic conceptions be found; it led both to deism and, more consonant with people's religious experience, panentheism. Now religion can come back to science with the gift of advanced panentheism as a theology which is compatible with scientific disciplines and can express scientists' own religious experience. In this way, science and panentheism have come full circle. The history of theology from deism onwards, in reactions to *Essays and Reviews*, *Lux Mundi*, modernism, and all the rest, is littered with the tragedy of theological defensiveness, entrenchment, and opposition to science, simply because theology itself was operating with a problematic doctrine of God (for some of this history and its nuances, see Bowler 2001). Theology now does not have to survive in this way or to perceive science as a threat. Scientists now do not have to reject theology for the wrong reason.

But that is just to put the benefit negatively. Panentheists are not just making up for the past; they have something positive to offer. They offer alternative and subtle doctrines of God of which scientists may not be aware (cf. McDaniel 2005: 39) and which provide a metaphysical 'fit' for those scientists who wish to integrate religious experience into their interpretations of reality (Clayton 2001c: 233–4), and who look for a metaphysic with moral 'edge'. Panentheism, in short, represents theology that is engaging and creative, and a metaphysic with rich and sensitive moral resources. And given the moral questions which now press so urgently on science and the contemporary world, that contribution of panentheism could be of paramount importance. 'Go for a walk. Get wet. Dig the earth.'

REFERENCES AND SUGGESTED READING

AUGUSTINE (1991). *Confessions*, trans. H. Chadwick. Oxford: Oxford University Press.

BORG, M. J. (1997). *The God We Never Knew: Beyond Dogmatic Religion to a More Authentic Contemporary Faith*. New York: HarperCollins.

——(2000). 'The God who is Spirit', in F. W. Schmidt (ed.), *The Changing Face of God*, Harrisburg, Pa.: Morehouse Publishing, 33–49.

——(2004). *The Heart of Christianity: Rediscovering a Life of Faith*. New York: HarperCollins.

BOWLER, P. J. (2001). *Reconciling Science and Religion: The Debate in Early-Twentieth-Century Britain*. Chicago and London: University of Chicago Press.

BRACKEN, J. A. (1991). *Society and Spirit: A Trinitarian Cosmology*. Cranbury, N.J.: Associated University Presses.

——(1992). 'The Issue of Panentheism in the Dialogue with the Non-Believer', *Studies in Religion*, 21: 207–18.

——(1995). 'Panentheism from a Trinitarian Perspective', *Horizons*, 22: 7–28.

——(1997). 'Panentheism from a Process Perspective', in J. A. Bracken and M. H. Suchocki (eds.), *Trinity in Process: A Relational Theology of God*, New York: Continuum, 95–113.

BRACKEN, J. A. (2004). 'Panentheism: A Field-Oriented Approach', in Clayton and Peacocke (2004), 211–21.

BRIERLEY, M. W. (2001). 'Introducing the Early British Passibilists', *Journal for the History of Modern Theology*, 8: 218–33.

——(2004a). 'Naming a Quiet Revolution: The Panentheistic Turn in Modern Theology', in Clayton and Peacocke (2004), 1–15.

——(2004b). 'Ripon Hall, Henry Major and the Shaping of English Liberal Theology', in M. D. Chapman (ed.), *Ambassadors of Christ: Commemorating 150 Years of Theological Education in Cuddesdon 1854–2004*, Aldershot: Ashgate, 89–155.

——(2005). 'Panentheism', in *The Encyclopedia of Christianity*, iv, Grand Rapids, Mich.: William B. Eerdmans Publishing Co.; Leiden: Brill, 21–5.

——(2006). 'Norman Pittenger (1905–1997) and Panentheism', *Theology*, 109.

CASE-WINTERS, A. (1990). *God's Power: Traditional Understandings and Contemporary Challenges*. Louisville, Ky.: Westminster/John Knox Press.

CLARKE, W. N. (1990). 'Charles Hartshorne's Philosophy of God: A Thomistic Critique', in S. Sia (ed.), *Charles Hartshorne's Concept of God: Philosophical and Theological Responses*, Dordrecht: Kluwer Academic Publishers, 103–23.

CLAYTON, P. (1997a). *God and Contemporary Science*. Edinburgh: Edinburgh University Press.

——(1997b). 'Pluralism, Idealism, Romanticism: Untapped Resources for a Trinity in Process', in J. A. Bracken and M. H. Suchocki (eds.), *Trinity in Process: A Relational Theology of God*, New York: Continuum, 117–45.

——(1998). 'The Case for Christian Panentheism', *Dialog*, 37: 201–8.

——(1999). 'The Panentheistic Turn in Christian Theology', *Dialog*, 38: 289–93.

——(2000a). 'On the Value of the Panentheistic Analogy: A Response to William Drees', *Zygon*, 35: 699–704.

——(2000b). *The Problem of God in Modern Thought*. Grand Rapids, Mich., and Cambridge: William B. Eerdmans Publishing Co.

——(2001a). 'In Whom We Have Our Being: Philosophical Resources for the Doctrine of the Spirit', in B. E. Hinze and D. L. Dabney (eds.), *Advents of the Spirit: An Introduction to the Current Study of Pneumatology*, Milwaukee: Marquette University Press, 173–207.

——(2001b). 'Panentheist Internalism: Living within the Presence of the Trinitarian God', *Dialog*, 40: 208–15.

——(2001c). 'Tracing the Lines: Constraint and Freedom in the Movement from Quantum Physics to Theology', in Russell *et al.* (2001), 211–34.

——(2002). 'The Impossible Possibility: Divine Causes in the World of Nature', in T. Peters, M. Iqbal, and S. N. Haq (eds.), *God, Life, and the Cosmos: Christian and Islamic Perspectives*, Aldershot: Ashgate, 249–80.

——(2003). 'God and World', in K. J. Vanhoozer (ed.), *The Cambridge Companion to Postmodern Theology*, Cambridge: Cambridge University Press, 203–18.

——(2004a). *Mind and Emergence: From Quantum to Consciousness*. Oxford: Oxford University Press.

——(2004b). 'Natural Law and Divine Action: The Search for an Expanded Theory of Causation', *Zygon*, 39: 615–36.

——(2004c). 'Panentheism in Metaphysical and Scientific Perspective', in Clayton and Peacocke (2004), 73–91.

——(2004d). 'Panentheism Today: A Constructive Systematic Evaluation', in Clayton and Peacocke (2004), 249–64.

——(2005). 'Kenotic Trinitarian Panentheism', *Dialog*, 44: 250–5.

—— (forthcoming). 'Toward a Theory of Divine Action that has Traction', in R. J. Russell and N. C. Murphy (eds.), *The Divine Action Debate: A Retrospective*, Vatican City State: Vatican Observatory Publications.

—— and PEACOCKE, A. R. (2004) (eds.). *In Whom We Live and Move and Have Our Being: Panentheistic Reflections on God's Presence in a Scientific World*. Grand Rapids, Mich., and Cambridge: William B. Eerdmans Publishing Co.

CONWAY MORRIS, S. (2003). *Life's Solution: Inevitable Humans in a Lonely Universe*. Cambridge: Cambridge University Press.

COWDELL, S. (2000). *A God for This World*. London and New York: Mowbray.

CROSS, F. L., and LIVINGSTONE, E. A. (2005) (eds.). *The Oxford Dictionary of the Christian Church*, 3rd edn., rev. Oxford: Oxford University Press.

DEANE-DRUMMOND, C. E. (2004). 'The Logos as Wisdom: A Starting Point for a Sophianic Theology of Creation', in Clayton and Peacocke (2004), 233–45.

DOMBROWSKI, D. A. (1996). *Analytic Theism, Hartshorne, and the Concept of God*. Albany, N.Y.: State University of New York Press.

FOX, M. (1974). 'Panentheistic Spirituality: Religious Education for the Future?', *Living Light*, 11: 357–67.

GREGERSEN, N. H. (2004). 'Three Varieties of Panentheism', in Clayton and Peacocke (2004), 19–35.

GRIFFIN, D. R. (2004). 'Panentheism: A Postmodern Revelation', in Clayton and Peacocke (2004), 36–47.

—— (2005). 'Panentheism's Significance in the Science-and-Religion Discussion', *Science and Theology News*, 5/9: 35 and 41.

HARRIS, H. A. (2004). 'On Understanding that the Struggle for Truth is Moral and Spiritual', in U. King and T. Beattie (eds.), *Gender, Religion and Diversity: Cross-Cultural Perspectives*, London and New York: Continuum, 51–64.

HARTSHORNE, C., and REESE, W. L. (2000). *Philosophers Speak of God*, 2nd edn. New York: Humanity Books.

JULIAN of NORWICH (1966). *Revelations of Divine Love*, trans. C. Wolters. Harmondsworth: Penguin Books.

KNIGHT, C. C. (2001). *Wrestling with the Divine: Religion, Science, and Revelation*. Minneapolis: Fortress Press.

—— (2004). 'Theistic Naturalism and the Word Made Flesh: Complementary Approaches to the Debate on Panentheism', in Clayton and Peacocke (2004), 48–61.

—— (2005). 'Naturalism and Faith: Friends or Foes?', *Theology*, 108: 254–63.

LEE, A. (1946). *Groundwork of the Philosophy of Religion*. London: Duckworth.

McDANIEL, J. B. (2005). 'Reconnecting the Dots: Religion, Science and Nature', *Science and Theology News*, 5/9: 38–9.

McFAGUE, S. (1982). *Metaphorical Theology: Models of God in Religious Language*. Philadelphia: Fortress Press.

MOLTMANN, J. (1981). *The Trinity and the Kingdom of God: The Doctrine of God*, trans. M. Kohl. London: SCM Press.

—— (1985). *God in Creation: An Ecological Doctrine of Creation*, The Gifford Lectures 1984–1985, trans. M. Kohl. London: SCM Press.

—— (2001). 'God's Kenosis in the Creation and Consummation of the World', in J. C. Polkinghorne (ed.), *The Work of Love: Creation as Kenosis*, Grand Rapids, Mich., and Cambridge: William B. Eerdmans Publishing Co.; London: SPCK, 137–51.

—— (2003). *Science and Wisdom*, trans. M. Kohl. London: SCM Press.

NIKKEL, D. H. (1995). *Panentheism in Hartshorne and Tillich: A Creative Synthesis*. New York: Peter Lang Publishing.

—— (2003). 'Panentheism', in *Encyclopedia of Science and Religion*, ii, New York: Macmillan Reference USA, 641–5.

PAGE, R. (2004). 'Panentheism and Pansyntheism: God in Relation', in Clayton and Peacocke (2004), 222–32.

PEACOCKE, A. R. (1993). *Theology for a Scientific Age: Being and Becoming—Natural, Divine and Human*, 2nd edn. London: SCM Press.

—— (1995). 'God's Interaction with the World: The Implications of Deterministic "Chaos" and of Interconnected and Interdependent Complexity', in Russell *et al.* (1995), 263–87.

—— (1998). 'Biological Evolution: A Positive Theological Appraisal', in Russell *et al.* (1998), 357–76.

—— (1999). 'The Sound of Sheer Silence: How Does God Communicate with Humanity?', in Russell *et al.* (1999), 215–47.

—— (2001). *Paths from Science towards God: The End of All Our Exploring*. Oxford: Oneworld Publications.

—— (2004a). 'Articulating God's Presence in and to the World Unveiled by the Sciences', in Clayton and Peacocke (2004), 137–54.

—— (2004b). *Creation and the World of Science: The Re-Shaping of Belief*, 2nd edn. Oxford: Oxford University Press.

PETERSON, G. R. (2001). 'Whither Panentheism?', *Zygon*, 36: 395–405.

POLKINGHORNE, J. C. (1994). *Science and Christian Belief: Theological Reflections of a Bottom-Up Thinker*, The Gifford Lectures for 1993–4. London: SPCK.

—— (1996). *Scientists as Theologians: A Comparison of the Writings of Ian Barbour, Arthur Peacocke and John Polkinghorne*. London: SPCK.

—— (2000). *Faith, Science and Understanding*. London: SPCK.

—— (2002). *The God of Hope and the End of the World*. London: SPCK.

—— (2004). *Science and the Trinity: The Christian Encounter with Reality*. London: SPCK.

RICKS, C. (1987) (ed.). *The Poems of Tennyson*, ii, 2nd edn. London: Longman.

RUSSELL, R. J. (1995). 'Introduction', in Russell *et al.* (1995), 1–31.

—— (2001). 'Introduction', in Russell *et al.* (2001), pp. i–xxvi.

—— CLAYTON, P., WEGTER-MCNELLY, K., and POLKINGHORNE, J. C. (2001) (eds.). *Quantum Mechanics: Scientific Perspectives on Divine Action*. Vatican City State: Vatican Observatory Publications; Berkeley: Center for Theology and the Natural Sciences.

—— MURPHY, N. C., and ISHAM, C. J. (1996) (eds.). *Quantum Cosmology and the Laws of Nature: Scientific Perspectives on Divine Action*, 2nd edn. Vatican City State: Vatican Observatory Publications; Berkeley: Center for Theology and the Natural Sciences.

—— —— MEYERING, T. C., and ARBIB, M. A. (1999) (eds.). *Neuroscience and the Person: Scientific Perspectives on Divine Action*. Vatican City State: Vatican Observatory Publications; Berkeley: Center for Theology and the Natural Sciences.

—— —— and PEACOCKE, A. R. (1995) (eds.). *Chaos and Complexity: Scientific Perspectives on Divine Action*. Vatican City State: Vatican Observatory Publications; Berkeley: Center for Theology and the Natural Sciences.

—— STOEGER, W. R., SJ, and AYALA, F. J. (1998) (eds.). *Evolutionary and Molecular Biology: Scientific Perspectives on Divine Action*. Vatican City State: Vatican Observatory Publications; Berkeley: Center for Theology and the Natural Sciences.

SOUTHGATE, C. (2005). 'A Test Case: Divine Action', in C. Southgate, J. H. Brooke, C. E. Deane-Drummond, P. D. Murray, M. R. Negus, L. Osborn, M. Poole, A. Robinson, J. Stewart, F. N. Watts, and D. Wilkinson, *God, Humanity and the Cosmos: A Companion to the*

Science–Religion Debate, 2nd edn., London and New York: T & T Clark International, 260–99.

TEMPLE, W. (1934). *Nature, Man and God: Being the Gifford Lectures Delivered in the University of Glasgow in the Academical Years 1932–1933 and 1933–1934.* London: Macmillan and Co.

THOMAS, O. C. (1999). 'Not Yet a Case for Christian Panentheism', *Dialog*, 38: 285–7.

WARD, J. S. K. (2004). 'The World as the Body of God: A Panentheistic Metaphor', in Clayton and Peacocke (2004), 62–72.

WHITTEMORE, R. C. (1956). 'Iqbal's Panentheism', *Review of Metaphysics*, 9: 681–99.

WILLIAMS, R. D. (2000). *On Christian Theology.* Oxford and Malden, Mass.: Blackwell Publishers.

—— (2005). 'Foreword', in *Sharing God's Planet: A Christian Vision for a Sustainable Future: A Report from the Mission and Public Affairs Council.* London: Church House Publishing, pp. vii–viii.

CHAPTER 38

....

PROBLEMS IN PANENTHEISM

....

OWEN C. THOMAS

INTRODUCTION

....

There are some serious problems in the understanding and interpretation of panentheism in what has become a fairly widespread movement that has gathered under this banner. These problems arise from the fact that panentheism is not one particular view of the relationship of the divine to the world (universe), but rather, a large and diverse family of views involving quite different interpretations of the key metaphorical assertion that the world is *in* God. This is indicated by the common locution among panentheists that the world is 'in some sense' in God, and by the fact that few panentheists go on to specify clearly and in detail exactly what sense is intended. As a result, panentheists do not offer a clear interpretation of the relationship of theology and science. This is not to say that there are not some particular versions of panentheism which are developed coherently, such as process panentheism, but that the panentheism movement *as a whole* does not present a coherent view.

Various attempts have been made to determine what this family of views has in common besides agreement on the metaphor, but again these determinations also vary rather widely. For example, in a recent symposium (henceforth referred to simply as 'the symposium') on panentheism one of the editors, Philip Clayton, in summarizing the interpretations offered, lists thirteen varieties of panentheism and also summarizes four quite different attempts to determine the characteristics of 'generic panentheism' (Clayton and Peacocke 2004: 250–4).[1] Clayton, perhaps

[1] Since this is the most recent symposium on panentheism and the only book-length treatment of it outside process theology, I will be referring to it regularly below by means of page numbers in parentheses.

somewhat uneasy about this, goes on to suggest that this broad variety of views 'may be even encouraging, insofar as they reflect the theological richness of the underlying notion' (p. 254). It may be noted, however, that the theological richness of a concept is mainly enhanced by its clarity rather than by its vagueness.

One source of this wide variety in the interpretations of panentheism is the even larger variety in the meanings of the key word 'in'. The *Oxford English Dictionary* devotes six full pages to the meanings of 'in', and specifies sixty-three distinct usages, many of which are employed by panentheists. These usages fall under the headings of preposition, adverb, adjective, and substantive: 'position or location', 'occupation', 'time', 'pregnant uses' (in himself), 'motion or direction', 'constructional' (believe in), and 'phrases' (insofar). The first usage listed, and probably the most prevalent in the symposium, is the preposition of 'position or location'. The other editor of the symposium, Arthur Peacocke, states that the 'in' of panentheism 'is clearly not intended in any locative sense.... It refers, rather, to an ontological relation so that the world is conceived as within the Being of God' (p. 145). Since the spatial metaphor is simply repeated in the statement of the ontological interpretation, however, without any indication of the ontology involved, this leaves it unexplained. Furthermore, the preposition of position or location seems to be 'clearly intended' by most of the authors in the symposium.

In another essay in the symposium, Clayton notes that theorists of emergence also use the terms 'in' and 'internal'. He states: 'What emergence actually offers... is the self-inclusion relation "⊂": "belongs to" or "is a member of", etc. This is a relation of logical inclusion rather than (primarily) one of location.... Emergence thus represents a powerful answer to misgivings about the preposition "in"' (p. 88). It is not clear, however, what this answer is and what misgivings it is supposed to address. What is clear is that other panentheists would probably not be satisfied with logical inclusion or class membership as the definition of the way in which the world is in God.

PROCESS PANENTHEISM

There is one major exception to this vagueness in the concept of panentheism, and that is the interpretation found in the process philosophy of Alfred North Whitehead and Charles Hartshorne that is expounded in detail and in highly novel and technical language. It might be called the original, or even the official, interpretation, since it was Hartshorne, who has been called 'the leading twentieth-century advocate of panentheism' (p. 3), who first popularized the notion. One symposium author, Celia E. Deane-Drummond, has described process panentheism as 'the measure of the "orthodoxy" of panentheism' (p. 234). Hartshorne describes Whitehead as 'the outstanding surrelativist or panentheist' (Hartshorne and Reese 1953: 273). Whitehead, who did not use the term, states his view as follows: 'God, as well as being primordial, is also consequent....The completion of God's nature into a fulness of physical feeling is derived from the objectification of the world in God...[the] prehension into God of each creature' (Whitehead 1978: 345).

Hartshorne has a somewhat different interpretation of panentheism. He states:

A supreme person [God] must be inclusive of all reality. We find that persons contain relations of knowledge and love to other persons and things, and since relations contain their terms, persons must contain other persons and things. If it seems otherwise, this is because of the inadequacy of human personal relations, which is such that the terms are not conspicuously and clearly contained in their objects....In God, terms of his knowledge would be absolutely manifest and clear and not at all 'outside' the knowledge or the knower. (Hartshorne 1948: 143–4)

Hartshorne explains why 'relations contain their terms' by a logical or definitional argument as follows: 'An individual relation is a *single* entity which is nothing without its terms, and hence its entire unitary activity must include that of each of its terms' (Hartshorne 1941: 238). However, the argument that relations contain their terms does not entail that one term in a relation contains the other term.

Thirty-five years later Hartshorne offered a different interpretation of inclusion, as follows: 'Can we define inclusion? I take my cue here from a formula of propositional logic. (There is a parallel formula in the logic of class inclusion.) P entails (logically includes) q if p & $q \equiv p$, or if asserting p is no less and no more than asserting p & q' (Hartshorne 1976: 247).[2]

There is another sense in process panentheism in which the world is in God. John B. Cobb Jr. puts it this way: 'The region of God [includes] the regions comprising the standpoints of all the contemporary occasions in the world' (Cobb Jr. 1965: 196). This is roughly equivalent to the assertion in classical Christian theism of divine omnipresence, that God is present at every point in space and time. It is also analogous to the version of panentheism in which God as infinite includes all finite realities. This latter view is mentioned in passing by two symposium authors (pp. 65, 146), and it is argued at some length by Clayton, who traces the history of the infinity and subjectivity of God to Hegel, who asserted that the truly infinite God 'must include the world within Godself' (p. 81). He neglects to mention, however, that this God cannot be understood as a subject or a person.

The interpretation of panentheism offered by Whitehead and Hartshorne is a very specific formulation that is analogous to knowing someone or something, and thus having them 'in mind'. It suggests that since we know and love our spouses, children, and friends, we have them in mind. Would we also be able to say that therefore they are in us? Probably not. This process interpretation of 'in', so far as I know, is not used by any other panentheists. At least it does not appear in any of the four lists of the characteristics of generic panentheism in the symposium. So what has been described as the original or official interpretation is not at all shared by other panentheists. The reason for this, apparently, is that the metaphor of human knowledge or love is not

---

[2] I am indebted to Donald Viney for this reference.

strong enough for the other panentheists, or that they are not persuaded by the concept of logical inclusion.

There is another problem in Whitehead's, and probably Hartshorne's, panentheism: namely, that God's knowledge of the world is neither complete nor of the contemporary world. The world is made up of actual entities or occasions of experience that begin with an initial aim supplied by God and proceed through a process of concrescence until they reach their 'satisfaction' and have 'perished'. It is only at this last stage that God 'prehends' or 'knows' them and incorporates them into the divine consequent nature (Neville 1980: 15–17). This may be another reason why other panentheists want to distinguish themselves from process panentheism. The main reason, however, why most other panentheists reject process panentheism is that in the latter the 'world' is held to be coeternal, and hence God is not the creator of the world, except in the very reduced sense that God supplies an 'initial aim' to each actual entity. Furthermore, the world, along with the 'eternal objects' and 'creativity', are coeternal with God, which leaves the totality unexplained. This is why Whitehead subtitles his main work 'An Essay in Cosmology' rather than cosmogony. Langdon Gilkey thus places Whitehead in the tradition of Platonic and Gnostic dualism, which suggests a tragic rather than a moral view of evil (Gilkey 1959: 49–51).[3]

Moreover, process panentheists tend to assert that the permanent correlation of God and world and the denial of creation out of nothing are *essential* to panentheism, and refer to other views as 'classical theism'. David Ray Griffin states:

This increased popularity [of panentheism] brings a danger that 'panentheism' will be appropriated for doctrines devoid of [the] promise [of panentheism]. There has been a tendency to extend the term to various doctrines that have modified classical theism sufficiently to say that the world is 'in' God, in the sense of affecting God, but that otherwise retain the defining characteristics of classical theism. (Griffin 2005: 35)

This further indicates the wide divergences among panentheists.

DIVINE IMMANENCE

Since panentheists usually criticize traditional or classical theism for attenuating divine immanence by asserting that God is outside the world, panentheism is often interpreted to mean primarily an intensification of divine immanence. One symposium author, Michael W. Brierley, states that one 'common panentheistic

---

[3] A summary of the criticisms of process theology has been supplied by Cobb and Griffin (1976: 184), to which should be added Neville (1980).

theme' is the use of the language of 'in and through', or the Lutheran eucharistic language of 'in, with, and under' to describe God's relation to the world (pp. 7–8). This implies, however, that God is in the world, rather than vice versa, and it also seems to be simply an emphasis on divine immanence, rather than a form of panentheism. Likewise, Peacocke in commending panentheism states: 'We...need a new model for expressing the closeness of God's presence to finite, natural events, entities, structures, and processes; and we need the divine to be as close to them as it is possible to imagine, without dissolving the distinction between Creator and what is created' (p. 145). Moreover, Clayton states that 'panentheists use [the 'in' metaphor] in two different directions—the world is in God and God is in the world.... [Therefore] it is not difficult to paraphrase the fundamental claim being made by the metaphor: the *inter*depend-ence of God and world' (p. 83). This implies that the world-in-God metaphor is not essential to panentheism, and what is essential is simply an emphasis on divine immanence. As John Cobb puts it in a review of the sypmposium, 'In this book the accent falls not on how all creatures are in God but on how God is in the world.... Certainly the two ideas are quite distinct, and it is possible, even common in the Christian tradition to affirm that God is in all things without affirming that all things are in God. Some of the essays in this book follow that line' (Cobb Jr. 2005: 241).

One symposium author, Niels Henrik Gregersen, however, argues that many, if not most, versions of classical theism have strong doctrines of divine immanence. He quotes Thomas Aquinas: 'God exists in everything...as an agent is present to that in which its action takes place.... So God must exist intimately in everything' (p. 23). Gregersen also notes that Thomas concludes with a panentheistic image: 'Immaterial things contain that in which they exist, as the soul contains the body. So God also contains things by existing in them. However, one does use the bodily metaphor and talk of everything being in God inasmuch as he contains them' (pp. 23–4).

Similarly Karl Barth, surely an exponent of classical theism, has an even stronger doctrine of divine immanence. After criticizing various versions of panentheism in Western history, he states:

Now the absoluteness of God strictly understood in this sense means that God has the freedom to be present with that which is not God, to communicate Himself and unite Himself with the other and the other with Himself, in a way which utterly surpasses all that can be effected in regard to reciprocal presence, communion and fellowship between other beings.... God...is free to be immanent, free to achieve a uniquely inward and genuine immanence of His being in and with the being which is distinct from Himself. (Barth 1957: 313)

So neither radical immanence nor panentheistic imagery are absent in classical theism. This leads to the question as to whether there is any real difference between the assertion that the world is in God and the doctrines in classical theism of the immanence and omnipresence of God. Because of the vagueness of the spatial metaphor, it is not clear that there *is* any difference.

Classical Theism

In this connection it may be noted that the symposium contains about thirty-five pejorative references to 'classical theism'. Brierley states, however, that 'classical theism has on occasion been made into a target of straw by panentheists', and lists three authors who have done so (pp. 13–14). He could also have listed my essay, which is a critique of an earlier essay by Clayton. This latter essay included an extended critique of what Clayton calls 'classical philosophical theism (CPT)'. He states that CPT is 'most often presupposed (usually unconsciously) by systematicians', that it is 'widely assumed by theologians today', and that 'its influence on theology continues to be immense' (Clayton 1998: 202b, 203b, 204a). My response was as follows: 'But Clayton offers no examples of [contemporary] theologians to support these claims.... Why not? Because there aren't any. I know of no major theologian in [the last] century who fits this caricature... except perhaps for a few neo-Thomists or conservative evangelical theologians' (Thomas 1999: 286b).

Soul–Body Metaphor

It was noted above that Thomas Aquinas employed the soul–body metaphor for the God–world relation, and it, along with mind–body, has become one of the main metaphors of contemporary panentheism. Hartshorne refers to God as the 'soul of the universe' (Hartshorne 1941: ch. 5). This metaphor, however, is not without its problems. For example, the soul is now understood to emerge in the body in the process of evolution, rather than to contain it. Moreover, the soul is usually understood to be in the body, rather than vice versa. Furthermore, since the body is essential to being human, the question arises as to whether or not the world is essential to or part of God, and thus divine and not creature. The mind–body metaphor is often used in this connection, but it may not be as illuminating as is supposed among panentheists, since it is often interpreted by contemporary philosophers as an anomaly that is unintelligible (see e.g. Shaffer 1967: v. 345a). Finally, the mind–body metaphor poses a problem for panentheists in regard to transcendence and immanence, both human and divine. Panentheists usually clarify this metaphor by explaining that God is more than the world even as we are more than our bodies, being essentially subjects or selves. This poses a dilemma for panentheists. If they emphasize the essential character of embodiment, they tend to divinize the world. If they stress the idea that we are essentially selves or subjects, they imply a dualistic anthropology. As a result of these problems, Peacocke comes out against this metaphor, and Gregersen concludes that the soul–body metaphor for the God–world relation 'no longer commends itself as an adequate contemporary model for the God–world relationship' (p. 20).

SOTERIOLOGICAL AND ESCHATOLOGICAL PANENTHEISM

There are two other interpretations of panentheism that are highly qualified or partial, in that not all of the world is in God or not at all times. One can be called soteriological panentheism, in that it draws on New Testament metaphors of persons being 'in Christ' or 'in the Spirit' or participating in the divine life. This being in God presumably applies only to Christians and not to others or the natural world. The other version can be called eschatological panentheism, in that it asserts that only in the final fulfilment will all the world be in God (pp. 21, 166–7). Most other panentheists would probably agree that these versions do not merit the title.

THE PROBLEM OF EVIL

Since, according to panentheism, every aspect of the world is in God, this must include evil in all its forms. Most panentheists are aware of this problem and attempt to resolve it. Peacocke argues that since all evil, both natural and human, is 'internal to God's own self', God 'can thereby transform it into what is whole and healthy... God heals and transforms from within' (pp. 151–2). Process panentheist Griffin, however, argues that while all evils are in God, they are not in God's essence, the primordial nature, but rather in God's experience of the world, the consequent nature. He concludes: 'There is evil only in God's experience, not in God's intentions. There is no moral evil in God' (p. 46). This may seem to contradict the previous approach, and in any case other panentheists may be concerned that the world is not in God's essence. Gregersen concludes that the process panentheistic solution to the problem of evil is inadequate. 'There seems to be no redemption possible for the tragically un[ful]filled aspirations of life, nor for the problem of the horrendous evils of wickedness....A soteriological deficit is obvious' (pp. 32–3).

DIVINE ACTION

Advocates of panentheism often state that classical theism cannot resolve the problem of divine action in the world, because in that view God acts from 'outside' the world, and therefore must intervene and disrupt the laws of nature, whereas in panentheism God acts from 'inside' the world, and this does not involve intervention in the closed causal nexus discovered by science. Peacocke states that in classical theism 'God can only exert influence "from outside" on events in the world. Such intervention, for that is what it would be, raises acute problems in the light of our contemporary scientific perception of the causal nexus of the world being a closed one' (p. 145). Clayton states in a previous essay that 'Panentheism's success turns ... not on a spatial concept of inside versus outside, but on its ability to give a more adequate account of divine agency than its competitors' (Clayton 1999: 293a).

Some questions arise here. First, if the world is in God, then is not God acting from outside the world? Second, it is not clear why it is easier for God to act in a closed causal nexus from inside rather than from outside. Third, what is the meaning of the assertion that the world is a closed causal nexus? Panentheists usually interpret this to mean that there are no gaps in the causal nexus, or that scientists do not appeal to non-natural causes to explain the processes of the world. This last, however, is simply the necessary methodological presupposition of all natural science: namely, that appeal to non-natural causes is ruled out. It is not an assertion that there are no non-natural causes, which would be a metaphysical, not a scientific, assertion. Furthermore, 'closed causal nexus' is not an adequate way to describe the world of quantum mechanics which lies at the base of the universe and which, according to the majority opinion, involves the *lack* of a closed causal nexus for all quantum events. John Polkinghorne states that 'individual quantum events are radically uncaused' (1988: 339).

Fourth, since panentheists often state that they reject a 'God of the gaps' approach (pp. 49, 98, 144), any assertion of God's action in a closed causal nexus must consider the extended, detailed, and persuasive argument of philosopher Thomas Tracy that divine action in the world requires gaps in the causal structure of the world. Tracy's argument proceeds by showing the insuperable difficulties of any view of divine action in a world of closed causal structures. If God makes a difference in the course of the world, then it can no longer be claimed that a complete account could be given of these events in terms of natural processes. If God makes a contribution to the causal nexus, then any account of the relevant events in terms of natural processes must contain gaps (Tracy 1995: 289–324).

Fifth, why is it that divine action from 'within' does not involve intervention? If it means that divine action does not modify the course of nature, how is this to be distinguished from the absence of divine action? Presumably divine action that is 'focally intended' (pp. 262–4) means that the world is different from what it would be in the absence of divine action, and this falls under the usual definition of intervention. Only process panentheists have an answer to this question: namely, that God is active in all events by supplying an initial aim for all actual occasions.

Clayton expands on the mind–body analogy to treat the issue of divine action in the world. 'The world is in some sense analogous to the body of God: God is analogous to the mind which indwells the body.... Call it the panentheistic analogy (PA). The power of this analogy lies in the fact that mental causation...is more than physical causation and yet still a part of the natural world' (pp. 83–4). (Note that the 'panentheistic analogy' means that God is in the world, rather than vice versa.) There is, however, another way to interpret this analogy. The mind acts to produce a state in the body which is different from the state of the body apart from the mind's action, and thus not according to biological laws. Therefore it is an intervention, and on this analogy, divine action will also be an intervention. Clayton states that 'one of our struggles is to understand what the minimal conditions are for asserting divine causal influence in the world'. He refers to the latter as 'focally intended divine actions' that make a difference in the way things happen in the world. He states that at least three authors in the symposium deny this possibility in an age of science (pp. 262–4).

It is often stated or implied in the symposium that if God intervened in the processes of nature and thus abrogated natural laws, science would be impossible. It should be noted, however, that human action, which is often used as an analogy of divine action, involves intervention and thus a modification of the world process, and scientists constantly have to take account of this problem. One of the most recent examples is that of research in extraterrestrial life by listening for radio signals from advanced civilizations. Many of the false identifications come from signals of human origin. But problems of this type have not brought science to an end. In the light of the considerations which have been mentioned, it is not clear that a panentheistic interpretation of divine action is the most fruitful one.

THEOLOGY AND SCIENCE

Finally, how does panentheism fare as a mediator between theology and science? Does it interpret this relationship more clearly and fruitfully than alternative views? Clayton lists as one of the reasons for adopting panentheism that it is 'more compatible than traditional theism with particular results in physics and biology or with common features shared across the scientific disciplines such as the structure of emergence' (p. 73), and this general view is echoed throughout the symposium. But one symposium author, Ruth Page, finds many problems in panentheism from the perspective of biology, including a focus on the emergence of complexity as too abstract, an avoidance of the fact of the loss of 90 per cent of all species in evolution,

an anthropocentrism which downgrades non-human species, and, thus, a triumphalist view of evolution (pp. 222–7).

What in fact is the panentheistic interpretation of the relationship of theology and
science? While most of the participants in the symposium assert that panentheism
involves the best understanding of the relation of theology and science, none of them
offers an explicit theory of this relationship in the symposium. The two editors,
however, suggest what seems to be an identification of divine action with the
processes of nature. Clayton states that in panentheism 'there would be no *qualitative*
or ontological difference between the regularity of natural law conceived as expressing the regular or repetitive operation of divine agency and the intentionality of
special divine actions' (p. 84). Peacocke states that 'The processes revealed by the
sciences are in themselves God acting as creator, and God is not to be found as some
kind of *additional* influence or factor added on to the processes of the world God is
creating.' He also states, however, that God 'infinitely transcends' the world. He
describes his view as 'naturalistic theism' (p. 144). Since science can be understood as
reflection on the processes of nature, and theology as reflection on divine action, this
interpretation seems to imply that theology and science are identical, or at least
closely integrated.

This last word points to the view of the relationship of theology and science
favoured by Ian Barbour in his typology of the relationship: namely, integration.
(Since the relationship of theology and science is one of the main themes of the
symposium, it is surprising that Barbour's well-known fourfold typology is not
mentioned.) Barbour suggests three versions of the integration type: natural theology, or the demonstration of the existence and attributes of God from our
experience of nature; the theology of nature, or the interpretation of nature in the
light of contemporary science; and systematic synthesis by means of a metaphysic
(Barbour 2000: 27–38). It is not clear which of these versions fits the view propounded by the editors.

Clarity about a relationship depends upon clarity about what is related, the terms,
or the *relata*. It is assumed in the symposium that they are Christian theology and
natural science, and more specifically natural-scientific theories and Christian theological doctrines in the form of statements. Then it is important to note that
integration is not a possible relationship between statements. (Barbour's first two
types, conflict (better contradiction) and independence, are possible relationships
between statements, but his latter two, dialogue and integration, are not.) Integration
means bringing items together into a single whole. It is a vague term open to a variety
of interpretations, and not very illuminating in regard to the relationship of theology
and science. Although individual panentheists do occasionally offer clear views of the
relationship of theology and science—for example, process panentheists—I must
conclude that panentheism as a movement does not.

Panentheism has become a fairly widespread movement in the last couple of
decades. Brierley states that 'a whole host of theologians identify themselves as
panentheists', that many 'others have been identified as panentheists', and that
'whole movements have been claimed for panentheism: Neoplatonism, Orthodox

Christianity, mysticism, and English modernism'. He reports that one enthusiast has declared that 'we are all panentheists now', and that panentheism has been described as a 'doctrinal revolution' (pp. 3–4). How can this widespread enthusiasm for panentheism be explained? We have seen that it is an extremely diverse movement, which ranges from the rigorous arguments of process philosophy to the simple emphasis on divine immanence, with the result that about all the movement has in common is the vague metaphor of the world being in God.

It seems clear that about all that panentheists have in common by means of their metaphor of the world in God is a strong emphasis on divine immanence. But this emphasis does not seem to be significantly different from the similar emphasis that we have found in classical theologians such as Thomas and Barth, or from the traditional doctrine of divine omnipresence, the presence of God at every point in space and time.

Moreover, we have seen that Clayton clearly implies that the world-in-God metaphor is not essential to panentheism. He states that the 'panentheistic analogy' is 'the mind which indwells the body', that the 'in' metaphor can refer to the world in God or God in the world, that the 'fundamental claim' of the panentheistic metaphor is 'the *inter*dependence of God and world' (p. 83), and that the success of panentheism does not depend on 'a special concept of inside versus outside' (Clayton 1999: 293a). Thus it would seem that the popularity of panentheism can hardly be explained by its unity or by the clarity and rigour of the arguments supporting it or by its self-evident validity.

INFLUENCE OF THE CURRENT ROMANTIC MOVEMENT

Therefore I am led to another explanation of the wide appeal of panentheism. I believe that it is a manifestation of the power of the current Romantic movement, which has influenced all areas of our life and culture, at least in the United States and England. This new Romantic movement began to emerge in the 1960s and is closely similar to the first Romantic movement of the late eighteenth and early nineteenth centuries in its favouring of vagueness, complexity, the irrational, the holistic, the apophatic, the mystical, and the inner. The first theological document of the first Romantic movement was the *Speeches on Religion* of 1799 by the Romantic theologian Friedrich Schleiermacher, who has been described as a panentheist for his strong emphasis on divine immanence (p. 4). The term 'panentheism' was coined in 1829 by the German philosopher Karl Christian Friedrich Krause, who was deeply influenced by Schelling, whose philosophy has been described as 'the epitome of German romantic philosophies' (Margoshes 1967: vii. 305b). In an essay arguing this thesis about the new Romantic movement I refer to the work of the historians Theodore Roszak and Sydney Ahlstrom, the British sociologists Bernice Martin, Christopher Booker, Frank Musgrove, and Colin Campbell, philosopher Edith Wyschogrod, and theologian Michael Ryan. I find evidence for the new Romantic movement also in contemporary culture, especially the cinema, the new consumerism, the neo-conservative movement, and in contemporary theology, especially the new emphasis on interiority and creation out of chaos, and in contemporary religion, especially the current spirituality movement (Thomas 2006). Romantic movements always favour organic metaphors over interpersonal ones, and therefore 'internal presence' over personal presence. Finally, since we have noted above the tendency to interpret panentheism in terms of the immanence of God, it should also be noted that the fundamental principle of Romanticism has been described as 'the coincidence of the finite and the infinite. In everything finite the infinite is present.' In other words, the radical immanence of God (Tillich 1967: 77). Therefore, I find that panentheism fits perfectly with the temperament of the current Romantic movement.

I will offer just one example of a contemporary theologian who is a panentheist (though not mentioned in the symposium) and also a perfect example of the current Romantic movement: namely, Catherine Keller. In her most recent book she describes her main theological proposal as 'apophatic panentheism' (Keller 2003: 219). She appeals to many authors and works that have been described as panentheist (Athanasius, Irenaeus, Cabbalah) and also to others claimed as panentheists who influenced the first Romantic movement (Nicholas of Cusa, Meister Eckhart, and Schelling), and many Romantic themes receive constant reiteration: depth, darkness, chaos, disorder, fluidity, and mystery. She states: 'The tehomic deity [a reference to *tehom*, or deep, in Genesis 1: 2] remains enmeshed in the vulnerabilities and poten-tialities of an indeterminate creativity. As *Tehom* it *is* that process; as deity it is *born from and suckles* that process' (Keller 2003: 226).

REFERENCES AND SUGGESTED READING

BARBOUR, I. G. (2000). *When Science Meets Religion.* New York: HarperCollins.

BARTH, K. (1957). *Church Dogmatics*, II/1, ed. G. W. Bromiley and T. F. Torrance. Edinburgh: T & T Clark.

CLAYTON, P. (1998). 'The Case for Christian Panentheism', *Dialog*, 37/3: 201–8.

—— (1999). 'The Panentheistic Turn in Christian Theology', *Dialog*, 38/4: 289–93.

—— and PEACOCKE, ARTHUR (2004) (eds.). *In Whom We Live and Move and Have Our Being: Panentheistic Reflections on God's Presence in a Scientific World.* Grand Rapids, Mich., and Cambridge: William B. Eerdmans.

COBB, J. B. Jr. (1965). *A Christian Natural Theology: Based on the Philosophy of Alfred North Whitehead.* Philadelphia: Westminster Press.

—— (2005). 'Review of Clayton and Peacocke', *Theology and Science*, 3/2: 240–2.

—— and GRIFFIN, D. R. (1976). *Process Theology: An Introductory Exposition.* Garden City, N.Y.: Seabury Press.

GILKEY, L. (1959). *Maker of Heaven and Earth: A Study of the Christian Doctrine of Creation*. Garden City, N.Y.: Doubleday.

GRIFFIN, D. R. (2005). 'Panentheism's Significance for the Science-and-Religion Discussion', *Science and Theology News*, 5/9: 35, 41.

HARTSHORNE, C. (1941). *The Vision of God and the Logic of Theism*. Chicago: Willett, Clark.

——(1948). *The Divine Relativity: A Social Conception of God*. New Haven: Yale University Press.

——(1976). 'Synthesis and Polyadic Inclusion', *Southern Journal of Philosophy*, 14/2: 245–55.

——and REESE, W. (1953). *Philosophers Speak of God*. Chicago: University of Chicago Press.

KELLER, C. (2003). *Face of the Deep: A Theology of Becoming*. London: Routledge.

MARGOSHES, A. (1967). 'Schelling, Friedrich Wilhelm Joseph', in Paul Edwards (ed.), *The Encyclopedia of Philosophy*, 8 vols., London and New York: Macmillan, vii: 305–9.

NEVILLE, R. C. (1980). *Creativity and God: A Challenge to Process Philosophy*. New York: Seabury Press.

POLKINGHORNE, JOHN (1988). 'The Quantum World', in Robert J. Russell, William R. Stoeger, SJ, and George V. Coyne, SJ (eds.), *Physics, Philosophy, and Theology: A Common Quest for Understanding*. Vatican City State: Vatican Observatory Publications; Berkeley: Center for Theology and Natural Sciences, 333–42.

SHAFFER, J. (1967). 'Mind–Body Problem', in Paul Edwards (ed.), *The Encyclopedia of Philosophy*, 8 vols., New York: Macmillan, viii: 336–46.

THOMAS, O. C. (1999). 'Not Yet a Case for Christian Panentheism', *Dialog*, 38/4: 285–7.

——(2006). 'On Doing Theology during a Romantic Movement', in Richard Valantasis (ed.), *The Subjective Eye: Essays in Culture, Religion, and Gender in Honor of Margaret R. Miles*, Eugene, Ore: Wipt & Stock, 136–56.

TILLICH, P. (1967). *Perspectives on 19th and 20th Century Theology*, ed. Carl E. Braaten. New York: Harper & Row.

TRACY, T. F. (1995). 'Particular Providence and the God of the Gaps', in Robert John Russell, Nancey Murphy, and Arthur R. Peacocke (eds.), *Chaos and Complexity: Scientific Perspectives on Divine Action*, Vatican City: Vatican Observatory Publications; Berkeley: Center for Theology and the Natural Sciences, 289–324.

WHITEHEAD, A. N. (1978). *Process and Reality: An Essay in Cosmology*, corrected edn, ed. David Ray Griffin and Donald W. Sherburne. New York: Free Press.

EVOLUTION, CREATION, AND BELIEF IN GOD

CHAPTER 39

EVOLUTION, RELIGION, AND SCIENCE

WILLIAM B. PROVINE

Understanding evolution does not undermine many beliefs in god: deism, gods that work through natural phenomena, gods invented from tortured arguments by theologians or academics, and many others. Understanding evolution is, nevertheless, the most efficient engine of atheism ever discovered by humans. It challenges the primary, worldwide, observable reason for belief in a deity: the feeling of intelligent design in biological organisms, including humans.

The first section of this chapter will document the widespread, worldwide sympathy with the feeling of intelligent design in animals and plants. The second section explains why and how the understanding of evolution in nature undermines and destroys the feeling of intelligent design and promotes atheism.

THE FEELING OF INTELLIGENT DESIGN

A dear friend of mine joined me for dinner one beautiful summer evening. It was dark when he left for home, and I had turned on the big sodium vapour lamp on the electric pole in the parking area. As we neared the pole, he said: 'Stop. Look at the moths.' They were flying high, near the light. He slipped his key chain from his pocket and rattled his keys. The moths instantly dropped like rocks from the light to the grass under the pole.

Pushed for an explanation of this neat trick, my friend, the world-renowned biologist Thomas Eisner, said: 'Will, I know your barn is filled with bats. Bats catch moths in the air by using sonar—high-pitched squeaks that bounce off objects back to the ear of the bat. Moths have developed a defence mechanism. When they hear sounds similar to the squeaks of bats, they immediately drop to the ground, as you saw. Bats cannot catch them there. My keys, when shaken, produce sounds very similar to those of bat sonar. Of course, you cannot hear the sounds because they are above the hearing level of humans. I happened to discover this one night when I was getting out my keys to open the door and the moths at the light in my yard dropped. So now I am just repeating the trick for you.'

Two astonishing adaptations: the sonar navigation system of bats and the ability of moths to hear the bat sonar system and escape to safety. In technical detail, these adaptations become even more impressive. They smack of intelligent design.

The feeling of intelligent design rarely exists by itself. We all yearn for moral guidance in human relationships, deep meaning in life, life after death, and human free will. The feeling of design in nature helps humans to believe that some intelligence greater than mere humans exists (not necessarily a personal god of any kind). This intelligence, when sensed, can help guide us in facing and solving the deep problems of life. The feeling of design is generally part of a whole world-view, though perhaps the most visible part.

That is why the feeling of design is so important. If one can gaze at a grasshopper or at another human and see the intelligent design, then all the other parts of one's world-view are in turn reinforced. The moral lessons of the feeling of design in biology are crucial to whole packages of beliefs.

Once the feeling of intelligent design disappears, however, then the most obvious and direct way of knowing that an intelligence is somehow responsible for what we can see with our eyes and hear with our ears also disappears. I will show how deeply this view is embedded in many world-views. Anyone wishing support for the feeling of intelligent design can just look around, since the vast majority of the great thinkers of all times have shared it, just as have the vast majority of humans, including the most reviled members of society.

ANTIQUITY OF THE FEELING
OF INTELLIGENT DESIGN

First I want to distinguish carefully between the Christian 'argument from design', which was so popular in England from the seventeenth to the mid-nineteenth century in England, and what I am calling the 'feeling of intelligent design'. The argument from design is a special logical argument. One sees an adaptation (or a constellation of them), and deduces directly the existence of the almighty Christian God. The feeling of design, on the other hand, has nothing special to do with Christianity, or even with a personal god or any god at all.

The feeling of design is just a feeling that biological organisms must have some kind of intelligence behind their being, with no necessary logical argument. You just look at a bird, or hear the music you love, or feel the touch of the person who loves you, and the feeling of design comes to you (from your cultural background).

Consider the following passage from a famous tale from Chuang Tzu:

Prince Huei's cook was cutting up a bullock. Every blow of his hand, every heave of his shoulders, every tread of his foot, every thrust of his knee, every whshh of rent flesh, every chhk of the chopper, was in perfect rhythm,—like the dance of the Mulberry Grove, like the harmonious chords of Ching Shou.

'Well done!' cried the Prince. 'Yours is skill indeed!'

'Sire', replied the cook laying down his chopper, 'I have always devoted myself to Tao, which is higher than mere skill. When I first began to cut up bullocks, I saw before me whole bullocks. After three years' practice, I saw no more whole animals. And now I work with my mind and not with my eye. My mind works along without the control of the senses. Falling back upon eternal principles, I glide through such great joints or cavities as there may be, according to the natural constitution of the animal. I do not even touch the convolutions of muscle and tendon, still less attempt to cut through large bones.

'A good cook changes his chopper once a year, because he cuts. An ordinary cook once a month, because he hacks. But I have had this chopper nineteen years, and although I have cut up many thousand bullocks, its edge is as if fresh from the whetstone. For at the joints there are always interstices, and the edge of a chopper being without thickness, it remains only to insert that which is without thickness into such an interstice. Indeed there is plenty of room for the blade to move about. I have kept my chopper for nineteen years as though fresh from the whetstone.

'Nevertheless, when I come upon a knotty part which is difficult to tackle, I am all caution. Fixing my eye on it, I stay my hand, and gently apply my blade, until with a hwah the part yields like earth crumbling to the ground. Then I take out my chopper and stand up, and look around, and pause with an air of triumph. Then wiping my chopper, I put it carefully away'.

'Bravo!' cried the Prince. 'From the words of this cook I have learned how to take care of my life.' (Lin Yutang 1983)

What a lovely story! The cook has a sense of Tao, the 'Way' of the world. He is following 'eternal principles'. The cook explains to the prince how he cuts up an ox; from the cook, the prince learns how to live his life. No force, no conscious or 'scientific' analysis is necessary. Tao cannot be approached directly. But it is there. Only by following Tao can a person be a fine butcher or a whole and morally upright person whose being is filled with meaning.

The passage from Chuang Tzu illustrates the feeling of design and its association with basic human needs and wishes. The feeling of order in nature followed by the cook is a guide to how humans should behave for the best society.

The most widely read book in English about the world's religions is by Huston Smith: *The World's Religions: Our Great Wisdom Traditions* (1991). In the last chapter he tries to determine if all wisdom traditions, despite their differences, share some common beliefs. For a moment, he imagines life as a great tapestry, which we face from the wrong side.

From a purely human standpoint the wisdom traditions are the species' most prolonged and serious attempts to infer from the maze on this side of the tapestry the pattern which, on its right side, gives meaning to the whole. As the beauty and harmony of the design derive from the way the parts are related, the design confers on those parts a significance that we, seeing only scraps of the design, do not normally perceive.... When we add to this the baseline they establish for ethical behaviour and their account of the human virtues, one wonders if a wiser platform for life has been conceived. (Smith 1991: 388–9)

In a later book, *Forgotten Truth: The Common Vision of the World's Religions* (Smith 1992), Smith argues again for the common base of intelligent design in the world, but points out clearly that modern science, evolutionary biology in particular, has limitations. Otherwise, intelligent design must disappear from the biological world.

Christianity, Judaism, Islam, and other religions with personal gods contain innumerable references to the natural world made by God. A believer can see the design and revel in it. Did philosophy also hold this tradition? Plato and Aristotle were the most famous Greek philosophers of antiquity and pre-dated the birth of Christ and Muhammad. Plato was mostly concerned with moral and political life. He wrote dialogues that he hoped would help bring about a better society, the most famous of which is his *The Republic*. Plato had so little interest in the natural world that Benjamin Harrington and Ernst Mayr, both distinguished scholars, have argued that Plato hindered the advance of science.

That is why one of Plato's late dialogues, the *Timaeus*, is of such great interest. The *Timaeus* is about the physical world, at least at first glance. Of course, we can guess that Plato had an ulterior motive for focusing upon the changing physical world in this dialogue. Plato wrote the *Timaeus* when he was over 70. He planned it as the first part of a trilogy: *Timaeus*, *Critias*, and *Hermocrates*. He broke off mid-sentence in writing *Critias*, and never wrote *Hermocrates*. The idea behind the trilogy was to show how one could see deep into the changing physical world and begin to understand the reality that lay behind it. So indeed the major focus of the trilogy was right back to politics, ethics, justice, and meaning in life. Timaeus speaks for Plato in this dialogue, and I will just use Plato's name. The dialogue is divided into three sections: works of reason, what comes about of necessity, and co-operation of reason and necessity.

The physical world is full of ordered things like time, stars, planets, the human soul, the structure of the human body, seeing, hearing, etc. How did this intelligent design originate in the physical world? Could it possibly come from just the mechanical workings of nature? Certainly not (we will address that in a moment). Intelligent design in the physical world must have come from an intelligent designer. Plato at first calls it the Demiurge. And in the first section, Plato describes a likely story about how the Demiurge used eternal ideas as models to design the physical universe and the animals and plants in it. The Demiurge is very murky, and by the end of the *Timaeus* Plato is using the lesser gods to do the intelligent designing. The important point is that things in this world giving the strong appearance of intelligent design, were exactly that: intelligently designed.

What can come about by mere necessity itself, by the action of physical nature without the influence of the intelligent designer? Chaos, but a chaos that has some order. Time existed, but did not pass regularly. Earth, air, fire, and water existed, but did not have their regular characters. Heavy things generally went down and light things went up. Nothing good came of this chaos. Everything was shaken and swayed unevenly. The four elements did not even get their characteristics until the gods designed regularities for them. Still, nature existed before the intelligent designers went to work on it. Moreover, and importantly, when the designers tried to design physical objects, they had to take into account the necessary. Designers cannot make just anything—only what is possible given the constraints of sheer necessity.

In the final section, Plato tells a likely story about how the gods put order into chaos, and then a detailed story about how the human body was designed in the face of necessity. The head is the most important part of a human, but the gods designed a body for it so that the head wouldn't just roll around on the ground and be unable to climb out of ditches. The designers put bone all around the brain, then flesh and sinews on top of that. But the human would live longer if the bone and flesh covering on the head was much stronger and deeper. Then the designers had to make a choice: let the humans live longer with their heads well protected, or live shorter lives but nobler ones with greater use of their minds. They chose the shorter and more intense life. Every decision made by the designers was always in the face of necessity.

How does any of this make a person into a better one, who might make a fine citizen or ruler? Plato thought you should start somewhere to understand reality. The organized things in the physical world must have been organized by intelligence, based upon the real intelligible world, not necessity by itself. Thus, by studying intelligent design in nature, one could begin an education that would end in thinking correctly about justice, morality, and political life.

As Plato's most famous student, Aristotle had some tough times breaking from his mentor. Aristotle rejected Plato's entire realm of the intelligible, all of Plato's unchanging Forms, or ideas. He also rejected all gods except for the 'Unmoved Mover', who started off motion. But he retained the essential contrast between the realms of 'reason' and 'necessity'. Aristotle's famous four causes are the key. His causes were (1) material—the stuff of which things are composed; (2) efficient—the direct cause that physically pushes something; (3) formal—the cause that suggests a design, especially easy to understand with biological organisms (acorns grow up into oak-trees, not palms); and (4) final—the hardest one to understand, but the cause that suggests a goal toward which the object develops, again understood best with biological organisms. Even mutilated embryos tend to develop toward an adult.

Aristotle wrote a huge amount, and much of it has been preserved; but we can turn, as in Plato, to a crucial place in his writing. In Book II, chapters 8 and 9, of his *Physics*, Aristotle explains how nature works. It works, coincidentally, like a craftsman (Solmsen 1960). Material and efficient causes are the necessity; formal and final causes are the reason. Material and efficient causes by themselves would never produce an organized being. The formal and final causes behave as a craftsman would, constructing intelligently designed organisms. No gods are required. Still, intelligent design appears in biological organisms because nature operates like a craftsman. When Aristotle sees what appears to be intelligent design in nature, it is indeed intelligent design.

Plato and Aristotle, despite all their differences, both see the fundamental difference between reason and necessity. For Plato, the great good of understanding nature was the road to understanding moral and political life. Aristotle argued that 'man is a political animal'. The only way to understand ethics was to understand that humans depended upon other humans, and to seek the goal of human moral behaviour required an accurate view of humans in relation to nature. Only humans fashioned by nature-as-craftsman could aspire to having an ethics, and be capable of following ethical goals. The examination of nature thus had deep consequences for understanding human political and ethical behaviour.

Plato's *Timaeus* was his only work to survive in the Latin West up to medieval times. Others re-entered the West by way of northern Africa and the Moors, along with many of the works of Aristotle. The medieval synthesis of Plato, Aristotle, and Christianity is hardly a surprise—the Christian idea of an all-powerful God who designed all the organized things of nature fitted well with the Platonic and Aristotelian notion of a split between reason and necessity. That is why St Augustine could describe himself as a Platonist, and St Thomas Aquinas could describe himself as an Aristotelian.

Creationists assert that nearly all the great scientists of the eighteenth century and earlier believed in the feeling of intelligent design, or were outright creationists themselves. This assertion is correct. The great *Dictionary of Scientific Biography* (Gillispie 1981) appeared in an eight-volume edition in 1981, a monumental ten-year effort by historians of science, covering all the most important dead (that was a requirement!) scientists. Of the more than 5,000 entries, the vast majority of those who died before 1850 thought that one deity or another created the natural world, which in any case exhibited intelligent design. A few prominent examples will have to suffice.

Galileo is a good example, because he fought vehemently with the Catholic Church. He made telescopes and discovered such things as craters on the Moon, the four moons of Jupiter, the unusual shape of Saturn (he could not see the rings with his telescope), and the moon-like phases of Venus. He saw sunspots clearly, and resolved the Milky Way into a vast number of hitherto unseen stars. He also supported the Copernican view that the Earth was a planet circling the Sun. The Catholic Church was wedded to the belief that the heavens were made by God to surround the Earth at the centre and to be unchanging. When Galileo offered a high prelate the chance to look through his telescope and see the truth of his assertions about the heavens, the prelate refused, and said that he did not wish to see the disgusting sight of nature contradicting reason. Who doesn't feel that way when a cherished view could be dashed so easily?

The Catholic Church forbade Galileo from publishing his ideas on the solar system, changing heavens, and dynamics. Galileo was combative and persisted, and was confined to house arrest the last ten years of his life. You might imagine that poor Galileo was an atheist or something.

Nothing could be further from the truth. In 1979, Pope John Paul II addressed the Vatican Academy of Sciences (in French) on science and religion. His comments were translated into English and published in *Science* magazine. The Pope quoted Galileo: 'Holy Scripture and nature proceed equally from the divine Word, the former as it were dictated by the Holy Spirit, the latter as a very faithful executor of God's orders.' And at the beginning of his *Siderious Nuncius* Galileo says, 'All of this has been discovered and observed these last days thanks to the "telescope" that I have invented, after having been enlightened by divine grace' (Pope John Paul II 1980: 1166). Galileo's theology was thus agreeable, according to Pope John Paul II, who said that Galileo's views on science and religion are almost exactly those of the Catholic Church today. Galileo thought he was discovering order put into nature by God, and whose hand he required to find it.

Born so small that his mother could fit him in a quart pot, Isaac Newton became a famous giant for his extensions of the inverse square law of gravitational attraction to planetary astronomy. In a short period of time, he deduced the orbit of the Moon around the Earth, the reason for tides on Earth, the orbits of the Earth and other planets around the Sun, and the paths of comets (some of which returned and some that didn't). Newton also did much work in optics. But he also got into a lot of controversies with Christians, who accused him of being a deist, meaning someone who could not actually see the handiwork of God in nature, but supposed God to be behind it. A deist was nearly, but not quite, an atheist. Newton fumed against these accusations in many places in his writings.

In the introduction to his great *Philosophiae naturalis principia mathematica* (fondly known as the *Principia*), Newton says something that earned him the opposition he later faced:

I wish we could derive the rest of the phenomena of Nature by the same kind of reasoning from mechanical principles, for I am induced by many reasons to suspect that they may all depend upon certain forces by which the particles of bodies, by some causes hitherto unknown, are either mutually repelled toward one another and cohere in regular figures, or are repelled and recede from one another. (Thayer 1953: 10–11)

It sounds like a revolutionary idea: mechanical principles with no god required. That is not what Newton meant, however. In Part III of the *Principia*, Newton describes the lovely, regular system of the Sun and the planets with all six (in his day) planets orbiting in the same direction and nearly in the same plane: 'This most beautiful system of the sun, planets, and comets could only proceed from the counsel and dominion of an intelligent and powerful Being.' He then writes pages about God and says that 'we know him only by his most wise and excellent contrivances of things and final causes'. And, perhaps most tellingly, 'Blind metaphysical necessity, which is certainly the same always and everywhere, could produce no variety of things. All that diversity of natural things which we find suited to different times and places could arise from nothing but the ideas and will of a Being necessarily existing' (Thayer 1953: 42–3).

Newton was also aware of the symmetries and organization of animals:

Atheism is so senseless and odious to mankind that it never had many professors. Can it be by accident that all birds, beasts, and men have their right side and left side alike shaped (except in their bowels); and just two eyes and no more on either side of the face; and just two ears on either side the head; and a nose with two holes; and either two forelegs or two wings or two arms on the shoulders, and two legs on the hips, and no more? Whence arises this uniformity in all their outward shapes but from the counsel and contrivance of an Author? (Thayer 1953: 65)

For Newton, all knowledge of the natural world was derived from the appearances of things, so if the handiwork of God were invisible, he would have trouble believing in him.

Often called the father of modern medicine, William Harvey discovered the motion of the heart and blood. His little book *De Motu Cordis* (Harvey 1928) contained some great experiments and reasoning. He measured carefully the volume of the left ventricle in cadavers and found that it held about 3 ounces. He estimated conservatively how much blood the left ventricle pumps out (it cannot come back because of the aortic valve) on each beat. He estimated a fourth, fifth, sixth, or even eighth of the 3 measured ounces. And he calculated that in one half-hour, the heart must be pumping out far more blood than the body contains. Thus the blood flow must go from the arteries to the veins and return to the heart. Every scientist must hope to invent such a neat and decisive experiment.

Yet Harvey believed, just like Newton and Galileo, that God, working through nature, created all the adaptations of the heart, blood vessels, and blood. He was working hand in hand with the assumptions of almost all other scientists of the time. Chinese or Indian scientists of the same period did not invoke the same kind of intelligence, but nevertheless, what one could see in organisms or the planetary system or the visible universe was a product of intelligence, not merely the grinding away of unintelligent nature.

In the history of Western civilization unbelievers have always been present: Democritus (or his tradition if he himself did not exist), Leucippus, Lucretius, village atheists in medieval villages all over Western Europe, La Mettrie in the eighteenth century, and the famous French atheists, such as LaPlace and Diderot in the Enlightenment period. They served as examples upon which other thinkers of the time could pile invective and convincing criticisms. How did those fools think that biological organisms could come about without intelligence? Can college students taking a view of no intelligence in the evolutionary process really be smarter than Plato, Aristotle, Galileo, Newton, or Harvey, or all put together?

Let's return to Huston Smith again, since he has studied so much about the religions of the world. Hope, he claims, is one of the hallmarks of religion around the world. All the religions he has studied give access to hope as one of the greatest fruits of humankind.

Hope is indispensable to human health—to psychological health most immediately, but because man is a psychosomatic whole, to physical health as well. Situated as we are in the Middle (hence middling) World, vicissitudes are a part of the human lot: external vicissitudes (hard times), and internal vicissitudes—the 'gravitational collapse' of the psyche that sucks us into depression as if it were a black hole. Against such vicissitudes hope is our prime recourse. (Smith 1992: 118)

Hope is dashed, according to Smith, by Darwinian-type evolution with no intelligent design. The adaptations are mere products with no values attached to them. Evolution gives no basis for life after death or ultimate meaning, the opposite of hope. Darwinian evolution is the antithesis of the best that religion has to offer. Either the collective wisdom of the ages, embodied in religion, is right about these fundamental questions, or ultimate hope is gone, and naturalistic evolution rules the playing field.

Smith is clearly appalled by modern evolutionary biology. 'Evolution proposes to be an explanatory theory. It is the claim that everything about man, his complete complement of faculties and potentials, can be accounted for by a process, natural selection, that works mechanically on chance variations' (Smith 1992: 130). This must be a wrong explanation, a wrong theory: 'Our personal assessment is that on no other scientific theory does the modern mind rest so much confidence on so little proportional evidence, on evidence, that is to say, which, in the ratio of the amount that would be needed to establish the theory in the absence of the will to believe, is so meager' (Smith 1992: 132). Evolution as evolutionists imagine it has no evidence to support it and is wrong. Smith does not deny the fossil record or long life of the Earth, nor the progression from bacteria to humans in its largest sense. How can we harmonize the coming to be of species with religions? 'In the celestial realm, the species are never absent: their essential forms or archetypes reside there from an endless beginning. As earth ripens to receive them, each in its turn drops to the terrestrial plane and, donning the world's fabric, gives rise to a new life form. The origin of species is metaphysical' (Smith 1992: 139).

Although saying utter nonsense about the origin of species, Smith has set out the issue just right. He properly focuses upon modern evolution as the difficult issue, if not the prime target, for contemporary religions. One or the other is true about the world. It simply cannot be mechanistic evolution. Giving up ultimate hope is just too much.

EVOLUTION AND GODS

Why is evolution so antagonistic to the feeling of intelligent design? I will detail only two reasons here, but many more could be added in a longer essay.

People from all over the world believe that a god or intelligent design account for the wonderful adaptations of animals and plants, including those of humans. The first thing that an evolutionist must understand is that these same adaptations virtually ensure the extinction of any species. The problem is that adaptation is to a particular environmental situation, with seasonal changes included. When the environment changes, the species in question must adapt to the new environment. Often a species is so successful that its very success uses up a necessary resource for the species, and it goes extinct precisely because of its great adaptations. For example, the giant sharks (*Carcharocles megalodon*), which flourished in the Miocene from 15 million years ago to about 1.7 million years ago, were up to 50 feet long with giant 6-inch teeth, and superbly adapted to eating small whales, whose bones can easily be found with the fossil shark teeth. *C. megalodon* was far more fearful than the shark in the movie *Jaws*, having jaws that can be seen in the Smithsonian Museum in a reconstruction (but with real fossil teeth in the outer layer!). Six humans can fit in this jaw. It could have chomped on a small car. The small whales that *C. megalodon* ate perished, some shark specialists think, because the giant sharks ate them to extinction; but in any case, when the primary food for these sharks disappeared, they went extinct too, with all their adaptations. This is the story of evolution in a nutshell.

The feeling of intelligent design disappears in the perspective of evolution. I asked David Raup and Jack Sepkoski, palaeontologists at the University of Chicago, to estimate for me how many species of vertebrates existed at the end of the Cretaceous, about 70,000,000 years ago. They said about 50,000. And how many exist now?, I asked. They said about 100,000. Then the crucial question: how many of the species of vertebrates 70,000,000 years ago gave rise to all that exist now? They said probably fewer than twenty, but at the outside, twenty-five. So, of the 50,000 or so species, all but twenty-five went extinct. And in the interval of 70 million years, most of the species of vertebrates that came into existence also became extinct. Even with all the exquisite adaptations that smack of an intelligent designer, these vertebrates were poor survivors.

What about natural selection? Natural selection is not a mechanism, does no work, does not act, does not shape, does not cause anything. Biologists are very lax in their language, and so was Charles Darwin. Natural selection is the *outcome* of a very complex process that basically boils down to heredity, genetic variation, ecology, and demographics (especially the overproduction of offspring, and consequent struggle). The adaptations that evolve we call 'naturally selected'. The process yields organisms with adaptations which help them to survive and flourish. The process also virtually guarantees extinction when the environment changes sufficiently, which it often does. The intelligent design apparent in the adaptations has no inkling of environmental change. The pattern of extinction, however, is precisely what one would expect of the causes of natural selection. If one gets to Carnegie Hall by 'practice, man, practice', then one understands natural selection as 'demographics, man, demographics', followed by extinction.

Every organism that has become extinct (about 99+ per cent of all species that have ever lived) was jam-packed with adaptations. Some of those adaptations became detriments to the organism when the environment changed and caused the organism to become extinct. The better an organism is adapted to a particular environment, the more certain it is that it will become extinct when the environment changes. Adaptations are hopelessly tied with extinction. The feeling of intelligent design in organisms must thus be tied to extinction, too. That is why evolutionists give up the feeling of intelligent design.

The second reason why understanding evolution precludes the feeling of intelligent design is that evolution also shows no hint of progress. At the same time as humans evolved, so did roundworms that infect and can only reproduce in humans. Roundworms from other species are more dangerous to us than our own roundworms, to which we are well adapted. Raccoons harbour a roundworm (*Baylisascaris procyonis*) special to them, as do cats, dogs, and most other ranging mammals.

Baylisascaris procyonis are up to 7–10 inches long, less than a quarter-inch in diameter, and reproduce only in raccoon intestines. They exist almost everywhere that raccoons live, and in some parts of the United States infect nearly 100 per cent of raccoons. The worms shed their eggs in the faeces of the raccoon, up to millions of eggs in a day. When the faeces dry, the worm eggs can last for up to six or seven years. Most often, raccoons infect each other, and they suffer little from the parasite.

Other animals can be infected, too, and the worm behaves very differently from in a raccoon. It cannot reproduce in any animal other than a raccoon. If its eggs are ingested by, say, a woodrat, the larvae leave the intestine and migrate to the other organs or to the nervous system of the woodrat. Soon the woodrat dies.

Birds are especially susceptible and die rapidly. Known hosts to *Baylisascaris procyonis*: humans, dogs, mice, squirrels, birds, guinea pigs, rabbits, prairie dogs, porcupines, woodchucks, squirrel monkeys, chinchillas, quails, and ostriches, along with some eighty other species. The larvae infect far more animals that have not yet been studied. Obviously, *Baylisascaris procyonis* did not have ostriches as a factor in its evolution. The behaviour pattern of the worms works well beyond the factors causing its adaptations.

Raccoons live near humans now, in sandboxes and house chimneys. Children are especially susceptible, because dried raccoon dung comes down chimneys as dust, or makes the sand in a sandbox look a bit darker. Just a hand in the mouth, only five to ten of the eggs or tiny larvae, can damage brains and perhaps kill. All damage is permanent and irreversible. In adults, the most frequent effect is problems with vision, and an optometrist or ophthalmologist can look into the eye and see the worms emerging from the optic nerve. Humans may also suffer from lack of coordination, liver enlargement, and nausea. Remember, the worm can be 7–10 inches long. In small children, degeneration of vision is often undetected, and neural problems usually occur before the child is taken to the doctor. If large numbers of worms were ingested, the symptoms can progress rapidly.

No effective treatment exists for the worms. The usual vermifuges don't work, as they do with our roundworms, the pinworms. The eggs are also resistant to most pesticides. When we were animal rehabilitators for New York State for eighteen years, my former wife and I raised more than twenty wild raccoon babies before we knew anything about *Baylisascaris procyunis*. We and our two boys were really fortunate not to get this nasty worm. Perhaps three or four people die per year in North

America from this worm, but many more suffer some brain or eye damage. You can read all about it at the Center for Disease Control in Atlanta, Georgia (Sorvillo *et al.* 2002).

Other diseases kill vastly more people. AIDS, caused by a virus, has infected close to 50 million people, especially in Africa and South-East Asia, and death rates will rise for many years to come, despite some expensive medicines that help keep the virus at low levels. Malaria kills almost as many people per year as AIDS, including recently the brilliant evolutionist William Hamilton. Schistosomiasis, a disease caused by worms, infects more than 200 million humans worldwide.

Each of these infectious agents has evolved as long as humans have existed. I can see no hierarchy whatsoever in the productions of evolution. Any deity that would work this way seems perfectly awful to me. The process that produced these very different pathogens and humans just happens, and speaking as if evolution 'cared' about its productions is unintelligible.

These two reasons to reject the feeling of intelligent design in biological organisms are just a sample of compelling reasons. The famous evolutionist George C. Williams has written an essay on the evolution of social behaviour, and concludes that social behaviour in animals is nothing less than ghastly, and any hope we have as humans to have a decent moral world is to fight fiercely against the selfishness that evolution has produced in us (Williams 1988).

Antony Flew, the well-known philosophical atheist, has given up his atheism at age 80 because of his growing feeling of intelligent design in biological organisms. I venture to guess that Flew's understanding of evolution is extremely weak, and that if he knew more about evolution, he would not make this move to intelligent design.

Certainly the most eminent evolutionists in the world overwhelmingly reject gods and the feeling of intelligent design. Greg Graffin sent a detailed questionnaire to every evolutionary biologist who belongs to an honorific national academy in any country in the world. He asked if they believed in theism, deism, or naturalism, and asked them to expand on this choice (Graffin 2003). Only five theists among the 149 (of 271 sent) who replied were found, and most did not hold to a strong theism. More were deists, but still the number was low. As a group, they strongly rejected belief in gods, intelligent design, life after death, and the supernatural. Really understanding evolution means that belief in a deity is nearly impossible.

You might object, however, that Pope John Paul II believed in evolution, and thus Catholics have no problems with evolution and their Christian God. Pope John Paul II accepted that human bodies and chimpanzees share a common ancestor. But he rejected roundly the thought that human minds share a common ancestor with chimpanzee minds:

If the human body takes its origin from pre-existent living matter the spiritual soul is immediately created by God.... Consequently, theories of evolution which, in accordance with the philosophies inspiring them, consider the mind as emerging from the forces of living matter, or as a mere epiphenomenon of this matter, are incompatible with the truth about man. (Pope John Paul II 1996: 2)

God gave humans an immortal soul and human free will sometime in recent human evolution; and this means that, from this point on, humans and chimpanzees never possibly shared a common ancestor. Pope John Paul II was a young-earth creationist of human souls; by holding this view, he graphically violated the overwhelming evidence that humans and chimpanzees do indeed share most of their DNA and share a common ancestor, minds and all. He also perpetuated the myth that Jesus came from a virgin, Mary. Some mammalian virgins, though not humans, can produce offspring, but they are always female. Thus the male complement of chromosomes of Jesus was either contributed by a human male (and Mary was not a virgin, my preferred explanation) or he carried a divine complement of chromosomes that turned him into a male. Evolutionary biology and typical views of Catholics are inconsistent, and it's a widespread myth that they fit together.

Creationists and believers in almost any religion are right that evolutionary biology should be feared and rejected. It is *the* great engine of atheism. Religious believers and theologians alike understand that serious evolutionists are probably atheists, even though evolutionists often try to sweep that fact under the rug. As a teacher of evolutionary biology, I have seen a minority of students every year move from weakly held theism to a naturalist evolutionist position. Strongly religious students deepen their faith from my evolution course; the course regularly ends with more creationists than when it began. Students who are already naturalists delight in what they find in evolution. This diversity, much like the products of evolution, offers a delight of its own.

REFERENCE AND SUGGESTED READING

GILLISPIE, CHARLES (1981) (ed.). *Dictionary of Scientific Biography.* New York: Scribner.

GRAFFIN, GREGORY W. (2003). *Monism, Atheism, and the Naturalist Worldview: Perspectives from Evolutionary Biology.* A Ph.D. dissertation submitted to Cornell University. For more information, see <http://www.cornellevolutionproject.org>.

HARVEY, WILLIAM (1928). *De Motu Cordis.* Florence: R. Lier.

LIN YUTANG (1983). *Famous Chinese Short Stories.* London: Dent.

POPE JOHN PAUL II (1980). 'Comments on evolution to the Vatican Academy of Sciences', *Science*, 210: 1166.

—— (1996). 'Message to Pontifical Academy of Sciences of October 22, 1996'. Official translation published in *L'Osservatore Romano*, 'Weekly Edition in English', 30 October 1996. Available online through the National Center for Science Education at <http://www.ncseweb.org/resources/articles/8712_message_from_the_pope_1996_1_3_2001.asp>, verified April 14, 2006.

SMITH, HUSTON (1991). *The World's Religions: Our Great Wisdom Traditions.* San Francisco: Harper SanFrancisco.

—— (1992). *Forgotten Truth: The Common Vision of the World's Religions.* San Francisco: Harper SanFrancisco.

SOLMSEN, FRIEDRICH (1960). *Aristotle's System of the Physical World: A Comparison with his Predecessors*. Ithaca, N.Y.: Cornell University Press.

SORVILLO, F. *et al.* (2002). 'Baylisascaris procyonis: An Emerging Helminthic Zoonosis', *Emerging Infectious Diseases*, 8/4: 355–9.

THAYER, HORACE (1953) (ed.). *Newton's Philosophy of Nature: Selections from his Writings*. New York: Hafner.

WILLIAMS, GEORGE C. (1988). 'Huxley's Evolution and Ethics in Sociobiological Perspective', *Zygon*, 23: 383–438.

CHAPTER 40

DARWINISM

ALISTER E. MCGRATH

One of the most vigorous debates within modern Christian thought concerns the implications of Darwinism for religious belief. It is a debate that is by no means limited to Christianity, as is evident from the generally hostile reaction to Darwinism within the Islamic world. So what is Darwinism? While the term is often used to refer specifically to the views set out by Charles Darwin in his *Origin of Species*, it is more widely used to refer to the theories that emerged from Darwin's work, as they have been received, developed, and modified. Although terms such as the 'neo-Darwinian synthesis' are often used to distinguish the present state of evolutionary theory from the earlier forms proposed by Darwin himself, I shall follow the widespread convention of using the word 'Darwinism' to define a family of theories.

In brief, Darwinism can be defined in terms of its core constructs—the 'minimal theory that evolution is guided in adaptively nonrandom directions by the nonrandom survival of small random hereditary changes' (Dawkins 2003: 81). This is often taken to be sufficient to distinguish it from the rival evolutionary paradigm associated with Jean-Baptiste de Lamarck (1744–1829), who proposed the theory that changes that are acquired during the lifetime of an organism are passed on to its offspring. The idea that phenotypic changes can be passed on to the genotype is now widely discredited as a mechanism for explaining biological evolution. It is, of course, important to note that a number of debates are under way within contemporary Darwinism. Stephen Jay Gould's notion of 'punctuated equilibrium', although clearly consistent with a broad construal of Darwinism, has met considerable resistance from those committed to more continuous modes of evolutionary development. Gould's critics argue that he placed too much emphasis on drift and historical contingency, and neglected the themes that adaptationists regard as significant in selectionist theories.

Discussions of the issues raised for religion in general, and Christianity in particular, by Darwinism tend to fall into two categories. The first adopt a broadly historical approach, documenting the religious response to Darwin's *Origin of Species* (1859) and *Descent of Man* (1871). The second kind are more thematic in approach, setting out the religious issues that arose from Darwinism, and assessing their significance. I shall consider both aspects of the question in this chapter.

THE HISTORICAL BACKGROUND

Darwin's radical theory of natural selection can be seen as the culmination of a long process of reflection on the origins of species. Among the studies which prepared the way for Darwin's theory, particular attention should be paid to Charles Lyell's *Principles of Geology* (1830). The prevailing popular understanding of the history of the Earth from its creation took the form of a series of catastrophic changes. Lyell argued for what he called 'uniformitarianism' (a term which was coined by James Hutton in 1795), in which the same forces that can now be observed at work within the natural world are argued to have been active over huge expanses of time in the past. Darwin's theory of evolution works on a related assumption: that forces which lead to the development of new breeds of plants or animals in the present operated over very long periods of time in the past.

The major rival to Darwin's theory was due to the eighteenth-century Swedish naturalist Carl von Linné (1707–78), more generally known by the Latinized form of his name, 'Linnaeus'. Linnaeus argued for the 'fixity of species'. In other words, the present range of species which can be observed in the natural world represents the way things have been in the past, and the way they will remain. Linnaeus's detailed classification of species conveyed the impression to many of his readers that nature was fixed from the moment of its origination. This seemed to fit in rather well with a traditional and popular reading of the Genesis creation accounts, and suggested that the botanical world of today more or less corresponded to that established in creation. Each species could be regarded as having been created separately and distinctly by God, and endowed with its fixed characteristics.

The main difficulty here, pointed out by Georges Buffon and others, was that the fossil evidence suggested that certain species had become extinct. In other words, fossils were found which contained the preserved remains of plants (and animals) which now had no known counterpart on the Earth. Did not this seem to contradict the assumption of the fixity of species? And if old species died out, might not new ones arise to replace them? Other issues seemed to cause some difficulty for the theory of special creation—for example, the irregular geographical distribution of species.

The publication of Charles Darwin's *Origin of Species* (1859) is rightly regarded as a landmark in nineteenth-century science. On 27 December 1831, *H.M.S. Beagle* set out from the southern English port of Plymouth on a voyage that lasted almost five years. Its mission was to complete a survey of the southern coasts of South America, and afterwards to circumnavigate the globe. The small ship's naturalist was Charles Darwin (1809–82). During the voyage, Darwin noted some aspects of the plant and animal life of South America, particularly the Galapagos Islands and Tierra del Fuego, which seemed to him to require explanation, yet which were not satisfactorily accounted for by existing theories. The opening words of *Origin of Species* set out the riddle that he was determined to solve:

When on board H.M.S. Beagle as naturalist, I was much struck with certain facts in the distribution of the organic beings inhabiting South America, and in the geological relations of the present to the past inhabitants of that continent. These facts, as will be seen in the latter chapters of this volume, seemed to throw some light on the origin of species—that mystery of mysteries, as it has been called by one of our greatest philosophers. (Darwin 1890: 1)

One popular account of the origin of species, widely supported by the religious and academic establishment of the early nineteenth century, held that God had somehow created everything more or less as we now see it. The success of the view owed much to the influence of William Paley (1743–1805), archdeacon of Carlisle, who compared God to one of the mechanical geniuses of the Industrial Revolution. God had directly created the world in all its intricacy. Paley accepted the viewpoint of his age—namely, that God had constructed (Paley preferred the word 'contrived') the world in its finished form, as we now know it. The idea of any kind of development seemed impossible to him. Did a watchmaker leave his work unfinished? Certainly not!

Paley argued that the present organization of the world, both physical and biological, could be seen as a compelling witness to the wisdom of a Creator God. Paley's *Natural Theology; or Evidences of the Existence and Attributes of the Deity, Collected from the Appearances of Nature* (1802) had a profound influence on popular English religious thought in the first half of the nineteenth century, and is known to have been read by Darwin. Paley was deeply impressed by Newton's discovery of the regularity of nature, which allowed the universe to be thought of as a complex mechanism, operating according to regular and understandable principles. Nature consists of a series of biological structures which are to be thought of as being 'contrived'—that is, constructed with a clear purpose in mind. Paley used his famous analogy of the watch on a heath to emphasize that contrivance necessarily presupposed a designer and constructor. 'Every indication of contrivance, every manifestation of design, which existed in the watch, exists in the works of nature.' Indeed, Paley argued, the difference is that nature shows an even greater degree of contrivance than the watch. Paley is at his best when he deals with the description of mechanical systems within nature, such as the immensely complex structure of the human eye and heart. Yet Paley's argument depended on a static world-view, and simply could not cope with the dynamic world-view underlying Darwinism.

Darwin knew of Paley's views and initially found them persuasive. However, his observations on the *Beagle* raised some questions. On his return, Darwin set out to develop a more satisfying explanation of his own observations and those of others.

Although Darwin appears to have hit on the basic idea of evolution through natural selection by 1842, he was not ready to publish. Such a radical theory would require massive observational evidence to be marshalled in its support.

Four features of the natural world seemed to Darwin to require particularly close attention, in the light of problems and shortcomings with existing explanations.

(1) The forms of certain living creatures seemed to be adapted to their specific needs. Paley's theory proposed that these creatures were individually designed by God with those needs in mind. Darwin increasingly regarded this as a clumsy explanation.

(2) Some species were known to have died out altogether—to have become extinct. This fact had been known before Darwin, and was often explained on the basis of 'catastrophe' theories, such as a 'universal flood', as suggested by the biblical account of Noah.

(3) Darwin's research voyage on the *Beagle* had persuaded him of the uneven geographical distribution of life forms throughout the world. In particular, Darwin was impressed by the peculiarities of island populations.

(4) Many creatures possess 'rudimentary structures', which have no apparent or predictable function—such as the nipples of male mammals, the rudiments of a pelvis and hind limbs in snakes, and wings on many flightless birds. How might these be explained on the basis of Paley's theory, which stressed the importance of the individual design of species? Why should God design redundancies?

These aspects of the natural order could all be explained on the basis of Paley's theory. Yet the explanations offered seemed cumbersome and strained. What was originally a relatively neat and elegant theory began to crumble under the weight of accumulated difficulties and tensions. There had to be a better explanation. Darwin offered a wealth of evidence in support of the idea of biological evolution, and proposed a mechanism by which it might work—*natural selection.*

The *Origin of Species* sets out with great care why the idea of 'natural selection' is the best mechanism to explain how the evolution of species took place, and how it is to be understood. The key point is that natural selection is proposed as nature's analogue to the process of 'artificial selection' in stockbreeding. Darwin was familiar with these issues, especially as they related to the breeding of pigeons. The first chapter of the *Origin of Species* therefore considers 'variation under domestication'— that is, the way in which domestic plants and animals are bred by agriculturists. Darwin notes how selective breeding allows farmers to create animals or plants with particularly desirable traits. Variations develop in successive generations through this process of breeding, and these can be exploited to bring about inherited character- istics which are regarded as being of particular value by the breeder. In the second chapter, Darwin introduces the key notions of the 'struggle for survival' and 'natural selection' to account for what may be observed in both the fossil record and the present natural world.

Darwin then argues that this process of 'domestic selection' or 'artificial selection' offers a model for a mechanism for what happens in nature. 'Variation under domestication' is presented as an analogue of 'variation under nature'. A process of 'natural selection' is argued to occur within the natural order which is analogous to a well-known process, familiar to English stockbreeders and horticulturalists: 'As man can produce and certainly has produced a great result by his methodical and unconscious means of selection, what may not nature effect?'

In the end, Darwin's theory had many weaknesses and loose ends. For example, it required that speciation should take place; yet the evidence for this was conspicuously absent. Darwin himself devoted a large section of the *Origin of Species* to detailing difficulties with his theory, noting in particular the 'imperfection of the geological record', which gave little indication of the existence of intermediate species, and the 'extreme perfection and complication' of certain individual organs, such as the eye. Nevertheless, he was convinced that these were difficulties which could be tolerated on account of the clear explanatory superiority of his approach. Yet even though Darwin did not believe that he had adequately dealt with all the problems which required resolution, he was confident that his explanation was the best available. Noting that no less an authority than Leibniz had once opposed the law of gravity as being 'subversive to religion', Darwin argued for the religious acceptance of his theory as a worthy successor:

There is grandeur in this view of life, with its several powers, having been originally breathed by the Creator into a few forms or into one; and that, whilst their planet has gone cycling on according to the fixed laws of gravity, from so simple a beginning endless forms most beautiful and most wonderful have been, and are being evolved. (Darwin 1890: 403)

Darwin's theories, as set out in the *Origin of Species* (1859) and the *Descent of Man* (1871), hold that all species—including humanity—result from a long and complex process of biological evolution. The religious implications of this will be clear. Traditional Christian thought regarded humanity as being set apart from the rest of nature, created as the height of God's creation, and alone endowed with the 'image of God'. Darwin's theory suggested that human nature emerged gradually, over a long period of time, and that no fundamental biological distinction could be drawn between human beings and animals in terms of their origins and development.

The popular account of the Darwinian controversy at this point focuses on the meeting of the British Association at Oxford on 30 June 1860. The British Association had always seen one of its most significant objectives as being to popularize science. As Darwin's *Origin of Species* had been published the previous year, it was natural that it should be a subject of discussion at the 1860 meeting. Darwin himself was in ill health, and was unable to attend the meeting in person. According to the popular legend, Samuel Wilberforce, bishop of Oxford, attempted to pour scorn on the theory of evolution by suggesting that it implied that humans were recently descended from monkeys. He was then duly rebuked by T. H. Huxley, who turned the tables on him, showing him up to be an ignorant and arrogant cleric. The classic statement of this legend dates from 1898, and takes the form of an autobiographical memoir from Mrs Isabella Sidgewick, published in *Macmillan's Magazine*:

I was happy enough to be present on the memorable occasion at Oxford when Mr Huxley bearded Bishop Wilberforce.... The Bishop rose, and in a light scoffing tone, florid and fluent, he assured us that there was nothing in the idea of evolution; rock pigeons were what rock pigeons had always been. Then, turning to his antagonist with a smiling insolence, he begged to know, was it through his grandfather or his grandmother that he claimed descent from a monkey? (Lucas 1979: 313–14)

The account, which dates from 1898, contradicts accounts published or in circulation closer to the meeting itself. The truth of the matter was that Wilberforce had written an extensive review of the *Origin of Species*, pointing out some serious weaknesses. Darwin regarded this review as significant, and modified his discussion at several points in response to Wilberforce's criticisms. The review shows no trace of 'ecclesiastical obscurantism'. Nevertheless, by 1900 the legend was firmly established, and went some way towards reinforcing the 'conflict', or 'warfare', model of the interaction of science and religion.

Since Darwin's time, there have been many developments which have led to modification and development of his ideas. These include the clarification of the mechanism of inheritance of acquired traits by Gregor Mendel (1822–84), the discovery of the gene by Thomas Hunt Morgan in 1926, and the clarification of the critical role of DNA in the transmission of genetic data, particularly through the establishment of its 'double helix' structure by James Watson and Francis Crick. On the basis of their research, Crick proposed what he called the 'Central Dogma' of a neo-Darwinian view of evolution: namely, that DNA replicates, acting as a template for RNA, which in turn acts as a template for proteins. The long, complex DNA molecule contains the genetic information necessary for transmission 'encoded' using the four nucleotide bases adenine (A), guanine (G), thymine (T) and cytosine (C) arranged in sequences of 'base pairs'.

Today, the term 'Darwinism' is generally used to mean the general approach to biological evolution set out in Darwin's canonical works, as developed and extended through clarification of the molecular basis of inheritance.

So what religious issues are raised by Darwinism? It will be evident from the historical account just presented that Darwin's account of the origin of species raises serious problems for a static understanding of the biological order. Paley's most noted critic in recent years is Richard Dawkins. In his *Blind Watchmaker* (1987), Dawkins relentlessly points out the failings of Paley's viewpoint and the explanatory superiority of Darwin's approach, especially as it has been modified through the neo-Darwinian synthesis. Dawkins argues that Paley's approach is based on a static view of the world, rendered obsolete by Darwin's theory. Dawkins himself is eloquent and generous in his account of Paley's achievement, noting with appreciation his 'beautiful and reverent descriptions of the dissected machinery of life'. Without in any way belittling the wonder of the biological 'watches' that so fascinated and impressed Paley, Dawkins argued that his case for God—though made with 'passionate sincerity' and 'informed by the best biological scholarship of his day'—is 'gloriously and utterly wrong'. The 'only watchmaker in nature is the blind forces of physics'. For Dawkins, Paley is typical of his age; his ideas are entirely understandable, given his historical location prior to Darwin. But nobody, Dawkins argues, could share these ideas now. Paley is obsolete.

This, then, is perhaps one of the most obvious religious issues raised by the rise of Darwinism—the undermining of an argument for the existence of God which had played a major role in British religious thought, both popular and academic, for more than a century. Of course, the argument could easily be restated in more appropriate forms—a development which took place during the second half of the nineteenth century, when many Christian writers stressed that evolution could be seen as the means by which God providentially directed what was now understood as an extended process, rather than a single event.

DARWINISM AS A UNIVERSAL THEORY

A significant debate, which emerged as important shortly after Darwin's death, was whether Darwinism was a domain-specific theory, limited to biology, or a universal theory capable of explaining many aspects of human cultural and intellectual development. Darwin himself was cautious on this matter, although there are points at which he seems to imply that there are parallels between biological and cultural evolution. Darwin's theory of natural selection began to transform the manner in which doctrinal development was conceptualized. If one could speak of evolution within the biological world, could not the same—or at least an analogous—process be discerned within the world of ideas? Darwinism rapidly began its subtle and pervasive transformation from a tool of biological explanation to a more general view of reality. Nineteenth-century cultural evolutionists—such as Sir Edward B. Tylor—were committed to a 'doctrine of progress', in which the human situation was confidently predicted to improve through the constant replacement of inferior beliefs by those which were considered to be superior.

Richard Dawkins, perhaps the most celebrated popularizer of Darwinian orthodoxy and most aggressive advocate of 'universal Darwinism', insists that, at least in two respects, humans do not conform to the mechanisms that shape the biosphere. In the first place, human beings have developed culture—something that he asserts has no direct counterpart within other evolved species. Secondly, and perhaps more significantly, Dawkins proposes an important—indeed, a decisive—distinction between humanity and every other living product of genetic mutation and natural selection. *We alone are able to resist our genes.* Whereas E. O. Wilson (1975) and others had insisted that human beings came within the scope of the methods of sociobiology or evolutionary psychology, Dawkins excludes them from its purview as a matter of principle.

Such a 'universal Darwinism' has met with considerable theological resistance. The idea that every aspect of human life and thought can be accounted for by such a reductionist approach is seen as eliminating the distinctiveness and integrity of human reasoning. This naturally leads to a consideration of one of the most significant areas of tension between Darwinism and traditional religious views—the place of humanity within the natural order.

DARWINISM AND THE NATURE OF HUMANITY

Traditional Christian theology regarded humanity as the height of God's creation, distinguished from the remainder of the created order by being created in the image of God. On this traditional reading of things, humanity is to be located within the created order as a whole, yet stands above it on account of its unique relationship to God, articulated in the notion of the *imago Dei*. Yet Darwin's *Origin of Species* posed an implicit, and *The Descent of Man* an explicit, challenge to this view. Humanity had emerged, over a vast period of time, from within the natural order.

If there was one aspect of his own theory of evolution which left Charles Darwin feeling unsettled, it was its implications for the status and identity of the human race. In every edition of the *Origin of Species*, Darwin consistently stated that his proposed mechanism of natural selection did not entail any fixed or universal law of progressive development. Furthermore, he explicitly rejected Lamarck's theory that evolution demonstrated an 'innate and inevitable tendency towards perfection'. The inevitable conclusion must therefore be that human beings (now understood to be participants within, rather than merely observers of, the evolutionary process) cannot in any sense be said to be either the 'goal' or the 'apex' of evolution. It was not an easy conclusion for Darwin, or for his age. The conclusion to the *Descent of Man* speaks of humanity in exalted terms, while insisting upon its 'lowly' biological origins:

Man may be excused for feeling some pride at having risen, though not through his own exertions, to the very summit of the organic scale; and the fact of his having thus risen, instead of having been aboriginally placed there, may give him hope for a still higher destiny in the distant future. But we are not here concerned with hopes or fears, only with the truth as far as our reason permits us to discover it; and I have given the evidence to the best of my ability. We must, however, acknowledge, as it seems to me, that man with all his noble qualities ... still bears in his bodily frame the indelible stamp of his lowly origin. (Darwin 1882: 619)

Most Darwinists would insist that it is a corollary of an evolutionary world-view that we must recognize that we are animals, part of the evolutionary process. Darwinism thus critiques the absolutist assumptions concerning the place of humanity within nature that lie behind 'speciesism'—a term introduced by Richard Ryder, and given wider currency by Peter Singer, currently of Princeton University. This has raised considerable difficulties beyond the realm of traditional religion, in that many political and ethical theories are predicated on the assumption of the privileged status of humanity within nature, whether this is justified on religious or secular grounds.

The question of the status of humanity is controversial within Darwinism itself. Evolutionary psychology has tended to emphasize how human habits, values, beliefs, and norms can be ascribed to essentially Darwinian processes. Others, including Richard Dawkins, have argued that humanity possesses the capacity to resist its genes. Understanding the evolutionary process is thus a means to prevent humanity from being shaped by its pressures. As Dawkins famously put it: 'We have the power to defy the selfish genes of our birth . . . We are built as gene machines and cultured as meme machines, but we have the power to turn against our creators. We, alone on earth, can rebel against the tyranny of the selfish replicators' (Dawkins 1989: 200–1).

DARWINISM AND THE REDUNDANCY OF GOD

The traditional Christian understandings of the notion of 'creation' prevalent within popular religious culture attributed the creation of the world, including humanity, to direct, special divine action. This notion of the special creation of each and every species underlies William Paley's celebrated *Natural Theology* (1802). Darwin, however, found this notion of special creation problematic on several grounds. What of vestigial or rudimentary organs? And what of the uneven geographical distribution of species? Darwinism holds that the origin of species is to be attributed to extended natural process of variation and selection, in which no divine intervention is required or presupposed.

For some, this implies that Darwinism is atheistic, on two counts: first, that it does not require divine action in order for it to take place; second, that the random nature of variation is inconsistent with the idea of divine creation and providence, which are linked with the ideas of design, purpose, and intentionality. Richard Dawkins is an excellent example of a Darwinian who argues that God has been rendered utterly superfluous by the theory of evolution. Many conservative Protestant writers agree, arguing that the role attributed to random events is inconsistent with the biblical material. Creationist writers often consider this one of the most important elements of their critique of Darwinism.

However, the force of this point is open to question. Benjamin B. Warfield, perhaps one of the most influential conservative Protestant theologians of the late nineteenth century, pointed out that evolution could easily be understood as a seemingly random process, which was nevertheless divinely superintended. God's providence was directing the evolutionary process towards its intended goals. More recently, other writers have proposed alternative mechanisms by which the divine superintendence of the evolutionary process could be conceptualized—for example, Arthur Peacocke's notion of 'top-down causality'.

Yet the most widely proposed mechanism which Christian writers have proposed to account for God's involvement in the evolutionary process is the classic notion of secondary causality, particularly as this was developed by Thomas Aquinas in the thirteenth century. For Aquinas, God's causality operates in a number of ways. While God must be considered capable of doing certain things directly, God delegates causal efficacy to the created order. Aquinas understands this notion of secondary causality to be an extension of, not an alternative to, the primary causality of God. Events within the created order can exist in complex causal relationships, without in any way denying their ultimate dependency upon God as final cause. The created order thus demonstrates causal relationships which can be investigated by the natural sciences. Those causal relationships can be investigated and correlated—for example, in the form of the 'laws of nature'—without in any way implying, still less necessitating, an atheist world-view. God creates a world with its own ordering and processes.

Yet, while Darwin's theory of evolution did not lead to the elimination of God, it highlighted a particular difficulty for Christian theology: how could the goodness of God be maintained, in the light of the wastefulness of the evolutionary process? Surely there was a more efficient, more humane way of achieving these goals? Darwin himself felt the force of this point. Paley's argument emphasized the wisdom of God in creation. But what, Darwin wondered, of God's goodness? How could the brutality, pain, and sheer waste of nature be reconciled with the idea of a benevolent God? In his 'Sketch of 1842', Darwin found himself pondering how such things as 'creeping parasites' and other creatures that lay their eggs in the bowels or flesh of other animals can be justified within Paley's scheme. How could God's goodness be reconciled with such less pleasant aspects of the created order?

There are indeed several important passages in Darwin's writings that can be interpreted to mean that Darwin ceased to believe in an orthodox Christian conception of God on account of his views on evolution. The problem is that there are other passages which variously point to Darwin maintaining a religious belief, or to his losing his faith for reasons other than evolutionary concerns. However, a note of caution must be injected: on the basis of the published evidence at our disposal, it is clear that Darwin himself was far from consistent in the matter of his religious views. It would therefore be extremely unwise to draw any confident conclusions about these issues. There can be no doubt that Darwin abandoned what we might call 'conventional Christian beliefs' at some point in the 1840s, although the dating of this must remain elusive. Yet there is a substantial theoretical gap between 'abandoning orthodox Christian faith' and 'becoming an atheist'. Christianity involves a highly specific conception of God; it is perfectly possible to believe in a god other than that of Christianity, or to believe in God and reject certain other aspects of the Christian faith. Indeed, the 'Victorian crisis of faith'—within which Darwin was both spectator and participant—can be understood as a shift away from the specifics of Christianity towards a more generic concept of God, largely determined by the ethical values of the day.

These points have been addressed by a number of writers, particularly John Haught. Recognizing the pain and apparent wastefulness of evolution, Haught (2005) argues that we must not limit our reflections to the apparent design of the present natural order, but also look forward to its transformation. 'Instead of focusing only on the fact of living design, which can be accounted for scientifically in terms of the Darwinian recipe, a revived natural theology will focus on nature's openness to the future.' Haught deploys a thoroughly trinitarian view of God in using the image of a 'self-emptying God' who 'participates fully in the world's struggle and pain'. In some way, we can think of evolution as being transformed by the notion of 'an incarnate God who suffers along with creation'. For Haught, this affirms that 'the agony of living beings is not undergone in isolation from the divine eternity, but is taken up everlastingly and redemptively into the very "life-story" of God'.

DARWINISM AND BIBLICAL INTERPRETATION

Many of the controversies concerning science and religion have focused on the issue of biblical interpretation. The Copernican controversy, for example, raised the question of whether the Bible actively promoted a geocentric view of the universe, or whether it had simply been interpreted in this way for so long that this impression had become widespread. A similar issue emerged with the debate concerning Darwinism.

It is important to note that Darwinism became of particular concern to Christians in cultures which had been particularly influenced by literal readings of the Book of Genesis. Such readings are known to have been widespread within popular Protestantism in Britain and the United States in the first half of the nineteenth century, even though more nuanced interpretative schemes had been proposed by Protestant academics in both countries. Despite these more sophisticated interpretations of the Genesis creation accounts, at the popular level it was widely assumed that the 'common-sense' reading of the Bible led to a six-day understanding of the creation of the world and humanity. Darwinism posed a significant challenge, both to this specific reading of the Book of Genesis, as well as to existing models of biblical interpretation in general. Were the six days of the Genesis creation account to be taken literally, as periods of 24 hours? Or as indefinite periods of time? And was it legitimate to suggest that vast periods of time might separate the events of that narrative? Or was the Genesis creation account to be interpreted as a historically and culturally conditioned narrative, reflecting ancient Babylonian myths, which could not be taken as a scientific account of the origins of life in general, and humanity in particular? The debates are many, and continue to this day.

Yet it is important to note that these challenges to existing biblical interpretations are to be set in the context of an ongoing dialogue between the community of faith and its foundational text. The history of Christian thought is characterized by a constant process of revisiting and re-evaluating existing interpretations of the Bible, which is precipitated and catalysed by many factors, including scientific advance.

CHRISTIAN RESPONSES TO DARWINISM

So what are the Christian responses to Darwinism? It is now a century and a half since the publication of Darwin's *Origin of Species*. During that time, at least four categories of response have emerged, which will be considered below.

Young Earth Creationism

This position represents the continuation of the 'common reading' of Genesis, which was widely encountered in popular and at least some academic writing before 1800. On this view, the Earth was created in its basic form between 6,000 and 10,000 years ago. Young earth creationists generally read the first two chapters of the book of Genesis in a way that allows for no living creatures of any kind before Eden, and no death before the Fall. Most young earth creationists hold that all living things were created simultaneously, within the time frame proposed by the Genesis creation accounts, with the Hebrew word *yom* ('day') meaning a period of 24 hours. The fossil records, which point to a much longer timescale and to the existence of extinct species, are often understood to date from the time of Noah's flood. This viewpoint is often, but not universally, stated in forms of a 144-hour creation and a universal flood. Representative young earth creationists include Henry Madison Morris and Douglas F. Kelly.

Old Earth Creationism

This view has a long history and is probably the majority viewpoint within conservative Protestant circles. It has no particular difficulty with the vast age of the world, and argues that the 'young earth' approach requires modification in at least two respects. First, the Hebrew word *yom* may need to be interpreted as an 'indefinite time participle' (not unlike the English word 'while'), signifying an indeterminate period of time which is given specificity by its context. In other words, the word 'day' in the Genesis creation accounts is to be interpreted as a long period of time, not a specific period of 24 hours. Second, there may be a large gap between Genesis 1: 1 and Genesis 1: 2. In other words, the narrative is understood not to be continuous, but to make way for the intervention of a substantial period of time between the primordial act of creation of the universe, and the emergence of life on Earth. This viewpoint is advocated by the famous *Schofield Reference Bible*, first published in 1909, although the ideas can be traced back to writers such as the earlier nineteenth-century Scottish divine Thomas Chalmers.

Intelligent Design

This movement, which has gained considerable influence in the United States in recent years, argues that the biosphere is possessed of an 'irreducible complexity', which makes it impossible to explain its origins and development in any other way than by positing intelligent design. Intelligent design does not deny biological evolution; its most fundamental criticism of Darwinism is teleological—that evolution has no goal. The intelligent design movement argues that standard Darwinism runs into significant explanatory difficulties, which can be adequately resolved only through the intentional creation of individual species. Its critics argue that these difficulties are overstated, or that they will in due course be resolved by future theoretical advances. Although the movement avoids direct identification of God with this intelligent designer (presumably for political reasons), it is clear that this assumption is intrinsic to its working methods. The movement is particularly associated with Michael Behe, author of *Darwin's Black Box* (1996) and William Dembski, author of *Intelligent Design: The Bridge between Science and Theology* (1999). Both Dembski and Behe are fellows of the Discovery Institute, a Seattle research institute.

Evolutionary Theism

A final approach argues that evolution is to be understood as God's chosen method for bringing life into existence from inorganic materials, and creating complexity within life. Whereas Darwinism gives a significant place to random events in the evolutionary process, evolutionary theism sees the process as divinely directed. Some evolutionary theists propose that each level of complexity is to be explained on the basis of 'God working within the system', perhaps at the quantum level. Others, such as Howard van Till (1986), adopt a 'fully-gifted creation' perspective, arguing that God built in the potential for the emergence and complexity of life in the initial act of creation, so that further acts of divine intervention are not required. Van Till argues that the character of divine creative action is not best expressed in terms of 'reference to occasional interventions in which a new form is imposed on raw materials that are incapable of attaining that form with their own capabilities', but rather by reference to 'God's giving being to a creation fully equipped with the creaturely capabilities to organize and/or transform itself into a diversity of physical structures and life-forms'. Variations on such approaches are found elsewhere, as in the writings of Denis O. Lamoureux and Arthur Peacocke.

How should we evaluate these positions and their underlying religious and scientific concerns? Perhaps the most important point to make is simply this: the natural sciences, including the various Darwinian paradigms, are patient of atheist, theist, and agnostic interpretations and accommodations—but demand and necessitate none of them. Darwinism can be 'spun' in ways that are seemingly totally consistent and seemingly totally opposed to Christian belief. Both 'Darwinism' and 'Christianity' designate a spectrum of possibilities, making determining their conceptual overlaps and tensions problematic, and critically dependent on definitional issues. Perhaps one of the more interesting paradoxes of the contemporary religious scene is that, for precisely these reasons, Darwinism has been prematurely and unnecessarily branded as 'atheist', on the left by writers such as Dawkins, and on the right by various American creationist individuals and organizations.

As Warfield uncontroversially pointed out, serious issues of biblical interpretation often underlie such controversies. Perhaps somewhat more controversially, but almost certainly correctly, Warfield also observed that scientific advance offered a means whereby the church could 'check out' its interpretation of the Bible, so avoiding being locked into an arcane or archaic biblical interpretation which, by force of tradition, was assumed to be the self-evident meaning of the Bible itself. It is for this reason that Darwinism has proved to be especially controversial within the conservative American Protestant constituency, whose general approach to biblical interpretation is often based on the seemingly straightforward (but actually nuanced and complex) notion of the 'plain sense of Scripture'. (Similar issues arise within the Islamic constituency concerning the interpretation of the Qu'ran.) For precisely the same reason, Darwinism has proved much more acceptable to Roman Catholic writers, on account of the notion of the magisterial interpretation of an occasionally opaque scripture.

Yet one point must be made in closing. When all is said and done, Darwinism is a scientific theory, provisional in its status and open to modification, correction, development, or even ultimate abandonment as the process of scientific advance continues. It may be the received scientific wisdom of our age; and no study of the history of science would be unwise enough to suggest that it is necessarily, and possibly uniquely, immune to the process of radical theory change that has characterized scientific advance in the past. So what will this debate look like a century from now? Will we have moved on, scientifically and theologically? I am no prophet and have no answers on this. It would be extremely unwise, however, to suppose that the present age has settled, or even begun to settle, the question of the relationship of religious faith to the scientific exploration of the biological origins of life.

References and Suggested Reading

Aunger, Robert (2000) (ed.). *Darwinizing Culture: The Status of Memetics as a Science.* Oxford: Oxford University Press.

Behe, Michael J. (1996). *Darwin's Black Box: The Biochemical Challenge to Evolution.* New York: Free Press.

Blackmore, Susan J. (1999). *The Meme Machine.* Oxford: Oxford University Press.

Boyd, Robert, and Richerson, Peter J. (1985). *Culture and the Evolutionary Process.* Chicago: University of Chicago Press.

Brooke, John Hedley (1991). *Science and Religion: Some Historical Perspectives.* Cambridge: Cambridge University Press.

Campbell, John Angus, and Meyer, Stephen C. (2003). *Darwinism, Design, and Public Education.* East Lansing, Mich.: Michigan State University Press.

Darwin, Charles (1882). *The Descent of Man,* 2nd edn. London: John Murray.

—— (1890). *The Origin of Species,* 6th edn. London: John Murray.

DAWKINS, RICHARD (1987). *The Blind Watchmaker: Why the Evidence of Evolution Reveals a Universe without Design*. New York: W. W. Norton.

—— (1989). *The Selfish Gene*, 2nd edn. Oxford: Oxford University Press.

—— (1995). *River Out of Eden: A Darwinian View of Life*. London: Phoenix.

—— (2003). *A Devil's Chaplain*. London: Weidenfeld & Nicolson.

DEMBSKI, WILLIAM A. (1998). *The Design Inference: Eliminating Chance through Small Probabilities*. Cambridge: Cambridge University Press.

—— (1999). *Intelligent Design: The Bridge between Science and Theology*. Downers Grove, Ill.: Intervarsity Press.

DENNEN, J. VAN DER, SMILLIE, DAVID, and WILSON, DANIEL R. (1999) (eds.). *The Darwinian Heritage and Sociobiology*. Westport, Conn.: Praeger.

DENNETT, DANIEL C. (1995). *Darwin's Dangerous Idea: Evolution and the Meaning of Life*. New York: Simon & Schuster.

DURANT, JOHN (1985). *Darwinism and Divinity*. Oxford: Blackwell.

ELDREDGE, NILES (1995). *Reinventing Darwin: The Great Debate at the High Table of Evolutionary Theory*. New York: John Wiley & Sons.

ELLEGÅRD, ALVAR (1990). *Darwin and the General Reader: The Reception of Darwin's Theory of Evolution in the British Periodical Press, 1859–1872*. Chicago: University of Chicago Press.

FYFE, AILEEN (1997). 'The Reception of William Paley's *Natural Theology* in the University of Cambridge', *British Journal for the History of Science*, 30: 321–35.

GIBERSON, KARL W., and YERXA, DONALD A. (2002). *Species of Origins: America's Search for a Creation Story*. Lanham, Md.: Rowman & Littlefield.

GILLESPIE, NEAL C. (1979). *Charles Darwin and the Problem of Creation*. Chicago: University of Chicago Press.

GOULD, STEPHEN JAY (2002). *The Structure of Evolutionary Theory*. Cambridge, Mass.: Belknap Press of Harvard University Press.

HARRISON, PETER (2002). 'Fixing the Meaning of Scripture: The Renaissance Bible and the Origins of Modernity', in Seán Freyne and E. J. van Wolde (eds.), *The Many Voices of the Bible*, London: SCM Press, 102–10.

HAUGHT, JOHN F. (2000). *God after Darwin: A Theology of Evolution*. Boulder, Colo.: Westview Press.

—— (2005). 'The Boyle Lecture 2003: Darwin, Design, and the Promise of Nature', *Science and Christian Belief*, 17: 5–20.

HULL, DAVID L. (1973). *Darwin and his Critics*. Cambridge, Mass.: Harvard University Press.

KELLY, ALFRED (1981). *The Descent of Darwin: The Popularization of Darwinism in Germany, 1860–1914*. Chapel Hill, N.C.: University of North Carolina Press.

KEYNES, RANDAL (2001). *Annie's Box: Charles Darwin, his Daughter and Human Evolution*. London: Fourth Estate.

LALAND, KEVIN N., and BROWN, GILLIAN R. (2002). *Sense and Nonsense: Evolutionary Perspectives on Human Behaviour*. Oxford: Oxford University Press.

LARSON, EDWARD J. (1989). *Trial and Error: The American Controversy over Creation and Evolution*. New York: Oxford University Press.

LeMAHIEU, D. L. (1976). *The Mind of William Paley: A Philosopher and his Age*. Lincoln, Nebr.: University of Nebraska Press.

LUCAS, JOHN R. (1979). 'Wilberforce and Huxley: A Legendary Encounter', *Historical Journal*, 22: 313–30.

McGRATH, ALISTER E. (2004). *Dawkins' God: Genes, Memes and the Meaning of Life*. Oxford: Blackwell.

MAYR, ERNST, and PROVINE, WILLIAM B. (1980). *The Evolutionary Synthesis: Perspectives on the Unification of Biology*. Cambridge, Mass.: Harvard University Press.

MIDGLEY, MARY (2002). *Evolution as a Religion: Strange Hopes and Stranger Fears*, 2nd edn. London: Routledge.

MOORE, JAMES R. (1979). *The Post-Darwinian Controversies: A Study of the Protestant Struggle to Come to Terms with Darwin in Great Britain and America, 1870–1900*. Cambridge: Cambridge University Press.

MORELAND, JAMES PORTER, and REYNOLDS, JOHN MARK (1999) (eds.). *Three Views on Creation and Evolution*. Grand Rapids, Mich.: Zondervan.

MORRIS, RICHARD (2001). *The Evolutionists: The Struggle for Darwin's Soul*. New York: W. H. Freeman.

NUMBERS, RONALD L. (1992). *The Creationists: The Evolution of Scientific Creationism*. Berkeley: University of California Press.

O'HEAR, ANTHONY (1997). *Beyond Evolution: Human Nature and the Limits of Evolutionary Explanation*. Oxford: Clarendon Press.

PALEY, WILLIAM (1802). *Natural Theology; or Evidences of the Existence and Attributes of the Deity, Collected from the Appearances of Nature*. London: Wilkes and Taylor.

—— (1849). 'Natural Theology', in *Works*, London: William Orr, 23–195.

PEACOCKE, ARTHUR (1986). *God and the New Biology*. London: Dent.

PENNOCK, ROBERT T. (2001). *Intelligent Design Creationism and its Critics: Philosophical, Theological, and Scientific Perspectives*. Cambridge, Mass.: MIT Press.

RICHARDS, R. J. (1987). *Darwin and the Emergence of Evolutionary Theories of Mind and Behaviour*. Chicago: University of Chicago Press.

ROBERTS, JON H. (1988). *Darwinism and the Divine in America: Protestant Intellectuals and Organic Evolution, 1859–1900*. Madison: University of Wisconsin Press.

RUSE, MICHAEL (1998). *Taking Darwin Seriously: A Naturalistic Approach to Philosophy*. New York: Prometheus Books.

—— (2003). *Darwin and Design: Does Evolution Have a Purpose?* Cambridge, Mass.: Harvard University Press.

SHANKS, NIALL (2004). *God, the Devil, and Darwin: A Critique of Intelligent Design Theory*. New York: Oxford University Press.

SHENNAN, STEPHEN (2002). *Genes, Memes and Human History: Darwinian Archaeology and Cultural Evolution*. London: Thames & Hudson.

VAN TILL, HOWARD (1986). *The Fourth Day: What the Bible and the Heavens are Telling Us about the Creation*. Grand Rapids, Mich.: Eerdmans.

WILSON, DAVID SLOAN (2002). *Darwin's Cathedral: Evolution, Religion, and the Nature of Society*. Chicago: University of Chicago Press.

WILSON, EDWARD O. (1975). *Sociobiology: The New Synthesis*. Cambridge, Mass.: Harvard University Press.

CHAPTER 41

GOD AND EVOLUTION

JOHN F. HAUGHT

INTRODUCTION

Any theology that hopes to connect with science in a truly candid way must remain firmly anchored to specific communities of faith and their own particular ideas of God. In this respect the physicist Steven Weinberg, no friend to theology, is correct in saying that for theists what is at stake in the conversation of science and religion is the existence of an 'interested' God, not an abstract and impersonal one (Weinberg 1993: 245). In the context especially (though not exclusively) of biblical faiths, it is the reality of a personal, caring, compassionate, and providential God that science sometimes seems to place in question, and it is on such an understanding of God that the debate about evolution must remain focused.

The theologian, one who reflects systematically on the meaning of religious faith, is commissioned to reflect on God and the world, including the story of life as it is now being brought to light by evolutionary science, from within the limited perspective of his or her own faith tradition. A neutral or dispassionate perspective is inconsistent with the idiosyncratic character of religious existence. Acknowledging such a limitation does not mean that theological reflection has no interreligious or ecumenical implications. It means only that each theology is obliged to begin its reflections on the world, and in the present essay the evolution of life, from a definite location along the broad spectrum of religious teachings and God-ideas, and not from some imagined universal perspective that overlooks differences in the worlds of faith.

As a Christian theologian (I am a Roman Catholic layperson) concerned about the relationship of science and religion, therefore, I am not qualified to say exactly what evolution means from the perspective of a Jew, a Muslim, a Hindu, a Buddhist, or a

Native American, although I can learn much from each. Rather, it is from within the Christian context that I feel called to explore, however tentatively, what evolution might mean when seen in the light of a Christian understanding of God. Moreover, it cannot be my method to speak here simply as a philosopher of religion, although such an approach has a legitimate function in considering whether theological ideas are at least consistent with reason.

On the issue of 'God and evolution' it is not helpful, therefore, to think of God only as a first cause, a final cause, or an intelligent designer, even though such attributes may be philosophically worth pondering. Rather, theologically speaking, my first task as I approach the topic of evolution must be to bring to the fore what is taken to be the revelatory image of God as given in the witness of my own religious tradition. Such an image is available to Christians in the picture of Jesus presented in the Gospels and interpreted by other biblical and classic texts throughout the centuries. The constant teaching of Christian tradition, ratified by numerous councils and creeds, is that Jesus, known as the Christ, is the face of God: that is, the normative manifestation in human history of the eternal divine mystery. Christians believe that the hidden God has become paradoxically manifest in the life, death, and resurrection of Jesus. And they are instructed never to take their eyes off what we may call the 'picture' of Jesus when they think about God, or when they worship and surrender their lives to the divine mystery. I would propose also that Christians never take their eyes off this picture when they look at evolution. Unfortunately, this has not always been the case.

Therefore, in the world of Christian thought and its engagements with Darwinian biology, the first question must be whether the idea of God as revealed in Christ—and not some episodically interventionist intelligent designer—is able to illuminate the story of life without contradicting or editing the scientific information pertaining to evolution. A serious theological discussion of the question of God after Darwin cannot start off by tailoring scientific ideas to fit theological habits of thought. It would be bad form for the theologian to favour or reject a particular scientific interpretation of evolution for theological reasons alone. It is, of course, appropriate that theology resist scientific or evolutionary *naturalism*, which is a world-view based on the belief that science, especially Darwinism, rules out the existence of God. But naturalistic metaphysics is not the same thing as science, and theology has no more business confusing evolutionary science with naturalism than do atheistic evolutionists (e.g. Dawkins 1986, 1995, 1996; Dennett 1995).

Since truth cannot contradict truth, the theological assumption must be that if there is scientific evidence for biological evolution, and a reasonable theory to explain it, it is wrong to dismiss these simply because they do not seem prima facie to fit inherited theological schemata. It may be that the theological mind-set, without surrendering the revelatory content that gives a tradition continuity in time, must undergo some change itself as a result of the encounter with science and other new experiences. History, in fact, offers numerous instances in which religious faiths have had to pass through the trauma of transformation in order to make sense of the world in each age, or simply to remain alive as complex organic systems adapting to new environments. In my view, evolution provides not only a challenge, but also a great opportunity for theological growth and renewal today (for more extended treatment see Haught 2000, 2003).

THE MEANING OF REVELATION

Christian theology, as I have already emphasized, is inseparable from the religious experience of the revelation of God in Jesus who is called the Christ. Revelation, as theologian H. Richard Niebuhr has said, is the 'gift of an image'. 'By revelation in our history we mean . . . that special occasion which provides us with an image by means of which all occasions of personal and common life become intelligible' (Niebuhr 1960: 80). One might also add that this image should help make more intelligible the enigmas of nature, including evolution, as well. For Christians the image is one in which God is disclosed in the person and career of Jesus in a surprising way, completely unexpected by God-seekers from all times and places. What stands out in this picture is the 'humility of God', or the 'divine descent' (Macquarrie 1978; Hallman 1991). As John Macquarrie writes, the image of a humble God whose very essence is self-giving love supplants decisively the idea of the divine potentate or engineer that countless people—and we can include here atheistic evolutionists, creationists, and intelligent design proponents—typically have in mind when they use the word 'God'.

That God should come into history, that he should come in humility, helplessness and poverty—this contradicted everything—this contradicted everything that people had believed about the gods. It was the end of the power of deities, the Marduks, the Jupiters . . . yes, and even of Yahweh, to the extent that he had been misconstrued on the same model. The life that began in a cave ended on the cross, and there was the final conflict between power and love, the idols and the true God. (Macquarrie 1978: 34)

I believe with Macquarrie that the image of God's humble descent is not optional but central to the Christian understanding of God. In my own conversation with evolutionists, many of whom think of God only as a cosmic engineer or designer, I have discovered that they are sceptical that Christians are permitted to believe in a self-effacing deity. One reason for such an impression is that religious thinkers themselves have often reduced the idea of God to an intelligent designer in their engagement with scientists. However, even the late Pope John Paul II, by no means a theological radical, testified in his encyclical *Fides et Ratio* (1998) to the pre-eminent theological importance of the shocking notion of a humble, *kenotic*, self-emptying God. The God of Christian faith and theology is one who undergoes a *kenosis* (that is, a pouring out) of the divine substance. This is the proper Christian understanding of God as manifested in the obedience and self-sacrifice of the Christ. Hence it is the chief purpose of Christian systematic theology, John Paul states, to highlight the divine *kenosis*, 'a grand and mysterious truth for the human mind, which finds it inconceivable that suffering and death can express a love which gives itself and seeks nothing in return' (1998: sect. 93). And so, it is not unlikely that the foremost stumbling block that scientifically enlightened people will have in relation to Christian faith is not the sad history of its conflicts with science, but the 'inconceivable' descent and humility that are fundamental to its understanding of God.

The first step in developing a Christian theology of evolution, therefore, is to reform our thoughts about God so as to make them correspond with the incredible idea of the divine self-outpouring. Only then should we inquire into the intelligibility of evolution for Christian faith. If the astonishing picture of God's descent becomes the lens through which we look at the life process, then evolution will not look the same as if we took as our point of departure the idea of God as an intelligent designer. Interestingly, this theological method can also lessen the temptation to deny the results of scientific research. It will still allow room for the belief that a deep divine wisdom and power underlie everything, but it will be a wisdom furled in humility and a power transformed by unquenchable love. It is my view that the two bewildering pictures—that of Christianity's self-abandoning God, on the one hand, and the scientific picture of life evolving gradually by natural selection, on the other—can be mapped onto each other without much difficulty, and in such a way as to reach a satisfactory resolution for both science and theology.

The image of a self-emptying God, according to the authoritative interpreters of Christian faith beginning with St Paul, can be appreciated only if the believer surrenders him or herself to it and becomes drawn into its transformative power. This may not be easy, but there can be no true encounter with the revelatory image without allowing oneself to be grasped by it. Yet, even though the image itself is inaccessible to philosophical or scientific understanding—since it fits no inherited conceptual scheme—it may have the effect, at least to the eyes of faith, of illuminating the totality of being in an entirely fresh way. This means that it should have the power to open up a whole new way of looking at the universe and the life process as well. It may even help make sense of something as enigmatic as evolution without requiring that we ignore, modify, or slant the data gathered by the fields of inquiry tributary to evolutionary theory (geology, palaeontology, anthropology, geography, anatomy, genetics, etc.). At the end of our theological reflections on evolution, of course, mystery will still remain; but perhaps it will be a mystery that, even in its inaccessibility, can suffuse the whole universe and the long story of life with a meaning that eludes creationism, scientism, and evolutionary naturalism.

DARWINISM, THEOLOGY, AND THE MEANING OF EVOLUTION

But hasn't Darwin already brought sufficient intelligibility to the life process precisely in his illuminating theory of evolution? What meaning could theology add to a theory that already makes sense of life and its diversity in terms of the notion of natural selection? What could possibly be left over, after evolutionary analysis, which a religious or Christian view of reality could illuminate any further? In its conversation with science, theology must give reasons for its conviction that evolutionary accounts do not provide exhaustive understanding of life. Theology cannot do the meticulous work proper to science; nor can it properly compete with scientific explanations. But it may at least demonstrate that a theological interpretation of the history of life adds a dimension of depth to our understanding of evolution that science cannot provide. In order to do so, however, it must first critically inquire whether the biological and evolutionary sciences can explain, even in principle, every dimension of the fascinating story of living beings in the universe.

For one reason or another it has become increasingly common for scientists to claim that life *can* be explained fully in purely naturalistic terms. For example, in his assessment of the Darwinian revolution, the renowned geneticist Francisco J. Ayala writes:

It was Darwin's greatest accomplishment to show that the directive organization of living beings can be explained as the result of a natural process, natural selection, without any need to resort to a Creator or other external agent. The origin and adaptation of organisms in their profusion and wondrous variations were thus brought into the realm of science. (Ayala 1994: 4)

There can be no doubt, of course, that science illuminates natural phenomena in ways that theology cannot. However, my concern here is whether Darwinian science leaves any explanatory role to theology at all. After Darwin, Ayala continues:

The origin and adaptive nature of organisms could now be explained like the phenomena of the inanimate world, as the result of natural laws manifested in natural processes. Darwin's theory encountered opposition in religious circles, not so much because he proposed the evolutionary origin of living things (which had been proposed many times before, even by Christian theologians), but because his mechanism, natural selection, *excluded* God as accounting for the obvious design of organisms. (Ayala 1994: 4–5, emphasis added)

It is the term 'excluded' that worries the theologian here. I am not certain in what exact sense Ayala intends to say that Darwinism *excludes* any appeal to theological explanation. Does he mean at every level of explanation, or just at the level of scientific understanding? If he means that science, methodologically speaking, must never invoke the idea of God, every good theologian will agree. And this is quite possibly all that Ayala means to say. However, some of Ayala's readers may take his words to mean that 'explanation' of life no longer requires the introduction of the concept of God at *any* level of understanding. If so, then such a claim goes beyond what science itself warrants. The view that science alone can provide complete explanation is not, logically speaking, a conclusion of science, but simply a belief characteristic of scientific naturalism. At the end of his essay Ayala can easily give the impression, perhaps without intending to do so, of subscribing to such a naturalistic interpretation of Darwin's revolution:

This is Darwin's fundamental discovery, that there is a process that is creative though not conscious. And this is the conceptual revolution that Darwin completed—that everything in nature, including the origin of living organisms, can be explained by material processes governed by natural laws. This is nothing if not a fundamental vision that has forever changed how mankind perceives itself and its place in the universe. (Ayala 1994: 4)

Thus Darwin has not only revolutionized human understanding of life; he has also altered for good the criteria of what should be allowed to pass as reasonable explanation of the life process. Modern physics had already assumed that impersonal, unintelligent causes are applicable to the understanding of non-living nature. But Darwin's revolution appears to have made lifeless and mindless causes exhaustively explanatory of life and mind as well, thus completing the scientific revolution (Ayala 1994: 4–5). The question remains, then, whether mindless causes, acceptable though they may be in scientific explanation, are enough to render any theological understanding of evolution completely pointless. I am still not sure what Ayala's answer to this question would be.

If Ayala leaves us hanging, however, other renowned evolutionists do not. In the 1970s Gavin de Beer's major article on evolution in the *Encyclopaedia Britannica* stated emphatically that Darwin's science rules out any role for God at all. 'Darwin', he contended, 'did two things: he showed that evolution was a fact *contradicting* scriptural legends of creation and that its cause, natural selection, was automatic *with no room for divine guidance or design*' (de Beer 1973–4: 23, emphasis added). Notice once again, however, that the idea of God here is reduced to that of a guider or designer. Subscribing to the belief that natural science alone can explain life, many other biologists and philosophers of biology today share the view that Darwinian science leaves no room at all for God understood as an intelligent designer (Cziko 1995; Dawkins 1995; Dennett 1995; Rose 1998; Wilson 1998; Ruse 2003). Furthermore, since in their view any intelligible role for God would be that of directly 'designing' nature's orderly patterns, they are confident in their assertion that it is Darwinian process *rather* than God that explains living phenomena in an ultimate way. Hence they consider it no longer acceptable to seek any deeper intelligibility or meaning in the story of life than what biology can provide in its surprisingly simple recipe for evolutionary diversity and complexity.

According to this recipe, three basic ingredients are necessary:

(1) Accidental or contingent occurrences: the improbable chemical events required for the origin of life, the random genetic mutations that make possible the diversity of life, and many other unplanned events in natural history that shape the course of evolution (for instance, mass extinctions caused by famines, ice ages, or meteorite impacts). It is the absence of purposive design and the presence of accident that seem to rule out the existence of God.

(2) The 'law' of natural selection along with the invariant rules of physics and chemistry. These unbreakable habits of nature operate so blindly and impersonally that they seem to undermine any reasonable trust that a compassionate divine providence could be operative in the universe at all, let alone in life.

(3) Deep time. Evolutionary science, in order to avoid mystery, magic, and miracle, sees evolutionary change as taking place very gradually, even if the long process is punctuated by occasional spasms of accelerated creativity. Evolution requires enough time for a sufficiently large number of minute random variations to supply the mindless process of natural selection with adaptable outcomes. The fact that so much time is required for this unwieldy epic to transpire, that so much death occurs along the way, and that so many mistakes and monstrosities appear, only to be

discarded, renders evolution all the more devoid of purpose apparently. Any creator who would 'fool around' so inefficiently for billions of years in order to produce living and thinking beings seems much less competent than the most mediocre of human engineers.

So, in the minds of many scientific thinkers since the time of Darwin the three-part recipe has been enough to cast doubt on the notion of God, and especially on that of divine intelligent design. It has also led many religious believers to reject Darwin's evolutionary ideas as incompatible with divine providence. A century and a half after Darwin, a great abyss still separates a sizeable proportion of evolutionists, on the one hand, from countless Christians (as well as Muslims and other religious believers), on the other. One significant outcome of this divide has been the increasing alienation of scientists from the world of religious faith, and a troublesome disillusionment with science among religious believers, especially in North America, though increasingly elsewhere as well.

However, this cleavage is not only unnecessary, it is a lost opportunity for both theological maturation and scientific flourishing. As I shall argue below, the revelatory image of God based on the picture of Jesus as the Christ can, for Christians at least, accommodate comfortably what I am calling the 'evolutionary recipe'. And it may help the believer not only to accept, but also to become enthusiastic about, exploring the wider world of life.

The revelatory image of God that I am working with includes at least two ideas. First, God is to be thought of as one who not only creates, but also makes and faithfully keeps promises, inviting (even commanding) people to hope. This God opens up the future even where there appear to be only dead ends. The opening of the future, however, is not to be thought of as an *ad hoc*, interventionist, readjustment of the laws of nature. Rather, it is a constant aspect of the way God relates to the world. God, therefore, may rightly be called the 'power of the future' (Pannenberg 1977: 58–9; Peters 2000).

Second, as I noted above, the revelatory image associated with the picture of Jesus as the manifestation of God is one in which the divine mystery gives itself away to the creation in humble and selfless love (Rahner 1984; 78–203). The image of a self-emptying God, as Eberhard Jüngel (1976) rightly observes, lies at the heart of Christian faith as well as the doctrine of the Trinity. Likewise, theologian Jürgen Moltmann, indebted to Jewish Cabbalistic thought as well as to biblical imagery, argues that the creation of the universe is the result of a divine self-limiting restraint:

(God's) ... creative activity outwards is preceded by his humble divine self-restriction. In this sense God's self-humiliation does not begin merely with creation. ... It begins beforehand, and is the presupposition that makes creation possible. God's creative love is grounded in his humble, self-humiliating love. This self-restricting love is the beginning of that self-emptying of God which Philippians 2 sees as the divine mystery of the Messiah. Even in order to create heaven and earth, God emptied Himself of his all-plenishing omnipotence, and as Creator took ... the form of a servant. (Moltmann 1985: 88)

This humility, therefore, is not an occasional, but instead a constant, aspect of God's constitution of the world as something distinct from its creative ground. In order to allow something other than God to exist at all, the Creator may be thought of as 'renouncing' the cruder kind of causation that deterministically manufactures entities in a completely finished or perfected state of being. The greatness of divine creativity consists in the fact that, unlike finite creators, it is able to call into being (from out of the future) a world that in some sense must make itself. God's humility-lined creativity in some sense is a 'letting be' of the world. Divine creative activity opens up space and time for creation to become itself, as something clearly distinct from (though not separate from) its creator.

There is no way, however, to understand this paradoxical divine mode of creativity by looking for clear analogies in the thought-world of science with its emphasis on efficient causation. Therefore, attempts to render divine action intelligible on the analogy of efficient causation or of what passes as causal in the natural sciences will, in my opinion, end up diminishing or obscuring the multifaceted way of influencing the world that theology must attribute to God. For that reason, theology must not apologize for its perpetual failure to arrive at complete intellectual clarity with respect to divine action and divine providence. Theology does not do justice to the power of divine action unless it employs a variety of images that cannot be smoothly mapped onto one another.

For theology to emulate science by seeking a reductive clarity about how God acts is to impoverish, rather than enrich, religious understanding. Consequently, by speaking of divine creativity as that of a humble 'letting be' of the world, I am using metaphorical language that communicates a dimension of divine creativity and providence that other theological expressions, such as 'ground of being' or 'power of the future' do not fully capture. Theology must avoid exclusive fixation on any of the metaphors it uses, or else it ends up shrinking the mystery of God. This ordinance means, therefore, that the expression 'letting be' must be balanced by images of divine immanence, lest God be thought of as uninvolved and apathetic. Theology need not be embarrassed that its subject-matter, especially the religious sense of divine action, must always be approached by a tentative and dialectical discourse that constantly allows itself to be relativized by appeal to a rich variety of symbols, analogies, and metaphors. It is a mark of the eminence, not the absence, of the divine that we need to clothe our fragile and finite words about it in a rich plurality of references.

The Main Focus of a Theology of Evolution

So, if God's influence on the world may be thought of, at least tentatively, as humbly 'letting be' and as 'the power of the future', how well does such a picture match up with the evolutionary recipe? In order to approach this question frankly, a Christian theology of evolution must begin by acknowledging that chance, law, and deep time are indeed fundamental facets of nature underlying evolutionary process. It is appropriate to engage evolutionary thought at the level of these fundamentals, since they are taken to be essential in almost all contemporary scientific interpretations of life. There may be disagreements among evolutionary biologists as to the exact proportion of causal efficacy to be attributed to each of the three foundational elements. For example, to Stephen Jay Gould contingency or happenstance is the main engine of evolutionary creation (Gould 1989; Gould and Lewontin 1979). For Richard Dawkins, selection is the chief cause of organic particularity and diversity (1986, 1995, 1996). And for both, time itself—or, more precisely, a great abundance of time—is also accorded an explanatory status. But all three factors are present, in one distribution or another, in these and most other contemporary scientific proposals about how life evolves.

It seems fitting, therefore, for a theology of evolution to focus on the religious meaning of this three-pronged suite of constituents, since it is just this complex that makes biological evolution work. Such a focus applies also to post-Darwinian interpretations of evolution that emphasize factors such as co-operation, self-organization, or ecological impacts on the shaping of life, influences that classical Darwinism did not notice or highlight (Depew and Weber 1995). No doubt the theory (or theories) of evolution will continue to evolve, but it seems unlikely that a scientific interpretation of life can ever dispense completely with any of the three fundamental explanatory elements—contingency, physical law (including selection), and deep time—as the main factors in evolution.

My point here, therefore, is that by identifying the three underlying cosmic features essential for evolution, biologists can scarcely expect to have closed the door on theology. For what remains remarkable, and hence in need of deep explanation even after Darwin, is the exquisite blend of the three features that makes our universe a *story* and not just a state. At heart, evolutionary science is, after all, the telling of a grand story, in any of the various versions now being debated. It turns out that the recipe for evolution is first of all the recipe for nature-as-narrative. So we cannot hope to get to the bottom of evolution unless we risk asking why the universe would have a narrative character in the first place. Many specific sciences can contribute to this discussion, of course, but alongside all of them so also can theology.

How so? Once again, contingency, predictability, and time are jointly the stuff of story. Any significant narrative requires, first, contingent events whose openness to unpredictable outcomes makes suspense and uncertainty possible; second, at least some constraints on contingency and uncertainty are essential to stiffen a story with the continuity and coherence that prevent it from collapsing into a jumble of disconnected moments; third, sufficient time is needed for the story to unfold, and a truly momentous story may require an enormous amount of time.

It is of great significance to theology that beneath the surface of all evolutionary occurrences there lies a narrative matrix that weaves openness to the future (made possible by contingency) and the element of reliability (lawfulness) onto an irreversible filament of time so as to give nature an irreversibly historical character. It is from this narrative womb that life and its evolution came to birth in our universe. Theology, unlike evolutionary science, does not take this remarkable narrative amalgam for granted. Instead, it roams beyond the borders of purely scientific inquiry to ask why the universe would be graced at all with such a potential for drama and adventure.

It is the *confluence* of contingency, necessity, and time into the stream of a still ongoing story that invites theological reflection. The deeper questions that evolution raises for theology, then, may be summarized succinctly as follows.

(1) Must we interpret contingency—that is, the undirected and spontaneous aspect of natural occurrences such as genetic mutations—as purely aimless accidents arising out of a dead and blind past? Or instead, may we understand them narratively as openings to an unpredictable cosmic future (Pannenberg 1993: 72–122), a future whose coming can conceivably give a meaning to events that science, including Darwinian biology, cannot possibly glimpse in the here and now?

(2) Does the invariance (or determinism) of natural selection signify a fundamentally impersonal and meaningless universe? Or can such rigid constraints as natural selection, along with other laws and constants of nature, be interpreted more generously as indispensable to any meaningful story? Is it possible that they are indicative of a divine principle that provides for structure and continuity to the history of the universe and life without forcing these to take any predetermined path toward the future (Pannenberg 1993: 72–122)? It would seem that unless nature is corralled by firm limits holding the sequence of events within finite boundaries, there would be no story for evolutionists to talk about at all, but instead a mere unravelling into an ultimate incoherence. Are blindness and impersonality the last words when it comes to understanding the rigorous mechanisms of evolution?

(3) And what sense can one make of the fact of deep time? The diversification of life—the origin of species—requires sufficient temporal amplitude for its experimental creativity to unfold. But is the passage of time simply a thrust forward into an ultimate void? Or is time a reality made possible by the coming of an inexhaustible and ultimately mysterious future rich with new potential (Moltmann 1996: 259–95; 1975: 48)? And does time lead inevitably to nothingness, or does it flow into an eternity where all events can be remembered and reordered into an unfathomable beauty (Whitehead 1978: 34–51, 346; 1967: 265)?

The present chapter cannot respond in detail to these three sets of questions, but when they are framed in this way, one can at least note that the conversation of theology with evolutionary science is not entirely new. The topics of contingency, predictability, and time have been perennial preoccupations of theologians. However, because it forces us to look at these three themes in a more intense way than ever, evolutionary biology has become an exciting new stimulus to theological reflection on the narrative character of the natural world. It may well be that a composite of contingency, necessity, and time is essential to any universe in which anything of real significance is working itself out. The following, then, are some very brief reflections on how theology, without in any way interfering with the investigations of scientists, may further develop such a proposal.

EVOLUTION IN THE LIGHT OF CHRISTIAN REVELATION

First, it is important for theology to recognize that chance, accident, or contingency is no illusion, but part of the fabric of a finite universe. Even the medieval theologian Thomas Aquinas theorized that a universe devoid of accidents would be so deterministic as not to have any autonomy at all in relation to its creator (Mooney 1996: 162). Rather than being embarrassed about the prevalence of accidents in evolution, theology can allow that these are completely consonant with the revelatory image of God as humble, self-giving love and provider of future fulfilment. If the character of God is such as that revealed in the way Jesus related to others, then the Creator's disposition is also one of respect for freedom and autonomy, including, we might assume, in the *entirety* of creation, not just the human part of it. God's relationship to the creation is apparently a kind of 'letting be'. This is not a posture of apathy or abdication, but one that providentially provides openings to a new future for the creation in such a way as to renounce any manipulative management of cosmic affairs. God acts by offering relevant possibilities that allow the world to become new. But, in a fashion consistent with profound love, God does not stamp these possibilities on the creation in a forceful way.

Theology—or at least Christian theology—understands God as relating to the universe in a manner analogous to the way in which Jesus related to others, even if this means submission to suffering and death. God, therefore, may be understood as infinitely humble, self-giving love that works by opening the creation to an ever new future and also by remaining faithfully present in the evolutionary struggle through the creative presence of the divine Spirit. God is not an engineer, magician, or conjuror. And God does not have to become one actor among others in the cosmic and evolutionary story in order to be profoundly effective. Rather, divine action with respect to evolution and cosmic process is to ground and sustain the narrative loom upon which an indeterminate and still unfinished cosmic drama may be woven. Because of the presence of the Spirit of God in nature, the narrative is not just the world's story, but in a very profound way God's as well. God's grounding and sustaining of the narrative structure of natural being is a much deeper kind of involvement in the world than could possibly be the case if divine action consisted essentially of engineering things in the immediate manner that evolutionary materialists, creationists, and intelligent design advocates consider most appropriate to a masterful God.

Unfortunately, it is not uncharacteristic of evolutionists today to remain fixated on the idea that God must be a designer or divine mechanic. So they typically assume that by destroying the plausibility of the notion of a direct intelligent design of natural phenomena by God, they have automatically debunked the idea of divine action completely. For example, the psychologist and science writer David Barash notes that some religious believers attribute the intricate design in life to an intelligent designer, since in their view 'only a designer could generate such complex, perfect wonders'. Then he goes on to say, 'in fact, the living world is shot through with imperfection. Unless one wants to attribute either incompetence or sheer malevolence to such a designer, this imperfection—the manifold design flaws of life—points incontrovertibly to a natural, *rather than* (my emphasis) a divine, process, one in which living things were not created *de novo*, but evolved' (Barsah 2005). Evolution, in this not uncommon kind of proclamation, is a purely natural process that excludes design, and apparently also any place at all for divine involvement.

Yet even prior to Darwin, Christian theologians were not always comfortable with the idea that God is a designer or engineer. The Catholic thinker Cardinal John Henry Newman speaks for many in saying that natural theology's divine designer is even contrary to Christianity (McGrath 2005: 29–30). For Newman it is not the business of theology to force a divine Mind to show up directly behind the data of science. It seems to me, likewise, that the revelatory image of God in Christianity is inconsistent with the idea that God directly and simply engineers or tinkers with creation. Instead, God is more concerned to enable the world to make itself— permitting it to experience many mistakes in the process. There is a profound respect on the part of God for the freedom of the creation, a selfless liberality that apparently desires to avoid any unmediated manipulation of things. So in the creation of life's diversity, it is not a matter of natural processes *rather* than God doing all the work, but of God creating through natural processes.

As for divine providence, we know from our limited human experience that genuine care for others does not control or compel, but instead provides sufficient scope for others to become themselves. So if there is any truth to the biblical and traditional conviction that God cares intimately for creation, then the world will be allowed to become something truly other than God. This means that creation will unfold as a narrative in which suspense, uncertainty, and—in the life process—contingency will be indispensable to the drama. A universe deeply loved by the Creator must always have some degree of indeterminacy and autonomy, or else it will lapse from story into stone. Thus the randomness in evolution is not contrary to trust in divine providence. Even during the universe's long pre-human run, and in its microcosmic make-up at present, there will be analogous room for spontaneity, a condition that even physics no longer frowns on. Without the narrative ingredient of contingency, the universe could never have been the great adventure that it is. Nor could it ever have been a universe at all, theologically speaking. It would have been nothing more than an ornament attached passively to the divine being, rather than a reality in its own right.

If nature is to be truly distinct from God, as it must be if one is to escape the clutches of pantheism, it has to be given leave to experiment with a wide, though not unlimited, range of possibilities of being, even if the allowance for contingency opens up the possibility of tragedy at the level of life. Any supposed theological alternative to an evolutionary world peppered with contingency might be one in which there is no suffering or death. But such a world would also be devoid of life, human freedom, and the possibility of surprising future outcomes. On the other hand, a theology that situates the world within the embrace of a humble and promising God, one who offers ever new possibilities to the creation, can hardly be taken aback that the cosmos did not spring into being full-blown. Instead, it should expect that the universe will take its time—deep time perhaps—in order to actualize itself in the presence of its selfless and ever faithful Creator. Such a creation would require an elaborate interweaving of contingency, law, and time in order to be a significant story. It seems to me that the scientific portrait of life emerging from this narrative matrix is completely consistent with a Christian understanding of God.

In summary, then, there can be no self-giving of God to the created world if that same world is not permitted in some sense to be self-organizing and even self-creating. Theologically, what this creative scenario requires is room for undirected (contingent) events constrained by lawful limits in the context of copious time—in other words, the narrative evolutionary recipe. It is hard to imagine how any other set of cosmic characteristics would be compatible with the Christian understanding of divine generosity and promise. The pattern of evolution seems to fit quite comfortably into a world-view that features at its centre the idea of a humble God who loves stories, and who offers an open future in which the story of this universe and perhaps others as well, can continue to unfold.

THE SUFFERING OF SENTIENT LIFE

But can theology make sense of the ruthless and often painful way in which natural selection brings evolution about? One answer, of course, is that evolution is incompatible with any theological sense of a compassionate God. Moreover, it may seem that evolutionary biology can make good sense of life's suffering without the help of theology. Suffering, from the point of view of natural science, is an adaptation which, like any others, enhances an organism's probability of surviving and reproducing. Darwin himself thought that suffering is 'well adapted to make a creature guard against any great or sudden evil' (Barlow 1958: 88–9). Even though suffering is as imperfect as other adaptations, it can often signal that an organism is in danger, and thus allow it to continue its existence long enough to have descendants.

But the enormous excess of suffering seems disproportionate to its informational value, and this is one of the reasons why Darwin eventually rejected the idea of divine design. He found it inconceivable that a benign providence would ever have deliberately designed wasps that would lay their eggs inside living caterpillars so that their larvae would not have to suffer the indignity of consuming undecayed flesh (Darwin [1859] 1995: 220). Other evolutionists are no less disturbed than Darwin by nature's allowing so much suffering (which in this essay I am not distinguishing from 'pain', as some authors do). Richard Dawkins complains about the 'pitiless indifference' of a Darwinian universe (1996: 133). Biologist George Williams refers to nature as a 'wicked old witch' (1995) for allowing so much pain and struggle. And Stephen Jay Gould declares that because of evolution's amorality, humans can no longer look to nature for any ethical guidance at all (2003: pp. xvi–xvii).

Such sensitivity seems entirely appropriate, an expression of deep humaneness and compassion. So in view of the evolutionists' own vehement ethical protests against pain and perishing, any theological response to evolutionary suffering must begin, not by trying to show how suffering may be consistent with some fixed idea of deity such as an intelligent designer, but by asking what the suffering of life might mean in view of the revelatory image of God that I have alluded to above. This is a properly theological, as distinct from philosophical, way of addressing the topic. Such a theology may begin by asking what the evolutionists' own protest against suffering tells us about the universe. After all, the protest is just as much a part of the cosmos as is the pain, and if the pain deserves to be explained, so does the protest. Yet it is precisely in giving a full accounting of the human outrage against suffering that purely evolutionary or, for that matter, sociocultural explanations will turn out to be insufficient, and a theological world-view will become relevant.

Explaining the protest against suffering in a *purely* naturalistic way, after all, would take all the starch out of the protest itself. We would not be obliged to take any protest against suffering seriously if that protest itself turns out to be merely one more evolutionary adaptation or, for that matter, just one cultural convention among others. In order to arouse a genuinely ethical resistance to unnecessary suffering across many generations, the evolutionists' disapproval of pain must be backed up by something more substantive and permanent than a mindless evolutionary past or by purely conventional cultural habits. Theologically, the disapproval of suffering is to be taken seriously, not only because the protest is the outcome of an evolutionary process or cultural sculpting of ethical life (which is true enough), but also because the human heart and mind have the backing of *an eternally established state of being* in which suffering and death are judged to have no final or ultimate status. To Christians this essential order of being is revealed in the compassionate personal God who humbly descends into the creation to the point of suffering and struggling along with nature, who seeks to defeat all suffering and death by raising the dead to new life, and who at the same time promises an ultimate redemption for the whole universe. In my view, the protest against pain can receive permanent and lasting legitimization only by firmly planting itself in such a ground.

Still, this leaves the question of why God would create a universe in which pain is possible at all. I have no better answer to this question than anyone else, but it may be helpful at least to situate the suffering of sentient life, and indeed the whole of biological evolution, within the narrative matrix made possible by the coalescence of contingency, law, and deep time as discussed earlier, and then consider what any cosmological alternatives to such a dramatic setting would be like. They might be painless, but would they be of any interest at all?

As noted above, any universe devoid of contingency, to start with, would be a world without suffering, but it would also be a world without a future and without freedom, since everything would be fixed in place perpetually. Likewise, a world devoid altogether of reliably functioning (impersonal) laws would not have enough internal continuity to give it any narrative coherence across the ages. Without fixed laws it would be a world without pain, since no stable entities, least of all living organisms, could exist. But such a world, it goes without saying, would be unbearably flat and uninteresting in comparison with the one we actually inhabit.

Finally, a world that is not endowed with ample temporal opportunity to unfold experimentally and incrementally would also be trivial and unremarkable in comparison with an evolutionary one. Once again, a timeless world might be one without suffering, but it could never have brought about complex forms of life or given rise to mind *from within itself*. Too short a period of time would have required magical assistance to bring about life's complexity, as Dawkins (1996) also emphasizes, and it would have interfered with the universe's own need to unfold in a truly autonomous way. These counterfactual worlds are much less attractive, at least to those who love life and adventure, than the Darwinian world that stretches out so grandly before us.

REFERENCES AND SUGGESTED READING

AYALA, FRANCISCO (1994). 'Darwin's Revolution', in J. H. Campbell and J. W. Schopf (eds.), *Creative Evolution?!*, Boston: Jones & Bartlett, 1–17.

BARASH, DAVID (2005). 'Does God Have Back Problems Too?', *Los Angeles Times*, 27 June.

BARLOW, NORA (1958) (ed.). *The Autobiography of Charles Darwin*. New York: Harcourt.

CZIKO, GARY (1995). *Without Miracles: Universal Selection Theory and the Second Darwinian Revolution*. Cambridge, Mass.: MIT Press.

DARWIN, CHARLES ([1859] 1995). *The Orgin of Species*. New York: Gramercy.

DENNETT, DANIEL C. (1995). *Darwin's Dangerous Idea: Evolution and the Meaning of Life*. New York: Simon & Schuster.

DAWKINS, RICHARD (1986). *The Blind Watchmaker*. New York: W. W. Norton & Co.

—— (1995). *River Out of Eden*. New York: Basic Books.

—— (1996). *Climbing Mount Improbable*. New York: W. W. Norton & Co.

DE BEER, GAVIN (1973–4). 'Evolution', in *The New Encyclopaedia Britannica*, 15th edn., Macropaedia London: Encyclopaedia Britannica, vii. 7–23.

DEPEW, DAVID J., and WEBER, BRUCE H. (1995). *Darwinism Evolving: Systems Dynamics and the Genealogy of Natural Selection*. Cambridge, Mass.: MIT Press.

GOULD, STEPHEN J. (1989). *Wonderful Life: The Burgess Shale and the Nature of History*. New York: W. W. Norton.

—— (2003). 'Introduction', in Carl Zimmer, *Evolution: The Triumph of an Idea—From Darwin to DNA*, London: Arrow Books, pp. ix–xvii.

GOULD, STEPHEN J. and LEWONTIN, R. C. (1979). 'The Spandrels of San Marco and the Panglossian Paradigm: A Critique of the Adaptationist Programme', *Proceedings of the Royal Society of London*, Series B, 205/1161: 581–98.

HALLMAN, JOSEPH M. (1991). *The Descent of God: Divine Suffering in History and Theology.* Minneapolis: Fortress Press.

HAUGHT, JOHN F. (2000). *God after Darwin: A Theology of Evolution.* Boulder, Colo.: Westview Press.

—— (2003). *Deeper than Darwin: The Prospect for Religion in the Age of Evolution.* Boulder, Colo.: Westview Press.

JÜNGEL, EBERHARD (1976). *The Doctrine of the Trinity: God's Being is in Becoming*, trans. Scottish Academic Press Ltd. Grand Rapids, Mich.: Eerdmanns.

MACQUARRIE, JOHN (1978). *The Humility of God.* Philadelphia: Westminster Press.

MCGRATH, ALISTER (2005). 'A Blast from the Past? The Boyle Lectures and Natural Theology', *Science and Christian Belief*, 17: 25–30.

MOLTMANN, JÜRGEN (1975). *The Experiment Hope*, ed. and trans. M. Douglas Meeks. Philadelphia: Fortress Press.

—— (1985). *God in Creation*, trans. Margaret Kohl. San Francisco: Harper & Row.

—— (1996). *The Coming of God: Christian Eschatology*, trans. Margaret Kohl. Minneapolis: Fortress Press.

MOONEY, CHRISTOPHER, SJ (1996). *Theology and Scientific Knowledge.* Notre Dame, Ind., and London: University of Notre Dame Press.

NIEBUHR, H. RICHARD (1960). *The Meaning of Revelation.* New York: Macmillan Co.

PANNENBERG, WOLFHART (1977). *Faith and Reality*, trans. John Maxwell. Philadelphia: Westminster Press.

—— (1993). *Toward a Theology of Nature: Essays on Science and Faith*, ed. Ted Peters. Louisville, Ky.: Westminster/John Knox Press.

PETERS, TED (2000). *God—the World's Future: Systematic Theology for a New Era*, 2nd edn. Minneapolis: Fortress Press.

POPE JOHN PAUL II (1998). *Encyclical Letter Fides et Ratio of the Supreme Pontiff to the Bishops of the Catholic Church on the Relationship of Faith and Reason.*<http://www.vatican.va/holy_father/john_paul_ii/encyclicals/documents/hf_jp_ii_enc_15101998_fides-et-ratio_en.html>

RAHNER, KARL, SJ (1984). *Foundations of Christian Faith*, trans. William Dych. New York: Crossroad.

ROSE, MICHAEL R. (1998). *Darwin's Spectre: Evolutionary Biology in the Modern World.* Princeton: Princeton University Press.

RUSE, MICHAEL (2003). *Darwin and Design: Does Evolution Have a Purpose?* Cambridge, Mass.: Harvard University Press.

WEINBERG, STEVEN (1993). *Dreams of a Final Theory: The Search for the Ultimate Laws of Nature.* New York: Pantheon.

WHITEHEAD, ALFRED NORTH (1967). *Adventures of Ideas.* New York: Free Press.

—— (1978). *Process and Reality*, corrected edn., ed. David Ray Griffin and Donald W. Sherburne. New York: Free Press.

WILLIAMS, GEORGE C. (1995). 'Mother Nature Is a Wicked Old Witch!', in Matthew H. Nitecki and Doris V. Nitecki (eds.), *Evolutionary Ethics*, Albany, N.Y.: State University of New York Press, 217–31.

WILSON, EDWARD O. (1998). *Consilience: The Unity of Knowledge.* New York: Knopf.

INTELLIGENT DESIGN
AND ITS CRITICS

CHAPTER 42

IN DEFENCE OF INTELLIGENT DESIGN

WILLIAM A. DEMBSKI

PRELIMINARY CONSIDERATIONS

Anyone new to the debate over intelligent design encounters many conflicting claims about whether it is science. A *Washington Post* front-page story (Slevin 2005) asserts that intelligent design is 'not science [but] politics'. In that same story, Barry Lynn, the director of Americans United for Separation of Church and State, claims that intelligent design is merely 'a veneer over a certain theological message', thus identifying intelligent design not with science, but with religion. In a related vein, University of Copenhagen philosopher Jakob Wolf (2004) argues that intelligent design is not science but philosophy (albeit a philosophy useful for understanding science). And finally, proponents of intelligent design argue that it is indeed science (e.g. Dembski 2002b: ch. 6). Who is right?

In determining how to answer this question, three points need to be kept in mind:

(1) Science is not decided by majority vote. Can the majority of scientists be wrong about scientific matters? Yes, they can. Historian and philosopher of science Thomas Kuhn, in his *Structure of Scientific Revolutions* (1970), documented numerous reversals in science where views once confidently held by the scientific community ended up being discarded and replaced. For instance, until the theory of plate tectonics was proposed, geologists used to believe that the continents were immovable (compare Kearey and Vine 1996 to Clark and Stearn 1960). Intelligent design is at

present a minority position within science. But that fact by itself does nothing to impugn its validity.

(2) Just because an idea has religious, philosophical, or political implications does not make it unscientific. According to the late evolutionist Stephen Jay Gould (1977a: 267), 'Biology took away our status as paragons created in the image of God.... Before Darwin, we thought that a benevolent God had created us.' Oxford University biologist Richard Dawkins (1986: 6) claims that 'Darwin made it possible to be an intellectually fulfilled atheist'. In his book *A Darwinian Left: Politics, Evolution, and Cooperation*, Princeton bioethicist Peter Singer (2000: 6) remarks that we must 'face the fact that we are evolved animals and that we bear the evidence of our inheritance, not only in our anatomy and our DNA, but in our behavior too'. Gould, Dawkins, and Singer are respectively drawing religious, philosophical, and political implications from evolutionary theory. Does that make evolutionary theory unscientific? No. By the same token, intelligent design's philosophical and theological implications do not render it unscientific.

(3) To call some area of inquiry 'not science' or 'unscientific', or to label it 'religion' or 'myth', is within contemporary Western culture a common manoeuvre for discrediting an idea. Physicist David Lindley (1993), for instance, in order to discredit cosmological theories that outstrip experimental data or verification, calls such theories 'myths'. Writer and medical doctor Michael Crichton (2003), in his Caltech Michelin Lecture, criticizes the Search for Extraterrestrial Intelligence (SETI) as follows:

SETI is not science. SETI is unquestionably a religion. Faith is defined as the firm belief in something for which there is no proof.... The belief that there are other life forms in the universe is a matter of faith. There is not a single shred of evidence for any other life forms, and in forty years of searching, none has been discovered. There is absolutely no evidentiary reason to maintain this belief. SETI is a religion.

Crichton's criticism, however, seems extreme. In the past, NASA has funded SETI research. And even if the actual search for alien intelligences has thus far proved unsuccessful, SETI's methods of search and the *possibility* of these methods proving successful validate SETI as a legitimate scientific enterprise.

WHAT IS INTELLIGENT DESIGN?

Intelligent design is the field of study that investigates *signs of intelligence*. It identifies those features of objects that reliably signal the action of an intelligent cause. To see what is at stake, consider Mount Rushmore. The evidence for Mount Rushmore's design is direct—eyewitnesses saw the sculptor Gutzon Borglum spend the better part of his life designing and fashioning this structure. But what if there were no direct evidence for Mount Rushmore's design? Suppose humans went extinct and aliens, visiting the Earth, discovered Mount Rushmore in substantially the same condition as now.

In that case, what about this rock formation would provide convincing circumstantial evidence that it was due to a designing intelligence, and not merely to wind and erosion? Designed objects like Mount Rushmore exhibit characteristic features or patterns that point to an intelligence. Such features or patterns constitute signs of intelligence. Proponents of intelligent design, known as *design theorists*, purport to study such signs formally, rigorously, and scientifically. In particular, they claim that a type of information, known as *specified complexity*, is a key sign of intelligence. An exact formulation of specified complexity first appeared in my book *The Design Inference* (Dembski 1998) and was then further developed in *No Free Lunch* (2002*b*).

What is specified complexity? Recall the novel *Contact* by Carl Sagan (1985). In that novel, radio astronomers discover a long sequence of prime numbers from outer space. Because the sequence is long, it is *complex*. Moreover, because the sequence is mathematically significant, it can be characterized independently of the physical processes that bring it about. As a consequence, it is also *specified*. Thus, when the radio astronomers in *Contact* observe specified complexity in this sequence of numbers, they have convincing evidence of extraterrestrial intelligence. Granted, real-life SETI researchers have thus far failed to detect designed signals from outer space. The point to note, however, is that Sagan based the SETI researchers' methods of design detection on actual scientific practice.

To employ specified complexity to detect design is to engage in effect-to-cause reasoning. As a matter of basic human rationality, we reason from causes to effects as well as from effects back to causes. Scientific experimentation, for instance, requires observation and the control of variables, and thus typically employs cause-to-effect reasoning: the experimenter, in setting up certain causal processes in an experiment, constrains the outcome of those processes (the effect). But, in many cases, we do not have control of the relevant causal processes. Rather, we are confronted with an effect, and must reconstruct its cause. Thus, an alien visiting Earth and confronted with Mount Rushmore would need to figure out whether wind and erosion could produce it or whether some additional factors might be required.

A worry now arises as to whether effect-to-cause reasoning leads to many absurd design hypotheses. Consider the 'Zeus hypothesis' in which lightning strikes are attributed to the divine intervention of the god Zeus (I'm indebted to Robert Pennock for this example). Such a hypothesis, though an example of effect-to-cause reasoning, would not be the conclusion of a design inference based on specified complexity. Individual lightning strikes are readily explained in terms of the laws of physics, with no need to invoke a designer. The only way that lightning strikes might require an intelligent design (ID) hypothesis is if they jointly exhibit some particularly salient pattern. Consider, for instance, the possibility that on a given day all, and only, those people in the United States who had uttered snide remarks about Zeus were hit by lightning and died. In that case, the joint pattern of lightning strikes would exhibit specified complexity, and the Zeus hypothesis might no longer seem altogether absurd.

To sum up, many special sciences already employ specified complexity as a sign of intelligence—notably forensic science, cryptography, random number generation, archaeology, and the search for extraterrestrial intelligence (Dembski 1998: chs. 1 and 2). Design theorists take these methods and apply them to naturally occurring systems (see Dembski and Ruse 2004: pt. IV). When they do, these same methods for identifying intelligence indicate that the delicate balance of cosmological constants (known as cosmological fine-tuning) and the machine-like qualities of certain tightly integrated biochemical systems (known as irreducibly complex molecular machines) are the result of intelligence, and are highly unlikely to have come about by purely material forces (like the Darwinian mechanism of natural selection and random variation). (For such design-theoretic arguments at the level of cosmology, see Gonzalez and Richards (2004); for such design-theoretic arguments at the level of biology, see Behe (1996).) In any event, it is very much a live possibility that design in cosmology and biology is scientifically detectable, thus placing intelligent design squarely within the realm of science.

The Charge of Creationism

Despite intelligent design's clear linkage, both methodologically and in content, with existing sciences that sift the effects of intelligence from undirected natural forces, critics of intelligent design often label it a form of creationism. Not only is this label misleading, but in academic and scientific circles it has become a term of abuse to censor ideas before they can be fairly discussed.

To see that the creationist label is misleading, consider that one can advocate intelligent design without advocating creationism. Creationism typically denotes a literal interpretation of the first two chapters of Genesis as well as an attempt to harmonize science with this particular interpretation (Morris 1975). It can also denote the view common to theists that a personal transcendent God created the world, a view taught by Judaism, Christianity, and Islam (Johnson 2004). In either case, however, creationism presupposes that the world came into being through a creative power separate from the world.

Intelligent design, by contrast, places no such requirement on any designing intelligence responsible for cosmological fine-tuning or biological complexity. It simply argues that certain finite material objects exhibit patterns that convincingly point to an intelligent cause. But the nature of that cause—whether it is one or many, whether it is a part of or separate from the world, and even whether it is good or evil—simply do not fall within intelligent design's purview. Thomas Aquinas, in his *Summa Contra Gentiles* (III. 38), put it this way (quoted from Pegis 1948: 454–5):

By his natural reason man is able to arrive at some knowledge of God. For seeing that natural things run their course according to a fixed order, and since there cannot be order without a cause of order, men, for the most part, perceive that there is one who orders the things that we see. But who or of what kind this cause of order may be, or whether there be but one, cannot be gathered from this general consideration.

Consistent with this statement, Aristotle, who held to an eternal uncreated world and to a purposiveness built into the world, would today hold to intelligent design, but not to creationism (see his *Physics* as well as his *Metaphysics* in McKeon 1941). The same is true for Antony Flew, who until recently was the English-speaking world's most prominent atheist. He now repudiates atheism, because he sees intelligent design as necessary to explain the origin of life (Associated Press 2004). Yet, in embracing an intelligence behind biological complexity, he does not hold to creationism (Habermas 2004).

Despite its constant repetition, the charge that intelligent design is a form of creationism is false. Robert Pennock (1999, 2001) and Barbara Forrest (see Forrest and Gross 2004), for instance, repeat this charge in virtually all of their writings that criticize intelligent design. Yet, as trained philosophers, they know that intelligent design is consistent with philosophical positions that hold to no doctrine of creation. Why, then, do they insist that intelligent design is creationism? The reason is that creationism has been discredited in the courts and among the scientific and academic élite. Thus, if the label can be made to stick, intelligent design will be defeated without the need to investigate its actual claims.

To see that 'creationism' is a question-begging label meant to stop the flow of inquiry before it can get started, consider that one of the most prominent critics of intelligent design, Kenneth Miller, has himself been called a creationist. In his book *Finding Darwin's God*, Miller is critical of intelligent design in biology. Nonetheless, in that book he argues for an intelligence or purposiveness that underlies the laws of physics (laws that are necessary for the universe to be life-permitting—see K. R. Miller 1999: 226–32). Miller's reward for proposing intelligent design at the level of physics and cosmology is to be called a creationist by University of California professor Frederick Crews. In reviewing Miller's book, Crews (2001) writes:

When Miller then tries to drag God and Darwin to the bargaining table [by finding design or purpose underlying the laws of physics], his sense of proportion and probability abandons him, and he himself proves to be just another 'God of the gaps' creationist. That is, he joins Phillip Johnson, William Dembski, and company in seizing upon the not-yet-explained as if it must be a locus of intentional action by the Christian deity.

Despite criticisms like this by Crews and others, mainstream physics is now quite comfortable with design in cosmology. Take the following remark by Arno Penzias, Nobel laureate and co-discoverer of cosmic background radiation (quoted in Margenau and Varghese 1992: 83): 'Astronomy leads us to a unique event, a universe which was created out of nothing, one with the very delicate balance needed to provide exactly the conditions required to permit life, and one which has an underlying (one might say "supernatural") plan.' Or consider the following insight by well-known astrophysicist and science writer Paul Davies (1988: 203): 'There is for me powerful evidence that there is something going on behind it all. . . . It seems as though somebody has fine-tuned nature's numbers to make the Universe. . . . The impression of design is overwhelming.' Elsewhere Davies (1984: 243) says: 'The laws [of physics] . . . seem to be the product of exceedingly ingenious design. . . . The universe must have a purpose.' Remarks like this by prominent physicists and cosmologists are now widespread.

Why should inferring design from the evidence of cosmology be scientifically respectable, but inferring design from the evidence of biology be scientifically disreputable, issuing in the charge of creationism? Clearly, a double standard is at work here. Design theorists argue that the evidence of biology confirms a design inference. But even if that confirmation were eventually overturned by new evidence, such a failure would constitute a failure of intelligent design as a scientific theory, and not a failure of intelligent design to qualify as a scientific theory, much less to deserve the label 'creationism'.

PROBLEMS WITH EVOLUTIONARY THEORY

Most scientific theories are imperfect, in the sense that what they claim about the natural world and what the natural world in fact displays do not match up perfectly. Newton's theory, for instance, predicts certain types of planetary orbits. Nevertheless, the perihelion of Mercury was found to violate this prediction—not by much, but enough to call Newton's theory into question. Ultimately, Einstein resolved this anomaly by replacing Newton's theory with his own theory of relativity.

The problem of theories not matching up with facts has been known since the time of the ancient Greeks, who described this problem in terms of 'saving the phenomena'. In other words, the task of science (known back then as 'natural philosophy') was to match up scientific theories with the phenomena (or appearances) of nature. The physicist Pierre Duhem (1969) even wrote a book on this topic. He also wrote another book (Duhem 1954) to describe what scientists do when their theories do not match up with the facts. In that case, according to Duhem, they have two options: one is simply to abandon the theory; the other, and by far the more common option, is to add auxiliary hypotheses to try to shore up the theory. Simply put, the second option is to put patches over those aspects of the theory that don't match up with the facts.

Which option is preferable? This is a judgement call. Is the mismatch so egregious and the patch so artificial that the theory cannot be reasonably salvaged? In that case, scientists prefer the first option, that of abandoning the theory. Has the theory proved itself useful in the past, and is the mismatch so minor, and the patch so unobtrusive, that the theory remains largely intact? In that case, scientists prefer the second option. The problem is, as Thomas Kuhn showed in his vastly influential *The Structure of Scientific Revolutions* (1970: ch. 10), there is no easy way to draw the line between these two options.

Scientists remain divided over what to do about the mismatches between contemporary evolutionary theory and the facts of biology. Nevertheless, the mismatches are there in plain view, as are the patches put on evolutionary theory in order to mitigate the mismatches. The best known mismatch is the overwhelming failure of the fossil record to match up with Darwin's expectation that living forms fall within one gigantic, gradually branching tree of life. In the sixth edition of Darwin's *Origin of Species*, there is exactly one diagram: one that depicts the evolution of organisms as a gradually branching tree (Darwin 1872: 90–1). Yet the fossil record is full of gaps that show no sign of being bridged by the mechanisms of evolutionary theory.

To see this, one does not need to look to the work of design theorists. Evolutionists have recognized the problem for some time now. For instance, Stephen Jay Gould (1977*b*), who until his death was the most prominent evolutionary theorist on the American side of the Atlantic, noted: 'The extreme rarity of transitional forms in the fossil record persists as the trade secret of paleontology. The evolutionary trees that adorn our textbooks have data only at the tips and nodes of their branches; the rest is inference, however reasonable, not the evidence of fossils.'

Gould's solution to this problem was to propose his idea of punctuated equilibrium, in which evolution takes place in isolated populations that are unlikely to be fossilized, with the result that the fossil record exhibits a pattern of sudden change, followed by stasis (see Eldredge and Gould 1973). But this patch has its own problems. For one, it does not address the mechanism of evolutionary change. Also, it is largely untestable, because all the interesting evolution happens where it is inaccessible to scientific observation.

There are many other mismatches between contemporary evolutionary theory and the facts of biology. Despite primitive Earth atmospheric simulation experiments, like the one by Stanley Miller (1953), the problem of life's origin remains completely unresolved in materialistic terms. Similarly, the challenge of irreducibly complex molecular machines raised by Michael Behe (1996) has resisted evolutionary explanations. Colorado State University biochemist Franklin Harold (2001: 205), citing Behe, writes: 'We should reject, as a matter of principle, the substitution of intelligent design for the dialogue of chance and necessity (Behe, 1996); but we must concede that there are presently no detailed Darwinian accounts of the evolution of any biochemical system, only a variety of wishful speculations.'

Or take the problem of 'junk' DNA. According to neo-Darwinian theory, the genomes of organisms are cobbled together over a long evolutionary history through a trial-and-error process of natural selection sifting the effects of random genetic errors. As a consequence, neo-Darwinism expects to find a lot of 'junk' DNA: that is, DNA that serves no useful purpose but is simply carried along for the ride because it is easier for cells to keep copying DNA that genetic errors render useless than to identify and eliminate such DNA from the genome.

The theory of intelligent design, on the other hand, in approaching organisms as designed systems, is less apt to dismiss seemingly useless DNA as junk. Instead, it encourages biologists to investigate whether systems that at first appear functionless might in fact have a function. And, as it is now turning out, seemingly useless 'junk' DNA is increasingly being found to serve useful biological functions. For instance, James Shapiro and Richard von Sternberg (2005) have recently provided a comprehensive overview of the functions of repetitive DNA—a classic type of 'junk' DNA. Similarly, Roy Britten (2004) has recently outlined the functions of mobile genetic elements—another class of sequences long thought to be simply parasitic junk.

Such mismatches between evolutionary theory and the facts of biology are significant for the public understanding of biology. Even without specialized biological knowledge, it is possible for laypersons to see that evolutionary theory, as taught in high school and college biology textbooks, is desperately in need of fuller treatment and more adequate discussion of alternatives. Right now, the basic biology textbooks from which most people in the English-speaking world receive their first serious exposure to evolutionary theory explain the origination of biological forms in terms of the neo-Darwinian mechanism of natural selection and random genetic errors. This mechanism, however, is now increasingly seen as inadequate to explain the diversity of biological forms, and not just by design theorists.

For instance, Lynn Margulis (in Margulis and Sagan 2002: 103), a biologist who is a member of the National Academy of Sciences, criticizes the neo-Darwinian theory as follows: 'Like a sugary snack that temporarily satisfies our appetite but deprives us of more nutritious foods, neo-Darwinism sates intellectual curiosity with abstractions bereft of actual details—whether metabolic, biochemical, ecological, or of natural history.' Robert Laughlin (2005: 168–9), a Nobel laureate physicist concerned with the properties of matter that make life possible, offers even stronger criticism:

Much of present-day biological knowledge is ideological. A key symptom of ideological thinking is the explanation that has no implications and cannot be tested. I call such logical dead ends antitheories because they have exactly the opposite effect of real theories: they stop thinking rather than stimulate it. Evolution by natural selection, for instance, which Charles Darwin originally conceived as a great theory, has lately come to function more as an antitheory, called upon to cover up embarrassing experimental shortcomings and legitimize findings that are at best questionable and at worst not even wrong. Your protein defies the laws of mass action? Evolution did it! Your complicated mess of chemical reactions turns into a chicken? Evolution! The human brain works on logical principles no computer can emulate? Evolution is the cause!

Note that neither Margulis nor Laughlin are advocates of intelligent design.

These criticisms cut to the very heart of contemporary evolutionary theory, and are directly pertinent to how evolution should be taught. According to Simon Conway Morris (2000: 1), 'When discussing organic evolution the only point of agreement seems to be: "It happened." Thereafter, there is little consensus, which at first sight must seem rather odd.' Odd indeed. Right now, basic biology textbooks reflect a 'consensus trance', giving the illusion that there is unanimity among biologists over how evolution occurred when in fact there is no such unanimity.

This 'consensus trance' needs to be broken, with scientific alternatives to conventional evolutionary theory welcomed into biology curricula. One such alternative theory is intelligent design.

METHODOLOGICAL MATERIALISM

Notwithstanding the previous discussion, critics of intelligent design argue that it is not a scientific theory. They do so, however, not by confronting the evidence and logic by which design theorists argue for their conclusions, but, rather, by definitional fiat. Essentially, they engage in conceptual gerrymandering, carefully defining science so that conventional evolutionary theory falls within the domain of science and intelligent design falls outside. The device by which they keep intelligent design at bay is a normative principle for science known as *methodological naturalism* or *methodological materialism*. ID's rejection of this principle is said to show that ID is committed to a form of supernaturalism. This, in turn, is supposed to make ID a form of religious belief. Barbara Forrest (in Forrest and Gross 2004) and Eugenie Scott (2005) make methodological materialism the centrepiece of their critique of ID.

The impression they give is that whereas conventional evolutionary theory is engaged in the hard work of real science, intelligent design appeals to the supernatural, and thus gives up on science, substituting magic for 'natural explanations'. But what are 'natural explanations'? What constitutes nature remains very much an open question. If one reviews the ID literature, one finds that early on there were quite a few references to 'the supernatural', but that by 2000 (especially with the *Nature of Nature* conference, organized by Baylor University's Michael Polanyi Center—see Dembski and Gordon 2000), references to the supernatural largely disappear. The reason for this is that the very term 'supernatural' concedes precisely the point at issue: namely, what nature is like and what are the causal powers by which nature operates.

Critics of intelligent design who hold to methodological materialism say that nature operates only by natural causes and is explained scientifically only through natural explanations. But what do they mean by 'nature'? Eugenie Scott (1998), director of the evolution watchdog group the National Center for Science Education (NCSE), explains how methodological materialism construes nature:

Most scientists today require that science be carried out according to the rule of *methodological materialism*: to explain the natural world scientifically, scientists must restrict themselves only to material causes (to matter, energy, and their interaction). There is a practical reason for this restriction: it works. By continuing to seek natural explanations for how the world works, we have been able to find them. If supernatural explanations are allowed, they will discourage—or at least delay—the discovery of natural explanations, and we will understand less about the universe.

Thus, for Scott, nature is 'matter, energy, and their interaction'. Accordingly, by natural explanations, Scott means explanations that resort only to such material causes. Yet, that is precisely the point at issue: namely, whether nature operates exclusively by such causes. If nature contains a richer set of causes than purely material causes, then intelligent design is a live possibility, and methodological materialism has misread physical reality. Note, also, that to contrast natural explanations with supernatural explanations further obscures this crucial point. 'Supernatural explanations' typically denote explanations that invoke miracles and cannot be understood scientifically. But explanations that call upon intelligent causes require no miracles and give no evidence of being reducible to Scott's trio of 'matter, energy, and their interaction'. Indeed, design theorists argue that intelligent causation is perfectly natural provided that nature is understood aright.

Scott's characterization of methodological materialism thus encounters two difficulties. First, if, as she suggests, methodological materialism is merely a working hypothesis that scientists employ because 'it works', then scientists are free to discard it when they deem that it is no longer working. Design theorists contend that for adequately explaining biological complexity, methodological materialism fails, and rightly needs to be discarded. Second, and more significantly, in defining science as the search for natural explanations, Scott presupposes precisely what must be demonstrated. If, by 'natural explanations', Scott simply means explanations that account for what is happening in nature, there would be no problem, and intelligent design would constitute a perfectly good natural explanation of biological complexity. But, clearly, this is not what she means.

Because so much of the debate over intelligent design's scientific status hinges on the role of methodological materialism in restricting the nature of nature, let us examine the nature of nature more closely. Nature, as conceived by Scott and most critics of intelligent design, consists of material entities ruled by fixed laws of interaction, often referred to as 'natural laws'. These laws can be deterministic or non-deterministic, which is why some scientists refer to nature as being governed by 'chance and necessity' (like Jacques Monod, 1972). Obviously, these laws of interaction rule out any form of intelligent agency acting in real time within nature. They operate autonomously and automatically: given certain material entities with certain energetic properties in certain spatio-temporal relationships, these entities will behave in certain prescribed ways.

An inescapable question now arises: How do we know that nature is in fact a set of material entities ruled by fixed laws of interaction? Equivalently, how do we know that everything that happens in nature can be accounted for in terms of antecedent material conditions and the material causes that act on them? Once the question is posed this way, it becomes an open question whether nature comprises a set of material entities ruled by fixed laws of interaction. In fact, it becomes a live possibility that nature, so conceived, is radically incomplete. My book *No Free Lunch* (Dembski 2002b: pp. xiii–xiv) summarizes what's at issue here as follows:

In arguing that naturalistic [materialistic] explanations are incomplete or, equivalently, that natural [material] causes cannot account for all the features of the natural world, I am placing natural causes in contradistinction to intelligent causes. The scientific community has itself drawn this distinction in its use of these twin categories of causation. Thus, in the quote earlier by Francisco Ayala, 'Darwin's greatest accomplishment [was] to show that the directive organization of living beings can be explained as the result of a natural process, natural selection, without any need to resort to a Creator or other external agent'. Natural causes, as the scientific community understands them, are causes that operate according to deterministic and nondeterministic laws and that can be characterized in terms of chance, necessity, or their combination (cf. Jacques Monod's *Chance and Necessity*). To be sure, if one is more liberal about what one means by natural causes and includes among natural causes telic processes that are not reducible to chance and necessity (as the ancient Stoics did by endowing nature with immanent teleology), then my claim that natural causes are incomplete dissolves. But that is not how the scientific community by and large understands natural causes.

Accordingly, to define science (in line with methodological materialism) as the search for natural explanations of natural phenomena is to affirm that such explanations exist for *all* natural phenomena. But how is this affirmation to be justified? Rather than justify it, methodological materialism begs the question. To see this, consider the following analogy from the game of chess. In chess, there are initially thirty-two pieces arranged on an eight-by-eight chessboard (see Figure 42.1).

Moreover, chess operates by certain fixed rules. For instance, bishops move diagonally, pawns only move forward and only take one square diagonally, etc. In this analogy, the chess pieces in their initial configuration correspond to the material entities that within methodological materialism constitute nature, and the rules of chess correspond to the laws of interaction that for methodological materialism govern nature.

Given the initial position of chess pieces and the rules of the game, we can ask whether the position shown in Figure 42.2 is possible. It turns out that it is not. There is no way to get from the first position to the second by the rules of chess.

So too, intelligent design purports to show that there exist configurations of material entities in biology (e.g. bacterial flagella, protein synthesis mechanisms, and complex organ systems) that cannot be adequately explained in terms of antecedent material conditions together with the law-governed processes (i.e. mechanistic evolutionary processes) that act on them. Granted, chess constitutes a toy example, whereas the biological examples that ID theorists investigate are far more complicated. Moreover, whereas chess operates according to precise mathematical rules, the laws of interaction associated with material entities are probabilistic, so the obstacles to producing complex biological configurations of material entities are not logical impossibilities, but empirical improbabilities. But the point of the analogy still holds. Whenever one has a theory about process—how one state is supposed, by some process, to transform into another—it is perfectly legitimate to ask whether the process in question is capable of accounting for the final state in terms of the initial state.

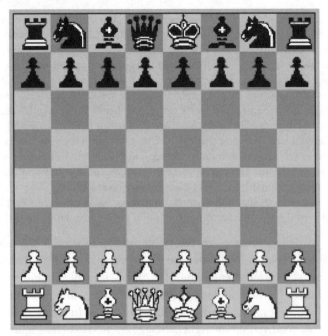

Fig. 42.1. Starting position for the game of chess.

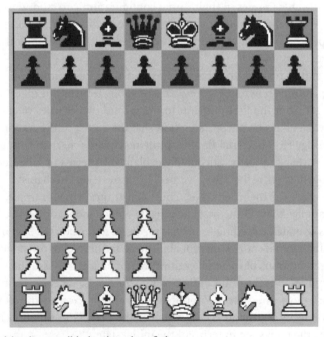

Fig. 42.2. Position inaccessible by the rules of chess.

It follows that the charge of supernaturalism against intelligent design cannot be sustained. Indeed, to say that rejecting naturalism entails accepting supernaturalism holds only if nature is defined as a closed system of material entities ruled by unbroken laws of material interaction. But this definition of nature begs the question. Nature is what nature is, not what we define it to be. To see this, consider the following riddle: How many legs does a dog have if one calls a tail a leg? The correct answer is four. Calling one thing another thing doesn't make it something else.

Likewise, defining nature as a closed system of material entities operating by fixed laws of interaction doesn't make it so. Nature is what nature is, and prescribing methodological materialism as a normative principle for science does nothing to change that. Intelligent design theorists argue that methodological materialism fundamentally distorts our understanding of nature. In assessing the validity of ID, the crucial thing is not whether they *are* right but whether they *might be* right. Given that they might be right, methodological materialism cannot be taken as a defining feature of science; much less should it be held dogmatically. To make methodological materialism a defining feature of science commits the pre-modern sin of forcing nature into a priori categories rather than allowing nature to speak for itself.

To sum up, methodological materialism presents us with a false dilemma: either science must be limited to 'natural explanations' (taken in a highly tendentious sense) or it must embrace 'supernatural explanations', by which is meant magic. But there is a third possibility: *neither materialism nor magic, but Mind*. Intelligent design theorists are not willing to concede the materialist claim that a designing intelligence (Mind) interacting with matter is 'supernatural'. Indeed, investigations by ID theorists are beginning to demonstrate that this interaction is perfectly natural—that nature cannot be properly understood apart from the activity of a designing intelligence (cf. Schwartz and Begley 2002).

THE CONTROVERSY SURROUNDING INTELLIGENT DESIGN

The controversy surrounding intelligent design occurs at many levels, but it is ultimately a scientific controversy within the scientific community. To be sure, there are educational, political, religious, and philosophical aspects to this controversy, but if there were no scientific controversy here, these other aspects would never have got off the ground.

There are a number of ways to see that this truly is a scientific controversy. One indicator is that design theorists are increasingly publishing research supporting intelligent design in the peer-reviewed mainstream scientific literature, especially in the biological literature (see Meyer 2004; Behe and Snoke 2004; Denton *et al.* 2002). A related indicator is that their work is increasingly being subjected to criticism within the mainstream scientific literature (see Thornhill and Ussery 2000; Schneider 2000; Lenski *et al.* 2003). And, most significantly, design theorists have a genuine programme of scientific research that they are now pursuing with increasing vigor (see the section on 'research themes' in Dembski 2002*a*).

To say that the intelligent design research programme is at odds with the conventional evolutionary theory is to offer a truism. Less obvious, perhaps, is that this controversy between competing theories is healthy for science, for it renders both intelligent design and evolutionary theory scientifically testable. Unfortunately, the way things stand now, given the artificial exclusion of intelligent design from scientific discussion (as by the device of methodological materialism), evolutionary theory has been rendered immune to scientific disconfirmation. In other words, it has become scientifically untestable.

Eshel Ben Jacob, a physicist who specializes in complex systems, is troubled by this state of affairs. He writes: 'Darwin, a free thinker who dared make far-reaching conclusions based on observations, would have been dismayed to see the petrified doctrine his brainchild has become. Must we admit that all organisms are nothing but watery Turing machines evolved merely by a sequence of accidents favoured by nature? Or do we have the intellectual freedom to rethink this fundamental issue?' (quoted in Dembski 2004, back cover).

Darwin's theory of descent with modification by means of natural selection acting on variations presents a non-teleological alternative to intelligent design. In fact, Darwin's *Origin of Species* can be viewed as a self-conscious response to the design argument. Contemporary evolutionary theory follows in this train. Hence Richard Dawkins (1986) gives his book *The Blind Watchmaker* the subtitle *Why the Evidence of Evolution Reveals a Universe without Design*. The study of biological origins is fundamentally incomplete so long as intelligent design is removed from scientific discussion. One can even go further and propose the following: evolutionary theory cannot be adequately understood apart from intelligent design as its proper foil and counterpart.

The integrity of current evolutionary theorizing depends on making room for intelligent design. Darwin himself would have agreed. In his *Origin of Species*, Darwin ([1859] 1964: 2) wrote: 'A fair result can be obtained only by fully stating and balancing the facts and arguments on both sides of each question.' When it comes to biological origins, intelligent design presents the facts and arguments for one side of this question. To pretend that there is no scientific controversy surrounding intelligent design is therefore itself unscientific.

References and Suggested Reading

Associated Press (2004). 'Famous Atheist Now Believes in God', 9 December; <http://abc news.go.com/US/wireStory?id=315976> (last accessed 25 March 2005).

Behe, Michael (1996). *Darwin's Black Box: The Biochemical Challenge to Evolution*. New York: Free Press.

—— and SNOKE, DAVID W. (2004). 'Simulating Evolution by Gene Duplication of Protein Features that Require Multiple Amino Acid Residues', *Protein Science*, 13: 2651–64.

BRITTEN, ROY J. (2004). 'Coding Sequences of Functioning Human Genes Derived Entirely from Mobile Element Sequences', *Proceedings of the National Academy of Sciences*, 101/48 (30 November): 16825–30.

CLARK, THOMAS H., and STEARN, COLIN W. (1960). *The Geological Evolution of North America*. New York: Ronald Press.

CONWAY MORRIS, SIMON (2000). 'Evolution: Bringing Molecules into the Fold', *Cell*, 100 (7 January): 1–11.

CREWS, FREDERICK C. (2001). 'Saving Us from Darwin, Part II', *The New York Review of Books*, (18 October); <http://www.nybooks.com/articles/article-preview?article_id=14622> (last accessed 25 March 2005).

CRICHTON, MICHAEL (2003). 'Aliens Cause Global Warming', Caltech Michelin Lecture (17 January); <http://www.crichton-official.com/speeches/speeches_quote04.html> (last accessed 15 March, 2005).

DARWIN, CHARLES ([1859] 1964). *On the Origin of Species*, 1st edn. Cambridge, Mass.: Harvard University Press; 6th edn. London: John Murray.

DAVIES, PAUL (1984). *Superforce: The Search for a Grand Unified Theory of Nature*. New York: Simon & Schuster.

—— (1988). *The Cosmic Blueprint: New Discoveries in Nature's Creative Ability to Order the Universe*. New York: Simon & Schuster.

DAWKINS, RICHARD (1986). *The Blind Watchmaker: Why the Evidence of Evolution Reveals a Universe without Design*. New York: Norton.

DEMBSKI, WILLIAM A. (1998). *The Design Inference: Eliminating Chance through Small Probabilities*. Cambridge: Cambridge University Press.

—— (2002a). 'Becoming a Disciplined Science: Prospects, Pitfalls, and Reality Check for ID', *Progress in Complexity, Information, and Design*, 1/4; <http://www.iscid.org/papers/Dembski_DisciplinedScience_102802.pdf> (last accessed 29 June, 2005).

—— (2002b). *No Free Lunch: Why Specified Complexity Cannot be Purchased without Intelligence*. Lanham, Md.: Rowman & Littlefield.

—— (2004) (ed.). *Uncommon Dissent: Intellectuals who Find Darwinism Unconvincing*. Wilmington, Del.: ISI Books.

—— and GORDON, BRUCE L. (2000). *The Nature of Nature: An Interdisciplinary Conference on the Role of Naturalism in Science*, hosted by Baylor University's Michael Polanyi Center, 12–15 April, 2000; conference schedule available online at <http://www.designinference.com/documents/2000.04.nature_of_nature.htm> (last accessed 29 June 2005).

—— and RUSE, MICHAEL (2004) (eds.). *Debating Design: From Darwin to DNA*. Cambridge: Cambridge University Press.

DENTON, M. J., MARSHALL, J. C., and LEGGE, M. (2002). 'The Protein Folds as Platonic Forms: New Support for the pre-Darwinian Conception of Evolution by Natural Law', *Journal of Theoretical Biology*, 219: 325–42.

DUHEM, PIERRE (1954). *The Aim and Structure of Physical Theory*, trans. P. P. Wiener. Princeton: Princeton University Press.

—— (1969). *To Save the Phenomena: An Essay on the Idea of Physical Theory from Plato to Galileo*, trans. E. Dolan and C. Maschler. Chicago: University of Chicago Press.

ELDREDGE, NILES, and GOULD, STEPHEN JAY (1973). 'Punctuated Equilibria: An Alternative to Phyletic Gradualism', in T. J. M. Schopf (ed.), *Models in Paleobiology*, San Francisco: Freeman, 82–115.

FORREST, BARBARA, and GROSS, PAUL (2004). *Creationism's Trojan Horse: The Wedge of Intelligent Design*. Oxford: Oxford University Press.

GONZALEZ, GUILLERMO, and RICHARDS, JAY WESLEY (2004). *The Privileged Planet: How Our Place in the Cosmos is Designed for Discovery.* Washington: Regnery.

GOULD, STEPHEN JAY (1977*a*). *Ever since Darwin: Reflections in Natural History.* New York: W. W. Norton.

——(1977*b*). 'Evolution's Erratic Pace', *Natural History*, 86/5: 12–16.

HABERMAS, GARY (2004). 'Atheist Becomes Theist: An Exclusive Interview with Former Atheist Antony Flew', *Philosophia Christi* (Winter); <http://www.biola.edu/antonyflew> (last accessed 29 June 2005).

HAROLD, FRANKLIN (2001). *The Way of the Cell: Molecules, Organisms and the Order of Life.* New York: Oxford University Press.

JOHNSON, PHILLIP (2004). 'Evolution as Dogma: The Establishment of Naturalism', in Dembski (2004), 23–40.

KEAREY, PHILIP, and VINE, FREDERICK J. (1996). *Global Tectonics.* Oxford: Blackwell Sciences.

KOESTLER, ARTHUR, and SMITHIES, J. R. (1969) (eds.). *Beyond Reductionism: New Perspectives in the Life Sciences*, Proceedings of the 1968 Alpbach Symposium. London: Hutchinson.

KUHN, THOMAS (1970). *The Structure of Scientific Revolutions*, 2nd edn. Chicago: University of Chicago Press.

LAUGHLIN, ROBERT B. (2005). *A Different Universe: Reinventing Physics from the Bottom Down* New York: Basic Books.

LENSKI, RICHARD E., OFRIA, CHARLES, PENNOCK, ROBERT T., and ADAMI, CHRISTOPH (2003). 'The Evolutionary Origin of Complex Features', *Nature*, 423 (8 May): 139–44.

LINDLEY, DAVID (1993). *The End of Physics: The Myth of a Unified Theory.* New York: Basic Books.

MARGENAU, HENRY, and VARGHESE ROY, (1992) (eds.). *Cosmos, Bios, and Theos.* La Salle, Ill.: Open Court.

MARGULIS, LYNN, and SAGAN, DORION (2002). *Acquiring Genomes: A Theory of the Origins of Species.* New York: Basic Books.

McKEON, RICHARD (1941) (ed.). *The Basic Works of Aristotle.* New York: Random House.

MEYER, STEPHEN C. (2004). 'The Origin of Biological Information and the Higher Taxonomic Categories', *Proceedings of the Biological Society of Washington*, 117/2: 213–39.

MILLER, KENNETH R. (1999). *Finding Darwin's God: A Scientist's Search for Common Ground between God and Evolution.* New York: HarperCollins.

MILLER, STANLEY (1953). 'A Production of Amino Acids under Possible Primitive Earth Conditions', *Science*, 117: 528–9.

MONOD, JACQUES (1972). *Chance and Necessity.* New York: Vintage.

MORRIS, HENRY (1975) (ed.). *Scientific Creationism.* San Diego: Creation-Life Publishers.

PEGIS, ANTON C. (1948) (ed.). *Introduction to St. Thomas Aquinas.* New York: Modern Library.

PENNOCK, ROBERT (1999). *Tower of Babel.* Cambridge, Mass.: MIT Press.

——(2001) (ed.). *Intelligent Design Creationism and its Critics.* Cambridge, Mass: MIT Press.

SAGAN, CARL (1985). *Contact.* New York: Simon Schuster.

SCHNEIDER, THOMAS D. (2000). 'Evolution of Biological Information', *Nucleic Acids Research*, 28/14: 2794–9.

SCHWARTZ, JEFFREY, and BEGLEY, SHARON (2002). *The Mind and the Brain: Neuroplasticity and the Power of Mental Force.* New York: HarperCollins.

SCOTT, EUGENIE C. (1998). ' "Science and Religion", "Christian Scholarship", and "Theistic Science": Some Comparisons', *Reports of the National Center for Science Education*, 18/2: 30–2.

——(2005). *Evolution vs. Creationism: An Introduction.* Berkeley: University of California Press.

SHAPIRO, JAMES A., and STERNBERG, RICHARD VON (2005). 'Why Repetitive DNA is Essential to Genome Function', *Biological Reviews*, 80: 1–24.

SINGER, PETER (2000). *A Darwinian Left: Politics, Evolution, and Cooperation*. New Haven: Yale University Press.

SLEVIN, PETER (2005). 'Battle on Teaching Evolution Sharpens', *Washington Post*, 14 March, A1.

THORNHILL, R. H., and USSERY, D. W. (2000). 'A Classification of Possible Routes of Darwinian Evolution', *Journal of Theoretical Biology*, 203: 111–16.

WOLF, JAKOB (2004). *Rosens Råb: Intelligent Design I Naturen, Opgør Med Darwinismen* (*The Cry of the Rose: Intelligent Design in Nature and the Critique of Darwinism*). Copenhagen: ANIS Publishers.

CHAPTER 43

THE PRE-MODERN SINS OF INTELLIGENT DESIGN

ROBERT T. PENNOCK

'To make methodological materialism a defining feature of science commits the premodern sin of forcing nature into a priori categories rather than allowing nature to speak for itself.' Do you consider this statement right or wrong? If wrong, why?

Dembski, 'The Vise Strategy'

ACTS OF GOD

In October 2004 a group of civic leaders in Dover, Pennsylvania, put in a bid to make it the next Dayton, Tennessee, the town made famous by the Scopes trial. The Dover school board voted to include intelligent design (ID) in its biology classes, and put sixty copies of the ID textbook *Of Pandas and People* (1993) in the library as instructional material. Eleven concerned parents sued the district, charging that it was unconstitutional to include creationism in the public schools. Through the month of October 2005, the *Kitzmiller* v. *Dover* case, regularly described as 'Scopes

Redux', was heard in the US District Court in Harrisburg. By happenstance, just days after the court portion of the trial ended, Dover residents went to the polls for a school board election. Eight of the incumbents were up for re-election, and every one lost to pro-science candidates. The next day televangelist Pat Robertson had a message for Dover. By voting to remove the ID supporters, the town had rejected God, he warned: 'I'd like to say to the good citizens of Dover. If there is a disaster in your area, don't turn to God, you just rejected Him from your city' (Associated Press 2005).

Robertson's comment was striking for two reasons. The first was his clear identification of intelligent design with God, something that the Thomas More Law Center attorneys defending the school board had tried to deny. The second was his suggestion, easily understood, given the recent devastation from Hurricane Katrina, that God might just send in a whirlwind. Hurricanes, tornadoes, and the like are indeed referred to as 'Acts of God'. But could it still be that people seriously propose that God might manipulate the world for such ends? Is it really possible to detect God's hand in the operations of the world?

That is the religious question that is at issue in the ID debate. ID theorists believe that they can do this, and claim to do it scientifically. ID creationists (IDCs) usually point to biology, but the patterns that they believe are signs of intelligence are also found in physics, chemistry, and meteorology. They point to the complexities of the eye, the immune system, and the bacterial flagellum, but for every system that serves a benign purpose, one finds those designed to cause suffering and devastation. No less intricate than the bacterial flagellum are the finely tuned parts of parasites that make them so effective in sucking the life from their hosts.

Such issues, known as the problem of evil, are made particularly poignant by the ID analysis of divine design. I will examine these and other theological problems shortly. But we should also face squarely the challenge posed in the epigraph. Are scientists really committing some pre-modern sin? Or might it be that IDCs should consider the beam in their own eye?

BEFORE THE ROOSTER CROWS

But wait. Why this talk of sin? Don't ID advocates object that ID is scientific, not religious? They say only that the complexities of the world are intelligently designed, but not that the designer is God. They claim ignorance of the designer's identity. In other, familiar, words: 'I don't know that man.' Before the rooster crows on any given morning, ID creationists will deny God in the public square more often than Peter. One could cite dozens of examples, but I'll mention just three general sorts of denials.

'ID is not Religion'

In a recent opinion editorial, Michael Behe, who was the main ID witness in the *Kitzmiller* trial, claimed: '[T]he theory of intelligent design is not a religiously based idea...
[I]ntelligent design itself says nothing about the religious concept of a creator' (Behe 2005). Behe's denials are hard to swallow, especially since just two paragraphs later he compares the ID argument to that of William Paley, whose watchmaker analogy is the most famous version of the design argument for the existence of God. Nor does ID stop with indirect philosophical arguments. In less guarded moments proponents of ID are more forthright that this is not just some generic higher power, but the God of the Bible.

Stephen Meyer, director of the Center for Science and Culture at the ID think tank Discovery Institute, speaks of ID as confirming 'the God Hypothesis'. Also citing Paley, he argues that the functional complexity of the world 'could not originate strictly through the blind forces of nature' (Meyer 1999: 3–4), and claims that ID theory supports 'a Judeo-Christian understanding of Creation' (Meyer 1999: 26) over all other metaphysical views.

For IDCs it is not Genesis that is the key, but the Gospel of John. As William Dembski, a Discovery Institute Senior Fellow also on the faculty at Southern Baptist Theological Seminary, explains, 'Intelligent design is the Logos of John's Gospel restated in the idiom of information theory' (Dembski 1999*b*: 84). And Phillip Johnson, godfather of the ID group, explains: 'Our strategy has been to change the subject a bit so that we can get the issue of intelligent design, which really means the reality of God, before the academic world and into the schools' (American Family Radio, 10 January 2003).

Protestations to the contrary notwithstanding, ID is religious to its core. The movement's goals, as adumbrated in 'The Wedge Strategy', a leaked internal manifesto from the Discovery Institute, are 'nothing less than the overthrow of materialism and its cultural legacies', and to re-establish 'a broadly theistic understanding of nature' and the proposition 'that human beings are created in the image of God' (Discovery Institute 1999).

In my own expert witness report for the *Kitzmiller* case, I documented ID's sectarian nature, further showing that it is religious simply by virtue of its essential appeal to an immaterial, supernatural designer (Pennock 2005). The definition of religion I assumed was the same as was critical in the ruling of the US Supreme Court in the 1987 *Edwards* v. *Aquillard* case, which found that creationism unconstitutionally endorsed religion 'by advancing the religious belief that a supernatural being created humankind'.

'ID is not Creationism'

Although they used the term in early writings, ID proponents today regularly deny that they are creationists. What exactly are they denying, and why? There are many different forms of creationism, but the generic notion is the rejection of the scientific account of evolution in favour of creation by some supernatural power or being. I have documented in detail elsewhere how ID is connected to creation science and other forms of creationism in its roots, strategies, and arguments (Pennock 1999), so here I will just let one ID proponent briefly tell the story. Nancy Pearcey, one of the authors of *Of Pandas and People*, describes how Johnson formulated ID as a strategy to unite various creationist factions.

[I]nstead of joining together to oppose the hegemony of the naturalistic worldview, Christians often got caught up in fighting each other. The bitterest debates were often not with atheistic evolutionists but among believers with conflicting scientific views: young-earth creationists, old-earth creationists, flood geologists, progressive creationists, 'gap' theorists, and theistic evolutionists. There were endless arguments over theological questions like the length of the creation 'days' and the extent of the Genesis flood.... It was Johnson himself, more than anyone else, who refocused the debate and brought about a rapprochement of the warring camps under the umbrella of the Intelligent Design movement. (Pearcey 2004: 173)

Pearcey herself was instrumental in helping to bring about this alliance, as were other authors of *Pandas*, including Meyer and Behe. Their negotiations were made easier by one important external factor.

Among the revelations of the *Kitzmiller* trial were details of the switch from the language of creation science to that of ID. The plaintiffs subpoenaed draft manuscripts of *Pandas*. In a series of drafts that changed titles from 'Creation Biology' (1983), to 'Biology and Creation' (1986), to 'Biology and Origins' (1987), and then several versions under the final title *Of Pandas and People* (1987), one could observe the terminological shift. Here is one example: 'Creation means that the various forms of life began abruptly through the agency of an intelligent creator with their distinctive features already intact—fish with fins and scales, birds with feathers, beaks, and wings, etc.' (quoted in Matzke 2005a). This sentence appears in the first four drafts through one version in 1987, but then in a fifth draft later in 1987 the wording changed abruptly: 'Intelligent design means that various forms of life began abruptly through an intelligent agency, with their distinctive features already intact—fish with fins and scales, birds with feathers, beaks, and wings, etc.' (quoted in Matzke 2005a).

What happened in 1987 that occasioned this linguistic fig leaf? That is when the *Edwards* case was decided, finding it unconstitutional to teach creation science in the public schools. In subsequent drafts leading up to the published text, 'creation science' became 'design theory', and 'creationists' became 'design proponents'. The new terms were substituted in an almost search-and-replace manner.

Barbara Forrest, an expert witness for the plaintiffs who examined the manuscripts, even turned up what is now humorously referred to as the 'Missing Link' between creationism and intelligent design—a sentence in the second 1987 draft that includes an accidental transitional form 'cdesign proponentsists' (Matzke 2005b).

'We're not saying it was God … could be extraterrestrials'

Despite the long paper trail demonstrating otherwise, creation scientists—that is, design proponents—still often deny God as the designer, claiming that it could have been extraterrestrial beings. If this is what they really hold, perhaps Robertson misspoke; were the voters of Dover rejecting not God, but ET?

However, if natural intelligent beings existed elsewhere in the universe, one would need to inquire about their origin as well. It will do no good to simply posit a prior natural designer. This tells us that ID proponents must assume that at least at one point in the chain the designer was supernatural. Thus, their public appeals to extraterrestrials are simply disingenuous.

FOOLISH, WICKED IDOLATERS

In seeking to detect divine action in the world, creationists are inspired by the words of Romans 1: 20: 'Ever since the creation of the world, his invisible attributes of eternal power and divinity have been able to be understood and perceived in what he has made' (New American Bible). Creationists of all stripes regularly cite this passage to justify their belief that God may be detected in creation. But this also reflects what can only be seen as a deep ambivalence towards science and scientists. They claim that ID scientifically demonstrates the glory of God, but they also believe that scientists are refusing, either through self-deception or wickedness, to recognize their discovery, and so they also cite Romans 1 to lay blame on those who claim not to see such design. Such unbelievers, as the continuation of the passage says, are 'without excuse'.

Johnson cited Romans when describing the success of ID in demonstrating what he took to be the essential Christian view regarding human origins:

I would say that the theistic and Biblical worldview has been tremendously validated. That is to say it's been validated in the sense that you do need a creator after all, but even more, what's been validated is the biblical view that it's a major part of the human project to get rid of the creator; because their deeds were evil, they did not want to honor god as God, and so instead they imagined various forms of idolatry and nature worship of which Darwinian evolution is just the most prevalent modern form. So, at this point, you say that not only has it been revealed that science points to the reality of a creator after all, but the enormously bad and self-deceptive thinking of the Darwinian evolutionist is something straight out of Romans 1. (Lawrence 1999)

Besides noting this as one more example of how ID is not just a religious but a biblical view, it is important to recognize what Johnson is implying here. Romans 1 says not only that unbelievers are without excuse, but that they are foolish and wicked haters of God—idolaters who 'deserve death' (Romans 1: 29–32). If one believes that these passages apply to 'the Darwinian evolutionist' in the way that Johnson implies, then it is but a small step to the views expressed by Robertson (Romans 1: 18).

Given that ID proponents think of scientists as foolish, wicked idolaters, it is ironic that they wish to have their own beliefs validated by science.

The Imprimatur of Science

IDCs regularly complain that their critics label ID as religious to discredit it. This may be true for William Provine and Richard Dawkins, the two atheist scientists who serve as the regular foils for ID, but they are hardly representative; the most consistent critics of ID value religion. The creationists' complaint thus reveals an interesting irony. Scientists may or may not be wicked fools, but creationists recognize that calling something 'scientific' carries a seal of trustworthiness by virtue of the unquestionable success of science. By comparison, claims of religious knowledge are not felt to carry the same weight; creationists are diffident about their faith. What is really going on is that ID proponents, in the same way as creation scientists before them, hope to have their particular religious beliefs legitimated by science. They value the imprimatur of science for its potential apologetic utility.

In a recent paper posted on the Internet, Dembski writes that ID 'is ultimately a scientific controversy within the scientific community. To be sure, there are educational, political, religious, and philosophical aspects to this controversy, but if there were no scientific controversy here, these other aspects would never have gotten off the ground' (Dembski 2005b). In fact, the opposite is true. I spent many years warning scientists about the ID movement, but the common reaction was to dismiss ID arguments as ignorant and unscientific. Few took IDCs seriously. If ID had not become a political and educational threat, scientists would have continued to ignore it.

For ID proponents to say that their view is a live controversy within science is self-deception of the first order. The ID leadership consists of a handful of individuals who have simply repeated old challenges to evolution and have failed to offer any positive evidence for their own view. IDCs often quote Charles Darwin's comment that 'A fair result can be obtained only by fully stating and balancing the facts and arguments on both sides of each question' (Darwin 1859, and 1871: 12). Americans are always sympathetic to appeals to fairness, but here the appeal is bogus. It is no more 'fair' to include arguments about whether evolution is true than to continue to consider whether the Earth revolves around the Sun. The scientific community has thoroughly examined ID and found it wanting. Dozens of recent statements from scientific organizations ranging from the American Association for the Advancement of Science to groups of Nobel laureates attest to this, but I'll mention just one representative example from the American Society of Agronomy:

Intelligent design is not a scientific discipline and should not be taught as part of the K–12 science curriculum. Intelligent design has neither the substantial research base nor the testable hypotheses as a scientific discipline. There are at least 70 resolutions from a broad array of scientific societies and institutions that are united on this matter.

Such unity is not significant just because of the sheer numbers; science is not, after all, decided by a vote, but by assessment of the evidence. What is significant is the common conclusion that ID has no evidence for its grand conclusions, and that it is not science. For ID proponents to pretend otherwise is certainly foolish. Theologically, one might also say that their programme is idolatrous, elevating science above faith. Others must judge whether it is also wicked.

QUEEN OF THE SCIENCES?

As revealed in the Wedge document, the ID movement aims to overturn the secular, scientific world-view. It hopes for a renewal of what is really the old medieval view of theology as queen of the sciences. ID proponents admit that for their view to prevail requires a change of 'the ground rules by which the natural sciences are conducted' (Dembski 2004b: 19). Their primary goal is to appeal to immaterial, transcendent causes, but to still call it 'science'. In my *Kitzmiller* report, I objected to this:

A famous philosopher posed the following question: If you call a tail a leg, how many legs does a dog have? The answer, he said, is *Four*; calling a tail a leg doesn't make it one. Calling intelligent design creationism an 'alternative scientific theory' and using scientific-sounding terminology does not make it a science now any more than it did for creation science. ID theory rejects both fundamental conclusions and basic methodological constraints of science. It posits an unnamed and undescribed supernatural designer as its sole explanatory principle. It provides no positive evidence for its extraordinary claims. And because it cannot stand on the evidential ground that science requires it tries to change the ground rules of science. Even by its own lights, ID theory is not science. (Pennock 2005)

Dembski was initially listed as an expert witness for the case, and so received my report. He has since taken my argument (and indeed my very phrasing, though usually without attribution) and tried to turn it around. In one instance he wrote that 'calling a tail a leg doesn't change that fact that a dog still only has four legs. Likewise, backing away from standard terminology and assigning to themselves other labels doesn't change the fact that most evolutionists are indeed Darwinists.' Why? Because, that supposedly is appropriate for anyone 'who holds that teleology ought to play no substantive role in evolutionary theory' (Dembski 2005a).

This is misleading in several ways. Contrary to Dembski's claim, evolutionary theory does include a substantive role for teleology; indeed, it is at the core of Darwin's law—namely, in the way in which adaptations are explained by natural selection. But this is a perfectly scientific notion of teleology, which is not what Dembski and company are speaking of. They have in mind a more cosmic teleology; they want to wedge supernatural purposes substantively into science. It is IDCs who define Darwinism in a non-standard, non-scientific way, by improperly building atheism into the very concept. Johnson even speaks of this strategically, as he did at D. James Kennedy's Coral Ridge Ministries: 'The objective, he said, is to convince people that Darwinism is inherently atheistic, thus shifting the debate from creationism vs. evolution to the existence of God vs. the non-existence of God. From there people are introduced to "the truth" of the Bible and then "the question of sin" and finally "introduced to Jesus"' (Boston 1999).

The question of sin and salvation is important to IDCs, as it was to creation scientists; they hold that if evolution is true, then there is no God, no basis for morality, and no purpose to life. Creationists even take the possibility of a metaphysical afterlife to be at stake. Johnson gave the ID answer when he was asked why he focused his attention upon Darwinism: 'I wanted to know whether the fundamentals of the Christian worldview were fact or fantasy. Darwinism is a logical place to begin because, if Darwinism is true, Christian metaphysics is fantasy' (quoted in 'Berkeley's Radical: An Interview with Phillip E. Johnson', 2002). In other words, should one take seriously the promise of a real life after death, or is this just fantasy? Without the hope or fear of an afterlife, people will behave, they seem to think, as if nothing really matters. I have previously addressed the issue of what creationists take to be at stake in the debate, and why their concerns are misplaced, and I will not rehearse those arguments again here (Pennock 1999: ch. 7). Suffice it to note that evolutionary biologists use exactly the same method as any other scientists, so to call Darwinism 'inherently atheistic' is a deceptive redefinition of terms. I will return to this issue in a moment, but let me first note another way in which Dembski attempts to turn the tables.

In Dembski's confidential briefing paper for the Thomas More Law Center attorneys, which he released after the trial, he suggested that they turn my point around in a slightly different way, and ask the following questions of me and the other expert witnesses during cross-examination:

Consider the following riddle (posed by Robert Pennock): if you call a tail a leg, how many legs does a dog have? Wouldn't you agree that the answer is four: calling a tail a leg doesn't make it one. Accordingly, wouldn't it be prejudicial to define nature as a closed system of material entities in which everything happens by material causation? Wouldn't you agree that nature is what nature is, and it is not the business of scientists to prescribe what nature is like in advance of actually investigating nature? (Dembski 2005c)

It is at this point that Dembski recommends that the attorneys pose his question about the supposed 'premodern sin of methodological materialism' quoted in the epigraph. No vise to the head is needed to make us forthrightly answer Dembski's question. Am I and are other scientists pre-modern sinners in his sense of the term?

Certainly not. Methodological naturalism does not define away any metaphysical possibilities or constrain the world; rather, it constrains science. Methodological naturalism is neutral with regard to supernatural possibilities. It takes a more humble view of what can be known. Science admits that it may miss true metaphysical facts about the world. Methodological naturalism does not claim access to all possible truths. Indeed, it expressly limits the purview of what can be known scientifically. If there are metaphysical truths beyond empirical test, then they are beyond science. While a few scientists may stake out a stronger metaphysical position, arguing that science provides an ultimate theory of everything, such *scientistic* (as opposed to scientific) views go beyond the evidence. Dembski's challenge improperly conflates methodological with metaphysical naturalism, and thereby equates science (not just evolution) with atheism.

Before continuing, this is a good place to dismiss a common red herring. Contrary to IDCs' assertions, science does not 'rule out all design'; in fact, science regularly allows us to draw inferences that, for instance, something was man-made. The constraints of methodological naturalism together with background knowledge allow archaeologists to draw all sorts of conclusions about peoples of the past. Forensic scientists may similarly find clues (e.g. fingerprints) that allow them to tell who pulled the trigger of the gun. It is telling that IDCs always crib from ordinary natural cases of agency when trying to motivate their inference to a transcendent world-maker. We see this too in the following common IDC argument, here made by Discovery Institute Fellow David DeWolf: '[W]hen we go to Mt. Rushmore we immediately recognise that what appear to be the faces of the Presidents are not the product of the random forces of erosion and rockslide' (DeWolf 1999). However, it is not science but ID that is caught in this landslide. This is a faulty analogy, for it is missing the key (non-random) process of natural selection. Moreover, the positive inference that we make that someone carved the faces on the mountain is a perfectly natural one. The Mount Rushmore case is a particularly poor example for ID, if only because it is so *unlike* cases of biological complexity.

Behe's notion of irreducible complexity and Dembski's account of specified complexity, upon which their design argument purportedly rests, have both been thoroughly critiqued and rebutted, and there is not space here to review the many problems with these general notions (see Pennock 2003). Instead, let us look at one of the few examples—a chessboard analogy—that Dembski offers to illustrate their supposed application (Dembski 2005*b*).

He shows a chessboard with all the pieces in standard starting positions, except that White's right four pawns (e2, f2, g2, and h2) are situated directly in front of the left four (at positions a3, b3, c3, and d3). He points out that one cannot get to this position from the standard starting position by the rules of chess. The purported lesson is that there may be material configurations that are not explainable in material terms (i.e. initial conditions and law-governed processes). Dembski concludes by once again trying to turn my argument around (again without attribution):

It follows that the charge of supernaturalism against intelligent design cannot be sustained. Indeed, to say that rejecting naturalism entails accepting supernaturalism holds only if nature is defined as a closed system of material entities ruled by unbroken laws of material inter-action. But this definition of nature begs the question. Nature is what nature is and not what we define it to be. To see this, consider the following riddle: how many legs does a dog have if one calls a tail a leg? The correct answer is four. Calling one thing another thing doesn't make it something else. (Dembski 2005*b*)

This is one of the most astonishing of many instances of creationists' creative redefinition. IDCs have insisted for years that no natural processes can produce biological complexity, and that design must therefore come from outside the system of nature; but they now want to say that they are not appealing to the supernatural by redefining the supernatural as part of nature. This is either a blatant self-contradiction, or else it is reconceiving God as a natural being. The former is a logical sin, and some would judge the latter a theological sin. We have already seen why methodological naturalism is not guilty. In fact, it is Dembski's argument that begs the question in at least four significant ways.

First, his chess analogy portrays the world as a game, which subtly presumes that there is a player who set the rules. The world does appear to be governed by rules, but these may or may not have been designed. Theologically we may hold that they are, but scientifically we are in no position to say either way; such possible transcendent purposes cannot just be read off the world. As I have argued before, this is a fundamental problem for the IDCs' supposed design inference (Pennock 1999: ch. 5).

Second, Dembski takes the canonical chess rules for granted, but there are any number of non-standard chess rules and non-standard starting positions that gamers and puzzlers also play which he fails to consider. How could Dembski know just by looking at the board that canonical chess rules were governing the pieces? Did the rule designer tell him? Instead of concluding that this is an inexplicable chess pattern, why not infer that another game is being played?

Third, nothing in the example demonstrates any need for a designer or a telic process, or even that these would be relevant. Why should we think that the pattern is purposeful at all? Did the designer say so? If the given pattern is supposed to be seen as a target, it is only because Dembski has drawn a circle around it himself. Again, he assumes what is at issue.

Fourth, Dembski's example assumes a hypothetical epistemological state that is completely at odds with our real one. Scientists start with knowledge neither of the rules (that is, the complete laws of nature) nor the starting position (that is, the initial conditions of the universe). These are what the scientist seeks to discover, to the degree that they are discoverable. Dembski's illustration assumes what is impossible for mere human beings—prior omniscience about the game and the board. But by this time, it should come as no surprise that IDCs presume such a God's-eye view.

I will make one final remark about Dembski's general challenge regarding the purported sin of science, which is to note how he misdefines methodological naturalism as a 'pre-modern' sin, no doubt for rhetorical purposes. If methodological naturalism is a sin at all, it is a modern sin, for it is the adoption of this methodology of science—of natural philosophy—that initiates and most characterizes the modern era. (IDCs do recognize this elsewhere, which is why the Wedge document speaks of their ultimate target as not just evolution or even just science, but 'modernism' itself.) The pre-modern sins are different; they are the sins of appeal to the occult and of claims to know the ways of the gods. As we have seen, these pre-modern sins are at the core of the ID argument.

Thunderbolts and Lightning

This takes us back to the meteorological cases with which we began and to the issue of divine action and responsibility. Can we really look up at the skies and see God's actions? Robertson seems to think that Dover citizens would be to blame for spitting in God's eye, but it would ultimately be God who is responsible for directing the winds. If we are to take ID programme seriously, we cannot stop with happy examples of benign designs; we must also confront the all-too-common cases where the complex designs of the world conspire, whether by pestilence or by storm, to inflict suffering. If a designer is to get the credit for the good, it also deserves the blame for the bad.

Dembski may be concerned about this possibility, for he tries to limit the scope of divine action:

A worry now arises whether effect-to-cause reasoning leads to many absurd design hypotheses. Consider the 'Zeus Hypothesis' in which lightning strikes are attributed to the divine intervention of the god Zeus (I'm indebted to Robert Pennock for this example). Such a hypothesis, though an example of effect-to-cause reasoning, would not be the conclusion of a design inference based on specified complexity. (Dembski 2005b)

Dembski fails to provide a reference (this passage, too, was taken from my expert report), so readers are not able to check and see that my point was somewhat different from the one he takes on. However, Dembski's passage usefully reveals some of the problems with his ID inference, so let us stick with it. Dembski continues his explanation:

Individual lightning strikes are readily explained in terms of the laws of physics, with no need to invoke a designer. The only way lightning strikes might require an ID hypothesis is if jointly they exhibit some particularly salient pattern. Consider, for instance, the possibility that on a given day all, and only, those people in the United States who had uttered snide remarks about Zeus were hit by lightning and died. In that case, the joint pattern of lightning strikes would exhibit specified complexity, and the Zeus Hypothesis might no longer seem altogether absurd. (Dembski 2005b).

IDCs regularly cite Dembski's specified complexity method of design detection as though it were a matter of mathematics or information theory, but in fact his design inference is a flawed and empty formalism (Pennock 2003). In the few cases when he gives a specific example of its supposed application, one quickly sees how substantive knowledge gets smuggled in. Invariably the examples trade on tacit natural assumptions; this case involves presuming that Zeus is peevish and vindictive, and treating him in human terms. Here, as elsewhere, to make his inference appear plausible, Dembski illicitly uses naturalized concepts. But how does Dembski know what a god takes to be snide, or that he would care, or that he would not be forgiving, and so on?

Robertson also believes that he can identify such a relevant specified pattern. The patterns he sees are no less precise than Dembski's speculations about what Zeus would take to be snide. But once one opens the door to the mysterious designs of an untestable supernatural being—be it Zeus, Yahweh, Allah, or Beelzebub—it is all too easy to read in a meaningful pattern. And what if there are apparent exceptions to some possible specification? Why was Mary struck even though no one heard her say anything snide about Zeus? She must have kept her snide thoughts to herself, or perhaps she did something else that rubbed Zeus the wrong way. Why wasn't John swept away with the others sinners in New Orleans? Perhaps he had done some good deed that deserved reward, or perhaps God simply had other plans for him. Dembski's supposed inference relies upon presuming that we can know the mind of God.

Another revealing point in this example is how Dembski conveniently speaks with hindsight, long after Benjamin Franklin's discoveries about the nature of lightning. But consider the situation beforehand. Specified patterns of lightning bolts equivalent to Dembski's example were readily apparent. Ironically, lightning seemed to be aimed rather pointedly at churches. Even individual bolts may exhibit a striking pattern; one eyewitness described how 'A bolt of lightning had struck the tower, partly melting the bell and electrocuting the priest; afterwards, continuing, it had shattered a great part of the ceiling, had passed behind the mistress, whom it deprived of sensibility, and after destroying a picture of the Savior hanging upon the wall, had disappeared through the floor' (quoted in Seckel and Edwards 1984). Or one might look to a broader pattern and note how it seemed that all church bell-ringers were being picked off one by one. In one thirty-three-year period in Germany, nearly 400 church towers were damaged by lightning, and 120 bell-ringers were killed (Seckel and Edwards 1984). Did God have designs on steeples and bell-ringers? Dembski apparently would have attributed this to Zeus.

Franklin's scientific work provided a natural explanation of lightning in terms of electrostatic discharge and also why it was likely to hit steeples, usually the tallest structures in a town. But Dembski cannot appeal retrospectively to Franklin's theory to save his design inference from absurdity. Indeed, he says that once one concludes design by identifying specified complexity, the inference can never be reversed. Incredibly, Dembski claims that his design inference gives no false positives. He adds that it would be worthless if it did (Dembski 1999a: 141).

The example illustrates the faulty reasoning of the ID approach: it tries to get a substantive conclusion—and not just a mundane one, but one about the designs of immaterial, transcendent powers—from gaps in scientific understanding. Behe (2005) summarizes the logic of their argument: 'in the absence of any convincing non-design explanation, we are justified in thinking that real intelligent design was involved in life'. Dembski simply hides this naïve negative argument within a formalism. Creationists have always proposed exactly the same fallacious argument, hoping to win by default without having to provide any positive evidence for a creator. Once one understands this, it is clear why the 'content' of 'ID theory', like that of creation science, is at base no more than a litany of supposed 'gaps', 'weaknesses', or 'problems' with evolution. ID is thus no more than what philosophers call an argument from ignorance.

Of course there are gaps in our understanding of evolution. There are any number of things that science cannot explain, and scientists are not shy about highlighting unanswered questions for which further research is needed. And there may be questions that science might never be in a position to answer; methodological naturalism is humble. Evolution is completely typical in not having all the answers; the same can be said for any theory in science. Meteorologists do not have a complete explanation of the incredible complexity of the weather system.

Should they therefore include ID as an alternative theory? Must we once again teach that God has a hand in thunderstorms and hurricanes, shooting arrows of lightning and directing the destructive wind any way it blows? Such an idea is very, very frightening.

BUILT ON SAND

The observation that ID is no more than an argument from ignorance brings us to a parallel theological problem. Theologians have long held that it is a serious mistake to posit God as plugging the holes in scientific knowledge, given that science regularly closes what might seem to be unfillable gaps. Nearly everyone who has looked closely at the ID arguments has criticized them on this ground. I discuss these issues in detail, especially regarding to how they relate to methodological naturalism in Pennock (2007).

I have already highlighted another major theological problem with ID: namely, the problem of evil. How can God be both omnipotent and omni-benevolent given such obvious suffering in the world? Blame for the cruelty and wastefulness of the biological world is laid much more directly upon a designer who, on the ID view, fashions each deadly 'irreducibly complex' pathogen and each 'finely tuned' debilitating parasite.

A third problem with ID that troubles theologians is its rejection of the view— theistic evolution—that evolution and faith are compatible. This mainstream Christian position makes it clear that IDCs are off base theologically in equating evolution and atheism. Indeed, some argue that Darwinism is a 'disguised friend' of Christian theism in that it highlights the sacramentality of nature and God's immanence in the world (Peacocke 2001). This view goes back to Darwin's own time, when Christians were early defenders of evolution (Livingstone 1987). But IDCs will have none of this. Johnson says that theistic evolution is a 'disastrous accommodation' to Darwinism which provides 'a veneer of biblical and Christian interpretation...to camouflage a fundamentally naturalistic creation story' (Johnson 2002*b*: 137). Theistic evolutionary views, he writes elsewhere, are 'bogus intellectual systems' that read the Bible 'figuratively rather than literally' (Johnson 1997: 111). Dembski is equally adamant, writing that '*Design theorists are no friends of theistic evolution*' (Dembski 1995: 3; emphasis original).

That the scientific community rejects ID is well known; what is less appreciated is that religious leaders are also dismissive, for reasons such as those noted above and more. Dembski admitted this years ago, writing: 'It is ironic that the design theorists have received an even cooler reception from the theological community than from the Darwinist establishment' (Dembski 1995: 1). Even evangelicals have not been as welcoming as one might have expected; Johnson says that he encounters 'bitter opposition' from professors at evangelical Christian schools (Christianbook.com 2000). If anything, theologians have become even more critical of ID over the years. Dembski recently noted that 'The president of the *Institute for Religion in an Age of Science*, Michael Cavanaugh, has now issued a formal warning about intelligent design, the Wedge, and Seattle's Discovery Institute, urging that people take seriously the threat to education and democracy that these pose' (Dembski 2004*a*). Cavanaugh (2004) minces no words: ID is 'totalitarian religious thought'.

Onward Christian Soldiers

Such quotations highlight one more sin of ID: namely, how it is reigniting old animosities between science and religion. This is the kind of view that led to the persecution of Galileo. Galileo himself thought that revealed truths in Scripture should not be interpreted to contradict the truths discovered by science, but ID proponents believe that one should *start* with 'what Christians know' (Plantinga 1997). IDCs regularly describe their movement as being in the front line of a 'culture war'. In their writings against evolution, they regularly use the metaphors of violence, war, and torture.

Dembski (2002) says that the 'Darwinian stranglehold' on public education will suffer a 'Taliban-style collapse', that his opponents have 'met their Waterloo' (Kern 2000), and so on. He recently recommended that biologists be subjected to what he called 'the Vise Strategy', to squeeze the truth out of them under oath, depicting this notion with 'humorous' images of people or, in one case, a stuffed Charles Darwin doll, with their heads being squashed in a vise. Johnson wonders whether atheists like Dawkins believe as strongly in their cause as religious believers, and whether he would have the requisite courage of his convictions comparable to that of the fundamentalists who blew up the World Trade Center.

A man who believes in something that is more important to him than life itself is potentially a dangerous man. He may do things that a person with more mundane purposes would never think of doing. This is true of secular as well as religious faiths. Consider, for example, the American Revolutionary War patriot Nathan Hale, who famously regretted that he had but one life to give for his country. Such a person might be capable of a suicide attack, given a sufficiently worthy end. (I would like to ask Dawkins if he might be capable of sacrificing his own life in an act of murderous violence if he were convinced that such an extreme measure was necessary to save science from being taken over by religious fundamentalists.) (Johnson 2002*a*)

Some ID leaders do claim something close to this level of zeal. In an unapologetic call to martyrdom, Dembski and Jay Wesley Richards, another Discovery Institute Fellow, write: 'this is our calling as Christian apologists, to bear witness to the truth, even to the point of death (be it the death of our bodies or the death of our careers)'. To be worthy apologists and to never give in to the ground rules set by the secular academy, they explain, is 'perhaps not a martyrdom where we spill our blood (although this too may be required)' (Dembski and Richards 2001: 15).

TROUBLING THEIR OWN HOUSE

This brings me to a few final considerations. IDCs want to find a special sort of purpose in the world. They claim to detect not just the natural kind of teleology found throughout evolutionary biology, but final causes of a more cosmic variety. They have confirmed 'the God hypothesis'. However, at the same time they claim that ID is not religion but science. In the *Kitzmiller* trial, defence attorneys tried to argue that ID was not inherently religious, but merely had religious 'implications'. But, as we have seen, much as its proponents deny God in the public square, ID is religious through and through. The 'implications' they take from their arguments emerge only because the religious elements are already part of their premises.

Those of us who have been critics of the movement might have viewed the ID proponents differently had they entered honestly into the long and honourable philosophical debate about the teleological argument. But ID is really a political movement that sees itself on the front line of the culture wars, and so is strategically deceptive, claiming to fight not in the name of God, but perhaps in the name of the Martians.

In the end, this debate between methodological naturalism and ID comes down to whether science should discuss plagues and hurricanes as natural disasters or as acts of God. I have argued that science is, and properly should remain, neutral with regard to divine possibilities. IDCs insist that their science can detect the workings of a transcendent mind in such complexities. But we have seen a few of the sins of their view. They are disturbing a delicate cessation of hostilities between science and religion. The lesson from one dramatic retelling of the Scopes trial might well be applied here: those who are tempted by the pre-modern sins of the ID movement shall inherit the wind.

REFERENCES AND SUGGESTED READING

Associated Press (2005). 'Pat Robertson warns Pa. town of disaster', *Seattle Post-Intelligencer*, 10 November.

BEHE, M. (2005). 'Design for Living'. *The New York Times*, 7 February.

'Berkeley's Radical: An Interview with Phillip E. Johnson'. (2002). *Touchstone: A Journal of Mere Christianity*.

BOSTON, ROB (1999). 'Missionary Man', *Church & State*, 52/4: 14–15.

CAVANAUGH, M. (2004). *Statement from Michael Cavavaugh*, from <www.creationismst rojanhorse.com#Cavanaugh_Endorsement>, retrieved 29 November 2005.

Christianbook.com (2000). *Interview: Phillip Johnson*, from <http://www.christianbook.com/Christian/Books/dpep/interview.pl/16559901?sku=22674>, retrieved 15 January 2005.

DARWIN, C. (1859 and 1871). *The Origin of Species and The Descent of Man*. New York: Random House.

DAVIS, P., and KENYON, D. H. (1993). *Of Pandas and People*. Dallas: Haughton Publishing Co.

DEMBSKI, W. A. (1995). 'What Every Theologian Should Know about Creation, Evolution, and Design', *Center for Interdisciplinary Studies Transactions*, 3/2: 1–8.
—— (1999a). *Intelligent Design: The Bridge between Science and Theology*. Downers Grove, Ill.: InterVarsity Press.
—— (1999b). 'Signs of Intelligence: A Primer on the Discernment of Intelligent Design', *Touchstone*, 12/4: 76–84.
—— (2002). 'Commentary on Eugenie Scott and Glen Branch's Guest Viewpoint: "Intelligent design" Not Accepted by Most Scientists'; <http://www.designinference.com/documents/2002.07.Scott_and_Branch.htm> retrieved 8 August 2002.
—— (2004a). 'Dealing with the Backlash against Intelligent Design'; from <http://www.designinference.com/documents/2004.04.Backlash.htm>, retrieved 29 May 2005.
—— (2004b). *The Design Revolution: Answering the Toughest Questions about Intelligent Design*. Downers Grove, Ill.: InterVarsity Press.
—— (2005a). 'Backing away from the term "Darwinism"', Web posting.
—— (2005b). 'In Defense of Intelligent Design', Web posting.
—— (2005c). *The Vise Strategy: Squeezing the Truth out of Darwinists*.
—— and RICHARDS, J. W. (2001). 'Introduction: Reclaiming Theological Education', in W. A. Dembski and J. W. Richards (eds.), *Unapologetic Apologetics: Meeting the Challenges of Theological Studies*, Downers Grove, Ill.: InterVarsity Press, 11–27.
DEWOLF, D. K. (1999). 'Teaching the Origins Controversy: A Guide for the Perplexed'; <http://www.discovery.org/crsc/articles/article6.html>, retrieved 13 October 1999.
Discovery Institute (1999). 'The Wedge Strategy 2002'; <www.stephenjaygould.org/ctrl/archive/wedge_document.html>, retrieved 5 August 1999.
JOHNSON, P. E. (1996). 'Starting a Conversation about Evolution'; <http://www.arn.org/docs/johnson/ratzsch.htm>, retrieved 9 January 2005.
—— (1997). *Defeating Darwinism*. Downers Grove, Ill.: InterVarsity Press.
—— (2002a). 'Recognizing the Power of Religion', *Touchstone*, 15/2: 12; <http://www.touchstonemag.com/docs/issues/15.2docs/15-2pg12.html>.
—— (2002b). *The Right Questions: Truth, Meaning and Public Debate*. Downers Grove, Ill.: Intervarsity Press.
KERN, L. (2000). *In God's Country*. Houston: Houston Press.
LAWRENCE, J. (1999). 'Communique Interview: Phillip E. Johnson', *Communique: A Quarterly Journal* Spring; <http://communiquejournal.org/q6_johnson.html>.
LIVINGSTONE, D. N. (1987). *Darwin's Forgotten Defenders: The Encounter between Evangelical Theology and Evolutionary Thought*. Edinburgh: Scottish Academy Press.
MATZKE, N. (2005a). 'I guess ID really was "Creationism's Trojan Horse" after all', Web posting.
—— (2005b). 'Missing Link discovered!'; <http://www2.ncseweb.org/wp/?p=80>, accessed 7 November 2005.
MEYER, S. C. (1999). 'The Return of the God Hypothesis', *Journal of Interdisciplinary Studies*, XI/1–2: 1–38.
PEACOCKE, A. R. (2001). 'Welcoming the "Disguised Friend"—Darwinism and Divinity', in R. T. Pennock (ed.) *Intelligent Design Creationism and its Critics: Philosophical, Theological and Scientific Perspectives*, Cambridge, Mass.: MIT Press, 471–86.
PEARCEY, N. (2004). *Total Truth: Liberating Christianity from its Cultural Captivity*. Wheaton, Ill.: Crossway Books.
PENNOCK, R. T. (1999). *Tower of Babel: The Evidence against the New Creationism*. Cambridge, Mass.: MIT Press.
—— (2003). 'Creationism and Intelligent Design', *Annual Review of Genomics and Human Genetics*, 4: 143–63.

PENNOCK, R. T. (2005). *Kitzmiller et al.* v. *Dover Area School District—Expert Report*; <http://www2.ncseweb.org/kvd/experts/pennock/pdf>.

——(2007). 'God of the Gaps: The Argument from Ignorance and the Limits of Methodological Naturalism', in A. J. Petto and L. R. Godfrey (eds.), *Scientists Confront Creationism: Creation Science, Intelligent Design and Beyond*, New York: W. W. Norton.

PLANTINGA, A. (1997). 'Methodological Naturalism?', *Perspectives on Science and Christian Faith*, 49/3: 143–54.

SECKEL, A., and EDWARDS, J. (1984). 'Franklin's UnHoly Lightning Rod', repr. in *ESS Journal*, 25 Nov. 2002; <http://www.esdjournal.com/articles/franklin/franklinrod.htm>.

THEOLOGIES OF EMERGENT COMPLEXITY AND THEIR CRITICS

CHAPTER 44

PHYSICS, COMPLEXITY, AND THE SCIENCE– RELIGION DEBATE

GEORGE F. R. ELLIS

PHYSICS AND THE EVERYDAY WORLD

The extraordinarily successful reductionist approach of present-day physics is based on the concept of an isolated system. Experiments carried out on such systems enable the physicist to isolate and understand the fundamental causal elements underlying physical reality. However, no real physical or biological system is in fact isolated, either physically or historically; biological systems are open systems (N. A. Campbell 1991), and in the real world context matters as much as laws (Bishop 2006). The physics approach tends to ignore three crucial features that enable the emergence of biological complexity out of the underlying physical substratum (Ellis 2004, 2005a; 2005b; Bishop 2005; Roederer 2005): namely, top-down action in the hierarchy of complexity, which affects both the operational context and the nature of constituent parts; the causal efficacy of stored information ('memory effects'); and the origin of biological information through evolutionary adaptation. These features enable the causal efficacy of emergent biological order, described by phenomenological laws of behaviour at each level of the hierarchy. Thus what occurs is contextual emergence of complexity (Bishop 2005): the higher-level laws emerge out of the underlying physics, which establishes a possibility landscape (Ellis 2004) delineating possible ways of creating biological functionality. However, the higher levels in the hierarchy of complexity have real autonomous causal powers, functionally independent of lower-level processes. The underlying physics both enables and constrains what is possible at the higher levels, creating the possibility space of outcomes, but it does not enable us to actually predict events in the everyday world around us (for example, future prices on the New York Stock Exchange), where human intentionality is causally effective. More than this, physics *per se* does not even uniquely causally determine the specific outcome of the higher-level functioning.

True complexity, with the emergence of higher levels of order and meaning, including life, occurs in modular hierarchical structures (Simon 1962; Booch 1994). They are structured, in that their physical nature reflects a precise ordering as in very large intricate networks—for example, the micro-connections in a computer chip or amongst neurons in the human brain. Such systems are complex not merely because they are complicated; 'order' means organization, in contrast to randomness or disorder. They are hierarchical, in that layers of emergent order and complexity build up on each other, with physics underlying chemistry, chemistry underlying biochemistry, and so on (Peacocke 1990; N. A. Campbell 1991). Each level is described in terms of concepts relevant to that level of structure (particle physics deals with quarks and gluons, chemistry with atoms and molecules, and so on), so a different descriptive language applies at each level. Thus we can talk of different levels of meaning embodied in the same complex structure, enabling emergent order: the higher levels display new properties not evident at the lower levels (N. A. Campbell 1991: 2–3). The essential key to understanding emergent properties is appropriate choice of higher-level concepts and associated variables. It is not possible to understand or explain the emergent properties in terms of the lower-level concepts and variables alone.

Bottom-up and Top-down Action

The first key issue underlying complex emergent behaviour is the occurrence of both bottom-up and top-down action in the hierarchy of structure and causation. What happens at each higher level is based on causal functioning at the level below; hence what happens at the highest level is based on physical functioning at the bottom-most level. The successive levels of order entail chemistry being based on physics, material science on physics and chemistry, geology on material science, and so on. This is the profound basis for reductionist world-views. In addition, however, higher-level structures together with the system's environment (which sets boundary conditions for physical variables) enable higher-level variables to influence lower-level variables by setting the context in which they function. This leads to downward causation (D. T. Campbell 1974) and contextual emergence (Bishop 2005). For example, when I move my arm, it moves because I have decided to move it; thus, in effect, my intention is causally effective in terms of instructing many millions of electrons and protons what to do. This is possible because the detailed physical structuring of the hierarchical system, in this case the physiology of the nervous system, provides the context in which the lower-level causality functions.

Top-down action affects the nature of causality significantly, both because inter-level feedback loops become possible, and because such effects modify the properties of the constitutive elements at the lower levels. For example, 'the emergence of the novel entity water obliges the two component elements to a relatedness (chemical bonding and the corresponding mixing of the electronic orbitals) that profoundly affects the properties of both hydrogen and oxygen' (Luisi 2002). A dramatic example is the properties of neutrons, which together with protons form atomic nuclei: they are unstable with a half-life of eleven minutes when unbound, but stable with a half-life of billions of years when bound into a nucleus. A change of context results in a major difference in local physical behaviour. At a much higher level of complexity, an individual human mind is crucially affected by the society in which it develops (Berger and Luckmann 1967); you cannot understand a mind in isolation, because the specific form of the modern mind has been determined largely by culture (Donald 2001). Thus complex systems are not just conglomerates of unchanged elementary constituents; rather, by their specific structuring, at all scales they profoundly affect the nature of the constituents out of which they are made.

Top-down action is prevalent in the real physical world and in biology. Potentials in the Schrödinger equation represent the summed effects of other particles and forces, and hence are the way in which the nature of both simple and complex structures can be described (from a particle in a box to the detailed structure of a computer or a set of brain connections). These potentials embody the ordered structure underlying causal relations (electrons flow in specific wires connecting specific components, neurons connect to specific other neurons, etc.). Top-down action is central to two main themes of molecular biology. First, the development of DNA codings (the particular sequence of base pairs in the DNA) occurs through an evolutionary process which results in adaptation of an organism to its ecological niche (N. A. Campbell 1991). A polar bear (*Ursus maritimus*) has genes for white fur in order to adapt to the polar environment, whereas a black bear (*Ursus americanus*) has genes for black fur in order to be adapted to the North American forest. The detailed DNA coding differs in the two cases because of the different environments in which the respective animals live. This is a classic case of top-down action from the environment to detailed biological microstructure—through the process of evolutionary adaptation, the environment (along with other causal factors) fixes the specific DNA coding (Roederer 2005). There is no way you could predict or explain this coding on the basis of biochemistry or microphysics alone. Also the central process of developmental biology, whereby positional information determines which genes get switched on and which do not in each cell, so determining their developmental fate, is a top-down process from the developing organism to the cell, based on the existence of gradients of positional indicators (morphogens) in the body

(Gilbert 1991; Wolpert 1998). Thus the crucial developmental mechanism determining the type of each cell in the body is controlled in an explicitly top-down way.

Concepts such as the plans for a jumbo jet, worked out on a rational basis through a process of computer-aided design, are causally effective in a top-down way because they determine the nature of physical objects in the world: they guide the manufacture of material objects. Socially devised rules and regulations (housing policy, health care systems, etc.) govern social relations and many resulting actions. The effectiveness of money, which can cause physical change in the world such as the construction of buildings, is based in social agreement. These are abstract variables based in social interaction over an extended period of time, and are neither the same as individual brain states, nor equivalent to an aggregate of current values of lower-level variables (although they may be represented by, and causally effective through, such states and variables). Causal models of the real world will be incomplete unless they include these various effects. Multiple top-down action from the mind co-ordinates action at lower levels in the body in a coherent way, and so gives the mind its causal effectiveness.

Feedback Systems, Goals, and Information

The second key issue underlying complex emergent behaviour (already alluded to above) is the existence of a hierarchy of goals that are causally effective, because they are the key to the functioning of feedback control systems, whereby the setting of goals results in specific actions taking place that aim to achieve those goals (Ashby 1958; Beer 1966, 1972). A comparator compares the system state with the goals, and sends an error message to the system controller if needed to correct the state by making it a better approximation to the goals. Examples are controlling the heat of a shower, the direction of an automobile, or the speed of an engine. Such systems damp out the effects of fluctuating initial data: they are designed precisely to give the same output whatever initial state occurs (within some limited domain that the system is designed to handle). The system output is determined by its goals rather than by the initial data. The series of goals in a feedback control system are clearly causally effective. They embody information about the system's desired behaviour or responses—indeed, living systems are goal seeking ('teleonomic'). A complete causal description of such systems must necessarily take such goals into account. The crucial issue now is what determines the goals: where do they come from? Two major cases need to be distinguished.

There are numerous systems in all living cells, plants, and animals that automatically, without conscious guidance, maintain homeostasis—they keep the structures in equilibrium through multiple feedback loops that fight intruders (the immune system), control energy and material flows, breathing, the function of the heart, etc. (Milsum 1966). They are affected through numerous enzymes, antibodies, and regulatory circuits of all kinds—for example, those that maintain body temperature and blood pressure. They have developed in the course of time through the adaptive processes of evolution, so are historically determined in a particular environmental context and are unaffected by individual history. Their existence is genetically determined, and embodies practical solutions to optimization problems faced by our animal and human ancestors.

However, at higher levels in humans and animals, important new features come into play; there are now individual behavioural goals that are not genetically determined. Many of them are conveyed to individuals through a variety of social mechanisms by which they become internalized (Berger 1963: ch. 5); others are learnt or consciously chosen. It is in the choice and implementation of such goals that explicit information processing comes into play. Information arrives from the senses and is analysed, sorted, and either discarded or stored in long-term and short-term memory, whence they help guide future behaviour. Thus humans are information-gathering and utilizing systems (Fernandez and Sole 2003; Hartle 2004). This is a highly non-linear process, which is non-local in both space and time. Conscious and unconscious processing of this information sets up the goal hierarchy, which then controls purposeful action in individual and social life. The goals may or may not be formulated explicitly.

At the highest level, the process of analysis and understanding is driven by the power of symbolic abstraction, codified into language embodying both syntax and semantics (Deacon 1997). This underpins other social creations, such as specialized roles in society and the monetary system, and higher-level abstractions such as mathematics, physical theories, philosophy, and legal systems—all encoded in symbolic systems. They gain their meaning in the context of a shared world-view and cognitive framework that are imparted to each individual by the society in which they live through many social processes (Berger and Luckmann 1967). Information is causally effective even though it is not a physical quantity but, rather, has an abstract nature (Roederer 2005). It is because of this effectiveness that it costs money to acquire information—it has an economic value. Not only that, but social constructions, too, are causally effective. The classic example of this is the laws of chess. Imagine someone coming from Mars and watching chess pieces moving. It is a very puzzling situation. Some pieces can only move in one way, and other pieces can only move in another way, so you imagine the Martian turning the board upside down or looking inside the rook, searching for a mechanism that causes these differences. But it is an abstraction, a social agreement, that is making the chess pieces move only in these ways. Such an agreement, reached by social convention over many hundreds of years, is not the same as any individual's brain state, though some people will try to tell you that it is. It is an abstraction that exists independently of any single mind, and that can be represented in many different ways. It is causally effective through the actions of individual minds, but none of them by itself created that abstraction or embodies it in its entirety.

Many other social constructions, including language, mathematics, and science, are equally causally effective. Non-physical entities such as the theory of thermodynamics and technology policy underlie the development and use of technology that enables transformation of the environment. They are created and maintained through social interaction and teaching, and are codified in books, and perhaps legislation. Thus, while they may be represented and understood in individual brains, their existence is not contained in any individual brain, and they are certainly not equivalent to brain states (electromagnetic theory, for example, is not the same as any individual's brain state). Rather, the latter serve as just one of many possible forms of embodiment of these features (they are also represented in books, journals, CDs, computer memory banks, diagrams, the spoken word, etc.). Indeed, they can be transformed between many such different representations precisely because they are independent of any single one of them. They are in fact abstract entities: equivalence classes of such representations. They are often socially agreed to, and exist in the context of a world of social constructions (Ellis 2004). They enable the active role of information in biological systems, which is always related to meaning, and enable information-driven interactions (Roederer 2005). Ethics is the subject shaping goals at the highest level of the causal hierarchy, which deals with life purpose and appropriate choice of lower-level goals. By determining the nature of lower-level goals chosen, and thence the nature of resulting actions, ethics is a set of abstract principles that are causally effective in the real physical world—indeed, they crucially determine what happens. Wars will be waged or not depending on ethical stances; large-scale physical devastation of the earth will result if thermonuclear war takes place.

The Causal Incompleteness of Physics

The higher-level feature of human consciousness is clearly causally effective in the world around us: we live in an environment dominated by manufactured objects that embody the outcomes of intentional design (buildings, motor cars, books, computers, clothes, teaspoons). The issue, then, is that the present-day subject of physics has nothing to say about the intentionality resulting in the existence of such objects. Thus it gives a causally incomplete account of the world (Ellis 2005a, 2005b). Even if we were to attain a 'theory of everything', such as string or M-theory (that is, a comprehensive theory of fundamental physics), this situation would remain unchanged: physics would still fail to comprehend human purpose, and hence would provide a causally incomplete description of the real world around us. This situation is characterized by the self-referential incompleteness of physics: there is no physics theory or experiment that can determine what will be the next experiment to be undertaken by the experimenter or theory to be created by the theorist.

There are three different aspects to this causal incompleteness of physics. First, as regards the present-day subject of physics, this is an incontrovertible statement of fact. There is no current physics theory or experiment that explains the nature of, or even the existence of, symphonies, football matches, teapots, or jumbo jet aircraft.

Secondly, one can ask if the present-day subject of physics could be extended to actually incorporate such features. The minimum requirement in order to have any hope of doing so would be to extend physical theory to include relevant higher-level variables, as happened in the past when the higher-level variables of entropy, specific heat, etc. were introduced into physics in order to explain the corresponding macroscopic physical effects. In the present case, the minimal need would be to include a function Ψ ('conscious intention'), to some degree dependent on lower-level variables, that would at least in principle be able to comprehend higher-level mental effects. One would then look for mathematical equations reliably predicting the evolution of this variable, or at least showing how it is related in principle to lower-level variables. I suspect that most physicists would regard this ambitious project as lying outside the proper scope of physics. It would in any case be too complex to be practicable. This is not to say that reductive bridge laws are not possible in general; it says only that they are not possible in the case of systems of the complexity of the human mind.

However, there is a third aspect—that of basic principle. Brains are networks of neural cells, so some claim that there's nothing in principle to stop us from fully understanding them. You just need to know enough about the state of the brain and the person's previous experience to apply physics and predict future behaviour. Predicting human intentionality is difficult only because we don't know enough about brains to make the calculation. The thing is doable in principle, though not in practice; physics causes all that happens, even if the outcome is not predictable in practice. In the end, physics is all there is, and by itself controls the outputs of the brain. Free will is an illusion. Despite its appeal to some, this kind of claim about the human mind is itself in fact an unprovable philosophical supposition about the nature of causation, with zero predictive ability (no observable consequences follow from it) and no experimental proof directly supporting it. On the contrary, every-day experience regarding our intentional actions suggests that this belief is wrong (Kane 1996; Pink 2004). Reductionism can explain many aspects of the way in which the brain works, but the claim here is that this *totally* explains consciousness: that is what is contentious. The key issue is whether the higher levels in the hierarchy of complexity have real autonomous causal powers, largely independent of the lower levels, and indeed controlling their context and hence their outcomes, or whether all the real causal powers reside at the lower levels and the higher levels dance to their algorithmic tune, merely appearing to have autonomy.

The claim made by assuming physical causal completeness is that for any specific physical system, physical laws alone give a unique outcome for each set of initial data. To see the improbability of this claim, one can contemplate what is required from this viewpoint when placed in its proper cosmic context. The implication is that the particles that existed at the time of decoupling of the cosmic background radiation in the early universe just happened to be placed so precisely as to make it inevitable that 14 billion years later, human beings would exist and Crick and Watson would discover DNA, Townes would conceive of the laser, and Witten would develop M-theory. Is it conceivable that truly random quantum fluctuations in the infla-tionary era can have had coded in them the future inevitability of the *Mona Lisa*, Nelson's victory at Trafalgar, Einstein's 1905 theory of relativity? Such later creations of the mind are clearly not random; on the contrary, they exhibit high levels of order embodying sophisticated understandings of painting, military tactics, and physics respectively, which cannot possibly have arisen directly from random initial data. This proposal simply does not account for the origin of such higher-level order.

Stars, galaxies, etc. represent the outworkings of the underlying physical laws, which lead to a specific class of structures (stars and star clusters, for example), with only detailed parameters determined by the initial conditions. Thus these kinds of outcomes are the almost inevitable outcome of physical forces, resulting from the nature of the possibility space, with just a few parameters set by the initial data. The higher-order meanings embodied in the mind and the resultant physical objects produced through its activity are not of this kind. For example, in the case of written text, almost all physically possible outputs are gibberish rather than text that has meaning. The possibility space of all written text does not specifically encode mathematical theorems or social theories—these certainly exist in this space, as they have indeed been written down, but as small islands of meaning in a vast sea of meaningless text, and no purely physics-based process has any way of telling which is which. Thus, if a purely physical evolution determines what happens, these meanings have to be already present in the initial data in some incipient or coded form, for they will not be probable outcomes of the way in which the possibility space is structured. But this cannot have been the case: there is no way these data could have been embedded there.

The later higher-level outcomes were not the consequences of specific aspects of the initial data, even though they arose out of them. There was no higher-level meaning somehow encoded in those initial data. Conditions at the time of decoupling of the cosmic background radiation in the early universe 14 billion years ago were such as to lead to the eventual development, first, of stars and planets which exist as attractors in the possibility space—their emergence is more or less inevitable, irrespective of the details of the initial data—and then much later, of life and ultimately minds that are autonomously effective, as they seem to be, due to the precise biological structuring of the brain, able to create higher-level order without any fine dependence on lower-level physical laws or initial data. The higher-level understandings in the mind were not specifically implied by the initial data in the early universe; neither were their physical outcomes such as television sets and cell phones. Reductive physics characterizes part of the causal nexus in operation in the workings of the brain—the bottom-up aspects—but not all of it. It cannot comprehend crucial top-down influences in operation, dependent on the higher-level processes just discussed; indeed, reductionism rules them out. Above all, we should not too hastily conclude that we can understand what is going on in the brain on the basis of physics alone until we properly understand the issues of consciousness and free will. Despite some extravagant claims made by a few adventurous souls, we actually don't have a clue as to how consciousness emerges from the underlying physics; we don't even know the appropriate questions to ask (Chalmers 1997).

Finally, we should recognize that the enterprise of science itself does not make sense if our minds cannot rationally choose between alternative theories on the basis of the available data, which is indeed the situation if one takes seriously the bottom-up mechanistic view that the mind simply dances to the commands of its constituent electrons and protons, algorithmically following the imperatives of Maxwell's equations and quantum physics. But a reasoning mind able to make rational choices is a prerequisite for physics to exist. On the reductionist view, rationality and reason are as illusory as free will; they are just the inevitable outcomes of microphysics. That viewpoint cannot account for the existence of physics as a rational enterprise. Just as there is a measurement problem underlying quantum theory—in essence, quantum theory does not seem able to describe the workings of the macroscopic measuring apparatus (Penrose 1989, 2004; Isham 1997)—so there is one underlying physics overall. In essence, physics does not seem able to account for the ability of the experimenter first to choose what to do, then to set up the apparatus as desired and to voluntarily carry out the appropriate series of measurements, and finally to rationally determine the scientific implications of the results. This means that physics today provides a causally incomplete theory of the world around us. It cannot describe all the causes acting to shape what happens in the real world.

Reductionism, Fundamentalism, and the Science and Religion Debate

The considerations above are sufficient to rebut simplistic reductionist views of humanity, and undermine the view that physicalist theories alone provide a sufficient metaphysical foundation for understanding the universe and humanity. These are key issues in the science and religion interaction, where many propose the view that physics is all there is, and that all the rest is an epiphenomenon with no meaning—a froth on the top of what is really going on.

This view is expressed in statements such as the following: under the headline 'Our life has no purpose', G. Monbiot (2005) says: 'Darwinian evolution teaches us that we are incipient compost: assemblages of complex molecules that—for no greater purpose than to secure sources of energy against competing claims—have developed the ability to speculate. After a few score years the molecules disaggregate and return to whence they came. Period.'

This kind of view, and others quoted below, deny the reality of all the higher levels of the hierarchy of complexity, even though they are irreducible to the lower levels and operate according to higher levels of emergent behaviour, with higher meaning demonstrably a causal factor. Indeed, the irony of it all is that Monbiot would not write that article unless he thought there was some meaning in doing so—his article is thus (like those of the extreme deconstructionists) an exercise in implicit self-contradiction. His scientifically based intellectual theories serve as a basis for dogmatic statements about the lack of meaning of life and the emptiness of religion, even though science has no ability to comment on high-level meaning because of the very limitations of its method. Thus he does not recognize the limits of science, and also fails to recognize the metaphysical issues underlying the existence of the universe. His statement is a form of scientific dogmatism which in the end demeans science as well as humanity. Such writings are plausibly one of the main reasons why science is regarded with such suspicion, and even rejected, by a large part of the general public (Ruse 2005).

Limits to Science

There are many limits to what science will ever be able to do that will never change—they are boundaries to its competence, because of its nature and its methods of investigation. There are many areas of concern to humans, of which only a subset are within the ambit of science. Outside this ambit are crucially important areas: in particular, ethics, aesthetics, metaphysics, and meaning. They are outside the competence of science because there is no scientific experiment which can determine any of them. Science can help illuminate some of their aspects, but is fundamentally unable to touch their core. This is not a 'God of the gaps' argument. It is about absolute boundaries to what science can ever do, because of the very nature of science.

There is a great deal of confusion about this, particularly in the case of ethics. This is outside the competence of science because there is no experiment which says that an act is good or bad. There are no units of good and bad, no measurements of so many milli-Hitlers. Ethics is simply an area that science cannot handle. It is true that science depends on and supports some basic virtues, such as respecting the data, telling the truth, and so on. But this does not begin to touch real ethical issues to do with the relative importance of ends and means, how to deal with conflicting interests, how to balance outcomes against principles, and so on. They simply do not help as regards real-world ethical dilemmas. However, sociobiology and evolutionary psychology produce arguments which claim to give complete explanations as to where our ethical views come from. There are many problems with those attempts, the first being that they do not explain ethics—they explain it away. If the true origin of our ethical beliefs lay in evolutionary biology, ethics would be completely undermined, because once you understood this, you would no longer necessarily believe that you had to follow its precepts. You could choose to buck the evolutionary imperative.

The second is that this looks at some of the causes in operation and ignores others; it simply leaves out of consideration two other important parts of the equation: namely, social effects embodied in culture (which some sociologists and anthropologists with equal fundamentalist vehemence claim are all that matter), and the key factor of personal choice. Above all, by fiat it denies the possibility of a realist ethics; but if morality is in any sense real—that is, if there are indeed standards of good and evil that are transcultural—then a realist ethics is strongly indicated (Murphy and Ellis 1996).

The third is that if you did follow those precepts, you would rapidly end up in very dangerous territory: namely, the domain of social Darwinism. That has been one of the most evil movements in the history of humanity, causing far more deaths than any other ideology has done (Weikart 2005). And the fact that one is able to say that it is evil shows that there are standards of ethics outside those provided by evolutionary biology. There is, of course, a substantial literature on the evolutionary rise of altruism, but as a historical fact the influence of evolutionary theory on ethics in practice has been to provide theoretical support for eugenics and social Darwinism, not for any movement of caring for others. The theoretical ideas in support of the evolutionary rise of altruism have had no discernible effect on public behaviour; and ultimately this is precisely because they explain ethics away, rather than providing a foundation for ethics.

Challenging evolutionary biologists who still maintain that their science can provide a basis for ethics, despite these arguments, is very simple. If a scientist says, 'Science can handle ethics', say to them: 'Tell me, what does science say should be done about Iraq today? And what does science say ethically about Israel and Palestine?' You will get a deafening silence, because the simple fact is that science cannot handle ethical questions. Ethical values, crucial for our individual and social lives, have to come from a value-based philosophical stance or a meaning-providing religious position. They cannot be justified by rationality alone, much less by science.

Similarly, aesthetics—the criteria of beauty—is also outside the boundaries of science. No scientific experiment can determine that something is beautiful or ugly, for these are not scientific concepts. The same is true for metaphysics and meaning. Thus there are major areas of life, incredibly important to humanity, which cannot be encompassed by science. They are the proper domain of philosophy, of religion, of art, and so on, but not of science.

Why are there these boundaries? Because experimental science deals with the generic, the universal, in very restricted circumstances. It works in circumstances so tightly prescribed that effects are repeatable. Most things which are of real value in human life are not repeatable. They are individual events which may have crucial meaning for individuals and for humanity in the course of history; but each occurs only once. So repeatable science does not encompass either all that is important or all that can reasonably be called knowledge.

Metaphysical Issues

There are major metaphysical issues underlying the existence of the universe—why it is the way it is, and in particular why the universe allows the existence of life. I will not pursue this complex argument here (see e.g. Ellis 2006b for an analysis). The point here is that much writing simply ignores the deep questions at the foundations of existence, which science cannot answer, yet nevertheless purports to give a definitive answer to the meaning of life. One can ignore these ultimate issues if one wants to, taking the nature of the universe and of the laws of physics for granted, and not needing explanation, but then precisely because one has done so, one is in no position to declaim on issues of higher meaning and purpose. One has simply chosen to exclude them from consideration a priori.

Fundamentalism as a Major Problem

Such exclusions are characteristics of scientific fundamentalism. I suggest that the essential nature of all fundamentalisms is a partial truth proclaimed as the whole truth. Only one viewpoint is allowed on any issue, all others are false. This dogmatism is combined with an inability to relate understanding to context. Admitting that what is important varies with context would undermine the fundamentalist's need to see the same single issue as dominant in every situation, come what may.

Perhaps its most obvious manifestation is in fundamentalist religion, where unquestionable revelations rule the day, and imply a total rejection of any competing views. It occurs in the social and human sciences—for example, in the work of the behaviourists and in the clash between the social sciences and biology (Pinker 2003). It occurs in the big divide between the sciences and the humanities, with extreme post-modernism on the one side and the scientific fundamentalists on the other. It occurs particularly in relation to the nature of humanity, with the understanding of evolution, on the one hand, and the mind–brain problem, on the other, being contested terrain. At issue here is the way we understand the nature of our existence: how we see ourselves and the meaning in our lives. Crucial consequences follow for how we treat people medically, individually, and politically. What is the essential nature of humanity in the light of modern biology: in particular, molecular biology and neuroscience? This is where there is real potential for conflict between science and religion, which is going to go on for a long time.

Part of the issue is how we make decisions and shape our understanding. Some science-based world-views claim in essence that reason is all that is needed for life, while emotion, faith, and hope simply get in the way of rationally desirable decisions. But this is a false view. It is not possible to reason things out and make decisions purely on a rational basis. First, emotions guide our actions at a more fundamental level than do rational decisions. They are a crucial feature of all human life (Damasio 1994/5). Secondly, in order to live our lives, we need faith and hope, because we always have inadequate information on which to base decisions. It is a part of daily life that when we make important decisions, like whom to marry, whether to take a new job, whether to move to a new city, they are always to a considerable degree guided by emotion, and in the end have to be concluded on the basis of partial information. Thus a lot of choices are based on faith and hope, faith about how things will be, hope that it will work out all right. This is true even in science. Scientists set up research groups to look at string theory or particle physics, in the belief that they will be able to obtain useful advances when their grant applications have been funded. They do not know that they will take those steps forward. It is a hope, based in belief. So embedded in the very foundations even of science there is a human structure of faith and hope. Furthermore, scientists actually carry out these enterprises because of the associated emotions and values that guide their actions— for example, the desire to understand is an emotion that underlies science.

Thirdly, we crucially need values to guide our rational decisions; but these cannot be arrived at rationally. Our minds act, as it were, as an arbiter between three tendencies guiding our actions: first, what rationality suggests is the best course of action—the cold calculus of more and less, the most beneficial choice economically; second, what emotion sways us to do—the way that feels best, what we would like to do; and third, what our values tell us we ought to do—the best option ethically, the right thing to do. These are distinct from each other, and in competition to gain the upper hand. Sometimes they may agree as to the best course of action, but often they will not. It is our personal responsibility to choose between them, making the best choice we can between these conflicting calls, with our best wisdom and integrity, and on the basis of the limited data available. This shows where value choices come in and help guide our actions. Rationality can help us decide which course of action will be most likely to promote specific ethical goals when we have made these value choices, but the choices themselves, the ethical system, must come from outside the pure rationality of rigorous proof, and certainly from outside science. As emphasized before, science cannot provide the basis of ethics. A deeper world-view is crucial here. It is essential to our well-being and proper fulfilment, because ethics and meaning are deeply intertwined. Humans have a great yearning for meaning, and ethics embodies those meanings and guides our actions in accordance with them.

The human mind and the question of consciousness are among the most serious potential points of tension between science and religion—and indeed, between science and the fullness of humanity. There are philosophers, psychologists, and neuroscientists who tell us that consciousness is an epiphenomenon: it is not real. We are not really conscious, but are machines driven by unconscious computations, so that what we think are conscious choices are not real. To me, this is the one real threat from the scientific side. Merlin Donald expresses this as follows:

Hardliners, led by a vanguard of rather voluble philosophers, believe not merely that consciousness is limited, as experimentalists have been saying for years, but that it plays no significant role in human cognition. They believe that we think, speak, and remember entirely outside its influence. Moreover, the use of the term 'consciousness' is viewed as pernicious because (note the theological undertones) it leads us into error. (Donald 2001: 28–9).

They support the downgrading of consciousness to the status of an epiphenomenon... A secondary byproduct of the brain's activity, a superficial manifestation of mental activity that plays no role in cognition. (Donald 2001: 36)

Dennett is actually denying the biological reality of the self. Selves, he says, hence self-consciousness, are cultural inventions. . . . The initiation and execution of mental activity is always outside conscious control. (Donald 2001: 40)

Consciousness is an illusion and we do not exist in any meaningful sense. But, they apologize at great length, this daunting fact Does Not Matter. Life will go on as always, meaningless algorithm after meaningless algorithm, and we can all return to our lives as if Nothing Has Happened. This is rather like telling you your real parents were not the ones you grew to know and love but Jack the Ripper and Elsa, She-Wolf of the SS. But not to worry... (Donald 2001: 45)

The practical consequences of this deterministic crusade are terrible indeed. There is no sound biological or ideological basis for selfhood, willpower, freedom, or responsibility. The notion of the conscious life as a vacuum leaves us with an idea of the self that is arbitrary, relative, and much worse, totally empty because it is not really a conscious self, at least not in any important way. (Donald 2001: 45)

But this is not in fact what is implied by the science, which has a long way to go before it properly understands the brain. It is important to reiterate here that, despite the enormous amount scientists know about neuroscience and its mechanisms, the neural correlates of consciousness, the different brain areas involved, and so on, we have no idea of how to solve the hard problem of consciousness (Chalmers 1997), however many of the hardliners deny even that there is such a problem.

As to the causal efficacy of consciousness, I take that as a given which underlies our ability to carry out science and to entertain philosophical and metaphysical questions. And as a consequence, ethical choices and decisions are real and meaningful. If theory denies this basic observational fact, then it is not in accord with the data and needs to be abandoned in favour of a more reliable theory.

Existence and Meaning

Thus there is a need for more humanist views to counter scientific fundamentalism and associated absolutist views:

- Promoting a consilience of very different world-views that are attempts to view important aspects of the same underlying reality through emphasizing humility: giving up the need to be right in favour of trying to see what is actually there from as many different viewpoints as possible, but all the time keeping in mind the need for evidence and testing of theory and the dangers of self-delusion.
- Emphasizing all the dimensions of humanity and the crucial role of an ethical system of values that cannot be derived from science alone, and that embody highest-level meaning.
- Helping develop world-views that can accommodate the pragmatic nature of science but also the kinds of deeper issues regarding existence and meaning that can be encountered in spiritual and religious world-views. Thus it can explore the deep nature of reality. Indeed, to attain a full depth of foundations for such views, one must look at the possibility of underlying purpose and meaning—a religious basis—being the foundation of all that is: that mind, rather than matter, is what ultimately underlies it all. This view can be expanded into a systematic view that makes overall sense with a kenotic ethical basis as a key element in the meaning of it all. (Ellis 1993; Murphy and Ellis 1996)

Overall, it can be important in emphasizing all the dimensions of humanity as well as what we can access by scientific observations and theories. This can be an important integrative factor, helping us fully to see ourselves and the universe in which we live. This affects our quality of life in a crucial way: it helps us to be fully human. The kinds of analysis of emergence and the limits of physics undertaken in the first part of this article can be helpful in this endeavour, providing support for understanding the reality of higher levels of meaning and affirming multiple levels and natures of existence (Ellis 2004, 2005*a*, 2005*b*).

References and Suggested Reading

ASHBY, R. (1958). *An Introduction to Cybernetics*. London: Chapman and Hall.

BEER, S. (1966). *Decision and Control*. New York: Wiley.

——(1972). *Brain of the Firm*. New York: Wiley.

BERGER, P. L. (1963). *Invitation to Sociology: A Humanistic Perspective*. Garden City, N.Y.: Doubleday.

——and LUCKMANN, T. (1967). *The Social Construction of Reality: A Treatise in the Sociology of Knowledge*. Garden City, N.Y.: Anchor.

BISHOP, R. C. (2005). 'Patching Physics and Chemistry Together', *Philosophy of Science* (forthcoming): <http://philsci-archive.pitt.edu/archive/00001880/>.

——(2006). 'The Hidden Premise in the Causal Argument for Physicalism', to appear in *Analysis*; available at <http://philsci-archive.pitt.edu/archive/00002415/01/Hidden_Premise.pdf>.

BOOCH, G. (1994). *Object-Oriented Analysis and Design with Applications*. Boston: Addison Wesley.

CAMPBELL, D. T. (1974). 'Downward Causation', in F. J. Ayala and T. Dobhzansky (eds.), *Studies in the Philosophy of Biology: Reduction and Related Problems*, Berkeley: University of California Press, 179–86.

CAMPBELL, N. A. (1991). *Biology*. Boston: Benjamin Cummings.

CHALMERS, D. (1997). *The Conscious Mind*. New York: Oxford University Press.

DAMASIO, A. R. (1994/5). *Descartes' Error: Emotion, Reason, and the Human Brain*. New York: Grosset/Putnam, 1994; New York: Harper Collins, 1995.

DEACON, T. (1997). *The Symbolic Species: The Co-evolution of Language and the Human Brain*. Harmondsworth: Penguin.

DONALD, M. (2001). *A Mind so Rare: The Evolution of Human Consciousness*. New York: Norton.

ELLIS, G. F. R. (1993). 'The Theology of the Anthropic Principle', in R. J. Russell, N. Murphy, and C. J. Isham (eds.), *Quantum Cosmology and the Laws of Nature*, Vatican City State: Vatican Observatory Publications; Berkeley: Center for Theology and the Natural Sciences, 367–406.

——(2004). 'True Complexity and its Associated Ontology', in J. D. Barrow, P. C. W. Davies, and C. L. Harper (eds.), *Science and Ultimate Reality: Quantum Theory, Cosmology and Complexity*, Cambridge and New York: Cambridge University Press, 607–36.

——(2005*a*). 'Physics, Complexity, and Causality', *Nature*, 435: 743.

——(2005*b*). 'Physics and the Real World', *Physics Today*, 58 (July): 49–54.

——(2006*a*). 'On the Nature of Emergent Reality', in P. Clayton and P. C. W. Davies (eds.), *The Re-emergence of Emergence*, Oxford: Oxford University Press, 79–107. Also at <http://www.mth.uct.ac.za/~ellis/emerge.doc>.

——(2006*b*). 'Issues in the Philosophy of Cosmology', in J. Butterfield and J. Earman (eds.), *Encyclopedia of the Philosophy of Science*, volume on physics (forthcoming); <http://www.mth.uct.ac.za/~ellis/enc2.pdf>.

FERNANDEZ, P., and SOLE, R. V. (2003). 'The Role of Computation in Complex Regulatory Networks'; <http://lanl.arXiv.org/abs/q-bio.MN/03110102>.

GILBERT, S. F. (1991). *Developmental Biology*. Sunderland, Mass.: Sinauer.

HARTLE, J. B. (2004). 'The Physics of Now'; <http://labl.arXiv.org/abs/gr-qc/0403001>.

ISHAM, C. J. (1997). *Lectures on Quantum Theory: Mathematical and Structural Foundations.* London: Imperial College Press.

KANE, R. (1996). *The Significance of Free Will*. New York: Oxford University Press.

LUISI, P. L. (2002). 'Emergence in Chemistry: Chemistry as the Embodiment of Emergence', *Foundations of Chemistry*, 4: 183–200.

MILSUM, J. H. (1966). *Biological Control Systems Analysis*. New York: McGraw-Hill.

MONBIOT, G. (2005). *Guardian Weekly*, 26 August–1 September, 14.

MURPHY, N., and ELLIS, G. F. R. (1996). *On The Moral Nature of the Universe: Cosmology, Theology, and Ethics*. Minneapolis: Fortress Press.

PEACOCKE, A. R. (1990). *An Introduction to the Physical Chemistry of Biological Organization*. Oxford: Clarendon Press.

PENROSE, R. (1989). *The Emperor's New Mind*. Oxford and New York: Oxford University Press.

——(2004). *The Road to Reality: A Complete Guide to the Laws of the Universe*. London: Jonathan Cape.

PINK, T. (2004). *Free Will: A Very Short Introduction*. Oxford: Oxford University Press.

PINKER, S. (2003). *The Blank Slate: The Modern Denial of Human Nature*. Harmondsworth: Penguin.

ROEDERER, J. G. (2005). *Information and its Role in Nature*. Heidelberg: Springer-Verlag.

RUSE, M. (2005). 'Evolving Problem', *Science and Spirit*, (September–October), 38–41.

SIMON, H. A. (1962). 'The Architecture of Complexity', *Proceedings of the American Philosophical Society*, 106: 467–82.

WEIKART, R. (2005). *From Darwin to Hitler*. New York and Houndmills: Palgrave Macmillan.

WOLPERT, L. (1998). *Principles of Development*. Oxford: Oxford University Press.

CHAPTER 45

EMERGENCE AND COMPLEXITY

NIELS HENRIK GREGERSEN

INTRODUCTION

In ordinary language 'emergence' refers to processes of coming forth from latency, or to states of things arising unexpectedly. As such, the term was used already by the father of chemistry, Robert Boyle, to refer to the rising of substances to the surface of liquids, and by Isaac Newton to designate the appearance of light refractions (*Oxford English Dictionary* (1989), v. 175–6).

However, it was only with the British emergentists of the 1920s (in particular, Samuel Alexander, C. Lloyd Morgan, and C. D. Broad) that emergence theory was formed as a meta-scientific interpretation of evolution in all its forms: cosmic, biological, mental, and cultural. Although emergentists differ in metaphysical orientation, they usually share three tenets. First, emergents are *qualitative novelties*, which should be distinguished from mere resultants that come about by a quantitative addition of parts. 'Weight', for example, is a resulting property of aggregating matter, whereas the liquidity and surface tension of water are new, emergent qualities in relation to the chemical compounds of hydrogen and oxygen. Secondly, nature is a *nested hierarchy of ontological levels*, so that the higher emergent levels (e.g. living organisms) include the lower levels (e.g. inorganic chemistry), on which they are based. Thirdly, emergentists are *explanatory holists*: that is, higher levels are not predictable from our knowledge of their constituent parts, and their operations are often in principle irreducible to the lower levels. In other words, bottom-up microphysical causation must be supplemented by various forms of top-down causation in order to account for the properties and functions of the higher levels.

Emergence theory was formed in the safe context of the Darwinian revolution in science. However, emergentism also developed well before revolution in the sciences of computation, information, and cybernetics in the wake of the work of Alan Turing, Claude Shannon, and John von Neumann in the 1940s and 1950s. In the reductionist climate in the philosophy of science between the 1930s and the 1960s, emergentist proposals were sidetracked. In their influential paper presenting the outline of the Covering Law model of scientific explanation, Carl G. Hempel and Paul Oppenheim (1948) redefined emergence as that which is unexpected relative to the present-day state of theories; future microstructure theories may thus explain the novel features in a fully reductive manner.

Since the 1970s, however, the idea of emergence has been rejuvenated by computer-aided studies of complexity. Emergentism has also been supported by more anti-reductionist trends within philosophy of science, trends that allow for a plurality of different explanations, each with its own explanatory domains. It is therefore appropriate to talk about a 're-emergence of emergence' (Clayton and Davies 2006).

Nevertheless, it should not be overlooked that the seminal idea of top-down causality has also been challenged by computer-aided studies of complexity. In the first section I therefore begin by locating the idea of emergence in a historical context. I aim to show that different claims of emergence (some weak, some strong) can, at least in part, be explained by the different research traditions within which the idea of emergence is placed. For even though the concepts of 'emergence' and 'complexity' often travel together, the framework of computational complexity remains a reductionist research programme, in which 'Its' are reduced to 'Bits'. Hence in the second section I present a short philosophical primer consisting of four types of emergence theory, two epistemological and two ontological. Only the ontological versions of emergence theory are committed to speaking of top-down causation. The question then is: what is it that emerges? Is it only properties, or is it also new causal capacities, or even new individuals that come forth? In the final section, I discuss what opportunities the idea of emergence may offer for religious reflection. Certainly no religious strings are attached to the idea of emergence. Theological reflection, however, may be propelled to rethink religious traditions, and to redescribe the world of creation in new terms, in the light of the sciences of complexity and emergence theory. In this context the question arises: to what degree can the idea of emergence also be applied to concepts of the divine? Is even deity an emergent property of the world, as suggested by Samuel Alexander (1920)? Or is God the atemporal ground of the emerging world, without himself falling prey to the temporal characteristics of emergence, as was supposed in classical atemporal theism? Or is there a need to speak theologically about God as both the atemporal ground of being *and* as having a life in and with time, 'in, with, and under' the emerging world, as in various forms of temporal theism?

EMERGENTISM AND COMPLEXITY THEORY: TWO RESEARCH TRADITIONS

Some of the seminal ideas of emergentism, especially the hierarchical and holist views of nature, are part of classic philosophical tradition. Already Plato observed that 'the whole is more than sum of the parts' (*Theaetetus* 203E; cf. *Laws* 903B–C). But also the Stoic materialists argued that reason needs to take into account the all-embracing world order, without which one cannot explain the 'sympathetic agreement, interconnection, and affinity of all things' (Cicero, *On the Nature of the Gods* II. 7. 19).

The notion of a *nested hierarchy* in nature also has important precursors in classic tradition. In the wake of Plotinus's philosophy, we find in Western tradition the influential notion of a *scala naturae*, a scale of being, as laid out in Arthur Lovejoy's study *The Great Chain of Being* ([1936] 1964: 88–92). The higher forms include the lower forms, so that, for instance, human beings include the features of animal and plant life, just as biological processes include the physical processes. According to the principle of plenitude (*principium plenitudinis*), we live in a 'filled world' in which all potentialities are represented. Here we have both hierarchy and holism. Still absent is the idea of evolutionary novelty: something more does not come out of something less. Rather, in Plotinus's philosophy of emanation, the occurrence of new things takes place as devolution, as a thinning out of the power of being that resides in the divine principle prior to the process of creation.

By contrast, emergentists subscribe to a robust scientific naturalism, according to which mental processes supervene on biological processes, and biological processes on physical processes. Emphatically, the British emergentists took leave of teleological causes in favour of efficient causes; they resisted vitalistic appeals to an *élan vital* (Bergson 1911) as well as theological explanations of biological particulars. As expressed by C. Lloyd Morgan, 'in a philosophy based on the procedure sanctioned by progress in scientific research, the advent of novelty of any kind is loyally to be accepted wherever it is found, without invoking any extra-natural Power (Force, Entelechy, Elan, or God) through the efficient Activity of which the observed facts may be explained' (Morgan 1923: 2).

Emergentism thus presents itself as a naturalistic research programme by presupposing an evolutionary monism, yet in a manner that contrasts with both physical reductionism and dualism. What is excitingly new about emergentism is its claim that naturalistic science is in need of a richer explanatory scheme. The combination of holism with novelty was probably not adumbrated before well into the nineteenth century. In his *System of Logic* of 1843, John Stuart Mill drew attention to two different sorts of compound causes (Bk. III. 6. 1–2; cf. 10. 4). Sometimes the joint effect of causes is simply the sum or average of their separate parts. For example, when a pistol projects a bullet towards the sky, the effect of the gunpowder is continuously counteracted by the bullet's gravity, so that in the end the bullet falls to the ground; most physical processes are of this mechanical type. But, in the case of 'heteropathic laws', the effects are entirely heterogeneous with respect to the causes. Mill found such entirely new properties already in chemical compounds: 'Not a trace of the properties of hydrogen or of oxygen is observable in those of their compound, water... nor is the colour of green vitriol a mixture of the colours of sulphuric acid and copper' (Mill [1843] 1970: 243). Mill's favourite examples, however, were the far more complex combinations of elements that constitute organized and sentient bodies, such as the capacity of a tongue to taste, which can not be deduced from elements such as gelatin and fibrin. Against this background Mill came to the general conclusion that '(t)he Laws of Life will never be deducible from the mere laws of the ingredients' ([1843] 1970: 245).

Mill's distinction between two types of joint causes was later rephrased by G. H. Lewes in *Problems of Life and Mind* (1877) as the distinction between *resultants* and *emergents*. The first simply come about through additions, while emergents constitute a new qualitative class of phenomena. Accordingly, C. D. Broad in *The Mind and its Place in Nature* (1925) argued against the homogenizing view of the material world entailed in mechanicism. While there is admittedly only 'one kind of stuff' at the fundamental level, Broad showed how matter is stratified in different kinds of order, so that irreducible properties and new effects arise. There are both 'intra-ordinal laws', which relate to the intrinsic properties of matter (e.g. at the atomic level), and 'trans-ordinal laws', which describe and predict how higher-order systems behave. 'A trans-ordinal law would be one which connects the properties of aggregates of adjacent orders. A and B would be adjacent, and in ascending order, if every aggregate of order B is composed of order A, and if it has certain properties which no aggregate of order A possesses and which cannot be deduced from the A properties' (Broad 1925: 77).

Observe that by speaking of trans-ordinal, or emergent, *laws*, Broad emphasized that emergent properties do not have a spurious *ad hoc* character. A trans-ordinal law can be investigated scientifically: 'once it has been discovered, it can be used like any other [law] to suggest experiments, to make predictions, and to give us practical control over external objects' (1925: 79). Once the capacity for sentience has emerged in animals, for example, new regularities of movement patterns can be investigated empirically.

Note also that emergents and resultants are not two classes of phenomena, for no emergents take place apart from resultant processes. As pointed out by C. Lloyd Morgan in *Emergent Evolution*, '[t]here may often be resultants without emergence; but there are no emergents that do not involve resultant effects' (Morgan 1923: 5). Thus no causal work at higher levels is possible without the causal plumbing going on at base level.

Emergentism here builds on a longer tradition of *organicism* in biology. Already Immanuel Kant (1724–1804) observed that it is characteristic for living organisms to coalesce 'into the unity of a whole' in such a way that the bodily parts 'are mutually cause and effect of one another' (*Critique of Judgment*, B 292). However, Kant's organicism was pre-evolutionary, and his position did not allow for living organisms to have causal laws superseding the mechanical laws of physics. For Kant organicism was a way of seeing the world innate to the human understanding, but not a discovery of new laws. While acknowledging that 'even the emergence of a blade of grass' cannot be accounted for by the laws of nature alone (B 338), Kant remained a Newtonian insofar as non-human nature is concerned.

Nonetheless, Kant had an important influence on later organicist programmes in biology. Embryologists such as Karl Ernst von Baer (1792–1876) argued that laws for embryonal growth exist, laws that are internal to the particular physiological structure of species, though always mediated by the interactions between the organism and its environment. Twentieth-century organicists, such as J. Needham and C. H. Waddington, have likewise pointed to the need for explanations of different levels: genes, cells, organs, organisms, and societies (Gilbert and Sarkar 2000).

Kant also inspired twentieth-century theories of self-organization. Stuart Kauffman, one of the pioneers in using computational models for explaining emergent properties in co-evolving systems, celebrates Kant as a forerunner of the idea of self-organizing biological systems (Kauffman 1993: 4 and 643). Kauffman also declares himself 'an unrepentant holist' (1998: 69). According to Kauffman, it is the general capacity of matter for autocatalytic behaviour at the chemical level that explains the emergence of life. Biology is rooted in the formative powers of thermodynamics, though he also insists that historical contingencies play an irreducible role in evolution (Kauffman 2000). Even though evolution itself may be like an 'incompressible algorithm' (1998: 23), the emergence of life is for Kauffman precisely an *expected* quality of physical self-organization. Working within the paradigm of computational complexity (CC), he aims to explain the evolutionary appearance of life through a bottom-up computational approach. For our purpose it is worth noting that the general strategy of CC on this point is at odds with the stronger emergentist claims of irreducibility. Thus, while emergentists work on the assumption of hierarchical levels, CC make use of a *flat theory design*: it is by the sheer reiteration of algorithmic rules that the complex patterns are formed on the computer screen.

In his book *Emergence* (1998), John Holland, one of the 'fathers' of evolutionary algorithms, therefore makes the uncompromising statement: 'we *can* reduce the behaviour of the whole to the lawful behaviour of its parts, if we take the non-linear interactions into account' (Holland 1998: 122). Non-linear interactions have traditionally posed problems for mathematical calculus, as is well known from the so-called three-body problem. But non-linear reactions can be computed rather effectively, as we know from fractal geometry, an important part of chaos theory. Moreover, historical trajectories can be modelled in the form of evolutionary algorithms that prescribe how to move forwards in a system, given this or that circumstance, and how to change the algorithm in the face of particular circumstances. CC thus inhabits an interesting middle stance between a physicalist reductionism and stronger programmes of emergentism. While acknowledging that mathematical *equations* are not able to explain the persistent emergence of complex structures, CC is interested in finding algorithmic *compressions* of evolutionary patterns and temporal sequences. Thus, the paradigm of CC sets before itself the aim of simulating the codes of naturally evolved phenomena in a bottom-up way. Simulating evolution means finding the minimal set of computational steps necessary for regenerating the logic of the complex phenomena under investigation. Mostly, however, these algorithmic *rules* (in contrast to *laws* of nature) are local in nature, and apply to contingent evolutionary conditions.

Now, what is the relation between the research programmes of emergentism and CC? First of all, both programmes study what we may term real-world complexity (RWC): that is, the production, stabilization, and development of higher-order systems. They also agree in criticizing the idea of a microphysical determination (Gregersen 2004), though for different reasons. While strong emergentists claim that a causal reduction is not possible, prime representatives of the complexity sciences believe that it is indeed possible to compress complex systems into relatively simple algorithms. Strong appeals to computational reduction can be found, for example, in Stephen Wolfram's *magnum opus, A New Kind of Science* (2002; see critique in Weinberg 2002).

Accordingly, emergence theory may be said to have its strength in proposing a richer causal model, which takes the ordered networks of RWC explanatorily seriously, while computational studies of complexity have the advantage of using a purely quantitative scientific approach even to complex domains of reality. These differences will inevitably lead to conflicts between emergentists who defend causal irreducibility and proponents of CC, unless the latter admit that CC offers only more or less helpful models of RWC without grasping the real causal processes. After all, the appearance of a virtual leaf on a computer screen (designed via fractal geometry) does not provide us with any empirical knowledge of the physical, genetic, and morphological causes behind the growth of tree leaves in the spring. On this analysis, only the strong proponents of CC (such as Wolfram or proponents of Artificial Intelligence) will be in conflict with strong emergentists (and vice versa), whereas strong emergentists can happily endorse any computational modelling of emergent phenomena as long as it is granted that computational models are only the first intimations of further inquiries into the causal interplay between higher- and lower-order levels.

There is also overlap with respect to the subject areas, though not identity between the study of complex systems and the study of emergent systems. Not all forms of complexity are ordered. The grains of sand on a beach are highly complex from a computational point of view (so many shapes and so many angles!), whereas they are uninteresting from an emergentist point of view. Complexity, but not emergence, is here at work. The freezing of water at low temperatures, by contrast, is caused by the structure of water molecules plus the temperature, and does not require particular complex constellations. Emergence, but not complexity, is here at work.

Furthermore, not all emergents arise from complexity, for quite a few emergent phenomena, such as consciousness, take place by leaving out information, or by exformation or de-complexification of neural firings (Nørretranders 1998). The sensation of something like 'scarlet red' is much simpler than an accurate description of the environmental, sensory, and neural processes involved in the production of that particular *qualia*. Thus we have here a series of arrows leading from complexity to emergence to simplicity. But in other cases the arrow goes the other way round, as when a fascination with scarlet red and other colours leads someone to become a painter and thereby participate in the wider cultural circulation of paintings, schools of painters, galleries, art buyers, and newspaper reviews. Here the arrow goes from low-degree complexity to emergence to high-degree complexity.

EPISTEMOLOGICAL AND ONTOLOGICAL VARIETIES OF EMERGENCE: EMERGENCE$_{1-4}$

Emergent phenomena are known within all the sciences. The most startling example is perhaps the fact that the entire world of classical physics continually comes into being on the basis of the microscopic processes described by quantum mechanics. If we stay within classical physics, we find many other cases of emergence. Think of the phenomenon of electromagnetism, where the magnetic effect appears only when the magnet is positioned in the direction of other metallic substances; it is only within such fields that the magnetic effect takes place. In biology, one might think of the formation of bird flocks, and the consequent migration patterns that move many kilos of atoms and molecules around the globe—phenomena that are unexplainable in terms of general physics. And in psychology we might think of the stunning fact of sentience and attention, without which goal-directed actions would not be possible. Nature is a continuous source of surprise.

Scientifically, however, it is not satisfying just to enumerate examples of emergence. For the scientific explanations will differ markedly from, say, the formation of ice crystals, which supervenes on a clearly defined subvenient base of chemistry, to the emergence of human language, which evolves through an intricate interplay between central nervous systems and historically evolved social institutions. Likewise, it is not philosophically sufficient to refer to mere 'family likenesses' between emergent phenomena. For the question is whether one can reach a workable typology of different forms of emergence.

Terrence Deacon (2003; also Chapter 50 with Ursula Goodenough below) has proposed a helpful logic between otherwise unrelated forms of emergence. Deacon suggests a typology of three forms of emergence, all of which have to do with the amplification of patterns or configurations. First we have the often trivial cases of *first-order emergence* through supervenience relations, in which the properties of the higher-order systems depend on their subvenient level, such as the viscosity or 'aquosity' of water, which depends on the chemical bonding between hydrogen and oxygen, the presence of many H_2O molecules, and the appropriate thermal conditions. A *second-order* variant of emergence takes place through chaos and self-organization, where the environmental conditions play a formative role in combination with the concrete history of the system. A simple example is the formation of snow crystals, a phenomenon caused not only by the first-order intrinsic physical properties of ice (which exhibits a preference for hexagonal structures) but also by the radial symmetry of heat dissipation, the unique history of temperature and humidity, and the initial form of the snow crystals. At this level the diachronic parameter of time begins to be important, since the growth of crystals happens by amplifying historical processes. More complex forms of second-order emergence involve the interaction between different components—for example, in autocatalytic reactions.

However, only at the level of *third-order emergence* do we observe an evolution in the stricter sense, insofar as systems begin to 'remember' their history by including their own prehistory in their organizational programme. The nucleic acids in genomes offer one example, but of course all cases of evolutionary learning can be cited, from intra-organismic co-ordination of parts to the emergence of inter-organismic communication. Evolutionary systems here begin to be self-referential, so that they maintain and produce themselves according to internal programmes that themselves are the results of a selective 're-entry' of past experiences. Here we have to do with 'self-referential self-organization, an autopoiesis of autopoieses' (Deacon 2003: 299). Thus the orders of emergence proceed logically from thermodynamics to morphodynamics to semeiodynamics. What matters is not only mass and energy, but also the informational or configurational capacities of matter.

All three variants of emergence are understood in full acknowledgement of bottom-up causality. What happens in self-organization and autopoiesis is propagated by the physical capacities of matter that afford such emergents. But once the emergent systems have been established, they are able to perform additional causal roles by constraining and channelling ('from above') what is dynamically possible ('from below'). Observe that Deacon's typology has itself the structure of a nested hierarchy. That is, the first-order emergence of supervenience and the second-order emergence of environmental conditions are present within the third-order emergence of autopoiesis (cf. Gregersen 1998).

One could add to Deacon's analysis the observation that the lower-order forms of emergence may also gain new qualities. Whereas water supervenes on the particular bonding between hydrogen and oxygen, the Principle of Multiple Realization applies to the supervenience relation between brain and mind. That is, different brains and brain states may produce the same content of consciousness—say, feeling a specific ache or seeing a specific colour. Physicalist supervenience may here be transformed into more holist forms of supervenience, where the effects of the particular token–token relations depend on the selective context (Murphy 1998; Gregersen 2000). Furthermore, the interaction with the environment takes a new form at the level of third-order emergence: a remembered and recognized environment 'influences' the organism in a different way from temperature 'influencing' the formation of snow crystals. In living organisms, the perception of the environment modulates and interprets the environment in accordance with the internal 'interests' of the system, whereas snow crystals are purely passive to the external influences. Thus the pheno-menon of *autopoiesis*, or self-productivity, significantly changes both the super-venience relations and the relation to the environment.

These points can, I assume, be taken as friendly amendments to Deacon's typology. It would also be in line with Deacon's position, stated in *The Symbolic Species* (1998) and elsewhere, to argue that the dynamics involved in semiotic processes differ markedly between, say, the internal response of an immune system to its environ-ment, the signals of birds to one another, and the new capacities of the human language. From an empirical point of view, these new *capacities* of emergent systems should be acknowledged scientifically, whether or not we are able to formulate them as emergent 'laws of nature' (cf. Cartwright 1989).

Now what is the status of emergent phenomena? What is actually emerging? Is it only *properties*, with no causal roles, or is it properties that involve new *causal capacities*, some of which may even be formulated in the form of *emergent laws*? Or again, can one legitimately speak of new *emergent individuals*? These questions about the status of the emergent phenomena are linked to the fundamental distinc-tion between 'weak' and 'strong' claims of emergence (O'Connor and Wong 2005).

According to *weak emergence*, systemic features at 'higher' (or more comprehen-sive) levels cannot be predicted by any finite knower from the standpoint of the pre-emergent stage, despite a knowledge of the empirical characters and governing laws concerning the ultimate constituents of that system. According to *strong emergence*, emergent phenomena obtain new causal capacities, which make it possible that higher-level or more comprehensive systems can exert a top-down, selective influ-ence on the lower-level (or local) constituents than would otherwise have obtained.

Weak emergence can also be termed *epistemological emergence*, since all the causal work is done at the base level, whereas strong emergence can be termed *ontological*, since what is causally effective must be deemed real. Note, however, that the weak emergentist can also accord a sort of reality to the higher-order qualities: namely, the reality of being an epiphenomenon that should nonetheless be taken with metaphysical seriousness, even if the emergent properties play no causal role of their own. In other words, not all reductionists are eliminative reductionists. Note too that the strong emergentist will admit that in many cases emergent properties are only epiphenomena, without any new causal capacities. One has to investigate on a case-by-case basis to determine which phenomena manifest weak and which manifest strong emergence.

On further reflection, the epistemological and ontological versions of emergence each come in two classes. Let me therefore suggest a typology of emergence$_{1-4}$ to supplement Deacon's typology, which focuses primarily on varieties of emergence$_{2,3}$. Emergence$_{1,2}$ are examples of 'weak emergence', while emergence$_{3-4}$ are examples of 'strong emergence'.

Emergence$_1$ is a purely logical or computational form of emergence, as investigated in strong programmes of CC. Examples are fractal patterns emerging on the com-puter screen, or cellular automata that are able to produce quite complex structures. Computational structures evolve by proceeding logically, step by step, on the basis of the initial design of the programme and the choice of the initial parameters. One 'runs a programme', as it is often aptly put. Emergence$_1$ is thus not necessarily related to the real physical world (apart from the fact that computer programs are implemented on the hardware of Newtonian machines).

Emergence$_2$ refers to cases where new physical properties appear, but are fully dependent on their subvenient physical bases. These properties are nothing but properties or attributes of the underlying level or levels, and ultimately properties of physics. A concrete example may be the emergence of water strictly based on the intrinsic chemical properties of H_2O. But this category would also apply to Deacon's example of the 'morphodynamical' formation of snowflakes. The difference is only that here the underlying physical base would be extended so as to encompass not only the intrinsic chemical properties but also their relations to an environment. This level of emergence can be modelled successfully in CC.

Emergence$_3$ refers to cases where new emergent properties, based on new spatial or hierarchical configurations, acquire new causal capacities in the context of relatively enduring higher-order systems. What is important for new causal capacities is that the emergent systems do not immediately fall prey to their changing environments (as snowflakes do). Causal capacities require the emergence of relatively stable systems that are able to follow their own programmatic 'ends' even under changing circumstances (say, the search for food when there is no food immediately available). It seems that this level is reached at least with living organisms that are able to respond selectively to their environments. I take it that the ambitious approaches in CC aim to model primarily this type of RWC, without thereby being able to conclude that RWC is identical with CC.

Emergence$_4$ refers to the very special cases in which new emergent properties, based on new spatial or hierarchical configurations, give rise to new causal capacities in the context of relatively enduring higher-order systems exhibiting not only self-referential but also self-reflective and unified features such as human consciousness. In this case one might speak of emergent individuals (Hasker 1999: 190).

This latter possibility is more controversial than emergence$_3$, since it assumes the existence of body–mind systems that have the form of self-conscious unity, a form that is usually ascribed only to human persons. Thus the philosopher William Hasker has proposed using the term 'emergent dualism' for the position that the human brain and its central nervous system actually produce the human mind, but that, once developed, the human mind acquires independent causal capacities due to its self-reflective ability and its unified field of consciousness. Even though Hasker has pointed to a possibility that is often overlooked in discussions of emergence, I don't find the term 'emergent dualism' fully appropriate for describing the specific capacities of human beings—say, in planning to go to the airport or to finish an article before a deadline.

Philip Clayton has opted for the more open term 'emergentist pluralism', which has the advantage of seeing the emergence of the human mind as continuous with other sorts of self-reproducing systems. Still, the human person constitutes a special case of emergence: emergence$_4$. Like Hasker, Clayton also rightly argues that the human person acts as a responsible person in relation to the environment and to him or herself. Thus human persons have the feature of being centres of intentions and autonomous actions; they are not only disparate fields seeking specific ends, such as food consumption: 'What emerges in the human case is a particular psycho-somatic unity, an organism that can do things both mentally and physically' (Clayton 2004: 148–9). Clayton therefore refers to the social and human sciences as the disciplines that offer the appropriate explanations of human actions. By implication, this view argues that human behaviour cannot be fully explained from a neural point of view.

THEOLOGICAL REFLECTIONS ON EMERGENCE AND COMPLEXITY

Already the early twentieth-century emergentists differed on their religious views. Morgan was a classical theist who understood God to be the immaterial source of all that exists (1923: 298–301). Alexander, on the other hand, saw God as the product of a developing world, all parts of which also sustain the reality of the deity (1920: ii. 394–6). Finally, other emergentists, such as Roy Wood Sellars (1922), wanted to stay within the confines of an unqualified naturalism.

However, what makes emergence and complexity open for a religious interpretation is the simple fact that an emergentist monism portrays nature as intrinsically inventive and filled with natural wonders, and that evolution again and again affords new centres of activity that can be seen as foci of divine interaction with an evolving world. As we will now see, some theological appropriations of emergence demand the stronger views of emergence$_{3,4}$, whereas other religious interpretations need only the weak versions of emergence$_{1,2}$.

In his Gifford Lectures on *Space, Time, and Deity*, Samuel Alexander developed a radical theory of an *emerging God*. Proceeding analogically from the bottom towards the top, Alexander posited a deity as a further step in the ladder of emergence. Deity has its roots in the strivings of spacetime towards ever more complex modes of realization: 'There is no intervention here, but only extension of a series whose principle is known, to another term. Even without the religious emotion, we could on purely speculative evidence postulate deity, on the ground of the general plan on which Space-Time works' (1920: ii. 381). This is a fully naturalized view of providence, for purpose is here rooted in the spacetime of matter, and deity is the quality or attribute that results from nature's upwards 'nisus'. The world is God's body, and God is said to be the 'possessor' of the emerging divine qualities. Accordingly, God's deity is lodged only in a portion of the big universe, for the divine qualities, in Alexander's view, supervene on the emergent qualities of life and mind: 'God includes the whole universe, but his deity, though infinite, belongs to, or is lodged in, only a portion of the universe' (ii. 357). Accordingly, spacetime is prior to the actuality of God, and 'God as an actual existent is always becoming deity but never attains it. He is the ideal God in embryo' (ii. 365).

Samuel Alexander's philosophical theology is an interesting case, since he was prepared to apply the idea of emergence without qualifications to God's nature. The very nature of God is emerging alongside with, and as a result of, the upwards drive of evolutionary history. But insofar as the universe is also understood as the body of God, God is coextensive with the universe, and may even be termed creative: 'as being the whole universe God is creative, but his distinctive character of deity is not creative but created' (ii. 397). While the specifics of Samuel Alexander's theology have not found many followers, the general form of his theological logic nonetheless reappears in other forms, most notably in Alfred North Whitehead's work (see below).

However, it is also possible to argue that God is the initiator of emergence, without falling prey to the confines of emerging realities. A first version of this view is classical *atemporal theism*. That God cannot change is an axiom inherited from ancient Greek philosophy, and in the Middle Ages it was the position shared by Jewish, Christian, and Muslim philosophical theologians such as Moses Maimonides, Thomas Aquinas, and Avicenna. Although God is the creator of a temporal world, God is assumed to be unimaginably beyond time and change.

In early modernity, this idea of a timeless God found a new shape in the wake of the scientific idea of the physical closure of the natural world. Whereas the timelessness of God was earlier professed to safeguard the otherness of God, the idea was now that, subsequent to creating the world, God could not interfere with it. The world was now conceived as a mechanistic world system, obeying the deterministic laws of physics. In its extreme form, atemporal theism grew into deism, which saw God only as the great initiator of the clockwork universe. More typically, God was seen not only as creating the universe, but also as sustaining the existence and order of the universe at every moment, including the laws of nature. This idea is usually referred to as uniformitarianism. In early nineteenth-century theology we find this uniformitarian position encapsulated in the work of the founder of modern theology, Friedrich Schleiermacher, who stated without equivocation that our 'absolute dependence on God coincides entirely with the view that all . . . things are conditioned and determined by the interdependence of Nature' ([1830] 1989: 170). Being preserved by God, and being part of the causal nexus of nature, is one and the same thing.

A more complex version of atemporal theism has recently been proposed by the physicist Paul Davies. He also takes his point of departure in a uniformitarian view of the God–world relationship, but he is far more interested in the new causal capacities of emergent phenomena. By focusing on the way in which the basic laws of physics, in combination with ever changing environments, seem to be fine-tuned for the emergence of organized complexity, he offers what he calls a 'modified uniformitarianism'. No divine supervision of the details of evolution is needed: 'By selecting judiciously, God is able to bestow a rich creativity on the cosmos, because the actual laws of the universe are able to bestow a remarkable capacity to canalize, encourage, and facilitate the evolution of matter and energy along pathways leading to greater organizational complexity' (1998: 158). Central to this view is a theistic inter-pretation of the so-called cosmic fine-tuning of laws and initial conditions for the production of a world consisting of life and consciousness. What is important is not only the laws of nature but also the balanced proportion of chance, which gives to creation its inherently emergent features. Emergent phenomena are not just pale manifestations of deterministic laws, for not only are some laws of nature probabilistic, but their outcomes depend on fragile environmental conditions. It is thus the felicitous interplay between necessity and chance that 'leads to the emergence of *a different sort of order*—the order of complexity—at the macro-, holistic level' (1998: 159).

Davies's position encourages him to be genuinely interested in emergent phe-nomena. God, in this view, is the creative mind who is both selecting the laws and making room for the appropriate portions of chance—for the purpose of emergent complexity. Davies thus proposes a sort of teleology in nature without a divine teleology tinkering with the detailed processes. God is instead like the inventor of a game such as chess, setting up the rules while leaving open the space for the self-development of the game of creation. There is, however, no explicit feedback from the world to God in Davies's model. One could argue, though, that his model would imply at least a certain order of succession in the divine mind as a result of God's coming to know what is coming to be in the world of creation. Furthermore, Davies is open to the additional possibility of God acting together with the natural capacities of the world, provided that one could develop a consistent theory for how God could do so without violating the laws and statistical distributions of nature.

Only the third theological model, temporal theism, to which we now turn, requires developing a theory of divine interaction with a developing world. During the twentieth century, various versions of temporal theism have been developed within philosophical theology. Process theology, mapped on the canvas of the philosophical theology proposed by the mathematician Alfred North Whitehead, was among the first to argue for a dipolar concept of God. According to process theology, God possesses not only a 'primordial nature', which is essentially unchanging, but also a 'consequent nature', which appears as a result of a consecutive divine absorption of temporal developments into the mind of God ([1929] 1978: 342–52). Other versions of the same idea have reappeared in many forms in later theology (both philosophical and doctrinal); and in the field of science-and-religion, temporal theism is the majority position. Temporal theism has a particular affinity to strong emergence. Though a temporal theism, understood as a passive divine responsiveness, is indeed possible under the premises of weak emergentism only, the view that God interacts with a developing world is particularly congenial to the idea of a God whose experience is growing with the emergent realities, in relation to which God is held to be actively involved.

In the work of the biochemist Arthur Peacocke we find a particularly clear way of combining temporal theism with strong emergence. According to Peacocke, God cannot know all future actualities. God can know the phase space of future possibil-ities, but not the exact route of emergent evolution. This limitation on divine knowledge may be seen by critics as an external limitation on God. But if God creates the world of nature by setting it free for fertile self-explorations, this limitation is not external to God but is instead rooted ultimately in the generosity of divine love. God's self-limitation, or kenosis, is motivated in the self-realization of divine love.

Now Peacocke combines this 'let go' view of an underlying divine creativity with a strong sense of a God who accompanies and actively responds to the world in a manner analogous to whole–part causation. Peacocke suggests that we understand God as the 'Circumambient Reality' of the universe. God, however, is not to be perceived as a far-away environment external to the universe because, being immanent in the world, God is ubiquitously present and incessantly active, working out the divine intentions 'in, with, and under' the nexus of nature as a whole. It is in this context that Peacocke uses the top-down (or 'whole–part') causality as an analogy to God's influence upon emergent evolution:

By analogy with the operation of whole–part influence in real systems, the suggestion is that, because the ontological gap between the world and God is located simply everywhere in space and time, God could affect holistically the state of the world-System. . . . God could cause particular patterns of events to occur which would express God's intentions. These latter would not otherwise have happened had God not so specifically intended. (Peacocke 2001: 110)

Observe that Peacocke is not arguing that there are in-principle gaps in scientific explanation. In general, he finds only two sources of ontological indeterminacy in the world of nature—quantum indeterminacy and consciousness—and in his view neither of them is sufficient for a theology of a transformative divine presence in nature. It is, rather, nature as a system of systems that exhibits cases of emergence and moves in the direction of ever greater complexity. Therefore Peacocke's theological explanation is related primarily to the world as a whole. In a manner analogous to the way in which higher-level systems (for example, a flock of migrating birds) exercise an informational influence on lower-level systems (the individual birds), God informs and reconfigures the world-as-a-whole. Since no information flows without some exchange of matter and energy, it is important for Peacocke that God's influence is seen not as an additional supernatural causality but rather as a causality always couched in, and hidden by, natural processes. The immanence of the transcendent God is thus quintessential to Peacocke's position. One would never be able to extract 'a divine factor' from the natural flow of information, for it is exactly in nature's operations that God is active. Thus the rational character of his theistic understanding of reality does not depend on gaps in scientific explanations, but appears rather as a meta-reflection on the trajectory of evolution as a whole. The epistemic support for his religious interpretation is cumulative and draws mainly on the outcomes rather than on specific causal loopholes in nature.

Peacocke has been criticized for hypostasizing 'the world-as-a-whole'. While insisting that the world does make up a 'system of systems', he admits that the causal work is actually done mostly in the interplay between distinctive, type-different systems (such as systems of sound and meaning in human language). He is therefore ready to argue, *ex hypothesi*, that relatively autonomous processes take off during history, processes that instantiate new modes of causality (Peacocke 1999).

For such a view one would have to accept a strong version of emergence, including not only Deacon's first- and second-order varieties of *supervenience* and *self-organization*, but also the third-order variety of *autopoiesis*, or self-productivity. As I have argued earlier, '[a]utopoietic theory does claim a process autonomy since type-different systems operate on the basis of their internal codes. Thus, the fact that type-different systems cannot be written together in a uniform causal scheme has an ontological basis in pluriform evolution itself' (Gregersen 1998: 363). For example, immune systems select antibodies to be cloned, birds and mammals produce species-specific warning signals, bird flocks produce population-specific cycles of migration, and the production of meaning in human language involves the selection of sounds to be uttered.

Observe that both emergence$_3$ and emergence$_4$ require here a correspondingly weaker notion of supervenience, since it is implied that supervenience relations cannot grasp the whole causal story enacted in the evolutionary processes. Supervenience may well explain the initial formation of higher-level systems (for example, the emergence of life and of central neural systems in natural history), but once these systems have appeared through evolution, they take on an autonomous causative role in co-determining the use of the available energy budget. Such a formative role of higher-order systems is exhibited both in the world of biology and in human experience.

Metaphysically, this view requires that information play a fundamental role in nature, not secondary to, but coterminous with, the mass-and-energy aspects of the physical world. Mass, energy, and information are three distinguishable, but not extractable, aspects of what we have been calling 'matter'. This view of reality also affords a richer theology, in which God may be actively present in the world as a resource of information, which then becomes manifest in the configurational powers of a self-complexifying world. Theologically, it is important to note that the idea of *auto*-poiesis or *self*-complexification does not mean that nature *eo ipso* operates without God. On the contrary, 'God could guide the process of emergence by introducing new information (formal causality) and by holding out an ideal or image that could influence development without altering the mechanical mechanisms of evolution or adding energy from the outside (final causality)' (Clayton 2002: 273).

Concepts of complexity and emergence may thus be used to map an understanding of divine action, yet without making God's action secondary in relation to the base of the material world. The critic may then ask, *How* exactly is all this possible? Neither Peacocke's more holistic account nor more pluralistic accounts like Clayton's or mine will be able to answer the 'how question' directly. For this question presupposes exactly what is denied in these theological proposals: namely, that God and nature can be conceived independently. Instead, on this view, God and nature are so intimately intertwined that the presence of the living God cannot be subtracted from the world of nature and still leave the world of nature as it is. Nature equals God-and-nature. For the same reason, the arguments that I have put forward are not part of a *natural theology* which seeks to argue for the 'existence of God' against the backdrop of emergence and complexity. Rather, I am concerned with examples of a *theology of nature* that represent a type of hypothetical reasoning: *if* nature includes cases of strong emergence (involving self-organization and autopoiesis), and *if* God is the creator at work in, with, and under creation, then there is a natural fit, and hence a strong coherence, between an emergentist view of nature and the tradition of temporal theism. Of course, it is possible to affirm strong emergence without affirming a religious explanation. At the level of ultimate explanations, however, both the theist and the non-theist will be asked how they propose to explain the fact that the laws of nature are as fertile as they are, and why evolution exhibits the upwards drive that we are able to observe (Gregersen 2003). No metaphysics can avoid the task of explaining the fact of nature's stunning capacity for self-complexification.

References and Suggested Reading

Alexander, Samuel (1920). *Space, Time, and Deity*, 2 vols. New York: The Humanitarian Press.

Bergson, Henri (1911). *Creative Evolution*, trans. Arthur Mitchell. New York: Henry Holt.

Broad, C. D. (1925). *The Mind and its Place in Nature*. London: Routledge & Kegan Paul.

Cartwright, Nancy (1989). *Nature's Capacities and their Measurement*. Oxford: Oxford University Press.

Clayton, Philip (2002). 'Divine Causes in the World of Nature', in Ted Peters, Muzzafar Iqbal, and Syed Nomanul Haq (eds.), *God, Life, and the Cosmos: Christian and Islamic Perspectives*, Aldershot: Ashgate, 249–80.

—— (2004). *Mind and Emergence: From Quantum to Consciousness*. Oxford: Oxford University Press.

—— and Davies, Paul (2006) (eds.). *The Re-Emergence of Emergence*. New York: Oxford University Press.

Davies, Paul (1998). 'Teleology without Teleology', in Robert John Russell, William R. Stoeger, SJ, and Fransisco J. Ayala (eds.), *Evolutionary and Molecular Biology: Scientific Perspectives on Divine Action*, Vatican City State: Vatican Observatory Publications; Berkeley: Center for Theology and the Natural Sciences, 151–62.

Deacon, Terrence W. (1998). *The Symbolic Species: The Co-Evolution of Language and Brain*. New York: W. W. Norton.

—— (2003). 'The Hierarchical Logic of Emergence: Untangling the Interdependence of Evolution and Self-Organization', in Bruce H. Weber and David J. Depew (eds.), *Evolution and Learning: The Baldwin Effect Reconsidered*, Cambridge, Mass.: MIT Press, 273–308.

Gilbert, Scott F., and Sarkar, Sahotra (2000). 'Embracing Complexity: Organicism for the 21st Century', *Developmental Dynamics*, 219: 1–9.

Gregersen, Niels Henrik (1998). 'The Idea of Creation and the Theory of Autopoetic Processes', *Zygon*, 33/3: 333–67.

—— (2000). 'God's Public Traffic: Holist versus Physicalist Supervenience', in Niels Henrik Gregersen, Willem B. Drees, and Ulf Görman (eds.), *The Human Person in Science and Theology*, Edinburgh: T & T Clark; Grand Rapids, Mich.: Eerdmans, 153–88.

—— (2003) (ed.). *From Complexity to Life: On the Emergence of Life and Meaning*. New York: Oxford University Press.

—— (2004). 'Complexity: What is at Stake for Religious Reflection', in Kees van Kooten Niekerk, and Hans Buhl (eds.), *The Significance of Complexity: Approaching a Complex World through Science, Theology, and the Humanities*, Aldershot: Ashgate, 135–65.

HASKER, WILLIAM (1999). *The Emergent Self*. Ithaca, N.Y.: Cornell University Press.

HEMPEL, CARL G., and OPPENHEIM, PAUL (1948). 'Studies in the Logic of Explanation', *Philosophy of Science*, 15: 567–79.

HOLLAND, JOHN (1998). *Emergence: From Chaos to Order*. Oxford: Oxford University Press.

KANT, IMMANUEL ([1790] 1952). *The Critique of Judgment*, trans. with Analytical Indexes by James Creed Meredith. Oxford: Oxford University Press.

KAUFFMAN, STUART (1993). *The Origins of Order: Self-Organization and Selection in Evolution*. New York: Oxford University Press.

—— (1998). *At Home in the Universe: The Search for Laws of Self-Organization and Complexity*. New York: Oxford University Press.

—— (2000). *Investigations*. New York: Oxford University Press.

LEWES, G. H. (1877). *Problems of Life and Mind*, ii. London: Kegan Paul.

LOVEJOY, ARTHUR ([1936] 1964). *The Great Chain of Being*. Cambridge, Mass.: Harvard University Press.

MILL, JOHN STUART ([1843] 1970). *A System of Logic Ratiocinative and Inductive Being a Connected View of the Principles of Evidence and the Methods of Scientific Investigation*. London: Longman.

MORGAN, C. LLOYD (1923). *Emergent Evolution*. London: Williams & Northgate.

MURPHY, NANCY (1998). 'Non-Reductive Physicalism: Philosophical Issues', in Warren S. Brown, Nancey Murphy, and H. Newton Malony (eds.), *Whatever Happened to the Soul?: Scientific and Theological Portraits of Human Nature*, Minneapolis: Augsburg Fortress, 127–48.

NØRRETRANDERS, TOR (1998). *The User Illusion: How to Cut Consciousness Down to Size*. New York: Penguin.

O'CONNOR, TIMOTHY, and WONG, HONG YU (2005). 'Emergent Properties', in Edward N. Zalta (ed.), *The Stanford Encyclopedia of Philosophy*, Summer 2005 edn; <http:/plato.stanford. edu/archives/sum2005/entries/propertiesemergent/>.

PEACOCKE, ARTHUR (1999). 'The Sound of Sheer Silence', in Robert John Russell, Nancey Murphy, Theo C. Meyering, and Michael A. Arbib (eds.), *Neuroscience and the Person: Scientific Perspectives on Divine Action*, Vatican City State: Vatican Observatory Publications; Berkeley: Center for Theology and the Natural Sciences, 215–48.

—— (2001). *Paths from Science Towards God: The End of all our Exploring*. Oxford: Oneworld Publications.

SCHLEIERMACHER, FRIEDRICH ([1830] 1989). *The Christian Faith*, ed. H. R. Mackintosh and J. S. Stewart. Edinburgh: T & T Clark.

SELLARS, ROY WOOD (1922). *Evolutionary Naturalism*. Chicago: Open Court.

WEINBERG, STEVEN (2002). 'Is the Universe a Computer?', *The New York Review of Books*, 49/16 (24 October): 43–7.

WHITEHEAD, ALFRED NORTH ([1929] 1978). *Process and Reality: An Essay in Cosmology*, corrected edn. ed. David Ray Griffin and Donald W. Sherburne. New York: Free Press.

WOLFRAM, STEPHEN (2002). *A New Kind of Science*. Champaign, Ill.: Wolfram Research.

CHAPTER 46

EMERGENCE, THEOLOGY, AND THE MANIFEST IMAGE

MICHAEL SILBERSTEIN

OVERVIEW

The first section discusses the relevance of emergence to theology. In the second section different varieties of emergence are characterized; it will be argued that only radical mereological/causal emergence and nomological emergence—both of which are forms of ontological emergence—have any real relevance for theology. The third section focuses on the promises and pitfalls of using ontological emergence to argue for the existence of God and divine action. Finally, the last section makes a variety of friendly suggestions to theologians, focusing on the direction of future work in emergence, theology, and the manifest image.

The Connections between Theology and Emergence

Theologians are hoping that emergence in one form or another will shed light on the nature and existence of God, divine action, the mind–body problem, and free will (Saunders 2000). Of course these concerns play themselves out in a variety of different ways in the hands of different theologians, but for the purposes of this chapter I will focus almost exclusively on one book, *Mind and Emergence: From Quantum to Consciousness* (2004) by Philip Clayton. The choice to focus on Clayton's book is due to its excellence and scope; there is no better attempt to argue from emergence to both the existence of God and the failure of physicalism in the theological literature. Clayton's conception of both God and the self are realist and robust in nature, and thus, while theologically sophisticated, also comes close to a widely held kind of theism and to the manifest view of human beings and their volitional powers. Clayton's book also considers an alternative connection between emergence and theology in terms of God as an emerging, evolving process or entity. Finally, Clayton's comprehensive work on the relationship between emergence and God is a perfect stalking horse to make some more general points about how the science and theology game ought to be played.

Clayton and many other theologians are hoping that emergence will at least help them establish that the following package of beliefs is coherent, logically consistent, and sufficiently probable.

(1) *Theism*   God exists and created, maintains, and acts in the world, though without recourse to 'miracles' (violating natural law) or 'intelligent design' (ID) of the sort defended by the ID movement. Furthermore, God may not be perfect, but, relatively speaking, he can be characterized by something approaching omnipotence, omniscience, omnipresence, and omni-benevolence—though Clayton is obviously sceptical and critical of 'perfect-being theology' and clearly rejects the claim that God is omnipotent.

(2) *The manifest image of personhood*   Short of positing immortal souls, there is some kind of robust 'ego-theory' of the self, understood as an agent possessing something approaching libertarian free will, though it is unclear whether Clayton fully embraces libertarianism. Mental causation *qua* mental is a real feature of the world, and thus physicalism and causal closure are false. Though some theologians would go further and add 'life after death', Clayton does not do so in *Mind and Emergence*.

Following the twentieth-century American philosopher Wilfrid Sellars, we can say that Clayton and some other theologians are hoping that emergence in some guise will help them to reconcile the 'manifest image' of God, creation, and self with the 'scientific image'. In the next section we will get clear on exactly what kind of emergence must obtain in the world if he is to have any shot at the goal of reconciliation.

ONTOLOGICAL VERSUS EPISTEMOLOGICAL EMERGENCE: A TAXONOMY

We can divide types of emergence into two basic categories: epistemological emergence and ontological emergence. (For detailed taxonomies of emergence see Silberstein 1999, 2001, 2002.) Historically there are two main construals of the problem of reduction and emergence: ontological and epistemological. The ontological construal of the question asks: is there some *robust* sense in which everything in the world can be said to be *nothing but* the fundamental constituents of reality (such as super-strings) or, at the very least, *determined by* those constituents? The epistemological construal of the question can be framed similarly: is there some *robust* sense in which our scientific theories/schemas (and our common-sense experiential conceptions) about the macroscopic features of the world can be inter-theoretically *reduced to* our scientific theories about the most fundamental features of the world such as quantum mechanics? Yet these two construals are inextricably related. For example, it seems impossible to justify ontological claims (such as the cross-theoretic identity of conscious mental processes with neurochemical processes) without appealing to epistemological claims (such as the attempted inter-theoretic reduction of folk psychology to neuroscientific theories of mind), and vice versa. That is, why believe that mental states are nothing but brain states if cognitive neuroscience does not eventually exhibit superior predictive or explanatory capacities with respect to folk psychology? We would like to believe that the unity of the world will be described in our scientific theories, and in turn, the success of those theories will provide evidence for the ultimate unity and simplicity of the world. Of course, things are rarely so straightforward.

Epistemological emergence involves relationships between representational items such as theories, concepts, models, frameworks, schemas, regularities, etc. The fact of epistemological emergence in any given case does not entail ontological emergence; there are many known cases of epistemological emergence with no obvious ontological significance. There are two main categories of epistemological emergence.

(1) *Predictive/explanatory emergence*   Wholes (systems) have features that cannot *in practice* be explained or predicted from the features of their parts, their mode of combination, and the laws governing their behaviour. In short, X bears predictive/explanatory emergence with respect to Y if Y cannot (reductively) predict/explain X. More specifically, in terms of types of inter-theoretic reduction, X bears predictive/explanatory emergence with respect to Y if Y cannot *replace* X, if X cannot be *derived* from Y, or if Y cannot be shown to be *isomorphic* to X. A lower-level theory Y (description, regularity, model, schema, etc.), can fail to predict or explain a higher-level theory X for purely epistemological reasons (i.e. conceptual, cognitive, or computational limits). If X is predictive/explanatory emergent with respect to Y for *all possible cognizers in practice*, then we might say that X is *incommensurable* with respect to Y. A paradigmatic and notorious example of predictive/explanatory emergence are chaotic, non-linear dynamical systems. The emergence exhibited in

chaotic systems (or models of non-linear systems exhibiting chaos) follows from their sensitivity to initial conditions, plus the fact that physical properties can be specified only to finite precision; infinite precision would be necessary to perform the required 'reduction', given this sensitivity. However, there is no reason to believe that chaotic systems provide evidence of ontological emergence (as defined below), or indefinite properties, or even dynamical indeterminism. For example, dynamical systems have attractors as high-level emergent features only in the sense that one cannot deduce them from equations for the system. Another example closer to home, McGinn (1999) and other 'mysterians' hold that folk psychology is predictive/explanatory emergent with respect to theories of neuroscience.

(2) *Representational/cognitive emergence*   Wholes (systems) exhibit features, patterns, or regularities that cannot be fully represented and understood using the theoretical and representational resources adequate for describing and understanding the features and regularities of their parts and reducible relations. X bears representational/cognitive emergence with respect to Y if X does *not* bear predictive/explanatory emergence with respect to Y but, nonetheless, X represents higher-level patterns or non-analytically guaranteed regularities that cannot be fully, properly, or easily represented or understood from the perspective of the lower-level Y. As long as X retains a significant *pragmatic* advantage over Y with respect to understanding the phenomena in question, then X is representational/cognitive emergent with respect to Y. Non-reductive physicalism (NRP) holds that folk psychology is representational/cognitive emergent with respect to theories of neuroscience. That is, NRP does not doubt that a future cognitive neuroscience will explain and predict conscious mental states and folk-psychological intentional capacities, but it does deny that cognitive neuroscience can fully subsume the explanatory power and counterfactual generalizations of folk psychology (Antony 1999).

Ontological emergence, in turn, involves relationships between items in the world such as entities, events, properties, laws, forces, etc. One can identify three main categories of ontological emergence.

(1) *Mereological or causal emergence*   This refers to cases in which wholes have *causally efficacious* properties that are not determined by the intrinsic, and perhaps even relational, physical properties of their most basic physical parts. Indeed, there might even be some cases in which entities are not even wholly composed of basic physical parts such as on some interpretations of quantum entanglement. The paradigm example of mereological emergence is quantum entanglement, or non-separability. Some claim that in the entangled state the particles that form such states cease to exist as individual entities (Humphreys 1997). Causal emergence need not be as radical as quantum entanglement, however. Any case in which global features or properties of the whole system—including contextual or environmental features of the system or irreducible relations between the parts—constrain, guide, change, or supersede the behaviour of the parts of that system may be said to possess causal emergence. These cases might include teleological and functional constraints as described by naturalistic formal and final causation, as well as various feedback and feed-forward mechanisms that perhaps impose such constraints such as self-organizing systems.

(2) *Radical mereological or causal emergence*   A particularly radical form of mereological emergence holds that what constitutes part and whole in any given system, and what fixes their respective causal capacities, are determined by environmental and contextual features of the system. For example, the Buddhist belief in 'interdependent arising'—the claim that the universe as a whole constitutes a system in which 'everything determines everything'—represents a very radical form of causal emergence. On this view, the whole universe is more ontologically fundamental than any particular subset or part of it; the universe is not made or composed of parts. Interdependent origination or arising says that nothing has independent, intrinsic existence; there are no intrinsic properties and no independently existing objects or things. There is mutual co-dependence between parts and wholes, which is to say that each determines and defines the other. On this view of the world, what are fundamental are a myriad of complex interrelations, not entities or things that compose larger things. Rather than causation as typically conceived, it is better to think of such radical determination relations as global or systemic acausal determination relations. Of course one need not adopt radical mereological emergence for the whole universe, but rather posit it only about particular sub-systems of the universe.

(3) *Nomological or strong emergence*   This refers to cases in which *higher-level causally efficacious* entities, properties, forces, potentials, laws, teleological organizing principles, etc. come into existence without in any way being necessitated or determined by lower-level features, and constrain, supersede, or change the behaviour of the latter. In such cases fundamental physical facts and laws would only provide at best a necessary condition for higher-level facts and laws. If, for example, brain states are necessary but not sufficient for mental states, and the latter causally affect the former, then mental states are nomologically emergent phenomena.

IMPLICATIONS OF THE THREE TYPES OF EMERGENCE

Non-radical mereological emergence is perhaps necessary, but certainly not sufficient, for nomological emergence. Quantum entanglement, for example, does not imply nomological emergence, since such quantum states are predicted by quantum mechanics itself. Furthermore, there is no obvious reason to think of entangled states as on a higher level than the level of particles that make up such states. Finally, if so-called downward causation implies that features of a *higher-level of reality* act downwardly on *lower-level features of reality* (such as mental states 'downwardly causing' brain states to change), then nomological emergence implies downward causation, whereas non-radical mereological emergence implies only violations of mereological or 'Humean supervenience'—the claim that the intrinsic and local physical properties of the basic parts of any given whole determine all the causal properties of that whole.

Of course, if radical mereological emergence is true globally, then both nomological emergence with its 'downward causation' and physicalism (the claim that the fundamental physical facts such as quantum states fix all the other higher-level facts) are wrong. Instead, reality as a whole is not divided up into a discrete hierarchy of levels, but is in fact an irreducibly relational, entangled, coupled, complex, self-organizing system.

Broadly, radical mereological emergence posits that the ordering on the complexity of structures, ranging from those of elementary physics to those of astrophysics and neurophysiology, is not discrete. Indeed, it holds that the interactions between such structures will be so entangled that any separation into levels will be conventional or contextual. The microscopic and the macroscopic may be only contextually or fictively separable from one another. On this view, the divisions and hierarchies between phenomena that are usually considered fundamental and emergent, simple and aggregate, kinematic and dynamic—and perhaps even what is considered physical, biological, and mental—are redrawn and redefined. Such divisions will be dependent on what question is being put to nature and what scale of phenomena is being probed. It is true that science is divided into hierarchical descriptions and theories; given radical mereological emergence, however, this might be only an epistemological artefact of scientific explanatory practice, not a fact about the world. The point is that neither nomological emergence nor physicalism can get off the ground if there is not some clear-cut intrinsic sense about what constitutes fundamental laws, entities, properties, etc.

It should be clear that the only kinds of emergence that might usefully figure in arguments against physicalism and causal closure[1] are nomological emergence and radical mereological emergence. Put another way, non-radical mereological emergence such as quantum entanglement is perfectly consistent with physicalism. Nomological emergence and certain brands of radical mereological emergence entail the falsity of physicalism and causal closure. It should also be clear that everything just said also applies to trying to argue from emergence divine action, free will, etc.

THE MOVE FROM ONTOLOGICAL EMERGENCE TO THE EXISTENCE OF GOD, MENTAL CAUSATION, AND FREE WILL

This section involves a close reading and a detailed critique of Clayton's excellent attempt to move from ontological emergence to the existence of God, mental causation, etc. (See Clayton (2004: 203 ff.) for a summary of his master argument.)

---

[1] I define causal closure to mean that, for any physical event e, if e has a cause at time t, then e has a wholly physical sufficient cause at t. Or, put differently, all physical events are determined, insofar as they are determined, by prior physical events and the physical laws that govern them.

Before we get into the argument itself, it should be kept in mind that without adverting to post-modern conceptions of rationality and/or faith alone, there are only two ways to justify belief in the existence of God as defined by Clayton and company: (1) to show that God fills some explanatory lacunae that cannot even in principle be filled by science because the universe really has such naturalistically unbridgeable gaps; or (2) to show that the existence of God is at the very least logically and physically consistent with what we take to be our best scientific theories about the world. All of the foregoing holds true also for justifying belief in human mental causation and/or free will.

What we will learn through the course of examining Clayton's argument is that option (1) is quite hard to defend and option (2) becomes harder to defend as natural science fills in explanatory gaps, although even at the ideal end-point of scientific investigation option (2), while uncomfortable, can be defended. We will also learn that when it comes to defending option (1), employing emergence as a premiss is a tricky proposition, because to be useful it can be neither a *limited* or merely 'in practice' 'God of the gaps'-type claim, nor an invocation of some additional natural feature of reality, such as new and novel laws of self-organization that make God's existence even less necessary from an explanatory perspective.

Clayton's master argument presented in standard argument form is as follows:

(P1) Ontological emergence of both the mereological and nomological variety is a real feature of the world from the realms of physics through psychology (emergent naturalism).
(P2) Mental phenomena are nomologically emergent, such that physicalism and causal closure fail to be true; that is, reasons are truly causes in this world (emergent naturalism).
(P3) Mental states or propositional attitudes such as beliefs and desires are best understood as states of emergent agents possessing libertarian-like free will. In short, folk psychology is a true theory.
(P4) Not even emergent naturalism can explain the following things, all of which stand in need of explanation, especially given that mind is a natural emergent:

  (A) what it means for a belief to be true and justified—the existence and semantics of propositional attitudes as given by folk psychology;
  (B) why the world is inherently rational and knowable by the exercise of human reason;
  (C) how and why the universe produces human agents with the capacity to understand the universe and form justified, true beliefs.
  (D) For good measure Clayton also throws in the following mysteries: why is there something rather than nothing? How is one to explain the existence of ethical facts, religious experience, and the meaning of life?

(P5) Nothing in the natural world, not even emergent naturalism, can possibly ever explain (A)–(D) above.
(P6) Thus, the fact of (A)–(D) above is either a brute fact or it has a non-natural explanation.
(P7) Given that it is highly improbable for (A)–(D) to be a brute fact, then the most likely non-natural explanation is that the universe and the agents in it were in some sense intentionally designed to have these features. It therefore follows that God is the one who designed the universe to have features (A)–(D).

Clayton's argument is a *kind* of argument from design. He uses ontological emergence to explain or justify naturalistically the existence of cognitive agents for which (*contra* physicalism and causal closure) reasons *qua* reasons are causes. Ironically, however, their existence in turn leads to explanatory gaps that Clayton claims can be filled only by God. First, Clayton's argument employs ontological emergence to show that mental causation (reasons as causes, etc.) is consistent with, and justified by, our best scientific theories: namely, those he alleges must appeal to ontological emergence to explain things such as consciousness and mental causation. Second, he uses the alleged limitations of ontological emergence to argue for the explanatory necessity of God.

The argument in a nutshell is that ontological emergence (*contra* physicalism and causal closure) provides for the reality and coherence of agents for whom reasons are causes; but the fact that such agents exist in turn begs the question of how (A)–(D) can be explained. Clayton claims that even ontological emergence cannot explain (A)–(D); for that, we need God. His argument embodies the tricky balancing act we spoke of earlier. Clayton needs ontological emergence to defeat physicalism and causal closure, thus producing cognitive agents for whom reasons as causes is plausible; but he must not grant so much universally explanatory ontological emergence that even (A)–(D) can be explained naturalistically. We shall have occasion to worry more about this knife-edged strategy as we proceed.

We will now turn to an evaluation of the truth-value or justification for each premiss in Clayton's argument and offer an analysis of its inductive strength. (P1) asserts the reality of both mereological and nomological emergence across the natural sciences. The latter, however, is much harder to prove; and as we know, only the latter constitutes a refutation of causal closure and physicalism. Clayton argues from enumerative induction to the conclusion of nomological emergence (see his ch. 3). Cases from the natural sciences that he canvasses include conductivity, chaotic hydrodynamics, Raleigh–Benard convection, autocatalysis, self-organization of various sorts such as the formation of snowflakes, Belousov–Zhabotinsky reactions, finite cellular automata, evolution, neural networks, ant colonies, the quantum Hall effect, quantum decoherence, and the Pauli Exclusion Principle. The only thing these myriad cases studies have in common is that they have all been tagged 'emergent phenomena' at one point or another in the literature; but the important question is what kind of emergence they represent. Unfortunately, none of these cases is an obvious candidate for nomological emergence or radical mereological emergence, and that is what Clayton needs to contradict causal closure and physicalism. Perhaps a case could be made that some of these examples *do* constitute nomological emergence, but Clayton does not attempt to make this case in his book. While I cannot possibly analyse each case in this chapter, what I can say is that many of them, such as finite cellular automata, obviously show no more than epistemological emergence. Such automata are ontologically deterministic, possess definite values, and in them each successive state, though unpredictable by us, follows logically from the rules and the antecedent states of the system. Many of the other cases, such as the quantum Hall effect, are good candidates for mereological emergence only; after all, as any condensed matter theorist would tell you (including Laughlin, Leggett, and Anderson), quantum mechanics does predict such effects.

In all fairness, Clayton recognizes that none of these cases constitutes prima facie evidence of nomological emergence. He also recognizes that it is nomological emergence that he needs to defeat physicalism and causal closure, but he goes no further (see Clayton 2004: 100, 129, 171–204). It is important to keep in mind the standard here: in order for an example of emergence to count against physicalism, it has to be the case that physics (or whatever the more fundamental physical theory happens to be) can not explain the example reductively *in principle*—at any rate, one must at least try to demonstrate that this is the case.

In (P2) Clayton asserts that mental phenomena such as consciousness and mental causation provide a much stronger case for nomological emergence. While he is certainly right that the natural sciences have as yet had very little success in predicting and explaining such phenomena, what we need are arguments that attempt to establish that the natural sciences must fail *in principle* in such a quest. Physicalism maintains that mental states are determined by brain states, not the other way around, and that someday we will be able to fill in the story by moving from the neural correlates of mental states (now allegedly being mapped by empirical investigation) to the complex causal neurochemical mechanisms underlying those mental states. In addition to enumerative induction based on past successes in the natural and neurosciences, physicalism believes the preceding claim based on several independent arguments for causal closure.

Thus, assuming causal closure, then according to physicalism, the only way in which mental states can be causal is if they are nothing but brain states. However the problem with the argument from enumerative induction based on past successes is as follows:

While there are some notable reductions (or partial reductions), for example, of thermodynamics to statistical mechanics, claims of reduction are usually accompanied by a great deal of hand waving. And while each successful reduction provides some reason in favour of physicalism, each failed reduction provides some reason against it. So this piecemeal approach is far from conclusive. (Loewer 2001: 49)

Loewer is a well-known advocate of physicalism, and his point is that, at this early stage in the evolution of science (especially cognitive neuroscience), neither side can hope to profit (or prophet!) much from arguments based on past successes or failures of inter-theoretic reduction. Thus, what Clayton still owes us are good reasons for thinking that consciousness and mental causation are irreducible phenomena *in principle* (that is, not determined by brain states) and that causal closure is most likely false. At the very least we need to be shown why the best arguments for causal closure are neither inductively strong nor deductively valid (see Silberstein 2006*b* for such an attempt). The assumption of causal closure is really the crucial issue with respect to physicalism, as it is the basis for most people's belief in physicalism.

(P3) asserts the folk-psychological picture of propositional attitudes, self, free will, and mental causation. Forms of scientific psychology such as the representational theory of mind (RTM), the computational theory of mind (CTM), and connection-ism—all of which purport to naturalize folk psychology while being realist about propositional attitudes—try to explain action and cognition without recourse to an irreducible self, libertarian free will, or irreducible intentionality. If Clayton thinks that these scientific ventures must fail *in principle*, then he needs to tell us why. Why, for example, should we believe that reasons are *irreducible* causes of action? After all, even the identity theorist does not deny that reasons are causes.

(P4) and (P5) assert that not even nomological emergence can explain the phe-nomena listed as (A)–(D) above. For example, Clayton says that nomological emergence cannot explain why the world is inherently rational and knowable by the exercise of human reason, or how and why the universe produces human agents with the capacity to understand the universe and form justified, true beliefs. The first task is to get clear on exactly what it is he is purporting cannot be explained by nomological emergence or any other natural feature of the world. An important issue here is: how is it that humans are so 'at home in the universe', such that both their common-sense and scientific theories are so successful, i.e. true? Of course, this presupposes the 'miracle argument' for scientific realism which I am perfectly happy to grant. Another issue here is why the universe is so induction-friendly. Finally, yet another related issue is the problem of intentionality—how natural language and thought manage to represent the world. Again, scientific psychology purports to answer the last question.

These are all certainly age-old philosophical and scientific conundrums; how hard or easy they are to resolve depends on one's background assumptions. For example, given only evolutionary psychology or neo-Darwinian mechanisms as the basis for explanation, then *perhaps* these questions cannot be answered. The problem with Clayton's strategy is that he never tells us *why* ontological emergence in its myriad forms *can* explain the emergence of the classical from the quantum, the emergence of life from physics, and the emergence of mind from biology, but *cannot* in principle explain how human reason reflects the structure of the world. Keep in mind that all of Clayton's alleged examples of ontological emergence come from science itself, and thus ontological emergence is part of the explanatory repertoire of science. Onto-logical emergence is not a failure of scientific explanation as such; it is a failure only of *reductive explanation*. Clayton's entire book is presumably a testament to how ontological emergence can fill in the explanatory gaps currently left unfilled by our best reductive scientific theories. Furthermore, if ontological emergence is a real, scientifically tractable feature of the world that explains the emergence of everything from basic physics, *including* conscious reasoning minds, then it is not clear what, if any, mysteries in *principle* are left to explain. In short, it is not clear that there is some *additional problem* of explaining the success of science. If there is some additional problem, it is not clear why ontological emergence and natural science cannot do the job. Is there any reason to doubt that in principle some day cognitive science could explain the success of scientific reasoning?

Per (P6) and (P7) we see that Clayton's argument is a kind of design argument or inference to the best explanation: even given ontological emergence, without God as designer it would be a miracle (i.e. highly unlikely) if the universe turned out to be induction-friendly, and we turned out to be smart enough to unravel its mysteries. However, Clayton never tells us *why* the induction-friendliness of the universe or the inductive powers of humans to know it are highly unlikely states of affairs to begin with, and the addition of ontological emergence to the scientific arsenal only makes it that much harder to make such a case. His argument is that given the choice between (A)–(D) being a brute fact and an explanation by design, probability theory says we must go with the latter. But even if we are somehow convinced that physicalism does not have the explanatory resources to explain (A)–(D), does Clayton's own onto-logical emergence not suggest that his argument embodies a false dichotomy? Furthermore, it is not obvious that (A)–(D) as a brute fact is less probable than (A)–(D) as designed by God when we consider *how probable or not the existence of God is in itself.* Why exactly is it more likely that God, as defined by Clayton, exists than it is that (A)–(D) are brute facts? Notice that when the question is put this way, the defender of the God hypothesis cannot invoke the fact of (A)–(D) as the answer.

Clayton (2004) seems to think that there is some kind of *noumenal* divide between the veil of human theories and conceptual schemas, on the one hand, and the nature of reality on the other, but he never tells us *why* he thinks this (p. 177). Certainly philosophers such as Putnam, Nagel, and Plantinga have worried about such *divides* over the years, but the worry is always relativized to a certain world-view, such as Putnam's foil of 'metaphysical realism' or Plantinga's foil of the starkly 'neo-Darwinian' world. But by hypothesis Clayton's world is a world of ontological emergence complete with allegedly irreducible self-organization, teleology, mental powers, and final causation. So while there may be *some things* that even ontological emergence cannot explain, there is no obvious reason to think that the induction-friendliness of the universe or the powers of human reason are among them.

My point, in a nutshell, is this: if we posit Clayton's ontological emergence, then we do not need God to explain the induction-friendliness of the universe or the powers of human reason, and if we posit the existence of God, then we do not need ontological emergence or any particular naturalistic story to explain anything at all *ultimately.* However, positing the existence of a God that is profoundly powerful, good, etc. raises far more worries about counterfactual worlds and possibilities than physics does. If such a God exists, then why would he even bother creating a dizzying and complex *physical world* of forces, causal mechanisms, atoms, cells, brains, etc.? Why not simply create immortal souls and the divine equivalent of the *Matrix*? If goodness, as opposed to, say, a good storyline is his bottom line, then why not make creation exactly as it *ought* to be from the beginning? God simply does not need ontological emergence or any other naturalistic process or mechanism. If it is none-theless his nature to create with such naturalistic tools, then obviously there are things, such as aesthetic considerations, which he values above divine providence, in which case the problem of evil rears its ugly head once again. If the end-point of the universe's evolution is morally and spiritually ideal, then why not just make the world like that from the beginning? Is doing so within God's power or not? Either way there is a problem for theism. If God can do this and does not, then we have the problem of evil. If on the other hand, God is only powerful enough to create the world and then merely co-evolve with it, that radically reduces God's explanatory power and providential capacities.

There are all kinds of trade-off problems here that God must face as well. For example, can God make a world that is both truly open cosmologically and open in terms of human actions, *yet* progressively evolving toward some fixed, divinely chosen, teleological end-point? Openness and a fixed teleological end-point seem to be at odds with one another unless the point of the universe is just to see what will happen, which is tantamount to some cosmic science experiment. Of course one can always get out of such worries by reducing God's power and/or goodness, but then one also reduces his explanatory power as well as his original appeal in terms of providence, meaning, teleology, etc.

Of course, perhaps God is just weak enough, and anti-providential enough to deflate the problem of evil, yet just powerful enough and good enough to explain all the gaps in our scientific theorizing; but we need some reason to believe this other than wishful thinking. Do we not now have a fine-tuning problem with respect to God himself? Why could God not have been otherwise, why *this* deity, not *that* one? Is it possible that in light of our experiences and investigations into reality we ought to rethink the axiomatic assumption of such a deity? After all, there are deeply religious, spiritual, and ethical traditions that disavow such an entity.

While Clayton's book does not focus on traditional fine-tuning worries, perhaps, as many have suggested, we need God to explain the following: the values of various physical constants, the fundamental physical laws, the initial and boundary conditions of the universe, etc. One must be cautious about fine-tuning arguments for a variety of reasons. First, no matter what one's ultimate explanatory schema is, at bottom it is going to contain brute axiomatic facts. That is, the ultimate explanatory schema will contain facts that themselves are *the fundamental features of reality*, and thus beyond explanation by definition from within that schema. However, one posits a multiverse, God, or fundamental laws of physics (a Theory of Everything), one is always free to ask: why *that* multiverse instead of some other? Why *that* or any other God? Why *those* fundamental laws of physics as opposed to some others? If one or more of the preceding posits are the fundamental axioms of reality, then there simply is no answering the preceding questions about them. Even if one posits a multiverse with all logical or physical possibilities actualized in it, or a timeless block world that never changes and has always existed, such posits would be the fundamental facts of the world. The point is that no allegedly *ultimate* explanatory schema can explain its most fundamental axioms, and that includes theism. It is also an interesting question just how many features of the world are brute facts. However, science by definition should never posit supernatural causes such as God as explanations, and it should only posit the multiverse if all else fails. Otherwise, science is a priori giving up the challenge of explanation and merely positing the phenomena itself to be explained as the answer.

Second, fine-tuning arguments all presuppose that the phenomena in need of explanation—such as the fine structure constant or the bio-friendliness of the universe—are highly improbable with respect to our best scientific theories, for that is what justifies positing God or the multiverse as explanations of the phenomena in question. The danger here is that unless the improbability claim is clearly an *in principle* one that will also apply to all future theories of science as well, then fine-tuning arguments become *in practice* merely 'God of the gaps'-type arguments. It is much too early in the history of science to know what, if any, features of the world will be unexplainable in principle short of invoking God or the multiverse. For example, physicists are hoping that future theories of quantum gravity, such as M-theory or loop quantum gravity, will allow us to derive the values of various physical constants. Just because there are many *logically possible* values for the various physical constants upon which life and mind depend, that does not imply that there are many *physically possible* values for those constants *given ultimate scientific knowledge*. From a God's-eye point of view, if you will, it may be that the values of various constants are highly probable or even fixed by other features of the world. Performing an optimistic meta-induction about the future, we can see how wrong we would have been in the past if we had given up on deriving the values of various physical constants and opted for God or a multiverse as an explanation. The discovery of quantum electrodynamics provides an excellent example of this. Could the values of the various constants have been different in a block world? No, in such a world everything is necessitated. Perhaps ours *is* a block world.

Third, even if fine-tuning arguments do prevail in the long run, there seems to be absolutely no explanatory reason to posit God in a multiverse, for what is then left for God to explain? Lastly, in Clayton's world of ontological emergence with irreducible teleology, self-organization, and naturalistic final causation, there might be no explanatory need for either God or the multiverse. Given cosmic self-organization, we have a false dilemma between God, on the one hand, and the multiverse on the other. It is precisely the point of self-organizing systems that they can give rise to novel complex behaviour with a high probability, without a central controller or without having to realize every posibility. Besides, invoking God as an explanation for anything just pushes the question back one (why God?) unless we are taking God as the ultimate brute fact. And it should be obvious that science cannot invoke God in this way unless all other natural explantory options fail in principle.

One response to all these problems is to forgo the conception of God as a transcendent supernatural creator in favour of a more naturalistic and process-theological conception of a God who emerges and unfolds along with the universe; this certainly seems like an obvious way to go, given Clayton's emergentist conception of the world. Clayton considers this possibility at some length in the last chapter of his *Mind and Emergence*, and largely rejects it for a variety of reasons. Indeed, given the conclusion of Clayton's master argument that a transcendent God is the best explanation for the existence of all the things that cannot in principle be explained naturalistically—not even with nomological emergence— he is forced to accept God as a transcendent creator. He fully appreciates that he is 'trading mind–body dualism for theological dualism' (2004: 185), a dualism between a supernatural and transcendent God, on the one hand, and a naturally emergent world, on the other. Much of this essay has focused on the worry that, for a variety of reasons, such a theological dualism is not a stable position. Again, one can try to combine the transcendent creator conception of God with the process conception of an emerging God, but (as is emphasized again in the conclusion) this also is not an obviously stable mixture.

Conclusion: Future Directions for Theology, the Manifest Image, and Emergence

Recall that there are only two ways to justify the belief in the existence of God as defined by Clayton and company: (1) to show that God fills some explanatory lacunae that cannot even in principle be filled by science; and/or (2) to show that the existence of God is at the very least logically and physically consistent with what we take to be our best scientific theories about the world. We learned that option (1) is quite hard to defend, because one has to demonstrate an *in principle* explanatory gap in present and future scientific explanations, which include all potential explanatory scientific resources, such as ontological emergence and the multiverse. It is much too early in the game to know if God's existence will be required to explain any phenomena. We need better arguments than only God can explain the success of science, fine-tuning, etc.

Employing emergence in defence of (1) is tricky, because one must thread the dilemma of mere epistemic emergence on the one side (i.e. a mere, in practice 'God of the gaps' claim) and, on the other, a nomological emergence that makes God unnecessary by filling in all the explanatory gaps. We saw that establishing the truth of nomological emergence or radical mereological emergence is necessary to defeat physicalism and causal closure, to ground 'downward' (mental) causation, and to open up the possibility of free will. We discovered that while epistemic emergence and non-radical mereological emergence are easy to establish, nomological emergence is quite hard to prove, insofar as it requires demonstrating an *in principle* explanatory or ontological gap in a lower-level scientific theory. In short, if nomological emergence were to be well established, it might obviate the explanatory need for God; and if it were not to be established, then we are stuck with causal closure and physicalism, the truth of which also negates the explanatory need for God. Of course, there might be just enough nomological emergence in the world to defeat physicalism, but not enough to eclipse God—one might call this theism's nomological emergence fine-tuning problem.

The best candidates for nomologically emergent phenomena are conscious mental processes. First, however, one must defeat the various arguments from causal closure and try to establish that not even future mind or brain science will be able to explain consciousness, action, or cognition without appeal to nomologically emergent mind (see Silberstein 2006a). Besides, if conscious mental processes are the only nomologically emergent phenomena in the world, then what we have established is not Clayton's monistic emergent naturalism but property dualism of the sort favoured by Chalmers (1996).

We have also learned that not only might nomological emergence negate the need for God, but the existence of God seriously calls into question the explanatory need for nomological emergence. Again, it is possible that God is just powerful enough to fill the explanatory gaps allegedly left by nomological emergence and other natural processes, but not so powerful as to be able to explain everything else that ontological emergence explains—a dilemma that I dubbed theism's divine fine-tuning problem. But even if we were to solve *both* of theism's fine-tuning problems, we would still be left with a version of the problem of evil, as well as various trade-offs such as openness versus progressive teleology. Nomological emergence is hard to establish, and if it were established, it might negate the explanatory need for God.

A bright spot for some theists may be that option (2) above is relatively easy to defend. Science will never be in a position to establish with near certainty that the existence of God and immortal souls with free will are logically or physically incompatible with scientific evidence. However, as science comes to explain more and more, option (2) becomes more and more uncomfortable—though never completely untenable. For example, even in a futuristic world in which reductive neuroscience can predict and explain the behaviour of an individual with great accuracy, substance dualism can still maintain that immortal souls bear the same kind of relationship to brains as light does to movie projectors. Once the projector (the brain) ceases to function, the light (the soul) is once again free to roam the netherworld. The free soul is no longer tethered to the Earth-bound brain with its sensory and cognitive processes. And regardless of the explanatory potency of any future neuroscience, one can always maintain parallelism or divine harmony as the explanation for the intimate relationship between brain and soul. For that matter, physicalism is logically compatible with theism—though it is hard to imagine why God would create such a world or what role he could possibly play in it.

Here are a couple of suggestions for those theologians who are not content with option (2). First, for now spend more time trying to defend the manifest image of the self and agent causation, and less time on directly God-related issues. Spend more time trying to refute empirical considerations and no-go arguments aimed against an irreducible self, mental causation, and libertarian free will. If the reigning paradigm consisting of physicalism, causal closure, and reductionism is to be seriously challenged, much more philosophical and empirical work needs to be done to establish the existence of nomological emergence, radical mereological emergence, or 'downward causation'. That is, one must make the case that there are deep, *in principle*, empirically grounded reasons for thinking that physicalism and causal closure are false. This is the work that Clayton is beginning to undertake in his important book; I am only admonishing him and others to keep going with this focus. If one cannot even establish the existence of mental *qua* mental causation, then what hope does one have of establishing the existence of God?

Second, if one wants to continue focusing on God-related issues directly, practise theology in the way suggested by Sir John Templeton:

Generate new spiritual information. First, the idea reflects a desire to avoid the stasis of closure. Why consider God only through a lens of fixed tradition without training the eye on anything new? In the quest of the spirit, why not look to the open and progressive model of science, which intrinsically abhors closure and for which the adventure of new discoveries is everything? Templeton's vision seeks to encourage people to cultivate a mindset of looking at the spiritual quest simultaneously as an adventure open to new insights from a wide variety of sources and as an endeavour to be taken seriously by using whatever methods of research might be fruitful. Sir John Templeton has described his vision as the 'humble approach'. (Harper 2005: pp. xv–xvi)

Clayton deserves much credit for adhering to Templeton's approach in his theological quest. One cannot practise the 'humble approach' to theology by axiomatically assuming the existence of a particular kind of deity and then bend over backwards to interpret all the empirical data to fit one's preconceived conception of God. Maybe no such entity meeting one's own preconceived description exists. According to humble empiricism, as we might call it, the question of the existence of such an entity is an empirical question to be resolved by science and direct mystical experience, not by wishful thinking, institutional authority, or 'sacred texts' that are not themselves justified by anything other than themselves. By analogy, it would be wrong for a humble empiricist in a philosophical or scientific venue to assume that physicalism is true, rather than letting open-minded empirical investigation resolve the matter.

References and Suggested Reading

Antony, L. (1999). 'Making Room for the Mental: Comments on Kim's "Making Sense of Emergence"', *Philosophical Studies*, 95/2: 37–43.

Chalmers, D. (1996). *The Conscious Mind: In Search of a Fundamental Theory*. Oxford: Oxford University Press.

Clayton, P. (2004). *Mind and Emergence: From Quantum to Consciousness*. Oxford: Oxford University Press.

Harper, C. (2005). *Spiritual Information: 100 Perspectives on Science and Religion*, ed. Charles L. Harper Jr., 3 vols. West Conshohocken, Pa.: Templeton Foundation Press.

Humphreys, Paul (1997). 'How Properties Emerge', *Philosophy of Science*, 64: 1–17.

Loewer, B. (2001). 'From Physics to Physicalism', in C. Gillett and B. Loewer (eds.), *Physicalism and its Discontents*, Cambridge: Cambridge University Press, 22–47.

McGinn, C. (1999). *The Mysterious Flame: Conscious Minds in a Material World*. New York: Basic Books.

Saunders, N. (2002). *Divine Action and Modern Science*. Cambridge: Cambridge University Press.

Silberstein, M. (1998). 'Emergence and the Mind–Body Problem', *Journal of Consciousness Studies*, 5/4: 464–82.

—— (1999). 'The Search for Ontological Emergence', *Philosophical Quarterly*, 49/195: 182–200.

—— (2001). 'Converging on Emergence: Consciousness, Causation and Explanation', *Journal of Consciousness Studies: The Emergence of Consciousness*, 8/9–10: 61–98.

—— (2002). 'Reduction, Emergence, and Explanation', in P. Machamer and M. Silberstein, (eds.), *The Blackwell Guide to the Philosophy of Science*, Oxford: Blackwell, 80–107.

—— (2006a). 'Enactive Cognition, Ontological Emergence and Mental Causation', in P. Clayton and P. Davies (eds.), *The Re-emergence of Emergence*, Oxford: Oxford University Press.

—— (2006b). 'Questioning Causal Closure and Physicalism', paper submitted to *Synthese*.

—— and HAWTHORNE, J. (1995). 'For Whom the Bell Arguments Toll', *Synthese*, 102: 99–138.

CHAPTER 47

THE HIDDEN BATTLES OVER EMERGENCE

CARL GILLETT

INTRODUCTION

Ontological reductionism has long dominated the sciences and intellectual life more broadly. It holds that a 'final theory' in physics would, in principle, suffice to explain all natural phenomena, and that, ultimately, the entities of such a theory, like quarks with their properties of spin, charm, and charge, are all that actually exists. Recently, however, a mounting challenge to this hegemonic reductionism has been focused around 'emergent' entities. On the one hand, philosophers and a range of writers in science-and-religion have provided new theoretical resources in anti-reductionist, 'emergentist' views of the structure of nature. On the other hand, a parade of eminent scientists, from disciplines as varied as condensed matter physics, evolutionary biology, the sciences of complexity, and cognitive science, have all argued that their empirical findings provide actual examples of 'emergence' in nature.

However, despite the range of evidence supporting 'emergence', the claims of 'emergentist' scientists, such as Philip Anderson, Robert Laughlin, Stuart Kauffman, James Crutchfield, and others, have made surprisingly little headway against reductionism. The reason for this, I will suggest, is that the deeper underpinnings of

I am grateful to Philip Clayton for comments and to the Templeton Foundation for their support for the 'Oxford Templeton Seminars on Science and Christianity' during which the research for this chapter was completed.

ontological reductionism are actually *metaphysical* in nature, and largely impervious to empirical data. One of my key contentions will thus be that, in addition to empirical research, a certain kind of theoretical task is now absolutely imperative to establishing the existence of 'emergence'. We once again need to consider metaphysical issues in the sciences, where the required 'metaphysics' is not the a priori, and purely reflective, investigation rightly disparaged by the Positivists, but is instead what has recently been dubbed the 'metaphysics of science': the careful, abstract examination of ontological issues as they arise within the sciences themselves and their findings, models, explanations, etc.[1] I will argue that recent work on 'emergence' in science-and-religion, by writers such as Arthur Peacocke, Philip Clayton, Nancey Murphy, and others, begins to supply the needed metaphysics of science, and thus has a significance that stretches far beyond science-and-religion—in effect making a key contribution to one of the great intellectual battles of our times.

I will illuminate the practical necessity of a metaphysics of science in the first section by highlighting the scientific findings that are replete with ontological concepts and by providing a 'minimal framework' in the metaphysics of science to articulate these concepts. I will then use this work to illuminate how ontological reductionism is driven by an implicit but widely held *metaphysical* argument. My work in the second section will then be to illuminate how writers on 'emergence' in science-and-religion, as well as in other areas, have all been reacting to the challenge of ontological reductionism by offering competing accounts in the metaphysics of sciences.[2] Using my minimal framework I will show there are at least *three* competing notions of emergence, each serving a very different position. I will also highlight a second important use of the ontological reductionist's metaphysical argument. For I will show how it has been used to argue for the *impossibility* of one of the three types of emergence and hence to force us into a dichotomy of options: accept either ontological reductionism or an 'anti-physicalism' which denies that the natural world is a comprehensive compositional hierarchy. Finally, and perhaps most importantly, I will show in the third section of the chapter that the reductionist's metaphysical argument is actually invalid, and that the third type of emergence is possible after all. By not explicitly addressing metaphysical issues in the sciences, I will argue that recent writers have consequently fallen into a false dichotomy of emergentist options. For these researchers have not appreciated that our problems arise from a *triad* of incompatible theses which actually allow a *tri*-chotomy of competing emergentist solutions. Most interestingly, I will suggest that this third

---

[1] By far the best existing introduction to this pioneering philosophical work on the metaphysics of science is provided by Brian Ellis (2002).

[2] I am only going to be concerned with ontological notions of emergence. For wider surveys covering both semantic/epistemic and ontological notions of emergence, see van Gulick (2001) or Silberstein (2002). For a review of uses of emergence in the rather different debates over consciousness, see Stephan (1999). Finally, for surveys of historical forms of emergentism, see Blitz's wide-ranging (1992), McLaughlin (1992), and Gillett (forthcoming *b*).

kind of emergence highlights a robust, emergentist version of 'non-reductive physicalism' that fully accommodates all our scientific evidence and offers special promise in understanding the 'emergence' identified by working scientists.

In concluding, I will briefly note the resources that recent accounts of 'emergence' provide for theologically important topics such as the nature of persons or divine action. However, the focus of my chapter will be resolutely upon the metaphysics of science. By the end of my work, I hope to have convinced the reader that this focus is an appropriate one not just for a chapter, but also for the broader field of science-and-religion in the coming decades. As we look more and more to the sciences for help in understanding ourselves, the universe, and even our conceptions of the divine, it becomes imperative that we explicitly address the ontology inherent in our scientific findings. Recent work in science-and-religion has blazed a pioneering trail in this regard, but my final conclusion will be that we must continue to press on with this work.

Ontological Reductionism and its Hidden Metaphysical Engine

Scientific Findings, the Structure of Nature, and the Metaphysics of Science

Perhaps the most stunning human phenomenon of the last two centuries has been the rise of the sciences and the staggering expansion of our explanations of all aspects of the natural world, from molecules to manic depression. These scientific investigations have given us a detailed, integrated account of such phenomena, which is often based on our understanding of the entities that compose them.[3] Consider, for instance, a familiar case from materials science and chemistry. We know that, under appropriate background conditions, when a cut diamond is dragged with pressure across a pane of glass, then the hardness of the diamond will produce a scratch in the glass. The sciences have explained why this occurs in terms of the parts of the diamond, their properties and powers, and the mechanisms they ground. Crudely put, carbon atoms have the properties of being tightly bonded to each other in a lattice structure and retain their relative positions even under high pressure; thus, when such bonded carbon atoms are pressed upon glass molecules, the latter are caused to change spatial position relative to each other and not the carbon atoms. A

---

[3] During my discussion I will use the term 'entity' to refer generically to power relations, properties, events, processes, and individuals.

range of causal powers, properties, and individuals bear compositional relations in this situation and, as a result, mechanisms involving carbon atoms together implement the mechanism of the diamond scratching the glass.

Similar cases, all centred on our understanding of compositional relations, are found across the full spectrum of scientific disciplines. The neurosciences are a useful case because they quickly show how comprehensive such compositional relations are. For it is increasingly clear that psychological states and behaviours are composed by brain areas, where the latter are themselves composed of collections of cells like neurons, which are in turn composed of cellular structures like ion channels, which are composed of complex biochemical molecules like protein 'sub-units', which are themselves composed of atoms, which are composed of the subatomic entities studied by physics. Such comprehensive compositional hierarchies are now to be found in most areas of the sciences. Nonetheless, despite their obvious importance, philosophers of science have paid relatively little attention to compositional relations, presumably due to the lingering influence of the Positivists' famous proscription of metaphysics. Happily, this situation has recently begun to change and we can use recent philosophical work on the metaphysics of science to sharpen our picture of such compositional relations.

To make our discussion manageable, and because they have been the primary focus of recent debates, I will focus simply upon giving a 'minimal' metaphysics of science framework for properties and their compositional relations. We therefore need an account of properties in the sciences, and we can find one in the 'causal theory of properties' developed by Rom Harré and E. H. Madden (1977), Sydney Shoemaker (1980), and others. I will use a variant of Shoemaker's account, according to which a property is individuated by the causal powers it *potentially* contributes to the individuals in which it is instantiated. Thus two properties are different when they contribute different powers under the same conditions. For reasons that will become apparent below, I will be concerned with properties I term 'causally efficacious'—that is, properties whose instantiation *actually* determines the contribution of causal powers to an individual. (In addition, I will take events to be the causal relata and to cause in virtue of their properties; though I shall often refer only to their properties for convenience.)

With the causal theory in hand, we can now return to our case of the diamond and the carbon atoms to illuminate the compositional relations between the properties of these individuals—or so-called realization relations. As I noted earlier, with the diamond the properties of the carbon atoms, such as their bonding and alignment, 'play the causal role' of the diamond's property of hardness; basically, the lower-level properties *together* result non-causally in the powers of the realized property. Thus, using such scientific cases and the causal theory of properties, we can formally frame the nature of the realization relations found throughout the sciences as follows:

(Realization) Property/relation instance(s) F_1-F_n realize an instance of a property G, in an individual *s*, *if and only if s* has powers that are individuative of an instance of G in virtue of the powers contributed by F_1-F_n to s or s's constituent(s), but not vice versa.[4]

Basically, realizer properties contribute powers that *determine* that the relevant individual has the powers individuative of the realized property. And we should carefully mark that this determination relation is not temporal in nature, since the upward determination involved in realization is instantaneous, does not involve wholly distinct entities, and does not involve the transfer of energy and/or the mediation of force. Realization is thus *not* a species of causal determination, since causal relations are temporally extended, do relate wholly distinct entities, and do involve the transfer of energy and/or the mediation of force. In contrast, realization is an example of what we might term '*non-causal*' determination, like the relations between the individuals bearing part–whole or constitution relations, or the implementation relations between processes.

Our account of realization in the sciences will allow us to illuminate a range of important issues, starting with two key theses that each have very strong scientific support and which are at the heart of recent debates. We have seen that scientific evidence apparently reveals the existence of a layered world wherein powers, properties, individuals, and mechanisms form a compositional hierarchy. And this implies that there are other entities than those studied by physics. The resulting thesis might seem too obvious to be worth framing, though we shall shortly see that it is highly contested; I will call it 'Higher Ontology', since it concerns the ontology of higher-level sciences:

(Higher Ontology) There are higher-level properties, and they are causally efficacious.

Higher Ontology expresses the claim that, in addition to the properties of spin, charm, and charge had by microphysical individuals such as quarks, the findings of the special sciences suggest that the world also contains higher-level properties such as being hard, being diabetic, having an action potential, believing that New Orleans is humid, and many more. In addition, we have seen that our slowly accumulating scientific evidence has made it steadily more plausible that the natural world is a *comprehensive* compositional hierarchy wherein *all* properties are realized by other properties, and ultimately by the properties of physics, and wherein *all* individuals are constituted by other individuals, and ultimately by the individuals of physics. Such evidence supports our second thesis, 'Physicalism':

(Physicalism) All individuals are constituted by, or identical to, microphysical individuals, and all properties are realized by, or identical to, microphysical properties.[5]

Physicalism simply makes the claim that the natural world is an *all-inclusive* system of compositional levels. At least initially, the truth of Physicalism and Higher

---

[4] I have defended this account at length in my 2002a, 2003a, and unpublished.

[5] Throughout the chapter the meanings of 'physical', 'microphysical', and 'non-physical' are those defended at length in Crook and Gillett (2001).

Ontology appear to go hand in hand, for each expresses one facet of the compositional hierarchy illuminated by the sciences over the last two centuries.

The Metaphysical Argument for Ontological Reductionism

On its face, ontological reductionism is an extremely radical view. It entails that, literally, there are no higher-level individuals like glaciers, cells, planets, or even people! Nor are there any of the characteristic higher-level properties of such individuals; thus there are literally no properties of being hard, being diabetic, having an action potential, or believing that New Orleans is humid. This is a shocking view that appears to fly in the face of the scientific evidence that supports the truth of *both* Physicalism *and* Higher Ontology. However, building upon the pioneering work of the philosopher Jaegwon Kim and using our minimal metaphysical framework, we can now illuminate why, over the last century, ontological reductionism has been so attractive, for there is a metaphysical argument that if Physicalism is taken to be true, then Higher Ontology ought not to be (see Kim 1997, 1998, 1999). This reasoning basically takes the form of an argument from simplicity, or ontological parsimony, and proceeds as follows. (Again, I will focus on properties, but we should note that there are similar arguments for individuals and processes.)

The reductionist will argue that it is ontologically profligate to take any realized property instance to determine the contribution of causal powers, and hence to be causally efficacious, *in addition* to its realizer property instances. For the reductionist claims that, given the nature of the realization relation, we can account for *all* the causal powers of individuals simply by using the contributions of powers by the realizer instances of these individuals (or by their constituents), rather than also as contributions from realized instances. But we cannot account for all causal powers of individuals simply as contributions by realized properties. If we assume that the causal powers of individuals are not overdetermined, then the critic, appealing to Occam's razor, argues that we should accept the existence of no more causally efficacious properties than we need to account for the causal powers of individuals. The proponent of this simple argument thus concludes that we should accept only that realizer property instances contribute powers to individuals, and hence only that realizer instances are causally efficacious. But it is also plausibly true that the only properties that exist make a difference to the causal powers contributed to individuals. We may thus further conclude that there are only realizer property instances. But if Physicalism is true, then all properties are realized by microphysical properties. Thus we must conclude that the only properties that exist are the fundamental microphysical realizer properties.

To capture the generality of the resulting reductionist challenge, I will refer to this reasoning as the 'Argument from Composition'. Such ontological arguments can easily be found just beneath the surface in many prominent defences of reductionism such as Steven Weinberg's *Dreams of a Final Theory* (1992). Once we appreciate the nature of the reductionist's primary argument, we can explain why such reductionist defences have succeeded in dismissing the many empirical examples of robust 'emergence', which, consequently, have made such little headway against reductionism. For, given the scientific evidence for Physicalism, the reductionist's metaphysical argument shows that we should not accept the existence of any higher-level property, and hence any robust form of 'emergence'; this argument is largely unaffected by any further empirical evidence we might identify.

We can now also clearly discern some of the damage done by the Positivists' proscription of metaphysics as an appropriate object of study for philosophers of science. Traditionally it was one task of philosophers, and especially philosophers of science, to carefully articulate and assess the metaphysical notions inherent in the sciences. Addressing the reductionist's metaphysical argument, at least initially, clearly involves a metaphysical account of the nature and implications of compositional relations like realization. But since philosophers of science have recently shunned such projects, the implicit arguments that drive one of the most important intellectual doctrines of our times, ontological reductionism, have thus gone critically unexamined. The philosophy of science bequeathed to us by the Positivists has arguably thus done a grave disservice to the sciences and to wider intellectual debates.

The Hidden Battles over the Varieties of Emergence

In this section I will show that recent emergentists are arguably unified in two respects: emergentists all use their accounts of emergence to react to the metaphysical arguments for ontological reductionism, and they are also united by a tacit commitment to the metaphysics of science as a key step in this project. I will nonetheless show that, although joined in these respects, emergentists are divided by disputes over their substantive ontological claims and defend a variety of incompatible, competing notions of emergence. Given the contested nature of various notions of emergence, there has been a corresponding tangle of terminology; for example, the terms 'Weak' and 'Strong' are used to refer to differing notions by writers. Elsewhere I have used the terms 'Weak', 'Ontological', and 'Strong' to distinguish the types of emergence. But to assuage concerns about usage, and to avoid confusion, I will here usually refer more neutrally to 'W-emergence', 'O-emergence', and 'S-emergence'.

More importantly, I will demonstrate that recent disputes are not merely about words, but reflect deeper ontological battles. In the first subsection we will see that 'W-emergence' is a tool of ontological reductionists which they use to ameliorate their own difficulties. In contrast, in the following subsection we will see that 'O-emergence' is used by anti-physicalists to frame their response to the challenge of ontological reductionism. Against this background I will then show in the third subsection how these two groups of emergentists, and many others, unite in accepting the validity of the Argument from Composition, and hence combine to attack the very possibility of what I term 'S-emergence'. We will thus see why many writers, including those in science and religion, have so often argued that there is a dichotomy of options in reductionism or anti-physicalism.

W-Emergence and Reductionist Apologetics

What I will term 'Weak' emergence, or more usually 'W-emergence', is a particular type of microphysically realized property that also has certain *epistemic* features. In particular, the law statements and/or explanations and/or theories that hold of a W-emergent property cannot be derived from the laws and/or explanations and/or theories of microphysics. Thus, although ontologically a W-emergent property is intimately related to microphysical properties (being realized by them), the law statements, explanations, and theories concerning this property nonetheless float free from the law statements, explanations, and theories concerning its realizer properties. The following criterion provides a useful target version of this concept:

(W-criterion) A property instance X, in an individual *s*, is W-emergent *only if* (i) X is a microphysically realized property instance; (ii) the law statements (and/or theories and/or explanations) true of X cannot be derived from the law statements (and/or theories and/or explanations) holding of the microphysical properties which realize X; and (iii) all microphysical events are determined, insofar as they are determined, by the laws of physics applying to simple systems.

I should mark that this only gives a *criterion* for this type of emergence, rather than telling us what is constitutive of it. For example, it offers only one necessary condition for a property to be W-emergent; plausibly there will be other conditions. My goal in offering this criterion, and also those for the other concepts of emergence, is simply to highlight the core features of the relevant kind of emergence.

W-emergence provides a vital tool to the ontological reductionist.[6] The primary obligation of the reductionist in supporting her position is to explain away the apparent scientific evidence for Higher Ontology outlined in part 1. Using W-emergence, the reductionist can argue that the appearance of causally efficacious higher-level properties is merely the result of the *epistemic* autonomy that characterizes W-emergent properties. The fact that the law statements (and/or theories and/or explanations) concerning W-emergent properties *epistemically* float free from law statements, etc. in physics grounds the ontological illusion that these layers of law statements (and/or theories and/or explanations) pick out distinct layers of entities. But, argues the reductionist, the Argument from Composition shows that there are

---

[6] A number of writers have tried to use W-emergence to pursue other goals, in particular the task of establishing 'S-emergence', whose nature I will examine shortly. Elsewhere I have argued at length that such uses of W-emergence are misguided (see Gillett 2002*b*).

really no such efficacious higher-level properties. Thus, using W-emergence, the reductionist can try to provide an 'apologetics' for her position.

The reductionist defender of W-emergence thus accepts both the validity and the soundness of the Argument from Composition, and hence abandons the truth of Higher Ontology, but sugars this bitter pill by pursuing the metaphysics of science to articulate the features of W-emergence. Recent explicit advocates of W-emergence, and just this type of apologetic strategy, have been the philosophers of science Jaegwon Kim (1999) and Mark Bedau (1997). We should note, however, that although W-emergence provides a way for reductionists to make their view more palatable, it nonetheless faces deep problems. For we do have powerful evidence supporting Higher Ontology from the sciences and our everyday lives. For instance, is there really no property of being diabetic, or any of the other properties illuminated by the special sciences? And are there truly no composite individuals—for example, do we really not exist?

O-Emergence and the Anti-Physicalist Solution

'O-emergent', or 'Ontologically' emergent, properties are starkly different from the kind of emergence trumpeted by ontological reductionists. For the central features of an O-emergent property are, first, that such a property is instantiated in higher-level individuals (whether physically constituted or non-physical individuals), and second, that an O-emergent property is an unrealized, i.e. uncomposed, and hence ontologically fundamental property, albeit a fundamental property found in a higher-level individual. We can capture the heart of this notion in the following criterion:

(O-criterion) A property instance X, in an individual s, is O-emergent *only if* (i) X is instantiated in s, where this individual is either constituted by microphysical individuals or is a non-physical individual; (ii) X is an unrealized property; and (iii) X is causally efficacious.[7]

This criterion has the merit of covering paradigm examples of O-emergence such as the accounts of Philip Clayton (1997, 2005), Tim O'Connor (1993), Tim Crane (2001), and William Hasker (1999), amongst others. Once again, by articulating the notion of O-emergence, these writers are implicitly pursuing the metaphysics of science. Philip Clayton nicely illuminates the significance of the O-emergent properties when, referring to O-emergence, he argues that 'theology does not need to embrace *either* a radical dualism of mind/soul and body *or* the physicalism that is widespread among scientists and philosophers today. The theory of *emergent properties* forms an attractive *via media* between these two poles of the discussion' (Clayton 1997: 247; emphasis original). We

---

[7] We need to mark the two types of individual—physically constituted and non-physical—because O-emergentists diverge over whether *both* properties *and* individuals emerge, hence generating fundamental non-physical individuals (Hasker), or whether properties *alone* emerge (Clayton, O'Connor, and others).

can appreciate Clayton's point when we remember that O-emergent properties are not realized microphysically; they are properties outside the system of compositional levels. Yet O-emergent properties are had by the individuals studied by the special sciences, whether these individuals are constituted or whether they are fundamental, non-physical individuals. These twin features of O-emergent properties allow them to support the truth of Higher Ontology in the face of the ontological reductionist's challenge.[8] For, being unrealized, O-emergent properties are uncomposed properties lying outside the compositional hierarchy forming much of the natural world, and are consequently unaffected by the Argument from Composition. The character of condition (ii) is thus used to secure the truth of condition (iii).

It is important to note, as we shall see in the next section, that O-emergentists thus accept the validity of the Argument from Composition. They accept that *if* its premises are true, then its conclusion must also be true, but they react to the Argument by denying the truth of one of its premises, viz. Physicalism, in order to preserve the truth of Higher Ontology. O-emergence is therefore a key weapon of what I will term 'anti-physicalism'. And we should mark that, like W-emergentists, O-emergentists face grave difficulties from our scientific evidence. In this case, the powerful scientific support for Physicalism is the problem. For example, we have good evidence that there are *only* four (possibly three) *fundamental*, i.e. uncomposed, forces, all of which are physical. But since they are physically uncomposed but have physical effects, if O-emergent properties exist, they would basically constitute further *non*-physical, *fundamental*, i.e. uncomposed, forces (see Gillett 2002*b* and 2003*c* for detailed versions of such criticisms). In this and many other respects, the anti-physicalism based on O-emergence thus faces grave conflicts with the range of scientific evidence presently supporting comprehensive composition in nature, and hence the truth of Physicalism.

The Unholy Alliance, its Dichotomy and (Charitable) Reinterpretations

The reductionists and anti-physicalists who are the primary proponents of W- and O-emergence respectively agree that the Argument from Composition is valid, though they disagree over how to react to this argument. As a result of this underlying agreement, W- and O-emergentists unite in what I will call the 'Unholy Alliance' which trenchantly rejects the possibility of 'S-emergence' (or what I have elsewhere termed 'Strong' emergence). It is now very simple to provide a criterion for S-emergence and to see why the Unholy Alliance rejects its very possibility, insofar as any property of which both Physicalism and Higher Ontology hold true is S-emergent. My criterion for this type of emergent property is thus simply:

---

[8] For an O-emergentist the term 'higher-level' in Higher Ontology should be read as meaning 'property of a constituted or non-physical individual', rather than 'realized'.

(S-criterion) A property instance X, in an individual *s*, is S-emergent *only if* (i) X is a realized property instance; and (ii) X is causally efficacious.

The theoretical qualms about S-emergence should be clear: the Argument from Composition putatively shows that S-emergence cannot exist! If one accepts that the Argument from Composition is valid, then a property cannot simultaneously be realized *and* be efficacious. The proponents of the Unholy Alliance have thus argued that S-emergent properties are like square circles: they simply cannot be—*whatever* our empirical evidence might superficially suggest.[9]

A number of writers in philosophy, in the field of science-and-religion, and especially in the sciences have apparently championed S-emergence over the last century. Indeed, there are clear reasons for the popularity of S-emergence, since it seeks to accommodate *all* of our scientific evidence—both the findings about the comprehensiveness of composition that support Physicalism and the evidence underlying Higher Ontology. S-emergence therefore potentially overcomes the difficulties facing both the reductionist's W-emergentism and the anti-physicalist's O-emergentism. A typical proponent of S-emergence, and perhaps its most sophisticated philosophical expositor, is Samuel Alexander (1920), an early emergentist who, in the 1920s, defended the efficacy (and hence the existence) of chemical, biological, and psychological properties, even though he explicitly took these properties to be comprehensively realized by microphysical properties. Alexander's idea was that the fundamental microphysical realizer properties were what we might term 'conditioned' in nature. That is, Alexander suggested that the microphysical realizers only contribute certain powers, and hence behave in particular manners, *when under certain conditions*—specifically, the condition of being in composites, i.e. realizing certain higher-level properties. Similar ideas are found in recent writers in science-and-religion, such as Arthur Peacocke (1990, 1995, 1999) and Nancey Murphy (Brown *et al.* 1998), and peppered across the work of scientific defenders of 'emergence' such as Philip Anderson (1972), Robert Laughlin (2005), James Crutchfield *et al.* (1986), and Stuart Kauffman (1995). For example, all of the latter scientists have suggested, in varying terms, that they have empirical evidence showing that the natural world is 'conditioned', where components behave in certain ways only under the condition of being in certain composites, thus apparently supporting S-emergence and explicitly arguing that ontological reductionism must be mistaken.

In response, members of the Unholy Alliance, who have long dominated theoretical discussions, routinely 'charitably reinterpret' the claims of these past and present proponents of S-emergence. Alexander's case is typical. Since the Argument from Composition putatively shows S-emergence to be untenable, recent writers all 'charitably reinterpret' Alexander as some *other* type of emergentist. Thus Kim (1999) and McLaughlin (1992) interpret Alexander as an O-emergentist who accepts Higher Ontology but abandons Physicalism, whilst O'Connor (1994) and Clayton

---

[9] Perhaps the strongest of these arguments is provided by Kim (1999), where he refers to S-emergence as 'synchronic emergence', and argues that it is quite literally impossible.

(2005) basically assume that Alexander must be a W-emergentist who discards Higher Ontology whilst endorsing Physicalism. Despite his explicit statements to the contrary, these members of the Unholy Alliance assume that Alexander just cannot be an S-emergentist (though what else he might be is obviously a little unclear!).[10] And similar 'charitable reinterpretations', whether overt or not, are meted out to the other defenders of S-emergence as well. In science and religion, Arthur Peacocke is consequently widely reinterpreted as holding the same view of the natural world as O-emergentists such as Philip Clayton, despite Peacocke's explicit commitment to the type of comprehensive compositional hierarchy in nature rejected by O-emergentism. Such charitable reinterpretation also extends to the many working scientists who apparently defend S-emergence. Regardless of their empirical findings, the Unholy Alliance tells us that, as with Alexander or Peacocke, we cannot take these scientists at their word; however much they protest, they must really be either O- or W-emergentists—either implicitly abandoning Physicalism or endorsing reductionism—since we know that S-emergence is impossible.

THE POSSIBILITY OF S-EMERGENCE AND THE BASIS OF THE FALSE DICHOTOMY

I will show in this section that the many scientists and others who have been attracted to S-emergence have wrongly been dismissed. I will argue in the first subsection that S-emergence is possible and that the Argument from Composition is invalid. In the second subsection I will then diagnose why writers have failed to appreciate the latter points. In particular, I will show that these mistakes arise from a deeper failure to appreciate that *three* theses, rather than two, are incompatible— thus obscuring the fact that we actually have a *tri*-chotomy of emergentist options, including S-emergence.

The Nature, and Possibility, of S-emergence

As we have seen, theoreticians and scientists have all thought that S-emergence is bound up with the fact that the fundamental ontological components are what I earlier termed 'conditioned' in nature. In examining whether S-emergence is possible, I will therefore follow the lead of these writers and explore a 'conditioned' scenario. That is, I assume, first, that microphysical properties are *conditioned*, and

---

[10] My forthcoming, *b*, provides a detailed account of Alexander's views and of his treatment by his critics.

hence *heterogeneous*, in their contributions of powers, because they contribute some powers only under certain conditions, which means that these properties have what Shoemaker (1980) terms 'conditional' powers. Second, I will assume that microphysical properties contribute such conditional powers when they realize certain higher-level properties, i.e. when they occur in certain composites.

In the resulting scenario, a realized property instance non-causally determines that one of its realizer properties contributes a certain power that it would not otherwise contribute. It is important to mark the *non-causal* nature of the determination exerted by the realized property in such a scenario, for this suggests that there will likely be no new ontologically fundamental, i.e. uncomposed, forces (or other properties). The relevant realized property instance 'H' is *not causing* a microphysical realizer property instance 'P' to contribute certain powers. Causal relations are typically mediated by forces and/or the transfer of energy; thus if H *causally* determined P's contribution of powers, then there might well be a new force. But in the scenario, H is exerting a *non-causal* determinative influence, and, as with part–whole or realization relations, this does not involve the exertion of a force and/or transfer of energy. As a consequence, H's non-causal determinative role will therefore not by itself produce any new, fundamental non-physical forces, causal powers, or properties.

Does Physicalism actually hold true in such a case? Given the points just noted, it appears that it might; but in order better to answer this question, let us now carefully examine the nature of the causal powers in the scenario. Although P's contributing some causal power is partially non-causally determined by realizing H, nonetheless this causal power is *still* a causal power of a microphysical property in P. Consequently, H is still apparently a realized property, for all of its causal powers result non-causally from powers contributed by microphysical properties such as P. In the scenario, all powers of individuals would therefore result from the powers contributed by microphysical properties, and all properties would still therefore be realized by microphysical properties. Thus Physicalism is true, even though the fundamental microphysical properties have a few conditional causal powers whose contribution is determined, in part, by realizing a certain higher-level property.

What of our other thesis, Higher Ontology? Would it also be true in such a case? It appears it would. A realized property instance, such as H, that partially non-causally determines the powers contributed by a microphysical property would be a necessary member of the set of properties which are only *jointly sufficient* for the contribution of a particular fundamental causal power, and hence any microphysical effects (call them 'E') that result from it. Given the latter point, realized properties would satisfy a widely held criterion for H to be one of the causes of E. It thus appears that Higher Ontology is true in the mooted scenario, for realized properties partially determine the contribution of powers to individuals, and hence are efficacious.

In our 'conditioned' scenario, through their partial non-causal determination of the powers contributed by the fundamental microphysical properties, realized properties would be causally efficacious, thereby supporting Higher Ontology, but *without adding* to the fundamental ontology of the universe, and hence *without violating* Physicalism. Our scenario thus demonstrates that there can be situations in which we ought to take *both* Physicalism *and* Higher Ontology to be true. In this 'conditioned' scenario we consequently have a promising abstract idea about the nature of S-emergence which is framed more precisely as follows:

A property instance X is S-emergent, in an individual *s*, *if and only if* (a) X is realized by other properties/relations and ultimately by microphysical properties; and (b) X partially non-causally determines the causal powers contributed by at least one of the fundamental microphysical properties/relations realizing X.

The latter is an actual account of S-emergence, for we have now seen that it satisfies the S-criterion. And we should carefully mark that this picture of S-emergence, and the scenario built upon it, also show that the Argument from Composition is plausibly *invalid*, insofar as the scenario shows that the Argument's conclusion may be false even while all of its premises are true. For a property can be efficacious whilst also being realized.

Though the scenario offers hope to S-emergentism, we should note that, like the other varieties of emergentism, this position also faces difficulties. First, although our brief discussion offers promise in the project of showing that S-emergence is coherent after all, a great deal of further theoretical work needs to be done in order to firmly establish such a possibility. Thus key ontological features of S-emergence need to be more carefully articulated, and a range of objections need to be addressed (see Gillett 2003*b*, 2003*d*, and forthcoming, *b*). Second, the S-emergentist also needs to provide empirical evidence that S-emergence actually exists in our world, rather than simply being some rarefied theoretical possibility. Again, providing such scientific support poses a large and very real challenge.

However, we should also note that S-emergentism has a special fund of resources to which it can look for help with these two tasks. All previous writers who have been attracted to S-emergence have been 'charitably reinterpreted', and hence had their ideas assessed under the assumption that they underpin *other* positions. We have now found that we need to re-examine these writers on their own terms, as offering defences of S-emergence. When reconsidered in this manner, their work potentially offers assistance with both tasks. Consider, for example, the contemporary scientists who have apparently defended S-emergence. A careful project of reassessment needs to be conducted on their empirical findings and wider claims. And when we apply our more adequate metaphysical framework, rather than the flawed prism of the Unholy Alliance, we may well find that these scientists have already provided empirical support for S-emergence.

The False Dichotomy Diagnosed, or an Incompatible Trio Exposed

Why have so many writers failed to appreciate both that the Argument from Composition is invalid and that S-emergence is possible? My diagnosis is that the failure to be explicit about metaphysical issues has blinded many to the role of a key thesis that exerts a powerful influence on their thinking. For much pioneering work in physics over the last century has been motivated by aspirations to provide a 'final theory': a simple set of laws referring only to the ontologically fundamental micro-physical entities, discovered by investigating simple systems of microphysical entities, and capable of explaining absolutely everything in the natural world. And this goal of finding a 'final theory' has been held for so long, and with such fervour, that it has hardened into a largely implicit metaphysical view of the microphysical entities themselves. Two assumptions about microphysical properties are thus now often tacitly presumed: (a) that the microphysical properties are largely *unconditioned* and *homogeneous* in their contributions of causal powers; and (b) that the contributions of powers by the microphysical properties are determined, insofar as they are determined, only by other microphysical properties. For in order to have the type of simple set of laws required by a 'final theory', it is all too easy to assume that (a) and (b) would both have to be true.

This metaphysical picture finds one of its most overt expressions in a thesis held by many philosophers that they appropriately term the 'Completeness of Physics':

(Completeness of Physics) All microphysical events are determined, insofar as they are determined, by prior microphysical events and the laws of physics.

The latter thesis is intended by philosophers to concern simple laws that refer directly only to microphysical entities, and that are discovered by studying simple, isolated systems of microphysical entities. As it is standardly interpreted, the Completeness of Physics thus basically encapsulates (a) by embodying the idea that microphysical properties are largely unconditioned in nature, because they are assumed to be fully described by the simple set of laws which is illuminated by studying simple, isolated microphysical systems. And the Completeness of Physics clearly entails a version of (b) by explicitly stating that microphysical events are determined only by other microphysical events and the laws of physics. Since the Completeness of Physics is an expression of (a) and (b), I will use it to illuminate the implications of this deeper metaphysical picture.

Our minimal framework for the metaphysics of science swiftly allows us to see that Physicalism, Higher Ontology, and the Completeness of Physics form an incompat-ible trio of theses—they cannot all be true together. For if a thesis like Completeness is true in combination with Physicalism, then not only do realized properties contribute no fundamental powers themselves, but realized properties also have no scope to determine contributions of powers by other properties either. As a result, realized properties cannot plausibly exist. Higher Ontology is therefore false when *both* Physicalism *and* the Completeness of Physics are true. And we can now provide a diagnosis of what has recently gone wrong in various debates, largely as a result of the failure to explicitly address ontological issues in the sciences.

Driven by their dreams of an 'ideal theory', contemporary writers have been so strongly in the *implicit* grip of assumptions like (a) and (b), and hence theses like the Completeness of Physics, that they have failed to recognize *explicitly* the role of such claims. Though using (a) and (b), or the Completeness of Physics, in their thinking, these writers have thought that Physicalism and Higher Ontology *alone* were in tension—thus wrongly thinking that the Argument from Composition was valid. But rather than two incompatible theses, we have found that we only get incompatibility with a *trio* of theses.

We can also see how the Unholy Alliance has, for similar reasons, fallen into the false dichotomy of thinking that one has to endorse either W-emergence or O-emergence. In fact, in response to the real incompatibility, there are obviously three distinct theses that one can reject, and hence a *tri*-chotomy of options. First, one may retain Physicalism and the Completeness of Physics, thus defending reductionism and W-emergence. Second, one can defend Higher Ontology (with or without the Completeness of Physics), thereby defending anti-physicalism and O-emergence. But there is also a third, previously hidden option. Rather than abandoning either Higher Ontology or Physicalism—both of which (as we have seen) have tremendous evidential support from the sciences—we may instead resolve the looming incompatibility by rejecting the Completeness of Physics.

Brief reflection on our earlier scenario for a 'conditioned' natural world shows that this is exactly what S-emergence involves. When the microphysical properties are 'conditioned' in nature, having powers they contribute only under the condition of realizing certain higher-level properties, then the Completeness of Physics will be false. For, most obviously, in such a case some microphysical events will be partially determined by realized S-emergent properties, and not just other microphysical properties. The result is an exciting third option embodying a robust 'non-reductive physicalism', which allows us to accommodate the scientific evidence for *both* comprehensive composition *and* the existence of higher-level entities. Furthermore, as Arthur Peacocke has argued at length, the understanding of the natural world provided by S-emergentism provides exciting new tools with which to tackle an array of key issues in Christian theology, such as God's interaction with the natural world, and many others (Peacocke 1990, 1995, 1999).

CONCLUSION: EXPLICIT ONTOLOGY FOR THE SCIENCES AND NEW HORIZONS FOR SCIENCE AND RELIGION

To conclude, recent work on emergence in science-and-religion, as well as other areas, is at the heart of perhaps the grandest intellectual battle of our times. Unfortunately, much of this debate has for too long been hidden from view, driven by tacit metaphysical assumptions and arguments. But, following the lead of pioneering work in science-and-religion, we have found that explicitly pursuing the metaphysics of science yields many benefits in this damaging situation. First, the metaphysics of science allows us to lay bare the mistakes and false dichotomies into which the debate has previously fallen through the Positivists' baleful influence. Second, the metaphysics of science provides tools which allow us to clarify and carefully distinguish the three metaphysical views of the natural world that presently do battle in ongoing debates. Perhaps most importantly, we can thus finally begin to see that it is also possible that we might live not in the reductionist's austere world, or the ontologically opulent world of the O-emergentist with its additional fundamental non-physical entities, but in what I have elsewhere termed a 'patchwork physicalist' world—one with a mosaic of efficacious entities and fundamental laws, at various levels in nature, firmly built around S-emergent properties.

W-, O-, and S-emergentism all still face ongoing difficulties, and our decisions between these competing views will ultimately be decided in large part by our empirical findings. But once their deeper ontological natures are revealed, as work on divine action in science and religion has recently demonstrated, each of these emergentist views offers a plethora of new resources to apply in understanding not only the natural world but also ourselves as a part of this world, and even our conceptions of the divine.[11] As we continue to learn about reality, the sciences and their findings play an ever increasing intellectual role in this project. In the future, the competing emergentist positions will consequently have a central role to play by clarifying our deepest problems and providing the competing visions by which we may seek to resolve them.[12]

REFERENCES AND SUGGESTED READING

ALEXANDER, S. (1920). *Space, Time and Deity: The Gifford Lectures at Glasgow*, 2 vols. Toronto: Macmillan.
ANDERSON, P. (1972). 'More is Different', *Science*, 177: 393–6.
BECKERMANN, A., FLOHR, H., and KIM, J. (1992) (eds.). *Emergence or Reduction?* New York: Walter de Gruyter.
BEDAU, M. (1997). 'Weak Emergence', *Philosophical Perspective*, 11: 375–99.
BLITZ, D. (1992). *Emergent Evolution*. New York: Kluwer.
BROWN, W., MURPHY, N., and MALONY, H. (1998). *Whatever Happened to the Soul?* Minneapolis: Augsburg Fortress Press.
CLAYTON, P. (1997). *God and Contemporary Science*. Grand Rapids, Mich.: Eerdmans.
—— (2005). *Mind and Emergence*. Oxford: Oxford University Press.
CRANE, T. (2001). *The Elements of Mind*. Oxford: Oxford University Press.
CROOK, S., and GILLETT, C. (2001). 'Why Physics Alone Cannot Define the "Physical"'. *Canadian Journal of Philosophy*, 31: 333–60.
CRUTCHFIELD, J., FARMER, J., PACKARD, N., and SHAW, R. (1986). 'Chaos', *Scientific American*, 255: 46–57.

[11] See e.g., the papers in Russell *et al.* 1995 and 1999. For the resources provided by O- and S-emergence for crafting our conceptions of ourselves, and the divine, see respectively Clayton (2005) and Peacocke (1990).

[12] The work of this paper draws in part from my 2002*b*, forthcoming, *a*, and forthcoming, *b*.

ELLIS, B. (2002). *The Philosophy of Nature*. Montreal: McGill University Press.

GILLETT, C. (2002*a*). 'The Dimensions of Realization: A Critique of the Standard View', *Analysis*, 62: 316–23.

—— (2002*b*). 'The Varieties of Emergence: Their Purposes, Obligations and Importance', *Grazer Philosophische Studien*, 65: 89–115.

—— (2003*a*). 'The Metaphysics of Realization, Multiple Realizability and the Special Sciences', *Journal of Philosophy*, 10: 591–603.

—— (2003*b*). 'Non-Reductive Realization and Non-Reductive Identity: What Physicalism Does Not Entail', in S. Walter and H.-D. Heckmann (eds.), *Physicalism and Mental Causation*, Exeter: Imprint Academic, 31–58.

—— (2003*c*). 'Physicalism and Panentheism: Good News and Bad News', *Faith and Philosophy*, 20: 3–23.

—— (2003*d*). 'Strong Emergence as a Defense of Non-Reductive Physicalism', *Principia*, 6 (special issue on emergence): 83–114.

—— (unpublished). 'Making Sense of Levels in the Sciences'.

—— (forthcoming, *a*). *Emergence: Philosophical and Scientific Perspectives*. Oxford: Oxford University Press.

—— (forthcoming, *b*). 'Samuel Alexander's Emergentism: Or, Higher Causation for Physicalists', *Synthese*.

HARRÉ, R., and MADDEN, E. H. (1977). *Causal Powers*. Oxford: Blackwell.

HASKER, W. (1999). *The Emergent Self*. Ithaca, N.Y.: Cornell University Press.

KAUFFMAN, S. (1995). *At Home in the Universe*. New York: Oxford University Press.

KELLERT, S. (1993). *In the Wake of Chaos*. Chicago: University of Chicago Press.

KIM, J. (1997). 'The Mind–Body Problem: Taking Stock After Forty Years', *Philosophical Perspectives*, 11: 185–207.

—— (1998). *Mind in a Physical World*. Cambridge, Mass.: MIT Press.

—— (1999). 'Making Sense of Emergence', *Philosophical Studies*, 95: 3–44.

LAUGHLIN, R. (2005). *A Different Universe: Reinventing Physics from the Bottom Down*. New York: Basic Books.

MCLAUGHLIN, B. (1992). 'The Rise and Fall of British Emergentism', in Beckermann *et al.* (1992), 49–93.

O'CONNOR, T. (1994). 'Emergent Properties', *American Philosophical Quarterly*, 31: 91–104.

PEACOCKE, A. (1990). *Theology for a Scientific Age*. Oxford: Blackwell.

—— (1995). 'God's Interaction with the World', in Russell *et al.* (1995), 263–87.

—— (1999). 'The Sound of Sheer Silence: How does God Communicate with Humanity?', in Russell *et al.* (1999), 215–48.

RUSSELL, J., and PEACOCKE, A. (1995) (eds.). *Chaos and Complexity: Scientific Perspectives on Divine Action*. Vatican City State: Vatican Observatory Publications.

——MURPHY, N., MEYERING, T., and ARBIB, A. (1999) (eds.). *Neuroscience and the Person: Scientific Perspectives on Divine Action*. Vatican City State: Vatican Observatory Publications.

SHOEMAKER, S. (1980). 'Causality and Properties', in P. Van Inwagen (ed.), *Time and Cause*, Dordrecht: Reidel, 109–36.

SILBERSTEIN, M. (2002). 'Reduction, Emergence and Explanation', in P. Machamer and M. Silberstein (eds.), *The Blackwell Guide to the Philosophy of Science*, Oxford: Blackwell, 80–107.

STEPHAN, A. (1999). 'Varieties of Emergentism', *Evolution and Cognition*, 5: 49–59.

VAN GULICK, R. (2001). 'Reduction, Emergence and other Options on the Mind/Body Problem', *Journal of Consciousness Studies*, 8: 1–34.

WEINBERG, S. (1992). *Dreams of a Final Theory*. New York: Random House.

FEMINIST APPROACHES

CHAPTER 48

GOING PUBLIC: FEMINIST EPISTEMOLOGIES, HANNAH ARENDT, AND THE SCIENCE-AND-RELIGION DISCOURSE

LISA L. STENMARK

That feminist epistemologies are relevant to the science-and-religion discourse (SRD) should be self-evident. After all, the contemporary SRD owes its existence in part to the insights of feminist epistemologies and science studies. Unfortunately, the contemporary SRD has largely failed to develop models that incorporate these insights into the methods and goals of the discourse, and instead continues to work out of old models which leads to a discourse that focuses on 'doctrines and discoveries': i.e. an emphasis on the truth claims of science and religion, as opposed to their practices, or viewing discourse as an attempt to reconcile truth claims. An emphasis on doctrines and discoveries not only undercuts the discourse we purport to engage in; it is antithetical to a central tenet of feminist epistemologies—and feminism in general—that we have to pay attention to the impact that our ideas and concepts have on people's lives. An emphasis on doctrines and discoveries undermines the hope that the SRD can embody a more just and a more inclusive society.

In this chapter, I will propose a model of the SRD as a 'public' discourse, a model which incorporates the insights of feminist epistemologies. Feminists have rejected objectivism, and have instead proposed an understanding of knowledge as a relationship between the knower and that which is known—a relational epistemology— and between the knower and society—a 'situated knower'. The assertion that *who* the knower is affects *what* is known has led feminists to argue, first, for the inclusion of women in various discourses, and later, for an even greater inclusivity, a 'borderlands epistemology'. Feminist epistemologies also recognize that knowledge has a real impact on and in the world, and that we can and should incorporate our values into knowledge and embody that knowledge in action.

In the second part, I will discuss the implications of this perspective for the SRD. I will argue that, to the extent that we seek agreement between science and religion, the objectivist notions of knowledge in the SRD may actually impede discourse, for these concepts are embedded in a world-view in which either science or religion must be subordinated to the other. I will conclude by offering a model of discourse based on the thought of Hannah Arendt. The model of the SRD as a 'public discourse' incorporates feminist insights about knowledge, diversity, and praxis,[1] and enables the SRD to move beyond doctrines and discoveries.

Feminist Epistemologies

The Situated Knower

The key critique of feminist epistemologies has been to challenge 'masculinist' conceptions of objectivity that posit knowledge as coming from a disinterested observer, the 'mind in the vat'. In this view, reliable knowledge is achieved through objective, dispassionate, and value-free inquiry, where knowers are isolated from the objects of knowledge, from subjective distortion, and from the influence of society or culture. This is the 'royal road' to truth, a universal method leading to a universal, value-free system of knowledge about life, the universe, and (pretty much) everything.

---

[1] Hannah Arendt was not a feminist, and was often at odds with feminism. Recently, however, feminists have begun to explore Arendt's ideas and criticisms, often finding them fruitful. For a specific discussion of feminist epistemologies, see Disch (1994). For a more general discussion of Arendt and feminism see Honig (1995).

Feminists have argued that far from being objective and value-neutral, objectivist epistemologies are the product of a specific world-view. Evelyn Fox Keller and others have argued that this is a *masculine* way of thinking, not because all men think like this, but because it relies on traits that have been *gendered* masculine—rationality, emotional distance, impassivity, dominance, and control (e.g. Keller 1985: 69). By extension, this excludes those things that are gendered feminine—bodies, emotions, passion—and thus largely excludes women. The crux of the argument is that notions of objectivity and universality—whether in science or in religion—are neither value-neutral nor universal: they project particular ways of thinking and embody particular values.

A feminist epistemology replaces the disinterested observer with the 'situated knower'. This entails a relational epistemology, which Keller describes as knowledge arising from a relationship between knower and known (1985: 117) and Hilary Rose (e.g. 1993, 1994) calls a relationship between the head, heart, and hand. We know the world because we are a part of the world. This is not radical relativism. There is a world that is known, but all of our knowledge of that world is embedded and embodied. The situated knower is embedded in the world, but also embodied in a particular body, a particular culture, and a particular location. All knowers have, therefore, particular concepts, values, and experiences.

This is not a rejection of objectivity as much as an acknowledgement that our subjective embodiments are part of knowledge. Subjectivity should be sought, not rejected, because feelings and values do not *distort* knowledge; they are in fact a *necessary* component of accurate knowledge. Knowledge comes not through the elimination of subjectivity but by examining and being *intentional* about our subjective preferences. As Sandra Harding asserts, objectivity is threatened not by subjectivity, but by 'coercive values—racism, classism, sexism—that deteriorate objectivity', while 'participatory values—anti-racism, anticlassism, antisexism'—decrease the cultural distortions in explanations and understandings. 'One can think of these participatory values as preconditions, constituents, or a reconception of objectivity' (1986: 249).

As Helen Longino puts it, when viewed from the perspective of social values, 'the idea of a value-free science is not just empty but pernicious' (1990: 191). Far from giving us an accurate picture of the world, Longino concludes that 'when purged of assumptions carrying social and cultural values' our knowledge is incapable of matching 'the beauty and power that characterize even the theories we do have' (1990: 219). For Longino, integrity happens only when values are a part of knowledge.

The location of situated knowers is not external to what is known, but provides a framework for knowledge, allowing us to place it in a meaningful pattern. Facts do not speak for themselves: we arrange them in a way that makes sense, in a way that tells a story (Haraway 1989). This narrative turn is important to feminist epistemologies, since it not only acknowledges the situated knower but also incorporates multiple perspectives.

Standpoint, Multiple Standpoints, and Borderlands Epistemology

Feminists argue that objectivist epistemologies lose their accuracy because they exclude different *kinds* of knowledge, as when, for example, Native American oral traditions are dismissed as 'myth' (Harding 1998; Deloria 1997). The exclusion of difference gives the appearance of universalism at the price of accuracy. Feminist perspectives reject assumptions based on homogeneity, arguing that exclusion distorts perception, and only multiple perspectives paint an accurate picture. To the extent that perspectives are excluded or dismissed, we end up with distorted truths and false universals that are, in fact, constructions.

The question is whether this exclusion has a real impact on the content of knowledge. Clearly, the notion of the situated knower implies that who is thinking makes a difference, although feminists have disagreed concerning the extent of the difference. Some have argued that science doesn't just happen to produce biased results, and religion doesn't just happen to produce abusive symbols, but that science and religion are each inextricably intertwined with masculinist models and ideas, so that they are themselves inherently biased against women. Science and religion as they exist must be discarded and replaced with a new kind of science and a new kind of religion, kinds that reflect totally new paradigms arising from women's perspectives.

Other feminists have argued that while existing science and religion are biased, this bias is external to the practices of religion and science. In this view, the solution to the problem is to include women in full participation in these discourses, to ask questions that start from women's lives, and to let those experiences and observations establish the framework, so that a whole new set of questions can be asked. This is more than just getting women to work in existing frameworks—a position more reflective of feminist empiricism (Harding 1986)—because the inclusion of women's perspectives actually changes the structure of the discourse. It is this position that I find most helpful for the SRD.

Initially, this approach meant a turn to a women's standpoint epistemology—giving a privileged position to women's experiences and perceptions. This privileged position was quickly criticized as simply replacing the universalized experiences of particular men—white, affluent, Northern, etc.—with the universalized experiences of particular women—white, affluent, Northern, etc. (Harding 1998; Collins 1990). Women of colour and white feminists have convincingly shown that in the same way that frameworks set by men can distort or obscure women's concerns, so the frameworks set by some women can distort or obscure the concerns of other women.

Feminists such as Harding responded, first by shifting the standpoint to, for example, women of colour or poor women in developing countries. But this, once again, re-created the problem. Moderate feminist epistemologies have largely abandoned a single standpoint altogether, in favour of multiple standpoints. Harding, for example, suggests that what is necessary is a kind of 'borderland epistemology'. Because different cultures are exposed to different parts of nature, each develop distinctive resources—metaphors, models, and language—which enable them to see their particular parts of the world in diverse ways. Thus, each culture has different resources for understanding, which generate both systematic knowledge and systematic ignorance. A single perspective produces less knowledge about the world, while the diverse resources of different cultures enable us to see more. A 'borderlands epistemology' can be expanded beyond our view of nature to, for example, our view of the divine. In both cases, the move is towards more and greater inclusion and a rejection of all totalizing theories.

What is important is that a borderland epistemology does not call for an integration of distinctive cultural understandings into a single, ideal knowledge system—this would be another totalizing theory! Any synthesis, any single theory—no matter how good it is—necessarily sacrifices the advantages of differing conceptual schemes. Instead, borderlands exist in many places at once, and we move from one perspective to another, learning what to value in Western science, what to value in Zen Buddhism, what to value in indigenous knowledge, etc. We learn which approaches provide the best set of maps for each particular journey.

Praxis

The 'situated knower' embodied in 'multiple standpoints' implies an additional element of a feminist epistemology, praxis, or a consideration of the ways in which thought, language, ideas, models, concepts, etc. affect our actions and communities. At its most basic level, this involves being critical of destructive modes of thought. Maria Mies, for example, argues that because scientists gain knowledge by forcibly removing objects from their environment, isolating them, and dissecting them, 'violence and force are therefore intrinsic methodological principles of the modern concept of science and knowledge' (Mies and Shiva 1988: 46–7).

The destructive impact of ideas is of particular importance to feminists, because such ideas and their negative consequences disproportionately affect women (e.g. Mies and Shiva 1988: 22). By privileging men's experiences, masculinist conceptions of nature, knowledge, the divine, sin, etc. not only exclude women; they contribute to and justify the subjugation of, and violence towards, women. Where these critiques merge is in ecofeminism, perhaps the most prolific and fruitful area of feminist science and religion, even though it is rarely acknowledged by the SRD.

Even outside ecofeminism, feminist epistemologies require that we focus on the impact of ideas, language, and concepts. Communities not only influence science and religion, but science and religion influence communities. If values are a part of knowledge, we have a responsibility to *choose* what values we express and embody in our knowledge of the world. We are not passive knowers, and cannot abdicate responsibility for choosing the course of knowledge, consistent with the values and commitments we express in the rest of our lives, such as justice, equality, and human flourishing. Further, we cannot abdicate our responsibility to act on what we know.

BEYOND DOCTRINES AND DISCOVERIES:
IMPLICATIONS OF FEMINIST
EPISTEMOLOGIES

Beyond Objectivity

Relational epistemologies are not just better as a matter of preference, expedience, or political correctness; they are better because they more closely approximate the world we seek to know. The SRD should reject any form of objectivism—and not just because it distorts our picture of the world. We should reject objectivism because concepts such as 'hard facts' and 'objective knowledge' are intertwined with an understanding of science and religion as distinct, and even hostile, discourses.

Early work in the contemporary SRD, such as Ian Barbour's *Myths, Models and Paradigms* (1974), recognized this, and arose out of developments in feminist episte-mologies and science studies which rejected such notions as a value-free science and detached objectivity. It would be difficult, if not impossible, to imagine a SRD without rejecting a fully objectivist epistemology. Yet there are ideas in the discourse, a focus on doctrines and discoveries—i.e. an emphasis on abstractions, an implicit privileging of science, the quest to reconcile truth claims (see below)—which betray a lingering reliance on objectivist epistemological assumptions. Old paradigms die hard!

The SRD exists because, among other things, feminists argued that 'facts' were not distinct from theory, and that 'objectivity'—in the sense of value-free thought or theories—did not exist. Thus Barbour argues, 'science is not as objective, nor religion as subjective' as philosophical models want to assume (1974: 5–6, 171). But assump-tions about objectivity and subjectivity remain embedded in the concepts we use and the way we use them—i.e. 'facts'.[2] These assumptions are in turn embedded in and reinforce a context in which there is a distinction, or even hostility, between science and religion. This not only precludes any meaningful relationship between religion and science; it is a world-view in which 'objective' thought is privileged and which therefore privileges scientific statements—which purport to be objective—over religious ones: when doctrines and discoveries clash, discoveries win.

To the extent that the categories and concepts we use to talk about science and religion presuppose a world of value-free facts and detached objectivity, religion and science cannot be in discourse. This is not to say that people in the SRD believe in objective notions of truth, that facts speak for themselves, or that science tells us the way things 'really' are (although, inevitably, there are those who do believe some or

---

[2] Alasdair MacIntyre points out that the concept of 'facts' developed in the eighteenth and nineteenth centuries to solve a particular problem—the gap between 'seems and is' (1984: 97). The context has changed, but 'facts' remain, with an implicit assertion that it is possible to strip away interpretation and theory—to detach oneself from any narrative—and confront facts as they are (1984: 84).

all of these things), but that we have failed to take into account that our concepts are historically and contextually situated, treating them instead as neutral. We may not believe in value- or context-free thought, but we speak—and act—as if we do.

The first way in which feminist epistemologies help the SRD is by offering an alternative to 'objective' knowledge that is *not* a kind of radical relativism wherein everything is constructed (interpreted as somehow 'not real') and wherein all knowledge consists simply of competing paradigms, with no means of evaluation. As noted above, feminist epistemologies are not relativistic, although the fear that relativism is the only alternative to objectivism might account for some of the reticence on the part of those in the SRD to fully explore the implications of feminist epistemologies. This perceived threat might also explain why the SRD continues to hold on to some forms of objectivism.

Feminist epistemologies reveal that the choice between objective fact and radical relativism is false. As Longino points out, the recognition that knowledge has a social component leads to relativism only in the context of an individualistic conception of knowledge (1990: 216). The choice between 'constructed' and 'facts' or 'truth' is tied to an older paradigm—one in which the SRD cannot exist.

The false dichotomy between facts and radical relativism is further illustrated by the feminist emphasis on knowledge as embodied, and the narrative shift in feminist epistemology. Consider the question 'Who am I?' The answer to that question involves a story. That story, like any story, is a mixture of fact and fiction. Things have happened to me: I *was* born in Sacramento, my father *was* in the military, I *did* live my first few years in Italy, etc. These are the 'facts'. In order to tell who I am, I recount these facts: I tell a story. But, as with any story, I elaborate on these facts, expanding on some, omitting others, adding commentary and colour. I can—I must!—take certain liberties with 'the facts' in order to express 'who' I am (and to keep my story from lasting as long as I have lived!). Storytelling involves tremendous freedom, as there is always another story to tell and another way to tell it. But, while there is a lot of leeway in the way I can tell my story, that leeway is not unlimited. I cannot, for example, claim to have fought with the Contras in Central America, even though I once had a vivid dream about doing this. At some point my story would be wrong, at some point I would simply be lying. Where that line is might be open to debate, but that line is still there.

Hannah Arendt makes a similar point in her discussion of the relationship between fact and opinion. She asserts that fact and opinion are not antagonistic—facts form opinions, and opinions can differ widely as long as they respect facts. Facts must be interpreted—'picked out of the chaos of other happenings...and then be fitted into a story that can be told only in a certain perspective' (1968b: 238)—but interpreting or rearranging is not the same as changing. What, Arendt asks, will historians say about the First World War? We don't know for certain; but they won't say that Belgium invaded Germany, because historians are powerless to eliminate German invasion.

It is not a leap to assert that a scientist constructs a theory by *emplotting* data. Bruno Latour (1999: 24–79) describes field scientists collecting samples to study the savannah of Boa Vista. First they tag and mark them, then more data are added—temperature, weather, colour, taste, smell. All of this is taken to the lab and reassembled so that scientists can study it. Throughout this process, scientists make contemporaries of things collected at different times; the specimens can be shuffled, rearranged, viewed from different angles. Eventually, patterns emerge. Through this collection and arrangement of samples, scientists extract a model from confusion.

There are a number of correlations between science and storytelling. Both scientific practice and storytelling arrange 'data' (experiences or samples), and in so doing create a configuration, drawing these disparate elements into a meaningful whole. Just as a story is not the same thing as the experience, so the reassembly of data and the pattern (the theory, model, etc.) that emerges are not the natural world, but a representation of it: the scientist substitutes a model for the 'real' thing, just as a storyteller substitutes the story for the actual experience. This is a dual move. Narratives provide distance, and we are removed from the immediacy of the experience or data so that we can examine it. But it is not objectivist, because narratives simultaneously draw us back into experience and data. Finally, our narrative configurations allow us to share experiences and data: scientists share data the way storytellers share experience and events.

Like storytelling, science is a creative act. No story is the only one that can be told, and there are always rival interpretations of data. But, while this is a rejection of objective, impartial knowledge and universals, it is also an assertion that not everything goes. There is still a world that must be acknowledged. There is leeway, but that leeway is not unlimited. Interpretation must always confront the data, and some theories will simply fit better than others. Interpreting and arranging events is not the same as changing them: no amount of interpretation can legitimately arrive at the assertion that Belgium invaded Germany, just as no legitimate interpretation can say that the universe revolves around the Earth. At some point, we are simply lying.

Rather than diminishing science, this understanding preserves science as a creative act, and the scientist as a responsible actor who is not shielded from the consequences of interpretation by the myth of objectivity. The best scientists do not get to be the best by being mouthpieces for nature, just as theologians do not simply repeat the 'word of God'. In this conception, the creative element of science is not sacrificed to the false god of objective knowledge or realism. Moreover, this understanding is more conducive to discourse, and thus to the roots of the SRD.

Beyond Agreement

In addition to moving beyond objectivism, feminist epistemologies suggest that it is necessary for the SRD to move beyond the goal of agreement. Implicit through much of the discourse is the assumption that science and religion should agree. Unfortunately, science and religion sometimes disagree. That they will disagree, and that those disagreements will be real and not reducible, is a fundamental assumption of feminist approaches, especially borderlands epistemologies. Thus, the SRD is confronted by the need to reduce the irreducible. This creates enormous pressure to shift to abstractions once again, a shift towards the truth claims of religion and science: doctrines and discoveries. This has the paradoxical effect of undermining discourse, because it assumes that these truth claims are separable from the practice of religion and science, that they are, in other words, timeless and universal. This is tied to an underlying objectivity, because it necessarily sees truth claims as independent of context, privileging those claims that purport to be true 'regardless'—in other words, claims which are 'scientific'. To the extent that it privileges universals and abstractions, such as 'Truth' and 'human nature', the SRD distorts reality by hiding particular experiences.

The goal of agreement—and its emphasis on doctrines and discoveries—is clearly bound to particular notions of 'science', 'religion', 'truth', and 'fact' that are, as I argued above, historically and contextually bound to a world-view that undermines or precludes meaningful discussion between science and religion. Because this understanding privileges detached 'objectivity', or a scientific approach, the best that can be achieved is either natural theology—in which the objective truths of science become the source of religious reflection, giving rise to a theology which necessarily shifts with shifting scientific interpretations—or ethics, in which religion tells us what we can or can't do with the results of scientific knowledge. While there is nothing inherently wrong with either natural theology or ethics, this is not what the SRD claims to do because, in each case, religion is subordinated to science, and true discourse is precluded.[3]

What feminist epistemologies have demonstrated is that detached objectivity relies on some purportedly independent—meaning universal—ground of justification. When religion and science conflict—and they *will* conflict—they each end up on their 'independent' ground. If one is religiously conservative, that ground will likely be some sort of revelation, or inerrancy, a position that clearly pits religion against science. Academic discourse fares no better, because here the neutral ground is reason and objectivity. Inevitably, then, objective, scientific discoveries challenge or change religious doctrines. Rarely, if ever, do religious doctrines—even long-established ones—change, much less challenge or even problematize, scientific discoveries. If this is doubted, try to imagine saying to a scientist, 'I'm sorry, I can't agree with you because that violates my religious beliefs' (or, 'millennia of religious tradition'). When doctrines and discoveries clash, discoveries win.

As is the case with objectivist epistemologies, the quest for agreement, although it may seem benign, actually hinders discourse. This desired agreement, even when we don't know the content, functions as a kind of totalizing or meta-discourse or 'universal truth'—and if science and religion can both agree, it must be the truth—

---

[3] The alternative is simply to reject science as not 'really' objective and to assert that religion has an objective basis. While the responses might be different, this one associated with conservative approaches, and the other with liberal ones, both accept the fundamental position that detached 'objective' knowledge is not only possible, it is better (see e.g. Murphy 1996).

that subsumes opinion. What we need is a new model of discourse, one that can accept, and perhaps embrace, a kind of endless deferment.

Going Public: Arendt and the SRD

While feminist epistemologies call into question some of the concepts and methods embedded in the current SRD, what is needed is a model that actually adopts feminist insights, moving beyond objectivist epistemologies and agreement—beyond doctrines and discoveries—and opening the door to a SRD that embodies a commitment to human flourishing. One way is to think of the SRD as a 'public' discourse. I will elaborate on this model, relying on three concepts from the work of Hannah Arendt: being in and for 'the world', engaging in a 'disputational friendship', and 'thinking in stories'.

'The world' is an important, complex, and controversial concept in Arendt's work, but it will be sufficient for this discussion to say that 'the world' describes the conditions necessary for the emergence of the public sphere. The world is a shared culture, both material and non-material, which provides that which is necessary for human personhood, freedom, and the exercise of collective power. These conditions include plurality, equality, and stability. While each of these has significance for the SRD, I will focus here on plurality.

Plurality is a basic human condition; it means that we are all the same, human, while we simultaneously maintain our distinctiveness. A world of plurality is one that we share, while maintaining our distinctive positions within it. Arendt refers to this as the 'space of appearance' (1956: 198). This plurality is linked to the capacity of individuals to act—it gives us the space to do something new—but it also makes *collective* action possible, which is to act together to change what exists and to *establish* something new. This is what Arendt means by power and political action: that *we* (not I) can create or change the world.

Plurality is also the condition that makes judgement possible. Judgement is the human capacity to evaluate the world and our environment as it appears to us. Because appearances can deceive, we need others to confirm our judgements. To be deprived of plurality is to be deprived of reality, because the reality of the world is guaranteed only by the presence of others which enables us to see from many sides at once (Arendt 1956: 199). Alone, we easily fall victim to the twists and turns of our own prejudices. The loss of plurality leads to the loss of judgement and our sense of reality.

The relationship of the SRD to the world is complex. To the extent that the SRD is multidisciplinary, and thus committed to plurality, it exists in, and is a manifestation of, the public realm. Further, the SRD needs the stability and openness of the world for its existence and success. This is true not only of the SRD, but also of the practices of religion and science as well.[4] Science and religion may need the world in different ways, but they both need it. Commitment to the world is thus in part a commitment to the conditions that make science and religion possible, as well as a commitment to maintaining the conditions of plurality, human freedom, and empowerment. Their role in building the world might be different—science produces stable artefacts, religion preserves myths and stories—but in each case that role is indispensable.

Religion and science each have what might be called a private dimension—what Arendt calls 'thinking'—but I am not concerned with that here. What I am concerned with is what happens when they engage in plural or public discourse, judgement. In private, science and religion are both in the 'truth claim' business, and can argue vehemently for 'the truth' or 'facts'. But, truth claims are disastrous to the public realm and to our sense of reality, because they are *not* a matter for discussion. All 'truth', once it is perceived as such, is beyond discussion, whether it is religious, scientific, or otherwise. The Earth revolves around the Sun whether one believes it or not; God exists no matter what one says. This is why it is so frustrating to talk to someone who is a 'true believer', whether religious, scientific, or otherwise. These conversations can feel violent because, as Arendt points out, truth has a coercive element and a despotic character (1968b: 241).

Given the difficulties that both 'truth' and 'fact' can present to the public realm, it seems that neither religion nor science should be public; nor is discourse possible between them. Arendt suggests a way out by arguing that the point of public discourse is not agreement, but disagreement. It is the disagreement between multiple perspectives that makes plurality, human spontaneity, and, ultimately, the exercise of collective power possible. True discourse is driven not by a commitment to the truth—although all parties to the discourse may be committed to the truth— but by a commitment to seek out others who disagree with us in order that we might maintain a sense of reality through multiple perspectives. It is driven by a commitment to a relationship, for the sake of the world. Arendt calls this a 'disputational friendship', which involves taking principled stands—which both science and religion want to do—without contravening the reality of plurality, which discourse and the world require.

A disputational friendship exists to preserve the world through plurality and discourse, because these make us human: 'However much we are affected by the things of the world, however deeply they may stir and stimulate us, they become human for us only when we can discuss them with our fellows' (1968a: 25). As a disputational friendship, the goal of the SRD would no longer be *harmonizing* religious doctrines and scientific discoveries, but *humanizing* the world.

---

[4] Although religion is more able to withstand the thinness of the world than science, which requires worldly objects, such as technology, labs, peer review, and the like. Religion can exist, even if only as a spiritual practice, even if the world has withered. Indeed, a shift from 'religion' to 'spirituality' might be a sign of diminished commitment to the world. But, while world loss severely contracts religion, it does not destroy it, as illustrated in the brilliant post-apocalyptic novel *Canticle for Liebowitz* (Miller 1976).

Seeing the SRD as a disputational friendship allows both religion and science—discourses that are in the truth and fact business—to take a stance on those principles without destroying plurality or having a coercive influence in the public sphere. This is not arguing simply for the sake of argument or a matter of dogmatic pronouncements, neither of which fosters or promotes relationship. Disputational friendship is critical thinking that takes sides for the world's sake—that is, for the sake of preserving plurality. It is 'partisanship for the world' which involves, among other things, abandoning scholarly argument in favour of critical models that provoke dispute (Arendt 1968a: 8).

Thinking in stories is just this kind of model. And it is a model of rationality which enables us to incorporate feminist insights and critiques. First, as I outlined above, stories present an alternative to the tired old dichotomy of objectivism and relativism. But they do much more. In particular, stories enable us to be for the world, because they situate us in the world. Behind theories, there are stories that contain their full meaning. No matter how abstract theories may seem, theories begin as experiences, and therefore they begin as stories. Stories situate theories within lived experience; they force theory to ground itself, to continually remind us about discrepancies between theory and the reality in which it should be embedded. In short, stories remind us that all theories—indeed, all statements—are embedded in a broader narrative. This does not make them wrong, but it does mean that detached neutrality is a fiction. Seen from the perspective of a commitment to the world, it is clear that the problem with detachment is that it cannot be *for* the world, because it is not *in* the world.

A second reason why thinking in stories provides a critical model for the SRD is the feminist assertion that objectivity and neutrality come from being embedded in the plurality of the world. Abstract thought does not eliminate prejudice: it reinforces it. The objectivity of disputational friendship requires thinking together and seeing from many sides. This approach does not seek an allegedly neutral 'any man', but instead approaches the discussion as *many* individual men and women. It takes divergent opinions into account.

This ability, and the impartiality that it implies, is the difference between thought, which is solitary, and judgement, which happens in public. Judgement is the ability to think from another perspective. The more views present in my mind, and the better I imagine those perspectives, 'the stronger will be my capacity for representative thinking and the more valid my final conclusions' (1968b: 241). The quality of my judgements depends on the quality and quantity of representation. Representational thinking is a Kantian 'enlarged mentality' (Arendt 1982: 42–3), which establishes a kind of critical distance, a stepping back and removing one's self from the immediacy of the moment, and assuming a kind of general standpoint. But Arendt worries that Kant's critical distance is too distant: despite his interest in the world, Kant was not a citizen of the world, but a *Weltbetrachter*, a world spectator (1982: 44).

Representative thinking—or thinking in stories—simultaneously distances us and draws us in. Arendt calls this the ability to 'train one's imagination to go visiting'

(1982: 43). 'Visiting' does not mean 'passively accepting another's thoughts or merely exchanging prejudices' (1982: 43); it involves training one's imagination to see the world from someone else's perspective—that is, not assuming another identity, but attempting to give voice to difference as authentically as possible.

Lisa Disch elaborates on Arendt's notion of visiting. On the one hand, it is not 'accidental tourism', which is the attempt to re-create all the comforts of home while one travels. It is also not assimilation, in which one loses one's identity in a new place. Neither of these respects and maintains plurality. Visiting means travelling to a new location and leaving behind what is familiar, but resisting the temptation to feel at home there. 'In order to tell yourself the story of an event from an unfamiliar standpoint, you have to position yourself there *as yourself*. That is, you can neither stand apart from nor identify with that position' (Disch 1994: 158). This involves thinking as oneself from an unfamiliar perspective and 'permitting oneself to experience the disorientation that is necessary to understanding just how the world looks different to someone else' (Disch 1994: 159). Visiting enables one to tell a story, even a familiar one, from an unfamiliar standpoint.

To the extent that the SRD emphasizes doctrines and discoveries, it cannot take plurality seriously. It leads to a tendency to either accidental tourism or assimilation. From the perspective of religion, accidental tourism is the attempt to bring all the comforts of our religious traditions along and impose those beliefs and methods on the practice of science. It can mean treating science like religion—or *a* religion—in one way or another. Assimilation, on the other hand, means leaving religious commitments and traditions behind and making one's self at home in the practice of science. Neither option maintains plurality; nor does it enable persons who are committed to a religious tradition to criticize science from the perspective of religion. In the case of tourism, we are outsiders with no right to complain (imagine one having a guest who did nothing but point out how one's culture was inferior!). In the case of assimilation, we lack the necessary distance to be critical.

A third reason why 'thinking in stories' provides a critical model for the SRD is that it resists definitive statements and is content to defer agreement endlessly. The visiting imagination does not privilege a particular position as an abstract position from nowhere on which all reasonable people, once they put aside their particulars, can agree. Visiting is the attempt to take up a number of different perspectives, never completely adopting one but, as Arendt says, exercising the ability 'to move from standpoint to standpoint' (1982: 43). It is this embrace of multiple perspectives that marks neutrality, not one supposedly neutral standpoint. The 'standpoint' of multiple perspectives never assumes that all relevant perspectives have been addressed, because there is always another story, always another way to see the world. Definitive agreement is not possible, because of the endless possibilities for stories.

The model of disputational friendship also opens the door to discourse between radically different cultures, because it assumes no common ground other than the commitment to openness and discourse. When we have no shared world—because, for example, we come from radically different cultures—the commitment to talk can be the beginning of a shared world. It is possible to start with nothing in common, although this would be quite rare, but human spontaneity, freedom, and creativity mean that we can create a shared world *ex nihilo*, as it were. Discourse as the attempt to build a world—not the attempt to find some shared belief, even if that shared belief is only abstractly possible—makes true discourse possible by maintaining difference and plurality.

It could be argued that if the goal is continually to disagree, that we have moved further from, not closer to, considerations of praxis. While the reasons why this is not the case can be complex, the brief answer is that discourse, while not itself action, makes action possible. In being for the world, the SRD makes it possible to judge and to act. Because the private activities of religion and science and the SRD have a world-building function, they maintain the conditions that make action and empowerment possible. Further, this model not only makes the SRD a public discourse, it implies a political dimension, because the public is the realm of politics, and because plurality in the presence of others and discourse are inextricably linked to politics. As Susan Bickford and others have noted, for Arendt, plurality, politics, and human identity are intertwined (Bickford 1995: 315).

Finally, while storytelling resists definitive endings, it does allow us to bring matters to a tentative conclusion and to make a collective judgement. If definitive agreement is a fiction—and discourses as disparate as feminist epistemologies, scientific method, and Christian assertions of human finitude suggest that it is— then waiting for it means that we will defer action indefinitely. Witness the debate on global warming. The current US administration claims we should not act until we have agreement from the entire scientific community on the nature or cause—or even existence!—of the problem. So, while there is a *general* consensus that global warming exists, and somewhat less agreement as to the cause, we defer action. Applying false and unattainable models of truth and agreement to public discourse is clearly not only destructive to discourse itself—as I have argued—it is conservative in the sense of preserving the *status quo* and resisting change. Clearly the 'best guess' and 'provisional endings' of a narrative approach are better for public discourse (judgement) and public policy (action).

In this model, the task of the SRD is not to achieve some sort of agreement, but to tell a story that invites criticism and which contains within it a multiplicity of perspectives. This is similar to Harding's borderlands epistemology, in that the goal is not a synthesis of perspectives, which leads to a new perspective, which itself obscures difference; the goal is resolutely to include other viewpoints, even those we radically disagree with, even those that make us uncomfortable (perhaps *especially* the ones that make us uncomfortable), for the sake of discourse. This goes beyond including women and people of colour; for the SRD this might mean taking such things as young earth creationism into account! The goal of the SRD is not—or should not be—agreement. Its goal should be preserving the conditions that make public judgement and public action possible: a commitment to the world, and to the conditions that preserve human freedom, empowerment, and action.

REFERENCES AND SUGGESTED READING

ARENDT, H. (1956). *The Human Condition.* Chicago: University of Chicago Press.

—— (1968a). *Men in Dark Times.* San Diego: Harcourt Brace & Company.

—— (1968b). 'Truth in Politics', in *Between Past and Future,* New York: Penguin Books, 227–64.

—— (1982). *Lectures on Kant's Political Philosophy.* Brighton: Harvester Press.

BARBOUR, I. (1974). *Myths, Models and Paradigms: A Comparative Study in Science and Religion.* New York: Harper & Row.

BICKFORD, S. (1995). 'In the Presence of Others: Arendt and Anzlaldúa on the Paradox of Public Appearance', in B. Honig (ed.), *Feminist Interpretations of Hannah Arendt,* University Park, Pa.: Pennsylvania State University Press, 312–35.

COLLINS, P. H. (1990). *Black Feminist Thought.* Boston: Unwin Hyman.

DELORIA, V. (1997). *Red Earth, White Lies: Native Americas and the Myth of Scientific Fact.* Golden, Colo.: Fulcrum Publishing.

DISCH, L. (1994). *Hannah Arendt and the Limits of Philosophy.* Ithaca, N.Y.: Cornell University Press.

HARAWAY, D. (1989). *Primate Visions.* New York: Routledge.

HARDING, S. (1986). *The Science Question in Feminism.* Ithaca, N.Y.: Cornell University Press.

—— (1998). *Is Science Multicultural?: Postcolonialisms, Feminisms, and Epistemologies.* Bloomington, Ind.: Indiana University Press.

HONIG, B. (1995) (ed.). *Feminist Interpretations of Hannah Arendt.* University Park, Pa.: Pennsylvania State University Press.

KELLER, E. F. (1985). *Reflections on Gender and Science.* New Haven: Yale University Press.

LATOUR, B. (1999). *Pandora's Hope: Essays on the Reality of Science Studies.* Cambridge, Mass.: Harvard University Press.

LONGINO, H. (1990). *Science as Social Knowledge.* Princeton: Princeton University Press.

MacINTYRE, A. (1984). *After Virtue: A Study in Moral Theology.* Notre Dame, Ind.: University of Notre Dame Press.

MIES, M., and SHIVA, V. (1988). *Ecofeminism.* Halifax, Nova Scotia: Fernwood Publications.

MILLER, W. M. JR. (1976). *A Canticle for Liebowitz.* New York: Bantam Books.

MURPHY, N. (1996). *Beyond Literalism & Fundamentalism: How Modern and Postmodern Philosophy Set the Theological Agenda.* Valley Forge, Pa.: Trinity Press.

ROSE, H. (1993). 'Hand, Brain, and Heart', *Signs: Journal of Women and Culture in Society,* 9/1: 73–90.

—— (1994). *Love, Power and Knowledge: Towards a Feminist Transformation of the Sciences.* Bloomington, Ind.: Indiana University Press.

....

CHAPTER 49

FEMINIST PERSPECTIVES IN MEDICINE AND BIOETHICS

....

ANN PEDERSON

For the last forty years or so, the dialogue between religion and science has wrestled with questions of meaning and purpose about human life. We wonder: Who are we? Where are we going? And why are we here? These questions are about as ancient as the humans who ask them; yet they have taken a new twist in the last few decades in light of research in evolutionary biology, genetics, cosmology, and biotechnology. How we answer these questions is critical, because the hazards we face for living a sustainable life together are multiplying. We enter and leave this world through tangled webs of technology, culture, and nature. The twentieth century brought us not only the benefits of technology but also its risks and costly price. While scientific and technological innovations multiply so rapidly that we cannot keep pace, not all the world has access to the benefits. The world divides into haves and have-nots. Some babies are born through *in vitro* fertilization, once considered unnatural but now commonplace; yet the infant mortality rate soars with mothers and babies dying of AIDS. For many, clean water, healthy food, and adequate shelter are not available. The high-end technological advances that many consider their right are simply prohibitive for most of the world's population, whose simple survival from day to day is precarious.

What does it mean to be a human person in a global community whose survival as a species is tenuous? Philip Hefner, a Lutheran theologian, writes that how we respond must be '*universal* in that answers to the basic questions must enable adequate and wholesome futures for the planet and all of its inhabitants; *particularly* in that they must be life-giving to all regions of the planet, to all sorts and conditions of persons upon the planet, as well as to its nonhuman sector' (Hefner 1993: 5; emphasis added). When answering such questions, we turn to telling stories to express our ultimate values. The way we know and make sense of our world is through interpreting narratives. The category of narrative is a natural fit for this chapter that emphasizes ethical concerns and feminist insights. Each of us is embedded in multiple stories: familial, cultural, social, political, economic, and religious. We tell stories that are evolutionary and religious in their epic scale. Since Newton and Descartes put their mark on the modern world, we have enlightened one another with rationales about our importance in the universe as a human species. Much of Christian theology has also reinforced this anthropocentric view that the natural world is valued only as a backdrop for humans on centre stage.

How does the dialogue between religion and science reinforce narratives that separate humans from the rest of nature and from each other? How can the dialogue between religion and science further the task of interpreting human identity so that our purpose as one human species is not only that we live but also that we live well? The answers depend in part on who we ask. Whose stories count? Is there only one Enlightenment narrative to which all must submit? Do multiple stories imply that all stories are true?

THE WELL-PEDIGREED
AND THE UNDOCUMENTED

Feminist thought challenges the dominant ways of answering both the starting points and the directions of these questions. While the Enlightenment is often held up as the 'problem child' of the modern world, feminists call for an examination and critique of its family members: e.g. colonialism, materialism, consumerism, etc. Enlightenment philosophical assumptions find expression in the economic, political, social, and religious schemas of the modern world. We must examine some Enlightenment philosophical presuppositions, and then move to their close sibling rivals. To accomplish this task, I will use the writings of Donna Haraway, a feminist philosopher of science.

Donna Haraway, a professor of History of Consciousness and Feminist Studies at the University of California, Santa Cruz, makes connections, crosses boundaries, and imagines new ways of thinking and doing. Haraway is a cultural critic and philosopher of science. She crafts creative metaphors to explore relationships between nature, culture, human, animal, and machine. In her most recent writing, she has moved from the image of cyborg to that of dog as a companion species. As she reflects on the relationship between dogs and humans, she illuminates the relationships of all companion species, co-constituted in their differences. In a word, she practises what she preaches.

Her doctorate from Yale was an interdisciplinary study between biology, philosophy, and history of science and medicine. She admits that keeping things neat and tidy runs against her feminist and Marxist grain. She muddies the waters of our traditional understandings. Binary dualisms are shattered. Boundaries between human/nature, technology/nature, and machine/human fuse together. Her work subverts, transgresses, and inspires. Consequently, I find that it provides a provocative platform for interpreting the multifarious relationships between religion and science.

The grand Western Enlightenment world-view creates separations—between nature and humans, male and female, public and private, secular and sacred, religion and science, technology and culture—that do not make sense for the majority of the world's inhabitants. These separations are helpful only for those who benefit from them, especially when the distinctions shift into hierarchical dualisms. We cannot avoid making distinctions, but we are responsible for how and why the world is divided and by whom. The religion-and-science dialogue cannot—nor should it—avoid these critical problems of unity amidst diversity, of how the many and the one are related, and how meaning between our similarities and differences is created. How do we as one human race deal with our differences? Why do they matter, and to whom? How we ask and answer these questions is a matter of life and death.

Humans are marked by a common narrative that speaks of our flesh and blood, life and death. From our common ground of finitude and flesh come particular stories that have often been left out, misplaced, misappropriated, and misunderstood by those of 'grander' narratives in the privileged Western world. Donna Haraway remarks: 'Anyone who has done historical research knows that the undocumented often have more to say about how the world is put together than do the well pedigreed' (Haraway 2003: 88). The well-pedigreed often dismiss the undocumented as alien, as foreigners. Who are the undocumented? Who are the well-pedigreed?

Feminists, among other liberationist and post-modern voices, challenge the way in which dominant voices have put the world together. Those who have had the good pedigrees have set the rules of the conversation, planned the publications, and worked together at conferences. Much excellent scholarship has come from such a dialogue, but this inter-'face' between religion and science has tended to reflect only the concerns of those participating. Other faces and voices haven't participated for a variety of reasons. Some simply weren't invited. For others, the dialogue didn't relate to their lives outside the dominant discourse. As individuals and groups have protested the *status quo* of the religion-and-science dialogue, the nature of the conversation has slowly changed.

Discourse has shifted from the academy to places of work, to faith communities, and to concerns about life and death. Multiple partners have changed the somewhat monogamous relationship of science and religion into one of plural partnerships that are fruitful and faithful. Scientists and theologians from other religious traditions, from developing countries, and different social classes are not just adding to, but are altering the heart of the discourse. While changes can often create dissonance and discord, these changes produce much needed novelty, transformation, and creativity.

I witnessed this productive and very dissonant process first hand in a religion-and-science conference in St Petersburg in 1999. Having been silenced for decades by the former government, Russians are not only re-entering but also relearning the way that religion relates to science. The conversations would at times turn to diatribes and monologues. The Russians were trying not only to come to terms with their own religion-and-science dialogue, but also with the pathologies of the Western religion-and-science dialogue. The discourse between the two, while productive, was often harsh, frustrating, and discordant. As the religion-and-science dialogue expands throughout the world, those in the Western world will also be changed and transformed.

Bound together on a planet that we share with the entire natural world, we need to find out what is at stake in our life together, and this finding out entails a moral imperative. Donna Haraway notes: 'The point is to make a difference in the world, to cast our lot for some ways of life and not others. To do that, one must be in the action, be finite and dirty, not transcendent and clean' (Haraway 2004: 236). The stories that unfold between religion and science will not come about as a result of a 'transcendent and clean' dialogue, but from a down-and-dirty working through of the challenges that we face not only to live, but also to live well on this planet. The interdisciplinary conversations between religion and science must start 'from below' in order to make a difference, a difference that matters to all.

Feminists claim that epistemology and ethics are not separate disciplines. What we know and what we do with what we know are related. Knowledge and practice, separated by Enlightenment short-sightedness, are joined in a partnership of inter-action. How we know and whom we know shape what we know. These ontological and epistemological questions of meaning, Donna Haraway suggests, can be answered only through practice, through engagement. 'Answers to these questions can only be put together in emergent practices; i.e., in vulnerable, on-the-ground work that cobbles together non-harmonious agencies and ways of living that are accountable both to their disparate inherited histories and to their barely possible but absolutely necessary joint futures. For me, that is what significant otherness signifies' (Haraway 2003: 7). In order to carry on a fruitful dialogue, religion and science need to find such emergent practices which begin with 'on-the-ground' work, joining together their respective harmonious and 'non-harmonious ways' of relating to each other. They need to meet each other in their 'disparate inherited histories'.

Dog-land: The Companion Species of Science and Religion

In her work *The Companion Species Manifesto* (2003), Haraway makes the world of dog-land a metaphor for these emergent practices. Otherness is not romanticized, tamed, or feared. When otherness is established and respected, then connections are created through the rigorous practice of agility training. Dog and human meet in a carefully constructed choreography of jumping hoops, winding through barriers, and running the course. Often nose to nose, and eye to eye, the dog and human work off each other. Practice might make perfect (or at least get close). This practice of agility training is similar to that of learning and practising improvisational skills in theatre and music. Partners work off each other. Spontaneous play occurs only after the basics are learned so thoroughly that they are intimately embodied in each member. How might these metaphors for emergent practices apply to the religion-and-science dialogue participants?

First, questions of meaning and value are approached through emergent practices and not through detached answers already presumed. Donna Haraway states that companion species are 'about a four-part composition, in which co-constitution, finitude, impurity, historicity, and complexity are what is' (Haraway 2003: 16). We compose the narratives along the way, face to face—a sort of incarnational fugue, if you will. Haraway comments about the demand upon the participants in agility training: 'In short, the major demand on the human is precisely what most of us don't even know we don't know how to do—to wit, how to see who the dogs are and hear what they are telling us, not in bloodless abstraction, but in one–one–one relationship, in otherness-in-connection' (Haraway 2003: 45). Much like the movements involved in agility training or in musical improvisation, the discourse between religion and science moves forward in rather unpredictable ways. All parties participate within their own roles, working off and listening to the other. So the task, according to Haraway, 'is to become coherent in an incoherent world, to engage in a joint dance of being that breeds respect and response in the flesh, in the run, on the course. And then to remember how to live like that at every scale, with all the partners' (Haraway 2003: 65). In a world where meaning slips and slides, where things fall apart, the task is to practise, 'in the flesh, in the run, on the course'. We live at every point, with all the partners we encounter, creating coherence from incoherence, and order from chaos. And we can expect such chaos when new partners change the dominant narratives. Incoherence for some becomes coherence for others. Nonetheless the task remains the same: to breed respect and response for all the partners involved.

We know the world through the relationships we create, with humans and non-humans. Haraway notes that distinctions between nature and culture, human and non-human, are often blurred. She draws on the philosophy of Alfred North Whitehead to describe these relationships and how they are rooted in our shared biology. She writes:

My love of Whitehead is rooted in biology, but even more in the practice of feminist theory as I have experienced it. This feminist theory, in its refusal of typological thinking, binary dualisms, and both relativisms and universalisms of many flavors, contributes a rich array of approaches to emergence, process, historicity, difference, specificity, co-habitation, co-constitution, and contingency. (Haraway 2003: 7)

No pure species exist. We come as mutts, not pure-breds.

Ontologies emerge from practices. We co-constitute each other in the relationships we form and practise. What we once divorced as dualistic opposites we now join together as partners. Haraway says that in dog-land, 'we are training each other in acts of communication we barely understand. We are, constitutively, companion species' (Haraway 2003: 2). The science-and-religion dialogue forms through complex layers and multiple partnerships. Trained in vastly different ways and often working in very dissimilar settings, participants in the science-and-religion dialogue must acknowledge their differences, and respect true otherness. Co-constitution doesn't happen in the abstract, in predetermined dialogues. The science-and-religion dialogue is fruitful when participants listen carefully to each other, pay close attention to what isn't always obvious, and commit to a partnership of ongoing collaboration.

If the pedigreed and pure-bred continue to reassure themselves that their story is the only reliable one, then they deny their rich and diverse biological lineage and separate themselves from all others. One can see how this works through history: American Indians are forced onto reservations; Africans become slaves of Euro-Americans; children become cheap labour for adults. The danger is that the stories we tell about our differences are used to justify the control and domination of others. In a similar manner, the pedigreed and pure-bred in the mainline science-and-religion dialogue can convince themselves that certain narratives are immune from the mongrel prejudices of culture, politics, race, class, gender, and sexual orientation. As if a kind of historical amnesia takes over, the dialogue partners have often proceeded without recognizing that their own cultural, political, and social contexts shape the dialogue itself.

Feminist theorists identify and clarify the ethical, political, and cultural layers of how we know and interpret the world around us. For Haraway and others, feminist methodologies are both practised and practical. 'None of this work is about finding sweet and nice—"feminine"—worlds and knowledges free of the ravages and productivities of power. Rather, feminist inquiry is about understanding how things work, who is in the action, what might be possible, and how worldly actors might somehow be accountable to and love each other less violently' (Haraway 2004: 7). In its first stages, feminist theorists were often privileged, heterosexual, white women working in the academy. They challenged the patriarchal, androcentric, sexist structures that limited, marginalized, and disempowered women. These dominant voices of feminism tended to define the categories, set the agendas, and publish the research. However, other women, those left out, those on the margins and edges of the *status quo*, began to challenge what it meant to be a woman and a feminist. The discourse of feminism changed; the lived experiences of marginalized women became the starting point for transforming feminist thought into a richer array of theories and practices.

Feminists have joined other diverse liberation movements. Womanist theologians like Jacquelyn Grant and Emilie Townes challenge the racism of white feminists and the sexism of black men. Ada Maria Isasi-Diaz, a Mujerista theologian and theorist, constructs new ways of interpreting personhood in light of the Latino/Hispanic experience. Other feminist philosophers of science like Sandra Harding challenge the heterosexist and colonialist Enlightenment narratives that formed the natural sciences. Liberationist voices challenge the powers that be, the powers that diminish and marginalize the weak, vulnerable, and the least in society.

A critical feminist methodology will help clarify how things work, discern the who and what of the science-and-religion dialogue, and then find ways for us to be more accountable and responsible to each other and to the world in which we live. So how do things work, and what do we do with what we know? I begin with specifics, with illustrative intersections and partnerships.

MEDICINE AS A MORAL ENTERPRISE

I live and work at the intersection of religion and science, specifically in a working group of physicians, nurses, chaplains, educators, and theologians. We are all part of the Section for Ethics and Humanities at the University of South Dakota School of Medicine. We are located in a mid-size community on the upper Great Plains of the Midwest in the United States. In a sense, we have become the well-pedigreed. We all have advanced degrees, have decent if not exorbitant salaries, and don't worry about where our next meal or clean water will come from. While members of our section have travelled and studied abroad, worked with the underserved, and are sensitive to the problematic nature of contemporary health care, we are all among the documented ones, the well-pedigreed. Each of us comes to the common task of educating medical students with different backgrounds and expertise. We all agree that what we know, how we know it, and what we do with that knowledge is a moral enterprise.

Yet our enterprise is fraught with the dangers of the dominant discourse of consumerism, individualism, and the language of rights. Many whom South Dakota physicians and other health care providers will serve are not those with pedigrees. The land of the upper Great Plains has been inhabited for centuries by its first natives, many of whom are now banished to reservations. American Indians, whose history has been stolen, defaced, and defiled, face hardships often brought on by those in the dominant, white culture. Other individuals in South Dakota face economic crises brought on by farming in the climate of corporate America. Small, rural towns are dying. Youth leave the state altogether or flock to the few urban centres. Many elderly men and women cannot afford medications, and the current religious climate prohibits adequate access to reproductive health care for women. Farmers and ranchers work with dangerous pesticides and chemicals. Those of us educating physicians for rural life must ask ourselves whether we are preparing them adequately for their work.

For some in the field of medicine, the Enlightenment myths still provide the safe separation of theory and practice. The narrative world of the patient does not impinge on the objectivity of the physician's care. While I find that some physicians and medical school faculty still reside in the sterile world of 'scientific objectivity', such a world-view can no longer provide a sanctuary from the tough moral issues that arise daily. How does one teach the practice of medicine as a moral discipline in a pluralistic, techno-scientific, messy, finite world? For example, internal medicine residents from different backgrounds and cultures will often have conflicting notions about how to inform a family whose relative is terminally ill. The use of life-extending technologies can also present religious and cultural conflicts for both patients and physicians. To acknowledge and work through such dilemmas, physicians need to be aware of how their own cultural and religious biases shape their medical practice. Epistemology and ethics have a messy relationship in the classrooms of health care clinics!

Biomedical ethics in the last forty years or so has been taught using a principled approach. Medical students should be able to recite by rote (or look at the placard in their briefcase) the four principles of bioethics: autonomy, beneficence, nonmaleficence, and justice. These principles were developed originally in order to help people wade through tough ethical decisions. Principles could provide objectivity. While principles worked in the abstract, they rarely took into account the messiness of clinical practice and public health care settings. Making decisions about, and with, patients face to face is very different from doing it in a sterile classroom.

Enter narrative. From the principled approach to bioethics made famous by Tom Beauchamp and James Childress, bioethicists have moved into the world of storytelling, of narrative. Karen Lebacqz, a theological ethicist, says that 'central to the narrative approach to bioethics, therefore, is (1) listening to the patient's story, (2) understanding what kind of story it is, and (3) responding with a story that fits the patient's own story' (Lebacqz 2004: 103). We know the world by the stories we tell. And much of the messy world of medicine cannot be fit into principles. Physicians who know more about patients' stories can provide richer and deeper analyses of the dilemmas that they face. In the teaching of bioethics, Rosemary Tong, a feminist bioethicist, explains this shift in pedagogy:

For a variety of reasons, mostly having to do with the fact that I started to see medicine in practice, I became convinced that the sweet reasonability of principlism did not fit the clinic nearly so well as it fit the classroom. I became interested in the bioethical theories that principlism had eclipsed and began to see in them what I sense principlism lacked. (Tong 2002: 418)

One can talk about autonomy in abstraction, but it is another thing to talk with a patient who is struggling with end-of-life decisions, or with a young woman trying to decide whether or not to terminate a pregnancy. Ambiguity and complexity become the fertile ground for teaching medicine as a moral discipline. The task of the educator is not to offer simplistic answers or moral absolutes to complex medical and ethical dilemmas. Instead, the educator must help the medical student to analyse and understand the complex layers involved in a patient's story.

Simply listening to the patient's story is not enough, however. While narrative has been the buzz-word in academic circles, one cannot simply reduce the world in all its dirty, messy relationships to a simple storyline. Stories create worlds, but the world also creates stories! Feminists insist that the category of narrative must be expanded beyond the individual to include the historical, contingent, cultural, and complex layers of the broader context. As a feminist educator, Rosemary Tong appreciated what narrative theory did for the teaching of bioethics. But it wasn't enough; it didn't go far enough to describe the who and what of the world and the meanings we construct along the way.

Tong draws upon the work of other feminists to illustrate why the category of narrative must be expanded. Differences are explored in order to create connections and, as Haraway notes, 'to make a difference in the world'. Tong notes that Uma Narayan, an Indian philosopher,

believes that it is Western intellectuals' guilt about their past role in oppression that explains their present reluctance to condemn actions, practices, and systems in developing nations they would immediately condemn were they to occur in the West.... Objecting to what amounts to a double-standard morality, Narayan stresses that she does not want Westerners to unreflectively respect her culture, but to insist with her that what is wrong about U.S. racist practices is precisely what is wrong about Indian racist practices. (Tong 2002: 427)

If the world is simply reduced to one's personal story, then the complex details of the bigger picture are ignored. Enlightenment individualism can reign again, this time in the form of cultural relativism. Tong, along with Haraway, urges us to avoid the pitfalls of universal absolutes and cultural relativisms. Both of these belong to the privileged standpoint that doesn't need to take others seriously. 'Clearly, contemporary bioethicists cannot afford to play with the bombs of relativism, let alone postmodernism, anymore. The heavy demands of living in a globalized world require us to find and use conceptual tools crafted to chisel a measure of unity within our diversity' (Tong 2002: 427).

The very conceptual tools that we use to shape our diverse stories can steer us to one story that we can all claim to tell: that we not only live, but also live well—faithfully and fruitfully as one species on this planet. We can begin our common human story by learning and telling the epic of evolution and other creation stories that give us clues as to who we are. Simply being a human person constitutes a rationale for constructing some kind of modest global story. If we only concentrate on our differences, we forget what we all share: our finitude and flesh. As Tong reminds us: 'Our common carnality and mortality invite us to acknowledge our shared human needs, lest we permit our diversity to reduce to rags a moral quilt that could have covered and comforted all of us' (Tong 2002: 431). We are moral and mortal.

The practice of medicine as a moral enterprise must face the limits of human life. Death cannot be avoided; financial and human resources are limited. All human persons have basic needs that must be met in order for them to survive. Medicine must be practised within sustainable limits, and basic human needs and access to public health should be the priorities for allocating resources. The language of rights gives way to that of responsibilities.

If feminists are right that the grand narratives of the Western Enlightenment trap us in lethal dualisms, then we must understand how the history and practice of medicine is shaped by such dualisms. For example, expensive technological innovations are often available only to the wealthy, not to those who are poor or earn minimum wage. Public health needs are ignored at the expense of private ones. To change the way in which medicine is practised requires different ways, new ways, of creating meaning and value. Haraway's notion of emergent practices offers a constructive approach for restructuring the way medicine is practised. Discussion of ambiguous and complicated medical and ethical problems can be a joint partnership between teacher and student, both parties vulnerable and accountable to each other. In turn, these emergent practices may act as models for understanding the relationship between physician and patient.

INCARNATION: THERE'S MORE TO THE STORY THAN MEETS THE EYE

To illustrate these emergent practices of medicine, I offer some reflections from my experience. I have taught graduate students in the nursing department in a private college, developed a class on death and dying for undergraduates, and worked with physicians at the University of South Dakota School of Medicine. Feminist thought has helped me understand the education of health care workers as an emergent partnership between teacher and student.

The Augustana College (a private, liberal arts, college of the Church) Nursing Department which offers a Master's degree in Community Health, focuses its mission on the needs and challenges that the communities face on the upper Great Plains. Advanced practice nurses begin their work with a course that forms community alliances. For example, a nurse might work with a group of senior citizens in his local congregation. Instead of telling the seniors what they need, the nurse spends time listening to and being with the community of seniors. This usually includes lengthy interviews to learn the participants' individual stories. Senior citizens in an urban area might have different concerns from those in isolated, rural ones. Once the nurse has discovered from the seniors how health is defined for them, then a protocol can be decided upon for meeting their health care needs. This kind of approach reverses a typical hierarchically based medical model of teaching nursing. The seniors are taken seriously as experts about their own health care and needs. The advanced practice nurse becomes a partner, not one who patronizes.

In a course on feminist and liberationist approaches to wellness and health care for the graduate education programme in nursing, I have used the writings of Terry Tempest Williams, a naturalist. Williams writes stories about the interdependent relationships between humans and the rest of nature. Her best-known book, *Refuge: An Unnatural History of Family and Place* (1992), intertwines a story about the death of a bird refuge with the death of her mother from ovarian cancer. The reader discovers how the causes of both deaths are related to one another. For Williams, narrative expands the imagination and opens eyes to bigger pictures and worlds. The main characters in her stories include, but are not limited to, humans.

Like Haraway's explanation of story, Williams's use of place is biological, historical, and cultural. And the lines blur. Williams's use of biological and ecological metaphors, like that of Haraway, calls for imaginative leaps and figurative dances:

Biologist Tim Clark says at the heart of good biology is a central core of imagination. It is the basis for responsible science. And it has everything to do with intimacy, spending time outside.... I believe that out of an erotics of place, a politics of place is emerging. Not radical, but conservative, a politics rooted in empathy in which we extend our notion of community, as Aldo Leopold has urged, to include all life forms—plants, animals, rivers, and soils. The enterprise of conservation is a revolution, an evolution of the spirit. (Williams 1994: 86–7)

Williams says we know what we love, and thus become responsible for those we love. Consequently, to live responsibly with others demands that we love them. Her broad understanding of community, which includes the entire natural order, is the context from which constructs of health and illness can be explored.

A few years ago in a course on end of life that I team-taught with an internal medicine physician, we asked our senior students to create final projects about someone they knew personally. One student chose to interview his grandmother who was in her eighties. She lived in a small rural town. To be elderly and live in such a community is truly to live on the margins. While death was hardly imminent, the student's grandmother had begun to experience dying in several ways. She was isolated and lonely. The economy was poor, and local families she knew had left or were leaving for other opportunities. As the community dissipated, the church and school grew smaller as well. She felt loss each time she went to church, each time she went to the small main street shopping area. And her friends were gone as well. All of the relationships that had sustained her well-being over time were slowly dying. Her notion of what she needed at this stage in her life didn't include technological innovations; instead she needed flesh—those lives around her that kept her going. For this elder on the prairie, loneliness was her diagnosis.

When teaching about the other end of life, I have discovered that there is more to the story than the embryo. While headlines trumpet the ethical issues surrounding beginnings of life, the lives of the women who bear the children are forgotten. Across the country, some physicians, pharmacists, and other health care providers refuse to perform abortions, prescribe or fill orders for contraception, and even oppose the teaching of abortion in medical schools. Religious beliefs are usually cited as the reasons. Yet even within each religious tradition, beliefs about abortion vary. In March 2006, the South Dakota Legislature passed a bill prohibiting all abortions with the exception of saving the life of the mother. According to the legislative agenda, a human being by definition begins at conception, and ending the life of the embryo by any means whatsoever would have been prohibited, even for the health of the mother. The implications are much more complicated than simply pro-life or pro-choice. For many women who are affected by infertility and live in rural areas, simply going to the local family doctor might initially be their only option. Fertility drugs are often prescribed by male physicians who don't explain their impact on the woman's physical and emotional well-being. For example, if a woman after using such drugs has multiple embryos and is not able to carry them all to term, she might face difficult and painful decisions about whether or not to terminate some of the embryos. Many family physicians and clergy are not equipped to help women cope with these difficult ethical dilemmas. Simple moral or religious pronouncements do not alleviate the complexity. Currently in South Dakota, the only *in vitro* fertilization clinic is at one end of the state. The life of such a woman gets very complicated. She often loses her local support systems. And, in some cases, local pharmacies and hospital systems may refuse medical assistance or care if they are opposed to such reproductive technologies. There is more to the story than the embryo.

The writings of John Lantos help students to know that there is often more than a simple, right answer to complex ethical dilemmas. Lantos, a professor of pediatrics and section chief of general pediatrics at the University of Chicago, knows at first hand the complexities of medicine as a moral discipline. In his most recent book, *The Lazarus Case: Life-and-Death Issues in Neonatal Intensive Care* (2001), he structures a narrative case study that sends the reader into the ambiguities of life and death in the NICU (Neonatal Intensive Care Unit) forestalling any easy assignment of blame. Lantos complicates the story from the beginning. Chicago is the cultural, social, political, and historical setting for the story. 'I see these cases as constituting a sort of cultural locator, an indication of where and how our culture tries to understand and to frame the tough issues raised by the double-edged sword of neonatal intensive care and, by analogy, other innovative medical interventions' (Lantos 2001: p. xiii). He looks for the patterns in the particular, like Haraway; furthermore, he refuses to separate narrative or 'fact' from the texture of the landscape in which decisions are made.

He begins with three visual landscape markers: the Sears Tower ('temple of retailing'), the Amoco Building (built on energy, oil, and antitrust laws), and the John Hancock Building (insurance premiums). All three temples, he explains, are 'at the center of all the business centers in the center of all the great American cities' (Lantos 2001: 1). These temples frame the context of the NICU—the modern temple of reproductive and life-extending medicine. Lantos relates how health care providers and family members struggle in the heated moment of crisis to discern what should be done. The luxury of waiting for the right answer to show up at the doorstep of the crisis is not available. The illusion of individual decisions made in the safe, unambiguous web of morality is exposed by Lantos. Moral decision making is woven into the multiple storylines of families, lawyers, health care professionals, and insurance companies. Moral decisions faced in the NICU are not abstract. They involve the flesh and blood of families, of the nurses in the NICU, of the doctors coming and going through the night.

Moral reflection begins with a particular type of personal suffering. Writing or reading about ethical dilemmas is an abstract exercise. Being there, in the night, was not. The babies and parents were there with me. I truly did not know, and neither did they, whether I was a savior or a torturer of babies, whether I offered hope or hubris, whether it was good to use my technology and skills or better to acquiesce gracefully. (Lantos 2001: 164)

BORDERLANDS

Medicine and morality are in the borderlands of flesh and blood, life and death, technology and nature, culture and biology. For the boundaries blur the borderlands, the familiar shifts into the foreign. Medical education must begin in the borderlands, at the edges where the once familiar Enlightenment, objective, narrative boundaries no longer exist. To practise medicine as a moral discipline is to enter deep into the flesh and blood of others, to make a difference in the world, and to cast lots for some ways of life and not others.

The dialogue between religion and science is a moral discipline, one that takes seriously making a difference in and for the world. Like the practice of medicine as a moral discipline, the dialogue between religion and science requires taking risks. As Haraway notes, answers to questions about meaning and purpose can only be found in emergent practices, i.e. in 'vulnerable, on-the-ground work' (Haraway 2003: 7). Such groundwork will require knowing and listening to all kinds of stories, particularly to those from the borderlands and margins. From these places and standpoints, the narratives of religion and science can be expanded to include all of life. From such incarnational practices, new narratives will emerge that help us as a human species not only to live, but to live well.

The science-and-religion dialogue does not exist apart from the embodiment of its participants. In these face-to-face encounters, we can assume that what seems lucid will blur, and what we thought was opaque will become clear. In the end, the dialogue will make a difference only when and if it begins from the experiences of those on the margins and edges. To respond to the questions about the meaning and purpose of human life requires that we engage the ambiguous, marvel at the mysterious, and hope with modesty. To experience moments of clarity and to envision moments of truth, one must accept the whirlwind, the busy buzzing craziness that life is. Human beings are always messy, complex, finite, and mysterious. Otherness is dizzying, often confusing, and even frightening. Haraway notes that, 'We also live with each other in the flesh in ways not exhausted by our ideologies. Stories are much bigger than ideologies. In that is our hope' (Haraway 2004: 17). The narrative of the science-and-religion dialogue can make a difference if the story moves beyond the ideologies of purity and pure-breds. We are, after all, mutts, mixed breeds, whose own stories are inherited from complex strands and layers.

References and Suggested Reading

HARAWAY, D. (2003). *The Companion Species Manifesto: Dogs, People, and Significant Otherness*. Chicago: Prickly Paradigm Press.

—— (2004). *The Haraway Reader*. New York and London: Routledge.

HEFNER, P. (1993). *The Human Factor: Evolution, Culture, and Religion*. Minneapolis: Fortress Press.

LANTOS, J. (2001). *The Lazarus Case: Life-and-Death Issues in Neonatal Intensive Care*. Baltimore and London: Johns Hopkins University Press.

LEBACQZ, K. (2004). 'Bioethics—Eleven Approaches', *Dialog*, 43/2: 100–6.

TONG, R. (2002). 'Teaching Bioethics in the New Millennium: Holding Theories Accountable to Actual Practices and Real People', *Journal of Medicine and Philosophy*, 27/4: 417–32.

WILLIAMS, T. (1992). *Refuge: An Unnatural History of Family and Place*. New York: Random House/Vintage Books.

—— (1994). *An Unspoken Hunger*. New York: Random House/Vintage Books.

HUMAN NATURE
AND ETHICS

CHAPTER 50

THE SACRED EMERGENCE OF NATURE

URSULA GOODENOUGH AND TERRENCE W. DEACON

REDUCTION AND EMERGENCE

Scientists have had spectacular success with reductionism. Take something like a human muscle, peer inside, and there are the muscle cells, contracting and relaxing in unison. Peer inside the cells and find actin and myosin polymers sliding past each other and generating the contractile forces in conjunction with ATP hydrolysis. Extract the polymers and discover that they're made of actin and myosin subunit proteins. Purify the subunits and learn that they're strands of amino acids that fold into shapes that allow them to interact with one another to generate force. Analyse the amino acids and encounter their component atoms and their bonding angles. Peer into the atoms, and it's a whole new world again. And finally, take a creature like an amoeba, find that the same kinds of actin and myosin proteins are propelling it along, and realize that our muscles are availing themselves of ancient evolutionary ideas.

Response to this success has been decidedly mixed. On the one hand, people slurp up the technologies and medicines that spin off from, and thereby validate, these reductionist understandings. On the other hand, they often decry the Humpty-Dumpty fragments that appear to be all that remains of their whole-egg world where the human is the point. And so we are awash in science wars and Darwin wars even as we are also awash in cell phones and Viagra. There's a lot of existential and religious havoc out there, and the situation doesn't seem to be improving.

This chapter offers some possible ways forward.

Whereas reductionism has yielded splendid results in science, there is an important sense in which it is artificial, and in this sense false. By starting from wholes and moving 'down' into parts, one is moving in the opposite direction from the way matters arise. To grasp how matters arise, one must run the muscle movie backwards, from the subatom to the atom to the amino acid to the protein to the polymer to the cell to the muscle to the contraction. To make such a movie, it is essential to begin with reductionist understandings—otherwise, there is no way to know what to put in the movie. But once the cast of characters is identified—once it is understood how proteins fold and myosin hydrolyses ATP and so on—it is possible to narrate such understandings in the correct temporal and spatial sequence, moving 'upwards' from one level to the next.

As scientists with casts of characters in hand engage in such 'upward' projects, they quickly arrive at an understanding that has in fact been around for some time (O'Conner and Wong 2002). Perhaps the most familiar phrase for stating this understanding is to say that 'the whole is greater than the sum of its parts'. A second phrasing is to say that as one moves 'up' in levels of scale, one encounters 'something more from nothing but' or, less euphoniously but more accurately, 'something else from nothing but'—since the point is not that one encounters something greater or something more, but that one encounters something else altogether. Importantly, this something else can, in turn, participate in generating a new something else at a different level of organization. That is, today's something else may be tomorrow's nothing but. The now widely adopted term to describe such dynamics is *emergence*.

Many are engaged in the religiopoetic project (Goodenough 2000*b*) of exploring the religious potential of our scientific understandings of Nature—an approach some are calling religious naturalism (Goodenough 2005*b*). In a book in this genre, *The Sacred Depths of Nature* (Goodenough 1998), the emergence concept is invoked both directly and indirectly, but the primary intent is to introduce the astonishing casts of characters revealed by reductionist approaches and to articulate religious responses to their foundational roles in the universe, in life, and in human mind and spirit.

In this chapter we assume familiarity with these characters, as in the foregoing examples of muscle → atoms and atoms → muscle, and will work from the emergentist perspective, running the movies in the emergent direction. We first give an overview of the emergentist view of nature, and then use these concepts to outline an emergentist view of the religious quest. We suggest that much—we would say most—of what religious persons seek is grounded in a thirst for the very emergent phenomena that in fact surround us. The concept of emergence, more than any other concept we have encountered, puts Humpty-Dumpty back together again in ways that are wonderfully resonant with our existential and religious yearnings.

EMERGENCE AS A GENERAL CONCEPT

Emergent properties arise as the consequence of relationships between entities. Robert Laughlin (2005) intriguingly suggests that emergent properties arise even at the level of relationships between subatomic entities—indeed, he suggests that the very 'laws' of nature may prove to be emergent—but since we are not trained in discourse at this level, we will begin with relationships between atoms.

Atoms interact with one another, and hence generate emergent outcomes, in accordance with two general features: their energy and their form. Thermodynamics describes the energy and the entropy parameters of an interaction, but critical as well are what we can loosely but conveniently call shape influences. Thermodynamically, two hydrogen atoms will interact to form H_2 because energy is released; NaCl will dissolve in water because energy is released and entropy increases. But there is more to chemistry than just thermodynamics. H_2 forms also because the electron shells of hydrogen atoms are conducive to a 'fit' between them, and the shapes of water molecules and their resultant dipolarity (charge separation) are conducive to separating and distributing the atoms of NaCl. These descriptors, and others offered below, may strike a scientifically challenged reader as too 'hard' and a scientifically sophisticated reader as too 'simplistic.' We ask you to bear with us, since what's important here are the concepts and not the details.

The key concept: if one starts with something like a water molecule, it is nothing but two hydrogen atoms and one oxygen atom, but each molecule has something-else properties that cannot be ascribed to hydrogen alone nor to oxygen alone. The interaction between the three atoms entails a reconfiguration of electron orbitals and generates a trapezoid-shaped entity that is more electrically positive on one facet and more negative on the opposite facet. Compared with hydrogen and oxygen atoms, a water molecule has unprecedented attributes, because the joining of these atoms has distorted the shapes of each and produced a composite shape with its own intrinsic properties. In chemistry, shape matters.

Now we can consider what happens when water molecules interact with one another. Here we encounter the interesting fact that it depends. Ice forms when the kinetic energy of the average molecule is low and the molecules' stickiness (capacity to form hydrogen bonds) overcomes their movement; liquid water forms when their movement is just sufficient to overcome the stickiness and allow them to slip over one another, forming hydrogen bonds with picosecond lifetimes; and steam forms when their relative velocities are high enough that collisions seldom allow sticking. The formation of each phase, and the transitions between phases, are generated by thermodynamics and shape, and the emergent outcomes are numerous. Thus ice displays buoyancy, crystalline organization, and hardness; water displays surface tension and viscosity. None of these properties is displayed by individual water molecules; what matter are dynamical regularities in the ways in which large numbers of these molecules interact with one another. And so we have here our first of countless examples in which a composite structure that is 'something more' (a water molecule, from hydrogen and oxygen) turns around and serves as a 'nothing but' (for the emergent properties of ice and liquid water). Higher-order properties have emerged by virtue of the regularities of interactions between their constituents.

All sorts of molecules besides water can adopt alternate phases, and in each case the resultant emergent properties are slightly different: the hardness of crystalline quartz is distinctive from that of ice, and the viscosity of liquid mercury is distinct from that of liquid water, because thermodynamic interaction effects and shape effects both contribute to generate slightly different emergent outcomes. That being said, phases of matter share characteristic features—hardness, viscosity— irrespective of their specific micro-properties, features arising from global regularities in the interactions that tend to wash out the details.

But this is not always the case. Sometimes one or more micro-details can get amplified.

Snowflakes illustrate this kind of amplification effect. A snowflake, usually a single planar hexagonal ice crystal, is again nothing but water, but as an initial crystalline 'seed' moves through a given set of humidity, temperature, and pressure conditions in its fall to earth, additional crystal growth is influenced both by the 'initial conditions' (the configuration of the seed) and the 'boundary conditions' (what kinds of crystalline structures most readily grow in that particular set of conditions). What makes snow-crystal growth special is that the pattern generated by one stage of growth serves as the initial condition for the next as the snowflake falls into a new set of atmospheric conditions. Consequently, each snowflake that reaches the earth displays a unique morphology that reflects its individual history, embodying all the initial and boundary conditions, all the constraints and possibilities, that it has both encountered and generated during its growth.

Unlike the emergence of material phases, where interactions tend toward the average over time, structural and thermodynamic effects become, in effect, multiplicative during snowflake formation. The heat released by each new water molecule's accretion to the growing crystal is dissipated throughout the crystal structure, meaning that the temperature of each part of the crystal, and thus the probability that new water molecules will attach to any given part, is biased by the just prior configuration as well as by external conditions. Hence the thermodynamic tendency to distribute heat evenly progressively exaggerates the subtle biases of different crystallization patterns, and these patterns in turn affect how heat will be distributed. Each bias amplifies the just previous bias of the other. This amplifying effect of complementary dynamics is designated *morphodynamic* emergence (form begetting form), in order to emphasize the critical role that both shape and dynamical regularity play in the process (Deacon 2006a).

Snowflake formation, and crystal growth in general, is said to be *autocatalytic*. A catalyst is any entity that influences both the probability that, and the rate at which, a given interaction will occur. Crystal growth is described as autocatalytic because the accretion of new molecules to the crystal lattice increases the surfaces available for subsequent accretions to occur, and thus accretion rates increase with each new accretion (as long as the supply of molecules and energetic conditions allow). In addition, the presence of the crystal lattice increases the probability that new molecules will link up to it in the same configuration.

More commonly we think of catalysts as molecules that increase the rate of some chemical reaction—e.g. the formation or breakage of chemical bonds to form new molecules—without themselves being chemically altered. Particularly interesting from an emergence perspective are cyclical chemical systems that generate catalysts during the course of forming new molecules, these being known as *autocatalytic cycles* (Kauffman 1996). A simple example might involve molecule A catalysing the chemical transformation of B into C, C then catalysing the transformation of D into E, and E then catalysing the transformation of F into A. As long as substrates B, D, and F are available to the system, catalysts A, C, and E will continue to be produced at ever greater rates as each traverse of the cycle generates more catalysts that can in turn catalyse the formation of more catalysts. An autocatalytic cycle, then, is a distinctive higher-order pattern of molecular interaction that alters the components that then alter the interaction pattern, the result being the self-amplification of both the dynamics and the relevant substrates. This recursive interaction pattern is made more probable by the complementary shapes of molecules, and the probability of interaction between such molecules is made more probable with each cycle as more are produced. As with the snowflake, micro-properties and interaction properties interact to reinforce each other, but strikingly different outcomes are possible once a system is circularly nested.

In the language of thermodynamics and morphodynamics, autocatalytic cycles are far-from-equilibrium dissipative systems that exhibit coherent behaviour by virtue of their dynamical regularity. They are usually transient in the non-living world because they, like all interacting systems, are dependent on initial and boundary conditions and proper energy/substrate flow, and these conditions are usually ephemeral. As we will see, life basically works by maintaining the conditions wherein such cycles can operate in a reliable fashion.

EMERGENCE AND THE NATURE OF LIFE

Emergence not only surrounds us in the non-living world; it is also the key dynamic of living organisms, as in, for example, the emergence of contractility (something else) from the interaction of myosin and actin polymers (nothing but). Life has a number of additional features, however, and these are perhaps best appreciated by considering ways that life may have emerged from non-life. All origin-of-life hypotheses are by definition speculative, since the default assumption is that the original lifeform is no longer extant but went on to evolve into the DNA-based, lipid-membrane-enclosed, protein-mediated, single-celled organisms that served as the common ancestors of all modern lifeforms. What is gained by exploring origin-of-life speculations is a grasp of what being alive entails.

Deacon (2006*b*) offers a scenario for the formation of hypothetical entities called *autocells* that display many, but not all, of the salient features of life by virtue of thermodynamics and morphodynamics alone. Their initiating feature is an auto-catalytic cycle that is enclosed in a container whose subunits are generated by the cycle, making it more likely that they assemble around the catalysts. The container can on occasion be disrupted by agitation, spilling its catalytic components, initiating new component production and then re-forming to enclose new substrates. Disrupted containers may also release subunits that reassemble into more than one autocell, or may break into partial autocells, each of which reassembles into a whole. In this way autocells are capable of self-replication. Moreover, a given autocell may capture novel substrates during disruption and reassembly that permit the emergence of novel and more complex autocatalytic cycles. Hence autocells are capable of evolution in this minimal sense. Critically, any autocells that evolve mechanisms for undergoing more efficient cycles would tend to self-replicate more often than others. Hence autocells would be subject to a form of natural selection.

And finally, autocells can be said to be 'end-directed' (toward a specific configuration), to have features with 'functions' (e.g. to maintain the autocell architecture and potential), and even to be *about* something (to the extent that their features exist 'with respect to' environmental factors conducive to making autocells with these same features). These are all facets of an emergence dynamics of a higher order than the thermodynamics and morphodynamics of autocell components, a property Deacon calls *teleodynamics*. The teleodynamic concept is in fact quite nuanced and not easily summarized, and readers are referred to Deacon (2006*a*) and Deacon and Jeremy Sherman (2006) for careful developments of this seminal idea.

The autocell, then, displays key features of life—substrate acquisition, self-propagation, evolution/natural selection, and end-directedness—without possessing a separate coding mechanism to specify these features. A critical difference that distinguishes all examples of modern lifeforms from autocells is the presence of such a coding mechanism in the system. The precursor to all modern organisms evolved some means of representing some of its structures and dynamical interaction patterns in separate molecular patterns, which would eventually evolve into the RNA/DNA-based coding systems that are now ubiquitous in life.

A coding mechanism is inherently just that—a mechanism, consisting of a set of markers, like an alphabet, coupled with a process that can interpret it. Its interpretation acquires significance to the extent that it codes *for* an entity—e.g. an autocell feature or an idea—in a way that promotes preservation both of that entity and of the code responsible for specifying it. These are the features of a *semiotic* system. When semiotic systems are copied, the capacity to generate more such entities is introduced; when semiotic systems change (mutate), the capacity to generate novel entities arises.

Setting aside consideration of the many complex steps that would attend the evolution of even a simple molecular coding system, we can instead focus on how this outcome would augment the teleodynamic process. In shifting from an autocell to an autocell containing an independent representation of its dynamical components, the process of autocell generation has undergone a division of labour: the autocell continues to operate in consonance with the thermodynamics, morphodynamics, and teleodynamics of its components, but its propagation and evolution are now also correlated with the genetic information that encodes those components. That is, an autocell with independently encoded features acquires some freedom from the constraints of its own dynamics. Internally coded structures are not inexorably tied to extrinsic conditions, allowing external materials to be used as generic building blocks and energy sources rather than specific structural precursors. By this means, an autocell dependent on encoded structures necessarily acquires an additional agenda: to provide conditions that permit the maintenance and propagation of this information as well as the system that contains it.

As we saw earlier, an autocell lacking encoded information still has the capacity to evolve if novel substrates happen to become incorporated into its container. The independent transmission of structural information provides an independent means by which structure can vary, since lineages can explore domains of possible structural and functional variants in a way that is unconstrained by what the environment offers in the way of variations. This is the power of representation. What continues through time is now not merely the pattern of a particular self-reproducing dynamics, but also the representation of potential alternative dynamics. With the addition of coding, autocells—and, by implication, living organisms—acquire a degree of freedom and autonomy from the specific configurations of matter and energy that constitute them, thereby generating a spectacular expansion in evolutionary potential.

With the addition of a metabolism that provides a continuous flow of materials and energy to maintain them in the non-equilibrium state necessary to incessantly run these processes, such entities can be said to be alive in the sense that biologists use the term. Life's emergent properties are no longer left to the vagaries of substrate diffusion and container disruption, but are now independently embodied in that emergent property called genetic information, which takes physical form in DNA- or RNA-based genomes in present-day organisms.

It is important to pause at this juncture and address a key issue. One way to read this account—a misreading, we will argue, but a common one—is that the genes are driving the system, that genes are 'selfish', that genes rule. Not only is this misreading inherently depressing, and religiously sterile; it also misses the point. Genomes are in fact the handmaidens of emergent properties, not the other way around. Natural selection doesn't 'see' genomes, and indeed, we saw natural selection operating in its canonical Darwinian fashion with our hypothetical autocells that have no genomes at all. The whole point of life is to generate emergent properties that, if successfully executed, have the additional feature of permitting transmission of genomes. Genomes represent a splendid convenience, allowing emergent properties, and hence organisms, to be generated ever more efficiently and with increasing levels of complexity. But they are useless unless they contribute to maintenance of the emergent dynamics that confer upon organisms the capacity to carry on. We can posit an autocell without an operational genome, but a genome that fails to specify an emergent teleodynamical system, such as is minimally present in an autocell, is dead on arrival.

So, a successful life outcome is to promote the transmission of information conducive to maintaining the emergent dynamical logic that gives it its meaning— that is, to promote the production of emergent outcomes (called *traits* in biology) that collectively make their own continuation more likely. It is traits that rule; genes follow in their wake. Traits common to all organisms include such non-depressing and religiously fertile capacities as end-directedness and identity maintenance; traits common to all animals include awareness and the capacity for pleasure and suffering; traits common to social beings include co-operation and meaning making; traits common to birds and mammals include bonding and nurturance; traits common to humans include language and its capacity to share subjective experiences, and thus to know love. Transmission of genomes is the steady background drumbeat; emergence is the music.

The How and the Why of Trait Generation

All modern organisms generate their traits in basically the same way, meaning that the common ancestor to all modern organisms, posited to have appeared more than 3 billion years ago, also employed these strategies. The core idea is that genes encode proteins that fold into useful shapes under the aegis of thermodynamics. Some of these proteins, such as actin and myosin, go on to mediate cell organization and behaviour; others, such as lactase, catalyse metabolism and hence mediate energy transduction; others, such as insulin, mediate cell-to-cell communication, where behaviour, metabolism, and communication are all emergent properties, the outcome of thermodynamics, morphodynamics, and teleodynamics, with countless variant manifestations.

Particularly interesting proteins mediate an emergent process called *regulation of gene expression*. Each gene comes equipped with an adjacent switching element, also made of DNA. When the switch is turned on, the gene is 'expressed' or 'transcribed', and the protein encoded by that gene is produced by the cell; when the switch is turned off, the protein is not made (the gene is 'not expressed' or 'not transcribed'). The switches are operated by proteins called 'transcription factors', some serving as activators and others as repressors of gene expression, that often act in concert on a given switching element—that is, the 'on' or 'off' command is the emergent outcome of complex interactions between these regulators and the components of the switch.

Each transcription factor is itself encoded by a gene that also has its own switch, which is subject to regulation by additional transcription factors. Hence the outcome is an elaborate system of feedback loops, where the genes can be considered as passive elements responding to protein commands. By virtue of these complex interaction dynamics, emergent patterns of gene expression are produced, resulting in emergent molecular and cellular interactions. In a real sense, then, proteins, and not genes, rule (though genes must be present to encode the proteins), and the emergent consequences of these protein interactions, with each other and with the genes, are the ultimate determinants of organism traits.

As different species' genomes have been sequenced, an initially surprising finding has been that complex organisms don't have all that many more genes than do simple organisms. A unicellular alga, for example, has about 17,000 genes, and a human has about 23,000 genes. Moreover, well over half of these genes encode 'housekeeping proteins'—actin, myosin, metabolic enzymes, and so on—that are present in both kinds of organisms. What has happened during evolution, then, is not so much the acquisition of new genes as changes in the patterns of expression of existing gene families such that novel combinations of proteins appear in a given cell at a given time, interact with one another, and generate novel emergent properties.

This strategy is particularly creative in multicellular organisms that begin as fertilized eggs and cleave to form embryos and eventually mature forms. Individual cells in the embryo set up novel patterns of gene expression such that cell A might produce and secrete hormone X and cell B might produce and display a receptor for hormone X. Hormone–receptor interactions then influence transcription factors such that genes are expressed in cell B that are not expressed at that stage in cell A, and the resultant proteins allow cell B to initiate a distinctive cell lineage whose activities go on to influence the gene expression patterns of cell C. And so on. In other words, embryogenesis can be thought of as a vast autocatalytic cycle—a metacatalytic cycle perhaps—made up of countless subcycles, feeding forwards and backwards in space and time under the aegis of protein–protein and protein–switch interactions, with genes obligingly responding when they are called upon to do so. A given gene may be switched on at the four-cell stage of development, switched off at the sixteen-cell stage, and then switched on again in one of the cells in a 256-cell embryo. The protein produced by this cell will encounter partner proteins that were not expressed at the four-cell stage, and their interactions will generate emergent properties in that cell's lineage only. The gene may be switched on again in primordial liver cells, again in concert with a distinctive set of partners, to generate novel liver-specific emergent outcomes. And so on. New kinds of embryos and hence new kinds of mature multicellular organisms result primarily from using the same old protein families in novel combinatorial patterns.

The important concept to grasp here is that the genome in no way represents a 'blueprint' for a multicellular organism—there exists no top-down design entity that can be analogized to an architect's blueprint. Nor is the organism assembled from pre-existing 'parts', like a house or a car. Rather, the organism literally builds itself, bottom-up, assembling tiny parts that modulate the assembly of the next set of tiny parts, where the same old protein families are used in novel combinatorial patterns along the way, all under the aegis of initial conditions and boundary conditions established and maintained by the information encoded in housekeeping genes. Thermodynamics, morphodynamics, and teleodynamics set up the constraints and possibilities, but organisms are not predetermined—even if they come into being in remarkably predictable ways. Their features predictably emerge because these emergent features are made almost inevitable by the hierarchy of biases of lower-order emergent features.

Particularly 'underdetermined' is the process of mammalian brain formation, albeit, again, features emerge in a predictable fashion—all gorilla brains, for example, are far more similar to one another than they are similar to the brains of any other species. While genes again switch on and off in various cell lineages at critical junctures during brain development, most of the action entails cell–cell interactions via protein receptors and hormones as the neurons move up into the cranium and establish connections with one other. Moreover, most of these hormones and receptors are not brain-specific: again they're the same old protein families put to use in a neurogenesis context. When one absorbs the fact that a mature mammalian brain may contain 100 billion neurons, each in synaptic communication with some 1,000 other neurons, all put together under the watch of a genome with some 20,000 genes, one comes to understand why it is so inaccurate to speak of a gene as being 'for' a particular mental capacity. True, a mutant gene encoding an aberrant protein may in some cases generate an aberrant brain function outcome, but this is not because that gene encodes that outcome; it's because the aberrant protein is defective in pointing neurogenesis in a particular emergent direction.

Embryogenesis occurs in environmental contexts—soils, ponds, nests, the uterus—and all brains, even clam brains, are capable of learning from experience. More generally, all creatures come into being and make a living in environmental contexts, where each ecosystem represents a rich interdigitation of the organic and inorganic, of organisms and planet. Genomes are transmitted to offspring when, and only when, all of this comes together. Life is not about survival of the fittest; it's about fitting in.

THE EMERGENT HUMAN

Deacon offers a bold emergentist claim in his book *The Symbolic Species* (1998): 'Biologically we are just another ape; mentally we are a whole new phylum of organism.' Our 'whole new' traits—symbolic languages, cultural transmission of ideas via languages, and generation of an autobiographical self—are of central importance to our lives and our religious lives, and much remains to be understood about how they operate from a reductionist perspective. At this juncture, however, the concept to take in is that these human-specific traits are quintessentially emergent: they are constructed bottom-up and then deeply influenced by environmental contexts; they make use of ancient protein families that are deployed in novel patterns and sequences. We are aware that this is a claim and that other concepts of human origins and essence are on offer, concepts that set humans apart; but from our perspective, the understanding that human-specific traits are emergent—something else popping through from all that has gone on before and continues to surround us—is fully consonant with what we now know about the course of natural history, and a deeply satisfying way to think about who we are.

What is particularly interesting about the course of human evolution is that it has entailed the co-evolution of three emergent modalities—brain, symbolic language, and culture—each feeding into and responding to the other two and hence generating particularly complex patterns and outcomes (Deacon 1998). While we don't know, and probably will never know, the actual details of the sequence, here is a plausible course of events.

- Initial social evolution of very simple symbol systems drives changes in children's brains that make the acquisition of symbolic abilities easier and easier, where the symbol systems, transmitted via culture, themselves evolve so as to be learnable by the evolving children's brains.
- These more complex symbolic abilities make possible the internalization of a 'public perspective': imagining an outside representation of self, one that can be superimposed onto real self-experience.
- This creates contexts for the juxtaposition of emotional states that would be otherwise mutually exclusive (e.g. love thine enemy).
- Out of this bisociation of emotional-cognitive experiences emerge unprecedented emotional experiences and perspectives.
- Hominid brains continually impacted by such unprecedented emotional states begin to evolve with respect to these states, developing modalities for integrating these experiences and cultural symbolic supports that both utilize and buffer them.
- Symbolic cognition, that is, precipitates a cascade of reorganizational cognitive and co-evolutionary events that eventually produce a brain with a capacity for the kind of mindfulness, intersubjective projection, aesthetic sensibility, and empathy that is now possible.

Absent from this account are such standard hominid milestones as upright posture, opposable thumbs, and tool use. While these showed up along the way and are important to who we are, their acquisition can be readily modelled using standard vertebrate embryology paradigms. By contrast, our emergent mentalities to date lack reductionist explanations, even if most neuroscientists are confident that such explanations will be forthcoming. As with life, moreover, analysing all the pieces is just the first step, not the final explanation. Making sense of brains will also entail an elaborate reconstruction to discern their emergent dynamics and what they entail.

Importantly, when the details become available, they will in fact have no impact on our experience of being self-aware beings, any more than our understanding of oxytocin's participation in romantic attachment impacts on our experience of being in love. Reductionist understandings of how minds work are fascinating, but they are also irrelevant to what it's like to be minded. While we don't know what it's like to be a bat, we know what it's like to be a human, and it entails a whole virtual realm that doesn't *feel* material at all. The beauty of the emergentist approach to mind is that it suggests that to experience our experience without awareness of its underlying mechanism is exactly what we should expect from an emergent property. The outcome has been given reverent names, like spirit or soul, names that conjure up the perceived absence of materiality. But we need not interpret this as evidence of some parallel transcendental immaterial world. We can now say that the experience of soul or spirit as immaterial is simply a reflection of the way the process of emergence progressively distances each new level from the details below.

We can now turn the page. What is the religious potential of the emergentist perspective?

WHAT IS MEANT BY RELIGIOUS?

In any such undertaking it is important to make clear what one means by religious, since every reader harbours a unique perspective on this adjective. We are not suggesting that our readers agree with or adopt our meanings, but only that they understand them.

We can begin by contrasting 'religious' with 'religion'. Loyal Rue proposes (1999) that a *religion* is a cultural entity, grounded in metanarratives indicating how things are and which things matter, which offers personal wholeness and social coherence to its adherents. In his recent book (2005) Rue argues persuasively that the common goal of a religion is to educate the emotions of its adherents such that the goals of the cultural tradition are realized.

A *religious* person—and we would say that all persons are religious—may or may not self-identify with a religion. Rather, a religious orientation encompasses three spheres of human experience (Goodenough 2003, 2005b): (1) The *interpretive* sphere (a.k.a. theological, philosophical, existential) describes responses to the Big Questions, such as, Why is there anything at all rather than nothing? Does the universe, or my life, have a Plan? a Purpose? How do I come to terms with death? Why is there evil and suffering? (2) The *spiritual* sphere describes such inward personal responses to existence as gratitude, awe, humility, reverence, assent, transcendence, and at-one-ness. (3) The *moral* sphere describes outward communal responses such as care, compassion, fair-mindedness, responsibility, trust, and commitment.

Religious naturalism, in this context, describes a person's interpretive, spiritual, and moral responses to our understandings of nature, in this case our emergentist understandings of nature, where these spheres are ordinarily not experienced as separate categories, but rather as an overall orientation. Since the authors are religious non-theists who do not self-identify with a given religion, our responses will be offered in this voice. (Willem Drees, Chapter 7 above, provides an excellent overview of the many forms taken by religious naturalism.) The responses are not intended to be comprehensive, but rather to suggest the religious potential of the emergentist world-view.

INTERPRETIVE RESPONSES

Creation and Purpose

All religious traditions offer ways to think about creation. A single Creator God is central to the Abrahamic faiths; creation myths are central to polytheistic faiths; and the Buddhist tradition includes the challenging concept of beginninglessness: all that is has always been, and was therefore never created. The emergence perspective offers us ways to think about creation, and creativity, that do not require a creator. Emergence can be thought of as nature's mode of creativity, giving rise to ever more complex outcomes by virtue of thermodynamics, morphodynamics, and teleodynamics.

In theistic traditions, creation is invariably coupled with purpose. There is some reason why creation occurred, and there exists a plan for what has been created, most prominently a plan for the human. It follows that a central focus of the theistic religious quest is to discern and attempt to act in accordance with that plan, finding guidance in texts and revelation and prayer. The emergence perspective, while not ruling out purpose or plan, is coherent without invoking either. Living beings, including humans, need not derive from, or be in the name of, some higher-order trajectory; instead, they can be understood to have followed trajectories made possible by planetary and ecological conditions and opportunities.

Jeffrey Dahms, Australian surgeon and naturalist philosopher, expresses these ideas as follows:

For me, the universe as a created object, no matter how subtly that concept is conceived, is unbearably depressing. Why would I want to live in a universe that is about me? For sure, a universe purposely built to generate complexity or life or primates or humanity is a cosmically impressive act, particularly if no tweaking is allowed after the initial roll of the dice.

But no thanks, I'd rather find another place to call home—a place that has no Purpose—a truly wild place—yes! A universe that spawns life and consciousness without instruction, just because it is ... Ah, now that is really something one can get excited about. It's the difference between inconceivably awesome cosmic skill and the purely magical. Naturalists are in love with the magic of the open, the magic of the possible. The very contingency of life is what makes it so unbearably sweet and precious. (Personal communication)

This stuff is clearly not for everyone (cf. Goodenough 2000a). For some, a self-creating universe generates angst and anomie; for others, orientation and excitement. For some, to think of purpose as an evolved emergent capacity is to destroy the very meaning of purpose; for others, purpose so understood is rendered meaningful, even sacred, for the first time. For some, to understand human-specific mental and emotional capacities as emergent is to violate their sense of self; others celebrate all that has gone on before us as setting the nothing-but stage for these spectacular something-else lives that we live. We count ourselves among the second group.

Contingency

Dahms ends his passage by invoking the contingency of life, and Stephen J. Gould (1989) has famously emphasized contingency as well, suggesting that if one were to 're-wind the tape of evolution', it is highly unlikely that the same kinds of organisms would have evolved a second time. Confucian scholar Michael Kalton (2000) lifts up the challenges, and the saliency, of the contingency concept:

The biocentered life orientation locates its center of value, meaning, and purpose squarely within the realm of the contingent, the very kind of irredeemable contingency identified with meaninglessness and absurdity within conventional frameworks. Indeed, contingency itself is a central element of its salvific message. Until we grasp our radical contingency, we have small chance of really understanding the nature of what is at stake.

Emergentism offers fresh ways to think about contingency. Whereas contingent is often understood to mean accidental or fortuitous, its etymology (*contigere*, to touch, meet) carries the sense of dependency, of something being conditional on something else, and this certainly maps onto the core understanding of the emergentist perspective. In the history of life, and indeed the history of the planet and the universe, things happen in the context of what has gone on before and the opportunities thereby generated—in Kauffman's (2002: 142) phrase, 'the adjacent possible'. Crystals build on crystals, traits build on traits, ideas build on ideas.

Once a planet with Earth's properties came into being, moreover, the emergence and evolution of life as we know it can even be considered as something expected. As developed by Richard Dawkins (2004), evolutionary history documents that the same *kinds* of emergent adaptations—e.g. vision, audition, flight, intelligence—keep popping through in diverse and independent lineages because niches that render such adaptations useful have been continuously available on this planet. Hence the same *kinds* of organisms, if not the same phenotypic details, would be expected with a Gouldian rewinding.

These perspectives go far to counteract the misunderstanding that evolutionary theory is about randomness. The variation itself is random, happily, since otherwise the possibilities for novel trajectories would never arise, but the emergent paths taken are contingent and, in an important sense, anticipated.

The Emergent Human

Evolutionary theory asks us to situate the human in the natural world, and this can generate cognitive dissonance given that our mental capacities would seem to place us 'above' the natural world and our cultures 'above' the natural order. The emergentist perspective allows us to see ourselves not as 'above' but rather as remarkably 'something else'. For all we know, and quite probably for all we will ever know, we are the only creatures in the universe who write psalms and sculpt marble and know how stars work. We inhabit a virtual reality of symbols and ideas (Goodenough 2005a), and we are uniquely endowed with the capacity to teach as well as to imitate.

Understanding the human as the emergent outcome of natural history, and in particular the outcome of a co-evolution of language, brain, and culture, allows us to understand who we are in exciting new ways. Deacon (2003: 306) lifts up one such perspective:

Human consciousness is not merely an emergent phenomenon; it epitomizes the logic of emergence in its very form. Human minds, deeply entangled in symbolic culture, have an effective causal locus that extends across continents and millennia, growing out of the experiences of countless individuals. Consciousness emerges as an incessant creation of something from nothing, a process continually transcending itself. To be human is to know what it feels like to be evolution happening.

SPIRITUAL RESPONSES

Enchantment

Dahms uses the word 'magical', a word often associated with supernatural miracles or the unexplainable. The emergentist perspective opens countless opportunities to encounter and celebrate the magical while remaining mindful of the fully natural basis of each encounter. There is a way in which the universe is re-enchanted each time one takes in its continuous coming into being, and there is a way in which our lives are re-enchanted each time we realize that we too are continually transcending ourselves.

Transcendence and Reverence

Transcendence is commonly used to denote a discontinuity, as in the 'top-down' agency of transcendent deity. But transcendence also aptly describes the phenomenon of emergence, where discontinuities (something elses) arise from, while remaining tethered to, their antecedents (nothing buts). This mode of understanding transcendence—Kalton (2000: 190) describes it as 'horizontal transcendence' (cf. Goodenough 2001)—facilitates deconstruction of the hubris that so often afflicts human sensibility, replacing it with reverence, the capacity to carry the sense that our context is vastly larger and more important than our selves (Woodruff 2001; Goodenough and Woodruff 2001). We can, in this framework, offer reverence not only to the living but also to the non-living. Here is Kalton again:

A reexamination of how we regard the 'non-living' aims to open the possibility of a mode of self-identification which transcends the boundary of biotic life. Once the boundary is down, an arena of immediate access to horizontal transcendence is created. What the poet Robinson Jeffers has referred to as 'the massive mysticism of stone' surrounds us, inviting us to discover the patterning that lives in geologic time or even cosmic time, substrate to patterns manifest in the rapid complexity of life time. What is it from which we have emerged, and to which we return at death? It cannot be less than us, for we are formed of it, belong to it, manifest it. (Kalton 2000: 199)

Gratitude

Gratitude is the most important facet of the spiritual life, allowing us to acknowledge and express our awe and our reverence. A universe that 'spawns because it is' generates our capacity to spawn because we are, inviting us to wrap our arms and minds and hearts around this astonishing whole to which we owe our lives and of which we are a part, and gasp our stammering gratitude.

MORAL RESPONSES

Emergent Morality

Probably the most commonly voiced concern about evolutionary theory is that it represents the slippery slope to moral relativism. Without some 'higher' authority, the argument goes, without some arbiter of plan and purpose, there is no ontological basis for human moral behaviour; therefore, anything goes. Many go further, arguing that the perceived moral decline in our culture has been caused by the dissemination of evolutionary understandings. Tom DeLay (R-TX) famously claimed that the Columbine massacres were a result of exposure to evolutionary theory, that 'our school systems teach the children that they are nothing but glorified apes who are evolutionized [sic] out of some primordial soup'.

Inherent in this mode of thinking is the notion that animals are 'brutes' and that therefore, if we 'evolutionized' from brutes, we are by nature brutes as well. Intensive studies of other primates refute this notion, however, instead documenting that these animals self-organize into highly effective and coherent social systems wherein are readily discerned versions of respect, friendship, co-operation and reciprocity, empathy, humour, loyalty, nurture, forgiveness, and so on (de Waal and Tyack 2003; van Schaik 2004). In primatologist Franz de Waal's words, the non-human apes are 'good-natured'. In this light, to be glorified apes is a promising moral starting point.

So in what ways, and by what means, have humans differentiated on the moral axis? We have considered this question in some depth (Goodenough 2003; Goodenough and Woodruff 2001; Goodenough and Deacon 2003). Basically, the idea is that during the brain/language/culture co-evolutionary trajectory outlined earlier, the pro-social emotions of our common ancestor with chimpanzees were not left in the evolutionary dustbin. Nor, however, are these emotions experienced as those common ancestors experienced them, nor as modern chimps experience them. Rather, they are experienced as humans experience things: via the cognitive-emotional juxtapositions that undergird our symbolic subjectivity. One's moral framework is not some instinct that just bubbles up. It is something that each of us constructs, amplifying and reconfiguring primate social emotions in the context of cultural stimuli and teachings. In Aristotle's words, 'We have the virtues neither by nor contrary to our nature. We are fitted by our nature to receive them.'

Importantly, the outcome of developing one's capacity for virtue is to experience pleasure and incur admiration: those who are courageous, reverent, fair-minded, and compassionate report deep satisfaction with these frames of mind—they are experienced as good, as beautiful—and they are held in high esteem by others. Geoffrey Miller (2000), in fact, goes on to argue that such developed traits are adaptive, in that they are substrates for sexual selection: persons who display these qualities may be more likely to be chosen as mates and to nurture their children with care and wisdom.

In any case, the emergentist perspective allows us to understand the human not as some discontinuous moral entity but as an emergent moral entity, expanding core primate capacities and sensibilities and celebrating their beauty and value in art, literature, and religious teachings. Indeed, the myths and metaphors that come to us from thousands of religious traditions convey timeless hopes and understandings of how best to be good.

Amorality

To look at the primates and lift up only their pro-social capacities is of course to tell only part of the story. Self-interest is central to the nature of all organisms (Goodenough 2003), and always lurking in the wings of primate self-interest are its 'darker' manifestations. It is here that the project of naturalizing morality encounters for many its insurmountable hurdle. When we remember that apes are also observed to injure and even kill one another, to use force in sex, to be cruel and rejecting, and to display robust xenophobia, and when we confront analogous behaviour in the human, as chillingly documented by Jonathan Glover (2001), gloom can descend.

A full consideration of the interplay between self-interest and pro-sociality, particularly as each plays out in its emergent manifestations, is well beyond the scope of this chapter, but a few observations are germane.

First, it is important to point out that the existence of self-interest, and its darker forms however defined, does not negate the existence of pro-sociality. Pro-social capacities are not just the absence of asocial capacities. They have emergent lives of their own.

We can then recall that primates, both non-human and human, most often engage in asocial behaviours when they are subjected to stress, and particularly to prolonged stress. Under these circumstances, we hunker down and engage in self-interested survival patterns, the default behaviour of all creatures, and these often take forms that are antithetical to pro-sociality.

One way to stack the deck in favour of morality, therefore, is to ameliorate the conditions wherein humans find themselves physically or emotionally impoverished, threatened, defeated, abused, humiliated, lonely, or insecure. Such conditions foster the dehumanization and demonization of those identified as the 'cause' of our frustrations, allowing them to become targets of exclusion and brutality (Glover 2001). Such conditions also render humans vulnerable to rigid fundamentalisms—many carrying morality labels—that activate our fear and greed in their promises of deliverance.

Ecomorality

Our ability to reconfigure the core social emotions also allows us to enlarge our moral vision such that we can come to care not just about family and troop and tribe but about conserving ecosystems and sustaining biodiversity. Ecomorality is a religious stance that flows effortlessly from emergentist understandings, asking not only for our allegiance but also for our continuous participation in protecting and celebrating that from which we have come.

REFERENCES AND SUGGESTED READING

DAWKINS, RICHARD (2004). *The Ancestor's Tale: A Pilgrimage to the Dawn of Evolution*. Boston: Houghton Mifflin.

DEACON, TERRENCE W. (1998). *The Symbolic Species*. New York: W. W. Norton.

——(2003). 'The Hierarchic Logic of Emergence: Untangling the Interdependence of Evolution and Self-Organization', in B. Weber and D. Depew (eds.), *Evolution and Learning: The Baldwin Effect Reconsidered*, Cambridge, Mass.: MIT Press, 273–308.

——(2006a). 'Emergence: The Hole at the Wheel's Hub', in Philip Clayton and Paul Davies (eds.), *Re-Emergence of Emergence*, Oxford: Oxford University Press, forthcoming.

——(2006b). 'Reciprocal Linkage between Self-organizing Processes is Sufficient for Self-reproduction and Evolvability', *Biological Theory* (in press).

—— and SHERMAN, JEREMY (2006). 'The Pattern which Connects Pleroma to Creatura: The Autocell Bridge from Physics to Life', *Biosemiotics* (in press).

DE WAAL, FRANZ B. M., and TYACK, PETER L. (2003) (eds.). *Animal Social Complexity*. Cambridge, Mass.: Harvard University Press.

GLOVER, JONATHAN (2001). *Humanity: A Moral History of the Twentieth Century*. New Haven: Yale University Press.

GOODENOUGH, URSULA (1998). *The Sacred Depths of Nature*. New York: Oxford University Press.

—— (2000a). 'Causality and Subjectivity in the Religious Quest', *Zygon*, 35: 725–34.

—— (2000b). 'Religiopoiesis', *Zygon*, 35: 352–5.

—— (2001). 'Vertical and Horizontal Transcendence', *Zygon*, 36: 21–31.

—— (2003). 'Religious Naturalism and Naturalizing Morality', *Zygon*, 38: 101–9.

—— (2005a). 'Reductionism and Holism, Chance and Selection, Mechanism and Mind', *Zygon*, 40: 369–80.

—— (2005b). 'Religious Naturalism', in B. Taylor and J. Kaplan (eds.), *Encyclopedia of Religion and Nature*, Bristol: Thoemmes Continuum, 1371–2.

—— and DEACON, TERRENCE W. (2003). 'From Biology to Consciousness to Morality', *Zygon*, 38: 801–19.

—— and WOODRUFF, PAUL (2001). 'Mindful Virtue, Mindful Reverence', *Zygon*, 36: 585–95.

GOULD, STEPHEN J. (1989). *Wonderful Life: The Burgess Shale and the Nature of History*. London: Hutchison Radius.

KALTON, MICHAEL (2000). 'Green Spirituality: Horizontal Transcendence', in M. E. Miller and P. Young-Eisendrath (eds.), *Paths of Integrity, Wisdom and Transcendence: Spiritual Development in the Mature Self*, London and Philadelphia: Routledge, 187–200.

KAUFFMAN, STUART A. (1996). *At Home in the Universe: The Search for Laws of Self-Organization and Complexity*. New York: Oxford University Press.

—— (2002). *Investigations*. New York: Oxford University Press.

LAUGHLIN, ROBERT B. (2005). *A Different Universe*. New York: Basic Books.

MILLER, GEOFFREY F. (2000). *The Mating Mind: How Sexual Choice Shaped the Evolution of Human Nature*. New York: Doubleday.

O'CONNER, TIMOTHY, and WONG, HONG YU (2002). 'Emergent Properties', in *Stanford Encyclopedia of Philosophy*; <http://plato.stanford.edu/entries/properties-emergent/>.

RUE, LOYAL D. (1999). *Everybody's Story: Wising Up to the Epic of Evolution*. Albany, N.Y.: State University of New York Press.

—— (2005). *Religion is Not about God: How Spiritual Traditions Nurture our Biological Nature*. New Brunswick, N.J.: Rutgers University Press.

VAN SCHAIK, CAREL (2004). *Among Orangutans: Red Apes and the Rise of Human Culture*. Cambridge, Mass.: Harvard University Press.

WOODRUFF, PAUL (2001). *Reverence: Renewing a Forgotten Virtue*. New York: Oxford University Press.

CHAPTER 51

SCIENCE, ETHICS, AND THE HUMAN SPIRIT

WILLIAM B. HURLBUT

He is before all things and in him all things hold together.

Colossians 1: 17

INTRODUCTION

Nearly 400 years ago, at the dawn of the Scientific Revolution, the French scientist and philosopher Blaise Pascal (1958: 44) lamented, 'When I commenced the study of man, I saw that these abstract sciences are not suited to man, and that I was wandering farther from my own state in examining them, than others in not knowing them.' Today many feel the same disquiet. The naturalistic methodologies of modern science have uncovered a vast wealth of understandings of the human condition that Pascal could not have imagined. These understandings have informed new naturalistic philosophical perspectives on the very nature of human nature, together with heretofore unimagined technological powers to manipulate it. Yet, there is a sense that what is most important and distinctly human has slipped through the fingers of the grasp of science. Somehow the new conceptions of human nature fall short of cohering with the fullness of life as actually lived and experienced. They fail to sustain our sense of irreducible personal identity, our capacity for free choice, and our commitment to ideals and transcendent truths. Likewise, they fail to provide an adequate account of our individual or collective proclivities for pride, perversity, and self-evident evil. Such aspects of our humanity, sometimes designated as 'spiritual', to a large extent lie outside the scope of naturalistic scientific methodologies, but are richly treated within the wisdom of our ancient religious traditions.

If, however, we relax naturalistic presuppositions that categorically exclude from initial consideration such spiritual aspects, it is possible to survey afresh the evolutionary panorama constructed by science and achieve a broader synthesis. We may then discern in the evolutionary emergence of human personhood and its moral sociality intimations of a richer conception both of human nature and of the natural order in and to which it has adapted. Human persons know one another with a personal kind of participatory knowing that at once transcends the naturalistic methodology of science and enables a knowing of aspects of reality that lie beyond the reach of science. When coupled with the perspective of Christian faith, such an expanded conception of the world culminates in an ethic that rises above mere moral duty and social utility to an alignment of life with the very source and power of self-giving love. Such love reflects the deepest nature and purpose of the cosmic order that evolved the capacity for its expression. No less than such an ideal will be needed to guide the increasing powers of our emerging science.

A CRITIQUE OF EVOLUTIONARY PSYCHOLOGY

Toward the middle of the last century, the eminent geneticist Theodosius Dobzhansky noted that 'nothing in biology makes sense except in the light of evolution'. In the decades that followed, it became increasingly evident that this perspective on biology in general also applies to the science of human life. Today, with the sequencing of the human genome and our deepening understanding of developmental biology, we are gaining a greater appreciation of the unbroken lineage and intricate interrelation of the whole of living nature. Together with the recognition of our ancient heritage of molecular and cellular mechanisms has come an understanding that our emotions, cognitive categories, and broader capacities of mind are likewise grounded in our evolutionary past. This insight has spawned the new multidisciplinary synthesis of evolutionary psychology that draws on genetics, neurobiology, anthropology, and sociology in the search for a general description and explanation of universal aspects of human nature.

As a methodological programme and theoretical framework that takes seriously the actual biological conditions of human life, its embedded existence within the ecology of nature, its embodied being and evolved form and functions, evolutionary psychology holds promise for increasing our scientific understanding of human nature. Indeed, no theory more certainly affirms that there is a human nature. At the same time, it calls into question our most fundamental natural intuitions and traditional notions concerning the meaning and purpose of human life, with consequent implications concerning the fundamental character of the natural world.

If we begin our discussion with a critical analysis of the presuppositions and implications of this theory, we can recognize its legitimate contributions, its limitations, and its non-scientific aspects. We can then return to reconsider the evolutionary panorama within a broader frame of human knowledge. Drawing on both science and Christian theology, we can seek a fuller, wider anthropology and cosmology that provide a richer, truer, and more integrated description of science, ethics, and the human spirit.

Evolutionary psychology is a theory about the origins of the human mind. It assumes that all human behaviour, like that of animals, is directed toward competitive advantage in the evolutionary struggle of life. A wide array of psychological mechanisms, ranging from perceptual categories to preferences in mate selection, have been preserved due to their adaptive advantages and these collectively form a central core of human nature that is transmitted through genetic heritage. Whereas Darwin had originally put an emphasis on the individual organism and its adaptations for survival, evolutionary psychology emphasizes the transmission of adaptive traits to successive generations. According to Wrangham and Peterson (1996: 22), 'the ultimate explanation of any individual's behaviour considers only how the behaviour tends to maximise genetic success: to pass that individual's genes into subsequent generations'.

This claim extends beyond the realm of behaviours that relate in an obvious way to survival and reproduction, to include the subtlest manifestations of aesthetic preference, religious practices, and moral judgements. In complex human culture, the ability to sustain co-operative community may be of great benefit to the individual's survival and reproduction. Mental structures that promote co-operation will be favoured traits. Cultures will organize around a set of beliefs and practices that enhance human survival and reproduction. Receiving, sustaining, and transmitting these beliefs then becomes essential to any society.

Sociobiologist E. O. Wilson sees religious beliefs in particular as providing a sense of 'sacredness' on which principles of social co-operation can be firmly constructed. As such, they form the grounding for moral precepts that in reality are 'principles of social contract hardened into rules and dictates' (Wilson 1998: 250). Yet Wilson is not arguing for the reality of religious belief as some kind of transcendent truth, only the utility of the belief in benefiting the individual and sustaining social unity. Indeed, Wilson claims that morality has no other demonstrable function than to keep human genetic material intact. As articulated by Robert Wright (1994: 146): 'people tend to pass the sort of moral judgments that help move their genes into the next generation . . . and there is definitely no reason to assume that existing moral codes reflect some higher truth apprehended via divine inspiration or detached philosophical inquiry.'

The objection may be raised that throughout nature, and in human life as well, there are numerous examples of self-sacrifice for the good of others. Evolutionary psychology acknowledges this, but discerns beneath the appearance of altruism a deeper genetic self-interest. The principle of 'inclusive fitness' explains altruism by its indirect benefit to our genetic heritage via our siblings, cousins, and other genetic relatives. Likewise, the concept of 'reciprocal altruism' proposes that with our remarkable capacity to form coalitions and carry an accounting sheet of social debits and credits, we gain individually from a co-operation that is of benefit to all. The category of 'reputational altruism' proposes that good deeds are deemed not to go unnoticed and therefore will redound to personal benefit through social approval and privileged status. All of these explanations of self-sacrifice, however, undercut the moral meaning of the term 'altruism'. As Robert Trivers comments, 'Models that attempt to explain altruistic behaviour in terms of natural selection are models designed to take the altruism out of altruism' (quoted in Ridley 1996: 132).

In considering the claims of evolutionary psychology, it is evident that they contradict the most fundamental beliefs of Christianity concerning personal, social, and spiritual reality. The very possibility of genuine altruism and a meaningful morality built on love evaporates when their sustaining source in transcendent truth is considered a mere functional fiction. These theories, with their depreciation of religious foundations, echo the earlier perspectives of the molecular biologist Jacques Monod: 'The ancient covenant is in pieces, man knows at last that he is alone in the universe's unfeeling immensity, out of which he emerged only by chance' (quoted in Barlow 1994: 197). Yet a closer examination of evolutionary psychology can illuminate where it missteps and perhaps opens up an avenue for a more thoughtful application of its fundamental insight, providing an approach more compatible with natural human experience and with the claims of Christian theology.

There are some obvious and commonly noted criticisms of evolutionary psychology. It is far from clear that our entire psychological nature is oriented simply toward the propagation of genes. Of course the genes must be preserved and life must be sustained; there can be nothing without these fundamental biological processes. But it is another thing to say that human life is ultimately about nothing more, in view of all its exuberant artistic expression and scientific curiosity. Likewise, to account for all of human nature in terms of earlier evolutionary mechanisms ignores the evident freedom and emergence in nature. Human consciousness, symbolic language, and moral community are not fully described or understood by anything that has come before. Indeed, much of the evidence used to support the theoretical assertions of evolutionary psychology is drawn from inappropriate extrapolation of research on animal behaviour to human circumstances. As Jerome Kagan (1998: 161) rather drolly states, 'there can be no mouse model for human pride, shame or guilt'.

Of much greater concern, however, is the fact that, instead of serving simply as a naturalistic methodology, evolutionary psychology goes beyond the scope of science to include assumptions that represent philosophical and theological views. These assumptions make evolutionary psychology into an extreme form of philosophical naturalism. It assumes no design to nature, no direction or purpose—and no intrinsic meaning to its process. But such assumptions preclude the very categories of values and virtues, at least as these terms are traditionally understood. The practical effect of this is to reduce all human behaviours to value-neutral adaptations having no genuine reference to transcendent truths, and to deny any ultimate significance to mind and moral culture. Good and evil are seen as functional fictions generated for social cohesion, and human freedom is reduced to a useful illusion for legislating responsible behaviour.

Although proponents of evolutionary psychology often disclaim the deeper implications of their ideas and call on us to rise above the process of our origins, their materialistic presuppositions allow little logic for either the freedom or the motivation to do so. Evolutionary psychology, which began as a provisional framework for scientific inquiry, has become the central principle of a secular ideology; what began as a methodological tool has become a metaphysical pessimism.

This pessimism prejudices the very approach of our scientific inquiry, and is blind to the evident testimony of the phenomenon of human life. It describes wholes only in terms of parts, the higher in terms of the lower, and matter and mechanism as somehow more fundamental than the mind that investigates them. In the end, these assumptions obscure the most startling fact of evolution: the emergence of human persons and the manifestation within matter of freedom, consciousness, and moral awareness. What is needed is to release evolutionary psychology from its limiting assumptions and engage its explanatory power in exploring how freedom and the capacity for moral awareness are anchored in and arise from basic biology, yet beckon beyond to matters of transcendent truth.

THE EVOLUTION OF FREEDOM

When we look back at the evolutionary process we are at once struck by both its continuity and its creativity. At every stage and level the unfolding of the diversity of forms and functions reveals new, previously unseen dimensions of nature, and so revises our understanding of the nature of nature. The very word for nature in Greek, *physis* (from which we get our words 'physics' and 'physician'), comes from the root *phuo*, 'to sprout', 'to grow', to become'; and in Latin, nature comes from *nasci*, which means 'to be born'.

Although evolutionary accounts often stress the contingency of development, it is more likely that the earliest phases of life were highly constrained by the requirements of carbon chemistry and the need for precise environmental conditions. Looking back over nearly 4 billion years of evolution, it is astonishing to realize that these simple, constrained early life forms were able to serve as the foundation for all subsequent life, with its vastly greater complexity and breadth of freedom.

At its most primary level, freedom within nature is prefigured as a widening range of both responses to environmental stress and options for change. In biology this capacity is first expressed at the level of mutation, and such a strategy works very well in rapidly reproducing organisms. A single bacterium, for example, which has a limited ability to adjust to a changing environment, can produce tens of thousands of varied offspring within a few hours. This allows not only adaptation within a changing environment, but also extension into a wider range of environments with different conditions.

In this way, even the earliest life forms exhibit a general characteristic of biological organisms: there is a constant dynamic balance between self-preservation and the production of variation. Life does not remain static and constrained; there is a continual thrusting outward, a probing for possibilities for ever further proliferation into new realms of opportunity. Moreover, species do not evolve only for simple adaptation, but for flexibility of response; they evolve for evolvability, favouring adaptations which themselves provide the platform for further flexible evolution. We can see this tendency continuing all the way up to and including the emergence of human beings.

While early life forms adapted by mutation that created variation within a population, more complex systems of adaptation evolved that allowed adjustment to changing environmental conditions by an individual organism. With the emergence of brains more than 500 million years ago, the primary capacities of selective perception and locomotion in simple organisms were transcended by programmes of integrated organismal response, innate reflex arcs of nerves and muscles triggered by external stimuli. In addition, whereas the oceans had provided a more or less stable chemical context and constant temperature, the ascent to dry land required more complex regulation of water and temperature, but opened a vast new range of opportunities for the extension of life. It also led to the refinement of integrated motor and endocrine programmes that formed the biological basis of the emotions.

The emergence of affective life aided immediate survival, but also pointed beyond it. Emotions had their evolutionary origins in the physiological processes of body regulation, such as postural and visceral changes that place the organism in a condition of readiness of response. For example, fear involves increased muscle tension and protective posture. As sensory perception and action became increasingly complex, organisms developed with a more integrated 'inner' sense of subjective feelings associated with emotions.

Within a rising scale of feeling and self-awareness, sustained patterns of response came to be motivated and co-ordinated by an inwardly felt sense of aversion, appetite, or desire. The philosopher of biology Hans Jonas considers this the essence of animal life: '[The animal] emancipates itself from its immersion in blind organic function and takes over an office of its own: its functions are the emotions. Animal being is thus essentially passionate being' (Jonas 1966: 106). The unconscious process of plant life becomes the immediate awareness and response of animal life.

In human beings these legacies of our animal ancestors are preserved, but they are largely subsumed under the control of voluntary actions guided by associative memory, analytical reason, and conscious aspiration, as a further extension of freedom within the phenomenon of life. Such refinement of internal mental capacities evolved in co-ordination with increasingly complex external morphology. Early in the evolutionary process the differentiation of the head region, with its organs of sensory perception and communication, was paralleled internally by cerebral structures capable of processing more complex impressions of the surrounding environment. In *Homo sapiens*, along with the transition to upright posture, came co-ordinated revisions of body form, increased range of motion, and a radical cerebral reorganization that made possible new relations with the world.

Our transformation to upright form is reflected in nearly every detail of our deep structure, both somatic and psychic. For example, the freeing of the upper limbs and the refinement of the 'tool of tools', as Aristotle called the hands, allowed the emergence of greater fine-motor control and the cerebral capabilities that could co-ordinate and sustain more complex actions on the world. Along with upright posture came the evolution of the highly flexible, furless canvas of self-presentation we call the *face*. With more than thirty finely tuned muscles of facial expression and vocal control, human beings are capable of a wide array of communicative expressions of emotions and intentions. Special ensembles of cells in the brain respond only to faces, and some respond to specific facial expressions and direction of gaze. Indeed, there is evidence that our very concept of person, of a distinct subjective locus of consciousness, with its intentions, hopes, and fears, is formed in a uniquely human extension of the brain structures that process facial and vocal expression. Furthermore, as sight replaced smell as the most prominent sense, it allowed rapid perception of objects and actions at great distances. The cerebral processing and storage of visual images led to the emergence of imagination and its creative powers. Sight allowed insight.

Imagination in human beings conferred extraordinary adaptive benefit. Here mutations of matter are transcended by permutations of mind, the self-generated production of possibilities independent of the constraints of immediate reality. The symbolic mind, which is capable of detaching image from object and thereby recombining images in new ways, can envision scenarios and sequences detached from immediate time and space to anticipate their implications and outcomes. This is yet another powerful form of freedom in which the organism can imagine possibilities and try them out (in a kind of dress rehearsal) without the expense of time and risk of resources in the process.

The human capacity for imagination, however, goes far beyond adaptive anticipation. Imagination is not mere memory or imitation, but envisioned creation. Forming mental images, maintaining them in the mind and achieving their realization, allows intention, planning, and implementation of ideals. The first recorded moment of true creativity occurred in our pre-human ancestors 1.5 million years ago: there, in the fossil record, the simple chipped tools found in layers representing a million years of hominid history are suddenly transcended by an artefact that bespeaks a cognitive leap, the production of the hand axe. 'These symmetrical implements, shaped from large stone cores, were the first to conform to a "mental template" that existed in the toolmaker's mind' (Tattersall 2000: 61). This is perhaps the first intentional innovation: the bringing into being of an imagined ideal.

Whereas most creatures exist in an unbroken immediacy of life, humans have the freedom to draw the past into the present from learning stored as memory, and the freedom to draw the future into the present through the creative imagination. Together with the ceaseless drive to organize the unexplained ('the cognitive impera- tive'), the capacities to calculate, extrapolate, and recombine are used to reconfigure that which is into that which could be. While most creatures are pushed by biological and ecological circumstances, we are pulled into the future by our dreams and images of fullest flourishing and envisioned ideals. The human ascent to a coherent moral ideal is the fullest extension, the culmination of the most fundamental drive in living nature. As Leon Kass writes, 'Desire, not DNA, is the deepest principle of life' (1994: 48).

The story of life began within the constraints of chemistry and has ascended to the open possibility of an imagined ideal. The strategies of life culminated in our human responsive ability to think and act. We are made not for a particular niche, but for unpredicted possibility, for comprehension and control, for flexibility and freedom. We have adapted for adaptability itself.

This entire phylogenetic progression reflects the interplay between the possibilities and potencies within living matter and the constraints and opportunities of the natural world. Although chance may generate the multitude of mutations and recombinations tossed up to the filter of natural selection, their preservation is not random or arbitrary. This is the insight expressed by Leon Kass (1985: 273) when he writes:

Ought we to be surprised, should we regard it as an accident, that, in a visible, odorous, and sounding world the powers of sight, or smell, or hearing once they appeared should have been preserved, magnified, perfected? Likewise with intellect. However accidentally intellect first appeared, is it surprising that it should have been preserved in a world of cause and effect, past and future, means and ends, all of which can be brought into consciousness and used to advantage in a being endowed with memory, a sense of time, self-awareness, and the ability to order means to ends in securing the future?

Thus the evolution of progressively higher degrees of freedom reflects a potential within nature; human freedom has evolved as a part of human nature in response to a world in which such freedom is possible and, indeed, promoted by adaptive advantage. This same pattern of freedom emerging within physical, and social constraints is evident in the development of each individual person.

EMERGENCE OF HUMAN PERSONHOOD

To understand the emergence of the human person, it is essential to recognize the inseparable relationship between body and mind, and the way they are formed in interaction with the environment. As Antonio Damasio has pointed out, 'The mind had to be first about the body, or it could not have been' (1994: p. xvi). Peripheral sensory perception of the body surface and proprioceptive awareness of the musculo-skeletal dynamics of body position and balance contribute to an inner web of self-awareness. Like a map suspended in mental space, it provides a constantly updated (but relatively stable) image of our state of being. This fundamental sense of self-awareness then serves as a standard against which any perturbation or alteration (experienced as sensory input) can be compared, thereby providing information and understanding of the world. Damasio explains that the body, as represented in the brain, provides 'the indispensable frame of reference for the neural processes that we experience as the mind; that our very organism rather than some absolute external reality is used as the ground reference for the constructions we make of the world around us and for the constructions of the ever-present sense of subjectivity' (1994: p. xvi).

The mind, then, is not an abstract neurologic function, but is an activity of the whole body. We know the world not as a separate reality, but with reference to ourselves. Even to the simplest sensory awareness we bring the entire weight of selective attention, contextual interpretation, and the images and ideals that guide our deepest goals. Likewise, rational process is deeply tied to the fundamental evaluations and adaptive responses encoded within our emotions. Furthermore, even as we are shaped by our perceptions, we also shape our perceptions, stretching forth as active agents, probing the world with our questions in a quest for its unifying principles and coherent order. This places the human person within a larger frame that beckons beyond biology to questions of the spiritual significance of life. And it raises the fundamental question: what kind of knowing is made possible by this inseparable psychophysical unity of the human person?

Our particular unique evolved form of human embodiment provides a common 'language' of mental categories, emotional responses, and shared needs that serve as the basis for intelligible communication. Our embodiment also provides a basis for the desires and intentions that shape our shared system of values. With increasing organismal complexity, the central values of evolutionary success, survival, and reproduction are served by pleasurable intermediate activities that become valued ends in themselves. The most obvious of these are the pleasures associated with eating and with sexuality. These basic biological drives, in turn, point beyond survival to a shared realm of more transcendent values. Notwithstanding the great variety of cultures and diversity of personalities, there is a central core of common humanity that forms our shared community of values and aspirations. Our unique human form and its concomitant capacities and inclination of mind make possible what Leon Kass describes as

a new world relation, one that admits of a knowing and accurate encounter with things, of a genuine and articulate communion and meaningful action between living beings, and of conscious delight in the order and variety of the world's many splendored forms—in short a world relation colored by a concern for the true, the good and the beautiful. (1994: 66)

All of these shared foundations of human existence, our particular form of embodied being and the common challenges of a similar environment in which our lives are embedded, provide the crucial underpinnings of human social life and culture. Human beings are, by their very nature, social beings. Even the way we enter the world, amid the dangers of childbirth, promotes dependence and co-operation. Furthermore, human infants are born at an early stage of neurological development. Their long period of childhood dependency assures that social stimulation plays a formative role in the maturation of the mind. This intricate social interplay, first established between mother and infant, is built on a remarkable set of anatomic and physiologic adaptations that make possible the unique human form of the capacity for *empathy*. The biological and psychological development of empathy is one of the most extraordinary stories in the history of life and has a profound bearing on crucial questions associated with the moral and spiritual meaning of the human person.

Simply defined, empathy is the ability to identify with and understand the situations, motives, and feelings of another. More than merely sympathy, it is a form of intersubjectivity in which the observer actually participates in the feelings of the other. This extraordinary capacity is built on a combination of evolutionarily ancient emotional responses and more recent anatomical and neurological innovations unique to primates and highly refined in human beings. The shared, bodily based quality of emotions, manifested and experienced in a similar way by all people, makes possible the process of empathy. By perceiving and understanding the bodily movements of others, we come to understand something of their inner life.

Neurologically, the process of empathy seems to work as follows. An emotional state such as anger, sadness, or surprise generates in an individual visible manifestations of body position and facial expression. Observing such postural and facial expressions subtly activates in the observer the same muscular movements and nervous system responses that together constitute the physical grounding of an inwardly felt subjective state that would be represented by such expressions. We experience this, for example, when we see someone grimace in pain. Such physical response in the observer translates into the corresponding emotional states, thus establishing an empathically shared psychophysiological state between the observer and the one being observed. The discovery of 'mirror neurons' in monkeys (and suggestions from neuroimaging studies of similar cells in humans) may provide a physiological basis for such an explanation. These cells fire not only when the individual *makes* certain motions (such as hand movements), but also when he *observes* others making the same motions. Such studies suggest that innate, hard-wired connections between the sensory, motor, and visceral components of emotions make possible a single psychophysiological state shared between individuals.

This empathic intersubjectivity provides the patterning for personal identity and the platform for cultural awareness, and may provide a solution to that most difficult of questions: how do we leap beyond our subjective solipsistic self into genuine society with others? Charles Taylor (1991: 33) writes: 'The genesis of the human mind is ... not "monological", not something each accomplishes on his or her own, but dialogical.' From earliest infancy there is an interactive empathic engagement between mother and child. Babies preferentially look at faces and follow the flow of the mother's emotional expressions and their vital association with the patterns of events. In a process that psychologist Daniel Sterns calls 'attunement' there is a continuing reciprocity of small, repeated exchanges (Goleman 1995: 100). These establish congruency of emotional connection, nurturing the ties of attachment and the non-verbal foundations upon which language will later be built. In a process of 'social referencing' that builds a common set of values, the infant will point or gaze at an object to establish joint attention and then observe the mother's reaction. The mother's spoken responses, which at first convey to the baby only feelings— the shared affective language of posture and prosody—begin to carry specific semantic content. A web of meaning is formed within this linguistic system of empathetically grounded symbolic gestures, the coded concepts on which all human cultures are constructed.

With language we weave an interpretive story of cultural meanings and values, rich with ideals and aspirations, a narrative by which we navigate the world. In a kind of 're-envoicement' the child begins to structure his understanding of the world, the very pattern of his thoughts, by the echo of the words of others. In this frame the social significance of self is placed within a pattern of moral meanings and transcendent truths. Slowly the child is entrained to the society in which he is born, raised to a realm of beliefs and hopes inaccessible to an isolated individual.

DEVELOPMENT OF MORAL AWARENESS

The child psychologist Jerome Kagan (1998) describes how moral awareness develops together with empathetically grounded sociality. In an orderly developmental progression, a child begins to crystallize a sense of self and other and discover the inner mental world of private beliefs and intentions. With conscious personal identity comes awareness of the distinct identity of others, and cognitive empathy, a willed and knowing stepping into the role of the other.

Between the second and third year of life children develop an appreciation of the symbolic categories of good and bad and learn to apply these to their own actions, thoughts, and feelings. The child's sensitivity about the propriety of his behaviour relates to a larger concern with the right order and relationship of things. Discordances such as broken toys and shirts missing buttons trouble the child, and he begins a lifelong search for a coherent and harmonious explanation of the larger order of the world. With a growing understanding of the relationship between present actions and future outcomes, a child begins to develop a conflict between acting on present desires and recognizing when to do so would have negative consequences on self and others. Before the age of 5, children have difficulty governing their actions, but by around six the sense of self-control, and therefore accountability, allows shame and guilt but also the happy sense of virtue, of consonant goodness of self.

Free choice becomes, increasingly, the central moral axis and, guided by the emotional pull of empathetic communion, leads to the poignant and powerful drama of the individual self in the quest for a sense of ethical worthiness that transcends any simple imperative of genetic propagation. As Richard Shweder (1999: 798) notes, 'The most powerful motive of human life is a desire to gain a feeling of virtue, the desire to be "good"—a desire unique among sentient animals.' Indeed, we are beings with an awareness of good and evil, of time and death. Who among us wants to have written on his or her tombstone 'successful hedonist' or 'prolific propagator of genes'?

It is the nearly universal human testimony that this drama of moral choice is experienced subjectively as an exercise of genuine freedom. Such a view is central to how we understand both ourselves and our sense of significance within the world. Without freedom, there is no meaningful creativity, conscious choice, or morality in the common, ordinary sense of such terms. Yet freedom is at once a powerful adaptive advantage and a potential liability; it must be governed or it dissolves into indecision or derails into destructiveness for individual and communal life. From the perspective of biology, the moral sense is an adaptation to guide and channel choice in conditions too complex or novel for the direct operation of our primary desires and fears or heritable patterns of control.

Guided by the deepest vision of life's meaning and purpose, moral action freely chooses to do that which promotes the flourishing and the freedom of life, serving both stable continuity and creative change. But there is also a sense in which moral action itself is fullness of life. It serves as both a means and an end. This is evident if one considers morality as a distillation of the true, the good, and the beautiful.

To see moral action as fullness of life leads to a view of morality that is more profound than mere social obligation or codified rules of behaviour. The moral sense provides the comprehensive frame for the alignment of life with its deepest source and ultimate significance. Moral actions unify and integrate our most primary biological needs, our call to honour and nobility of spirit, and our fullest sense of the meaning and purpose of human life. Moral ideals form the foundation of values, sustain social communion, and provide the coherent vision that constitutes and animates the cohesion of culture.

Indeed, the moral sense provides the very infrastructure of personal identity and unifies reason, emotion, and intuition. It is to the mental life what the skeletal system is to the physical frame. It is the central core of the cognitive convergence we call consciousness. Taylor (1991: 31) writes, 'Reasoning in moral matters is always reasoning with someone.' The very root of our word conscience in Latin means 'joint or mutual knowledge'! Our modern word 'consciousness' has the same root; its current meaning was originally included in the term conscience, meaning 'consciousness of moral sense'. Looked at in light of the neurophysiology and psychology of empathy, it is clear why these words must be related. Human consciousness is intrinsically interpersonal and moral, for as we develop our sense of self out of empathic interaction with others, we must also develop our sense of how to relate to those others. As Pope John Paul II said, 'When we look into the eyes of another person we know we have encountered a limit to our self-will.'

We know that limit, but often we deny it. Even as empathy contributes to human community and co-operation by sensitizing individuals to the emotions and needs of others, it also enables the perpetration of harmful or hurtful actions. Other human beings are at once both our companions and our competitors. Group life implies previously unimagined freedoms, but also the opportunity for exploitation. The insight into the other person enabled by the capacity for empathy can be used for open communication or for calculated deception. The natural struggle of life may engage our empathic capacities for the purposes of intimidation, dominance, and exploitation. Thomas Hobbes in *Leviathan* wrote, 'I put for a general inclination of all mankind, a perpetual and restless desire of power that ceases only in death' (1651: pt. 1, ch. 4). But from the perspective of evolutionary psychology, such behaviour is a normal and expected consequence of the genetic competition, with little room for altruism, except in limited contexts to benefit one's group. Nevertheless, altruism, as E. O. Wilson (1975: 98) has written, is 'the central theoretical problem of sociobiology'. He goes on to say that the phenomenon of apparent altruism has not diminished in the scope of nature, but has increased and become 'the culminating mystery of all biology'.

Yet, in spite of so much in experience that is selfish, still at the level of everyday life, amid the hustle of the street and haggling of the marketplace, there is everywhere, in greater or lesser ways, the evidence of love. Many people, perhaps most, in some way give the energy and effort of their lives from a belief in love and a desire to make a positive contribution in the world. The natural sentiment and hope is that love is real. If we are cynically on guard about the lack of love as we seek to protect ourselves from exploitation, nevertheless we still remain alert for the signs of its grace, and its generosity and joy. Wherever it appears, it can still reorient and realign our lives, not with obligation but with inspiration.

People long to be loved, not conditionally for their usefulness or attractiveness, but unconditionally. But such love is costly to maintain. Where it is mutual, there must be reciprocity of trust, both parties risking vulnerability to betrayal as they open their innermost selves to one another. Without such trust and risk they cannot come to know the full personal reality of each other with self-revealing intimacy. Indeed, the intimate relationship depends on both qualities of faith and self-opening, and the relationship is itself a reality that cannot be known in any other way. This is a participatory type of knowing that is different from the knowing of empirical science; and what it knows within the relational reality of love cannot be fully known through the methods of science.

Such altruistic love is recognized as a moral ideal, and serves the highest human flourishing, as is evident wherever such sacrifice is absent as in circumstances of conflict without conciliation, bitterness without forgiveness, and misfortune without mercy. It is the kind of love that positively promotes the good and freely does unto others what we would wish for ourselves, even as we remain imperfect in our love. Of crucial importance is the question of how to sustain such love in the face of violation of it, and what the source would be for the conviction and the motivation to keep believing in and acting on love as the most primary reality. There is a clue in the role of central cultural narratives; for the moral sense is sustained as well as formed in community.

CENTRAL NARRATIVES OF CULTURE

Within every culture the moral distils into an account of origins and ultimate ends that organizes and bestows meaning, aligning life with the deepest principles and purposes of being. And generally these accounts distil as sacred story, binding belief with the compelling power of transcendent truths. In the words of sociologist Christian Smith (2003: 67), 'the human animal is a moral, believing animal—inescapably so. And the cultural frameworks within which the morally oriented believings of the human animal make sense are most deeply narrative in form.' He goes on to say that all such views, including our contemporary scientific accounts, reach beyond the fact/faith distinction and must 'ultimately rest on empirically unverifiable belief commitments and suppositions'.

The notion that morality should be linked to transcendent truths is understandable in our experience of life. To be effective without external coercion, moral norms must be believed; and to be believed they must be understood to be true as well as useful. As soon as moral obligations are not 'real', they are irrelevant. Without a reference to truth, moral behaviour rapidly loses its utility as rigid rules become tyranny or individual self-aggrandizement becomes anarchy. Even proponents of evolutionary psychology recognize the necessity to maintain at least the social fiction of an objective basis for morality. As expressed by Michael Ruse, 'It is important... that biology not simply put moral beliefs in place but also put into place a way of keeping them up. What this means is, that even though morality may not be objective in the sense of referring to something "out there", it is an important part of the experience of morality that we think it is' (quoted in Matteo 1999: 42). Any purely instrumental notion of the nature of morality or its basis, however, is corrosive to self-understanding and incompatible with the way in which morals operate within the community. For the stubborn evidentiary fact remains that, as Jerome Kagan (1998: 158) expresses it, 'Although evolutionary biologists insist that the appearance of humans was due to a quirky roll of the genetic dice, our species refuses to act as if good and evil are arbitrary choices bereft of natural significance.'

The issue for moral motivation, then, is one of faith in the transcendent truths of a central cultural narrative, and the extent to which the truths expressed within that narrative sustain a balance of flourishing and freedom. Increasingly, evolutionary science is a source for the modern central narratives of the origin and nature of humankind. When that source is evolutionary psychology, it has the remarkable characteristic of not only failing to provide transcendent truths, but of expressly denying that they exist. Even if a culture built around such a narrative attempted to formulate values as a utilitarian ethics based on natural laws of genetic flourishing, it would lack any basis to motivate its values with the compelling power of a transcendent referent. Without the binding power of moral truth, it would be vulnerable to the tendency for individuals to prefer their own alternative values, based on the more primary persuasions of their immediate appetites and competitive ambitions.

We must ponder the relative fitness of such a cultural construction, because in the coming era of biomedical intervention in human life the stakes will be higher than they have ever been. Brought forth in an evolutionary trajectory of ever increasing freedom, we have ascended to a comprehension and control of the natural world that places in our hands the very powers that have formed us. In the nearly 5,000 generations our species has been on the Earth, never has the exploratory edge of human experience been at once so open and so full of danger. Liberated from the basic struggles for survival, we are opened to the prideful ambitions of technological self-transformation that could shatter the fragile balance of our physical and psychological functioning.

It is not hard, however, to recognize the deep source of the pessimism of evolutionary psychology. I have in this essay emphasized the positive development of human life and our capacities for love. But the evolutionary panorama also presents the spectacle of unspeakable suffering that is inseparably woven into the entire process of predation and natural catastrophe. A comprehensive picture of the world must face the problem that is posed by such suffering for any notion of transcendent goodness.

Torn between the pull of pride, the private lures and longings of self-will, and the aspirations of the communal ideal, the fundamental question arises, 'In whose image are we made'? Pascal recognized the full significance of this question and warned that those who sought God apart from Christ, who went no further than nature, would fall into atheism. The natural world, with its strife and struggle, poses a question that it cannot answer: How can there be both suffering and love? Yet with this question the deepest meaning of the material world is opened to understanding. All of creation, and its evolutionary ascent to mind and moral awareness, may be recognized as a kind of 'living language' in a drama of the deepest spiritual significance. To the participatory knowing of faith, the entire cosmic order of time and space and material being may be seen as an arena for the revelation of Love, for the creation of a creature capable of ascending to an apprehension of its Creator; but more profoundly, for the reaching down, the compassionate condescension of Love himself.

There within the human form with its capacity for empathy, for genuine communication, moral truth was revealed in matter; the true Image of God was borne within a body. In the face of Jesus was made evident the face of Love, and most specifically in his suffering on the cross. Those who looked upon him felt his pain, yet recognized his righteousness and knew the injustice of his plight; his was the ultimate, defining act of altruism. In this the transcendent was revealed in and through the immanent;

nature and God were reconciled, and the cosmos was restored in its intelligibility. The fullness of Love was revealed in human form.

In that moment of human history, all of creation was lifted to another level of meaning. All of the evolutionary struggle, the seeming futility of suffering and sacrifice and death itself was raised to the possibility of participation in a higher order of being. Christian faith is a faith in the God whose very nature is Love, an affirmation that reaches beyond all suffering to the ultimate goodness of life. It is here that, while decisively denying the pessimism, cynicism, and amoral implications of evolutionary psychology, Christianity may at once affirm the reality and positive significance of the material world and its evolutionary process. In the emergence of moral nature, humanity, as the culmination of creation, is called into communion with the very life of God, the life of Love.

REFERENCES AND SUGGESTED READING

Barlow, C. (1994) (ed.). *Evolution Extended.* Cambridge, Mass.: MIT Press.

Damasio, Antonio (1994). *Descartes' Error.* New York: Grosset/Putnam.

Goleman, Daniel (1995). *Emotional Intelligence: Why it can Matter more than IQ.* New York: Bantam Books.

Hobbes, T. (1651). *Leviathan.* New York: Touchstone.

Jonas, H. (1966). *The Phenomenon of Life: Toward a Philosophical Biology.* New York: Harper & Row.

Kagan, Jerome (1998). *Three Seductive Ideas.* Cambridge, Mass.: Harvard University Press.

Kass, Leon (1985). *Towards a More Natural Science.* New York: Free Press.

—— (1994). *The Hungry Soul: Eating and the Perfecting of our Nature.* New York: Macmillan.

Matteo, Anthony (1999). 'The Light of Reason: Evolutionary Psychology and Ethics', *Science and Spirit*, 10/2: 14–15, 42.

Pascal, Blaise (1958). *Pensées*, intro T. S. Eliot. New York: E. P. Dutton & Co.

Ridley, M. (1996). *The Origins of Virtue.* New York: Viking.

Shweder, Richard (1999). 'Humans Really Are Different', *Science*, 283: 798–9.

Smith, Christian (2003). *Believing, Moral Animals.* New York: Oxford University Press.

Tattersall, I. (2000). 'Once We Were Not Alone', *Scientific American*, 282/1: 56–62.

Taylor, Charles (1991). *The Ethics of Authenticity.* Cambridge, Mass.: Harvard University Press.

Wilson, E. O. (1975). *Sociobiology: The New Synthesis.* Cambridge, Mass.: Harvard University Press.

—— (1998). *Consilience.* New York: Alfred A. Knopf.

Wrangham, R., and Peterson, D. (1996). *Demonic Males.* Boston: Houghton Mifflin.

Wright, Robert (1994). *The Moral Animal.* New York: Random House.

PART VI

VALUES ISSUES IN RELIGION AND SCIENCE

CHAPTER 52

THEOLOGY, ECOLOGY, AND VALUES

CELIA DEANE-DRUMMOND

INTRODUCTION

The nexus between ecology and religious reflection has been on the agenda for discussion among theologians since the 1970s, which were characterized by an increased social, public, and political awareness of environmental questions arising specifically from human activities. For the purposes of this chapter, ecology is understood to include ecological science, concerned with the interrelationships between different species, as well as broader impacts on ecosystems through processes such as climate change, agriculture, and biotechnology. These applied sciences might seem less glamorous than more abstract sciences but are, I suggest, highly significant in terms of their impact on the social conditions of the present and the future health of planet Earth, its image gloriously captured from outer space. The intention of this chapter is to engage with the values that arise within ecology, to offer an illustrative analysis of current ecological challenges, and to highlight debates within Christian theological ethics as to how best to respond to such challenges. I will argue for a virtue ethics approach to complex questions in ecology, drawing specifically on the classical tradition of Thomas Aquinas and including the virtues of wisdom, prudence, justice, and fortitude alongside the theological virtues of faith, hope, and charity. Finally, I will offer pointers towards the way the field might develop in the future.

There are, in the first place, a number of questions that ecological science raises for theology, and for all monotheistic religions that give a high place to humanity. In the first place, it challenges the notion that humanity is the 'crown of creation', in that humanity is one species among many in a complex ecosystem. It also challenges theology to reflect not just on human history and human redemption but also, in common with all the biological sciences, on the significance of the material world as created by God. Moreover, there are practical challenges emerging within environmental science that impinge not just on particular cultures and particular societies, but also have global consequences. Such pressing concerns invite serious questions about the future of the cosmos and humanity's place in it.

Scientific investigation has been particularly successful since the dawn of experimental science in the seventeenth century largely because it has specialized into subfields that can focus on particular problems at hand. Such specialization offers the strength of expert knowledge, but such expertise has its drawbacks, for knowledge becomes more fragmented and it is hard to view problems from a more holistic perspective. It is here that non-expert, public knowledge can make a contribution, and social scientists have found ways of demonstrating that new questions arise in such contexts (Deane-Drummond and Szerszynski 2003). Ecological science has the advantage of exploring interrelationships in a way that is lost in many other sciences, and has sometimes been called a subversive science. Contemporary theological thought, similarly, has also fragmented. Yet because theology is acutely aware of its historical roots, it is permissible to return to those classical modes of thought less subject to fragmentation. Of course there is no need to pretend that we can merely return to a pre-modern golden age. Rather, some facets of that earlier thinking might be useful in navigating territory where contemporary thinking has failed either to convince or prove effective (Deane-Drummond 2000).

There are considerable debates in the field as to how far and to what extent a recovery of classic thinking is useful, permissible, or helpful. Many ecofeminists and process theologians, for example, believe that the classical approach is too contaminated with patriarchal or monotheistic thinking. While wanting to take into account these concerns, I argue for the relevance of a classical approach, revised in the context of contemporary scientific discourse. There is some suspicion among those in the science-and-religion field that it is impermissible to deal with subjects that are overtly 'political'; as it somehow prevents a dispassionate and objective approach. Yet science has its own philosophy, and politics, and particular commitments, as does theology. Some authors have argued for the inclusion of politics in theological reflection on the natural world (Scott 2003). This is not simply an aping of 'liberation theology', including 'nature' under the category of the oppressed, though certainly arguments might be made for this by liberation theologians (see e.g. Boff 1997). Rather, it is the recognition that our human sphere is necessarily inclusive of political communities, and it is within the scope of theological reflection, at least, to critique and consider what might be the most appropriate response. Ecotheology needs to resist attachment to a 'green' political agenda, offering instead a theological critique of political science. I suggest further that to miss out on reflection on political decision making is a mistake, for it amounts to a failure of nerve, a failure of theology to embed itself in secular society.

VALUES IN ECOLOGICAL SCIENCE

One of the core values in ecological science is the challenge to anthropocentrism: that is, the view that humanity is the centre of the universe. Some have argued that Lovelock's Gaia hypothesis presents an even deeper challenge in this respect (Primavesi 2003; Midgley 2001). Lovelock's hypothesis addresses the interaction between the biota and global geochemistry. Lovelock's hypothesis argues for the homeostatic regulation of the earth's atmosphere by the biota in order to keep environmental conditions constant (Lovelock 1987). The values implicit in this hypothesis are more commonly interpreted in terms of co-operation and symbiosis (Galleni and Scalfari 2005; Midgley 2001). The theory of Gaia can be interpreted along more Darwinian lines, which includes the idea that the dominant species are the most influential in setting the environmental state. In this respect Gaia is rather more ambiguous as an ethical paradigm than some of its proponents allow (Deane-Drummond 2004a: 162–85). Insofar as it raises the importance of co-operation and symbiosis, it reminds humanity of the varied ways in which species relate to each other.

Alongside a paradigm of interrelatedness, there is an equally strong notion of 'balance' that has served ecology. Of course, it is a disputed point as to what extent the idea of the 'balance of nature' arose from ecology; more likely it was already embedded in cultural assumptions about the natural world, and ecology just reinforced this notion (Worster 1977). This idea of balance continues to dominate the discourse among many contemporary Christian theologians, who find it congenial in relation to connectivity and ideal community, serving as an ethical pointer (Northcott 1996). Yet ecologists themselves in their current practice are now less attuned to this notion. Instead, the notion of flux, unstable equilibrium, openness to external influences, disturbance from internal and external forces, including humanity, is much more to the foreground (Pickett, Kolasa, and Jones 1994: 160–1). This shift illustrates the difficulties of using ecological science as a natural basis for values, for other values surface that undermine the original notions (Deane-Drummond 2004a: 36–8).

The concept of balance cannot be dispensed with altogether, since just as Newtonian mechanics proved accurate at the broadest level but was eventually supplemented by Einstein's ideas, so too the concept of ecological balance can persist in the broadest sense, even if flux is a more accurate description of what actually happens at other scales. Balance is also crucial in consideration of global processes such as climate change, but here too, research shows that such balances are subject to disturbance, often anthropogenic in origin (Houghton 2004). The concept of eco-systems being in fragile balance, subject to disturbance, is important to *inform* the way humanity thinks about ethical conduct in relation to the environment; that is, we need to be aware that this is the case, but it should not become paradigmatic as a *value term* for ethical behaviour. Why not? I am reluctant to adopt a strong philosophy of naturalism: that is, deriving how we *ought* to behave from the way things are. This is not the same as saying that we can ignore ecological science; rather, such aspects of the way ecology *is* serve to help us interpret how to treat the environment even if, to some extent, this knowledge is provisional. If flux became a value, this would lead to chaos, and shows how inappropriate it is to derive or read off values from the natural world. There need, in other words, to be other reasons why humanity has decided that balance serves the common good and reflects the goodness of creation understood as God's gift.

A case might be made for ecological wisdom, arising out of a perceived sense of ecological interrelationships. In one sense natural wisdom exists, arising from the way in which creatures are so marvellously attuned to their environment and converge into particular patterns during the course of evolution (Deane-Drummond 2006c). Of course the meaning of wisdom is itself somewhat illusive. Wisdom as a particularly human capacity is, according to Jeffrey Schloss, that ability to critique culture in the light of human evolutionary history. He suggests, 'Wisdom is living in a way that corresponds to how things are' (Schloss 2000: 156). The way things are includes ecological insights and knowledge. Wisdom, then, would be living in accordance with our knowledge of such sciences, but in such a way that gives meaning to existence. A deeper meaning to wisdom comes through theological reflection.

Wonder (Deane-Drummond 2006c), like wisdom, is also a virtue, or a particular habit of mind, which is common to many of the sciences, including ecology. Wonder among ecologists largely springs from observation of the marvellous fecundity and diversity of life on this planet. Wilson contends that humans possess an innate tendency to focus on life and lifelike processes, a tendency that he calls *biophilia* (Wilson 1984: 1). He argues that modern biology is a genuinely new way of looking at the world that happens to be in tune with this tendency. It is through such a search that he believes we discover the core of wonder, due to the rich abundance of life (Wilson 1984: 10).

Rachel Carson (1965), a biologist writing in the era before E. O. Wilson developed his sociobiology, argued passionately for a wondrous appreciation of the natural world. However, this appreciation came from a fundamental anxiety about what might happen to the world if humans continued to despoil the environment. Although Wilson used the language of wonder to express his forays into natural history, he appealed to baser aspects of human nature as a basis for environmental protection. The natural world was valuable for its instrumental use to humans, so it could be viewed as an untapped source of new pharmaceuticals, crops, fibres, and so on (Wilson 1996: 174). Carson believed that the despoliation of nature was not in humanity's best interest. She believed that wonder is a prelude to care for the earth, and in this her views are directly contradictory to those of Wilson (Carson 1965).

CHALLENGES OF ECOLOGICAL SCIENCE

The challenges posed by our knowledge of ecology, understood broadly to include the environmental sciences, poses pressing issues for national and global economies. Stenmark (2002) argues that environmental policies are influenced by particular environmental philosophies, so it is incorrect to assume that different philosophies converge when applied to policy making. The following examples are illustrative of the kind of issues that need to be addressed at national and global levels, and raise important religious as well as philosophical issues. This discussion will be brief and illustrative, using a few case studies.

The projected rate of climate change in this century is far greater than anything experienced by the Earth in the last 10,000 years, mostly caused by human activity through the burning of fossil fuels (Houghton 2004). There have always been sceptics who challenge the detailed predictions anticipated by climate change models. The *extent* of change anticipated is debated, rather than whether there is any change taking place at all. Accumulated scientific evidence points to increasing unpredictability. The overall consensus is, in spite of the uncertainties, that most of the observed global warming over the last fifty years is due to changes in greenhouse gas concentrations.

The impacts of climate change on the ecology of both human and non-human communities come about through things like rising sea levels and an increase in number and frequency of climate extremes; floods and droughts are becoming more common due to a more intense hydrological cycle (IPCC 2001). The changes that have a 90–99 per cent probability, if trends continue, include higher maximum temperatures and more hot days as well as higher minimum temperatures and fewer cold days and frost days over nearly all land areas; reduced diurnal temperature range and increase in heat index over most land areas; and more intense precipitation events over many areas. Those countries that are the most subject to change are in the poorer subtropical parts of the world. Low-lying lands, such as Bangladesh, and island communities are the most vulnerable to sea-level rises, which could lead to swamping of entire communities (Houghton 2004: 150–2). Such devastating loss would lead to a huge number of environmental refugees, estimated to be 150 million by 2050 if present rates of change continue. The resulting social problems inevitably raise the issue of environmental injustice: namely, the proportionally greater impact of climate change on the poorer communities of the world, which are, in relative terms, contributing significantly less to the anthropogenic effects on climate change.

The colonization of land by human populations has always led to some loss of biodiversity, even among prehistoric peoples. Since 1600 researchers have identified more than a thousand recorded cases of extinctions of plant and animal species (Gaston and Spicer 2004: 135). Ecological knowledge helps to define biological diversity, though the species that have so far been identified represent only a small fraction of the actual diversity, especially in areas such as the tropical rainforest.

Conservation is, at best, a safety measure for known species; there are others that are disappearing before they can even be identified. Species may be lost through direct exploitation by humans, and in areas such as the tropical rainforest it is difficult to estimate original wildlife diversity because there are virtually no regions entirely free from human exploitation. The targets for human exploitation in this context include bush meat, as well as fuel wood, which overall exceeds replacement planting (Gaston and Spicer 2004: 116–17).

After direct exploitation, loss of habitat is the second most common reason for species extinction, correlated with population density and clearance of land, as well as climate change as mentioned above. Although habitat loss has been a feature of human history since the dawn of civilization, its extent has increased rapidly with the march of technology.

A third cause of species loss is introduced species. About 400,000 species have been introduced over time (Pimentel 2001). While some may have been intentional, others have come through transport routes between human populations. In about 10 per cent of cases, introduced species have effects on existing ecology, leading to negative impacts, including species extinction. Endemic species in lakes and islands are particularly vulnerable to predation and parasitism by introduced species. For example, the accidental introduction of the brown tree snake *Bioga irregularis* to the island of Guam in 1950 resulted in the loss either directly or indirectly of twelve species of an original fauna of twenty-two native birds, reduction in the remaining forest species to a mere remnant, and the loss of over a third of native reptiles (Fritts and Rodda 1998). Ecologists also recognize that due to the interconnection between species, extinction of one species almost inevitably leads to extinction of another, leading to what are termed 'extinction cascades' (Gaston and Spicer 2004: 129).

The relationship between biodiversity and ecosystem function is a major research area within ecology (Loreau *et al.* 2001). In some cases a species may be lost without any effect, known as 'redundancy'. In other cases there are crucial species whose loss disrupts the ecosystem, an effect known as 'rivet popping'. In a third 'idiosyncratic' scenario it is not possible to predict accurately the impact of the loss of a particular species, because each species has complex and varied roles. An ecosystem with a large number of species is more likely to show species 'redundancy'. Yet the value of conserving biodiversity is not simply related to ecosystem function. The benefits of biodiversity relate to (a) direct uses, instrumental uses for human benefit; (b) indirect use, achieved through biogeochemical cycling, necessary for the maintenance of global ecology. So-called 'redundant' species may also be indirectly useful as food sources for other species that are of instrumental use. Other indirect uses might include tourism and recreation. There are also (c) four values that do not relate to present usefulness. Future generations may have (i) instrumental or other uses for biodiversity, known as (ii) bequest value. There is also (iii) existence value (Gaston and Spicer 2004: 103–5), which is the value of species as such for human beings. Finally, the value of species in and of themselves apart from human valuation leads to the notion of (iv) intrinsic value.

As well as climate change and biodiversity, the impact of genetic biotechnology has local, national, and international ramifications. Environmental risk factors associated with genetic engineering are related to the power of the technology to bring about irreversible changes in the hereditary material of life forms. Genetically engineered rapeseed and maize, for example, which are resistant to specific herbicides, have increased production over the last ten years since their first release, with the majority of the crop grown in the USA, though a significant portion is grown in Canada, Argentina, and parts of Europe. In many cases there were no public consultations prior to the release, which may be one reason why there has been a backlash in Europe and increasing resistance in the USA. While the companies concerned deny any risk to the environment, the *science* suggests the opposite, since rapeseed can cross-fertilize with wild mustard plants.

The increased dependence of the farmers on herbicides for weed control encourages an equal dependence on the *hybrid seed* sold by the same company. Hybrid seeds do not breed true, so the traditional method of saving seed for the next year's crop is possible only with loss of yield. The impacts of these stipulations are of particular significance for farmers in the poorer communities in the world, where recycling seed is seen as an integral part of their lifestyle. Another indirect risk of GM crops, which is also characteristic of conventional plant breeding, is the overall loss in genetic diversity. Wild strains have a much greater variability, which protects them naturally from pests and disease. When a crop is genetically engineered (or bred) the resultant uniformity brings the desired increase in yield, but also carries a greater vulnerability to disease. In order to find new sources of variation, researchers have sought wild, genetically diverse strains that are on the whole confined to the poorer Southern continents. Patent payments are more often than not accrued by the company developing the technology (Deane-Drummond 2004*b*).

REDISCOVERING CHRISTIAN VALUES

Christian approaches to ecology and environmental concern are often on the defensive, responding to the charge that Christianity itself has exacerbated the ecological crisis by fostering the notion of human dominion of the earth (White 1967). There are three possible responses to this accusation. The first response is apologetic: to deny that dominion as set forth in the Book of Genesis means domination, arguing that it should instead be thought of in terms of stewardship (Berry 2003: 183–200). The second response is to admit the anthropocentric tendencies in the biblical text, but to seek to recover other strands in the biblical tradition which are more accommodating to environmental care, such as the affirmation of creation in the Book of Job (Habel 2000).

The third response is to admit to flaws in the tradition and seek other paradigms outside that tradition, such as in feminism, pantheism, or the Gaia hypothesis, in order to generate a basis for value (Hessel and Ruether 2000; Primavesi 2003). The association of experimental scientific activity with a devaluation of the natural world, 'the death of nature' (e.g. Merchant 1982), has tended to alienate still further those in the scientific world from those engaged in ecofeminist discourse. Some authors retain aspects of the tradition, but weave them into new models, such as the idea of the Earth as the body of God or the Earth as sacred, highlighting God's immanence in a way that was, arguably, neglected in the past (McFague 1993; Grey 2003; see also Edwards 2001). Another common stance of many ecotheologians emphasizes the ideal of the Earth as the home for all species, including humans. The difficulty, of course, is that this view romanticizes the extent to which humans can genuinely feel at home, given the reality of suffering and loss currently experienced across a multitude of species and cultures (Conradie 2005a). Ernst Conradie also argues that ecotheologians have leaned too far in the direction of stressing the immanence of God in creation, in such a way that there has been a failure to rediscover the meaning of God's transcendence in a global, ecological context. Others believe that the focus needs to be on the future of the Earth, a recovery of eschatological traditions in line with process thought, rather than on either stewardship or re-sacralizing nature (Haught 1993: 100–10). There are also Eastern Orthodox theologians, who insist that the early patristic traditions neither need to be dispensed with nor lead inevitably to the kind of exploitative anthropocentrism that is implied by critics of tradition. Orthodox liturgy is a cosmic celebration in which each species is valued in itself and in which humanity's role is that of priest (Theokritoff 2005). In addition, I have argued that the notions that God creates through Wisdom, and that Wisdom is a key theological category, are ones that can serve to inform a theological basis for ethics (e.g. Deane-Drummond 2000, 2004d).

How far and to what extent the notion of stewardship is a helpful model for human valuing from a Christian perspective has been the subject of intense debate. Proponents argue that stewardship, if properly conceived, addresses the important issue of human responsibility. It leads to a model understandable at a secular level, and results in some broad guidelines for policy making (Berry 2003: 183–200). A number of authors are critical of this approach, not least because it seems to reinforce, albeit obliquely, the managerial style of human relatedness to the Earth (Conradie 2005b; Haught 1993; Deane-Drummond 2004a: 34–5). A second option argues for citizenship, that we are all fellow citizens and companions on Earth with other creatures (Page 1996). The difficulty in this case is how to connect this theme with specifically theological insights. It becomes more explicitly theological in association with the themes of covenant theology and the ideal of the Sabbath (Murray 1992; Ruether 2000; Deane-Drummond 2004c). A third, more directly theological option is that of servanthood; Christ as model of servant then becomes the model for how Christians need to relate to the Earth (Wirzba 2003). In this case the difficulty is how servanthood connects to secular policy making: for example, what meaning might it have for those in government trying to set policy where there are numerous and competing demands for value?

My own preference is to opt for a virtue ethic approach,[1] one that is philosophically informed but also leans towards a classic virtue ethic which draws specifically on the work of Thomas Aquinas (e.g. Deane-Drummond 2000, 2003, 2004a–d, 2006b). The particular virtues relevant to consider are wisdom, prudence, justice, temperance and fortitude, alongside the more theological virtues of faith, hope, and charity. In addition, prudence and temperance need to come alongside charity and beauty, expressed through wonder. Of course, in arguing for the practice of prudence in ecological decision making, I am presupposing that the world in all its diversity is understood in the Christian tradition as created by God, is good, that it is a gift, and, moreover, that life is worth preserving. Prudence as a virtue is, I suggest, a faculty that helps to *distinguish* between different goods where there are competing demands for attention. The good to be aimed at is the *common good*, encompassing all peoples as situated in an ecological context; in other words, it is not simply individualistic, but looks to the good of the individual alongside the wider good of the community as a whole.

Prudence, or *practical wisdom* for Thomas Aquinas, is the 'mother' of all the other cardinal virtues. Unlike contemporary secular accounts of prudence, which associate it with political expediency and caution, classical approaches to prudence link it with goodness. Prudence includes deliberation, judgement, and action; it is 'wisdom in human affairs', rather than absolute wisdom. Its orientation is practical, rather than simply theoretical. Prudence does not define what the good is; rather, it *facilitates what makes for right choices* by so acting in different virtues in accordance with the ultimate orientation towards the good. It also demands openness to others in a way that prevents prudence from being individualistic, for taking counsel is the first part in the overall process of prudential activity. Prudential understanding avoids a rule-based approach, so it allows some flexibility when it comes to practical decision making. Aquinas lists the various components of prudence as memory, insight or intelligence, teachableness, acumen (*solertia*), reasoned judgement, foresight, circumspection, and caution. (Aquinas, *Summa Theologiae* 2a2ae, qu. 49). There are also three types of prudence—individual, domestic, and political—a fact of particular significance in an ecological ethics that includes decision making at all these levels. This is important to stress, since virtue ethics has more commonly been associated with individualism; but if we return to a Thomistic approach to prudence, it opens up an approach to decision making at broader, communal levels in a way that counters this criticism.

Aquinas also distinguishes between imprudence, negligence, and sham prudence. One mode of imprudence is thoughtlessness, acting too quickly without sufficient time for consideration. Examples relevant to the ecological sphere might be a rush to introduce a new species, without due consideration of possible negative effects. Or it

---

[1] There are other authors who have argued for the relevance of virtues for environmental ethics; see Bouma-Prediger 2001, and van Wensveen 2000.

might be a politically short-sighted approach to the impact of global warming, along with a failure to address the long-term strategy needed to address climate change. Aquinas named negligence as a distinctive vice: namely, one of *not* choosing the good, a failure in implementation of prudence (Aquinas, *Summa Theologiae*, 2a2ae, qu. 54.1, 54.2). Negligence would also apply where there is a failure to protect species loss, in spite of known causes leading to its destruction. Sham prudence is prudence directed to the wrong end, such as material gain, set up as the final goal of the human life. Biotechnology that is operative purely on the basis of lucrative patents, regardless of the possible effects on either the human community or biological diversity, would be a good example of sham prudence, where the benefits are confined to the producers. Another form of sham prudence comes through using false or deceitful means: namely cunning, to attain either a good or an evil end. In this case one might cite the example of an environmental gloss given to a particular project in order to win political and public support, when the intentions are confined to baser motives.

Given this understanding of prudence, such a task would be almost impossible given the propensity to err (sin) in all its varied facets without Aquinas's belief that perfect prudence is possible only through the grace of God. The form of prudence to which the Christian aspires is prudence in the light of the three theological virtues of faith, hope, and charity. Commentators on prudence in Aquinas seem to have ignored the possibility of prudence acting under the influence of God's grace; it has proved to be an embarrassment to modern readers. Debate exists, therefore, in how far it is appropriate to enlarge an understanding of virtues as guided by the Holy Spirit, rather than simply contain this within humanistic categories of knowing. This also opens up debate about the relationship between nature and grace, and how far a Thomistic approach was sufficiently aware of the organic unity of the human person. In such a context it is preferable, perhaps, to draw more on Karl Rahner's understanding of the way that grace perfects nature (1954: 297–317), rather than simply to adopt a neo-Thomistic understanding that grace simply 'added' to natural capacities. Yet, even with this modification, we can retain a form of grace-laden prudence as one that also relates to wisdom, since it can be more readily identified as a theological category. While prudence is in one sense a form of wisdom, known as practical wisdom, wisdom in itself moves beyond even the scope of prudence delineated above.

The ability of prudence to be still, to deliberate well, is a quality desperately needed in the frenzied search for new methods and techniques in technology that are considered to have particular usefulness for humanity in spite of environmental impacts. Prudence demands full encounter with experience, including the experience of ecological and environmental sciences, taking time to perceive what is true in the natural world. Such close attention to reality as perceived in the scientific world involves a kind of studied attention. Prudence invokes not just contemplation of the world, but positive action as well, action that has in mind the goodness of God. Consequentialist approaches to ecological ethics have sought to frame decisions in terms of costs and benefits, or risks, more often than not expressed in terms of the precautionary principle. While prudence would include some perception of risks where they are known, the ability to have accurate foresight depends on how far such decisions promote the overall goal of prudence towards goodness, understood as the common good in the light of God's goodness.

A discussion of the significance of the virtue of prudence would not be complete without mention of the three other cardinal virtues of justice, temperance, and fortitude. These virtues express important values that are richly embedded in Christian teaching. In secular debate, justice is often split off from a consideration of virtue ethics, as it is more commonly associated with rule-based ethics. Consideration of justice, like the communal and political dimensions of prudence, is equally important in that it counters some of the criticisms of virtue ethics that it is too focused on individual tendencies. Yet locating justice as a virtue deepens the practice of justice beyond simply following the rules. Though related to the law-like approaches, it moves beyond them again, in that it considers the attitude of the agents. Justice is concerned broadly with the idea that each is given her or his due. Unlike many other virtues, justice specifically governs relationships with others. Justice is located in the will, rather than the emotions, maintaining right relations between individuals and between them and others in one's own and other communities. By calling justice 'a habit', Aquinas locates justice within the parameters of virtue, even while recognizing the importance of law through his threefold distinction of natural law, positive law, and eternal law. In other words, for every principle of justice there is a corresponding virtue or habit that facilitates this practice.

The 'natural' does not always satisfy the demands of justice, but human will establishes positive laws commensurate with natural justice. In Aquinas natural law does not so much read off human values from nature, but acknowledges the importance of living in relation to how things are as creatures—in this sense it is connected to Schloss's interpretation of wisdom. In Aquinas the alignment in natural law also, and importantly, has to take its bearing from the eternal law, which serves to define what the good means. Legal justice, or general or contributive justice between the individual and state, can be compared with distributive justice between the state and the individual, and commutative justice between individuals. Considerable debate exists in the literature as to how far an environmental ethic that draws on natural law should be secular—that is, detached from theological norms—or theocentric—that is, related to considerations about who God is through the notion of the eternal law. While it is possible to argue for the value of a natural law ethic independent of Thomistic notions of the eternal law, my suggestion is that disembedding this concept from his theology amounts to a reading of Aquinas that takes us too far from his original intentions.[2]

The question arises, What is due, and to whom? How do we work out what is due, especially in relation to other creatures? Aquinas's specific writing on ethics might encourage a lack of respect for non-human creatures, as, like other authors of his

---

[2] For a valuable discussion of natural law in Aquinas and argument for retaining his theocentric approach, see e.g. Porter (2005).

time, he describes animals in terms that merely express their usefulness to humanity.[3] He was writing when animals were often treated disrespectfully, though it is important to recognize that he did allow for a degree of possible emotion in animals that pointed in an alternative direction. His ontology presents us with what Judith Barad describes as a 'glaring inconsistency' (1995: 145), in that in spite of his somewhat negative specific ethical approach to animals, his ontology points in the opposite direction, and suggests a strong continuity between plants, animals, and humans as *creatures* made by God and sharing in divine providence. Thomas's views on creatures as related in a hierarchical manner also clearly require some modification in the light of our current understanding of evolutionary theory. Such adjustment would not do disservice to his basic method, for he relied on science that was current at the time of writing. Hence it is not necessary to adopt Aquinas's views on nature as a hierarchical chain of being in order to draw out different dimensions of his discussion of virtues. One might even go further and suggest that he was remarkably positive about the possibility of emotion in animals, in a way that softened his hierarchical approach. The fact that he saw some humans as lower than 'the beasts' in terms of their behaviour also shows that his division between humans and animals was not as strong as might be supposed. Certainly, simple retention of a hierarchical chain of being is both unnecessary and unrealistic today. It is also important to note that even in Aquinas's scheme, justice, like wisdom, cannot flourish unless it is rooted and informed by the grace-laden virtue of charity, given by the Holy Spirit. Justice is more accurately described in Aquinas's thought as theocentric, rather than anthropocentric. This is important, for it opens up the possibility of alternative interpretations of justice that are more inclusive of the moral standing of non-human creatures (Deane-Drummond 2004a: 54–85). Debates about ecological justice, economic justice, and environmental justice, in terms of both principles and virtues are important, but outside the scope of this present chapter.[4]

Fortitude, for Aquinas, springs from prudence and justice, insofar as it presumes a correct evaluation of things. Fortitude is the ability to stand firm in the face of difficult circumstances, to be willing to suffer for the good. Fortitude is necessary in order to preserve the good that is perceived by prudence and established by justice. Fortitude is linked with the fourth cardinal virtue, temperance. Popular understanding of this word might imply some sense of restriction, or just not acting in excess. Classical understanding pointed to a real awareness and sense of the ordered unity of the human person. Out of this inner unity a serenity of spirit arises. For such a turning in towards self to be temperance, it must be *selfless*, rather than *selfish*.

---

[3] Andrew Linzey, e.g., is particularly vitriolic in his disdain for Aquinas's animal ethics in terms of human usefulness (see Linzey 1996: 29–33). I suggest that this dismissal of the usefulness of Aquinas's position is incorrect in its alignment with Descartes, and is unaware of the context of Aquinas's specific reason for rejecting vegetarianism, which Aquinas would have associated with Manichaeanism: i.e. a negative attitude towards the material world. In other words, humans are permitted to enjoy flesh foods on account of their goodness as material, created reality (Deane-Drummond 2004a: 67–77).

[4] For more discussion, see Deane-Drummond 2006a.

Intemperance is the misdirection of the desires for self-preservation in actions that are selfish and ultimately self-destructive. Temperance is also associated with humility, the ability to be assertive of the self in a way that is genuinely directed towards self-preservation and is a truthful estimation of one's worth, as opposed to pride, which is a false estimation of one's own worth. Temperance has relevance in consideration of the range of goods that are parodied as needs, rather than wants, in Western consumer society, ultimately leading to excessive waste, with associated environmentally destructive consequences. Pride and covetousness creep in when estimates of the benefits of new biotechnologies overreach their genuine possibility and value.

I resist the possible criticism that virtues are too anthropocentric to be helpful in developing an ecological ethic. Unless we look to ourselves and are critical of what we might become or have become, then ethics simply directed towards nature becomes a false dawn. It emerges, further, from a disjointed sense of who we are as persons, a projection of self into the cosmos as in deep ecology, without any sense of what that self might be like. The Word made Flesh excludes any Manichaean understanding of the body or creation as evil. The Incarnation is a fundamental affirmation of the value of the natural world, while Christology is a reminder also of humanity's tendency for sinfulness alongside creaturely suffering.[5] For Aquinas the essence of all things is that they arise from the creative activity of God; his approach is fundamentally theocentric, rather than anthropocentric. Wisdom is the virtue that can express the mediation between scientific reasoning and faith, between the discoveries of the creaturely world and the hoped for future of the created order as participation in God. While prudence is wisdom in human affairs, wisdom pure and simple surpasses this, as it aims directly at God, at knowledge of divine things (Deane-Drummond 2000). Wisdom, moreover, is rooted in charity, in alignment with the goodness and beauty of God.

Yet I would also suggest that in order to appreciate and love the natural world in appropriate ways we need to be tuned into the world as *created*, to appreciate its magnificent wisdom and beauty as mirrors of the wisdom and beauty of God. Taking wisdom and beauty as motifs need not be troublesome in the light of suffering, and some might say natural evil, for in theological terms wisdom includes the wisdom of the cross, and beauty includes the radical beauty of Christ, crucified and risen, showing a depth to beauty beyond the superficial value of mere attractiveness (Deane-Drummond 2006*b*). This appreciation of beauty and wisdom in the natural world is not so much a finding of natural values in ecology as a *basis* for ecological ethics, but rather a realization that the world is charged with the goodness of God, dimly reflected in spite of the mystery of human and non-human suffering. It leads for non-believers to the secular experience of wonder. For Christians, however, that wondering is caught up in a journey into wonder, one that is celebrated both

---

[5] For more discussion of the importance of Christology for an understanding of the relationships between God as Creator, wisdom, and humanity, see Deane-Drummond (2004*d*).

individually and in community (Deane-Drummond 2006c). This restful celebration of God's goodness is one that is in tune with the theological notion of the Sabbath. The Sabbath rest is not simply passive, but also active, since it points to transformation of the way in which the human community is situated in the context of the wider community of creation (Deane-Drummond 2004c). It is through living out of such a context that the gifts of wisdom, prudence, justice, fortitude, and temperance become established, received through the Holy Spirit in the context of Christian liturgical life and practice. Virtues, in other words, which can be learned through education and family life, become reinforced and transformed into gifts in the context of Christian community, but such community needs to be reminded continually of its dependence on both the grace of God and the wider biotic community in which it is situated.

FUTURE DIRECTIONS FOR THEOLOGY AND ECOLOGY

Given the above, what might be future directions for relating theology and ecology in so far as they bear on the theme of values in science and religion? The following are working proposals.

(a) Our knowledge of the natural world in all its complexity is always provisional. Hence what once might have been perceived as a 'redundant' species may, over time, prove to have more value for the total functioning of an ecosystem, perhaps at a global level, than originally thought. We cannot afford, therefore, to label species as 'redundant', just as geneticists first labelled non-coding DNA as 'junk' but subsequently found it to be critical in regulatory functions. The virtue of humility is relevant in this context, as well as wisdom, for both remind the human community of how little we know.

(b) Given the immensity of the tasks of species conservation and stabilization of climate change, as well as biotechnological challenges, it might be easy to give up or resist action. Yet prudence at an individual and political level is a reminder that negligence is not an option; rather, humans are called to act as responsibly as they can within the limitations of existing knowledge. How can the value of hope embedded in the Christian tradition find expression in such a way that it makes sense to those both outside and inside the Christian community? Questions here relate to how such natural hints of wonder and wisdom in ecology relate to the Christian tradition, beyond the discussion of natural and revealed theology.

(c) Both ecosystems and climate systems are always subject to disturbance by human activity. We inevitably disturb local and global ecology, since such disturbance has been an integral aspect of human civilization and living since the dawn of human history. The question that arises is, *What disturbances might be acceptable, or not?* Christian ethics cannot endorse all that might be possible, but rather needs to incorporate that aspect of prudence known as foresight in order to have a clearer idea of what is permissible in relation to the common good. In addition, the wisdom to perceive where resources might usefully be focused in research requires more than just expert knowledge, for some areas might be more fruitful in terms of either knowledge or application than others. While we have to give up on the notion that we can 'manage' the earth community successfully, we must not give up on the idea that our actions will have a significant impact on future generations (Muers 2004). Moreover, there is a requirement to develop those dimensions of justice that are attuned to the non-human world alongside and in companionship with the needs of future generations of persons.

(d) There is also a requirement for a review of the broader agenda for Christian theology in different Christian traditions and contexts in the light of ecology, including a review of Christian theological traditions of God's transcendence/immanence, anthropology, eschatology, the relationship between creation and redemption, and the relationship between Christology and Pneumatology (Conradie 2005*b*).

(e) Competing demands in the human community, all pressing for particular rights—including, for example, environmental justice for those in the poorer regions of the world subject to the effects of climate change—engage the human community at a local as well as a global level. The engagement with issues and problems cannot afford to remain parochial; it needs to be bold enough to reach out to those human communities unlike the ones in which we are situated in order to find some common ground and work for common solutions.

(f) While I have argued for a particular version of virtue ethics that makes sense in the context of a Christian community, there is need for dialogue with those committed to other religious traditions, as well as with those outside institutional religious practices. A religious instinct is still present in public debates in terms of implicit religious values, even if these are not always fully articulated. Further work is required in order to tease out more aspects of public responses and engage further with other religious traditions. Common values that lend themselves to such a dialogue include wisdom, prudence, justice, temperance, fortitude, beauty, wonder, charity, hope, and faith.

References and Suggested Reading

Aquinas, Thomas (1973). *Summa Theologiae*, xxxvi: *Prudence*, trans. T. Gilby. London: Blackfriars.

Barad, J. (1995). *Aquinas on the Nature and Treatment of Animals*. San Francisco: Scholars International Press.

Berry, R. J. (2003). *God's Book of Works*. London: T & T Clark/Continuum.

Boff, L. (1997). *Cry of the Earth, Cry of the Poor*. Maryknoll, N.Y.: Orbis.

BOUMER-PREDIGER, S. (2001). *For the Beauty of the Earth*. Grand Rapids, Mich.: Baker Academic.

CARSON, R. (1965). *A Sense of Wonder*. New York: Harper & Row.

CONRADIE, E. (2005a). *An Ecological Christian Anthropology*. Basingstoke: Ashgate.

——(2005b). 'Towards an Agenda for Ecological Theology: An Intercontinental Dialogue', *Ecotheology*, 10/2: 281–343.

DEANE-DRUMMOND, C. (2000). *Creation through Wisdom*. Edinburgh: T & T Clark.

——(2003). 'Wisdom, Justice and Environmental Decision-Making in a Biotechnological Age', *Ecotheology*, 8/2: 173–92.

——(2004a). *The Ethics of Nature*. Oxford: Blackwell.

——(2004b). 'Genetic Interventions in Nature: Perspectives from a Christian Ethic of Wisdom', in Denis Edwards and Mark Worthing (eds.), *Biodiversity and Ecology*, Adelaide: ATF Press, 30–44.

——(2004c). 'Living from the Sabbath: Developing an Ecological Theology in the Context of Biodiversity', in Denis Edwards and Mark Worthing (eds.), *Biodiversity and Ecology*, Adelaide: ATF Press, 1–13.

——(2004d). 'The Logos as Wisdom: A Starting Point for a Sophianic Theology of Creation', in Philip Clayton and Arthur Peacocke (eds.), *In Whom We Live and Move and Have Our Being*, Grand Rapids, Mich.: Eerdmans, 233–45.

——(2006a) 'Environmental Justice and the Economy: A Christian Theologian's View', *Ecotheology*, 11/3, in press.

——(2006b). 'Where Streams Meet? Ecology, Wisdom and Beauty in Bulgakov, von Balthasar and Aquinas', in Hubert Meisinger, Willem B. Drees, and Zbigniew Liana (eds.), *Wisdom or Knowledge?*, Issues in Science and Theology (IST), 4. London: Continuum, 108–26.

——(2006c). *Wonder and Wisdom: Conversations in Science, Spirituality and Theology*. London: Darton, Longman and Todd.

——and SZERSZYNSKI, B. (2003). *Re-Ordering Nature*. London: T & T Clark/Continuum.

EDWARDS, D. (2001) (ed.). *Earth Revealing: Earth Healing*. Collegeville, Minn.: Liturgical Press.

FRITTS, T. H., and RODDA, G. H. (1998). 'The Role of Introduced Species in the Degradation of Island Ecosystems: A Case History of Guam', *Annual Review of Ecology and Systematics*, 29: 113–40.

GALLENI, L., and SCALFARI, F. (2005). 'Teilhard de Chardin's Engagement with the Relationship between Science and Theology in Light of Discussions about Environmental Ethics', *Ecotheology*, 10/2: 196–214.

GASTON, K. J., and SPICER, J. I. (2004). *Biodiversity*, 2nd edn. Oxford: Blackwell.

GREY, M. (2003). *Sacred Longings*. London: SCM Press.

HABEL, N. (2000). *The Earth Bible*, i. Sheffield: Sheffield Academic Press.

HAUGHT, J. F. (1993). *The Promise of Nature*. Mahwah, N.J.: Paulist Press.

HOUGHTON, J. (2004). *Global Warming: The Complete Briefing*, 3rd edn. Cambridge: Cambridge University Press.

IPCC (2001). *Climate Change 2001: Synthesis Report of the Third Assessment of the Intergovernmental Panel on Climate Change*. Cambridge: Cambridge University Press.

LINZEY, A. (1996). 'Animal Rights', in P. Clarke and A. Linzey (eds.), *Dictionary of Ethics, Theology and Society*, London: Routledge, 29–33.

LOREAU, M., NAEEM, S., INCHAUSTI, P., BENGTSSON, J., GRIME, J., HECTOR, A., HOOPER, D. U., HUSTON, M. A., RAFFAELLI, D., SCHMID, B., TILMAN, D., and WARD, D. A. (2001). 'Biodiversity and Ecosystems Functioning: Current Knowledge and Future Challenges', *Science*, 294: 804–8.

LOVELOCK, J. (1987). *Gaia: A New Look at Life on Earth*, 2nd edn. Oxford: Oxford University Press.

McFAGUE, S. (1993). *The Body of God*. Minneapolis: Fortress Press.

MERCHANT, C. (1982). *The Death of Nature*. London: Wildwood House.

MIDGLEY, M. (2001). *Gaia*. London: Demos.

MUERS, R. (2004). 'Pushing the Limit: Theology and the Responsibility to Future Generations', *Studies in Christian Ethics*, 16/2: 36–51.

MURRAY, R. (1992). *The Cosmic Covenant*. London: Sheed & Ward.

NORTHCOTT, M. (1996). *Christianity and Environmental Ethics*. Cambridge: Cambridge University Press.

PAGE, R. (1996). *God and the Web of Creation*. London: SCM Press.

PICKETT, S. T. A., KOLASA, J., and JONES, C. G. (1994). *Ecological Understanding*. London and Boston: Academic Press.

PIMENTEL, D. (2001). 'Agricultural Invasions', in S. A. Levin (ed.), *Encyclopedia of Biodiversity*, i, San Diego: Academic Press, 71–85.

PORTER, J. (2005). *Nature as Reason: A Thomistic Theory of Natural Law*. Grand Rapids, Mich.: Eerdmans.

PRIMAVESI, A. (2003). *Gaia's Gift*. London: Routledge.

RAHNER, K. (1954). *Theological Investigations*, i: *God, Christ, Mary and Grace*, trans. Cornelius Ernst. London: Darton, Longman and Todd.

RUETHER, R. (2000). 'Conclusion: Ecojustice at the Center of the Church's Mission', in D. T. Hessel and R. R. Ruether (eds.), *Christianity and Ecology*, Cambridge, Mass.: Harvard University Press, 603–14.

SCHLOSS, J. (2000). 'Wisdom Traditions as Mechanisms for Organismal Integration', in W. S. Brown (ed.), *Understanding Wisdom*, Philadelphia: Templeton Foundation Press, 153–91.

SCOTT, P. (2003). *A Political Theology of Nature*. Cambridge: Cambridge University Press.

STENMARK, M. (2002). *Environmental Ethics and Policy Making*. Basingstoke: Ashgate.

THEOKRITOFF, E. (2005). 'Creation and Priesthood in Modern Orthodox Thinking', *Ecotheology*, 10/3: 344–63.

VAN WENSVEEN, L. (2000). *Dirty Virtues*. New York: Humanity Books.

WHITE, L. (1967). 'The Historical Roots of Our Ecologic Crisis', *Science*, 155: 1203–7.

WILSON, E. O. (1984). *Biophilia*. Cambridge, Mass.: Harvard University Press.

—— (1996). *In Search of Nature*. Washington: Island Press/Shearwater Books.

WIRZBA, N. (2003). *The Paradise of God*. New York: Oxford University Press.

WORSTER, D. (1977). *Nature's Economy*. New York: Cambridge University Press.

CHAPTER 53

ENVIRONMENTAL ETHICS AND RELIGION/ SCIENCE

HOLMES ROLSTON III

What to make of who we are, where we are, what we ought to do? These perennial questions are familiar enough; what is recently extraordinary is how the science–religion dialogue reframes these old questions with an *on Earth* dimension. What to make of Earth, the home planet? Earth is proving to be a remarkable planet, and humans have deep roots in and entwined destinies with this wonderland Earth. Simultaneously, however, humans are remarkable on this remarkable planet, a wonder on wonderland Earth. But the foreboding challenge is that these spectacular humans, the sole moral agents on Earth, now jeopardize both themselves and their planet. Science and religion are equally needed, and strained, to bring salvation (to use a religious term), to keep life on Earth sustainable (to use a more secular, scientific term).

WHERE ON EARTH ARE WE?

Where are we located? Where are we on Earth? What's the 'set-up'? Environmentalists claim that an organism's surroundings, from the skin out, are as telling as organismic identity, from the skin in. They insist on knowing locations, habitats, niches. This locating of ourselves has been escalating from ecosystems to bioregions and continents, and lately, increasingly requires a global view. At cosmological scales there is deep space and time; at evolutionary scales Earth is a marvellous planet, a wonderland lost in this deep spacetime. Alas too, at anthropocentric and economic scales, this Earth is in deep jeopardy.

Astronaut Edgar Mitchell, a rocket scientist, reports being earthstruck:

Suddenly from behind the rim of the room, in long, slow-motion moments of immense majesty, there emerges a sparkling blue and white jewel, a light, delicate sky-blue sphere laced with slowly swirling veils of white, rising gradually like a small pearl in a thick sea of black mystery. It takes more than a moment to fully realise this is Earth-home. (Quoted in Kelley 1988: at photograph 42)

That indicates what many scientists think in their more philosophical moments: we are located on a sparkling planetary jewel. Almost everyone on Earth has been moved by those photographs from space. The mystery, however, is not only the surrounding black space in which we are located but how there comes to be this spectacular home planet. These are limit questions that reach toward religious answers: the meaning and significance of life on Earth.

Science has found some surprising things about that surrounding black space; this too is part of the set-up. Astrophysics and nuclear physics, combining quantum mechanics and relativity theory, have described a universe 'fine-tuned' for life, originating some 15 billion years ago in a 'big bang'. Startling interrelationships are required for these creative processes to work; these are gathered under the concept of the 'Anthropic Principle', which might better have been termed a 'biogenic principle' (Leslie 1990; Barrow and Tipler 1986; Barr 2003).

The native-range Earth world stands about midway between the infinitesimal and the immense on the natural scale. The mass of a human being is the geometric mean of the mass of Earth and the mass of a proton. A person contains about 10^{28} atoms, more atoms than there are stars in the universe. In astronomical nature and micro-nature, at both ends of the spectrum of size, nature lacks the complexity that it demonstrates at the meso-levels, found at our native ranges on Earth. On Earth, the surprises compound.

Earth is a kind of providing ground for life, a planet with promise. Located at a felicitous distance from the Sun, Earth has liquid water, atmosphere, a suitable mix of elements, compounds, minerals, and an ample supply of energy. Geological forces generate and regenerate landscapes and seas—mountains, canyons, rivers, plains, islands, volcanoes, estuaries, continental shelves. The Earth-system is a kind of cooking pot sufficient to make life possible. Spontaneously, natural history organizes itself.

The system proves to be pro-life; the story goes from zero to 5 million species (more or less) in 5 billion years, passing through 5 billion species (more or less) which have come and gone en route, impressively adding diversity and complexity to simpler forms of life. Prokaryotes dominated the living world more than 3 billion years ago; there later appeared eukaryotes, with their well-organized nucleus and cytoplasmic organelles. Single-celled eukaryotes evolved into multi-celled plants and animals with highly specialized organ systems. First there were cold-blooded animals at the mercy of climate, later warm-blooded animals with more energetic metabolisms. From small brains emerge large central nervous systems. Instinct evolves into acquired learning. Parental care develops, animal societies, and at length humans with their capacities for language and culture. Palaeontology reports this in increasing detail, and taking the broad view, prima facie the most plausible account seems to find some programmatic evolution toward value.

Does the Earth set-up make life probable, even inevitable? Biologists spread themselves across a spectrum thinking that natural history is random, contingent, caused, unlikely, likely, determined, open. Many hold that we need to put some kind of an arrow on evolutionary time. Simon Conway Morris is recently the most vigorous palaeontologist arguing that human life has appeared only on Earth but did so here as a law of the universe: We are 'inevitable humans in a lonely universe'. 'The science of evolution does not belittle us. . . . Something like ourselves is an evolutionary inevitability' (Conway Morris 2003: p. xv). Christian de Duve, Nobel laureate, argues that 'Life was bound to arise under the prevailing conditions. . . . I view this universe [as] . . . made in such a way as to generate life and mind, bound to give birth to thinking beings' (de Duve 1995: pp. xv, xviii).

On opposing accounts, the history of life is a random walk with much struggle and chance (famously in Monod 1972; repeatedly in Gould 1989). Evolutionary history can seem tinkering and make-shift. Natural selection is thought to be blind, both in the genetic variations bubbling up without regard to the needs of the organism, some few of which by chance are beneficial, and also in the evolutionary selective forces, which select for survival, without regard to advance. Many evolutionary theorists doubt that the Darwinian theory predicts the long-term historical innovations that have in fact occurred (Maynard Smith and Szathmáry 1995: 3).

Whether evolutionary theory explains this or not, those who want a comprehensive view insist that it would be a quite anomalous result if there had appeared novel kinds steadily over many millennia but only by drifting into them. The natural history suggests a creative genesis of life. Earth history is the story of how significant values are generated and endure through a context of suffering, stress, perpetual perishing, and regeneration. At this point we reach the equally debated 'value' question. Are we humans located on this Earth in value isolation from the nature out of which we have come? Or is our location, with our evidently unique capacities for valuing, part of a more comprehensive community of value?

Western science accompanied the Enlightenment; and, with its legacies in Cartesian mind–body dualism, the prevailing account found that nature is value-free. That seemed plausible looking at Sun, Moon, and stars, or rock strata, or atoms—in the physical sciences. Biologists, confronted by life in its relentless vitality, were never so easily persuaded of this. But life did increasingly seem to be a matter of biochemistry. Value, on the Enlightenment account, appeared only in psychology, with the experience of felt interests, and this flowered in human life. With the coming of ecological science, prodding by a revisiting of the issue in evolutionary science, and with the coming of cybernetic interpretations of genetics, the value issue has been thrown into new light. We were not so 'enlightened' as we supposed.

'Value' is a frequently encountered term in evolutionary biology and in ecosystem science—and this despite the 'value-free' science—Enlightenment humanism. 'An ability to ascribe value to events in the world, a product of evolutionary selective processes, is evident across phylogeny. Value in this sense refers to an organism's facility to sense whether events in its environment are more or less desirable' (Dolan 2002: 1191). Adaptive value, survival value, is the basic matrix of Darwinian theory. An organism is the loci of values defended; life is otherwise unthinkable. Such organismic values are individually defended; but, ecologists insist, organisms occupy niches and are networked into biotic communities. At this point ethicists, looking over the shoulders of biologists describing this display of biodiversity, millions of species defending their kind over millennia, begin to wonder whether there may be goods (values) in nature which humans ought to consider. Animals, plants, and species, integrated into ecosystems, may embody values that, though non-moral, count morally when moral agents encounter these.

Here Judaeo-Christian monotheists will invoke the Genesis accounts of a good creation—and may begin to wonder how so-called enlightened biology ever got into the mode of a value-free nature. They cheer rather for the 'conservation biologists', delighted that within academic biology the growth of groups such as the Society for Conservation Biology has been spectacular, and that conservation biology is regularly featured in such publications as *Science* and *BioScience*. Biblical writers already had an intense sense of the worth (value) of creation. Nature is a wonderland. 'Praise the Lord from the earth, you sea monsters and all deeps, fire and hail, snow and frost, stormy wind fulfilling his command! Mountains and all hills, fruit trees and all cedars! Beasts and all cattle, creeping things and flying birds!' (Psalm 148: 7–10; RSV). 'The hills gird themselves with joy, the meadows clothe themselves with flocks, the valleys deck themselves with grain, they shout and sing for joy' (Psalm 65: 12–13).

Encountering the vitality in their landscapes, the Hebrews formed a vision of creation, cast in a Genesis parable: the brooding Spirit of God animates the Earth, and Earth gives birth. 'The earth was without form and void, and darkness was upon the face of the deep; and the Spirit of God was moving over the face of the waters. And God said, "Let there be..."' (Genesis 1: 2–3). 'Let the earth bring forth living things according to their kinds' (Genesis 1: 11, 24). 'Let the waters bring forth swarms of living creatures' (Genesis 1: 20). Earth speciates. God, say the Hebrews, reviews this display of life, finds it 'very good', and bids it continue. 'Be fruitful and multiply and fill the waters in the seas, and let birds multiply on the earth' (Genesis 1: 22). In current scientific vocabulary, there is dispersal, conservation by survival over generations, and niche saturation up to carrying capacity.

Anciently, the Hebrews marvelled over the 'swarms' (= biodiversity) of creatures Earth brings forth in Genesis 1. These were brought before man to name them (a taxonomy project!). Classically, theologians spoke of 'plenitude of being'. Contemporary biologists concur that Earth speciates with marvellous fecundity; the systematists have named and catalogued a far vaster genesis of life than any available to ancient or medieval minds. Contemporary biologists, almost without exception, urge the conservation of this richness of biodiversity. Learn it from a conservation biology textbook, or learn it from the Bible, science and religion have a common and urgent agenda.

Value in nature is recognized again when the fauna are included within the Hebrew covenant. 'Behold I establish my covenant with you and your descendants after you, and with every living creature that is with you, the birds, the cattle, and every beast of the earth with you' (Genesis 9: 5). In modern terms, the covenant was both ecumenical and ecological. It was 'theocentric', theologians might insist; but if so, it was also less 'anthropocentric' and more 'biocentric' than traditional Jews or Christians realized. Noah with his ark was the first Endangered Species Project. The science is rather archaic, but the environmental policy ('Keep them alive with you' (Genesis 6: 19)) is something the US Congress reached only with the Endangered Species Act in 1973.

Biologists find biological creativity indisputable, whether or not there is a Creator. Biologists have no wish to talk theologians out of genesis. Whatever one makes of God, biological creativity is indisputable. There is creation, whether or not there is a Creator, just as there is law, whether or not there is a Lawgiver. Biologists are not inclined, nor should they be as biologists, to look for explanations in supernature, but biologists nevertheless find a nature that is super! Superb! Biologists may be taught to eliminate from nature any suggestions of teleology, but no biologist can doubt genesis.

Somewhat ironically, just when humans, with their increasing industry and technology, seemed further and further from nature, having more knowledge about natural processes and more power to manage them, just when humans were more and more rebuilding their environments, thinking perhaps to escape nature, the natural world has emerged as a focus of concern. Nature remains the milieu of culture—so both science and religion have discovered. In a currently popular vocabulary, humans need to get themselves 'naturalized'. Using another metaphor, nature is the 'womb' of culture, but a womb that humans never entirely leave. Almost like God—to adopt classical theological language—nature is 'in, with, and under us'.

WHO ON EARTH ARE WE?

We next encounter the question of the human place on Earth. Humans are part of, yet apart from, nature; we evolved out of nature, yet we did evolve *out of* it. Yes, transcend wild spontaneous nature though we may, we still require an earthen life support system; but humans are something special on Earth. With us the black mystery compounds again. Humans are a quite late and minor part of the world in evolutionary and ecological senses. They are one more primate among hundreds, one more vertebrate among tens of thousands, one more species among many millions.

But there is also a way in which this 'last' comes to be 'first'. Humans can seem minuscule at astronomical levels; they can seem ephemeral on evolutionary scales. Humans do not live at the range of the infinitely small, or at that of the infinitely large, but humans on Earth do seem, at ecological ranges, to live at the range of the infinitely complex, evidenced both in the biodiversity made possible by genetics and in the cultural history made possible by the human mind. If we ask where are the 'deep' thoughts about this 'deep' nature, they are right here. *Homo sapiens* is the first and only part of the world free to orient itself with a view of the whole, to seek wisdom about who we are, where we are, where we are going, what we ought to do.

We humans are the most sophisticated of known natural products. In our 150 pounds of protoplasm, in our 3-pound brain is more operational organization than in the whole of the Andromeda galaxy. On a cosmic scale, humans are minuscule atoms. Yet the brain is so curiously a microcosm of this macrocosm. Not only evolutionary biologists, but also astrophysicists, are studying their own origins (since our elements were made in the stars). They are an end of what they are watching the beginnings of, one of the consequences of the stellar chemistry, which now can reflect over this world. We humans too are 'stars' in the show. In that sense, the most significant thing in the known universe is still immediately behind the eyes of the astronomer!

Animal brains are already impressive. In a cubic millimetre (about a pinhead) of mouse cortex there are 450 metres of dendrites and 1–2 kilometres of axons; each neuron can synapse on thousands of others. But this cognitive development has reached a striking expression point in the hominid lines leading to *Homo sapiens*, going from about 300 to 1,400 cubic centimetres of cranial capacity in a few million years. The human brain has a cortex 3,000 times larger than that of the mouse. Our protein molecules are 97 per cent identical to those in chimpanzees, only 3 per cent different. But we have three times their cranial cortex. The connecting fibres in a human brain, extended, would wrap around the Earth forty times.

The human brain is of such complexity that descriptive numbers are astronomical and difficult to fathom. A typical estimate is 10^{13} neurons, each with several thousand synapses (possibly tens of thousands). Each neuron can 'talk' to many others. This network, formed and re-formed, makes possible virtually endless mental activity. The result of such combinatorial explosion is that the human brain is capable of forming more possible thoughts than there are atoms in the universe.

Some trans-genetic threshold seems to have been crossed. Humans have made an exodus from determination by genetics and natural selection and passed into a mental and social realm with new freedoms. Richard Lewontin, Harvard biologist, puts it this way:

Our DNA is a powerful influence on our anatomies and physiologies. In particular, it makes possible the complex brain that characterizes human beings. But having made that brain possible, the genes have made possible human nature, a social nature whose limitations and possible shapes we do not know except insofar as we know what human consciousness has already made possible. . . . History far transcends any narrow limitations that are claimed for either the power of the genes or the power of the environment to circumscribe us. . . . The genes, in making possible the development of human consciousness, have surrendered their power both to determine the individual and its environment. They have been replaced by an entirely new level of causation, that of social interaction with its own laws and its own nature. (Lewontin 1991: 123)

The genes outdo themselves. Mind of the human kind seems to require incredible opening up of new possibility space.

J. Craig Venter and more than 200 co-authors, reporting on the completion of the Celera Genomics version of the Human Genome Project, caution in their concluding paragraph:

In organisms with complex nervous systems, neither gene number, neuron number, nor number of cell types correlates in any meaningful manner with even simplistic measures of structural or behavioural complexity.... Between humans and chimpanzees, the gene number, gene structural function, chromosomal and genomic organization, and cell types and neuroanatomies are almost indistinguishable, yet the development modifications that predisposed human lineages to cortical expansion and development of the larynx, giving rise to language culminated in a massive singularity that by even the simplest of criteria made humans more complex in a behavioural sense.... The real challenge of human biology, beyond the task of finding out how genes orchestrate the construction and maintenance of the miraculous mechanism of our bodies, will lie ahead as we seek to explain how our minds have come to organize thoughts sufficiently well to investigate our own existence. (Venter *et al.* 2001: 1347–8)

The surprise is that human intelligence becomes reflectively self-conscious and builds cumulative transmissible cultures. An information explosion gets pinpointed in humans. Humans alone have 'a theory of mind'; they know that there are ideas in other minds, making linguistic cultures possible. Our ideas and our practices configure and re-configure our own sponsoring brain structures. In the vocabulary of neuroscience, we have 'mutable maps'. For example, with the decision to play a violin well, and resolute practice, string musicians alter the structural configuration of their brains to facilitate fingering the strings with one arm and drawing the bow with the other (Elbert *et al.*, 1995). With the decision to become a taxi driver in London, and long experience driving about the city, drivers likewise alter their brain structures, devoting more space to navigation-related skills than non-taxi drivers have (Maguire *et al.* 2000: 4398). Similarly with other decisions to learn. The human brain is as open as it is wired up. Our minds shape our brains.

In the most organized structure in the universe, molecules, trillions of them, spin round in and generate the unified, centrally focused experience of mind. These events have 'insides' to them, subjective experience. There is 'somebody there', already in the higher animals, but this becomes especially 'spirited' in human persons. The peculiar genius of humans is that, superposed on biology, we become, so to speak, 'free spirits', not free from the worlds of either nature or culture, but free *in* those environments.

That humans are embodied spirits, capable of thinking about themselves and what they can and ought to do, is really beyond dispute. The act of disputing it, verifies it. The self-actualizing characteristic of all living organisms doubles back on itself in this reflexive animal with the qualitative emergence of what the Germans call *Geist*, what existentialists call *Existenz*, what philosophers and theologians often call 'Spirit'. An object, the brained body, becomes a spirited subject.

There is a 'massive singularity' in humans (Venter *et al.* 2001). This cybernetic, cognitive emergence does not 'reduce' well; rather it tends to 'expand'. The past is not a good guide to what the future holds when there is this massive singularity. That brings, again, paradox and dialectic. Are we part of nature, or apart from nature? Yes and no. Nature hardly seems up to the guidance of the child she has delivered.

For some, that is cause for freedom and relief. Humans are self-defining animals. They do not need to consult nature, but are intellectually and morally free to do their own thing. But it also seems fitting that humans be defined in their place. Otherwise, we cancel all promise of showing a systematic unity between human life and cosmic or earthen nature. It is one thing to be set free in the world, another to be set adrift in it. So that—if you like—has now become the main agenda: the place of this spirit waked up in nature. What does human uniqueness imply for human responsibility? Once again, science and religion are equally challenged, and stressed, to answer.

WHAT ON EARTH ARE WE DOING?

That brings us to ask what we are now doing? We start with three graphs as icons of the contemporary scene, indicating population, consumption, and distribution.

People are a good thing, people with energy at their service are fortunate, people need goods and services for an abundant life. On this both scientists and theologians agree; scientists have celebrated how applied science has given us better things for better living; Christians and other believers have shown great social concern for taking care of people, their physical as well as their spiritual needs. But when we join the first two graphs with the third, troubles loom. What we have been doing is rapidly escalating the human population, rapidly escalating energy consumption, typical of consumerism generally, and the distribution is quite disproportionate.

Human numbers are escalating around the world, much more so in lesser developed nations than in developed ones. About 20 per cent of the world's population in the developed nations (the Group of 7 (now 8), the big nations of North America, Europe, and Japan, 'the North') produces and consumes about 80 per cent of the world's goods and services. Conversely, 80 per cent of the people in the world produce and consume about 20 per cent of these goods and services (the G77 nations, once 77 but now including some 128 lesser developed nations, often south of the industrial north). Capitalism has become the dominant global economic system; coupled with science and technology it makes possible a growth in consumerism. But on its present course it is making the rich richer proportionately to any trickle-down benefits to the poor, evidenced in the 20–80 differential. For every dollar of economic

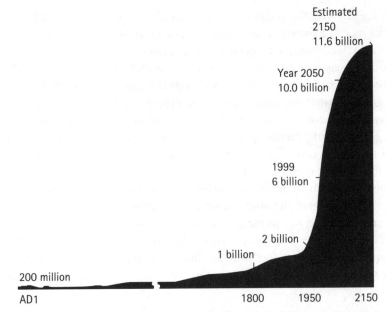

Fig. 53.1. World population growth data from (2000) *Statistical Abstract of the United States*, 120th edn. (2000).

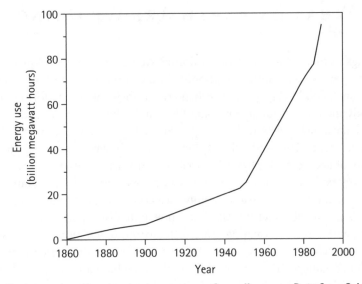

Fig. 53.2. Energy consumption: inanimate energy use from all sources. Data from Cohen (1995) and World Resources Institute (1994).

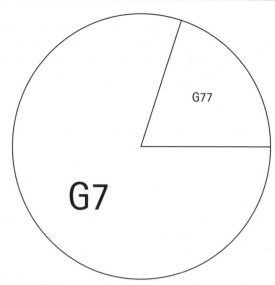

Fig. 53.3. G7 versus G77 nations. The pie chart summarizes population and production data in World Bank (2001). The G7 group has now become the G8 (United States, United Kingdom, France, Germany, Canada, Russia, Japan, and Italy).

growth per person in the South, twenty dollars accrue in the North. Free trade moves capital and goods across national boundaries in international markets, but the labour also required for production is confined within nations, which means that capital can relocate production seeking the cheapest labour.

The shadow side of this is a degrading environment in both developed and developing nations. Since the coming of science with its technology, since the invention of motors and gears in the mid-nineteenth century, giving humans orders of magnitude more power to transform the landscape, since the coming of modern medicine, there have been unprecedented changes in world population, in agricultural production, in industrial production, in transportation and communication, in economic systems, in military commitments—all these literally altering the face of the planet.

Humans now control 40 per cent of the planet's land-based primary net productivity: that is, the basic plant growth which captures the energy on which everything else depends (Vitousek *et al.* 1986). If the human population doubles, the capture will rise to 60–80 per cent. In another survey, researchers found the proportions of Earth's terrestrial surface altered as follows: (1) little disturbed by humans, 51.9 per cent; (2) partially disturbed, 24.2 per cent; (3) human-dominated, 23.9 per cent. Factoring out ice, rock, and barren land, which support little human or other life, the percentages become (1) little disturbed, 27.0 per cent; (2) partially disturbed, 36.7 per cent; (3) human-dominated, 36.3 per cent (Hannah *et al.* 1994). Most terrestrial nature is dominated or partially disturbed (73.0 per cent). There is the sea; Earth might as well have been called Aqua, since oceans cover 70 per cent of the planet. But the sea too is increasingly affected. Over 90 per cent of the world's fisheries have been depleted; coastal development and pollution have caused sharp declines in ocean health. Increasingly less habitat remains for forms of life that cannot be accommodated in the nooks and crannies of a human-dominated world. This is producing an escalating extinction rate, comparable to that of the great catastrophic extinctions of the palaeontological past—with the difference that after natural extinctions there is re-speciation, whereas human extinctions shut down the speciating processes.

Humans do not use the lands they have domesticated effectively. A World Bank study found that 35 per cent of the Earth's land has now become degraded (Goodland 1992). In developed nations there has been much progress cleaning up air and water, but still, in the United States almost half the population lives in areas that do not meet national air-quality standards. Global warming threatens to disrupt not only fragile semi-arid areas but equally long-established agricultural patterns.

A central problem is that many environmental problems result from the incremental aggregation of actions that are individually beneficial. Coupled with a long lag time for environmental problems to become manifest, this masks the problem in both nature and human nature. A person may be doing what, taken individually, would be a perfectly good thing, a thing he or she has a right to do were he alone, but which, taken in collection with thousands of others doing the same thing, becomes a harmful thing. A good thing escalates into a bad thing. This is Garrett Hardin's tragedy of the commons (Hardin 1968). Pursuit of individual advantage destroys the commons.

Biologists may continue here with a more troubling concern. Theologians have classically found in humans a tendency to self-interest, to selfishness, to sin, and now the biologists concur. Indeed, the biologists may claim that humans are innately 'selfish' by Darwinian natural selection. The nature inherited in human nature is self-interested, and this, in an environmental crisis, may prove self-defeating. Theologians and biologists alike find too much in human nature that is irrational, blind. Although the conservation biologists celebrate Earth's biodiversity, the sociobiologists (and, later, evolutionary psychologists) worry that the human disposition to survive, a legacy of our evolutionary heritage, has left humans too locally short-sighted to deal with the environmental crisis at the global level.

Humans are not genetically or psychologically equipped to deal with collective issues that upset individual goals (Ehrlich and Ehrlich 2004). Biologists hold that we are naturally selected to look out for ourselves and our families, perhaps also to co-operate in tribes or for reciprocal benefits. Beyond that, humans are not biologically capable of more comprehensive vision, of considering the interests of others in foreign nations or in future generations, if this is at expense to our own interests (Sober and Wilson 1998; D. S. Wilson 2002). Humans are not rational in any 'absolute' or even 'global' sense; they bend their reason to serve their interests, competitively against others—other nations, tribes, or neighbours—when push comes to shove. Hence the escalating violence and terrorism in today's world, the perpetrators often as not claiming their cause in the name of some faith. Humans inherit Pleistocene urges, such as an insatiable taste for sugar, salt, and fats, traits once adaptive, but which today make obesity a leading health problem. Our global environmental problems result from such insatiable consumptive Pleistocene urges. Too often the religions remain tribal; God is for me, for my kind, my nation; love your neighbour and hate your enemies. Can enlightened science or enlightened religion get us past this legacy? Were we not just celebrating the genius of the human mind, transcending genetics, making exoduses out of previous determinisms, opening up promising new possibility spaces? Even 'Enlightenment science' and 'Enlightenment religion', with all their focus on the (Western) human powers and achievements, may now need transcending, leaving behind a debilitating anthropocentric humanism, embracing a more inclusive vision of the goodness of the whole community of life on Earth.

Science and religion are equally challenged by this environmental crisis, each to re-evaluate the natural world, and each to re-evaluate its dialogue with the other. Both are thrown into researching fundamental theory and practice in the face of an upheaval unprecedented in human history, indeed in planetary history. Life on Earth is in jeopardy owing to the behaviour of one species, the only species that is either scientific or religious, the only species claiming privilege as the 'wise species', *Homo sapiens*. Facing the next century, Earth, the planet of promise, is a planet in peril. Science and religion will both be required for our salvation. We will need to mix science and conscience.

What on Earth Ought We to Do?

Science and Conscience

Scientists and ethicists alike have traditionally divided their disciplines into the realm of the *is* and the realm of the *ought*, continuing the Cartesian tradition, later further elaborated by David Hume and G. E. Moore. By this division, no study of nature can tell humans what ought to happen, on pain of committing the naturalistic fallacy. Ecologists who claim to know what we ought to do, sociobiologists who claim that humans can only be selfish or tribal, or theologians who claim to base ethics on ecology may be violating the long-established taboo against mixing facts and values. Recently, this neat division has been challenged by ecologists and conservation biologists and their philosophical interpreters.

Still, there is ambiguity: ecology reframes ways that we think about nature, but leaves deeper questions unanswered. Biologists may describe the nature out of which we have evolved; they may move us to regard such nature with care and concern; they may caution us about legacies in our human nature. But have we, with our complex minds generating culture, not emerged into new ethical possibilities? What seem always to remain after science are the deeper value questions. After four centuries of Enlightenment and Western science, and with due admiration for impressive successes, the value questions in today's world are as urgent, sharp, and painful as ever. There is no scientific guidance of life.

Science could be part of the problem, not part of the solution. Science can, and often does, serve noble interests. But science can, and often does, become self-serving, a means of perpetuating injustice, of violating human rights, of making war, of degrading the environment. Science is used for Western dominion over nature. Science is equally used for Western domination of other nations. The values surrounding the pursuit of science, as well as those that govern the uses to which science is put, are not generated out of the science, not even ecological science, much less the rest of science.

Where science seeks to control, dominate, manipulate either persons or nature, or both, it blinds quite as much as it guides. Nothing in science ensures against philosophical confusions, against rationalizing, against mistaking evil for good, against loving the wrong gods. The whole scientific enterprise of the last four centuries could yet prove demonic, a Faustian bargain; and as good an indication as any of that is our ecological crisis (Rasmussen 1996).

Ecology as a science has to join with human ecology, where the religious dimension is more evident. Perhaps we ought not to focus on the ecological *science* that biblical writers might have known, but rather on the *human* ecology into which they had insight. Emphasize the *human*, not the *ecology*, side of the relationship. We need to regain their insights into human nature more than into nature. True, one cannot know the right way for humans to behave if one is ignorant of how human behaviours result in this or that causal outcome in the natural systems about which one is concerned—for example, whether letting the land lie fallow one year in seven is adequate to restore its productivity. But if humans by nature are prone to exploit, the rich gaining power over the poor, then does society need, after seven times seven years, to declare a sabbatical, resetting land-holding patterns more equitably? Or to find other ways to ensure fair access to resources?

Religions are about that gap between *is* and *ought*, and how to close that gap. This often requires revealing how human nature functions and dysfunctions, and how to re-form, or redeem, this 'fallen' nature. Whatever biology discovers about our nature, in religion God is redeeming humans. Humans must repair their broken wills, discipline innate self-interest, and curb corrupt social forces. What it means to be blessed, what it means to be wicked: these are theological questions. One is not going to get much help here from ecology or from elsewhere in biology, any more than from astrophysics or soil chemistry.

However much ecology reframes nature for our re-evaluation, the deeper evaluative questions are still left open. In that sense, science cannot teach us what we most need to know about nature: that is, how to value it. Ecologists may be able to tell us what our options are, what will work and what will not, what is the minimum baseline health of landscapes. But ecologists have no special competence in evaluating what rebuilding of nature a culture desires, and how far the integrity of wild nature should be sacrificed to achieve this. A people on a landscape will have to make value judgements about how much original nature they have, or want, or wish to restore, and how much culturally modified nature they want, and whether it should be culturally modified with more or less natural patterns remaining. Ought we to give priority to 'sustainable development', as the World Commission on Environment and Development (1987) recommends? Or to a 'sustainable biosphere', as the Ecological Society of America (Risser *et al.* 1991) recommends? There is nothing in ecology *per se* that gives ecologists any authority or skills to make these further decisions. Although ecological science cum conservation biology seems to couple the concerns of biology and religion congenially, we still have to be cautious and worry about that naturalistic fallacy.

This mix of science and conscience requires caring for people and caring for nature, and a fundamental tension in environmental ethics is whether and how far our ethics is human-centred, anthropocentric, and how far it is biocentric, respecting the comprehensive community of life on Earth. Maybe, to put it provocatively, religion cum science will move us to care for people, the science cum religion will move us to care for nature. Religion is for and about people caring for people; in environmental concerns such caring needs to be well-informed scientifically. Science, at least natural science, is about nature, describing how nature works; science has been doing this in ways that reveal a wonderland Earth. That prompts us to wonder whether caring for such nature is a religious concern. But this, again, proves a half-truth, mistaken if taken for the whole.

Caring for People

How *nature* works is the province of the physical and biological sciences. How *human nature* works is the province of religion, perhaps also of human sciences such as psychology and sociology. How human nature ought to work, how it can be reformed to work as it ought, is the province of ethics and religion. If we emphasize the *human*, not the *ecology*, side of the relationship, we recognize that religion has a vital role to play. We need human ecology, humane ecology. Religious ethicists can with considerable plausibility make the claim that neither technological development, nor conservation, nor a sustainable biosphere, nor sustainable development, nor any other harmony between humans and nature can be gained until persons learn to use the Earth both justly and charitably. Those twin concepts are not found either in wild nature or in any science that studies nature. They must be grounded in some ethical authority, and this has classically been religious.

The Hebrews were given a blessing with a mandate. The land flows with milk and honey (assuming good land husbandry) if and only if there is obedience to Torah. Abraham said to Lot, 'Let there be no strife between me and you, and between your herdsmen and my herdsmen' (Genesis 13: 8), and they partitioned the common good equitably among themselves. The righteous life depicted in the Hebrew Bible is about a long life on Earth, sustainable until the third and fourth generations. Whatever it has to say about heaven, or life after death, the Bible is also about keeping this earthly life divine, godly, or at least human, humane, or 'righteous' and 'loving'.

Any people who cope on a landscape for centuries will have some store of ecological wisdom, but that is not what we really turn to classical religious faiths to learn. We turn to religions to deal with the disvalues in humans—their irrationality, their greed, their short-sightedness, in short, their sinfulness. Religions save, they regenerate; they hold forth an *ought* to guide what *is*. Humans sin, unlike the fauna and flora. Religion is for people, and not for nature; nor does salvation come naturally—even the earthly good life is elusive.

Ultimately such salvation is beyond the natural; perhaps it is supernatural by the grace of the monotheist God, perhaps in some realization of depths underlying the natural, such as Brahman or *sunyata*. Meanwhile, whatever the noumenal ultimate, humans reside in a phenomenal world, which they must evaluate, and in which they must live, hopefully redeemed or enlightened by their faiths.

Much concern has come to be focused on environmental justice; the way people treat each other is related to the way they treat nature. If humans have a tendency by nature to exploit, they will as soon exploit other people as nature (revealed by the 80 per cent–20 per cent chart and the environmental degradation sketched above). These are the underlying theological and ethical issues underlying global capitalism, consumerism, and nationalism. The combination of escalating populations, escalating consumption, global capitalism, struggles for power between and within nations, militarism, results in environmental degradation that seriously threatens the welfare of the poor today and will increasingly threaten the rich in the future. The four critical items on our human agenda are population, development, peace, and the environment. All are global; all are local; all are intertwined; in none have we modern humans anywhere yet achieved a sustainable relationship with our Earth. Our human capacities to alter and reshape our planet are already more profound than our capacities to recognize the consequences of our activity and deal with it collectively and internationally.

Caring for Nature

What we want is not just 'riches', but a 'rich life', and appropriate respect for the biodiversity on Earth enriches human life. Humans belong on the planet; they will increasingly dominate the planet. But we humans, dominant though we are, want to be a part of something bigger. Environmental justice needs to be eco-justice, as with the World Council of Churches' emphasis on 'justice, peace, and the integrity of creation'. Contemporary ethics has been concerned to be inclusive. Environmental ethics is even more inclusive. It is not simply what a society does to its poor, its blacks, slaves, children, minorities, women, handicapped, or future generations that reveals the character of that society, but also what it does to its fauna, flora, species, ecosystems, and landscapes. Whales slaughtered, wolves extirpated, whooping cranes and their habitats disrupted, ancient forests cut, Earth threatened by global warming—these are ethical questions intrinsically, owing to values destroyed in nature, as well as also instrumentally, owing to human resources jeopardized. Humans need to include nature in their ethics; humans need to include themselves in nature.

Ecologists may insist at this point that environmental science sometimes does inform an environmental evaluation in subtle ways. Consider some of the descriptive categories used of ecosystems: the *order, stability, complexity,* and *diversity* in these biotic *communities.* Ecologists describe their *interdependence,* or speak of their *health* or *integrity,* perhaps of their *resilience* or *efficiency.* Biologists describe the *adapted fit* that organisms have in their niches, the roles they play. Biologists may describe an ecosystem as *flourishing,* as *self-organizing,* perhaps as *dynamically developing,* or *regenerating.* Strictly interpreted, these are just descriptive terms; yet often they are already quasi-evaluative terms.

Other ecologists challenge such a positive account of ecosystems (Pickett *et al.* 1992). Disturbance interrupts the orderly succession of ecosystems, producing patchwork and even chaotic landscapes. Over decades and centuries, ecosystems change. Over millennia, one ecosystem evolves into another. Always, though, evolution and ecology both require organisms selected to be good, adapted fits, each in the niche it inhabits. Misfits go extinct, and easily disrupted ecosystems collapse and are replaced by more stable ones. There are ordered regularities (seasons returning, the hydrologic cycle, acorns making oak-trees, squirrels feeding on the acorns) mixed with episodic irregularities (droughts, fires, lightning killing an oak, mutations in the acorns).

The rains come; leaves photosynthesize; insects and birds go their way; earthworms work the soil; bacteria break down wastes that are recycled; coyotes have their pups and hunt rabbits; and on and on. Natural systems (such as the Serengeti plains of Africa) were often sustained in the past for long periods, even while they were gradually modified. R. V. O'Neill *et al.* conclude that those who see stability and those who see change are looking at two sides of one coin: 'In fact, both impressions are correct, depending on the purpose and time-space scale of our observations' (O'Neill *et al.* 1986: 3). 'The dynamic nature of ecosystems', concludes Claudia Pahl-Wostl, is 'chaos and order entwined' (Pahl-Wostl 1995).

But this dynamic openness is also welcome. Ecosystems are equilibrating systems composed of co-evolving organisms, with checks and balances pulsing over time. Many general characteristics are repeated; many local details vary. Patterns of growth and development are orderly and predictable enough to make ecological science possible—and also to make possible an environmental ethics respecting these creative, vital processes. Natural selection means changes, but natural selection fails without order, without enough stability in ecosystems to make the mutations selected for dependably good for the time being. A rabbit with a lucky genetic mutation that enables it to run a little faster has no survival advantage to be selected for unless there are coyotes reliably present to remove the slower rabbits.

Ecosystems have to be more or less integrated (in their food pyramids, for example), relatively stable (with more or less dependable food supplies, grass growing again each spring for the rabbits), and with persistent patterns (the hydrologic cycle watering the grass), or nothing can be an adapted fit, nor can adaptations evolve. Ecosystems get tested over thousands of years for their resilience. This is true even though ecosystems are continually changing and though from time to time natural systems are upset (when volcanoes erupt, tsunamis destroy whole regions, or catastrophic epidemics break out). Then organisms have to adapt to altered circumstances, and, as new interdependencies and networks appear, the integrity of ecosystems has to become re-established.

Evolutionary biologists add that their science has made quite commanding discoveries about the comprehensive history of life on Earth: that is, about what these dynamic changes and upsets in ecosystems have produced. There is something awesome about an Earth that begins with zero and runs up toward 5–10 million species in several billion years, setbacks and upsets notwithstanding. The long evolutionary history fact of the matter seems valuable; it commands respect, as biologists recognize, even reverence, as theologians claim. When one celebrates the biodiversity and wonders whether there is a systemic tendency to produce it, biology and theology become natural allies. Perhaps this alliance can help humans to correct the misuses to which science has been put—with more respect and reverence for life.

Though biologists (in their philosophical moments) are typically uncertain as to whether life has arrived on Earth by divine intention, they are almost unanimous in their respect for life and seek biological conservation on an endangered planet. Earth's impressive and unique biodiversity, evolved and created in the context of these ecosystems, warrants wonder and care. There is but one species aware of this panorama of life, a species at the same time jeopardizing this garden Earth.

Asking about respect for creation, critics of Western monotheism may reply that the problem is the other way round. Judaeo-Christian religion has not adequately cared for nature because it saw nature as the object of human dominion. Famously, historian Lynn White, though himself a Christian, laid much of the blame for the ecological crisis on Christianity, an attack published in *Science*, the leading journal of the American Association for the Advancement of Science (White 1967). God's command in Genesis 1 for humans to 'have dominion' flowered in medieval Europe, licensed the exploitation of nature, and produced science and technology that have resulted in an ecological crisis. Ecofeminists, post-modernists, and proponents of Asian faiths have joined in such criticism. Equally of course White was attacking science for buying into a secular form of the dominion hypothesis, but the original authorization, so he claimed, was religious. After the Fall, and the disruption of garden Earth, nature too is corrupted, and life is even more of a struggle than before. Nature needs to be redeemed by human labour.

Theologians have replied that appropriate dominion requires stewardship and care (Birch, Eakin, and McDaniel 1990; Cobb 1972; DeWitt 1998; Nash 1991). Adam and Eve are also commanded to 'till the garden and keep it' (Genesis 2: 15), a more positive sense of dominion. Adapting biblical metaphors for an environmental ethic, humans on Earth are, and ought to be, prophets, priests, and kings—roles unavailable to non-humans. Humans should speak for God in natural history, should reverence the sacred on Earth, and should rule creation in freedom and in love.

The same Genesis stories teach the human fall into sin driven by desire to be like God, in tension with being made in the image of God. Humans covet, worship false gods; they corrupt their faiths, they rationalize in self-deception. Faiths must be ever reforming; humans need their prophets and priests to constrain their kings. The righteous, the humane life balances all three dimensions. Christianity has indeed often been too anthropocentric, just as Christians have often been too self-centred. The need for repentance is perennial (Rasmussen 1996).

Here, religious persons, as prophets and priests, can bring a perspective of depth to nature conservation, one that science can help launch but cannot complete. With too much kings' dominion (those escalating control, consumption, exploit-ation concerns, also fuelled by science), we lose the world we seek to gain. Monotheists will see in forest, sky, mountain, and sea the presence and symbol of forces in natural systems that transcend human powers and human utility. They will find in encounter with nature forces that are awe-inspiring and overpowering, the signature of time and eternity. Although nature is an incomplete revelation of God's presence, it remains a mysterious sign of divine power. In the teachings of Jesus, the birds of the air neither sow nor reap yet are fed by the heavenly Father, who notices the sparrows that fall. Not even Solomon is arrayed with the glory of the lilies, though the grass of the field, today alive, perishes tomorrow (Matthew 6: 26–30). There is in every seed and root a promise. Sowers sow, the seed grows secretly, and sowers return to reap their harvests. God sends rain on the just and the unjust. 'A generation goes, and a generation comes, but the earth remains forever' (Ecclesiastes 1: 4).

Theologians claim that humans are made in the image of God. Biologists find that, out of primate lineages, nature has equipped *Homo sapiens*, the wise species, with a conscience. Ethicists, theologians, and biologists in dialogue wonder if conscience is not less wisely used than it ought to be when, as in classical Enlightenment ethics, it excludes the global community of life from consideration, with the resulting paradox that the self-consciously moral species acts only in its collective self-interest toward all the rest. Biologists may find such self-interest in our evolutionary legacy; but now, superposing ethics on biology, an *is* has been transformed into an *ought*. Ecologists and religious believers join to claim that we humans are not so 'enlightened' as once supposed, not until we reach a more inclusive ethic.

In a new century, a new millennium, with science flourishing as never before, we face a crisis of the human spirit. Central to these misgivings is the human relation to nature. In other centuries, critics complained that humans were alienated from God. In the most recent century, with its World Wars, East versus West, North versus South, critics worried about our alienation from each other. In this new century, critics are more likely to worry that humans are alienated from their planet. One may set aside cosmological questions, but we cannot set aside global issues, except at our peril. We face an identity crisis in our own home territory, trying to get the human spirit put in its natural place.

Several billion years' worth of creative toil, several million species of teeming life, have been handed over to the care of this late-coming species in which mind has flowered and morals have emerged. Ought not those of this sole moral species do something less self-interested than count all the produce of an evolutionary ecosystem as resources to be valued only for the benefits they bring? Such an attitude today hardly seems biologically informed (even if it claims such a tendency as our inherited Pleistocene urge), much less ethically adequate for an environmental crisis where humans jeopardize the global community of life. Nor does it seem very godly. Ecologists and theologians agree: humans need a land ethic. In the ancient biblical world, Palestine was a promised land. Today and for the century hence, the call is to see Earth as a planet with promise. That might be the God's-eye view.

Even secular naturalists may be drawn toward respect, even reverence for nature. Stephen Jay Gould, for example, found on Earth 'wonderful life', if also 'chance riches' (Gould 1989, 1980), and he was moved, among the last words he wrote, to call the earthen drama 'almost unspeakably holy' (Gould 2002: 1342). Edward O. Wilson, a secular humanist, ever insistent that he can find no divinity in, with, or under nature, still exclaims: 'The biospheric membrane that covers the Earth, and you and me, . . . is the miracle we have been given' (E. O. Wilson 2002: 21).

In the midst of its struggles, life has been ever 'conserved', as biologists find; life has been perpetually 'redeemed', as theologians find. To adapt a biblical metaphor, the light shines in the darkness, and the darkness has not overcome it (see John 1: 5). Science and religion join to celebrate this saga of life perennially generated and regenerated on this planet, this pearl in a sea of black mystery. We are then indeed enlightened; yet deep tragedy looms as we humans jeopardize life on Earth.

References and Suggested Reading

BARR, S. M. (2003). *Modern Physics and Ancient Faith.* Notre Dame, Ind.: University of Notre Dame Press.

BARROW, J. D., and TIPLER, F. J. (1986). *The Anthropic Cosmological Principle.* New York: Oxford University Press.

BIRCH, C., EAKIN, W. and McDANIEL, J. (1990) (eds.). *Liberating Life: Contemporary Approaches to Ecological Theology.* Maryknoll, N.Y.: Orbis Books.

COBB, J. B. Jr. (1972). *Is It Too Late?: A Theology of Ecology.* Beverly Hills, Calif.: Bruce.

COHEN, J. E. (1995). 'Population Growth and Earth's Carrying Capacity', *Science*, 269: 341–6.

CONWAY MORRIS, S. (2003). *Life's Solution: Inevitable Humans in a Lonely Universe.* Cambridge: Cambridge University Press.

DE DUVE, C. (1995). *Vital Dust: The Origin and Evolution of Life on Earth.* New York: Basic Books.

DEWITT, C. B. (1998). *Caring for Creation: Responsible Stewardship of God's Handiwork.* Grand Rapids, Mich.: Baker Books; Washington: The Center for Public Justice.

DOLAN, R. J. (2002). 'Emotion, Cognition, and Behaviour', *Science*, 298: 1191–4.

EHRLICH, P. R., and EHRLICH, A. H. (2004). *One with Nineveh: Politics, Consumption and the Human Future*. Washington: Island Press.

ELBERT, T., *et al.* (1995). 'Increased Cortical Representation of the Fingers of the Left Hand in String Players', *Science*, 270: 305–7.

GOODLAND, R. (1992). 'The Case that the World has Reached Limits', in R. Goodland, H. E. Daly, and S. E. Serafy (eds.), *Population, Technology, and Lifestyle*, Washington: Island Press, 3–22.

GOULD, S. J. (1980). 'Chance Riches', *Natural History*, 89/11: 36–44.

—— (1989). *Wonderful Life: The Burgess Shale and the Nature of History*. New York: Norton.

—— (2002). *The Structure of Evolutionary Theory*. Cambridge, Mass.: Harvard University Press.

HANNAH, L., LOHSE, D., HUTCHINSON, C., CARR, J. L., and LANKERANI, A. (1994). 'A Preliminary Inventory of Human Disturbance of World Ecosystems', *Ambio*, 23: 246–50.

HARDIN, G. (1968). 'The Tragedy of the Commons', *Science*, 169: 1243–8.

KELLEY, K. W. (1988) (ed.). *The Home Planet*. Reading, Mass.: Addison-Wesley.

LESLIE, J. (1990) (ed.). *Physical Cosmology and Cosmology*. New York: Macmillan.

LEWONTIN, R. C. (1991). *Biology as Ideology: The Doctrine of DNA*. New York: HarperCollins.

MAGUIRE, E. A., *et al.* (2000). 'Navigation-Related Structural Change in the Hippocampi of Taxi Drivers', *Proceedings of the National Academy of Sciences of the United States of America*, 97/8: 4398–4403.

MAYNARD SMITH, J., and SZATHMÁRY, E. (1995). *The Major Transitions in Evolution*. New York: W. H. Freeman.

MONOD, J. (1972). *Chance and Necessity*. New York: Random House.

NASH, J. A. (1991). *Loving Nature: Ecological Integrity and Christian Responsibility*. Nashville: Abingdon Cokesbury.

O'NEILL, R. V. *et al.* (1986). *A Hierarchical Concept of Ecosystems*. Princeton: Princeton University Press.

PAHL-WOSTL, C. (1995). *The Dynamic Nature of Ecosystems: Chaos and Order Entwined*. New York: John Wiley.

PICKETT, S. T. A. *et al.* (1992). 'The New Paradigm in Ecology: Implications for Conservation Biology above the Species Level', in P. L. Fiedler, and S. K. Jain, (eds.), *Conservation Biology*, New York: Chapman & Hall, 65–88.

RASMUSSEN, L. L. (1996). *Earth Community Earth Ethics*. Maryknoll, N.Y.: Orbis Books.

RISSER, P. G., LUBCHENCO, J., and LEVIN, S. A. (1991). 'Biological Research Priorities: A Sustainable Biosphere', *BioScience*, 47: 625–7.

SOBER, E., and WILSON, D. S. (1998). *Unto Others: The Evolution and Psychology of Unselfish Behaviour*. Cambridge, Mass.: Harvard University Press.

Statistical Abstract of the United States, 120th edn. (2000). Washington: US Census Bureau.

United States Congress (1969). *National Environmental Policy Act*. 83 Stat. 852. Public Law 91–190.

VENTER, J. C. *et al.* (2001). 'The Sequence of the Human Genome', *Science*, 291 (16 Feb.): 1304–51.

VITOUSEK, P. M., EHRLICH, P. R., EHRLICH, A. H., and MATSON, P. A. (1986). 'Human Appropriation of the Products of Biosynthesis', *BioScience*, 36: 368–73.

WHITE, L. JR. (1967). 'The Historical Roots of our Ecological Crisis', *Science*, 155: 1203–7.

WILSON, D. S. (2002). *Darwin's Cathedral: Evolution, Religion, and the Nature of Society*. Chicago: University of Chicago Press.

WILSON, E. O. (2002). *The Future of Life*. New York: Alfred A. Knopf.

World Bank (2001). *World Development Report 2000/2001.* New York: Oxford University Press.

World Commission on Environment and Development (1987). *Our Common Future.* Oxford: Oxford University Press.

World Resources Institute (1994). *World Resources 1994–95.* New York: Oxford University Press.

CHAPTER 54

BIOTECHNOLOGY AND THE RELIGION– SCIENCE DISCUSSION

RONALD COLE-TURNER

INTRODUCTION

Ever since the rise of modern biotechnology in the mid-1960s, religious institutions and theologians have explored its most interesting features and problems. Even so, it is often said that technology outpaces morality, and that human power exceeds wisdom. Having created biotechnology, human beings are losing control of their own creation. Biotechnology seems to drive itself, as if by some technological imperative that compels us to solve problems even if our solutions create greater problems. To some extent, this complaint points out the deficiencies in religion's engagement with biotechnology: deficiencies of rigour, consistency, and of course universality. When, for example, a religious body takes a position on biotechnology, at best the position reflects committee compromise. Rarely does the wider member-ship know or agree with the position, and the general public is hardly ever affected. Whatever else their virtues might be, few would claim that today's religious and moral assessments of biotechnology are up to the task of guiding tomorrow's achievements.

The complaint that morality lags behinds technology, however, is based in a deeper set of apprehensions and anxieties about modernity, technology in general, and the various forms of biotechnology in particular. In the 1970s, some were concerned that human beings were 'taking evolution into their own hands'. In a lecture given in 1942, C. S. Lewis warned that 'what we call Man's power over Nature turns out to be a power exercised by some men over other men with Nature as its instrument' (for more information, see <http://users.ox.ac.uk/~jrlucas/lewis.html>). In this simple statement, Lewis points out the central ambiguity of technology: if technology controls nature, who or what controls technology?

If for no other reason, anxiety about biotechnology moves some to press religion into public service as a kind of technology braking mechanism. When religion is forced into that role, its contributions are easily reduced to simplistic cries of 'playing God' or to a narrowly focused defence of human embryos. Given the assumption that many still hold that religion and science are inevitable rivals, religion seems perfectly suited for the task of stopping technology. So suited, religion is cast forever again in its perennial role, as in some renaissance myth.

On closer examination, however, biotechnology is more than embryos, and religion's stake in biotechnology is broader and far more positive than is often thought. For the most part, the religions of the world are strongly in favour of biotechnology in general, particularly as it applies to human health and nutrition. Common to the religions are certain foundational attitudes, such as compassion, altruism, and a commitment to healing. To the extent that biotechnology is consistent with these attitudes, it has the blessing and approval of nearly all people of religious faith. The blessing pronounced by religion upon technology is conditional, depending not so much on the quality of the biotechnology as on the moral maturity and spiritual wisdom of its makers and users (for specific virtues applied to biotechnology, see Deane-Drummond 2004: 127).

While giving attention to the embryo question and indeed to the meaning of 'playing God', this chapter surveys more generally some religiously significant aspects of recent genetics and biotechnology. The first section considers what genetics suggests about human nature. The next section, on biotechnology, looks first at work on plants and animals, but moves quickly to human applications, from gene therapy to cloning and stem cells, asking about the moral implications. The final section offers a theological interpretation of genetics and biotechnology, reflective of Christianity but intended for wider readership. This section looks at the question of the human embryo and its role in research, the theological implications of technological transformation of the human self, and how we are to understand in religious terms our new role in creation.

Genetics and its Implications for Human Self-Understanding

Little separates the sciences of genetics and developmental biology from biotechnology. Science leads to new technology, and technology is the method of science. The modern revolution in the science of genetics is possible because of the power to manipulate genes and other molecules. Added to this is heavy reliance in science and biotechnology on computers to absorb all the data ('bioinformatics'), to test for the presence of particular DNA sequences (gene chips), and perhaps most interestingly, to image molecular biology on the screen. Taken together, these technologies make it possible to model the complex and precise three-dimensional structures of proteins and to observe the interior workings of the living cell, contributing to understanding and to further control. Without separating science from technology, however, it is possible to focus for the moment on the science, on what we are learning more than on what we are doing, and to consider its significance for religion.

A central achievement of recent genetics is the Human Genome Project, which presents the entire DNA sequence for human beings. The achievement is all the more significant because the genomes of many other organisms have been sequenced, allowing scientists to compare among closely or distantly related species. Using this information, researchers are probing the basis of disease in hopes of finding new pharmaceutical products and new understandings of human development and illness. Of greater theological interest, the completion of the human genome, together with contemporary neuroscience, opens the way to greater understanding of the uniqueness of the human species and of differences within our species. Genetics confirms the view that all human beings have descended from one ancestral population, and thus are one race or species. In the past, perceptions of differences have divided ethnic groups. Some will claim that new findings bolster ancient divisions and fuel new ones. It is true that human traits, including cognitive abilities, have their basis in genetic inheritance, and therefore that variation in traits is based in part on genetic variation as well as on environmental factors. No evidence suggests, however, that traits such as intelligence vary genetically between human groups or 'races' (Graves 2001: 157–72).

Nevertheless, what genetics adds to neuroscience is an understanding of differences within the human species. Genetics is the science of differences and offers a method for separating genes and environment as factors explaining human variation. This allows researchers to tease out how genetic variation results in observable variation in human traits and, in time, to identify exactly where relevant genetic variations are located in the human DNA sequence. So far, researchers have found consistently that while genes almost always play a role in human differences, environment also plays a determining role. Even so, these findings raise the potential for undesirable genetic discrimination or élitist attitudes.

While religions can help to counter racist interpretations of genetics, theology today must itself come to terms with the simple fact that we live with growing knowledge of human genetic variation, which of course is expressed not just in differences in skin pigmentation but quite probably in traits ranging from cognitive abilities to mood (anxiety or depression) to religiosity (Hamer 2004). Keeping in mind that the differences are within, not between, ethnic groups, how do we come to a religious understanding of difference? To take just one example, what theological explanation is there for different levels of religious responsiveness? In Christianity, one traditional explanation of difference is that God chooses or elects only some for salvation. A rival theological explanation is that all humans are equally free (after environment is taken into account, of course), but that only some choose to respond religiously, for which there can be no cause without destroying freedom. Genetics, however, may suggest that neither explanation is correct and that the difference in religious response is explained at least in part by differences in our genes.

Does this exclude free will or the authenticity of human religious response? Human freedom and responsibility depend upon the successful function of many genes that give rise to highly complex neurological processes. Philosophical and theological interpretations of neuroscience have explored at length the question of free will and determinism. What is often set aside in those discussions is the contribution of genetics as the study of difference—in this case, differences between individual human beings in their capacities for freedom, moral conscience, and responsibility. In law, individuals are rarely exonerated on the basis of psychological factors, but it is known that some individuals lack moral capacity, and therefore culpability. This forces us to ask not just whether freedom is compatible with the constraints of genetic/neurological processes in which they are lodged, but whether some individuals are more or less free than others. Too subtle for legal defence, slight variations in human moral capacities appear to arise from simple variations in DNA, either in genes themselves or in DNA that regulates their expression. One way in which this variation is encountered is in the debate over attention deficit disorder, a condition for which medical remedies (usually Ritalin or Strattera) are widely used in the United States and some other countries. Some argue that the use of drugs is unwarranted, or at least excessive, and that all people, with few exceptions, should be held accountable equally for the coherence of their decisiveness and for their success in achieving their aims. Others point to successful use of these drugs as evidence not only that genetic/neurological factors bear on attentiveness but that human beings vary in their capacities.

While theology cannot ignore these findings, neither should it rush to interpret when the science is still at an early stage. There can be little doubt that we will learn much more in coming decades about the neurological significance of genetic variation. For now, it can be said that freedom and responsibility are dependent upon many factors, which vary in subtle ways from individual to individual, affecting the scope of freedom but not undermining it conceptually, and only rarely incapacitating it functionally. The genes an individual inherits interact with the environment over the lifespan, resulting in a composite biological reality that shapes and limits the individual at each moment. Within the unique shape and limits of the biological individual, there is an exercise of freedom and responsibility, and perhaps an expression of religious response. These human possibilities arise within the matrix of genes/environment interaction, which is different in each individual. In the future, just as medical therapy is tailored to the genes of the individual, spiritual or religious care will be individualized, not just to a unique environment or personal history but to genetic factors unique to the individual.

Human beings differ from each other, but share a surprising number of genes with other species, from closely related primates such as chimpanzees to distantly related insects or even bacteria. DNA sequences that evolved hundreds of millions of years ago are conserved through evolution and shared in widely diverse living species, often performing basic biological tasks in the cells of complex organisms just as they did in the single-cell organisms. A new view of life emerges, in which human beings are linked in highly detailed ways in a web of organic commonality. In contrast to a traditional view of fixed species, biologists today see species as fluid. To the dynamic fluidity of nature must now be added the powers of technology to move DNA sequences across lines and to implant cells from one species into another, creating chimeras. Human neural stem cells have been implanted into the foetal brains of Old World monkeys (Ourednik *et al.* 2001), and the resulting growth of monkey and human brain cells together in the developing brain of the monkey poses new questions, not just of safety or the meaning of species, but of how we should respond if non-human animals with human brain cells begin to exhibit what might be considered human traits (Greene *et al.* 2005).

Recent science and technology demand that we ask in a new way: what makes us unique? At the level of DNA, surprisingly little; yet somehow small genetic difference give rise to huge functional and cultural differences, with traits such as language playing a key role. Traits as complex as language depend on many genes, of which many are shared with chimpanzees, for instance. No single gene makes us human. Slight evolutionary variations in shared genes, however, may play a decisive role in uniquely human traits, such as complex language (Enard *et al.* 2002).

So far, genetics has not offered much help in explaining how even simple multicellular organisms develop from a single cell (the fertilized egg) through embryonic stages to adulthood. Biological development depends on genes, but genes alone cannot explain how cells with identical DNA take on specialized structures and functions in various parts of the body. Cells seem to know where they are in the developing body, based on information not coded in genes but necessary nonetheless for development. No complete explanation is currently available, but we know that genes alone do not explain development (Keller 2002). The fields of embryology and cell development are poised for substantial progress, driven in large part by the prospect of medical applications coming from the isolation of human embryonic stem cells. Researchers will learn in detail how the embryo develops and how cells differentiate, and a new view of the developing embryo may emerge.

BIOTECHNOLOGY: MORAL CONCERNS

Depending on how the term is defined, 'biotechnology' goes back about 10,000 years when our ancestors first engaged in selective breeding of plants and animals. Today's wheat, rice, and corn bear little resemblance to their wild cousins. Thousands of years of seed selection have produced heartier plants and bigger yields. Yeast was used to ferment beverages and produce bread, while animals ranging from dairy herds to herding dogs were bred. By the 1800s, selective breeding was so refined that Charles Darwin drew upon it for the core metaphor for evolution, 'natural selection'. In the 1900s, techniques were further refined. Hybrids (crosses between various strains) were created, resulting in the so-called green revolution, increasing yields further but introducing new problems to global agriculture. For millennia, some have suggested that these breeding techniques should be applied to human beings, and early twentieth-century eugenicists made various attempts along these lines, with tragic results.

Traditional selective breeding could only select traits already found in a breeding population. Recombinant DNA technology, by contrast, can transfer traits found almost anywhere by transferring their genes from one species to another. With powerful new tools, biotechnology seeks to develop plants that can produce more food, survive drought or salt water, self-fertilize, resist pathogens in the field or rot in storage, produce novel proteins to enhance nutrition, or produce pharmaceutical compounds. Similar work is being done in agricultural animals. Laboratory animals (rodents) are routinely bred to mimic human beings who have diseases. Almost without exception, these modified plants and animals are patented and sold as valuable commodities.

Global controversy has followed these developments, which raise possible health concerns, especially in light of the difficulty in distinguishing between modified and unmodified food in the food supply. Others object to patenting as commodification of life, an economic form of reductionism that is exploitative of traditional farmers in developing countries. Loss of genetic diversity in human food sources, with growing risk of vulnerability to pathogens, is a concern, especially in light of possible cloning of livestock. All this is debated against the backdrop of global controversy on trade protection for agriculture. Religious questions, such as whether human beings have a right to act upon other species in these ways (for instance, to create them in order to model human diseases), are insufficiently addressed.

When applied to human beings, these same technologies take on quite different forms and meanings. As the relationship between DNA sequences and some human diseases becomes clearer, it is possible to predict the future development of disease by testing DNA. With growing frequency, genetic testing is done during pregnancy, and in 1–2 per cent of cases, problematic results are reported to prospective parents, who must then face the prospect that their unborn child may not survive pregnancy or may be born with severe health problems. In a very few situations, it is possible to use prenatal genetic information to benefit the child. In nearly all other cases, morally and theologically difficult challenges arise for mothers and couples, especially over the use of abortion to prevent the birth of a child with disease (Cole-Turner and Waters 1996).

The goal of human genetics, of course, is not simply to predict diseases, but to treat them by interrupting or modifying the DNA that causes them. This idea, 'gene therapy', is powerfully attractive but surprisingly difficult, with minimal success as of 2005, following fifteen years of serious effort. Modifying DNA in the cells of the body, in sufficient quantity at the right location and in a way that remains stable over years in the tissue with the disease, is challenging in part because cells have evolved for billions of years to resist such modifications.

Without exception, ethicists and theologians who have pondered the idea of human gene therapy have endorsed it, provided of course that it can be done safely and effectively, and as long as it is confined to 'therapy'. Safety and effectiveness have not yet been achieved, but already there is growing concern that it will be impossible to hold the line at therapy. Commercially, at least, far more attractive applications lie in what might be called 'enhancement'. The boundary between 'therapy' and 'enhancement' is notoriously difficult to define, even though most agree that it is grounded in a clear intuition: it is right to treat a disease, but not to 'improve' human beings through medicine in ways that are not related to disease. Vaccines, of course, improve or enhance the immune system, but in a way related to disease, so nearly everyone accepts their use. But using medicine and biotechnology to make children taller, more athletic, or smarter, borders on illegitimate enhancement, even though the economic incentives for doing so might be great. Whether the public will accept these uses is unclear, but widespread use of cosmetic medicine leads many to assume not just acceptance but eagerness.

Gene 'therapy', also known as 'somatic cell modification', targets DNA only in limited cells of the body of the patient. Somatic modification is contrasted with 'germ line' modification, which affects all of the cells, including germ cells (sperm or egg) passed to future generations. 'Germ line' or inheritable genetic modification, some-times called 'designer babies', is more controversial than somatic modifications. Safety concerns are far more complex, in part because the modification might affect the health of generations in the distant future, in part because any modification affects all the cells of the body over the entire lifespan of development, starting with the embryo.

Even if safety concerns can be resolved, philosophical and religious objections remain. It has been said repeatedly that germ line modification violates human dignity, perhaps by bringing into existence persons whose genetic identity is deter-mined in part by the decisions of others. The objection based on dignity, however, is not developed in a complete or compelling way, because the key term 'dignity' lacks definition, and because it is unclear just how germ line modification (and not other medical interventions, for instance on infants) might violate it. Presumably, human dignity is the worth possessed intrinsically by every member of the human species, but not shared with other species. Recent science, which tends to dismiss any qualitative difference between human and non-human life, threatens the concept of dignity, while technology (according to the argument) threatens its reality. Appeals to dignity tend to be made by bioethical conservatives, drawing upon a tradition of Kantian philosophy according to which human beings are ends, never means to be used in service of another end, no matter how worthy (Kass 2002).

In place of dignity, theologians sometimes appeal to notions of the image of God, mentioned in Hebrew Scriptures but not defined. While it seems obvious that the idea of the image of God gave rise to the concept of dignity, the first term is irreducibly relational, while the second is not. That is, in theology, human beings possess worth, not intrinsically (or because of some biologically based superiority over other species) but because of a relationship with God. For Christians, the God–human relationship is centred not primarily in creation but in the Incarnation, God's presence in creation in human form. Theologians who argue that human germ line modification would violate the image of God in humanity are faced with the same challenges as their secular counterparts: how are 'image of God' or 'dignity' to be defined, and how does germ line modification, but not other medical treatments, threaten it?

In addition to possible violations of human dignity, another argument against human germ line modification is that it diminishes freedom. This argument nicely exploits an irony. Advocates of human germ line modification defend it by appealing to the freedom of parents to use technology to acquire the sort of child they want (Stock 2002). Secular bioethics, with its insistence on patient autonomy, seems to support their argument, at least on the surface. But opponents have countered that the freedom of the parents comes at the cost of the freedom of the child, who enters the world with genes determined in some respect by parents. Two versions of this objection can be made. The stronger claim is that such genetic modification actually produces a human life with impaired freedom by interfering with the genetic-neurological-volitional pathways and diminishing the capacity of the child to act freely, perhaps by giving the child a selected propensity toward a behaviour of interest to parents. But of course, the capacity for freedom in every human being is limited by the genetic-neurological pathways, which in some sense are fixed or determined by nature at conception; so advocates of this claim must show how technological determination impairs freedom but natural determination does not, or how having a selected propensity results in less freedom than having some other natural propensity. A more modest claim is that while there is no actual impairment, the relationship between engineering and the engineered child is forever warped, improperly asymmetrical, and incapable of the normal development necessary for the developing freedom of the child (Habermas 2003).

In contrast to philosophers and bioethicists, theologians who address germ line modification tend to challenge the use of freedom as the starting point for assessment. If germ line modification is permitted, it is not because parents are free, but because they have responsibilities (Wheeler 2003; Moraczewski 2003). Acting within the scope of parental responsibility, parents might rightly choose some narrowly defined form of germ line modification. This argument, which amounts to a limited approval of germ line modification, can be found as early as 1970 in the writings of

Paul Ramsey: 'The notation to be made concerning genetic surgery, or the introduction of some anti-mutagent chemical intermediary, which will eliminate a genetic defect before it can be passed on through reproduction, is simple. Should the practice of such medical genetics become feasible at some time in the future, it will raise no moral questions at all' (Ramsey 1970: 44). A similar assessment is offered by Pope John Paul II, whose *Donum vitae* offers qualified permission for something like germ line modification: '*Medical research must refrain from operations on live embryos, unless there is a moral certainty of not causing harm to the life or integrity of the unborn child and the mother, and on condition that the parents have given their free and informed consent to the procedure*' (Congregation for the Doctrine of the Faith 1987; italics original). This endorsement is limited, additionally, to what is clearly therapy, avoiding anything that might be called enhancement. But both statements grant that it might be permissible to use future techniques to modify an embryo (with germ line effects) to prevent a disease, in order to allow the embryo to develop and live, but not to enhance physical or cognitive traits. Religious acceptance of germ line modification for therapeutic purposes is not particularly controversial. Theologians should, however, go further in offering support for at least some forms of enhancement (perhaps to prevent cancer), while at the same time helping prepare future parents to use these technologies with restraint and wisdom, avoiding uses that control or limit future children.

By most accounts, germ line modification will occur in time, but not for several decades at least. Already at hand, however, is a technology that combines genetics with the reproductive technique of *in vitro* fertilization (IVF). This technology tests embryos genetically before they are implanted—hence its name, pre-implantation genetic diagnosis (PGD). Using PGD, couples at risk for genetic disease create multiple embryos. Each is allowed to divide to about eight cells. One cell is carefully removed from the embryo, and its DNA is tested for specific diseases. Embryos clear of disease are implanted, usually two at a time. Many see this procedure, expensive as it is, as morally (and not just psychologically) superior to traditional conception, followed by prenatal testing and a possible abortion. In one sense, PGD moves the date of the test from the fourteenth week or so of pregnancy to the four days after conception, before pregnancy. All opposed to IVF, such as the Catholic Church, reject PGD; but even some theologians and church leaders who permit IVF have their concerns over PGD, which puts human beings in the role of deciding over human life, determining which embryos live and which do not, all the while creating embryos with the knowledge that some will be destroyed. Others are concerned that PGD will be used to select for other traits, unrelated to health. Already it has been used to select embryos as suitable donors to save the life of an older sibling. Some see this as the unacceptable use of human life, a utilitarian assault on dignity, while others see it as grounded clearly in the parental duty to love and protect children.

Alongside these developments, advances in embryology and cell biology have widened the scope of biotechnology and its potential for human impact. The report of the birth of Dolly the cloned sheep in 1997, followed in 1998 by the derivation of human embryonic stem cells, launched a new era in developmental biology, accompanied by some of the most intense public debates in the history of biotechnology. With Dolly came the prospect of mammalian cloning, understood here as the beginning of a new individual with the nuclear DNA of another. When Dolly was first introduced, many thought that human cloning would follow quickly, and that within perhaps a decade, human cloned children would be produced. Cloning of human beings—indeed, cloning of mammals generally, and primates in particular— has turned out to be more difficult than was first thought.

Stem cells are cells that have the potential to divide and develop into cells with specialized functions, such as skin or bone or brain. Stem cells in the adult human body are partway along a pathway of development, not fully mature developmentally but already committed to becoming a certain category of cell. Finding these cells in the adult body is difficult but immensely promising medically, because they could be used to help the body heal itself by regenerating damaged or diseased tissues. For years it has been understood that in the embryo, there are stem cells that developmentally are just at the starting point of differentiation, with potential to develop into all the cell types of the human body. In 1998 researchers isolated these cells and grew them in the laboratory, destroying donated embryos in the process.

The central moral issue raised by human embryonic stem cell research is whether the human embryo may under any circumstances be used either as an object of research or as a means to benefit another. Research using cells from adults or even from dead embryos, aborted foetuses, or placentas raises no such objections, because nothing is destroyed. Research or therapy that destroys the human embryo raises profound concerns for some. Three major positions can be distinguished. The first and most restrictive position, held by the Catholic Church and by others, rejects any use of human embryos, even if they are going to be destroyed anyway. The middle position allows, and in some cases publicly funds, research using embryonic stem cells if the embryos are left over from reproductive clinics and are destined to be destroyed. The third position allows and may fund research that includes the creation of new embryos specifically for research purposes, including the use of nuclear transfer or cloning as the method of creation. The conflict between these positions is not likely to be resolved any time soon. Different Christian denominations have endorsed all three positions, and individual theologians have advocated restrictions or conditional support for research.

Biotechnology and Theology

The final section of this chapter considers three theological themes: the human embryo, human self-transformation via technology, and the human role in creation. The question of the human embryo was introduced in the previous section in the context of human embryonic stem cell research. Recent research sets before us the possibility that the embryo may be used as an object of research. Furthermore, now that it is possible to use nuclear transfer or cloning to create an entity very much like an embryo, we are faced with a new question. In the past, the embryo was the result of conception, which occurred inside a woman's body. Now, not only may the process of conception occur in a dish in a laboratory, but conception itself is no longer necessary for the origination of the new entity. Furthermore, the potential of these cloned entities to develop normally is not yet understood, but it is generally believed by experts in the field that these cloned entities are significantly lacking in normal potential. If these entities come into existence without conception, if they are located not in the womb but in the lab, and if they lack normal potential for development, are they embryos in the normal and normative sense? This is the new question before us.

If they are embryos, what else must be regarded as an embryo? If they are not embryos, what are they? What are the minimal criteria that must be met for a cell or cluster of cells to be regarded as an embryo? In many respects, this is a scientific question, and the very research that fuels the controversy also promises to shed new light on the fundamentals of developmental biology and embryology. But this is also a question of definition, a matter of where and how to make distinctions within the data, and thus a question for the philosophy or even the theology of biology as much as for biology itself.

These questions now take their place next to the familiar question of the ontological and moral status of the embryo. When is the embryo ontologically a human life or a human person? When does it acquire moral status comparable to that of a child? In this familiar form of the embryo question, philosophical and theological judgements rebound upon science, not only limiting what may be done in research but perhaps shedding light on what it is that the researcher confronts when looking at the early stages of the developing organism. But the ontological and moral claims that philosophy and theology bring to embryology must be informed by embryology itself, not a simple task in a field of such rapid expansion of knowledge. At the very least, today's research should challenge theological and philosophical opinions now known to be based on incorrect views of embryo development. For example, it is now clear that genes alone do not account for the development of the embryo, and therefore that fertilization is a necessary but not a sufficient condition for all the requirements for personhood.

As far back as we know, the developing embryo has inspired awe and respect. Among the religions of the world, however, Christianity is uniquely concerned about the embryo. When Christians first separated from Judaism, they not only objected to abortion, but equated it with murder (*Didache* 2. 2). The difference probably lies in the exclusively Christian belief that salvation depends upon Christ becoming human, first by becoming an embryo. Even so, the history of Christianity contains a wide and conflicting range of theological interpretations of the ontological and moral status of the embryo (Jones 2004; Ford 2002).

Today's embryology, despite recent breakthroughs in genetics and developmental biology, is not yet close to a comprehensive explanation of the extraordinarily complex pathways from fertilized egg to foetus to newborn to adult. Development from fertilization to birth is a pathway marked by many milestones, none of which can support the conclusion that what a moment ago was not a person has now become a person. While advances in embryology should inform theological understanding, it must also be acknowledged that too often the religious debate is not about the facts but about cultural entailments and commitments, having more to do with birth control, reproductive medicine, and abortion than with either theology or science. Also at stake are future technologies of human modification, which will likely focus on the embryo as a locus of intervention. Protecting the embryo is one way to resist these technologies, while permitting use of the embryo in research opens the path to some of them. One side in the embryo debate opposes human attempts at biotechnological self-modification, while the other is open to consider such things. At stake are not just two views of the embryo, but two views of human nature, its malleability, and the theological legitimacy of human self-modification. Christian churches have taken positions at both ends of this debate.

If anything, however, too much focus on the embryo question has diverted attention away from considerations of the likely long-term consequences of these fields of research. The future of PGD and the possibilities of human germ line modification need more reflection. Stem cell research promises to open a new field often called 'regenerative medicine'. By injecting stem cells into patients, the regenerative medicine of the future might offer new tools, more creatively powerful than surgery, to help the body regenerate tissues or organs damaged by injury or disease. Quite possibly the brain itself, or significant regions or types of cells in the brain, might be regenerated enough to offset the effects of some diseases. In time it might be possible to reinvigorate the human body, organ by organ, perhaps with genetically modified stem cells, and so forestall ageing and possibly even natural death itself. Whereas medicine so far has made it possible for human beings to live out their full lifespan, the medicine of the future might extend the human lifespan, delaying degenerative diseases and the processes of ageing while enhancing cognitive ability or other capacities.

While promising, these developments will raise new problems. What will be the economic impact of profound shifts in patterns of human ageing or competitiveness? If it becomes possible to extend the human lifespan, if we learn to regenerate portions of the body, or if we discover how to make our offspring more talented or intelligent, how will we distribute these 'benefits' justly, if indeed we regard them as benefits at all? Should we permit PGD or use germ line modification to alter these traits in our offspring? Or should we permit the use of gene therapy, perhaps combined with stem cell therapy, to modify such traits in ourselves? Already we are faced with widespread use of prescription drugs to modify some psychological traits, such as anxiety or mood. Such self-transformation, now and in the future, may in time become a replacement for what might be seen as less effectual methods of religious transformation.

If so, then some might justify the technologies of human self-transformation by claiming that their only crime is that they actually deliver what religion only promises. Whether the objective is health or long life or a renewed self, technology cannot be condemned by religion for meeting religion's goals with technological efficiency. If technology offers a new self in an efficient time frame and perhaps at less cost than religion, then why not use technology? An answer is that religious *means* are essential to the full meaning of religious ends (Cole-Turner 1999). A new self through prayer is not the same as a new self through technology.

This does not mean, however, that we pit religion against technology, portraying technical means of self-modification as rival graces of false gods. Religious means and technological powers can complement each other without losing their distinctness. The biblical tradition of the Abrahamic faiths insists that all creaturely powers are mere instruments in the hands of the one true and all-powerful God, whose works are often hidden in the actions of intermediaries, including technology. So in truth technology is not a rival god (although some may embrace it as such), but a tool in the hands of the real God. If we approach all of the powers and processes of creation with this expectation, and if we actively add new and future technologies to the list of God's agents, then we can use them without excluding religion. But if so, then our use of these technologies, while warranted religiously, is limited morally. We may use them only in ways that are consistent with what we honestly take to be God's will and purposes. Our moral standard for assessing technology is based in our beliefs about God's will and purpose in creation and in its future transformations, including our individual creatureliness and its transformation.

A more subtle, and therefore greater, danger in technological self-transformation lies in technology's offer of control of the self *versus* religion's demand of surrender of the self. Religion offers a way to change the self, but requires first the giving of control of the self to God, with the promise that the surrendered self is truly and finally free. The danger of the technologies of self-transformation for the religious person is that they leave the self unchanged. They are self-asserting rather than self-transforming, enhancing the ego rather than surrendering it to a greater reality and purpose. The threat is subtle, because technology offers ways to change various features of the self, but it promises paradoxically to keep the self in control of the process. Technology invites us to see ourselves as a list of personality traits or habits or failures that can be alienated from the core self and thereby changed at the command of the self. These changeable parts now include dimensions of the self, such as mood or focus, which are hardly detachable from the self. The problem surfaces in a simple way when a person forgets to take medication intended to help with focus on daily tasks. Underneath lies a deeper enigma of the incoherence of the self, which sincerely wants and yet does not want to change. The danger of technology is that it offers the illusion of a managed grace whereby the self can fix itself up without changing, smoothing off rough spots while remaining in control, when in reality it is not in control.

Here is not just a rivalry between religion and technology but a diametrically opposite tendency, not of an intellectual but of a moral and spiritual sort. And exactly at this point, the warnings of the ancient Hebrew prophets are most relevant. Idols may be nothing but delusions, names for powers that are nothing more than tools in the hands of God. But when they are taken for powers independent of God, indeed powers of our own making, and when we further believe that we manipulate them and their saving powers, we run the risk of self-destruction. We believe we have created the means to expand our control of the self when in truth we have increased only the power of the self to control, leaving the self unchanged yet self-changing, uncontrollable yet more controlling. Technology is not out of control because it is a real power, but because we cannot control what is supposed to control it: namely, ourselves. Collectively, we lack the social means; but, more to the point, individually, we lack the inner resources to use these powers responsibly and wisely.

No wonder some stand back from the whole field with fear and condemnation, while others warn that technology too often goes too far, and that when it does, it amounts to 'playing God'. The phrase 'playing God' rightly calls attention to the human tendency toward hubris that readily accompanies success in technology. Succeeding in one thing, we think too quickly that we can succeed in all things, and that our success is unambiguously good. We overestimate our abilities and, most of all, our moral maturity, refusing to see our own egocentrism and blind spots. The theologian Paul Ramsey captures the rhetorical power of the phrase in his comment that 'Men ought not to play God before they learn to be men, and after they have learned to be men they will not play God' (Ramsey 1970: 138). The phrase 'playing God' challenges us to consider how biotechnology changes fundamentally the relationship between human beings and living nature, including human nature. Of course, the phrase goes too far in suggesting that we might be taking control of evolution or that life has become a plaything in our hands. More precisely, there are two connotations of 'playing God' that should be rejected. First, it implies that technological modification of nature is illegitimate because creation is completed, static, and fixed in place by God. Second, it suggests that some specific parts of the creation, such as the human embryo, are God's domain and wholly off limits to technology.

An alternative, and more positive, way to connect God and biotechnology is to speak of biotechnology as 'co-creation'. 'Co-creation' recognizes that we are creatures in a vast creative process, which we are only beginning to understand, but which we can already join through the power of our technology. The vision of 'co-creation' is that through biotechnology we can accompany God's own continuing creative work, which because of technology can operate now at a wholly new level. Thanks to biotechnology, the God who created over aeons through natural processes now has new tools for creation. Compared to 'playing God', the notion of 'co-creation' is more in keeping with the scriptural tradition, which insists that whether or not we honour God, our technology is a tool in the hands of a sovereign God, who deserves our service and whose holiness we must imitate. Even so, 'co-creation' lacks a critical principle, and so implies that all technology is equally good in contributing to creation. In this respect, 'playing God' and 'co-creation' suffer from the same deficit. Neither can distinguish good uses of technology from bad. 'Playing God' rejects everything, and 'co-creation' rejects nothing.

By itself the term 'co-creation' fails to criticize technology or the technologist, and even suggests that God and humanity are equal partners. This concern led to the suggestion by Philip Hefner (1989) that a qualifier be inserted. A human being is a 'created co-creator', a junior partner, late on the scene, and very much part of the systems we tweak. Even as junior partners, however, it is not clear that 'co-creation' takes human fallibility and self-centredness with the sort of seriousness that their potential for damage rightly requires. 'Co-creators' are morally and intellectually fallible human beings whose qualification for doing anything *with* God is first of all to be transformed *by* God. 'Playing God' suggests we are incapable of being transformed into instruments of God's service, while 'co-creation' suggests that we do not need to be transformed.

Whether in the end our power exceeds our wisdom may prove to be true of our time and our civilization, with its unprecedented powers over nature. If we fail as a technologically advanced culture, the fault will lie not in the sophistication of our technology or even in our theology and ethics, so much as in our character. Of course, we face questions to which we do not readily know the right answer, and rigorous theological and ethical analysis is needed. However, it is the moral and spiritual character of the technologist and, far more often, of the consumer of technology that lies at the heart of the problem. Technology demands not just theological subtlety and ethical clarity, but moral maturity. For this, not just the intellectual but all the institutional, public, communal, and character-forming resources of religion are needed more than ever.

REFERENCES AND SUGGESTED READING

COLE-TURNER, RONALD (1999). 'Do Means Matter?', in E. Parens (ed.), *Enhancing Human Traits: Ethics and Social Implications*, Washington: Georgetown University Press, 151–61.

—— and WATERS, BRENT (1996). *Pastoral Genetics: Theology and Care at the Beginning of Life*. Cleveland: Pilgrim Press.

Congregation for the Doctrine of the Faith (1987). 'Respect for Human Life (*Donum vitae*)'. Vatican.

DEANE-DRUMMOND, CELIA E. (2004). *The Ethics of Nature*. Oxford: Blackwell.

ENARD, WOLFGANG, *et al.* (2002). 'Molecular Evolution of *FOXP2*, a Gene Involved in Speech and Language', *Nature*, 418: 869–72.

FORD, NORMAN M. (2002). *The Prenatal Person: Ethics from Conception to Birth*. Oxford: Blackwell.

GRAVES, JOSEPH L., JR. (2001). *The Emperor's New Clothes: Biological Theories of Race at the Millennium*. New Brunswick, N.J.: Rutgers University Press.

GREENE, MARK, *et al.* (2005). 'Moral Issues of Human-Non-Human Primate Neural Grafting', *Science*, 309: 385–6.

HABERMAS, JÜRGEN (2003). *The Future of Human Nature*, trans. William Rehg, Max Pensky, and Hella Beister. Cambridge: Polity.

HAMER, DEAN (2004). *The God Gene: How Faith is Hardwired into our Genes*. New York: Doubleday.

HEFNER, PHILIP (1989). 'The Evolution of the Created Co-Creator', in Ted Peters (ed.), *Cosmos as Creation: Science and Theology in Consonance*, Nashville: Abingdon, 211–33.

JONES, DAVID ALBERT (2004). *The Soul of the Embryo: An Enquiry into the Status of the Human Embryo in the Christian Tradition*. London: Continuum.

KASS, LEON (2002). *Life, Liberty and the Defense of Dignity: The Challenge for Bioethics*. San Francisco: Encounter Books.

KELLER, EVELYN FOX (2002). *Making Sense of Life: Explaining Biological Development with Models, Metaphors, and Machines*. Cambridge, Mass.: Harvard University Press.

LEWIS, C. S. (1962). *The Abolition of Man*. New York: Collier Books.

MORACZEWSKI, ALBERT (2003). 'The Moral Tradition of the Catholic Church', in Audrey R. Chapman and Mark S. Frankel (eds.), *Designing our Descendants: The Promises and Perils of Genetic Modifications*, Baltimore: Johns Hopkins University Press, 199–211.

OUREDNIK, VACLAV, et al., (2001). 'Segregation of Human Neural Stem Cells in the Developing Primate Forebrain', *Science*, 293: 1820–4.

RAMSEY, PAUL (1970). *Fabricated Man: The Ethics of Genetic Control*. New Haven: Yale University Press.

STOCK, GREGORY (2002). *Redesigning Humans: Our Inevitable Genetic Future*. Boston: Houghton Mifflin.

WHEELER, SONDRA (2003). 'A Theological Appraisal of Parental Power', in Audrey R. Chapman and Mark S. Frankel (eds.), *Designing our Descendants: The Promises and Perils of Genetic Modifications*, Baltimore: Johns Hopkins University Press, 238–51.

....

CHAPTER 55

RELATIONS BETWEEN *HOMO SAPIENS* AND OTHER ANIMALS: SCIENTIFIC AND RELIGIOUS ARGUMENTS

....

NANCY R. HOWELL

Science and theology artfully negotiate the relationship between *Homo sapiens* and other animals, speaking sometimes about the continuity of humans with other animals and other times about the distinguishing characteristics of humans and animals. At stake for both science and religion is the compelling question, What makes humanity unique among living beings? The problematic issues for both fields are discernment and expression of what constitutes similarity and difference.

The contention here is that science and theology in dialogue recommend a nuanced understanding of both similarity and difference in the relationship of *Homo sapiens* to other animals. The chapter begins by identifying anthropomorphism, the projection of human traits onto animals, as a pivotal issue around which scientists debate human uniqueness and the relationship of humans to other animals. Similarly the chapter reviews the historical anthropocentric focus of Christian theology, which Rosemary Radford Ruether describes as humanocentrism, 'making humans the norm and crown of creation in a way that diminishes the other beings in the community of creation' (1983: 20). Contemporary animal science, however, poses some challenges for anthropomorphism and anthropocentrism, and ethologist Marc Bekoff informs the essay with observations about how the best practices in science undertake comparison of humans and animals and understand the relationship of *Homo sapiens* and other animals. Finally, the chapter concludes with concrete proposals to shift the anthropocentric focus of theology by adopting updated science to inform Christian thought and emphasizing some existing theological options.

SIMILARITY AND DIFFERENCE: ANTHROPOMORPHISM AND SCIENCE

Science cautions against drawing too close an analogy between *Homo sapiens* and other animals and follows a methodological principle that resists anthropomorphism. The prohibition of anthropomorphism guards against inappropriate projection of human qualities, emotions, and motivations onto non-human animals. Sociobiology best exemplifies resistance to anthropomorphism by rather strictly attributing animal behaviour to genetic predispositions and survival-based actions, which favour reproduction and well-being of species.

Some scientists, however, argue for carefully qualified forms of anthropomorphism when the method enhances the process of learning and discovery. For example, Gordon Burghardt (1985: 917) argues for critical anthropomorphism, which depends upon diverse forms of information ranging from descriptive anecdote and imaginative identification with the animal to prior experimentation and observation. The claims of critical anthropomorphism are, of course, subject to the rigours of science, which entail testable hypotheses, reliable predictions, and replicable results (Bekoff 2002: 49; de Waal 1996: 64).

Other scientists propose appropriate anthropomorphism moderated by an animal-centred rather than an anthropocentric orientation towards observation and description of animal behaviour. Marc Bekoff (2002: 48) advocates biocentric anthropomorphism, which makes other animal emotions and behaviours more accessible to human observation and scientific interpretation. Bekoff's biocentric anthropomorphism does not permit scientists and other humans simply to collapse or assign identity to human and other animal behaviour and feelings, but our modes of expression, understanding, and language are human and, therefore, must depend on analogy between *Homo sapiens* and other animals. Biocentric anthropomorphism does require that the observed animals and the interpretations of their behaviour and emotions remain focused on the animals' points of view.

Frans de Waal similarly advocates a fresh examination of anthropomorphism. In the book *Good Natured: The Origins of Right and Wrong in Humans and Other Animals*, de Waal observes how difficult anthropomorphism is to avoid even for scientists most dedicated to preserving anthropomorphism as a methodological ideal. De Waal observes that sociobiologists appear comfortable attributing negative motivations and behaviours to non-human animals, and he writes that 'current scientific literature routinely depicts animals as "suckers," "grudgers," and "cheaters" who act "spitefully," "greedily," and "murderously." There is really nothing lovable about them!' (1996: 18). If kinder emotions and behaviours appear among animals, de Waal notes that some scientists use quotation marks to qualify altruistic behaviours or qualify the behaviours with negative terms, such as *nepotism* rather than 'love for kin' to note positive relations with family members (1996: 18). The same scientists, who betray their own anthropomorphism by attributing negative human characteristics to other animals, criticize de Waal and other scientists for ascribing reconciliation or friendship to non-human animals. De Waal wants us to see that non-human animals capable of negatively valued behaviours, emotions, and motivations are likewise capable of positive actions and motivations, and obscuring similarities between human and other animal behaviour results from manipulation of language and values (1996: 19). Nevertheless, de Waal is clear that science must wrestle with human language, the only common expression at our disposal, because we use human language precisely to describe behaviours in other animals that are not identical with human behaviours and motivations, even though the language reminds us of human interactions, emotions, and intentions (1996: 63).

De Waal's extensive discussion of anthropomorphism is developed in *The Ape and the Sushi Master: Cultural Reflections of a Primatologist*, where he posits the problem of anthropo-denial and proposes the method of animal-centric anthropomorphism. Anthropo-denial is a facile and rigid rejection of anthropomorphism, which overlooks the possibility that some appropriate analogies between *Homo sapiens* and other animals might actually generate reliable scientific knowledge (2001: 69). Both anthropomorphism and anthropo-denial entail risk, which de Waal expresses by addressing why scientists cannot quite rid themselves of anthropomorphism: 'Isn't it partly because, even though anthropomorphism carries the risk that we overestimate animal mental complexity, we are not entirely comfortable with the opposite either, which is to deliberately create a gap between ourselves and other animals?' (2001: 68). Make no mistake; de Waal is not an opponent of unsupportable, simplistic, or naïve anthropomorphism, which results from 'insufficient information or wishful thinking' (2001: 68). The issue with anthropo-denial and naïve anthropomorphism is that our thinking is self-referential or anthropocentric, serving human purposes and biases and reflecting little knowledge of the other animals themselves. In light of tendencies towards anthropocentric anthropomorphism, de Waal proposes a more scientifically credible form of anthropomorphism, which he labels animal-centric anthropomorphism. Animal-centric anthropomorphism works within accepted information about animals and requires the scientist to adopt the observed animal's point of view (2001: 77). Properly conceived, anthropomorphism supports scientific method and experimentation by permitting human identification with or shared characteristics with animals to generate hypotheses and predictions subject to testing (2001: 78).

Before leaving discussion of how the scientific community understands the relationship between *Homo sapiens* and other animals, genetic and evolutionary bases for analogy should be mentioned alongside the issues of anthropomorphism in terms of behaviour, emotions, and motivations. While genetic and evolutionary connections account for expected similarities in behaviour, emotions, and motivations, genetic and evolutionary similarities between humans and other animals also form the basis for research and medical achievement. Animal testing and experimentation are common techniques for establishing the safety and effectiveness of commercial products, therapeutic drugs, and medical procedures. In addition, the physiological and functional analogy of humans and other animals supports using animals as effective teaching models for students and research models for disease and treatment.

Understanding other animals as analogous to humans for the sake of education and experimentation invites contention in the scientific community similar to the debates about anthropomorphism. Jane Goodall, for example, asserts that research using animals must be held accountable to knowing the whole animal subject. To accept that chimpanzees are appropriate analogues of human physiology and immune systems grounds medical research (in such diseases as hepatitis and AIDS), but morally obligates researchers to concede similarities in brain and central nervous systems between humans and chimpanzees. Goodall is compelled to ask, 'If physiological similarities between chimpanzees and man [*sic*] mean that a disease pattern is likely to follow a similar course in our two species and be affected by similar preventative or curative agents, is it not logical to infer that similarities in the central nervous systems of chimpanzees and ourselves may have led to corresponding similarities in cognitive abilities?' (in Goodall and Berryman 1999: 214). Goodall contends that scientists must count the mental and physical costs of research exacted from chimpanzees. In addition to the point that similarity suggests both promise and concern about animal experimentation, Roger Fouts illustrates the point that similarity may not guarantee successful results or applications to human curative therapies. Fouts reports that AIDS research found human studies much more effective than chimpanzee models for understanding the disease and discovering genetic bases for AIDS resistance. In AIDS research, the fundamental differences in human and chimpanzee immune systems prevented animal research from generating useful information (in Fouts and Mills 1997: 362).

Negotiating similarity and difference in the sciences entails discernment about appropriate and demonstrable shared characteristics of humans and other animals. The quality of scientific investigation and theory depends on avoiding the extremes of naïve anthropomorphism and anthropo-denial. Generally the concern is not to decide whether humans and other animals are related, but how humans and other animals are related.

SIMILARITY AND DIFFERENCE: ANTHROPOCENTRISM IN CHRISTIAN THEOLOGY

The conventions of Western science are situated in a world-view and history shaped and inhabited by Christian thought. Because the scope of the essay prevents explorations of multiple religious traditions, Christian theology offers a reasonable example of the ambivalence in religious traditions on the question of the relatedness of *Homo sapiens* and other animals. Christian theology suggests grounds for affirming the relationship of humans and other animals at the same time as anthropocentric Christian doctrines preserve human uniqueness. Certainly, Stephen Jay Gould's *Ever since Darwin* asserts that 'we are so tied to our philosophical and religious heritage that we still seek a criterion for strict division between our abilities and those of chimpanzees' (1977: 51).

Supported by companion philosophical influences, Christian thought, not surprisingly, established a history exploring the natures and relationship of God and humans, so that Christian anthropocentrism is a matter of neglect of other animals as well as a product of human arrogance intentionally guarding human superiority over and difference from other animals. John Cobb and Charles Birch's *The Liberation of Life* characterizes the dominant Christian model of humans, other animals, and nature by highlighting a few central Christian themes. First, Christian doctrine entails recognition that humans are part of a larger creation, including other animals and living creatures, but emphasizes the distinction of humans as creatures made in the image of God. Second, as a consequence of defining humans in relation to the soul and a Fall that distorted the image of God, Christian thought tends to envision a destiny for humans ultimately different from the destiny of other animals. Third, Christian theology, easily hospitable to Cartesian dualism and a mechanistic world-view, further separates humans from other animals and nature—sometimes characterizing nature as a mere stage for the human historical drama (Birch and Cobb 1981: 99).

Moving away from biblical theological statements that animals are valued by God, founders of Christian faith addressed the riddle of the image of God, generally arguing from a strict demarcation between humans and other animals. Engaging a philosophical debate of his time, St Augustine, for example, supported anthropocentrism and argued for human superiority over other animals: 'Among the many ways in which it can be shown that human beings surpass animals in reason, this is obvious to all: beasts [*beluae*] can be tamed by human beings, but human beings cannot be tamed by beasts' (Augustine, n.d.; Clark 1991: 68; her translation from *83 Questions on Various Topics*, Corpus Christianorum, ser. Latina 44A: 20). Augustine further argued that the lack of reason in animals absolved humans of any responsibility with regard to animal suffering (Birch and Cobb 1991: 147).

One feature of Augustine's reasoning entailed defining the *imago Dei* based on human reason as a distinction from other animals. Starting with the Genesis text, Augustine affirmed that humans are created in the image of God and are instructed to have dominion over animals, and then concluded that the image of God must reside in a part of human nature not shared with animals (Augustine 1982: 3, 20, 30). Because humans share embodiment with other animals, and especially because God is not embodied, the image of God cannot be found in the body and must reside in the reason or rational soul. In *The Literal Meaning of Genesis*, Augustine wrote: 'If, therefore, He Himself formed man from the earth and the beasts from the earth, what is the basis of man's greater dignity except that he was created in the image of God? This was not, however, in his body but in his intellect... And yet he does have in his body also a characteristic that is a sign of this dignity in so far as he has been made to stand erect' (1982: 6, 12, 22). Augustine concluded that the pre-eminence of humans in creation is due to human endowment with the *imago Dei* or rational soul, which surpasses the intellect of other animals.

St Augustine, as Gillian Clark observes, did not discuss animals in and of themselves, but mentioned animals, as his arguments required, to establish larger points about human nature, morality, and spirituality or about God. Clark characterizes the shape of Augustine's thought with regard to animals: 'So the implications of the rule of reason are, according to Augustine, that animals, animal behaviour, and animal suffering are all for the physical or spiritual benefit of human beings; that God's providence is concerned (within limits) for the physical survival of animals, but animals are not in spiritual contact with God, lack knowledge, and cannot experience happiness' (1991: 78). The shape of Augustine's arguments and theology clearly exemplifies anthropocentrism in Christian thought, and significantly so because of Augustine's continued influence on Christian theology and ethics.

Thomas Aquinas's theology is equally formational in defining human relationships with other animals in the Christian tradition. St Thomas, like St Augustine, upheld a strict, anthropocentric distinction between humans and other animals. Thomas's world-view conceived creation as hierarchical, reflecting relative positions of creatures by virtue of spiritual, rational being or material being. The composite of spirit and body defines humans as closer participants in the divine life than animals, who are not in the image of God.

As Paul Badham notes, however, Thomas followed Aristotle in asserting that animals have souls. The word *animal* itself establishes, by definition, that other animals have souls because the Latin *anima* translates as 'soul'. But the souls of other animals must not be confused with human souls, because the human immortal, rational soul is distinct from the mortal, sentient soul of other animals (Badham 1991: 181). Badham represents the importance of Thomas's distinction between human and animal souls with the following quotation from the *Summa Theologiae*:

Aristotle established that understanding, alone among the acts of the soul, took place without a physical organ... So it is clear that the sense-soul has no proper activity of its own, but every one of its acts is of the body-soul compound. Which leaves us with the conclusion that since souls of brute animals have no activity which is intrinsically of soul alone, they do not subsist... Hence though man is of the same generic type as other animals, he is a different species. (1970: 17)

The difference between humans and other creatures is upheld in the threefold understanding of the soul, which attributes a vegetative soul to creatures that grow and develop (such as plants), a sense soul, or sensitive soul, to creatures that move and feel (such as animals), and a rational soul to creatures that reason, which is reserved for humans. The rational soul of humans is immortal, surviving death because it is immaterial; the animating sense soul is mortal, dying with the animal because it is physical (Badham 1991: 182).

Dorothy Yamamoto interprets St Thomas's theological understanding of animals as contributing to the maintenance of social, moral, and political boundaries. At one level, animals function to mirror moral truths to humans. While God embodies moral ideals in animals, the irrational animals possess no critical awareness of the morality they mirror. The moral truths divinely imprinted or coded in animal bodies must be interpreted by human intellect. Yamamoto observes that this physically encoded morality establishes a role for animals in Thomas's hierarchy, which simultaneously establishes the *status quo* as the divinely created order:

Thus, if the template of animal society is held up to the human one, it can be seen that the precepts extracted from the former will be those that favour order, stability—in other words, the *status quo*.... So it can be seen that if animals are presented to humans as social exemplars a heavily weighted message is likely to emerge. It is one which will privilege things as they are and will censure nonconformity or attempts to change old practices for new ones. (1991: 82–3)

In Thomism, the absolute difference between humans and other animals demarcates a borderline useful for distinguishing some humans from others in the social order and hierarchical *status quo*. Because animals operate by instinct and humans think and act by reason, according to some Thomistic and medieval thinkers, the place of some humans (women and Jews, for example) in the social order could be determined by similar distinctions (Yamamoto 1991: 86).

The ambivalence of the Christian tradition towards animals is best expressed by exceptions to the rule of anthropocentrism, and three widely recognized exceptions occur in Chrysostom, Saint Francis of Assisi, and Albert Schweitzer. Birch and Cobb's *The Liberation of Life* reminds us that the Christian tradition is not limited to anthropocentrism. Chrysostom advocated Christian gentleness toward animals. Saint Francis, according to some interpretations of legend and biography, expressed extraordinary regard for nature, including animals. Albert Schweitzer's reverence for life, though flawed and incomplete as an ethic perhaps, resists valuing animals exclusively in proportion to the standard of humanity (Birch and Cobb 1981: 148–9).

The relationship of *Homo sapiens* and other animals in Christian thought generally sets human nature, uniqueness, and interests at the centre of reflection, with animals a more marginal concern. The result of anthropocentrism is a variety of ethical perspectives assessing different values for humans and animals and different values among humans. Christian perspectives attentive to the relationship of humans and other animals, especially establishing a caring or non-hierarchical relationship, are considered outside the mainstream.

SIMILARITY AND DIFFERENCE: SCIENCE LESSONS FOR THEOLOGY

Building on ideas originally presented to the American Academy of Religion in 2004, the final sections of the chapter advocate critical reflection on the meaning of similarity and difference, and propose a theological alternative to anthropocentrism. Because Marc Bekoff's writing is careful in interpreting similarity and difference, his scientific perspective is the framework that urges a theology of human and animal relationship responsive to studies of animal behaviour, emotions, and motivations.

The value of Marc Bekoff's observations lies in careful attention to the unnoticed, ordinary behaviours of dogs, which can be a window to the occurrence of important social behaviours, such as fair play. Bekoff's eye for complex canid social behaviours challenges assumptions about animals and humans common in earlier scientific research and still present in much theological scholarship. The purpose of my theological reflection is to explore assumptions about animals and humans, perhaps demonstrating that centred focus on animals might be fruitful for theological reflection. The following argument is that lessons in continuity, comparison, variation, and uniqueness from Bekoff's scholarship recommend theological reflection about method, personhood, God, and justice.

Marc Bekoff's empathy for dogs and other animals gives him a discerning eye for the complexity and importance of behaviours which are unnoticed, ordinary, and unremarkable from the standpoint of most theological scholarship. Theologians are virtual novices in the realm of the ordinary, and I suggest that when theologians begin the project of theologizing the ordinary, we must form partnerships with scholars such as Bekoff, whose expertise can tutor us in seeing and interpreting beauty in the world made invisible by theological commentary. To focus the experiment in theologizing the ordinary, I enter dialogue with Bekoff on the research question: What difference does difference make? I propose that Bekoff's observations and reflections instruct us in vital lessons about similarity, difference, continuity, and uniqueness. The resulting focus moves animals from the margins to the centre of theological reflection.

The first lesson is that *all similarities and differences between species (especially between humans and other animals) must be understood in light of the scientific basis for continuity.* Bekoff writes, 'Although there are numerous differences between humans and other animals, in many important ways "we" (humans) are very much one of "them" (animals), and "they" are very much one of "us"' (2002: 142). As news reports have made listeners aware, comparative studies of genes and proteins show continuity and similarity between humans and chimpanzees. In addition to genetics, evolution is an important theoretical basis for asserting the continuity of humans and other animals. In his consideration of emotions in animals, for example, Bekoff proposes the unlikelihood that human love emerged in nature 'with no evolutionary precursors, no animal lovers' (2002: 20). Further, Bekoff reminds us that the continuity or similarity of other animals with humans provides the rationale for some scientific and medical research involving animals as objects of study: ' "We" versus "them" dualisms do not work. The similarities rather than the differences between humans and other animals drive much research in which animals' lives are compromised. If "they" who are used in research are so much like "us," then much more work needs to be done to justify certain research practices' (2002: 55). Bekoff contends that the justifications for some research programmes are deficient because the argument for continuity is paired with human/animal dualism that objectifies animals.

A similar irony is the speciesist linear hierarchy of species, which ranks some species as 'higher' and others as 'lower'. The hierarchy in some ways reflects the continuity of animal species, yet neglects a careful understanding of evolutionary continuity and intra-species diversity, which, in fact, contradict a simple linear hierarchy and undermine simple determinations of the value of individuals and species (2002: 54).

If I interpret Bekoff appropriately, I conclude that genotypic and phenotypic evidence and evolutionary continuity require theology to take account of a more complex profile of the animal continuum, which includes humans. As Bekoff (citing Patrick Bateson) suggests, continuity between humans and other animals must be supported by empirical evidence, but without specific evidence ruling out certain similarities, science (and, I add, theology) cannot assume that particular continuities do not exist (2002: 95).

A second lesson is that *any comparisons between species or observations of charac-teristics within species must be described and evaluated within the species context* (2002: p. xx). As Bekoff notes, even scientists tend to make comparisons of animal behav-iour and abilities without situating the comparisons within the particular animals' or species' habitat. For example, to paraphrase Bekoff, observations support proposals that some chimpanzees have a sense of self, dogs plan for the future, and many animals experience emotions, pleasure, and pain (2002: 86). However, Bekoff writes, 'In addition to learning about the cognitive abilities of animals, some researchers are interested in making comparisons between the cognitive abilities or cognitive "levels" of animals and humans' (2002: 86). Bekoff asserts that such comparisons are not always helpful in learning about animals or humans because the abilities of all animals, including humans, are matters of appropriate fit and adaptation to the species' context. Bekoff concludes:

I am not sure that it is very useful to claim that a chimpanzee can reach the 'intellectual' level of a two-and-a-half-year-old human infant. Neither will we learn much by continuing to rear chimpanzees as if they are human. These so-called cross-fostering studies tell us little if anything about the behaviour of normal chimpanzees and raise numerous ethical questions.

Each organism does what it needs to do in its own world, and surely a young human (or most humans at any age) could not survive in the world of a chimpanzee. (2002: 86)

The questionable importance of comparisons of human and other animal behaviour suggests biases about human superiority. Bekoff offers a second example, which begins with an anecdote about the cleverness of a dog Skipper to retrieve a stick floating downstream by running ahead of the stick to catch it. Noting that young children might not possess the cognitive abilities to anticipate and intercept the stick similarly, Bekoff comments, 'While there may be other explanations for Skipper's behaviour, I am not sure what I would discover if I were told that children of a certain age usually develop the same ability that Skipper displayed and that Skipper was as smart as a child of that age, but no smarter' (2002: 86). The comparisons that we tend to make often select intelligence as the point of reference, and perhaps misinterpret the real significance of differences in behaviour. One final reference to Bekoff again emphasizes the importance of understanding behaviour in context: 'To claim that variations in the behaviour of different species are due to members of one species being less intelligent than members of another species shifts attention away from the various needs of the organisms that may explain the behavioural differences. Dogs are dog-smart and monkeys monkey-smart. Each does what is required to survive in its own world' (2002: 91).

The lesson that theologians take from Bekoff is that we must beware of comparisons between humans and other animals that neglect contextual awareness of behavioural characteristics. Behaviours are situated in ecological and evolutionary contexts.

The third lesson is that *accounts of animal behaviour should remember that individuals within a species exhibit variations in behaviour and personality.* Bekoff's research is attuned to individuals, which means that he is interested in individual behavioural variation and in the evolution of behavioural variation (2002: p. xviii). While speciesism is content to characterize individuals and their relationship to other animals by a species label, Bekoff's non-speciesism recognizes that individual differences within a species should not be dismissed. Attention to individuals may be especially important when similarities in behaviours cross species lines because, as Bekoff writes, 'it is possible that individual members of different species may be "equivalent" with respect to various traits or that individuals of a given species may possess characteristics that are exclusively theirs' (2002: 54). Individual variations suggest that ecology plays a role in behavioural variation because genetics alone is insufficient to explain variation that may be more appropriately linked to ecology or social factors (2002: 61).

Bekoff encourages awareness that variations in intelligence and adaptability should be expected among individuals of a single species (2002: 91). The problem is that generalizations about intelligence and cognition may reflect more about the limitations of observations and research than about the cognitive limitations of animals. Scientific conclusions, Bekoff notes, 'are based on small data sets from a small number of individuals who may have been exposed to a narrow array of behavioral challenges' (2002: 98). Primatologist Barbara Smuts likewise notes that the limitations of observers have much to do with perceived limitations of animals, and Bekoff cites her reflection that 'limitations most of us encounter in our relations with other animals reflect not their shortcomings, as we so often assume, but our own narrow views about who they are and the kinds of relationships we can have with them' (2002: 99).

Common methods in so-called objective behavioural sciences advocate treating unnamed animals as objects of study and discourage attention to individual personalities (including references to animals using grammar reserved for persons), yet Bekoff, Jane Goodall, and some other ethologists argue that naming individual animals is appropriate and no less effective than numbering animal subjects (Bekoff 2002: 45–7). I would guess, in following Bekoff's thinking, that naming animals and attending to individual personalities might be a more effective methodological approach when attention to individual behavioural variations and social and family relationships are central to the research project at hand.

Just as he points us to the diversity within animal species behaviour, so Bekoff similarly reminds us of the diversity in human behaviour, personhood, and morality, which makes comparison of species with humans even more complicated (2002: 15, 122). The lesson for theologians is that both our method and our constructions require attention to difference in very particular and concrete details as a guard against inappropriate generalizations about all animals and species, including humans.

A fourth lesson is that *human uniqueness must make room for dog uniqueness (and, of course, dolphin uniqueness, elephant uniqueness, chimpanzee uniqueness, etc.).* Human uniqueness is something of a moving target. Some claims about human uniqueness have been slowly dismantled by observations of diverse animal abilities with tools, language, culture, aesthetics, and reason; yet Bekoff speculates that such concepts as contemplation of mortality may still be defensible as uniquely human behaviour (2002: 13). When we claim that animals use tools or language, we do not necessarily mean that animals and humans are identical, which is a claim empirically unsupportable, but we mean that humans cannot be absolutely separated from animals by evidence of specified behaviours, and that animals have complex languages or communication within their social groups, although the language is not human language (2002: 138).

Further, even where similarities appear, the 'uniquenesses' of species must be acknowledged. For example, similar emotions in animal species may entail difference. Bekoff calls for more attention to research in species differences in expression of emotions and experiences of feeling: 'Even if joy and grief in dogs are not the same as joy and grief in chimpanzees, elephants, or humans, this does not mean that there is no such thing as dog-joy, dog-grief, chimpanzee-joy, or elephant-grief. Even wild animals (for example, wolves), and their domesticated relatives (dogs), may differ in the nature of their emotional lives' (2002: 119).

Unique behaviour, emotions, contexts, and social interactions are grounds for acknowledging that the word 'unique' is an appropriate adjective for all animals. Bekoff asks, 'Are humans unique? Yes, but so are other animals. The important question is *"What differences make a difference?"'* (2002: 138). The lesson for theologians is twofold: (a) theology needs revision in its claim that only human animals are unique, and (b) theology needs a method of reflection that attends carefully to differences among animals.

Similarity and Difference: Theological Responses to Science

In the remainder of the chapter I will sketch some existing options in theology that promise constructive dialogue with the lessons generated by Bekoff's scientific reflections on animals.

First, Sallie McFague has already proposed a methodological option in theology that calls for greater awareness of the intrinsic value of animals. Attention epistemology, as McFague defines it, is 'a rather abstract term for a very concrete and basic phenomenon: the kind of knowledge that comes from paying close attention to something other than oneself' (1993: 49). Attention epistemology means setting aside assumptions about human uniqueness and superiority in order to place other creatures in focus. The assumptions of attention epistemology are that animals (and plants and elements of nature) have intrinsic value and unique perspective. In an extended description of attention epistemology, McFague writes:

An attention epistemology is central to embodied knowing and doing, for it takes with utmost seriousness the differences that separate all beings: the individual unique site from which each is in itself and for itself. Embodiment means paying attention to differences, and we can learn this lesson best perhaps when we gauge our response to a being very unlike ourselves, not only to another human being (who may be different in skin color or sex or economic status), but to a being who is *indifferent* to us and whose existence we cannot absorb into our own—such as a kestrel (or turtle or tree). If we were to give such a being our attention, we would most probably act differently than we presently do toward it—for from this kind of knowing-attention to the other in its own, other, different embodiment—follows a doing appropriate to what and who that being is. (1993: 50–1)

Attention epistemology works against the dominant tendency in theology to generalize about nature, and challenges us to look deeply and empirically at the unique value in and differences among species and individuals.

Attention epistemology is evident in Bekoff's approach to research, which takes the animal's point of view. When Bekoff uses the phrase 'minding animals', in part he means 'caring for other animal beings, respecting them for who they are, appreciating their own worldviews, and wondering what and how they are feeling and why' (2002: p. xvi). Bekoff argues that field study requires 'taking the animal's point of view' in order to make sense of animal behaviour, emotions, and purposes (2002: 60).

Both McFague and Bekoff conclude that the observer is transformed by close attention to another creature, and that the transformation is evident in the evocation of compassion or love, which generates urgency for justice and advocacy (Bekoff 2002: 135–6; McFague 1993: 50). Attention epistemology conforms to the orientation of Bekoff's approach, about which he claims, 'My research has taken me in many different directions. Most important, it has led me deep into the minds, hearts, spirits, and souls of many other animals. It has also led me deeply into my own mind, heart, spirit, and soul. Animals have been my teachers and healers' (2002: 9). The transformation that occurs in humans who decentre themselves in relation to animals is not a sentimentality for similar creatures, an infatuation with the exotic, or a self-aggrandizement from charitable openness, but the transformative incarnation, if you will, of the different other for the difference made in human knowing and doing.

A second option is to develop a theology of nature that includes the personhood of non-human animals. Philosopher Alfred North Whitehead influences a number of theologians who consider animals to have continuity with humanity. Whitehead encouraged scholars not to make judgements about similarities and differences between humans and other animals apart from empirical evidence (Cobb 1965: 58). In Whitehead's *Modes of Thought*, the continuity of other animals and humans is based on the concepts of novelty, language, and religion. While acknowledging that humans have a more complex relationship with novelty, Whitehead noted that animal intelligence responds to 'conventional novelty with conventional devices' (1938: 35). Humans have conventional moments, too, but with humanity, Whitehead asserts that nature crossed a boundary permitting beings in nature to entertain unrealized possibility: 'In this way, outrageous novelty is introduced, sometimes beatified, sometimes damned, and sometimes literally patented or protected by copyright' (1938: 36). Concerning language, Whitehead observed that humans and other animals engage in communication (at least, in the embryonic form of speech) that 'varies between emotional expression and signalling' (1938: 52). While understanding that religion is comprised of ritual, emotion, belief, and rationalization, Whitehead attributed ritual and emotion to animals including humans, but he reserved belief and rationalization for human development of religion (1926: 20–1).

Given the continuities and differences among humans, Whitehead suggested that the more advanced capacities for freedom and creativity in humans and vertebrates and the presence of a central organizing principle to co-ordinate organic and social relationships in humans and other animals are grounds for extending the definition of *person* to include at least some animals other than humans. Whitehead defined persons as individuals whose life history of experience is co-ordinated by a 'presiding occasion of experience' (1978: 107). The presiding occasion of experience is the natural phenomenon that Whitehead called the psyche or soul, which means that Whitehead included non-human animals among persons, who by definition are endowed with souls that preside over behaviour (Cobb 1965: 48).

Like Whitehead, Bekoff hopes to convince us that non-human animals should also be designated as persons. Bekoff defines persons using several criteria: 'being

conscious of one's surroundings, being able to reason, experiencing various emotions, having a sense of self, adjusting to changing situations, and performing various
cognitive and intellectual tasks' (2002: 14). Bekoff notes that humans vary considerably in their abilities to meet the criteria, yet we still appropriately consider humans
(such as infants) who cannot meet all the criteria to be persons (2002: 14). Claiming
that humans have nothing to lose by sharing personhood with animals, Bekoff
suggests that animals as persons have much to gain. Calling animals persons
'would mean that animals would come to be treated with respect and compassion
that is due them, that their interests in not suffering would be given equal consideration with those of humans' (2002: 15). Bekoff's research programme promises the
empirical evidence that Whitehead required and theologians need to develop an
adequate theology of and for animals—a theology of animal personhood.

A third option in theology is to develop explicitly a panentheistic concept of God
that takes account of the rich and complex diversity of experiences in the animal
world. A panentheistic world-view understands that God's experience encompasses
the world's experience. Sallie McFague proposes that God's radical transcendence
and immanence are expressed in the panentheistic metaphor describing the universe
as God's body, which entails all bodies, all embodiments (1993: 134).

Panentheism suggests a creative reciprocity between cosmic experiences and divine
experiences, and the attendant world-view holds that God acts creatively and persuasively in the world and that the experiences of the world are creative events in the
body of God (Whitehead 1978: 348). The experiences that make up the world take on
a sacramental character when the world's embodiments directly constitute God's
embodiment. In technical Whiteheadian language, God's inclusive experience of the
experiences of the world is called 'intensity', a term which refers generally to the
ability to entertain the variety, depth, and breadth of experiences without loss of
personal integrity (1978: 83).

We might then imagine two ways to contribute experience to God's body. One way
points toward complex individuals, such as humans, whose freedom and creativity
enable them to contribute rich and intense experience to God's experience. The first
way inclines theologians and philosophers to maintain gradations of value in nature
that give greater importance to creatures who individually contribute rich experience
to God's body. However, if closer examination of animal behaviour leads to appreciation of the intensity of non-human individuals, then hierarchical interpretations
of experience may give way to genuine recognition of the intrinsic and sacramental
value of non-human animal behaviour and experience. The second way points
toward diversity and community, the volume of life in total, as the truly inclusive
source of intensity and rich experience in the body of God. With the second way, no
experience is unimportant in contributing to the intensity of divine experience. As an
ecological interpretation of depth and breadth of experience, the second approach
values all experience—human and non-human, animal and plant, living and non-
living—as sacred in the experience, body, and being of God.

In dialogue with Bekoff (and other ethologists), theological perspectives can be
enriched by deeper understanding of the rich diversity of animal behaviour, experi-

ence, and emotions, which characterize individuals, species, and animal life. With theological imagination, we might expect that a better understanding of the beauty and intensity of animal experience (including varieties of difference and similarity) might generate and support a deeper and more interesting concept of God.

A fourth option for theology is to continue our work to establish justice and compassion. Theology is not always tolerant of difference, and the problem of diversity is addressed with repeated and recognizable habits of thought. *Mujerista* theologian Ada María Isasi-Díaz criticizes traditional theology in terms of its habits of mind that define difference 'as absolute otherness, mutual exclusion, categorical opposition' (1996: 80). Making one group of humans the norm against which other persons are measured, traditional theology is essentialist with regard to difference. Defining difference in essentialist terms 'expresses a fear of specificity and a fear of making permeable the boundaries between oneself and the others, between one's ideas and those of others' (1996: 80). Cuban American theologian Luis G. Pedraja adds one further characteristic of theological habits of mind: traditional theology creates hierarchies that place some groups of humans closer to God, hence justifying superiority over and domination of groups who are deemed inferior (2003: 120).

To be clear about the context of Isasi-Díaz's and Pedraja's observations, I must note that their characterization of theology addresses human differences—the difference between dominant culture and Latino/a culture brought to light by consciousness of marginality and by engagement of *mulatez* and *mestizaje* diversity within Latino/a culture. Isasi-Díaz asserts that one challenge to traditional theology is the *mujerista* theologian's claim that embracing diversity is a moral obligation for theology (1996: 80).

For some, the connection between Latino/a theology and concern for animals may not be apparent. However, I have cited Isasi-Díaz and Pedraja to demonstrate that theological injustice is committed toward diverse humans and other animals when we fail to attend to intrinsic value and particularity. Theologians use the same habits of mind to justify exclusion, dehumanization, and exploitation of all not-quite-human beings and animals. Just as Isasi-Díaz challenges theology to attend to specificity for the sake of justice toward the Latino/a community, so Bekoff challenges us to attend to the particularity of individuals and species, so that our awareness of the remarkable behaviour and intrinsic value of animals might convince us that justice toward animals is not an extraordinary expectation.

A deep sense of relationship attends both particularity and value. The tendency in some historical theology is toward universal and homogeneous interpretations of nature and humanity, and the result is neglect of the particular. Resistance toward universalism appears in contemporary theology that forces the issues of gender, race, and class into central place and that decentres the theological imagination away from humans and toward creation (for example, in ecological theologies). What theologians have learned, however, with a good deal of struggle, is that even general theories about gender, race, and nature are insufficient.

Theology cannot be based on some abstracted concept of woman because the ethnic, cultural, social, religious, national, and class contexts of diverse women demand a more sophisticated and plural interpretation of gender. Likewise, a theology that generalizes about issues of race still dangerously holds dominant (white) racial motifs as normative until the cultural, social, and economic diversity among races and within races informs deeper and broader developments in theological anthropology. Similarly, then, theology might expect to discover that a broad theology of creation or nature is inadequate to interpret the particularity of species and within species. Ultimately, theological reflections on humanity and nature are distorted and even unjust when particularity is neglected.

Rosemary Ruether names the connection between particularity and value, and the following remarks repeat and extend a quotation cited early in the chapter:

> Women must also criticize humanocentrism, that is, making humans the norm and crown of creation in a way that diminishes the other beings in the community of creation. This is not a question of sameness but of recognition of value, which at the same time affirms genuine variety and particularity. It reaches for a new mode of relationship, neither a hierarchical model that diminishes the potential of the 'other' nor an 'equality' defined by a ruling norm drawn from the dominant group; rather a mutuality that allows us to affirm different ways of being. (1983: 20)

Attending to particularity is a part of attributing appropriate value to persons and relationships. Forms of instrumental value tend to interpret groups or species as valuable for the sake of their usefulness to dominant groups or individuals. Instrumental value may limit relationship to subject–object interaction. Intrinsic value recognizes that Washoe the chimpanzee and Jethro the dog have value in and for themselves, without regard to their utility for human purposes; but intrinsic value is rendered invisible by theological reflection that understands Washoe and Jethro as functionally and objectively indistinguishable from others of their species or from non-human animals in general. Theological imagination accountable to empirical evidence and observations is adequate only when the particularity and intrinsic value—the diverse personalities, emotions, cultures, behaviours, motivations, and uniqueness—of other animals inform how we think about relationships among animals and between humans and specific animals.

In conclusion, an empirical and specific understanding of similarities and differences blurs boundaries and eliminates borders. To speak of humankind and otherkind (language I have used in other writing) perpetuates the idea of unsubstantiated, absolute difference. Hierarchical and value-burdened categories of the unique Self and the Other create too much separation and too much temptation to exploit and marginalize the Other—other animals and other humans. Theology must recast uniqueness, equating uniqueness with the intrinsic value and differences evident in all species and individuals, as well as the cosmic community embodied in the divine.

References and Suggested Reading

Aquinas, Thomas (1970). *Summa Theologiae*, ii, trans. Timothy Suttor. London: Eyre & Spotttiswoode.

Augustine (1982). *The Literal Meaning of Genesis*, in J. H. Taylor (trans.), Ancient Christian Writers, 41–2. New York and Ramsey, N.J.: Newman Press.

—— (n.d.). '83 Questions on Various Topics', in G. Clark (trans.), Corpus Christianorum, Series Latina, 44A: 20.

Badham, P. (1991). 'Do Animals Have Immortal Souls?', in A. Linzey and D. Yamamoto (eds.), *Animals on the Agenda*, Urbana, Ill., and Chicago: University of Illinois Press, 181–9.

Bekoff, M. (2002). *Minding Animals: Awareness, Emotions, and Heart*. Oxford: Oxford University Press.

Birch, C., and Cobb, John B. Jr. (1981). *The Liberation of Life: From the Cell to the Community*. Cambridge: Cambridge University Press.

Burghardt, G. M. (1985). 'Animal Awareness: Current Perceptions and Historical Perspective', *American Psychologist*, 40: 905–19.

Clark, G. (1991). 'The Fathers and the Animals: The Rule of Reason?', in A. Linzey and D. Yamamoto (eds.), *Animals on the Agenda*, Urbana, Ill., and Chicago: University of Illinois Press, 67–79.

Cobb, J. B. Jr. (1965). *A Christian Natural Theology: Based on the Thought of Alfred North Whitehead*. Philadelphia: Westminster Press.

de Waal, F. (1996). *Good Natured: The Origins of Right and Wrong in Humans and Other Animals*. Cambridge, Mass.: Harvard University Press.

—— (2001). *The Ape and the Sushi Master: Cultural Reflections of a Primatologist*. New York: Perseus Books Group, Basic Books.

Fouts, R., with Mills, S. T. (1997). *Next of Kin: My Conversations with Chimpanzees*. New York: Avon Books, Inc.

Goodall, J., with Berryman, P. (1999). *Reason for Hope: A Spiritual Journey*. New York: Warner Books.

Gould, S. J. (1977). *Ever since Darwin: Reflection in Natural History*. New York: W. W. Norton.

Isasi-Díaz, A. M. (1996). *Mujerista Theology*. Maryknoll, N.Y.: Orbis Books.

McFague, S. (1993). *The Body of God: An Ecological Theology*. Minneapolis: Fortress Press.

Pedraja, L. G. (2003). *Teología: An Introduction to Hispanic Theology*. Nashville: Abingdon Press.

Ruether, R. R. (1983). *Sexism and God-Talk: Toward a Feminist Theology*. Boston: Beacon Press.

Whitehead, A. N. (1926). *Religion in the Making*. New York: Macmillan Company.

—— (1938). *Modes of Thought*. New York: Macmillan Company.

—— (1978). *Process and Reality*, corrected edn., ed. D. R. Griffin and D. W. Sherburne. New York: Free Press.

Yamamoto, D. (1991). 'Aquinas and Animals: Patrolling the Boundary?', in A. Linzey and D. Yamamoto (eds.), *Animals on the Agenda*, Urbana, Ill., and Chicago: University of Illinois Press, 80–9.

....

CHAPTER 56

CONCLUDING REFLECTIONS: DOVER BEACH REVISITED

....

MARY MIDGLEY

ON BEING EXILES IN OUR OWN LAND

....

Matthew Arnold, writing sadly of the receding Sea of Faith, gave his image a vast and deadly application:

> The world, which seems
> To lie before us like a land of dreams
> So various, so beautiful, so new,
> Hath *really* neither joy, nor love, nor light,
> Nor certitude, nor peace, nor help for pain
>
> *Dover Beach*, 4. 2–6; (emphasis mine)

No joy, no love, no light? As Arnold then saw it, the loss of Christian metaphysics drained away all the normal meaning from life, leaving us desperately trying to make sense of a dead, empty world by pulling on our own bootstraps.

This passage was actually the beginning, not the end, of Arnold's long spiritual quest. But it is this part of his message that has been most clearly remembered, because later theorists have developed and preached the same vision. Like Arnold, they have enforced it by strong ontological language. They have dismissed our natural sense of connection with the world around us as simply an illusion. Thus Jacques Monod, following Sartre, gives us this: 'Man must at last wake out of his millennary dream and discover his total solitude, his fundamental isolation. He must realise that, like a gypsy, he lives on the border of an alien world, a world that is deaf to his music' (Monod 1972: 160).

But who, we might ask, are the mysterious figures that Monod calls 'we', these creatures living quite outside the normal system of nature? They are not much like ourselves. In fact, Monod's whole vision is just one more optional imaginative picture, a fantasy as much at odds with modern science as it is with common sense.

That nihilistic message—which has had enormous influence during the past century—draws its power from exploiting the unreal split that Descartes introduced between human minds and the rest of nature. It assumes that the natural world is— as seventeenth-century physicists believed—alien to us because it is just a mass of inert solid particles bouncing off each other in accord with a few simple laws. Animals are unconscious automata. Life itself is quite alien to nature. We ourselves are unearthly, supernatural entities, God's colonists sent to supervise the Earth.

The Newtonian age could use this crude dualism because the Christian God still kept the two elements together. Later, as church influence waned, the faults of the system became obvious. But, instead of rethinking our relation to the outside world from scratch, as was then needed, people saved time by cobbling together parts of their existing ideas. Perhaps largely due to the Industrial Revolution, they often dealt with dualism by simply dropping half the subject-matter. They threw out Descartes's Ghost and kept his Machine. Thus they could turn dualism into materialism and leave the inert, life-free physical world to manage on its own.

It has turned out, however, that this doesn't really work. The inert Machine and the active Ghost were carefully designed to fit each other. Neither can be used effectively alone. The idea of a machine without a maker is a very obscure one, and it cannot be clarified by simply nominating Natural Selection to fill that post (see Goodwin 1994; Rose 1997). Moreover, the inert particle model itself no longer makes any sense because it is contrary to modern physics. On the human side, too, a world of objects without active subjects does not make much sense. Eventually, therefore, behaviourism collapsed, leaving us with the 'problem of consciousness' which is now giving us so much trouble. The behaviourists, misguidedly obsessed with parsimony, had tried to explain human life entirely from the outside, ignoring the inner experience that lies at the heart of all human action. In their ambition to simplify, they had cut psychology off from a crucial part of its subject-matter—from the very thing that it was trying to explain.

WHAT IS REALISM?

Besides religious matters, however, there is another pile of difficulties here that are often discussed today—problems that arise when people try to restore various parts of the huge inner landscape that Arnold felt he had lost and Monod thought he had demolished. They concern, for instance, the status of subjective experience and the nature of moral values. Some philosophers now lump all these problems together under the heading of 'realism'. This language suggests that they are all ontological problems, questions about the reality of doubtful entities trying to get themselves admitted to the charmed realm of things known to exist—trying, in fact, to get classed as physical objects. The question then approximates to what might be called the Loch Ness Monster model, asking whether certain things—souls, gods, moral values—can actually be found in the physical world around us.

Put this way, the topic naturally invites anyone with an Enlightenment background to reach for Occam's razor, laying the burden of proof on the candidate entity—an immigrant begging for citizenship and most unlikely to deserve it. But to make any headway with these questions, what we actually need is a quite different model. We need the one which already comes to our minds when somebody says, 'There are no real scholars any longer', or perhaps, 'no real shamans'. Or again, when we say such things as, 'I don't believe in progress' or 'in basic instincts' or 'in the intentionalist fallacy' or 'in the hidden hand of the market'. Or indeed, 'in the death of the author'.

Here it is obvious that the question is not about admitting a particular item into the inventory of things in the world. In the first set of remarks, the word 'real' has the same sort of sense that it has when we talk of real cream or real coffee. It contrasts proper examples of a certain class with faulty ones. For this use we always need a clear idea of the particular kind of wrongness or falsity to which we are opposing it (granulated coffee, synthetic cream, imitation eyelashes . . .). This is indeed probably the clearest and most satisfactory use of the word 'real' because it brings out the conceptual background underlying the use of that particular term. That background becomes still more obvious in the second set of examples, because there the intention is clearly not to banish particular objects from the world but to question the use of certain *concepts*—certain patterns of thought, certain ways of dividing and naming our experience—and to suggest that we must find better ones to replace them.

The Loch Ness Monster approach is surely ill-omened, because, quite apart from religion, it has already discredited itself by failing in two other areas which are of the first philosophical importance. About subjective experience—a topic that the great empiricists, from Locke to William James, rightly thought central—it imposed a paralysing, indiscriminate inattention during most of the twentieth century. Essentially, conscious minds were deemed not to be real since they were not part of the physical world, so consciousness itself could never be mentioned. This behaviourist frost is now beginning to thaw, but it has left us in the shocking chaos that surrounds the 'problem of consciousness' today.

Similarly in the case of ethics, this archaic ontological thinking paralysed thought by ruling that values were radically shut off from facts. Since value-judgements were not statements about physical facts, they could have no real subject-matter at all. These judgements had no logical relation to the rest of thought and could never be explained. So, for half a century, English-speaking moral philosophers solemnly maintained that rational thought about how moral problems relate to the world was simply impossible, because such thought would involve a 'naturalistic fallacy'.

THE FORGOTTEN EARTH

What is wrong with this counter-productive way of thinking is not just its neglect of mind but, more deeply, its narrow, arbitrary conception of *facts* and *matter.* Thirty years back, I wrote of it in *Beast and Man*:

The really monstrous thing about Existentialism is its proceeding as if the world contained only dead matter (things) on the one hand and fully rational, educated human beings on the other—as if there were no other life-forms. The impression of *desertion* or *abandonment* which Existentialists have is due, I am sure, not to the removal of God, but to this contemptuous dismissal of almost the whole biosphere—plants, animals, and children. Life shrinks to a few urban rooms; no wonder it becomes absurd. (Midgley 1978: 18–19)

Arnold's impression of desolation on his stony beach and Monod's strange claim to the marginal, outlawed condition of a gypsy are, I think, simply expressions of the impoverished status to which the isolated, cerebral, seventeenth-century Self had sunk in Enlightenment thinking once it lost its celestial background. Deprived of heaven, this Cartesian ghost belonged nowhere. It certainly could not think of itself as at home on the Earth, which it had long regarded as simply the opposite of heaven (see Midgley 2003: ch. 19). It had been separated most carefully from the human body, and thus from its natural context in the rich life surrounding us.

Christian thinking had denounced all reverence for the natural world as pagan, and secular thinkers still avoided it as idle and superstitious. Indeed, *reverence* itself tended to be viewed in enlightened circles as a dangerous state of mind, not suitable for an independent, autonomous, rational individual.

In consequence, human spiritual capacities—which are, after all, as real as our other faculties and as much a part of the natural world as frogs and daffodils—were viewed with deep suspicion; their cultivation was discouraged. The result was that this impressive, totally independent entity MAN was left with nothing to worship but himself. This of course he eagerly did, building technological shrines to his own glory. The heathen in his blindness bowed down to wood and stone—steel and glass, plastic and rubber and silicon—of his own devising, and saw them as embodying the final truth. This has gone on until he began to find, to his surprise, that they are fast exhausting the resources of the Earth he lives on.

WHY GAIA HELPS

It was at this point that James Lovelock suggested that we might instead start trying to understand the Earth itself and our own relation to it in different terms. If (he said) we begin to attend properly to our own planet, to see that it is not an inert heap of resources but a self-maintaining whole, a vast, complex living system—if we begin to grasp our own total dependence as a minuscule part of it—we might indeed also sensibly view it with reverence rather than taking it for granted as exploitable. In order to make this huge shift of perspective clear, he called this living earth Gaia, after the Greek earth-goddess, the deeply revered primal mother of gods and men.

This move suddenly opened the window which, for a century, had been firmly closed between modern scientific thinking and the spiritual world that our ancestors, and most other human cultures, had always assumed was there around them. Many people were alarmed, and the official scientific world, in particular, turned its back, firmly refusing, at first, to listen to the unwelcome message at all (Lovelock 2000: ch. 9). In time, however, it became clear that, scientifically speaking, Lovelock was actually correct. Living things have indeed played a crucial part in moderating the Earth's atmosphere and climate in a way that has made their own survival possible. Without their saving action, Earth would indeed quickly have become a dead, airless planet like Venus and Mars, and there would certainly be no researchers here to speculate about why this had happened.

In principle, then, scientists saw that life was indeed not an alien import but actually a crucial component of the earth-system. Geology and biology were not really separate studies but ought to be continuous with each other. And universities, conceding this, did indeed begin to combine the two in departments of Earth Science where scholars now study the interaction. But the one thing that most of them still will not willingly do is to use the name 'Gaia' in this context. They see that this would mean acknowledging something rather more revolutionary than a mere link between two departments of study. It would affect our own status, which would be highly unsettling. They therefore still prefer at present to dismiss thoughts about Gaia by calling them 'flaky', 'New Age', or 'Californian'.

Certainly the idea was taken up in those quarters. But of course that alone is not a sensible reason for dismissing it. Lovelock himself was disturbed about these distortions, and for a time he even considered dropping the name Gaia, replacing it by the term 'geophysiology' (Lovelock 1991: 6, 11, 31). In the end, however, he decided against this change because he thought that, in spite of their difficulties, the wider applications of the idea were still essential to it.

I am well aware that the term [life] itself is metaphorical and that the earth is not alive in the same way as you or me, or even as a bacterium. At the same time I insist that Gaia theory is real science and no mere metaphor... [The metaphor is necessary because] ... real science is riddled with metaphor. Science grows from imaginary models in the mind and is sharpened by measurements that check the fit of the models with reality. (Lovelock 1991: 31, 6, 11)

The sense in which the totality of life is alive must obviously be rather different from the sense that fits the individual organisms within it. Its implications are wider. As Lovelock points out, this idea inevitably bears on matters that go far beyond the borders of science, but that does not mean that scientists must not envisage it: 'For me, Gaia is a religious as well as a scientific concept, and in both spheres it is manageable.... God and Gaia, theology and science, even physics and biology are not separate but a single way of thought' (Lovelock 1988: 206–7, 212).

These notions are now becoming quite widespread. The idea that the biosphere is in some sense a living whole looks increasingly plausible as we are increasingly forced to recognize that this whole is *sick* and in trouble. For it is surely only living things that can be ill or well. Gaian thinking thus calls for a sudden return to realism (in the perfectly ordinary sense). It makes possible a saner, more rational, and more usable view of our own situation—a view that finally explodes dualism and exposes the hollowness of the Monodian melodrama.

For all purposes, we need to grasp finally that we are not machines or ghosts or loose combinations of those entities vacationing on this planet. We are earthly creatures who are thoroughly *at home* here—part of the system, native to the place, living in the only dwelling we could ever have, and thus responsible for its upkeep. Mass migration to space is as idle a fantasy as any wish fulfilment that religious people may have dreamed up in the past.

This insight needs no ontological excesses. Gaia is not an extra entity, but the earthly biosphere that has actually produced us. Occam's razor can, I think, be safely sheathed.

Parsimony and Pluralism

Why, however, was that razor so eagerly brought out in the first place? It was meant, of course, to provide that most laudable feature, conceptual parsimony. William of Occam was indeed right to say that varieties of entities should not be multiplied beyond necessity—that we should not say that something exists unless we have a clear idea of what it is and what work it does in the world. We do not want idle fifth wheels on our cart such as phlogiston or the 'animal spirits'. But how do we decide just which explanatory entities are really needed? What constitutes necessity? Not all these ideas are fifth wheels. Some may be babies which it would be false economy to throw out with their bathwater.

It is often said that science easily resolves this problem of selection, because science deals only in entities that it can prove to be present by experiment. But scientific practice does not really go on in this restricted way, least of all in that holy of holies, theoretical physics. Debates about the physical workings of the universe do not usually revolve around whether a particular entity exists but around its function, the reason for invoking it, the work that it does in the larger conceptual structure. Current attempts to find the Higgs boson may look superficially like expeditions to find the Loch Ness Monster, but they are actually something very different. What is really going on here is a comparison between various possible conceptual systems— large-scale ways of interpreting the whole. That comparison will be settled mainly by their coherence with other ideas and with the great mass of physical facts. Particular experiments play only a very small part in this process.

Throughout these large inquiries new items constantly crop up that are far distant from any possible experience—strings, super-strings, branes, dark matter, dark energy, eleven-dimensional space. Sometimes indeed experiments are eventually found that seem to confirm their presence. But the acceptance of these results always depends on a great structure of interpretation, an edifice of conceptual investigation, not the experimental kind.

In this area, perhaps the most remarkable anti-Occamist flight in recent times has been the Many Worlds Interpretation of quantum mechanics. This is a theory that accounts for the uncertainty belonging to quantum events by providing enough new universes to accommodate both alternative possible outcomes simultaneously on each occasion. It posits that an infinite number of new universes—not just new worlds—comes into existence every moment, so that both possibilities are always provide for. As Roger Penrose puts it:

Not just an observer but the entire universe that he inhabits splits in two (or more) at each 'measurement' that he makes of the world. Such splitting occurs again and again. . . . So that these universe 'branches' proliferate wildly. Indeed, every alternative possibility would co-exist in some vast superposition. This is hardly the most economical of viewpoints. . . . It seems to me that the many-worlds view introduces a multitude of problems of its own without really touching upon the *real* puzzles of quantum measurement. (Penrose 1989: 382)

This is a discreet understatement. Actually the extravagance of the supposition surely dwarfs almost anything that the various religions have ever been bold enough to put forward, simply because of its scale. How is such grandiloquence to be distinguished from mere empty talk? *In what context* is all this supposed to be happening? *Where*, so to speak, do we put all these universes? I doubt, in fact, whether it is actually possible to *believe* in this indefinitely accelerating creation of endless wholes in any normal sense, because it is impossible to imagine it. But today a fair number of reputable scientists do entertain this idea, along with other strange visions, as a workable background for scientific thought—a pattern that may eventually throw up further ideas that they can use to make sense of experience. Indeed, Penrose him-self—though he has other objections to it—remarks later that 'despite . . . the multi-tude of problems and inadequacies that it presents us with, [the many-worlds theory] cannot be ruled out as a possibility' (Penrose 1989: 560).

Thus, in the world of theoretical physics, other considerations often prevail over parsimony. Occam certainly does not always rule. Ways of thinking are considered legitimate if they apparently explain the data more effectively than other available ways, even if they do it more circuitously.

This may indeed be one reason why physicists are, today, often more receptive to religious thinking than are biologists or social scientists. They may be more aware that it is necessary to try out many different ways of thinking and sometimes to use different ones together—that problems often do not admit of a single satisfactory solution—that, in fact, the world we are trying to understand is often a great deal more complex than our current thinking allows. Or, as the great biologist J. B. S. Haldane put it, this world is probably not just much queerer than we suppose but much queerer than we *can* suppose. Our faculties, said Haldane, may simply not be adequate to begin to represent its queerness.

Because this complexity is now admitted, we now know that it is vanishingly unlikely that a single way of thinking will ever explain the world's workings completely. No one pattern of thought—not even in physics—is so 'fundamental' that all others will eventually be reduced to it. For most important questions in human life, a number of different conceptual toolboxes always have to be used together. And there is no single law showing us how we should combine them.

The more profound thinkers among today's physicists have, I think, indeed understood the need for this kind of pluralism, a need which Heisenberg made clear half a century back (Heisenberg 1958). They have swung right away from the position that their forefathers held at the dawn of modern science. Seventeenth-century thinkers were convinced that the world was ultimately simple in a way that really would allow it to be understood completely by a single pattern of thought, a pattern that would essentially reduce to something closely related to mathematics. And they set about developing physics as the candidate for that sublime role.

Since then, a depressing series of disappointments, diversions, and surprises has interrupted the ideal straightforward progress that was hoped for. The most traumatic of these has, of course, been the sudden collapse, a century back, of the Newtonian paradigm which was expected to support the whole enterprise. That shock has indeed led many physicists to adopt a more realistic, more pluralistic approach to the capacities of their own discipline. Unluckily, however, many scholars in other disciplines seem not to have noticed this explosion. They still back a 'physicalist' view of the world which assumes that this single ideal, universal explanation will eventually be found.

THERE IS INNER COMPLEXITY, TOO

How does this point about pluralism bear on our questions about religion?

Here, what we are trying to understand and explain is the whole vast landscape of the religious experience and thinking of humankind. Confronting that range, William James famously found huge 'varieties of religious experience' to occupy him, even within the Western Christian tradition alone. He explained that, in order to do justice to that range, he had to leave aside other traditions, and also the public aspect of religious practice in the West itself. Even so, he found enough complexities to fill a packed volume of 500 pages, which he concluded on an unpretentious and tentative note:

Meanwhile the practical needs and experiences of religion seem to me sufficiently met by the belief that, beyond each man and in a fashion continuous with him there exists a larger power which is friendly to him and to his ideals. All that the facts require is that the power should be both other and larger than our conscious selves. Anything larger will do if only it be large enough to trust for the next step. It need not be infinite, it need not be solitary. (James 1960: 500; cf. 141)

In writing that 'anything larger will do', James was not, of course, endorsing some kind of mindless subjectivism. He was not saying that each of us can pick any symbolic vision at random and use it to 'construct' a private world-view which will be as 'valid' as any other. He is taking the observer's point of view on the range of human spirituality and pointing out the extreme diversity of perceptions—the differences of temper and standpoint that underlie various beliefs. He was saying that those differences must inevitably produce different responses and would do so even if what confronts them outside is actually the same for all.

EMPIRICIST VISIONS

James is connecting the possible plurality of gods with the more general plurality of aspects that pervades all life—the irreducible richness and complexity of the real world, constantly exceeding our powers of understanding. For James, empiricism is not (as it was for Hobbes and Hume) one more reductive method, a rival simplification brought forward to outdo the order imposed by rationalistic systems. It is something much more ambitious—an attempt to stick close to experience, to do as much justice as possible to life's actual richness, while still finding enough order to make life navigable. As he said, 'A simple conception is an equivalent for the world only so far as the world is simple—the world meanwhile, whatever simplicity it may harbour, being also a mightily complex affair' (see James 1956: 70, 'The Sentiment of Rationality').

When we are trying to understand religious experience, the workings of our inner life make this complexity even more mysterious. But (he said) we do have ways of resolving our doubts here:

Looking back on my own experiences, they all converge towards a kind of insight to which I cannot help ascribing some kind of metaphysical significance. The keynote of it is invariably a reconciliation. It is as if the opposites of the world, whose contradictoriness and conflict makes all our difficulties and troubles, were melted into unity. Not only do they, as contrasted species, belong to one and the same genus, but *one of the species*, the nobler and better one, *is itself the genus, and so soaks up and absorbs its opposite into itself.* (James 1956: 374; emphasis original)

This account surely reflects James's own experiences in the grim struggles with depression that he endured in early life, before finally reaching a sane and usable balance. But it seems also a fair description of what goes on in any of us when we grapple with conflicts between deep attitudes and the different sorts of thinking that go with them—including conflicts between what are viewed as purely intellectual 'paradigms'.

WHAT SORT OF GOD?

What all this surely shows is that the emphasis here needs to be on conceptual schemes rather than entities. What comes first is not an exercise in metaphysics but a choice between wider patterns of thought and the attitudes that underlie them. The metaphysics must follow later. We cannot start from the question of whether a particular entity, such as God or the soul, exists.

This is clear in the case of God if we notice how much his situation differs from that of Bigfoot or the Loch Ness Monster. In a seriously religious person's world God is not an optional component. That person's entire world is shot through with divinity. It is a world in which God is not one item, but a basic principle determining the whole structure. As Wittgenstein pointed out, different attitudes to life as a whole do not alter facts, but they do profoundly alter a person's world—the enclosing context which gives those facts their meaning: 'The effect must be that it becomes an altogether different world. It must, so to speak, wax or wane as a whole. The world of the happy man is a different one from the world of the unhappy man' (Wittgenstein 1961: 6.43).

Or, as Francis Thompson put it:

> Does the fish soar to find the ocean,
>   The eagle plunge to find the air—
> That we ask of the stars in motion
>   If they have rumour of thee there?
>
> Thompson, *The Kingdom of God*

God, then, is not just one item among others, but the principle at the root of the structure. That principle may or may not be conceived in personal form, and the gap between these two approaches remains one of our most disturbing mysteries. For Taoism and Confucianism the central principle is quite impersonal. And Christian thought too contains a strong mystical tradition, drawing on Neoplatonist sages such as Plotinus, which altogether rejects the personal approach, crying out with Meister Eckhart, 'Pray God to be rid of God.' Buddhism too dismisses the idea of a central, personal God, though it does have a special role for the Buddha's personality and allows for minor gods who may be part of the whole.

By contrast, the mainstream of Christian, Judaic, and Islamic thought is strongly personal, attached to the God of Abraham, Isaac, and Jacob. That God is, of course, also believed to be universal and immanent—not a fully human god like Homer's Olympian deities, but a creative power underlying the structure of the cosmos. It has never been easy, however, to combine these aspects. The difficulty of doing so has surely been central to our culture's recent increasing unease about the whole concept of God, an unease that was already rampant in Arnold's time. And in this friction between the two aspects of deity, the personal aspect is the one that, on the whole, modern thought has found it hardest to assimilate.

An interesting indication of this has been the recent change in the fortunes of the word 'spiritual'. That word, which was long banished from intellectual circles along with the rest of the religious vocabulary, seems lately to have been paroled and returned to circulation in a favourable sense, and is used to describe a wide range of matters pertaining to the inner life. It is clear that many people outside the churches are no longer happy to dismiss all such topics briskly as superstitious. And many of them are friendly to the idea of an immanent God or life force, though they can no longer bring the idea of a fully personal deity into focus.

IS HEAVEN ELSEWHERE?

Of course, this kind of position is not new in Western thought. There are, however, still special difficulties about accepting it in a Western context, because the idea of a personal God has traditionally been conceived in a way that still shapes the imagery, which is profoundly important here. In the Judaeo-Christian tradition, God has been seen as an absolute ruler and one especially interested in punishment. This imagery now strikes many of us as alien and objectionable. That shift in perception may not actually be an unmixed blessing. The current adulation of 'celebrities' suggests that the habit of misplaced worship may have been redirected rather than cured. But in liberal politics this disenchantment with monarchy cannot now be reversed.

Another difficulty is the local—indeed, parochial—character of the biblical God, especially the God of the Old Testament. The many justifications of tribal warfare and genocide given there are not easily forgotten. The New Testament is indeed less pugnacious, more universal in its message, but it contains many stories of miracles, stories which grate against today's scientific outlook. And the most important of these miracles—Christ's incarnation and resurrection—also raise a wider difficulty of perspective.

The Gospel treats these events as central to the whole history of creation; yet in the much vaster, much less tidy, and anthropocentric universe that we now inhabit, they tend now to strike us as local. In fact, much of the difficulty centres round the changes in our notion of the cosmos itself that have followed the Copernican revolution, changes which not only disturb the traditional placing of God in the sky but perhaps afflict the whole notion of seeing him as a person in a way that continues the earlier trend.

Since persons—people—are particular beings, it is natural to place them *somewhere*, and this placing affects their character. Traditionally, the most obvious place for divine persons was either above us or below us—in the sky or under the earth. Thus there were sky-gods and earth-gods, or rather, quite often, sky-gods and earth-goddesses. Jehovah, like Zeus, was originally a sky-god, wielding his thunderbolt from above, and in his early days he seems (like other chief gods) to have had colleagues and a family to balance his special interests. But after King Hezekiah, who was anxious to unify his distracted realm, threw out these other gods and made Jehovah an absolute ruler (2 Kings 18: 4), this balance was disturbed.

No doubt it was intended that Jehovah should absorb all their former symbolism into himself, but that did not happen. Instead, he treated the deities themselves as vanquished enemies, and the provinces of life that they had formerly represented as conquered territory. Milton, who loved both worlds, feelingly displayed this debacle in his 'Hymn on the Morning of Christ's Nativity'. He saw that, in the new monotheistic world, nymphs and agricultural deities could expect no mercy. As Milton saw, in spite of the deeper and more universalistic insights of the Christian tradition, Jehovah still remained a partial arbiter—essentially a patriarch, ruling human earthly life from outside. From his base above us in heaven, he used the earth simply as a trying-ground for souls and a convenient container for the punishments of hell that lay at its centre. After the Last Judgement, when he would no longer need this earth, he would burn it up. For human souls, therefore, the life that really counted was always the next one— their destiny in heaven or hell, not on the planet where they were actually living.

This earth-despising bias is something that we need to be aware of, because it is still with us today. It did not evaporate along with the political influence of the churches. It continued to determine the status, not just of God but, more crucially, of MAN as well. The confidence with which thinkers like Bacon undertook to remodel nature did not fade along with the habit of backing it by biblical texts. Instead, that confidence remained, even without the texts, as a core support of the Scientific and Industrial Revolutions that followed. Anthropocentrism, which in many ways took over from theistic religion, freely presented modern MAN with the news that the physical world had no independent value and was therefore simply his oyster. That (it seems) is how we have got into the mess in which we find ourselves today. And the point of Gaian imagery is precisely to correct that illusion.

DIFFICULTIES OF DEICIDE

Thus the figure of a personal God that has been handed down by our tradition did indeed have plenty wrong with it. No wonder many people have vowed to destroy it altogether. The more cheerful of them, indeed, have often thought that this would be quite easy, that they could amputate the useless organ without trouble.

Nietzsche, however, working at the centre of the operation, never made that mistake. When he first launched his project of eliminating the divine Lawgiver, he was racked with guilt. He wrote, 'Doubt devours me. I have killed the Law, and now the Law haunts me as a cadaver haunts a living person. *If I am not more than the law, then I am among the damned souls the most damned*' (quoted by Reinhardt 1964: 5). And in *The Joyful Wisdom*, where there is much discussion of the project, he developed that thought further in the Madman's speech:

We have killed him—you and I. We are all his murderers!...Do we not hear the noise of the grave-diggers who are burying God? Do we not smell the divine putrefaction—for even Gods putrefy! God is dead. God remains dead! And we have killed him! How shall we console ourselves, the most murderous of all murderers? . . . Is not the magnitude of this deed too great for us? Shall we not ourselves have to become gods, merely to seem worthy of it? (Nietzsche 1960: Book III, para. 125; emphasis original)

This indeed is Nietzsche's solution to the problem. Humans who have gone beyond religion have (he says) no choice except to transform themselves, by efforts hitherto unheard of, into a quite new species of being who can perform all that was formerly expected of deity. They must take on God's functions, becoming a kind of god: namely, *Supermen*. Each of them must become his own lawgiver, acquiring by independent thought the authority needed to back his moral precepts. (Since this is a men-only story, there is of course no question of writing 'his or her own lawgiver' here.) Moreover, they must recognize that the 'death of God' which makes this effort necessary is not an accident. It is something for which they themselves are responsible, something that is still being brought about by their own choice. That death is not (as people often seem to imply) a historical event like the extinction of the dinosaurs, which just happened to occur in the late nineteenth century. It is a choice that still constantly has to be made—a choice between two different ways of seeing the world.

Since Nietzsche wrote, this vision of turning ourselves into gods—this conception of autonomy as a spiritual regeneration by will-power that shall make each of us into a final authority and the only proper object of our own worship—has been much admired and widely developed. While it has certainly made possible certain good qualities belonging to our individualistic age, I think we can see by now that it cannot do the work he expected of it. The choice of moral solitude, indeed of moral solipsism—a choice that he constantly glorified as a proof of heroic courage—can just as easily be a sign of weakness, a fear of other people. And the vices that are encouraged by this kind of inward-looking self-importance certainly have not turned out to be less dangerous than those that were associated with lazy-minded conformity.

What is the right course here? It seems reasonable to ask why Nietzsche insisted on finding a new absolute monarch rather than questioning the institution of monarchy itself. Why did he want to refill what he took to have been God's legislative function rather than re-examine it? It was already a bad idea to suppose that the reasons for doing right come from an infallible authority even when that authority was supposed to be divine. It is still a bad idea when that authority is oneself. In fact, the whole demand for an authority here is surely a misguided form of foundationalism, as unhelpful in morals as it is in the theory of knowledge. If there is indeed a God-shaped hole, this cannot be the right way to fill it.

WHAT IT MEANS TO BE SOCIAL

The reason why we are not Nietzschean supreme gods is that we do not live alone, 'creating values' in a moral vacuum. We make our choices in a social context. Though we are indeed each 'autonomous' in the sense of relying on our own conscience, this conscience is not isolated from the consciences of others. It is part of a wider workshop which is, as it were, the joint conscience of our society. We need constantly to take part in its discussions. Of course we do indeed sometimes have to make the kind of sharp protests there which chiefly engaged Nietzsche. But much more often, what we are doing is engaging in an ongoing conversation—hearing, welcoming, and developing the suggestions made by others and adding our own items to the mix. The mantra on Dover Beach is never 'Cogito ergo sum', always 'Cogitamus, ergo sumus'. If there is any salvation, we will have to find it together.

Arnold indeed made this other-regarding aspect explicit at the outset of his lamentation, crying out:

> Ah love, let us be true
> To one another!

And it is the key to all his later work on the problem. He insists that our relation to other people is not just an irrelevant comfort, a consolation, a distraction from the chilling truth of atheism. To the contrary, it is the guide to a deeper and fuller truth beyond atheism. It is a light that is very hard to follow but that can, if followed, show us how the world is meaningful after all.

This is the sense of his startling remark that '*God . . .* is really, at bottom, a deeply moved way of saying *conduct* or *righteousness*' (Arnold 1873: 46). This is not just a reductive move, because what he means by conduct and righteousness is so tremendous. He is proposing that, if we take other people really seriously, if we try our hardest to understand them and to treat them properly, we can, in that way, gradually become aware—through them—of 'an enduring Power, not ourselves, that makes for righteousness' (Arnold 1873: 323).

Michael McGhee, in a most illuminating discussion, calls this 'a modest non-theistic transcendentalism' and connects it with various Buddhist philosophical positions on the matter. He comments:

In Arnold we have the beginnings of the notion that 'conduct' or 'righteousness' increasingly opens us to realities that were formerly and otherwise concealed. . . . There are, on this view, realities that are concealed from the unregenerate consciousness. . . . Arnold is invoking the idea of a gradually revealed *given* that cannot be traced back to our choice or construction, though we do indeed construct a great deal around it when we reflect upon it theoretically. (McGhee 2000: 132, 142)

This suggestion is not fishy epistemologically. We already know that moral changes can alter people's cognitive capacities—that people's power of understanding each other's feelings and motives depends quite as much on the understanders' virtues and sensibilities as it does on their IQs. And we can also understand that there is always a special difficulty when the less tries to understand the greater. As McGhee puts it:

The 'not fully grasped object of the speaker's consciousness' is just this obscurely present given that opens out progressively to conduct and, of course, *towards* it. The use of God-language is therefore to be understood in terms of assertions that can be verified in experience. . . .

Arnold's account of theological language makes its use no more than a cultural contingency, a means by which the intimation of a progressive disclosure of reality is articulated. This is not a revisionist voluntarism that makes belief a *commitment* that saves us from futility.... What he put in the place of 'belief' or faith is the notion of a revelatory *life*. (McGhee 2000: 144)

This seems not to be far from William James's idea that 'beyond each man and... continuous with him there exists a larger power which is friendly to him and to his ideals... [a power] both other and larger than our conscious selves' (James 1960: 500; cf. 141).

At this point, however, modern thought tends to get stuck on the question of *otherness*, insisting on being told whether the power is actually a part of each self—perhaps, as James suggests, 'a larger and more godlike self'—or whether it is, as Arnold puts it, simply 'not ourselves'. But is the estuary part of the river or part of the ocean? What makes us think that we can cut each self off from all the others like this and map them so exactly as to draw lines here? What stops us thinking about the properties of the whole of life to which they belong?

This dichotomy, this atomistic view of individuals, seems to me to show far more confidence than is justified in our knowledge of our own inner selves. Certainly William James's suggestion does depict a self much larger and more mysterious than any that modern psychology recognizes. But since modern psychology has resolutely turned its back on any problems that might call on it to recognize such vastness and mystery, this does not necessarily mean very much. Heraclitus, by contrast, remarked that 'you would not find out the boundaries of the soul, even by travelling along every path; so deep a measure does it have' (fragment 232, in Kirk, Raven, and Schofield 1983: 203). Carl Jung, too, suggested such an approach to the psyche. It seems reasonable to suggest that, to make any sense of these difficult questions, we had better follow their example.

REFERENCES AND SUGGESTED READING

ARNOLD, MATTHEW (1873). *Literature and Dogma: An Essay Towards a Better Apprehension of the Bible.* London: Smith, Elder & Co.

GOODWIN, BRIAN (1994). *How The Leopard Changed Its Spots.* London: Weidenfeld & Nicolson.

HEISENBERG, WERNER (1958). *Physics and Philosophy.* London: Penguin.

JAMES, WILLIAM (1956). *The Will to Believe and Other Essays in Popular Philosophy.* New York: Dover.

—— (1960). *The Varieties of Religious Experience.* Glasgow: Collins, Fontana.

KIRK, G. S., RAVEN, J. E., and SCHOFIELD, M. (1983). *The Presocratic Philosophers.* Cambridge: Cambridge University Press.

LOVELOCK, JAMES (1988). *The Ages of Gaia.* Oxford: Oxford University Press.

—— (1991). *Gaia: The Practical Science of Planetary Medicine.* London: Gaia Books Ltd.

—— (2000). *Homage to Gaia: The Life of an Independent Scientist.* Oxford: Oxford University Press.

McGHEE, MICHAEL (2000). *Transformations of Mind: Philosophy as Spiritual Practice.* Cambridge: Cambridge University Press.

MIDGLEY, MARY (1978). *Beast and Man.* Ithaca, N.Y.: Cornell University Press.

—— (2003). *The Myths We Live By.* London: Routledge.

MONOD, JACQUES (1972). *Chance and Necessity,* trans. Austryn Wainhouse. Glasgow: Fontana, William Collins Sons.

NIETZSCHE, FRIEDRICH (1960). *Joyful Wisdom,* trans. Thomas Common. New York: Frederick Ungar Publishing Co.

PENROSE, ROGER (1989). *The Emperor's New Mind.* Oxford: Oxford University Press.

ROSE, STEVEN (1997). *Lifelines.* Harmondsworth: Penguin.

REINHARDT, KURT (1960). Introduction to Nietzsche, *Joyful Wisdom.* New York: Frederick Ungar Publishing Co.

WITTGENSTEIN, LUDWIG (1961). *Tractatus Logico-Philosophicus,* trans. D. F. Pears and B. F. McGuinness. London: Routledge.

Name Index

Includes all referenced authors and Organizations

Subject Index